Pharmaceutical Biotechnology

Daan J. A. Crommelin
Robert D. Sindelar · Bernd Meibohm
Editors

Pharmaceutical Biotechnology

Fundamentals and Applications

Sixth Edition

Springer

Editors
Daan J. A. Crommelin
Utrecht University
Utrecht, The Netherlands

Bernd Meibohm
College of Pharmacy
University of Tennessee Health Science Center
Memphis, TN, USA

Robert D. Sindelar
Faculty of Pharmaceutical Sciences
The University of British Columbia (UBC)
Vancouver, BC, Canada

ISBN 978-3-031-30022-6   ISBN 978-3-031-30023-3   (eBook)
https://doi.org/10.1007/978-3-031-30023-3

Editorial Contact: Charlotte Nunes

This Springer imprint is published by the registered company Springer Nature Switzerland AG
The registered company address is: Gewerbestrasse 11, 6330 Cham, Switzerland

Paper in this product is recyclable.

Preface

Over the past 35 years, the share of biotechnologically derived drug products in the arsenal of medicinal products has been growing steadily. These drug products include proteins, such as monoclonal antibodies, antibody fragments, endogenous or modified hormones and growth factors, as well as antisense oligonucleotides, mRNA vaccines, DNA preparations for gene therapy and cell therapies. Drug products such as epoetin-α (Epogen®, Eprex®, Procrit®), interferons-α (Intron®A, Roferon®A) and -β (Avonex®, Rebif®, Betaseron®), etanercept (Enbrel®), infliximab (Remicade®), adalimumab (Humira®), bevacizumab (Avastin®) and trastuzumab (Herceptin®) are all examples of highly successful biotech drugs that have revolutionized the pharmacotherapy of previously unmet medical needs. The fast development and introduction of the mRNA vaccines to fight the COVID-19 pandemic showed the world the lifesaving power of this class of biotechnological products. Biotech drugs have a major socioeconomic impact. They present the opportunity to introduce new first-in-class medicines for challenging diseases. Also, they are rapidly trending toward becoming among the most prescribed drugs. In 2021, six of the ten top selling drugs in the world were biotechnologically derived drug products. The biotechnology market size is expected to reach ~$1345 billion USD by 2030 owing to increasing prevalence of chronic illnesses.

The techniques of biotechnology are a driving force of modern drug discovery as well. Due to the rapid growth in the importance of biopharmaceuticals and the techniques of biotechnology to modern medicine and the life sciences, the field of pharmaceutical biotechnology has become an increasingly important component in the education of today's and tomorrow's pharmacists and pharmaceutical scientists. We believe that there is a critical need for an introductory textbook on Pharmaceutical Biotechnology that provides well-integrated, detailed coverage of both the relevant science and the clinical application of pharmaceuticals derived by biotechnology.

Previous editions of the textbook *Pharmaceutical Biotechnology: Fundamentals and Applications* have provided a well-balanced framework for education in various aspects of pharmaceutical biotechnology, including production, quality, dosage forms, administration, economic and regulatory aspects and therapeutic applications. With more than three-million-chapter downloads, the fifth edition of the textbook has achieved widespread distribution as a key educational resource for the field of pharmaceutical biotechnology. Rapid growth and advances in the field, however, made it necessary to revise this textbook in order to continue providing up-to-date information and introduce readers to cutting-edge knowledge of this field.

This sixth edition of the textbook *Pharmaceutical Biotechnology: Fundamentals and Applications* builds on the successful concept used in the preceding editions, and further expands its availability as electronic versions of the full book as well as individual chapters are readily available and downloadable through online platforms.

The textbook is structured into three sections. An initial basic science and general features section on biologics 'Pharmaceutical Biotechnology: The science, techniques, and important concepts' comprises the first 12 chapters introducing the reader to key concepts at the foundation of the technology relevant for biologics, including molecular biology, production and analytical procedures, formulation development, pharmacokinetics and pharmacodynamics, immunogenicity, and chapters dealing with regulatory, economic and pharmacy practice considerations and with evolving new technologies and applications. The second section 'Pharmaceutical Biotechnology: Oligonucleotides, genes and cells/cell subunits, vaccines— The science, techniques, and clinical use' (Chaps. 13, 14, and 15) discusses the rapidly expanding field of nucleotide- and cell-based therapies. It includes both the fundamentals and the clinical applications. Finally, the third section 'Pharmaceutical Biotechnology: The protein products of biotechnology and their clinical use' covers various therapeutic classes of protein biologics.

All chapters of the previous edition were revised, and some were completely overhauled. The section on monoclonal antibodies is differentiated into a section on general considerations for this important class of biologics and sections focused on their application in oncology, transplantation, inflammation and neurological diseases in order to allow for a comprehensive discussion of the substantial number of approved antibody drugs.

In accordance with previous editions, the new edition of *Pharmaceutical Biotechnology: Fundamentals and Applications* will have as a primary target students in undergraduate and professional pharmacy programs as well as graduate students in the pharmaceutical sciences. An additional important audience are pharmaceutical scientists in industry and academia, particularly those who have not received formal training in pharmaceutical biotechnology and are inexperienced in this field.

We are convinced that this sixth edition of *Pharmaceutical Biotechnology: Fundamentals and Applications* makes an important contribution to the education of pharmaceutical scientists, pharmacists and other healthcare professionals as well as serving as a ready resource on biotechnology. By increasing the knowledge and expertise in the development, application and therapeutic use of 'biotech' drugs, we hope to help facilitate a widespread, rational and safe application of this important and rapidly evolving class of therapeutics.

Utrecht, The Netherlands   Daan J. A. Crommelin
Vancouver, BC, Canada   Robert D. Sindelar
Memphis, TN, USA   Bernd Meibohm
December 2023

Contents

Abbreviations

11β-HSD1	11β-Hydroxysteroid dehydrogenase type 1
2-MOE	2′-O-methoxyethyl
5-HT	Serotonin
a	Allometric coefficient
A	Adenine
A	Amyloidosis
AAV	Adeno-associated virus
ABE	Adenine base editors
ABO	ABO blood group system
ABW	Adjusted body weight
ACER	Average cost-effectiveness ratio
ACR	American College of Rheumatology
Ad	Adenovirus
AD	Alzheimer's disease
AD	Psoriasis, atopic dermatitis
ADA	Adenosine deaminase
ADA	Antidrug antibody
ADAR	Adenosine deaminase acting on RNA
ADA-SCID	Severe combined immunodeficiency due to adenosine deaminase deficiency
ADC	Antibody–drug conjugate
ADCC	Antibody-dependent cell cytotoxicity
ADCC	Antibody-dependent cell-mediated cytotoxicity
ADCC	Antibody-dependent cellular cytolysis
ADCP	Antibody-dependent cell-mediated phagocytosis
ADCP	Antibody-dependent cellular phagocytosis
ADME	Absorption, distribution, metabolism, and elimination
ADME	Absorption, distribution, metabolism, and excretion
ADPR	Adenosine diphosphate ribose
ADPRP	Adenosine diphosphate ribose-2′-phosphate
AdV	Adenovirus
AEX	Anion exchange
AF4	Asymmetrical flow field-flow fractionation
AID	Automated insulin delivery
AIDS	Acquired immunodeficiency syndrome
AKT	Protein kinase B
ALK	Anaplastic lymphoma kinase
ALL	Acute lymphocytic leukemia
Alpha-gal	Galactose-alpha-1,3-galactose
ALT	Alanine transaminase
AM	Adrenomedullin receptors

AMD	Age-related macular degeneration
AML	Acute myeloid leukemia
AMR	Antibody-mediated rejection
ANC	Absolute neutrophil count
AOSD	Adult-onset Still's disease
APC	Activated protein C
APC	Antigen-presenting cell
APE1	Apurinic/apyrimidinic endonucleases from humans
API	Active pharmaceutical ingredient
APP	Amyloid precursor protein
APRIL	A proliferation-inducing ligand; CD256
ARCA	Anti-reverse cap analog
ARF	Alternative reading frame
ART	Antiretroviral therapy
ART	Assisted reproductive technologies
AS	Ankylosing spondylitis
ASCVD	Atherosclerotic cardiovascular disease
ASGPR	Asialoglycoprotein receptor
ASO	Antisense oligonucleotides
AST	Aspartate aminotransferase
ATF	Alternating tangential flow
ATMP	Advanced therapy medicinal product
ATP	Adenosine triphosphate
AUC	Analytical ultracentrifuge
AUC	Area under the concentration–time curve
AUC_{0-24}	Area under the plasma concentration-time curve from time 0–24 h
AWP	Actual wholesale price
b	Allometric exponent
BAFF	B-cell activating factor
BBB	Blood-brain barrier
BC	Breast cancer
BCG	Bacille Calmette-Guérin (live attenuated tuberculosis vaccine)
BCMA	B-cell maturation antigen
BCR	B-cell receptor
BDD-FVIII	B-domain deleted factor VIII
BE	Base editors
BHK	Baby hamster kidney
BICLA	British Isles Lupus Assessment Group-based Composite Lupus Assessment
BID	Twice a day
BiTE	Bispecific T-cell engager
BLyS	B-lymphocyte stimulator
BMI	Backgrounded membrane
BMI	Body mass index
BPCI	Biologics Price Competition and Innovation (Act)
BRAF	B-Raf proto-oncogene, serine/threonine kinase
BSE	Bovine spongiform encephalopathy
BSE/TSE	Bovine/transmittable spongiform encephalopathy, mad cow disease
C	Cytosine
C1q	Complement component 1q
cADPR	Cyclic adenosine diphosphate ribose
CALD	Cerebral adrenoleukodystrophy
CAPEX	Capital expenditures

CAPS	Cryopyrin-associated periodic syndrome
CAR	Chimeric antigen receptor
CAR	Coxsackie and adenovirus receptor
CARPA	Complement activation related pseudoallergy
CAR-T	Chimeric antigen receptor (CAR) T-cell therapy
CAR-T-cells	Chimeric antigen receptor T cells
CAS	Chemical abstract service
CAT	Committee for advanced therapies
CBA	Cost–benefit analysis
CBE	Cytosine base editors
CBER	Center for biologics evaluation and research
CCR	Chemokine receptors
CD	Circular dichroism (spectroscopy)
CD	Cluster of differentiation
CD	Crohn's disease
CDAI	Crohn's Disease Activity Index
CDC	Center for Disease Control and Prevention US
CDC	Complement-dependent cytotoxicity
cDNA	Copy DNA
CDR	Complementarity-determining regions
CDRP	Calcitonin gene-related peptide receptor
CE	Capillary electrophoresis
CEA	Carcinoembryonic antigen
CEA	Cost-effectiveness analysis
CE-SDS	Capillary electrophoresis sodium dodecyl sulfate
CEX	Cationic exchange (chromatography)
CF	Cystic fibrosis
CFA	Corifollitropin alfa
CFTR	Cystic fibrosis transmembrane conductance regulator
CFU	Colony-forming unit
CG	Chorionic gonadotropin
CGD	Chronic granulomatous disease
CGE	Capillary gel electrophoresis
CGM	continuous glucose monitor
cGMP	Current good manufacturing practice
CGRP	Calcitonin gene-related peptide
CHMP	Committee for medicinal products for human use
CHO cells	Chinese hamster ovary cells
CHO	Chinese hamster ovary
CI	Checkpoint inhibitors
CI	Confidence interval
CID	Collision-induced dissociation
CIP	Clean-in-place
circRNA	Circular RNA
CIU	Chronic idiopathic urticaria
CK	Chemokines
CKD	Chronic kidney disease
CL	Clearance
$CL_{extravasation}$	Transfer clearance from the vascular to the interstitial space
CLL	Chronic lymphocytic leukemia
CL_{lymph}	Transfer clearance from the interstitial space to the lymphatic system
CLR	Calcitonin-like receptor

CLss	Clearance at steady state
CLtot	Total clearance
CM	Chronic migraine
CMA	Cost-minimization analysis
CMA	Critical material attribute
C_{max}	Maximum concentration
CMC	Chemical, manufacturing, and controls
CML	Chronic myeloid leukemia
CMV	Cytomegalovirus
CNS	Central nervous system
COCs	Combined oral contraceptives
COI	Cost of Illness
COP	Cyclic olefin polymer
COS	Controlled ovarian stimulation
COVID-19	Coronavirus disease 2019
CP	Capacitance manometer
CP	Carboplatin and paclitaxel
Cp	Plasma concentration
CPB	Carboplatin, paclitaxel, and bevacizumab
CpG	Cytosine-phosphodiester-guanine
CPP	Cell penetrating peptide
CPP	Critical process parameters
cPPT	Central polypurine tract
CQA	Critical quality attribute
CRC	Colorectal carcinoma
CRI	Chronic renal insufficiency
CRISPR	Clustered regularly interspaced short palindromic repeats
CRISPR-Cas	CRISPR-associated protein 9
CRP	C-reactive protein
crRNA	Complementary sequence RNA
CRS	Cytokine release syndrome
CRSwNP	Chronic rhinosinusitis with nasal polyps
CRT	Calreticulin
CsA	Cyclosporin A
CSD	Cortical spreading depression/depolarization
CSF	Cerebrospinal fluid
CSF	Colony-stimulating factor
CSII	Continuous subcutaneous insulin infusion
CSU	Chronic spontaneous urticaria
CT	Cholera toxin
CTA	Clinical trial authorization
CTL	Cytotoxic T lymphocytes
CTLA-4	Cytotoxic T-lymphocyte-associated antigen 4
CTP	Carboxy-terminal peptide
C_{trough}	Concentration immediately prior to next treatment
CUA	Cost-utility analysis
CVD	Cardiovascular disease
CXC	Chemokine (of the CXC chemokine family)
CXCR4	C-X-C motif chemokine receptor 4
CXCR5	C-X-C chemokine receptor type 5
CYP	Cytochrome P450
CYP3A4/5	Cytochrome P450-A4 and P450-A5
CYP450	Cytochrome P450

CZE	Capillary zone elecrophoresis
DC	Dendritic cells
dCas9	"Dead" Cas9 enzyme
DCC	Dual-chamber cartridge
DC-chol	3-β[N(NV,NV-dimethylaminoethane)-carbamoyl] cholesterol
DDC	Human dopa decarboxylase
DDI	Drug-drug interaction
DIGE	Differential gel electrophoresis
DIRA	Deficiency of the interleukin-1 receptor antagonist
DIY	Do it yourself
DLBCL	Diffuse large B-cell lymphoma
DM	Diabetes mellitus
DM1	Emtansine
DM4	Ravtansine
DMARDs	Disease-modifying antirheumatic drugs
DMD	Duchenne's muscular dystrophy
DMSO	Dimethyl sulfoxide
DMT	Dimethoxytrityl
DNA	Deoxynucleic acid
DNase I	Deoxyribonuclease I
DO	Dissolved oxygen
DOTMA	2, 3-dioleyloxypropyl-1-trimethyl ammonium bromide
DP	Drug product
DS	Drug substance
DSB	Double-strand break
DSC	Differential scanning calorimetry
dsDNA	Double-stranded DNA
DSP	Downstream process
dsRNA	Double-stranded RNA
DTT	Dithiothreitol
DXd	Deruxtecan
EACA	Epsilon amino caproic acid
EASI	Eczema Area and Severity Index
EBV	Epstein-Barr virus
EC50	Concentration of the drug that produces half of the maximum effect
ECD	Extracellular domain
ECP	Eosinophil cationic protein
EDD	Electron detachment dissociation
EDSS	Expanded disability status scale
EDTA	Ethylenediaminetetraacetic acid
EGF	Epidermal growth factor
EGFR	Epidermal growth factor receptor
EGPA	Eosinophilic granulomatosis with polyangiitis
EI	Electrospray ionization
ELISA	Enzyme-linked immune sorbent assay
EM	Episodic migraine
EMA	European Medicines Agency
Emax	Maximum (achievable) effect
EMBL	European Molecular Biology Laboratory
env gene	Envelope gene
EPAR	European Public Assessment Report
EpCAM	Epithelial cell adhesion molecule
EPO	Epoetin alfa, erythropoietin

EPOR	Erythropoietin receptor
ER	Endoplasmic reticulum
ErbB	Member of the ErbB family of proteins: ErbB1, ErbB2, ErbB3, and ErbB4
ERK	Extracellular signal-regulated kinase
ESA	Erythropoietin-stimulating agents
ESC	Embryonic stem cell
ESC	Enhanced stabilization chemistry
ESCF	Epidemiologic study of cystic fibrosis
ESKD	End-stage kidney disease
ESR	Erythrocyte sedimentation rate
ETD	Electron transfer dissociation
EU	European Union
EV	Extracellular vesicles
EZS	Electrical zone sensing
F	Bioavailability
Fab	Antigen-binding fragment
FACS	Fluorescence-activated cell sorting
F-actin	Filamentous actin
FBS	Fetal bovine serum
Fc receptor	Fragment crystallizable receptor
Fc	Constant fragment, fragment crystallizable, immunoglobulin constant region
FCAS	Familial cold autoinflammatory syndrome
FcRn	Neonatal Fc receptor
FcγR	Fcγ receptor
FDA	(US) Food and Drug Administration
FDC	Food, Drug, and Cosmetic (Act)
FeNO	Fractional exhaled nitric oxide
FEV1	Forced expiratory volume in 1 s
FEV_1	Mean forced expiratory volume in 1 s
FFF	Field-flow fractionation
FGF	Fibroblast growth factor
FGF-2	Fibroblast growth factor 2
FH	Familial hypercholesterolemia
FIH	First-in-human
FIM	Flow imaging microscopy
FISH	Fluorescence in situ hybridization
FL	Follicular lymphoma
FLT3	FMS-like tyrosine kinase 3
FMEA	Failure mode effects analysis
FMF	Familial Mediterranean fever
FN	Febrile neutropenia
FP	Forward primer
FPI	Full prescribing information
FSC	Forward scatter
FSGS	Focal segmental glomerulosclerosis
FSH	Follicle-stimulating hormone
FSH-CTP	FSH-C-terminal peptide
FTET	Frozen-thawed embryo transfer
FTIR	Fourier transform infrared spectroscopy
Fv	Variable fragment
FVC	Forced vital capacity

FVIII, FIX, etc.	Factor VIII, factor IX, etc.
G	Guanine
GAD	Glutamic acid decarboxylase
gag	Group-specific antigen gene
GalNAc	N-acetyl galactosamine
Gas	Gs alpha subunit
GC	Gastric cancer
GCA	Giant cell arteritis
G-CSF	Granulocyte colony-stimulating factor
G-CSFR	Granulocyte colony-stimulating factor receptor
gCV	Geometric coefficient of variation
GD2	Disialoganglioside expressed on tumors of neuroectodermal origin
GHBP	Growth hormone-binding proteins
GHD	Growth hormone deficiency
GHR	hGH receptor
GHRH	Growth hormone-releasing hormone
GI	Gastrointestinal tract
Gla	γ-Carboxyglutamic acid
GlcNAc	N-acetylglucosamine
GLP($-$1RA)	Glucagon-1-like peptide (receptor agonist)
GLP-1	Glucagon-like peptide 1
GM	Gluteus maximus site
GM-CSF	Granulocyte macrophage colony-stimulating factor
GMP	Good manufacturing practices
GNA	Glycol nucleic acid
GnRH	Gonadotropin-releasing hormone
GOI	Gene of interest
GPP	Generalized pustular psoriasis
GSD II	Glycogen storage disease II
GSK-3β	Glycogen synthase kinase-3β
GST	Glutathione S-transferase
GTAC	Gene therapy advisory committee
GTMP	Gene therapy medicinal product
GTP	Guanosine triphosphate
GvHD	Graft-versus-host disease
HACA	Human antichimeric antibodies
HAE	Hereditary angioedema
HAHA	Human antihuman antibodies
HAMA	Human anti-mouse antibodies
HAT (Medium)	Hypoxanthine-aminopterin-thymidine (medium)
Hb	Hemoglobin
HBsAg	Hepatitis B surface antigen
HBV	Hepatitis B virus
HC	Heavy chain
hCG	Human chorionic gonadotropin
HCP	Host cell protein
HDI	HER2 dimerization inhibitor
HDL	High density lipoproteins
HDR	Homology-directed repair
HDX-MS	Hydrogen-deuterium exchange
HEK	Human embryonic kidney
HEK293	Human embryonic kidney cells 293

HER1	ErbB1 transmembrane tyrosine kinase inhibitor; also called human epidermal growth factor receptor1
HER2	ErbB2 transmembrane tyrosine kinase inhibitor; also called human epidermal growth factor receptor 2
HF	Hollow fiber
hFVIII-SQ	SQ form of human coagulation factor VIII
(h)G-CSF	(human) Granulocyte colony stimulating factor
HGF	Hematopoietic growth factor
hGH	Human growth hormone
HGNC	HUGO Gene Nomenclature Committee
HGPRT	Hypoxanthine-guanine phosphoribosyltransferase
Hib	Haemophilus influenzae type b
HIC	Hydrophobic interaction chromatography
HIDS/MKD	Hyperimmunoglobulin D syndrome/mevalonate kinase deficiency
HiSCR	Hidradenitis suppurativa clinical response
HIV	Human immunodeficiency virus
HLA	Human leukocyte antigen
HLH	Hemophagocytic Lymphohistiocytosis
HMA	Heads of medicines agencies
HMGB1	High-mobility group box 1
hMSC	Human mesenchymal stromal cell
HMWP	High-molecular-weight protein
HOS	Higher order structures
HPAEC-PAD	High-performance anion-exchange chromatography with pulsed amperometric detection
HPLC	High performance liquid chromatography
HPRT	Hypoxanthine guanine phosphoribosyl transferase
HP-SEC	High performance size exclusion chromatography
HPV	Human papillomavirus
HR	Hazard ratio
HS	Hidradenitis suppurativa
HSA	Human serum albumin
HSC	Hematopoietic stem cell
HSR	Hypersensitivity reactions
HSV-1	Herpes simplex virus type 1
HSV-TK Mut2	Herpes simplex I virus thymidine kinase
i.v.	Intravenous
IBD	Inflammatory bowel disease
IC50	Concentration at 50% of maximum inhibition
ICD	Immunogenic cell death
ICER	Incremental cost-effectiveness ratio
ICH	International Conference of Harmonization or International Council for Harmonization of Technical Requirements for Pharmaceuticals for Human Use
ICI	Immune checkpoint inhibitor
icIEF	Imaged capillary isoelectric focusing
ICS	Inhaled corticosteroid
ICSI	Intracytoplasmic sperm injection
ICUR	Incremental cost-utility ratio
IDO	Indoleamine 2,3-dioxygenase
IEC	Ion-exchange chromatography
IEF	Isoelectric focusing
IER	Institute for Clinical and Economic Review

IFN	Interferon
IFNAR (1)	Interferon-alpha/beta receptor (alpha chain)
Ig	Immunoglobulin
IGA	Investigator's Global Assessment
IgE	Immunoglobulin E
IGF	Insulin-like growth factor
IGF-1	Insulin-like growth factor 1
IgG	Immunoglobulin G
IgG1	Immunoglobulin G1
IgG2	Immunoglobulin G2
IgG3	Immunoglobulin G3
IgG4	Immunoglobulin G4
IHC	Immunohistochemistry
IL	Interleukin
IL-1Ra	Il-1 receptor antagonist
IL-5	Interleukin-5
IL-6	Interleukin-6
ILD	Interstitial lung disease
IM	Intramuscular
IMS-MS	Ion-mobility spectrometry-MS
INN	International non-proprietary name
IO	Immune oncology
IP	Intraperitoneal
IPN	Intravenous parenteral nutrition
IP-RP	Ion-pair reversed-phase
iPSC	Induced pluripotent stem cell
IRA	Inflation Reduction Act
irAEs	Immune-related adverse events
IRES	Internal ribosomal entry site
IRR	Infusion-related reaction
irRC	Immune-related response criteria
irTCP	Immune-related thrombocytopenia
ISS	Idiopathic short stature
ITP	Immune thrombocytopenia purpura
ITR	Inverted terminal repeats
IU	International units
IV	Intravenous
IVF	In vitro fertilization
IVIG	Intravenous immunoglobulin
IVT mRNA	In vitro transcribed messenger RNA
IVT	In vitro transcription
JAK	Janus kinase
JAKi	Janus kinase inhibitors
JAK-STAT	Janus activated kinase/signal transducer and activator of transcription
JCV	John Cunningham virus
JIA	Juvenile idiopathic arthritis
k_a	Absorption rate constant
k_{app}	Apparent absorption rate constant
K_d	Binding affinity constant
K_D	Equilibrium dissociation rate constant
kDa	Kilodalton
kGy	Kilogray

Km	Michaelis–Menten constant
KRAS	Family member of the RAS proto-oncogene protein family
LA	License application
LAG-3	Lymphocyte activation gene-3
LC	Light chain
LC	Liquid chromatography
LC/MS/MS	Liquid chromatography tandem mass spectrometry
LC/MS/MS	Liquid chromatography/mass spectrometry/mass spectrometry
LC-DMB	Liquid chromatography-1,2-diamino-4,5-methylenedioxybenzene-2HCl
LDL(R)	Low-density lipoprotein (receptor)
LFA	Lymphocyte function-associated antigen
LFA-3/Fc	Human leukocyte function antigen-3 fusion protein
LH	Luteinizing hormone
LHRH	Luteinizing hormone-releasing hormone
LIF	Leukemia inhibitory factor
LN	Lupus nephritis
LNA	Locked nucleic acid
LNGFR	Low-affinity nerve growth factor receptor
LNP	Lipid nanoparticle
LO	Light obscuration
LON	Late-onset neutropenia
LPLD	Lipoprotein lipase deficiency
LPS	lipopolysaccharide
LRP	Low-density lipoprotein receptor-related protein
LSM	Least square mean
LTR	Long-terminal repeats
M cell	Microfold cell
M	Mass spectrometry
MA	Marketing authorization
MAA	Marketing authorization application
mAb	Monoclonal antibody
mAbs	Monoclonal antibodies
MACE	Major adverse cardiovascular events
MACI	Matrix-assisted chondrocyte implantation
MACS	Magnetic activated cell sorting
MAdCAM-1	Mucosal addressing cell adhesion molecule-1
MAGE-A3	Melanoma-associated antigen 3
MALDI	Matrix-assisted laser desorption ionization
MALS	Multi-angle lights scattering
MAPK	Mitogen-activated protein kinase
MBP	Maltose-binding protein
MCB	Master cell bank
MCS	Multiple cloning site
M-CSF	Macrophage colony-stimulating factor
MDA	Melanoma differentiation-associated gene
MDCK	Madin Darby canine kidney
MDR	Multidrug resistance
MDS	Myelodysplastic syndrome
MDSCs	Myeloid-derived suppressor cells
MEK	Mitogen-activated protein kinase
MET	MET receptor tyrosine kinase
MHC	Major histocompatibility complex
MHRA	Medicines and Healthcare products Regulatory Agency (UK)

miRNA	microRNA
MIU	Million international units
mL	Milliliter
MLD	Metachromatic leukodystrophy
MM	Multiple myeloma
MMAD	Mass median aerodynamic diameter
MMAE	Monomethyl auristatin E
MMAF	Monomethyl auristatin F
MMDs	Monthly migraine days
MMEJ	Microhomology-mediated end-joining
MMF	Mycophenolate mofetil
MMR	Measles-mumps-rubella (combination vaccine)
MOA	Mechanisms of action
MoA	Mode of action
MoMLV	Moloney murine leukemia virus
mPEG	Methoxy-polyethylene glycol
MPS	Mononuclear phagocyte system
MRI	Magnetic resonance imaging
mRNA	Messenger ribonucleic acid
MS	Mass spectrometer/metry
MS	Multiple sclerosis
MS4A	Membrane-spanning 4-domains A4A
MSC	Mesenchymal stromal cell
MSX	Methionine sulfoximine
MT	Methyltransferase
MTX	Methotrexate
mVar	Murine variable
Mw	Molecular weight
MWS	Magnesium wasting syndrome
MWS	Muckle-Wells syndrome
MxA protein	Myxovirus resistance protein 1
n	Hill coefficient
NA	Information not available or not found
NAADP	Nicotinic acid adenine dinucleotide phosphate
NAD	Nicotinamide adenine dinucleotide
NBB	Nucleotide-based biologic
NCA	National competent authority
nCas9	Nickase
NCBI	National Center for Biotechnology Information
ncRNA	Noncoding RNA
NDA	New drug application
NETD	Negative electron transfer dissociation
NF-κB	Nuclear factor kappa B
NGS	Next generation sequencing
NHEJ	Nonhomologous end-joining
NHL	Non-Hodgkin's lymphoma
NIH	National Institute of Health
NK cell	Natural killer cell
NLRP-3	Nucleotide-binding domain, leucine-rich-containing family, pyrin domain-containing 3
NMOSD	Neuromyelitis optica spectrum disorder
NMR	Nuclear magnetic resonance spectroscopy
NMSC	Non-melanoma skin cancer

NNT	Number needed to treat
NO	Nitric oxide
NOAEL	No observable adverse effect level
NOMID	Neonatal-onset, multisystem, inflammatory disorder
NONMEM	Nonlinear mixed effect modeling
NOR	Normal operating range
NOS	Not otherwise specified
NPH	Neutral Protamine Hagedorn
NPL	Neutral protamine lispro
NR	Information not reported
nr-axSpA	Non-radiographic axial spondyloarthritis
NRP	Neuropilin
NRS	Numerical rating scale
NSAIDs	Nonsteroidal anti-inflammatory drugs
NSCLC	Non-small-cell lung cancer
NTA	Nanoparticle tracking analysis
OAS	Oligoadenylate synthetase
OBDS	On-body delivery systems
OBRR	Office of Blood Research and Review
OCS	Oral corticosteroids
OCTGT	Office of Cellular, Tissues, and Gene Therapies
OHSS	Ovarian hyperstimulation syndrome
OLE	Open label extension
ON	Oligonucleotide
OpenAPS	Open artificial pancreas system
OPEX	Operating expenditures
ORF	Open reading frame
ORi	Origin of replication
OS	Overall survival
OVRR	Office of Vaccines Research and Review
P	Pharmacokinetic parameter
PAI-1	Plasminogen activator inhibitor-1
PAM	Protospacer adjacent motif
PAMAM	Polyamidoamine
PAMP	Pathogen-associated molecular pattern
PAR	Proven acceptable range
PASI	Psoriasis Activity and Severity Index
PBAE	Poly(β-amino ester) polymers
PBD	Pyrrolobenzodiazepine
PBPK	Physiologically based pharmacokinetic
PCR	Polymerase chain reaction
PCSK	Proprotein convertase subtilisin/kexin
PCSK9	Proprotein convertase subtilisin/kexin type 9
PCV	Pneumococcal conjugate vaccine
PD	Pharmacodynamics
PD-1	Programmed cell death protein 1
PDA	Parenteral drug association
PDGFRα	Platelet-derived growth factor receptor α
PD-L1	Programmed death ligand 1
pDMAEMA)	Poly[(2-dimethylamino) ethyl methacrylate
PDMDD	Pharmacodynamic-mediated drug disposition
pDNA	Plasmid DNA
PE	Prime editors/editing

PedACR	Pediatric American College of Rheumatology Criteria
PEG	Polyethylene glycol
Peg-MGDF	Pegylated megakaryocyte growth and development factor
pegRNA	Engineered prime editing guide RNA
PEI	Polyethyleneimine
PEPT1	Human peptide transporter 1
PEPT2	Human peptide transporter 2
PFS	Prefilled syringe
PFS	Progression-free survival
PHS	Public Health Service
pI	Isoelectric point
PI3K	Phosphatidylinositol 3-kinase
PIDs	Primary immune deficiencies
PIGF	Phosphatidylinositol-glycan biosynthesis class F protein
pit-hGH	Pituitary hGH
pJIA	Polyarticular JIA
PK	Pharmacokinetic
PK/PD (modeling)	Pharmacokinetic-pharmacodynamic (modeling)
PKA	Protein kinase A
PLGA	Polylactic acid–polyglycolic acid
PlGF	Placenta growth factor
PLL	Poly(L-lysine)
PML	Progressive multifocal leukoencephalopathy
PNA	Peptide nucleic acid
pol	Polymerase gene
Poly(A)	Poly(adenosine monophosphate)
Poly-IC	Polyinosinic and polycytidylic acid
Poly-ICLC	Polylysine-complexed poly-IC
PRCA	Pure red cell aplasia
PRES	Posterior reversible encephalopathy syndrome
PRP	Platelet-rich plasma
PRR	Pattern recognition receptor
PS	Phosphorothioate
PSA	Prostate-specific antigen
PsA	Psoriatic arthritis
PSCs	Pluripotent stem cells
PSEN1 or 2	Presenilin 1 or 2
PSO	Plaque psoriasis
PTLD	Posttransplant lymphoproliferative disease/disorder
PTM	Posttranslational modification
PVDF	Polyvinylidene difluoride
PWS	Prader-Willi syndrome
Q2W	Biweekly
Q3W	Every 3 weeks
QALY	Quality-adjusted life-year
QbD	Quality by design
QC	Quality control
QCM	Quartz crystal microbalance
QD	Once a day
QLT	Quarterly
QM	Monthly
qPCR	Quantitative polymerase chain reaction
QW	Every week

RA	Rheumatoid arthritis
rAAT	Alfa1-antitrypsin
RAMP	Receptor activity-modifying protein
RANKL	Receptor activator of nuclear factor kappa-B ligand
RAS	RAS proto-oncogene protein family
Rb	Retinoblastoma
RBC	Red blood cell
RCL	Replication-competent lentivirus
RCP	Receptor component protein
RCT	Randomized clinical trial
rDNA	Recombinant DNA
RECIST	Response evaluation criteria in solid tumors
REMS	Risk evaluation and mitigation strategy
RES	Reticuloendothelial system
rhEPO	Recombinant human erythropoietin
rhGH	Recombinant human growth hormone
rhIFNα-2b	Recombinant human interferon α-2b
rhIFNβ	Recombinant human interferon β
RIC	Radio immune conjugate
RISC	RNA-induced silencing complex
RIT	Radioimmunotherapeutic
RLD	Reference listed drug
RMM	Resonance mass measurement
RNA	Ribonucleic acid
RNAi	RNA interference
RNP	Ribonucleoprotein
ROW	Other countries (rest of world)
RP	Recurrent pericarditis
RP	Reverse primer
RPE	Retinal pigmented epithelial cells
RP-HPLC	Reverse phase high-performance liquid chromatography
RPLS	Reversible posterior leukoencephalopathy syndrome
RR	Risk ratio
RSV	Respiratory syncytial virus
RSV	Rous sarcoma virus
RT	Reverse transcriptase
RT-PCR	Reverse transcriptase polymerase chain reaction
RWD	Real-world data
s.c.	Subcutaneous
sALCL	Systemic anaplastic large cell lymphoma
saRNA	Self-amplifying RNA
SARS-CoV	Severe acute respiratory syndrome coronavirus
SBS	Short bowel syndrome
SC	Subcutaneous
SCC	Single-chamber cartridge
SCF	Stem cell factor
scFv	Single-chain variable fragment
SCNT	Somatic cell nuclear transfer
SCTMP	Somatic cell therapy medicinal products
SD	Standard deviation
SDR	Specificity determining residues
SDS	Sodium dodecyl sulfate
SDS-PAGE	Sodium dodecyl sulfate–polyacrylamide gel electrophoresis

SEC	Size exclusion chromatography
SELEX	Systematic evolution of ligands by exponential enrichment
SEM/EDX	Scanning Electron Microscopy (SEM) with Energy Dispersive X-Ray Analysis
SFS	Sustained follicle stimulant
SGA	Small for gestational age
sgRNA	Single guide RNA
SHOX	Short stature homeobox-containing gene
SID-1	Systemic RNA Interference Deficiency-1
SIN	Self-inactivating
SIP	Steam-in-place
siRNA	Short interfering RNA
SIRS	Systemic inflammatory response syndrome
sJIA	Systemic JIA
SLAMF7	Signaling lymphocytic activation molecule family member 7
SLE	Systemic lupus erythematosus
SLP	Synthetic long peptide
SMA	Spinal muscular atrophy/dystrophy
SMase	Sphingomyelin phosphodiesterase
SmPC	Summary of product characteristics
SOCS	Suppressors of cytokine signaling
SOP	Standard operating procedure
SPR	Surface plasma resonance
SRIF	Somatotropin release-inhibiting factor
SRP	Signal recognition particle
SSC	Side scatter
SSc-ILD	Systemic sclerosis-associated interstitial lung disease
ssRNA	Single-stranded RNA
STC	Standard template chemistry
SUB	Single use bioreactor
SUF	Single use fermentor
SvP	Sub-visible particles
T	Thymidine
T1/2	Half-life
$t_{1/2}$	Terminal half-life
T1DM	Type 1 diabetes
T2DM	Type 2 diabetes
TALEN	Transcription activator-like effector nucleases
T-ALL	T-cell acute lymphoblastic leukemia
TB	Tuberculosis
Tc	Collapse temperature or temperature condensor
TC	Trigeminal complex
TCR	T-cell receptor
TCS	Topical corticosteroids
T-DM1	Trastuzumab emtansine
T-DXd	Trastuzumab deruxtecan
Te	Eutectic temperature
TEAE	Treatment-emergent adverse event
$T_{early\ 1/2\ max}$	Time to early half maximum concentration
TEP	Tissue-engineered products
TETA	Triethylenetetramine
TF	Tissue factor
TFF	Tangential flow filtration

Tfh	T-follicular helper cell
TG	Trigeminal ganglion
Tg'	Glass transition temperature of frozen liquid
TGF	Transforming growth factor
TGF-beta	Transforming growth factor beta
TGVS	Trigeminovascular system
Th	T helper (T-cell, cytokine)
TID	Thrice a day
TIGIT	T-cell immunoreceptor with Ig and ITIM domains
TILs	Tumor-infiltrating lymphocytes
TIM-3	T-cell immunoglobulin domain and mucin domain-3
TIW	Three times a week
TKY2i	Tyrosine kinase 2 inhibitors
TLR	Toll-like receptor
t_{max}	Time of the maximum concentration
Tmax	Time to peak drug concentration
TMDD	Target-mediated drug disposition
TME	Tumor microenvironment
TNBC	Triple-negative breast cancer
TNF	Tissue necrosis factor
TNF-alpha	Tumor necrosis factor-alpha
TNFRSF17	Tumor necrosis factor receptor (TNFR) superfamily 17
TNFα	Tumor necrosis factor alpha
TNFβ	Lymphotoxin-α
TNSALP	Tissue nonspecific alkaline phosphatase
TO%	Total relative uptake
TOP I	DNA topoisomerase 1
Tp	Temperature sublimating product
TP	Therapeutic protein
t-PA	Tissue plasminogen activator
TPO	Thrombopoietin
TPOR	Thrombopoietin receptor
TPP	Therapeutic protein product
TRA	Thrombopoietin receptor agonist
tracrRNA	Trans-activating RNA
TRAPS	Tumor necrosis factor receptor-associated periodic syndrome
Treg	Regulatory T cell
tRNA	Transfer RNA
Trop-2	Tumor-associated calcium signal transducer 2
Trop-2	Trophoblast cell surface antigen 2
TRPM6	Transient receptor potential cation channel, under family M member 6
Ts	Shelf temperature
TS	Turner syndrome
TSE	Transmissible spongiform encephalopathies
TSH	Thyroid-stimulating hormone
TSLP	Thymic stromal lymphopoietin
TSS	Transcription start sites
TTP	Time to progression
TTR	Transthyretin
UC	Ulcerative colitis
UF/DF	Ultrafiltration/diafiltration
UGT1A1	Uridine diphosphate glucuronosyltransferase 1A1

UPLC	Ultra performance liquid chromatography
US	United States
USP	Upstream processing
UV	Ultraviolet
UV	Uveitis
UVPD	Ultraviolet photodissociation
V1	Volume of distribution of central compartment
V2	Volume of distribution of peripheral compartment
Vc	Volume of central compartment
VCAM-1	Vascular cell adhesion molecule-1
Ve	Excluded volume
VEGF	Vascular endothelial growth factor
VEGF-R	Vascular endothelial growth factor receptor
VH	Heavy chain variable domain
VL	Light chain variable domain
VL	Vastus lateralis
V_L, V_H	Variable light and variable heavy chain of monoclonal antibody
VLP	Virus-like particles
VLS	Vascular leak syndrome
Vmax	Maximum clearance rate
Vss	Volume of distribution at steady state
VSV	Vesicular stomatitis virus
VSVg	Vesicular stomatitis virus G
VWF	Von Willebrand factor
Vz	Volume of distribution during the terminal phase
W	Body weight
WCB	Working cell bank
WCE	Weak cation exchange
WHO	World Health Organization
WPRE	Woodchuck hepatitis posttranslational regulatory element
WT	Wild type
XNA	Xeno nucleic acids
X-SCID	X-linked severe combined immunodeficiency
ZAP-70	Zeta-chain-associated protein kinase 70
ZFNs	Zinc-finger nucleases
α2-PI	α2-Plasmin inhibitor
Ψ	Packaging signal

Part I

Pharmaceutical Biotechnology: The Science, Techniques, and Important Concepts

Molecular Biotechnology: From DNA Sequence to Therapeutic Protein

Olivier G. de Jong and Ronald S. Oosting

Introduction

Proteins are already used for more than 100 years to treat or prevent diseases in humans. It started in the early 1890s with "serum therapy" for the treatment of diphtheria and tetanus by Emile von Behring and others. The antiserum was obtained from immunized rabbits and horses. Behring received the Nobel Prize for Medicine in 1901 for this pioneering work on passive immunization. A next big step in the development of therapeutic proteins was the use of purified insulin isolated from pig or cow pancreas for the treatment of diabetes type I in the early 1920s by Banting and Best (in 1923 Banting received the Nobel Prize for this work). Soon after the discovery of insulin, the pharmaceutical company Eli Lilly started large-scale production of the pancreatic extracts for the treatment of diabetes. Within 3 years after the start of the experiments by Banting, already enough animal-derived insulin was produced to supply the entire North American continent. Compare this to the present time-to-market of a new drug (from discovery to approval) of 10–15 years (Van Norman 2016). Note: Both the European Medicines Agency (EMA) and the US Food and Drug Administration (FDA) have accelerated approval procedures to speed up the market introduction of new drugs with "meaningful advantages" over the available therapies.

Thanks to advances in biotechnology (e.g., recombinant DNA technology, hybridoma technology), we have moved almost entirely away from animal-derived proteins to proteins with the complete human amino acid sequence.

Such therapeutic human proteins are less likely to cause side effects and to elicit immune responses. Banting and Best were very lucky. They had no idea about possible sequence or structural differences between human and porcine/bovine insulin. Nowadays, we know that porcine insulin differs only by one amino acid from the human sequence and bovine insulin differs by three amino acids (see Fig. 1.1). Thanks to this high degree of sequence conservation, porcine/bovine insulin can be used to treat human patients. In 1982, human insulin became the first recombinant human protein approved for sale in the USA (also produced by Eli Lilly) (cf. Chap. 16). Since then, a large number of biopharmaceuticals have been developed. There are now several hundreds of human proteins marketed for a wide range of therapeutic areas.

O. G. de Jong (✉)
Department of Pharmaceutics, Utrecht Institute for Pharmaceutical Sciences, Utrecht University, Utrecht, The Netherlands
e-mail: o.g.dejong@uu.nl

R. S. Oosting
Utrecht University, Utrecht, The Netherlands
e-mail: r.s.oosting@amrif.eu

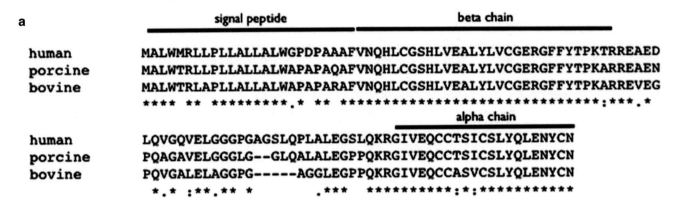

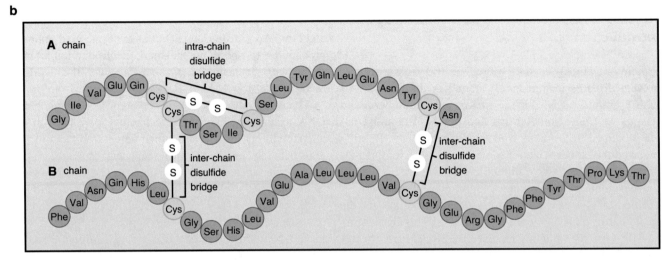

Fig. 1.1 (a) Multiple alignment (http://www.ebi.ac.uk/Tools/msa/clustalw2) of the amino acid sequences of human, porcine, and bovine prepro-insulin.(*): identical residue. (b) Schematic drawing of the structure of insulin. The alpha and beta chain are linked by two disulfide bridges. Both the one-letter and three-letter codes for the amino acids are used in this figure: alanine (Ala, A), arginine (Arg, R), asparagine (Asn, N), aspartic acid (Asp, D), cysteine (Cys, C), glutamic acid (Glu, E), glutamine (Gln, Q), glycine (Gly, G), histidine (His, H), isoleucine (Ile, I), leucine (Leu, L), lysine (Lys, K), methionine (Met, M), phenylalanine (Phe, F), proline (Pro, P), serine (Ser, S), threonine (Thr, T), tryptophan (Trp, W), tyrosine (Tyr, Y), and valine (Val, V) (Figure b is taken from Wikipedia)

Pharmaceutical Biotechnology, Why This Book, Why This Chapter?

In this book, we define pharmaceutical biotechnology as all technologies needed to produce biopharmaceuticals (other than (non-genetically modified) animal- or human blood-derived medicines). Attention is paid both to these technologies and the products thereof. Biotechnology makes use of findings from various research areas, such as molecular biology, biochemistry, cell biology, genetics, bioinformatics, microbiology, bioprocess engineering, and separation technologies. Progress in these fields has been and will remain a major driver for the development of new biopharmaceuticals. Biopharmaceuticals form a fast-growing segment in the world of medicines opening new therapeutic options for patients with severe diseases. This success is also reflected by the fast growth in global sales.

Until recent years, biopharmaceuticals were primarily proteins, but therapeutic DNA or RNA- based molecules (think about gene therapy products, DNA/ RNA vaccines, and RNA interference-based products; Chaps. 13, 14, and 15) have now become part of our therapeutic arsenal. Examples of this development include patisaran (Onpattro®): the first FDA-approved small interfering RNA (siRNA)-based drug to treat transthyretin-mediated amyloidosis, and onasemnogene abeparvovec (Zolgensma®): an FDA-approved adeno-associated virus vector-based gene therapy that delivers a fully functional copy of the SMN1 gene to motor neuron cells of spinal muscular atrophy (SMA) patients. The most prominent examples are the RNA-based vaccines that were brought to the market at an unprecedented speed and scale in response to the coronavirus disease 2019 (COVID-19) pandemic, including the mRNA vaccines tozinameran (produced by Pfizer-BioNTech, Comirnaty®) and elasomeran/imelasomeran (produced by Moderna,

Spikevax®) as well as the adenoviral (DNA-based) vaccines produced by AstraZeneca (COVID-19 vaccine ChAdOx1-S [recombinant], Covishield®/Vaxzevria®) and Janssen (COVID-19 vaccine Ad26.COV2-S [recombinant], Jcovden®)" that contain the gene of the SARS-CoV-2 Spike protein. These topics will be discussed in more detail in Chaps. 14 and 15.

Therapeutic proteins differ in many aspects from classical, small molecule drugs. They differ in size, composition, production, purification, contaminations, side effects, stability, formulation, regulatory aspects, etc. These fundamental differences justify paying attention to therapeutic proteins as a separate family of medicines. These general aspects are discussed in the first 15 chapters of this book. After those general topics, the different subfamilies of biopharmaceuticals are dealt with in detail. This first chapter should be seen as a chapter where many of the basic elements of the selection, design, and production of biopharmaceuticals are touched upon. For further detailed information, the reader is referred to relevant literature (see Recommended Reading at the end of this chapter) and other chapters in this book.

From an In Silico DNA Sequence to a Therapeutic Protein

We will discuss now the steps and methods needed to select, design, and produce a recombinant therapeutic protein (see Fig. 1.2). We will not discuss in detail the underlying biological mechanisms. The reader is referred to Box 1.1, for a short description of the central dogma of molecular biology, which describes the flow of information from DNA via RNA into a protein.

Selection of a Therapeutic Protein

The selection of what protein should be developed for a treatment of a particular disease is often challenging, with lots of uncertainties. This is why most big pharmaceutical companies only become interested in a certain product when there is some clinical evidence that the new product actually works and that it is safe. This business model gives opportunities for startup biotech companies and venture capitalists to engage in this important early development process.

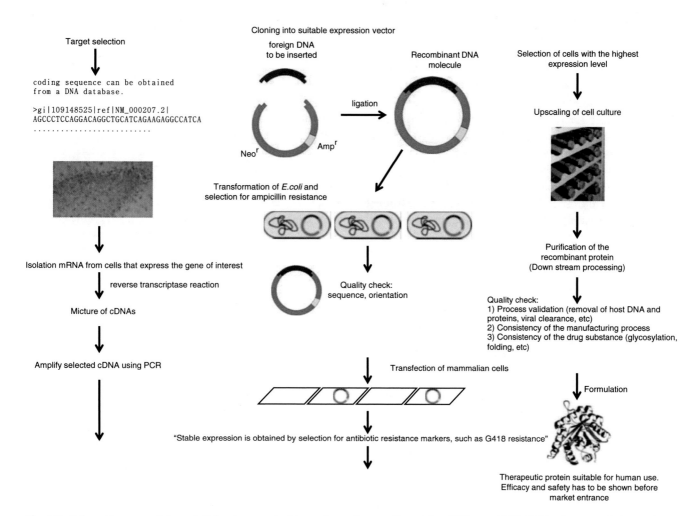

Fig. 1.2 Schematic representation of all the steps required to produce a therapeutic protein. *cDNA* copy DNA, *PCR* polymerase chain reaction

Sometimes the choice for a certain protein as a therapeutic drug is simple. Think, for instance, about replacement of endogenous proteins such as insulin and erythropoietin for the treatment of diabetes type I and anemia, respectively. For many other diseases, it is much more difficult to identify an effective therapeutic protein or target. For instance, an antibody directed against a growth factor receptor on a tumor cell may look promising based on in vitro and animal research but may be largely ineffective in human cancer patients.

It is beyond the scope of this chapter to go further into the topic of therapeutic protein and target discovery. For further information, the reader is referred to the large number of scientific papers on this topic, as can be searched using PubMed (http://www.ncbi.nlm.nih.gov/pubmed).

In the rest of this chapter, we will mainly focus on a typical example of the steps in the molecular cloning process and production of a therapeutic protein. At the end of this chapter, we will briefly discuss the cloning and large-scale production of monoclonal antibodies (see also Chap. 8).

Molecular cloning is defined as the assembly of recombinant DNA molecules (most often from two different organisms) and their replication within host cells.

DNA Sequence
The DNA, mRNA, and amino acid sequence of every protein in the human genome can be obtained from publicly available gene and protein databases, such as those present at the National Center for Biotechnology Information (NCBI) in the USA and the European Molecular Biology Laboratory (EMBL). Their websites are http://www.ncbi.nlm.nih.gov/ and http://www.ebi.ac.uk, respectively.

DNA sequences in these databases are always given from the 5′ end to the 3′ end and protein sequences from the amino- to the carboxy-terminal end (see Fig. 1.3). These databases also contain information about the gene (e.g., exons, introns, and regulatory sequences, see Box 1.1 for explanations of these terms) and protein structure (domains, specific sites, posttranslational modifications, etc.). The

presence or absence of certain posttranslational modifications determines what expression hosts (e.g., *Escherichia coli* (*E. coli*), yeast, or a mammalian cell line) may be used (see below).

Selection of Expression Host
Recombinant proteins can be produced in bacteria, yeasts, plants (e.g., rice and tomato), mammalian cells, and even by transgenic animals. All these expression hosts have different pros and cons (see Chap. 4 for more details).

Most marketed therapeutic proteins are produced in cultured mammalian cells. In particular Chinese hamster ovary (CHO) cells are used (more than 70% of marketed proteins are produced in CHO cells). However, for the production of therapeutic glycoproteins, in particular those with complex human glycosylation patterns, the most commonly used cell lines are Human Embryonic Kidney 293 (HEK293) and HT-1080 fibrosarcoma cell lines (Lalonde 2017). Various derivatives of the HEK293 cell line have been produced to increase protein production as well as scalability. For example, 293-F and 293-H cells are commercially generated HEK293 cell lines that can be grown in suspension, allowing fast growth and large-scale production. Another example is the adherent HEK293T cell line which, due to the permanent integration of a viral protein that facilitates plasmid replication, is capable of maintaining a high number of plasmid DNA copies resulting in increased levels of protein expression. On first sight, mammalian cells are not a logical choice. They are much more difficult to culture than, for instance, bacteria or yeast. On average, mammalian cells divide only once every 24 h, while cell division in *E. coli* takes ~20 min and in yeast ~1 h. In addition, mammalian cells need expensive growth media and in many cases bovine (fetal) serum as a source of growth factors (see Table 1.1 for a comparison of the various expression systems and Table 4.1). Since the outbreak of the bovine or transmissible spongiform encephalopathy epidemic (BSE/TSE, better known as mad cow disease) in cattle in the United Kingdom, the use of bovine serum for the production of therapeutic proteins is consid-

Fig. 1.3 DNA sequences are always written from the 5′ → 3′ direction and proteins sequences from the amino-terminal to the carboxy-terminal. (**a**) Nucleotide code, (**b**) amino acid code, see legend to Fig. 1.1 or Table Fig. 2.3

a
```
>gi|109148525|ref|NM_000207.2| Homo sapiens insulin (INS), transcript
variant 1, mRNA
5'AGCCCTCCAGGACAGGCTGCATCAGAAGAGGCCATCAAGCAGATCACTGTCCTTCTGCCATGGCCCTGT
GGATGCGCCTCCTGCCCCTGCTGGCGCTGCTGGCCCTCTGGGGACCTGACCCAGCCGCAGCCTTTGTGAAC
CAACACCTGTGCGGCTCACACCTGGTGGAAGCTCTCTACCTAGTGTGCGGGGAACGAGGCTTCTTCTACAC
ACCCAAGACCCGCCGGGAGGCAGAGGACCTGCAGGTGGGGCAGGTGGAGCTGGGCGGGGGCCCTGGTGCAG
GCAGCCTGCAGCCCTTGGCCCTGGAGGGGTCCCTGCAGAAGCGTGGCATTGTGGAACAATGCTGTACCAGC
ATCTGCTCCCTCTACCAGCTGGAGAACTACTGCAACTAGACGCAGCCCGCAGGCAGCCCCACACCCGCCGC
CTCCTGCACCGAGAGAGATGGAATAAAGCCCTTGAACCAGCAAAA 3'
```

b
```
>gi|4557671|ref|NP_000198.1| insulin preproprotein [Homo sapiens]
(NH2)MALWMRLLPLLALLALWGPDPAAAFVNQHLCGSHLVEALYLVCGERGFFYTPKTRREAEDLQVGQV
ELGGGPGAGSLQPLALEGSLQKRGIVEQCCTSICSLYQLENYCN-(COOH)
```

Table 1.1 Pros and cons of different expression hosts

	Prokaryotes	Yeast	Mammalian cells
	E. coli	*Pichia pastoris, Saccharomyces cerevisiae*	e.g. CHO or HEK293 cells
+	Easy manipulation Rapid growth Large-scale fermentation Simple media High yield	Grows relatively rapidly Large-scale fermentation Performs some posttranslational modifications	May grow in suspension, perform all required posttranslational modifications
–	Proteins may not fold correctly or may even aggregate (inclusion bodies) Almost no posttranslational modifications	Posttranslational modifications may differ from humans (especially glycosylation)	Slow growth Expensive media Difficult to scale up Dependence on serum (BSE)

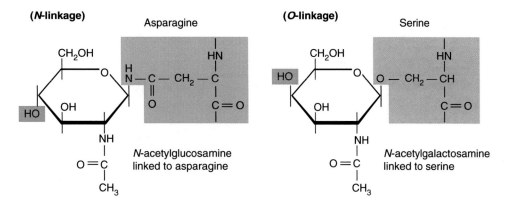

Fig. 1.4 Glycosylation takes place either at the nitrogen atom in the side chain of asparagine (N-linked) or at the oxygen atom in the side chain of serine or threonine (O-linked). Glycosylation of asparagine takes place only when this residue is part of an Asn-X-Ser or Ans-X-Thr sequence (X can be any residue except proline). Not all potential sites are glycosylated. Which sites become glycosylated depends also on the protein structure and on the cell type in which the protein is expressed

ered a safety risk by the regulatory authorities (such as the EMA in Europe and the FDA in the USA). To minimize the risk of transmitting TSE via a medicinal product, bovine serum has to be obtained from animals in countries with the lowest possible TSE risk, e.g. the USA, Australia, and New Zealand. However, because of the inherent risk of using animal-derived products, serum-free culture media containing recombinant growth factors are increasingly used.

The main reason why mammalian cells are used as production platform for therapeutic proteins is that in these cells posttranslational modification (PTM) of the synthesized proteins most closely resembles the human situation. An important PTM is the formation of disulfide bonds between two cysteine moieties. Disulfide bonds are crucial for stabilizing the tertiary structure of a protein. Wild-type *E. coli* is only able to produce disulfide bonds in the periplasm, the space between the inner cytoplasmic membrane and the bacterial outer membrane. Cytosolic expression of recombinant protein in *E. coli* is needed for high level expression and proper folding. By knocking out two major reducing enzymes: thioredoxin reductase and glutathione reductase, *E. coli* strains have been produced that are able to express disulfide

bonds. Unfortunately, the yield of recombinant disulfide bonded proteins by such strains is usually low. However, new developments in this area are very promising (e.g. co-expression of the disulfide bonded protein with sulfhydryl oxidase and disulfide bond isomerase (Gaciarz et al. 2017)) and may ultimately result in an *E. coli*-based expression system suited for high level expression of therapeutic proteins with disulfide bonds.

Another important PTM of therapeutic proteins is glycosylation. Around 70% of all marketed therapeutic proteins, including monoclonal antibodies, are glycosylated. Glycosylation is the covalent attachment of oligosaccharides to either asparagine (N-linked) or serine/threonine (O-linked) (see Fig. 1.4). The oligosaccharide moiety of a therapeutic protein affects many of its pharmacological properties, including stability, solubility, bioavailability, in vivo activity, pharmacokinetics, and immunogenicity. Glycosylation differs between species, between different cell types within a species, and even between batches of therapeutic proteins produced in cell culture. N-linked glycosylation is found in all eukaryotes (and also in some bacteria, but not in wildtype *E. coli* (Nothaft and Szymanski 2010)) and takes

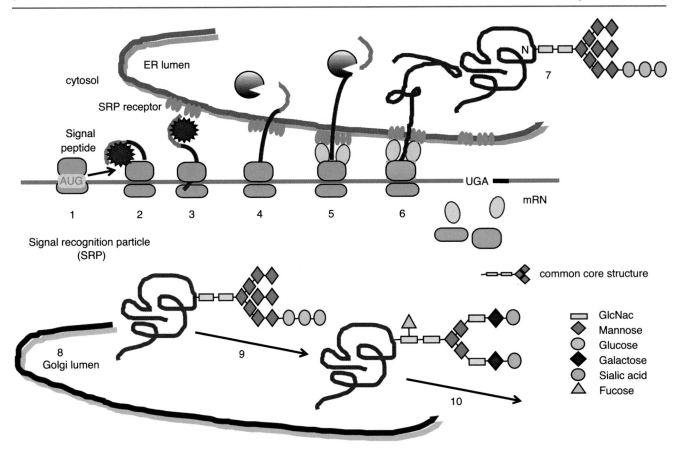

Fig. 1.5 Schematic drawing of the N-linked glycosylation process as it occurs in the endoplasmic reticulum (ER) and Golgi system of an eukaryotic cell. (*1*) The ribozyme binds to the mRNA and translation starts at the AUG start codon. The first ~20 amino acids form the signal peptide. (*2*) The signal recognition particle (SRP) binds the signal peptide. (*3*) Next, the SRP docks with the SRP receptor to the cytosolic side of the ER membrane. (*4*) The SRP is released and (*5*) the ribosomes dock onto the ER membrane. (*6*) Translation continues until the protein is complete. (*7*) A large oligosaccharide (activated by coupling to dolichol phosphate) is transferred to the specific asparagine (N) residue of the growing polypeptide chain. (*8*) Proteins in the lumen of the ER are transported to the Golgi system. (*9*) The outer carbohydrate residues are removed by glycosidases. Next, glycosyltransferases add different carbohydrates to the core structure. The complex type carbohydrate structure shown is just an example out of many possible varieties. The exact structure of the oligosaccharide attached to the peptide chain differs between cell types and even between different batches of therapeutic proteins produced in cell culture. (*10*) Finally, secretory vesicles containing the glycoproteins are budded from the Golgi. After fusion of these vesicles with the plasma membrane, their content is released into the extracellular space

place in the lumen of the endoplasmic reticulum and the Golgi system (see Fig. 1.5). All N-linked oligosaccharides have a common pentasaccharide core containing three mannose and two *N*-acetylglucosamine (GlcNAc) residues. Additional sugars are attached to this core. These maturation reactions take place in the Golgi system and differ between expression hosts. In yeast, the mature glycoproteins are rich in mannose, while in mammalian cells much more complex oligosaccharide structures are possible. To make yeast more suitable for expressing human glycosylated proteins, yeast strains with humanized glycosylation pathways have been developed. Unfortunately, the protein yields with these strains were too low and the level of glycosylation too heterogenous (Wells and Robinson, 2017). O-linked glycosylation takes place solely in the Golgi system. Interestingly, over recent years efforts have been made in the development of bacterial protein production systems that include both

N-linked and O-linked glycosylation by introducing yeast and/or human-derived glycosylation genes (Secreters, 2022). *E. coli*-based systems do not express endogenous glycosylation machinery. This is an advantage as there is no competition with non-human glycosylation processes, like the high-mannose structures that are observed in yeast. However, as is also the case for yeast, production levels of glycosylated proteins in bacteria require significant improvement and optimization before they are economically competitive with current eukaryotic cell based protein production systems.

CopyDNA
The next step is to obtain the actual DNA that codes for the protein. This DNA is obtained by reverse-transcribing the mRNA sequence into copyDNA (cDNA). To explain this process, it is important to discuss first the structure of a mammalian gene and mRNA.

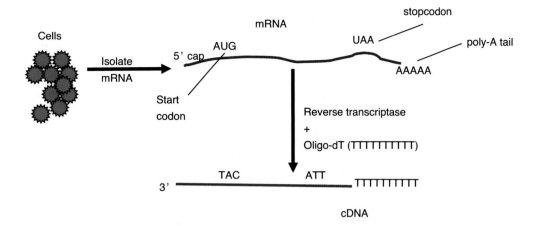

Fig. 1.6 Reverse transcriptase reaction

Most mammalian genes contain fragments of coding DNA (exons) interspersed by stretches of DNA that do not contain protein-coding information (introns). Messenger RNA synthesis starts with the making of a large primary transcript. Then, the introns are removed via a regulated process, called splicing. The mature mRNA contains only the exon sequences. Most mammalian mRNAs contain also a so-called poly-A "tail," a string of 100–300 adenosine nucleotides. These adenines are coupled to the mRNA molecule in a process called polyadenylation. Polyadenylation is initiated by binding of a specific set of proteins at the polyadenylation site at the end of the mRNA. The poly-A tail is important for transport of the mRNA from the nucleus into the cytosol, for translation, and it protects the mRNA from degradation.

An essential tool in cDNA formation is reverse transcriptase (RT). This enzyme was originally isolated from retroviruses. These viruses contain an RNA genome. After infecting a host cell, their RNA genome is reverse-transcribed first into DNA. The finding that RNA can be reverse-transcribed into DNA by RT is an important exception of the central dogma of molecular biology (as discussed in Box 1.1).

To obtain the coding DNA of the protein, one starts by isolating (m)RNA from cells/tissue that expresses the protein. Next, the mRNA is reverse-transcribed into copyDNA (cDNA) (see Fig. 1.6). The RT reaction is performed in the presence of an oligo-dT (a single-stranded oligonucleotide containing ~18 thymidines). The oligo-dT binds to the poly-A tail and reverse transcriptase couples deoxyribonucleotides complementary to the mRNA template, to the 3′ end of the growing cDNA. In this way, a so-called library of cDNAs is obtained, representing all the mRNAs expressed in the starting cells or tissue.

The next step is to amplify specifically the cDNA for the protein of interest using the polymerase chain reaction (PCR, see Fig. 1.7). A PCR reaction uses a (c) DNA template, a forward primer, a reverse primer, deoxyribonucleotides (dATP, dCTP, dGTP, and dTTP), Mg^{2+}, and a thermostable DNA polymerase. DNA polymerase adds free nucleotides

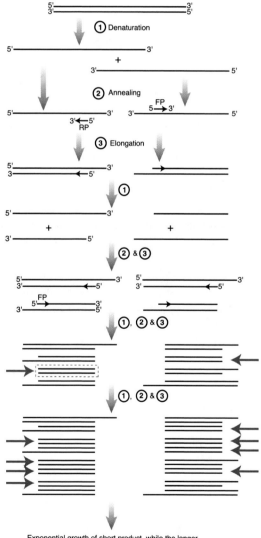

Fig. 1.7 The PCR process. (*1*) DNA is denatured at 94–96 °C. (2) The temperature is lowered to ±60 °C. At this temperature, the primers bind (anneal) to their target sequence in the DNA. (3) Next, the temperature is raised to 72 °C, the optimal temperature for Taq polymerase. Four cycles are shown here. A typical PCR reaction runs for 30 cycles. The *arrows* point to the desired PCR product

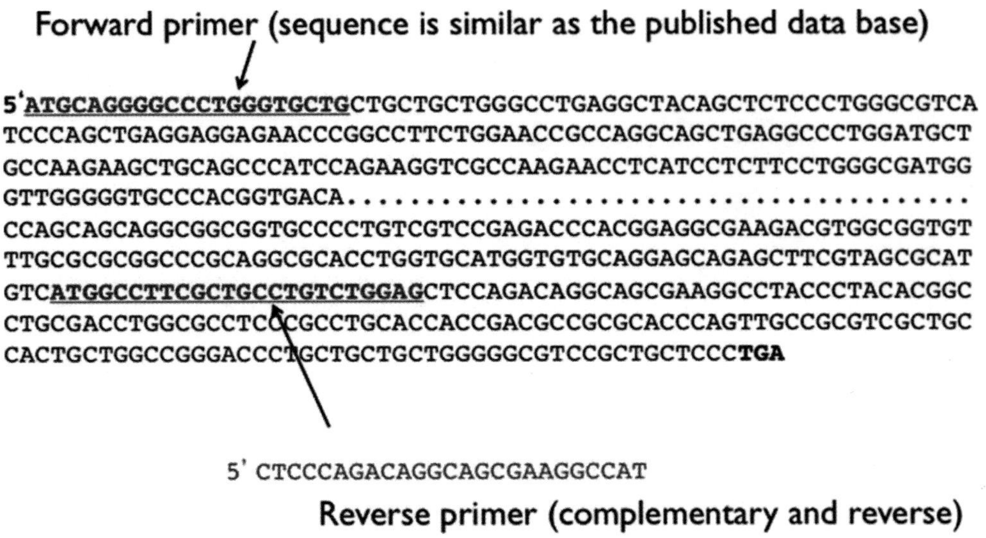

Fig. 1.8 PCR primer design (see Question 4)

Forward primer (sequence is similar as the published data base)

5'<u>ATGCAGGGGCCCTGGGTGCTG</u>CTGCTGCTGGGCCTGAGGCTACAGCTCTCCCTGGGCGTCA
TCCCAGCTGAGGAGGAGAACCCGGCCTTCTGGAACCGCCAGGCAGCTGAGGCCCTGGATGCT
GCCAAGAAGCTGCAGCCCATCCAGAAGGTCGCCAAGAACCTCATCCTCTTCCTGGGCGATGG
GTTGGGGGTGCCCACGGTGACA. . . . .
CCAGCAGCAGGCGGCGGTGCCCCTGTCGTCCGAGACCCACGGAGGCGAAGACGTGGCGGTGT
TTGCGCGCGGCCCGCAGGCGCACCTGGTGCATGGTGTGCAGGAGCAGAGCTTCGTAGCGCAT
GTC<u>ATGGCCTTCGCTGCCTGTCTGGAG</u>CTCCAGACAGGCAGCGAAGGCCTACCCTACACGGC
CTGCGACCTGGCGCCTCCCGCCTGCACCACCGACGCCGCGCACCCAGTTGCCGCGTCGCTGC
CACTGCTGGCCGGGACCCTGCTGCTGCTGGGGGGCGTCCGCTGCTCCC**TGA**

5' CTCCCAGACAGGCAGCGAAGGCCAT

Reverse primer (complementary and reverse)

only to the 3' end of the newly forming strand. This results in elongation of the new strand in a 5' → 3' direction. DNA polymerase can add a nucleotide only to a preexisting 3'-OH end, and therefore it needs a primer at which it can add the first nucleotide. PCR primers are single-stranded oligonucleotides around 18–30 nucleotides long, flanking opposite ends of the target DNA (see Fig. 1.8). The PCR is usually carried out for 30 cycles. Each cycle consists of three stages: a denaturing stage at ~94 °C (the double-stranded DNA is converted into single-stranded DNA), a primer annealing stage at ~60 °C (the optimal annealing temperature depends on sequences of the primers and template), and an extension stage commonly at 72 °C. Temperatures may vary among polymerases. Theoretically, the amount of DNA should double during each cycle. A 30-cycle-long PCR should therefore result in a 2^{30} fold (~10^9) increase in the amount of DNA. In practice this is never reached. In particular at later cycles the efficiency of the PCR reaction reduces, as specific contents of the reaction buffer such as nucleotides or primers are used up and their concentrations become a limiting factor. After a PCR reaction, the PCR product is then isolated and the other components of the PCR reaction are removed (buffers, nucleotides, enzymes, and primers and other impurities). A common approach to clean PCR products is by using silica membranes that bind DNA in high-salt containing media. As these membranes do not efficiently bind DNA molecules smaller than 100 nucleotides, primers are immediately removed. After binding the DNA product, the membranes can then be treated with various washing buffers and the PCR product can subsequently be eluted (isolated) from the membrane using a low-salt elution buffer. Alternatively, the PCR product can be loaded onto an agarose gel, followed by electrophoresis alongside molecular weight standards. The specific PCR product can then be isolated from the agarose gel and purified using silica membranes or precipitation pro-

Fig. 1.9 A hot spring in Yellowstone National Park. In hot springs like this one, Archaea, the bacterial source of thermostable polymerases, live

tocols. This approach is required when the PCR reaction yields products from off-target sequences, or when incomplete/duplex products are formed, as the molecular weight standards can be used to isolate the PCR product with the correct size.

PCR uses a thermostable DNA polymerase. These polymerases were obtained from Archaea living in hot springs such as those occurring in Yellowstone National Park (see Fig. 1.9) and at the ocean bottom. DNA polymerases make mistakes. When the aim is to clone and express a PCR product, a thermostable DNA polymerase should be used with 3' → 5' exonuclease "proofreading activity." Innovations in this area still continue. The reader is referred to websites of companies, such as New England Biolabs, Roche, Invitrogen, and others, to look for the latest additions to the PCR toolbox.

Box 1.1 The Central Dogma of Molecular Biology

The central dogma of molecular biology was first stated by Francis Crick in 1958 and deals with the information flow in biological systems and can best be summarized as "DNA makes RNA makes protein" (this quote is from Marshall Nirenberg who received the Nobel Prize in 1968 for deciphering the genetic code). The basis of the information flow from DNA via RNA into a protein is pairing of complementary bases; thus, adenine (A) forms a base pair with thymidine (T) in DNA or uracil (U) in RNA and guanine (G) forms a base pair with cytosine (C).

To make a protein, the information contained in a gene is first transferred into a RNA molecule. RNA polymerases and transcription factors (these proteins bind to regulatory sequences on the DNA, such as promoters and enhancers) are needed for this process. In eukaryotic cells, genes are built of exons and introns. Intron sequences (the term intron is derived from intragenic region) are removed from the primary transcript by a highly regulated process which is called splicing. The remaining mRNA is built solely of exon sequences and contains the coding sequence or sense sequence. In eukaryotic cells, transcription and splicing take place in the nucleus.

The next step is translation of the mRNA molecule into a protein. This process starts by binding of the mRNA to a ribosome. The mRNA is read by the ribosome as a string of adjacent 3-nucleotide-long sequences, called codons. Complexes of specific proteins (initiation and elongation factors) bring aminoacylated transfer RNAs (tRNAs) into the ribosome-mRNA complex. Each tRNA (via its anticodon sequence) base pairs with its specific codon in the mRNA, thereby adding the correct amino acid in the sequence encoded by the gene. There are 64 possible codon sequences. Sixty-one of those encode for the 20 possible amino acids. This means that the genetic code is redundant (see Table 1.2). Translation starts at the start codon AUG, which codes for methionine and ends at one of the three possible stop codons: UAA, UGA, or UAG. The nascent polypeptide chain is then released from the ribosome as a mature protein. In some cases, the new polypeptide chain requires additional processing to make a mature protein.

Table 1.2 The genetic code

1st base		2nd base				3rd base
		U	C	A	G	
U	Phe	Ser	Tyr	Cys	U
	Phe	Ser	Tyr	Cys	C
	Leu	Ser	Stop	Stop	A
	Leu	Ser	Stop	Trp	G
C	Leu	Pro	His	Arg	U
	Leu	Pro	His	Arg	C
	Leu	Pro	Gln	Arg	A
	Leu	Pro	Gln	Arg	G
A	Ile	Thr	Asn	Ser	U
	Ile	Thr	Asn	Ser	C
	Ile	Thr	Lys	Arg	A
	Met	Thr	Lys	Arg	G
G	Val	Ala	Asp	Gly	U
	Val	Ala	Asp	Gly	C
	Val	Ala	Glu	Gly	A
	Val	Ala	Glu	Gly	G

Synthetic DNA Production

There are additional ways to acquire DNA sequences of potential genes of interest, alongside obtaining the sequence from a host cell. The last decades have seen a rapid development in gene synthesis techniques and commercial availability. Nowadays, there is a large number of companies that offer synthesis of synthetic double-stranded DNA (dsDNA) molecules ranging in size from 100 to 3000 basepairs. Although there are some limitations to the techniques used to synthesize and validate synthetic sequences, such synthetic dsDNA molecules can be designed and produced to contain the sequences of interest. The first synthesis of a DNA molecule was the sequence of a 77 nucleotide tRNA sequence in the late 1960s, which took more than 5 years to complete (Agarwal et al. 1970). These days, column- and microarray-based phosphoramidite synthesis chemistry allows for fast high-throughput synthetic dsDNA production. Worldwide many different companies exist that are able to supply DNA sequences to a researcher within a manner of days to weeks. The time required depends on the length of the sequence.

As mentioned, these synthetic DNA molecules are generally generated through phosphoramidite synthesis chemistry (Fig. 1.10). In this technique, DNA molecules attached to a solid matrix are elongated with one nucleotide at a time. This process starts with a chemically modified nucleoside (for context, a nucleoside is a nucleotide missing a phosphate group), a so-called dimethoxytrityl (DMT)-protected nucleoside, attached to a solid surface. The DMT group is then removed using trichloroacetic acid resulting in a free hydroxyl group, allowing for the coupling of an additional nucleoside. When another DMT-protected nucleoside is added, this results in the coupling of a single nucleoside at a time, since no other nucleosides can be added to the chain until the DMT group of the newly added nucleoside is removed. By selectively adding nucleoside phosphoramidites (A, C, G, and T) in the right order, a molecule with the

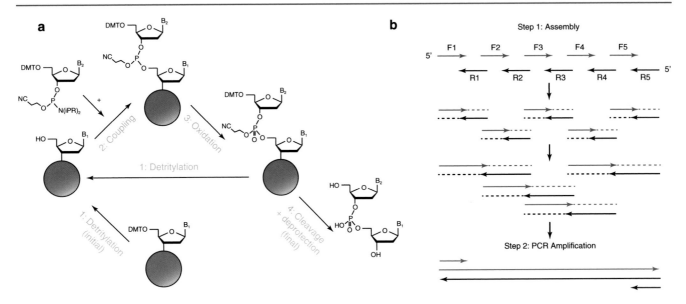

Fig. 1.10 Production of synthetic DNA fragments by phosphoramidite synthesis chemistry. (**a**) The phosphoramidite oligonucleotide synthesis starts with a single DMT-modified nucleoside attached to a solid surface. Per synthesis cycle, a single DMT-modified nucleoside is added, followed by an oxidation and a detritylation step. Once the oligonucleotide sequence is completed, it can be uncoupled from the solid surface and its final protective groups can be removed. (**b**) To produce longer double-stranded DNA fragments, a variety of shorter oligonucleotides with overlapping sequences are mixed in a one-pot reaction. These oligonucleotides are taken from either strand of the double-stranded DNA sequence (*F:* forward orientation, *R:* reverse orientation on the complementary strand) with partially overlapping sequences at the 5'- and 3' ends. In this reaction, the oligonucleotides progressively generate full-length molecules via a process called assembly PCR. After this process, the full-length product is amplified by PCR amplification using primers that target the 5'- and 3' ends of the complete construct

desired sequence of choice can be generated. In between each single nucleoside elongation step, several additional chemical reactions are performed to stabilize the chain (capping), and generating a phosphotriester linkage (oxidation), thereby forming an oligonucleotide structure. After the desired oligonucleotide sequence is produced, it can be cleaved from the solid matrix and the final protective groups can be removed. For an in-depth review on the process, the reader is referred to the review from Hughes and Ellington (2017).

This technique does have limitations and chance of generating error-free full-length oligonucleotide chains decreases as the oligonucleotide chain length increases. Therefore, fragments are generally synthesized with a length of up to 100 nucleotides. In order to generate dsDNA fragments up to 3000 base pairs, a collection of these shorter single stranded oligonucleotides is designed that have short overlapping sequences with oligonucleotides on the opposite strand. These overlapping oligonucleotides are then mixed together and assembled in a PCR-based assembly method (for a thorough explanation of this process, see the review from Czar et al. 2009). To save time and resources, these oligonucleotides are often assembled in a single, so-called "one-pot" reaction, rather than assembling these oligonucleotides sequentially. Finally, after this assembly procedure is completed, products may then be amplified through traditional PCR reactions, as described above.

Alternative Sources for Cloned DNA Sequences

With the expansion of both commercial and academic research toward the development of biologicals, various companies, as well as non-profit initiatives for academic research purposes, directly supply plasmids with gene open reading frame sequences, as well as services that provide sequences for specific gene knockdown or knockout constructs. In many cases, a range of plasmid characteristics are available, such as suitability for lentiviral production, transient transfection, or compatibility with advanced cloning techniques such as gateway cloning (explained in this chapter, section "Cloning PCR Products into an Expression Vector"). Whereas commercial purchases of such sequences may be costly, the amount of time, effort, and personnel costs that would otherwise be invested in the generation of such sequences should also be taken into consideration.

Alongside commercial services, there are also several non-profit plasmid repositories such as the European Plasmid Repository (EPA), DNASU, or Addgene. These repositories allow sharing of DNA plasmids for nonprofit research purposes. The most rapidly growing nonprofit repository is Addgene (www.addgene.org), which was founded in 2004 with the mission to "Accelerate research and discovery by improving access to useful research materials and information." This repository allows research labs to submit and request plasmid sequences, and currently lists over 100,000 plasmids from over 5000 nonprofit research labs. These plasmids do not only include sequences of genes but also include a variety of research tools such as plasmids for gene expression, gene knockdown, genomic engineering, viral vectors, and fluorescent proteins. Whereas most of the plasmids that are available on Addgene are currently solely purposed for academic research, there is a growing collection of plasmids and tools that are becoming available for commercial purposes.

Box 1.2 Plasmids and Enzyme Toolbox

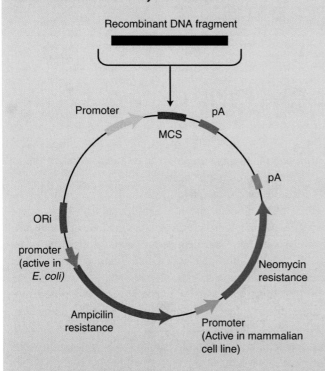

Schematic drawing of an expression plasmid for a mammalian cell line

Plasmids are self-replicating circular extrachromosomal DNA molecules. The plasmids used nowadays in biotechnology are constructed partly from naturally occurring plasmids and partly from synthetic DNA. The figure above shows a schematic representation of a plasmid suitable for driving protein expression in a mammalian cell. The most important features of this plasmid are:

1. An origin of replication (ORi). The ORi allows plasmids to replicate separately from the host cell's chromosome.
2. A multiple cloning site (MCS). The MCS contains recognition sites for a number of restriction enzymes. The presence of the MCS in plasmids makes it relatively easy to transfer a DNA fragment from one plasmid into another.
3. Antibiotic-resistance genes. All plasmids contain a gene that makes the recipient *E. coli* resistant to an antibiotic, in this case resistant to ampicillin. Other antibiotic-resistance genes that are often used confer resistance to tetracycline and Zeocin®. The expression plasmid also contains the neomycin resistance gene. This selection marker enables selection of those mammalian cells that have been transfected with the plasmid, or that have integrated the DNA sequence into their chromosome. The protein product of the neomycin resistance gene inactivates the toxin G418 or Geneticin®.
4. Promoter to drive gene expression. Many expression vectors for mammalian cells contain the cytomegalovirus (CMV) promoter, which is taken from the cytomegaloma virus and is constitutively active. To drive recombinant protein expression in other expression hosts, other plasmids with other promoter sequences have to be used.
5. Poly (A) recognition site (pA). This site becomes part of the newly produced mRNA and binds a protein complex that adds a poly-A tail to the 3′ end of the mRNA. Expression vectors that are used to drive protein expression in *E. coli* do not contain a poly-A recognition site.

Besides the use of multiple promoters for the expression of multiple genes (as is shown in the figure above), there are additional methods to express multiple genes from a single plasmid. Multicistronic vectors contain mRNA with multiple coding areas (open reading frames). Additional features that are commonly included in multicistronic vectors are internal ribosomal entry sites (IRES) and 2A sequences.

An *IRES* is a secondary RNA structure that allows for initiation of translation of a protein on a secondary site of an mRNA molecule. This means that both a gene of interest (GOI) and an antibiotic-resistant gene can be translated from a single RNA molecule, which is expressed from a single promoter. It should be noted that the translation of a protein following an IRES is substantially lower than the expression of a GOI directly at the start of an mRNA. However, an advantage of this approach is that, since both the GOI and the antibiotic resistance gene are under the same promoter, loss of the transcription of the GOI mRNA will also result in the loss of the antibiotic resistance gene.

2A peptides are ~20 amino acid self-cleaving peptide sequences that are found in a variety of viral genomes. 2A sequences consist of amino acid sequences that "self-cleave" due to the inability of the ribosome to make a peptide bond between two specific amino acids during translation as a result of a challenging amino acid sequence. When this cleavage occurs, the ribosome will, however, not terminate protein translation. This means that, while the ribosome translates a single open reading frame from an mRNA molecule, the end product is cleaved into multiple separate proteins. As all proteins are derived from the same translation process, expression levels are less variable than those expressed from an IRES.

Molecular biology enzyme toolbox

DNA polymerase produces a polynucleotide sequence against a nucleotide template strand using base-pairing interactions (G against C and A against T). It adds nucleotides to a free 3'OH, and thus it acts in a 5' → 3' direction. Some polymerases also have 3' → 5' exonuclease activity (see below), which mediates proofreading.

Reverse transcriptase (RT) is a special kind of DNA polymerase, since it requires an RNA template instead of a DNA template.

Restriction enzymes are endonucleases that bind specific recognition sites on DNA and cut both strands.

Restriction enzymes can either cut both DNA strands at the same location (blunt end) or they can cut at different sites on each strand, generating a single-stranded end (better known as a sticky end).

Examples:

HindIII	5'A^{a}AGCTT	XhoI	5'C^{a}TCGAG
	3'TTCGA^{a}A		3'GAGCT^{a}C
KpnI	5'GGTAC^{a}C	EcoRV	5'GAT^{a}ATC
	3'C^{a}CATGG		3'CTA^{a}TAG
NotI	5'GC^{a}GGCCGC	PacI	5'TTAAT^{a}TAA
	3'CGCCGG^{a}CG		3'AAT^{a}TAATT

^{a} Location where the enzyme cuts

DNA ligase joints two DNA fragments. It covalently links the 3'-OH of one strand with the 5'-PO$_4$ of the other DNA strand. The linkage of two DNA molecules with complementary sticky ends by ligase is much more efficient than blunt-end ligation.

Alkaline phosphatase. A ligation reaction of a blunt-end DNA fragment into a plasmid also with blunt ends will result primarily in empty plasmids, being the result of self-ligation. Treatment of a plasmid with blunt ends with alkaline phosphatase, which removes the 5'-PO$_4$ groups, prevents self-ligation.

Exonucleases remove nucleotides one at a time from the end (exo) of a DNA molecule. They act, depending on the type of enzyme, either in a 5' → 3' or 3 → 5' direction and on single- or double-stranded DNA. Some polymerases also have exonuclease activity (required for proofreading). Exonucleases are used, for instance, to generate blunt ends on a DNA molecule with either a 3' or 5' extension.

Cloning DNA Sequences into an Expression Vector

There are several ways to clone a PCR product. One of the easiest ways is known as TA cloning (see Fig. 1.11). TA cloning makes use of the property of Taq polymerase to add a single adenosine to the 3′ end of a PCR product. Such a PCR product can subsequently be ligated (using DNA ligase, see Box 1.1) into a plasmid with a 5′ thymidine overhang (see Box 1.2 for a general description of expression plasmids). A disadvantage of TA and blunt-end cloning is that directional cloning is not possible, so the PCR fragment can be cloned either in the sense or antisense direction (see Fig. 1.11). An additional issue of TA cloning is that Taq polymerase does not contain a so-called proofreading domain. This proofreading domain searches for, and excises, mismatched nucleotides that are incorporated during the polymerase chain reaction. As a result, this polymerase has an error rate of 0.23%, meaning that on average per PCR cycle 2.3% of 1000 nucleotide base pair products will contain one mutation. As polymerases with proofreading domains have a 10–50 times lower error rate, their use is generally preferred. However, PCR products obtained with a DNA polymerase with proofreading activity have a blunt end, and thus they do not contain the 3′ A overhang. Fortunately, such PCR fragments can easily be A-tailed by incubating for a short period with Taq polymerase and dATP.

Blunt PCR products (meaning PCR products that are double-stranded DNA fragments that do not have single strand nucleotides at the end, see Box 1.2), can directly be cloned into a linearized plasmid with 2 blunt ends. One downside of this approach is that plasmids with two blunt ends can self-ligate without incorporating genomic inserts. To prevent this, the blunt ends of the destination vector can be dephosphorylated (using a phosphatase, see Box 1.1). As ligation of DNA fragments is based on joining a PO_4 (phosphate) group on the 5′ end of a DNA molecule to a 3′ OH (hydroxyl) group, enzymatically removing the 5′ PO_4 groups will make self-ligation impossible. However, ligation can still occur with other DNA fragments that still contain their 5′ PO_4 groups, such as a DNA insert of interest that has not been dephosphorylated (see Fig. 1.12). PCR products do not contain 5′ PO_4 groups. However, 5′ PO_4 groups can be added by T4 kinase enzymes in a process called phosphorylation. Whereas we describe dephosphorylation in the context of blunt-end ligations, it can also be applied for ligation reactions DNA overhangs. It should be noted that the efficiency of blunt-end PCR cloning is much lower than cloning with restriction enzymes

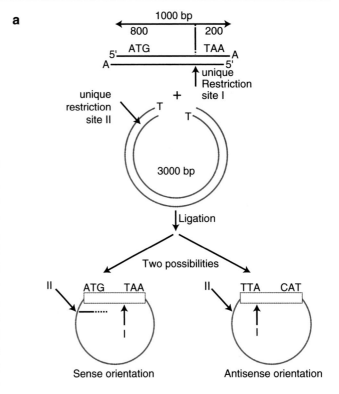

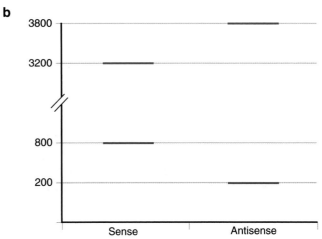

Fig. 1.11 Cloning of a PCR product via TA cloning (**a**). This cloning strategy makes use of the property of Taq polymerase to add an extra A to the 3′ end of the PCR product. To determine the orientation of the insert, the plasmid is cut by enzymes I and II (enzyme I cuts in the insert and enzyme II cuts in the plasmid). On the basis of the obtained fragment size (as determined by agarose electrophoresis), the orientation of the insert can be deduced (**b**)

that leave DNA overhangs (see Box 1.2), since ligation with blunt ends is substantially less efficient than ligation with DNA overhangs.

As a solution to address the lower efficiency of blunt-end cloning, PCR products can also be cloned by adding unique

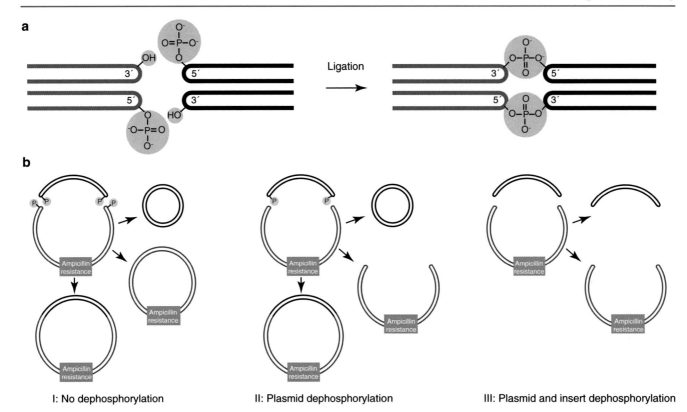

Fig. 1.12 Phosphorylation in DNA ligation. (**a**) Ligation of double-stranded DNA fragments occurs through the enzymatic formation of covalent phosphodiester linkages, from a 3′hydroxyl group and a 5′phosphate group. (**b**) Ligation of two compatible DNA ends (including blunt overhangs) requires the presence of a phosphate group on the 5′ end of at least one of the DNA fragments. I: When a destination plasmid (blue) is restricted with blunt ends (or compatible overhangs) on both restriction sites, it can self-ligate. II: When only the DNA insert (black) is phosphorylated, a circular plasmid can only be formed when the DNA insert is incorporated. III: without the presence of any 5′ phosphate groups, no ligation can occur. Keep in mind that ampicillin resistance genes are not properly expressed from linearized DNA fragments

recognition sites of restriction enzymes to both ends of the PCR product. This can be done by incorporating these sites at the 5′ end of the PCR primers. After the PCR reaction, the PCR product can then be isolated. Whereas this approach requires some additional effort, it is a popular cloning technique as it allows direct incorporation of PCR products into most plasmids, as any recognition site can be placed at the 5′- and 3′ ends of PCR products (see Fig. 1.13). It should be noted that for some restriction enzymes up to 6 extra nucleotides (a so-called "leader sequence") may need to be added in front of the recognition sequence in the primers, as some restriction enzymes do not cut efficiently directly at the end of a linear piece of DNA.

In some cases, it might be desirable to clone DNA sequences into destination vectors without the incorporation or use of restriction enzyme sequences. This could be the case when multiple DNA sequences need to be fused directly together

without the incorporation of additional sequences, for example, to avoid interfering with the open reading frame of a gene. For such purposes, Gibson Assembly Cloning was developed (Gibson et al. 2009). In this process, overlapping DNA fragments are joined in an enzymatic reaction (see Fig. 1.14). To generate these overlapping sequences, long PCR primers are designed to create overlapping DNA ends of the fusion sites of the destination vector and the insert. As a result, the products of these PCR reactions will now have overlapping double-stranded DNA ends. To be able to anneal these sequences, both fragments are treated with a T5 Exonuclease enzyme. This enzyme creates single-strand DNA 3′ overhangs by removing nucleotides from the 5′ end of DNA. These single-stranded overhangs can then anneal with their complementary overhangs, after which a proof-reading polymerase can fill in any missing nucleotides in the annealed regions. Lastly, a DNA ligase can then covalently link these DNA fragments

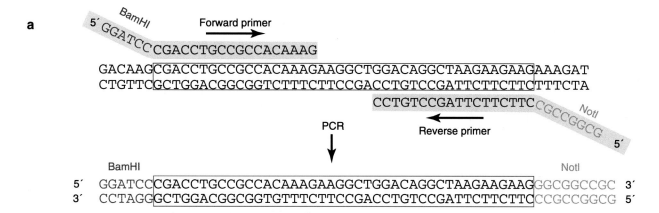

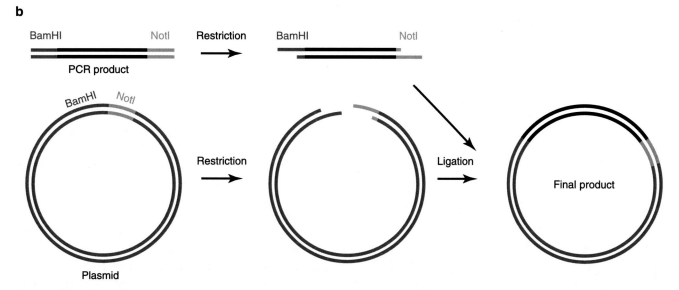

Fig. 1.13 PCR-based restriction enzyme cloning (**a**) Restriction sites can added to both the 5′- and 3′ends on sequences of interest through PCR by incorporating recognition sequences for restriction enzymes at the 5′ end of PCR primers (yellow). Hereafter, PCR products can be digested with restriction enzymes, and ligated into a destination plasmid that has been digested with the same restriction enzymes. In this illustration two separate restriction enzymes, BamHI (red) and NotI (green), are used as an example (**b**)

together (see Fig. 1.12). As this technique can be used to join almost any sequence, does not require the incorporation of additional sequences, and can be used to fuse multiple DNA fragments in one reaction its use has become progressively popular. Whereas the approach sounds complicated, online tools for computer-generated primer design and the availability of commercial reagent mixes for single-tube reactions have made use of this technique increasingly accessible.

Another technique that does not require the use of restriction enzymes is Gateway Cloning. Gateway cloning is based on recombination reactions by enzymes used by bacteriophages to integrate their genome into bacteria. Interestingly,

these bacteriophages use two separate enzymatic reactions to excise and reintegrate their genome from the bacterial genome, as a mechanism to defend themself from bacterial genomic defense mechanisms. These mechanisms have been adapted to efficiently exchange genomic sequences between plasmids, as is explained in Fig. 1.15. First, the gene of interest is flanked with attB sequences by PCR, in a similar process as is shown in Fig. 1.13. This sequence is then transferred into a donor vector that contains attP sequences by the BP clonase enzyme (originally responsible for integration of phage genomic sequences into the bacterial genome), which switches out sequences that are located between attB

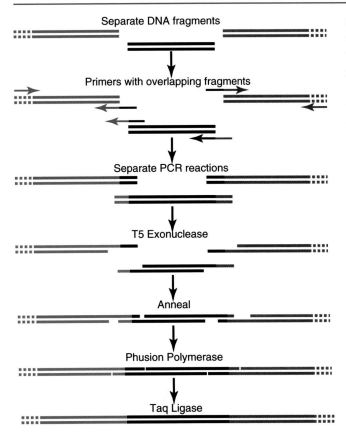

Separate DNA fragments

↓

Primers with overlapping fragments

↓

Separate PCR reactions

↓

T5 Exonuclease

↓

Anneal

↓

Phusion Polymerase

↓

Taq Ligase

Fig. 1.14 Gibson assembly cloning. In this example, three different DNA fragments (blue, black, and red) are directly fused together through Gibson cloning. Overlapping DNA sequences at the fusion sites are generated by PCR. A T5 exonuclease enzyme then removes nucleotides from the 5' end of all fragments, generating single-stranded DNA overhangs. These overhangs are compatible with their adjacent fragments due to the overlapping sequences previously attached by PCR. After temperature-based annealing of the fragments, a proofreading polymerase is added to fill any gaps (Phusion polymerase), and finally a ligase is added to covalently bind the fragments together

sequences and attP sequences. In this process, the sequence between the attP sequences, a gene encoding for the CcdB toxin (*ccdB*) which is toxic for *E. coli*, is removed from the donor vector and replaced with the gene of interest. In this recombination process the attB and attP sequences are also altered, resulting in the formation of attR and attL sequences respectively. This donor vector also contains a kanamycin resistance gene, which remains unaffected by the BP clonase recombination. After this reaction, the reaction mix is introduced into *E. coli* by a process called transformation, followed by kanamycin antibiotic selection. Since expression of

the *ccdB* gene is toxic for *E. coli,* only the bacteria that take up plasmids that have undergone a successful recombination will survive. *E. coli* that did not take up any plasmids are removed by the kanamycin antibiotic selection. As a final step, the gene of interest can then be transferred into any destination vector containing attR sequences by the LR clonase enzyme (originally responsible for excision of the phage genomic sequences from the bacterial genome), which switches out sequences that are located between attL sequences and attR sequences. These destination vectors also contain a *ccdB* sequence, to once again remove plasmids that did not undergo a successful recombination after *E. coli* transformation. As the destination vector contains and ampicillin resistance gene, ampicillin selection will result in the removal of *E. coli* that did not take up any plasmids, or took up any potentially remaining original donor vector (as it contains a kanamycin resistance gene instead). The expression vectors that are finally formed can have a wide variety of functions, such as protein expression in cell culture, bacterial protein production, and even production of lentivirus that contain the gene of interest. For an in-depth review on Gateway cloning and its applications, the reader is referred to the review of Hartley (Hartley et al, 2003). Whereas Gateway cloning is more complicated than restriction enzyme-based techniques, there are some advantages to this technique. The BP- and LR clonase steps are fast, efficient, and even allow for the cloning of multiple fragments in a row in a single reaction, by using modified versions of attB, attL, attP, and attR sequences. Moreover, there is a high variety of Gateway destination vectors available, suitable for a vast number of applications and target organisms. Once a gene of interest has been successfully cloned into a donor vector, it can easily be transferred to any of these vectors with high speed and efficiency.

After the gene of interest has been cloned into its destination vector, the plasmid is introduced into *E. coli* by transformation. There are several ways to transform *E. coli*. Most used are the calcium chloride method (better known as heat shock) and electroporation (the bacteria are exposed to a very high electric pulse). Whatever the transformation method, channels in the membrane are opened through which the plasmid can enter the cell. Next, the bacteria are plated onto an agar plate with an antibiotic. Only bacteria that have taken up the plasmid with an antibiotic-resistant gene, and thus produce a protein that degrades the antibiotic, will survive. As bacteria will not properly express this antibiotic resistance gene

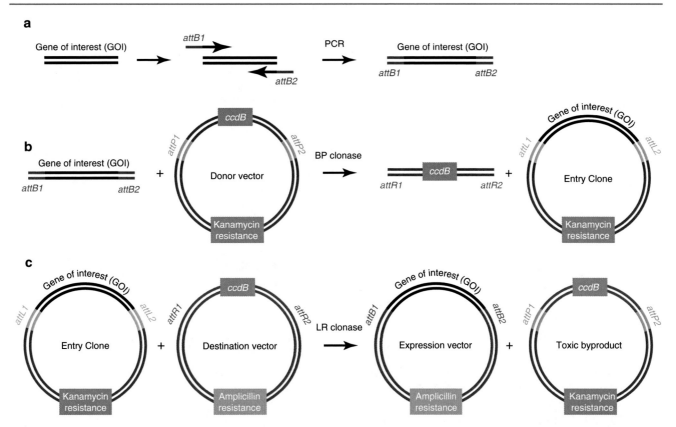

Fig. 1.15 Gateway cloning. First, flanking attB sequences are added around a gene of interest by PCR (**a**). Then, the gene of interest is transferred into a donor vector, containing attP sites, through recombination by BP clonase (recombining attB and attP sequences, resulting in the formation of attL and attR sites), generating a so-called entry clone (**b**). Lastly, the gene of interest can then be placed into any destination vector containing attR sites using LR clonase (recombining attL and attR sequences, resulting in the reformation of attB and attP sites) (**c**). As *ccdB* expresses a toxic protein for *E. coli*, plasmids that did not undergo proper recombination, or toxic byproduct plasmids that are created in the LR clonase reaction, are not replicated in *E. coli* after transformation. Ampicillin selection after the LR clonase recombination ensures that only expression vector plasmids, and not entry clone plasmids, are replicated by the *E. coli*

from a linearized DNA product, only bacteria that have been transfected with a successful ligation product survive. After an overnight incubation at 37 °C, the agar plate will contain a number of colonies. The bacteria in each colony are the descendants of one bacterium and as such all contain the same cloned plasmid copy. Subsequently, aliquots of a number of these colonies are grown overnight in liquid medium at 37 °C. From these cultures, plasmids can be isolated. The next steps will be to determine whether the obtained plasmid preparations contain an insert, and if so, to determine what the orientation is of the insert relative to the promoter that will drive the recombinant protein expression. The orientation can, for instance, be determined by cutting the obtained plas-

mids with a restriction enzyme that cuts only once somewhere in the plasmid and with another enzyme that cuts once somewhere in the insert. On the basis of the obtained fragment sizes (determined via agarose gel electrophoresis using appropriate molecular weight standards), the orientation of the insert in the plasmid can be determined (see Fig. 1.11).

As already discussed above, DNA polymerases make mistakes, and therefore, it is crucial to determine the nucleotide sequence of the cloned PCR fragment. DNA sequencing is a very important method in biotechnology (the developments in high-throughput sequencing have enabled the sequencing of many different genomes, including that of humans) and is therefore further explained in Box 1.3.

Box 1.3 DNA Sequencing

Technical breakthroughs in DNA sequencing, the determination of the nucleotide sequence, permit the sequencing of entire genomes, including the human genome. It all started with the sequencing in 1977 of the 5386-nucleotide-long single-stranded genome of the bacteriophage φX174.

Basic method: Chain-termination method

The most frequently used method for DNA sequencing is the chain-termination method, also known as the dideoxynucleotide method, as developed by Frederick Sanger in the 1970s.

The method starts by creating millions of copies of the DNA to be sequenced. This can be done by isolating plasmids with the DNA inserted from bacterial cultures or by PCR. Next, the obtained double-stranded DNA molecules are denatured, and the reverse strand of one of the two original DNA strands is synthesized using DNA polymerase, a DNA primer complementary to a sequence upstream of the sequence to be determined with normal deoxynucleotidetriphosphates (dNTPs), and *di*deoxyNTPs (ddNTPs) that terminate DNA strand elongation. The four different ddNTPs (ddATP, ddGTP, ddCTP, or ddTTP) miss the 3′OH group required for the formation of a phosphodiester bond between two nucleotides and are each labeled with a different fluorescent dye, each emitting light at different wavelengths. This reaction results in different reverse strand DNA molecules extended to different lengths. Following denaturation and removal of the free nucleotides, primers, and the enzyme, the resulting DNA molecules are separated on the basis of their molecular weight with a resolution of just one nucleotide (corresponding to the point of termination). The presence of the fluorescent label attached to the terminating ddNTPs makes a sequentially read out in the order created by the separation process possible. See also the figures below. The separation of the DNA molecules is nowadays carried out by capillary electrophoresis. The available capillary sequencing systems are able to run in parallel 96 or 384 samples with a length of 600–1000 nucleotides. With the more common 96 capillary systems, it is possible to obtain around six million bases (Mb) of sequence per day.

Advanced methods: High-throughput sequencing

The capillary sequencing systems are hardly used anymore. They are now replaced by alternative systems with a much higher output and at the same time a very strong reduction in the costs. These new high-throughput sequencing technologies parallelize the sequencing process and produce thousands or millions of sequences concurrently. It is beyond the scope of this chapter to go further into the technology of these (extreme) high-throughput sequencing technologies. A good starting point to read more about modern DNA sequencing and the bioinformatics involved are articles at Wikipedia (https://en.wikipedia.org/wiki/DNA_sequencing).

It is important to mention here that all sequencing technologies produce raw data that need to be assembled into longer sequences such as complete genomes (sequence assembly). Of course, this is all done by dedicated software. Very challenging are DNA fragments with repetitive sequences that often prevent complete genome assemblies because they occur in many places of the genome. As a consequence, many sequences may not be assigned to particular chromosomes.

a DNA synthesis in the presence of dNTPs and fluorescently labeled ddNTPs (T,C,A,or G)

```
Target sequence
3' ---GGGTCCAGTGGCAGAGGATTCCGCC
5' ---CCCAGG  →
        ---CCCAGGT
        ---CCCAGGTC
        ---CCCAGGTCA
        ---CCCAGGTCAC
        ---CCCAGGTCACC
        ---CCCAGGTCACCG
```
(primer extension)

b Separation of the synthesized DNA molecules by capillary electrophoresis and read out of fluorescence

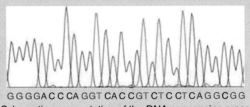

G G GGA C C CA GGT CA C CGT CT C CT CA G G C G G
Schematic representation of the DNA sequencing process

Schematic representation of the basic DNA sequencing process

Transfection of Host Cells and Recombinant Protein Production

Introducing DNA into a mammalian cell is called transfection (and as already mentioned above, transformation in *E. coli*). There are several methods to introduce DNA into a mammalian cell line. Most often, the plasmid DNA is complexed to cationic lipids (as in Lipofectamine, for example) or polymers (such as polyethyleneimines (PEI)) and then pipetted on to the cells. Next, the positively charged aggregates bind to the negatively charged cell membrane and are subsequently endocytosed (see Fig. 1.16). Then, the plasmid DNA has to escape from the endosome and has to find its way into the

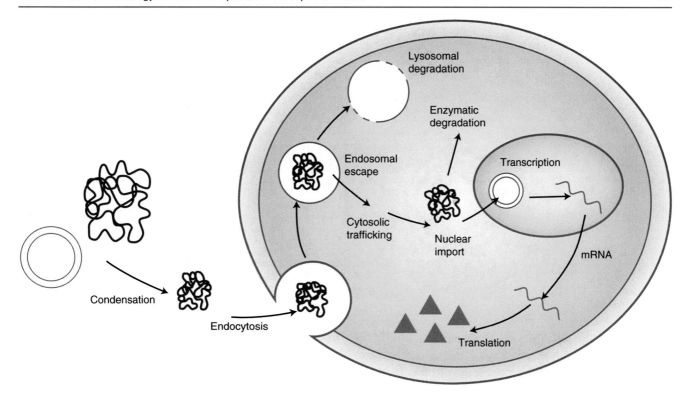

Fig. 1.16 Carrier-mediated transfection of mammalian cells

nucleus where mRNA synthesis can take place. This is actually achieved during cell division when the nuclear membrane is absent. Another way to introduce DNA into the cytosol is through electroporation. During electroporation, an electric pulse is applied to the cells, which results in the formation of small pores in the plasma membrane. Through these pores, the plasmid DNA can enter the cells (Chap. 14).

Transfection leads to transient expression of the introduced gene. The introduced plasmids are rapidly diluted as a consequence of cell division or even degraded. However, it is possible to stably transfect cells leading to long expression periods. Then, the plasmid DNA has to integrate into the chromosomal DNA of the host cell. To accomplish this, a selection gene is normally included into the expression vector, which gives the transfected cells a selectable growth advantage. Only those cells that have integrated the selection marker (and most likely, but not necessary, also the gene of interest) into their genome will survive. There are various selection markers that are used for mammalian cell lines. The most common selection marker is the neomycin resistance gene (Neo^r). This gene codes for a protein that neutralizes the toxic drug Geneticin, also known as G418. Other commonly used selection markers are resistance genes against puromycin, blasticidin, hygromycin, and Zeocin®. These selection antibiotics can also be used in combination, allowing for the simultaneous selection of multiple constructs. The entire selection process takes 1–4 weeks, depending on the chosen selection antibiotic, and results in a

tissue culture dish with several colonies. Each colony contains the descendants of 1 stably transfected cell. Then, the cells from individual colonies have to be isolated and further expanded. The next step will be to quantify the recombinant protein production of the obtained cell cultures and to select those with the highest yields.

Transfection of mammalian cells is a very inefficient process (compared to transformation of *E. coli*) and needs relatively large amounts of plasmid DNA. Integration of the transfected plasmid DNA into the genome is a very rare event. As a typical example, starting with 10^7 mammalian cells, one obtains usually not more than 10^2 stably expressing clones. If higher efficiency is required more efficient techniques, such as the use of lentiviral vectors, can be employed.

Cell Culture

A big challenge is to scale up cell cultures from lab scale (e.g. a 75 cm^2 tissue culture bottle) to a large-scale production platform (such as a bioreactor). Mammalian cells are relatively weak and may easily be damaged by stirring or pumping liquid in or out of a fermenter (shear stress). In this respect, *E. coli* is sturdier, and this bacterium can therefore be grown in much larger fermenters.

A particular problem is the large-scale culturing of adherent (versus suspended) mammalian cells. One way to grow adherent cells in large amounts is on the surface of small beads. After a while the surface of the beads will be completely covered (confluent) with cells, and then it is neces-

Fig. 1.17 Cell culturing in roller bottles

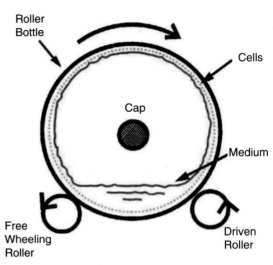

sary to detach the cells from the beads and to re-divide the cells over more (empty) beads and to transfer them to a bioreactor compatible with higher working volumes. To loosen the cells from the beads, usually the protease trypsin is used. It is very important that the trypsinization process is well timed: if it is too short, many cells are still on the beads, and if it is too long, the cells will lose their integrity and will not survive this treatment.

Some companies have tackled the scale-up problem by "simply" culturing and expanding their adherent cells in increasing amounts of roller bottles. These bottles revolve slowly (between 5 and 60 revolutions per hour), which bathes the cells that are attached to the inner surface with medium (see Fig. 1.17). See Chap. 4 in this book for more in-depth information or reviews by Kim et al. (2012) and Kunert and Reinhart (2016).

Purification and Downstream Processing

Recombinant proteins are usually purified from cell culture supernatants or cell extracts by filtration and conventional column chromatography, including affinity chromatography (see Chap. 4).

The aim of downstream processing (DSP) is to purify the therapeutic protein from (potential) endogenous and extraneous contaminants, such as host cell proteins, DNA, and viruses.

It is important to mention here that slight changes in the purification process of a therapeutic protein may affect its activity and the amount and nature of the co-purified impurities. This is one of the main reasons (in addition to differences in expression host and culture conditions) why follow-on products (after expiration of the patent) made by a different company will never be identical to the original preparation and that is why they are not considered a true generic product as is the case with small, low-molecular-weight drug molecules (see Chap. 11). A generic drug must contain the same active ingredient as the original drug and in the case of a therapeutic protein this is almost impossible and that is why the term "biosimilar" was invented.

Although not often used for the production of therapeutic proteins, recombinant protein purification may be simplified by linking it with an affinity tag, such as the his-tag (6 histidines). His-tagged proteins have a high affinity for Ni^{2+}-containing resins. There are two ways to add the 6 histidine residues. The DNA encoding the protein may be inserted into a plasmid already encoding a his-tag. Another possibility is to perform a PCR reaction with a regular primer and a primer with at its 5′ end 6 histidine codons (CAT or CAC) (see Fig. 1.18). To enable easy removal of the his-tag from the recombinant protein, the tag may be followed by a suitable amino acid sequence that is recognized by an endopeptidase.

In *E. coli*, recombinant proteins are often produced as a fusion protein with another protein such as thioredoxin, maltose-binding protein (MBP), and glutathione S-transferase (GST). These fusion partners may improve the proper folding and solubility of the recombinant protein and act as affinity tags for purification. For a review on recombinant protein expression in *E. coli*, see Rosano et al. (2019).

a

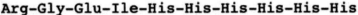

Arg-Gly-Glu-Ile-His-His-His-His-His-His

Recognition site for the protease Factor Xa   His-tag binds to Ni²⁺

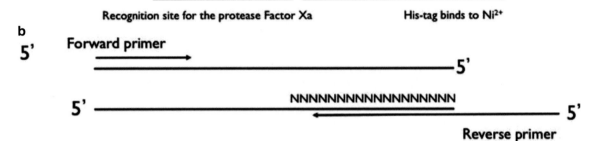

b

5'   Forward primer
          →
                                                                    5'

     5' —————————————————————— 5'
                    NNNNNNNNNNNNNNNNNNNN
                         ←
                                         Reverse primer

    XXX XXX XXX XXX XXX XXX ILE GLU GLY ARG HIS HIS HIS HIS HIS HIS **stop**
5'NNN NNN NNN NNN NNN NNN ATT GAA GGA CGT CAT CAT CAT CAT CAT CAT TAA

REVERSE PRIMER:
5' TTA ATG ATG ATG ATG ATG ATG ACG TCC TTC AAT NNN NNN NNN NNN NNN NNN

Fig. 1.18 (**a**) Schematic drawing of his-tagged fusion protein. (**b**) Design of the primers needed to generate his-tag at the carboxy-terminal end of a protein

Monoclonal Antibodies

So far, we discussed the selection, design, and production of a protein starting from a DNA sequence in a genomic database. There is no database available of the entire repertoire of human antibodies. Potentially, there are millions of different antibodies possible, and our knowledge about antibody–antigen interactions is not large enough to design a specific antibody from scratch.

Many marketed therapeutic proteins are monoclonal antibodies (cf. Chaps. 8, 23, 24, 25 and 26). We will focus here on the molecular biological aspects of the design and production of (humanized) monoclonal antibodies in cell culture (primarily CHO cells are used). For a description of the structural elements of monoclonal antibodies, we refer to Chap. 8, Figs. 8.1 and 8.2.

The classic way to make a monoclonal antibody starts by immunizing a laboratory animal with a purified human protein against which the antibody should be directed (see Fig. 1.19). In most cases, mice are used. The immunization process (a number of injections with the antigens and an adjuvant) will take several weeks. Then the spleens of these mice are removed and lymphocytes are isolated. Subsequently, the lymphocytes are fused with a myeloma cell using polyethylene glycol (PEG). The resulting hybrid-

oma cell inherited from the lymphocytes the ability to produce antibodies and from the myeloma cell line the ability to divide indefinitely. To select hybridoma cells from the excess of nonfused lymphocytes and myeloma cells, the cells are grown in HAT selection medium. This culture medium contains hypoxanthine, aminopterin, and thymidine. The myeloma cell lines used for the production of monoclonal antibodies contain an inactive hypoxanthine-guanine phosphoribosyltransferase (HGPRT), an enzyme necessary for the salvage synthesis of nucleic acids. The lack of HGPRT activity is not a problem for the myeloma cells because they can still synthesize purines de novo. By exposing the myeloma cells to the drug aminopterin also de novo synthesis of purines is blocked and these cells will not survive anymore. Selection against the unfused lymphocytes is not necessary, since these cells, like most primary cells, do not survive for a long time in cell culture. The hybridoma cells, however, will be able to survive, as they will both survive the aminopterin selection and will be able to survive in the HAT medium due to the enzymatic armamentarium that they have inherited from the lymphocytic cells. After PEG treatment, the cells are diluted and divided over several dishes. After approximately 2 weeks, individual clones are visible. Each clone contains the descendants of one hybridoma cell and will produce one particular type of antibody (that is why they

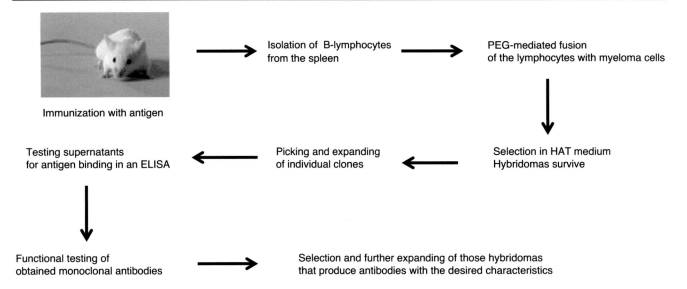

Immunization with antigen → Isolation of B-lymphocytes from the spleen → PEG-mediated fusion of the lymphocytes with myeloma cells

Testing supernatants for antigen binding in an ELISA ← Picking and expanding of individual clones ← Selection in HAT medium Hybridomas survive

Functional testing of obtained monoclonal antibodies → Selection and further expanding of those hybridomas that produce antibodies with the desired characteristics

Fig. 1.19 The making of a mouse monoclonal antibody

are called monoclonal antibodies). The next step is to isolate hybridoma cells from individual clones and grow them in separate wells of a 96-well plate. The hybridomas secrete antibodies into the culture medium. Using a suitable test (e.g., an ELISA), the obtained culture media can be screened for antibody binding to the antigen. The obtained antibodies can then be further characterized using other tests. In this way, a mouse monoclonal antibody is generated.

These mouse monoclonal antibodies cannot be used directly for the treatment of human patients. The amino acid sequence of a mouse antibody is too different from the sequence of an antibody in humans and thus will elicit an immune response. To make a mouse antibody less immunogenic, the main part of its sequence must be replaced by the corresponding human sequence. Initially, human-mouse chimeric antibodies were made. These antibodies consisted of the constant regions of the human heavy and light chain and the variable regions of the mouse antibody. Later, so-called humanized antibodies were generated by grafting only the complementarity-determining regions (CDRs), which are responsible for the antigen-binding properties, of the selected mouse antibody onto a human framework of the variable light (V_L) and heavy (V_H) domains. The humanized antibodies are much less immunogenic than the previously used chimeric antibodies. To even further reduce immunogenicity, specificity determining residues (SDR) grafting is used (Kashmiri et al. 2005). SDR stands for 'specificity determining residues'. From the analysis of the 3D structure of antibodies, it appeared that only ~30% of the amino acid residues present in the CDRs are critical for antigen binding. These residues, which form the SDR, are thought to be unique for a given antibody.

Humanization of a mouse antibody is a difficult and tricky process. It usually results in a reduction of the affinity of the antibody for its antigen. One of the challenges is the selec-tion of the most appropriate human antibody framework. This framework determines basically the structure of the antibody and thus the orientation of the antigen recognition domains in space. Sometimes it is necessary to change some of the residues in the human antibody framework to restore antigen binding. To further enhance the affinity of the humanized antibody for its antigen, mutations within the CDR/SDR sequences are introduced. How this in vitro affinity maturation is done is beyond the scope of this chapter.

So far, the generation of a humanized antibody has been described in a rather abstract way. How is it done in practice? First, the nucleotide sequence of each of the V_L and V_H regions is deduced (contains either the murine CDRs or SDRs). Next, the entire sequence is divided over four or more alternating oligonucleotides with overlapping flanks (see Fig. 1.20). These relatively long oligonucleotides are made synthetically. The reason why the entire sequence is divided over four nucleotides instead of over two or even one is that there is a limitation to the length of an oligonucleotide that can be synthesized reliably (a less than 100% yield of each coupling step (nucleotides are added one at the time) and the occurrence of side reactions make that oligonucleotides hardly exceed 100 nucleotide residues).

To the four oligonucleotides, a heat-stable DNA polymerase and the 4 deoxyribonucleotides (dATP, dCTP, dGTP, and dTTP) are added, and the mixture is incubated at an appropriate temperature. Then, 2 primers, complementary to both ends of the fragment, are added, which enable the amplification of the entire sequence. The strategy to fuse overlapping oligonucleotides by PCR is called PCR sewing.

Finally, the PCR product encoding the humanized V_L and V_H region is cloned into an expression vectors carrying the respective constant regions and a signal peptide. The signal peptide is required for glycosylation. Subsequently, the expression constructs will be used to stably transfect CHO or

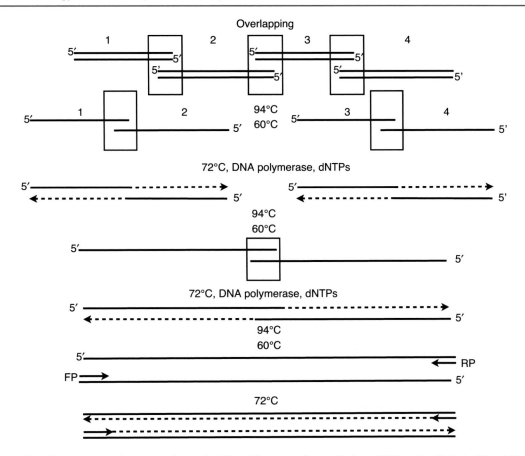

Fig. 1.20 The making of a sequence containing the humanized V_L or V_H region of an antibody by PCR sewing. Both the V_L and V_H regions contain three highly variable loops (known as complementarity-determining regions 1, 2, and 3). The V_L and V_H regions are approximately 110 amino acids in size. Four alternating oligonucleotides with overlapping flanks are incubated together with a DNA polymerase and deoxynucleotides. DNA polymerase fills in the gaps. The sequences of these oligonucleotides are based upon the original mouse CDR/SDR sequences inserted into a human V_L or V_H framework. Next, the entire sequence is PCR amplified using end primers (*FP* forward primer, *RP* reverse primer). The resulting PCR fragments will be around 330 base pairs in size. Note: an alternative for "PCR sewing" is the Gibson Assembly cloning strategy (see Fig. 1.14)

HEK293 cells. The obtained clones will be tested for antibody production, and clones with the highest antibody production capacity will be selected for further use.

Yields

To give an idea about the production capacity needed to produce a monoclonal antibody, we will do now some calculations. The annual amount needed for the most successful therapeutic monoclonal antibodies is around 1000 kg. In cell culture, titers of up to 13 g/L are reached in fed-batch cultures (see Carrara et al. 2021, for more details on manufacturing of therapeutic antibodies), and the yield of the DSP is around 80%. Thus, to produce 1000 kg monoclonal antibody, one needs at least 100,000 liter of cell culture supernatant. In Chap. 4, more details can be found regarding protein manufacturing.

Conclusion

Thanks to advances in many different areas, including molecular biology, bioinformatics, and bioprocess engineering, we have moved from an animal–/human-derived therapeutic protein product toward in vitro-produced therapeutic proteins with the fully human sequence and structure. Importantly, we now have access to large amounts of high-quality therapeutic proteins. Of course, there will always be a risk for (viral) contaminations in the in vitro-produced therapeutic protein preparation, but this risk is much smaller than when the protein has to be isolated from a human source. Examples from the past include the transmission of hepatitis B and C and human immunodeficiency virus (HIV) via blood-derived products and the transmission of Creutzfeldt-Jakob disease from human growth hormone preparations from human pituitaries.

As basic knowledge in molecular biology and engineering keeps on growing, the efficiency of the cloning and production process will increase in parallel.

Self-Assessment Questions

Questions

1. A Researcher Wanted to Clone and Subsequently Express the Human Histone H4 Protein in *E. coli*

She obtained the sequence below from the NCBI, as shown below. The start and stop codons are underlined.

>gi|29,553,982|ref.|NM_003548.2| Homo sapiens histone cluster 2, H4a (HIST2H4A), mRNA.

AGAAGCTGTCTATCGGGCTCCAGCGGTC<u>ATG</u>TCCGG
CAGAGGAAAGGGCGGAAAAGGCTTAGGCAAAGGG.
GGCGCTAAGCGCCACCGCAAGGTCTTGAGAGACAA
CATTCAGGGCATCACCAAGCCTGCCATTCGGCGTC.
TAGCTCGGCGTGGCGGCGTTAAGCGGATCTCTGGC
CTCATTTACGAGGAGACCCGCGGTGTGCTGAAGGT.
GTTCCTGGAGAATGTGATTCGGGACGCAGTCACCTA
CACCGAGCACGCCAAGCGCAAGACCGTCACAGCC.
ATGGATGTGGTGTACGCGCTCAAGCGCCAGGGGCG
CACCCTGTACGGCTTCGGAGGC<u>TAG</u>GCCGCCGCTC.
CAGCTTTGCACGTTTCGATCCCAAAGGCCCTTTTTA
GGGCCGACCA.

(a) Is *E. coli* a suitable expression host for the H4 protein?

(b) Design primers for the amplification of the coding sequence of the H4 protein by PCR.

(c) To ease purification the researcher decided to add an affinity tag (Trp-Ser-His-Pro-Gln-Phe-Glu-Lys) to the carboxy-terminal end of the H4 protein. PCR was used to clone this tag in frame with the H4 protein.

What was the sequence of the primers she probably used?

To answer this question, make use of the Table 1.3 below.

(d) And finally, she decided to optimize the codon usage for expression in *E. coli*. What is coding optimalization? What is its purpose?

(e) Design a strategy/method to optimize the codon usage of the H4 protein.

(f) The human H4 mRNA differs from most other human mRNAs by lacking a poly-A tail (instead the H4 mRNA is protected by a palindromic termination element), and thus the cDNA encoding this protein cannot be obtained by a reverse transcriptase reaction using an oligo-dT as primer. Describe a method to obtain the H4 cDNA.

2. Ampicillin, G418, and HAT medium are used to select for transformed *E. coli*, transfected mammalian cells, and hybridomas, respectively. Describe shortly the mechanism underlying the three mentioned selection strategies.

3. *E. coli* does not take up plasmid DNA spontaneously. However, the so-called chemical competent *E. coli* is able to take up plasmids following a heat shock (30 s 42 °C, followed by an immediate transfer to 0 °C). These competent bacteria can be obtained by extensive washing with a 100 mM $CaCl_2$ solution.

Transformation of competent *E. coli* of good-quality results in ±10^8 colonies/μg of supercoiled plasmid DNA. The bacteria in each colony are the descendants of one bacterium that had initially taken up one plasmid molecule.

Calculate the transformation efficiency defined as the number of plasmids taken up by the competent bacteria divided by the total number of plasmids added. Make the calculation for a plasmid of 3333 base pairs (the MW of a nucleotide is 300 g/mol and the Avogadro constant is 6×10^{23} molecules/mol).

4. Is the reverse primer sequence in Fig. 1.18 correct?

Table 1.3 The genetic code

			2nd	base			
			U	C	A	G	
		U	Phe	Ser	Tyr	Cys	U
			Phe	Ser	Tyr	Cys	C
			Leu	Ser	Stop	Stop	A
1			Leu	Ser	Stop	Trp	G	3
s		C	Leu	Pro	His	Arg	U	r
t			Leu	Pro	His	Arg	C	d
			Leu	Pro	Gln	Arg	A	
b			Leu	Pro	Gln	Arg	G	b
a		A	Ile	Thr	Asn	Ser	U	a
s			Ile	Thr	Asn	Ser	C	s
e			Ile	Thr	Lys	Arg	A	e
			Met	Thr	Lys	Arg	G	
		G	Val	Ala	Asp	Gly	U	
			Val	Ala	Asp	Gly	C	
			Val	Ala	Glu	Gly	A	
			Val	Ala	Glu	Gly	G	

Answers

1. (a) Information about the protein structure can be obtained from http://www.expasy.org/. The H4 protein does not contain disulfide bridges and is unglycosylated. It is therefore likely that *E. coli* is able to produce a correctly folded H4 protein.

   (b) PCR primers are usually around 18–20 nucleotides long. The sequences of the forward and reverse primer are *ATG* TCC GGC AGA GGA AAG (identical to the published sequence) and *CTA* GCC TCC GAA GCC GTA (complementary and reverse), respectively.

   (c) The forward primer will be as above. At the 5′ end of the reverse primer, additional sequences must be added. First, the DNA sequence encoding the affinity tag Trp-Ser-His-Pro-Gln-Phe-Glu-Lys must be determined using the codon usage table: TCG CAC CCA CAG TTC GAA AAG. It is important to place the tag in front of the stop codon (TAG). The sequence of the reverse primer will then be 5′- CTA CTT TTC GAA CTG TGG GTG CGA CCA GCC TCC GAA GCC GTA CAG- 3′.

   (d) For most amino acids more than one codon exists (see the codon usage table).

   Differences in preferences for one of the several codons that encode the same amino acid exist between organisms. In particular in fast-growing organisms, such as *E. coli*, the optimal codons reflect the composition of their transfer RNA (tRNA) pool. By changing the native codons into those codons preferred by *E. coli*, the level of heterologous protein expression may increase. Alternatively, and much easier, one could use as expression host an *E. coli* with plasmids encoding extra copies of rare tRNAs.

   (e) The H4 protein is 103 amino acids long. The easiest way to change the sequence at many places along the entire length of the coding sequence/mRNA is by designing four overlapping oligonucleotides.

   Next, the four overlapping oligonucleotides must be "sewed" together by a DNA polymerase in the presence of dNTPs. Finally, by the addition of two flanking primers, the entire, now optimized, sequence can be amplified.

   (f) An oligo-dT will not bind to the mRNA of H4, and therefore one has to use a H4-specific primer. One could use for instance the reverse primer as designed by Question 1(b).

2. (a) *Selection of transformed bacteria using ampicillin.* The antibiotic ampicillin is an inhibitor of transpeptidase. This enzyme is required for the making of the bacterial cell wall. The ampicillin resistance gene encodes for the enzyme beta-lactamase, which degrades ampicillin.

   (b) *Selection of stably transfected mammalian cells using G418.* Most expression plasmids for mammalian cells contain as selection marker the neomycin resistance gene (Neo^r). This gene codes for a protein that neutralizes the toxic drug Geneticin, also known as G418. G418 blocks protein synthesis both in prokaryotic and eukaryotic cells. Only cells that have incorporated the plasmid with the Neo^r gene into their chromosomal DNA will survive.

   (c) *Selection of hybridomas using HAT medium.* HAT medium contains hypoxanthine, aminopterin, and thymidine. The myeloma cell lines used for the production of monoclonal antibodies contain an inactive hypoxanthine-guanine phosphoribosyltransferase (HGPRT), an enzyme necessary for the salvage synthesis of nucleic acids. The lack of HGPRT activity is not a problem for the myeloma cells because they can still synthesize purines de novo. By exposing the myeloma cells to the drug aminopterin also de novo synthesis of purines is blocked and these cells will not survive anymore. Selection against the unfused lymphocytes is not necessary, since these cells, like most primary cells, do not survive for a long time in cell culture.

3. First, calculate the molecular weight of the plasmid: $3333 \times 2 \times 300 = 2 \times 10^6$ g/mol. $\rightarrow 2 \times 10^6$ g plasmid $= 6 \times 10^{23}$ molecules. $\rightarrow 1$ g plasmid $= 3 \times 10^{17}$ molecules. $\rightarrow 1$ µg $(1 \times 10^{-6}$ g$) = 3 \times 10^{11}$ molecules.

   1 µg gram plasmid results in 10^8 colonies. Thus, only one in 3000 plasmids is taken up by the bacteria.

4. No, it should read "5′CTCCAGACAGGCAGCGAAG GCCAT".

Recommended reading and references

Agarwal KL, Büchi H, Caruthers MH, Gupta N, Korana HG, Kleppe K, Kumar A, Ohtsuka E, Rajbhandary UL, Van de Sande JH, Sgamarella V, Weber H, Yamada T (1970) Total synthesis of the gene for an alanine transfer ribonucleic acid from yeast. Nature 227(5253):27–34

Brekke OH, Sandlie I (2003) Therapeutic antibodies for human diseases at the dawn of the twenty-first century. Nat Rev Drug Discov 2(1):52–62

Carrara SC, Ulitzka M, Grzeschik J, Kornmann H, Hock B, Kolmar H (2021) From cell line development to the formulated drug product: the art of manufacturing therapeutic monoclonal antibodies. Int J Pharm 594:120164

Czar MJ, Anderson C, Bader JS, Peccoud J (2009) Gene synthesis demystified. Trends Biotechnol 27(2):63–72

EvaluatePharma® World Preview 2018(2018) Outlook to 2024 www. evaluate.com/PharmaWorldPreview2018 pdf (June 2018)

Gaciarz A, Khatri NK, Velez-Superbie ML, Saaranen MJ, Uchida Y, Keshavarz-Moore E, Ruddock LW (2017) Efficient soluble expression of disulfide bonded proteins in the cytoplasm of Escherichia coli in fed batch fermentations on chemically defined minimal media. Microbiol Cell Fact 16:108

Gibson DG, Young L, Chuang RY, Venter JC, Hutchison CA, Smith HO (2009) Enzymatic assembly of DNA molecules up to several hundred kilobases. Nat Methods 6(5):343–345

Hartley JL (2003) Use of the gateway system for protein expression in multiple hosts. Curr Protoc Protein Sci:5.17.1–5.17.10. https://doi.org/10.1002/0471140864.ps0517s30

Hughes RA, Ellington AD (2017) Synthetic DNA synthesis and assembly: putting the synthetic in synthetic biology. Cold Spring Harb Perspect Biol 9(1):a023812

Kashmiri SV, De Pascalis R, Gonzales NR, Schlom J (2005) SDR grafting-a new approach to antibody humanization. Methods 36(1):2–34

Kim JY, Kim YG, Lee GM (2012) CHO cells in biotechnology for production of recombinant proteins: current state and further potential. Appl Microbiol Biotechnol 93:917–930

Kircher M, Kelso J (2010) High-throughput DNA sequencing–concepts and limitations. BioEssays 32(6):524–536

Kunert R, Reinhart D (2016) Advances in recombinant antibody manufacturing. Appl Microbiol Biotechnol 100:3451–3461

Lalonde ME, Durocher Y (2017) Therapeutic glycoprotein production in mammalian cells. J Biotechnol 251:128–140

Leader B, Baca QJ, Golan DE (2008) Protein therapeutics: a summary and pharmacological classification. Nat Rev Drug Discov 7(1):21–39

Lodish H, Berk A, Kaiser CA, Krieger M, Scott MP (2007) Molecular cell biology, 6th edn. WH. Freeman & CO, New York

Nothaft H, Szymanski CM (2010) Protein glycosylation in bacteria: sweeter than ever. Nat Rev Microbiol 8(11):765–778

Rosano GL, Morales ES, Ceccarelli EA (2019) New tools for recombinant protein production in Escherichia coli: a 5-year update. Protein Sci 28(8):1412–1422

SECRETERS-European Union's Horizon 2020 Programme (2022) microbial protein cell factories fight back? Trends Biotechnol 40(5):576–590

Strohl WR, Knight DM (2009) Discovery and development of biopharmaceuticals: current issues. Curr Opin Biotechnol 20(6):668–672

Van Norman GA (2016) Drugs, devices, and the FDA: part 1: an overview of approval processes for drugs. JACC Basic Transl Sci 1(4):277–287

Wells EA, Robinson AS (2017) Cellular engineering for therapeutic protein production: product quality, host modification, and process improvement. Biotechnol J 12(16001015):1–12

Biophysical and Biochemical Characteristics of Therapeutic Proteins

2

Wim Jiskoot and Daan J. A. Crommelin

Introduction

For a recombinant human protein to become a therapeutic product, its biophysical and biochemical characteristics must be well understood. These properties serve as a basis for understanding the behavior of the protein under various circumstances, e.g., for establishing the range of conditions to properly purify the protein and to stabilize it during production, storage, and shipping.

Protein Structure

Primary Structure

Most proteins that are developed for therapy perform specific functions by interacting with other small and large molecules, e.g., cell-surface receptors (mostly proteins), nucleic acids, carbohydrates, and lipids. The functional properties of proteins are derived from their folding into distinct three-dimensional structures. Each protein fold is based on its specific polypeptide sequence in which different natural amino acids are connected through peptide bonds in a specific way. This alignment of the 20 amino acid residues, called a primary sequence, has in general all the information necessary for folding into a distinct tertiary structure comprising different secondary structures such as α-helices and β-sheets (see

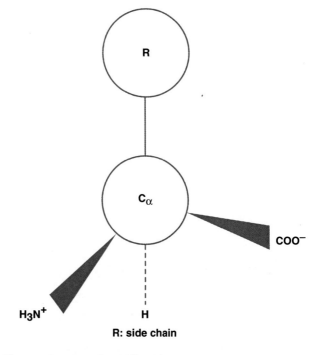

Fig. 2.1 Structure of L-amino acids

below). Because the 20 amino acids possess different side chains, polypeptides with widely diverse properties are obtained.

All of the natural amino acids consist of a C_α carbon to which an amino group, a carboxyl group, a hydrogen, and a side chain are covalently attached. All natural amino acids, except glycine (having a proton as the side chain), are chiral and have an L-configuration (Fig. 2.1). In a polypeptide, these amino acids are joined by condensation to yield peptide bonds consisting of the C_α-carboxyl group of an amino acid joined with the C_α-amino group of the next amino acid (Fig. 2.2).

The condensation gives an amide (NH) group at the N-terminal side of C_α and a carbonyl (C=O) group at the C-terminal side. These groups, as well as the side chains, play important roles in protein folding. Owing to their ability

Sadly, Wim Jiskoot passed away by the time of publication of this sixth edition.
This text is a revised and abbreviated version of the chapter by Tsutomu Arakawa and John S. Pilo in the fourth, edition of this book. The discussion of techniques to physicochemically characterize protein structures was taken out and forms a separate chapter, Chap. 3.

W. Jiskoot (Deceased)
Division of BioTherapeutics, Leiden Academic Center for Drug Research (LACDR), Leiden University, Leiden, The Netherlands

D. J. A. Crommelin (✉)
Department of Pharmaceutics, Utrecht Institute for Pharmaceutical Sciences (UIPS), Utrecht University, Utrecht, The Netherlands
e-mail: d.j.a.crommelin@uu.nl

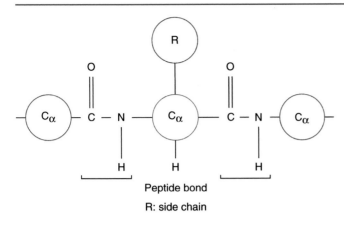

Fig. 2.2 Schematical structure of two sequential peptide bonds

to form hydrogen bonds, they make major energetic contributions to the formation of two important secondary structures, α-helix and β-sheet. The peptide bonds between various amino acid residues are very much equivalent, however, so that they do not determine which part of a sequence should form an α-helix or β-sheet. Sequence-dependent secondary structure formation is determined by the side chains.

The 20 natural amino acids commonly found in proteins are shown in Fig. 2.3. They are described by their full names and three- and one-letter codes. Their side chains are structurally different in such a way that at physiological pH values, aspartic, and glutamic acid are negatively charged and lysine and arginine are positively charged. At pH 7.4, a minor

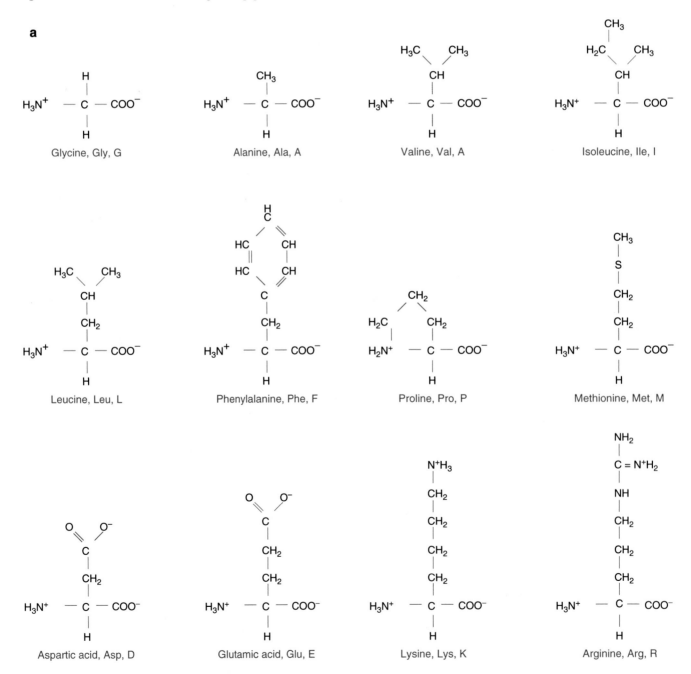

Fig. 2.3 (**a and b**) Chemical structure of the 20 natural amino acids, which are the building blocks commonly found in proteins

b

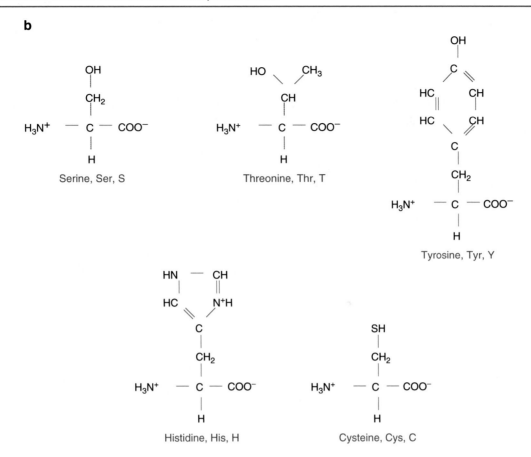

Fig. 2.3 (continued)

fraction of the histidine side chains is positively charged (pKa = 6). Tyrosine and cysteine are protonated and uncharged at physiological pHs, but become negatively charged above pH 10 and 8, respectively.

Polar amino acids consist of serine, threonine, asparagine, and glutamine, as well as cysteine, while nonpolar amino acids consist of alanine, valine, phenylalanine, proline, methionine, leucine, and isoleucine. Glycine behaves neutrally while cystine, the oxidized form of two cysteines (i.e., a Cys-Cys, or disulfide bridge), is characterized as hydrophobic. Although tyrosine and tryptophan often enter into polar interactions, they are better characterized as nonpolar, or hydrophobic, as described later.

These 20 amino acids are incorporated into a unique sequence based on the genetic code, as the example in Fig. 2.4 shows. This is an amino acid sequence of human granulocyte-colony-stimulating factor (G-CSF), which selectively regulates the proliferation and maturation of neu-

trophils. Although several properties of this protein depend on the location of each amino acid and hence the location of each side chain in the three-dimensional structure, some properties can be estimated simply from the amino acid composition, as shown in Table 2.1.

Using the pK$_a$ values of these side chains, one amino terminus and one carboxyl terminus, one can calculate total charges (positive plus negative charges) and net charges (positive minus negative charges) of a protein as a function of pH, i.e., a titration curve. Since cysteine can be oxidized to form a disulfide bond or can be in a free form, accurate calculation above pH 8 requires knowledge of the status of cysteinyl residues in the protein. The titration curve thus obtained is only an approximation, since some charged residues may be buried and the effective pKa values depend on the local environment of each residue. Nevertheless, the calculated titration curve gives a first approximation of the overall charged state of a protein at a given pH and hence its

Fig. 2.4 Amino acid sequence of human granulocyte-colony-stimulating factor

```
TPLGPASSLPQSFLLKCLEQVRKIQGDGAALQEKLCATYK    40
LCHPEELVLLGHSLGIPWAPLSSCPSQALQLAGCLSQLHS    80
GLFLYQGLLQALEGISPELGPTLDTLQLDVADFATTIWQQ   120
MEELGMAPALQPTQGAMPAFASAFQRRAGGVLVASHLQSF   160
LEVSYRVLRHLAQP
```

Table 2.1 Amino acid composition and physicochemical parameters of granulocyte-colony-stimulating factor

Parameter	Value
Molecular weight	18,673
Total number of amino acids	174
1 μg	53.5 pmol
Molar extinction coefficient	15,820
1 A (280)	1.18 mg/mL
Isoelectric point	5.86
Charge at pH 7	−3.39

Amino acid	Number	% by weight	% by frequency
A Ala	19	7.23	10.92
C Cys	5	2.76	2.87
D Asp	4	2.47	2.30
E Glu	9	6.22	5.17
F Phe	6	4.73	3.45
G Gly	14	4.28	8.05
H His	5	3.67	2.87
1 Me	4	2.42	2.30
K Lys	4	2.75	2.30
L Leu	33	20.00	18.97
M Met	3	2.11	1.72
N Asn	0	0.00	0.00
P Pro	13	6.76	7.47
Q Gln	17	11.66	9.77
R Arg	5	4.18	2.87
S Ser	14	6.53	8.05
T Thr	7	3.79	4.02
V Val	7	3.71	4.02
W Trp	2	1.99	1.15
Y Tyr	3	2.62	1.72

solubility (Cf. Fig. 5.7, Chap. 5). Other molecular parameters, such as isoelectric point (pI, where the net charge of a protein becomes zero), molecular weight, extinction coefficient, partial specific volume, and hydrophobicity, can also be estimated from the amino acid composition, as shown in Table 2.1.

The primary structure of a protein, i.e., the sequence of the 20 amino acids, can lead to the three-dimensional structure because the various amino acids have diverse physicochemical properties. As an example, Fig. 2.5 shows a cartoon of the three-dimensional structure of filgrastim (recombinant human G-CSF). Each type of amino acid has the tendency to be more preferentially incorporated into certain secondary structures. The frequencies with which each amino acid is found in α-helix, β-sheet, and β-turn, secondary structures that are discussed later in this chapter, can be calculated as an average over a number of proteins whose three-dimensional structures have been solved. These frequencies are listed in Table 2.2. The β-turn has a distinct configuration consisting of four sequential amino acids, and there is a strong preference for specific amino acids in these four positions. For example, asparagine has an overall high frequency of occurrence in a β-turn and is most frequently observed in the first and third positions of a β-turn. This characteristic of asparagine is consistent with its side chain being a potential site of N-linked glycosylation (see below). Furthermore, effects of glycosylation on the biological and physicochemical properties of proteins are extremely important. However, their contribution to structure is not readily predictable.

Another property of amino acids, which impacts protein folding, is the hydrophobicity of their side chains. Although nonpolar amino acids are basically hydrophobic, it is important to know how hydrophobic they are. This property has been determined by measuring the partition coefficient or solubility of amino acids in water and organic solvents and normalizing such parameters relative to glycine. Relative to the side chain of glycine, a single hydrogen, such normalization shows how strongly the side chains of nonpolar amino acids prefer the organic phase to the aqueous phase. A representation of such measurements is shown in Table 2.3. The values indicate that the free energy increases as the side chain of tryptophan and tyrosine is transferred from an organic solvent to water and that such transfer is thermodynamically unfavorable. Although it is unclear how comparable the hydrophobic property is between an organic solvent and the interior of protein molecules, the hydrophobic side chains favor clustering together, resulting in a core structure with properties similar to those of an organic solvent. These hydrophobic characteristics of nonpolar amino acids and hydrophilic characteristics of polar amino acids generate a partition of amino acyl residues into a hydrophobic core and a hydrophilic surface, which contributes to the conformational stability of folded proteins.

Secondary Structure

Immediately evident in the primary structure of a protein is that each amino acid is linked by a peptide bond (Fig. 2.2). The amide, NH, is a hydrogen donor and the carbonyl, C=O, is a hydrogen acceptor, and they can form a stable hydrogen bond when they are positioned in an appropriate configuration of the polypeptide chain. Such structures of the polypeptide chain are called secondary structure. Two main structures, α-helix and β-sheet, accommodate such stable hydrogen bonds. Besides α-helices and β-sheets, loops and turns are common secondary structures found in proteins.

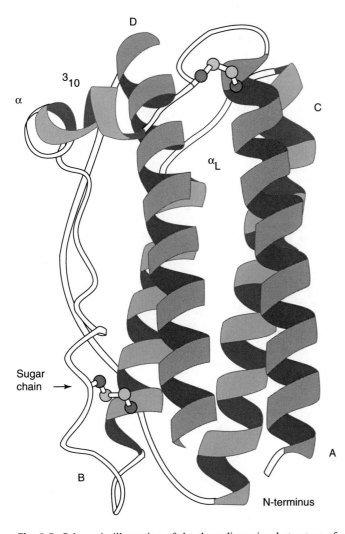

Fig. 2.5 Schematic illustration of the three-dimensional structure of filgrastim (recombinant human G-CSF). Filgrastim is a 175-amino acid protein. Its four antiparallel alpha helices (**a**, **b**, **c**, and **d**) and short 3-to-10 type helix (3_{10}) form a helical bundle. The two biologically active sites (α and $α_L$) are remote from modifications at the N-terminus of the α-helix and the sugar chain attached to loops **c–d**. Note: filgrastim is not glycosylated; the sugar chain is included to illustrate its location in endogenous G-CSF

Table 2.2 Frequency of occurrence of 20 amino acids in α-helix, β-sheet, and β-turn

α-Helix		β-Sheet		β-Turn		β-Turn position 1		β-Turn position 2		β-Turn position 3		β-Turn position 4	
Glu	1.51	Val	1.70	Asn	1.56	Asn	0.161	Pro	0.301	Asn	0.191	Trp	0.167
Met	1.45	Lie	1.60	Gly	1.56	Cys	0.149	Ser	0.139	Gly	0.190	Gly	0.152
Ala	1.42	Tyr	1.47	Pro	1.52	Asp	0.147	Lys	0.115	Asp	0.179	Cys	0.128
Leu	1.21	Phe	1.38	Asp	1.46	His	0.140	Asp	0.110	Ser	0.125	Tyr	0.125
Lys	1.16	Trp	1.37	Ser	1.43	Ser	0.120	Thr	0.108	Cys	0.117	Ser	0.106
Phe	1.13	Leu	1.30	Cys	1.19	Pro	0.102	Arg	0.106	Tyr	0.114	Gln	0.098
Gln	1.11	Cys	1.19	Tyr	1.14	Gly	0.102	Gln	0.098	Arg	0.099	Lys	0.095
Trp	1.08	Thr	1.19	Lys	1.01	Thr	0.086	Gly	0.085	His	0.093	Asn	0.091
Ile	1.08	Gln	1.10	Gln	0.98	Tyr	0.082	Asn	0.083	Glu	0.077	Arg	0.085
Val	1.06	Met	1.05	Thr	0.96	Trp	0.077	Met	0.082	Lys	0.072	Asp	0.081
Asp	1.01	Arg	0.93	Trp	0.96	Gln	0.074	Ala	0.076	Tyr	0.065	Thr	0.079
His	1.00	Asn	0.89	Arg	0.95	Arg	0.070	Tyr	0.065	Phe	0.065	Leu	0.070
Arg	0.98	His	0.87	His	0.95	Met	0.068	Glu	0.060	Trp	0.064	Pro	0.068
Thr	0.83	Ala	0.83	Glu	0.74	Val	0.062	Cys	0.053	Gln	0.037	Phe	0.065
Ser	0.77	Ser	0.75	Ala	0.66	Leu	0.061	Val	0.048	Leu	0.036	Glu	0.064
Cys	0.70	Gly	0.75	Met	0.60	Ala	0.060	His	0.047	Ala	0.035	Ala	0.058
Tyr	0.69	Lys	0.74	Phe	0.60	Phe	0.059	Phe	0.041	Pro	0.034	Ile	0.056
Asn	0.67	Pro	0.55	Leu	0.59	Glu	0.056	Ile	0.034	Val	0.028	Met	0.055
Pro	0.57	Asp	0.54	Val	0.50	Lys	0.055	Leu	0.025	Met	0.014	His	0.054
Gly	0.57	Glu	0.37	Ile	0.47	Ile	0.043	Trp	0.013	Ile	0.013	Val	0.053

Taken and edited from Chou PY, Fasman GD (1978) Empirical predictions of protein conformation. Ann Rev. Biochem 47: 251–276 with permission from Annual Reviews, Inc.

Table 2.3 Hydrophobicity scale: transfer free energies of amino acid side chains from organic solvent to water

Amino acid side chain	Cal/mol
Tryptophan	3400
Norleucine	2600
Phenylalanine	2500
Tyrosine	2300
Dihydroxyphenylalanine	1800
Leucine	1800
Valine	1500
Methionine	1300
Histidine	500
Alanine	500
Threonine	400
Serine	−300

Taken from Nozaki Y, Tanford C (1971) The solubility of amino acids and two glycine peptides in aqueous ethanol and dioxane solutions. Establishment of a hydrophobicity scale. J Biol Chem 246:2211–2217 with permission from American Society of Biological Chemists

α-Helix

The α-helix is a right-handed helix that makes one turn per 3.6 residues. The overall length of α-helices can vary widely. Figure 2.6 shows an example of a short α-helix. In this case, the C=O group of residue 1 forms a hydrogen bond with the NH group of residue 5 and the C=O group of residue 2 forms a hydrogen bond with the NH group of residue 6. All the hydrogen bonds are aligned along the helical axis. Since peptide NH and C=O groups both have electric dipole moments pointing in the same direction, they will add to a substantial dipole moment throughout the entire α-helix,

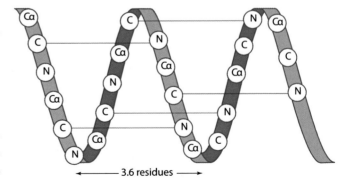

Fig. 2.6 Schematic illustration of the structure of an α-helix

Fig. 2.7 Helical wheel analysis of erythropoietin sequence, from His94 to Ala111, with amino acid residues indicated in one-letter code; open circle = hydrophobic side chain, open rectangle = hydrophilic side chains) (Elliott S, personal communication, 1990)

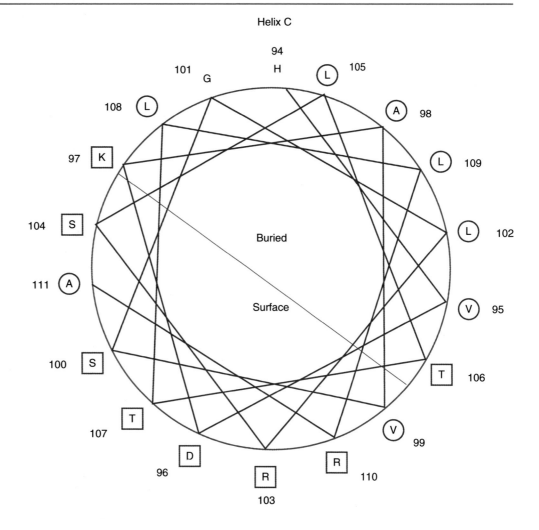

with the negative partial charge at the C-terminal side and the positive partial charge at the N-terminal side.

The side chains project outward from the α-helix. This projection means that all the side chains surround the outer surface of an α-helix and interact both with each other and with side chains of other regions within the protein that come in contact with these side chains. These interactions, so-called long-range interactions, can stabilize the α-helical structure and help it to act as a folding unit. Often an α-helix serves as a building block for the three-dimensional structure of globular proteins by bringing hydrophobic side chains to one side of a helix and hydrophilic side chains to the opposite side of the same helix. Distribution of side chains along the α-helical axis can be viewed using the helical wheel. Since one turn in an α-helix is 3.6 residues long, each residue can be plotted every 360°/3.6 = 100° around a circle (viewed from the top of the α-helix), as shown in Fig. 2.7. Such a plot shows the projection of the position of the residues onto a plane perpendicular to the helical axis.

One of the helices in erythropoietin is shown in Fig. 2.7, using an open circle for hydrophobic side chains and an open rectangle for hydrophilic side chains. It becomes immediately obvious that one side of the α-helix is highly hydrophobic, suggesting that this side forms an internal core, while the other side is relatively hydrophilic and is hence most likely exposed to the surface. Since many biologically important proteins function by interacting with other macromolecules, the information obtained from the helical wheel is extremely useful. For example, mutations of amino acids in the solvent-exposed side may lead to identification of regions responsible for biological activity, while mutations in the internal core may lead to altered protein stability.

β-Sheet

The second major secondary structural element found in proteins is the β-sheet. In contrast to the α-helix, which is built up from a continuous region with a peptide hydrogen bond linking every fourth amino acid, the β-sheet comprises peptide hydrogen bonds between different regions of the polypeptide that may be far apart in sequence. β-strands can interact with each other in one of the two ways shown in Fig. 2.8, i.e., either parallel or antiparallel. In a parallel β-sheet, each strand is oriented in the same direction with peptide hydrogen bonds formed between the strands, while in antiparallel β-sheets, the polypeptide sequences are oriented in the opposite direction. In both structures, the C=O and NH groups project into opposite sides of the polypeptide chain, and hence, a β-strand can interact from either side of that particular chain to form peptide hydrogen bonds with adjacent strands. Thus, more than two β-strands can contact each other either in a parallel or in an antiparallel manner, or

even in combination. Such clustering can result in all the β-strands lying in a plane as a sheet. The β-strands that are at the edges of the sheet may have unpaired alternating C=O and NH groups.

Side chains project perpendicularly to this plane in opposite directions and can interact with other side chains within the same β-sheet or with other regions of the molecule, or may be exposed to the solvent.

However, in almost all known protein structures, β-strands are right-handed twisted. This way, the β-strands adapt into widely different conformations. Depending on how they are twisted, all the side chains in the same strand or in different strands do not necessarily project in the same direction.

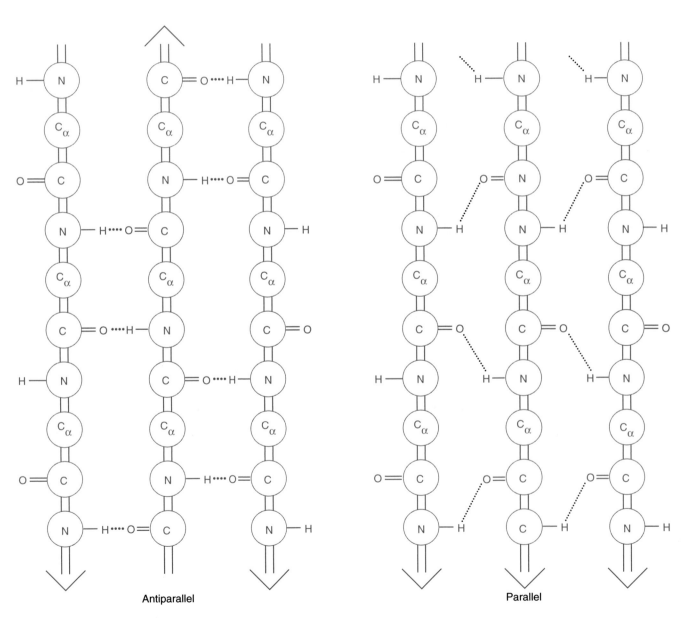

Fig. 2.8 Schematic illustration of the structure of an antiparallel (*left side*) and a parallel (*right side*) β-sheet. *Arrow* indicates the direction of amino acid sequence from the N-terminus to C-terminus

Loops and Turns

Loops and turns serve to connect other secondary structure elements, such as α-helices and β-strands. They are composed of an amino acid sequence, which is usually hydrophilic and exposed to the solvent. These regions consist of β-turns (reverse turns), short hairpin loops, and long loops. Loops and turns usually cause a change in direction of the polypeptide chain, allowing it to fold back to create a compact three-dimensional structure. Turns are short, typically consisting of four amino acid residues, and are stabilized by hydrogen bonds. Loops can be longer and are typically unstructured. Many hairpin loops are formed to connect two antiparallel β-strands.

The amino acid sequences which form β-turns are relatively easy to predict, since turns must be present periodically to fold a linear sequence into a globular structure. Amino acids found most frequently in the β-turn are usually not found in α-helical or β-sheet structures. Thus, proline and glycine represent the least-observed amino acids in these typical secondary structures. However, proline has an extremely high frequency of occurrence at the second position in the β-turn, while glycine has a high preference at the third and fourth position of a β-turn.

Although loops are not as predictable as β-turns, amino acids with high frequency for β-turns also can form a long loop. Loops are an important secondary structure, since they form a highly solvent-exposed region of the protein molecules and allow the protein to fold onto itself.

Tertiary Structure and Quaternary Structure

The spatial arrangement of the various secondary structures in a protein results in its three-dimensional structure. Many proteins fold into a fairly compact, globular structure. Larger proteins usually are folded into several structural domains. Examples are immunoglobulins and Factor VIII.

The folding of a protein molecule into a distinct three-dimensional structure determines its function. Enzyme activity requires the exact coordination of catalytically important residues in the three-dimensional space. Binding of antibodies to antigens and binding of growth factors and cytokines to their receptors all require a distinct, specific surface for high-affinity binding. These interactions do not occur if the tertiary structures of antibodies, growth factors, and cytokines are altered.

A unique tertiary structure of a protein can often result in the assembly of the protein into a distinct quaternary structure consisting of a fixed stoichiometry of protein chains within the complex. Assembly can occur between identical or between different polypeptide chains. Each molecule in the complex is called a subunit. For instance, actin and tubulin self-associate into F-actin and microtubule; hemoglobin is a tetramer consisting of two α- and two β-subunits; among the cytokines and growth factors, interferon-γ is a homodimer, while platelet-derived growth factor is a homodimer of either A or B chains or a heterodimer of the A and B chain. The formation of a quaternary structure occurs via noncovalent interactions and may be stabilized through disulfide bonds between the subunits, such as in the case of the two heavy chains and two light chains of immunoglobulins.

Forces

Interactions occurring between chemical groups in proteins are responsible for formation of their specific secondary, tertiary, and quaternary structures. Either repulsive or attractive interactions can occur between different groups. Repulsive interactions consist of steric hindrance and electrostatic effects. Like charges repel each other and bulky side chains, although they do not repel each other, cannot occupy the same space. Folding is also against the natural tendency to move toward randomness, i.e., increasing entropy. Folding leads to a fixed position of each atom and hence a decrease in entropy. For folding to occur, this decrease in entropy, as well as the repulsive interactions, must be overcome by attractive interactions, i.e., hydrophobic interactions, hydrogen bonds, electrostatic attraction, and van der Waals interactions. Hydration of proteins, discussed in the next section, also plays an important role in protein folding.

These interactions are all relatively weak and can be easily broken and formed. Hence, each folded protein structure arises from a fine balance between these repulsive and attractive interactions. The stability of the folded structure is a fundamental concern in developing protein therapeutics.

Hydrophobic Interactions

The hydrophobic interaction reflects a summation of the van der Waals attractive forces among nonpolar groups in the protein interior, which change the surrounding water structure necessary to accommodate these groups if they become exposed. The transfer of nonpolar groups from the interior to the surface requires a large decrease in entropy, so that hydrophobic interactions are essentially entropically driven. The resulting large positive free energy change prevents the transfer of nonpolar groups from the largely sheltered interior to the more solvent-exposed exterior of the protein molecule. Thus, nonpolar groups preferentially reside in the protein interior, while the more polar groups are exposed to the surface and surrounding environment. The partitioning of different amino acyl residues between the inside and outside of a protein correlates well with the hydration energy of their side chains, that is, their relative affinity for water.

Hydrogen Bonds

The hydrogen bond is ionic in character since it depends strongly on the sharing of a proton between two electronegative atoms (generally oxygen and nitrogen atoms). Hydrogen bonds may form either between a protein atom and a water molecule or exclusively as protein intramolecular hydrogen bonds. Intramolecular interactions can have significantly more favorable free energies (because of entropic considerations) than intermolecular hydrogen bonds, so the contribution of all hydrogen bonds in the protein molecule to the stability of protein structures can be substantial. In addition, when the hydrogen bonds occur in the interior of protein molecules, the bonds become stronger due to the hydrophobic environment.

Electrostatic Interactions

Electrostatic interactions occur between any two charged groups. According to Coulomb's law, if the charges are of the same sign, the interaction is repulsive with an increase in energy, but if they are opposite in sign, it is attractive, with a lowering of energy. Electrostatic interactions are strongly dependent upon distance, according to Coulomb's law, and inversely related to the dielectric constant of the medium. Electrostatic interactions are much stronger in the interior of the protein molecule because of a lower dielectric constant. The numerous charged groups present on protein molecules can provide overall stability by the electrostatic attraction of opposite charges, for example, between negatively charged carboxyl groups and positively charged amino groups. However, the net effects of all possible pairs of charged groups must be considered. Thus, the free energy derived from electrostatic interactions is actually a property of the whole structure, not just of any single amino acid residue or cluster.

Van der Waals Interactions

Weak van der Waals interactions exist between atoms (except the bare proton), whether they are polar or nonpolar. They arise from net attractive interactions between permanent dipoles and/or induced (temporary and fluctuating) dipoles. However, when two atoms approach each other too closely, the repulsion between their electron clouds becomes strong and counterbalances the attractive forces.

Hydration

Water molecules are bound to proteins internally and externally. Some water molecules occasionally occupy small internal cavities in the protein structure and are hydrogen bonded to peptide bonds and side chains of the protein and often to a prosthetic group, or cofactor, within the protein. The protein surface is large and consists of a mosaic of polar and nonpolar amino acids, and it binds a large number of water molecules, from the surrounding environment, i.e., it is hydrated. As described in the previous section, water molecules trapped in the interior of protein molecules are bound more tightly to hydrogen-bonding donors and acceptors because of a lower dielectric constant.

Solvent around the protein surface clearly has a general role in hydrating peptide and side chains but might be expected to be rather mobile and nonspecific in its interactions. Well-ordered water molecules can make significant contributions to protein stability. One water molecule can hydrogen bond to two groups distant in the primary structure on a protein molecule, acting as a bridge between these groups. Such a water molecule may be highly restricted in motion and can contribute to the stability, at least locally, of the protein, since such tight binding may exist only when these groups assume the proper configuration to accommodate a water molecule that is present only in the native state of the protein. Such hydration can also decrease the flexibility of the groups involved.

There is also evidence for solvation over hydrophobic groups on the protein surface. So-called hydrophobic hydration occurs because of the unfavorable nature of the interaction between water molecules and hydrophobic surfaces, resulting in the clustering of water molecules. Since this clustering is energetically unfavorable, such hydrophobic hydration does not contribute to the protein stability. However, this hydrophobic hydration facilitates hydrophobic interaction. This unfavorable hydration is diminished as the various hydrophobic groups come in contact either intramolecularly or intermolecularly, leading to the folding of intrachain structures or to protein–protein interactions.

Both the loosely and strongly bound water molecules can have an important impact, not only on protein stability but also on protein function. For example, certain enzymes function in nonaqueous solvent provided that a small amount of water, just enough to cover the protein surface, is present. Bound water can modulate the dynamics of surface groups, and such dynamics may be critical for enzyme function. Dried enzymes are, in general, inactive and become active after they absorb 0.2 g water per gram protein. This amount of water is only sufficient to cover surface polar groups, yet may give sufficient flexibility for function.

Evidence that water bound to protein molecules has a different property from bulk water can be demonstrated by the presence of nonfreezable water. Thus, when a protein solution is cooled below −40 °C, a fraction of water, ~0.3 g water/g protein, does not freeze, and can be detected by high-resolution nuclear magnetic resonance (NMR). Several other techniques also detect a similar amount of bound water. This unfreezable water reflects the unique property of bound water that prevents it from adopting an ice structure.

Proteins are dissolved under physiological conditions or in test tubes in aqueous solutions containing not only water but also other solution components, e.g., salts, metals, amino acids, sugars, and many other (minor) components (cf. Chaps. 4 and 5). These components also interact with the protein surface and affect protein folding and stability. For examples, sugars and amino acids are known to enhance folding and stability of the proteins, as described later.

Posttranslational Modifications

In eukaryotic cells, the amino acid sequence of a protein is synthesized in the ribosomes. Subsequently, so-called post-translational modification processes in the endoplasmatic reticulum and the Golgi body of the cell may change the "aminoacid-only" structure. For instance, sugar groups (glycosylation), phosphate groups (phosphorylation), sulfate groups (sulfation) can be enzymatically attached to the primary amino acid structure of the protein. Disulfide bridge formation is a posttranslational modification as well. For therapeutic proteins that undergo posttranslational modifications, glycosylation and disulfide bridging are the most relevant ones. An important family of glycosylated and disulfide bridge-carrying molecules is the monoclonal antibody family (Chap. 8). Other examples of highly glycosylated proteins (also called glycoproteins) are follicle stimulating hormone (Chap. 19), erythropoeitin (Chap. 17), and Factor VIII (Chap. 18). On the other hand, proteins such as insulins, human growth hormone, and interferon alfa lack sugar chains. But, all those proteins contain disulfide bonds.

Protein disulfide bonds are essential for the proper folding, functioning, and stability of proteins. Oxidative enzymatic steps in the endoplasmatic reticulum lead to cystine bridge formation. When a protein product goes through redox cycling, disulfide bonds may open up and form again. Then, disulfide bond shuffling may result in improper protein structures and covalent aggregates.

Glycosylation is a process where many different enzymes generate often complex sugar structures. During glycosylation, various sugars can be attached via glycosidic linkages in chains of different lengths and complexity. An important prerequisite for glycosylation is that the protein has sites for N-linked (via asparagine) or O-linked (usually via serine or threonine) glycosylation. Some sugar chains are unbranched, others are branched chains. Common units found in sugar chains of glycoproteins are mannose, galactose, xylose, sialic acid (derivatives of N-acetylneuraminic acid, negatively charged), and N-acetylgalactosamine, N-acetylglucosamine. Figure 2.9a shows examples of glycosylation patterns found for the monoclonal antibody rituximab.

Glycosylation plays a prominent role in the folding process of proteins and their stability, e.g., against aggregation. Moreover, therapeutic activity may be affected as, e.g., with monoclonal antibodies, where the antibody-dependent cellular cytotoxicity (ADCC) depends on the glycosylation pattern. Finally, glycosylation, in particular the presence of sialic acid affects PK profiles. High sialic

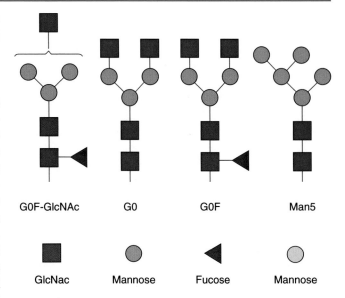

Fig. 2.9 Glycosylation pattern of the monoclonal antibody rituximab (Rituxan®/Mabthera®) (From: Schiestl et al. (2011) Acceptable changes in quality attributes of glycosylated biopharmaceuticals. Nature Biotechnology 29: 310–312)

acids densities cause prolonged circulation times (cf. Chap. 19, Fig. 19.3).

Standard prokaryotic cells such as *E. coli* do not perform glycosylation reactions. That may be different for engineered prokaryotes (cf. Chap. 4). Proteins expressed in eukaryotic cells can attach sugar chains to their primary structure. They undergo varying degrees of glycosylation depending on the host cell used and cell culture conditions. Moreover, downstream processing conditions may affect (and are sometimes exploited to adjust) the glycosylation profile of the final drug substance. Mammalian cells have an advantage over yeast or plant production cells. Their sugar chain composition closest resembles the "natural" one. The sugar chains produced in yeast are rich in mannose; their sialic acid content is low. This will increase the clearance rate of these glycoproteins upon injection. Plant-derived glycoproteins are rich in fucose and xylose levels. Those structures are believed to increase immunogenicity. As with yeast-derived glycoproteins, they have a relatively low sialic acid content, which affects the pharmacokinetic profile, i.e., accelerated clearance from the blood.

It is important to realize that, unlike the synthesis of a protein's unique primary structure, posttranslational modifications result in multiple chemical species, i.e., they introduce heterogeneity in the chemical structure of a protein. For instance, glycosylation variants within one batch of a glycoprotein (e.g., follicle-stimulating hormone, FSH, monoclonal antibodies, cf. Fig. 2.9) yield a mixture of various molecular species within one product. This heterogeneity complicates protein characterization but can be identified by analytical techniques such as isoelectric focusing and mass spectrometry (see Chap. 3).

Protein Folding

Proteins become functional only when they assume a distinct tertiary structure. Many physiologically and therapeutically important proteins present their surface for recognition by interacting with molecules such as substrates, receptors, signaling proteins, and cell-surface adhesion macromolecules. Upon biosynthesis in vivo, proteins fold in a spatially organized environment that is composed of ribosomes, ribosome-associated enzymes, chaperones, and a highly concentrated macromolecule solution (200–400 mg/mL). The nascent peptide chain often may fold co-translationally. The interplay of codon usage and tRNA abundance plays an important role in introducing appropriate translation pauses in order to regulate the in vivo speed of protein synthesis to achieve correct folding.

The majority of the current therapeutic proteins require several posttranslational modifications, such as disulfide formation and glycosylation, in order to be functional (see above). This is why mammalian cell line expression platforms such as CHO and HEK293 are best suited for the expression of proteins that require posttranslational modifications for their activity. Several E. coli strains have been engineered to translocate recombinant protein to the periplasmic space around the E. coli cell membrane, which harbors enzymes that can introduce or break disulfide bonds. However, when recombinant proteins are produced in E. coli, they often form inclusion bodies into which they are deposited as insoluble proteins. Therefore, an in vitro process is required to refold insoluble recombinant proteins into their native, physiologically active state. This is usually accomplished by solubilizing the insoluble proteins with detergents or denaturants, followed by the purification and removal of these reagents concurrent with refolding the proteins (see Chap. 4).

Unfolded states of proteins are usually highly stable and soluble in the presence of denaturing agents. Once the proteins are folded correctly, they are also relatively stable. During the transition from the unfolded form to the native state, the protein must go through a multitude of other transition states in which it is not fully folded, and denaturants or solubilizing agents are at low concentrations or even absent.

The refolding of proteins can be achieved in various ways. The dilution of proteins at high denaturant concentration into aqueous buffer will decrease both denaturant and protein concentration simultaneously. The addition of an aqueous buffer to a protein-denaturant solution also causes a decrease in concentrations of both denaturant and

protein. The difference in these procedures is that, in the first case, both denaturant and protein concentrations are the lowest at the beginning of dilution and gradually increase as the process continues. In the second case, both denaturant and protein concentrations are highest at the beginning of dilution and gradually decrease as the dilution proceeds. Dialysis or the diafiltration of protein in the denaturant against an aqueous buffer resembles the second case, since the denaturant concentration decreases as the procedure continues. In this case, however, the protein concentration remains practically unchanged. Refolding can also be achieved by first binding the protein in denaturants to a solid phase, i.e., to a column matrix, and then equilibrating it with an aqueous buffer. In this case, protein concentrations are not well defined. Each procedure has advantages and disadvantages and may be applicable to one protein, but not to another.

If proteins in the native state have disulfide bonds, cysteines must be correctly oxidized. Such oxidation may be done in various ways, e.g., air oxidation, glutathione-catalyzed disulfide exchange, or mixed-disulfide formation followed by reduction and oxidation or by disulfide reshuffling.

Protein folding has been a topic of intensive research since Anfinsen's demonstration that ribonuclease can be refolded from the fully reduced and denatured state in in vitro experiments. This folding can be achieved only if the amino acid sequence itself contains all information necessary for folding into the native structure. This is the case, at least partially, for many proteins. However, a lot of other proteins do not refold in a simple one-step process. Rather, they refold via various intermediates which are relatively compact and possess varying degrees of secondary structures, but which lack a rigid tertiary structure. Intrachain interactions of these preformed secondary structures eventually lead to the native state. However, the absence of a rigid structure in these preformed secondary structures can also expose a cluster of hydrophobic groups to those of other polypeptide chains, rather than to their own polypeptide segments, resulting in intermolecular aggregation. High efficiency in the recovery of native protein depends to a large extent on how this aggregation of intermediate forms is minimized. The use of chaperones or polyethylene glycol has been found quite effective for this purpose. The former are proteins, which aid in the proper folding of other proteins by stabilizing intermediates in the folding process and the latter serves to solvate the protein during folding and diminishes interchain aggregation events.

Protein folding is often facilitated by cosolvents, such as polyethylene glycol or glycerol. As described above, proteins are functional and highly hydrated in aqueous solutions. True physiological solutions, however, contain not only water but also various ions and low- and high-molecular-weight solutes, often at very high concentrations. These ions and other solutes play a critical role in maintaining the functional structure of the proteins. When isolated from their natural environment, the protein molecules may lose these stabilizing factors and hence must be stabilized by certain compounds, often at high concentrations. These solutes are also used in vitro to assist in protein folding and to help stabilize proteins during large-scale purification and production as well as for long-term storage. These solutes encompass sugars, amino acids, inorganic and organic salts, and polyols. They may not strongly bind to proteins, but instead typically interact weakly with the protein surface to provide significant stabilizing energy without interfering with their functional structure.

Self-Assessment Questions

Questions

1. What is the net charge of granulocyte-colony-stimulating factor at pH 2.0, assuming that all the carboxyl groups are protonated?
2. Based on the above calculation, do you expect the protein to unfold at pH 2.0?
3. Are hydrophilic and hydrophobic amino acids ad random distributed in an alfa-helix primary sequence of a folded protein?
4. (a) What are the different types of forces that stabilize the secondary, tertiary, and, eventually, quaternary structure of a protein?
   (b) Why are solutes changing these folded structures?

Answers

1. Based on the assumption that glutamyl and aspartyl residues are uncharged at this pH, all the charges come from protonated histidyl, lysyl, arginyl residues, and the amino terminus, i.e., 5 His +4 Lys + 5 Arg + N-terminal = +15.

2. Whether a protein unfolds or remains folded depends on the balance between the stabilizing and destabilizing forces. At pH 2.0, extensive positive charges destabilize the protein, but whether such destabilization is sufficient or insufficient to unfold the protein depends on how stable the protein is in the native state. The charged state alone cannot predict whether a protein will unfold.

3. An α-helix serves as a building block for the three-dimensional structure of globular proteins by bringing hydrophobic side chains to one side of a helix and hydrophilic side chains to the opposite side of the same helix.

4. (a) These forces are covalent forces (disulfide bonds), hydrophobic interactions, hydrogen bonds, electrostatic interactions, van der Waals interactions, and hydration forces.
   (b) These components interact with the protein surface and affect protein folding and stability. For example, sugars and amino acids are known to enhance folding and stability of the proteins, as described below. Another example is buffers that change the pH and by that the charge on the proteins and by that electrostatic interactions, or (high) concentrations of surfactants that denature (unfold) proteins.

Further Reading

Buxbaum E (2015) Fundamentals of protein structure and function, 2nd edn. Springer, New York

Creighton TE (ed) (1989) Protein structure: a practical approach. IRL Press, Oxford

Gregory RB (ed) (1994) Protein-solvent interactions. Marcel Dekker, New York

Schulz GE, Schirmer RH (eds) (1979) Principles of protein structure. Springer, New York

Shirley BA (ed) (1995) Protein stability and folding. Humana Press, Totowa

Whitford D (2005) Proteins: structure and function. John Wiley, Hoboken, NJ

Stability and Characterization of Protein- and Nucleotide-Based Therapeutics

3

Atanas V. Koulov

Introduction

One of the main tasks in the development of biologicals is the detailed characterization of the recombinant protein or nucleotide-based drug candidate. Gaining intimate knowledge of the molecular characteristics of the protein is required for understanding and controlling the manufacturing process and also the stability of the molecule. The latter is critical for developing a stable and fit-for-purpose drug product, as well as defining an appropriate control strategy for monitoring the stability during long-term storage. Protein therapeutics were the central feature in this chapter. However, a section on nucleotide-based pharmaceuticals is added in this sixth edition.

Analytical Toolbox: General Overview

The large diversity of possible protein modifications necessitates the use of a broad array of analytical approaches. Naturally, the largest share of analytical methods comes from well-established and traditional technologies of separation science, such as chromatography and electrophoresis. Over the last couple of decades, mass spectrometry (MS) approaches have also gained vast popularity, largely owed to the very rapid development of the technology in this field. Unlike traditionally used separation technology approaches, MS allows for elucidating the structure of protein modifications and in turn monitoring specific molecular modifications (e.g., oxidation at a specific amino acid residue), as opposed to measuring global (population) changes in the protein structure.

Chromatography

Chromatography techniques are extensively used in biotechnology, not only in protein purification procedures (see Chap. 4) but also in assessing the integrity of the product. Routine procedures are highly automated so that comparisons of similar samples can be made. An analytical chromatographic system consists of an autosampler, which will take a known amount (usually a known volume) of material for analysis and automatically places it in the solution stream (mobile phase) headed toward a separation column used to fractionate the sample. Another part of this system is a pump module, which provides a reproducible flow rate. In addition, the pumping system can provide a gradient, which changes the properties of the mobile phase such as pH, ionic strength, and polarity. A detection system (or possibly multiple detectors in series) is located at the outlet of the column. This measures the relative amount of protein exiting the column. Coupled to the detector is a data acquisition system. This takes the signal from the detector and integrates it into a value related to the amount of material (see Fig. 3.1). When the protein solution emerges from the column, the signal begins to increase, and as the protein passes through the detector, the signal subsequently decreases. The area under the peak of the signal is proportional to the amount of material that has passed through the detector. By analyzing known amounts of protein, a peak area versus amount of protein plot can be generated and this may be used to estimate the amount of this protein in the sample under other circumstances. Another benefit of this integrated chromatography system is that low levels of components which appear over time can be estimated relative to the major, desired protein being analyzed. This is a particularly useful function when the long-term stability of the product is under evaluation.

During the more than 100 years of history of chromatography, a large variety of separation modes has been developed and many of these are actively used today for characterization of proteins. Proteins and peptides can be chromatographically separated based on their polarity

A. V. Koulov (✉)
Clear Solutions Laboratories AG, Basel, Switzerland
e-mail: atanas.koulov@clearsolutions-labs.com

© The Author(s), under exclusive license to Springer Nature Switzerland AG 2024
D. J. A. Crommelin et al. (eds.), *Pharmaceutical Biotechnology*, https://doi.org/10.1007/978-3-031-30023-3_3

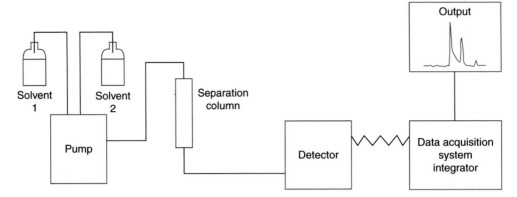

Fig. 3.1 Components of a typical chromatography station. The pump combines solvents one and two in appropriate ratios to generate a pH, salt concentration, and/or hydrophobic gradient. Proteins that are fractioned on the column pass through a detector, which measures their occurrence. Information from the detector is used to generate chromatograms and estimate the relative amount of each component

(reversed-phase chromatography, hydrophobic interaction chromatography (HIC), hydrophilic interaction chromatography), charge distribution (ion exchange chromatography), size (size exclusion chromatography), etc. In addition, mixed-mode chromatography (using columns with both hydrophobic and charged groups, i.e., a combination of ion-exchange and HIC) and two-dimensional chromatographic approaches (using a sequential combination of separation modes e.g., reversed phase and ion exchange chromatography) are regularly used when characterizing proteins and peptides.

Electrophoresis

Generally speaking, the family of electrophoretic techniques separates proteins in an electrical field, based on their charge-to-mass ratio. The charge of the protein depends on the presence of acid and basic amino acids (cf. Chap. 2) and can be controlled by the pH of the solution in which the protein is separated. The farther away the pH of the solution is from the pI value of the protein, that is, the pH at which it has a net charge of zero, the greater is the net charge and hence the greater is its charge to mass ratio. Alternatively, additives such as sodium dodecyl sulfate (SDS) may impart an overwhelming negative charge to the protein molecules. This phenomenon forms the basis of the SDS-PAGE (sodium dodecyl sulfate polyacrylamide gel electrophoresis) technique that is/was extensively used to determine the molecular weight of proteins.

Throughout the twentieth century, gel electrophoresis was one of the main methods of choice for characterization of proteins. Since the advent of capillary electrophoresis in the 1980s, significant improvements were achieved in the electrophoretic separation of proteins. Today, capillary electrophoresis is the main electrophoretic method used in the biopharmaceutical analytics.

Mass Spectrometry

MS is a technique in which ions of the various species present in the sample are generated using different ionization techniques and where their molecular masses are measured with high accuracy. This technique is one of the most impactful analytical methods in the current biopharmaceutical analytical practice. While this method was used in the past to analyze small, relatively volatile molecules, the molecular weights of highly charged proteins with masses of over 100 kilodaltons (kDa) can now be accurately determined. Together with the rapid development of informatics and MS analytical instrumentation incorporating different ionization and detection modes, a large number of different variants of MS have been developed and are currently in use.

One of the main advantages of MS is its ability to determine molecular masses with unparalleled accuracy. This attribute has enabled measuring posttranslational modifications with mass differences of only 1 Da and specific modifications that arise during stability studies. For example, an increase in mass of 16 Da suggests that an oxygen atom has been added to the protein as happens when a methionyl residue is oxidized to a methionyl sulfoxide residue. The molecular mass of peptides obtained after proteolytic digestion and separation by high-performance liquid chromatography (HPLC) indicates from which region of the primary structure they are derived. Such an HPLC chromatogram is called a "peptide map." An example is shown in Fig. 3.2. This is obtained by digesting a protein with pepsin and by subsequently separating the digested peptides by reverse HPLC. This highly characteristic pattern for a protein is called a "protein fingerprint." Peaks are identified by elution times on HPLC. If peptides have molecular masses differing from those expected from the primary sequence, the nature of the modification to that peptide can be implicated. Moreover, molecular mass estimates can be made for peptides obtained from unfractionated proteolytic digests.

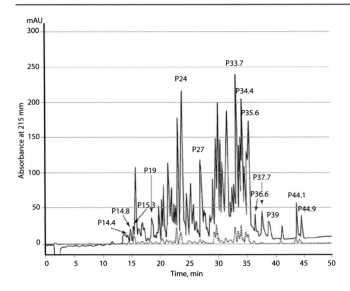

Fig. 3.2 Peptide map of a pepsin digest of recombinant human β-secretase. Each peptide is labeled by elution time in HPLC

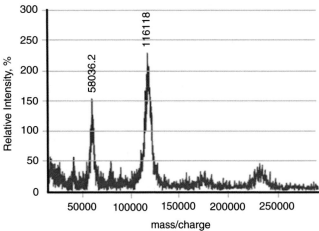

Fig. 3.3 MALDI-mass analysis of a purified recombinant human β-secretase. Numbers correspond to the singly charged and doubly charged ions

Molecular masses that differ from expected values indicate that a part of the protein molecule has been altered, e.g., that a different glycosylation pattern occurs, that a different or degraded amino acid has been found, or that the protein under investigation still contains contaminants.

Another way of using MS as an analytical tool is in the sequencing of peptides. A recurring structure, the peptide bond, in peptides tends to yield fragments of the mature peptide, which differ stepwise by an amino acyl residue. The difference in mass between two fragments indicates the amino acid removed from one fragment to generate the other. Except for leucine and isoleucine, each amino acid has a different mass and hence a sequence can be read from the mass spectrometer. Stepwise removal can occur from either the amino terminus or carboxy terminus. In addition, subsequent fragmentation of the individual peptides (MS^2) yields a highly regular fragmentation pattern, which enables sequencing of the parent peptide.

By changing three basic components of the mass spectrometer, i.e., the ion source, the analyzer, and the detector, different types of measurement may be undertaken. Typical ion sources that volatilize proteins are electrospray ionization, fast atom bombardment, and liquid secondary ionization. Common analyzers include quadrupole, magnetic sector, and time-of-flight, electrostatic sector, quadrupole ion trap, and ion cyclotron resonance instruments. The function of the analyzer is to separate the ionized biomolecules based on their mass-to-charge ratio. The detector measures a current whenever impinged upon by charged particles. Electrospray ionization (EI) and matrix-assisted laser desorption (MALDI) are two sources that can generate high-molecular-weight volatile proteins. In the former method, droplets are generated by spraying or nebulizing

the protein solution into the source of the mass spectrometer. As the solvent evaporates, the protein remains behind in the gas phase and passes through the analyzer to the detector. In MALDI, proteins are mixed with a matrix, which vaporizes when exposed to laser light, thus carrying the protein into the gas phase. An example of MALDI-mass analysis is shown in Fig. 3.3, indicating the singly charged ion (116,118 Da) and the doubly charged ion (58,036.2) for a purified protein. Since proteins may carry multiple charges, a number of components are observed representing mass-to-charge forms, each differing from the next by one charge. By imputing various charges to the mass-to-charge values, a molecular mass of the protein can be estimated. The latter step is empirical since only the mass-to-charge ratio is detected and not the net charge for that particular particle.

Spectroscopic and Other Techniques for Studying Higher Order Structure

A variety of spectroscopic techniques have found broad use for studying protein structure. These techniques differ significantly by the information content provided and by the amount of expert knowledge required for operation and data interpretation. As a rule: the higher the information content (spatial resolution of the structure) provided by a given method, the more laborious, complex to operate and interpret the data obtained it is.

Circular dichroism (CD) is a method that utilizes the property of proteins as chiral molecules to differently absorb the right- and left-handed polarized light across the UV and visible parts of the spectrum. CD is used extensively for probing the secondary and tertiary structure of proteins.

Fourier transformed infrared spectroscopy (FTIR) is used to measure the absorption of infrared (IR) light by proteins due to vibrational transitions of various functional groups. In this technique, the absorption of IR light over a broad wavelength range is measured simultaneously, using a device called interferometer. Different factors (such as hydrogen bonding, redox state, bond angles, and conformation) can influence the absorption of the vibrating group, which is why FTIR spectroscopy is widely used to probe a protein structure. The repeat units in proteins give rise to several characteristic absorption bands, of which the Amide I (1700–1600 cm^{-1}) bands are perhaps the most useful, due to their sensitivity to the protein secondary structure.

Fluorescence is another widely used spectroscopic technique to study protein conformation: its secondary, tertiary, and quaternary structure. In this technique, the fluorescence of the two main fluorophores (Tyr and Trp) present in proteins is used. Trp fluorescence is of particular interest due to the peculiar properties of this fluorophore. One of the two overlapping transitions in Trp is highly sensitive to hydrogen bonding of the indole's—NH group, which in practical terms gives rise to a high sensitivity of the Trp fluorescence spectrum to exposure to water. This property of Trp fluorescence is used to indirectly measure the protein structure as a function of Trp exposure to the aqueous environment.

Some MS techniques have been developed with the specific goal to study the higher order protein structure. The most prominent example is hydrogen-deuterium exchange MS (HDX-MS). In this technique, the different exchange rates of the amide hydrogens over the peptide backbone of the protein are measured using a highly specialized LC–MS system after exposing the protein to deuterium (D$_2$O) for a brief period of time, followed by quenching of the exchange and a rapid peptide mapping measurement (enzymatically digesting the protein into peptide fragments and measuring the deuteration levels of the individual peptides). Kinetic experiments allow measuring the rates of deuteration of the different peptides, which largely depend on the local structural environment—the level of exposure to the aqueous environment, as well as the conformational flexibility of the given peptide. The results from HDX-MS experiments are visualized using protein maps indicating the rates of exchange in the different protein regions. These maps are extremely useful in understanding the protein dynamics, flexibility, and accessibility.

Ion mobility, MS (IMS-MS) is another MS technique to study the higher order structure of a protein. Briefly, this technique measures the differential mobility of different protein species in the gas phase in an electric field. This method is mostly used to measure the aggregation state of protein mixtures.

Nuclear magnetic resonance (NMR) is a technique that has made major contributions to elucidate the 3D structure of proteins and is becoming more and more popular with the concomitant development of the analytical instrumentation. Very briefly, this technique measures the magnetic properties of atomic nuclei (more specifically the interaction of the magnetic moment of an atomic nucleus with an external magnetic field) which strongly depends of their local environment. Thus, when employing various experimental strategies including 2-D (two dimensional), NMR approaches (using measurements of different nuclei) in principle allows the determination of the 3D (three dimensional) structure of a protein at atomic resolution. One advantage that NMR has over other high-resolution techniques such as X-ray diffraction (see below) is that one can directly observe protein dynamics in kinetic experiments in solution.

X-ray diffraction is still considered the ultimate technique for studying the structure of proteins. This technique uses the phenomenon of diffraction of a monochromatic X-ray beam by protein crystals. In a protein crystallography experiment, the diffraction pattern of the protein crystal is captured from many different orientations of the crystal. From the diffraction patterns obtained (intensity and location of the resulting spots), the positions of the atoms in the molecule can be determined which in turn allows for a calculation of a molecular model of the protein in crystal form, often at atomic resolution.

In the next sections of this chapter, the wide-ranging arsenal of analytical methods to separate and characterize various protein structural modifications is presented in the specific context of these protein modifications. More precisely, applicable analytical techniques are discussed in the context of specific protein attribute(s) altered by a given modification. For example, analytical methods to characterize protein charge heterogeneity are discussed in the context of the modifications which introduce change in protein charge, etc.

Protein Stability: What Can Go Wrong and How to Measure It?

All levels of structural organization of proteins (see Chap. 2) are susceptible to damage as a consequence of physical or chemical stress (Table 3.1). Different modifications of the protein structure may be manifested as changes in various attributes (properties) of the protein. This is why assessing the stability of protein therapeutics is a complex and multifaceted task. In the following sections of this chapter, the most common structural modifications of proteins are presented together with the typical analytical approaches currently applied to measure these modifications.

Table 3.1 Common protein modifications and corresponding methods of analysis

Protein Modification	Typical causes and important factors	Physical property affected	Method of analysis
Oxidation	Light, metal ions, peroxides	Hydrophobicity	RP-HPLC, HIC, and mass spectrometry
Cys		Hydrophobicity	
Disulfide			
Intrachain			
Interchain			
Met, Trp, Tyr			
Fragmentation	pH, sequence (nearest AA neighbor)	size	Size-exclusion chromatography, SDS-PAGE
N to O migration		Hydrophobicity	RP-HPLC inactive in Edman reaction
Ser, Thr		Chemistry	
α-Carboxy to β-carboxy migration		Hydrophobicity	RP-HPLC inactive in Edman reaction
Asp, Asn		Chemistry	
Deamidation	pH, sequence (nearest AA neighbor), HOS	charge	Ion-exchange chromatography
Asn, Gln			
Acylation		Charge	Ion-exchange chromatography mass spectrometry
α-Amino group, ε-amino group			
Esterification/carboxylation		Charge	Ion-exchange chromatography mass spectrometry
Glu, asp, C-terminal			
Secondary structure changes Aggregation		Hydrophobicity	RP-HPLC
		Size	Size-exclusion chromatography
		Sec/tert structure	CD
		Sec/tert structure	FTIR
		Sec/tert structure	Fluorescence Light scattering
			Analytical ultracentrifugation, AF4

Protein Modifications Introducing Changes in Charge Heterogeneity

Deamidation and Isomerization

Some of the most common and most significant modifications in terms of impact on the properties of protein biopharmaceuticals are the deamidation of asparagine (Asn) and isomerization of aspartate (Asp). The mechanism of deamidation involves the formation of a cyclic imide intermediate (succinimide), which in turn hydrolyzes spontaneously to a mixture of isoaspartic/aspartic acid at an approximate ratio of 3:1. This reaction may be accompanied by further racemization of the isoaspartyl and aspartyl residues via the succinimide intermediate (see Fig. 3.4). Typically, the succinimide intermediate is short-lived at neutral pH, but in some cases may be stabilized.

Isomerization of Asp to isoAsp and deamidation of Asn occur frequently in biotherapeutics. In some cases, these modifications may be benign, but in others they may result in severe consequences for the product, for example, in cases when the complementarity-determining regions (CDR) regions of mAbs are affected. The biological activity of these molecules may be altered. The most important factors that influence deamidation and isomerization rates are tempera-ture, pH, local protein structure, and flanking aminoacyl residues. All chemical reactions mentioned above either result in changes in the charge of the affected protein (deamidation and succinimide formation) or in changes of the surface charge distribution (isomerization). Whereas deamidation results in an increase of acidic species, succinimide formation contributes to an increase of basic variants of the protein.

In should be noted that deamidation can also occur in glutamine (Gln) residues. However, the rates of Gln deamidation are much slower than Asn deamidation rates.

Pyro-Glu Formation

Another common modification that results in a change in the protein charge heterogeneity is the formation of pyroglutamate (pyro-Glu). Cyclization of the N-terminal glutamate (Glu) to pyroGlu (see Fig. 3.5) may occur either enzymatically or spontaneously. PyroGlu formation typically results in protein species with a higher pI than the main isoform.

Glycation

Glycation of proteins is the addition of reducing sugars (e.g., glucose or lactose) to the primary amine of lysine residues. It typically occurs during manufacturing in glucose-containing

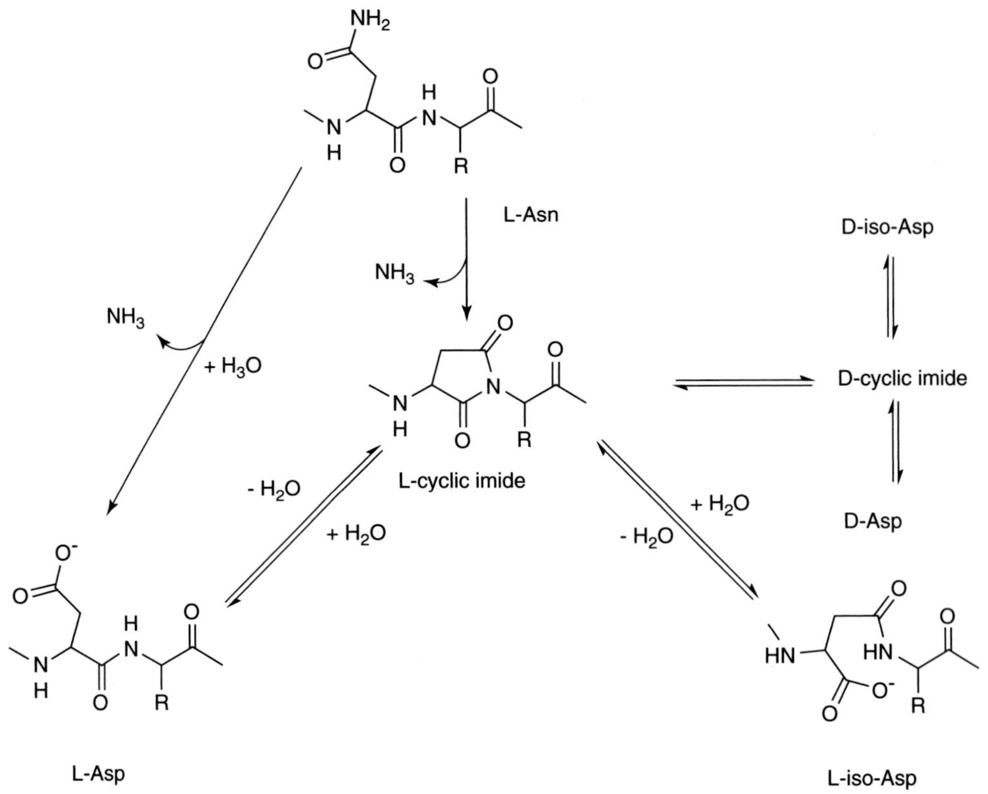

Fig. 3.4 Deamidation, isomerization, and succinimide formation

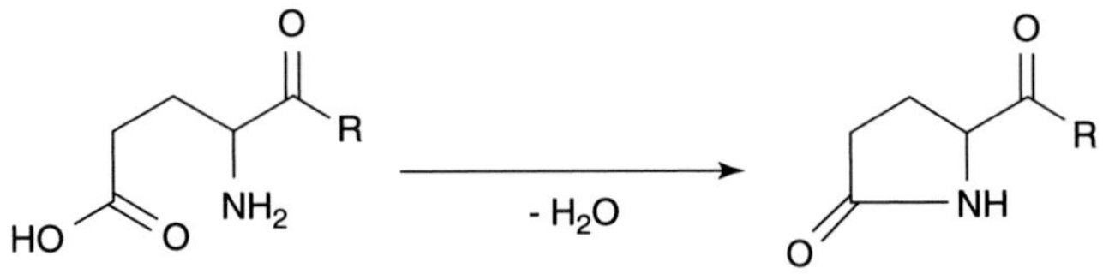

Fig. 3.5 Pyro-Glu formation

culture media. The glycation of proteins results in the increase of acidic protein variants.

Additional Modifications Inducing Changes in Protein Charge Heterogeneity

All protein charge modifications mentioned above occur as a result of chemical instability of proteins, i.e., under various types of physico-chemical stress: extreme pH, high temperature, etc. Other modifications that may result in the formation of protein charge variants occur as a result of enzymatic reactions—typically during the fermentation process while manufacturing biopharmaceuticals. Such modifications are, for example, the formation of C-terminal Lys variants and the sialylation of proteins.

Measuring Changes in Protein Charge Heterogeneity

A number of different protein modifications occurring either during long-term storage or during fermentation (upstream processing) may result in changes of the protein charge het-

erogeneity profile (e.g., deamidation, isomerization, glycation, etc.—see previous section). Because these modifications commonly occur simultaneously, in practice the resulting charge heterogeneity patterns of proteins are often relatively complex. Thus, characterization of the various species underlying the complex protein charge heterogeneity may require the application of different analytical approaches.

There are two main groups of techniques commonly used to measure changes in the charge profile distribution of protein therapeutics: electrophoretic techniques (IEF/icIEF, CZE) and chromatographic techniques (IEC), see below.

Ion-Exchange Chromatography (IEC)

A group of methods that has traditionally been applied for the assessment of charge heterogeneity and still finds a very broad use in this context is the group of analytical ion exchange chromatographic techniques (IEC or IEX). The commercial availability of a variety of stationary phases (chromatographic columns) for separation of charge variants using HPLC provides a choice of separation modes (anionic or cationic, strong or weak) and the opportunity for a very good separation and fractionation of variants that are difficult to separate by other techniques.

This technique takes advantage of the electric charge properties of proteins. Some of the amino acyl residues are negatively charged and others are positively charged. The net charge of the protein can be modulated by the pH of its environment relative to the pI value of the protein. At a pH value lower than the pI, the protein has a net positive charge, whereas at a pH value greater than the pI, the protein has a net negative charge. IEC utilizes various resins (chromatographic stationary phases), containing functional groups with either positive or negative charges (anion- or cation-exchange chromatography, correspondingly), depending on the pI of the separated protein. Positively charged proteins bind to negatively charged matrices and negatively charged proteins bind to positively charged matrices. Proteins bound to the chromatographic column are displaced (eluted) from the resin either by increasing the salt concentration of the mobile phase (screening the protein-column charge-charge interactions), or changing the pH of the mobile phase (effectively changing the charge of the protein). Proteins or protein variants with different net charges are separated from one another during elution with the change in the gradient (salt or pH). The choice of the charged resin and elution conditions are dependent upon the protein of interest.

Figure 3.6 shows an example separation of a monoclonal antibody using cationic exchange chromatography (CEX). Acidic isoforms elute before (left side of the chromatogram) and basic isoforms after (right side of the chromatogram) the main isoform. Upon exposure to low pH and elevated temperature stress conditions, significant changes in the charge heterogeneity of this protein can be observed. One finds a large decrease of the main isoform, accompanied by a decrease in the basic and increase in the acidic charge isoforms. Due to the complexity of the possible reactions mentioned in the previous section, it is difficult to assess what are the specific changes underlying the re-distribution of charge variants using this chromatogram alone. For this purpose, typically, it is necessary to fractionate the individual peaks (or groups of peaks) and subject them to further analyses (typically MS) in order to establish unequivocally the specific sequence modifications (see section MS).

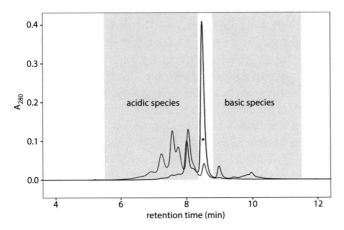

Fig. 3.6 Charge heterogeneity profile of a MAB using cationic exchange chromatography: control (black line) and stressed (blue line) acidic species formed during incubation at 40 °C and pH 4.5. The main peak is indicated by a star

Isoelectric Focusing (IEF/cIEF)

Another family of analytical methods to separate proteins based on their electric charge properties is isoelectric focusing (IEF). IEF techniques rely on separating proteins based on their isoelectric point (pI). In a first run, a pH gradient is established within a polyacrylamide gel (or a capillary in cIEF) using a mixture of small-molecular-weight ampholytes with varying pI values. After introduction of the protein sample and application of an electric field, all proteins or protein species/variants migrate within the pH gradient to the pH where their corresponding net charge is zero (their apparent pI). This technique is very useful for separating protein charge variants, such as deamidated or glycated species, from the native protein.

IEF, or its capillary configuration: Imaged capillary isoelectric focusing (icIEF) has the advantages that it can be applied to a broad variety of molecules and that it typically requires minimal method development efforts. icIEF has found particularly broad use in the biotech industry as a "platform" or "generic" charge heterogeneity assessment method due to its relative ease of use, minimal sample requirements, and its broad applicability.

Capillary Zone Elecrophoresis (CZE)

Another method which has gained an increased presence over the last decades is capillary zone electrophoresis (CZE). Rather than separating the proteins in a matrix, as in polyacrylamide gel electrophoresis through which the proteins migrate, in CZE the proteins are free in solution in an electric field within the confines of a capillary tube with a diameter of 25–50 μm. After passing through the capillary tube, proteins encounter an UV absorbance or fluorescence detector which can be used to quantify the proteins. The movement of one protein relative to another is a function of the molecular mass and the net charge on the protein. The latter can be influenced by pH and analytes in the solution. Typically, various additives or capillary coatings (e.g., epsilon amino caproic acid (EACA), or triethylenetetramine (TETA)) are used to suppress the interaction of proteins with the capillary wall, as well as the electro-osmotic flow.

CZE offers several advantages over other analytical methods described here to assess charge heterogeneity, such as the relatively easy implementation and low development efforts required. This makes CZE very suitable as a platform method. In addition, it offers a robust and rapid separation, which makes it amenable to high-throughput applications.

All analytical methods for measuring protein charge heterogeneity described in this chapter have advantages and disadvantages. The specific application of a given method depends on the given protein or protein mixture measured (e.g., the column, selection of method conditions,) and on the type of modification of the protein structure. Surface modifications, for example, are easily detected by IEC,

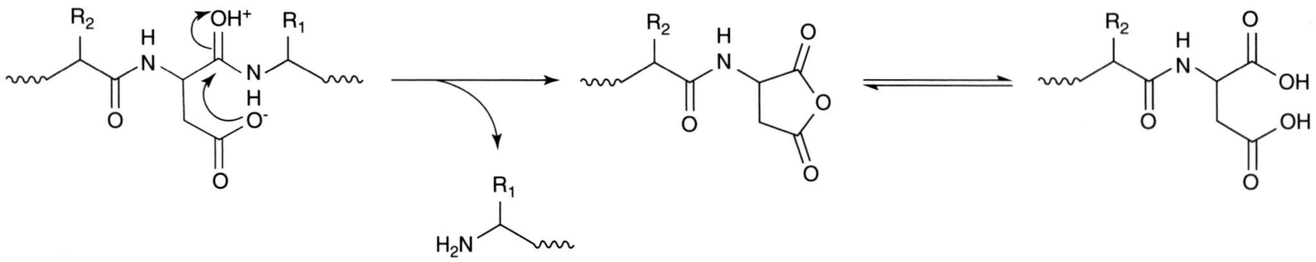

Fig. 3.7 Fragmentation of a polypeptide chain at an "Asp-X" site

whereas modifications buried within the structure of the protein may be detected better by IEF and CZE. Additional considerations for method selection depend on the specific purpose of the analyses. In cases where the characterization goal is to simply measure the changes under given stress condition, an IEF or a CZE measurement may be sufficient. However, one disadvantage of these electrophoretic methods is the inability for direct fractionation of molecular variants and online coupling (hyphenation) to MS for measurements of changes in the primary sequence. Thus, if the goal of the investigation is to understand the chemical nature of these changes and additional measurements (e.g., MS) may be required, IEC may be preferred. In practice, often a combined approach is used—e.g. electrophoretic methods may be used to measure charge heterogeneity routinely, while complementary IEC methods are used for fractionation of specific variants when needed.

Protein Modifications Introducing Changes in Size

Proteins can undergo a number of changes that affect their size. These changes may be either covalent modifications such as fragmentation of the polypeptide chain and intermolecular disulfide scrambling or noncovalent modifications such as protein aggregation and particle formation. Often, such changes dramatically affect protein biopharmaceuticals' potency and safety profiles.

Protein Fragmentation

Despite the fact the peptide bond as such is remarkably stable, fragmentation of the polypeptide backbone of recombinant proteins is a commonly observed modification. The reason for this apparent contradiction is that often, the adjacent amino acid side chains or local structure flexibility may contribute significantly to rendering a site susceptible to fragmentation. Neighboring amino acid side chains (most commonly Asp, Ser/Thr, Cys/Cys-Cys, Asn) may result in fragmentation which may occur via distinct mechanisms (see details below). In addition, flexible regions, such as disordered loops or the IgG hinge regions, for example, may be

particularly prone to fragmentation. Various additional factors and conditions (such as pH, temperature, and the presence of radicals) may contribute to increased fragmentation of the polypeptide chain of a protein as well. One of the most common examples of polypeptide chain fragmentation in recombinant proteins is fragmentation at an Asp-X site, where X is any amino acid residue (shown in Fig. 3.7). In this example, at low pH a nucleophilic attack of the ionized side-chain carboxylate of the Asp on the protonated carbonyl carbon of the peptide bond takes place, followed by release of the C-terminal peptide, cyclization, and a further hydrolysis of the unstable aspartic anhydride to an Asp residue.

There are many other reported mechanisms of fragmentation of the polypeptide chain, which involve various amino acid residues. Depending on the specific mechanism, various factors affect the rate of fragmentation. In the example above, low pH and small amino acid residues at position X favor the fragmentation reaction. Typically, the structural context (three-dimensional structure of the protein) also influences fragmentation with more disordered regions being more susceptible.

Protein Aggregation

The term "protein aggregation" refers to the process of agglomeration of two or more protein molecules, but it is typically distinct from functional protein–protein binding or quaternary structure formation. This term is too general for practical use, as it encompasses a vast diversity of different molecular phenomena. Protein aggregates can be reversible or irreversible, soluble or insoluble, covalent or non-covalent, etc. Furthermore, protein aggregates are typically present as a continuum of species spanning an enormous size range: from a few nm to >100 μm. Despite various attempts to categorize protein aggregation in a systematic fashion, to date there is still no sufficient clarity and agreement on nomenclature.

Protein aggregation may take place as a result of a variety of phenomena, such as local unfolding or perturbation of the protein secondary, tertiary, or quaternary structure. This general mechanism is typically evoked in cases when the system (e.g., protein solution) receives an excess of energy, such as during thermal stress. The physical stability of a protein is expressed as the difference in free energy, ΔG_U, between the native and denatured states. Thus, protein molecules are in equilibrium between the above two states. As long as this unfolding is reversible and ΔG_U is positive, it does not matter how small the ΔG_U is. In many cases, this reversibility does not hold. This is often seen when ΔG_U is decreased by heating. Most proteins denature (i.e., unfold) upon heating and subsequent aggregation of the denatured molecules results in irreversible denaturation. Thus, reversible unfolding is made irreversible by aggregation:

$$\text{Native state} \overset{\Delta G_U}{\Leftrightarrow} \text{Denatured state} \overset{k}{\Rightarrow} \text{Aggregated state}$$

Therefore, any stress that decreases ΔG_U and increases k will cause the accumulation of irreversibly inactivated forms of the protein. Such stresses may include chemical modifications as described above. Protein aggregation may occur as a result of oxidative processes such as disulfide scrambling or the chosen physical conditions, such as pH, ionic strength, protein concentration, and temperature. Development of a suitable formulation that prolongs the shelf life of a recombinant protein is essential when it is to be used as a human therapeutic (cf. Chap. 5).

Despite the variety of molecular mechanisms via which protein aggregation may ensue, the final outcome is the generation of protein species larger than the original molecule. Thus, all techniques for measuring protein aggregation are size based.

Measuring Changes in Protein Size

Similar to all other analytical techniques, the size-based protein characterization methods each have their advantages and disadvantages. Thus, some find more extensive use in characterizing small protein fragments and others in measuring large molecular weight species and proteinaceous particles. Because of the huge diversity of protein aggregation mechanisms and products, naturally there is a large array of analytical techniques, which have been developed to study specifically protein aggregation and assess its various aspects. Some techniques are suitable for measuring and characterizing small aggregates (oligomers), whereas others are useful for measuring exclusively high-molecular weight species (protein particles). Some techniques are specifically utilized to evaluate covalent aggregates, whereas others are used for noncovalent aggregates. However, in reality, nearly all of the common size-based protein characterization analytical techniques find dual use – for assessment of fragmentation and aggregation alike. This section will present only the most commonly used techniques, without delving into details and discussing the highly specialized technologies with very limited, niche use.

Size-Exclusion Chromatography

As the name implies, this technique separates proteins based on their size or molecular weight or shape. The matrix consists of very fine beads containing cavities and pores accessible to molecules of a certain size or smaller, but inaccessible to larger molecules. The principle of this technique is the distribution of molecules between the volume of solution within the beads and the volume of solution surrounding the beads (cf. Fig. 4.6). Small molecules have access to a larger volume than large molecules. As the solution flows through the column, molecules can diffuse back and forth, depending upon their size, in and out of the pores of the beads. Smaller molecules can reside within the pores for a finite period, whereas larger molecules, unable to enter these spaces, continue along in the fluid stream. Intermediate-sized molecules spend an intermediate amount of time within the pores. They can be fractionated from large molecules that cannot access the matrix space at all and from small molecules that have free access to this volume and spend most of the time within the beads.

Size-exclusion chromatography can be used to estimate the mass of proteins by calibrating the column with a series of globular proteins of known mass. However, the separation depends on molecular shape (conformation) as well as mass and highly elongated proteins—proteins containing flexible, disordered regions—and glycoproteins will often appear to have masses as much as two to three times the true value. Other proteins may interact weakly with the column matrix and be retarded, thereby appearing to have a smaller mass. Size-exclusion chromatography can be used to measure both protein fragmentation and protein aggregation, with the latter being more the common application. Sedimentation (ultracentrifugation), light scattering, or MS methods are preferred for accurate mass measurement.

Polyacrylamide Gel Electrophoresis ((SDS)-PAGE)

One of the earliest methods for analysis of proteins is polyacrylamide gel electrophoresis (PAGE). Polyacrylamide gels are used as a sieve. By adjusting the concentration of acrylamide that is used in these gels, one can control the migration rate of material within the gel. The more acrylamide, the more hindrance for the protein to migrate in an electrical field. The polyacrylamide gel provides not only a separation matrix but also a matrix to hold the proteins in place until they can be detected with suitable reagents.

The direction and speed of migration of the protein in a gel depend on the pH of the gel. If the pH of the gel is above its pI value, then the protein is negatively charged and hence migrates toward the positive electrode. The higher the pH of the gel, the faster the migration. This type of electrophoresis is called native gel electrophoresis.

The addition of a detergent, sodium dodecyl sulfate (SDS), to the electrophoretic separation system allows for the separation to take place primarily as a function of the size of the protein. Dodecyl sulfate ions form complexes with proteins, resulting in an unfolding of the proteins, and the amount of detergent that is complexed is proportional to the mass of the protein. The larger the protein, the more detergent that is complexed. Dodecyl sulfate is a negatively charged ion. When proteins are in a solution of SDS, the net effect is that the own charge of the protein is overwhelmed by that of the dodecyl sulfate complexed with it, so that the proteins take on a net negative charge proportional to their mass. Polyacrylamide gel electrophoresis in the presence of sodium dodecyl sulfate is commonly known as SDS-PAGE. All the proteins take on a net negative charge, with larger proteins binding more SDS but with the charge to mass ratio being fairly constant among the proteins. Since all proteins have essentially the same charge to mass ratio, how can separation occur? This is done by controlling the concentration of acrylamide in the path of proteins migrating in an electrical field. The greater the acrylamide concentration, the more difficult it is for large protein molecules to migrate relative to smaller protein molecules. This is sometimes thought of as a sieving effect, since the greater the acrylamide concentration, the smaller the pore size within the polyacrylamide gel. Indeed, if the acrylamide concentration is sufficiently high, some high-molecular-weight proteins may not migrate at all within the gel. Since in SDS-PAGE the proteins are denatured, their hydrodynamic size, and hence the degree of retardation by the sieving effects, is directly related to their mass. Proteins containing disulfide bonds will have a much more compact structure and higher mobility for their mass unless the disulfides are reduced prior to electrophoresis.

An example of SDS-PAGE is shown in Fig. 3.8. Here, SDS-PAGE is used to monitor the polypeptide chains of three different monoclonal antibodies and their various fragmentation products.

As described above, native gel electrophoresis and SDS-PAGE are quite different in terms of the mechanism of protein separation. In native gel electrophoresis, the proteins are in the native state and migrate on their own charges. Thus, this electrophoresis can be used to characterize proteins in the native state. In SDS-PAGE, proteins are unfolded and migrate based on their molecular mass.

Detection of Proteins Within Polyacrylamide Gels

Although the polyacrylamide gels provide a flexible support for the proteins, with time the proteins will diffuse and spread within the gel. Consequently, the usual practice is to fix the proteins using fixing solutions (rendering the proteins insoluble) or trap them at the location where they migrated to.

There are many methods for staining proteins in gels, but the two most common and well-studied methods are either

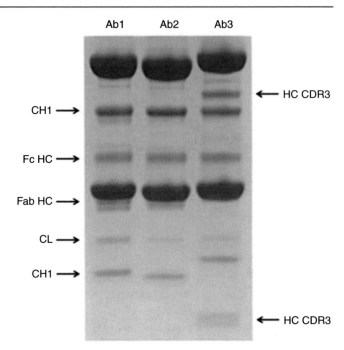

Fig. 3.8 Fragmentation of three (Ab1, Ab2, and Ab3) monoclonal antibodies monitored using SDS-PAGE *CH1* heavy chain constant domain 1, *Fc HC* heavy chain Fc domain, *Fab HC* heavy chain Fab domain, *CL* light chain constant domain, *HC* heavy chain, *CDR3* complementarity-determining region 3 (Taken from Vlasak and Ionescu, 2011– permission requested 23 Dec 2022)

staining with a dye called Coomassie blue or by a method using silver. The latter method is used if a very low limit of detection (LOD) needs to be achieved. The principle of developing the Coomassie blue stain is the hydrophobic interaction of a dye with the protein. Thus, the gel takes on a color wherever a protein is located. Using standard amounts of proteins, the amount of protein or contaminant may be estimated. Quantification using the silver staining method is more complex.

Blotting Techniques

Blotting methods form an important niche in the analytical toolbox of biotech products. Typically, they are used to detect very low levels of unique molecules in a milieu of proteins, nucleic acids, and other cellular components. For example, they can be used to detect components from the host cells used for the production of recombinant proteins (cf. Chap. 4). When blotting, biomolecules are transferred to a synthetic membrane ("blotting"), and this membrane is then probed with specific reagents to detect the molecule of interest. Membranes used in protein blots are made of a variety of materials including nitrocellulose, nylon, and polyvinylidene difluoride (PVDF), all of which avidly bind protein.

Liquid samples can be analyzed by methods called dot blots or slot blots. A solution containing the biomolecule of interest is filtered through a membrane, which captures the biomolecule. Often, the sample is subjected to some type of fractionation, such as polyacrylamide gel electrophoresis, prior to the blotting step. An early technique, Southern blotting, named after the discoverer, E.M. Southern, is used to detect DNA fragments. When this procedure was adapted to RNA fragments and to proteins, other compass coordinates were chosen as labels for these procedures, i.e., northern blots for RNA and western blots for proteins. Western blots involve the use of labeled antibodies to detect specific proteins.

Following polyacrylamide gel electrophoresis, the transfer of proteins from the gel to the membrane can be accomplished in a number of ways. Originally, blotting was achieved by capillary action. The transfer of proteins to the membrane can occur under the influence of an electric field, as well. The electric field is applied perpendicularly to the original field used in separation so that the maximum distance the protein needs to migrate is only the thickness of the gel, and hence, the transfer of proteins can occur very rapidly. This latter method is called electroblotting.

Once the transfer has occurred, the next step is to identify the presence of the desired protein. In addition to various colorimetric staining methods, the blots can be probed with reagents specific for certain proteins, as for example, antibodies to a protein of interest. This technique is called immunoblotting. In the biotechnology field, immunoblotting is used as an identity test for the product of interest. An antibody that recognizes the desired protein is used in this instance. Second, immunoblotting is sometimes used to show the absence of host proteins. In this instance, the antibodies are raised against proteins of the organism in which the recombinant protein has been expressed. This latter method can attest to the purity of the desired protein.

The antibody reacts with a specific protein on the membrane only at the location of that protein because of its specific interaction with its antigen. When immunoblotting techniques are used, methods are still needed to recognize the location of the interaction of the antibody with its specific protein. A number of procedures can be used to detect this complex (see Table 3.2).

The antibody itself can be labeled with a radioactive marker such as ^{125}I and placed in direct contact with X-ray film. After exposure of the membrane to the film for a suitable period, the film is developed and a photographic negative is made of the location of radioactivity on the membrane. Alternatively, the antibody can be linked to an enzyme that, upon the addition of appropriate reagents, catalyzes a color or light reaction at the

site of the antibody. These procedures entail purification of the antibody and specifically label it. More often, "secondary" antibodies are used. The primary antibody is the one that recognizes the protein of interest. The secondary antibody is then an antibody that specifically recognizes the primary antibody. Quite commonly, the primary antibody is raised in rabbits. The secondary antibody may then be an antibody raised in another animal, such as a goat, which recognizes rabbit antibodies. Since this secondary antibody recognizes rabbit antibodies in general, it can be used as a generic reagent to detect rabbit antibodies in a number of different proteins of interest that have been raised in rabbits. Thus, the primary antibody specifically recognizes and complexes a unique protein, and the secondary antibody, suitably labeled, is used for detection (see also section "ELISA" and Fig. 3.9).

The secondary antibody can be labeled with a radioactive or enzymatic marker group and used to detect several different primary antibodies. Thus, rather than purifying a number of different primary antibodies, only one secondary antibody needs to be purified and labeled for recognition of all primary antibodies. Because of their wide use, many common secondary antibodies are commercially available in kits containing the detection system and follow routine, straightforward procedures.

In addition to antibodies raised against the amino acyl constituents of proteins, specific antibodies can be used which recognize unique posttranslational components in proteins, such as phosphotyrosyl residues, which are important during signal transduction, and carbohydrate moieties of glycoproteins.

Figure 3.9 illustrates the above-mentioned detection methods that can be used on immunoblots. The primary antibody, or if convenient, the secondary antibody, can have an appropriate label for detection. They may be labeled with a radioactive tag as mentioned previously. Second, these antibodies can be coupled with an enzyme such as horseradish peroxidase (HRP) or alkaline phosphatase (AP). Substrate is added and is converted to an insoluble, colored product at the site of the protein-primary antibody-secondary antibody-HRP product. An alternative substrate can be used which yields a chemiluminescent product. A chemical reaction leads to the production of light that can detected by photographic or X-ray film. The chromogenic and chemiluminescent detection systems have comparable sensitivities to radioactive methods. The former detection methods are dis-

Table 3.2 Detection methods used in blotting techniques (further explained in the text below)

1. Antibodies are labeled with radioactive markers such as ^{125}I

2. Antibodies are linked to an enzyme such as horseradish peroxidase (HRP) or alkaline phosphatase (AP). On incubation with substrate, an insoluble colored product is formed at the location of the antibody. Alternatively, the location of the antibody can be detected using a substrate which yields a chemiluminescent product, an image of which is made on photographic film

3. Antibody is labeled with biotin. Streptavidin or avidin is added to strongly bind to the biotin. Each streptavidin molecule has four binding sites. The remaining binding sites can combine with other biotin molecules which are covalently linked to HRP or to AP

Fig. 3.9 Common immunoblotting detection systems used to detect antigens (protein of interest). Ag, on membranes. *Ab* antibody, *E* enzyme, such as horseradish peroxidase or alkaline phosphatase, *S* substrate, *P* product, either colored and insoluble or chemiluminescent, *B* biotin, *SA* streptavidin

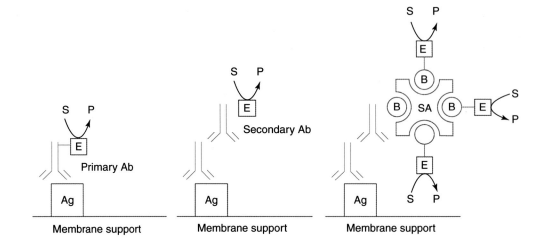

placing the latter method, since problems associated with handling radioactive material and radioactive waste solutions are eliminated.

As illustrated in Fig. 3.9, streptavidin, or alternatively avidin, and biotin can play an important role in detecting proteins on immunoblots. This is because biotin forms very tight complexes with streptavidin and avidin. Second, these proteins are multimeric and contain four binding sites for biotin. When biotin is covalently linked to proteins such as antibodies and enzymes, streptavidin binds to the covalently bound biotin, thus recognizing the site on the membrane where the protein of interest is located.

Capillary Electrophoresis Sodium Dodecyl Sulfate (CE-SDS)

Today, the traditional slab gel SDS-PAGE is increasingly being replaced by capillary electrophoresis sodium dodecyl sulfate (CE-SDS) due to the improved convenience and possibilities for automation, superior separation and reproducibility of this newer technique. In CE-SDS, the separation is carried out in a capillary in the presence of a sieving matrix. Whereas the basic electrophoretic separation principle of CE-SDS is the same as the one of SDS-PAGE, there are also some significant differences. Unlike SDS-PAGE, where only cross-linked polyacrylamide is used as a sieving matrix, in CE-SDS a variety of linear or slightly branched polymers may be used for the same purpose (e.g., linear polyacrylamide, polyethylene oxide, polyethylene glycol, dextran, and pullulan). This contributes to the method's flexibility. Another aspect where CE-SDS differs is the detection mode. In this technique, the laborious step of post-separation staining of the SDS-PAGE gels is eliminated and replaced by online UV or highly sensitive fluorescence detection. The elimination of the staining/destaining step as well as the need for scanning of the gels in CE-SDS (online detection generates quantitative electropherograms), together with the CE instrument design contributes to faster and more reproduc-

ible analysis, as well as amenability to automation. Altogether, CE-SDS is considered a superior method, demonstrating better accuracy, linearity, and precision than SDS-PAGE, which is why the latter has been effectively replaced by CE-SDS in the current pharmaceutical analytical practice.

Figure 3.10 shows an example of reduced and nonreduced CE-SDS separations of a monoclonal antibody.

An interesting new trend in the CE instrument development that has emerged recently—the development of new systems based on a microchip technology—has promised to revolutionize these analyses even further. These new chip- or cartridge-based configurations of CE-SDS (also IEF) enable the fully automatic separation of a larger number of samples and improve the ease of use even further. They achieve even faster and higher throughput separations, while maintaining the advantages of CE-SDS over SDS-PAGE.

Asymmetric Field Flow Field Fractionation

Protein aggregation is a process that can produce species spanning a vast size range, stretching from a few nanometers (oligomeric species) to hundreds of micrometers (visible particles). Measuring the various species across this continuum is impossible using one single technique. Whereas the methods described above (SEC, electrophoresis) are limited by the corresponding separation matrices (columns, gels) and can measure protein aggregates up to a certain range (depending on the protein size typically large oligomeric species), for quantifying larger protein species the application of other techniques is required. One of the best techniques for measuring high-order protein aggregates is asymmetric field flow fractionation (AF4). Although this technique was discovered in the 1960s and its application for separation of proteins developed in the 1980s, it has only gained popularity in the last decade.

AF4 is based on the migration of analytes in a mobile phase flowing through a channel with a semipermeable bot-

Fig. 3.10 CE-SDS reduced and nonreduced separations of a monoclonal antibody, showing the intact MAB, the heavy (HC), and light chains (LC) as well as various fragments, nonglycosylated form and incompletely reduced recombinant rmAb. Taken from Salas-Solano, O. and Felten, 2008

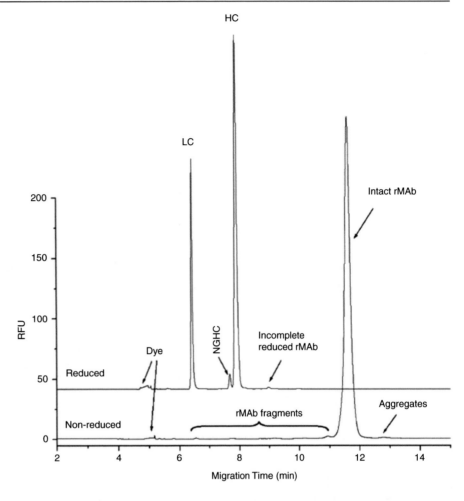

tom wall. During the separation, as the analytes advance through the channel they are subjected to an asymmetric field, generated by the application of a flow perpendicular to the sample flow. This leads to the differential migration of the analytes – smaller species eluting faster due to their faster lateral diffusion and larger species eluting slower. Thus, a separation of aggregates of various sizes is achieved without using a stationary phase. The lack of a stationary phase is an advantage as it eliminates the filter effect of columns, column frits and gels, which often leads to exclusion of the large aggregate/particle species from separation altogether. A second advantage is its very wide size range of separation—it can separate aggregates ranging from several nanometers to hundreds of nanometers and even micrometers.

Techniques for Measuring Sub-visible Particles

Protein aggregate species of tens of nanometers and larger are commonly termed "sub-visible particles" (SvP). Because the sub-visible particles are a critical quality attribute of protein therapeutics (cf. Chap. 7), their accurate and precise measurements are of high importance for the development of biotherapeutics. Due to the broad size-range span of these species, the simultaneous application of several techniques is

required in order to measure all applicable species (see Table 3.3).

The traditional and "gold standard" method for measuring particles in the micrometer size-range is light obscuration (LO). This method uses a flow cell through which the sample is led. A laser illuminates the flow cell. A particle passing with the liquid flow casts a shadow over the photodiode detector, which is registered and quantified via the resulting current drop.

A newer technique is flow imaging microscopy (FIM). In this technique, instead of photodiode detector, a high-speed camera is used to capture the images of all individual particles imaged via a microscope. This invention allows for studying the morphology of the particles detected and potentially provides the option to draw conclusions about their composition and origin.

Other methods for particle characterization include some newly emerged techniques, such as resonance mass measurement (RMM) or nano tracking analysis (NTA). Both of these techniques allow measuring the concentrations of submicrometer particles, which is their major application. Nanoparticle tracking analysis uses single-particle tracking to calculate the diffusion coefficient of each individual par-

Table 3.3 Analytical techniques commonly used for measuring and characterizing sub-visible particles (reproduced with modifications from Ríos Quiroz et al. 2016)

	Method principle and data analysis	Size (μm)												Optimal sample concentration (part./ mL)
		0.03	0.05	0.20	0.30	0.50	0.60	0.80	1.00	2.00	5.00	10.00	25.00	
NTA	*Tracking of Brownian motion of individual particles:* Hypothetical hard spheres that diffuse at the same speed of the tracked particles are assumed. The hydrodynamic diameter is obtained according to the 2D-modified Stokes-Einstein equation. For count determinations the averaged particle abundance (average number of particles per frame) is divided by the estimated volume of the sample chamber.	▓	▓	▓	▓	▓	▓	▓						3×10^8 -1×10^9 ~20-70 centers per frame
RMM	*Changes in frequency due to added mass:* Shifts in frequency with respect to sensor baseline resonance are converted into buoyant mass using the sensor-specific sensitivity. Sensitivity is obtained using size standards as calibrators. Knowing the fluid's and particle's density, buoyant mass is converted into dry mass. Assuming a sphere shape, particle diameter is calculated. Concentration is obtained relating the number of events (particles) registered with the volume of sample dispensed		▓	▓	▓	▓	▓	▓	▓	▓	▓			$< 8 \times 10^6$
EZS	*Changes in resistance due to volume displacement:* The impedance pulses generated as particles are pumped through an orifice in a glass tube are individually analyzed by the instrument electronic components. As the electrical current is constrained in the aperture orifice each pulse is directly proportional to the volume that the particle displaced and its size. Concentration is obtained relating the number of events (particles) registered with the volume of sample dispensed					▓	▓	▓	▓	▓	▓	▓	▓	~ 2×10^5 Coincidence <5%
FIM	*Image analysis of single particles:* Digital images of the particles in the sample are captured and analyzed by the instrument software. Following background comparison, intensity values are assigned to each activated pixel. Adjoining pixels below 96% of the maximum brightness is grouped as particles. Internal algorithms are used to generate morphological descriptors per each particle. Concentration is obtained relating the number of events (particles) registered with the volume of sample dispensed.								▓	▓	▓	▓	▓	$< 9 \times 10^4$
LO	*Drop in current due to light obscuration:* A calibration curve size vs. voltage is defined using calibration size standards. Particle size is obtained by direct interpolation in the calibration curve of the voltage recorded when a particle blocks the sensor. Concentration is obtained relating the number of events (particles) registered with the volume of sample dispensed.									▓	▓	▓	▓	$< 1.8 \times 10^4$

NTA nanoparticle tracking analysis, *RMM* resonance mass measurement, *EZS* electrical zone sensing, *FIM* flow imaging microscopy, *LO* light obscuration

ticle and in turn—its size. The concentrations of particles in solution are then inferred from the small subset measured. Due to the unique capability of this technique to measure particles >30 nm, it finds extensive use in vaccine development and recombinant virus characterization. RMM also offers some unique features, namely the ability to distinguish between particles with different densities. The latter is very useful in discriminating proteinaceous particles from silicone oil droplets (often present in biopharmaceutical drug products in prefilled syringes or cartridges), for example.

The availability of the various SvP methods allows for coverage of the entire particle size range. However, as with all analytical methodologies, an important consideration when using different SvP characterization methods in parallel is to recognize the specific advantages and shortcomings that apply to each of them.

Protein Modifications Introducing Changes in Hydrophobicity

Protein Oxidation

Oxidation is a common degradative pathway for proteins. It often has profound effects on their physico-chemical properties. Such major property changes may in turn result in alteration of the biological functions of the affected protein, such as loss of binding, reduction of enzymatic activity, and unexpectedly rapid clearance. Thus, monitoring protein oxidation is very critical for the successful development of biopharmaceuticals.

Protein oxidation may occur during all stages of protein manufacturing, processing, and storage, whenever the proteins may be exposed to oxidative agents. The latter may include peroxides, transition metal ions, exposure to light, etc.

Whereas theoretically all amino acids can be oxidized, in practice the most commonly oxidized amino acid residues are Trp, Met, Tyr, His, Phe, and Cys.

Tryptophan residues are particularly susceptible to oxidation due to the relatively high reactivity of the aromatic indole with reactive oxygen species. Tryptophan oxidation typically requires some level of exposure of the Trp residues, which a commonly buried in the three-dimensional structure of proteins. However, when oxidized, tryptophan residues may convert to a large variety of products (see Fig. 3.11), all of which with properties very different from the original Trp. The most common pathway for Trp oxidation includes the formation of N-formylkinurenine.

Another commonly oxidized amino acid is methionine. The sulfur atom in the Met residue can accept either one or two oxygen atoms leading to the formation of sulfoxide or sulfone, correspondingly (see Fig. 3.12). Due to the typically high surface exposure of Met, this modification is relatively common.

Measuring Changes in Protein Hydrophobicity

Most oxidative modifications of proteins result in some changes of the polarity of the affected residues. In the Met and Trp oxidation examples shown here, the resulting products differ from the original residues by their relative hydrophobicity. Thus, protein oxidation is commonly detected and quantified with analytical methods utilizing polarity-based separation. The most common technique using this separation principle is reversed-phase chromatography.

Reversed-Phase High-Performance Liquid Chromatography

Reversed-phase high-performance liquid chromatography (RP-HPLC) takes advantage of the hydrophobic properties of proteins. The functional groups on the column matrix may contain from one to up to 18 carbon atoms in a hydrocarbon chain. The longer this chain, the more hydrophobic is the matrix. The hydrophobic patches of proteins interact with the hydrophobic chromatographic matrix. Proteins are then eluted from the matrix by increasing the hydrophobic nature of the solvent passing through the column. Acetonitrile is a solvent commonly used, although other organic solvents such as ethanol also may be employed. The solvent is made acidic by the addition of trifluoroacetic acid, since proteins have increased solubility at pH values further removed from their pI. A gradient with increasing concentration of hydrophobic solvent is passed through the column. Different proteins have different hydrophobicities and are eluted from the column depending on the "hydrophobic potential" of the solvent.

This technique can be very powerful. It may detect the addition of a single oxygen atom to the protein, as is the case when a methionyl residue is oxidized or when the hydrolysis of an amide moiety on a glutamyl or asparaginyl residue occurs. Disulfide bond formation or shuffling also changes the hydrophobic characteristic of the protein. Hence, RP-HPLC can be used not only to assess the homogeneity of the protein but also to follow degradation pathways occurring during long-term storage.

RP-HPLC does not always provide sufficient resolution for separation of oxidized species of an intact protein, particularly when larger and more complex proteins are analyzed (such as monoclonal antibodies). In such cases, various methods can be applied to solve this problem. For example, the intact protein can be digested into subdomains, such as the Fab and Fc fragments in the case of a MAB, or even smaller fragments (cf. Chap. 8). This latter approach typically employs more frequently cutting enzymes, such as trypsin, or Lys-C, in order to generate a large number of small peptide fragments, which can be better separated on a RP-HPLC from their oxidized isoforms (see Fig. 3.13).

Such RP-HPLC separations of proteolytic digests of recombinant proteins typically yield complex and unique

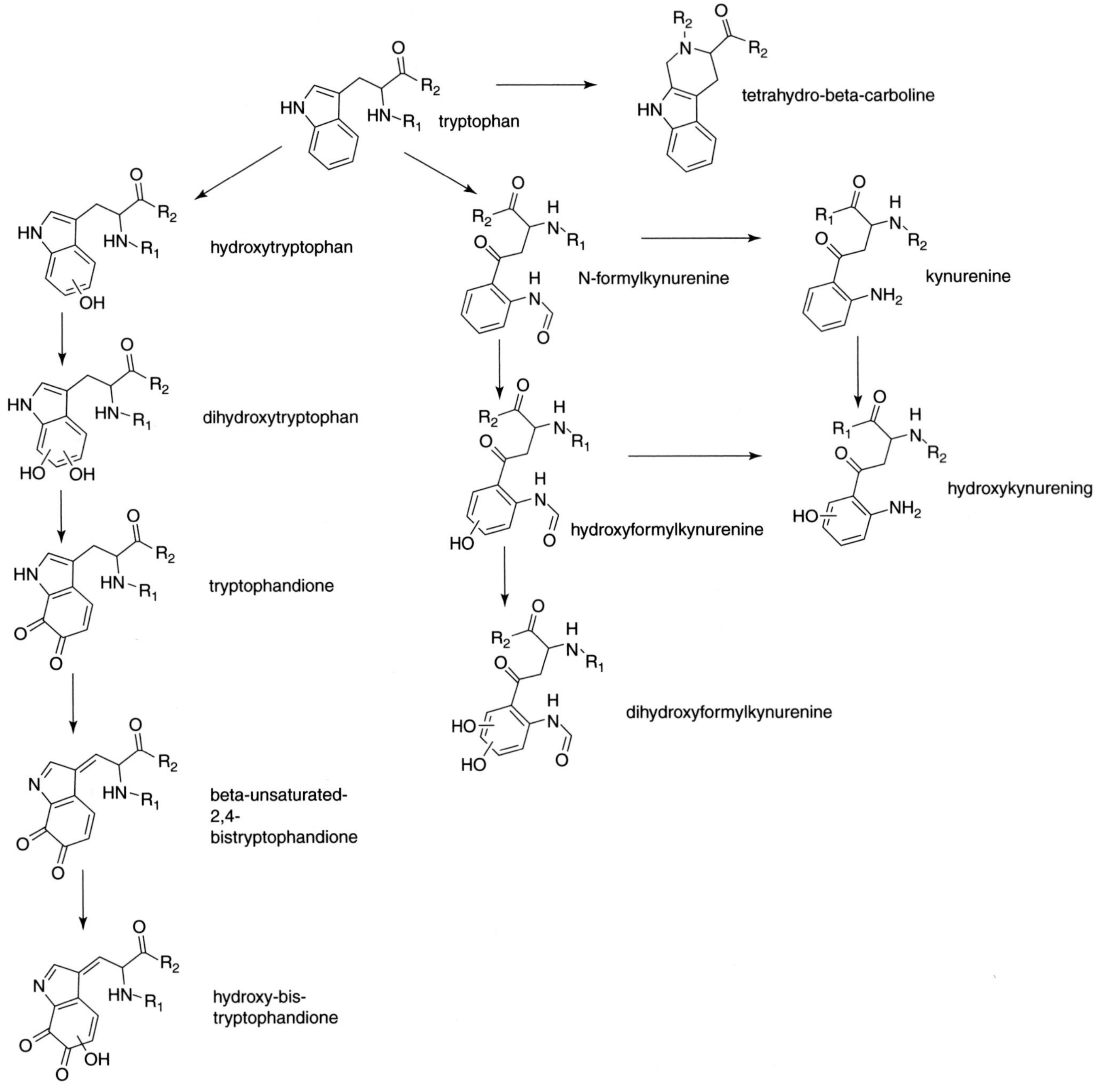

Fig. 3.11 Possible oxidation products of tryptophan

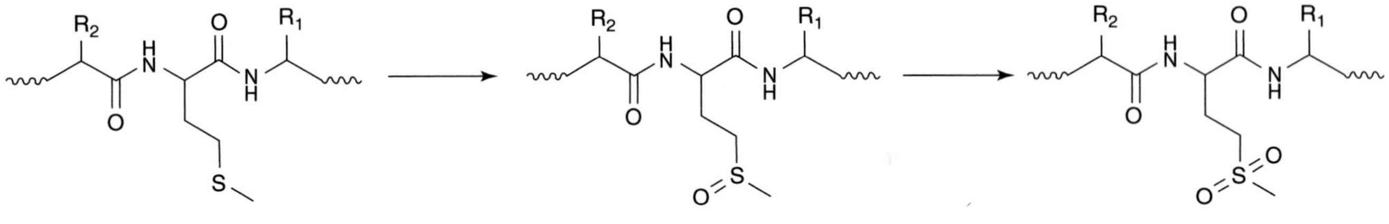

methionine methionine sulfoxide methionine sulfone

Fig. 3.12 Oxidation of a methionine-containing peptide

Fig. 3.13 A RP-HPLC peptide map of a MAB showing the oxidation of individual methionine residues (indicated by arrows). Top trace shows the peptide map of a nonoxidized protein sample, middle trace—sample oxidized using H_2O_2 and bottom—sample to which H_2O_2 and an antioxidant have been added (Courtesy of Folzer et al. 2015)

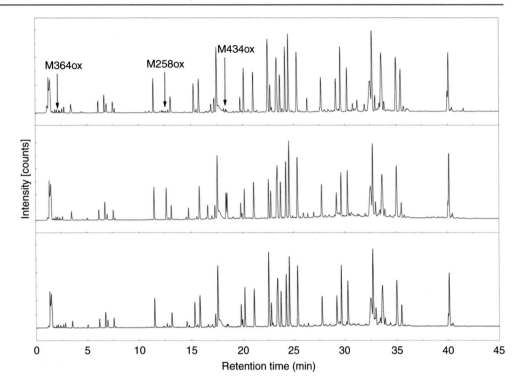

separation patterns ("peptide maps"), which are often used as a method to identify a protein. Several different proteases, such as trypsin, chymotrypsin, and other endoproteinases, are used for these identity tests (see below under "Mass Spectrometry").

Hydrophobic Interaction Chromatography

A companion to RP-HPLC is HIC. In principle, this latter method is normal-phase chromatography, i.e., here an aqueous solvent system rather than an organic one is used to fractionate proteins. The hydrophobic characteristics of the solution are modulated by inorganic salt concentrations. Ammonium sulfate and sodium chloride are often used, since these compounds are highly soluble in water. In the presence of high salt concentrations (up to several molar), proteins are attracted to hydrophobic surfaces on the matrix of resins used in this technique. As the salt concentration decreases, proteins have less affinity for the matrix and eventually elute from the column. This method lacks the resolving power of RP-HPLC, but is gentler, since low pH values or organic solvents as used in RP-HPLC can be detrimental to some proteins.

Two-Dimensional (Hyphenated) Techniques

Some analytical techniques can be combined (hyphenated) to achieve additional functionality. Two prominent examples are discussed below.

2-Dimensional Gel Electrophoresis and Differential Gel Electrophoresis

Isoelectric focusing and SDS-PAGE can be combined into a procedure called 2D gel electrophoresis. Briefly, proteins are first fractionated by isoelectric focusing based upon their pI values. They are then subjected to SDS-PAGE run perpendicular to the first dimension and fractionated based on the molecular weights of the proteins. These separations produce a gel on which each protein appears as a separate spot, corresponding to a specific molecular weight and pI value combination. This setup allows for separating very complex protein mixtures (e.g., extracted from cells or tissues) and is commonly used in the proteomic field (cf. Chap. 9). Another situation where 2D gel electrophoresis is regularly applied concerns profiling of host cell proteins.

Another 2-dimensional gel technique is differential gel electrophoresis (DIGE). This technique is essentially 2D gel electrophoresis where 2 or 3 different samples are separated simultaneously. The proteins in the samples are labeled with differently colored fluorescent dyes (typically Cy2, Cy3, and Cy5, which are charge- and mass-matched). The proteins co-migrate on the gel and are typically detected simultaneously using a multichannel scanner or a camera. The overlay of the different channels allows for identifying/visualization of individual proteins being over- or underrepresented in the different samples, which is otherwise very difficult to find out with complex protein mixtures.

2-Dimensional Chromatography

One important strategy for improving the selectivity (specificity) of chromatographic separations is the coupling of two or more columns—for example, an ion exchange column, directly followed by a reversed phase column. This strategy allows for the separation of highly complex analyte mixtures, such as the mixtures of peptides generated during shotgun proteomic experiments. In these experiments, multiprotein (up to thousands of proteins) mixtures are digested using a protease (typically trypsin) to produce an even richer mixture of peptides. These peptide mixtures are separated on a 2D chromatographic system in order to reduce the complexity and are commonly analyzed using an MS as a detector. MS/MS fragmentation of the individual peptides allows for the sequencing and identification of each peptide and correspondingly—protein, thus permitting the semi-quantitative analysis of the original protein mixtures. Such approaches are very commonly used in host cell protein (HCP) profiling and biomarker research.

Modifications to the Higher Order Structure of Proteins

All levels of protein structural organization can be altered as a consequence of physical or chemical stress. These alterations can be manifested in a large diversity of protein modifications, each changing the physico-chemical and biological properties of the protein, see Table 3.1.

In addition to covalent modifications (modifications to the primary structure/amino-acid sequence) described in earlier sections, the higher order structure of proteins can undergo changes as well. Such changes can be relatively minor, such as alterations of the quaternary structure (subunit configuration) of a protein complex due to an incorrect disulfide bridge, or more substantial like perturbation of the tertiary structure of a given domain. Some of these possible modifications are described in the following section.

Measuring Changes in Higher Order Structure of Proteins

A large number of analytical techniques can be used to measure the structural organization of proteins. However, all these methods differ significantly from each other by the level of information content they provide and the ease of use. Typically, the most accessible and easy to use methods provide a relatively low information content, whereas higher resolution methods (providing specific information about the structure of separate domains and even functional groups and atoms) require highly specialized and expensive equipment and dedicated, highly trained specialized personnel.

Lower Resolution Techniques: CD, FTIR, Fluorescence

Most of the spectroscopic techniques used for characterization of the higher order protein structure (see before) are relatively easy to use, although interpretation of the experimental results typically requires expert knowledge. Very often in biotech development, spectroscopic measurements such as CD, FTIR, or fluorescence spectroscopy are applied in order to compare protein therapeutic products from different manufacturing batches (typically after changes introduced into the manufacturing process), asking the specific question whether significant differences (alterations of the secondary or tertiary structure) are present between the different batches. In these, so called "comparability studies" (cf. Chap. 12), the first goal is to identify if such changes are present at all. To answer this question, it is sufficient to overlay the CD, FTIR, or fluorescence spectra from the different batches and look for any differences. However, one common downside of CD, FTIR, and fluorescence spectroscopy is the fact that if differences are seen, it is difficult to judge to which region of the molecule these differences are related. The reason for this is that all of the above-mentioned techniques provide a summary/population information for all of the spectroscopically active functional groups in the molecule (chromophores in the case of CD, fluorophores in the case of fluorescence and amide absorption bands in case of FTIR) and do not provide specific spacial information for individual groups. This means that using the results from such experiments one cannot pinpoint where exactly in the structure of the molecule the detected changes are located. To answer that question, additional analytical work employing higher resolution techniques (see below) is required.

One additional complication resulting from the fact that the methods mentioned above measure the overall molecular population present in the test solution is the fact that the LOD of specific structural changes is relatively high. More specifically, if a given structural modification has occurred only within a small portion of the overall population (for example, let us say in only 5% of the molecules), this change is unlikely to give sufficiently strong spectral signals to modify the overall (summary) spectrum collected in the experiment. Thus, the techniques described above are typically useful for the detection of gross modifications of the secondary, tertiary (CD, FTIR, and fluorescence), and quaternary (fluorescence) structure.

Higher Resolution Techniques

In contrast to the spectroscopic methods described in the previous section, some techniques are capable of providing specific spatial information for specific domains, functional groups or even individual atoms in proteins. The degree of structural detail available varies from method to method and typically, the higher the information content (structural details)—the more complex and specialized the method.

An increasingly popular technique for higher order structures (HOS) determination is HDX-MS (see Analytical Toolbox). This method typically provides much higher spatial resolution than the spectroscopic techniques mentioned above, although not as high as X-ray crystallography or NMR (see below). What is typically achieved using HDX-MS is at peptide-level information, allowing to map individual subdomains according to their mobility and solvent accessibility. These maps (using 3D molecular models) can be extremely useful in understanding structural alterations limited to small regions of the protein. Compared to the higher resolution methods described below, HDX-MS is more accessible (both in terms of instrumentation and also expertise), and it is not limited by the protein size, crystallographic properties, or required sample amounts, which are some of the downsides for NMR and X-ray crystallography.

X-ray crystallography is the ultimate spatial resolution method. Using this technique, it is quite common to determine protein structures at an atomic resolution. This technique is indispensable in research focused on enzymes or specific protein-binding sites. A number of X-ray crystallographic structure analyses of Ab-Ag complexes, for example, or drug–target molecule complexes have been very illuminating and were critical with respect to advancing drug discovery. In drug development, this technique typically does not find as broad use, due to the huge efforts required. Despite advances in this field, including the use of robotics and machine learning approaches, it is not uncommon to take year(s) for solving a given structure, not to mention the fact that solving some structures is currently impossible due to the fact that some proteins are exceedingly difficult to crystalize.

Another very high-resolution technique is NMR (see Analytical Toolbox). One very significant advantage that this method offers is the fact that experiments can be carried out in solution, meaning that often important aspects of the protein dynamics can be interrogated, a feature not available with other high-resolution techniques. Until recently, major limitations of this technique came from the need to label the proteins prior to analyses using stable isotopes and also the size limit to the proteins analyzed. More recent advances in the protein NMR field led to the possibility to obtain 2D ^{13}C NMR methyl fingerprint data for structural mapping of an intact MAB at natural isotopic abundance, which significantly ameliorated the shortcomings mentioned above.

One technique which has undergone a development boom in the twenty-first century is Cryo Electron Microscopy (Cryo EM). This technique also allows looking at native structures and complexes in some cases to near-atomic resolution. The huge impact of this technique on biological research and studies of protein complexes was recognized in 2017, when the co-discoverers of the method received the Nobel Prize for Chemistry (Henderson 2018).

Stability and Characterization of Nucleotide-Based Pharmaceuticals

Nucleotide-based therapeutic and preventative pharmaceuticals have gained tremendous popularity during the last decade. One of the remarkable success stories during the COVID-19 pandemic in 2020 has been the rapid development of mRNA-based vaccines, but there are other classes of nucleotide-based pharmaceuticals, featuring significant structural diversity: oligonucleotides and (oligo)nucleotide-protein conjugates including anti-sense oligonucleotides (ASO), small interfering RNA (siRNA), and gene therapeutics (viral and non-viral). They are discussed in more detail in Chaps. 13, 14 and 15. Naturally—being biological molecules—nucleotide-based pharmaceuticals exhibit a variety of liabilities in terms of their stability and similarly to protein-based pharmaceuticals they can degrade under various types of physical and chemical stress. However, there are several unique features of nucleotide-based pharmaceuticals, which have a distinct impact of their stability and in turn—efficacy.

One particular feature of nucleotide-based pharmaceuticals that differentiates them from protein-based pharmaceuticals is the fact that the various structural classes of the former vary significantly in stability. For example, DNA is significantly more stable than RNA and double-stranded RNA and is more stable than single-stranded RNA. The significant influence of structure (primary and higher order structures) on stability results in differential importance of the various structural modifications, as well as the corresponding analytical approaches that need to be selected for monitoring and quality control.

Another unique feature of nucleotide-based pharmaceuticals (most relevant to RNA-based molecules) is the use of covalent modifications to regulate their stability. One prominent example amongst many others is the use of pseudouridine in mRNA-based pharmaceuticals such as some of the approved COVID-19 vaccines marketed today. Whereas certain chemical modifications do help to protect nucleotide-based pharmaceuticals from chemical degradation, in some cases, the influence of these modifications on the interactions with the immune system (in vivo stability) is of even higher relevance. For example, nucleotide-based pharmaceuticals can trigger immune response by activation of the Toll-like receptor (TLR) system, an innate anti-microbial defense mechanism. This may result in rapid clearance and poor efficacy, which can be counteracted using chemical modifications.

Finally, an important distinction between protein- and nucleotide-based pharmaceuticals is that most often than not proteins preserve their potency even when modified, unless the modifications reside in locations critical for their biological activity such as binding sites or catalytic centers, or change their pharmacokinetic profile. In contrast, even a single chemical modification may render nucleotide-based pharmaceutical molecules completely inactive. A good example is a phosphodiester bond cleavage in an mRNA molecule, resulting in an incomplete message and therefore—complete lack of efficacy.

Chemical Stability of Nucleotide-Based Pharmaceuticals

There are several pathways through which nucleotide-based pharmaceuticals can undergo degradation in vitro (Fig. 3.14). However, the most impactful degradation mechanisms from a practical point of view are hydrolysis or transesterification and oxidation, whereas depurination, depyrimidination, and deamination are much less common.

Hydrolytic Degradation

Hydrolysis of nucleotide-based pharmaceuticals predominantly affects the backbone phosphodiester bonds. Importantly, phosphodiester bonds in RNA molecules are much more labile than DNA, due to presence of a free 2′ OH group on the ribose. Mechanistically, the 2′ OH group initiates a nucleophilic attack on the phosphate ester bond leading to a cleavage at the P-5′ O ester bond. This reaction can be catalyzed by nucleases, but can also occur in the absence of enzymes, catalyzed by Brønsted acids and bases (Fig. 3.15). Interestingly, the susceptibility of RNA molecules to hydrolytic degradation is highly dependent on their higher order structure with the resulting differences in base stacking being considered as most impactful structural factor determining hydrolytic susceptibility.

It appears that nuclease-mediated hydrolysis of RNA is most relevant to mRNA-based pharmaceuticals, likely due to their low level of modification, as compared to siRNA, which are typically heavily modified, or to ASO, which most often feature moderate modifications (Chaps. 13 and 14).

Oxidation

Another common degradative mechanism relevant to nucleotide-based therapeutics is oxidation. Despite being less common in practical terms, oxidation remains an important stability-relevant factor for nucleotide-based therapeutics. Via a variety of reactive oxygen species oxidation may affect the electron-rich purine and pyrimidine nucleobases, leading to base cleavage, strand scission, or adduct formation (Fig. 3.14).

Measuring Chemical Modifications of Nucleotide-Based Pharmaceuticals

Traditional and still relevant techniques used for the analysis of nucleic acids are ultraviolet-visible (UV/Vis) spectrophotometry, fluorescence spectrometry, electrophoresis, polymerase chain reactions (PCR), and next-generation sequencing (NGS). However, in quality control protocols of nucleotide-based pharmaceuticals separation-based technologies, such as HPLC- and CZE-based techniques, are preferred.

The majority of the modifications in the chemical structure of nucleotide-based pharmaceuticals described in the

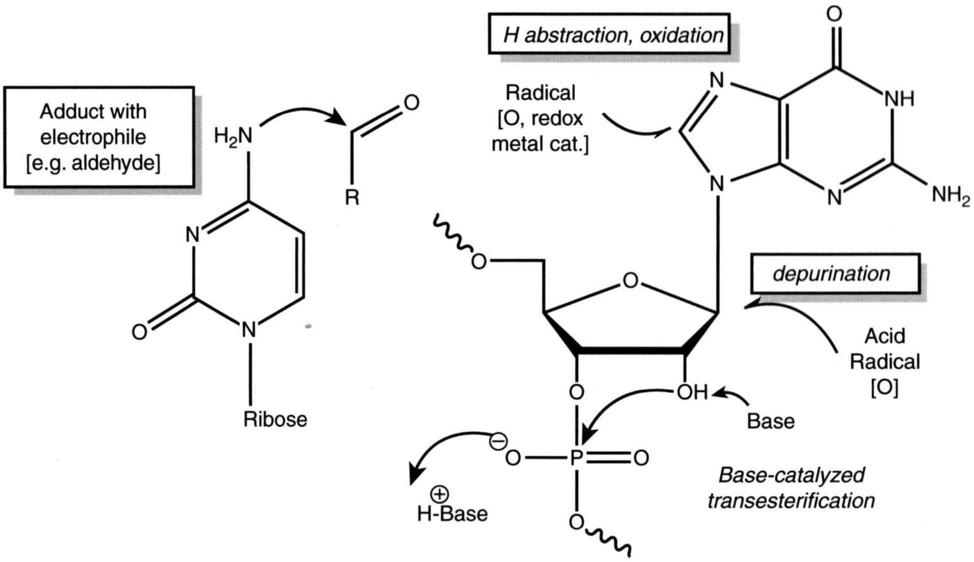

Fig. 3.14 Schematic of theoretical mRNA degradation mechanisms and their sites on the mRNA molecule. [O]: oxidant. The R (lipid)-aldehyde forming the lipid-mRNA adduct is the result of lipid oxidation reactions of the ionizable lipid present in many mRNA delivery systems, i.e., lipid nanoparticles, LNPs (cf. Chap. 5). Courtesy of Oude Blenke et al. 2022

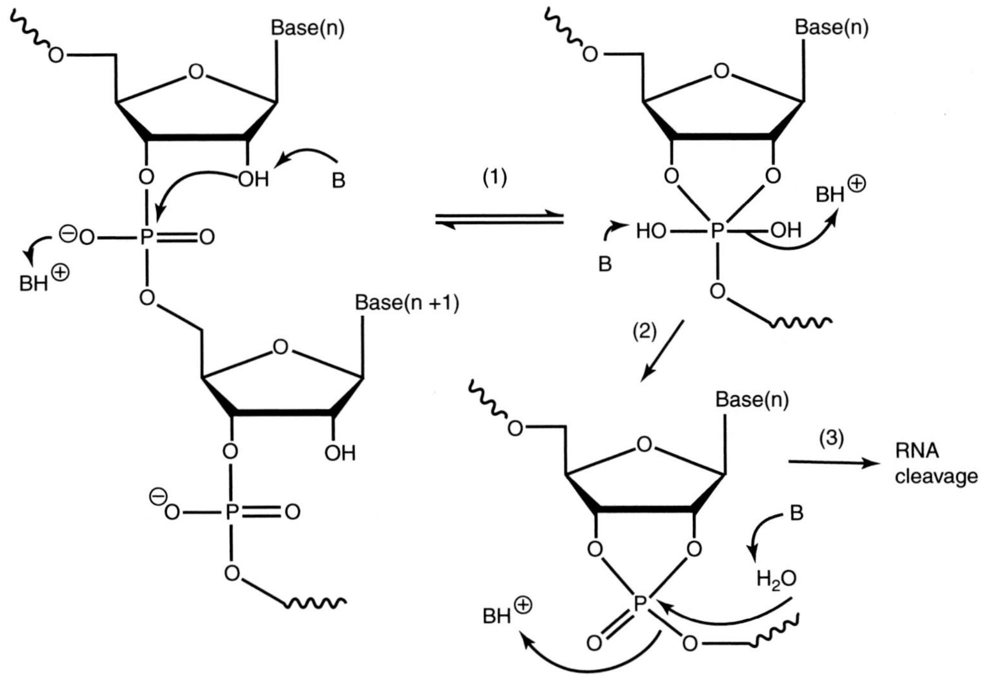

Fig. 3.15 Detailed mechanism of the degradative transesterification reaction in mRNA strands leading to strand cleavage. B denotes a brønsted base, and BH+ is the corresponding conjugate acid. Courtesy of Oude Blenke et al. (2022)

previous section result into a change of the charge—hydrophobicity profile of these molecules. Thus, nearly universally, variants or hyphenated versions of reversed-phase and ion exchange chromatography (with the former being more prevalent) are used to measure these molecular modifications.

RP-HPLC analysis of proteins was reviewed in more detail earlier in this chapter. However, as nucleotide-based pharmaceuticals are highly charged macromolecules, specific conditions need to be created for their analyses by RP-HPLC. For example, in a version of the method called ion-pair reversed-phase (IP-RP) liquid chromatography, lipophilic cations that ion-pair with the negatively charged backbone of the nucleic acids are added to the mobile phase, thus enhancing the hydrophobicity-based separation of these species.

Hydrophilic interaction liquid chromatography (HILIC) is another separation technology commonly used in the analyses and characterization of nucleotide-based therapeutics. In HILIC, contrary to RP-HPLC, analytes are eluted in order of their increased polarity from a polar stationary phase using an aqueous eluent.

The traditionally used agarose gel electrophoresis to separate molecular variants of nucleic acids has been replaced by capillary electrophoresis (capillary gel electrophoresis (CGE) and capillary zone electrophoresis (CZE)). These electrophoretic techniques are more sensitive, have higher resolution and higher throughput than the agarose gel-based instrument, and are thereby much better suited to be used in the pharmaceutical development phase and for product quality control. These techniques can be applied to characterize

both size variants, e.g., hydrolytic degradants or other modifications, such as nucleobase modifications.

As mentioned earlier, a peculiar feature of the nucleotide-based therapeutics is the severe impact of a single modification on the biological activity of the molecule. Hence, the need for high-resolution characterization techniques is much higher, and they are much more widely used than in pharmaceutical protein development. High-resolution techniques are, for example, MS or other sequencing techniques, such as reverse-transcriptase quantitative polymerase chain reaction (RT-qPCR) or NGS.

In MS analyses of nucleotide-based pharmaceuticals a variety of activation methods are applied: collision-induced dissociation (CID), ultraviolet photodissociation (UVPD), electron transfer dissociation (ETD), negative electron transfer dissociation (NETD), electron detachment dissociation (EDD), to name a few. Similarly to protein analytics, MS methods for nucleotide-based pharmaceutical analyses are commonly hyphenated with various chromatographic separation techniques.

Higher Order Structure Modifications in Nucleotide-Based Pharmaceuticals

The role of the secondary and tertiary structure of nucleic acids on the biological activity and stability of nucleotide-based therapeutics, in particular mRNA, has been an area of developing appreciation over the last couple of decades. Interestingly, and highly relevant for the stability of pharmaceuticals, the RNA folding appears dynamic, rather representing molecular ensembles than static structures, and highly dependent on the conditions such as pH, ionic strength, temperature, the presence or absence of molecular partners. Because of the high complexity and dynamic nature of the higher order structure of RNA-based pharmaceuticals, the development of these agents employs a plethora of sophisticated computational analyses. An important consideration in maintaining and monitoring the higher order structure of nucleotide-based pharmaceuticals such as mRNA products, is that it may, to a considerable extent, determine the chemical stability of these products as noted.

Stability of Supramolecular Assembles in Nucleotide-Based Pharmaceuticals

Nucleotide-based pharmaceuticals are rarely used as "naked" nucleic acid. This is driven by the need for intracellular delivery of the active ingredient and also to ensure its stability. There are various delivery vehicles widely used in practice today. These encompass viral capsids and a variety of lipid complexes. Naturally, it is of considerable importance to monitor the integrity and various attributes of these vehicles. For example: the assembly stoichiometry and protein modifications of viral capsids or virus-like particles, the composition and stoichiometry (including encapsulation efficiency) of lipid-nucleotide complexes, as well as the size distribution of these nanoparticles, e.g., viruses or lipid complexes (cf. Chap. 5).

Measuring Higher Order Structure Modifications in Nucleotide-Based Pharmaceuticals

All analytical methods reviewed in the sections on protein therapeutics above are used to various degrees also for characterization of the higher order structure of nucleotide-based therapeutics. For example, the techniques CD, FTIR, fluorescence spectroscopy, ion mobility, and MS can be readily utilized for characterization and monitoring purposes. Naturally, the same limitations mentioned above apply to a large extent to the field of nucleotide-bases pharmaceuticals. Being population techniques, i.e., measuring the average value of an attribute across a molecular population, as opposed to being able to discern individual molecules, these methods possess a limited ability to detect individual modifications. As mentioned previously, such ability is more important in the field of nucleotide-based therapeutics, as individual modifications can render a molecule inactive.

The same limitation applies to some of the high-resolution techniques available today such as NMR and X-ray crystallography and cryo EM. Not as relevant to protein-based pharmaceuticals, other considerations become important when applying these methods. For example, NMR is a solu-

tion technique that can capture the dynamics of the RNA molecules, which is not the case in X-ray crystallography. On the other hand, NMR is typically limited to about 100 kDa-sized molecules, whereas X-ray crystallography is not. X-ray crystallography in turn is limited to relatively rigid structures, which is a disadvantage when studying RNA-based products.

In terms of measuring the size distribution of nanoparticle-based pharmaceuticals, e.g., virus- or lipoplex-based, all techniques described in Table 3.3 are fully applicable, including the previously discussed considerations. It needs to be mentioned that because of their size, nanoparticles are often characterized using light-scattering techniques that do suffer some deficiencies dictated by the measurement principle, e.g., measurement results are disproportionally skewed toward larger sizes and heavy use of mathematical models.

Biological Activity (Potency) Assays/ Bioassays

Binding Assays (ELISA), Surface Plasmon Resonance (SPR)

Immunoassays

Enzyme-linked immunosorbent assay (ELISA) provides a means to quantitatively measure extremely small amounts of proteins. This procedure utilizes the fact that plastic surfaces are able to adsorb low but detectable amounts of proteins. Typically, antibodies against a certain protein of interest are allowed to adsorb to the surface of microtitration plates. Each plate may contain up to 96 or 384 wells so that multiple samples can be assayed. After incubating the antibodies in the wells of the plate for a specific period, excess antibody is removed and residual protein-binding sites on the plastic are blocked by incubation with an inert protein. Several microtitration plates can be prepared at one time since the antibodies coating the plates retain their binding capacity for an extended period. During the ELISA, the sample solution containing the protein of interest is incubated in the wells and the protein (Ag) is captured by the antibodies coating the well surface. Excess sample is removed and other antibodies which now have an enzyme (E) linked to them are added to react with the bound antigen.

The format described above is called a sandwich assay since the antigen (protein of interest) is located between the antibody on the titer well surface and the antibody containing the linked enzyme. Figure 3.16 illustrates a number of formats that can be used in an ELISA. A suitable substrate is added and the enzyme linked to the antibody–antigen–antibody well complex converts this compound to a colored product. The amount of product obtained is proportional to the enzyme adsorbed in the well of the plate. A standard

Fig. 3.16 Examples of several formats for ELISA in which the specific antibody is adsorbed to the surface of a microtitration plate. See Fig. 3.9 for abbreviations used. The antibody is represented by the Y-type structure. The product *P* is colored and the amount generated is measured with a spectrometer or plate reader

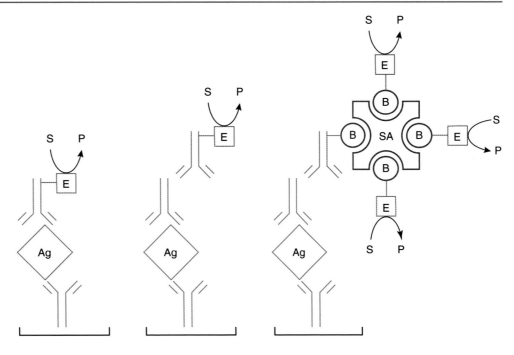

curve can be prepared if known concentrations of antigen (protein of interest) are tested in this system and the amount of antigen in unknown samples can be estimated from this standard curve. A number of enzymes can be used in ELISAs. However, the most common ones are horseradish peroxidase and alkaline phosphatase. A variety of substrates for each enzyme is available; they yield colored products catalyzed by the antibody-linked enzyme. Absorbance of the colored product solutions is measured on plate readers, instruments that rapidly measure the absorbance in all 96 wells of the microtitration plate. Data processing can be automated for rapid throughput of information. Note that detection approaches partly parallel those discussed in the section on "Blotting." The above ELISA format (sandwich assay) is only one of many different ELISA set ups. For example, the microtitration wells may be coated directly with the antigen rather than having a specific antibody attached to the surface. The concentration of antigen is established by comparison with known quantities of antigen (protein of interest) used to coat individual wells.

Another approach is to use—this time subsequent to the binding of the antigen (protein of interest) either directly to the surface or to an antibody on the surface—an antibody specific to the antibody binding the protein antigen, that is, a secondary antibody (cf. Fig. 3.9). This latter, secondary, antibody contains the linked enzyme used for detection. As already discussed in the section on "Blotting," the advantage of this approach is that such antibodies can be obtained in high purity and with the desired enzyme linked to them from commercial sources. Thus, a single source of enzyme-linked antibody can be used in assays for different protein antigens. Should a sandwich assay be used, then antibodies from dif-

ferent species need to be used for each side of the sandwich. A possible scenario is that rabbit antibodies are used to coat the microtitration wells; mouse antibodies, possibly a monoclonal antibody, are used to complex with the antigen; and then, a goat anti-mouse immunoglobulin containing linked HRP or AP is used for detection purposes.

As with immunoblots discussed above, streptavidin or avidin can be used in these assays if biotin is covalently linked to the antibodies and enzymes.

SPR-Based Binding Assays

Surface plasmon resonance techniques are based on the excitation of free electrons (called surface plasmons when excited) by polarized light from a metal film at an interface with a medium with a different refractive index. Binding of molecules to this interfacial layer results in shifts in their reflection curves. Since the refractive index changes are linearly proportional to the number of molecules bound, this technique can be used to calculate a number of binding parameters such as the equilibrium association constant (K_A), equilibrium dissociation constant (K_D), as well as the concentration of a protein in solution. In practice, these measurements are carried out using commercially available SPR chips, which are typically covalently derivatized with antibodies or antigens to which the protein of interest can bind. The solution with the protein of interest flows over the chip at a defined rate and from the SPR signals the characteristics mentioned above can be calculated.

Today, both ELISA- and SPR-based binding assays are extensively used in the development of biopharmaceuticals. Although neither method has undergone fundamental changes over the last decade or so, one recent improvement which has been broadly implemented is the automation of these assays. Since these techniques typically require a significant hands-on time for analysis, recent advances as the introduction of robotic systems have increased the throughput significantly.

Cell-Based Activity Assays

Paramount to the development of a protein therapeutic is to have an assay that identifies its biological function. Chromatographic and electrophoretic methodologies can address the purity of a biotherapeutic and be useful in investigating stability parameters. However, it is also essential to ascertain whether the therapeutic protein has adequate bioactivity. Typically, bioassays (potency assays) are carried out in vitro by monitoring the biological response in cells when the therapeutic protein is added to the system. These biological responses need to reflect the mode of action (MoA) of the therapeutic. There is a wide and ever-increasing variety of cell-based assays.

Common cell-based bioassays are

(a) Cell proliferation/anti-proliferation assays in which the proliferation (or reduction of proliferation) of the cells in culture is measured as a response to the drug,

(b) Cytotoxicity assays in which cell death occurring as a response to the drug is measured,

(c) Adhesion assays in which the influence of the drug on the adhesion properties of the cells are measured,

(d) Kinase receptor activation assays in which the phosphorylation of a tyrosine kinase as a response to the drug as a ligand is measured by capture-ELISA after cell lysis,

(e) Cellular response to the biopharmaceutical is often monitored via the activation of a specific cellular signaling cascade,

(f) Antibody-dependent cell-mediated cytotoxicity (ADCC) assays are assays in which cell lysis by immune system effector cells upon their activation by a therapeutic antibody bound to a receptor of the target cell is measured,

(g) Complement-dependent cytotoxicity (CDC) assays are assays in which the lysis of the target cell by components of the complement system are measured after complement activation by the therapeutic antibody bound to a cellular receptor,

(h) Cytokine release assays measure the release of cytokines by the target cell as a response to the protein therapeutic.

An interesting new approach to cell-based potency assays is the use of reporter genes. In the reporter gene assays, the activation of a gene regulatory element (as a response to a signaling cascade activation) is monitored using a common reporter gene, for example, luciferase. The assay readout is the expression of the reporter gene as a response to the drug. The advantage of this approach is the potential for using it as a generic (platform) approach to a number of different bioassays, thus simplifying the bioassay development, cell banking, etc.

Concluding Remarks

With the ever-increasing clinical application and variety of (human) recombinant proteins and nucleotide-based therapeutics, the need for characterization of their structure, function, and purity has also increased. Naturally, the analytical technology has also undergone a rapid evolution. Today, a large array of techniques is used to characterize the primary, secondary, tertiary, and quaternary structure of proteins and to determine the quality, purity, and stability of the recombinant drug product.

The information provided in this chapter offers only a general guidance to the process and tools for analytical characterization of various protein modifications. In reality, this process is rarely straightforward and is often convoluted by a number of factors. For example, a lot of different protein modifications occur simultaneously, thus inducing simultaneous changes in a number of molecular attributes. Often the results of these simultaneous changes may be masked. For example, it is possible that no changes in the IEX charge heterogeneity profile of a protein molecule are detected when simultaneous succinimide formation (inducing the formation of more basic species) and deamidation (inducing the formation of acidic species) occur, in spite of a significant redistribution of various charged species.

Many protein modifications result in changes in more than one molecular attribute. Thus, more than one analytical technique should be used to characterize these modifications. For example, pyro-Glu formation may be measured using either charge-based techniques (e.g., IEX) or techniques measuring changes in polarity (e.g., RP-HPLC). Moreover, the extent to which the molecular properties of a protein are modified as a result of a given modification, depends on the structural context in which this modification occurs (overall molecular charge, hydrophobicity, size, etc.). For example, clipping of a small fragment of a very large protein may not be easy to detect using size-exclusion chromatography, due to the resolution limits of this technique.

Because of the complexity of the interplay between various potential protein modifications, the definition of the characterization and quality control strategies is an important intellectual challenge for the analytical experts in the therapeutic protein development team, and critical for the success of every development program.

The stability and characterization of nucleotide-based pharmaceuticals are discussed in a separate section of this chapter. This is a family of DNA- or RNA-based molecules. Molecular weight and nature of the units, e.g., the presence of modified nucleotides, can vary widely with siRNA as an example of a small and plasmid DNA or mRNA as large molecules. Nucleotide-based molecules share many physicochemical features with proteins which is reflected in the use of a similar "analytical toolbox." A basic difference between pharmaceutical proteins and nucleotide-based pharmaceuticals is that activity and safety of proteins is often preserved when minor degradation reactions occur, whereas activity totally disappears with one break in the chain or a degradation reaction. This has a clear impact on formulation design and shelf-life conditions.

Self-Assessment Questions

Questions

1. What are the most common chemical modifications resulting in changes in protein charge heterogeneity?
2. What are the three most common techniques for measuring charge heterogeneity of proteins?
3. What is the transfer of proteins to a membrane such as nitrocellulose or PDVF called?
4. What are the two most commonly oxidized amino acids in protein biotherapeutics? How can one detect these oxidized amino acids in molecular structure?
5. In a 2-dimensional electrophoresis analysis of a protein mixture, the first method of separation is an IEF run, followed by a SDS-PAGE run in a perpendicular direction to the first run? If one runs the SDS-PAGE analysis first, followed by the IEF run, would one get a similar result?
6. List three techniques for separating proteins based on molecular size.
7. Which technique provides the ultimate (atomic) resolution of the structure of a protein?
8. What are limitations of an ELISA when analyzing a protein product?

Answers

1. Deamidation and isomerization.
2. Ion exchange chromatography, isoelectric focusing, and capillary zone electrophoresis.
3. This method is called blotting. If an electric current is used, then the method is called electroblotting.
4. Methionine and tryptophan. MS, RP-HPLC, and HIC are preferred analytical techniques to detect oxidation of proteins.
5. No. In the SDS-PAGE analysis, the protein is denatured (unfolds), and SDS interacts with the protein giving it a uniform negative charge that masks the amino acids charges in the protein. The pI of the protein(s) cannot be determined in a subsequent IEF run.
6. Size-exclusion chromatography, SDS-PAGE, asymmetric field flow field fractionation (AFFF).
7. X-ray crystallography.
8. The ELISA (sandwich assay set up) may measure degradation products and/or product-related variants with similar affinity to intact molecules.

Acknowledgments This chapter is a re-work of part (including a number of figures) of the chapter "*Biophysical and Biochemical Analysis of Recombinant Proteins,*" by Tsutomu Arakawa and John Philo, which appeared in the previous editions of this book—cf. fourth edition, 2013. Despite the significant changes made to the original chapter, it still contains fragments of the excellent original text by Arakawa and Philo.

The author thanks Abbas Razvi and Marigone Lenjani for providing chromatograms and electropherograms shown in this chapter.

Further Reading

Arbogast LW, Brinson RG, Marino JP (2015) Mapping monoclonal antibody structure by 2D 13C NMR at natural abundance. Anal Chem 87(7):3556–3561

Berkowitz SA, Engen JR, Mazzeo JR, Jones GB (2012) Analytical tools for characterizing biopharmaceuticals and the implications for biosimilars. Nat Rev. Drug Discov 11(7):527–540

Bilikallahalli K, Muralidhara BK, Baid R, Bishop SM, Huang M, Wang W, Nema S (2016) Critical considerations for developing nucleic acid macromolecule based drug products. Drug Discov Today 21(3):430–444

Boo SH, Kim YK (2020) The emerging role of RNA modifications in the regulation of mRNA stability. Exp Mol Med 52:400–408

Butler JE (ed) (1991) Immunochemistry of solid-phase immunoassay. CRC Press, Boca Raton

Cavanagh J, Fairbrother WJ, Skelton NJ (2007) Protein NMR spectroscopy principles and practice book, 2nd edn. Elsevier, Amsterdam

Chirino AJ, Mire-Sluis A (2004) Characterizing biological products and assessing comparability following manufacturing changes. Nat Biotechnol 22(11):1383–1391

Coligan J, Dunn B, Ploegh H, Speicher D, Wingfield P (eds) (1995) Current protocols in protein science. Wiley, New York

Crabb JW (ed) (1995) Techniques in protein chemistry VI. Academic, San Diego

Domon B, Aebersold R (2006) Mass spectrometry and protein analysis. Science 312(5771):212–217

Du Y, Walsh A, Ehrick R, Xu W, May K, Liu H (2012) Chromatographic analysis of the acidic and basic species of recombinant monoclonal antibodies. MAbs 4(5):578–585

Dunbar BS (1994) Protein blotting: a practical approach. Oxford University Press, New York

Fekete S, Guillarme D, Sandra P, Sandra K (2016) Chromatographic, electrophoretic, and mass spectrometric methods for the analytical characterization of protein biopharmaceuticals. Anal Chem 88(1):480–507

Folzer E, Diepold C, Bomans K, Finkler C, Schmidt R, Bulau P, Huwyler J, Mahler H-C, Koulov AV (2015) Selective oxidation of methionine and tryptophan residues in a therapeutic IgG1 molecule. J Pharm Sci 104:2824–2831

Gahoual R, Beck A, Leize-Wagner E, François YN (2016) Cutting-edge capillary electrophoresis characterization of monoclonal antibodies and related products. J Chromatogr B Analyt Technol Biomed Life Sci 1032:61–78

Hames BD, Rickwood D (eds) (1990) Gel electrophoresis of proteins: a practical approach, 2nd edn. IRL Press, New York

Henderson R (2018) From electron crystallography to single particle CryoEM (Nobel lecture). Angew Chem Int Ed Engl 57:10804

Jiskoot W, Crommelin DJA (eds) (2005) Methods for structural analysis of protein pharmaceuticals. AAPS Press, Arlington

Kahle J, Wätzig H (2018) Determination of protein charge variants with (imaged) capillary isoelectric focusing and capillary zone electrophoresis. Electrophoresis 39:2492

Kaltashov IA, Bobst CE, Abzalimov RR, Wang G, Baykal B, Wang S (2012) Advances and challenges in analytical characterization of biotechnology products: mass spectrometry-based approaches to study properties and behavior of protein therapeutics. Biotechnol Adv 30(1):210–222

Kulkarni JA, Witzigmann D, Thomson SB et al (2021) The current landscape of nucleic acid therapeutics. Nat Nanotechnol 16:630–643

Liu H, Gaza-Bulseco G, Faldu D, Chumsae C, Sun J (2008) Heterogeneity of monoclonal antibodies. J Pharm Sci 97(7):2426–2447

McEwen CN, Larsen BS (eds) (1998) Mass spectrometry of biological materials, 2nd edn. Dekker, New York

Moritz B, Schnaible V, Kiessig S, Heyne A, Wild M, Finkler C, Christians S, Mueller K, Zhang L, Furuya K, Hassel M, Hamm M, Rustandi R, He Y, Solano OS, Whitmore C, Park SA, Hansen D, Santos M, Lies M (2015) Evaluation of capillary zone electrophoresis for charge heterogeneity testing of monoclonal antibodies. J Chromatogr B Analyt Technol Biomed Life Sci 983–984:101–110

Müllertz A, Perrie Y, Rades T (eds) (2016) Analytical techniques in the pharmaceutical sciences. springer, New York

Oude Blenke E, Örnskov E, Schöneich C, Nilsson GA, Volkin DB, Mastrobattista E, Almarsson O, Crommelin DJA (2022) The storage and in-use stability of mRNA vaccines and therapeutics: not a cold case. J Pharm Sci 112:386. https://doi.org/10.1016/j.xphs.2022.11.001

Pace CN, Grimsley GR, Scholtz JM, Shaw KL (2014) Protein stability. Wiley

Parr MK, Montacir O, Montacir H (2016) Physicochemical characterization of biopharmaceuticals. J Pharm Biomed Anal 130:366–389

Pogocki D, Schöneich C (2000) Chemical stability of nucleic acid-derived drugs. J Pharm Sci 89(4):443–456

Ponniah G, Nowak C, Neill A, Liu H (2017) Characterization of charge variants of a monoclonal antibody using weak anion exchange chromatography at subunit levels. Anal Biochem 520:49–57

Rathore D, Faustino A, Schiel J, Pang E, Boyne M, Rogstad S (2018) The role of mass spectrometry in the characterization of biologic protein products. Expert Rev. Proteomics 15(5):431–449

Ríos Quiroz A, Lamerz J, Da Cunha T, Boillon A, Adler M, Finkler C, Huwyler J, Schmidt R, Mahler H-M, Koulov AV (2016) Factors governing the precision of subvisible particle measurement methods–a case study with a low-concentration therapeutic protein product in a prefilled syringe. Pharm Res 33:450–461

Salas-Solano O, Felten O (2008) Capillary electrophoresis methods for pharmaceutical analysis. Sep Sci Technol 9:401

Santos IC, Brodbelt JS (2021) Recent developments in the characterization of nucleic acids by liquid chromatography, capillary electrophoresis, ion mobility, and mass spectrometry (2010–2020). J Sep Sci 44(1):340–372

Shirley BA (ed) (1995) Protein stability and folding. Humana Press, Totowa

Schoenmaker L, Witzigmann D, Kulkarni JA, Verbeke R, Kersten G, Jiskoot W, Crommelin DJA (2021) mRNA-lipid nanoparticle COVID-19 vaccines: structure and stability. Int J Pharm 601:120586

Strege MA, Lagu AL (2004) Capillary electrophoresis of proteins and peptides. Humana Press, Totowa

Thorpe R, Wadhwa M, Mire-Sluis A (1997) The use of bioassays for the characterisation and control of biological therapeutic products produced by biotechnology. Dev Biol Stand 91:79–88

Vicens Q, Kieft JS (2022) Thoughts on how to think (and talk) about RNA structure. Proc Natl Acad Sci U S A 119(17):e2112677119

Vlasak J, Ionescu R (2011) Fragmentation of monoclonal antibodies. MAbs 63:253–263

Vlasak J, Ionescu R (2008) Heterogeneity of monoclonal antibodies revealed by charge-sensitive methods. Curr Pharm Biotechnol 9:468–481

Wei H, Mo J, Tao L, Russell RJ, Tymiak AA, Chen G, Iacob RE, Engen JR (2014) Hydrogen/deuterium exchange mass spectrometry for probing higher order structure of protein therapeutics: methodology and applications. Drug Discov Today 19(1):95–102

Wild D (ed) (2013) The immunoassay handbook theory and applications of ligand binding, ELISA and Related techniques, 4th edn. Elsevier, New York

Zhang J, Fei Y, Sun L et al (2022) Advances and opportunities in RNA structure experimental determination and computational modeling. Nat Methods 19:1193–1207

Production and Purification of Recombinant Proteins

4

Alfred Luitjens and Emile van Corven

Introduction

The growing therapeutic use of proteins increases the need for practical and economical processing techniques. As a result, biotechnological production methods have advanced significantly over the past three to four decades. Also, single-use production technology that has the potential to mitigate many of the economical and quality issues arising from manufacturing these products has evolved rapidly (Hodge 2004; Luitjens et al. 2012), as did the development of continuous manufacturing processes (Zhou et al. 2021; Khanal and Lenhoff 2021).

When producing proteins for therapeutic use, a number of challenges arise related to the manufacturing, purification, and characterization of the products. Biotechnological products for therapeutic use have to meet strict specifications especially when used via the parenteral route. The regulatory agencies both in Europe (EMA: European Medicines Agency) and in the United States of America (FDA: Food and Drug Administration) play a pivotal role in providing legal requirements and guidelines (www.ICH.org, www.FDA.gov).

In this chapter, the focus is on the technical aspects of production (upstream processing) and purification (downstream processing) of recombinant therapeutic proteins. However, a majority of the techniques discussed can also be applied to vaccines and viral vector production. For further details, the reader is referred to the literature mentioned.

Upstream Processing

Expression Systems

General Considerations

Expression systems for proteins of therapeutic interest include both pro- and eukaryotic cells (bacteria, yeast, fungi, plants, insect cells, and mammalian cells) and transgenic animals. The choice of a particular system will be determined to a large extent by the nature and origin of the desired protein, the intended use of the product, the amount needed, and the cost.

In principle, any protein can be produced using genetically engineered organisms, but not every type of protein can be produced by every type of cell. In most cases, the protein is foreign to the host cells, and although the translation of the genetic code can be performed by the cells, the post-translation modifications of the protein might be different compared to the native protein.

About 5% of the mammalian proteome is thought to comprise enzymes performing over 200 types of post-translation modifications of proteins (Walsh 2006). These modifications are species and/or cell-type specific. The metabolic pathways that lead to these modifications are genetically determined by the host cell. Thus, even if the cells can produce the desired post-translation modifications, these modifications, such as glycosylation, might still be different from that of the native protein. Correct N-linked glycosylation of therapeutically relevant proteins is important for full biological activity, immunogenicity, stability, targeting, and pharmacokinetics. Prokaryotic cells, such as bacteria, are sometimes capable of producing N-linked glycoproteins. However, the observed N-linked structures differ from the structures found in eukaryotes (Dell et al. 2011). Yeast cells are able to produce recombinant proteins such as albumin, and yeast has been engineered to produce glycoproteins with human-like glycan structures including terminal sialylation (reviewed by Celik and Calik 2011).

A. Luitjens
Batavia Biosciences, Leiden, The Netherlands
e-mail: a.luitjens@bataviabiosciences.com

E. van Corven (✉)
Corven Holding, Utrecht, The Netherlands

© The Author(s), under exclusive license to Springer Nature Switzerland AG 2024
D. J. A. Crommelin et al. (eds.), *Pharmaceutical Biotechnology*, https://doi.org/10.1007/978-3-031-30023-3_4

Still, most products on the market and currently in development use cell types that are, if possible, closely related to the original protein-producing cell type. Therefore, for human-derived proteins, typically mammalian cells are chosen for production as prokaryotic cells are less effective in producing post-translational modifications. Those are often essential when it comes to complex protein structures such as monoclonal antibodies. However, driven by the increasing demand for inexpensive products, especially for costly antibody therapies, two trends opened new opportunities to produce antibody fragments in *E. coli*; (1) generation of engineered *E. coli* strains and (2) new knowledge in using biologically functional antibody fragments. Based on this, various *E. coli*-derived antibody fragments have been approved by the regulatory bodies (e.g., ranibizumab, certolizumab pegol). Therefore, although still to be further developed, bacteria and yeast will likely keep on playing a role as future production systems given their ease of growing in large-scale fermenters and lower cost of large-scale manufacturing relative to mammalian cells.

Generalized features of proteins expressed in different biological systems are listed in Table 4.1 (see also Walter et al. 1992; Yao et al. 2015). However, it should be kept in mind that there are exceptions to this table for specific product/expression systems.

The Research Cell Bank (RCB) and the Master Cell Bank (MCB)

To develop a stable mammalian cell line that produces a therapeutic protein, the expression vector containing the gene of interest (GOI) and selection marker is transfected into the host cell line. The choice of the selection marker depends on the cell type and selection protocol used. By using a selection marker, it is possible to perform a selection between therapeutic protein producing cells and nonproducing cells. For selection, often methotrexate (MTX), methionine sulf-

oximine (MSX) or zeozin are used. Subsequently, the protein-producing cells undergo a cloning process, usually by using a combination of limiting dilutions and cell imaging. This is how a protein producing cell line is generated from one cell line increasing its homogeneity. An important aspect in the selection of the producing cell line for manufacturing is the stability of the producing clone. The GOI of interest must be expressed during at least 40 to 60 doublings to ensure that the GOI is stably expressed at the end of the production process and beyond. Of the selected cell clone, usually the highest producer, a Research Cell bank (RCB) is prepared. The RCB is transferred to a Good Manufacturing Practice (GMP) facility, where the Master Cell Bank (MCB) is produced. The MCB is extensively tested for many safety and other quality parameters and forms the basis for the development of the manufacturing process for the product used in clinical studies and for the market (Matasci et al. 2008; Lai et al. 2013).

Transgenic Animals

Foreign genes can be introduced into animals like mice, rabbits, pigs, sheep, goats, and cows through nuclear transfer and cloning techniques. Using milk-specific promoters, the desired protein can be expressed in the milk of the female offspring. During lactation, the milk is collected, the milk fats are removed, and the skimmed milk is used as the starting material for the purification of the protein.

The advantage of this technology is the relatively cheap method to produce the desired proteins in vast quantities when using larger animals such as cows. Disadvantages are the long lead time to generate a herd of transgenic animals and concerns about the health of the animal, food safety, and ethics (see: report Bundesministerium für Gesundheit, Familie und Jugend, Sektion IV http://www.science-art.at/uploads/media/report_transgenic_animals_02.pdf). Some proteins expressed in the mammary gland leak back into the

Table 4.1 Generalized features of proteins of different biological origin

Protein feature	Prokaryotic bacteria	Eukaryotic yeast	Eukaryotic mammalian cells	Eukaryotic Plant cells	Transgenic animals
Concentration	High	Medium-high	High	Low	Medium-high
Molecular weight	Low	High	High	High	High
S-S bridges	Limitation	No limitation	No limitation	No limitation	No limitation
Secretion	No	Yes/no	Yes	Yes/no	Yes
Aggregation state	Inclusion body	Singular, native	Singular, native	Singular, native	Singular, native
Folding	Risk of misfolding	Correct folding	Correct folding	Correct folding	Correct folding
Glycosylation (human-like)	Limited	Limited	Possible	Limited	Possible
Contamination risk	Possible (endotoxin)	Low	Possible (virus, prion, oncogenic DNA)	Low	Very possible (virus, prion, and endotoxin)
Cost to manufacture	Low-medium	Low-medium	High	High[a]	Medium-high

[a] Due to current limited scalability (Shukla et al. 2017)

circulation and cause serious negative health effects. An example is the expression of erythropoietin in cows. Although the protein was well expressed in the milk, it caused severe health effects and these experiments were stopped.

The purification strategies and purity requirements for proteins from milk can be different from those derived from bacterial or mammalian cell systems. Often the transgenic milk containing the recombinant protein also contains significant amounts of the nonrecombinant counterpart. To separate these closely related proteins poses a purification challenge. The "contaminants" in proteins for oral use expressed in milk that is otherwise consumed by humans are known to be safe for consumption.

The transgenic animal technology for the production of pharmaceutical proteins has progressed within the past few years. The FDA and EMA approved recombinant antithrombin III (ATryn®, GTC Biotherapeutics) produced in the milk of transgenic goats, as well as recombinant human C1 esterase inhibitor (Ruconest®, Pharming Group N.V.) produced in the milk of transgenic rabbits. More details about this technology are presented in Chap. 9.

Plants

Therapeutic proteins can also be expressed in plants and plant cell cultures. For instance, human albumin has been expressed in potatoes and tobacco. Whether these production vehicles are economically feasible has yet to be established. The lack of genetic stability of plants was sometimes a drawback. Stable expression of proteins in edible seeds has been obtained. For instance, rice and barley can be harvested and easily kept for a prolonged period of time as raw material sources. Especially for oral therapeutics or vaccines, this might be the ideal solution to produce large amounts of cheap therapeutics, because the "contaminants" are known to be safe for consumption. However, challenges are the presence of high endotoxin levels, a relatively low expression level of the product, and secretion of proteases limiting the shelf life of plant extracts (Shukla et al. 2017). A better understanding of the plant molecular biology together with more sophisticated genetic engineering techniques and strategies to increase yields and optimized glycan structures resulted in an increase in the number of products in development including late-stage clinical trials (reviewed by Orzaez et al. 2009; Peters and Stoger 2011). Biosafety concerns (such as pollen contamination and immunogenicity of plant-specific glycans) and costly downstream extraction and purification requirements, however, have hampered moving therapeutic protein production in plants from the laboratory to industrial size application (Yao et al. 2015).

More details about the use of plant systems for the production of pharmaceutical proteins are presented in Chap. 9.

Cultivation Systems

The remainder of this chapter will focus on mammalian cell-based expression systems. Non-mammalian expression systems will be discussed only briefly.

General

In general, cells can be cultivated in vessels containing an appropriate liquid growth medium in which the cells are either immobilized and grow as a monolayer, attached to microcarriers, in suspension, or entrapped in matrices. The culture method will determine the scale of the separation and purification methods. Production-scale cultivation is commonly performed in fermenters, used for bacterial and fungal cells, or bioreactors, used for mammalian and insect cells. Bioreactor systems can be classified into four different types:

- Stirred tank (Fig. 4.1a)
- Airlift (Fig. 4.1b)
- Fixed bed (Fig. 4.1c)
- Membrane bioreactors (Fig. 4.1d)

Because of its reliability and experience with the design and scaling up potential, the stirred tank is still the most commonly used bioreactor. This type of bioreactor is not only used for suspension cells like CHO, HEK293, and PER. C6® cells but it is also used for production of adherent cells like Vero and MDCK cells. In the latter case, the production is performed on microcarriers (Van Wezel et al. 1985).

Bioreactor Processes

The kinetics of cell growth and product formation will not only dictate the type of bioreactor used but also how the growth process is performed. Three types of bioreactor processes are commonly employed and discussed below:

- Batch
- In a batch process, the bioreactor is filled with the entire volume of medium needed during the cell growth and/or production phase. No additional supplements are added to increase the cell growth or production during the process. Waste products, such as lactate and ammonium, and the product itself accumulate in the bioreactor. The product is harvested at the end of the process. Maximum cell density and product yields will be lower compared to a fed-batch process.
- Fed-batch
- In a fed-batch process, a substrate is supplemented to the bioreactor. The substrate consists of the growth-limiting nutrients that are needed during the cell growth phase and/ or during the production phase of the process. Like the batch process, waste products accumulate in the bioreactor. The product is harvested at the end of the process. With the

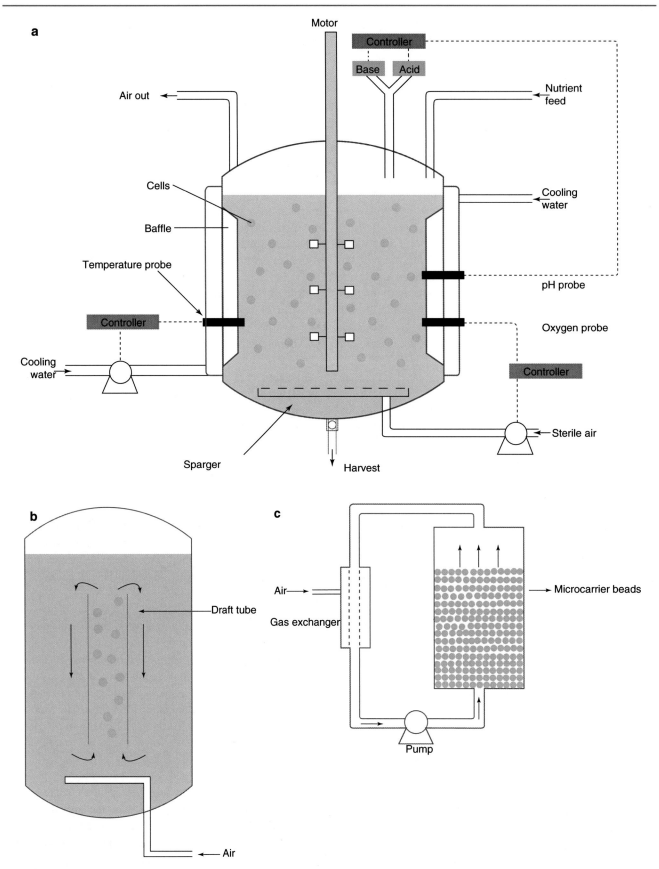

Fig. 4.1 (**a**) Schematic representation of stirred-tank bioreactor. (**b**) Schematic representation of airlift bioreactor. (**c**) Schematic representation of fixed-bed stirred-tank bioreactor. (**d**) Schematic representation of hollow fiber perfusion bioreactor. All schematics are adapted from Klegerman and Groves (1992)

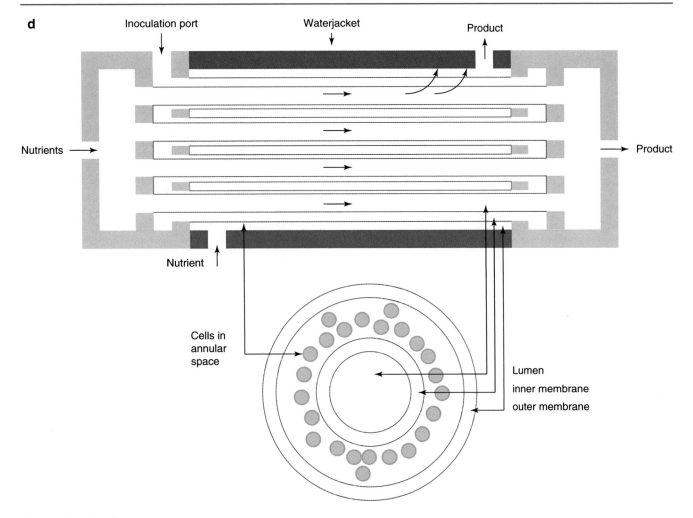

d

Inoculation port   Waterjacket   Product

Nutrients →   → Product

Nutrient ↑

Cells in
annular
space

Lumen
inner membrane
outer membrane

Fig. 4.1 (continued)

fed-batch process, higher cell densities and product yields can be reached compared to the batch process due to the extension of production time that can be achieved compared to a batch process. The substrate used is highly concentrated and can be added to the bioreactor at certain points in time or as a continuous feed. The fed-batch mode is currently widely used for the production of proteins. The process is well understood and characterized.

• Perfusion
• In a perfusion process, the media and waste products are continuously exchanged and the product is harvested throughout the culture period. A membrane device is used to retain the cells in the bioreactor, and waste medium is removed from the bioreactor by this device (Fig. 4.2). To keep the medium level constant in the bioreactor, fresh medium is supplemented to the bioreactor. By operating in perfusion mode, the level of waste products will be kept constant and one generates a stable environment for the cells to grow or to produce (see below). With the perfusion process, much higher cell

densities can be reached and therefore higher productivity (Compton and Jensen 2007).

• Chemostat

• The chemostat process is a perfusion process with the goal to keep all process parameters like growth and dead rate constant by controlling the addition of nutrients and removal of waste medium and metabolites. As a chemostat process is difficult to manage, it is very unlikely that this process is used for commercial manufacturing.

In all these four protocols, the cells go through four distinctive phases:

1. Lag Phase
   In this phase, the cells are adapting to the conditions in the bioreactor and do not yet grow.
2. Exponential Growth Phase
   During this phase, cells grow in a more or less constant doubling time for a fixed period. However, under the right

Fig. 4.2 Schematic representation of perfusion device coupled to a stirred-tank bioreactor (*ATF*: alternating tangential flow)

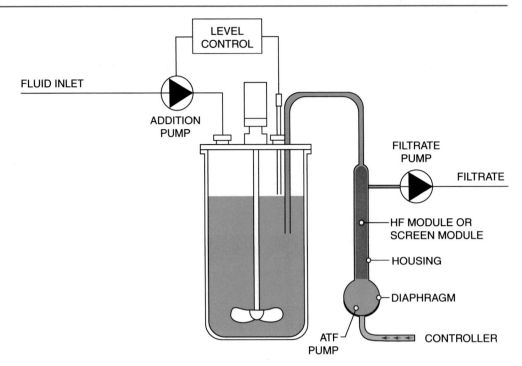

process conditions, mammalian cell doubling time is dependent on the cell type used and will usually vary between 20 and 40 h. Plotting the natural logarithm of cell concentration against time produces a straight line. Therefore, the exponential growth phase is also called the log phase. The growth phase will be affected by growth conditions such as temperature, pH, oxygen pressure, and external forces like stirring and baffles that are inserted into the bioreactor. Furthermore, the growth rate is affected by the supply of sufficient nutrients, buildup of waste products, etc.

3. Stationary Phase

   In the stationary phase, the growth rate of the cells slows down since nutrients are depleted and/or buildup of toxic waste products like lactate and ammonium. In this phase, constant cell concentrations are found because a balance between cell growth and cell death has been reached.

4. Death Phase

   Cells die due to depletion of nutrients and/or presence of high concentrations of toxic by-products such as lactate and ammonium.

   Examples of animal cells that are commonly used to produce recombinant proteins of clinical interest are Chinese Hamster Ovary cells (CHO), immortalized human embryonic retinal cells (PER.C6® cells), baby hamster kidney cells (BHK), lymphoblastoid tumor cells (interferon production), melanoma cells (plasminogen activator), and hybridized tumor cells (monoclonal antibodies).

   The cell culture has to be free from undesired microorganisms that may destroy the cell culture or present hazards to the patient by producing endotoxins. Therefore, strict measures are required for both the production procedures

and materials used (WHO 2010; Berthold and Walter 1994) to prevent a possible contamination with extraneous agents such as viruses, bacteria, and mycoplasma. Furthermore, strict measures are needed, especially with regard to the raw materials used, to prevent contaminations with transmissible spongiform encephalopathies (TSEs).

Cultivation Medium

In order to achieve optimal growth of cells and optimal production of recombinant proteins, it is of great importance not only that conditions such as stirring, pH, oxygen pressure, and temperature are chosen and controlled appropriately but also that a cell growth and protein production medium with the proper nutrients are provided for each stage of the production process.

The media used for mammalian cell culture are complex and consist of a mixture of diverse components, such as sugars, amino acids, electrolytes, vitamins, fetal calf serum (caveat see below), and/or a mixture of peptones, growth factors, hormones, and other proteins (see Table 4.2). Many of these ingredients are preblended either as concentrate or as homogeneous mixtures of powders. To prepare the final medium, components are dissolved in purified water before sterilization. The preferred method for sterilization is heat ($\geq$15 min at 121 °C). However, most components used in cell culture medium cannot be sterilized by heat, therefore filtration is used. Then, the medium is filtrated through 0.1 μm (to prevent mycoplasma and bacterial contamination) or 0.2 μm filters (to prevent bacterial contamination). Some supplements, especially fetal bovine serum, contribute considerably to the

Table 4.2 Major components of growth media for mammalian cell structures

Type of nutrient	Example(s)
Sugars	Glucose, lactose, sucrose, maltose, dextrins
Fat	Fatty acids, triglycerides
Water (high quality, sterilized)	Water for injection
Amino acids	Glutamine
Electrolytes	Calcium, sodium, potassium, phosphate
Vitamins	Ascorbic acid,—tocopherol, thiamine, riboflavine, folic acid, pyridoxin
Serum (fetal calf serum, "synthetic" serum)	Albumin, transferrin
Trace minerals	Iron, manganese, copper, cobalt, zinc
Hormones	Growth factors

presence of contaminating proteins and may seriously complicate purification procedures. Moreover, the composition of serum is variable. It depends on the individual animal, season of the year, processing differences between suppliers, etc. The use of serum may introduce adventitious material such as viruses, mycoplasma, bacteria, and fungi into the culture system (Berthold and Walter 1994). Furthermore, the possible presence of prions that can cause transmissible spongiform encephalitis almost precludes the use of materials from animal origin. However, if use of this material is inevitable, one must follow the relevant guidelines in which selective sourcing of the material is the key measure to safety (EMA 2011). A measure to prevent the contaminations mentioned above is gamma irradiation of the fetal bovine serum at 25 kGy and use sourcing from countries that have a TSE/BSE free status (Australia, New Zealand, Tasmania, USA, etc.). Many of these potential problems when using serum in cell culture media led to the development of chemically defined medium, free from animal components and material derived from animal components. These medium formulations were not only developed by the suppliers. There is the trend that the key players in the biotech industry develop their own chemically defined medium for their specific production platforms. The advantage of this is that manufacturers are less dependent on medium suppliers and have full knowledge on the composition of the medium used for their products. The chemically defined media have been shown to give satisfactory results in large-scale production settings for monoclonal antibody processes. However, hydrolysates from non-animal origin, such as yeast and plant sources, are more and more used for optimal cell growth and product secretion (reviewed by Shukla and Thömmes 2010).

Downstream Processing

Introduction

Recovering a biological reagent from a cell culture supernatant (downstream processing, DSP) is one of the critical parts of the manufacturing procedure for biotech products, and purification costs typically outweigh those of the upstream part of the production process. For the production of mono-

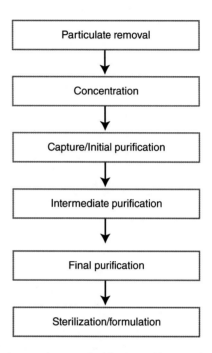

Fig. 4.3 Basic operations required for the purification of a biopharmaceutical macromolecule. For monoclonal antibody processes the concentration occurs within the capture step. Final purification is often called "polishing"

clonal antibodies, protein A resin and virus removal by filtration can account for a significant part, e.g., 40%, of the cost (Gottschalk 2006; Sinclair et al. 2016).

In the 1980s and early 1990s of the twentieth century, the protein of interest was produced in bioreactors at low concentrations (e.g., 10–200 mg/L). At most concentrations, up to 500–800 mg/L could be reached (Berthold and Walter 1994). Developments in mammalian cell culture technology through application of genetics, proteomics, medium compositions, and increased understanding of bioreactor technology resulted in product titers well above 1 g/L. Product titers above 20 g/L are also reported (Monteclaro 2010). These high product titers pose a challenge to the downstream processing unit operations (Shukla and Thömmes 2010).

With low-yield processes, a concentration step is often required to reduce handling volumes for further purification. Usually, the product subsequently undergoes a series of purification steps (Figs. 4.3 and 4.4). The first step in a purifica-

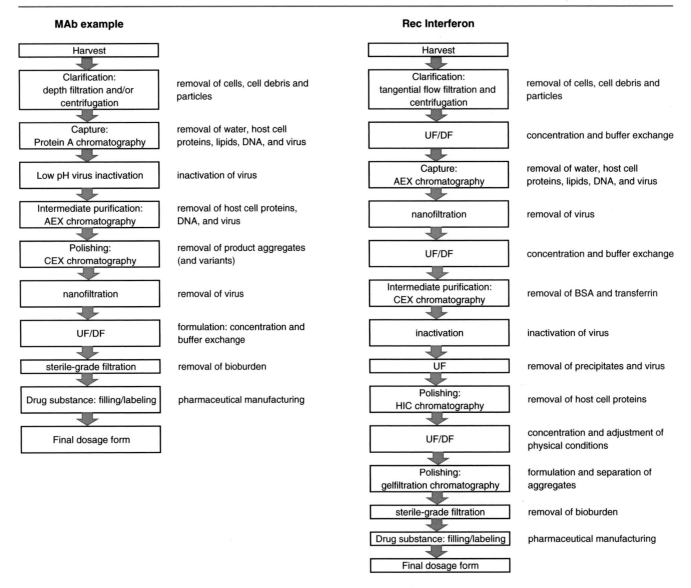

Fig. 4.4 Downstream processing of a monoclonal antibody (mAb) and a glycosylated recombinant interferon, describing the purpose of the inclusion of the individual unit operations. (*F* filtration, *TFF* tangential flow filtration, *UF* ultrafiltration, *DF* diafiltration, *A* adsorption Rec Interferon adapted from Berthold and Walter 1994). For mAbs, the sequence of anion exchange chromatography *AEX*, cationic exchange chromatography *CEX*, and nanofiltration steps can change. Also, instead of ion exchange ligands, hydrophobic interaction chromatography *HIC* or mixed mode ligands are used (Shukla et al. 2017)

tion process is to remove cells and cell debris from the process fluids ("clarification"). This process step is normally performed using centrifugation and/or depth filters. Depth filters are often used in combination with filter aid/diatomaceous earth. The next step in the purification process is often a capture step. Subsequent steps remove the residual bulk contaminants, and a final step removes trace contaminants and sometimes variant forms of the molecule. Alternatively, the reverse strategy, where the main contaminants are captured and the product is purified in subsequent steps, might result in a more economic process, especially if the product is not excreted from the cells. In the case where the product is excreted into the cell culture medium, the product will

generally not represent more than 1–5% of total cellular protein, and a specific binding of the cellular proteins in a product-specific capture step will have a high impact on the efficiency of that step. If the bulk of the contaminants can be removed first, the specific capture step will be more efficient and smaller in size and therefore more economical. Subsequent unit operation steps, e.g., with chromatography columns, can be used for further purification.

After purification, additional steps are performed to bring the desired product into a formulation buffer in which the product is stabilized and can be stored for the desired time until further process steps are performed. Before storage of the final bulk drug substance, potential residual amounts of adventitious agents, e.g., bacteria, fungi, have to be removed, usually by 0.2 µm filtration. Formulation aspects will be dealt with in Chap. 5).

As for the (upstream) production process, when designing a DSP protocol, the possibility for scaling up should be carefully considered. A process that has been designed for small quantities is often not suitable for a large-scale production for technical, economical, and safety reasons. During the development of a DSP protocol two stages can be defined: *design* and *scale-up*.

As mentioned above, separating the impurities from the product protein requires a series of purification steps (*process design*), each removing some of the impurities and bringing the product closer to its final specification. In general, the starting feedstock contains cell debris and/or whole-cell particulate material that must be removed. Defining the major contaminants in the starting material is helpful in the DSP design. This includes detailed information on the source of the material, e.g., bacterial or mammalian cell culture, and major contaminants that are used or produced in the upstream process such as albumin, serum, or product analogs. Moreover, the physico-chemical characteristics of the product versus the known contaminants (stability, isoelectric point, molecular weight, hydrophobicity, density, and specific binding properties) largely determine the process design. Processes used for the production of therapeutics in humans should be safe, reproducible, robust, and produced at the desired cost of goods. The DSP steps may expose the protein molecules to high physical stress, e.g., high temperatures and extreme pH, which can alter the protein properties possibly leading to loss in efficacy. Any therapeutic protein that is used by injection must be safe. The endotoxin concentration must be below a certain level depending on the product. Compendial limits are stated in the European Pharmacopoeia and the US Pharmacopeia: e.g., less than 0.2 and 5 endotoxin units per kg body mass for intrathecal and intravenous application, respectively. Aseptic techniques have to be used wherever possible throughout the DSP. This necessitates procedures to be performed with clean air and microbial control of all materials and equipment used. During validation of the purification process, one must also demonstrate that potential viral contaminants are inactivated and removed (Walter et al. 1992). The purification matrices should be at least sanitizable or, if possible, steam-sterilizable. For depyrogenation, the purification material must withstand either extended dry heat at ≥180 °C or treatment with 1–2 M sodium hydroxide. If any material in contact with the product inadvertently releases compounds, these leachables must be analyzed and their removal by subsequent purification steps must be demonstrated during process validation, or it must be demonstrated that the leachables are below a toxic level. In the last decade, the increased use of plastic film-based single-use production technology, e.g., sterile single-

use bioreactor bags, bags to store liquids and filter housings, brought these aspects more to the forefront. Suppliers have reacted by providing a significant body of information regarding leachables and biocompatibility for typical solutions used during processing. The problem of leachables is especially hampering the use of affinity chromatography (see below) in the production of pharmaceuticals for human use. The removal of any leached ligands well below a toxic level has to be demonstrated. Because leached affinity ligands will bind to the product, the removal might be cumbersome.

Scale-up is the term used to describe a number of processes employed in converting a laboratory procedure into an economical, industrial process. During the scale-up phase, the process moves from the laboratory to the pilot plant and finally to the production plant. The objective of scale-up is to reproducibly manufacture a product of high quality at a competitive price. Since the costs of DSP can be as high as 50–80% of the total cost of the final drug product, practical and economical ways of purifying the product should be used. Superior protein purification methods (Van Reis and Zydney 2007; Baumann and Hubbuch 2017; Tripathi and Shrivastava 2019) hold the key to a strong market position.

Basic unit operations required for a purification process used for macromolecules from biological sources are shown in Fig. 4.3.

As mentioned before, the design of the DSP protocol is highly product dependent. Each product requires a specific multistage purification procedure. The basic scheme as represented in Fig. 4.3 becomes complex. Two typical examples of a process flow for downstream processing are shown in Fig. 4.4. These schemes represent the processing of a typical monoclonal antibody (about 150 kDa) and another glycosylated protein, recombinant interferon (about 28 kDa) produced in mammalian cells. The aim of the individual unit operations are described in the figure as well.

A number of purification methods are available to separate proteins on the basis of a variety of different physico-chemical criteria such as size, charge, hydrophobicity, and solubility (Table 4.3). Detailed information about some separation and purification methods commonly used in purification schemes is provided below.

Centrifugation

Recombinant protein products in a cell harvest must be separated from suspended particulate material, including whole cells, lysed cell material, and fragments of broken cells generated when cell breakage has been necessary to release intracellular products. Most DSP flow sheets will, therefore, include at least one unit operation for the removal ("clarification") of particulates. Most frequently used methods are cen-

Table 4.3 Frequently used separation processes and their physical basis

Separation technique	Mode/principle	Separation based on
Filtration	Microfiltration	Size
	Ultrafiltration	Size
	Nanofiltration	Size
	Dialysis	Size
	Charged membranes	Charge
	Depth filtration	Size
Centrifugation	Isopycnic banding	Density
	Nonequilibrium setting	Density
Extraction	Fluid extraction	Solubility
	Liquid/liquid extraction	Partition, change in solubility
Precipitation	Fractional precipitation	Change in solubility
Chromatography	Ion exchange	Charge
	Gel filtration	Size
	Affinity	Specific ligand-substrate interaction
	Hydrophobic interaction	Hydrophobicity
	Adsorption	Covalent/noncovalent binding

trifugation and filtration techniques. However, the expense and effectiveness of such methods are highly dependent on the physical nature of the particulate material, of the product and the scale of the unit operation. Various clarification technologies are summarized in Turner et al. (Turner et al. 2018).

Besides the use of centrifugation for removal of cells and cell debris (clarification), centrifugation can also be used to separate subcellular particles and organelles, for example, particles produced when cells are disrupted by mechanical procedures. However, subcellular particles and organelles are difficult to separate either by using one fixed centrifugation step (or by filtration), but they can be isolated efficiently by centrifugation at different speeds. For instance, nuclei can be obtained by centrifugation at 400 × g for 20 min, while plasma membrane vesicles are pelleted at higher centrifugation rates and longer centrifugation times (fractional centrifugation). In many cases, however, total biomass can easily be separated from the medium and classified by a simple centrifugation step, e.g., in a continuous disc-stack centrifuge. Buoyant density centrifugation can be useful for separation of particles as well. This technique uses a viscous fluid with a continuous gradient of density in a centrifuge tube. Particles and molecules of various densities within the density range in the tube will cease to move when the isopycnic region has been reached. Both techniques of continuous (fluid densities within a range) and discontinuous (blocks of fluid with different density) density gradient centrifugation are used in buoyant density centrifugation on a laboratory scale. However, for application on an industrial scale, continuous centrifuges, e.g., tubular bowl centrifuges, are only used for discontinuous buoyant density centrifugation of protein products. This type of industrial centrifuge is mainly applied to recover precipitated proteins or contaminants.

Filtration

Filtration is often applied at various stages during downstream processing. The most successful setups being normal flow depth filtration, membrane filtration, and tangential flow filtration (TFF, also referred to as "cross flow"). Separation is achieved based on particle size differences. Below the main types of filtration are described.

Depth Filtration

Depth filters are often applied in the clarification of cell harvest to remove cells and cell debris. Depth filters consist of a complex porous matrix of materials, often including charged components and filter aids such as diatomaceous earth, enabling cellular debris and other contaminants to be retained at both the surface and internal layers of the depth filter (Turner et al. 2018). Issues at large manufacturing scale are usually the large membrane area needed to prevent clogging/fouling, and the large hold up volumes. For large harvest volumes depth filters are also used in combination with centrifugation.

Membrane Filtration

While depth filters retain contaminants within the filter structure, membrane filters have defined pore size ranges (e.g., in the micrometer or nanometer range) that trap supra-pore size particles on the membrane surface while allowing passage of smaller particles. The main membrane filters are either used in a dead-end mode in which the retained particles collect on the surface of the filter media as a stable filter cake that grows in thickness and increases flow resistance, or in a tangential flow mode in which the high shear across the membrane surface limits fouling, gel layer formation and concentration polarization. Important applications of membrane filters within pharmaceutical processes are described below.

Tangential Flow Filtration

Tangential flow filtration (TFF) is often used for the concentration and buffer exchange of purified product and sometimes within clarification processes. Depending upon the molecules/particles to be separated or concentrated, ultrafilter or micro membranes are used. Mixtures of molecules of highly different molecular dimensions are separated by passage of a dispersion under pressure across a membrane with a defined pore size. In general, ultrafiltration achieves little purification of protein product from other molecules with a comparable size, because of the relatively large pore size distribution of the membranes. As mentioned, this technique is widely used to concentrate macromolecules, and also to change the aqueous phase (e.g., re buffer components) in which the particles are dispersed or in which molecules are dissolved (diafiltration) to one required for the subsequent purification steps.

Sterilizing-Grade Filtration

Bioburden reduction filters are an essential part of most pharmaceutical processes. These dead-end filters consist of a membrane with an average pore size between 0.1 and 0.45 μm and a narrow size distribution. They are very effective in the removal of bioburden, and as such used at various steps in the purification process, e.g., at hold steps, and at the final steps to produce drug substance or drug product.

Virus Filtration

As mentioned later in this chapter, removal of potentially contaminating viruses is essential in a pharmaceutical process. Nanofiltration is an elegant and effective technique and the validation aspects of this technology are well described (PDA technical report 41, Sofer et al. 2005). Filtration through 15 nm pore membranes can remove even the smallest nonenveloped viruses such as bovine parvovirus (Maerz et al. 1996). Nanofilters are a significant contributor to the costs of the purification process.

Charged Membranes

A type of membrane that is increasingly used within the biopharmaceutical industry is the charged membrane (Zhou and Tressel 2006; Etzel and Arunkumar 2017). As for ion exchange chromatography (see below), negatively (sulphonic, S) or positively (quaternary ammonium, Q) charged ligands are attached to the multilayer membranes, enabling the removal of residual impurities such as host cell DNA, viruses or host cell proteins from the recombinant protein product. In contrast to ion exchange chromatography, the open structure of the charged membranes enables relatively

high diffusion rates of product/contaminants, thus a fast process step. A downside is the lower capacity. Charged membranes are often used in a flow-through mode in e.g. monoclonal antibody production processes, as such replacing the Q-based chromatography columns.

Extraction

Extraction, including liquid–liquid extraction, is a technique often used in the chemical industry, but rarely used for biopharmaceuticals. Liquid–liquid extraction basically separates molecules on solute affinity due to differences in the molecule's physical–chemical properties in a mixture of two immiscible phases (reviewed by Dos Santos et al. 2017). Traditionally, the phases consist of an aqueous and an organic phase. Upon phase separation, the target molecules are extracted to one of the two phases allowing its concentration and sometimes purification. Due to the possible impact of organic solvents on the structure and biological activity of biopharmaceuticals as well as the environmental impact, this traditional extraction method is rarely used anymore.

To overcome the main concerns, aqueous two-phase systems are developed. The compounds enabling separation of biopharmaceuticals encompass polymers, salts, surfactants, amino acids, and ionic liquids. Compared to chromatography, the operational costs of the two-phase systems are relatively low, scale up is straightforward, and the technique can be easily integrated in the early steps of a downstream process. However, two-phase systems are rarely applied in biopharmaceutical processes due to in general relatively low recovery values and limited purification abilities (reviewed by Dos Santos et al. 2017). A better understanding of the partitioning processes may reduce these limitations in the future.

Precipitation

The solubility of a particular protein depends on the physicochemical environment, for example, pH, ionic species, and ionic strength of the solution (see also Chap. 5). A slow continuous increase of the ionic strength (of a protein mixture) will selectively drive proteins out of solution. This phenomenon is known as "salting out." A wide variety of agents, with different "salting-out" potencies are available. Chaotropic series with increasing "salting-out" effects of negatively (I) and positively (II) charged molecules are given below:

I. SCN^-, I^-, CLO_4^- NO_3^-, Br^-, Cl^-, CH_3COO^-, PO_4^{3-}, SO_4^{2-}
II. Ba^{2+}, Ca^{2+}, Mg^{2+}, Li^+, Cs^+, Na^+, K^+, Rb^+, NH_4^+

Ammonium sulfate is highly soluble in cold aqueous solutions and is frequently used in "salting-out" purification.

Another method to precipitate proteins is to use water-miscible organic solvents (change in the dielectric constant). Examples of precipitating agents are polyethylene glycol and trichloroacetic acid. Under certain conditions, chitosan and nonionic polyoxyethylene detergents also induce precipitation (Cartwright 1987; Homma et al. 1993; Terstappen et al. 1993). Cationic detergents have been used to selectively precipitate DNA.

Precipitation is a scalable, simple, and relatively economical procedure for the recovery of a product from a dilute feedstock. It has been used for the isolation of proteins from culture supernatants. Unfortunately, with most bulk precipitation methods, the gain in purity is generally limited and product recovery can be low. Moreover, extraneous components are introduced which must be eliminated later. Finally, large quantities of precipitates may be difficult to handle. Despite these limitations, recovery by precipitation has been used with considerable success for some products.

Chromatography

In preparative chromatography systems, molecular species are primarily separated based on differences in distribution between two phases: one is the stationary phase (mostly a solid phase) and the other moves. This mobile phase may be liquid or gaseous (see also Chap. 3). Nowadays, almost all stationary phases (fine particles providing a large surface area) are packed into a column. The mobile phase is passed through by pumps. DSP protocols usually have at least two to three chromatography steps. Chromatographic methods used in purification procedures of biotech products are listed in (Table 4.3) and are briefly discussed in the following sections.

Chromatographic Stationary Phases

Chromatographic procedures often represent the rate-limiting step in the overall downstream processing. An important primary factor governing the rate of operation is the mass transport into the pores of conventional packing materials. Adsorbents employed include inorganic materials such as silica gels, glass beads, hydroxyapatite, various metal oxides (alumina), and organic polymers (cross-linked dextrans, cellulose, agarose). Separation occurs by differential interaction of sample components with the chromatographic medium. Ionic groups such as amines and carboxylic acids, dipolar groups such as carbonyl functional groups, and hydrogen bond-donating and bond-accepting groups control the interaction of the sample components with the stationary phase, and these functional groups slow down the elution rate if interaction occurs.

Chromatographic stationary phases for use on a large scale are evolving over time. An approach to the problems associated with mass transport in conventional systems is to use chromatographic particles that contain some large "through pores" in addition to conventional pores (see Fig. 4.5). These flow-through or "perfusion chromatography" media enable faster convective mass transport into particles and allow operation at much higher speeds without loss in resolution or binding capacity (Afeyan et al. 1989; Fulton 1994).

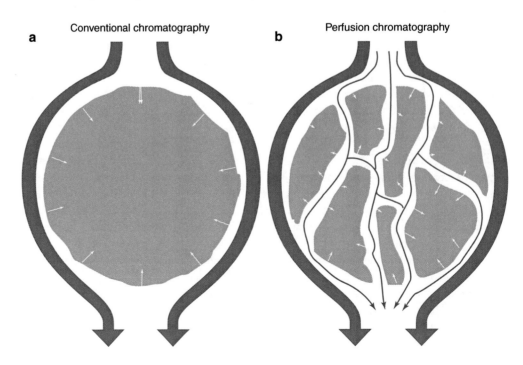

Fig. 4.5 The structure of conventional chromatographic particles (**a**) and the perfusion of flow through chromatographic particles (**b**) (Adapted from Fulton 1994)

a Conventional chromatography

b Perfusion chromatography

The ideal stationary phase for protein separation should possess a number of characteristics, among which are high mechanical strength, high porosity, no nonspecific interaction between protein and the support phase, high capacity and mass transfer rate, biocompatibility, and high stability of the matrix in a variety of solvents. The latter is especially true for columns used for the production of pharmaceuticals that need to be cleaned, depyrogenized, disinfected, and sterilized at regular intervals.

In production environments, chromatography columns that operate at relatively low back pressure are often used. These can be made of stainless steel. But the low back pressure allows the introduction of disposable (plastic) columns in a GMP manufacturing environment. Unlike conventional stainless steel, plastic columns are less sensitive to, e.g., salt corrosion. A disadvantage can be leaching of plastic components into the product stream. Disposable plastic columns permit the efficient separation of proteins in a single batch, making this an attractive unit operation in a manufacturing process. A new development is the use of chromatography equipment with fully disposable flow paths that resists almost all chemicals used in protein purification including disinfection and sterilization media.

Adsorption Chromatography

In adsorption chromatography (also called "normal phase" chromatography), the stationary phase is more polar than the mobile phase. The protein of interest selectively binds to a static matrix under one condition and is released under a different condition. Adsorption chromatography methods enable high ratios of product load to stationary phase volume. Therefore, this principle is economically scalable.

Ion-Exchange Chromatography

Ion-exchange chromatography can be a powerful step early in a purification scheme. It can be easily scaled up. Ion-exchange chromatography can be used in a negative mode, i.e., the product flows through the column under conditions that favor the adsorption of contaminants to the matrix, while the protein of interest does not bind (Tennikova and Svec 1993). The type of the column needed is determined by the properties of the proteins to be purified, e.g., isoelectric point and charge density. Anion exchangers bind negatively charged molecules and cation exchangers bind positively charged molecules. In salt-gradient ion-exchange chromatography, the salt concentration in the perfusing elution buffer is increased continuously or in steps. The stronger the binding of an individual protein to the ion exchanger, the later it will appear in the elution buffer. Likewise, in pH-gradient chromatography, the pH is changed continuously or in steps. Here, the protein binds at 1 pH and is released at a different pH. As a result of the heterogeneity in glycosylation, e.g., a varying number of

sialic acid moieties, cf. Chap. 19, glycosylated proteins may elute over a relatively broad pH range (up to 2 pH units).

In order to simplify purification, a specific amino acid tail can be added to the protein at the gene level to create a "purification handle." For example, a short tail consisting of arginine residues allows a protein to bind to a cation exchanger under conditions where almost no other cell proteins bind. However, this technique is useful for laboratory-scale isolation of the product and generally not at production scale due to regulatory problems related to the removal of the arginine or other specific tags from the protein.

(Immuno) Affinity Chromatography

Affinity Chromatography

Affinity chromatography is based on highly specific interactions between an immobilized ligand and the protein of interest. Affinity chromatography is a very powerful method for the purification of proteins. Under physiological conditions, the protein binds to the ligand. Extensive washing of this matrix will remove contaminants, and the purified protein can be recovered by the addition of ligands competing for the stationary phase binding sites or by changes in physical conditions (such as low or high pH of the eluent) that greatly reduce the affinity. Examples of affinity chromatography include the purification of glycoproteins, which bind to immobilized lectins, and the purification of serine proteases with lysine binding sites, which bind to immobilized lysine. In these cases, a soluble ligand (sugar or lysine, respectively) can be used to elute the required product under relatively mild conditions. Another example is the use of the affinity of protein A and protein G for antibodies. Protein A and protein G have a high affinity for the Fc portions of many immunoglobulins from various animals. Protein A and G matrices can be commercially obtained with a high degree of purity. Protein A resins are often used in the capture of biotherapeutic monoclonal antibodies at large scale, and these resins are also one of the most expensive parts of the production process. In the last decade the amino acid composition has been modified to generate Protein A ligands that are more resistant to hydroxide, allowing better cleaning of the resin. Also, the coupling chemistry has been improved to allow re-use of the resin for over a hundred cycles, and by that reducing the cost of goods.

Since the elution of antibodies from Protein A occurs at low pH, acid-sensitive antibodies, especially IgG2, IgG4 and bispecific antibodies, cannot be purified using Protein A since they tend to aggregate. Recently a Protein A ligand was developed (Z-Ca) which is dependent on calcium ions for its binding to antibodies. Elution was achieved by depletion of calcium with a sodium chloride-containing buffer close to neutral pH (Schwartz et al. 2022).

For the purification of, e.g., hormones or growth factors, the receptors or short peptide sequence that mimic the binding site of the receptor molecule can be used as affinity ligands. Some proteins show highly selective affinity for certain dyes commercially available as immobilized ligands on purification matrices. When considering the selection of these ligands for pharmaceutical production, one must realize that some of these dyes are carcinogenic and that a fraction may leach out during the process.

An interesting approach to optimize purification is the use of a gene that codes not only for the desired protein but also for an additional sequence that facilitates recovery by affinity chromatography. At a later stage, the additional sequence is removed by a specific cleavage reaction. As mentioned before, this is a complex process that needs additional purification steps.

In general, the use of affinity chromatography in the production process for therapeutics may lead to complications during validation of the removal of free ligands or protein extensions. Consequently, except for monoclonal antibodies where affinity chromatography is part of the purification platform at large scale, this technology is rarely used in the industry.

Immunoaffinity Chromatography

The specific binding of antibodies to their epitopes is used in immunoaffinity chromatography (reviewed by Abi-Ghanem and Berghman 2012). This technique can be applied for purification of either the antigen or the antibody. The antibody can be covalently coupled to the stationary phase and act as the "receptor" for the antigen to be purified. Alternatively, the antigen, or parts thereof, can be attached to the stationary phase for the purification of the antibody. Advantages of immunoaffinity chromatography are its high specificity and the combination of concentration and purification in one step.

A disadvantage associated with immunoaffinity methods is the sometimes very strong antibody-antigen binding. This requires harsh conditions during elution of the ligand. Under such conditions, sensitive ligands could be harmed (e.g., by denaturation of the protein to be purified). This can be alleviated by the selection of antibodies and environmental conditions with high specificity and sufficient affinity to induce an antibody-ligand interaction, while the antigen can be released under mild conditions. Another concern is the disruption of the covalent bond linking the "receptor" to the matrix. This would result in the elution of the entire complex. Therefore, in practice, a further purification step after affinity chromatography as well as an appropriate detection assay (e.g., Enzyme-Linked Immuno Sorbent Assay, ELISA) is almost always necessary. On the other hand, improved coupling chemistry that is less susceptible to hydrolysis has been developed to prevent leaching.

Scale-up of immunoaffinity chromatography is often hampered by the relatively large quantity of the specific "receptor" (either the antigen or the antibody) that is required and the lack of commercially available, ready-to-use matrices. The use of immunoaffinity in pharmaceutical processes will have major regulatory consequences since the immunoaffinity ligand used will be considered by the regulatory bodies as a "second product" and will be subjected to basically the same regulatory scrutiny as the drug substance. Moreover, immunoaffinity ligands can have a significant effect on the final costs of goods.

Examples of proteins of potential therapeutic value that have been purified using immunoaffinity chromatography are interferons, urokinase, epoetin, interleukin-2, human factor VIII and X, and recombinant tissue plasminogen activator.

Hydrophobic Interaction Chromatography

Under physiological conditions, most hydrophobic amino acid residues are located inside the protein core, and only a small fraction of hydrophobic amino acids is exposed on the "surface" of a protein. Their exposure is suppressed because of the presence of hydrophilic amino acids that attract large clusters of water molecules and form a "shield." High salt concentrations reduce the hydration of a protein, and the surface-exposed hydrophobic amino acid residues become more accessible. Hydrophobic interaction chromatography (HIC) is based on noncovalent and nonelectrostatic interactions between proteins and the stationary phase. HIC is a mild technique, usually yielding high recoveries of proteins that are not damaged, are folded correctly, and are separated from contaminants that are structurally related. HIC is ideally placed in the purification scheme after ion-exchange chromatography, where the protein usually is released in high ionic strength elution media (reviewed by Chen et al. 2015).

Gel-Permeation Chromatography

Gel-permeation or size-exclusion chromatography, also known as gel filtration, separates molecules according to their shape and size (see Fig. 4.6). Inert gels with narrow pore-size distributions in the size range of proteins are available. These gels are packed into a column. The protein mixture is loaded on top of the column and the proteins diffuse into the gel. The smaller the protein, the more volume it will have available in which to disperse. Molecules that are larger than the largest pores are not able to penetrate the gel beads and will therefore stay in the void volume of the column. When a continuous flow of buffer passes through the column, the larger proteins will elute first and the smallest molecules last. Gel-permeation chromatography is a good alternative to membrane diafiltration for buffer exchange at almost any purification stage, and it is often used in labora-

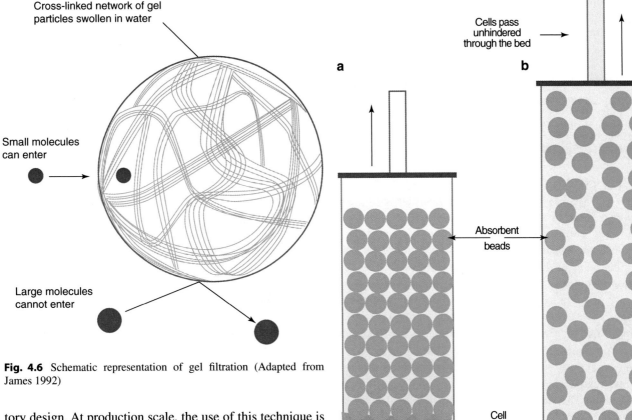

Cross-linked network of gel
particles swollen in water

Small molecules
can enter

Large molecules
cannot enter

Fig. 4.6 Schematic representation of gel filtration (Adapted from James 1992)

Cells pass
unhindered
through the bed

a

b

Absorbent
beads

Cell
cake

Particulate
containing
process fluid

Fixed bed

Expanded bed

Fig. 4.7 Comparison between (**a**) a packed bed and (**b**) an expanded bed (Adapted from Chase and Draeger 1993)

tory design. At production scale, the use of this technique is usually limited, mainly because it is a slow process and only relatively small sample volumes can be loaded on a large column (up to one-third of the column volume in the case of "buffer exchange"). It is therefore best avoided or used late in the purification process when the protein is available in a highly concentrated form. Gel filtration is sometimes used as the final step in the purification to bring proteins in the appropriate buffer used in the final formulation. In this application, its use has little if no effect on the product purity characteristics.

Expanded Beds

As mentioned before, purification schemes are based on multistep protocols. This not only adds greatly to the overall production costs but can also result in significant loss of product. Therefore, there still is an interest in the development of new methods for simplifying the purification process. Adsorption techniques are popular methods for the recovery of proteins, and the conventional operating format for preparative separations is a packed column (or fixed bed) of adsorbent. Particulate material, however, can be trapped near the bed, which results in an increase in the pressure drop across the bed and eventually in clogging of the column. This can be avoided by the use of precolumn filters (e.g., 0.2 μm pore size) to save the column integrity. Another solution to this problem may be the use of expanded beds (Chase and Draeger 1993; Fulton 1994), also called fluidized beds (see

Fig. 4.7). In principle, the use of expanded beds enables clarification, concentration, and purification to be achieved in a single step. The concept is to employ a particulate solid-phase adsorbent in an open bed with upward liquid flow. The hydrodynamic drag around the particles tends to lift them upward, which is counteracted by gravity because of a density difference between the particles and the liquid phase. The particles remain suspended if particle diameter, particle density, liquid viscosity, and liquid density are properly balanced by choosing the correct flow rate. The expanded bed allows particulates (e.g., cells and cell debris) to pass through, whereas molecules in solution are selectively retained (e.g., by the use of ion-exchange or affinity adsorbents) on the adsorbent particles. Feedstocks can be applied to the bed without prior removal of particulate material by centrifugation or filtration, thus reducing process time and costs. Fluidized beds have been used previously for the industrial-scale recovery of antibiotics such as streptomycin and novobiocin (Fulton 1994; Chase 1994). Stable, expanded beds can be obtained using simple equipment adapted from that used for conventional, packed bed adsorption and chromatography processes. Ion-exchange adsorbents are likely to be chosen for such separations.

Single-Use Systems

In the past two decades, the development of single-use production systems has been boosting. This is reflected by the growing number of single-use systems available for mammalian cell culture and microbial cultures (see below). Single-use systems are currently not only developed for culturing but also for downstream unit operations such as the filtration (depth, membrane) and chromatography steps. It is currently possible to produce proteins with only single-use systems.

Single-use bioreactors for mammalian cell culture and protein production applications are characterized by a low power input, low mixing capabilities, limited oxygen transfer, restrictive exhaust capacity, and limited foam management. Therefore, transferring these single-use bioreactors into single-use fermenters that can be used for microbial production is a challenge as microbial fermentation requires high mixing capabilities and has higher oxygen transfer needs. The present generation of single-use fermenters is only used in the production of the least challenging 5% of microbial fermentations (Jones 2015).

Single-use bioreactors are used for the manufacturing of products in development and on the market. Shire (Dublin, Ireland) was the first company that used single-use bioreactors up to 2000 L for the manufacturing of one of its products. The advantages of the single-use technology are:

- Cost-effective manufacturing technology
- By introducing single-use systems, the design is such that all items not directly related to the process can be removed from the culture system, such as clean-in-place (CIP) and steam-in-place (SIP) systems that are critical within a stainless-steel plant. Furthermore, a reduction in capital costs (CAPEX) is achieved by introducing single-use systems. In a case study that compares the costs for a single-use versus multiuse stainless steel 2 × 1000 L new facility, the single-use facility reduces CAPEX significantly, while operating costs (OPEX) are increased. Overall, these studies show that investing in a flexible

single-use facility is beneficial compared to a fixed stainless steel facility (Eibl and Eibl 2011; Goldstein and Molina 2016). It must be noted that investment decisions on new production facilities must be taken before the product is accepted by regulatory bodies.

- Increases the number of GMP batches
- By introducing single-use systems, it is possible to increase the number of GMP batches that can be produced within a manufacturing campaign since time-consuming cleaning and sterilization of the equipment is not needed anymore. Thus the turnover time needed from batch to batch is shortened.
- Provides flexibility in GMP facility design
- When stainless steel systems are used, changes to the equipment might have an impact on the design of the stainless steel tanks, piping, etc. These equipment changes will influence the overall validation status of the facility. By using single-use systems, equipment changes can easily be incorporated as the setup of the single-use process is flexible. As with the stainless steel systems, in case a change will influence the validated process, the validated status of the process must be reconsidered and a revalidation might be needed.
- Speeds up implementation and time to market
- Due to the great flexibility of the single-use systems, the speed of product to market is less influenced by process changes that might be introduced during the different development stages of the production process than 'traditional' stainless-steel equipment. However, the process needs to be validated before market introduction. When changes are introduced after process validation, a revalidation might be needed. Here again, there is no difference in this respect to the traditional stainless-steel setup.
- Reduces water and wastewater costs
- Since the systems are single-use, there will be a great reduction in the total costs for cleaning. Not only through a reduction in water consumption but also a reduction in the number of hours needed to clean systems and to set them up for the next batch of product.
- Reduces validation costs
- No annual validation costs for cleaning and sterilization are needed anymore when single-use systems are used.

A disadvantage of the single-use system is that the operational expenses will increase and storage facilities for single-use bags and tubing are needed. Moreover, the dependence of the company on one supplier of single-use systems is a factor to consider. This clearly became evident during the recent worldwide outbreak of the corona virus. Finally, leachables and extractables from the single use plastics may end up in your product, causing potential safety and efficacy issues.

The advantages of the stainless steel bioreactors are obvious as this traditional technology is well understood and controlled, although the stainless steel pathway had and still has major disadvantages such as expensive and inflexible design, installation and maintenance costs combined with significant expenditures of time in facilities and equipment qualification and validation efforts. For very large volume products, stainless steel is still the most economically viable option due to limited scalability of current single-use bioreactors (i.e., 2000 L max).

Continuous Manufacturing

Most biopharmaceuticals are currently produced in batches. In general, a harvest from one bioreactor is processed in one purification train to produce the drug substance. Continuous processes are designed to achieve higher product quality at lower cost. The basic design is to run the bioreactor process in a continuous mode.

In a batch process, the cell concentration increases during culture, until it reaches a maximum level, followed by a decline in cell density due to cell lysis. The product will be harvested at a predefined viable cell density to prevent that cell death will have an effect on product quality. In a bioreactor process in a continuous mode the cell concentration is kept at a relatively steady level, not for, e.g., 2 weeks as for an average batch process for mAbs, but for many weeks. This is achieved by connecting the bioreactor to a filtration device in perfusion mode (e.g., ATF, see Fig. 4.2) removing product continuously, adding fresh medium, and by bleeding the bioreactor, i.e., removal of excess cells. The continuously removed product is subsequently loaded on the capture column, which can run in a semi-continuous mode. The reader is referred to the literature for more information (Konstantinov and Cooney 2015; Sao Pedro et al. 2021). In principle, continuous processes can only be operated by automated systems where the desired values for critical quality attributes CQAs are achieved by adjusting critical process parameters (CPPs) in real time (see Quality by Design section). This greatly enhances process robustness, the productivity, and equipment utilization. The overall result of such an automated continuous process is a lower cost of manufacturing. The interest in the biopharmaceutical industry is growing, but the adaption rate is still low due to a number of challenges: (1) the different single unit operations run in continu-

ous mode need to be carefully tuned; (2) a higher level of process understanding is required, also in an earlier stage of the product development process compared to batch processes; (3) a thorough understanding about the behavior of cells in perfusion mode is essential, which requires advanced analytical tools that can be applied as in process control assays to ensure that the CPPs are maintained (Dream et al. 2018).

Contaminants

Parenteral product purity mostly is $\geq 99\%$ (Berthold and Walter 1994; ICH 1999a). Purification processes should yield potent proteins with well-defined characteristics for human use from which "all" contaminants have been removed to a major extent. The purity of the drug protein in the final product largely depends on the applied purification technology.

Table 4.4 lists potential contaminants and product variants that may be present in recombinant protein products from bacterial and mammalian sources. These contaminants can be host-related, process-related, and product-related. In the following sections, special attention is paid to the detec-

Table 4.4 Potential contaminants/variants in recombinant protein products derived from bacterial and mammalian hosts

Origin	Contaminant
Host-related	Viruses
	Bacteria (mycoplasma)
	Host-derived proteins and DNA
	Endotoxins (from gram-negative bacterial hosts)
Product-related	Glycosylation variants
	Amino acid substitution and deletion
	Denatured protein (loss of secondary, tertiary, quaternary structure)
	Oxidized variants
	Conformational isomers
	Dimers and aggregates
	Disulfide pairing variants
	Succinimide formation
	(De)amidated species
	Protein fragments
Process-related	Viruses
	Bacteria
	Cell culture medium components
	Purification reagents
	Metals
	Column materials/leachables
	Leachables from single-use system (tubes, bags, etc.)

tion and elimination of contamination by viruses, bacteria, cellular DNA, and undesired proteins.

Viruses

Endogenous and adventitious viruses, which require the presence of living cells to propagate, are potential contaminants of animal cell cultures and, therefore, of the final drug product. If present, their concentration in the purified product will be very low and it will be difficult to detect them. Viruses such as retrovirus can be visualized by (nonsensitive) electron microscopy. For retroviruses, a highly sensitive RT-PCR (reverse-transcriptase polymerase chain reaction) assay is available, but for other viruses, a sensitive in vitro assay might be lacking. The risks of some viruses (e.g., hepatitis virus) are known (Walter et al. 1991; Marcus-Sekura 1991), but there are other viruses whose risks cannot be properly judged because of lack of solid experimental data. Some virus infections, such as parvovirus, can have long latent periods before their clinical effects become apparent. Long-term effects of introducing viruses into a patient treated with a recombinant protein should not be overlooked. Therefore, it is required that products used parenterally are free from viruses. The specific virus testing regime required will depend on the cell type used for production (Löwer 1990; Minor 1994).

Viruses can be introduced by nutrients, by an infected production cell line, or they are introduced (by human handling) during the production process. The most frequent source of virus introduction is animal serum. In addition, animal serum can introduce other unwanted agents such as bacteria, mycoplasmas, prions, fungi, and endotoxins. Appropriate screening of cell banks and growth medium constituents for viruses and other adventitious agents should be strictly regulated and supervised (Walter et al. 1991; FDA 1993; ICH 1999b; WHO 2010). Validated, orthogonal methods (cf. Chap. 5) to inactivate and remove possible viral contaminants during the production process are mandatory for licensing of therapeutics derived from mammalian cells

or transgenic animals (EMA 1996; ICH 1999b). Viruses can be inactivated by physical and chemical treatment of the product. Heat, irradiation, sonication, extreme pH, detergents, solvents, and certain disinfectants can inactivate viruses. These procedures can be harmful to the product as well and should therefore be carefully evaluated and validated (Walter et al. 1992; ICH 1999b). As mentioned in the filtration section, removal of viruses by nanofiltration is an elegant and effective technique and the validation aspects of this technology are well described (PDA technical report 41, Sofer et al. 2005). A significant log reduction of even the smallest nonenveloped viruses such as bovine parvovirus can be obtained by filtration through 15 nm membranes (Maerz et al. 1996). Another common, although less robust, method to remove viruses in antibody processes is by ion-exchange chromatography and Q-charged membranes (Zhou and Tressel 2006). A number of methods for removing or inactivating viral contaminants are mentioned in Table 4.5.

In general, a protein production process using mammalian cells should contain 2 or more orthogonal virus reduction steps. As mentioned, virus validation studies need to be performed on the developed production process and they should show sufficient removal or inactivation of spiked model viruses before the start of clinical studies. The choice of viruses to be spiked depends upon the production cell line, the ease of growing model viruses to high titers, and should include various types of virus (large vs small, enveloped vs nonenveloped, DNA vs RNA-based). These types of studies are performed in specialized laboratories.

Bacteria

Bacterial contamination may be a problem for cells in culture or during pharmaceutical purification. Usually, the size of bacteria allows simple filtration over 0.2 or 0.45 μm (or smaller) filters for adequate removal. Special attention is given to potential contaminations with mycoplasma, a genus of bacteria having no cell wall around their cell membrane. Some mycoplasma species are pathogenic to humans, and

Table 4.5 Methods for reducing or inactivating viral contaminants

Category	Types	Example
Inactivation	Heat treatment	Pasteurization
	pH extremes	Low pH
	Radiation	UV-light
	Dehydration	Lyophilization
	Cross linking agents, denaturing or disrupting agents	β-Propiolactone, formaldehyde, NaOH, organic solvents (e.g., chloroform), detergents (e.g., Na-cholate)
	Neutralization	Specific, neutralizing antibodies
Removal	Chromatography	Ion-exchange, immuno-affinity chromatography
	Filtration	Nanofiltration, Q-charged membranes
	Precipitation	Cryo-precipitation

hundreds of mycoplasma species infect animals (Larsen and Hwang 2010). Testing for mycoplasma is a regulatory requirement for human biopharmaceuticals.

In order to further prevent bacterial contamination during production, the used raw materials have to be sterilized, if feasible, preferably at 121 °C or higher, and the products are manufactured under strictly aseptic conditions wherever possible. Production most often takes place in so-called clean rooms where the chances of environmental contamination are reduced through careful control of the environment, for example, filtration of air. Additionally, antibiotic agents can be added to the culture media in some cases, but have to be removed further downstream in the purification process. The use of beta-lactam antibiotics such as penicillin is strictly prohibited due to oversensitivity of some individuals to these compounds. Because of the persistence of antibiotic residues, which are difficult to eliminate from the product, appropriately designed manufacturing plants and extensive quality control systems for added reagents (medium, serum, enzymes, etc.) permitting antibiotic-free operation are preferable.

Pyrogens

Pyrogens are compounds that induce fever. Humans are sensitive to pyrogen contamination at very low concentrations (picograms per mL). Exogenous pyrogens (pyrogens introduced into the body, not generated by the body itself) can be derived from bacterial, viral, or fungal sources. Bacterial pyrogens are mainly endotoxins shed from gram-negative bacteria. They are lipopolysaccharides, and Fig. 4.8 shows the basic structure. The conserved structure in the full array of thousands of different endotoxins is the lipid-A moiety. Another general property shared by endotoxins is their high, negative electrical charge. Their tendency to aggregate and to form large units with M_W of 10^4 to over 10^6 Daltons in

water, and their tendency to adsorb to surfaces indicate that these compounds are amphipathic in nature. Sensitive tests to detect and quantify pyrogens are commercially available.

They are stable under standard autoclaving conditions but break down when heated in the dry state. For this reason, equipment and container are treated at temperatures above 160 °C for prolonged periods e.g., 30 min dry heat at 250 °C. Removal is complicated because pyrogens vary in size and chemical composition. Pyrogen removal of recombinant products derived from bacterial sources should be an integral part of the preparation process. Ion exchange chromatographic procedures (utilizing its negative charge) can effectively reduce endotoxin levels in solution.

Excipients used in the protein formulation should be essentially endotoxin-free. For solutions "water for injection" (compendial standards) is (freshly) distilled or produced by reverse osmosis. The aggregated endotoxins cannot pass through the reverse osmosis membrane. Removal of endotoxins immediately before filling the final container can be accomplished by using activated charcoal or other materials with large surfaces offering hydrophobic interactions. Endotoxins can also be inactivated on utensil surfaces by oxidation, e.g., peroxide, or dry heating, e.g., 30 min dry heat at 250 °C.

Cellular DNA

The application of continuous mammalian cell lines for the production of recombinant proteins might result in the presence of oncogene-bearing DNA fragments in the final protein product (Walter and Werner 1993; Löwer 1990). A stringent purification protocol that is capable of reducing the DNA content and fragment size to a safe level is therefore necessary (Berthold and Walter 1994; WHO 2010; ICH 2017). A number of approaches are available to validate that the purification process removes cellular DNA and RNA. One

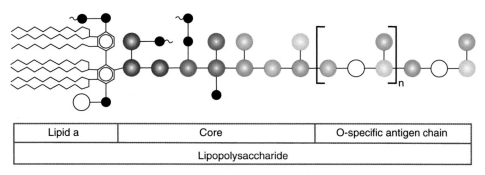

Fig. 4.8 Generalized structure of endotoxins. Most properties of endotoxins are accounted for by the active, insoluble "lipid A" fraction being solubilized by the various sugar moieties (circles with different colors). Although the general structure is similar, individual endotoxins vary according to their source and are characterized by the O-specific antigenic chain (adapted from Groves (1988))

such approach involves incubating the cell line with radiola-beled nucleotides and determining radioactivity in the puri-fied product obtained through the purification protocol. Other methods are dye-binding fluorescence-enhancement assays for nucleotides and PCR-based methods. If the pres-ence of nucleic acids persists at significant levels in a final preparation, then additional steps must be introduced in the purification process. The question about a safe level of nucleic acids in biotech products is difficult to answer. Transfection with so-called naked DNA is very difficult and a high concentration of DNA is needed. Nevertheless, it is agreed for safety reasons that final product contamination by nucleic acids should not exceed 100 pg or 10 ng per dose depending on the type of cells used to produce the pharma-ceutical (WHO 2010; European Pharmacopoeia 2011).

Protein Contaminants and Product Variants

As mentioned before, minor amounts of host-, process-, and product-related protein contaminants will likely be present in biotech products. These types of contaminants are a poten-tial health hazard because, if present, they may be recognized as antigens by the patient receiving the recombinant protein product. On repeated use, the patient may show an immune reaction caused by the contaminant, while the protein of interest is performing its beneficial function. In such cases, the immunogenicity may be misinterpreted as being due to the recombinant protein itself. Therefore, one must be very cautious in interpreting safety data of a given recombinant therapeutic protein. Some contaminants may also affect effi-cacy of the product, for example, if they bind to an epitope important for the product to exert its function. Hence, careful control is needed.

Generally, the sources of host- and process-related pro-tein contaminants are the cell culture medium used and the host proteins of the cells. Among the host-derived contami-nants, the host species' version of the recombinant protein could be present (WHO 2010). As these proteins are similar in structure, it is possible that undesired proteins are co-purified with the desired product. For example, urokinase is known to be present in many continuous cell lines. The syn-thesis of highly active biological molecules such as cyto-kines by hybridoma cells might be another concern (FDA 1990). Depending upon their nature and concentration, these cytokines might enhance the antigenicity of the product.

"Known" or expected contaminants should be moni-tored at the successive stages in a purification process by suitable in-process controls, e.g., sensitive immunoassay(s). Tracing of the many "unknown" cell-derived proteins is more difficult. When developing a purification process, other less-specific analytical techniques such as SDS-PAGE (sodium dodecyl sulfate—polyacrylamide gel elec-

trophoresis) are usually used in combination with various staining techniques.

Product-related contaminants may pose a safety issue for patients. These contaminants can, for example, be aggre-gated, deamidated or oxidized forms of the product. And, importantly, one has to keep in mind that recombinant pro-teins produced in cells are inherently variable in structure, for example, at the level of glycosylation. Such molecules are generally considered product variants. Some of these contaminants/variants are described in the following paragraphs.

N- and C-Terminal Heterogeneity

A major problem connected with the production of biotech products is the problem associated with the amino (NH2)-terminus of the protein, e.g., in *E. coli* systems, where pro-tein synthesis always starts with methylmethionine. Obviously, it has been of great interest to develop methods that generate proteins with an NH2-terminus as found in the authentic protein. When the proteins are not produced in the correct way, the final product may contain several methionyl variants of the protein in question or even contain proteins lacking one or more residues from the amino terminus. This is called the amino-terminal heterogeneity. This heterogene-ity can also occur with recombinant proteins such as interferon-alpha that are susceptible to proteases secreted by the host or introduced by serum-containing media. These proteases can clip off amino acids from the C-terminal and/or N-terminal of the desired product (amino- and/or carboxy-terminal heterogeneity). Amino- and/or carboxy-terminal heterogeneity is not desirable since it may cause difficulties in purification and characterization of the proteins. In case of the presence of an additional methionine at the N-terminal end of the protein, its secondary and tertiary structure can be altered. This could affect the biological activity and stability and may make it immunogenic. Moreover, N-terminal methionine and/or internal methionine is sensitive to oxida-tion (Sharma 1990).

C-terminal lysine clipping is often observed in monoclo-nal antibodies produced in mammalian cells. This does not have to be an issue, since the C-terminal lysine is clipped off rapidly in the blood upon injection in humans. The glutamine on the N-terminus of monoclonal antibodies can be con-verted in pyro-glutamate, increasing the acidity of the anti-body. These types of post-translational modifications should be controlled within a certain range to ensure a robust pro-duction process.

Conformational Changes/Chemical Modifications
Although mammalian cells are able to produce proteins structurally equal to endogenous proteins, some caution is needed. Transcripts containing the full-length coding sequence could result in conformational isomers of the pro-

tein because of unexpected secondary structures that affect translational fidelity (Sharma 1990). Another factor to be considered is the possible existence of equilibria between the desired form and other forms such as dimers. The correct folding of proteins after biosynthesis is important because it determines the specific activity of the protein). Therefore, it is important to determine if all molecules of a given recombinant protein secreted by a mammalian expression system are folded in their native conformation. Apart from conformational changes, proteins can undergo chemical alterations, such as proteolysis, deamidation, and hydroxyl and sulfhydryl oxidations during the purification process (cf. Chaps. 2 and 3) These alterations can result in (partial) denaturation of the protein. Vice versa, denaturation of the protein may cause chemical modifications as well, e.g., as a result of exposure of sensitive groups.

Glycosylation (Cf. Chap. 2)

Many therapeutic proteins produced by recombinant DNA technology are glycoproteins of which the majority are monoclonal antibodies. The presence and nature of oligosaccharide side chains in proteins affect a number of important characteristics, such as the serum half-life, solubility, and stability of the protein, and sometimes even the pharmacological function (Cumming 1991). Darbepoetin, a second-generation, genetically modified erythropoietin, has a carbohydrate content of 50% compared to 40% for the native molecule, which increases the in vivo half-life after intravenous administration from 8 h for erythropoietin to 25 h for darbepoetin (Sinclair and Elliott 2005). Antibody-dependent cell cytotoxicity (ADCC) is dependent on the degree of fucosylation of the antibody product (Hossler et al. 2009; reviewed by Krasnova and Wong 2016). As a result, the therapeutic profile may be "glycosylation" dependent. As mentioned previously, protein glycosylation is not determined by the DNA sequence. It is an enzymatic modification of the protein after translation and depends on the metabolic state of the cell (Hossler et al. 2009). Although mammalian cells are very well able to glycosylate proteins, it is hard to fully control glycosylation. Carbohydrate heterogeneity can occur through the size of the chain, type of oligosaccharide, and sequence of the carbohydrates. This has been demonstrated for a number of recombinant products including monoclonal antibodies, interleukin-4, chorionic gonadotropin, erythropoietin, and tissue plasminogen activator. Carbohydrate structure and composition in recombinant proteins may differ from their native counterparts because the enzymes required for synthesis and processing vary among different expression systems, e.g., glycoproteins from insect cells are frequently smaller than the same glycoproteins expressed in mammalian cells or even from one mammalian system to another.

Proteolytic Processing

Proteases play an important role in processing, maturation, modification, or isolation of recombinant proteins. Proteases from mammalian cells are involved in secreting proteins into the cultivation medium, e.g., by cleaving off a signal peptide. Proteases are released if cells die and undergo lysis during production in the bioreactor and at harvest. It is therefore important to control growth and harvest conditions in order to minimize this effect. Another source of proteolytic attack is found in the components of the medium in which the cells are grown. For example, serum contains a number of proteases and protease zymogens that may affect the secreted recombinant protein. If present in small amounts and if the nature of the proteolytic attack on the desired protein is identified, appropriate protease inhibitors to control proteolysis could be used. It is advised to document the integrity of the recombinant protein basically after each purification step.

Proteins become much more susceptible to proteases at elevated temperatures. Purification strategies should be designed to carry out all the steps at 2–8 °C (Sharma 1990) if proteolytic degradation occurs. Alternatively, Ca^{2+} complexing agents (e.g., citrate) can be added as many proteases depend on Ca^{2+} for their activity. From a manufacturing perspective, however, cooling large-scale downstream process unit operations, although not impossible, is a complicating and expensive factor.

Bacteria: Protein Inclusion Body Formation

In bacteria, soluble proteins can form dense, finely granular inclusions within the cytoplasm. These "inclusion bodies" often occur in bacterial cells that overproduce proteins by plasmid expression. The protein inclusions appear in electron micrographs as large, dense bodies often spanning the entire diameter of the cell. Protein inclusions are probably formed by a buildup of amorphous protein aggregates held together by covalent and noncovalent bonds. The inability to measure inclusion body proteins directly may lead to the inaccurate assessment of recovery and yield and may cause problems if protein solubility is essential for efficient, large-scale purification (Berthold and Walter 1994). Several schemes for recovery of proteins from inclusion bodies have been described. The recovery of proteins from inclusion bodies requires cell breakage and inclusion body recovery. Dissolution of inclusion proteins is the next step in the purification scheme and typically takes place in extremely dilute solutions, thus increasing the volumes of the unit operations during the manufacturing phases. This can make process control more difficult if, for example, low temperatures are required during these steps. Generally, inclusion proteins dissolve in denaturing agents such as sodium dodecyl sulfate

(SDS), urea, or guanidine hydrochloride. Because bacterial systems generally are incapable of forming disulfide bonds, a protein containing these bonds has to be refolded under oxidizing conditions to restore these bonds and to generate the biologically active protein. This so-called renaturation step is increasingly difficult if more S-S bridges are present in the molecule and the yield of renatured product could be as low as only a few percent. Once the protein is solubilized, conventional chromatographic separations can be used for further purification of the protein.

Aggregate formation at first sight may seem undesirable, but there may also be advantages as long as the protein of interest will unfold and refold properly. Inclusion body proteins can easily be recovered to yield proteins with >50% purity, a substantial improvement over the purity of soluble proteins (sometimes below 1% of the total cell protein). Furthermore, the aggregated forms of the proteins are more resistant to proteolysis because most molecules of an aggregated form are not accessible to proteolytic enzymes. Thus, the high yield and relatively cheap production using a bacterial system can offset a low-yield renaturation process. For a nonglycosylated, simple protein molecule, this production system is still used.

Quality by Design

The current expectations of regulatory agencies, particularly in implementing the twenty-first century's risk-based GMPs, is to employ the principles of risk analysis, design space (see below), control strategy and quality by design (QbD). Implementing QbD should result in a manufacturing process that consistently delivers a high-quality product. Furthermore, it ensures that the critical sources of variability are identified and controlled through appropriate control strategies. A detailed end-to-end assessment of the product, its manufacturing process and raw materials will result in the definition of:

1. Critical Quality Attributes (CQAs)

    The definition of a CQA according to ICH Q8 (R2) (2009) is as follows: "a physical, chemical, biological, or microbiological property or characteristic that should be within an appropriate limit, range, or distribution to ensure the desired product quality." The CQAs of biologics are basically assessed by measuring their impact on safety and efficacy.

2. Critical Process Parameters (CPPs)

    CPPs are according to ICH Q8(R2) (2009) "process parameters whose variability have an impact on critical quality attributes". They are identified by sound scientific judgment and based on prior knowledge, development, scale-up, and manufacturing experience. CPPs should be controlled and monitored to confirm that the product quality is comparable to or better than historical data from development and manufacturing. Quality attributes that should be considered in defining CPPs are, for example, purity, qualitative and quantitative impurities, microbial quality, biological activity, and content.

3. Critical Material Attributes (CMAs)

    CMAs are materials used in the process that affect the quality attributes. They are judged as described above for the CPPs. CMAs should be controlled and monitored by validated incoming goods assays.

4. Control Strategy

    The control strategy for the product is defined by controlling CPPs and the CMAs. Based on the risks related to the CPPs/CMAs, an appropriate control strategy should be designed. A proper control strategy will decrease the probability/likelihood of out-of-range CQA and increase the detectability of CPP/CMA failure. During the lifecycle of the product, the control strategy should be adjusted based on new knowledge. The control strategy will be assessed by means of a risk assessment, e.g., failure mode effects analysis, FMEA.

Above-mentioned analysis must be performed during all stages of process development. However, the starting point for a QbD exercise is to study the (potential) CQAs that are defined in early-stage discovery. The analysis should continue during early and late development and commercial scale manufacturing. Prior knowledge, analytical development, comparability studies, and (non-)clinical study results contribute to the understanding of CQA. By performing this analysis during all stages, the QbD principles will be continuously updated as they are based on increased know-how during product development and commercial scale manufacturing.

Although not obligatory, the authorities encourage to implement a design space in the processes. ICH Q8 defines the design space of a process as follows: "the multidimensional combination and interaction of input variables (e.g., material attributes) and process parameters that have been demonstrated to provide assurance of quality. Working within the design space is not considered as a change. Movement out of the design space is considered to be a change and would normally initiate a regulatory post-approval change process. The design space is proposed by the manufacturer. The advantage of the design space that it is usually broader than the operating ranges."

Based on these assessments the CPPs and CMAs are specified for:

(a) Normal Operating Ranges (NOR)

    A process range that is representative of historical variability in the manufacturing process

(b) Acceptable Operating Ranges

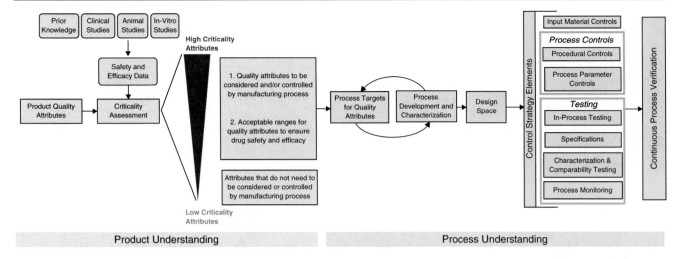

Fig. 4.9 The overall approach for A-Mab product realization illustrates a sequence of activities that starts with the design of the molecule and spans the development process ultimately resulting in the final process and control strategy used for commercial scale manufacturing (adapted from Berridge et al. 2009)

A range that is specified in the manufacturing batch record

(c) Proven Acceptable Range (PAR)

A characterized range of the process which will result in producing a product meeting the relevant quality criteria.

Company representatives of the Biotechnology Industry were brought together in 2008 helping to advance the principles which are contained in ICH Q8 (R2), Q9, and Q10, focusing on the principles of quality by design. The outcome of this collaboration resulted in a unique document: "A-Mab: A case study in BioProcess Development" (Berridge et al. 2009). The case study is a must read for people involved in the biotechnology industry. Figure 4.9 below shows the overall approach for A-Mab product realization.

Quality Assurance and Quality Control

It is important that the GMP-grade manufacturing process consistently yields products with high purity and quality. Both FDA and EMA demand rationalized and strict documentation of manufacturing controls. This includes documentation and/or in process testing at each step of the manufacturing process to ensure that the final product meets the predefined specifications for safety and efficacy. A thorough understanding of both the cell cultivation and protein generation as well as the purification process is essential to minimize biological variability of the recombinant protein product that might affect its safety and efficacy. This variability may be undetected, even with the most sensitive and sophisticated analytical methods.

As mentioned in earlier paragraphs of this chapter, quality by design should be used to develop the manufacturing process. The contaminants paragraph mentions critical components of the manufacturing process. For each analytical method used in the manufacturing process a Standard Operating Procedure (SOP) should be implemented. These methods and the manufacturing process should be validated, this to unsure that the process adequately characterizes the final protein product, and can consistently remove extraneous compounds such as column leachables, endotoxins, residual host cell proteins and DNA, bacteria, viruses, chemicals used in the process. The level of validation depends on the stage of product development. In the early stages (e.g., for product for preclinical studies, or phase 1/2 clinical studies), it is critical to fully validate the safety methods and validate your process for virus removal. The product used for phase 3 clinical studies and market-scale manufacturing require full validation of all in-process, release and stability-indicating methods as well a full process validation. For more information, please check the guidelines on the FDA and EMA websites.

Commercial-Scale Manufacturing and Innovation

A major part of the recombinant proteins on the market consists of monoclonal antibodies produced in mammalian cells. Pharmaceutical production processes have been set up since the early 1980s of the twentieth century. These processes essentially consist of production in stirred tanks bioreactors, clarification using centrifugation, and membrane technology, followed by protein A capture, low-pH virus inactivation, cation-exchange and anion-exchange chromatography (or an alternative chromatographic ligand), virus filtration, and UF/DF for product formulation (Shukla and Thömmes 2010).

Such platform processes run consistently at very large scale, e.g., multiple 10,000 L bioreactors and higher volumes. Product recovery is generally very high (>70%). Since product titers in the bioreactors have increased to a level where further increases have no or a minimal effect on the cost of goods, the focus of process development in companies with these large-scale manufacturing plants working at full capacity is shifting to understanding the process fundamentals of the current platform (Kelley 2009). However, it is also anticipated that the monoclonal antibody demands for some disease indications may decrease due to the introduction of more efficacious products such as antibody–drug conjugates (e.g. Adcetris®, Seattle Genetics) and increased competition with biosimilar products (e.g., Celltrion's Remsina®/Infectra® as biosimilar of the Johnson & Johnson blockbuster Remicade®), and the introduction of new products with (much) smaller market sizes, including those used in personalized medicines approaches. A lower demand together with the increase in recombinant protein titers and yields will lead to a decrease in bioreactor size, an increase in the need for flexible facilities, and faster turnaround times leading to a growth in the use of disposables and other innovative technologies as discussed above (Shukla and Thömmes 2010). Such innovative technologies and capabilities encompass process intensification, in which production is intensified by using highly concentrated product and reactants, and in which process steps are combined into single units. Innovation is also seen in the introduction of continuous processing strategies in the pharmaceutical industry, as well as steps toward fully automated facilities, enabling a fast response to capacity demands at lower costs and higher quality. Facilities will become modular and mobile, allowing standardized "plug and play" manufacturing systems to be configured, assembled and relocated quickly. A further introduction of process analytical technology (PAT) is expected, allowing in-line process monitoring and real time drug product release. This includes development of software enabling multivariate data analysis, predictive models and closed feedback control loops (BioPhorum Operations Group 2017).

Self-Assessment Questions

Questions

5. Name the 4 expression systems mentioned in this chapter?
6. What is the main reason to use eukaryotic mammalian cells as expression system?
7. What are the main reasons for manufacturing companies to change from stainless steel system to single-use systems?
8. Which bioreactor processes are generally used for the production of biopharmaceuticals?

9. Membrane filters are frequently used within the purification process of biotech products. Name 4 different membrane filter types.
10. Compared to other chromatographic methods, what is in general the most significant advantage and disadvantage of affinity purification chromatography?
11. Name at least 6 different product-related variants.
12. What is the difference between a NOR and PAR? Which of these 2 parameters gives most flexibility in a process?
13. What are the major safety concerns in the purification of cell-expressed proteins?
14. Glycosylation may affect several properties of the protein. Mention at least three possible effects in case of changing a glycosylation pattern.
15. What is in general the expectation of the size of future GMP manufacturing facilities? What is the reasoning behind this?

Answers

1. Prokaryotic bacteria, eukaryotic yeast, eukaryotic mammalian cells, and eukaryotic plant cells.
2. For biopharmaceutical products used in human health care the glycosylation process is the most important reason. The glycosylation pattern should be human-like which is possible with the eukaryotic mammalian cell system.
3. The main reasons are the speed to market, possibility to increase the number of batches produced per year in a manufacturing facility, providing flexibility in facility design, reduction of water consumption, and reduced validation costs.
4. The main bioreactor processes are batch, fed-batch and perfusion.
5. Sterilizing-grade filters, tangential flow filters, virus removal filters, charged filters.
6. Advantage: high degree of purity can be obtained; disadvantage: usually very costly, and extra regulatory burden due to characterization of affinity ligand.
7. Glycosylation variants, amino acid substitution and deletion, denatured protein, oxidized variants, conformational isomers, dimers and aggregates, disulfide paring variants, succinimide formation, (de)amidated variants, protein fragments.
8. The normal operating range (NOR) is a process range that is representative of historical variability in the manufacturing process, while a proven acceptable range (PAR) is a characterized range of the process which will result in producing a product meeting the relevant quality criteria. The PAR gives most flexibility since it allows operation beyond the NOR.
9. Removal of viruses, bacteria, protein contaminants, and cellular DNA.

10. Solubility, pKa, charge, stability, and biological activity.
11. GMP manufacturing facilities will become smaller, modular, and mobile. Rationale: manufacturing volumes will become smaller due to process intensification and the generation of products with a smaller market capture.

References

Abi-Ghanem DA, Berghman LR (2012) Immunoaffinity chromatography: a review. IntechOpen, London. Cdn.intechopen.com/pdfs/33050pdf

Afeyan N, Gordon N, Mazsaroff I, Varady L, Fulton S, Yang Y, Regnier F (1989) Flow-through particles of the high-performance liquid chromatographic separation of biomolecules, perfusion chromatography. J Chromatogr 519:1–29

Baumann P, Hubbuch J (2017) Downstream process development strategies for effective bioprocesses: trends, progress, and combinatorial approaches. Eng Life Sci 17:1142–1158

Berridge J, Seamon K, Venugopal S (2009) A-Mab: a case study in BioProcess Development, version 2.1. http://c.ymcdn.com/sites/www.casss.org/resource/resmgr/imported/A-Mab_Case_Study_Version_2-1.pdf

Berthold W, Walter J (1994) Protein purification: aspects of processes for pharmaceutical products. Biologicals 22:135–150

BioPhorum Operations Group Ltd (2017) Biomanufacturing technology roadmap. BioPhorum, London. www.biophorum.com/wp-content/uploads/2017/07/SupplyPartMgmnt.pdf

Cartwright T (1987) Isolation and purification of products from animal cells. Trends Biotechnol 5:25–30

Celik E, Calik P (2011) Production of recombinant proteins by yeast cells. Biotechnol Adv 30:1108. https://doi.org/10.1016/j.biotechadv.2011.09.011

Chase HA (1994) Purification of proteins by adsorption chromatography in expanded beds. Trends Biotechnol 12:296–303

Chase H, Draeger N (1993) Affinity purification of proteins using expanded beds. J Chromatogr 597:129–145

Chen T, Zhang K, Gruenhagen J, and Medley CD (2015) Hydrophobic interaction chromatography for antibody drug conjugate drug distribution analysis. American Pharmaceutical Review. www.americanpharmaceuticalreview.com/Featured-Articles/177927

Compton B, Jensen J (2007) Use of perfusion technology on the Rise–New modes are beginning to gain ground on Fed-Batch strategy. Genet Eng Biotechnol News 27:48

Cumming DA (1991) Glycosylation of recombinant protein therapeutics: control and functional implications. Glycobiology 1:115–130

Dell A, Galadari A, Sastre F, Hitchen P (2011) Similarities and differences in the glycosylation mechanisms in prokaryotes and eukaryotes. Int J Microbiol 2010:148178

Dos Santos NV, De Carvalho Santos-Ebinuma V, Pessoa A Jr, Brandao Pereira JF (2017) Liquid-liquid extraction of biopharmaceuticals from fermented broth: trends and future prospects. J Chem Technol Biotechnol 93:1845. https://doi.org/10.1002/jctb.5476

Dream R, Herwig C, Pelletier E (2018) Continuous manufacturing in biotech processes–challenges for implementation. ISPE, Tampa, FL

Eibl R, Eibl D (eds) (2011) Single-use technology in biopharmaceutical manufacture. Wiley, Hoboken

EMA (2011) Note for guidance on minimising the risk of transmitting animal spongiform encephalopathy agents via human and veterinary medicinal products (EMA/410/01 rev.3)

Etzel MR, Arunkumar A (2017) Charged ultrafiltration and microfiltration membranes for antibody purification. In: Gottschalk U (ed) Process scale purification of antibodies. Wiley, Hoboken. https://doi.org/10.1002/9781119126942.ch12

European Agency for the Evaluation of Medicinal Products (EMA) (1996) Note for guidance for virus validation studies: design, contribution and interpretation of studies validating the inactivation and removal of viruses (CPMP/BWP/268/95)

European Pharmacopoeia (2011) 5.2.3. Cell substrates for the production of vaccines for human use, 7th edn. Council of Europe, Strasbourg

FDA, Center for Biologics Evaluation and Research (1990) Cytokine and growth factor pre-pivotal trial information package with special emphasis on products identified for consideration under 21 CFR 312 subpart E. Bethesda, Maryland

FDA, Office of Biologicals Research and Review (1993) Points to consider in the characterization of cell lines used to produce biologicals, Bethesda, Maryland

Fulton SP (1994) Large scale processing of macromolecules. Curr Opin Biotechnol 5:201–205

Goldstein A, Molina O (2016) Implementation strategies and challenges: single use technologies. PepTalk Presentation

Gottschalk U (2006) The renaissance of protein purification. BioPharm Int 19:S8–S9

Groves MJ (1988) Parenteral technology manual: an introduction to formulation and production aspects of parenteral products. Interpharm Press, Buffalo

Hodge G (2004) Disposable components enable a new approach to biopharmaceutical manufacturing. BioPharm Int 15:38–49

Homma T, Fuji M, Mori J, Kawakami T, Kuroda K, Taniguchi M (1993) Production of cellobiose by enzymatic hydrolysis: removal of β-glucosidase from cellulase by affinity precipitation using chitosan. Biotechnol Bioeng 41:405–410

Hossler P, Khattak SF, Li ZJ (2009) Optimal and consistent protein glycosylation in mammalian cell culture. Glycobiology 19:936–949

ICH (International Conference on Harmonization) (1999a) Topic Q6B Specifications: test procedures and acceptance criteria for biotechnology/biological products

ICH (International Conference on Harmonization) (1999b) Topic Q5A on viral safety evaluation of biotechnology products derived from cell lines of human or animal origin

ICH (International Conference On Harmonization) (2009) Topic Q8 (R2) Pharmaceutical development

ICH (International Conference on Harmonization) (2017) Guideline M7 on assessment and control of DNA reactive (mutagenic) impurities in pharmaceuticals to limit potential carcinogenic risk

James AM (1992) Introduction fundamental techniques. In: James AM (ed) Analysis of amino acids and nucleic acids. Butterworth-Heinemann, Oxford, pp 1–28

Jones N (2015) Single-use processing for microbial fermentations. BioProcess Int 13:56–62

Kelley B (2009) Industrialization of mAb production technology. MAbs 1:443–452

Khanal O, Lenhoff M (2021) Developments and opportunities in continuous biopharmaceutical manufacturing. MAbs 13:e1903664

Klegerman ME, Groves MJ (1992) Pharmaceutical biotechnology. Interpharm Press, Buffalo

Konstantinov KB, Cooney CL (2015) White paper on continuous bioprocessing, may 20–21, 2014 continuous manufacturing symposium. J Pharm Sci 104:813–820

Krasnova L, Wong CH (2016) Understanding the chemistry and biology of glycosylation with glycan synthesis. Annu Rev Biochem 85:599–630

Lai T, Yang Y, Kong Ng S (2013) Advances in mammalian cell line development technologies for recombinant protein production. Pharmaceuticals 6:579–603

Larsen B, Hwang J (2010) Mycoplasma, Ureaplasma, and adverse pregnancy outcomes: a fresh look. Infect Dis Obstet Gynecol 2010:1–7

Löwer J (1990) Risk of tumor induction in vivo by residual cellular DNA: quantitative considerations. J Med Virol 31:50–53

Luitjens A, Lewis J, Pralong A (2012) Single-use biotechnologies and modular manufacturing environments invite paradigm shifts in bioprocess development and biopharmaceutical manufacturing. In: Subramanian G (ed) Biopharmaceutical production technology, vol 1&2. Wiley, Weinheim, pp 817–857

Maerz H, Hahn SO, Maassen A, Meisel H, Roggenbuck D, Sato T, Tanzmann H, Emmrich F, Marx U (1996) Improved removal of viruslike particles from purified monoclonal antibody IgM preparation via virus filtration. Nat Biotechnol 14:651–652

Marcus-Sekura CJ (1991) Validation and removal of human retroviruses. Center for biologics evaluation and research. FDA, Bethesda

Matasci M, Hacker DL, Baldi L, Wurm M (2008) Recombinant therapeutic protein production in cultivated mammalian cells: current status and future prospects. Drug Discov Today 5:37–42

Minor PD (1994) Ensuring safety and consistency in cell culture production processes: viral screening and inactivation. Trends Biotechnol 12:257–261

Monteclaro F (2010) Protein expression systems, ringing in the new. Innov Pharm Technol 12:45–49

Orzaez D, Granell A, Blazquez MA (2009) Manufacturing antibodies in the plant cell. Biotechnol J 4:1712–1724

Peters J, Stoger E (2011) Transgenic crops for the production of recombinant vaccines and anti-microbial antibodies. Hum Vaccin 7:367–374

Sao Pedro MN, Silva TC, Patil R, Ottens M (2021) White paper on high-throughput process development for integrated continuous biomanufacturing. Biotechnol Bioeng 118:3275–3286

Schwartz H, Gomis Fons J, Isaksson M, Scheffel J, Andersson N, Andersson A, Castan A, Solbrand A, Hober S, Nilsson B, Chotteau V (2022) Integrated continuous biomanufacturing on pilot scale for acid-sensitive monoclonal antibodies. Biotechnol Bioeng 119:2152. https://doi.org/10.1002/bit.28120

Sharma SK (1990) Key issues in the purification and characterization of recombinant proteins for therapeutic use. Adv Drug Deliv Rev 4:87–111

Shukla AA, Thömmes J (2010) Recent advances in large-scale production of monoclonal antibodies and related proteins. Trends Biotechnol 28:253–261

Shukla AA, Wolfe LS, Mostafa SS, Norman C (2017) Evolving trends in mAb production processes. Bioeng Transl Med 2:58–69

Sinclair AM, Elliott S (2005) Glycoengineering: the effect of glycosylation on the properties of therapeutic proteins. J Pharm Sci 94:1626–1635

Sinclair A, Brower M, Lopes AG, Pollard D, Abe Y (2016) Standardized economic cost modeling for next-generation MAb production. BioProcess Int 14:14–23

Sofer G, Brorson K, Abujoub A, Aranha H, Burnouf T, Carter J, Jocham UE, Jornitz M, Korneyeva M, Krishnan M, Marcus-Sekura C (2005) Technical report No. 41, virus filtration. PDA J Pharm Sci Technol 59(S-2):1–42

Tennikova T, Svec F (1993) High performance membrane chromatography: highly efficient separation method for proteins in ion-exchange, hydrophobic interaction and reversed phase modes. J Chromatogr 646:279–288

Terstappen G, Ramelmeier R, Kula M (1993) Protein partitioning in detergent-based aqueous two-phase systems. J Biotechnol 28:263–275

Tripathi NK, Shrivastava A (2019) Recent developments in bioprocessing of recombinant proteins: expression hosts and process development. Front Bioeng Biotechnol 7:420. https://doi.org/10.3389/fbioe.2019.00420

Turner R, Joseph A, Titchener-Hooker N, Bender J (2018) Manufacturing of proteins and antibodies: chapter downstream processing technologies, harvest operations. Adv Biochem Eng Biotechnol 165:95–114. https://doi.org/10.1007/10_2016_54

Van Reis R, Zydney A (2007) Bioprocess membrane technology. J Membr Sci 297:16–50

Van Wezel AL, Van der Velden-de Groot CA, De Haan HH, Van den Heuvel N, Schasfoort R (1985) Large scale animal cell cultivation for production of cellular biologicals. Dev Biol Stand 60:229–236

Walsh C (2006) Post-translational modification of proteins: expanding nature's inventory, vol xxi. Roberts and Co, Englewood, p 490

Walter J, Werner RG (1993) Regulatory requirements and economic aspects in downstream processing of biotechnically engineered proteins for parenteral application as pharmaceuticals. In: Kroner KH, Papamichael N, Schütte H (eds) Downstream processing, recovery and purification of proteins, a handbook of principles and practice. Verlag, Muenchen

Walter J, Werz W, McGoff P, Werner RG, Berthold W (1991) Virus removal/inactivation in downstream processing. In: Spier RE, Griffiths JB, MacDonald C (eds) Animal cell technology: development, processes and products. Butterworth-Heinemann, Linacre House, Oxford, pp 624–634

Walter K, Werz W, Berthold W (1992) Virus removal and inactivation, concept and data for process validation of downstream processing. Biotech Forum Europe 9:560–564

WHO (World Health Organization) (2010) Recommendations for the evaluation of animal cell cultures as substrates for the manufacture of biological medicinal products and for the characterization of cell banks. Technical report series, proposed replacement of 878, annex 1 (not yet published)

Yao J, Weng Y, Dickey A, Tressel KY (2015) Plants as factories for human pharmaceuticals: applications and challenges. Int J Mol Sci 16:28549–28565

Zhou JX, Tressel T (2006) Basic concepts in Q membrane chromatography for large scale antibody production. Biotechnol Prog 22:341–349

Zhou H, Fang M, Zheng X, Zhou W (2021) Improving an intensified and integrated continuous bioprocess platform for biologics manufacturing. Biotechnol Bioeng 118:3618. https://doi.org/10.1002/bit.27768

Formulation of Biologics Including Biopharmaceutical Considerations

5

Daan J. A. Crommelin, Andrea Hawe, and Wim Jiskoot

Introduction

This chapter provides an introduction to the process of formulating biologics. In this formulation process, a drug *substance* (DS), often called active pharmaceutical ingredient (API), is turned into a drug *product* that can be administered to the patient. It addresses questions regarding the tests to be run, e.g., to characterize the DS and to ensure its stability, and the choice of excipients and delivery system. It also discusses biopharmaceutical issues such as the route and rate of administration.

The text concentrates on formulating proteins used in therapy, but the same principles also apply to other biologics, such as vaccines and nucleotide-based products e.g., oligonucleotides, as discussed in Chaps. 13 and 15 of this book.

Formulating a biologic is not a one-step, routine process with fixed strategies. Several different, sometimes overlapping, phases can be recognized during the product development process, as depicted in Fig. 5.1. In the formulation development process one starts with preformulation activities and ends up—after months/years of running stability tests—with late-stage fine tuning of the

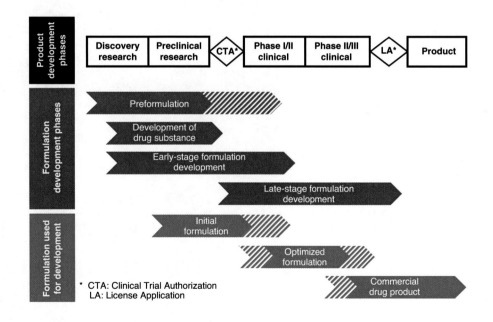

Fig. 5.1 Diagram of a formulation development process

Sadly, Wim Jiskoot passed away by the time of publication of this sixth edition.

D. J. A. Crommelin (✉)
Department of Pharmaceutics, Utrecht Institute for Pharmaceutical Sciences (UIPS), Utrecht University, Utrecht, the Netherlands
e-mail: d.j.a.crommelin@uu.nl

A. Hawe
Coriolis Pharma Research GmbH, Martinsried, Germany
e-mail: Andrea.hawe@coriolis-pharma.com

W. Jiskoot (Deceased)
Leiden Academic Center for Drug Research (LACDR), Leiden University, Leiden, The Netherlands

© The Author(s), under exclusive license to Springer Nature Switzerland AG 2024
D. J. A. Crommelin et al. (eds.), *Pharmaceutical Biotechnology*, https://doi.org/10.1007/978-3-031-30023-3_5

selected, optimized product composition and dosage form. Therefore, the formulation used in the preclinical and clinical development phases may change as in a learning process: from 'initial formulation' to 'commercial drug product'.

Points to Consider in the Process of Formulating a Therapeutic Protein

Protein Structure and Protein Stability

Table 5.1 lists 'points to consider' when formulating a protein. An early and deep understanding of the structural properties of the protein at hand such as its primary structure, higher-order structures, molecular weight, isoelectric point, post-translational modifications, hydrophobicity, and its physical (unfolding and aggregation) and chemical stability (cf. Fig. 5.2 and Fig. 2.1) as function of its direct environment (e.g., pH, ionic strength) will speed up the formulation process. This basic information helps to design a product that is stable not only on the shelf, but also under real-life conditions e.g., during manufacturing, transportation, handling (e.g., dilution in an intravenous infusion bag) and administration. Table 5.2 (adapted from Hawe et al. 2012) shows various stress factors a product can encounter.

Table 5.1 Points for consideration in the formulation process of pharmaceutical proteins

Factor	Description/attributes/examples
API or drug substance	Type of protein, physico-chemical properties, e.g., molecular weight, pI, hydrophobicity, solubility, post-translational modifications, PEG-ylation, physical and chemical stability and concentration, available amount, purity
Clinical factors	Patient population (e.g., age and concomitant medication), self-administration versus administration by professional, compatibility with infusion solution, indication (e.g., one-time application or chronical application)
Route of administration	Subcutaneous, intravenous, intramuscular, intravitreal, intra-articular, intradermal, pulmonal
Dosage form	Single- or multi-dose, prefilled syringe, dual chamber cartridge, pen cartridge, liquid, lyophilizate, frozen liquid, API concentration, injection volume, injection rate, controlled delivery/release
Primary packaging	Glass, polymers, rubber, silicone oil, metals, leachables (anti-oxidants, plasticizers, etc.)
Excipients	Pharmaceutical quality, safety record (for intended administration route and dose), manufacturer, tested for critical impurities, stability
Analytical methods	Characterization of API, stability-indicating assays, quality control assays and assays for excipient content and stability
Intellectual property	Possible restrictions to the use of the API or the choice of excipients e.g., when considering to develop a biosimilar product

Adapted from Weinbuch et al. (2018)
API active pharmaceutical ingredient, *PEG* polyethylene glycol.

Fig. 5.2 Chemical and physical reactions jeopardizing protein stability (cf. Table 3.1)

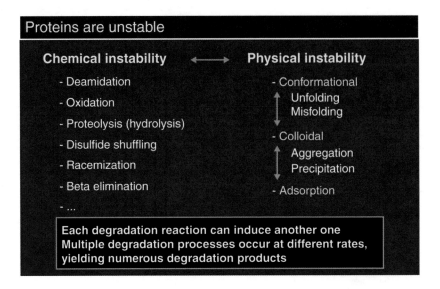

Table 5.2 Stress factors a therapeutic protein may encounter

Stress factor	When encountered/examples
Elevated temperature, temperature excursions	Production (upstream and downstream processing); improper shipment; storage or handling deviations
Freezing, freeze-thawing	Storage of frozen (bulk) material; accidental freezing during storage or shipment; lyophilization
Mechanical stress	Production (e.g., pumping, stirring, filtration)
Light	Production; shipment; storage; handling
Oxidative stress	Production (exposure to oxygen); exposure to peroxide or metal ion impurities in excipients; shipment (cavitation)
pH changes	Production (downstream processing); freezing; formulation; dilution in infusion liquids; administration
Interfaces	Air-water interface; filters; primary packaging material; infusion bags and administration lines; particulate impurities
X-ray	Air freight transportation

Adapted from Hawe et al. (2012)

Table 5.3 Examples of protein degradation products and techniques to analyze them

Type of degradation product	Examples of analytical techniques
Soluble aggregates (dimers, trimers, oligomers) and fragments	Size-exclusion HPLC/UPLC, AF4, analytical ultracentrifugation, SDS-PAGE, CE-SDS
Nanometer-sized aggregates	Dynamic light scattering; nanoparticle tracking analysis; AF4; Taylor dispersion analysis; turbidimetry/nephelometry; static light scattering; microfluidic resistive pulse sensing; oil-immersion flow imaging microscopy; RMM
Micrometer-sized aggregates	Light obscuration; light microscopy; flow imaging microscopy; BMI; electric zone sensing; fluorescence microscopy; turbidimetry/nephelometry; particle identification techniques, e.g., Raman microscopy; FTIR microscopy; SEM-EDX
Visible particles	Visual inspection; (semi-)automated visual inspection
Conformational changes	Circular dichroism; infrared, intrinsic fluorescence, extrinsic fluorescence spectroscopy and secondary-derivative UV spectroscopy
Chemical changes	Reversed-phase HPLC/UPLC; (HPLC-)mass spectrometry; ion-exchange chromatography; (capillary) isoelectric focusing

AF4 asymmetrical flow field-flow fractionation, *BMI* backgrounded membrane imaging, *CE-SDS* capillary electrophoresis sodium dodecyl sulfate, *FTIR* Fourier-transform infrared spectroscopy, *HPLC* high performance liquid chromatography, *RMM* resonant mass measurement, *SEM/EDX* Scanning Electron Microscopy (SEM) with Energy Dispersive X-Ray Analysis, *UPLC* ultra-performance liquid chromatography (adapted from Hawe et al. 2012)

Analytical Toolbox

The need for an 'analytical toolbox' with stability-indicating orthogonal and complementary analytical techniques (see Textbox 5.1) to characterize a protein in various stages of formulation development in as much detail as possible is evident. Table 5.3 (adapted from Hawe et al. 2012) lists selected analytical methodologies that are being used to monitor protein stability. Textbox 5.2. discusses the differences in uniformity of small molecule APIs and large protein APIs such as monoclonal antibodies.

Box 5.1 Orthogonal Vs Complementary Analytical Techniques

Orthogonal analytical techniques are combinations of techniques that monitor the same (similar) properties of a protein (in its formulation) with a different measurement principle (cf. Table 5.3).

For example,

- Size-exclusion chromatography (SEC), asymmetrical flow field flow fractionation (AF4) and analytical ultracentrifugation (AUC)
- Near-UV circular dichroism (CD) and intrinsic fluorescence spectroscopy
- Far-UV circular dichroism (CD) and Fourier transform infrared (FTIR) spectroscopy
- Light obscuration (LO), Flow-Imaging microscopy, backgrounded membrane imaging and electric zone sensing

Complementary analytical techniques are techniques that measure different properties of a protein (in its formulation) with a different measurement principle (cf. Table 5.3).

For example when monitoring aggregation of the protein,

- Size-exclusion chromatography (SEC) for oligomers and dynamic light scattering for nanometer size aggregates
- Analytical ultracentrifugation (AUC) for oligomers and nanoparticle tracking analysis (NTA) for nanometer size aggregates,
- Light obscuration or flow imaging microscopy for micrometer size aggregates, visual inspection for larger, visible particles
- Techniques for sizing (see above) and methods for conformational analysis (CD, FTIR, fluorescence spectroscopy)

Techniques for size (e.g., SEC), charge (e.g., ion-exchange chromatography) and hydrophobicity (RP-HPLC).

Box 5.2 Uniformity of Small Molecule Pharmaceuticals Versus Biologics

Small therapeutic molecules such as paracetamol/acetaminophen (consisting of one defined structure of the API) are 98%+ pure. For relatively small, nonglycosylated proteins, such as insulin-derivatives, active ingredient purity levels are still high i.e., in the 95%+ range. However, monoclonal antibody products with molecular weights of 150,000 Da are mixtures of related protein molecules. Even if it is stated for a product that the purity (e.g., based on monomer content by HP-SEC) is 95%+, the monomer fraction may be composed of a mixture of related molecules, isoforms.

For example, in one batch heavy chain Asp at positions 388 and 393 are partly deamidated and the C-terminal lysine composition and glycosylation profile vary (e.g., the cation-exchange HPLC tracing in Fig. 5.3). All pharmacological testing in the development phase of the API has been done with these various modifications present, so they are inherent to the product. This heterogeneity can be reproduced from batch to batch only if well-controlled manufacturing protocols are in place.

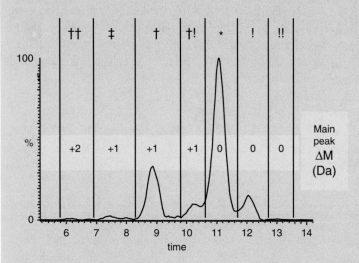

	Isoform Modification	ΔM (Da)
††	2 x Deamidation LC-N30	+2
‡	IsoAspartate HC-N55	+1
†	Deamidation LC-N30	+1
†!	Deamidation LC-N30 IsoAspartate HC-D102	+1
*	Relatively unmodifiedt	0
!	IsoAspartate HC-D102	0
!!	2 x IsoAspartate HC-D102	0

Fig. 5.3 Weak cation exchange (WCX) chromatography outcomes coupled directly to Orbitrap MS (mass spectrometer) for trastuzumab. Assignment of isoforms (Bailey et al. 2018). *HL* heavy chain, *LC* light chain

These analytical techniques will provide the necessary data and guide the formulator through the subsequent stages of the development process ending up with the marketed drug product. More information about protein (in)stability and analytical methodology can be found in Chaps. 2 and 3 of this book and in articles/books by Manning et al. (1989, 2010), Jiskoot and Crommelin Jiskoot and Crommelin 2005; Kamerzell et al. 2011; Zölls et al. 2012, Roesch et al. 2021; Hawe et al. 2012; Weinbuch et al. 2018).

Physical and Chemical Stability

Already in an early stage, data is collected on physical stability (colloidal and conformational stability) and on chemical stability of the API and formulations. Monitoring and controlling aggregate formation is particularly important, because protein aggregates are readily formed under a variety of conditions and have been associated with enhanced risk of immunogenicity of therapeutic proteins (cf. Chap. 7). It should be emphasized that aggregation of proteins can happen at concentrations much below their solubility and at temperatures far below their denaturation temperature. Proteins in solution have an increased tendency to aggregate upon mechanical stress and interaction with interfaces. Therefore, protein stability should not only be studied under quiescent conditions, but tests should be performed on protein stability simulating relevant stress factors a protein can be exposed to, in order to ensure the robustness of a formulation. This includes stress factors (1) under manufacturing conditions including fill and finish e.g., contact to tubing, pipes, columns, filters and pumps, (2) during shipment e.g. agitation/shaking of vials and (3) during handling in the clinic/by the patient, e.g. accidental freezing, dilution in carrier solutions, contact to syringes, filters, infusion bags/lines and the infusion itself.

For shelf-life assessment of the drug product typically 'real time' data at 2–8 °C is collected in combination with data from forced degradation studies. These include collecting stability information under stress conditions such as exposure to elevated temperatures e.g., 25 °C and 40 °C, to mechanical stress (cavitation, shear, interfacial effects), freeze-thawing and/or to light. This information is used to assess the overall robustness of a formulation during manufacturing. The ICH (International Council for Harmonization of Technical Requirements for Pharmaceuticals for Human Use) guideline Q5C provides global information about accelerated stability testing of biological products but does not outline exact conditions for forced degradation studies, except light stress. Table 5.4 lists a typical set of stress test conditions that are used in practice. With respect to temperature stress, one should realize that protein degradation typically does not follow Arrhenius kinetics; this is due to the complexity of the different degradation reactions that may run parallel to each other (Manning et al. 2010). Therefore, accelerated degradation studies at elevated temperatures can never replace real-time experiments (Hawe et al. 2012). Typically, a shelf life of at least 18–24 months for the drug product in its final primary packaging container (e.g., vial, syringe, cartridge pen, autoinjector) is desired. Various strategies are available to increase a protein's shelf life to or beyond the preferred 24 months range. These follow later in this chapter.

Primary Packaging

The vast majority of therapeutic proteins is parenterally administered by injection. Various primary packaging materials are available to the formulator, such as vials, cartridges and syringes (see Fig. 5.4 and Sacha et al. 2010; Sacha et al. 2015). The choice of the primary packaging

Table 5.4 Examples of accelerated stability and forced degradation studies

Type of stress	Examples of stress conditions	Anticipated instability types
Temperature	Real time (2–8 °C; up to several years) Accelerated (e.g., 25 °C, 40 °C, up to several months)	Aggregation, conformational changes, chemical changes
Mechanical	Shaking (50–500 rpm, hours–days) stirring, 50–500 rpm, hours–days) shipment studies	Aggregation, adsorption, conformational changes
Freeze-thawing	Freeze-thawing, (1–5 cycles, from 25 °C to −20 °C or − 80 °C)	Aggregation, conformational changes
Light	E.g., 1.2 million lux hours and an integrated near ultraviolet energy of not less than 200 Wh/m^2 (ICH Q1B)	Chemical changes, aggregation, conformational changes
pH	Exposure to low pH (<3) or high pH (>10)	Aggregation, conformational changes, chemical changes
Oxidation	H_2O_2 (0.01–3%, 1–7 days, 2–8° to 40 °C depending on the molecules); bivalent metals e.g., Cu^{2+}, Fe^{2+}	Chemical changes, aggregation, conformational changes
Humidity[a]	0–100% relative humidity	Aggregation, conformational changes, chemical changes

Adapted from Weinbuch et al. (2018)
[a] Specifically for freeze dried products
ICH International Council for Harmonization of Technical Requirements for Pharmaceuticals for Human Use

material depends on a number of factors. For chronic therapy, the subcutaneous route of administration is often preferred as the patient can self-administer the drug. Convenience of use then excludes vial containers. One can choose among pen injectors and prefilled syringes (cf. Fig. 5.4). Pen injectors are cartridge-based syringes for multidose administration. They are typically used when frequent subcutaneous injections of variable doses of the drug are required, as with insulin. The patient has to insert the needle her/himself. Prefilled syringes for subcutaneous administration are gaining increased popularity, especially in combination with an autoinjector to facilitate controlled and reproducible self-administration.

Typically, vials and the barrel of prefilled syringes consist of glass. Glass has the advantage of transparency, which allows visual inspection of the injected solution. Prefilled glass syringes are coated with silicone oil that acts as a lubricant to help moving the plunger. Fully polymer-based syringes offer an alternative option. A disadvantage of prefilled glass syringes is that a small fraction of the silicone oil coating can be released in the solution in the form of (subvisible) oil droplets. Proteins can adsorb to these silicone oil droplets, which potentially leads to aggregation. Adding nonionic surfactants to the formulation can prevent protein adsorption and aggregation. Furthermore, in the manufacturing process of glass syringes tungsten, which has been associated with protein aggregate formation as well, may end up in the product.

On storage of glass vials and syringes, the glass surface can release (heavy) metal ions; organic compounds may leach out of the polymer-based materials as used in vial stoppers, syringe plungers and barrels of polymer-based syringes. The formulator should collect information on these leachables and take proper action when necessary, such as changing packaging material or its vendor, add coatings or adjust formulation characteristics such as the pH.

All these issues stimulate the development of alternative packaging material (nonglass) made of polymer e.g., COP (cyclic olefin polymer) and alternative lubricants replacing silicone oil vials/syringes (Yoneda et al. 2021).

Higher gauge (G)—i.e., thinner–needles are preferred by patients as the pain sensation during injection decreases with a decreasing needle diameter. However, these narrow needles will negatively affect the 'syringeability': the force required to inject a certain volume at a certain rate through a needle with a certain length. The narrower the needle, the more force is needed.

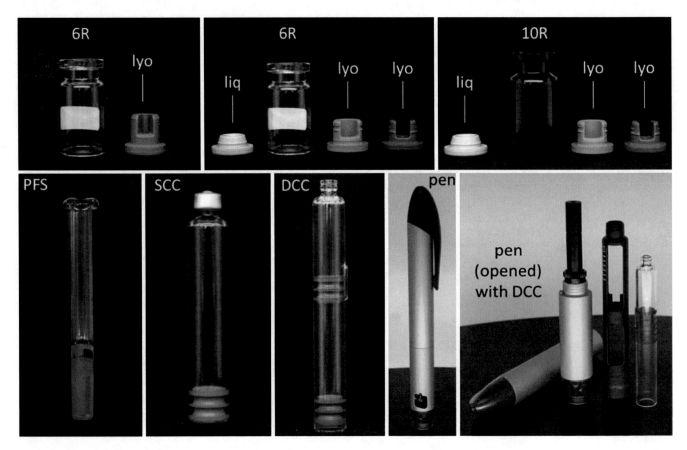

Fig. 5.4 Examples of primary packaging materials and a pen device (pen): 6-mL (6R) vials and 10-mL (10R) vial with corresponding stoppers for liquid (liq) and lyophilized (lyo) formulations; *PFS* (empty) prefilled syringe, *SCC* single-chamber cartridge, *DCC* dual-chamber cartridge. Scaling of the pictures is not identical. Photos taken by Matthias Wurm

Monoclonal antibody (mAb) formulations for subcutaneous injections often need high concentrations of the API (up to 300 mg mAb/mL) in the formulated drug product because of restrictions to the subcutaneous injection volume (1–2 mL) and the high mAb doses needed. This may lead to viscous protein solutions and, consequently, syringeability problems and increased tendencies to aggregate formation. Careful selection of the excipients used in the mAb protein products may ensure both low viscosity and colloidal stability (Jiskoot et al. 2022).

Formulation Development of Marketed Products

A marketed drug product may undergo changes in its formulation and the way it is used. For instance, the company may introduce a new route of application, a new dose, a new type of primary packaging material, a change in route of administration, a change in master cell bank, or new column material in the purification process. Such changes may affect protein structure and its stability profile and thereby the safety and efficacy profile of the product. Chapter 7 (immunogenicity) mentions an example of a formulation change of an epoetin product that caused a dramatic increase in the incidence of anti-drug antibody induced pure red cell aplasia, a serious adverse effect. Dependent on the proposed change, regulatory bodies may request new data to exclude changes in structure and shelf life of the new product and ensure that protein efficacy and safety have not changed. These comparability studies may include clinical studies, if analytical, preclinical and pharmacokinetic studies are considered insufficient to prove the claim of 'comparability' (cf. Chap. 11).

Excipients

In a protein formulation one finds, apart from the API, a number of excipients that are selected to serve different purposes. This selection process should be carried out with great care to ensure therapeutically efficacious and safe products. The nature of the protein (e.g., stability) and its therapeutic use (e.g., multiple injection systems for chronic use) can make these formulations complex in terms of excipient profile and manufacturing (freeze-drying, aseptic preparation). Table 5.1 mentions clinical factors, route of administration and dosage form as points to consider when designing the formulation. For example, the choice of an intravenously administered product (hospital setting) versus a subcutaneously administered product (self-administration) impacts the selection of excipients.

Both the choice of the excipient and its concentration are important. For instance, low concentrations of polysorbates may stabilize the protein (see below), while higher concentrations may reduce the thermal stability of a protein. On the other hand, too low concentrations of polysorbates may result in particle formation during storage caused by protein instability or polysorbate degradation when the cleaved fatty acids (from the polysorbate) are no longer solubilized by the remaining polysorbate. Moreover, polysorbate degradation may lead to a surfactant concentration that is insufficient to stabilize the protein.

Table 5.5 lists components commonly found in presently marketed formulations. Clearly, an excipient can have different functions. For instance, sugars may be added for achieving isotonicity, as conformation stabilizer in liquid products, and as bulking agent and cryo/lyoprotectant in freeze-dried products. Kamerzell et al. (2011) discuss in detail the role of different classes of excipients in protein formulations and their mechanism of action.

In the following sections the reasons for including excipients from this list to protein products are discussed in more detail.

Solubility Enhancement
Approaches to enhance protein solubility include the selection of the proper pH (see below) and ionic strength condi-

Table 5.5 Common excipients in protein drug products

Excipient class	Function	Examples
Buffers	pH control, tonicity	Histidine, phosphate, acetate, citrate
Salts	Tonicity, stabilization, viscosity reduction	Sodium chloride
Sugars[a], polyols	Tonicity, stabilization, cryoprotection, lyoprotection[b], bulking agent[b], reconstitution improvement[b]	Sucrose, trehalose, mannitol, sorbitol
Surfactants	Adsorption prevention, solubilization, stabilization, reconstitution improvement[b]	Polysorbate 20, polysorbate 80, poloxamer 188
Amino acids	Stabilization, viscosity reduction, tonicity, pH control, bulking agent[b]	Arginine, glycine, succinate, histidine
Anti-oxidants	Oxidation prevention	Methionine, sodium edetate
Preservatives[c]	Bacterial growth prevention	m-cresol, benzyl alcohol, phenol

[a] Only non-reducing sugars
[b] For freeze-dried products
[c] Multi-dose containers (adapted from Weinbuch et al. 2018)

tions. Addition of amino acids, such as arginine, or surfactants can also help to increase the solubility. The mechanism of action of these solubility enhancers depends on the type of enhancer and the protein involved and is not always fully understood. As an example, Fig. 5.5 shows the dramatic effect of the arginine concentration on the apparent solubility of tissue plasminogen activator (alteplase).

Protection against Adsorption, Interfacial Stress and Aggregation in the Bulk Solution

Most proteins are prone to adsorb to interfaces. Upon adsorption, proteins tend to expose hydrophobic sites, nor-

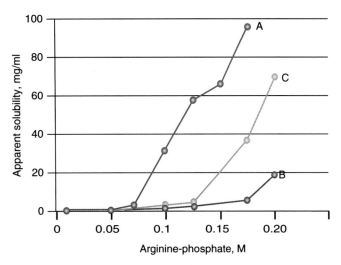

Fig. 5.5 Effect of arginine concentration on the apparent solubility of (**a**) type I and (**b**) type II alteplase, and (**c**) a 50:50 mixture thereof at pH 7.2 and 25 °C (Adapted from Nguyen and Ward (1993))

mally present in the core of the native protein structure when an interface is present. Examples of interfaces are the formulation liquid–air interface, the liquid–container wall interface, or interfaces formed between the liquid and utensils used to administer the drug (e.g., infusion bags and lines, syringes, needles). Importantly, adsorbed, partially unfolded protein molecules not only present a loss of API but also may form aggregates, leave the surface, return to the aqueous phase, and form larger aggregates. Figure 5.6 (adapted from Sediq et al. 2016) shows a schematic representation of this mechanism of aggregation at a solid surface. A similar situation may occur at gas-liquid interfaces. For some proteins the reconstitution protocol for the freeze-dried cake explicitly stipulates to swirl the vial instead of shaking it to avoid protein exposure to large liquid-air interfaces.

Many protein formulations include a surfactant to reduce protein adsorption. Surfactants readily adsorb to hydrophobic interfaces with their own hydrophobic groups and render this interface hydrophilic by exposing their hydrophilic groups to the aqueous phase. Protein accumulation at the interface is suppressed and thereby aggregate formation. The most commonly used surfactants for parenteral use are polysorbate 20 and 80. Poloxamer 188 is also used and is gaining importance because of issues with polysorbate degradation (Martos et al. 2017). Furthermore, 2-hydroxypropyl-beta-cyclodextrin can prevent adsorption and is accepted as an excipient for parenteral use as well. Human serum albumin prevents adsorption, but is nowadays rarely used because of potential infectious content and interference with analytical characterization of the API.

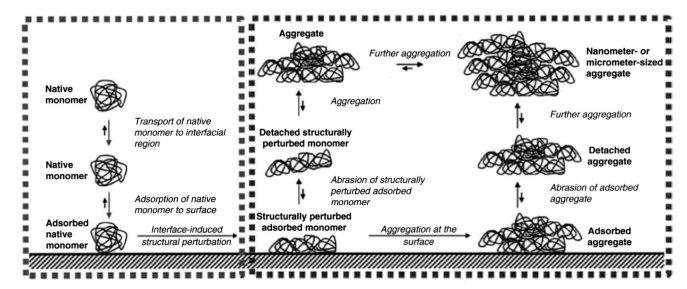

Fig. 5.6 Schematic representation of the suggested mechanism of stirring-induced protein aggregation. The left part (framed in blue) depicts the process of protein adsorption onto solid surfaces with potential perturbation of the native structure of the protein on adsorption. This process is followed by aggregation at the surface and in the bulk (framed in red). Contact sliding results in abrasion of the adsorbed protein layer, leading to renewal of the surface for adsorption of a fresh protein layer. Addition of surfactants, such as polysorbate 20, and avoidance of contact stirring will inhibit the steps shown as blue and red arrows, respectively. (Adapted from Sediq et al. (2016)

Apart from interface-induced aggregation, aggregates may be formed in the bulk of a solution because of colloidal and/or conformational instability (Chi et al. 2003). Sugars, selection of a proper pH value and buffer components may mitigate the tendency to this bulk aggregation.

These excipients (sugars and polyhydric alcohols) may not be inert; they may influence protein stability. For example, sugars and polyhydric alcohols can stabilize the protein structure through the principle of "preferential exclusion". They enhance the interaction of the solvent with the protein and are themselves excluded from the protein surface layer; the protein is preferentially hydrated. This results in an increased conformational stability of the protein.

Glucose may perfectly act as a conformational stabilizer, but may induce chemical instability through the Maillard reaction. Primary amino groups of the proteins react with the reducing sugar, resulting in brownish/yellow solutions. Sucrose should not be used below pH 6 because of hydrolysis, leading to formation of fructose and glucose; the latter being a reducing sugar. Trehalose and polyols such as mannitol and sorbitol may be used as well, also at low pH.

Buffer Selection

Buffer selection is an important part of the formulation process, because of the pH dependence of protein solubility, as illustrated in Fig. 5.7. Moreover, both the pH and the buffer species itself can have profound effects on the physical (aggregation) and chemical stability of proteins (Zbacnik et al. 2017) (cf. Fig. 5.8). Buffer systems regularly encountered in protein formulations are phosphate, citrate, histidine, succinate, glutamate and acetate. Highly concentrated protein solutions (protein concentration >50 mg/mL) may not need a buffer as they have sufficient intrinsic buffer capacity. Even short, temporary pH changes can cause protein aggregation. These conditions can occur, for example, during elution of a monoclonal antibody from a protein A column at low pH (see Chap. 4) or during the freezing step in a freeze-drying process, when one of the buffer components is crystallizing and the other is not. For instance, in a sodium phosphate buffer, Na_2HPO_4 crystallizes faster than NaH_2PO_4. This causes a pronounced drop in pH during the freezing step (see below). In the presence of high concentrations of a sugar, which is typically added as lyo- and cryoprotectant during lyophilization, the effect of pH changes is less pronounced. Other buffer components do not crystallize but form amorphous systems, and then pH changes are negligible.

Protection Against Oxidation

Methionine, cysteine, tryptophan, tyrosine, and histidine are amino acid residues that are readily oxidized (see Chap. 3). As these amino acid residues occur in almost all proteins,

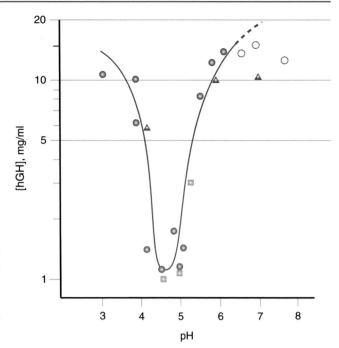

Fig. 5.7 A plot of the solubility of various forms of hGH (human growth hormone) as a function of pH. Samples of hGH were either recombinant hGH (circles), Met-hGH (triangles), or pituitary hGH (squares). Solubility was determined by dialyzing an approximately 11 mg/mL solution of each protein into an appropriate buffer for each pH. Buffers were citrate, pH 3–7, and borate, pH 8–9, all at 10 mM buffer concentrations. Concentrations of hGH were measured by UV absorbance as well as by RP-HPLC, relative to an external standard. The closed symbols indicate that precipitate was present in the dialysis tube after equilibration, whereas open symbols mean that no solid material was present, and thus the solubility is at least this amount (From Pearlman and Bewley (1993) adapted)

oxidative degradation is a regular threat to the stability of proteins. The sensitivity of an amino acid residue towards oxidation depends on its position within the protein, as this determines its accessibility for oxidative reagents. Replacement of oxygen by inert gases (e.g., argon) in the vials or minimizing the headspace, such as in prefilled syringes, helps reducing oxidative stress. Moreover, one may consider the addition of antioxidants, such as methionine, which competes with methionine residues for oxidation. Interestingly, some antioxidants can accelerate protein oxidation (Vemuri et al. 1993). Ascorbic acid, for example, can act as an oxidant in the presence of trace amounts of heavy metals which may be present as impurities. To reduce the catalytic activity of heavy metals, one may consider introducing chelators such as EDTA (ethylenediaminetetraacetic acid) (Kamerzell et al. 2011).

Preservation

Proteins may be marketed in containers designed for multiple injections. After administering the first dose, contamina-

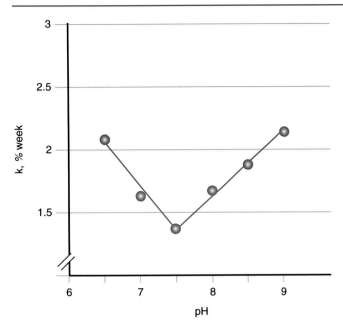

Fig. 5.8 pH stability profile (at 25 °C) of monomeric recombinant α1-antitrypsin (rAAT) by size exclusion-HPLC assay, k degradation rate constant. Monomeric rAAT decreased rapidly in concentration both under acidic and basic conditions. Optimal stability occurred at pH 7.5 (Adjusted from Vemuri et al. (1993))

tion with microorganisms may occur. Therefore, formulations in multi-dose containers must contain a preservative. Common antimicrobial agents include phenol, meta-cresol, benzyl alcohol, and chlorobutanol (Kamerzell et al. 2011). These preservative molecules can interact with the protein, which may compromise both the activity of the protein and the effectivity of the preservative. An example is the well-established interaction between insulin and phenols (Chap. 16). A caveat is the incompatibility of polysorbates and m-cresol, which may lead to precipitation.

Tonicity Adjustment
For proteins, the regular rules apply for adjusting the tonicity of parenteral products. Formulation excipients, such as buffers and amino acids, contribute to the tonicity. Disaccharides, polyols, and sodium chloride are commonly added to reach isotonicity.

Protection against Freezing and Drying
Cryoprotectants (mainly sugars: sucrose, trehalose and sugar alcohols: mannitol, sorbitol) are excipients that protect a protein during freezing or in the frozen state. This is relevant for the development of frozen liquid formulations. The key stabilizing mechanisms are "preferential exclusion." As explained above, these additives ("water structure promoters") enhance the interaction of the solvent (water) with the protein and are themselves excluded from the protein surface layer; the protein is preferentially hydrated.

Lyoprotectants protect the protein in the lyophilized state (e.g., sugars). The key mechanisms of protection described for lyoprotectants are (cf. Mensink et al. 2017): (1) the "water replacement theory': replacement of water as stabilizing agent by forming hydrogen bonds with the protein and (2) the "vitrification theory": formation of a glassy amorphous matrix keeping protein molecules separated from each other.

When stored in the presence of reducing sugars, such as glucose and lactose, the Maillard reaction (see above) may occur also in the dried state and the cake color turns yellow brown. Therefore, reducing sugars should not be used as lyoprotectant and nonreducing sugars such as sucrose (above pH 6) or trehalose are preferred.

Viscosity Reduction
For high concentration protein formulations, the challenge of an exponential rise in viscosity with increasing protein concentration needs to be tackled by the formulator. High viscosity is a result of protein–protein interactions. Depending on the underlying mechanism, e.g., excluded volume repulsion, hydrophobic interactions or electrostatic interactions, the ionic strength (mainly via NaCl), formulation pH, or the addition of amino acids, e.g., arginine, lysine or proline are commonly applied strategies to reduce viscosity.

Freeze-Drying of Proteins

The abundant presence of water in liquid protein formulations promotes chemical and physical degradation processes. This explains why proteins in solution often do not meet the preferred stability requirements for industrially produced pharmaceutical products (>2 years), even when kept permanently under refrigerator conditions (the "cold chain").

Freeze-drying may provide the required stability (Constantino and Pikal 2004). During freeze-drying, water is removed through sublimation. The freeze-drying process consists of three steps: (1) freezing (if required, this includes an annealing step), (2) primary drying, and (3) secondary drying. Figure 5.9 shows what happens with chamber pressure and temperature over time during these stages.

Although aimed to improve protein stability, freeze-drying may cause irreversible damage to the protein. This is particularly true when applying improper lyophilization process conditions (see below) and/or without using proper excipients (see above). Table 5.5 lists excipients typically encountered in successfully freeze-dried protein products.

Freezing
In the freezing step (see Fig. 5.9), the temperature of the solution (typically in vials) is lowered. Ice crystal formation

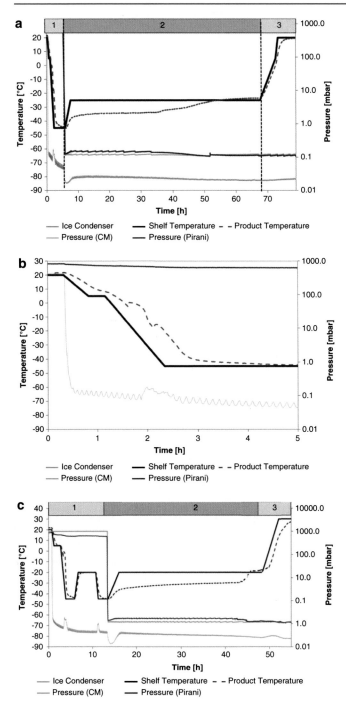

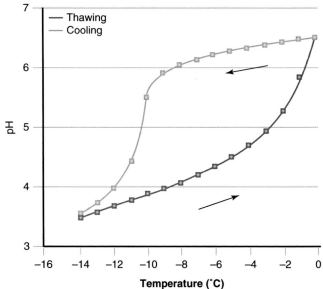

Fig. 5.10 Thawing/cooling; thawing, cooling. The effect of freezing on the pH of a citric acid–disodium phosphate buffer system (Cited in Pikal 1990, adapted)

Fig. 5.9 (**a**) Example of freeze-drying showing shelf temperature, product temperature, pressure (by Pirani and CP and ice condenser temperature). 1 = freezing stage, 2 = primary drying stage, 3 = secondary drying stage. See the text for explanation CP and Pirani measurement. (**b**) Zoom in on the freezing stage. (**c**) Similar to Fig. 9a, but now with annealing step in the freezing stage (temperature rise from −45 °C to −20 °C and drop again to −45 °C)

does not start right at the thermodynamic or equilibrium freezing point, but supercooling occurs. That means that ice crystallization often only occurs when reaching temperatures of −15 °C or lower. During ice crystallization, the tempera-

ture temporarily rises in the vial because of the generation of crystallization heat. During the cooling stage, the concentration of the protein and excipients increases because of the growing ice crystal mass at the expense of the liquid aqueous phase. This can cause precipitation of one or more of the excipients, which may result in pH shifts (see above and Fig. 5.10) and ionic strength changes. It may also induce protein denaturation because protein molecules are getting in close contact in the concentrated nonfrozen phase. Cooling of the vials is done through lowering the shelf temperature. Selecting the proper cooling scheme for the shelf—and consequently the vials—is important, as it dictates the degree of supercooling and ice crystal size. Small crystals form during fast cooling; large crystals form at lower cooling rates. Small ice crystals are required for porous cakes and fast sublimation rates.

Ice nucleation is a random and stochastic approach, which may lead to inhomogeneity in pore size and cake structure. Controlled nucleation can be applied to assure homogenous freezing at a defined temperature. This results in reduced primary drying times and shorter reconstitution times of highly concentrated lyophilized protein formulations (Geidobler and Winter 2013).

When choosing the freezing temperature, it is important to assure that the product is fully frozen. Crystallizing compounds (e.g., NaCl, mannitol) need to be cooled below the eutectic temperature (Te) and compounds forming amorphous structures (e.g., sugars) below the glass transition temperature of the maximally freeze-concentrated solution (Tg'). In the amorphous phase, the viscosity changes dra-

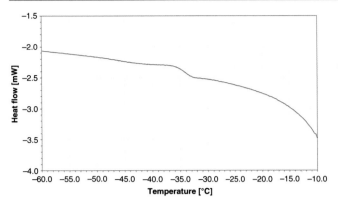

Fig. 5.11 Differential scanning calorimetry trace of a 5% sucrose solution with a Tg' at about −35 °C

matically in the temperature range around Tg': A "rubbery" state exists above and a glass state below the Tg'.

Lyophilized formulations can be either amorphous (e.g., sugar-based) or (partially) crystalline (e.g., mannitol-based). To assure that bulking agents such as mannitol crystallize, an annealing step can be included to promote crystallization during freezing: The frozen mass is then kept at a temperature above the Tg' for some time and then refrozen, before moving towards primary drying. At the start of the primary drying stage, no "free and fluid" water should be present in the vials. Minus forty to minus fifty degrees Celsius is a typical freezing temperature range where sublimation is initiated through chamber pressure reduction.

Primary Drying
In the primary drying stage (see Fig. 5.9), sublimation of the water mass in the vial starts by lowering the pressure. The water vapor condenses on a condenser, with a (substantially) lower temperature than the shelf with the vials (typically −80 °C). Sublimation costs energy (about 2500 kJ/g ice). The supply of heat from the shelf to the vial prevents the vial temperature to drop. Thus, the shelf is heated during this stage.

The pressure (vacuum) and the heat supply rate control the resulting product temperature. It is important to keep the product temperature below the Tg' (e.g., determined by differential scanning calorimetry (DSC)) or the collapse temperature (Tc) (e.g., determined by freeze-drying microscopy). At and above the collapse temperature, the material softens and cannot support its own structure anymore (see below). Collapse causes a strong reduction in sublimation rate and poor cake formation, resulting in nonelegant cake appearance and long reconstitution times. Although cake appearance is negatively affected, collapse may result in products with acceptable protein stability (Schersch et al. 2010; Patel et al. 2017). An example of a DSC scan providing information on the Tg' is presented in Fig. 5.11.

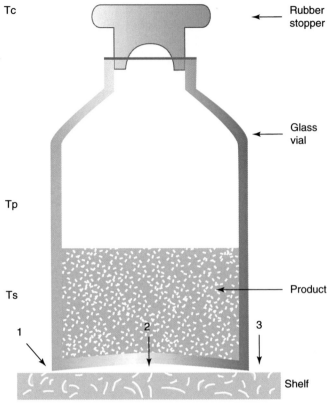

Fig. 5.12 Heat transfer mechanisms during the freeze-drying process: (1) Direct conduction via shelf and glass at points of actual contact. (2) Gas conduction: contribution heat transfer via conduction through gas between shelf and vial bottom. (3) Radiation heat transfer. Ts shelf temperature, Tp temperature sublimating product, Tc temperature condensor. Ts > Tp > Tc

Heat reaches the vial through (1) direct shelf–vial contact (conductance), (2) radiation, and (3) gas conduction (Fig. 5.12). Gas conduction depends on the pressure: if one selects relatively high gas pressures, heat transport increases because of a high conductivity. However, it reduces mass transfer, because of a low driving force: the pressure between equilibrium vapor pressures at the interface between the frozen mass/dried cake and the chamber pressure (Pikal 1990). During the primary drying stage, one transfers heat from the shelf through the vial bottom and the frozen mass to the interface frozen mass/dry powder, to keep the sublimation process going.

During this drying stage, the product temperature should never reach Te or Tg'/Tc, as the cake may collapse. Typically, a safety margin of 2–5 °C is used. For highly concentrated protein formulations, Tc is typically higher than Tg', and the products can be dried at product temperatures above Tg' (below Tc) without resulting in collapse. Therefore, knowledge of Te and Tg' is of great importance to develop a rationally designed freeze-drying protocol.

Heat transfer resistance decreases during the drying process by the reduction of the transport distance as the interface retreats. With the mass transfer resistance (transport of water vapor), however, the opposite occurs. Mass transfer resistance increases during the drying process, as the dry cake becomes thicker. Therefore, parameters such as chamber pressure and shelf temperature are not necessarily constant during the primary drying process. They should be carefully chosen and adjusted as the drying process proceeds.

When all frozen or "unbound" water (i.e., not bound to protein or excipients) has been removed, the primary drying step has finished (Fig. 5.9). The primary drying process can be monitored by following individual vials (e.g., product temperature of individual vials) or batch methods, e.g., comparative pressure measurements, manometric temperature measurement/pressure rise test, mass spectrometry to monitor gas composition in the chamber, or tunable diode laser absorption spectroscopy to provide information on the sublimation rate.

The end of the primary drying stage can be measured by thermocouples within the product vials: the end of primary drying is reached when product temperature and shelf temperature become equal, or when the partial water pressure drops (Pikal 1990).

Comparative pressure measurement with a Pirani gauge and a capacitance manometer (CP) is another approach next to thermocouples to detect the end of primary drying. This is based on the gas composition dependent reading of a Pirani gauge, whereas a capacitance manometer reading is not influenced by the gas composition. Consequently, the Pirani gauge (which is typically calibrated under nitrogen or air) shows higher pressure values than the CP in the presence of water vapor in the chamber, i.e., as long as primary drying is progressing. The end of primary drying, i.e., when sublimation of "unbound" water has finished, is indicated when the pressure measured by the Pirani gauge and the CP show the same signal.

Secondary Drying

In the secondary drying stage, the temperature is slowly increased to remove "bound" water; the chamber pressure is still kept low. The temperature should stay all the time below the collapse/eutectic temperature, which continues to rise as the residual water content drops. Typically, the secondary drying step ends when the product has been kept at 20–40 °C for several hours. The residual water content of the product (e.g., determined by a Karl Fischer assay) is a critical, end point-indicating parameter. Values as low as 1% residual water in the cake have been recommended; however, this needs to be evaluated on a product-specific basis and certain products may require higher or lower residual moisture contents to result in a stable product. Fig. 5.13 (Pristoupil 1985; Pikal 1990) exemplifies the decreasing stability of freeze-dried hemoglobin with increasing residual water content.

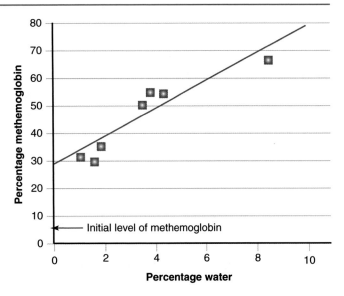

Fig. 5.13 The effect of residual moisture on the stability of freeze-dried hemoglobin (−6%) formulated with 0.2 M sucrose; decomposition to methemoglobin during storage at 23 °C for 4 years (Adapted from Pikal (1990). Data reported by Pristoupil et al. (1985))

Points to Consider in the Process of Formulating a Nucleotide-Based Biologic (NBB)

Nucleotide-based biologics appear in many molecular sizes and shapes. From oligonucleotides via different RNA (ribonucleic acid) structures such as siRNA (small interfering RNA) and mRNA (messenger RNA) to DNA (deoxynucleic acid) used in gene therapeutics. Oligonucleotides and siRNA API are chemically synthesized and, strictly spoken, do not fall under the FDA definition of biologics. However, their formulation issues are similar to larger, natural source derived biologics and will therefore be dealt with in this chapter.

The nucleotide-building blocks (cf. Chap. 13) are hydrophilic and negatively charged. The linkages between these building blocks are sensitive to enzymatic degradation. In Chap. 13, molecular approaches to improve their stability are discussed. In this chapter, the focus is on the formulation and design of delivery systems for NBBs.

Formulation

Table 5.6 lists a number of marketed NBB. First, note that the (soluble) excipients are selected from the same list as used with therapeutic proteins. Second, all NBBs are parenterally administered, just like therapeutic proteins. Third, siRNA and mRNA products are formulated as lipid nanoparticles (LNP).

Table 5.6 Examples of marketed nucleotide based biologics

			Formulation	Route of administration
Oligonucleotide	Nusinersen		Phosphate buffer, NaCl, KCl, CaCl, MgCl	Intrathecal
	Mipomersen			Subcutaneous
	Pegatanib sodium		Phosphate buffer, NaCl	Intravitreal
siRNA	Patiseran		Phosphate buffer, NaCl Lapid nanoparticles[a]	Intravenous
mRNA	Covid-19 vaccine Moderna		Tris buffer, NaAc, sucrose Lapid nanoparticles[a]	Intramuscular
	Covid-19 vaccine Pfizer BioNTech		Phosphate or Tris buffer, phosphate NaCl Lapid nanoparticles[a]	Intramuscular

[a] *LNP* Lipid nanoparticles

Lipid Nanoparticles (LNP): Structure

LNP are NBB-containing lipid structures in the 100 nm range. Typically, LNP are built of four essential building blocks. First, a cationic ionizable lipid, usually a tertiary amine group coupled to two lipid chains; it binds to the negatively charged NBB and this complex forms part of the core of the NLP. Second, a neutral lipid or a combination of neutral lipids. E.g., a phospholipid and/or cholesterol, collectively referred to as helper lipids. They are part of the core but also form the outer lipid layer of the particle. Third, PEGylated lipid molecules positioned on the outside as well for stabilization of the LNP during manufacturing and upon storage and aiding in the disposition of the LNP-payload in vivo. As fourth component, water can be found throughout the core of the LNP (Schoenmaker et al. 2021). High-quality lipids should be used as lipid induced oxidation and lipid adduct formation may jeopardize safety and potency of the NBB product (Oude Blenke et al. 2023).

Lipid Nanoparticles: Their Role in NBB In Vivo Performance and Stability

LNP serve a dual purpose. They provide a carrier for the NBB upon injection by screening the mRNA from ribonuclease abundantly present in the extracellular space. Second, the negative charge and hydrophilicity of NBB interfere with membrane passage and consequently intracellular delivery. The siRNA component of the commercial product patiseran is transported by the LNP to the liver hepatocytes where it inhibits a specific mRNA molecule to be translated. With the mRNA COVID-19 vaccines the lipid components presumable play an adjuvating role. "Administration of LNP-formulated RNA vaccines i.m. results in transient local inflammation that drives recruitment of neutrophils and antigen presenting cells (APCs) to the site of delivery. Recruited APCs are capable of LNP uptake and protein expression and can subsequently migrate to the local draining lymph nodes where T cell priming occurs" (EMA 2020).

Apart from their pharmacological effect, LNP formation has a stabilizing effect on the shelf life of the encapsulated NBB. In particular, the long mRNA single-stranded molecules with typically 3000 nucleotides are highly sensitive to activity loss. One break in the strand leads to inactivity. The LNP structure provides some protection. However, at present, long-term stability can only be achieved by storing the LNP at subzero temperatures and shelf life at room temperature is limited to a number of days (Oude Blenke et al. 2023).

Handling of Pharmaceutical Proteins Post-Production

In the formulation process, the manufacturer will expose the protein product under development to several stress factors as mentioned in Table 5.4 to decide on the final formulation composition and shelf life. In addition, studies are performed by the developer to test the stability of the drug product during the actual planned clinical application, e.g., studies on stability upon dilution in carrier solutions, in infusion lines, or hold time studies in bags/syringes. The composition of the final formulation of the protein product will reflect the outcome of those (laboratory) stress experiments in combination with real-time stability testing: optimum formulation conditions for chemical and physical stability of the protein will be chosen. In spite of all these efforts, pharmaceutical proteins and polynucleotides remain sensitive to "real life" handling and may readily show degradation reactions that obviously affect both efficacy and safety. Therefore, the manufacturer—together with the regulatory authorities—creates a package insert text that points out to health care professionals and patients the conditions that should be maintained for the product, e.g., storage temperature window, avoidance of shaking/shear, exposure to light. As an example, the package insert of trastuzumab states: "Swirl the vial gently to aid reconstitution. DO NOT SHAKE." Trastuzumab may be sensitive to shear-induced stress, e.g., agitation or rapid expulsion from a syringe (FDA Trastuzumab 1998). Other examples are the mRNA-LNP vaccines (discussed above). These are delicate dosage forms and need careful handling following detailed instructions provided in the insert (EMA 2021).

Surprisingly, little information is available on the actual storage and administration practices for pharmaceutical proteins in hospitals. Anecdotal information of exposing protein or nucleotide solutions to high shear conditions (shaking, use of pneumatic tube transport) and the use of incorrect administration techniques can be found in the literature. This is also true for the patient's home setting. Real world handling data are available of a group of rheumatoid arthritis patients of whom only a minority stored their protein product (anti-TNF-alpha therapy) in the prescribed temperature window, i.e., 2–8 °C. Both freezing to −20 °C and storing above 25 °C occurred (Vlieland et al. 2016). The consequences of this behavior for the stability of this protein, in particular aggregate or particle formation, may be an increased chance of formation of antidrug antibodies (Vlieland et al. 2018). Considering the clinical importance of these medicines and the high prices paid, health care professionals and patients should be aware of the importance of' 'Good Handling Practices for Biologicals' (Nejadnik et al. 2018).

Delivery of Proteins: Routes of Administration

The Parenteral Route of Administration

Parenteral administration is here defined as administration via those routes where a needle is used, including intravenous (IV), intramuscular (IM), subcutaneous (SC), intracutaneous, intraperitoneal (IP), and intravitreal injections. Chapter 6 provides more information on the pharmacokinetic behavior of recombinant proteins. It suffices here to state that the blood half-life of biotech products can vary over a wide range. For example, the blood circulation half-life of tissue plasminogen activator is a few minutes, whereas monoclonal antibodies have half-lives of a few days to weeks.

A simple way to expand the mean residence time for short half-life proteins is to switch from IV to IM or SC administration. One should realize that by doing that, changes in disposition may occur with a significant impact on the bioavailability (slower uptake in the blood compartment and lower extent of absorption) and therapeutic performance of the biologic. For instance, the extent of absorption of SC administered protein injections (compared to IV administration) may be as low as 30% (Kinnunen and Mrsny 2014; Bittner et al. 2018). This change in disposition is caused by various factors such as: (1) the prolonged residence time at the IM or SC site of injection compared to IV administration, differences in protein environment upon injection e.g., enhanced exposure to degradation reactions (peptidases), and (2) differences in disposition.

Regarding point 1: For instance, diabetics can become "insulin resistant" through high tissue peptidase activity (Maberly et al. 1982). Other factors that can contribute to absorption variation are differences in level of activity of the muscle at the injection site as well as massage and heat at the injection site. The state of the tissue, for instance, the occurrence of pathological conditions, may be important as well.

Regarding point 2: Upon administration, the protein may reach the blood through the lymphatics or enter the blood circulation through the capillary wall at the site of injection (Figs. 5.14 and 5.15). The fraction of the administered dose taking this lymphatic route depends on the molecular weight of the protein (Supersaxo et al. 1990). Lymphatic transport takes time (hours), and uptake in the blood circulation is

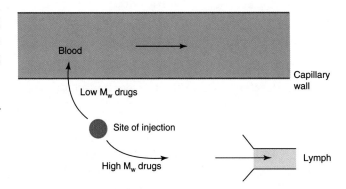

Fig. 5.14 Routes of uptake of SC- or IM-injected drugs

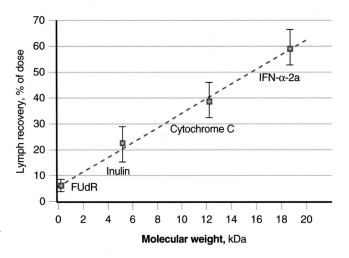

Fig. 5.15 Correlation between the molecular weight (Mw) and the cumulative recovery of rIFN alpha-2a (Mw 19 kDa), cytochrome c (MW 12.3 kDa), insulin (MW 5.2 kDa), and FUDR (MW 256.2 Da) in the efferent lymph from the right popliteal lymph node following SC administration into the lower part of the right hind leg of sheep. Each point and bar shows the mean and standard deviation of three experiments performed in separate sheep. The line drawn is the best fit by linear regression analysis calculated with the four mean values. The points have a correlation coefficient r of 0.998 ($p < 0.01$) (Adapted from Supersaxo et al. (1990))

highly dependent on the injection site. On its way to the blood, the lymph passes through draining lymph nodes. There, contact is possible between lymph contents and cells of the immune system, such as macrophages and B and T lymphocytes residing in the lymph nodes.

Other Routes of Administration

For several reasons, e.g., ease of administration, patient friendliness and cost, alternative administration routes to the parenteral route would be welcome for the successful systemic delivery of recombinant proteins. This is particularly true for the oral route. Nature, unfortunately, does not allow us to use the oral route of administration for therapeutic proteins if a high (or at least constant) bioavailability is required. The two main reasons (A and B) for this failure of uptake are: (A) protein degradation in the GI (gastrointestinal track) and (B) poor permeability of the wall of the GI tract in case of a passive transport process.

Regarding point A: Protein degradation. The human body has developed a very efficient system to break down proteins in our food to amino acids or di- or tripeptides. These building blocks for body proteins are actively absorbed for use in newly formed proteins. The stomach secretes pepsins, a family of aspartic proteases. They are particularly active between pH 3 and 5 and lose activity at higher pH values. Pepsins are endopeptidases capable of cleaving peptide bonds distant from the ends of the peptide chain, preferentially peptide bonds between two hydrophobic amino acids. Other endopeptidases are active in the gastrointestinal tract at neutral pH values, e.g., trypsin, chymotrypsin, and elastase. They have different, complementary peptide bond cleavage characteristics. Exopeptidases, proteases degrading peptide chains from their C- or N-terminus, are present as well. In the intestinal epithelium, brush border and cytoplasmic proteases of the enterocytes continue to cut proteins into fragments down to amino acids, di- and tripeptides.

Regarding point B: Permeability. High-molecular-weight molecules with a hydrophilic "coat" such as therapeutic proteins do not readily penetrate the intact and mature epithelial barrier of the intestinal lumen. Active transport of intact proteins over the GI-epithelium has not been described. This leaves diffusion to and partitioning into the enterocyte membrane as the sole pathway for mass transfer. Diffusion coefficients and partition coefficients are low for intact therapeutic proteins leading to very low extent of uptake. Some oral vaccines, however, are available on the market, as discussed in Chap. 15. For those products, only a (small) fraction of the antigen must reach its target site to illicit an immune response.

Some proteins are administered locally for local therapy. For instance, ranibizumab (Fab fragment), aflibercept (a recombinant fusion protein consisting of vascular endothelial growth factor (VEGF)-binding portions from the extracellular domains of human VEGF receptors 1 and 2, which are fused to the Fc portion of human IgG1), and bevacizumab are proteins to block neovascularization in the retina. When injected in the vitreous cavity of the eye, they interact with vascular endothelial growth factor (VEGF) and slow down wet, age-related macular degeneration. Another example of administering a biologic near its site of action is dornase alfa (Pulmozyme®). It is taken via inhalation to break down DNA in sputum of cystic fibrosis patients (cf. Chap. 21).

In 2017, an oral formulation of semaglutide, a glucagon-like peptide 1 receptor agonist (GLP-1RA), was approved for treating diabetes type 2 patients. Native GLP-1RA is a 30 amino acid peptide; semaglutide (Rybelsus®) is a modified form of native human GLP to enhance stability and blood circulation time (cf. liraglutide/semaglutide, below). Co-formulated with an absorption enhancer its bioavailability is still low: 0.4–1% (FDA 2017) (Cf. Chap. 16).

Apart from the oral route, the eye and lungs (as mentioned above when discussing local therapy), the nose, rectum, oral cavity (buccal absorption) and skin have been studied as potential sites of application. Table 5.7 lists the potential pros and cons for the different relevant routes. Moeller and Jorgensen (2009) and Jorgenson and Nielsen (2009) describe "the state of the art" in more detail. The nasal, buccal, rectal, and transdermal routes all are of little clinical relevance if systemic action is required. In general, bioavailability is too low and varies too much. The pulmonary route may be the exception to this rule.

FDA approved the first pulmonary insulin formulation (Exubera®) in January 2006. However, the supplier took it off the market in 2008 because of poor market penetration. Inhalation technology plays a critical role when considering the prospects of the pulmonary route for the systemic delivery of therapeutic proteins. Dry powder inhalers and nebulizers are the delivery systems considered and tested. The fraction of protein that is ultimately absorbed depends on (1) the fraction of the inhaled/nebulized dose that is actually leaving the device, (2) the fraction that is actually deposited in the lung, and (3) the fraction that is being absorbed, i.e., total relative uptake (TO%) = % uptake from device × % deposited in the lungs × % actually absorbed from the lungs. For insulin, TO% is estimated to be about 10% (Patton et al. 2004). The fraction of insulin that is absorbed from the lung is about 20%. The reproducibility of the blood glucose response to inhaled insulin was equivalent to SC-injected insulin. These figures demonstrate that insulin absorption via the lung may be a possible route, but the fraction that reaches the blood circulation is small and commercial success failed to materialize until now. Later, Afrezza® entered the market as a pulmonary insulin product (Cf. Chap. 16).

Table 5.7 Alternative routes of administration to the IV, IM, and SC route for biopharmaceuticals. + = relative advantage, – = relative disadvantage Oral + easy to access, proven track record with "conventional" medicines, sustained/controlled release possible—negligible bioavailability for proteins

Nasal
+ easily accessible, fast uptake, proven track record with a number of "conventional" medicines, probably lower proteolytic activity than in the GI tract, avoidance of first pass effect, spatial containment of absorption enhancers is possible
– reproducibility (in particular under pathological conditions), safety (e.g., ciliary movement), negligible bioavailability for proteins
Pulmonary
+ relatively easy to access, fast uptake, proven track record with "conventional" medicines, substantial –in the 10% range- fractions of insulin are absorbed, lower proteolytic activity than in the GI tract, avoidance of hepatic first pass effect
– reproducibility (in particular under pathological conditions, smokers/nonsmokers), safety (e.g., immunogenicity), presence of macrophages in the lung with high affinity for particulates
Rectal
+ easily accessible, partial avoidance of hepatic first pass, probably lower proteolytic activity than in the upper parts of the GI tract, spatial containment of absorption enhancers is possible, proven track record with a number of "conventional" drugs
– negligible bioavailability for proteins
Buccal
+ easily accessible, avoidance of hepatic first pass, probably lower proteolytic activity than in the lower parts of the GI tract, spatial containment of absorption enhancers is possible, option to remove formulation if necessary
– negligible bioavailability of proteins, no proven track record yet
Transdermal
+ easily accessible, avoidance of hepatic first pass effect, removal of formulation if necessary is possible, spatial containment of absorption enhancers, proven track record with "conventional" medicines, sustained/controlled release possible—negligible bioavailability of proteins
intravitreal + direct access to vitreous, delivery close to the target site—not suitable for systemic effects

Table 5.8 Communication between cells: chemical messengers

Endocrine hormones:
A hormone secreted by a distant cell to regulate cell functions distributed widely through the body. The bloodstream plays an important role in the transport process
Paracrine-acting mediators:
The mediator is secreted by a cell to influence surrounding cells, short-range influence
Autocrine-acting mediators:
The agent is secreted by a cell and affects the cell by which it is generated, (very) short-range influence

Delivery of Proteins by the Parenteral Route: Approaches for Rate-Controlled Delivery

Presently used therapeutic proteins widely differ in their pharmacokinetic characteristics (see Chap. 6). If they are recombinant counterparts of endogenous agents such as insulin, tissue plasminogen activator, growth hormone, epoetin, interleukins, or factor VIII, it is important to realize why, when, and where—by which cells—they are secreted. Cells can communicate with each other through the endocrine, paracrine and/or autocrine pathway leading to secretion of mediator molecules (Table 5.8).

The presence of these mediators may activate a complex cascade of events that needs to be carefully controlled. Therefore, key issues for their therapeutic success are (1) access to target cells, (2) retention at the target site, and (3) proper timing of delivery.

In particular, for paracrine- and autocrine-acting proteins, such as tumor necrosis factor and interleukin-2, severe side effects were reported upon parenteral (IV or SC) administration (see Chap. 22). The occurrence of these side effects lim-

its the therapeutic potential of these compounds. Therefore, the delivery of these proteins at the proper site, rate, and dose is crucial for their therapeutic success.

Various technologies similar to those used for 'small, low-molecular-weight' medicines may achieve rate control. E.g., for insulin one can choose from a spectrum of options (see Chap. 16). Moreover, continuous/"smart" infusion systems are on the market for insulin, see below.

Mechanical Pumps

In general, proteins are parenterally administered as an aqueous solution. Only recombinant vaccines and a number of insulin formulations are (colloidal) dispersions. For continuous and controlled administration of these solutions pump systems are used: continuous infusion'. These pumps typically deliver the protein formulation via the intravenous route. However, patients may receive subcutaneously up to 25 mL over prolonged periods (up to 1 h) with a pump system (20% immune globulin solution; Hizentra®, cf. Jiskoot

Table 5.9 Controlled release and input systems for parenteral delivery

- Continuous infusion with pumps. Input: Preset with limited variability. E.g., elastomer or spring-driven pumps.
- Continuous infusion with pumps. Input: Variable, controlled by health care professional or patient: e.g., mechanically/electronically driven (smart) pumps.
- Rate control through implants, biodegradable polymer-based microspheres. Input, limited control
- Rate control through a closed-loop approach/feedback system
- Biosensor–pump combination.

et al. 2022). Table 5.9 lists some of the technologically feasible options. They are briefly touched upon below.

Pumps can be chosen in various sizes/prices, being portable or not, for inside/outside the body, with/without sophisticated rate control software. A pump system needs constant attention as it may fail because of power failure (batteries serve as backup power supply), problems with the syringe, accidental needle withdrawal, leakage of the catheter, and problems at the injection or implantation site. Moreover, long-term protein drug stability may become a problem. The protein should be stable at 37 °C or ambient temperature (for internal/implanted and external devices, respectively) between two refills.

Controlled administration of a drug does not necessarily imply a constant input rate. Pulsatile or variable-rate delivery is the desired mode of input for a number of proteins. For these biologics, the applied pumps should provide options for a flexible input rate. Insulin is a prime example of a therapeutic protein, where there is a need to adjust the input rate to the needs of the body (Cf. Chap. 16). Today, by far, the most experience has been gained with pump systems with adjustable input rates in an ambulatory setting with this protein drug. Even with high-tech pump systems, the patient still has to collect data to adjust the pump rate. This implies invasive sampling of body fluids on a regular basis, followed by calculations and setting of the required input rate. Progress made in developing the concept of closed-loop systems integrating these three actions: monitoring, calculating and choosing the rate of administration (a "natural" biofeedback system is discussed under "Biosensor-Pump Combinations."

Biodegradable Microspheres

Polylactic acid–polyglycolic acid (PLGA)-based delivery systems are being used extensively for the delivery of therapeutic peptides, in particular for luteinizing hormone-releasing hormone (LHRH) agonists, such as leuprolide in the therapy of prostate cancer. The first LHRH agonist-controlled release formulations were implants, rods containing leuprolide with dosing intervals of 1–3 months. Later, microspheres loaded with leuprolide entered the market with dosing intervals up to 6 months. Considerations to design these controlled release systems are (1) the drug has to be highly potent (only a small dose is required over the dosing interval) and stable in the dosage form until release, (2) a

sustained presence in the body is required, and (3) no adverse reactions at the injection site should occur. Only two such microsphere products for sustained delivery of therapeutic proteins instead of peptides made it to the market. Nutropin Depot® released recombinant human growth hormone over prolonged periods (monthly injection). Introduced in 1999, it was taken off the market in 2004 because of low perceived added therapeutic value, manufacturing problems, and costs. A glucagon-like protein-1 (GLP-1, 39 amino acids, exanitide) slow-release formulation (Bydureon™) based on PLGA microspheres for once-a-week administration to type II diabetics was released in 2012.

Biosensor-Pump Combinations

If input rate control is desired to stabilize a certain body function, then this function or a suitable biomarker should be monitored. An algorithm converts this data into a drug-input rate and corresponding pump settings. If there is a known relationship between plasma level of the biomarker and pharmacological effect, these systems contain (Fig. 5.16):

1. A biosensor, measuring the (plasma) level of the biomarker
2. An algorithm, to calculate the required input rate for the delivery system
3. A pump system, able to administer the drug at the required rate over prolonged periods

Before considering building a fully integrated closed-loop delivery system for a biological, a number of points have to be checked. A simple relationship between plasma level and therapeutic effect does not always exist (see Chap. 6). There are many exceptions known to this rule; for instance, "hit and run" drugs can have long-lasting pharmacological effects after only a short exposure time. Also, drug effect–blood level relationships may be time dependent, as in the case of downregulation of relevant receptors on prolonged stimulation. Finally, if circadian rhythms exist, these will be responsible for variable PK/PD (pharmacokinetic/pharmacodynamic) relationships as well.

If PK/PD relationships can be established, as with insulin in diabetics, then integrated biosensor–pump combinations are available that act as biofeedback systems. The biosensor

Fig. 5.16 Therapeutic system with closed control loop (Adapted from Heilman (1984). (1) A biosensor, measuring the plasma level of the protein. (2) An algorithm, to calculate the required input rate for the delivery system. (3) A pump system, able to administer the drug at the required rate over prolonged periods

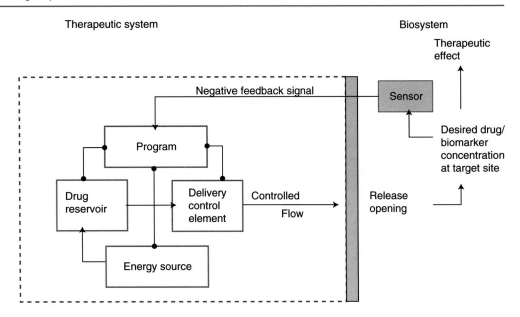

measures interstitial fluid glucose levels and sends the outcome (wireless) to a therapy management algorithm. This software program adjusts the insulin pump settings to deliver an appropriate dose of insulin for basal glucose levels. However, the patient still has to inject a bolus before meals or adjust the dose before exercise. Biosensor stability, robustness, absence of histological reactions, and handling postprandial highs are still challenges in the design of fully integrated closed loop systems for chronic use. Leelarathna et al. (2021) describe the state of the art in this fast-moving field (see Chap. 16).

Delivery of Proteins by the Parenteral Route: Half-Life Extension by Modification of the API

Chemical modifications can change protein characteristics. For example, insulin half-life can be prolonged by exploiting the long circulation time of serum albumin and its high binding affinity for fatty acids such as myristic acid. In insulin detemir (Levemir®) lysine replaces the C-terminal threonine of insulin and myristic acid is chemically coupled through this lysine. After subcutaneous injection, the myristic acid–insulin combination reaches the blood circulation and binds to albumin. Thereby the half-life of insulin is prolonged from less than 10 min to over 5 h. A similar approach is used with glucagon-1-like peptide receptor agonist (GLP-1RA (7–37)) for the treatment of diabetes type 2. Conjugating myristic acid to GLP-1RA (amino acids 7–37) (liraglutide marketed as Victoza®) increases the plasma half-life from 2 min to over 10 h. In semaglutide, a similar approach was followed with a stearic acid conjugated to the GLP-1RA (Ozempic®). A human growth hormone-lipid anchor combination

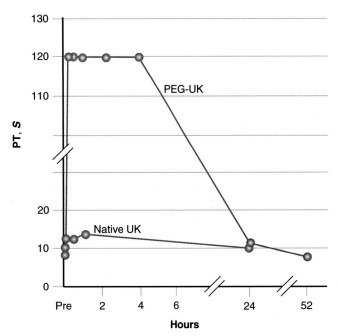

Fig. 5.17 Influence of chemical grafting of polyethylene glycol (PEG) on the ability of urokinase (UK) to affect the prothrombin time in seconds (PT) in vivo in beagles as function of time after a single administration (through Tomlinson (1987)), adapted

extended plasma half-life for hGH from a few hours to a few days (somapacitan-beco) (cf. Chap. 16).

Another chemical modification approach that has been very successful in prolonging plasma circulation times and dosing intervals is the covalent attachment of polyethylene glycol (PEG) to proteins. Figure 5.17 shows an example of this approach. Commercially successful products that were developed later are PEGylated interferon alfa-2a and -2b (see Fig. 22.3c), PEGylated hG-CSF (human granulocyte

Table 5.10 Examples of PEGylated biologicals

Protein Name	PEGylation	Product Name
IFN α-2a IFN α-2b	Branched, 40 kDa Linear, 12 kDa	Pegasys® PegIntron®
Interferon β	Linear, 20 kDa	Plegridy® (mPEGIFN β1a)
Erythropoietin	60 kDa	Mircera® (mPEG epoetinβ)
G-CSF	Linear, 20 kDa	Neulasta® (pegfilgrastim)
Adenosine deaminase	Linear, 5 kDa	Adagen® (pegademase)
Asparaginase	Linear, 5 kDa	Oncaspar® (pegaspargase)
rhGH analog-antagonist	Linear, 5 kDa	Somavert® (pegvisomant)
rhFactor VIII	Linear, 40 kDa	Adynovate® (octocog alfa pegol)
B-domain deleted FVIII	40 kDa	Jivi® (damoctocog alfa pegol)
B-domain deleted FVIII	40 kDa	Esperoct® (turoctocog alfa pegol)
Factor IX	40 kDa	Rebinyn, Refixia (nonacog beta pegol)
Fab' fragment	40 kDa	Cimzia® (cetolizumab pegol)
RNA aptamer	40 kDa	Macugen® (pegaptanib)
hGH	40 kDa	Skytrofa® (lonapegsomatropin-tcgd)

colony-stimulating factor, filgrastim; see Chap. 17) and a PEGylated Fab' fragment as in cetolizumab, see Chap. 24 and Table 5.10.

Furthermore, genetic modification of active proteins has been successful in extending plasma half-life in a number of cases. Both serum albumin and antibodies are long circulating physiological proteins, see e.g., Chap. 8). Fusing active proteins with human serum albumin or Fc-parts from monoclonal antibodies (e.g., etanercept, aflibercept) prolongs their plasma half-life. Attaching 28 carboxy-terminal residues of human chorionic gonadotropin to the terminal ends of human growth hormone protein has a similar effect (somatrogron, Chap. 20). This concept was also followed to prolong the dosing interval for follicle stimulating hormone in designing corifollitropin alfa, cf. Chap. 19. With erythropoietin another approach was chosen. Darbepoetin alfa (Aranesp) is a hyperglycosylated form of epoetin. It has the same mechanism of action, but has a threefold longer half-life compared to epoetin.

Concluding Remarks

In order to successfully formulate a biologic API and turn it into a medicinal product that can be administered to a patient, the first requirement is an in depth understanding of the chemical and physical characteristics of the molecule in question, including its stability under the preferred storage conditions. A set of stability indicating, complementary, and orthogonal analytical techniques (see Textbox 5.1 as an example for protein drugs) should be available to successfully achieve the defined target product profile and select the proper formulation composition and the packaging material for a stable product (freeze-dried or not).

In spite of considerable efforts made over almost 100 years, the parenteral route is the preferred to administer protein-based medicines for systemic delivery to the patient. All other routes of administration failed so far.

Self-Assessment Questions

Questions

1. A protein, which is poorly water soluble around its pI (pH 7.4), has to be formulated for subcutaneous administration. What conditions would one select to produce a water-soluble, injectable solution?
2. Why are many proteins to be used in the clinic formulated in freeze-dried form? Why is, as a rule, the presence of a lyoprotectant required? Why is it important to know the glass transition temperature of the maximally freeze concentrated solution or eutectic temperature of the system?
3. Why should glucose be avoided as an excipient in protein formulations?
4. Why is the oral bioavailability of therapeutic proteins close to zero?
5. What is the impact on the pharmacokinetics when changing from the IV to the SC route of administration of a therapeutic protein?
6. A company decides to explore the possibility to develop a biofeedback system for a therapeutic protein. What information should be available for estimating the chances for success?
7. What is the function of a preservative? For which type of protein formulations are they required? Example? Potential disadvantage?
8. Why do many protein formulations contain a surfactant? Example? Potential disadvantage?

Answers

1. Both solubility and stability should be considered. As both the aqueous solubility and the stability will be pH dependent, information on the solubility and stability as function of pH should be collected. The pH should be controlled by using a buffer. If needed, other excipients can be added to improve both the physical and the chemical stability of the protein and to achieve isotonicity.

2. Chemical and physical instability of proteins in aqueous media is usually the reason to stabilize the protein in a freeze-dried form.

   Freeze-drying is the preferred drying technology, as other drying techniques do not result in rapidly reconstitutable dry forms for the formulation and are difficult to perform under GMP conditions for large-scale manufacturing and/or because elevated temperatures necessary for drying jeopardize the integrity of the protein. Lyoprotectants protect the proteins from degradation during the freeze-drying process and in the dried state.

   The collapse temperature (Tc) should not be exceeded (stay a few degrees below Teu or Tg'), as otherwise collapse of the cake occurs. Collapse slows down the freeze-drying process rate, collapsed material does not rapidly dissolve upon adding water for reconstitution, and the cake appearance is not acceptable anymore.

3. Because the protein will degrade through the Maillard reaction, as glucose is a reducing sugar. This is true for both liquid and lyophilized formulations.

4. Because of the hostile environment in the GI tract regarding protein stability and the poor absorption characteristics of proteins (high molecular weight/often hydrophilic).

5. Both the extent and rate of uptake into the blood circulation are affected. When changing from IV to SC administration, the AUC (area under the curve) and the absorption rate are reduced.

6. Information that should be available.

   - The desired pharmacokinetic profile (e.g., information on the PK/PD relationship/circadian rhythm).
   - Chemical and physical stability of the protein on long-term storage at body/ambient temperature.
   - Availability of a biosensor system (stability in vivo, precision/accuracy).
   - Availability of a reliable pump system (see Table 5.9).

7. A preservative is included to neutralize contaminations in containers. A multidose formulation needs a preservative. Examples of preservatives used in protein formulations are phenol, meta-cresol, benzyl alcohol, and chlorobutanol. A disadvantage of preservatives is that they may interact with the protein, affecting their own and/or the protein's performance, and can interfere with the analytical methods due to UV absorbance at 280 nm.

8. Surfactants reduce adsorption to interfaces and by doing so prevent aggregation. Examples of commonly used surfactants are polysorbate 80 and 20, as well as poloxamer 188. A disadvantage of surfactants is that too high concentrations may cause denaturation (unfolding) of the protein. A disadvantage of polysorbates is that they can hydrolyze, which reduces their stabilizing potential and may lead to insoluble degradation products (a.o. fatty acids). Oxidation can occur for both polysorbates and poloxamer 188.

References

Bailey AO, Han G, Phung W, Gazis P, Sutton J, LS JJ, Sandoval W (2018) Charge variant native mass spectrometry benefitsmass precision and dynamic range of monoclonal antibody intact mass analysis. MAbs 10:1214–1225. https://doi.org/10.1080/19420862.2018.152113

Bittner B, Richter W, Schmidt J (2018) Subcutaneous administration of biotherapeutics: an overview of current challenges and opportunities. BioDrugs 32:425–440. https://doi.org/10.1007/s40259-018-0295-0

Chi E, Krishnan S, Randolph TW, Carpenter JF (2003) Physical stability of proteins in aqueous solution: mechanism and driving forces in nonnative protein aggregation. Pharm Res 20:1325–1336

Constantino HR, Pikal MJ (2004) Lyophilization of biopharmaceuticals. AAPS Press, Arlington

EMA (2020) Assessment report Comirnaty. https://www.ema.europa.eu/en/documents/assessment-report/comirnaty-epar-public-assessment-report_en.pdf. Accessed 10 Mar 2022

EMA (2021) Annex 1 Summary of Product Characteristics Comirnaty. https://www.ema.europa.eu/en/documents/product-information/comirnaty-epar-product-information_en.pdf. Accessed 10 Mar 2022

FDA (1998) Insert trastuzumab. https://www.accessdata.fda.gov/drugsatfda_docs/label/2010/103792s5250lbl.pdf. Accessed 10 Mar 2022

FDA (2017) Rybelsus (semaglutide) tablets, for oral use. https://wwsemiglutidew.accessdata.fda.gov/drugsatfda_docs/label/2019/213051s000lbl.pdf. Accessed 10 Mar 2022

Geidobler R, Winter G (2013) Controlled ice nucleation in the field of freeze-drying: fundamentals and technology review. Eur J Pharm Biopharm 85:214–222

Hawe A, Wiggenhorn M, van de Weert M, Garbe JHO, Mahler H-C, Jiskoot W (2012) Forced degradation of therapeutic proteins. J Pharm SciJ Pharm Sci 101:895–913

Heilmann K (1984) Therapeutic systems. Rate controlled delivery: concept and development. Thieme, Stuttgart

Jiskoot W, Crommelin DJA (2005) Methods for Structural Analysis of Protein Pharmaceuticals. AAPS Press, Arlington VA

Jiskoot W, Hawe A, Menzen T, Volkin DB, Crommelin DJA (2022) Ongoing challenges to develop high concentration monoclonal antibody-based formulations for subcutaneous administration:

quo Vadis? J Pharm Sci 111:861–867. https://doi.org/10.1016/j.xphs.2021.11.008

Jorgensen J, Nielsen HM (eds) (2009) Delivery technologies for biopharmaceuticals: peptides, proteins, nucleic acids and vaccines. Wiley, Chichester

Kamerzell TJ, Esfandiary R, Joshi SB, Middaugh CR, Volkin DB (2011) Protein–excipient interactions: mechanisms and biophysical characterization applied to protein formulation development. Adv Drug Deliv Rev 63:1118–1159

Kinnunen HM, Mrsny RJ (2014) Improving the outcomes of biopharmaceutical delivery via the subcutaneous route by understanding the chemical, physical and physiological properties of the subcutaneous injection site. J Control Release 182:22–32

Leelarathna L, Choudhary P, Wilmot E (2021) Hybrid closed-loop therapy: where are we in 2021? Diabetes Obes Metab 23:655–660. https://doi.org/10.1111/dom.14273

Maberly GF, Wait GA, Kilpatrick JA, Loten EG, Gain KR, Stewart RDH, Eastman CJ (1982) Evidence for insulin degradation by muscle and fat tissue in an insulin resistant diabetic patient. Diabetologia 23:333–336

Manning MC, Patel K, Borchardt RT (1989) Stability of proteins. Pharm Res 6:903–918

Manning MC, Chou DK, Murphy BM, Payne RW, Katayama DS (2010) Stability of protein pharmaceuticals: an update. Pharm Res 27:544–575

Martos A, Koch W, Jiskoot W, Wuchner K, Winter G, Friess W, Hawe A (2017) Trends on analytical characterization of polysorbates and their degradation products in biopharmaceutical formulations. J Pharm Sci 106:1722–1735

Mensink MA, Frijlink HW, van der Voort MK, Hinrichs W (2017) How sugars protect proteins in the solid state and during drying (review): mechanisms of stabilization in relation to stress conditions. Eur J Pharm Biopharm 114:288–295

Moeller EH, Jorgensen L (2009) Alternative routes of administration for systemic delivery of protein pharmaceuticals. Drug Discov Today Technol 5:89–94

Nejadnik MR, Randolph TW, Volkin DB, Schöneich C, Carpenter JF, Crommelin DJA, Jiskoot W (2018) Post-production handling and administration of protein pharmaceuticals and potential instability issues. J Pharm Sci 107(8):2013–2019. https://doi.org/10.1016/j.xphs.2018.04.005

Nguyen TH, Ward C (1993) Stability characterization and formulation development of alteplase, a recombinant tissue plasminogen activator. In: Wang YJ, Pearlman R (eds) Stability and characterization of protein and peptide drugs. Case histories. Plenum Press, New York, pp 91–134

Oude Blenke E, Örnskov E, Schöneich C, Nilsson G, Volkin D, Mastrobattista E, Almarsson O, Crommelin DJA (2023) The storage and in-use stability of mRNA vaccines and therapeutics: not a cold case. J Pharm Sci 112(2):386–403

Patel S, Nail S, Pikal M, Geidobler R, Winter G, Hawe A, Davagnino J, Rambhatla GS (2017) Lyophilized drug product cake appearance: what is acceptable? J Pharm Sci 106:1706–1721

Patton JS, Bukar JG, Eldon MA (2004) Clinical pharmacokinetics and pharmacodynamics of inhaled insulin. Clin Pharmacokinet 43:781–801

Pearlman R, Bewley TA (1993) Stability and characterization of human growth hormone. In: Wang YJ, Pearlman R (eds) Stability and characterization of protein and peptide drugs. Case histories. Plenum Press, New York, pp 1–58

Pikal MJ (1990) Freeze-drying of proteins. Part I: process design. BioPharm 3:18–27

Pristoupil TI (1985) Haemoglobin lyophilized with sucrose: effect of residual moisture on storage. Haematologia 18:45–52

Roesch A, Zölls S, Stadler D, Helbig C, Wuchner K, Kersten G, Hawe A, Jiskoot W, Menzen T (2021) Particles in biopharmaceutical formulations, part 2: an update on analytical techniques and applications for therapeutic proteins, viruses, vaccines and cells. J Pharm Sci 111:933. https://doi.org/10.1016/j.xphs.2021.12.011

Sacha GA, Saffell-Clemmer W, Abram K, Akers MJ (2010) Practical fundamentals of glass, rubber, and plastic sterile packaging systems. Pharm Dev Technol 15(1):6–34. https://doi.org/10.3109/10837450903511178

Sacha G, Rogers JA, Miller RL (2015) Pre-filled syringes: a review of the history, manufacturing and challenges. Pharm Dev Technol 20:1–11. https://doi.org/10.3109/10837450.2014.982825

Schersch K, Betz O, Garidel P, Muehlau S, Bassarab S, Winter G (2010) Systematic investigation of the effect of lyophilizate collapse on pharmaceutically relevant proteins I: stability after freeze-drying. J Pharm SciJ Pharm Sci 99:2256–2278

Schoenmaker L, Witzigmann D, Kulkarni YA, Verbeke R, Kersten G, Jiskoot W, Crommelin DJA (2021) mRNA-lipid nanoparticle COVID-19 vaccines: structure and stability. Int J Pharm 601:120586. https://doi.org/10.1016/j.ijpharm.2021.120586

Sediq AS, van Duijvenvoorde RB, Jiskoot W, Nejadnik MR (2016) Subvisible particle formation during stirring. J Pharm SciJ Pharm Sci 105:519–529

Supersaxo A, Hein WR, Steffen H (1990) Effect of molecular weight on the lymphatic absorption of water-soluble compounds following subcutaneous administration. Pharm Res 7:167–169. https://doi.org/10.1023/A:1015880819328

Tomlinson E (1987) Theory and practice of site-specific drug delivery. Adv Drug Deliv Rev 1:87–198

Vemuri S, Yu CT, Roosdorp N (1993) Formulation and stability of recombinant alpha1-antitrypsin. In: Wang YJ, Pearlman R (eds) Stability and characterization of protein and peptide drugs. Plenum Press, New York, pp 263–286

Vlieland ND, Gardarsdottir H, Bouvy ML, Egberts TCG, van den Bemt BJF (2016) The majority of patients do not store their biologic disease-modifying antirheumatic drugs within the recommended temperature range. Rheumatology 55:704–709. https://doi.org/10.1093/rheumatology/kev394

Vlieland ND, Nejadnik MR, Gardarsdottir H, Romeijn AS, Sediq S, Bouvy ML, Egberts ACG, van den Bemt BJF, Jiskoot W (2018) The impact of inadequate temperature storage conditions on aggregate and particle formation in drugs containing tumor necrosis factor-alpha inhibitors. Pharm Res 35:42. https://doi.org/10.1007/s11095-017-2341-x

Weinbuch D, Hawe A, Jiskoot W, Friess W (2018) In: Mahler HC, Warne NW (eds) Challenges in protein product development. AAPS advances in the pharmaceutical sciences series. Springer, New York, pp 3–22

Yoneda S, Torisu T, Uchiyama S (2021) Development of syringes and vials for delivery of biologics: current challenges and innovative. Expert Opin Drug Deliv 18:459–470. https://doi.org/10.1016/j.xphs.2021.12.011

Zbacnik TJ, Holcomb RE, Katayama DS, Murphy BM, Payne RW, Coccaro RC, Evans GJ, Matsuura JE, Henry CS, Manning MC (2017) Role of buffers in protein formulations. J Pharm SciJ Pharm Sci 106:713–733

Zölls S, Tantipolphan R, Wiggenhorn M, Winter G, Jiskoot W, Friess W, Hawe A (2012) Particles in therapeutic protein formulations—part I. overview of analytical methods. J Pharm SciJ Pharm Sci 101:914–935

Further Reading

Carpenter JF, Manning MC (2002) Rational design of stable protein formulations—theory and practice. Springer, New York

Mahler H-C, Jiskoot W (2012) Analysis of aggregates and particles in protein pharmaceuticals. Wiley, Hoboken, NJ

Mahler HC, Warne NW (2018) Challenges in protein product development. AAPS advances in the pharmaceutical sciences series. Springer, New York

Moeller EH, Jorgensen L (2009) Alternative routes of administration for systemic delivery of protein pharmaceuticals. Drug Discov Today Technol 5:89–94

Pharmacokinetics and Pharmacodynamics of Therapeutic Proteins and Nucleic Acids

6

Bernd Meibohm

Introduction

The rational use of drugs and the design of effective dosage regimens are facilitated by the appreciation of the central paradigm of clinical pharmacology that there is a defined relationship between the administered dose of a drug, the resulting drug concentrations in various body fluids and tissues, and the intensity of pharmacologic effects caused by these concentrations (Meibohm and Derendorf 1997). This dose–exposure–response relationship and thus the dose of a drug required to achieve a certain effect are determined by the drug's pharmacokinetic and pharmacodynamic properties (Fig. 6.1).

Pharmacokinetics describes the time course of the concentration of a drug in a body fluid, preferably plasma or blood, which results from the administration of a certain dosage regimen. It comprises all processes affecting drug absorption, distribution, metabolism, and excretion. Simplified, pharmacokinetics characterizes *"what the body does to the drug."* In contrast, pharmacodynamics characterizes the intensity of a drug effect or toxicity resulting from certain drug concentrations in a body fluid, usually at the assumed site of drug action. It can be simplified to *what the drug does to the body* (Fig. 6.2) (Holford and Sheiner 1982; Derendorf and Meibohm 1999).

The understanding of the dose–concentration–effect relationship is crucial to any drug—including peptides and proteins—as it lays the foundation for dosing regimen design and rational clinical application. General pharmacokinetic and pharmacodynamic principles are to a large extent equally applicable to protein- and nucleic acid-based therapeutics as they are to traditional small molecule-based therapeutics. Deviations from some of these principles and additional challenges with regard to the characterization of the pharmacokinetics and pharmacodynamics of therapeutic proteins, however, arise from some of their specific properties:

(a) Their definition by the production process in a living organism rather than a chemically exactly defined structure and purity as it is the case for small-molecule drugs
(b) Their structural similarity to endogenous structural or functional proteins and nutrients
(c) Their intimate involvement in physiologic processes on the molecular level, often including regulatory feedback mechanisms
(d) The analytical challenges to identify and quantify them in the presence of a myriad of similar molecules
(e) Their large molecular weight and macromolecule character

This chapter highlights some of the major pharmacokinetic properties and processes relevant for the majority of proteins and provides examples of well-characterized pharmacodynamic relationships for protein drugs. It also briefly discusses the pharmacokinetic characteristics of nucleotide-based therapeutics as emerging new group of biotechnology products, even though they are usually chemically synthesized rather than produced in living organisms. The clinical pharmacology of monoclonal antibodies (mAbs), including special aspects in their pharmacokinetics and pharmacodynamics, is discussed in further detail in Chap. 8. For a more general discussion on pharmacokinetic and pharmacodynamic principles, the reader is referred to several textbooks and articles that review the topic in extensive detail (see Suggested Reading).

B. Meibohm (✉)
Department of Pharmaceutical Sciences, University of Tennessee
Health Science Center, College of Pharmacy, Memphis, TN, USA
e-mail: bmeibohm@uthsc.edu

© The Author(s), under exclusive license to Springer Nature Switzerland AG 2024
D. J. A. Crommelin et al. (eds.), *Pharmaceutical Biotechnology*, https://doi.org/10.1007/978-3-031-30023-3_6

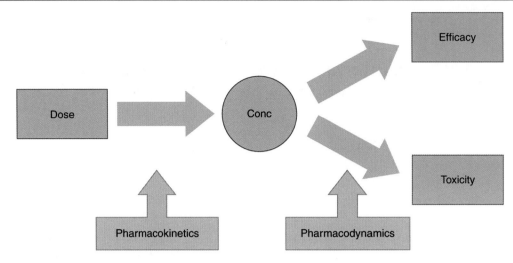

Fig. 6.1 The central paradigm of clinical pharmacology: the dose–concentration–effect relationship

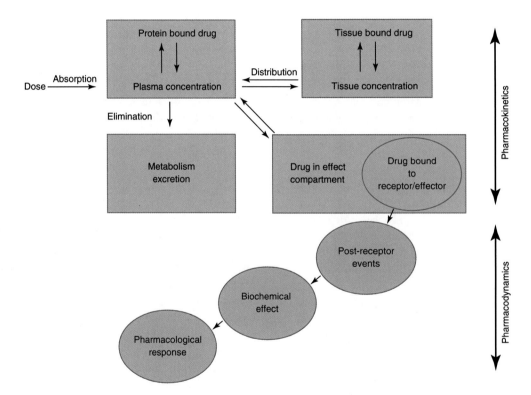

Fig. 6.2 Physiological scheme of pharmacokinetic and pharmacodynamic processes

Pharmacokinetics of Therapeutic Proteins

The in vivo disposition of protein drugs may often be predicted to a large degree from their physiological function (Tang and Meibohm 2006). Small proteins and peptides, for example, which frequently have hormone activity, usually have short elimination half-lives, which is desirable for a close regulation of their endogenous levels and thus function. Insulin, for example, shows dose-dependent elimination with a relatively short half-life of 26 and 52 min at 0.1 and 0.2 U/ kg, respectively. Contrary to that, large proteins that have transport tasks such as albumin or long-term immunity functions such as immunoglobulins have elimination half-lives of several days, which enables and ensures the continuous maintenance of physiologically necessary concentrations in the bloodstream (Meibohm 2006). This is, for example, reflected by the elimination half-life of antibody drugs such as the anti-epidermal growth factor receptor antibody cetuximab, an IgG1 chimeric antibody for which a half-life of approximately 7 days has been reported (Herbst and Langer 2002).

Absorption of Therapeutic Proteins

Enteral Administration

Therapeutic proteins, unlike conventional small-molecule drugs, are generally not therapeutically active upon oral administration (Fasano 1998; Mahato et al. 2003; Tang et al. 2004). The lack of systemic bioavailability is mainly caused by two factors: (1) high gastrointestinal enzyme activity and (2) low permeability through the gastrointestinal mucosa. In fact, the substantial peptidase and protease activity in the gastrointestinal tract make it the most efficient body compartment for protein metabolism. Furthermore, the gastrointestinal mucosa presents a major absorption barrier for water-soluble macromolecules such as proteins (Tang et al. 2004). Thus, although various factors such as permeability, stability, and gastrointestinal transit time can affect the rate and extent of orally administered proteins, molecular size is generally considered the ultimate obstacle (Shen 2003).

Since oral administration is still a highly desirable route of delivery for protein drugs due to its convenience, cost-effectiveness, and painlessness, numerous strategies to overcome the obstacles associated with oral delivery of proteins have recently been an area of intensive research. Suggested approaches to increase the oral bioavailability of protein drugs include encapsulation into micro- or nanoparticles thereby protecting proteins from intestinal degradation (Lee 2002; Mahato et al. 2003; Shen 2003; Verma et al. 2021). Other strategies are chemical modifications such as amino acid backbone modifications and chemical conjugations to improve the resistance to degradation and permeability of the protein drug (Diao and Meibohm 2013). Coadministration of protease inhibitors has also been suggested for the inhibition of enzymatic degradation (Pauletti et al. 1997; Mahato et al. 2003). More details on approaches for oral delivery of therapeutic proteins are discussed in Chap. 5.

Parenteral Administration

Most protein drugs are currently formulated as parenteral formulations because of their poor oral bioavailability. Major routes of administration include intravenous (IV), subcutaneous (SC), and intramuscular (IM) administration. In addition, other nonoral administration pathways are utilized, including nasal, buccal, rectal, vaginal, transdermal, ocular, and pulmonary drug delivery (see Chap. 5).

IV administration of proteins offers the advantage of circumventing presystemic degradation, thereby achieving the highest concentration in the biological system. Therapeutic proteins given by the IV route include, among many others, the tissue plasminogen activator (t-PA) analogues alteplase and tenecteplase, the recombinant human erythropoietin epoetin-α, and the granulocyte colony-stimulating factor filgrastim (Tang and Meibohm 2006).

IV administration as either a bolus dose or constant rate infusion, however, may not always provide the desired concentration–time profile depending on the biological activity of the product. In these cases, IM or SC injections may be more appropriate alternatives. For example, luteinizing hormone-releasing hormone (LH-RH) in bursts stimulates the release of follicle-stimulating hormone (FSH) and luteinizing hormone (LH), whereas a continuous baseline level will suppress the release of these hormones (Handelsman and Swerdloff 1986). To avoid the high peaks from an IV administration of leuprorelin, an LH-RH agonist, a long-acting monthly depot injection of the drug is approved for the treatment of prostate cancer and endometriosis (Periti et al. 2002). A study comparing SC versus IV administration of epoetin-α in patients receiving hemodialysis reported that the SC route can maintain the hematocrit in a desired target range with a lower average weekly dose of epoetin-α compared to IV (Kaufman et al. 1998). In addition, SC injections have become increasingly popular as they allow self-administration by the patient, especially with the introduction of microneedles and pen devices, and thus not only circumvent the need to intravenous access but also have increased patient acceptance and overall lower administration cost. Thus, numerous mAbs that had initially been brought to market as IV dosage forms have more recently been extended to also offer SC dosage forms. One of the related challenges beyond dosage volume limitations and resulting formulation issues is the translation of body weight-based dosing for the IV dosage form to fixed dosing as required by the SC route (Yapa et al. 2016).

Potential limitations of SC and IM administration, however, are the presystemic degradation processes frequently associated with these administration routes, resulting in a reduced systemic bioavailability compared to IV administration. No correlation between the molecular weight of a therapeutic protein and its systemic bioavailability has so far been described in any species (Richter et al. 2012), and clinically observed bioavailability seems to be product-specific based on physicochemical properties and structure. For many subcutaneously administered mAbs; however, SC bioavailability is in the range of 50–80% (Ryman and Meibohm 2017).

Bioavailability assessments for therapeutic proteins may be challenging if the protein exhibits the frequently encountered nonlinear pharmacokinetic behavior. Classic bioavailability assessments comparing systemic exposures quantified as area-under-the-concentration-time curve (AUC) resulting from extravascular versus IV administration assume linear pharmacokinetics, i.e., a drug clearance independent of concentration and the administration pathway. As this is not the case for many therapeutic proteins, especially those that undergo target-mediated drug disposition (see respective section in this chapter), bioavailability assessments using the classic approach can result in substantial bias (Limothai and Meibohm 2011). Potential approaches suggested to minimize or overcome these effects include bioavailability assessments at doses at which the target- or receptor-mediated processes are saturated or to compare concentration–time profiles with similar shape and magnitude for extravascular and IV administration by modulating the input rate in the IV experiment.

The pharmacokinetically derived apparent absorption rate constant k_{app} for protein drugs administered via these administration routes is the combination of absorption into the systemic circulation and presystemic degradation prior to entering the blood stream, i.e., the sum of a true first-order absorption rate constant k_a and a first-order degradation rate constant. The true absorption rate constant k_a can then be calculated as

$$k_a = F \cdot k_{app}$$

where F is the systemic bioavailability compared to IV administration. A rapid apparent absorption, i.e., large k_{app}, can thus be the result of a slow true absorption and a fast presystemic degradation, i.e., a low systemic bioavailability (Colburn 1991).

Other potential factors that may limit the rate and/or extent of uptake of proteins after SC or IM administration include variable local blood and lymph flow, injection trauma, and limitations of uptake into the systemic circulation related to effective capillary pore size, diffusion, and convective transport.

Several therapeutic proteins including anakinra, etanercept, insulin, and pegfilgrastim but also mAbs such as adalimumab, omalizumab, or alirocumab are administered as SC injections. Following a SC injection, therapeutic peptides and proteins may enter the systemic circulation either via blood capillaries or through lymphatic vessels (Porter and Charman 2000). There appears to be a defined relationship between the molecular weight of the protein and the proportion of the dose absorbed by the lymphatics (see Fig. 5.15) (Supersaxo et al. 1990). In general, peptides and proteins larger than 16 kDa are predominantly absorbed into the lymphatics, whereas those under 1 kDa are mostly absorbed into the blood circulation. While diffusion is the driving force for the uptake into blood capillaries, transport of larger proteins through the interstitial space into lymphatic vessels is mediated by convective transport with the interstitial fluid following the hydrostatic and osmotic pressure differences between vascular and interstitial space (see paragraphs on distribution). The fraction of insulin (5.2 kDa), for example, that has been described to undergo absorption through the lymphatic system is approximately 20% (see Chap. 16), while this fraction is approaching 100% for mAbs (150 kDa).

For mAbs and fusion proteins with antibody Fc fragment, interaction with the neonatal Fc receptor (FcRn) has also been identified as a potential absorption process (Roopenian and Akilesh 2007). In this context, FcRn prevents the mAb or fusion protein from undergoing lysosomal degradation (see Chap. 8 for details) and thereby increases systemic bioavailability but may also facilitate transcellular transport from the absorption site into the vascular space. The contribution of this pathway to overall absorption, however, is limited.

Since lymph flow and interstitial convective transport are substantially slower than blood flow and diffusion processes, larger proteins taken up into lymphatic vessels usually show a delayed and prolonged absorption process after SC administration that can even become the rate-limiting step in their overall disposition. For mAbs, for example, the time of the maximum concentration (t_{max}) was substantially delayed after SC administration, ranging from 1.7 to 13.5 days, with frequent values around 6–8 days. A related model-based analysis suggests that lymphatic flow rate is the most influential factor for t_{max} of SC administered mAbs (Zhao et al. 2013).

Preferential uptake into lymphatic vessels after SC administration is of particular importance for those agents that target lymphoid cells (i.e., interferons and interleukins). Studies with recombinant human interferon α-2a (rhIFN α-2a) indicate that following SC administration, high concentrations of the recombinant protein are found in the lymphatic system, which drains into regional lymph nodes (Supersaxo et al. 1988). Due to this targeting effect, clinical studies show that palliative low-to-intermediate-dose SC recombinant interleukin-2 (rIL-2) in combination with rhIFN α-2a can be administered to patients in the ambulatory setting with efficacy and safety profiles comparable to the most aggressive IV rIL-2 protocol against metastatic renal cell cancer (Schomburg et al. 1993).

Beyond molecular weight and size, charge has also been described as an important factor in the SC absorption of proteins: While the positive and negative charges from collagen and hyaluronan in the extracellular matrix seem to be of similar magnitude, additional negative charges of proteoglycans may lead to a negative interstitial charge (Richter et al. 2012). This negative net charge and the associated ionic interactions with SC-administered proteins result in a slower transport for more positively rather than negatively charged proteins, as could be shown for several mAbs (Mach et al. 2011).

Distribution of Therapeutic Proteins
Distribution Mechanisms and Volumes

The rate and extent of protein distribution is largely determined by the molecule size and molecular weight, physicochemical properties (e.g., charge, lipophilicity), binding to structural or transport proteins, and their dependency on active transport processes to cross biomembranes. Since most therapeutic proteins have high molecular weights and are thus large in size, their apparent volume of distribution is usually small and limited to the volume of the extracellular space due to their limited mobility secondary to impaired passage through biomembranes (Zito 1997). In addition, there is a mutual exclusion between therapeutic proteins and the structural molecules of the extracellular matrix. This fraction of the interstitial space that is not available for distribution is expressed as the excluded volume (Ve). It is dependent on the molecular weight and charge of the macromolecule and further limits extravascular distribution. For albumin (MW 66 kDa), the Ve has been reported as ~50% in muscle and skin (Ryman and Meibohm 2017). Active tissue uptake and binding to intra- and extravascular proteins, however, can substantially increase the apparent volume of distribution of protein drugs, as reflected by the relatively large volume of distribution of up to 2.8 L/kg for interferon β-1b (Chiang et al. 1993).

In contrast to small-molecule drugs, protein transport from the vascular space into the interstitial space of tissues is largely mediated by convection rather than diffusion, following the unidirectional fluid flux from the vascular space through paracellular pores into the interstitial tissue space (Fig. 6.3). The subsequent removal from the interstitial space is accomplished by lymph drainage back into the systemic circulation (Flessner et al. 1997). This underlines the unique role that the lymphatic system plays in the disposition of therapeutic proteins as already discussed in the section on absorption. The fact that the transfer clearance from the vascular to the interstitial space is smaller than the transfer clearance from the interstitial space to the lymphatic system results in lower protein concentrations in the interstitial space compared to the vascular space, thereby further limiting the apparent volume of distribution for therapeutic proteins. For endogenous and exogenous immunoglobulin G antibodies, for example, the tissue:blood concentration ratio is in the range of 0.1–0.5, i.e., antibody concentrations are substantially lower in the tissue interstitial fluid than in plasma (Ryman and Meibohm 2017). For brain tissue, the ratio is even in the range of 0.01 or lower, but may be higher in cases of compromised blood-brain barrier (Kingwell 2016).

Another, but much less prominent pathway for the movement of protein molecules from the vascular to the interstitial space is transcellular migration via endocytosis (Baxter et al. 1994; Reddy et al. 2006).

Besides the size-dependent sieving of macromolecules through the capillary walls, charge may also play an important role in the biodistribution of proteins. It has been suggested that the electrostatic attraction between positively charged proteins and negatively charged cell membranes might increase the rate and extent of tissue distribution. Most cell surfaces are negatively charged because of their abundance of glycosaminoglycans in the extracellular matrix.

After IV administration, proteins usually follow a biexponential plasma concentration–time profile that can best be described by a two-compartment pharmacokinetic model (Meibohm 2004). A biexponential concentration–time profile has, for example, been described for clenoliximab, a macaque-human chimeric mAb specific to the CD4 molecule on the surface of T lymphocytes (Mould et al. 1999). Similarly, secukinumab, a human mAb that binds and neutralizes interleukin 17A for the treatment of psoriasis, exhibited biphasic pharmacokinetics after IV administration (Bruin et al. 2017). The central compartment in this two-compartment model represents primarily the vascular space and the interstitial space of well-perfused organs with permeable capillary walls, including the liver and the kidneys. The peripheral compartment is more reflective of concentration–time profiles in the interstitial space of slowly equilibrating tissues.

The central compartment in which proteins initially distribute after IV administration has thus typically a volume of distribution equal to or slightly larger than the plasma volume, i.e., 3–8 L. The total volume of distribution frequently comprises with 14–20 L not more than two to three times the

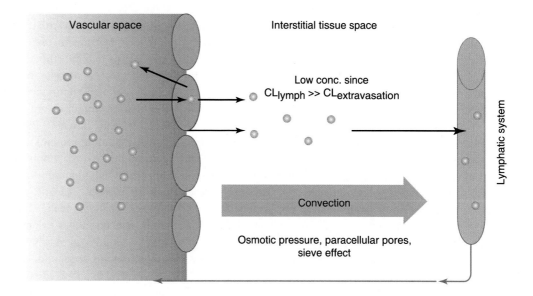

Fig. 6.3 Distribution mechanisms of therapeutic proteins: convective extravasation rather than diffusion as major distribution process. $CL_{extravasation}$ transfer clearance from the vascular to the interstitial space, CL_{lymph} transfer clearance from the interstitial space to the lymphatic system

initial volume of distribution (Colburn 1991; Dirks and Meibohm 2010). An example for such a distribution pattern is the t-PA analog tenecteplase. Radiolabeled ^{125}I-tenecteplase was described to have an initial volume of distribution of 4.2–6.3 L and a total volume of distribution of 6.1–9.9 L with liver as the only organ that had a significant uptake of radioactivity. The authors concluded that the small volume of distribution suggests primarily intravascular distribution for tenecteplase, consistent with the drug's large molecular weight of 65 kDa (Tanswell et al. 2002).

Epoetin-α, for example, has a volume of distribution estimated to be close to the plasma volume at 0.056 L/kg after an IV administration to healthy volunteers (Ramakrishnan et al. 2004). Similarly, volume of distribution for darbepoetin-α has been reported as 0.062 L/kg after IV administration in patients undergoing dialysis (Allon et al. 2002), and distribution of thrombopoietin has also been reported to be limited to the plasma volume (~3 L) (Jin and Krzyzanski 2004).

It should be stressed that pharmacokinetic calculations of volume of distribution may be problematic for many therapeutic proteins (Tang et al. 2004; Straughn 2006). Noncompartmental determination of volume of distribution at steady state (V_{ss}) using statistical moment theory assumes first-order disposition processes with elimination occurring from the rapidly equilibrating or central compartment (Perrier and Mayersohn 1982; Straughn 1982; Veng-Pedersen and Gillespie 1984). These basic assumptions, however, are not fulfilled for numerous therapeutic proteins, as proteolysis and receptor-mediated elimination in peripheral tissues may constitute a substantial fraction of the overall elimination process. If therapeutic proteins are eliminated from slowly equilibrating tissues at a rate greater than their distribution process, substantial error in the volume of distribution assessment may occur. A simulation study could show that if substantial tissue elimination exists, a V_{ss} determined by noncompartmental methods will underestimate the "true" V_{ss}, and that the magnitude of error tends to be larger the more extensively the protein is eliminated by tissue routes (Meibohm 2004; Straughn 2006; Tang and Meibohm 2006).

These challenges in characterizing the distribution of therapeutic proteins can only be overcome by determining actual protein concentrations in the tissue by biopsy or necropsy or via biodistribution studies with radiolabeled compound and/or imaging techniques.

Biodistribution studies are imperative for small organic synthetic drugs, since long residence times of the radioactive label in certain tissues may be an indication of tissue accumulation of potentially toxic metabolites. Because of the possible reutilization of amino acids from protein drugs in endogenous proteins, such a safety concern does not exist for therapeutic proteins. Therefore, biodistribution studies for protein drugs are usually only performed to assess drug tar-

geting to specific tissues or to detect the major organs of elimination.

If a biodistribution study with radiolabeled protein is performed, either an external label such as ^{125}I can be chemically coupled to the protein if it contains a suitable amino acid such as tyrosine or lysine or internal labeling can be used by growing the production cell line in the presence of amino acids labeled with ^{3}H, ^{14}C, ^{35}S, etc. The latter method, however, is not routinely used because of the prohibition of radioactive contamination of fermentation equipment (Meibohm 2004). Moreover, internally labeled proteins may be less desirable than iodinated proteins because of the potential reutilization of the radiolabeled amino acid fragments in the synthesis of endogenous proteins and cell structures. Irrespective of the labeling method, but more so for external labeling, the labeled product should have demonstrated physicochemical and biological properties identical to the unlabeled molecule (Bennett and McMartin 1978).

Protein Binding of Therapeutic Proteins

Another factor that can influence the distribution of therapeutic proteins is binding to endogenous protein structures. Physiologically active endogenous peptides and proteins frequently interact with specific binding proteins involved in their transport and regulation. Furthermore, interaction with binding proteins may enable or facilitate cellular uptake processes and thus affect the drug's pharmacodynamics.

It is a general pharmacokinetic principle, which is also applicable to proteins, that only the free, unbound fraction of a drug substance is accessible to distribution and elimination processes as well as interactions with its target structures at the site of action, for example, a receptor or ion channel. Thus, protein binding may affect the pharmacodynamics but also disposition properties of therapeutic proteins. Specific binding proteins have been identified for numerous protein drugs, including recombinant human DNase for use as mucolytic in cystic fibrosis (Mohler et al. 1993), growth hormone (Toon 1996), and recombinant human vascular endothelial growth factor (rhVEGF) (Eppler et al. 2002).

Protein binding not only affects the unbound fraction of a protein drug and thus the fraction of a drug available to exert pharmacological activity, but many times it also either prolongs protein circulation time by acting as a storage depot or it enhances protein clearance. Recombinant cytokines, for example, may after IV administration encounter various cytokine-binding proteins including soluble cytokine receptors and anti-cytokine antibodies (Piscitelli et al. 1997). In either case, the binding protein may either prolong the cytokine circulation time by acting as a storage depot or it may enhance the cytokine clearance.

Growth hormone, as another example, has at least two binding proteins in plasma (Wills and Ferraiolo 1992). This protein binding substantially reduces growth hormone elimination with a tenfold smaller clearance of total compared to free growth hormone but also decreases its activity via reduction of receptor interactions.

Ectodomain shedding is another source of binding proteins circulating in plasma where the extracellular domain of a membrane-standing receptor is cleaved and released into the circulation (Hayashida et al. 2010). For therapeutic proteins targeting these receptors, the shed ectodomain constitutes a binding reservoir that by being in the vascular space is often more easily accessible than the intact membrane-standing receptor on target cells in the extravascular space. Thus, shed antigen can limit the disposition of a therapeutic protein and can inactivate a fraction of the administered therapeutic protein by preventing it from accessing its intended target (Ryman and Meibohm 2017). Different patients may have vastly different shed antigen concentrations and thus different effects, as shown for CD52, the target for the mAb alemtuzumab (Albitar et al. 2004).

Apart from this specific binding, peptides and proteins may also be nonspecifically bound to plasma proteins. For example, metkephamid, a met-enkephalin analog, was described to be 44–49% bound to albumin (Taki et al. 1998), and octreotide, a somatostatin analog, is up to 65% bound to lipoproteins (Chanson et al. 1993).

Distribution Via Receptor-Mediated Uptake

Aside from physicochemical properties and protein binding of therapeutic proteins, site-specific receptor-mediated uptake can also substantially influence and contribute to the distribution of therapeutic proteins, as well as to elimination and pharmacodynamics (see section on "Target-Mediated Protein Metabolism").

The generally low volume of distribution should not necessarily be interpreted as low tissue penetration. Receptor-mediated specific uptake into the target organ, as one mechanism, can result in therapeutically effective tissue concentrations despite a relatively small volume of distribution. Nartograstim, a recombinant derivative of the granulocyte colony-stimulating factor (G-CSF), for example, is charac-

terized by a specific, dose-dependent, and saturable tissue uptake into the target organ bone marrow, presumably via receptor-mediated endocytosis (Kuwabara et al. 1995).

Elimination of Therapeutic Proteins

Therapeutic proteins are generally subject to the same catabolic pathways as endogenous or dietetic proteins. The end products of protein metabolism are thus amino acids that are reutilized in the endogenous amino acid pool for the de novo biosynthesis of structural or functional proteins in the human body (Meibohm 2004). Detailed investigations on the metabolism of proteins are relatively difficult because of the myriad of potential molecule fragments that may be formed and are therefore generally not conducted. Nonmetabolic elimination pathways such as renal or biliary excretion are negligible for most proteins. If biliary excretion occurs, however, it is generally followed by subsequent metabolic degradation of the compound in the gastrointestinal tract.

Proteolysis

In contrast to small-molecule drugs, metabolic degradation of therapeutic proteins by proteolysis can occur unspecifically nearly everywhere in the body. Due to this unspecific proteolysis of some proteins already in blood as well as potential active cellular uptake, the clearance of protein drugs can exceed cardiac output, i.e., >5 L/min for blood clearance and >3 L/min for plasma clearance (Meibohm 2004). The clearance of proteins in this context describes the irreversible removal of active substance from the vascular space, which includes besides metabolism also cellular uptake. Thus, intracellular uptake is per se more an elimination rather than a distribution process (Tang and Meibohm 2006). The metabolic rate for protein degradation generally increases with decreasing molecular weight from large to small proteins to peptides (Table 6.1), but is also dependent on other factors such as size, charge, lipophilicity, functional groups, and glycosylation pattern as well as secondary and tertiary structure.

Proteolytic enzymes such as proteases and peptidases are ubiquitous throughout the body within subcellular compart-

Table 6.1 Molecular weight as major determinant of the elimination mechanisms of peptides and proteins

Molecular weight (kDa)	Elimination site	Predominant elimination mechanisms
<0.5	Blood, liver	Extracellular hydrolysis Passive lipoid diffusion
0.5–1	Liver	Carrier-mediated uptake Passive lipoid diffusion
1–60	Kidney	Glomerular filtration and subsequent degradation processes (see Fig. 6.4)
50–200	RES, endothelial cells (skin, muscle, gut), liver	Receptor-mediated endocytosis Pinocytosis
200–400	Immune system	Opsonization
>400	Phagocytic cells	Phagocytosis

Based on Meijer and Ziegler (1993) and Eigenmann et al. (2017)
Other determining factors are size, charge, lipophilicity, functional groups, sugar recognition, vulnerability for proteases, aggregation to particles, formation of complexes with opsonization factors, etc. As indicated, mechanisms may overlap. Endocytosis may occur at any molecular weight range; RES reticuloendothelial system

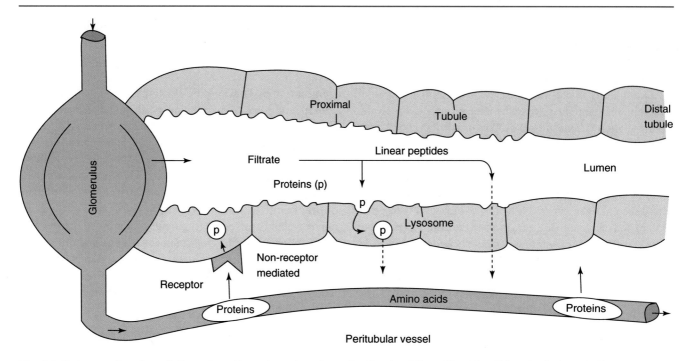

Fig. 6.4 Pathways of renal metabolism of peptides and proteins: glomerular filtration followed by either (I) intraluminal metabolism or (II) tubular reabsorption with intracellular lysosomal metabolism and (III) peritubular extraction with intracellular lysosomal metabolism. (Modified from Maack et al. (1985))

ments such as lysosomes. Thus, intracellular uptake is the rate limiting step for nonspecific, proteolytic clearance of therapeutic proteins. This uptake occurs either by pinocytosis, a fluid-phase endocytosis, or by a receptor-mediated endocytosis. Pinocytosis is a relatively unspecific and inefficient endocytic process by endothelial cells lining the blood and lymphatic vessels. In pinocytosis, protein molecules are taken up into cells by forming invaginations of cell membrane around extracellular fluid droplets that are subsequently taken up as membrane vesicles. Due to the large surface area of endothelial cells in the body (>1000 m²), the process can despite its inefficiency substantially contribute to the elimination of therapeutic proteins. Nonspecific proteolytic degradation following pinocytotic uptake is thus not limited to a specific organ but occurs throughout the body, particularly in those organs and tissues rich in capillary beds with endothelial cells. Thus, the skin, muscles, and the gastrointestinal tract are the major elimination organs for the nonspecific proteolytic degradation of many therapeutic proteins, including immunoglobulin G-based therapeutics such as mAbs and antibody derivatives. In addition, the phagocytic cells of the reticuloendothelial system have been identified as a major contributor to the unspecific proteolytic degradation of many therapeutic proteins (Ryman and Meibohm 2017). Receptor-mediated endocytosis processes are more relevant for specific organs and tissues and will be discussed in the subsequent sections.

While peptidases and proteases in the gastrointestinal tract and in lysosomes are relatively unspecific, soluble pep-

tidases in the interstitial space and exopeptidases on the cell surface have a higher selectivity and determine the specific metabolism pattern of an organ. The proteolytic activity of subcutaneous and particularly lymphatic tissue, for example, results in a partial loss of activity of SC compared to IV administered interferon-γ.

Gastrointestinal Protein Metabolism

As pointed out earlier, the gastrointestinal tract is a major site of protein metabolism with high proteolytic enzyme activity due to its primary function to digest dietary proteins. Thus, gastrointestinal metabolism is one of the major factors limiting systemic bioavailability of orally administered protein drugs. The metabolic activity of the gastrointestinal tract, however, is not limited to orally administered proteins. Parenterally administered proteins may also be metabolized in the endothelial cells lining the vast capillary beds of the gastrointestinal tract as well as in resident phagocytic cells. At least 20% of the degradation of endogenous albumin, for example, has been reported to take place in the gastrointestinal tract (Colburn 1991).

Renal Protein Metabolism

The kidneys are a major site of protein metabolism for smaller-sized proteins that undergo glomerular filtration. The size-selective cutoff for glomerular filtration is approximately 60 kDa, although the effective molecule radius based on molecular weight and conformation is probably the limiting factor (Edwards et al. 1999). Glomerular filtration is most efficient, however, for proteins smaller than 30 kDa (Kompella and Lee 1991). Peptides and small proteins

(<5 kDa) are filtered very efficiently, and their glomerular filtration clearance approaches the glomerular filtration rate (GFR, ~120 mL/min in humans). For molecular weights exceeding 30 kDa, the filtration rate falls off sharply. In addition to size selectivity, charge selectivity has also been observed for glomerular filtration where anionic macromolecules pass through the capillary wall less readily than neutral macromolecules, which in turn pass through less readily than cationic macromolecules (Deen et al. 2001).

The importance of the kidneys as elimination organ could, for example, be shown for interleukin-2, macrophage colony-stimulating factor (M-CSF), and interferon-α (McMartin 1992; Wills and Ferraiolo 1992).

Renal metabolism of peptides and small proteins is mediated through three highly effective processes (Fig. 6.4). As a result, only minuscule amounts of intact protein are detectable in urine.

The first mechanism involves glomerular filtration of larger, complex peptides and proteins followed by reabsorption into endocytic vesicles in the proximal tubule and subsequent hydrolysis into small peptide fragments and amino acids (Maack et al. 1985). This mechanism of elimination has been described for IL-2 (Anderson and Sorenson 1994), IL-11 (Takagi et al. 1995), growth hormone (Johnson and Maack 1977), and insulin (Rabkin et al. 1984).

The second mechanism entails glomerular filtration followed by intraluminal metabolism, predominantly by exopeptidases in the luminal brush border membrane of the proximal tubule. The resulting peptide fragments and amino acids are reabsorbed into the systemic circulation. This route of disposition applies to small linear peptides such as glucagon and LH-RH (Carone and Peterson 1980; Carone et al. 1982). Studies implicate the proton-driven peptide transporters PEPT1 and especially PEPT2 as the main route of cellular uptake of small peptides and peptide-like drugs from the glomerular filtrates (Inui et al. 2000). These high-affinity transport proteins seem to exhibit selective uptake of di- and tripeptides, which implicates their role in renal amino acid homeostasis (Daniel and Herget 1997).

For both mechanisms, glomerular filtration is the dominant, rate-limiting step as subsequent degradation processes are not saturable under physiologic conditions (Maack et al. 1985; Colburn 1991). Under pathologic conditions or very high doses of the therapeutic protein, however, renal tubular reuptake processes may be overwhelmed, resulting in dose-dependent increases in urinary excretion of filtered proteins, as observed for the humanized mAb Fab fragment (48 kDa) idarucizumab. The likely underlying mechanism is temporary saturation of the promiscuous endocytic receptors, megalin and cubilin, on the apical membrane of renal tubular cells that facilitate endocytic uptake of proteins from the tubular lumen (Glund et al. 2018).

Due to this limitation of renal elimination, the renal contribution to the overall elimination of proteins is dependent on the proteolytic degradation of these proteins in other body regions. If metabolic activity for these proteins is high in other body regions, there is only minor renal contribution to total clearance, and it becomes negligible in the presence of unspecific degradation throughout the body. If the metabolic activity is low in other tissues or if distribution to the extravascular space is limited; however, the renal contribution to total clearance may approach 100%.

The involvement of glomerular filtration in the renal metabolism of therapeutic proteins implies that the pharmacokinetics of therapeutic proteins below the molecular weight or hydrodynamic volume cutoff size for filtration will be affected by renal impairment. Indeed, it has been reported that the systemic exposure and elimination half-life increases with decreasing glomerular filtration rate for recombinant human interleukin-10 (18 kDa), recombinant human growth hormone (22 kDa), and the recombinant human IL-1 receptor antagonist anakinra (17.3 kDa). Consistent with these theoretical considerations is also the observation that for mAbs (150 kDa) such as rituximab, cetuximab, bevacizumab, trastuzumab and elotuzumab, no effect of renal impairment on their disposition has been reported (Meibohm and Zhou 2012; Berdeja et al. 2016).

The third mechanism of renal metabolism is peritubular extraction of peptides and proteins from post-glomerular capillaries with subsequent intracellular metabolism. Experiments using radioiodinated growth hormone (^{125}I-rGH) have demonstrated that while reabsorption into endocytic vesicles at the proximal tubule is still the dominant route of disposition, a small percentage of the hormone may be extracted from the peritubular capillaries (Krogsgaard Thomsen et al. 1994). Peritubular transport of proteins and peptides from the basolateral membrane has also been shown for insulin (Nielsen et al. 1987).

Hepatic Protein Metabolism

Aside from nonspecific proteolytic clearance via endothelial cells and the reticuloendothelial system, as well as renal and gastrointestinal metabolism, the liver may also play a major role in the metabolism of some therapeutic proteins, especially for larger proteins. Exogenous as well as endogenous proteins undergo proteolytic degradation to dipeptides and amino acids that are reused for endogenous protein synthesis. Proteolysis usually starts with endopeptidases that attack in the middle part of the protein, and the resulting oligopeptides are then further degraded by exopeptidases. The rate of hepatic metabolism is largely dependent on the specific amino acid sequence of the protein (Meibohm 2004).

The major prerequisite for hepatic protein metabolism in the liver as in any other cells in the body is the active uptake of proteins into the different liver cell types as these protein molecules are unable to passively cross cell membranes due to their high molecular weight and charge. Uptake of larger peptides and proteins can either be facilitated through pino-cytosis as described above or by receptor-mediated endocytosis.

Receptor-mediated endocytosis is usually a clathrin-mediated endocytosis process via relatively unspecific, pro-miscuous membrane receptors (McMahon and Boucrot 2011). In receptor-mediated endocytosis, circulating proteins are recognized by specific membrane-standing receptor proteins. The receptors are usually integral membrane glyco-proteins with an exposed binding domain on the extracellular side of the cell membrane. Many different receptor systems use this same clathrin-mediated endocytosis process. After the binding of the circulating protein to the receptor, the complex is already present or moves to clathrin-coated pit regions in the cell membrane, and the membrane invaginates and pinches off to form an endocytotic coated vesicle that contains the receptor and ligand. This process is referred to as internalization of the drug-receptor complex. The vesicle coat consists of proteins (clathrin, adaptin, and others), which are then removed by an uncoating adenosine triphos-phatase (ATPase). The vesicle parts, the receptor, and the ligands dissociate and are targeted to various intracellular locations. Some receptors, such as the low-density lipopro-tein (LDL), asialoglycoprotein, and transferrin receptors, are known to undergo recycling. Since sometimes several hun-dred cycles are part of a single receptor's lifetime, the associ-ated receptor-mediated endocytosis is oftentimes of high capacity. Other receptors, such as the interferon receptor, undergo degradation. This degradation leads to a decrease in the concentration of receptors on the cell surface (receptor downregulation). Others, such as insulin receptors, for exam-ple, undergo both recycling and degradation (Kompella and Lee 1991).

For glycoproteins, receptor-mediated endocytosis through sugar-recognizing C-type lectin receptors is an efficient hepatic uptake mechanism if a critical number of exposed sugar groups (mannose, galactose, fucose, N-acetylglucosamine, N-acetylgalactosamine, or glucose) is exceeded (Meijer and Ziegler 1993). Important C-type lectin receptors in the liver are the asialoglycoprotein receptor on hepatocytes and the mannose and fucose receptors on Kupffer and liver endothelial cells (Smedsrod and Einarsson 1990; Bu et al. 1992). MAb-based therapeutics usually consist of a heterogeneous mixture of differ-ent glycoforms based on the glycan chains attached to amino acid Asn297 on each heavy chain. Some of these glycoforms that have a high content of mannose (Man5, Man8, Man9) have been described to exhibit a three times faster clearance compared to other glycan structures, presumably via interaction with the hepatic mannose receptor (Falck et al. 2021). Similarly, the asialoglycoprotein receptor recognizes glycosylated proteins with terminal galactose and galactose derivatives and has been implicated in the rapid clearance of erythropoietin, reteplase, lan-oteplase, and clotting factor VIII (Lunghi et al. 2022).

The low-density lipoprotein receptor-related protein (LRP) is a member of the LDL receptor family responsible for endocytosis of several important lipoproteins, proteases, and protease-inhibitor complexes in the liver and other tis-sues (Strickland et al. 1995).

Uptake of proteins by liver cells is followed by transport to an intracellular compartment for metabolism. Proteins internalized into vesicles via an endocytotic mechanism undergo intracellular transport toward the lysosomal com-partment near the center of the cell. There, the endocytotic vehicles fuse with or mature into lysosomes, which are spe-cialized acidic vesicles that contain a wide variety of hydro-lases capable of degrading all biological macromolecules. Proteolysis is started by endopeptidases (mainly cathepsin D) that act on the middle part of the proteins. Oligopeptides—as the result of the first step—are further degraded by exo-peptidases. The resulting amino acids and dipeptides reenter the metabolic pool of the cell. The hepatic metabolism of glycoproteins may occur more slowly than the naked protein because protecting oligosaccharide chains need to be removed first. Metabolized proteins and peptides in lyso-somes from hepatocytes, hepatic sinusoidal cells, and Kupffer cells may be released into the blood. Degraded pro-teins in hepatocyte lysosomes can also be delivered to the bile canaliculus and excreted by exocytosis.

Besides intracellular degradation, a second intracellular pathway for proteins is the direct shuttle or transcytotic path-way (Kompella and Lee 1991). In this case, the endocytotic vesicle formed at the cell surface traverses the cell to the peribiliary space, where it fuses with the bile canalicular membrane, releasing its contents by exocytosis into bile. This pathway bypasses the lysosomal compartment com-pletely. It has been described for polymeric immunoglobulin A but is not assumed to be a major elimination pathway for most protein drugs.

Target-Mediated Protein Metabolism

Therapeutic proteins frequently bind with high affinity to membrane-associated receptors on the cell surface if the receptors are the target structure to which the therapeutic protein is directed. This binding can lead to receptor-mediated uptake by endocytosis and subsequent intracellular lysosomal metabolism. The associated drug disposition behavior in which the binding to the pharmacodynamic tar-get structure affects the pharmacokinetics of a drug com-pound is termed "target-mediated drug disposition" (Levy 1994).

For conventional small-molecule drugs, receptor binding is usually negligible compared to the total amount of drug in the body and rarely affects their pharmacokinetic profile. In contrast, a substantial fraction of a therapeutic protein can be bound to its pharmacologic target structure, for example, a receptor. Target-mediated drug disposition can affect distri-bution as well as elimination processes. Most notably,

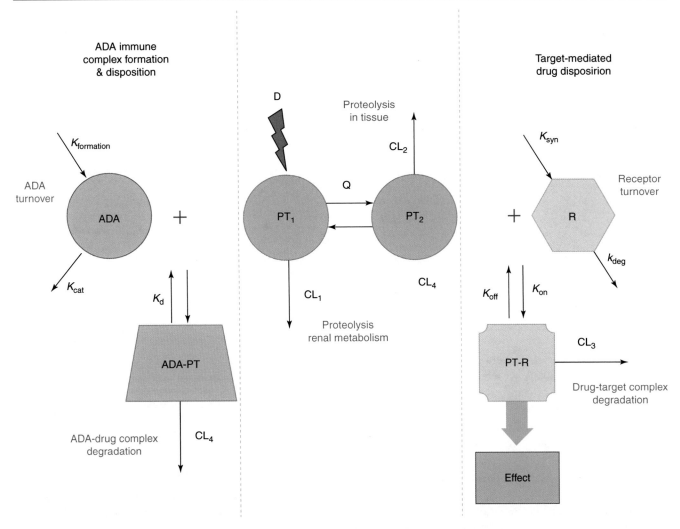

Fig. 6.5 Example of multiple clearance pathways affecting the pharmacokinetics of a typical therapeutic protein. Depicted is a two-compartment pharmacokinetic model with intravenous administration of a dose (*D*), concentrations of the therapeutic protein in the central (*PT$_1$*) and peripheral (*PT$_2$*) compartment, and interdepartmental clearance Q. The pharmacokinetic model includes two clearance pathways, one from the central compartment (*CL$_1$*) representative of, for example, renal metabolism or proteolytic degradation through the reticuloendothelial system and a second proteolytic degradation pathway from the peripheral compartment (*CL$_2$*) representative of, for example, proteolytic degradation through a receptor-mediated endocytosis pathway. Added to these two clearance pathways is on the right side a target-mediated disposition pathway that constitutes interaction of the therapeutic protein with its pharmacologic target receptor, which is in a homeostatic equilibrium of synthesis and degradation (synthesis rate k_{syn} and degradation rate constant k_{deg}). The dynamic equilibrium for the formation of the resulting therapeutic protein-receptor complex (*PT-R*) is determined through the association rate constant k_{on} and the dissociation rate constant k_{off}. The formation of PT-R does not only elicit the pharmacologic effect but also triggers degradation of the complex. Thus, target binding and subsequent PT-R degradation constitute an additional clearance pathway for the therapeutic protein (*CL$_3$*). The left side of the graphic depicts the effect of an immune response to the therapeutic protein resulting in antidrug antibody (*ADA*) formation. Again, the circulating concentration of the ADA is determined by a homeostatic equilibrium between its formation rate ($k_{formation}$) and a catabolic turnover process (rate constant k_{cat}). The ADA response results in the formation of immune complexes with the drug (*ADA-PT*). Dependent on the size and structure of the immune complexes, endogenous elimination pathways though the reticuloendothelial system may be triggered, most likely via Fcγ-receptor mediated endocytosis. Thus, immune complex formation and subsequent degradation may constitute an additional clearance pathway (*CL$_4$*) for therapeutic proteins. (From Chirmule et al. (2012))

receptor-mediated protein metabolism is a frequently encountered elimination pathway for many therapeutic proteins (Meibohm 2004).

Receptor-mediated uptake and metabolism via interaction with these generally high-affinity, low-capacity binding sites is not limited to a specific organ or tissue type. Thus, any tis-

sue including the therapeutically targeted cells that express receptors for the drug can contribute to the elimination of the therapeutic protein (Fig. 6.5) (Zhang and Meibohm 2012).

Since the number of protein drug receptors is limited, receptor-mediated protein metabolism can usually be saturated within therapeutic concentrations, or more specifically

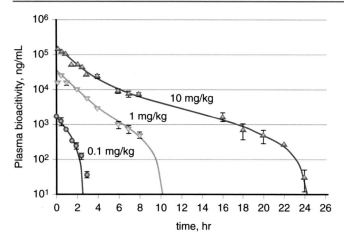

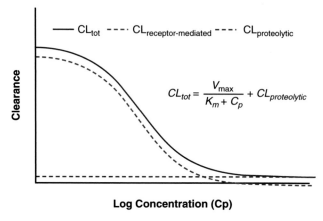

Fig. 6.6 Nonlinear pharmacokinetics of macrophage colony-stimulating factor (M-CSF), presented as measured (*triangles* and *circles*; mean ± SE) and modeled plasma bioactivity–time curves (*lines*) after intravenous injection of 0.1 mg/kg (*n* = 5), 1.0 mg/kg (*n* = 3), and 10 mg/kg (*n* = 8) in rats. Bioactivity is used as a substitute for concentration. (From Bauer et al. (1994), with permission from American Society for Pharmacology and Experimental Therapeutics)

Fig. 6.7 Conceptualization of the concentration-dependent changes in clearance for a therapeutic protein that undergoes receptor- (or target-) mediated elimination displayed in a semi-logarithmic plot. The therapeutic protein is assumed to be eliminated by two parallel clearance processes, one linear, nonsaturable process with relatively low efficiency such as nonspecific proteolytic clearance ($CL_{proteolytic}$), and a second nonlinear, saturable process characterized by Michaelis–Menten-type kinetics and high efficiency at low concentrations such as a receptor- or target-mediated clearance process ($CL_{receptor-mediated}$ = V_{max}/[K_m + Cp]). The total clearance (CL_{tot}) for the therapeutic protein is the sum of the clearances for both pathways. At low concentration, the total clearance is fast and dominated by the target mediated elimination, and the contribution of the nonspecific proteolytic pathway is limited to a low level. With increasing drug plasma concentrations, the receptor-mediated elimination pathway becomes increasingly saturated once the drug concentrations are in the range of or larger than the K_m value for this pathway. Consequently, the total clearance progressively decreases. At very high drug concentrations relative to K_m, the receptor-mediated clearance asymptotically reaches 0, and the total clearance is only determined by the nonspecific proteolytic clearance. V_{max} maximum clearance rate [amount/time]; K_m Michaelis–Menten constant: concentration [amount/volume] at 50% of V_{max}, Cp plasma concentration

at relatively low molar ratios between the protein drug and the receptor (Mager 2006). As a consequence, the elimination clearance of these protein drugs is not constant but dose- and concentration-dependent and decreases with increasing dose or concentration. Thus, receptor-mediated elimination constitutes a major source for nonlinear pharmacokinetic behavior of numerous protein drugs, i.e., systemic exposure to the drug increases more than proportional with increasing dose (Tang et al. 2004).

Recombinant human macrophage colony-stimulating factor (M-CSF), for example, undergoes besides linear renal elimination a nonlinear elimination pathway that follows Michaelis–Menten kinetics and is linked to a receptor-mediated uptake into macrophages. At low concentrations, all M-CSF elimination pathways are active and unsaturated, while at high concentrations nonrenal elimination pathways are saturated resulting in nonlinear pharmacokinetic behavior (Fig. 6.6) (Bauer et al. 1994).

The concentration-dependent change in clearance for therapeutic proteins undergoing receptor-mediated elimination is conceptualized in Fig. 6.7. Nonlinearity in pharmacokinetics resulting from target-mediated drug disposition has also been observed for numerous mAbs, for instance for the anti-EGFR chimeric mAb cetuximab in patients with head-and-neck cancer (Dirks et al. 2008) and the antiproprotein convertase subtilisin/kexin type 9 (PCSK9) mAb evolocumab in patients with hypercholesterolemia (Gibbs et al. 2017). For cetuximab, increasing concentrations lead to the saturation of the available EGFR molecules expressed in the vascular space, the primary distribution space of the mAb, thereby saturating this target-mediated clearance pathway. Similarly, for evolocumab, increasing doses and their corre-

sponding concentrations led to the saturation of the available target PCSK9 in liver, kidneys, and small intestine, resulting in an over-proportional increase in exposure with increasing doses.

Modulation of Protein Disposition by the FcRn Receptor
Immunoglobulin G (IgG)-based mAbs and their derivatives as well as albumin conjugates constitute important classes of therapeutic proteins with many members currently being under development or in therapeutic use. Interaction with the neonatal Fc receptor (FcRn) constitutes a major component in the drug disposition of IgG molecules (Roopenian and Akilesh 2007). FcRn has been well described in the transfer of passive humoral immunity from a mother to her fetus by transferring IgG across the placenta via transcytosis. More importantly, interaction with FcRn in a variety of cells, including endothelial cells and monocytes, macrophages, and other dendritic cells, protects IgG from lysosomal catabolism, and thus constitutes a salvage pathway for IgG molecules that have been internalized in these cell types. This is facilitated by intercepting IgG in the endosomes via a pH-dependent binding process and recy-

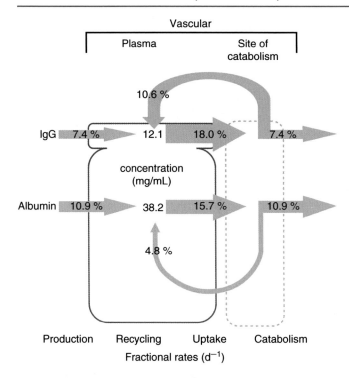

Fig. 6.8 Effect of FcRn-mediated recycling on IgG and albumin turnover in humans expressed as fractional rates. Shown are homeostatic plasma concentrations (12.1 and 38.2 mg/mL), fractional catabolic rates (7.4 and 10.9%/day), the FcRn-mediated fractional recycling rates (10.6 and 4.8%/day), and the fractional production rates (7.4 and 10.9%/day). The figure is to scale: areas for plasma amounts and arrow widths for rates. (From Kim et al. (2007), with permission from Elsevier)

cling it to the systemic circulation (Wang et al. 2008). The interaction with the FcRn receptor thereby prolongs the elimination half-life of IgG, with a more pronounced effect the stronger the binding of the Fc fragment of the antibody is to the receptor: Based on the affinity of this binding interaction, human IgG1, IgG2, and IgG4 have a half-life in humans of 18–21 days, whereas the less strongly bound IgG3 has a half-life of only 7 days, and murine IgG in humans with only very weak binding has a half-life of 1–2 days (Dirks and Meibohm 2010).

Similar to IgG, FcRn is also involved in the disposition of albumin molecules. The kinetics of IgG and albumin recycling are illustrated in Fig. 6.8. For IgG1, approximately 60% of the molecules taken up into lysosomes are recycled, for albumin 30%. As FcRn is responsible for the extended presence of IgG, albumin, and other Fc- or albumin-conjugated proteins in the systemic circulation, modulation of the interaction with FcRn allows to deliberately control the half-life of these molecules (Kim et al. 2007).

Immunogenicity and Protein Pharmacokinetics

The antigenic potential of therapeutic proteins may lead to antibody formation against the therapeutic protein during chronic therapy. This is especially of concern if

animal-derived proteins are applied in human clinical studies but also if human proteins are used in animal studies during preclinical drug development. Chapter 7 discusses in detail the phenomenon of immunogenicity and its consequences for the pharmacotherapy with therapeutic proteins.

The formation of antidrug antibodies (ADA) against a therapeutic protein may not only modulate or even obliterate the biological activity of a protein drug but may also modify its pharmacokinetic profile. In addition, ADA–drug complex formation may lead to immune complex-mediated toxicity, particularly if the complexes get deposited in a specific organ or tissue. Glomerulonephritis has, for example, been observed after deposition of ADA–protein drug complexes in the renal glomeruli of Cynomolgus monkeys after intramuscular administration of recombinant human interferon-γ. Similar to other circulating immune complexes, ADA–protein drug complexes may trigger the regular endogenous elimination pathways for these complexes, which consist of uptake and lysosomal degradation by the reticuloendothelial system. This process has been primarily described for the liver and the spleen and seems to be mediated by Fcγ receptors.

The ADA formation may either lead to the formation of neutralizing or non-neutralizing ADA. Neutralizing ADA bind at or near the target-binding domain of the therapeutic protein and interfere with its ability to bind to its target receptor, thereby reducing its biologic activity. Non-neutralizing ADA bind to regions of the therapeutic protein that are more distant to the target-binding domain and do not interfere with its target binding. Independent on whether ADA are neutralizing or non-neutralizing, they can both modulate the therapeutic protein's pharmacokinetics: Clearing ADA increase the clearance of the therapeutic protein, whereas sustaining ADA decrease the clearance of the therapeutic protein (Fig. 6.9). For clearing ADA, the immune complex formation triggers elimination via the reticuloendothelial system, which constitutes an additional elimination pathway for the protein (Fig. 6.5). This increase in clearance for the protein results in a decreased systemic exposure and reduced elimination half-life, which ultimately leads to reduced activity also for non-neutralizing ADA. A clearing effect of ADA is often observed for large therapeutic proteins such as mAbs (Richter et al. 1999).

For sustaining ADA, the immune complex formation does not trigger the regular endogenous elimination processes, but serves as a storage depot for the protein, thereby reducing its clearance, increasing its systemic exposure, prolonging its half-life, and thereby increasing its activity in case of non-neutralizing ADA. This behavior has often been described for small therapeutic proteins where the immune complex formation, for example, prevents glomerular filtration and subsequent tubular metabolism. The elimination half-life of

Fig. 6.9 Effect of antidrug antibody (ADA) formation on the pharmacokinetics and pharmacodynamics of therapeutic proteins. *CL* clearance

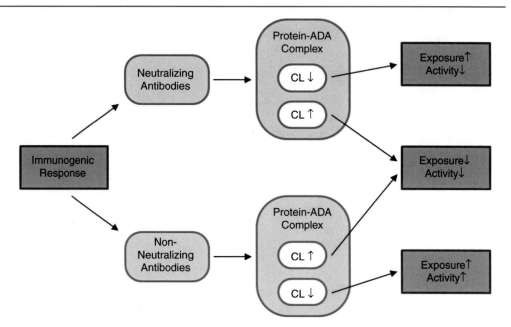

the therapeutic protein is then often increased to approach that of IgG (Chirmule et al. 2012).

Whether ADA–protein drug complex formation results in clearing or sustaining effects seems to be a function of its physicochemical and structural properties, including size, antibody class, ADA-antigen ratio, characteristics of the antigen, and location of the binding epitopes. For example, both an increased and decreased clearance is possible as ADA effect for the same protein, dependent on the dose level administered. At low doses, protein–antibody complexes delay clearance because their elimination is slower than the unbound protein. In contrast, at high doses, higher levels of protein–antibody complex result in the formation of larger aggregates, which are cleared more rapidly than the unbound protein.

The enhancement of the clearance of the cytokine interleukin-6 (IL-6) via administration of cocktails of three anti-IL-6 mAbs was suggested as a therapeutic approach in cytokine-dependent diseases like multiple myeloma, B-cell lymphoma, and rheumatoid arthritis (Montero-Julian et al. 1995). The authors could show that, while the binding of one or two antibodies to the cytokine led to stabilization of the cytokine, simultaneous binding of three anti-IL-6 antibodies to three distinct epitopes induced rapid uptake of the complex by the liver and thus mediated a rapid elimination of IL-6 from the systemic circulation.

It should be emphasized that ADA formation is a polyclonal and usually relatively unspecific immune response to the therapeutic protein, with formation of different antibodies with variable binding affinities and epitope specificities, and that this ADA formation with its multiple-involved antibody species is different in different patients. Thus, reliable prediction of ADA formation and effects remains elusive at the current time (Chirmule et al. 2012).

The immunogenicity of therapeutic proteins is also dependent on the route of administration. Extravascular injection is known to stimulate antibody formation more than IV application, which is most likely caused by the increased immunogenicity of protein aggregates and precipitates formed at the injection site. Further details on these aspects of immunogenicity are discussed in Chap. 7.

Species Specificity and Allometric Scaling
Proteins often exhibit distinct species specificity with regard to structure and activity. Proteins with identical physiological function may have different amino acid sequences in different species and may have no activity or be even immunogenic if used in a different species. The extent of glycosylation of a protein molecule is another factor of species differences, e.g., for interferon-α or erythropoietin, which may not only alter its efficacy and immunogenicity (see Chap. 7) but also the drug's clearance.

Projecting human pharmacokinetic behavior for therapeutic proteins based on data in preclinical species is often performed using allometric approaches. Allometry is a methodology used to relate morphology and body function to the size of an organism. Allometric scaling is an empirical technique to predict body functions based on body size. Allometric scaling has found wide application in drug development, especially to predict pharmacokinetic parameters in humans based on the corresponding parameters in several animal species and the body size differences among these species and humans. Multiple allometric scaling approaches have been described with variable success rates, predominantly during the transition from preclinical to clinical drug development (Dedrick 1973; Boxenbaum 1982). In the most frequently used approach, pharmacokinetic parameters between different species are related via body weight using a power function:

$$P = a \cdot W^b$$

where P is the pharmacokinetic parameter scaled, W is the body weight in kg, a is the allometric coefficient, and b is the allometric exponent. a and b are specific constants for each parameter of a compound. General tendencies for the allometric exponent are 0.75 for biological rates (i.e., clearance, flow rates), 1 for volumes of distribution, and 0.25 for half-lives. More recently, allometric approaches are being complemented by physiologically based pharmacokinetic modeling.

For most traditional small-molecule drugs, allometric scaling is often imprecise, especially if hepatic metabolism is a major elimination pathway and/or if there are interspecies differences in metabolism. For peptides and proteins, however, allometric scaling has frequently proven to be much more precise and reliable if their disposition is governed by relatively unspecific proteolytic degradation pathways. The reason is probably the similarity in handling peptides and proteins among different mammalian species (Wills and Ferraiolo 1992). Clearance and volume of distribution of numerous therapeutically used proteins like growth hormone or t-PA follow a well-defined, weight-dependent physiologic relationship between lab animals and humans. This allows relatively precise quantitative predictions for their pharmacokinetic behavior in humans based on preclinical findings (Mordenti et al. 1991).

Figure 6.10, for example, shows allometric plots for the clearance and volume of distribution of a P-selectin antagonist, P-selectin glycoprotein ligand-1, for the treatment of P-selectin-mediated diseases such as thrombosis, reperfusion injury, and deep vein thrombosis. The protein's human pharmacokinetic parameters could accurately be predicted

using allometric power functions based on data from four species: mouse, rat, monkey, and pig (Khor et al. 2000).

More recent work on scaling the pharmacokinetics of mAbs has suggested that allometric scaling from one nonhuman primate species, in this case the Cynomolgus monkey, using an allometric exponent of 0.85 might be superior to traditional allometric scaling approaches (Deng et al. 2011). Especially allometric extrapolation of pharmacokinetic parameters from mice to humans for mAbs has been challenging because murine FcRn has a substantially higher binding affinity to human IgG molecules compared to human FcRn (Ober et al. 2001). Thus, preclinical pharmacokinetic experiments in mice with humanized or human mAbs result in arbitrarily low and thus overly optimistic nonspecific proteolytic clearance values in mice that when allometrically scaled largely underestimate human clearance. Transgenic animal models such as the Tg32 and Tg276 mouse models that express the human instead of the murine FcRn are increasingly used to circumvent this problem but are also challenged by other resulting effects such as arbitrarily low endogenous immunoglobulin levels (Ko et al. 2021).

In any case, successful allometric scaling seems so far largely limited to unspecific protein elimination pathways. Once interactions with specific receptors are involved in drug disposition, for example, in receptor-mediated processes or target-mediated drug disposition, then allometric approaches oftentimes have large prediction error margins or even fail to scale drug disposition of therapeutic proteins across species due to differences in binding affinity and specificity, as well as expression and turnover kinetics of the involved receptors and targets in different species. In this situation, it becomes especially important to only consider for scaling preclinical pharmacokinetic data from "relevant" animal species for which the therapeutic protein shows

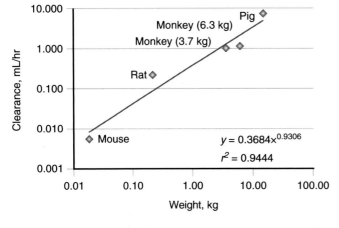
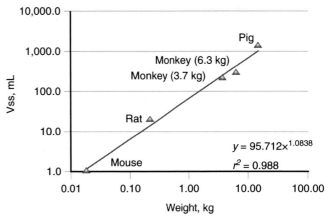

Fig. 6.10 Allometric plots of the pharmacokinetic parameter clearance and volume of distribution at steady state (V_{ss}) for the P-selectin antagonist rPSGL-Ig. Each data point within the plot represents an averaged value of the respective pharmacokinetic parameter in one of five species: mouse, rat, monkey (3.7 kg), monkey, (6.3 kg), and pig, respectively. The *solid line* is the best fit with a power function to relate pharmacokinetic parameters to body weight. (Khor et al. 2000; with permission from American Society for Pharmacology and Experimental Therapeutics)

cross-reactivity between animal and human receptors or targets. Dong et al. (2011) provided practical examples how unspecific and receptor-mediated elimination pathways for the same therapeutic protein can independently be scaled to improve human clearance predictions.

It needs to be emphasized that allometric scaling techniques are useful tools for predicting a dose that will assist in the planning of dose-ranging studies, including first-in-human studies, but are not a replacement for such studies. The advantage of including such dose prediction in the protocol design of dose-ranging studies is that a smaller number of doses need to be tested before finding the final dose level. Interspecies dose predictions simply narrow the range of doses in the initial pharmacological efficacy studies, the animal toxicology studies, and the human safety and efficacy studies. More recently, physiologically-based pharmacokinetic modeling has become more widely used to make more mechanistically based and accurate predictions of human pharmacokinetic behavior of therapeutic proteins (Diao and Meibohm 2013; Glassman and Balthasar 2016).

Chemical Modifications for Optimizing the Pharmacokinetics of Therapeutic Proteins

In recent years, approaches modifying the molecular structure of therapeutic proteins have repeatedly been applied to affect the immunogenicity, pharmacokinetics, and/or pharmacodynamics of protein drugs (Kontermann 2012). These approaches include the addition, deletion, or exchange of selected amino acids within the protein's sequence, synthesis of truncated proteins with a reduced amino acid sequence, glycosylation or deglycosylation, and covalent linkage to polymers (Veronese and Caliceti 2006). The latter approach has been used for several therapeutic proteins by linking them to polyethylene glycol (PEG) molecules of various chain lengths in a process called PEGylation (Caliceti and Veronese 2003).

The conjugation of high polymeric mass to protein drugs is generally aimed at preventing the protein being recognized by the immune system as well as reducing its elimination via glomerular filtration or proteolytic enzymes, thereby prolonging the oftentimes relatively short elimination half-life of endogenous proteins. Conjugation of protein drugs with PEG chains increases their molecular weight, but because of the attraction of water molecules by PEG even more their hydrodynamic volume, this in turn results in a reduced renal clearance and restricted volume of distribution. PEGylation can also shield antigenic determinants on the protein drug from detection by the immune system through steric hindrance (Walsh et al. 2003). Similarly, amino acid sequences sensitive toward proteolytic degradation may be shielded against protease attack. By adding a large, hydrophilic molecule to the protein, PEGylation can also increase drug solubility (Molineux 2003).

PEGylation has been used to improve the therapeutic properties of numerous therapeutic proteins including interferon-α, asparaginase, and filgrastim. The therapeutic application of L-asparaginase in the treatment of acute lymphoblastic leukemia has been hampered by its strong immunogenicity with allergic reactions occurring in 33–75% of treated patients in various studies. The development of pegaspargase, a PEGylated form of L-asparaginase, is a successful example for overcoming this high rate of allergic reactions toward L-asparaginase using PEG conjugation techniques (Graham 2003). Pegaspargase is well-tolerated compared to L-asparaginase, with only 3–10% of the treated patients experiencing clinical allergic reactions.

Pegfilgrastim is the PEGylated version of the granulocyte colony-stimulating factor filgrastim, which is administered for the management of chemotherapy-induced neutropenia. PEGylation minimizes filgrastim's renal clearance by glomerular filtration, thereby making neutrophil-mediated clearance the predominant route of elimination. Thus, PEGylation of filgrastim results in so-called self-regulating pharmacokinetics since pegfilgrastim has a reduced clearance and thus prolonged half-life and more sustained duration of action in a neutropenic compared to a normal patient because only few mature neutrophils are available to mediate its elimination (Zamboni 2003).

The hematopoietic growth factor darbepoetin-α is an example of a chemically modified endogenous protein with altered glycosylation pattern. It is a glycosylation analog of human erythropoietin, with two additional N-linked oligosaccharide chains (five in total) (Mould et al. 1999). The additional N-glycosylation sites were made available through substitution of five amino acid residues in the peptide backbone of erythropoietin, thereby increasing the molecular weight from 30 to 37 kDa. Darbepoetin-α has a substantially modified pharmacokinetic profile compared to erythropoietin, resulting in a threefold longer serum half-life that allows for reduced dosing frequency. More details on hematopoietic growth factors, including erythropoietin and darbepoetin-α, are provided in Chap. 17.

Pharmacodynamics of Therapeutic Proteins

Therapeutic proteins are usually highly potent compounds with steep dose–effect curves as they are targeted therapies toward a specific, well-described pharmacologic structure or mechanism. Thus, a careful characterization of the concentration–effect relationship, i.e., the pharmacodynamics, is especially desirable for therapeutic proteins (Tabrizi and Roskos 2006; Mould and Meibohm 2016). Combination of pharmacodynamics with pharmacokinetics by integrated

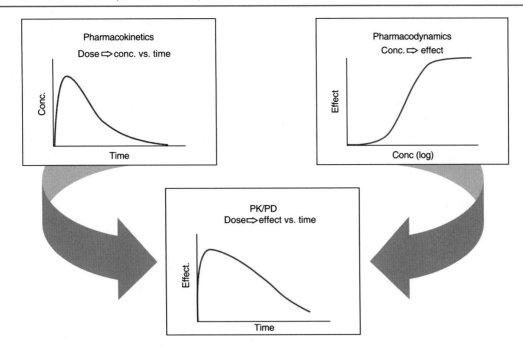

Fig. 6.11 General concept of PK/PD modeling. Pharmacokinetic-pharmacodynamic (PK/PD) modeling combines a pharmacokinetic model component that describes the time course of drug in plasma and a pharmacodynamic model component that relates the plasma concentration to the drug effect in order to describe the time course of the effect intensity resulting from the administration of a certain dosage regimen. (From Derendorf and Meibohm (1999))

pharmacokinetic-pharmacodynamic modeling (PK/PD modeling) adds an additional level of complexity that allows furthermore characterization of the dose–exposure–response relationship of a drug and a continuous description of the time course of effect intensity directly resulting from the administration of a certain dosage regimen (Fig. 6.11) (Meibohm and Derendorf 1997; Derendorf and Meibohm 1999).

PK/PD modeling is a technique that combines the two classical pharmacologic disciplines of pharmacokinetics and pharmacodynamics. It integrates a pharmacokinetic and a pharmacodynamic model component into one set of mathematical expressions that allows the description of the time course of effect intensity in response to administration of a drug dose. This so-called integrated PK/PD model allows deriving pharmacokinetic and pharmacodynamic model parameters that characterize the dose–concentration–effect relationship for a specific drug based on measured concentration and effect data. In addition, it allows simulation of the time course of effect intensity for dosage regimens of a drug beyond actually measured data, within the constraints of the validity of the model assumptions for the simulated condition. Addition of a statistical model component describing inter- and intraindividual variation in model parameters allows expanding PK/PD models to describe time courses of effect intensity not only for individual subjects but also for whole populations of subjects.

Integrated PK/PD modeling approaches have widely been applied for the characterization of therapeutic proteins (Tabrizi and Roskos 2006). Embedded in a model-informed drug development paradigm (EfpiaMidWorkgroup et al. 2016), modeling and simulation based on integrated PK/PD does not only provide a comprehensive summary of the available data but also enables to test competing hypotheses regarding processes altered by the drug, allows making predictions of drug effects under new conditions, and facilitates to estimate inaccessible system variables (Meibohm and Derendorf 1997; Mager et al. 2003; Liu et al. 2021).

Mechanism-based PK/PD modeling appreciating the physiological events involved in the elaboration of the observed effect has been promoted as superior modeling approach as compared to empirical modeling, especially because it does not only describe observations but also offers some insight into the underlying biological processes involved and thus provides flexibility in extrapolating the model to other clinical situations (Levy 1994; Derendorf and Meibohm 1999; Suryawanshi et al. 2010). Since the molecular mechanism of action of a therapeutic protein is generally well understood, it is often straightforward to transform this available knowledge into a mechanism-based PK/PD modeling approach that appropriately characterizes the real physiological process leading to the drug's therapeutic effect.

The relationship between exposure and response may be either simple or complex, and thus obvious or hidden. However, if no simple relationship is obvious, it would be misleading to conclude a priori that no relationship exists at all rather than that it is not readily apparent (Levy 1986).

The application of PK/PD modeling is beneficial in all phases of preclinical and clinical drug development and has been endorsed by the pharmaceutical industry, academia, and regulatory agencies (Peck et al. 1994; Lesko et al. 2000; Sheiner and Steimer 2000; Meibohm and Derendorf 2002; Lesko 2007; Zhu et al. 2019). Thus, PK/PD concepts and model-informed drug development play a pivotal role especially in the drug development process for biologics, and their widespread application supports a scientifically driven, evidence-based, and focused product development for therapeutic proteins (Zhang et al. 2008; Mould and Meibohm 2016).

While a variety of PK/PD modeling approaches has been employed for biologics, we will in the following focus on five classes of approaches to illustrate the challenges and complexities, but also opportunities to characterize the pharmacodynamics of therapeutic proteins:

- Direct link PK/PD models
- Indirect link PK/PD models
- Indirect response PK/PD models (also referred to as turnover models)
- Cell life span models
- Complex response models

It should not be unmentioned, however, that PK/PD models for therapeutic proteins are not only limited to continuous responses as shown in the following, but are also used for binary or graded responses. Binary responses are responses with only two outcome levels where a condition is either present or absent, e.g., dead versus alive. Graded or categorical responses have a set of predefined outcome levels, which may or may not be ordered, for example, the categories "mild," "moderate," and "severe" for a disease state. Lee et al. (2003), for example, used a logistic PK/PD modeling approach to link cumulative AUC of the anti-TNF-α protein etanercept with a binary response, the American College of Rheumatology response criterion of 20% improvement (ARC20) in patients with rheumatoid arthritis (Lee et al. 2003).

Direct Link PK/PD Models
The concentration of a therapeutic protein is usually only quantified in plasma, serum, or blood, while the magnitude of the observed response is determined by the concentration of the protein drug at its effect site, the site of action in the target tissue (Meibohm and Derendorf 1997). Effect site concentrations, however, are usually not accessible for measurement, and plasma, serum, or blood concentrations are usually used as their substitute. The relationship between the drug concentration in plasma and at the effect site may either be

constant or undergo time-dependent changes. If equilibrium between both concentrations is rapidly achieved or the site of action is within plasma, serum, or blood, there is practically a constant relationship between both concentrations with no temporal delay between plasma and effect site. In this case, measured plasma concentrations can directly serve as input for a pharmacodynamic model (Fig. 6.12). The most frequently used direct link pharmacodynamic model is a sigmoid E_{max} model:

$$E = \frac{E_{max} \cdot Cp^n}{EC_{50}^n + Cp^n}$$

with E_{max} as maximum achievable effect, Cp as drug concentration in plasma, and EC_{50} the concentration of the drug that produces half of the maximum effect. The Hill coefficient n is an empirical shape factor that allows for an improved fit of the relationship to the observed data. As represented by the equation for the sigmoid E_{max} model, a direct link model directly connects measured concentration to the observed effect without any temporal delay (Derendorf and Meibohm 1999).

A direct link model was, for example, used to relate the serum concentration of the antihuman immunoglobulin E (IgE) antibody CGP 51901 for the treatment of seasonal allergic rhinitis to the reduction of free IgE via an inhibitory E_{max} model (Fig. 6.13) (Racine-Poon et al. 1997). It should be noted that the peak and trough concentrations and effects

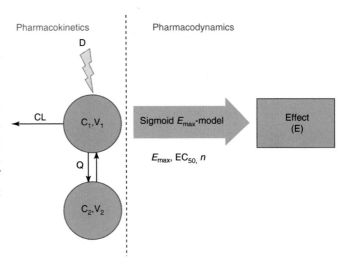

Fig. 6.12 Schematic of a typical direct link PK/PD model. The PK model is a typical two-compartment model with a linear elimination clearance from the central compartment (*CL*) and a distributional clearance (*Q*). C_1 and C_2 are the concentrations in the central and peripheral compartments, and V_1 and V_2 are their respective volumes of distribution. The effect (*E*) is directly linked to the concentration in the central compartment C_1 via a sigmoid E_{max} model. The sigmoid E_{max} relationship is characterized by the pharmacodynamic parameters E_{max}, the maximum achievable effect, EC_{50}, the concentration of the drug that produced half of the maximum effect, and the Hill coefficient n as via the sigmoid E_{max} equation

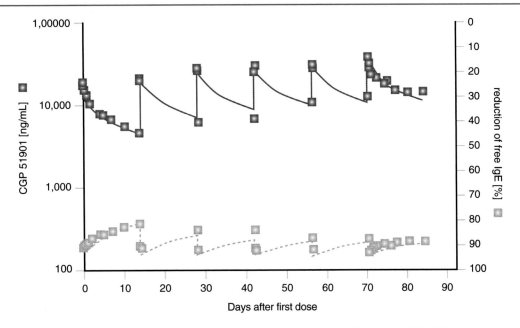

Fig. 6.13 Observed (■) and model-predicted (——) serum concentration of the antihuman IgE antibody CGP 51901 and observed (■) and model-predicted (----) reduction of free IgE in one representative patient, given six IV doses of 60 mg biweekly. The predictions were modeled with a direct link PK/PD model. (Modified from Racine-Poon et al. (1997); with permission from Macmillan Publishers Ltd.)

are directly related and thus occur at the same times, respectively, without time delay. Similarly, a direct link model was used to relate the effect of recombinant interleukin-10 (IL-10) on the ex vivo release of the pro-inflammatory cytokines TNF-α and interleukin-1β in LPS-stimulated leukocytes (Radwanski et al. 1998). In the first case, the site of action and the sampling site for concentration measurements of the therapeutic protein were identical, i.e., in plasma, and so the direct link model was mechanistically well justified. In the second case, the effect was dependent on the IL-10 concentration on the cell surface of leukocytes where IL-10 interacts with its target receptor. Again sampling fluid and effect site were in instant equilibrium.

Indirect Link PK/PD Models

The concentration–effect relationship of many protein drugs, however, cannot be described by direct link PK/PD models, but is characterized by a temporal dissociation between the time courses of plasma concentration and effect. In this case, plasma concentration maxima occur before effect maxima; effect intensity may increase despite decreasing plasma concentrations and may persist beyond the time when drug concentrations in plasma are no longer detectable. The relationship between measured concentration and observed effect follows a counterclockwise hysteresis loop. This phenomenon can either be caused by an indirect response mechanism (see next section) or by a distributional delay between the drug concentrations in plasma and at the effect site.

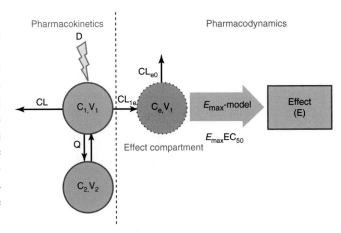

Fig. 6.14 Schematic of a typical indirect link PK/PD model. A hypothetical effect compartment is linked to the central compartment of a two-compartment pharmacokinetic model. The concentration in the effect compartment (C_e) drives the intensity of the pharmacodynamic effect (E) via an E_{max} relationship. CL_{1e} is the transfer clearance from the central to the effect compartment, CL_{e0} the equilibrium clearance for the effect compartment. All other PK and PD parameters are identical to those used in Fig. 6.12

The latter one can conceptually be described by an indirect link model, which attaches a hypothetical effect compartment to a pharmacokinetic compartment model (Fig. 6.14). The effect compartment addition to the pharmacokinetic model does not account for mass balance, i.e., no actual mass transfer is implemented in the pharmacokinetic part of the PK/PD model. Instead, drug transfer with respect to the effect compartment is defined by the time course of the effect itself (Sheiner et al. 1979, Holford and Sheiner 1982). The effect-compartment approach, however, is necessary, as the effect site can be viewed as a small part of a pharmacokinetic compartment that from a pharmacokinetic point of view cannot be distinguished from other tissues within that compartment. The concentration in the effect compartment represents the active drug concentration at the effect site that is slowly equilibrating with the plasma and is usually linked to the effect via an E_{max} model.

Although this PK/PD model is constructed with tissue distribution as the reason for the delay of the effect, the distribution clearance to the effect compartment can be interpreted differently, including other reasons of delay, such as transduction processes and secondary post-receptor events.

Human regular U-500 insulin has recently been developed for insulin-resistant and high-dose insulin-treated patients to provide the ability of administering large doses (500 U/mL) at one-fifth the volume of that of the previously highest concentrated dosage form, human regular U-100 insulin. In order to explore the effect-time course after administration of once-daily, twice-daily, and thrice-daily administration of U-500 insulin, a PK/PD model was developed based on single-dose euglycemic clamp studies in healthy individuals and patients with type I diabetes. Insulin concentrations were related to glucose infusion rate as measure of pharmacodynamics effect via an effect compartment model (de la Pena et al. 2014). Model-based simulations of the different administration frequencies at steady state (Fig. 6.15) suggest that BID and TID dosing may provide adequate insulin action throughout the day, but QD dosing

leads to fluctuations in effect that may increase the risk for hypoglycemia and may thus not be adequate for use as basal insulin therapy.

Indirect Response PK/PD Models

The effect of most therapeutic proteins, however, is not mediated via a direct interaction between drug concentration at the effect site and response systems but frequently involves several transduction processes that include at their rate-limiting step the stimulation or inhibition of a physiologic process, for example, the synthesis or degradation of a molecular response mediator like a hormone or cytokine. In these cases, the time courses of plasma concentration and effect are also dissociated resulting in counterclockwise hysteresis for the concentration–effect relationship, but the underlying cause is not a distributional delay as for the indirect link models, but a time-consuming indirect response mechanism (Meibohm and Derendorf 1997).

Indirect response models generally describe the effect on a representative response parameter via the dynamic equilibrium between increase or synthesis and decrease or degradation of the response, with the former being a zero-order and the latter a first-order process (Fig. 6.16). The response itself can be modulated in one of four basic variants of the model. In each variant, the synthesis or degradation process of the response is either stimulated or inhibited as a function of the effect site concentration. A stimulatory or inhibitory E_{max} model is used to describe the drug effect on the synthesis or degradation of the response (Dayneka et al. 1993; Sharma and Jusko 1998; Sun and Jusko 1999).

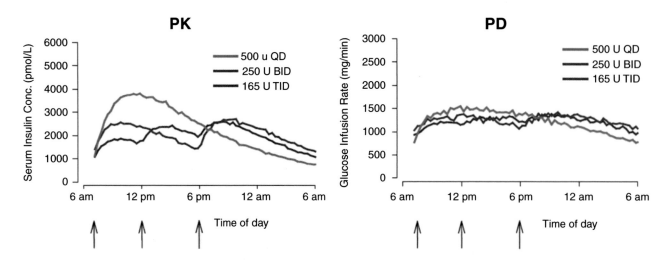

Fig. 6.15 PK/PD model-based simulations of different dosing regimens of U-500 insulin during 24 h at steady-state: 500 U QD (*green*), 250 U BID (*blue*), 165 U TID (*red*). Arrows represent dose administration times: for QD at 7 am, BID at 7 am and 6 pm, and TID at 7 am, 12 noon, and 6 pm. The PK panel on the left shows the resulting serum insulin concentration–time profiles, the PD panel on the right side the time course of the glucose infusion rate needed to maintain euglycemia. (From de la Peña et al. (2014))

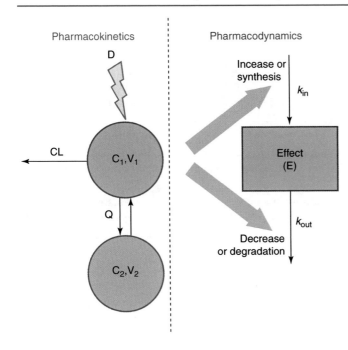

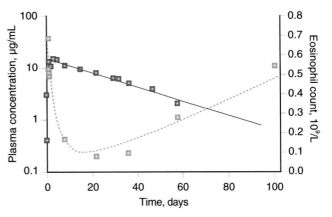

Fig. 6.17 Model-predicted and observed plasma concentration (observed, *blue squares*; predicted, *solid blue line*) and eosinophil count (observed, *orange squares*; predicted, *dashed orange line*) following SC administration of 1 mg/kg of the anti-IL-5 humanized monoclonal antibody SB-240563 in a Cynomolgus monkey. A mechanism-based indirect response PK/PD model was used to describe eosinophil count as a function of SB-240563 plasma concentration. The reduction in eosinophil count in peripheral blood (as effect E) was modeled as a reduction of the recruitment of eosinophils from the bone, i.e., an inhibition of the production rate k_{in} using the indirect response model of subtype I (see Fig. 6.16). (Zia-Amirhosseini et al. (1999); with permission from American Society for Pharmacology and Experimental Therapeutics)

Indirect response model subtypes

Subtype I:
inhibition of synthesis (k_{in})

$$\frac{dE}{dt} = k_{in}\cdot\left[1 - \frac{C_1}{EC_{50}+C_1}\right] - k_{out}\cdot E$$

Subtype II:
inhibition of degradation (k_{out})

$$\frac{dE}{dt} = k_{in} - k_{out}\cdot\left(1 - \frac{C_1}{EC_{50}+C_1}\right)\cdot E$$

Subtype III:
stimulation of synthesis (k_{in})

$$\frac{dE}{dt} = k_{in}\cdot\left[1 + \frac{E_{max}\cdot C_1}{EC_{50}+C_1}\right] - k_{out}\cdot E$$

Subtype IV:
stimulation of degradation (k_{out})

$$\frac{dE}{dt} = k_{in} - k_{out}\cdot\left(1 + \frac{E_{max}\cdot C_1}{EC_{50}+C_1}\right)\cdot E$$

Fig. 6.16 Schematic of a typical indirect response PK/PD model. The effect measure (E) is maintained by a dynamic equilibrium between an increase or synthesis and a decrease or degradation process. The former is modeled by a zero-order process with rate constant k_{in}, the latter by a first-order process with rate constant k_{out}. Thus, the rate of change in effect (dE/dt) is expressed as the difference between synthesis rate (k_{in}) and degradation rate (k_{out} times E). Drug concentration (C_1) can stimulate or inhibit the synthesis or the degradation process for the effect (E) via an E_{max} relationship using one of four subtypes (model I, II, III or IV) of the indirect response model. The pharmacokinetic model and all other PK and PD parameters are identical to those used in Fig. 6.12

As indirect response models appreciate the underlying physiological events involved in the elaboration of the observed drug effect, their application is often preferred in PK/PD modeling as they have a mechanistic basis on the molecular and/or cellular level that often allows for extrapolating the model to other clinical situations.

An indirect response model was, for example, used in the evaluation of SB-240563, a humanized mAb directed towards IL-5 in monkeys (Zia-Amirhosseini et al. 1999). IL-5 appears to play a significant role in the production, activation, and maturation of eosinophils. The delayed effect of SB-240563 on eosinophils is consistent with its mechanism of action via binding to and thus inactivation of IL-5. It was modeled using an indirect response model with inhibition of the production of response (eosinophil count) (Fig. 6.17). The obtained low EC_{50} value for reduction of circulating eosinophils combined with a long terminal half-life of the therapeutic protein of 13 days suggests the possibility of an infrequent dosing regimen for SB-240563 in the pharmacotherapy of disorders with increased eosinophil function, such as asthma.

Indirect response models were also used for the effect of growth hormone on endogenous IGF-1 concentration (Sun et al. 1999), for the effect of epoetin-α on two response parameters, free ferritin concentration, and soluble transferrin receptor concentration (Bressolle et al. 1997) as well as for the effect of the alirocumab, a proprotein convertase subtilisin kexin type 9 (PCSK9) targeting mAb, on low-density lipoprotein cholesterol (Nolain et al. 2022). Similarly, a modified indirect response model was used to relate the concentration of the humanized anti-factor IX antibody SB-249417 to factor IX activity in Cynomolgus monkeys as well as humans (Benincosa et al. 2000; Chow et al. 2002). The drug effect in this model was introduced by interrupting the natural degradation of factor IX by sequestration of factor IX by the antibody.

Cell Life Span Models

A sizable number of therapeutic proteins exert their pharmacologic effect through direct or indirect modulation of blood and/or immune cell types. For these kinds of therapeutics, cell life span models have been proven useful to capture their exposure–response relationship and describe and predict drug effects (Perez-Ruixo et al. 2005). Cell life span models are mechanism-based, physiologic PK/PD models that are established based on the sequential maturation and life span-driven cell turnover of their affected cell types and progenitor cell populations. Cell life span models are especially widely used for characterizing the dose–concentration–effect relationship of hematopoietic growth factors aimed at modifying erythropoiesis, granulopoiesis, or thrombopoiesis (Perez-Ruixo et al. 2005; Agoram et al. 2006). The fixed physiologic time span for the maturation of precursor cells is the major reason for the prolonged delay between drug administration and the observed response, i.e., change in the cell count in peripheral blood. Cell life span models accommodate this sequential maturation of several precursor cell populations at fixed physiologic time intervals by a series of transit compartments linked via first- or zero-order processes with a common transfer rate constant.

A cell life span model was, for example, used to describe the effect of a multiple dose regimen of erythropoietin 600 IU/kg given once weekly by SC injection (Ramakrishnan et al. 2004). The process of erythropoiesis and the applied PK/PD approach including a cell life span model are depicted in Figs. 6.18 and 6.19, respectively. Erythropoietin is known to stimulate the production and release of reticulocytes from the bone marrow. The erythropoietin effect was modeled as stimulation of the maturation of two progenitor cell populations (P1 and P2 in Fig. 6.18), including also a feedback inhibition between erythrocyte count and progenitor proliferation. Development and turnover of the subsequent populations of reticulocytes and erythrocytes was modeled, taking into account their life spans as listed in Fig. 6.18. The hemoglobin concentration as pharmacodynamic target parameter was calculated from erythrocyte and reticulocyte counts and hemoglobin content per cell. Figure 6.20 shows the resulting time courses in reticulocyte count, erythrocyte count, and hemoglobin concentration.

Complex Response Models

Since the effect of most therapeutic proteins is mediated via complex regulatory physiologic processes including feedback mechanisms and/or tolerance phenomena, some PK/PD models that have been described for protein drugs are much more sophisticated than the four classes of models previously discussed.

One example of such a complex modeling approach is the cytokinetic model used to describe the effect of pegfilgrastim on the granulocyte count in peripheral blood (Roskos et al. 2006; Yang 2006). Pegfilgrastim is a PEGylated form of the human granulocyte colony-stimulating factor (G-CSF) analog filgrastim. Pegfilgrastim, like filgrastim and G-CSF, stimulates the activation, proliferation, and differentiation of neutrophil progenitor cells and enhances the functions of mature neutrophils (Roskos et al. 2006). Pegfilgrastim is mainly used as supportive care to ameliorate and enhance recovery from neutropenia secondary to cancer chemotherapy regimens. As already discussed in the section on PEGylation, pegfilgrastim follows target-mediated drug disposition with saturable receptor-mediated endocytosis by neutrophils as major elimination pathway and a parallel first-order process as minor elimination pathway (Fig. 6.21). The clearance for the receptor-mediated pathway is determined by the absolute neutrophil count (ANC), the sum of the peripheral blood band cell, and segmented neutrophil populations.

A maturation-structured cytokinetic model of granulopoiesis was established to describe the relationship between pegfilgrastim serum concentration and neutrophil count (Fig. 6.21). The starting point is the production of metamyelocytes from mitotic precursors. Subsequent maturation stages are captured as band cells and segmented neutrophils in the bone marrow. Each maturation stage is modeled by three sequential transit compartments. Pegfilgrastim concentrations are assumed to increase ANC by stimulating mitosis and mobilization of band cells and segmented neutrophils from the bone marrow into the systemic circulation. Pegfilgrastim also promotes rapid margination of peripheral blood neutrophils, i.e., adhesion to blood vessels; this effect is modeled as an expansion of neutrophil dilution volume.

Figure 6.22 shows observed and modeled pegfilgrastim concentration time and ANC time profiles after escalating single SC dose administration of pegfilgrastim. The presented PK/PD model for pegfilgrastim allowed determining its EC_{50} for the effect on ANC. Based on this EC_{50} value and the obtained pegfilgrastim plasma concentrations, it was concluded that a 100 µg/kg dose was sufficient to reach the maximum therapeutic effect of pegfilgrastim on ANC (Roskos et al. 2006; Yang 2006).

Fig. 6.18 Process of erythropoiesis. Erythropoietin stimulates the proliferation and differentiation of the erythrocyte progenitors (*BFU* burst-forming unit erythroid, *CFUe* colony-forming unit erythroid), as well as the erythroblasts in the bone marrow. The life spans (τ) of the various cell populations are indicated at the right. (From Ramakrishnan et al. (2004), with permission from John Wiley & Sons, Inc. Copyright American College of Clinical Pharmacology 2004)

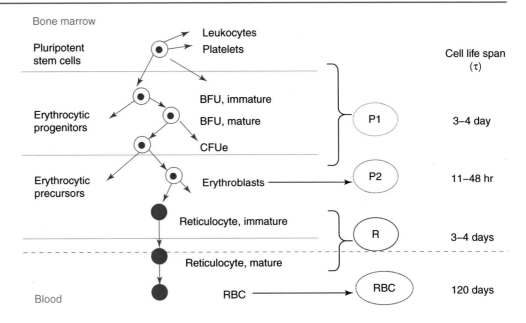

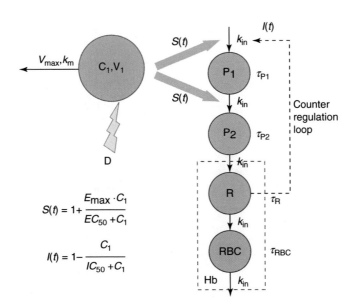

Fig. 6.19 A PK/PD model describing the disposition of recombinant human erythropoietin and effects on reticulocyte count, red blood cell count, and hemoglobin concentration. The PK model is a one-compartment model with Michaelis–Menten type elimination (K_m, V_{max}) from the central compartment. The PD model is a cell life span model with four sequential cell compartments, representing erythroid progenitor cells (P_1), erythroblasts (P_2), reticulocytes (R), and red blood cells (*RBC*). τ_{P1}, τ_{P2}, τ_R, and τ_{RBC} are the corresponding cell life spans, k_{in} the common zero-order transfer rate between cell compartments. The target parameter hemoglobin in the blood (*Hb*) is calculated from the reticulocyte and red blood cell count and the hemoglobin content per cell. The effect of erythropoietin is modeled as a stimulation of the production of both precursor cell populations (P_1 and P_2) in the bone marrow with the stimulation function $S(t)$. E_{max} is the maximum possible stimulation of reticulocyte production by erythropoietin, EC_{50} the plasma concentration of erythropoietin that produced half-maximum stimulation. A counter-regulatory feedback loop represents the feedback inhibition of reticulocytes on their own production by reducing the production rate of cells in the P_1 compartment via the inhibitory function $I(t)$. IC_{50} is the reticulocyte count that produced half of complete inhibition. (Modified from Ramakrishnan et al. (2004))

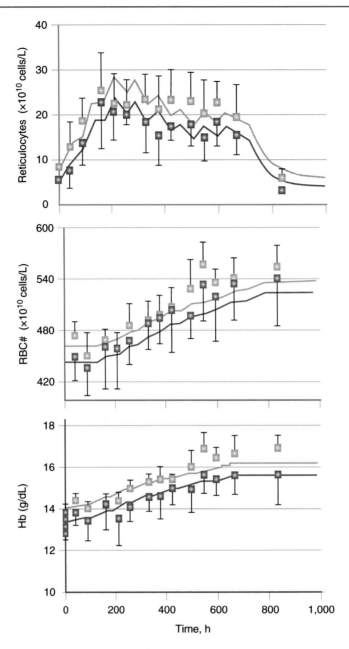

Fig. 6.20 Reticulocyte, red blood cell (*RBC*), and hemoglobin (*Hb*) time courses after multiple SC dosing of 600 IU/kg/week recombinant human erythropoietin. *Orange* and *blue squares* represent data for males and females, whereas the *orange* and *blue lines* for the reticulocytes are model fittings. The *lines* in the RBC and Hb panels are the predictions using the model-fitted curves for the reticulocytes and the life span parameters. (From Ramakrishnan et al. (2004), with permission from John Wiley & Sons, Inc. Copyright American College of Clinical Pharmacology 2004)

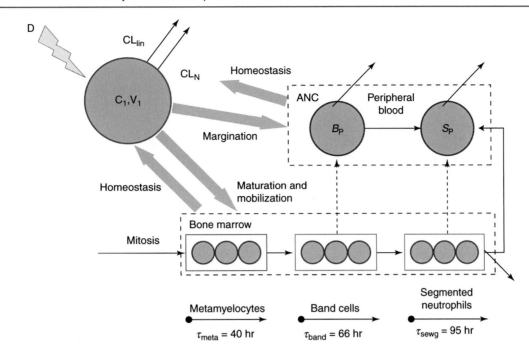

Fig. 6.21 A PK/PD model describing the granulopoietic effects of pegfilgrastim. The PK model is a one-compartment model with two parallel elimination pathways, a first-order elimination process (CL_{lin}), and a neutrophil-mediated elimination process (CL_N). C_1 and V_1 are the concentrations in the PK compartment and the corresponding volume of distribution. The PD model is a cytokinetic model similar to the cell life span model in Fig. 6.19. Three maturation stages of neutrophils and their respective life spans (t_{meta}, t_{band}, t_{seg}) are included in the model, metamyelocytes, band cells, and segmented neutrophils. Each maturation stage is modeled by three sequential transit compartments. Serum concentrations of pegfilgrastim stimulate mitosis and mobilization of band cells and segmented neutrophils in bone marrow, decrease maturation times for postmitotic cells in marrow, and affect margination of the peripheral blood band cell (B_P) and segmented neutrophil (S_P) populations, the sum of which is the total absolute neutrophil count (*ANC*). Changes in neutrophil counts in peripheral blood provide feedback regulation of pegfilgrastim clearance. (Modified from Roskos et al. (2006))

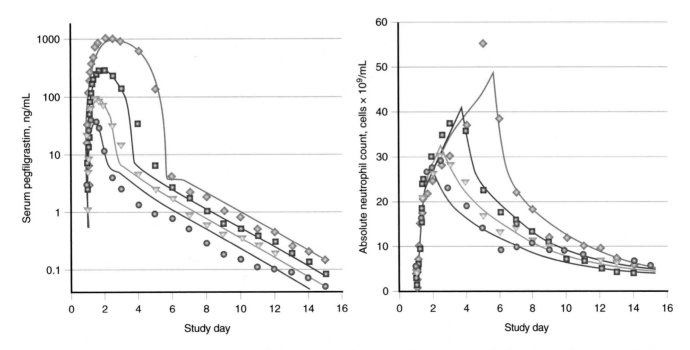

Fig. 6.22 Pegfilgrastim concentration time course and absolute neutrophil count (ANC) time profiles in healthy subjects after a single SC administration of 30, 60, 100, and 300 μg/kg pegfilgrastim (*n* = 8/dose group). Measured data are presented by symbols as mean. Lines represent modeled time courses based on the cytokinetic PK/PD model presented in Fig. 6.22. (From Roskos et al. (2006), with permission from John Wiley & Sons, Inc. Copyright American College of Clinical Pharmacology 2006)

Pharmacokinetics of Nucleic Acid-Based Therapeutics

Therapeutic drugs derived from nucleic acids have in the past decade become a rapidly evolving modality in our armamentarium of pharmacologic interventions to treat human disease. In contrast to conventional drugs and therapeutic proteins that generally target proteins in the body, nucleic acid-based therapeutics target in most instances gene expression. Especially four platform technologies have emerged and are increasingly used: (1) chemically modified antisense oligonucleotides (ASOs); (2) N-acetylgalactosamine (GalNAc)-modified short interfering RNA (siRNA) conjugates; (3) lipid nanoparticle (LNP)-based RNA systems; and (4) adeno-associated virus (AAV) vectors (Kulkarni et al. 2021). LNP- and AAV-based systems have recently gained popularity as basis for COVID-19 vaccines and are extensively discussed in Chap. 15. AAV-based technology is also used in approved and investigational gene therapy approaches and is discussed in Chap. 14. Thus, the following section focuses on the pharmacokinetics of ASOs and siRNA.

ASOs are short synthetic single-stranded nucleic acid polymers, usually 15–30 nucleotides in length with a molecular weight of 6–8 kDa that hybridize with cellular RNA using classic Watson-Crick base pairing to modulate gene expression post-transcriptionally. Precise molecular sequence design provides ASOs with high therapeutic potential and specificity compared to other nucleic acid-based drugs. They typically mediate their effects via modulation of pre-RNA splicing, RNA degradation, or regulation of protein translation. In contrast to ASOs, artificial siRNA-based gene regulation relies on sequence complementarity of 7–8 nucleotides of the target mRNA 3′ untranslated region such that a single siRNA may interact with multiple mRNAs with different affinities.

Several classes of chemical modifications were employed over the past decades to reduce the susceptibility of nucleic acid-based drugs to nuclease degradation. First-generation ASOs employed phosphorothioate linkages instead of phosphate linkages in the ASO backbone. The sulfur substitution dramatically modifies the pharmacokinetics of oligonucleotides by stabilizing them against nuclease digestion and increasing nonspecific plasma protein binding, resulting in a prolonged residence time in tissues and cells, improved tissue distribution, and reduced urinary excretion (Geary et al. 2001).

Second generation ASOs further increased nuclease resistance by chemical modification of the ribose close to the 3′- and 5′- end of the molecule, including 2′-O-methoxyethyl (2-MOE) and 2′-fluoro modifications for ASOs and siRNAs. Especially gapmer technology with a central unmodified deoxynucleotide region for optimal RNAse H1 activity and terminal chemical modifications on both, the 3′ and the 5′ ends, has been proven to improve potency and nuclease stability for ASOs (Kulkarni et al. 2021).

siRNAs are large hydrophilic molecules that consist of two complementary strands of RNA that form a double-helical structure of 19–21 base pairs with a molecular mass of ~14 kDa. Their polyanionic backbone and hydrophilic character prevent passive intracellular uptake and therefore require specialized delivery solutions (Migliorati et al. 2022). Similar to ASOs, siRNA also requires chemical modification of the backbone structure to ensure stability in the circulation. To increase target organ accumulation in the liver, siRNA has been conjugated to a triantennary GalNAc moiety that targets the asialoglycoprotein receptor (ASGPR). ASGPR is primarily expressed by hepatocytes and thus allows efficient targeting to this cell population in the liver (Kulkarni et al. 2021).

Pharmacokinetic properties of ASO- and siRNA-based drugs are largely driven by the chemical structure of their backbone and are thus sequence independent within a chemical class. This similarity has been observed in preclinical models as well as in humans. Due to their similar molecular structure, many common class-wide characteristics with regard to pharmacokinetics and drug disposition can be identified across these platform technologies, although each of the currently approved ASO- or siRNA-based therapeutics has unique features (Park et al. 2016; Weidolf et al. 2021).

Administration and Absorption of Nucleic Acid-Based Therapeutics

Similar to proteins, nucleic acids are due to their large molecular size and high molecular charge not orally bioavailable to a relevant degree. Together with their limited stability in the gastrointestinal tract due to nuclease digestion, the resulting oral bioavailability is in the range of 1–3%. Thus, nucleic acid-based therapeutics require parental administration. Most approved ASOs and siRNAs are administered by the IV or SC route. Localized administration by intravitreal or intrathecal administration has shown to be efficacious as well.

Mipomersen, an approved ASO for the treatment of familial hypercholesterolemia, is administered by SC administration. It primarily targets ApoB mRNA in the liver to induce its liver-based degradation. Similarly, inotersen is administered subcutaneously for the treatment of polyneuropathy in patients with hereditary transthyretin-mediated amyloidosis, thereby inhibiting the formation of protein deposits predominantly formed by the liver (Migliorati et al. 2022). Givorisan, inclisiran, and lumasiran are FDA-approved subcutaneously administered GalNAc-conjugated siRNAs for the treatment of acute hepatic porphyria, hypercholesterolemia, and primary hyperoxaluria type 1, respectively (Kulkarni et al.

2021). For nusinersen, an ASO for the treatment of spinal muscle atrophy, circumventing the blood-brain barrier that is usually impenetrable for ASOs, is crucial to achieve its therapeutic target in the brain. This was accomplished by creating a dosage form for intrathecal administration (Luu et al. 2017).

After SC administration of a 2′-MOE gapmer-modified ASO, bioavailability is close to 100% (Geary et al. 2015). Peak plasma concentrations occur 2–4 h after SC administration (Migliorati et al. 2022). For other less metabolically stabilized ASO structures, subcutaneous bioavailability may remain below 40%.

Distribution and Tissue Uptake of Nucleic Acid-Based Therapeutics

ASOs and siRNA generally have multiphasic distribution profiles after intravenous administration, with a rapid initial distribution (3–4 h) and a long terminal half-life that may reach several weeks for second-generation ASOs (Weidolf et al. 2021). As ASOs and siRNA are generally considered to undergo distribution rate-limited elimination, the terminal half-life in tissues is considered to be in parallel with the terminal half-life in plasma (Geary et al. 1997).

After intravenous administration, ASOs are detected in nearly all tissues and organs except for the brain and testes, suggesting significant transport barriers in these tissues. Major accumulation of ASOs occurs in liver and kidneys, and to a lesser extent in spleen, bone marrow, adipose tissue, and lymph nodes, which seems to be independent of ASO sequence. This may, however, limit their viability against diseases with targets in other tissues, such as in the heart or skeletal muscle (Weidolf et al. 2021).

Cellular uptake as prerequisite for distribution and metabolism is predominantly facilitated by clathrin-mediated endocytosis This internalization into endosomes is followed by subcellular endosome release and trafficking. As ASOs have different intracellular target locations (e.g., RNAase H-mediated mRNA degradation in the cytoplasm (ribosomes) or nucleus, vs. exon skipping in the spliceosome in the nucleus), endosomal release and intracellular trafficking are key determinants of the pharmacologic activity of ASO drugs (Juliano 2018).

For siRNA conjugates, clathrin-dependent receptor-mediated endocytosis via ASGPR has been utilized to target and enrich siRNA in hepatocytes. While hepatocytes constitute 80% of the liver volume, only 15% of the amount of unconjugated nucleic acid drugs taken up by the liver reaches hepatocytes, while majority ends up in Kupffer cells and endothelial cells (Wang et al. 2019). Thus, GalNAc conjugation is necessary for efficient drug delivery to hepatocytes.

As previously discussed in this chapter, ASGPR is a promiscuous C-type lectin receptor that recognizes and facilitates the intracellular update of glycans with end-standing galactose or GalNAc. It is primarily expressed on the sinusoidal membrane of hepatocytes and possesses a high internalization and recycling rate for efficient substrate delivery. Thus, it is an ideal conduit to target hepatocytes with ASGPR-recognized glycans, thereby circumventing broad uptake of siRNA to various organs and tissues throughout the body. Thus, siRNA molecules conjugated with triantennary GalNAc sugar moieties achieve substantially higher potency and in vivo efficacy when targeting gene expression mechanisms in hepatocytes (Kulkarni et al. 2021).

Chemical modifications of the phosphorothioate backbone structure have altered protein binding and organ distribution. Chemically modified ASOs and conjugated siRNA are highly bound to plasma proteins, with more than 85% for nusinersen, inotersen, and mipomersen (Crooke and Geary 2013) as well as approved siRNA conjugates (McDougall et al. 2022). This high binding was present in humans as well as mice, rats, and monkeys. Major binding proteins seem to be β_2-macroglobulin and albumin. The high plasma protein binding is a major determinant in the pharmacokinetics of chemically modified ASOs and siRNA. High plasma protein binding, for example, protects ASOs from renal filtration, as based on their molecular weight below the filtration cutoff of the kidneys their unbound fraction undergoes renal filtration. Thus, plasma protein binding severely restricts the renal elimination of ASOs, so that urinary excretion of intact compound is only a minor elimination pathway (Crooke et al. 1996).

Metabolism of Nucleic Acid-Based Therapeutics

Unmodified RNA oligonucleotides are rapidly degraded in biological matrices, thereby limiting their utility as therapeutics. To enable the development of oligonucleotides into viable therapeutic agents, a variety of chemical modifications were necessary to increase their in vivo stability as outlined above. Endonuclease- and exonuclease-mediated degradation occurs in the blood stream and in target cells. For most nucleic acid-based drugs, endonuclease activity occurs first, cleaving the oligonucleotide into fragments. Then, exonuclease activity further degrades the fragments.

While ASOs and siRNA are relatively rapidly removed from blood, predominantly by distribution and uptake into tissues, their residence time in tissues was found to be relatively long and dependent on their chemical modification (Geary et al. 2001). ASOs are cleared from tissues by nuclease-mediated metabolism, with half-lives of up to several weeks (Geary et al. 2015). GalNAc-conjugated siRNA metabolism generally results in loss of one, two or all three GalNAc sugar chains, followed by excretion of parts of, or the full triannetary molecule (Weidolf et al. 2021). Knowledge on the metabolizing enzymes is still evolving and may require further investigation.

Plasma concentration–time profiles are predominantly characterized by rapid distribution after intravenous administration, followed by slow redistribution and tissue elimination in the terminal phase. Thus, terminal half-lives in plasma mirror those in tissues and are driven by slow tissue elimination. The partition ratios between liver and plasma in the post-distribution phase are approximately 5000:1 for 2′MOE-modified ASOs and are species independent. Thus, whole-organ pharmacokinetics after 24 h is thought to present intracellular exposure, as only very little ASO remained bound to extracellular components by 24 h after injection, and post-distribution plasma concentrations can be used as surrogate for tissue exposure in all species, including humans (Geary et al. 2015). Similarly, siRNA levels in hepatocytes are considered better predictors for pharmacologic activity of GalNAc-conjugated siRNA than plasma concentrations (Migliorati et al. 2022).

Givosiran as siRNA conjugate consists of a 21-base sense strand and a 23-base antisense strand that are fully modified with 16 nucleotides containing a 2′-F substitution and the remaining nucleotides being 2′-OMe substituted. Six of its backbone linkages distributed at the ends of the strands are PS-modified. The combination of chemical modification of the backbone and conjugation to tri-GalNAc allows for sufficient metabolic stability and high uptake into the liver for efficient treatment of the approved indication acute hepatic porphyria (Zhang et al. 2021).

So far, little is known on the specific nucleases involved in ASO and siRNA metabolism, such as identity, specificity, capacity, potential competition with endogenous RNA, and interspecies differences. Given the structure and function of RNA are conserved across species, the function and substrate specificity of nucleases are also likely to be conserved, and significant species differences in metabolite structures are not expected (Weidolf et al. 2021).

Excretion of Nucleic Acid-Based Therapeutics
Urinary excretion is a major route of excretion for ASOs, regardless of sequence or chemical structure, with the majority being shorter length metabolites rather than unchanged parent drug (Agrawal et al. 1995). Metabolites are assumed to undergo cellular release via membrane leakage, vesicle release, or endosome release. Only a minor fraction of the dose is excreted into feces although enterohepatic recirculation has been suggested (Dvorchik 2000). For phosphorothioate-modified ASOs, the increased plasma protein binding substantially decreased their renal excretion. For fomivirsen, for example, the excretion of the parent compound was only 16% in urine and 3% in feces. For phosphorothioate ASOs with 2′-MOE gapmer technology, renal excretion of parent drug remains even lower (<1–3%). Metabolite levels remain low in the systemic circulation, but

higher percentages are recovered in the urine, likely because of lower plasma protein binding than the parent compound (Weidolf et al. 2021).

Limited assessments of the effects of renal impairment on plasma exposures for ASOs suggest that mild or moderate renal impairment showed no effect on the plasma exposure, but end-stage renal disease may result in a mild increase (34%) (Wang et al. 2023).

Conclusion

The pharmacokinetic and pharmacodynamic characteristics of proteins and nucleic acid-based drugs form the basis for their therapeutic application. Appreciation of the pharmacokinetic and pharmacodynamic differences between therapeutic biologics and traditional small-molecule drugs will empower the drug development scientist as well as the healthcare provider to handle, evaluate, and apply these compounds in an optimal fashion during the drug development process as well as during applied pharmacotherapy. Rational, scientifically based drug development and pharmacotherapy based on the use of pharmacokinetic and pharmacodynamic concepts will undoubtedly propel the success and future of protein- and nucleic acid-based therapeutics and might ultimately contribute to provide the novel medications that may serve as the key for the aspired "precision medicine" in the healthcare systems of the future (Dugger et al. 2018).

Self-Assessment Questions

Questions

1. What are the major elimination pathways for protein drugs after administration?
2. Which pathway of absorption is rather unique for proteins after SC injection?
3. What is the role of plasma-binding proteins for natural proteins?
4. How do the sugar groups on glycoproteins influence hepatic elimination of these glycoproteins?
5. In which direction might elimination clearance of a protein drug change when antibodies against the protein are produced after chronic dosing with the protein drug? Why?
6. What is the major driving force for the transport of proteins from the vascular to the extravascular space?
7. Why are therapeutic proteins generally not active upon oral administration?

8. Many therapeutic proteins exhibit Michaelis–Menten type, saturable elimination kinetics. What are the underlying mechanisms for this pharmacokinetic behavior?
9. Explain counterclockwise hysteresis in plasma concentration-effect plots.
10. Why is mechanism-based PK/PD modeling a preferred modeling approach for therapeutic proteins?
11. What are common chemical modifications in ASOs and siRNA to increase their stability toward degradation by nucleases?

Answers

1. Proteolysis, glomerular filtration followed by intraluminal metabolism or tubular reabsorption with intracellular lysosomal degradation, and receptor-mediated endocytosis followed by metabolism in the skin, muscle, liver, and possibly other organs and tissues.
2. Biodistribution from the injection site into the lymphatic system.
3. Plasma proteins may act as circulating reservoirs for the proteins that are their ligands. Consequently, the protein ligands may be protected from elimination and distribution. In some cases, protein binding may protect the organism from undesirable, acute effects; in other cases, receptor binding may be facilitated by the binding protein.
4. In some cases, the sugar groups are recognized by hepatic receptors (e.g., mannose by the mannose receptor), facilitating receptor-mediated uptake and metabolism. In other cases, sugar chains and terminal sugar groups (e.g., terminal sialic acid residues) may shield the protein from binding to receptors and hepatic uptake.
5. Clearance may increase or decrease by forming antibody–protein complexes. A decrease of clearance occurs when the antibody–protein complex is eliminated slower than free protein. An increase of clearance occurs when the protein–antibody complex is eliminated more rapidly than the unbound protein, such as when reticuloendothelial uptake is stimulated by the complex.
6. Protein extravasation, i.e., transport from the blood or vascular space to the interstitial tissue space, is predominantly mediated by fluid convection. Protein molecules follow the fluid flux from the vascular space through pores between adjacent cells into the interstitial space. Drainage of the interstitial space through the lymphatic system allows therapeutic proteins to distribute back into the vascular space.
7. The gastrointestinal mucosa is a major absorption barrier for hydrophilic macromolecule such as proteins. In addition, therapeutic peptides and proteins are degraded by the extensive peptidase and protease activity in the gastrointestinal tract. Both processes minimize the oral bioavailability of therapeutic proteins.

8. Receptor-mediated endocytosis is the most frequent cause of nonlinear pharmacokinetics in therapeutic proteins. Its occurrence becomes even more prominent if the therapeutic protein undergoes target-mediated drug disposition, i.e., if the receptor-mediated endocytosis is mediated via the pharmacologic target of the therapeutic protein. As the binding to the target is usually of high affinity, and the therapeutic protein is often dosed to saturate the majority of the available target receptors for maximum pharmacologic efficacy, saturation of the associated receptor-mediated endocytosis as elimination pathway is frequently encountered.
9. Counterclockwise hysteresis is an indication of the indirect nature of the effects seen for many protein drugs. It can be explained by delays between the appearance of drug in plasma and the appearance of the pharmacodynamic response. The underlying cause may either be a distributional delay between the drug concentrations in plasma and at the effect site (modeled with an indirect link PK/PD model) or by time-consuming post-receptor events that cause a delay between the drug-receptor interaction and the observed drug effect, for example, the effect on a physiologic measure or endogenous substance (modeled with an indirect response or turnover PK/PD model).
10. Therapeutic proteins are often classified as "targeted therapies," where the drug compound acts on one specific, well-defined response pathway. This well-documented knowledge on the mechanism of action can relatively easily be translated into a mechanism-based PK/PD modeling approach that incorporates the major physiological processes relevant for the pharmacologic effect. The advantage of mechanism-based as compared to empirical PK/PD modeling is that mechanism-based models are usually more robust and allow more reliable simulations beyond the actually measured data.
11. Substitution of a sulfur atom for a nonbridging oxygen in the phosphate backbone of an oligonucleotide leads to a phosphorothioate linkage between nucleotides that is resistant to nuclease degradation. In addition, select nucleotides are modified at the 2′ position of the ribose, with, for example, a methoxyethyl group (2'-MOE) for ASOs or a fluoride (2'-F) for siRNA.

References

Agoram B, Heatherington AC, Gastonguay MR (2006) Development and evaluation of a population pharmacokinetic-pharmacodynamic model of darbepoetin alfa in patients with nonmyeloid malignancies undergoing multicycle chemotherapy. AAPS J 8(3):E552–E563

Agrawal S, Temsamani J, Galbraith W, Tang J (1995) Pharmacokinetics of antisense oligonucleotides. Clin Pharmacokinet 28(1):7–16

Albitar M, Do KA, Johnson MM et al (2004) Free circulating soluble CD52 as a tumor marker in chronic lymphocytic leukemia and its implication in therapy with anti-CD52 antibodies. Cancer 101(5):999–1008

Allon M, Kleinman K, Walczyk M et al (2002) Pharmacokinetics and pharmacodynamics of darbepoetin alfa and epoetin in patients undergoing dialysis. Clin Pharmacol Ther 72(5):546–555

Anderson PM, Sorenson MA (1994) Effects of route and formulation on clinical pharmacokinetics of interleukin-2. Clin Pharmacokinet 27(1):19–31

Bauer RJ, Gibbons JA, Bell DP, Luo ZP, Young JD (1994) Nonlinear pharmacokinetics of recombinant human macrophage colony- stimulating factor (M-CSF) in rats. J Pharmacol Exp Ther 268(1):152–158

Baxter LT, Zhu H, Mackensen DG, Jain RK (1994) Physiologically based pharmacokinetic model for specific and nonspecific monoclonal antibodies and fragments in normal tissues and human tumor xenografts in nude mice. Cancer Res 54(6):1517–1528

Benincosa LJ, Chow FS, Tobia LP, Kwok DC, Davis CB, Jusko WJ (2000) Pharmacokinetics and pharmacodynamics of a humanized monoclonal antibody to factor IX in cynomolgus monkeys. J Pharmacol Exp Ther 292(2):810–816

Bennett HP, McMartin C (1978) Peptide hormones and their analogues: distribution, clearance from the circulation, and inactivation in vivo. Pharmacol Rev 30(3):247–292

Berdeja J, Jagannath S, Zonder J et al (2016) Pharmacokinetics and safety of Elotuzumab combined with Lenalidomide and dexamethasone in patients with multiple myeloma and various levels of renal impairment: results of a phase Ib study. Clin Lymphoma Myeloma Leuk 16(3):129–138

Boxenbaum H (1982) Interspecies scaling, allometry, physiological time, and the ground plan of pharmacokinetics. J Pharmacokinet Biopharm 10(2):201–227

Bressolle F, Audran M, Gareau R, Pham TN, Gomeni R (1997) Comparison of a direct and indirect population pharmacodynamic model: application to recombinant human erythropoietin in athletes. J Pharmacokinet Biopharm 25(3):263–275

Bruin G, Loesche C, Nyirady J, Sander O (2017) Population pharmacokinetic modeling of Secukinumab in patients with moderate to severe psoriasis. J Clin Pharmacol 57(7):876–885

Bu G, Williams S, Strickland DK, Schwartz AL (1992) Low density lipoprotein receptor-related protein/alpha 2-macroglobulin receptor is an hepatic receptor for tissue-type plasminogen activator. Proc Natl Acad Sci U S A 89(16):7427–7431

Caliceti P, Veronese FM (2003) Pharmacokinetic and biodistribution properties of poly(ethylene glycol)-protein conjugates. Adv Drug Deliv Rev 55(10):1261–1277

Carone FA, Peterson DR (1980) Hydrolysis and transport of small peptides by the proximal tubule. Am J Phys 238(3):F151–F158

Carone FA, Peterson DR, Flouret G (1982) Renal tubular processing of small peptide hormones. J Lab Clin Med 100(1):1–14

Chanson P, Timsit J, Harris AG (1993) Clinical pharmacokinetics of octreotide. Therapeutic applications in patients with pituitary tumours. Clin Pharmacokinet 25(5):375–391

Chiang J, Gloff CA, Yoshizawa CN, Williams GJ (1993) Pharmacokinetics of recombinant human interferon-beta ser in healthy volunteers and its effect on serum neopterin. Pharm Res 10(4):567–572

Chirmule N, Jawa V, Meibohm B (2012) Immunogenicity to therapeutic proteins: impact on PK/PD and efficacy. AAPS J 14(2):296–302

Chow FS, Benincosa LJ, Sheth SB et al (2002) Pharmacokinetic and pharmacodynamic modeling of humanized anti-factor IX antibody (SB 249417) in humans. Clin Pharmacol Ther 71(4):235–245

Colburn W (1991) Peptide, peptoid, and protein pharmacokinetics/pharmacodynamics. In: Garzone P, Colburn W, Mokotoff M (eds) Petides, peptoids, and proteins. Harvey Whitney Books, Cincinnati, OH, pp 94–115

Crooke ST, Geary RS (2013) Clinical pharmacological properties of mipomersen (Kynamro), a second generation antisense inhibitor of apolipoprotein B. Br J Clin Pharmacol 76(2):269–276

Crooke ST, Graham MJ, Zuckerman JE et al (1996) Pharmacokinetic properties of several novel oligonucleotide analogs in mice. J Pharmacol Exp Ther 277(2):923–937

Daniel H, Herget M (1997) Cellular and molecular mechanisms of renal peptide transport. Am J Phys 273(1 Pt 2):F1–F8

Dayneka NL, Garg V, Jusko WJ (1993) Comparison of four basic models of indirect pharmacodynamic responses. J Pharmacokinet Biopharm 21(4):457–478

de la Pena A, Ma X, Reddy S, Ovalle F, Bergenstal RM, Jackson JA (2014) Application of PK/PD modeling and simulation to dosing regimen optimization of high-dose human regular U-500 insulin. J Diabetes Sci Technol 8(4):821–829

Dedrick RL (1973) Animal scale-up. J Pharmacokinet Biopharm 1(5):435–461

Deen WM, Lazzara MJ, Myers BD (2001) Structural determinants of glomerular permeability. Am J Physiol Ren Physiol 281(4):F579–F596

Deng R, Iyer S, Theil FP, Mortensen DL, Fielder PJ, Prabhu S (2011) Projecting human pharmacokinetics of therapeutic antibodies from nonclinical data: what have we learned? MAbs 3(1):61–66

Derendorf H, Meibohm B (1999) Modeling of pharmacokinetic/pharmacodynamic (PK/PD) relationships: concepts and perspectives. Pharm Res 16(2):176–185

Diao L, Meibohm B (2013) Pharmacokinetics and pharmacokinetic-pharmacodynamic correlations of therapeutic peptides. Clin Pharmacokinet 52(10):855–868

Dirks NL, Meibohm B (2010) Population pharmacokinetics of therapeutic monoclonal antibodies. Clin Pharmacokinet 49(10):633–659

Dirks NL, Nolting A, Kovar A, Meibohm B (2008) Population pharmacokinetics of cetuximab in patients with squamous cell carcinoma of the head and neck. J Clin Pharmacol 48(3):267–278

Dong JQ, Salinger DH, Endres CJ et al (2011) Quantitative prediction of human pharmacokinetics for monoclonal antibodies: retrospective analysis of monkey as a single species for first-in-human prediction. Clin Pharmacokinet 50(2):131–142

Dugger SA, Platt A, Goldstein DB (2018) Drug development in the era of precision medicine. Nat Rev Drug Discov 17(3):183–196

Dvorchik BH (2000) The disposition (ADME) of antisense oligonucleotides. Curr Opin Mol Ther 2(3):253–257

Edwards A, Daniels BS, Deen WM (1999) Ultrastructural model for size selectivity in glomerular filtration. Am J Phys 276(6 Pt 2):F892–F902

EfpiaMidWorkgroup MSF, Burghaus R et al (2016) Good practices in model-informed drug discovery and development: practice, application, and documentation. CPT Pharmacometrics Syst Pharmacol 5(3):93–122

Eigenmann MJ, Fronton L, Grimm HP, Otteneder MB, Krippendorff BF (2017) Quantification of IgG monoclonal antibody clearance in tissues. MAbs 9(6):1007–1015

Eppler SM, Combs DL, Henry TD et al (2002) A target-mediated model to describe the pharmacokinetics and hemodynamic effects of recombinant human vascular endothelial growth factor in humans. Clin Pharmacol Ther 72(1):20–32

Falck D, Thomann M, Lechmann M et al (2021) Glycoform-resolved pharmacokinetic studies in a rat model employing glycoengineered variants of a therapeutic monoclonal antibody. MAbs 13(1):1865596

Fasano A (1998) Novel approaches for oral delivery of macromolecules. J Pharm Sci 87(11):1351–1356

Flessner MF, Lofthouse J, Zakaria el R (1997) In vivo diffusion of immunoglobulin G in muscle: effects of binding, solute exclusion, and lymphatic removal. Am J Phys 273(6 Pt 2):H2783–H2793

Geary RS, Leeds JM, Henry SP, Monteith DK, Levin AA (1997) Antisense oligonucleotide inhibitors for the treatment of cancer: 1. Pharmacokinetic properties of phosphorothioate oligodeoxynucleotides. Anticancer Drug Des 12(5):383–393

Geary RS, Yu RZ, Levin AA (2001) Pharmacokinetics of phosphorothioate antisense oligodeoxynucleotides. Curr Opin Investig Drugs 2(4):562–573

Geary RS, Norris D, Yu R, Bennett CF (2015) Pharmacokinetics, biodistribution and cell uptake of antisense oligonucleotides. Adv Drug Deliv Rev 87:46–51

Gibbs JP, Doshi S, Kuchimanchi M et al (2017) Impact of target-mediated elimination on the dose and regimen of Evolocumab, a human monoclonal antibody against Proprotein convertase Subtilisin/Kexin type 9 (PCSK9). J Clin Pharmacol 57(5):616–626

Glassman PM, Balthasar JP (2016) Physiologically-based pharmacokinetic modeling to predict the clinical pharmacokinetics of monoclonal antibodies. J Pharmacokinet Pharmacodyn 43(4):427–446

Glund S, Gan G, Moschetti V et al (2018) The renal elimination pathways of the dabigatran reversal agent Idarucizumab and its impact on dabigatran elimination. Clin Appl Thromb Hemost 24(5):724–733

Graham ML (2003) Pegaspargase: a review of clinical studies. Adv Drug Deliv Rev 55(10):1293–1302

Handelsman DJ, Swerdloff RS (1986) Pharmacokinetics of gonadotropin-releasing hormone and its analogs. Endocr Rev 7(1):95–105

Hayashida K, Bartlett AH, Chen Y, Park PW (2010) Molecular and cellular mechanisms of ectodomain shedding. Anat Rec (Hoboken) 293(6):925–937

Herbst RS, Langer CJ (2002) Epidermal growth factor receptors as a target for cancer treatment: the emerging role of IMC-C225 in the treatment of lung and head and neck cancers. Semin Oncol 29(1 Suppl 4):27–36

Holford NH, Sheiner LB (1982) Kinetics of pharmacologic response. Pharmacol Ther 16(2):143–166

Inui K, Terada T, Masuda S, Saito H (2000) Physiological and pharmacological implications of peptide transporters, PEPT1 and PEPT2. Nephrol Dial Transplant 15(Suppl 6):11–13

Jin F, Krzyzanski W (2004) Pharmacokinetic model of target-mediated disposition of thrombopoietin. AAPS PharmSci 6(1):E9

Johnson V, Maack T (1977) Renal extraction, filtration, absorption, and catabolism of growth hormone. Am J Phys 233(3):F185–F196

Juliano RL (2018) Intracellular trafficking and endosomal release of oligonucleotides: what we know and what we Don't. Nucleic Acid Ther 28(3):166–177

Kaufman JS, Reda DJ, Fye CL et al (1998) Subcutaneous compared with intravenous epoetin in patients receiving hemodialysis. Department of Veterans Affairs Cooperative Study Group on erythropoietin in hemodialysis patients. N Engl J Med 339(9):578–583

Khor SP, McCarthy K, DuPont M, Murray W, Timony G (2000) Pharmacokinetics, pharmacodynamics, allometry, and dose selection of rPSGL-Ig for phase I trial. J Pharmacol Exp Ther 293(2):618–624

Kim J, Hayton WL, Robinson JM, Anderson CL (2007) Kinetics of FcRn-mediated recycling of IgG and albumin in human: pathophysiology and therapeutic implications using a simplified mechanism-based model. Clin Immunol 122(2):146–155

Kingwell K (2016) Drug delivery: new targets for drug delivery across the BBB. Nat Rev Drug Discov 15(2):84–85

Ko S, Jo M, Jung ST (2021) Recent achievements and challenges in prolonging the serum half-lives of therapeutic IgG antibodies through fc engineering. BioDrugs 35(2):147–157

Kompella U, Lee V (1991) Pharmacokinetics of peptide and protein drugs. In: Lee V (ed) Peptide and protein drug delivery. Marcel Dekker, New York, pp 391–484

Kontermann R (2012) Therapeutic proteins: strategies to modulate their plasma half-lives. Wiley, Weinheim

Krogsgaard Thomsen M, Friis C, Sehested Hansen B et al (1994) Studies on the renal kinetics of growth hormone (GH) and on the GH receptor and related effects in animals. J Pediatr Endocrinol 7(2):93–105

Kulkarni JA, Witzigmann D, Thomson SB et al (2021) The current landscape of nucleic acid therapeutics. Nat Nanotechnol 16(6):630–643

Kuwabara T, Uchimura T, Kobayashi H, Kobayashi S, Sugiyama Y (1995) Receptor-mediated clearance of G-CSF derivative nartograstim in bone marrow of rats. Am J Phys 269(1 Pt 1):E1–E9

Lee HJ (2002) Protein drug oral delivery: the recent progress. Arch Pharm Res 25(5):572–584

Lee H, Kimko HC, Rogge M, Wang D, Nestorov I, Peck CC (2003) Population pharmacokinetic and pharmacodynamic modeling of etanercept using logistic regression analysis. Clin Pharmacol Ther 73(4):348–365

Lesko LJ (2007) Paving the critical path: how can clinical pharmacology help achieve the vision? Clin Pharmacol Ther 81(2):170–177

Lesko LJ, Rowland M, Peck CC, Blaschke TF (2000) Optimizing the science of drug development: opportunities for better candidate selection and accelerated evaluation in humans. J Clin Pharmacol 40(8):803–814

Levy G (1986) Kinetics of drug action: an overview. J Allergy Clin Immunol 78(4 Pt 2):754–761

Levy G (1994) Mechanism-based pharmacodynamic modeling. Clin Pharmacol Ther 56(4):356–358

Limothai W, Meibohm B (2011) Effect of dose on the apparent bioavailability of therapeutic proteins that undergo target-mediated drug disposition. AAPS J 13(S2)

Liu Q, Ahadpour M, Rocca M, Huang SM (2021) Clinical pharmacology regulatory sciences in drug development and precision medicine: current status and emerging trends. AAPS J 23(3):54

Lunghi B, Morfini M, Martinelli N et al (2022) The Asialoglycoprotein receptor minor subunit gene contributes to pharmacokinetics of factor VIII concentrates in hemophilia a. Thromb Haemost 122(5):715–725

Luu KT, Norris DA, Gunawan R, Henry S, Geary R, Wang Y (2017) Population pharmacokinetics of Nusinersen in the cerebral spinal fluid and plasma of pediatric patients with spinal muscular atrophy following intrathecal administrations. J Clin Pharmacol 57(8):1031–1041

Maack T, Park C, Camargo M (1985) Renal filtration, transport and metabolism of proteins. In: Seldin D, Giebisch G (eds) The kidney. Raven Press, New York, pp 1773–1803

Mach H, Gregory SM, Mackiewicz A et al (2011) Electrostatic interactions of monoclonal antibodies with subcutaneous tissue. Ther Deliv 2(6):727–736

Mager DE (2006) Target-mediated drug disposition and dynamics. Biochem Pharmacol 72(1):1–10

Mager DE, Wyska E, Jusko WJ (2003) Diversity of mechanism-based pharmacodynamic models. Drug Metab Dispos 31(5):510–518

Mahato RI, Narang AS, Thoma L, Miller DD (2003) Emerging trends in oral delivery of peptide and protein drugs. Crit Rev Ther Drug Carrier Syst 20(2–3):153–214

McDougall R, Ramsden D, Agarwal S et al (2022) The nonclinical disposition and pharmacokinetic/Pharmacodynamic properties of N-Acetylgalactosamine-conjugated small interfering RNA are highly predictable and build confidence in translation to human. Drug Metab Dispos 50(6):781–797

McMahon HT, Boucrot E (2011) Molecular mechanism and physiological functions of clathrin-mediated endocytosis. Nat Rev Mol Cell Biol 12(8):517–533

McMartin C (1992) Pharmacokinetics of peptides and proteins: opportunities and challenges. Adv Drug Res 22:39–106

Meibohm B (2004) Pharmacokinetics of protein- and nucleotide-based drugs. In: Mahato RI (ed) Biomaterials for delivery and targeting of proteins and nucleic acids. CRC Press, Boca Raton, pp 275–294

Meibohm B (2006) Pharmacokinetics and pharmacodynamics of biotech drugs. Wiley, Weinheim

Meibohm B, Derendorf H (1997) Basic concepts of pharmacokinetic/pharmacodynamic (PK/PD) modelling. Int J Clin Pharmacol Ther 35(10):401–413

Meibohm B, Derendorf H (2002) Pharmacokinetic/pharmacodynamic studies in drug product development. J Pharm Sci 91(1):18–31

Meibohm B, Zhou H (2012) Characterizing the impact of renal impairment on the clinical pharmacology of biologics. J Clin Pharmacol 52(1 Suppl):54S–62S

Meijer D, Ziegler K (1993) Biological barriers to protein delivery. Plenum Press, New York

Migliorati JM, Liu S, Liu A et al (2022) Absorption, distribution, metabolism, and excretion of US Food and Drug Administration-approved antisense oligonucleotide drugs. Drug Metab Dispos 50(6):888–897

Mohler M, Cook J, Lewis D et al (1993) Altered pharmacokinetics of recombinant human deoxyribonuclease in rats due to the presence of a binding protein. Drug Metab Dispos 21(1):71–75

Molineux G (2003) Pegylation: engineering improved biopharmaceuticals for oncology. Pharmacotherapy 23(8 Pt 2):3S–8S

Montero-Julian FA, Klein B, Gautherot E, Brailly H (1995) Pharmacokinetic study of anti-interleukin-6 (IL-6) therapy with monoclonal antibodies: enhancement of IL-6 clearance by cocktails of anti-IL-6 antibodies. Blood 85(4):917–924

Mordenti J, Chen SA, Moore JA, Ferraiolo BL, Green JD (1991) Interspecies scaling of clearance and volume of distribution data for five therapeutic proteins. Pharm Res 8(11):1351–1359

Mould DR, Meibohm B (2016) Drug development of therapeutic monoclonal antibodies. BioDrugs 30(4):275–293

Mould DR, Davis CB, Minthorn EA et al (1999) A population pharmacokinetic-pharmacodynamic analysis of single doses of clenoliximab in patients with rheumatoid arthritis. Clin Pharmacol Ther 66(3):246–257

Nielsen S, Nielsen JT, Christensen EI (1987) Luminal and basolateral uptake of insulin in isolated, perfused, proximal tubules. Am J Phys 253(5 Pt 2):F857–F867

Nolain P, Djebli N, Brunet A, Fabre D, Khier S (2022) Combined semi-mechanistic target-mediated drug disposition and pharmacokinetic-Pharmacodynamic models of Alirocumab, PCSK9, and low-density lipoprotein cholesterol in a pooled analysis of randomized phase I/II/III studies. Eur J Drug Metab Pharmacokinet 47(6):789–802

Ober RJ, Radu CG, Ghetie V, Ward ES (2001) Differences in promiscuity for antibody-FcRn interactions across species: implications for therapeutic antibodies. Int Immunol 13(12):1551–1559

Park J, Park J, Pei Y, Xu J, Yeo Y (2016) Pharmacokinetics and biodistribution of recently-developed siRNA nanomedicines. Adv Drug Deliv Rev 104:93–109

Pauletti GM, Gangwar S, Siahaan TJ, Jeffrey A, Borchardt RT (1997) Improvement of oral peptide bioavailability: Peptidomimetics and prodrug strategies. Adv Drug Deliv Rev 27(2–3):235–256

Peck CC, Barr WH, Benet LZ et al (1994) Opportunities for integration of pharmacokinetics, pharmacodynamics, and toxicokinetics in rational drug development. J Clin Pharmacol 34(2):111–119

Perez-Ruixo JJ, Kimko HC, Chow AT, Piotrovsky V, Krzyzanski W, Jusko WJ (2005) Population cell life span models for effects of drugs following indirect mechanisms of action. J Pharmacokinet Pharmacodyn 32(5–6):767–793

Periti P, Mazzei T, Mini E (2002) Clinical pharmacokinetics of depot leuprorelin. Clin Pharmacokinet 41(7):485–504

Perrier D, Mayersohn M (1982) Noncompartmental determination of the steady-state volume of distribution for any mode of administration. J Pharm Sci 71(3):372–373

Piscitelli SC, Reiss WG, Figg WD, Petros WP (1997) Pharmacokinetic studies with recombinant cytokines. Scientific issues and practical considerations. Clin Pharmacokinet 32(5):368–381

Porter CJ, Charman SA (2000) Lymphatic transport of proteins after subcutaneous administration. J Pharm Sci 89(3):297–310

Rabkin R, Ryan MP, Duckworth WC (1984) The renal metabolism of insulin. Diabetologia 27(3):351–357

Racine-Poon A, Botta L, Chang TW et al (1997) Efficacy, pharmacodynamics, and pharmacokinetics of CGP 51901, an anti-immunoglobulin E chimeric monoclonal antibody, in patients with seasonal allergic rhinitis. Clin Pharmacol Ther 62(6):675–690

Radwanski E, Chakraborty A, Van Wart S et al (1998) Pharmacokinetics and leukocyte responses of recombinant human interleukin-10. Pharm Res 15(12):1895–1901

Ramakrishnan R, Cheung WK, Wacholtz MC, Minton N, Jusko WJ (2004) Pharmacokinetic and pharmacodynamic modeling of recombinant human erythropoietin after single and multiple doses in healthy volunteers. J Clin Pharmacol 44(9):991–1002

Reddy ST, Berk DA, Jain RK, Swartz MA (2006) A sensitive in vivo model for quantifying interstitial convective transport of injected macromolecules and nanoparticles. J Appl Physiol 101(4):1162–1169

Richter WF, Gallati H, Schiller CD (1999) Animal pharmacokinetics of the tumor necrosis factor receptor-immunoglobulin fusion protein lenercept and their extrapolation to humans. Drug Metab Dispos 27(1):21–25

Richter WF, Bhansali SG, Morris ME (2012) Mechanistic determinants of biotherapeutics absorption following SC administration. AAPS J 14(3):559–570

Roopenian DC, Akilesh S (2007) FcRn: the neonatal fc receptor comes of age. Nat Rev Immunol 7(9):715–725

Roskos LK, Lum P, Lockbaum P, Schwab G, Yang BB (2006) Pharmacokinetic/pharmacodynamic modeling of pegfilgrastim in healthy subjects. J Clin Pharmacol 46(7):747–757

Ryman JT, Meibohm B (2017) Pharmacokinetics of monoclonal antibodies. CPT Pharmacometrics Syst Pharmacol 6(9):576–588

Schomburg A, Kirchner H, Atzpodien J (1993) Renal, metabolic, and hemodynamic side-effects of interleukin-2 and/or interferon alpha: evidence of a risk/benefit advantage of subcutaneous therapy. J Cancer Res Clin Oncol 119(12):745–755

Sharma A, Jusko W (1998) Characteristics of indirect pharmacodynamic models and applications to clinical drug responses. Br J Clin Pharmacol 45:229–239

Sheiner LB, Steimer JL (2000) Pharmacokinetic/pharmacodynamic modeling in drug development. Annu Rev Pharmacol Toxicol 40:67–95

Sheiner LB, Stanski DR, Vozeh S, Miller RD, Ham J (1979) Simultaneous modeling of pharmacokinetics and pharmacodynamics: application to d-tubocurarine. Clin Pharmacol Ther 25(3):358–371

Shen WC (2003) Oral peptide and protein delivery: unfulfilled promises? Drug Discov Today 8(14):607–608

Smedsrod B, Einarsson M (1990) Clearance of tissue plasminogen activator by mannose and galactose receptors in the liver. Thromb Haemost 63(1):60–66

Straughn AB (1982) Model-independent steady-state volume of distribution. J Pharm Sci 71(5):597–598

Straughn AB (2006) Limitations of noncompartmental pharmacokinetic analysis of biotech drugs. In: Meibohm B (ed) Pharmacokinetics and pharmacodynamics of biotech drugs. Wiley, Weinheim, pp 181–188

Strickland DK, Kounnas MZ, Argraves WS (1995) LDL receptor-related protein: a multiligand receptor for lipoprotein and proteinase catabolism. FASEB J 9(10):890–898

Sun YN, Jusko WJ (1999) Role of baseline parameters in determining indirect pharmacodynamic responses. J Pharm Sci 88(10):987–990

Sun YN, Lee HJ, Almon RR, Jusko WJ (1999) A pharmacokinetic/pharmacodynamic model for recombinant human growth hormone effects on induction of insulin-like growth factor I in monkeys. J Pharmacol Exp Ther 289(3):1523–1532

Supersaxo A, Hein W, Gallati H, Steffen H (1988) Recombinant human interferon alpha-2a: delivery to lymphoid tissue by selected modes of application. Pharm Res 5(8):472–476

Supersaxo A, Hein WR, Steffen H (1990) Effect of molecular weight on the lymphatic absorption of water-soluble compounds following subcutaneous administration. Pharm Res 7(2):167–169

Suryawanshi S, Zhang L, Pfister M, Meibohm B (2010) The current role of model-based drug development. Expert Opin Drug Discovery 5(4):311–321

Tabrizi M, Roskos LK (2006) Exposure-Reponse relationships for therapeutic biologics. In: Meibohm B (ed) Pharmacokinetics and pharmacodynamics of biotech drugs. Wiley, Weinheim, pp 295–330

Takagi A, Masuda H, Takakura Y, Hashida M (1995) Disposition characteristics of recombinant human interleukin-11 after a bolus intravenous administration in mice. J Pharmacol Exp Ther 275(2):537–543

Taki Y, Sakane T, Nadai T et al (1998) First-pass metabolism of peptide drugs in rat perfused liver. J Pharm Pharmacol 50(9):1013–1018

Tang L, Meibohm B (2006) Pharmacokinetics of peptides and proteins. In: Meibohm B (ed) Pharmacokinetics and pharmacodynamics of biotech drugs. Wiley, Weinheim, pp 17–44

Tang L, Persky AM, Hochhaus G, Meibohm B (2004) Pharmacokinetic aspects of biotechnology products. J Pharm Sci 93(9):2184–2204

Tanswell P, Modi N, Combs D, Danays T (2002) Pharmacokinetics and pharmacodynamics of tenecteplase in fibrinolytic therapy of acute myocardial infarction. Clin Pharmacokinet 41(15):1229–1245

Toon S (1996) The relevance of pharmacokinetics in the development of biotechnology products. Eur J Drug Metab Pharmacokinet 21(2):93–103

Veng-Pedersen P, Gillespie W (1984) Mean residence time in peripheral tissue: a linear disposition parameter useful for evaluating a drug's tissue distribution. J Pharmacokinet Biopharm 12(5):535–543

Verma S, Goand UK, Husain A, Katekar RA, Garg R, Gayen JR (2021) Challenges of peptide and protein drug delivery by oral route: current strategies to improve the bioavailability. Drug Dev Res 82(7):927–944

Veronese FM, Caliceti P (2006) Custom-tailored pharmacokinetics and pharmacodynamics via chemical modifications of biotech drugs. In: Meibohm B (ed) Pharmacokinetics and pharmacodynamics of Boptech drugs. Wiley, Weinheim, pp 271–294

Walsh S, Shah A, Mond J (2003) Improved pharmacokinetics and reduced antibody reactivity of lysostaphin conjugated to polyethylene glycol. Antimicrob Agents Chemother 47(2):554–558

Wang W, Wang EQ, Balthasar JP (2008) Monoclonal antibody pharmacokinetics and pharmacodynamics. Clin Pharmacol Ther 84(5):548–558

Wang Y, Yu RZ, Henry S, Geary RS (2019) Pharmacokinetics and clinical pharmacology considerations of GalNAc(3)-conjugated antisense oligonucleotides. Expert Opin Drug Metab Toxicol 15(6):475–485

Wang Y, Diep JK, Yu RZ et al (2023) Assessment of the effect of organ impairment on the pharmacokinetics of 2'-MOE and Phosphorothioate modified antisense oligonucleotides. J Clin Pharmacol 63(1):21–28

Weidolf L, Bjorkbom A, Dahlen A et al (2021) Distribution and biotransformation of therapeutic antisense oligonucleotides and conjugates. Drug Discov Today 26(10):2244–2258

Wills RJ, Ferraiolo BL (1992) The role of pharmacokinetics in the development of biotechnologically derived agents. Clin Pharmacokinet 23(6):406–414

Yang BB (2006) Integration of pharmacokinetics and pharmacodynamics into the drug developmetn of pegfilgrastim, a pegylated protein. In: Meibohm B (ed) Pharmacokinetics and pharmacodynamics of biotech drugs. Wiley, Weinheim, pp 373–394

Yapa SW, Roth D, Gordon D, Struemper H (2016) Comparison of intravenous and subcutaneous exposure supporting dose selection of subcutaneous belimumab systemic lupus erythematosus phase 3 program. Lupus 25(13):1448–1455

Zamboni WC (2003) Pharmacokinetics of pegfilgrastim. Pharmacotherapy 23(8 Pt 2):9S–14S

Zhang Y, Meibohm B (2012) Pharmacokinetics and pharmacodynamics and therapeutic peptides and proteins. In: Kayzer O, Warzecha H (eds) Pharmaceutical biotechnology: drug discovery and clinical Apllications. Wiley, Weinheim, pp 337–368

Zhang L, Pfister M, Meibohm B (2008) Concepts and challenges in quantitative pharmacology and model-based drug development. AAPS J 10(4):552–559

Zhang MM, Bahal R, Rasmussen TP, Manautou JE, Zhong XB (2021) The growth of siRNA-based therapeutics: updated clinical studies. Biochem Pharmacol 189:114432

Zhao L, Ji P, Li Z, Roy P, Sahajwalla CG (2013) The antibody drug absorption following subcutaneous or intramuscular administration and its mathematical description by coupling physiologically based absorption process with the conventional compartment pharmacokinetic model. J Clin Pharmacol 53(3):314–325

Zhu H, Huang SM, Madabushi R, Strauss DG, Wang Y, Zineh I (2019) Model-informed drug development: a regulatory perspective on Progress. Clin Pharmacol Ther 106(1):91–93

Zia-Amirhosseini P, Minthorn E, Benincosa LJ et al (1999) Pharmacokinetics and pharmacodynamics of SB-240563, a humanized monoclonal antibody directed to human interleukin-5, in monkeys. J Pharmacol Exp Ther 291(3):1060–1067

Zito SW (1997) Pharmaceutical biotechnology: a programmed text, vol 10. Technomic Pub. Co., Lancaster, PA, p 328

Further Reading

General Pharmacokinetics and Pharmacodynamics

Bonate PL (2011) Pharmacokinetic-pharmacodynamic modeling and simulation. Springer, New York

Derendorf H, Meibohm B (1999) Modeling of pharmacokinetic/pharmacodynamic (PK/PD) relationships: concepts and perspectives. Pharm Res 16(2):176–185

Derendorf H, Schmidt S (2019) Rowland and Tozer's clinical pharmacokinetics and pharmacodynamics: concepts and applications. Lippincott, Baltimore

Gabrielsson J, Hjorth S (2012) Quantitative pharmacology. Swedish Academy of Pharmaceutical Sciences, Stockholm

Gibaldi M, Perrier D (1982) Pharmacokinetics. Marcel Dekker Inc., New York

Holford NH, Sheiner LB (1982) Kinetics of pharmacologic response. Pharmacol Ther 16(2):143–166

Huang SM, Lertora JJ, Atkinson AJ (2012) Principles of clinical pharmacology. Academic, San Diego

Pharmacokinetics and Pharmacodynamics of Proteins

Diao L, Meibohm B (2013) Pharmacokinetics and pharmacokinetic-pharmacodynamic correlations of therapeutic peptides. Clin Pharmacokinet 52(10):855–868

Diao L, Meibohm B (2015) Tools for predicting the PK/PD of therapeutic proteins. Expert Opin Drug Metab Toxicol 11(7):1115–1125

Kontermann R (2012) Therapeutic proteins: strategies to modulate their plasma half-lives. Wiley, Weinheim

Meibohm B (2006) Pharmacokinetics and pharmacodynamics of biotech drugs. Wiley, Weinheim

Mould DR, Meibohm B (2016) Drug development of therapeutic monoclonal antibodies. BioDrugs 30(4):275–293

Ryman JT, Meibohm B (2017) Pharmacokinetics of monoclonal antibodies. CPT Pharmacometrics Syst Pharmacol 6(9):576–588

Tang L, Persky AM, Hochhaus G, Meibohm B (2004) Pharmacokinetic aspects of biotechnology products. J Pharm Sci 93(9):2184–2204

Pharmacokinetics of Nucleic Acid-Based Therapeutics

Migliorati JM et al (2022) Absorption, distribution, metabolism, and excretion of US Food and Drug Administration-approved antisense oligonucleotide drugs. Drug Metab Dispos 50(6):888–897

Weidolf L et al (2021) Distribution and biotransformation of therapeutic antisense oligonucleotides and conjugates. Drug Discov Today 26(10):2244–2258

Immunogenicity of Therapeutic Proteins

Theo Rispens, Wim Jiskoot, and Grzegorz Kijanka

Introduction

Proteins were first used at the end of the nineteenth century when antisera from animals were introduced for the treatment of serious complications of infections, such as diphtheria and tetanus. However, because such antisera were loaded with proteins foreign to the patients' immune system, they often led to serious and sometimes even fatal side effects. Persons who had been treated in general had a warning in their passports or identification cards to alert physicians for a possible anaphylactic reaction after rechallenge with an antiserum.

In the 1920s, porcine and bovine insulin products were introduced to treat diabetes. Many patients receiving these insulins developed antibodies neutralizing the drug (antidrug antibodies, ADA). At first, this had been again explained by the animal origin of the products. However, reduction of immunogenicity of these products following improvements in the production methods and increasing purity strongly indicated that animal origin of protein was not the only factor leading to their immunogenicity. In the second half of the twentieth century, a number of human proteins from natural sources such as plasma derived clotting factors and growth hormone produced from pituary glands of cadavers were introduced into the clinic. These products were given mainly to children with an innate deficiency who therefore lacked the natural immune tolerance. Therefore, their immune response was also interpreted as a response to foreign proteins. The correlation between the factor VIII gene defect and level of deficiency with the immune response in hemophilia patients confirmed this explanation (Fakharzadeh and Kazazian 2000). However, similar to the animal derived insulins, improved purification protocols led to reduced immunogenicity levels for some of the human protein products.

The true breakthrough in availability of therapeutic proteins occurred in the 1980s when recombinant DNA technologies allowed large-scale production of highly purified proteins. In 1982, human insulin was marketed as the first recombinant DNA-derived protein for human use. Since then, dozens of recombinant proteins have been introduced and some of these products, such as the interferons and the epoetins, are among the most widely used drugs in the world. However, although these proteins were developed as close copies of human endogenous proteins, nearly all of them induce ADA, sometimes even in a majority of patients (Table 7.1). Importantly, several of these products are used in patients who do not have an innate deficiency and, presumably, are immune tolerant to the protein.

The initial assumption was that the production by recombinant technology in nonhuman host cells and the downstream processing modified the proteins and the immunological response was the classical response to a foreign protein. However, in some cases, the antibody response to human homologs might be based on circumventing B-cell tolerance. This phenomenon is not yet completely understood but appears to be different from the immune response against foreign antigens used in vaccines. On the other hand, for therapeutic monoclonal antibodies, there are usually determinants present in the molecule that are foreign to the recipient (see section below, "Issues specifically related to monoclonal antibodies").

Sadly, Wim Jiskoot passed away by the time of publication of this sixth edition.

T. Rispens (✉)
Department of Immunopathology, Sanquin Research, Amsterdam, The Netherlands
e-mail: t.rispens@sanquin.nl

W. Jiskoot (Deceased)
Leiden Academic Center for Drug Research (LACDR), Leiden University, Leiden, The Netherlands

G. Kijanka
Division of BioTherapeutics, Leiden Academic Centre for Drug Research (LACDR), Leiden University, Leiden, The Netherlands

Table 7.1 Nonexhaustive list of recombinant proteins showing immune reactions upon administration

Drug	Immunogenicity rate	Reference
Growth hormone	7–22%	Rougeot et al. (1991)
Factor VIII	3–35%	Oldenburg et al. (2015)
Insulin	14–44%	Fineberg et al. (2007)
Interferon beta	2–47%	Bertolotto et al. (2004)
Monoclonal antibodies	0–89%	Harding et al. (2010)
Interleukin-2	20–100%	Prümmer (1997)
Alfa-galactosidase A	5–91%	Deegan (2012)

The clinical manifestations of both types of reaction are very different. The typical vaccine-like response to foreign proteins occurs within days to weeks and often a single injection is sufficient to induce a substantial ADA response. In general, high levels of neutralizing antibodies are induced, and a rechallenge leads to a booster reaction, indicating a memory response.

However, the development of an immune response against certain recombinant human proteins may require months, sometimes years of chronic treatment. Moreover, secreted antibodies often do not neutralize the injected therapeutic protein and sometimes even do not manifest any apparent clinical effects. Additionally, in some cases, these ADA, especially when produced in low or moderate amounts, tend to disappear shortly after the treatment has been stopped and sometimes even during treatment (Perini et al. 2001). This response does not appear to generate immunological memory. In patients treated with recombinant human interferon β (rhIFNβ) or recombinant human growth hormone, the therapy often can be paused to allow the ADA levels to decline and then restarted without boosting of ADA titers (Schellekens and Casadevall 2004; Perini et al. 2001).

In contrast, the ADA response against therapeutic monoclonal antibodies, such as adalimumab and infliximab, follows a more classical pattern (Brandse et al. 2016; Bartelds et al. 2011), with ADA detected as early as 2 weeks after start of treatment. Moreover, the vast majority of these ADA is neutralizing (van Schie et al. 2015a; van Schie et al. 2017), and intermittent infliximab treatment is associated with more immunogenicity (Han and Cohen 2004).

The Immune Response

The therapeutic proteins currently available cover the whole spectrum, from completely foreign (e.g., bacterial derived asparaginase) to completely human (e.g., recombinant human interferon α-2b (rhIFNα-2b)) as well as unnatural proteins, such as fusion proteins (e.g., etanercept), truncated antibody (like) proteins (e.g., Fab fragments, nanobodies), and PEGylated proteins (e.g., PEGylated rhIFNα).

Foreign proteins elicit antibodies by the classical pathway that includes ingestion and cleaving of the proteins into peptides by macrophages and dendritic cells, presentation of peptides by the MHC-II system and activation of B-cells and boosting, and affinity maturation and isotype switching of the B-cells by helpe CD4[+] T-cells. Furthermore, memory B-cells are induced (see Chap. 15 for details). This situation is encountered with, e.g., proteins in case of genetic defects. Also, monoclonal antibodies, regardless of their degree of humanization, will always contain foreign determinants (see section: Issues specifically related to monoclonal antibodies).

It is much less clear how B-cell tolerance is circumvented. One of the theories trying to explain antibody formation against a self(-like) therapeutic protein was that aggregated protein induced exclusively T-cell independent B-cell activation via crosslinking of B-cell receptors by protein aggregates resembling bacterial or viral structures (Moore and Leppert 1980; Bachmann et al. 1993). Such structures present repetitive epitopes in a highly regular matter allowing rapid recognition as foreign substance and response without engaging CD4[+] T-cells. However, since this mechanism was first proposed, numerous preclinical and clinical observations have indicated involvement of CD4[+] T-cells in the antibody response against therapeutic self(-like) proteins. During CD4[+] T-cell-independent responses mostly low affinity IgMs are produced, but in patients and in animal models, mostly IgGs have been observed. Efficient isotype switching is one of the hallmarks of a CD4[+] T-cell response. However, production of non-neutralizing antibodies in part of the patients suggests impaired affinity maturation.

Another indication of a CD4[+] T-cell-dependent mechanism is the observation that certain HLA alleles correlate with a higher probability of ADA formation against, e.g., rhIFNβ, erythropoietin, and anti-TNF mAb (Barbosa et al. 2006; Moss et al. 2013; Praditpornsilpa et al. 2009; Fijal et al. 2008). Moreover, in several animal models, CD4[+] T-cell depletion resulted in almost complete abolishment of antibody production (reviewed by Jiskoot et al. (2016)). Nevertheless, full explanations for immunogenicity occurrence in only part of the patients receiving the protein, the frequent formation of non-neutralizing antibodies, and the apparent lack of immunological memory are still missing.

Factors Influencing Antibody Formation to Therapeutic Proteins

Figure 7.1 depicts the different factors that influence immunogenicity (Schellekens 2002a; Hermeling et al. 2004).

Structural Factors

The similarity of a therapeutic protein's sequence to that of the human counterpart is one of the factors determining the probability of an ADA response against the protein drug. However, the degree of nonself necessary to induce a vaccine-type response is highly dependent on the protein involved and the site of the divergence from the natural sequence of the endogenous protein. For instance, single mutations in insulin can lead to a new epitope and a response, whereas other mutations have no influence at all; and consensus IFNα, in which more than 10% of the amino acids diverge from the nearest naturally occurring IFNα subtype, is not more immunogenic than the IFNα-2 homologue. Moreover, some unnatural proteins, such as etanercept, are relatively poorly immunogenic.

Glycosylation is another important structural factor for the immunogenicity of therapeutic proteins. There is little evidence that modified glycosylation, e.g., by expressing human glycoproteins in plant cells or other nonhuman eukaryotic hosts, may induce an immune response (Singh 2011). However, the level of glycosyl-ation has a clear effect. For instance, rhIFNβ produced in *E. coli* (nonglycosylated) is much more immunogenic than rhIFNβ produced in mammalian cells. This may be explained by the higher hydrophobicity, causing higher aggregation levels in the nonglycosylated *E. coli*-derived product.

Furthermore, in certain populations, preexisting IgE antibodies to nonhuman glycans present on cetuximab resulted in severe anaphylactic reactions (Chung et al. 2008).

Impurities

Impurities are considered to be important risk factors for the immunogenicity of therapeutic protein products. Among the clinically relevant impurities, protein aggregates have received most attention; the presence of aggregates is widely accepted as one of the most important risk factors for immunogenicity. Aggregation may be triggered by a variety of factors, such as thermal stress, pH shift, agitation, freeze-thawing, and UV irradiation. Importantly, aggregates may be induced at every step of the production process, storage, shipment, or even during product administration (cf. Chap. 5). Aggregation is believed to be one of the main causes of immunogenicity of, e.g., human growth hormone or rhIFNβ-1b (Moore and Leppert 1980; Barnard et al. 2013). However, it has to be noted that aggregation is not the only risk factor and products containing small amounts of aggregates might also be very immunogenic.

Fig. 7.1 Factors influencing immunogenicity (Schellekens 2002a)

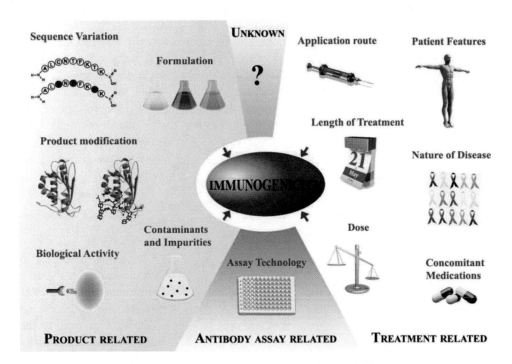

Chemical modification of human proteins, e.g., oxidation, might lead to formation of neo-epitopes, which might be recognized by the patient's immune system and trigger ADA. Moreover, as demonstrated in numerous animal studies, antibodies induced by chemically modified proteins can be cross-reactive with unmodified protein (Jiskoot et al. 2016).

Besides aggregates and chemically modified proteins, substances derived from the production process, such as host cell components, resins from chromatographic columns, enzymes used to activate the product, and monoclonal antibodies used for affinity purification, may end up in the final product. Impurities may also be introduced by components of the formulation, may leak from the container and sealing of the product, or may be introduced during the fill and finish steps. These impurities may be immunogenic by themselves and thereby elicit an immune response different than that to the therapeutic protein. Moreover, endotoxin from bacterial host cells and G-C-rich bacterial host cell DNA are examples of impurities that can act as adjuvants. Antibodies induced by impurities may lead to general immune reactions such as skin reactions, allergies, anaphylaxis, and serum sickness. Antibodies to impurities also raise quality issues concerning the product and therefore need to be monitored.

Formulation

Human therapeutic proteins are often highly biologically active and the doses may be at the µg level, making it a technological challenge to formulate a stable product with a reasonable shelf-life and to avoid the formation of aggregates and other product modifications (cf. Chap. 5) The importance of formulation in avoiding immunogenicity is highlighted in two historical cases. In the case of rhIFNα-2a, a large difference in immunogenicity was noted among different formulations (Fig. 7.2). A freeze-dried formulation, containing human serum albumin as a stabilizer that according to its instructions could be kept at room temperature, was

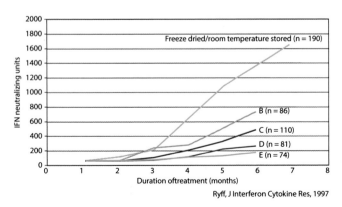

Fig. 7.2 Immunogenicity differences between rhIFNα-2a formulations in patients

particularly immunogenic. It appeared that at room temperature rhIFNα-2a became partly oxidized. This led to the formation of aggregates, which most likely were responsible for the immune response (Ryff 1997). Interestingly, studies in preclinical animal models have confirmed that aggregates induced by oxidation are particularly immunogenic degradants (Jiskoot et al. 2016).

In the second case, an ADA-mediated severe form of anemia (pure red cell aplasia; PRCA) occurred after the formulation of an epoetin-α product was changed (Casadevall et al. 2002). Human serum albumin was replaced by glycine and polysorbate 80. How this formulation change led to a higher incidence of immunogenicity is still not certain, but it has been postulated that the new formulation is slightly less stable, resulting in aggregate formation when the product is not appropriately handled (Schellekens and Jiskoot 2006).

Route of Administration

Historically, the subcutaneous route was considered to be the most and intravenous the least immunogenic among administration routes used for protein drugs regardless of other circumstances (Schellekens 2002b). However, direct, head-to-head comparison of administration routes in patients is challenging, as the treatment regimen and/or the formulation have to be adjusted to compensate for altered pharmacokinetics and usually decreased bioavailability upon subcutaneous injection (Richter et al. 2012), all of which may influence the risk of an immune response. Subcutaneous formulations usually differ from intravenous ones because subcutaneous administration generally requires lower injection volumes. Overall, the influence of the administration route on immunogenicity seems to be strongly product dependent. Whereas for some products the subcutaneous route indeed seems to be associated with higher immunogenicity risk as compared to the intravenous route, for others the influence of administration route seems to be very minor and immunogenicity risk seems to be similar for different routes (Hamuro et al. 2017). Nevertheless, immunogenicity can be seen after any route of application.

Dose

The effect of the dose is not quite clear. There are studies with the lowest incidence of ADA formation in the highest dose group (Maini et al. 1999; Food and Drug Administration 2002; Ross et al. 2000). However, such data should be interpreted with caution. In the highest dose group, the higher drug levels in the circulation may interfere with the assay, or the measured ADA level may be lower by increased immune complex formation. Nevertheless, high-dose drug exposure has been exploited to overcome immunogenicity issues, in particular for FVIII therapy (Hausl et al. 2005).

Patient Features

Several patient related factors may influence the incidence of ADA formation. For example, many of the patients receiving, e.g., monoclonal antibodies are immune compromised by diseases such as cancer or by immune suppressive treatment. These patients are less likely to produce ADA than patients with a normal immune status. An opposite effect can happen in patients suffering from chronic infections (like hepatitis): these patients may be more prone to development of ADA, as observed for rhIFNα-2 (Antonelli and Dianzani 1999).

There are several indications that circumventing B-cell tolerance is HLA-type dependent. This may explain, at least in part, why some individuals or populations do and others do not produce ADA when treated with the same product. For, instance, for infliximab associations between HLA-type and ADA formation were reported (Billiet et al. 2015).

In many patients treated with therapeutic proteins, ADA are found before the start of the treatment (Gorovits et al. 2016). These so called preexisting ADA might be formed against structures, e.g., common glycans, which closely resemble epitopes present in the protein drug. In many of these patients the preexisting ADA don't seem to have any impact on the therapy's safety or efficiency. However, since some patients who tested positive for preexisting ADA might be more susceptible to acute side effects or development of high titers of ADA, preexisting ADA are considered as an immunogenicity risk factor.

Assays for Antibodies

Assays are an important factor influencing the reported incidence of ADA induction by therapeutic proteins. In the published studies with rhIFNα-2 in patients with viral infections the incidence of ADA induction varied from 0% to more than 60% positive patients. A similar variation has been seen with rhIFNβ (van Beers et al. 2010). This variation must be assay related. Evaluations of the performance of different test laboratories with blind panel testing showed a more than 50-fold difference in titers found in the same sera. Thus, any reliable comparison between different groups of patients when looking for a clinical effect of ADA or studying factors influencing immunogenicity can only be done if the antibody quantification is done with in a well-validated assay in the same laboratory.

This situation existed when evaluating antibody formation against TNF inhibitors (infliximab, adalimumab), resulting in incidence rates varying from 0 to 87% in case of adalimumab (Vincent et al. 2013). Several factors account for these large discrepancies between studies, but one factor in particular was found to be of prime importance: drug interference. In particular for therapeutic monoclonal antibodies, which are usually dosed at high concentrations, serum samples will often contain considerable amounts of drug that may interfere with the detection of ADA. Accordingly, many efforts have been made to design assays that overcome this drug interference, resulting in assays that are "drug-tolerant" to variable degrees (Bloem et al. 2015).

The relationship between drug concentrations and ADA is important not only for the detection of ADA but also for the impact of ADA on efficacy. Often the main consequence of ADA formation is the reduction of "active" drug concentrations. Thus, small amounts of ADA without a noticeable effect on the drug concentrations may be detected with a modern drug-tolerant assay but have no impact on efficacy (van Schouwenburg et al. 2013a).

Recommendations for Antibody Assays

There is a lack of standardization of assay methodology, and only a few reference and/or standard antibody preparations are available. Nevertheless, a number of papers appeared with recommendations for setting up and validating immunoassays for immunogenicity testing (Mire-Sluis et al. 2004; Shankar et al. 2008, 2014). A brief discussion of these recommendations follows below.

A single assay may not be sufficient to evaluate the immunogenicity of a new protein drug, and a number of assays may have to be used in conjunction. Most antibody assay strategies are based on a two-tier approach: a screening + confirmation assay to identify the ADA positive sera followed by further characterization, e.g., establishing whether the antibodies are neutralizing, their titer, affinity, and isotype. Especially quantification may be useful, given that ADA levels can vary widely across individuals and the potential clinical impact of an ADA response usually relates to the extent of antibody formation.

In general, the screening assay is a binding assay, often a bridging type of assay (either ELISA or ECL; see Chap. 3) with the radio-immune-precipitation methodology as an alternative. Screening assays are designed for optimal sensitivity to avoid false negatives, and often, the cut point (i.e., threshold positive/negative) for the assay is set at a 5% false positive level by using a panel of normal human sera and/or untreated patient sera representative of the groups to be treated. The results would have to be evaluated in conjunction with a confirmatory assay that evaluates those samples found positive in the screening assay. The confirmatory assay may be the same screening assay, in which excess unlabeled drug is added to evaluate if the signal is reduced. Moreover, a stricter confirmatory cut point is defined to make sure that only "true" positive samples are identified as such.

One issue that complicates cut-point determination is the possible existence of preexisting antibodies. In rare cases (e.g., cetuximab, see above), these might have clinical consequences. More often, preexisting antibodies towards a protein drug may be detected in a (small) fraction of individuals without a clear clinical impact (Xue and Rup 2013). This needs to be dealt with on a case-by-case basis (Gorovits et al. 2016). One common type of preexisting antibodies that is found in most rheumatoid arthritis patients but also in a few percent of the general population is the so-called "rheumatoid factor," low-affinity antibodies binding to the Fc portion of human IgG. There is no evidence for these antibodies being relevant in immunogenicity assessment and measures should be taken to avoid their measurement (van Schie et al. 2015b). Antibodies to PEG groups attached to proteins are another commonly observed phenomenon. Their accurate measurement has been challenging, hampering the evaluation of their potential risk (Schellekens et al. 2013; Krishna et al. 2015). Recent technological developments have resulted in assays and standards that appear to more reliably detect anti-PEG antibodies, and studies suggest a high incidence in the general population (Ehlinger et al. 2019). The clinical significance of these antibodies is uncertain. Anti-PEG antibodies emerging after protein drug exposure, but not preexisting anti-PEG antibodies, have been associated with enhanced clearance for pegloticase (Lipsky et al. 2014).

The assay for neutralizing antibodies is in general a modification of the potency assay for the therapeutic protein product. The potency assay is in most cases an in vitro cell-based assay. A predefined amount of product is added to the serum and a reduction of activity evaluated in the bioassay. An appropriate alternative is the competitive ligand binding assay, which evaluates reduction in target binding (Finco et al. 2011). The latter type of assays are much easier to set up and validate and may be preferred unless there is a risk of antibodies formed to the drug that neutralize its activity via a mechanism other than preventing target binding.

Issues Specifically Related to Monoclonal Antibodies

The first generation of monoclonal antibodies was of murine origin. They induced an immune response in the majority of patients as foreign proteins should trigger a classical vaccine-type immune response. This so-called HAMA response (human antibodies to murine antibodies) was a major restriction in the clinical success of these murine antibodies. Over the years, however, methods were introduced in different stages to humanize monoclonal antibodies (cf. Chaps. 1 and 8). Recombinant DNA technology was used to exchange the murine constant parts of the immune globulin chains with their human counterparts resulting in chimeric monoclonal antibodies. The next step was to graft murine complementarity determining regions (CDR's), which determine the specificity, into a human immune globulin backbone creating humanized monoclonal antibodies. And the final step was the development of transgenic animals, phage display technologies, and other developments allowing the production of human monoclonal antibodies. The assumption that human monoclonal antibodies would have no immunogenicity proved to be wrong. Although humanization has reduced the immunogenicity, even completely human monoclonal antibodies have been shown to induce antibodies, as illustrated in Fig. 7.3.

As discussed, multiple factors may cause the immunogenicity of human therapeutic proteins including monoclonal

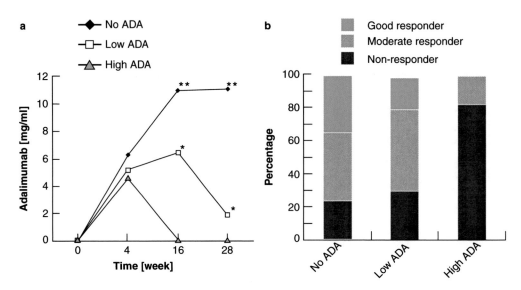

Fig. 7.3 Impact of immunogenicity on the bioavailability and treatment efficacy of fully human monoclonal antibody adalimumab. (**a**) The concentration of adalimumab in blood is altered by the presence of ADA. Low ADA titer corresponds to 12–100 AU/mL, high ADA titer corresponds to >100 AU/mL of ADA. *$p < 0.05$, **$p < 0.01$ (**b**) In patients with high ADA titers treatment is hampered to a higher extent than in patients with moderate or no ADA. (Figure adapted from Bartelds et al. (2011))

antibodies. One of them is aggregation. In fact, in the classical studies of B-cell tolerance performed more than 40 years ago aggregated immunoglobulin preparations were used to break tolerance (Weigle 1971). More recently, several pre-clinical studies have indicated that aggregates enhance the immunogenicity of monoclonal antibodies (reviewed by Jiskoot et al. 2016). However, also mAb products containing a very low number of aggregates may be highly immunogenic. In contrast to other recombinant human proteins, monoclonal antibodies, even when fully human, may expose foreign epitopes in their complementarity-determining regions (CDR). Analysis of several monoclonal antibodies indeed confirmed that, in most cases, foreign CD4+ T cells epitopes are found primarily within the CDR sequence (Harding et al. 2010).

Monoclonal antibodies have properties which may contribute to their immunogenicity. They can activate T-cells by themselves and may boost the immune response by their Fc functions such as macrophage activation and complement activation. Indeed, removal of N-linked glycosyl chains from the Fc part of the immunoglobulin may reduce Fc function and lead to a diminished immunogenicity (Liu 2015).

The molecule to which an antibody binds also influences its immunogenicity. Monoclonal antibodies targeting cell bound antigens generally induce a higher level of ADA than those targeting soluble targets (Harding et al. 2010). Monoclonal antibodies directed to antigens on immune cells with the purpose of inducing immune suppression also suppress an immunological response.

Although for a number of therapeutic proteins more injections and higher doses are associated with a higher immune response, this may not apply to therapeutic antibodies In fact, lower doses and episodic treatment are usually associated with less antibody formation (Han and Cohen 2004). Although this may in part be the result of ADA assays not detecting antibodies in the presence of high drug concentra-tions (see previous section), transient antibody formation has been reported for both infliximab and adalimumab, using drug-tolerant assays to measure the ADA response (van Schouwenburg et al. 2013b; van de Casteele et al. 2013). A phenomenon called "high-dose tolerance" might be operational, where the immune system is overwhelmed by soluble antigen.

Another important aspect when studying the immunogenicity of monoclonal antibodies is timing of the blood sampling of patients. These products may have a relative long half-life (several weeks) and the circulating product may interfere with the detection of induced antibodies and may lead to false negative results. Sampling sera up to 20 weeks after the patient has received the last injection may be necessary to avoid the interference of circulating monoclonal antibodies. Also, natural antibodies, soluble receptors and immune complexes may interfere with assays and lead to either false positive or false-negative results.

Clinical Effects of Induced Antibodies

Despite the methodological drawbacks, the list of protein products with clinically relevant immunogenic side effects is growing (Schellekens 2002a; Malucchi and Bertolotto 2008). The most common consequence is loss of efficacy. Sometimes this loss can be overcome by increasing the dose or changing to another product.

The most dramatic and undisputed complication occurs when the antibodies to the product cross neutralize an endogenous factor with an important biological function. This has been described for a megakaryocyte growth and differentiation factor which induced antibodies cross reacting with endogenous thrombopoietin (see Table 7.2). Volunteers and patients in a clinical trial developed severe thrombocytopenia and needed platelet transfusions. Because of this compli-

Table 7.2 Examples of clinical consequences of immunogenicity of therapeutic proteins (adopted from Schellekens (2002a) and Malucchi and Bertolotto (2008))

Clinical consequence	Drug
Loss of efficacy	• Animal and recombinant human (rh) insulin • Factor VIII (both natural and rh) • Rh interferon alpha 2 • Rh interferon beta • Rh interleukin 2 • Human chorionic gonadotropin • Monoclonal antibodies
Neutralization of endogenous protein	• Rh megakaryocyte-derived growth factor • Epoetin
No apparent biological consequence	• Rh growth hormone • Rh insulin • Monoclonal antibodies
General immune effects: (i) Allergy (ii) Serum sickness (iii) Anaphylaxis (iv) Injection site reactions	• Various therapeutic proteins

cation, the product was withdrawn from further development.

Another example is the upsurge of PRCA (see above) associated with a formulation change of epoetin-α marketed outside the USA. The antibodies induced by the product neutralized the residual endogenous erythropoietin in these patients resulting in a severe anemia which could only be treated with blood transfusions.

Antibodies can also influence the side effects of therapeutic proteins. The consequences are dependent on the cause of the side effects. If the adverse effects are the results of the intrinsic activity of the products, antibodies may reduce the side effects as it is the case with rhIFNα-2. Sometimes the mitigation of the side effects is even the first clinical sign of the induction of antibodies.

With some products the side effects are caused by the ADA formation. This is in general the case when the product is administered in relatively high doses, like with some monoclonal antibodies. Symptoms caused by immune complexes like delayed type hypersensitivity and serum sickness are related to the level of antibodies induced.

The general effects caused by an immune reaction to a therapeutic protein such as acute anaphylaxis, hypersensitivity, skin reaction, and serum sickness are relatively common when large amounts of nonhuman proteins are administered. These effects are relatively rare for modern biotechnology-derived products, which are highly purified human proteins administered in relatively low amounts. However, these side effects caused by an immune response are currently still relatively common during treatment with high doses of monoclonal antibodies.

Immunogenicity Risk Assessment and Reduction

The occurrence of immunogenicity is probably seldom the result of a single risk factor. Rather, coincidence of several factors is believed to be needed to trigger immunogenicity. For example, the higher immunogenicity of interferon beta 1b (Betaferon) than interferon beta 1a (Avonex) is most likely a combined result of differences in treatment regimen and product quality (Bertolotto et al. 2004). Moreover, the immune mechanisms leading to antibody induction by therapeutic proteins are still not completely understood. As a consequence, it is impossible to fully predict the immunogenicity of a new product in a patient population, let alone in individual patients. However, immunogenicity risk assessment

and potential mitigation strategies are required by authorities for all new products. Most commonly used tool types for assessing relative immunogenicity risk are summarized in Table 7.3.

Tools to Assess Immunogenicity

According to current knowledge, CD4+ T-cells are key players in immunogenicity. Therefore, one of the most commonly used approaches to assess immunogenicity risk is an in silico analysis of CD4+ T-cell epitopes, i.e., short peptides within the protein sequence capable of binding to MHC II molecules. Multiple algorithms have been developed for this purpose (reviewed by Jawa et al. 2013) Generally speaking, the more high-affinity CD4+ T-cell epitopes a protein contains, the higher the expected risk of immunogenicity. Recent developments in understanding of the immune system allowed improving in silico prediction algorithms. Instead of assessing the overall/total number or CD4+ T-cell epitopes, nowadays epitopes for effector and regulatory CD4+ T-cells can be discriminated (Cousens et al. 2014). Combined analysis of effector and regulatory CD4+ T-cell analysis has been suggested to be more reliable than analysis of total CD4+ T-cell epitopes. However, as these approaches rely exclusively on the primary sequence of the protein and do not take into consideration other factors such as protein folding, post-translational modifications, presence of impurities, and complexity of the interaction between different immune cells, their predictive power is limited especially for self(−like) proteins.

Another approach is represented by a number of in vitro CD4+ T-cell stimulation and MHC II binding tests. Although, similarly to in silico tools, these assays do not fully represent the complexity of the immune system, they do allow assessing the impact of formulation related factors on immunogenicity risk. Thus, they can be used for validation of in silico results and to measure relative immunogenicity of different protein variants or different formulations of the same protein.

A third type of immunogenicity assessment tool is a set of in vivo models ranging from wild type and transgenic mice to nonhuman primates (Jiskoot et al. 2016; Brinks et al. 2013). The main benefit of in vivo models is that complexity of the tested immune system is similar to that of patients. Thus, they might be used not only to assess the immunogenicity but also to study immune mechanisms and to provide insight into the possible clinical effects of immunogenicity.

Table 7.3 Tools used for assessing protein immunogenicity (adopted from Brinks et al. (2013) and Kijanka et al. (2012))

Assessment tool		Pros	Cons
In Silico	• CD4⁺ T-cell epitope prediction • Regulatory T-cell epitope prediction	1. Fast and cheap 2. Useful as a first step of immunogenicity assessment 3. Allows design or selection of less immunogenic protein variant	1. Tendency to be over predictive 2. Results based exclusively on primary sequence 3. The quality of results depends on degree of understanding of studied process 4. Translational power limited
In vitro	• Dendritic cell uptake and activation • CD4⁺ T-cell activation • MHC II binding	1. Relatively fast and cheap 2. Enable studying formulated products (assessing factors other than primary sequence) 3. Some assays may allow studying biological effect	1. Quality of results strongly dependent on assay format 2. Often, samples from large set of donors required 3. Focus on isolated immune cells 4. Translational power limited
In vivo	• Nontransgenic rodents • Transgenic rodents • Nonhuman primates	1. (Usually) Presence of complete, functional immune system 2. Immune processes similar to those in patients 3. Various factors influencing immunogenicity can be studied 4. In some models, studying biological effect possible	1. Time-consuming and expensive 2. Translational power depends on protein and animal model 3. Immune system of used model differs from that of patient to various extents (nontransgenic mice vs transgenic mice)

Reducing Immunogenicity

Several strategies are being applied to reduce immunogenicity. At an early stage of development, one may alter the amino acid sequence of the lead candidate molecule in order to remove potentially immunogenic or aggregate prone sequences (Griswold and Bailey-Kellogg 2016). Furthermore, pharmacokinetic properties of the protein can relate to its immunogenicity. PEGylation is one strategy to prolong half-life of proteins and may or may not favorably impact immunogenicity (see above). IgG Fc fusion proteins also have in general a long half-life, and multiple examples of therapeutic proteins with low reported immunogenicity exist including etanercept and abatacept. Once the drug substance has been chosen, the immunogenicity maybe reduced by redesigning the formulation. As discussed previously, improvement in quality of the first therapeutic products resulted in considerable reduction of immunogenicity. The treatment regimen can also have a profound impact on the observed levels of immunogenicity. For instance, avoidance of intermittent use was found to reduce immunogenicity of infliximab. Furthermore, many protein therapeutics, especially mAbs, are used in combination with immunosuppressant drugs. Medicines such as methotrexate, rapamicin, or aziothioprine can strongly inhibit formation of antibodies, but patients might be more prone to infections or malignancy (de Mattos et al. 2015). For a number of proteins, targeting the antibody-producing cells have been explored to reduce the ADA response, including the use of rituximab, an anti-CD20 B cell-depleting antibody in the case of enzyme replacement therapy with alglucosidase alfa (Li et al. 2021). Another strategy is an induction of specific tolerance toward the protein. As discussed above, administration of high dose(s) of factor VIII may induce tolerance in hemophilia patients with antibodies to factor VIII (Aledort 1994).

Conclusions

The most important points of this chapter are summarized in the following bullet points:

- The immunogenicity of therapeutic proteins is a commonly occurring phenomenon.
- The clinical consequences can vary.
- Validated detection methods are essential to study the immunogenicity of therapeutic proteins.
- Correctly predicting immunogenicity in patients based on physico-chemical characterization and animal studies is difficult.
- There is still a lot to be learned about why and how patients produce antibodies to therapeutic proteins.

The growing awareness of the importance of immunogenicity of therapeutic proteins is illustrated by the adoption of a standard requirement in regulatory dossiers for new proteins and biosimilars to evaluate their immunogenicity in clinical trials (cf. Chap. 11).

Self-Assessment Questions

Questions

1. Which factors contribute to unwanted immunogenicity of therapeutic proteins?
2. What are possible clinical consequences of antibody formation against biopharmaceuticals in patients?
3. Why do aggregates of recombinant human proteins induce antibodies that cross-react with the (nonaggregated) drug?
4. Explain the fundamental difference between (a) antibody formation in children with growth hormone deficiency treated against recombinant human growth hormone and (b) antibody formation against rh erythropoietin in patients with chronic renal failure.
5. Give an example of a case that demonstrates that the formulation of a biopharmaceutical can affect the immune response.
6. Give at least 3 approaches that can be followed to reduce the immunogenicity of a biopharmaceutical.
7. Why is standardization of assays for detection of antidrug antibodies important?
8. Why are antidrug-antibody titers against a monoclonal antibody more difficult to determine accurately than antibodies against interferon?

Answers

1. See Fig. 7.1.
2. Reduction of therapeutic efficacy, (seldom) enhancement of efficacy, anaphylactic reactions, cross-reactivity with endogenous protein.
3. Aggregates can circumvent B-cell tolerance against native (like) epitopes (repetitive epitopes); the more "native like" the aggregate, the more likely cross-reactivity with the monomer will occur.
4. (a) is the classical immune response versus (b) circumventing B-cell tolerance.
5. The examples given in the text re erythropoietin and interferon α.
6. Design another formulation, remove aggregates, change the glycosylation pattern of the protein, use amino acid mutants, use human(ized) versions of the proteins or select another route of administration. NB some of these approaches will lead to a new bioactive drug molecule and that has implications for the way authorities will judge the procedure to be followed for obtaining marketing approval.
7. Different assay formats and blood sampling schedules give different answers and thus hamper direct comparison between studies. Therefore, it is difficult to compare the results obtained with different products that are tested for immunogenicity in different labs.
8. Monoclonal antibodies are often administered in high doses and have a long circulation time (days/weeks). This will likely cause interference with the assay by the circulating drug (resulting in false negatives or underestimation of antibody titers). Another possibility for interference is the occurrence of cross reactivity of the reagents in the test for the induced antibodies and the original drug–antibody. With interferon a different situation is encountered: interferons are rapidly cleared and administered in low doses (microgram range); therefore, interferons will less likely interfere with the measurement of anti-IFN antibodies.

References

Aledort L (1994) Inhibitors in hemophilia patients: current status and management. Am J Hematol 47:208–217
Antonelli G, Dianzani F (1999) Development of antibodies to interferon beta in patients: technical and biological aspects. Eur Cytokine Netw 10:413–422
Bachmann MF, Rohrer UH, Kündig TM, Bürki K, Hengartner H, Zinkernagel RM (1993) The influence of antigen organization on B cell responsiveness. Science 262:1448–1451
Barbosa MD, Vielmetter J, Chu S, Smith DD, Jacinto J (2006) Clinical link between MHC class II haplotype and interferon-beta (IFN-β) immunogenicity. Clin Immunol 118:42–50
Barnard JG, Babcock K, Carpenter JF (2013) Characterization and quantitation of aggregates and particles in interferon-β products: potential links between product quality attributes and immunogenicity. J Pharm Sci 102:915–928
Bartelds GM, Krieckaert CL, Nurmohamed MT, van Schouwenburg PA, Lems WF, Twisk JW, Dijkmans BA, Aarden L, Wolbink GJ (2011) Development of antidrug antibodies against adalimumab and association with disease activity and treatment failure during long-term follow-up. JAMA 305:1460–1468
Bertolotto A, Deisenhammer F, Gallo P, Sölberg SP (2004) Immunogenicity of interferon beta: differences among products. J Neurol 251:II15–II24
Billiet T, Vande Casteele N, Van Stappen T, Princen F, Singh S, Gils A, Ferrante M, Van Assche G, Cleynen I, Vermeire S (2015) Immunogenicity to infliximab is associated with HLA-DRB1. Gut 64:1344–1345
Bloem K, van Leeuwen A, Verbeek G, Nurmohamed MT, Wolbink GJ, van der Kleij D, Rispens T (2015) Systematic comparison of drug-tolerant assays for anti-drug antibodies in a cohort of adalimumab-treated rheumatoid arthritis patients. J Immunol Methods 418:29–38
Brandse JF, Mathôt RA, van der Kleij D, Rispens T, Ashruf Y, Jansen JM, Rietdijk S, Löwenberg M, Ponsioen CY, Singh S, van den Brink GR, D'Haens GR (2016) Pharmacokinetic features and presence of antidrug antibodies associate with response to infliximab induction therapy in patients with moderate to severe ulcerative colitis. Clin Gastroenterol Hepatol 14:251–258
Brinks V, Weinbuch D, Baker M, Dean Y, Stas P, Kostense S, Rup B, Jiskoot W (2013) Preclinical models used for immunogenicity prediction of therapeutic proteins. Pharm Res 30:1719–1728
Casadevall N, Nataf J, Viron B, Kolta A, Kiladjian JJ, Martin-Dupont P, Michaud P, Papo T, Ugo V, Teyssandier I, Varet B, Mayeux P (2002) Pure red-cell aplasia and antierythropoietin antibodies in

patients treated with recombinant erythropoietin. N Engl J Med 346:469–475

Chung CH, Mirakhur B, Chan E, Le QT, Berlin J, Morse M, Murphy BA, Satinover SM, Hosen J, Mauro D, Slebos RJ, Zhou Q, Gold D, Hatley T, Hicklin DJ, Platts-Mills TA (2008) Cetuximab-induced anaphylaxis and IgE specific for galactose-α-1,3-galactose. N Engl J Med 358:1109–1117

Cousens L, Najafian N, Martin WD, De Groot AS (2014) Tregitope: immunomodulation powerhouse. Hum Immunol 75:1139–1146

de Mattos BR, Garcia MP, Nogueira JB, Paiatto LN, Albuquerque CG, Souza CL, Fernandes LG, Tamashiro WM, Simioni PU (2015) Inflammatory bowel disease: an overview of immune mechanisms and biological treatments. Mediat Inflamm 2015:1–11

Deegan PB (2012) Fabry disease, enzyme replacement therapy and the significance of antibody responses. J Inherit Metab Dis 35:227–243

Ehlinger C, Spear N, Doddareddy R, Shankar G, Schantz A (2019) A generic method for the detection of polyethylene glycol specific IgG and IgM antibodies in human serum. J Immunol Methods 474:112669

Fakharzadeh SS, Kazazian HH Jr (2000) Correlation between factor VIII genotype and inhibitor development in hemophilia a. Semin Thromb Hemost 26:167–172

Fijal B, Ricci D, Vercammen E, Palmer PA, Fotiou F, Fife D, Lindholm A, Broderick E, Francke S, Wu X, Colaianne J, Cohen N (2008) Case–control study of the association between select HLA genes and anti-erythropoietin antibody-positive pure red-cell aplasia. Pharmacogenomics 9:157–167

Finco D, Baltrukonis D, Clements-Egan A, Delaria K, Gunn GR 3rd, Lowe J, Maia M, Wong T (2011) Comparison of competitive ligand-binding assay and bioassay formats for the measurement of neutralizing antibodies to protein therapeutics. J Pharm Biomed Anal 54:351–358

Fineberg SE, Kawabata TT, Finco-Kent D, Fountaine RJ, Finch GL, Krasner AS (2007) Immunological responses to exogenous insulin. Endocr Rev 28:625–652

Food and Drug Administration (2002) Adalimumab. 2002. Pharmacology reviews: adalimumab product approval information—licensing action. http://www.accessdata.fda.gov/drugsatfda_docs/nda/2008/125057s110TOC.cfm. Accessed on 4 Nov 2017

Gorovits B, Clements-Egan A, Birchler M, Liang M, Myler H, Peng K, Purushothama S, Rajadhyaksha M, Salazar-Fontana L, Sung C, Xue L (2016) Pre-existing antibody: biotherapeutic modality-based review. AAPS J 18:311–320

Griswold KE, Bailey-Kellogg C (2016) Design and engineering of deimmunized biotherapeutics. Curr Opin Struct Biol 39:79–88

Hamuro L, Kijanka G, Kinderman F, Kropshofer H, Bu DX, Zepeda M, Jawa V (2017) Perspectives on subcutaneous route of administration as an immunogenicity risk factor for therapeutic proteins. J Pharm Sci 106:2946–2954

Han PD, Cohen RD (2004) Managing immunogenic responses to infliximab: treatment implications for patients with Crohn's disease. Drugs 64:1767–1777

Harding FA, Stickler MM, Razo J, DuBridge RB (2010) The immunogenicity of humanized and fully human antibodies: residual immunogenicity resides in the CDR regions. MAbs 2:256–265

Hausl C, Ahmad RU, Sasgary M, Doering CB, Lollar P, Richter G, Schwarz HP, Turecek PL, Reipert BM (2005) High-dose factor VIII inhibits factor VIII-specific memory B cells in hemophilia A with factor VIII inhibitors. Blood 106:3415–3422

Hermeling S, Crommelin DJ, Schellekens H, Jiskoot W (2004) Structure-immunogenicity relationships of therapeutic proteins. Pharm Res 21:897–903

Jawa V, Cousens LP, Awwad M, Wakshull E, Kropshofer H, De Groot AS (2013) T-cell dependent immunogenicity of protein therapeutics: preclinical assessment and mitigation. Clin Immunol 149:534–555

Jiskoot W, Kijanka G, Randolph TW, Carpenter JF, Koulov AV, Mahler HC, Joubert MK, Jawa V, Narhi LO (2016) Mouse models for assessing protein immunogenicity: lessons and challenges. J Pharm Sci 105:1567–1575

Kijanka G, Jiskoot W, Sauerborn M, Schellekens H, Brinks V (2012) Immunogenicity of 12 therapeutic proteins. In: Hovgaard L, Frokjaer S, van de Weert M (eds) Pharmaceutical formulation development of peptides and proteins. CRC Press, Boca Raton, FL, pp 297–322

Krishna M, Palme H, Duo J, Lin Z, Corbett M, Dodge R, Piccoli S, Myler H, Pillutla R, Desilva B (2015) Development and characterization of antibody reagents to assess anti-PEG IgG antibodies in clinical samples. Bioanalysis 7:1869–1883

Li C, Desai AK, Gupta P, Dempsey K, Bhambhani V, Hopkin RJ, Ficicioglu C, Tanpaiboon P, Craigen WJ, Rosenberg AS, Kishnani PS (2021) Transforming the clinical outcome in CRIM-negative infantile Pompe disease identified via newborn screening: the benefits of early treatment with enzyme replacement therapy and immune tolerance induction. Genet Med 23:845–855

Lipsky PE, Calabrese LJ, Kavanaugh A, Sundy JS, Wright D, Wolfson M, Becker MA (2014) Pegloticase immunogenicity: the relationship between efficacy and antibody development in patients treated for refractory chronic gout. Arthritis Res Ther 16:R60

Liu L (2015) Antibody glycosylation and its impact on the pharmacokinetics and pharmacodynamics of monoclonal antibodies and fc-fusion proteins. J Pharm Sci 104:1866–1884

Maini R, St Clair EW, Breedveld F, Furst D, Kalden J, Weisman M, Smolen J, Emery P, Harriman G, Feldmann M, Lipsky P (1999) Infliximab (chimeric anti-tumour necrosis factor α monoclonal antibody) versus placebo in rheumatoid arthritis patients receiving concomitant methotrexate: a randomised phase III trial. Lancet 354:1932–1939

Malucchi S, Bertolotto A. (2008) Clinical aspects of immunogenicity to biopharmaceuticals. In: Weert MV, Møller EH. Immunogenicity of biopharmaceuticals Springer, New York, NY; 47: 27–56

Mire-Sluis AR, Barrett YC, Devanarayan V, Koren E, Liu H, Maia M, Parish T, Scott G, Shankar G, Shores E, Swanson SJ, Taniguchi G, Wierda D, Zuckerman LA (2004) Recommendations for the design and optimization of immunoassays used in the detection of host antibodies against biotechnology products. J Immunol Methods 289:1–16

Moore WV, Leppert P (1980) Role of aggregated human growth hormone (hGH) in development of antibodies to hGH. J Clin Endocrinol Metab 51:691–697

Moss AC, Brinks V, Carpenter JF (2013) Review article: Immunogenicity of anti-TNF biologics in IBD–the role of patient, product and prescriber factors. Aliment Pharmacol Ther 38:1188–1197

Oldenburg J, Lacroix-Desmazes S, Lillicrap D (2015) Alloantibodies to therapeutic factor VIII in hemophilia a: the role of von Willebrand factor in regulating factor VIII immunogenicity. Haematologica 100:149–156

Perini P, Facchinetti A, Bulian P, Massaro AR, Pascalis DD, Bertolotto A, Biasi G, Gallo P (2001) Interferon-Beta (INF-Beta) antibodies in interferon-beta1a- and interferon-beta1b-treated multiple sclerosis patients. Prevalence, kinetics, cross-reactivity, and factors enhancing interferon-beta immunogenicity in vivo. Eur Cytokine Netw 12:56–61

Praditpornsilpa K, Kupatawintu P, Mongkonsritagoon W, Supasyndh O, Jootar S, Intarakumthornchai T, Pongskul C, Prasithsirikul W, Achavanuntakul B, Ruangkarnchanasetr P, Laohavinij S, Eiam-Ong S (2009) The association of anti-R-HuEpo-associated pure red cell aplasia with HLA-DRB1*09-DQB1*0309. Nephrol Dial Transplant 24:1545–1549

Prümmer O (1997) Treatment-induced antibodies to interleukin-2. Biotherapy 10:15–24

Richter WF, Bhansali SG, Morris ME (2012) Mechanistic determinants of biotherapeutics absorption following sc administration. AAPS J 14:559–570

Ross C, Clemmesen KM, Svenson M, Sørensen PS, Koch-Henriksen N, Skovgaard GL, Bendtzen K (2000) Immunogenicity of interferon-beta in multiple sclerosis patients: influence of preparation, dosage, dose frequency, and route of administration. Ann Neurol 48:706–712

Rougeot C, Marchand PM, Dray F, Girard F, Job JC, Pierson M, Ponte C, Rochiccioli P, Rappaport R (1991) Comparative study of biosynthetic human growth hormone immunogenicity in growth hormone deficient children. Horm Res 35:76–81

Ryff JC (1997) Clinical investigation of the immunogenicity of interferon-alpha 2a. J Interf Cytokine Res 17:S29–S33

Schellekens H (2002a) Bioequivalence and the immunogenicity of biopharmaceuticals. Nat Rev Drug Discov 1:457–462

Schellekens H (2002b) Immunogenicity of therapeutic proteins: clinical implications and future prospects. Clin Ther 24:1720–1740

Schellekens H, Casadevall N (2004) Immunogenicity of recombinant human proteins: causes and consequences. J Neurol 251:II4–II9

Schellekens H, Jiskoot W (2006) Erythropoietin-associated PRCA: still an unsolved mystery. J Immunotoxicol 3:123–130

Schellekens H, Hennink WE, Brinks V (2013) The immunogenicity of polyethylene glycol: facts and fiction. Pharm Res 30:1729–1734

Shankar G, Devanarayan V, Amaravadi L, Barrett YC, Bowsher R, Finco-Kent D, Fiscella M, Gorovits B, Kirschner S, Moxness M, Parish T, Quarmby V, Smith H, Smith W, Zuckerman LA, Koren E (2008) Recommendations for the validation of immunoassays used for detection of host antibodies against biotechnology products. J Pharm Biomed Anal 48:1267–1281

Shankar G, Arkin S, Cocea L, Devanarayan V, Kirshner S, Kromminga A, Quarmby V, Richards S, Schneider CK, Subramanyam M, Swanson S, Verthelyi D, Yim S (2014) Assessment and reporting of the clinical immunogenicity of therapeutic proteins and peptides—harmonized terminology and tactical recommendations. AAPS J 16:658–673

Singh SK (2011) Impact of product-related factors on immunogenicity of biotherapeutics. J Pharm Sci 100:354–387

van Beers MM, Jiskoot W, Schellekens H (2010) On the role of aggregates in the immunogenicity of recombinant human interferon beta in patients with multiple sclerosis. J Interf Cytokine Res 30:767–775

van de Casteele N, Gils A, Singh S, Ohrmund L, Hauenstein S, Rutgeerts P, Vermeire S (2013) Antibody response to infliximab and its impact on pharmacokinetics can be transient. Am J Gastroenterol 108:962–971

van Schie KA, Hart MH, de Groot ER, Kruithof S, Aarden LA, Wolbink GJ, Rispens T (2015a) The antibody response against human and chimeric anti-TNF therapeutic antibodies primarily targets the TNF binding region. Ann Rheum Dis 74:311–314

van Schie KA, Wolbink GJ, Rispens T (2015b) Cross-reactive and pre-existing antibodies to therapeutic antibodies–effects on treatment and immunogenicity. MAbs 7:662–671

van Schie KA, Kruithof S, van Schouwenburg PA, Vennegoor A, Killestein J, Wolbink G, Rispens T (2017) Neutralizing capacity of monoclonal and polyclonal anti-natalizumab antibodies: the immune response to antibody therapeutics preferentially targets the antigen-binding site. J Allergy Clin Immunol 139:1035–1037

van Schouwenburg PA, Rispens T, Wolbink GJ (2013a) Immunogenicity of anti-TNF biologic therapies for rheumatoid arthritis. Nat Rev Rheumatol 9:164–172

van Schouwenburg PA, Krieckaert CL, Rispens T, Aarden L, Wolbink GJ, Wouters D (2013b) Long-term measurement of anti-adalimumab using pH-shift-anti-idiotype antigen binding test shows predictive value and transient antibody formation. Ann Rheum Dis 72:1680–1686

Vincent FB, Morand EF, Murphy K, Mackay F, Mariette X, Marcelli C (2013) Antidrug antibodies (ADA) to tumour necrosis factor (TNF)-specific neutralising agents in chronic inflammatory diseases: a real issue, a clinical perspective. Ann Rheum Dis 72:165–178

Weigle WO (1971) Recent observations and concepts in immunological unresponsiveness and autoimmunity. Clin Exp Immunol 9:437–447

Xue L, Rup B (2013) Evaluation of pre-existing antibody presence as a risk factor for posttreatment anti-drug antibody induction: analysis of human clinical study data for multiple biotherapeutics. AAPS J 15:893–896

Further Reading

Brinks V, Jiskoot W, Schellekens H (2011) Immunogenicity of therapeutic proteins: the use of animal models. Pharm Res 28:2379–2385

Ducret A, Ackaert C, Bessa J, Bunce C, Hickling T, Jawa V, Kroenke MA, Lamberth K, Manin A, Penny HL, Smith N, Terszowski G, Tourdot S, Spindeldreher S (2011) Assay format diversity in preclinical immunogenicity risk assessment: toward a possible harmonization of antigenicity assays. MAbs 14:1993522

Filipe V, Hawe A, Schellekens H, Jiskoot W (2010) Aggregation and immunogenicity of therapeutic proteins, in aggregation of therapeutic proteins. Wiley, Hoboken, NJ

Hermeling S, Crommelin DJA, Schellekens H, Jiskoot W (2007) Immunogenicity of therapeutic proteins. In: Gad SC (ed) Handbook of pharmaceutical biotechnology. Wiley, Hoboken, NJ, pp 911–931

Moussa EM, Panchal JP, Moorthy BS, Blum JS, Joubert MK, Narhi LO, Topp EM (2016) Immunogenicity of therapeutic protein aggregates. J Pharm Sci 105:417–430

Schellekens H (2010) The immunogenicity of therapeutic proteins. Discov Med 9:560–564

Schellekens H, Crommelin D, Jiskoot W (2007) Immunogenicity of antibody therapeutics. In: Dübel S (ed) Handbook of therapeutic antibodies. Wiley, Weinheim, pp 267–276

Monoclonal Antibodies: From Structure to Therapeutic Application

8

Rong Deng, Junyi Li, C. Andrew Boswell, Amita Joshi, and Chunze Li

Introduction

The field of therapeutic monoclonal antibodies (mAbs) had its origins when Köhler and Milstein presented their murine hybridoma technology in 1975 (Kohler and Milstein 1975). This technology provides a reproducible method for producing mAbs with unique target selectivity in almost unlimited quantities. In 1984, both scientists received the Nobel Prize for their scientific breakthrough, and their work is viewed as a key milestone in the history of mAbs as therapeutic modalities and their other applications. Although it took some time until the first therapeutic mAb received FDA approval in 1986 (Orthoclone OKT3, Chap. 25), mAbs are now the standard of care in several disease areas. In particular, in oncology (Chap. 23), inflammatory diseases (Chap. 24), transplantation (Chap. 25), infectious diseases, and neurological diseases (Chap. 26), patients now have novel life-changing treatment alternatives for diseases that had very limited or nonexistent medical treatment options before the emergence of mAbs. Up to Q1 2022, more than 117 mAbs and mAb derivatives, including fusion proteins and mAb fragments, are available for a variety of indications (Table 8.1). The majority of approved biologic therapies are mAbs, antibody–drug conjugates (ADCs), antibody fragments, and Fc fusion proteins. Technological evolutions have subsequently allowed much wider application of mAbs thanks to the ability to generate mouse/human chimeric, humanized, and fully human mAbs from antibodies (Abs) of pure murine origin. In particular, the reduction of the xenogenic portion of the mAb structure decreased the immunogenic potential of murine mAbs, allowing their wider application. mAbs are generally well-tolerated drugs because of their target selectivity, thus avoiding unnecessary exposure to, and consequently activity in, nontarget organs. This is particularly apparent in the field of oncology where mAbs like rituximab, trastuzumab, bevacizumab, cetuximab and immune-oncology mAbs like atezolizumab, pembrolizumab, and nivolumab can offer a more favorable risk–benefit profile compared to common chemotherapeutic treatment regimens for some hematologic cancers and solid tumors.

The advent of mAbs not only resulted in new drugs but also triggered the development of an entirely new business model for drug research and development and the founding of hundreds of biotech companies focused on mAb development. Furthermore, the ability to selectively target disease-related molecules with mAbs helped to launch a new era of targeted medicine and set new standards for successful drug research and development. The term "translational medicine" was coined to describe the use of biochemical, biological, and (patho)physiological understanding to find novel interventions to treat disease. During this process, biomarkers (e.g., genetic expression levels of marker genes, protein expression of target proteins, or molecular imaging) can be used to gain deeper understanding of the biological activities of drugs in a qualitative and, most importantly, quantitative sense, essentially encompassing the entire field of pharmacokinetics and pharmacodynamics (PK/PD). The application of these scientific methods, together with the principle of molecular-targeted medicine and the favorable PK and safety of mAbs, may at least partly explain the higher success rates of biotechnologically derived products in reaching the market compared to chemically derived small-molecule drugs.

This chapter addresses the following topics: Antibody structure and classes, currently approved mAb based therapies, mechanisms of action, clinical development, and drug properties. In this sense, this chapter provides a general introduction to Chaps. 23, 24, 25, and 26, where the currently marketed mAbs and mAb derivatives including antibody fragments, fusion proteins, and ADCs are discussed in

R. Deng · J. Li · A. Joshi · C. Li (✉)
Clinical Pharmacology, Genentech Inc.,
South San Francisco, CA, USA
e-mail: rdeng@gene.com; li.junyi@gene.com;
joshi.amita@gene.com; li.chunze@gene.com

C. A. Boswell
Preclinical and Translational Pharmacokinetics, Genentech Inc.,
South San Francisco, CA, USA
e-mail: boswell.charles@gene.com

Table 8.1 Therapeutic biologics (monoclonal antibodies, antibody fragments, fusion proteins and antibody conjugates) approved up to Q1 2022[a]

Drug name	Therapeutic area	Type	Antibody composition	Target Receptor/antigen/activity	Target Type	PK Behavioral[b]	Elimination/terminal half-life[c]	Clearance	Vss[d]	Route/dosing highlights
Anifrolumab	Autoimmune	MAB	Human IgG1	IFN a, b, ω receptor 1	Soluble	Nonlinear	NR	0.193 L/day	6.23 L	IV infusion, flat mg dose
Bimekizumab	Autoimmune	MAB	Humanized IgG1	IL-17A and IL-17F	Soluble	Linear	23 days	0.337 L/day	11.2 L	SC flat mg dose
Caplacizumab	Autoimmune	Nanobody	Humanized nanobody	von Willebrand factor	Soluble	Nonlinear	Target level dependent	NR	6.33 L	IV and SC, flat mg dose
Emapalumab	Autoimmune	MAB	Human IgG1	IFNr	Soluble	Nonlinear: Slightly more than proportional from 1–3 mg/kg; less than dose proportional from 3–10 mg/kg	22 in HV, 2.5–18.9 days in HLH patients	0.007 L/h	Central: 4.2 L, peripheral: 5.6 L	IV infusion, mg/kg dose
Risankizumab	Autoimmune	MAB	Humanized IgG1	IL-23 p19	Soluble	Linear	28 days	0.31 L/day	11.2 L	SC flat mg dose
Tralokinumab	Autoimmune	MAB	Human IgG4	IL-13	Soluble	Linear	22 days	0.149 L/day	4.2 L	SC flat mg dose
Tildrakizumab	Autoimmune	MAB	Humanized IgG1	IL23 p19	Soluble	Linear	23 days	0.32 L/day	10.8 L	SC flat mg dose
Abciximab[e,f]	Cardiovascular/metabolism	Fragment	Chimeric Fab: +D31mVar-hIgG1	CD41	Cell bound	Linear	0.29 days	0.068 L/h/kg	0.12 L/kg	IV bolus mg/kg doses
Alirocumab	Cardiovascular/metabolism	MAB	hIgG1	PCSK9	Soluble	Nonlinear	17–20 days	NR	Central Vd: 0.04–0.05 L/kg	SC flat mg dose
Dulaglutide	Cardiovascular/metabolism	Fusion protein	Disulfide-linked homodimer GLP-1 + modified hIgG4H (Fc)	GLP-1 R	Cell bound	NR	5 days	0.75 mg dose: 0.111 L/h; 1.5 mg dose: 0.107 L/h	0.75 mg dose: 19.2 L (range 14.3–26.4) 1.5 mg: 17.4 L (range 9.3–33)	SC flat mg dose
Evolocumab	Cardiovascular/metabolism	MAB	hIgG2	PCSK9	Soluble	Nonlinear	11–17 days	12 mL/h after 420 mg IV	3.3 L after 420 mg IV	SC flat mg dose
Idarucizumab	Cardiovascular/metabolism	Fragment	Humanized IgG1 (Fab)	Dabigatran and its acylglucuronide metabolites	NR	NR	Initial: 47 min (gCV 11.4%); Terminal: 10.3 h (gCV 18.9%)	47.0 mL/min	8.9 L	IV infusion flat gram doses
Alprolix	Hemophilia	Fusion protein	Human coagulation Factor IX + hIgG1 (Fc)	NR	NR	NR	50 IU/kg: 86 h 100 IU/kg: 91 h	50 IU/kg dose: 3.3 mL/h/kg 100 IU/kg: 2.6 mL/h/kg	50 IU/kg: 327 mL/kg 100 IU/kg: 236 mL/kg	IV infusion IU/kg doses

Drug	Indication	Type	Format	Target	Soluble/cell bound	Linearity	Half-life	Clearance	Volume of distribution	Individualized per label
Antihemophilic factor (recombinant), Fc fusion protein	Hemophilia	Fusion protein	B-domain deleted human coagulation Factor VIII + hIgG1 Fc	NR	NR	NR	1–5 years: 12.7 h; 6–11 years: 14.9 h; 12–17 years: 16.4 h; Adults: 19.7 h	1–5 years: 3.60 mL/h/kg; 6–11 years: 2.78 mL/h/kg; 12–17 years: 2.66 mL/h/kg; Adults: 2.06 mL/h/kg	1–5 years: 58.6 mL/kg; 6–11 years: 52.1 mL/kg; 12–17 years: 60.3 mL/kg; Adults: 49.5 mL/kg	
Emicizumab-kxwh	Hemophilia A	Bispecific	Humanized bi-specific IgG4	Factor IXa and factor X	Soluble	Linear withing dose range of 0.3–3.0 mg/kg	27.8±8.1 days	CL/F: 0.24 L/Day (95% CI: 0.22–0.26)	V/F: 11.4 L (95% CI: 10.6–12.1)	SC injection mg/kg dose
Abatacept[h]	Inflammation	Fusion protein	hCTLA-4 ECD + hFc (hinge)	CD80, CD86	Cell bound	Linear	13.1 days	0.22 mL/h/kg	0.07 L/kg	IV infusion, mg/kg dose
Alefacept[i]	Inflammation	Fusion protein	LFA-3 + hIgG1 (Fc)	CD2	Cell bound	Nonlinear	11.3 days	0.25 mL/h/kg)	IV: 94 mL/kg; IM: F = 63%	IV infusion, flat mg dose
Etanercept[n]	Inflammation	Fusion protein	TNF receptor + hIgG1 (Fc)	TNFα	Soluble, cell bound	Linear	4 days	120 mL/h; CL/F: 160 mL/h	F: 58% Vd: 6–11 L	SC, flat mg dose
Rilonacept	Inflammation	Fusion protein	hIgG1	IL-1beta	Soluble	NR	7.6 days	CL/F: 0.866 L/day	Vz/F: 9.73 L	SC, flat mg dose
Adalimumab[j]	Inflammation/autoimmunity	MAB	hIgG1	TNFα	Soluble, cell bound	Linear	14.7–19.3 days	9–12 mL/h	5.1–5.75 L, SC: F = 64%	SC injection flat mg dose
Belimumab	Inflammation/autoimmunity	MAB	hIgG1	BLyS	Soluble, cell bound	Linear	19.4 days	215 mL/day	5.29 L	IV infusion mg/kg dose
Benralizumab	Inflammation/autoimmunity	MAB	Humanized IgG1	IL-5Rα	Cell bound	Linear within dose range of 20–200 mg	15 days	70 kg patient: 0.29 L/day	70 kg patient: 5.7 L; SC: F ≈ 58%	SC flat mg dose
Brodalumab	Inflammation/autoimmunity	MAB	hIgG2	hIL-17RA	Cell bound	Nonlinear	NR	CL/F: 3.0 ± 3.5 L/day	Vz/F: 8.9 ± 9.4 L	IV infusion mg/kg dose
Canakinumab	Inflammation/autoimmunity	MAB	hIgG1	IL-1β	Soluble	Linear	26 days	0.174 L/day	6.01 L, F = 70%	SC flat mg dose
Certolizumab pegol	Inflammation/autoimmunity	Fragment	Humanized IgG1 (Fab) conjugated with PEG2MAL40K	TNFα	Soluble, cell bound	Linear	14 days	9.21–14.38 mL/h	6–8 L, F = 76–88%	SC flat mg dose
Daclizumab HYP	Inflammation/autoimmunity	MAB	Humanized IgG1	CD25	Cell bound	Linear	22 days	CL: 0.24 L/day CL/F: 0.274 L/day	6.41 L	SC flat mg dose
Dupilumab[k]	Inflammation/autoimmunity		hIgG4	IL-4Rα common subunit IL-4R and IL-13R complexes	Cell bound	Nonlinear	NR	0.126 L/day	4.8 ± 1.3 L	SC, flat mg dose with loading dose
Eculizumab	Inflammation/autoimmunity	MAB	Humanized IgG2/4 k	Complement protein C5	Soluble	Linear	8–14.8 days	22 mL/h	7.7 L	IV infusion flat mg dose
Efalizumab[l]	Inflammation/autoimmunity	MAB	Humanized IgG1	CD11a	Cell bound, internalized	Nonlinear	NR	6.6 mL/kg/day	58 mL/kg SC, F = 50%	SC mg/kg dose
Golimumab	Inflammation/autoimmunity	MAB	hIgG1	TNFα	Soluble, cell bound	Linear	2 weeks	4.9–6.7 mL/day/kg	Vd = 58–126 mL/kg, F = 53%	SC flat mg dose

(continued)

Table 8.1 (continued)

Drug name	Therapeutic area	Type	Target			PK				
			Antibody composition	Receptor/antigen/activity	Type	Behavioral[b]	Elimination/terminal half-life[c]	Clearance	Vss[d]	Route/dosing highlights
Infliximab[o]	Inflammation/autoimmunity	MAB	Chimeric: mVar-hIgG1	TNFα	Soluble, cell bound	Linear	7.7–9.5 days	9.8 mL/h	NR	IV infusion mg/kg dose
Ixekizumab	Inflammation/autoimmunity	MAB	Humanized IgG4	IL-17A	Soluble, membrane bound	Linear	13 days	0.39 L/day	7.11 L	SC flat mg dose with loading dose
Mepolizumab	Inflammation/autoimmunity	MAB	Humanized IgG1, kappa	IL-5	Soluble	Linear	20 days (range 16–22 days)	70 kg patient: 0.28 L/day	70 kg patient: Vc = 3.6 L	SC flat mg dose
Natalizumab	Inflammation/autoimmunity	MAB	Humanized IgG4k	Alfa 4 integrins	Cell bound	NR	11 days	16 mL/h	NR	IV infusion flat mg dose
Ocrelizumab	Inflammation/autoimmunity	MAB	Humanized IgG1, glycosylated	CD20	Cell bound	Nonlinear, linear over 400–2000 mg dose range	26 days	CL linear: 0.17 L/day	Vc 2.78 L, Vp 2.68 L	IV infusion flat mg doses
Omalizumab	Inflammation/autoimmunity	MAB	Humanized IgG1, kappa	IgE	Soluble	Linear	26 days (asthma) 24 days (CIU)	2.4 mL/day/kg (asthma); 3.0 mL/day/kg (CIU)	78 ± 32 mL/kg SC: $F = 62\%$	Asthma: SC, mg dose based on body weight and pretreatment IgE level; CIU: Flat mg dose
Reslizumab	Inflammation/autoimmunity	MAB	Humanized IgG4kappa	IL-5	Soluble	Linear	24 days	7 mL/h	5 L	IV infusion mg/kg doses
Secukinumab	Inflammation/autoimmunity	MAB	hIgG1kappa	IL-17A	Soluble, cell bound	Linear over 25–300 mg dose range	22–31 days	0.14–0.22 L/day	Vz: 7.10–8.60 L	SC flat mg dose with loading dose
Siltuximab	Inflammation/autoimmunity	MAB	Chimeric: Mouse-human IgG1	IL-6	Soluble and cell bound	Linear over a 2.8–11 mg/kg dose range	20.6 days range 14.2–29.7 days	0.23 L/day	70 kg male subject: 4.5 L	IV infusion mg/kg doses
Tocilizumab[s]	Inflammation/autoimmunity	MAB	Humanized IgG1	IL-6R	Soluble, cell bound	Nonlinear	Up to 13 days	12.5 mL/h	6.4 L	IV infusion mg/kg doses
Ustekinumab[p]	Inflammation/autoimmunity	MAB	hIgG1	IL-12, IL-23	Soluble	Linear	14.9–45.6 days	0.19 L/day	4.62 L	SC flat mg dose
Vedolizumab	Inflammation/autoimmunity	MAB	Humanized IgG1	α4β7	Cell bound	Linear + nonlinear	25 days at 300 mg	Linear CL: 0.157 L/day	5 L	IV infusion flat mg dose
Rituximab	Inflammation/autoimmunity, oncology	MAB	Chimeric: mVar-hIgG1	CD20	Cell bound	Linear	RA: 18 days (5.17–77.5 days); NHL: 22 days (6.1–52 days); CLL: 32 days (14–62 days)	RA: 0.335 L/day	3.1 L	IV infusion mg/m2 doses
Atoltivimab, maftivimab and odesivimab	Infectious disease	MAB	Human IgG1 mixture (3 mAbs)	Ebola virus	Cell bound	Linear	A: 21.2 days; M: 22.3 days; O: 25.3 days	A: 3.08 mL/day/kg; M: 2.78 mL/day/kg; O 2.02 mL/day/kg	A: 58.2 mL/kg; M: 57.6 mL/kg; O: 56.0 mL/kg	IV infusion, flat mg dose

Drug	Category	Type	Structure	Target	Form	Kinetics	Half-life	Clearance	Volume	Administration
Bezlotoxumab	Infectious disease	MAB	hIgG1	C. difficile toxin B	Cell bound	Linear	19 days	0.317 L/day	7.33 L	IV infusion mg/kg dose
Casirivimab + imdevimab	Infectious disease	MAB	Mixture of 2 human IgG1	SARS-CoV-2	Soluble virus	Linear	Casirivimab: 31.8 (±8.35) day; Imdevimab: 26.9 (±6.80) day	NR	NR	IV and SC flat mg dose
Ibalizumab	Infectious disease	MAB	Humanized IgG4	CD4 receptor	Cell bound	Nonlinear	2.7–64 h with dose from 0.3–25 mg/kg	0.36–9.54 mL/h/kg	4.8 L	IV infusion flat mg dose
Obiltoxaximab[g]	Infectious disease	MAB	Chimeric: mVar-hIgG1κ	PA component B. anthracis toxin	Cell bound	Linear over 4–16 mg/kg dose range	15–23 days	NR	NR	IV infusion mg/kg dose
Palivizumab	Infectious disease	MAB	Humanized IgG1	RSV F protein	Cell bound	NR	20 days	NR	NR	IM mg/kg doses
Regdanvimab	Infectious disease	MAB	Human IgG1	SARS-CoV-2	Soluble virus	Linear	17 days	0.20 mL/h/kg	83 mL/kg	IV infusion, flat mg dose
Asfotase alfa	Metabolic disease	Fusion protein / enzyme	Human tissue nonspecific alkaline phosphatase (TNSALP) catalytic domain + hIgG1 (fc)	Hydrolyzes phospho-monoesters	NR	Linear	5 days	NR	NR	SC mg/kg dose
Evinacumab	Metabolic disease	MAB	Human IgG4	Angiopoietin-like protein 3	Soluble	Nonlinear	Elimination/half-life HL is not a constant change with serum concentration.	NR	4.8 L	IV infusion, mg/kg dose
Romosozumab	Metabolic disease	MAB	Humanized IgG2	Sclerostin	Soluble	Nonlinear	12.8 days at 3 mg/kg	0.38 mL/h/kg at 3 mg/kg	3.92 L	SC flat mg dose
Aducanumab	Neurology	MAB	Human IgG1	Amyloid beta	Soluble/insoluble form	Linear	24.8 (range: 14.8–37.9) days	0.0159 (range: 0.0156–0.0161) L/h	9.63 L (range: 9.48–9.79)	IV infusion, mg/kg dose
Eptinezumab	Neurology	MAB	Humanized IgG1	CGRP	Soluble	Linear	27 days	0.006 L/h	Vc: 3.7 L	IV infusion, flat mg dose
Erenumab	Neurology	MAB	Human IgG2	CGRP	Cell bound	Nonlinear	Effective half-life at 70–140 mg: 28 days	NR	3.86 (±0.77) L at 140 mg, F = 82%	SC flat mg dose
Fremanezumab	Neurology	MAB	Humanized IgG2	CGRP	Cell bound	Linear	31 days	0.141 L/day	6 L	SC flat mg dose
Galcanezumab	Neurology	MAB	Humanized IgG4	CGRP	Cell bound	Linear	27 days	0.008 L/h	7.3 L	SC flat mg dose
Inebilizumab	Neurology	MAB	Humanized IgG1	CD19	Cell bound	NR	18 days	0.19 L/day	Vc: 2.95 L; Vp: 2.57 L	IV infusion, flat mg dose
Satralizumab	Neurology	MAB	Humanized IgG2	IL-6 R	Cell bound	Nonlinear	60–240 mg: 30 days (22–37 days)	Linear clearance: 0.0601 L/day	Vc: 3.46 L, Vp: 2.07 L	SC flat mg dose

(continued)

Table 8.1 (continued)

Drug name	Therapeutic area	Type	Antibody composition	Target Receptor/antigen/activity	Target Type	PK Behavioral[b]	Elimination/terminal half-life[c]	Clearance	Vss[d]	Route/dosing highlights
Alemtuzumab[a]	Oncology	MAB	Humanized IgG1	CD52	Soluble	Nonlinear	11 h after single dose, 12 days following multiple doses	NR	0.18 L/kg	IV infusion flat mg dose
Amivantamab	Oncology	Bispecific	Human IgG1 bispecific	EGFR, cMet	Cell bound	Linear	11.3 (±4.53) days	360 (±144) mL/day	5.13 (±1.78) L	IV infusion, flat mg dose
Atezolizumab	Oncology	MAB	Nonglycosylated humanized IgG1	PD-L1	Soluble, cell bound	Linear over 1–20 mg/kg	27 days	0.2 L/kg	6.9 L	IV infusion flat mg dose
Avelumab	Oncology	MAB	hIGg1 lambda	PD-L1	Soluble, cell bound	Linear over 10–20 mg/kg	6.1 days at 10 mg/kg	0.59 L/day	4.72 L	IV infusion mg/kg dose
Belantamab mafodotin	Oncology	ADC	Humanized IgG1 ADC	BCMA	Cell bound	Linear	Intact ADC: First dose: 12 days, 14 days at SS	Intact ADC: First dose: 0.7 L/day, 0.9 L/day at SS	Intact ADC: 11 L	IV infusion, flat mg dose
Bevacizumab	Oncology	MAB	Humanized IgG1	VEGF	Soluble	Linear	20 days	0.207–0.262 L/day	2.66–3.25 L	IV infusion mg/kg dose
Blinatumomab	Oncology	Bispecific	Antibody fragment	CD19	Cell bound	Linear	2.1 h	3.11 L/h	4.35 L	IV infusion flat mcg dose
Brentuximab vedotin[f]	Oncology	ADC	Chimeric mVar-hIgG1	CD30	Cell bound	Linear over 1.2–2.7 mg/kg	4.43 days	1.76 L/day	8.21 L	IV infusion mg/kg dose
Catumaxomab (EU only)	Oncology	Bispecific	Chimeric: Rat-mouse tri-functional MAB	EpCAM and CD3	Cell bound	NR	2.5 days (range 0.7–17)	NR	NR	IP infusion flat mg dose
Cemiplimab	Oncology	MAB	Human mAb	PD-1	Cell bound	Linear	20.3 days	CL0: 0.29 L/day; CLss 0.2 L/day	5.3 L	IV infusion flat mg dose
Cetuximab	Oncology	MAB	Chimeric: mVar-hIgG1	EGFR	Soluble	Nonlinear	4.8 days MRT: 12.6 days	0.02–0.08 L/h/m2	2–3 L/m2	IV infusion mg/m2 dose
Daratumumab	Oncology	MAB	hIgG1 kappa	CD38	Cell bound	Nonlinear	18 ± 9 days (mono) 23 ± 12 days (combo)	171.4 ± 95.3 mL/day	Vc: 4.7 ± 1.3 L (mono) 4.4 ± 1.5 L (combo)	IV infusion mg/kg dose
Denosumab	Oncology	MAB	hIgG2	RANKL	Cell bound	Linear	28 days	NR	F = 62%	SC injection flat mg dose
Dinutuximab	Oncology	MAB	Chimeric: mVar-hIgG1κ	GD2 glycolipid	Cell bound	NR	10 days	0.21 L/day	Vdss: 5.4 L	IV infusion mg/m2/day dose
Dostarlimab	Oncology	MAB	Humanized IgG4	PD-1	Soluble	Linear	25.4 days	0.007 L/h	5.3 L	IV infusion, flat mg dose
Durvalumab	Oncology	MAB	hIgG1κ	PD-L1	Soluble, cell bound	Linear over dose range of 3–20 mg/kg	17 days	CLss: 8.24 mL/h (clearance decreases over time)	5.6 L	IV infusion mg/kg dose

Name		Modality	Structure	Target		Kinetics	Half-life	Clearance	Volume	Administration
Elotuzumab	Oncology	MAB	Humanized IgG1	CD319 (SLAMF7)	Cell bound	Nonlinear	NR	NR	NR	IV infusion mg/kg dose with loading dose
Enfortumab vedotin	Oncology	ADC	Human IgG1+ MMAE	Nectin-4	Cell bound	Linear	Intact ADC: 3.4 days; MMAE: 2.4 days	Intact ADC: 0.1 L/h; MMAE: 2.7 L/h	Intact ADC: 11 L	IV infusion, mg/kg dose
Gemtuzumab ozogamicin	Oncology	ADC	Humanized IgG4k (hP67.6) + N-acetyl-gamma-calicheamicin (cytotoxin)	CD33	Cell bound	Nonlinear	Terminal plasma t1/2 hP67.6 mAb: After first dose = 62 h, after second dose = 90 h	Terminal plasma CLss hP67.6 mAb: After first dose = 0.35 L/h, after second dose = 0.15 L/h	V1 = 6.31 L, V2 = 15.1 L for hP67.6 mAb	IV infusion mg/m2 dose
Ibritumomab tiuxetan	Oncology	MAB	Murine IgG1+ tiuxetan (chelator)	CD20	Cell bound, stable	NR	47 h	NR	NR	IV injection over 10 min, mCi/kg dose
Inotuzumab ozogamicin	Oncology	ADC	Humanized IgG4 kappa + N-acetyl-gamma-calicheamicin (cytotoxin)	CD22	Cell bound	Nonlinear	12.3 days	CLss, 0.0333 L/h	12 L for ADC	IV infusion mg/m2 dose
Ipilimumab	Oncology	MAB	hIgG1	CTLA-4	Cell bound	Linear	15.4 days	16.8 mL/h	7.21 L	IV infusion mg/kg dose
Isatuximab	Oncology	MAB	Chimeric IgG1	CD38	Cell bound	Nonlinear from 1–20 mg/kg and linear from 5–20 mg/kg Q2W	At steady state ~12 days (99% eliminate in~2 months)	NR	8.13 L	IV infusion, mg/kg dose
Loncastuximab tesirine	Oncology	ADC	Humanized IgG1 ADC	CD19	Cell bound	NR	Intact ADC: 20.8 (±7.06) days	Intact ADC: First dose: 0.499 L/ day, 0.275 L/day at SS	Intact ADC: 7.11 L	IV infusion, mg/kg dose
Mogamulizumab	Oncology	MAB	Humanized IgG1	CCR4	Cell bound	Linear	17 days	12 mL/h	3.6 L	IV infusion, mg/kg dose
Moxetumomab pasudodox	Oncology	Fusion protein	Murine IgG1 dsFv immunotoxin	CD22	Cell bound	Linear	1.4 h (±0.35)	Initial: 25 (±29) L/h, SS: 4 (±4.4) L/h	6.5 (±2.4) L	IV infusion, mg/kg dose
Naxitamab	Oncology	MAB	Humanized IgG1	GD2	Cell bound	NR	8.2 days	NR in label	NR in label	IV infusion, mg/kg dose
Necitumumab	Oncology	MAB	hIgG1 kappa	EGFR	Cell bound	Nonlinear	14 days after 800 mg on days 1 and 8 of each 21-day cycle	14.1 mL/h after 800 mg on days 1 and 8 of each 21-day cycle	7.0 L after 800 mg on days 1 and 8 of each 21-day cycle	IV infusion flat mg dose

(continued)

Table 8.1 (continued)

Drug name	Therapeutic area	Type	Antibody composition	Target Receptor/antigen/activity	Target Type	PK Behavioral[b]	Elimination/terminal half-life[c]	Clearance	Vss[d]	Route/dosing highlights
Nivolumab	Oncology	MAB	hIgG4 kappa	PD-1	Cell bound	Linear	25 days	8.2 mL/h	6.8 L	IV infusion flat mg doses
Obinutuzumab	Oncology	MAB	Humanized IgG1 with reduced fucose content	CD20	Cell bound	Linear within recommended dose range	CLL patients: 25.5 days NHL patients: 35.3 days	CLL: 0.11 L/day; NHL: 0.08 L/day (both values after TMDD saturation)	CLL: 4.1 L NHL: 4.3 L	IV infusion flat mg doses
Ofatumumab	Oncology	MAB	hIgG1κ	CD-20	Cell bound	Nonlinear	14 days	0.01 L/h	1.7–5.1 L	IV infusion flat mg dose
Olaratumab	Oncology	MAB	hIgG1	PDGFR-α	Cell bound	NR	11 days (range 6–24 days)	0.56 L/day	7.7 L	IV infusion mg/kg doses
Panitumumab[u]	Oncology	MAB	hIgG2	EGFR	Cell bound	Nonlinear	7.5 days	4.9 mL/day/kg	82 mL/kg	IV infusion mg/kg dose
Pembrolizumab	Oncology	MAB	hIgG4 kappa	PD-1	Cell bound	Linear over 2–10 mg/kg dose range	22 days	252 mL/day after first dose, 195 mL/day at steady state (geometric mean)	6.0 L	IV infusion flat mg dose
Pertuzumab[v]	Oncology	MAB	Humanized IgG1	HER2	Cell bound, shed	Linear over 2–25 mg/kg dose range	18 days	0.235 L/day	5.57 L	IV infusion flat mg doses
Polatuzumab Vedotin	Oncology	ADC	Humanized IgG1 ADC	CD79b	Cell-bound	Linear	Intact ADC: ~12 days; MMAE: ~4 days	Intact ADC: 0.9 L/day	Intact ADC: 3.15 L	IV infusion, mg/kg dose
Ramucirumab	Oncology	MAB	hIgG1	VEGFR2	Cell bound	Linear	14 days	0.015 L/h	NR	IV infusion mg/kg doses
Sacituzumab govitecan	Oncology	ADC	Humanized IgG1 ADC	TROP-2	Cell bound	NR	Sacituzumab govitecan-hziy: 13.1 h; free SN-38: 19.7 h	0.14 L/h	2.96 L	IV infusion, mg/kg dose
Tafasitamab	Oncology	MAB	Humanized IgG1	CD19	Cell bound	NR	17 days (95% CI: 15–18 days)	0.41 L/day	9.3 L (95% CI: 8.6–10 L)	IV infusion, flat mg dose
Tebentafusp-tebn	Oncology	Bispecific	Bispecific fusion protein	gp100 peptide presented by HLA	Cell bound	Linear	7.5 h (range: 6.8–7.5 h)	16.4 L/d	7.56 L	IV infusion, flat mg dose
Tisotumab vedotin	Oncology	ADC	Human IgG1 ADC	Tissue factor	Cell bound	Linear	Intact ADC: 4.04 days; unconjugated MMAE8: 2.56 days	Intact ADC: 1.54 L/day; unconjugated MMAE8: 45.9 L/day	Intact ADC: 7.83 L	IV infusion, mg/kg dose

Drug	Therapeutic area	Type	Format	Target	Solubility	Linearity	Half-life	Clearance	Volume	Administration
Tositumomab	Oncology	MAB radiolabeled	mIgG2α	CD20	Cell bound	Nonlinear	NR	68.2 mL/h	NR	IV infusion flat mg doses
Trastuzumab[w]	Oncology	MAB	Humanized IgG1	HER2	Cell-bound, shed	Nonlinear	NR	0.173–0.283 L/day (breast cancer) 0.189–0.337 L/day (gastric cancer)	6 L (breast cancer) 6.6 L (gastric cancer)	IV infusion, mg/kg dose
Trastuzumab deruxtecan	Oncology	ADC	Humanized IgG1 + Dxd	Her2	Cell-bound shed	Linear	Intact ADC: 5.7–5.8 days; DXd: 5.5–5.8 day	Intact ADC: 0.42 L/day; DXd: 19.6 L/h	Intact ADC: Vc 2.78 L	IV infusion, mg/kg dose
Trastuzumab emtansine	Oncology	ADC	Humanized IgG1 + emtansine (cytotoxin)	HER2	Cell bound, shed	Linear at dose ≥ 2.4 mg/kg	4 days for ADC	0.68 L/day for ADC	3.13 L for ADC	IV infusion mg/kg dose
Ziv-aflibercept	Oncology	Fusion protein	Human VEGF receptors 1 and 2 ligand-binding ECD + hIgG1 (Fc)	hVEGF-A, hVEGF-B, hPIGF	Soluble	Linear	6 days (range 4–7 days)	NR	NR	IV infusion mg/kg doses
Aflibercept	Ophthalmology	Fusion protein	hVEGF receptors 1 and 2 ECDs + human IgG1 (fc)	VEGF-A, PIGF	Soluble	NR	5–6 days after IV	NR	6 L after IV	Intravitreal flat mg dose
Brolucizumab	Ophthalmology	Fragment	Humanized scFv	VEGF	Soluble	NR	4.4 (±2.0) days	NR	NR	IV bolus, flat mg dose
Ranibizumab	Ophthalmology	Fragment	Humanized IgG1κ	VEGF	Soluble	NR	9 days	NR	NR	Intravitreal injection, flat mg doses
Teprotumumab	Ophthalmology	MAB	Human IgG1	IGF-1 R	Cell bound	Linear	20 (±5) days	0.27 (±0.08) L/day	Central: 3.26 (±0.87) L, peripheral: 4.32 (±0.67) L	IV infusion, mg/kg dose
Burosumab	Rare disease (X-linked hypophosphatemia)	MAB	Human IgG1	FGF23	Soluble	Linear	19 days	0.29 L/day	8 L	SC mg/kg dose
Lanadelumab	Rare disease (hereditary angioedema attacks)	MAB	Human IgG1	Plasma Kallikrein	Soluble	Linear	~2 weeks	0.67–0.81 L/day	14.1–16.6 L	SC flat mg dose
Crizanlizumab	Sickle cell disease	MAB	Humanized IgG2	CD62 (aka P-selectin)	Cell bound	Linear	10.6 days in HV; 7.6 day in sickle cell disease patients	11.7 mL/h in HV; NR in patients	4.26 L	IV infusion, mg/kg dose
Basiliximab[x]	Transplantation	MAB	Chimeric: mVar-hIgG1	CD25	Cell bound	NR	4.1 days	75 mL/h	5.5–13.9 L	IV bolus or infusion flat mg dose

(continued)

Table 8.1 (continued)

Drug name	Therapeutic area	Type	Antibody composition	Target Receptor/antigen/activity	Target Type	PK Behavioral[b]	Elimination/terminal half-life[c]	Clearance	Vss[d]	Route/dosing highlights
Belatacept	Transplantation	Fusion protein	hCTLA-4 ECD + hFc (hinge)	CD80, CD86	Cell bound	Linear	9.8 days	0.49 mL/h/kg	0.11 L/kg	IV infusion mg/kg dose
Daclizumab	Transplantation	MAB	Humanized IgG1	CD25	Cell bound	Linear	20 days	15 mL/h	5.9 L	IV infusion mg/kg dose
Muromonab-CD3[y]	Transplantation	MAB	mIgG2α	CD3	Cell bound	NR	0.75 days	NR	NR	IV bolus flat mg dose

Abbreviations: *ADC* antibody–drug conjugate, *CD* cluster of differentiation, *CDR* complementarity determining region, *CLL* chronic lymphocytic leukemia, *CTLA* cytotoxic T lymphocyte-associated antigen, *CLss* clearance at steady state, *ECD*, extracellular domain, *EGFR* epidermal growth factor receptor, *F* bioavailability, *Fab* antigen-binding fragment, *Fc* constant fragment, *HER2* human epidermal growth factor receptor 2, *Ig* immunoglobulin, *IL* interleukin, *IV* intravenous, *LFA* lymphocyte function-associated antigen, *MAB* monoclonal antibody, *MS* multiple sclerosis, *mVar* murine variable, *NA* information not available or not found, *NHL* Non-Hodgkin lymphoma, *NR* information not reported, *PD-1* programmed cell death 1 receptor, *PDGFR* prostaglandin, *PIGF* phosphatidylinositol-glycan biosynthesis class F protein, *RA* rheumatoid arthritis, *RANKL* receptor activator of nuclear factor kappa-B ligand, *RSV* respiratory syncytial virus, *SC* subcutaneous, *TNSALP* tissue nonspecific alkaline phosphatase, *TNF* tumor necrosis factor, *V1* volume of distribution of central compartment, *V2* volume of distribution of peripheral compartment, *Vc* volume of central compartment, *VEGF* vascular endothelial growth factor, *Vss* volume of distribution at steady state, *Vz* volume of distribution during the terminal phase

[a]Source: Prescribing information (not shown) and/or additional references as shown

[b]Where PK are nonlinear, parameters are reported at usual clinical dose

[c]Terminal t1/2 reported for linear PK only

[d]Volume or apparent volume of distribution at steady state, unless otherwise indicated

[e,f]Kleiman et al. (1995), Mager et al. (2003)

[g]CDER (2015)

[h]Hervey and Keam (2006)

[i]Weisman et al. (2003)

[j]In 2011, manufacturing, distribution and sales of Amevive were halted during a supply disruption. According to the manufacturer, Astellas Pharma US, Inc., the decision to cease Amevive sales was neither the result of any specific safety concern nor of any FDA-mandated or voluntary product recall

[k]Kovalenko et al. (2016), Kovarik et al. (2001)

[l]Withdrawn from market

[m]Joshi et al. (2006)

[n]Lee et al. (2003)

[o]Cornillie et al. (2001)

[p]Zhu et al. (2009)

[q]Morris et al. (2003)

[r]Younes et al. (2010)

[s]Gibiansky et al. (2012)

[t]Wiseman et al. (2001)

[u]Ma et al. (2009)

[v]Garg et al. (2014)

[w]Kirschbrown et al. (2017)

[x]Kovarik et al. (2001)

[y]Hooks et al. (1991)

the context of their therapeutic applications. Efalizumab (anti-CD11a), a mAb marketed as an antipsoriasis drug in the US and EU, was chosen to illustrate the application of PK/PD principles in the drug development process.

Antibody Structure and Classes

Antibodies, or immunoglobulins (Igs), are roughly Y-shaped molecules or combinations of such molecules. There are five major classes of Ig: IgG, IgA, IgD, IgE, and IgM. Table 8.2 summarizes the characteristics of these molecules, particularly their structure (monomer, dimer, hexamer, or pentamer), molecular weight (ranging from ~150 to ~1150 kDa), and functions (e.g., activate complement, FcγR binding). Among these classes, IgGs and their derivatives form the framework for the development of therapeutic antibodies, though IgM antibodies recently emerged as a potential therapeutic modality with a few examples currently being investigated clinically (Keyt et al. 2020). Figure 8.1 depicts the general structural components of IgG and a conformational structure of efalizumab. An IgG molecule has four peptide chains, including two identical heavy (H) chains (50–55 kDa) and two identical light (L) chains (25 kDa), which are linked via disulfide (S–S) bonds at the hinge region. The first ~110 amino acids of both chains form the variable regions (V_H and V_L) and are also the antigen-binding regions. Each V domain contains three short stretches of peptide with hypervariable sequences (HV1, HV2, and HV3), known as complementarity determining regions (CDRs), i.e., the region that binds antigen. The remaining sequences of each light chain consist of a single constant domain (C_L). The remainder of each heavy chain contains three constant regions (C_{H1}, C_{H2}, and C_{H3}). Constant regions are responsible for effector recognition and binding. IgGs can be further divided into four subclasses (IgG1, IgG2, IgG3, and IgG4). The differences among these subclasses are also summarized in Table 8.2.

Murine, Chimeric, Humanized, and Fully Human mAbs

The first therapeutic mAbs were murine mAbs produced via hybridomas; however, these murine mAbs easily elicited formation of neutralizing human antimouse antibodies (HAMA) (Kuus-Reichel et al. 1994). With the advancement of technology, murine mAbs have been engineered further to produce chimeric (mouse CDR, human Fc), humanized, and fully human mAbs (Fig. 8.2). Murine mAbs, chimeric mAbs, humanized mAbs, and fully human mAbs have 0%, ~60–70%, ~90–95%, and ~100% sequence similarity to human mAbs, respectively. Decreasing the xenogenic portion, which is the different peptide sequence from the human pro-

tein, of the mAb potentially reduces the immunogenic risks of generating anti-drug antibodies (ADAs), which could impact the PK, PD, safety or efficacy of the mAbs. Muromonab-CD3 (Orthoclone OKT3), a first-generation mAb of murine origin, has shown efficacy in the treatment of acute transplant rejection and was the first mAb licensed for use in humans. It is reported that 50% of the patients who received OKT3 produced HAMA after the first dose. HAMA interfered with OKT3's binding to T cells, thus decreasing the therapeutic efficacy of the mAb (Norman et al. 1993). Later, molecular cloning and the expression of the variable region genes of IgGs facilitated the generation of engineered antibodies. A second generation of mAbs, chimeric mAbs, consists of human constant regions and mouse variable regions. The antigen specificity of a chimeric mAb is the same as the parental mouse antibody; however, the human Fc region renders a longer in vivo half-life than the parent murine mAb, and similar effector functions as a human antibody. Currently, there are 11 chimeric mAbs, fragments, and ADCs on the market (abciximab, basiliximab, cetuximab, dinutuximab, infliximab, obiltoxaximab, rituximab, siltuximab, isatuximab and brentuximab vedotin, and catumaxomab). These mAbs can still induce human antichimeric antibodies (HACA). For example, about 61% of patients who received infliximab had a HACA response associated with shorter duration of therapeutic efficacy and increased risk of infusion reactions (Baert et al. 2003).

The development of ADAs appears to be different across indications. For example, 6 of 17 patients with systemic lupus erythematosus receiving rituximab developed high-titer HACA (Looney et al. 2004), whereas only 1 of 166 B cell depleted lymphoma patients developed HACA (McLaughlin et al. 1998). Humanized mAbs contain significant portions of human sequence except the CDR which is still of murine origin. There are more than 45 humanized mAbs (including ADCs) on the market (see Table 8.1). The incidence of ADAs (in this case, human antihuman antibodies or HAHAs) was greatly decreased for these humanized mAbs. Trastuzumab has a reported HAHA incidence of ~0.1% (1 of 903 cases) (FDA 2006a), but another humanized mAb daclizumab had a HAHA rate as high as 34% (FDA 2005b).

Another way to achieve full biocompatibility of mAbs is to develop fully human antibodies, which can be produced by two approaches: through phage-display library or by using transgenic animals, such as the XenoMouse® or Trianni Mouse™ (Trianni 2018; Weiner 2006). Adalimumab is the first licensed fully human mAb generated using a phage-display library. Adalimumab was approved in 2002 and 2007 for the treatment of rheumatoid arthritis (RA) and Crohn's diseases, respectively (FDA 2007). However, despite its fully human antibody structure, the incidence of HAHA was about 5% (58 of 1062 patients) in three randomized clinical trials with adalimumab (Cohenuram and Saif 2007; FDA 2007).

Table 8.2 Important properties of endogenous immunoglobulin subclass (Goldsby et al. 1999; Kolar and Capra 2003)

Property		IgA		IgG				IgM	IgD	IgE
		IgA1	IgA2	IgG1	IgG2	IgG3	IgG4			
Serum concentration in adult (mg/mL)		1.4–4.2	0.2–0.5	5–12	2–6	0.5–1	0.2–1	0.25–3.1	0.03–0.4	0.0001–0.0002
Molecular form		Monomer, dimer		Monomer				Pentamer, hexamer	Monomer	Monomer
Functional valency		2 or 4		2				5 or 10	2	2
Molecular weight (kDa)		160 (m), 300 (d)	160 (m), 350 (d)	150	150	160	150	950(p)	175	190
Serum half-life (days)		5–7	4–6	21–24	21–24	7–8	21–24	5–10	2–8	1–5
% Total Ig in adult serum		11–14	1–4	45–53	11–15	3–6	1–4	10	0.2	50
Function	Activate classical complement pathway	−		+	±	++	−	+++	−	−
	Activate alternative complement pathway	+	−	−	−	−	−	−	−	−
	Cross placenta	−		+	±	+	+	−	−	−
	Present on membrane of mature B cell	−		−	−	−	−	+	−	+
	Bind to fc receptors of phagocytes	−		++	±	++	+	+	−	−
	Mucosal transport	++		−	−	−	−	+	−	−
	Induces mast cell degranulation	−		−	−	−	−	−	+	−
Biological properties		Secretory Ig, binds to polymeric Ig receptor		Placental transfer, secondary antibody for most response to pathogen, binds macrophage, and other phagocytic cells by Fcγ receptor				Primary antibody response, some binding to polymeric Ig receptor, some binding to phagocytes	Mature B cell marker	Allergy and parasite reactivity, binds FcεR on mast cells and basophiles

Panitumumab is the first approved fully human mAb generated using transgenic mouse technology. HAHA responses have been reported as less than 1% by an acid dissociation bridging enzyme-linked immunosorbent assay (ELISA) in clinical trial after chronic dosing with panitumumab to date (Cohenuram and Saif 2007; FDA 2015). Of note, typically ADAs are measured using ELISA, and the reported incidence rates of ADAs for a given mAb can be influenced by the sensitivity and specificity of the assay. Additionally, the observed incidence of ADA positivity in an assay may also be influenced by several other factors, including sample handling, timing of sample collection, concomitant medications, and underlying disease. For these reasons, comparison of the incidence of ADAs to a specific mAb with the incidence of ADAs to another product may be misleading.

Key Structural Components of mAbs

Proteolytic digestion of antibodies releases different fragments termed Fv (fragment variable), Fab (fragment antigen binding), and Fc (fragment crystallization) [reviewed by Wang et al. (2007)]. These fragments can also be generated by recombinant engineering. Treatment with papain generates two identical Fab's and one Fc. Pepsin treatment generates a F(ab')2 and several smaller fragments. Reduction of F(ab')2 will produce two Fab fragments. The Fv consists of the heavy chain variable domain (V_H) and the light chain variable domain (V_L) held together by strong noncovalent interaction. Stabilization of the Fv by a peptide linker generates a single-chain Fv (scFv).

a

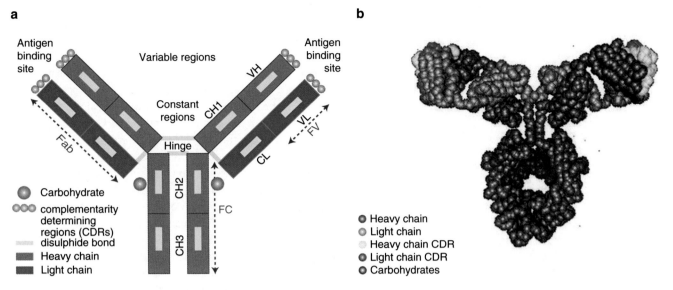

b

Fig. 8.1 (**a**) IgG1 antibody structure. Antigen is bound via the variable range of the antibody, whereas the Fc part of the IgG determines the mode of action (also called effector function). (**b**) Example of a conformational structure: efalizumab (anti-CD11a). *H chain* heavy chain consisting of VH, CH1, CH2, CH3; *L chain* light chain consisting of VL, CL; *VH, VL* variable heavy and light chain; *CHn, CL* constant heavy and light chain; *Fv* variable fraction; *Fc* crystallizable fraction; *Fab* antigen-binding fraction. (http://people.cryst.bbk.ac.uk/~ubcg07s/gifs/IgG.gif)

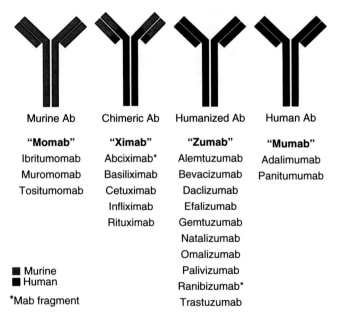

Fig. 8.2 Different generations of therapeutic antibodies. The well-known stem-mab was discontinued and the new International Nonproprietary Names (INN) was approved and adopted by WHO in October 2021. Details about the new INN mAb nomenclature scheme can be found here (Balocco et al. 2022)

Modifying Fc Structures

The Fc regions of mAbs play a critical role not only in their function but also in their disposition in the body. mAbs can elicit effector functions, including antibody-dependent cellular cyto-

toxicity (ADCC), antibody-dependent cellular phagocytosis (ADCP) and complement-dependent cytotoxicity (CDC) (refer to section mAb therapeutic mechanisms of action [MOA]), following interaction between their Fc regions and different Fcγ receptors (FcγR) and complement fixation (C1q, C3b). The CH2 domain and the hinge region joining CH1 and CH2 have been identified as the crucial regions for binding to FcγR (Presta 2002; Presta et al. 2002). Engineered mAbs with enhanced or decreased ADCC, ADCP, and CDC activity have been produced by manipulation of the critical Fc regions. Umana et al. (Umana et al. 1999) engineered an antineuroblastoma IgG1 with enhanced ADCC activity compared with wild type (WT). Shields et al. (2001) demonstrated that selected IgG1 variants with improved binding to FcγRIIIA showed enhanced ADCC by peripheral blood monocyte cells and natural killer cells. These findings indicate that Fc-engineered antibodies may have important applications for improving therapeutic efficacy. It was found that the *FCGR3A* gene dimorphism generates two allotypes, FcγRIIIa-158 V and FcγRIIIa-158F, and the polymorphism in FcγRIIIA is associated with favorable clinical response following rituximab administration in non-Hodgkin's lymphoma patients (Cartron et al. 2004; Dall'Ozzo et al. 2004). Obinutuzumab, an anti-CD20 mAb with enhanced effector functions as compared to rituximab, was approved for the treatment of patients with previously untreated chronic lymphocytic leukemia (CLL) and patients with follicular lymphoma (FL) who relapsed after, or are refractory to, a rituximab-containing regimen. The efficacy of antibody-interleukin 2 fusion protein (Ab-IL2) was improved by reducing its interaction with Fc receptors

(Gillies et al. 1999). In addition, the Fc portion of mAbs also binds to the neonatal Fc receptor, FcRn (named based on its discovery in neonatal rats), an Fc receptor belonging to the major histocompatibility complex structure, which is involved in IgG transport and clearance (Junghans 1997). Engineered mAbs with a decreased or increased FcRn-binding affinity have been investigated for its potential to modify the pharmacokinetic behavior of mAbs (see the section on Clearance for details).

Antibody Derivatives: F(Ab')2, Fab, Antibody Drug Conjugates and Fusion Proteins

The fragments of antibodies [Fab, F(ab')2, and scFv] have a shorter half-life compared with the corresponding full-sized antibodies (Wu and Sun 2014). A single-chain fragment variable (scFv) can be further engineered into a dimer (diabody, ~60 kDa), or trimer (triabody, ~90 kDa). Two diabodies can be further linked together to generate a bispecific tandem diabody (tandab). A single Fab can be fused to a complete Fc engineered to form a single arm mAb, which is monovalent. Figure 8.3 illustrates the structure of different antibody fragments. Of note, abciximab, idarucizumab, and ranibizumab are three Fabs approved by FDA. Abciximab is a chimeric Fab used to prevent blood clotting and has a 20–30 min half-life in serum and 4-h half-life in platelets (Schror and Weber 2003). Ranibizumab, administered via intravitreal injection, was approved for the treatment of macular degeneration in 2006 and exhibits a vitreous elimination half-life of 9 days (Albrecht and DeNardo 2006).

The half-life of Fc fragments is more similar to that of full-sized IgGs (Lobo et al. 2004). Therefore, Fc portions of IgGs have been used to form fusions with molecules such as cytokines, growth factor enzymes, or the ligand-binding region of receptor or adhesion molecules to improve their half-life and stability. There are 11 Fc fusion proteins currently on the market [abatacept, aflibercept, alefacept, alprolix, antihemophilic factor (recombinant) Fc fusion protein, belatacept, dulaglutide, etanercept, rilonacept and ziv-aflibercept, moxetumomab pasudodox]. Etanercept, a dimeric fusion molecule consisting of the TNF-α receptor fused to the Fc region of human IgG1, has a half-life of approximately 70–100 h (Zhou 2005), which is much longer than that of the TNF-α receptor itself (30 min to ~2 h) (Watanabe et al. 1988).

Antibodies and antibody fragments can also be linked covalently with cytotoxic radionuclides or small-molecule drugs to form radioimmunotherapeutic (RIT) agents or ADCs (Fig. 8.4), respectively. In each case, the antibody is used as a delivery mechanism to selectively target the cytotoxic moiety to tumors (Girish and Li 2015; Prabhu et al. 2011). For both ADCs and RIT agents, the therapeutic strategy involves selective delivery of a cytotoxin (drug or radionuclide) to tumors via the antibody. As targeted approaches, both technologies exploit the overexpression of target on the surface of the cancer cells and thereby minimize damage to normal tissues. Such approaches are anticipated to minimize the significant side effects encountered when cytotoxic small-molecule drugs or radionuclides are administered as single agents, thus leading to enhanced therapeutic windows.

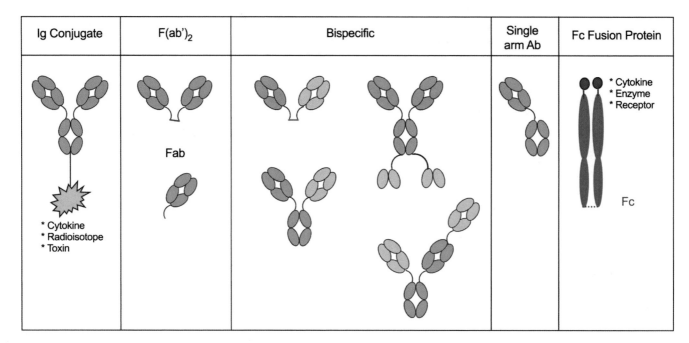

Fig. 8.3 Schematic representation of antibody derivatives: Ig conjugate, F(ab')2, Fab, bispecfic Ab, single arm Ab, and Fc fusion proteins. For Ig conjugate, the toxin can link to Ig at different sites

Fig. 8.4 Schematic representation of ADC structures. ADCs are a heterogeneous mixture with different drug-to-antibody ratio (DAR) species, with individual molecules exhibiting a range of DARs. (Adapted with permission from Kaur et al., Mass Spectrometry of Antibody-Drug Conjugates in Plasma and Tissue in Drug Development. In Characterization of Protein Therapeutics Using Mass Spectrometry, 2013. Guodong Chen, Ed., Springer Press, New York, NY, pp. 279–304)

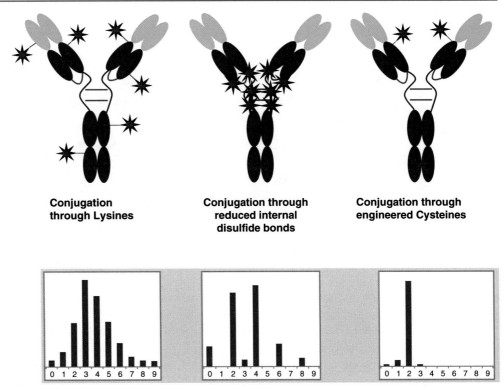

Conjugation through Lysines

Conjugation through reduced internal disulfide bonds

Conjugation through engineered Cysteines

Drug-to-Antibody Ratio Distribution

However, important distinctions exist between these two therapeutic modalities. For example, ADCs usually require internalization into the endosomes and/or lysosomes for efficacy, while RIT agents are often able to emit beta or gamma radiation, even from the cell surface, to achieve cell killing following direct binding to membrane antigens. Furthermore, RIT can deliver high levels of radiation even with very low doses of radioimmunoconjugate.

Currently, there are 11 approved ADCs (Table 8.1) and many more are currently being investigated in early-stage and late-stage clinical trials. All 11 ADCs are approved for oncology indications, with six of them against hematological tumors and five against solid tumors. Among the 11 approved ADCs, six of them use humanized immunoglobulin G1 (IgG1) antibodies except brentuximab vedotin (chimeric IgG1), gemtuzumab ozogamicin (humanized IgG4), inotuzumab ozogamcin (humanized IgG4), enfortumab vedotin (fully human IgG1), and tisoumab vedotin (fully human IgG1). Two of the 11, trastuzumab emtansine and belantamab mafodotin utilize noncleavable linkers with the remaining nine adopting cleavable linkers. Seven unique cytotoxic payloads are utilized in the 11 approved ADCs. The strategy of conjugating a potent, nonspecific payload to an antibody can dramatically improve the therapeutic index of drugs whose cytotoxicity would otherwise be intolerable, allowing them to be used therapeutically. Following the great success of these ADCs, numerous innovative approaches (e.g., site-specific conjugation or novel payloads) have been implemented to further improve the therapeutic window, resulting in the "next-generation" ADCs.

The only current radioimmunotherapeutic agents licensed by the FDA are ibritumomab tiuxetan and tositumomab plus ^{131}I-tositumomab, both for non-Hodgkin's lymphoma. Both intact murine mAbs bind CD20 and carry a potent beta particle-emitting radioisotope (^{90}Y for ibritumomab/tiuxetan and ^{131}I for tositumomab). In the case of ibritumomab, the bifunctional chelating agent, tiuxetan, is used to covalently link the radionuclide to the mAb ibritumomab. However, another approved anti-CD20 mAb, rituximab, is included in the dosing regimen as a nonradioactive predose to improve the biodistribution of the radiolabeled mAb. Despite impressive clinical results, radioimmunotherapeutic mAbs have not generated considerable commercial success; various financial, regulatory, and commercial barriers have been cited as contributing factors to this trend (Boswell and Brechbiel 2007; Girish and Li 2015; Prabhu et al. 2011).

mAb Therapeutic Mechanisms of Action (MOAs)

The pharmacological effects of antibodies are first initiated by the specific interaction between antibody and antigen. mAbs generally exhibit exquisite specificity for the target antigen. The binding site on the antigen, called the epitope, can be linear or conformational and may comprise continuous

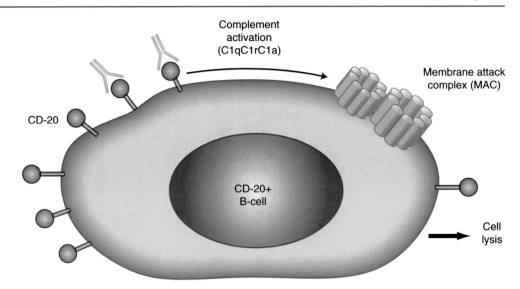

Fig. 8.5 An example of complement-dependent cytotoxicity (CDC), using a B cell lymphoma model, where the monoclonal antibody (mAb) rituximab binds to the receptor and initiates the complement system, also known as the "complement cascade." The end result is formation of a membrane attack complex (MAC), which leads to cell lysis and death

or discontinuous amino acid sequences. The epitope is the primary determinant of the antibody's modulatory functions, and depending on the epitope, the antibody may exert antagonist or agonist effects, or it may be nonmodulatory. The epitope may also influence the antibody's ability to induce ADCC and CDC. mAbs exert their pharmacological effects via multiple mechanisms that include direct modulation of the target antigen, CDC and ADCC, ADCP, apoptosis, delivery of a radionuclide or cytotoxic drug to target cells and multitarget engagement or immune cell activation using bispecific constructs.

Direct Modulation of Target Antigen

Examples of direct modulation of the target antigen include anti-TNFα, anti-IgE, and anti-CD11a therapies that are involved in blocking and removal of the target antigen. Most mAbs act through multiple mechanisms, such as antagonism of a soluble ligand or receptor, blockage of cell–cell interaction and agonism of a receptor, and may exhibit cooperativity with concurrent therapies.

Complement-Dependent Cytotoxicity (CDC)

The complement system is an important part of the innate (i.e., nonadaptive) immune system. It consists of many enzymes that form a cascade with each enzyme acting as a catalyst for the next. CDC results from interaction of cell-bound mAbs with proteins of the complement system. CDC is initiated by binding of the complement protein, C1q, to the Fc domain. The IgG1 and IgG3 isotypes have the highest CDC activity, while the IgG4 isotype lacks C1q binding and complement activation (Presta 2002). Upon binding to immune complexes, C1q undergoes a conformational change, and the resulting

activated complex initiates an enzymatic cascade involving complement proteins C2 to C9 and several other factors. This cascade spreads rapidly and ends in the formation of the membrane attack complex (MAC), which inserts into the membrane of the target cell and causes osmotic disruption and lysis of the target. Figure 8.5 illustrates the mechanism for CDC with rituximab (a chimeric mAb that targets the CD20 antigen) as an example.

Antibody-Dependent Cellular Cytotoxicity (ADCC)

ADCC is a mechanism of cell-mediated immunity whereby an effector cell of the immune system actively lyses a target cell that has been bound by specific antibodies. It is one of the mechanisms through which antibodies, as part of the humoral immune response, can act to limit and contain infection. Classical ADCC is mediated by natural killer (NK) cells, monocytes, or macrophages, but an alternate ADCC is used by eosinophils to kill certain parasitic worms known as helminths. ADCC is part of the adaptive immune response due to its dependence on a prior antibody response. The typical ADCC involves activation of NK cells, monocytes, or macrophages and is dependent on the recognition of antibody-coated infected cells by Fc receptors on the surface of these cells. The Fc receptors recognize the Fc portion of antibodies such as IgG, which bind to the surface of a pathogen-infected target cell. The Fc receptor that exists on the surface of NK cell is called CD16 or FcγRIII. Once bound to the Fc receptor of IgG, the NK cell releases cytokines such as IFN-γ and cytotoxic granules like perforin and granzyme that enter the target cell and promote cell death by triggering apoptosis. This is similar to, but independent of, responses by cytotoxic T cells. Figure 8.6 illustrates the mechanism for ADCC with rituximab as an example.

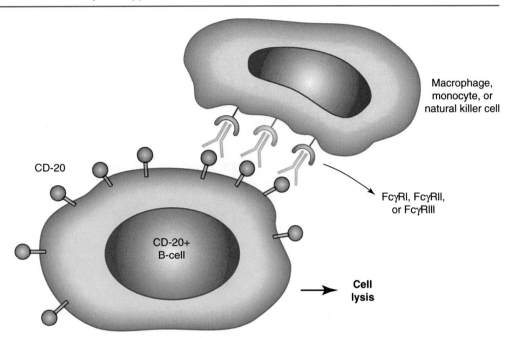

Fig. 8.6 An example of antibody-dependent cellular cytotoxicity (ADCC) and antibody-dependent cellular phagocytosis (ADCP). The monoclonal antibody (mAb) rituximab targets the CD20 antigen, which is expressed on a significant number of B cell malignancies. The Fc fragment of the mAb binds the Fc receptors found on effector cells such as monocytes, macrophages, and NK cells. These cells in turn either engulf the mAb-bound tumor cell (ADCP) or release cytotoxic agents such as perforin and granzymes, leading to destruction of the tumor cell (ADCC)

Macrophage, monocyte, or natural killer cell

CD-20

FcγRI, FcγRII, or FcγRIII

CD-20+ B-cell

Cell lysis

Antibody-Dependent Cellular Phagocytosis (ADCP)

ADCP is an immune effector function in which cells or particles opsonized with antibodies are engulfed by phagocytic effector cells, such as macrophages, following interactions between the Fc region of antibodies and Fcγ receptors on effector cells. In vivo, ADCP can be mediated by monocytes, macrophages, neutrophils, and dendritic cells via all three types of activating Fcγ receptors: FcγRI, FcγRIIa, and FcγRIIIa. Studies have shown that mAbs against tumor antigens induce phagocytosis of cancer cells in vitro, promote macrophage infiltration into tumors, and elicit macrophage-mediated destruction of tumors in mice (Weiskopf and Weissman 2015). ADCP is an important MOA of several antibody therapies for cancer, such as rituximab, obinutuzumab, and ocrelizumab. Engagement of Fcγ receptors expressed on phagocytic effector cells with antibodies bound to target cells triggers a signaling cascade leading to the engulfment of the antibody-opsonized tumor cells. Upon full engulfment, a phagosome is formed, which fuses with lysosomes, leading to acidification and digestion of the tumor cells. Figure 8.6 illustrates the mechanism for ADCP with rituximab as an example.

Apoptosis

mAbs achieve their therapeutic effect through various mechanisms. In addition to the above-mentioned effector functions, they can have direct effects in producing apoptosis or programmed cell death, which is characterized by nuclear

DNA degradation, nuclear degeneration and condensation, and the phagocytosis of cell remains. For example, Cetuximab induces cell cycle arrest and apoptosis in tumor cells by blocking ligand binding (Li et al. 2005) and receptor dimerization (Patel et al. 2009).

Targeted Delivery of Cytotoxic Drugs Via ADCs

ADCs achieve their therapeutic effect through selectively delivering a potent cytotoxic agent to tumor cells (Girish and Li 2015). The mAb component enables the ADC to specifically bind to targeted cell surface antigens overexpressed on the tumor cells. After binding to the cell surface antigen, the ADC is internalized by the tumor cell, where it undergoes lysosomal degradation, leading to the release of the cytotoxic agent (Fig. 8.7, an example of major MOA for ADC). Targeted delivery of cytotoxic drugs to tumors enables ADCs to potentially harness and improve their antitumor effect while minimizing their impact on normal tissues, thereby enhancing the benefit-risk profile.

Mechanism of Actions for Bispecific Constructs

Bispecific antibodies can achieve their therapeutic effect through two different MOAs. One category of the bispecific antibodies can function to bridge patient's own immune cells (e.g., through CD3 + T cell or CD16 + NK cell) to attack target-positive tumor cells. This type of bispecific constructs have one arm directed against an antigen on the immune effector cells and the other arm directed

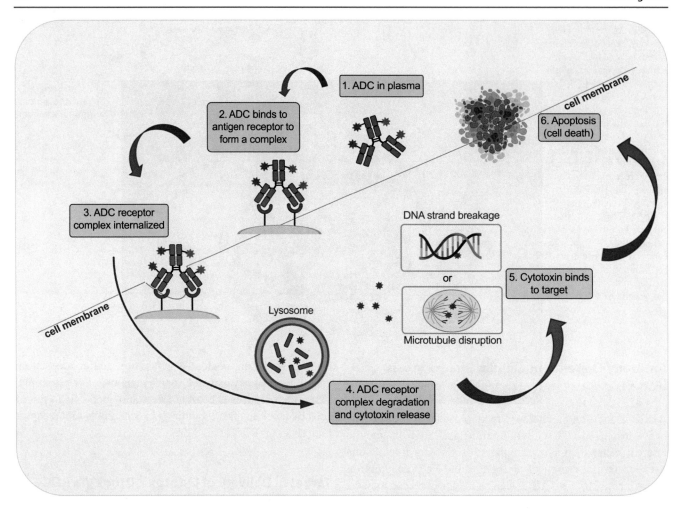

Fig. 8.7 An example of major MOAs for targeted delivery of cytotoxic drugs via ADC. An ADC can also work through monoclonal antibody or toxin-related mechanism

against a target cell surface antigen overexpressed by tumor cells (Mandikian et al. 2018). Simultaneous engagement of both arms results in formation of an immunologic synapse between the target tumor cell and effector cell, which leads to killing of the target tumor cells (e.g., through granzyme- and perforin-induced cell lysis). In contrast, the other category of the bispecific antibodies does not require to bridge the two target cells simultaneously for efficacy, but acts through engaging multiple targets. Bispecific antibodies in this category may target two soluble cytokines, bind different epitopes of the same tumor cell or viral antigen, or engage different targets to mimic the function of an endogenous protein. Such an approach may have advantages over combination therapy with multiple antibodies (FDA 2021).

Translational Medicine/Development Process

The tight connection of basic to clinical sciences is an essential part of translational medicine, which aims to *translate* the knowledge of basic science into practical therapeutic applications for patients. This knowledge transfer is often referred to as the process of moving *from-bench-to-bedside*, emphasizing the transition of scientific advancements into clinical applications. This framework of translational medicine is applied during the discovery and drug development process of a specific mAb against a certain disease. It includes major steps such as identifying an important and viable pathophysiological target antigen to modify the disease in a beneficial way, producing mAbs with structural ele-

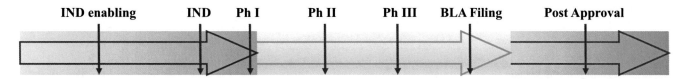

| IND enabling | IND | Ph I | Ph II | Ph III | BLA Filing | Post Approval |

Preclinical
- Mechanism of action
- In vitro and in vivo efficacy
- Safety pharmacology
- Toxicology/Toxicokinetics
  (single/repeated dose in relevant
  species)
- Tissue cross-reactivity
- Local tolerance
- PK support for molecule selection
- Assay support for PK/PD
- PK/PD support for
  Dose/Route/Regimen
- Assay development

Preclinical
- General Toxicity/Toxicokinetics
- Reproductive Toxicity/Toxicokinetics
- Immunotoxicity

Clinical
- Safety and efficacy
- Characterization of the concentration-
  effect relationship
- Material comparability studies
- Mechanistic modeling approach
- Immunogenicity: anti-drug antibodies
  and neutralizing antibodies
- Proof of mechanism
- Proof of concept
- Population pharmacokinetics/predictions

- Post-marketing
  commitment
- Drug-drug interaction
  studies
- Dose optimization
- Pharmacovigilance

Fig. 8.8 Flowchart depicting high level PK/PD/toxicology study requirements during preclinical and clinical drug product development

ments providing optimal PK, preclinical safety and efficacy testing in relevant models, and finally clinical trials in patients. An overview of the development phases of the molecules comprising the preclinical activities is outlined in Fig. 8.8. Furthermore, the critical components of the entire mAb development process are explained in detail from a PK/PD perspective below.

Preclinical Safety Assessment of mAbs

Preclinical safety assessment of mAbs encounters unique challenges, as many of the classical evaluations employed for small molecules are not appropriate for protein therapeutics in general and mAbs in particular. For example, in vitro genotoxicology tests such as the Ames and chromosome aberration assays are generally not conducted for mAbs given their limited interaction with nuclear material and the lack of appropriate receptor/target expression in these systems. As mAb binding tends to be highly species specific, suitable animal models are often limited to nonhuman primates, and for this reason, many common in vivo models, such as rodent carcinogenesis bioassays and some safety pharmacology bioassays, are not viable for mAb therapeutic candidates. For general toxicology studies, cynomolgus and rhesus monkeys are most commonly employed and offer many advantages given their close phylogenetic relationship with humans; however, due to logistics, animal availability, ethical constraints, and costs, group sizes tend to be much smaller than typically used for lower species, thus limiting statistical power. In some cases,

alternative models are employed to enable studies in rodents. Rather than directly testing the therapeutic candidate, analogous mAbs that can bind to target epitopes in lower species (e.g., mice) can be engineered and used as a surrogate mAb for safety evaluation (Clarke et al. 2004). Often the antibody framework amino acid sequence is modified to reduce antigenicity thus enabling longer term studies (Albrecht and DeNardo 2006; Cohenuram and Saif 2007; Weiner 2006). Another approach is to use transgenic models that express the human receptor/target of interest (Bugelski et al. 2000), although results must be interpreted with caution as transgenic models often have altered physiology and typically lack historical background data for the model (Boswell et al. 2013). To address development issues that are specific to mAbs and other protein therapeutics, the International Conference of Harmonization (ICH) has developed guidelines specific to the preclinical evaluation of biotechnology-derived pharmaceuticals (ICH 1997a, b).

For general safety studies, species selection is an important consideration given the exquisite species specificity often encountered with mAbs. Model selection needs to be justified based on appropriate expression of the target epitope, appropriate binding affinity with the therapeutic candidate, and appropriate biologic activity in the test system. To aid in the interpretation of results, tissue cross-reactivity studies offer the ability to compare drug localization in both animal and human tissues. For mAb therapeutic candidates, a range of three or more dose levels are typically selected to attain pharmacologically relevant serum concentrations, to approximate levels anticipated in the clinic, and to provide information at doses higher than anticipated in the clinic. For most indications, it is important to include dose levels that allow identification of a no observable adverse effect level (NOAEL). If feasible, the highest dose should fall within the range where toxicity is anticipated; although, in practice, many mAbs do not exhibit toxicity, and other factors limit the maximum dose. To best reflect human exposures, doses are often normalized and selected to match and exceed anticipated human therapeutic exposure in plasma, serum, or blood based upon the exposure parameters, area under the concentration–time curve (AUC), maximum concentration (C_{max}), or concentration immediately prior to next treatment (C_{trough}). The route of administration, dosing regimen, and dosing duration should be selected to best model the anticipated use in clinical trials (ICH 1997a, b).

To adequately interpret nonclinical study results, it is important to characterize ADA responses. For human mAbs, ADA responses are particularly prominent in lower species but also evident in nonhuman primates albeit to a lesser degree, making these species more viable for chronic toxicity studies. ADAs can impact drug activity in a variety of ways. Neutralizing ADAs are those that bind to the therapeutic in a manner that prevents activity, often by inhibiting direct binding to the target epitope. Non-neutralizing antibodies may also indirectly impact drug activity, e.g., rapid clearance of drug–ADA complexes can effectively reduce serum drug concentrations. In situations where prominent ADA responses are expected, administration of high-dose multiples of the anticipated clinical dose may overcome these issues by maintaining sufficient circulating concentrations of active drug when supported with sufficient safety margin. To properly interpret study results, it is important to characterize ADA incidence and magnitude, as the occurrence of ADA responses could mask toxicities. Alternatively, robust ADA responses may induce significant signs of toxicity, such as infusion-related anaphylaxis, that may not be predictive of human outcome where ADA formation is likely to be less of an issue. If ADA formation is clearly impacting circulating drug levels, ADA-positive individual animals are often removed from consideration when evaluating PK parameters to better reflect the anticipated PK in human populations.

Pharmacokinetics

A thorough and rigorous PK program in the early learning phase of preclinical drug development can provide a linkage between drug discovery and preclinical development. PK information can be linked to PD by mathematical modeling, which allows characterizing the time course of the effect intensity resulting from a certain dosing regimen. Antibodies often exhibit PK properties that are complex and different than those typically associated with small-molecule drugs (Meibohm and Derendorf 2002). The PK of ADCs is more complex due to the presence of both an antibody component as well as a small-molecule component. In the following sections, the basic characteristics of mAb and ADC PK are summarized in contrast to small-molecule drugs.

The PK of antibodies are very different from that of small molecules, as summarized in Table 8.3. Precise, sensitive, and accurate bioanalytical methods are essential for PK interpretation. However, for mAbs, the immunoassays and bioassay methodologies are often less specific than assays used for small-molecule drugs (e.g., LC/MS/MS). Compared to mAbs and small-molecule drugs, the assays for ADCs are much more complex given it has both small- and large-molecule drug components. Multiple analytes (e.g., conjugate, total antibody, and unconjugated cytotoxic drug) using diverse assay approaches (e.g., immunoassays, LC/MS/MS, or hybrid immunoaffinity LC/MS/MS) are typically measured in preclinical and clinical studies to characterize the PK properties of an ADC (Gorovits et al. 2013; Kaur et al. 2013).

mAbs are handled by the body very differently than are small molecules. In contrast to small-molecule drugs, the typical metabolic enzymes and transporter proteins, such as cytochrome P450 and multidrug resistance (MDR) efflux pumps, are not involved in the disposition of mAbs. Consequently, drug–drug interactions (DDI) at the level of these drug-metabolizing enzymes and transporters are not complicating factors in the drug development process of mAbs and in general do not need to be addressed by in vitro and in vivo studies. Because of their large molecular weight, intact mAbs are not usually cleared by the kidneys; however, renal clearance processes may play an important role in the elimination of molecules of smaller molecular weight such as Fab's and chemically derived small-molecule drugs. Renal elimination (i.e., glomerular filtration) occurs for non-branched molecules such as Fab's (50 kDa) and one-armed antibodies (100–150 kDa) but not for branched molecules such as (scFv)2Fc (100 kDa) and intact IgG (150 kDa), consistent with molecular shape being a stronger determinant of renal filtration of biologics than molecular weight (Rafidi et al. 2021). The PK of ADCs is more complex with combined characteristics of small-molecule drugs and large-molecule biotherapeutics. There are two alternative clearance pathways for ADCs: proteolytic degradation and deconjugation (Kamath and Iyer 2015, 2016). Similar to mAbs, ADC clearance through proteolytic degradation is driven primarily by catabolism mediated by target-specific or nonspecific cellular uptake followed by lysosomal degradation. Deconjugation clearance is usually mediated by enzymatic or chemical cleavage (e.g., maleimide exchange) of the linker leading to the release of the cytotoxic drug from the ADC (Shen et al. 2012). The released cytotoxic drug is

Table 8.3 Comparison of the pharmacokinetics between small-molecule drugs, monoclonal antibodies, and antibody drug conjugates (Kamath 2016; Lobo et al. 2004; Mould et al. 1999; Mould and Sweeney 2007; Roskos et al. 2004)

Small-molecule drugs	Monoclonal antibodies	Antibody drug conjugates
High potency and low specificity	Low potency and high specificity	High potency and high specificity
PK usually independent of PD	PK usually dependent of PD	Same as mAb
Binding generally nonspecific (can affect multiple enzymes)	Binding very specific for target protein or antigen	Same as mAb
Linear PK at low doses (usually therapeutic doses); nonlinear PK at high doses (after saturation of metabolic enzymes)	Nonlinear PK at low doses; linear PK at high doses after saturation of target	Same as mAb
Relatively short $t_{1/2}$ (h)	Long $t_{1/2}$ (days or weeks)	Long $t_{1/2}$ of antibody; sustained delivery of small molecule (formation rate limited)
Oral delivery often possible	Need parenteral dosing. Subcutaneous (SC) or intramuscular (IM) is possible	Need parenteral dosing. SC or IM has not been tested
Metabolism by cytochrome P450 or other phase I/phase II enzymes	Catabolism by proteolytic degradation	Catabolism by proteolytic degradation; small molecule component can undergo excretion unchanged or metabolism by cytochrome P450 enzymes or other phase I/phase II enzymes
Renal clearance often important	No renal clearance of intact antibody. May be eliminated by damaged kidneys. Antibody fragment might be eliminated by renal clearance	Combination of mAb and small molecule; released small molecule can be cleared renally and/or hepatically
High volume of distribution due to binding to tissues	Distribution usually limited to blood and extracellular space	Same as mAb
No immunogenicity	Immunogenicity may be seen	Same as mAb
Narrow therapeutic window	Large therapeutic window	Depends on ADC molecule

expected to undergo metabolic enzyme and/or transporter-mediated clearance mechanisms consistent with small molecules. The different *ADME* (*A*bsorption, *D*istribution, *M*etabolism, and *E*limination) processes comprising the PK of mAbs are discussed separately to address their individual specifics.

Absorption

Most mAbs are not administered orally because of their limited gastrointestinal stability, lipophilicity, and size, all of which result in insufficient resistance against the hostile proteolytic gastrointestinal milieu and very limited permeation through the lipophilic intestinal wall. Therefore, intravenous (IV) administration is still the most frequently used route, which allows for immediate systemic delivery of a large volume of drug product and provides complete systemic availability. Subcutaneous (SC) or intramuscular (IM) administration, however, may offer a number of benefits over IV administration. Being less invasive and with a much shorter injection duration (2–8 min versus 30–90 min for IV infusion), and commonly with a fixed dose, SC or IM dosing is expected to offer more convenience to patients compared to IV infusion. Additionally, IV infusion is typically administered in a hospital or physician's office; SC or IM administration may allow self or healthcare professional-assisted home administration. Of note, 22 of the 117 FDA-approved

mAb or mAb-derived therapies listed in Table 8.1 are administered by an extravascular route, either SC or IM. Aflibercept and ranibizumab are administered via intravitreal injection. Port Delivery System with ranibizumab (PDS) every 24 weeks (Q24W) demonstrated noninferior and equivalent efficacy to monthly ranibizumab (Holekamp et al. 2022). Recently, oral delivery of mAb has been studied in preclinical species using an orally dosed gastric auto-injector with promising PK profiles (Abramson et al. 2022).

The absorption mechanisms of SC or IM administration are poorly understood. However, it is believed that the absorption of mAbs after IM or SC injection is likely via lymphatic drainage due to its large molecular weight, leading to a slow absorption rate (see Chap. 6). The bioavailability of mAbs after SC or IM administration has been reported to be around 50–100% with maximal plasma concentrations observed 1–8 days following administration (Lobo et al. 2004). For example, following an IM injection, the bioavailability of alefacept was ~60% in healthy male volunteers; its C_{max} was threefold (0.96 versus 3.1 μg/mL) lower, and its T_{max} was 30 times longer (86 versus 2.8 h) than a 30-min IV infusion (Vaishnaw and TenHoor 2002). Interestingly, differences in PK have also been observed between different sites of IM dosing. PAmAb, a fully humanized mAb against *Bacillus anthracis* protective antigen, has significantly different pharmacokinetics between IM-GM (gluteus maximus site) and IM-VL (vastus lateralis site) injection in healthy volunteers (Subramanian et al. 2005). The bioavailability of PAmAb is 50–54% for IM-GM injection and 71–85% for IM-VL injection (Subramanian et al. 2005). Of note, mAbs appear to have greater bioavailability after SC administration in monkeys than in humans (Oitate et al. 2011). The mean bioavailability of adalimumab is 52–82% after a single 40 mg SC administration in healthy adult subjects, whereas it was observed to be 94–100% in monkeys. Similarly, the mean bioavailability of omalizumab is 66–71% after a single SC dose in patients with asthma versus 88–100% in monkeys (Oitate et al. 2011).

Although SC administration of mAbs initially used low-volume injections (1–2 mL), in recent years, larger volume injections (>2 mL) have been used, with and without permeation enhancers. SC injections of a viscous (5 cP) placebo buffer, characteristic of a high-concentration mAb formulation, at volumes of up to 3.5 mL had acceptable tolerability in healthy adult subjects at injection rates up to 3.5 mL/min (Dias et al. 2015a, b). Due to relatively large therapeutic doses (several hundred milligram), dosing volumes of high-concentration mAbs may still be too large to facilitate a painless SC injection. Without co-injection of a permeation enhancer, large volume injections may produce swelling at the injection site, particularly in the thigh and arm. A permeation enhancer, such as recombinant human hyaluronidase (rHuPH20), reduces this swelling. Hyaluronidase is a 61-kD naturally occurring enzyme that temporarily degrades hyaluronan in the skin and increases dispersion of the mAb over a greater area. Co-formulation of rHuPH20 with therapeutic proteins allows SC administration of larger injection volumes and potentially enhances absorption of the therapeutic protein into the systemic circulation (Frost 2007). Trastuzumab and rituximab have both been co-formulated with rHuPH20 to facilitate large-volume SC injections (Bittner et al. 2012). Trastuzumab is available as a 5-mL SC injection to be administered over 2–5 min, while rituximab is available at 11.7 or 13.4 mL injection volumes to be administered over 5 or 7 min, respectively. The same strategy was used for SC fixed dose combination of trastuzumab and pertuzumab, which is available as a 15 mL SC injection to be administered over approximately 8 min for the initial dose and a 10 mL SC injection to be administered over approximately 5 min for the maintenance dose (FDA 2020b). SC delivery devices such as prefilled syringe (PFS), autoinjectors, on-body delivery systems (OBDS), and infusion pump have also been used to facilitate the SC dosing (Bittner et al. 2018).

Distribution

After reaching the bloodstream, mAbs undergo biphasic elimination from serum, beginning with a rapid distribution phase. The volume of distribution of the rapid-distribution compartment is relatively small, approximating plasma volume. It is reported that the volume of the central compartment (Vc) is about 2–3 L, and the steady-state volume of distribution (Vss) is around 3.5–7 L for mAbs in humans (Lobo et al. 2004; Roskos et al. 2004). The small Vc and Vss for mAbs indicate that the distribution of mAbs is restricted to the blood and extracellular spaces, which is in agreement with their hydrophilic nature and their large molecular weight, limiting access to the intracellular compartment surrounded by a lipid bilayer. Small volumes of distributions are consistent with relatively small tissue:blood ratios for most antibodies typically ranging from 0.1 to 0.5 (Baxter et al. 1994, 1995; Berger et al. 2005). For example, the tissue-to-blood concentration ratios for a murine IgG1 mAb against the human ovarian cancer antigen CA125 in mice at 24 h after injection are 0.44, 0.39, 0.48, 0.34, 0.10, and 0.13 for the spleen, liver, lung, kidney, stomach, and muscle, respectively. Total (lump sum) murine tissue concentrations of an antiglycoprotein D IgG1 expressed as percentages of serum concentrations were 12–14% in lung, 9–14% in heart, 10–11% in kidneys, 3–4% in muscle, 9–11% in skin, 4–6% in intestines, 10–13% in spleen, 8–9% in liver, 4–5% in stomach, 6–8% in lymph nodes, 4–5% in fat, and 0.9% in brain. Expressing these data in a more pharmacologically relevant manner, tissue interstitial fluid concentrations as a percentage of serum concentrations were 17–26% in lung, 33–69% in heart, 28–36% in kidneys, 34–35% in muscle, 24–29% in skin, 15–24% in intestines, 12–21% in stomach, 12–18% in lymph nodes, 7–8% in fat, and 1.7–2.0% in brain (Rafidi et al. 2022). Brain and cerebrospinal fluid are anatomically protected by blood–brain barriers. Although the blood–brain barrier was believed to be impaired in certain neurodegenerative disease states, recent work has brought this into question (Bien-Ly et al. 2015). Therefore, both compartments are very limited distribution compartments for mAbs (Cf. Chap. 26). For example, endogenous IgG levels in cerebrospinal fluid were shown to be in the range of only 0.1–1% of their respective serum levels (Wurster and Haas 1994; Yadav et al. 2017; Yu et al. 2014).

It has been repeatedly noted that the reported Vss obtained by traditional noncompartmental or compartmental analysis may be not correct for mAbs that primarily undergo catabolism within tissue (Lobo et al. 2004; Straughn et al. 2006; Tang et al. 2004). The rate and extent of mAb distribution will be dependent on the kinetics of mAb extravasation within tissue, distribution within tissue, and elimination from tissue. Convection, diffusion, transcytosis, binding, and catabolism are important determining factors for antibody distribution (Lobo et al. 2004). Therefore, Vss might be substantially greater than the plasma volume in particular for those mAbs demonstrating high binding affinity in the tissue. Different research groups have reported effects of the presence of specific receptors (i.e., antigen sink) on the distribution of mAbs (Bumbaca et al. 2012; Danilov et al. 2001; Kairemo et al. 2001). Danilov et al. (2001) found in rats that an anti-PECAM-1 (anti-CD31) mAb showed tissue-to-blood concentration ratios of 13.1, 10.9, and 5.96 for the lung, liver, and spleen, respectively, 2 h after injection. Therefore, the true Vss of the anti-PECAM-1 is likely to be 15-fold greater than plasma volume.

Another complexity to consider is that tissue distribution via interaction with target proteins (e.g., cell surface proteins) and subsequent internalization of the antigen-mAb complex may be dose dependent. For the murine analog mAb of efalizumab, M17, a pronounced dose-dependent distribution was demonstrated by comparing tissue-to-blood concentration ratios for liver, spleen, bone marrow, and lymph node after a tracer dose of radiolabeled M17 and a high-dose treatment (Coffey et al. 2005). The tracer dose of M17 resulted in substantially higher tissue-to-blood concentration ratios of 6.4, 2.8, 1.6, and 1.3 for the lung, spleen, bone marrow, and lymph node, respectively, in mice at 72 h after injection. In contrast, the saturation of the target antigen at the high-dose level reduced the tissue distribution to the target independent distribution and resulted consequently in substantially lower tissue-to-blood concentration ratios (less than 1).

FcRn may play an important role in the transport of IgGs from plasma to the interstitial fluid of tissue. Recently, the data from Yip et al. increased understanding of FcRn's role in antibody PK and catabolism at the tissue level (Yip et al. 2014). They reported that distribution of the wild-type IgG and the variant with enhanced binding for FcRn were largely similar to each other in mice, but vastly different for the low-FcRn-binding variant due to its very low systemic exposure and widespread catabolism, particularly in liver and spleen. Ferl et al. (2005) reported that a physiologically based pharmacokinetic (PBPK) model, including the kinetic interaction between the mAb and the FcRn receptor within intracellular compartments, could describe the biodistribution of an anti-CEA mAb in a variety of tissue compartments such as plasma, lung, spleen, tumor, skin, muscle, kidney, heart, bone, and liver. FcRn was also reported to mediate the crossing of placental barriers by IgG (Junghans 1997) and the vectorial transport of IgG into the lumen of intestine (Dickinson et al. 1999) and lung (Spiekermann et al. 2002).

Clearance

Antibodies are mainly cleared by catabolism and broken down into peptide fragments and amino acids, which can be recycled—to be used as energy supply or for new protein synthesis. Due to the small molecular weight of antibody fragments (e.g., Fab and Fv), elimination of these fragments is faster than for intact IgGs, and they can be filtered through the glomerulus and reabsorbed and/or metabolized by proximal tubular cells of the nephron (Lobo et al. 2004). Murine

monoclonal anti-digoxin Fab, F(ab')2, and IgG1 have half-lives of 0.41, 0.70, and 8.10 h in rats, respectively (Bazin-Redureau et al. 1997). Several studies reported that the kidney is the major route for the catabolism of Fab and elimination of unchanged Fab (Druet et al. 1978; McClurkan et al. 1993).

Typically, IgGs have serum half-lives of approximately 21 days, resulting from CL values of about 3–5 mL/day/kg, and Vss's of 50–100 mL/kg. The exception is IgG3, which has a half-life of only 7 days. The half-life of IgG is much longer than that of other Igs (IgA, 6 days; IgE, 2.5 days; IgM, 5 days; IgD, 3 days). The FcRn receptor has been demonstrated to be a primary determinant of the disposition of IgG antibodies (Ghetie et al. 1996; Junghans 1997; Junghans and Anderson 1996). FcRn, which protects IgG from catabolism and contributes to the long plasma half-life of IgG, was first postulated by Brambell in 1964 (Brambell et al. 1964) and cloned in the late 1980s (Simister and Mostov 1989). FcRn is a heterodimer comprising of a β_2m light chain and a MHC class I-like heavy chain. The receptor is ubiquitously expressed in cells and tissues. Several studies have shown that IgG CL in β_2m knockout mice (Ghetie et al. 1996; Junghans and Anderson 1996), and FcRn heavy chain knockout mice (Roopenian et al. 2003) is increased 10–15-fold, with no changes in the elimination of other Igs. Figure 8.9 illustrates how the FcRn receptor protects IgG from catabolism and contributes to its long half-life. The FcRn receptor binds to IgG in a pH-dependent manner: binding to IgG at the acidic pH (6.0) of the endosome and releasing IgG at physiological pH (7.4). The unbound IgG proceeds to the lysosome and undergoes proteolysis.

It has been demonstrated that IgG half-life is dependent on its affinity to FcRn receptors. The shorter half-life of IgG3 was attributed to its low binding affinity to the FcRn receptor (Junghans 1997; Medesan et al. 1997). Murine mAbs have serum half-lives of 1–2 days in human. The shorter half-life of murine antibodies in human is due to their low binding affinity to the human FcRn receptor. It is reported that human FcRn binds to human, rabbit, and guinea pig IgG, but not to rat, mouse, sheep, and bovine IgG; however, mouse FcRn binds to IgG from all of these species (Ober et al. 2001). Interestingly, human IgG1 has greater affinity to murine FcRn (Petkova et al. 2006), which indicates potential limitations of using mice as preclinical models for human IgG1 pharmacokinetic evaluations. Ward's group confirmed that an engineered human IgG1 had disparate properties in murine and human systems (Vaccaro et al. 2006). Engineered IgGs with higher affinity to FcRn receptor have a two- to threefold longer half-life compared with WT in mice and monkeys (Hinton et al. 2006; Petkova et al. 2006). Two engineered human IgG1 mutants with enhanced binding affinity to human FcRn show a considerably extended half-life compared with WT in hFcRn transgenic

Fig. 8.9 Schematic disposition pathway of IgG antibodies via interaction with FcRn in endosomes. (*1*) IgGs enter cells by receptor-mediated endocytosis by binding of the Fc part to FcRn. (*2*) The intracellular vesicles (endosomes) fuse with lysosomes containing proteases. (*3*) Proteases degrade unbound IgG molecules, whereas IgGs bound to FcRn are protected. (*4a*) The intact IgG bound to FcRn is transported back to the cell surface and (*4b*) released back to the extracellular fluid

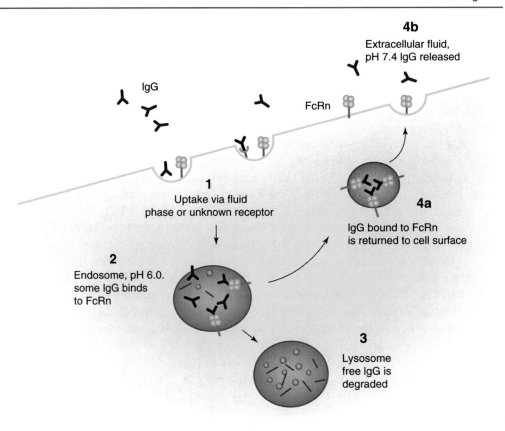

mice (4.35 ± 0.53, 3.85 ± 0.55 days versus 1.72 ± 0.08 days) (Hinton et al. 2006; Petkova et al. 2006). It was found that the half-life of IgG1 FcRn mutants with increasing binding affinity to human FcRn at pH 6.0 is about 2.5-fold longer than the WT antibody in monkey (838 ± 187 h versus 336 ± 34 h). It has been reported that the motavizumab FcRn affinity-enhancing Fc mutant (M252Y/S254T/T256E (YTE)) was able to extend serum half-life in humans by two to fourfold compared with the WT motavizumab IgG1 (~74–100 days vs ~19–34 days) (Robbie et al. 2013).

Dose-proportional, linear CL has been observed for mAb against soluble antigens with low endogenous levels (such as TNF-α, IFN-α, VEGF, and IL-5). For example, linear PK has been observed for a humanized mAb directed to human interleukin-5 following IV administration over a 6000-fold dose range (0.05–300 mg/kg) in monkeys (Zia-Amirhosseini et al. 1999). The CL of rhumAb against VEGF after IV dosing (2–50 mg/kg) ranged from 4.81 to 5.59 mL/day/kg and did not depend on dose (Lin et al. 1999). The mean total serum CL and the estimated mean terminal half-life of adalimumab were reported to range from 0.012 to 0.017 L/h and 10.0–13.6 days, respectively, for a 5-cohort clinical trial (0.5–10 mg/kg), with an overall mean half-life of 12 days (den Broeder et al. 2002). However, mAbs against soluble antigens with high endogenous levels (such as IgE) exhibit nonlinear PK. The PK of omalizumab, an mAb against IgE, is linear only at doses greater than 0.5 mg/kg (FDA 2006b; Petkova et al. 2006).

Elimination of mAbs may also be impacted by interaction with the targeted cell-bound antigen, and this phenomenon was demonstrated by dose-dependent clearance and half-life. At low dose, mAbs may show a shorter half-life and a faster clearance due to receptor-mediated elimination. With increasing doses, receptors become saturated, the half-life gradually increases to a constant, and the CL gradually decreases to a constant. The binding affinity (K_d), antigen density, and antigen turnover rate may influence the receptor-mediated elimination. Koon et al. found a strong inverse correlation between CD25 cell expression and the apparent half-life of daclizumab (a mAb specifically binding to CD25) (Koon et al. 2006). It has been shown that the PK of murine antihuman CD3 antibodies may be determined by the disappearance of target antigen (Meijer et al. 2002). In monkeys and mice, clearance of SGN-40, a humanized anti-CD40 mAb, was much faster at low dose, suggesting nonlinear PK (Kelley et al. 2006). In addition, Ng et al. demonstrated that an anti-CD4 mAb (TRX-1) had ~five-fold faster CL at 1 mg/kg dose compared with 10 mg/kg dose (37.4 ± 2.4 versus 7.8 ± 0.6 mL/day/kg) in healthy volunteers (Ng et al. 2006). They also found that receptor-mediated CL via endocytosis became saturated at higher doses; nonspecific clearance of TRX-1 contributed 8.6, 27.1, and 41.7% of total CL when dose was 1, 5, and 10 mg/kg, respectively.

In addition to FcRn and antigen–antibody interaction, other factors may also contribute to mAb elimination (Lobo et al. 2004; Roskos et al. 2004; Tabrizi et al. 2006):

1. *Immunogenicity* The elimination of mAbs in humans may increase with increasing level of immunogenicity (Tabrizi et al. 2006; Ternant and Paintaud 2005).

2. *Degree and the nature of glycosylation* The impact of glycosylation on the pharmacokinetics and effector functions of therapeutic IgG1 monoclonal antibodies has been previously reviewed (Putnam et al. 2010).

3. *Susceptibility to proteolysis* Gillies and coworkers improved the circulating half-life of antibody-interleukin 2 immunocytokine by twofold compared with wild type (1.0 h versus 0.54 h) by increasing the resistance to intracellular degradation (Gillies et al. 2002).

4. *Charge* Deliberate modification of the isoelectric point (pI) of an antibody by approximately one pI unit or more can lead to noticeable differences in the PK of an intact antibody (Bumbaca Yadav et al. 2015; Igawa et al. 2010; Li et al. 2014). Using a humanized anti-IL-6 receptor IgG1 as an example, Igawa et al. showed that lowering the pI point from 9.2 to 7.2 by engineering the V region reduced the IgG elimination in cynomolgus monkeys (Igawa et al. 2010). In contrast, minor changes in the nature of ionic charge resulting in pI differences of less than approximately one pI unit are not expected to affect the biological function of mAbs, including tissue retention and whole blood clearance (Boswell et al. 2010b; Khawli et al. 2010).

5. *Effector function* Effector functions, such as interactions with FcγR, can also regulate elimination and PK of mAbs (Mahmood and Green 2005). Mutation of the binding site of FcγR, for example, had dramatic effects on the clearance of an Ab-IL-2 fusion protein (Gillies et al. 1999).

6. *Concomitant medications* Methotrexate reduced adalimumab apparent CL after single dose and multiple dosing by 29 and 44%, respectively, in patients with RA (FDA 2007). In addition, azathioprine and mycophenolate mofetil were reported to reduce CL of basiliximab by approximately 22 and 51%, respectively (FDA 2005a). The effects of small-molecule drugs on the expression of Fcγ receptors could explain this finding. It has also been shown that methotrexate affects expression of FcγRI on monocytes significantly in RA patients (Bunescu et al. 2004).

7. *Off-target binding* Although specificity to their targets is a major characteristic of mAbs, they may have off-target binding that may result in atypical PK, such as faster CL and larger volume distribution. An antirespiratory syncytial virus mAb, A4b4, developed by affinity maturation of palivizumab, had poor PK in rats and cynomolgus monkeys due to broad nonspecific tissue binding and sequestration (Wu et al. 2007). The rapid elimination of a humanized antihuman amyloid beta peptide mAb, anti-Aβ Ab2, in cynomolgus monkeys was linked to off-target binding to cynomolgus monkey fibrinogen (Vugmeyster et al. 2011). In addition, a humanized anti-fibroblast growth factor receptor 4 mAb had a rapid CL in mice that was attributable to binding to mouse complement component 3 (Bumbaca et al. 2011). Other examples of mAbs with off-target effects include mAbs targeting Factor IXa/X (Sampei et al. 2013), interleukin-21 receptor (Vugmeyster et al. 2010). It is important to eliminate mAbs with higher risk of failure at the discovery stage, to increase the success rate. As PK of these therapeutic proteins might be influenced by a large number of both specific and nonspecific factors, Dostalek et al. have proposed multiple pharmacokinetic de-risking tools for selection of mAb lead candidates (Dostalek et al. 2017).

8. *Body weight, age, disease state, and other demographic factors* Individual characteristics can also change mAb PK (Mould and Sweeney 2007; Ryman and Meibohm 2017) (see Population Pharmacokinetics section).

9. *Albumin* Albumin levels are often an indicator of disease status and a significant covariate correlating with clearance for several mAbs, including infliximab, pertuzumab, trastuzumab emtansine (Lu et al. 2014), and bevacizumab (Dirks and Meibohm 2010). It is believed that albumin, which binds to FcRn at different sites than IgG, is an indicator of increased protein turnover (Dirks and Meibohm 2010; Fasanmade et al. 2010). Nevertheless, the correlation between the levels of albumin and the CL of pertuzumab and bevacizumab was moderate and dose modification was not recommended (Dirks and Meibohm 2010). Regardless, it has been suggested that serum albumin levels are a predictive factor for PK of infliximab and clinical response to the drug in patients with ulcerative colitis (Fasanmade et al. 2010).

10. *Disease state* It has been reported that disease state can impact mAb PK. Lower exposure and faster CL for trastuzumab (Han et al. 2014; Yang et al. 2013), bevacizumab (Han et al. 2014), pertuzumab (Kang et al. 2013), and trastuzumab emtansine (Chen et al. 2017) in patients with gastric cancer (GC) versus breast cancer (BC) have been reported. Steady-state trastuzumab trough concentration (C_{trough}) in patients with metastatic GC is 24–63% lower than in BC (Yang et al. 2013). The underlying mechanism for faster CL of mAbs in GC is unknown and warrants further research. Population PK analyses of ofatumumab were performed for various diseases with varying CD20 B-cell counts and indicated that target-mediated CL in CLL is greater than that in RA and FL, which is consistent with the higher B-cell count seen in CLL (Struemper et al. 2014). Diabetic comorbidity resulted in 28.7% higher CL/F for ustekinumab (Zhu et al. 2009). Infliximab CL is 40–50% higher in inflammatory bowel disease patients, which is likely due to

protein losing enteropathy (Fasanmade et al. 2009). Recently, it has been observed that mAbs in immune-oncology, such as pembrolizumab (Li et al. 2017; Turner et al. 2018) and nivolumab (Bajaj et al. 2017), have time-dependent CL. Naturally, patients with elevated catabolic rates of endogenous proteins arising from cancer-associated cachexia could also tend to have faster CL of exogenously administered biologics as well (Castillo et al. 2021; Fearon et al. 2012).

In summary, the association between disease factors and PK complicates the interpretation of the exposure-efficacy analyses for mAbs and ADCs in cancer patients, as only one-dose level is usually tested in the pivotal study. Although correction methods can be applied, the effect of disease severity on treatment exposure may result in an overestimation of exposure–response relationships, i.e., visually a steep trend is seen when the true relationship is flat (Liu et al. 2015; Wang 2016).

Therapeutic mAb–Drug Interactions

mAbs and other therapeutic proteins are increasingly combined with small-molecule drugs to treat various diseases. Assessment of the potential for PK- and/or PD-based mAb–drug interactions is frequently incorporated into the drug development process (Girish et al. 2011). The DDI risk assessment should consider the interplay between the hypothesized mechanism for the interaction, disease, and its severity, biological product type and clearance pathways (FDA 2020a; Schrieber et al. 2019). The exposure and response of concomitantly administered drugs can be altered by mAbs (mAb as perpetrator), and other drugs can affect the PK and PD of therapeutic mAbs (mAb as victim).

Several different mechanisms have been proposed for mAb–drug interactions (FDA 2020a). Various cytokines and cytokine modulators can influence the expression and activity of cytochrome P450 (CYP) enzymes and drug transporters (Lee et al. 2010). Therefore, if a therapeutic mAb is a cytokine or cytokine modulator, it can potentially alter the systemic exposure and/or clinical response of concomitantly administered drugs that are substrates of CYPs or transporters (Huang et al. 2010), particularly those with narrow therapeutic windows. T-cell engaged bispecific antibody treatment, which can induce cytokine release syndrome, could have such an effect on substrates of CYPs or transporters. For example, an increase in cyclosporin A (CsA) trough level was observed when given in combination with muromonab (Vasquez and Pollak 1997). Similarly, basiliximab has

been shown to increase CsA and tacrolimus level when used in combination (Sifontis et al. 2002). In diseases states such as infection or inflammation, cytokines or cytokine modulators can also normalize previously changed activity of CYPs or transporters, thereby altering the exposure of co-administered drugs. Examples include tocilizumab co-administered with omeprazole and tocilizumab co-administered with simvastatin.

At present, in vitro and preclinical systems have shown limited value in predicting a clinically relevant effect of cytokine-mediated therapeutic protein (TP)-DDI, and clinical evidence is preferred for informing the evolving risk assessment for TP-DDI (Huang et al. 2010; Slatter et al. 2013). To determine the necessity for a dedicated clinical DDI study, a four-step approach was proposed by the IQ Consortium/FDA TP-DDI workshop (San Diego, 2012) in assessing TP-DDI risk for cytokines or cytokine modulators on CYP enzymes. This includes stepwise investigations of: (1) the disease effect on cytokine levels and CYP expression; (2) TP mechanism and its impact on cytokine-mediated DDI; (3) DDI liability of the concurrently used small-molecule drugs; and (4) the above overall driving force in determining appropriate clinical TP-DDI strategies (Kenny et al. 2013). To date, a few dedicated clinical DDI studies have been performed for mAbs that specifically target cytokines or cytokine receptors, e.g., tocilizumab (Schmitt et al. 2011), sirukumab (Zhuang et al. 2015), daclizumab HYP (Tran et al. 2016), and dupilumab (Davis et al. 2018). However, the overall impact of these cytokine-blocking mAbs on PK of the CYP substrates (mAb as a perpetrator) were minimal (no effect, e.g., daclizumab HYP, ustekinumab, and dupilumab) or moderate (18–57% reduction in AUC for CYP 2C19 or 3A4 substrates, e.g., tocilizumab and sirukumab) and have not been implicated in dose justification for the relevant concurrent medicines. Understanding the cytokine-time profile following administration of the mAb can inform the DDI risk assessment or need for a dedicated clinical DDI study.

mAb–drug interactions can also occur when a therapeutic mAb is administered with a concomitant drug that can alter the formation of ADAs. This may in turn alter mAb clearance from the systemic circulation. For example, methotrexate (MTX) reduced the apparent CL of adalimumab by 29 and 44% after single and repeated dosing (FDA 2007). MTX also had a similar effect on infliximab (Maini et al. 1998). PD-based interactions can result from alteration of target biology, such as the site of expression, relative abundance of expression, and the pharmacology of the target (Girish et al. 2011). Small molecules may have indirect impact on antibody PK by altering the target-mediated clearance, which is relevant to the target expression level. A mAb could have PK DDI with another mAb through the same mechanism. Examples include efalizumab in combination with triple immune-suppressant therapy (Vincenti et al. 2007) and anakinra in combination with etanercept (Genovese et al. 2004).

To date, evidence of therapeutic mAb–drug interactions via nonspecific clearance appears to be limited, although downregulation of Fcγ receptors by MTX is observed in patients with RA. It is possible that changes in Fcγ receptors can affect mAb clearance in the presence of MTX (Girish et al. 2011). Co-administered medications that compromise the function of the FcRn can affect mAbs that interact with the FcRn. It has been reported that rozanolixizumab, a FcRn inhibitor, can reduce human serum IgG concentration (Kiessling et al. 2017).

The DDI potential for an ADC should be evaluated for the antibody and the small-molecule drug components independently. ADCs can interact with drugs or mAbs via the mechanisms described above through the antibody components. However, evidence of ADC–drug or ADC–mAb interaction appears to be limited. Lu et al. reported lack of interaction between trastuzumab emtansine (T-DM1) and pertuzumab in patients with HER2-positive metastatic breast cancer (Lu et al. 2012). Similarly, no interaction was observed between T-DM1 and paclitaxel or T-DM1 and docetaxel (Lu et al. 2012). In many cases, the systemic concentration of the released payload might be too low to act as a perpetrator. It might be necessary to evaluate the circulating payload (administered as an ADC) as a victim, which may impact the overall ADC safety profile, as these payloads are highly potent and typically have a very narrow therapeutic index. A data-driven and risk-based decision tree was published to guide the assessment of the circulating payload of an ADC as a victim (Li et al. 2022). Recently, PBPK modeling has been used as a tool to predict small-molecule component-based drug interactions for ADCs (Chen et al. 2015; Li et al. 2020). A PBPK modeling approach, in conjugation with a supported clinical DDI study from brentuximab vedotin, has been successfully used for the DDI predictions to inform polatuzumab vedotin prescribing information (Samineni et al. 2020).

With the theoretical potential for, and current experiences with, mAb–drug interactions, a question and risk-based integrated approach depending on the mechanism of the mAbs and patient population have been progressively adopted during drug development to address important questions regarding the safety and efficacy of mAb and drug combinations (Girish et al. 2011). Various in vitro test systems have been used to provide some insight into the mAb–drug interactions, such as isolated hepatocytes and liver microsomes. However, the interpretation of these in vitro data is difficult. More importantly, prospective predictions of drug interactions based on in vitro findings have not been feasible for mAbs. Therefore, clinical methods are primarily used to assess mAb–drug interactions. Three common methods used

are population PK, clinical cocktail studies, and less frequently, dedicated drug interaction studies. Details of various strategies used in the pharmaceutical industry were reviewed in a 2011 AAPS white paper (Girish et al. 2011).

Prediction of Human PK/PD Based on Preclinical Information

Prior to a first-in-human (FIH) clinical study, a number of preclinical in vivo and in vitro experiments are conducted to evaluate the PK/PD, safety, and efficacy of a new drug candidate. However, the ultimate goal is at all times to predict how these preclinical results on PK, safety, and efficacy data translate into a given patient population. Therefore, the objective of translational research is to predict PK/PD/safety outcomes in a target patient population, acknowledging the similarities and differences between preclinical and clinical settings.

Over the years, many theories and approaches have been proposed and used for scaling preclinical PK data to humans (Fig. 8.10). Allometric scaling, based on a power–law relationship between size of the body and physiological and anatomical parameters, is the simplest and most widely used approach (Dedrick 1973; Mahmood 2005, 2009). More recently, experimental efforts have been dedicated to accurate measurement of physiological parameters that are required for calculating drug concentrations at site of action and for physiologically based models (Boswell et al. 2012, 2010a, 2014). Physiologically based PK modeling (Shah and Betts 2012), species-invariant time method (Dedrick approach) (Oitate et al. 2011), and nonlinear mixed effect modeling based on allometry (Jolling et al. 2005; Martin-Jimenez and Riviere 2002) have also been used for interspecies scaling of PK. While no single scaling method has been shown to definitively predict human PK in all cases, especially for small-molecule drugs (Tang and Mayersohn 2005), the PK for mAbs can be predicted reasonably well, especially for mAb at doses where the dominant clearance route is likely to be independent of concentration. Most therapeutic mAbs bind to nonhuman primate antigens more often than to rodent antigens, due to the greater sequence homology observed between nonhuman primates and humans. The binding epitope, in vitro binding affinity to antigen, binding affinity to FcRn, tissue cross-reactivity profiles, and disposition and elimination pathways of mAbs are often comparable in nonhuman primates and humans. It has been demonstrated that clearance and distribution volume of mAbs with linear PK in humans can be reasonably projected based on data from nonhuman primates alone, with a fixed scaling exponent ranging from 0.75 to 0.9 for CL and a fixed

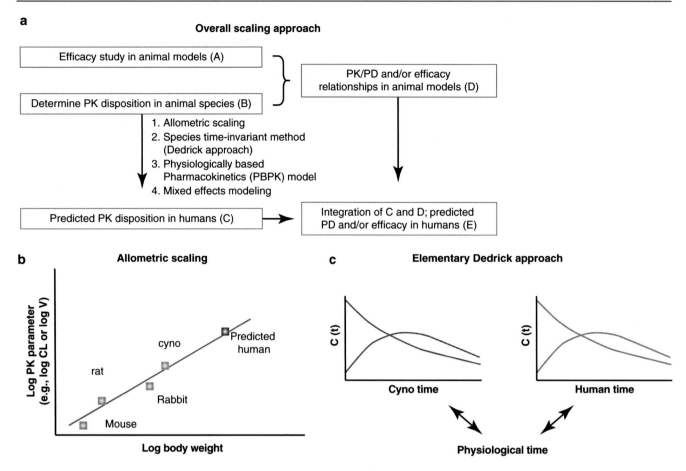

Fig. 8.10 PK/PD scaling approach from preclinical studies to humans. (**a**) Overall scaling approach. (**b**) Allometric scaling. (**c**) Elementary Dedrick approach

scaling exponent 1 for volume of distribution (Deng et al. 2011; Dong et al. 2011; Ling et al. 2009; Oitate et al. 2011; Wang and Prueksaritanont 2010). Recently, based on a data set of 11 ADCs, Li et al. demonstrate that scaling using a binding species such as cynomolgus monkey is pharmacologically appropriate for therapeutic ADCs that demonstrate linear CL (Li et al. 2019). Allometric scaling of CL using cynomolgus monkey alone with an exponent of 1.0 provided a good estimate of human CL for both ADC conjugate and total antibody analytes (Li et al. 2019). For mAbs or ADCs that exhibited nonlinear PK, the best predictive performance was obtained above doses that saturated the target of the mAb or ADC (Dong et al. 2011). Pharmacokinetic prediction for low doses of a mAb or ADC with nonlinear elimination remains challenging and will likely require further exploration of species difference in target expression level, target antibody binding and target kinetics, as well as strategic in vivo animal PK studies, designed with relevant dose ranges (Deng et al. 2011; Dong et al. 2011; Li et al. 2019). Immunogenicity is an additional challenge for the prediction of mAb PK. Alterations in the PK profile due to

immune-mediated clearance mechanisms in preclinical species cannot be scaled up to humans, since animal models are not predictive of human immune response to human mAbs. Thus, either excluding ADA-positive animals from PK scaling analysis or using only the early time points prior to ADAs observation in ADA-positive animals has been a standard practice in the industry.

Due to its complexity, any extrapolation of PD to humans requires more thorough consideration than for PK. Little is known about allometric relationships in PD parameters. It is expected that the physiological turnover rate constants of most general structures and functions among species should obey allometric principles, whereas capacity and sensitivity tend to be similar across species (Mager et al. 2009). Through integration of PK/PD modeling and interspecies scaling, PD effects in humans may be predicted if the PK/PD relationship is assumed to be similar between animal models and humans (Duconge et al. 2004; Kagan et al. 2010). For example, a PK/PD model was first developed to optimize the dosing regimen of a mAb against EGF/r3 using tumor-bearing nude mice as an animal model of human disease (Duconge et al. 2004). This PK/PD model was subsequently integrated with allometric scaling to calculate the dosing schedule required in a potential clinical trial to achieve a specific effect (Duconge et al. 2004).

In summary, species differences in antigen expression level, antigen–antibody binding and antigen kinetics, differences in FcRn binding between species, the immunogenicity, and other factors must be considered during PK/PD scaling of a mAb from animals to humans.

Role of PK/PD in Clinical Development of Antibody Therapeutics

Drug development has traditionally been performed in sequential phases, divided into preclinical as well as clinical phases I–IV. During the development phases of the molecules, the safety and PK/PD characteristics are established in order to select a compound for development and define a dosing regimen. This information-gathering process has been characterized as two successive learning–confirming cycles (Sheiner and Wakefield 1999; Sheiner 1997).

The first cycle (phases I and IIa) comprises learning about the dose regimen that is tolerated in healthy subjects and confirming that this dose regimen has shown drug-target engagement, acceptable tolerability and measurable clinical benefits in the targeted patients. An affirmative answer at this first cycle provides the justification for a larger and more costly second learn–confirm cycle (phases IIb and III), where the learning step is focused on defining the drug benefit/risk profile, whereas the confirm step is aimed at demonstrating acceptable benefit/risk in a large patient population (Meibohm and Derendorf 2002).

The drug development process at the clinical stage provides several opportunities for integration of PK/PD concepts. Clinical phase I dose escalation studies provide, from a PK/PD standpoint, the unique opportunity to evaluate the dose–concentration–effect relationship for therapeutic and toxic effects over a wide range of doses up to or even beyond the maximum tolerated dose under controlled conditions (Meredith et al. 1991). PK/PD evaluations at this stage of drug development can provide crucial information regarding the potency and tolerability of the drug in vivo and the verification and suitability of the PK/PD relationship established during preclinical studies.

Collecting robust data on the PK of the drug and PD or disease biomarkers that are indicative of drug pharmacology and disease progression/improvement is key to informing dose selection. Tocilizumab, omalizumab, and evolocumab are examples of mAbs that utilized PK/PD or disease biomarker data to facilitate dose selection for pivotal trials, final doses, and/or label revisions. In general, the strategy includes (1) understanding the PK profile and selecting clinical doses in the linear range, if possible; (2) identifying biomarkers having profiles correlated to clinically meaningful endpoints for PK/PD or exposure-response analyses; and (3) leveraging modeling and simulation approaches to predict the clinical outcome under different regimen scenarios, which is essential to determine the dose regimen for a pivotal trial or the final dose regimen on the label.

In the case of omalizumab, a dosing table for asthma patients was developed to select the dose based on an individual's pretreatment serum IgE level and body weight. The dosing table was designed to achieve a serum-free IgE level associated with clinical improvement (Hochhaus et al. 2003). PK/PD modeling and simulation approaches were subsequently used to revise and expand the dosing table (Honma et al. 2016; Lowe et al. 2015). In the case of evolocumab, a high-level summary of the development program and dosing strategy follows.

Evolocumab is a recombinant, human IgG2 mAb that specifically binds to human proprotein convertase subtilisin/kexin type 9 (PCSK9). It prevents PCSK9 from interacting with the low-density lipoprotein receptor (LDLR), thus upregulating LDLR, increasing uptake of circulating LDL-cholesterol (LDL-C), and reducing LDL-C concentration in plasma (Page and Watts 2015). Evolocumab is used as an adjunct to diet and maximally tolerated statin therapy for the treatment of adults with heterozygous familial hypercholesterolemia or clinical atherosclerotic cardiovascular disease. Its use is also indicated as an adjunct to diet and other LDL-lowering therapies for the treatment of patients with homozygous familial hypercholesterolemia.

The PK of evolocumab following multiple SC doses was evaluated in a Phase I study in subjects on a stable dose of statin over a dose range of 14–420 mg of evolocumab weekly, every 2 weeks, or every 4 weeks. Multiple doses of evolocumab resulted in nonlinear PK for the lower doses (up to 140 mg SC). Dose regimens of 140 mg and greater led to linear PK and concentrations associated with near complete suppression of PCSK9 (CDER 2014). Dose-dependent decreases in LDL-C levels were seen following treatment with evolocumab (CDER 2014), and this PD readout is also indicative of a meaningful clinical readout. There was a clear exposure–response relationship between evolocumab trough concentrations and LDL-C response (CDER 2014). These PK/PD data and the exposure–response relationship were used to support the final approved dose and dosing regimen.

Efalizumab Case Study (Raptiva®)

In the following sections, the recombinant humanized IgG1 mAb efalizumab is provided as a more detailed case study to understand the various steps during the development of therapeutic antibodies for various indications. Raptiva® received approval for the treatment of patients with psoriasis in more than 30 countries, including the United States and the European Union (FDA 2004). However, it was withdrawn from the market when the use of efalizumab was found to be associated with an increased risk of progressive multifocal leukoencephalopathy (PML).

A summary of the preclinical program, the overall PK/PD data from multiple clinical studies, and the selection of the subcutaneous doses of efalizumab for the treatment of psoriasis will be discussed. Psoriasis is a chronic skin disease characterized by abnormal keratinocyte differentiation and hyperproliferation and by an aberrant inflammatory process in the dermis and epidermis. T cell infiltration and activation in the skin and subsequent T cell-mediated processes have been implicated in the pathogenesis of psoriasis (Krueger 2002). Efalizumab is a targeted inhibitor of T cell interactions (Werther et al. 1996). An extensive preclinical research program was conducted to study the safety and MOA of efalizumab. Multiple clinical studies were also conducted to investigate the efficacy, safety, PK, PD, and MOA of efalizumab in patients with psoriasis.

Preclinical Program of Efalizumab

A thorough and rigorous preclinical program provides a linkage between drug discovery and clinical development. At the preclinical stage, activities may include the evaluation of in vivo potency and intrinsic activity, the identification of bio-/surrogate markers, understanding of MOA, and characterization of nonclinical PK/PD, as well as dosage form/regimen selection and optimization. The role of surrogate molecules in assessing ADME of therapeutic antibodies is important as the antigen specificity of humanized mAbs limits their utility in studies with rodents. Surrogate rodent mAbs (mouse/rat) provide a means of gaining knowledge of PK and PD in a preclinical rodent model, facilitating dose optimization in the clinic.

In the case of efalizumab, to complete a more comprehensive safety assessment, a chimeric rat antimouse CD11a antibody, muM17, was developed and evaluated as a species-specific surrogate molecule for efalizumab. muM17 binds mouse CD11a with specificity and affinity similar to that of efalizumab to its human target antigen. In addition, pharmacological activities of muM17 in mice were demonstrated to be similar to those of efalizumab in humans (Clarke et al. 2004; Nakakura et al. 1993).

The preclinical ADME program for efalizumab consisted of PK, PD (CD11a down-modulation and saturation) and toxicokinetic data from PK, PD, and toxicology studies with efalizumab in chimpanzees and with muM17 in mice. The use of efalizumab in the chimpanzee and muM17 in mice for PK and PD and safety studies was supported by in vitro activity assessments. The preclinical data were used for PK and PD characterization, PD-based dose selection, and toxicokinetic support for confirming exposure in toxicology

studies. Together, these data supported both the design of the preclinical program and its relevance to the clinical program.

The observed PD as well as the MOA of efalizumab and muM17 is attributed to binding CD11a present on cells and tissues. The binding affinities of efalizumab to human and chimpanzee CD11a on CD3 lymphocytes are comparable, supporting the use of chimpanzees as a preclinical model for human responses. CD11a expression has been observed to be greatly reduced on T lymphocytes in chimpanzees and mice treated with efalizumab and muM17, respectively. Expression of CD11a is restored as efalizumab and muM17 are eliminated from the plasma.

The disposition of efalizumab and of the mouse surrogate muM17 is mainly determined by the combination of both specific interactions with the ligand CD11a and by their IgG1 framework. The disposition is governed by the species specificity of the antibody for its ligand CD11a, the amount of CD11a in the system, and the administered dose. Binding to CD11a serves as a major pathway for clearance of these molecules, which leads to nonlinear PK depending on the relative amounts of CD11a and efalizumab or muM17 (Coffey et al. 2005).

Based on the safety studies, efalizumab was considered to be generally well tolerated in chimpanzees at doses up to 40 mg/kg/week IV for 6 months, providing an exposure ratio of 339-fold based on cumulative dose and 174-fold based on the cumulative AUC, compared with a clinical dose of 1 mg/kg/week. The surrogate antibody muM17 was also well tolerated in mice at doses up to 30 mg/kg/week SC. Overall, efalizumab was considered to have an excellent nonclinical safety profile, thereby supporting the use in adult patients. There was no signal for PML in the nonclinical studies, which subsequently led to withdrawal of efalizumab from the market.

Clinical Program of Efalizumab: PK/PD Studies, Assessment of Dose, Route, and Regimen

Efalizumab PK and PD data were available from ten studies in which more than 1700 patients with psoriasis received IV or SC efalizumab. In the phase I studies, PK and PD parameters were characterized by extensive sampling during treatment; in the phase III trials, steady-state trough levels were measured once or twice during the first 12-week treatment period for all the studies and during extended treatment periods for some studies. Several early phase I and II trials examined IV injection of efalizumab, and dose-ranging findings from these trials have served as the basis for SC dosing levels used in several subsequent phase I and all phase III trials.

IV Administration of Efalizumab

The PK of mAbs varies greatly, depending primarily on their affinity for and the distribution of their target antigen (Lobo et al. 2004). Efalizumab exhibited concentration-dependent nonlinear PK after administration of single IV doses of 0.03, 0.1, 0.3, 0.6, 1.0, 2.0, 3.0, and 10.0 mg/kg in a phase I study. This nonlinearity is directly related to specific and saturable binding of efalizumab to its cell surface receptor, CD11a, and has been described by a PK/PD model developed by Bauer et al. (Bauer et al. 1999), which was expanded to a PK/PD/ efficacy model by Ng et al. (Ng et al. 2005). The PK profiles of efalizumab following single IV doses with observed data and model predicted fit are presented in Fig. 8.11. Mean CL decreased from 380 to 6.6 mL/kg/day for doses of 0.03 mg/kg–10 mg/kg, respectively. The volume of distribution of the central compartment (Vc) of efalizumab was 110 mL/kg at 0.03 mg/kg (approximately twice the plasma volume) and decreased to 58 mL/kg at 10 mg/kg (approximately equal to plasma volume), consistent with saturable binding of efalizumab to CD11a in the vascular compartment. Because of efalizumab's nonlinear PK, its half-life ($t_{1/2}$) is dose dependent.

In a phase II study of efalizumab, it was shown that at a weekly dosage of 0.1 mg/kg IV, patients did not maintain maximal downmodulation of CD11a expression and did not maintain maximal saturation. Also, at the end of 8 weeks of efalizumab treatment, 0.1 mg/kg/week IV, patients did not have statistically significant histological improvement and did not achieve a full clinical response. The minimum weekly IV dosage of efalizumab tested that produced histological improvements in skin biopsies was 0.3 mg/kg/week, and this dosage resulted in submaximal saturation of CD11a-binding sites but maximal down-modulation of CD11a expression. Improvements in patients' psoriasis were also observed, as determined by histology and by the Psoriasis Area and Severity Index (PASI) (Papp et al. 2001).

Determination of SC Doses

Although efficacy was observed in phase I and II studies with 0.3 mg/kg/week IV efalizumab, dosages of 0.6 mg/kg/week and greater (given for 7–12 weeks) provided more consistent T lymphocyte CD11a saturation and maximal PD effect. At dosages ≤0.3 mg/kg/week, large between-subject variability was observed, whereas at dosages of 0.6 or 1.0 mg/kg/week, patients experienced better improvement in PASI scores, with lower between-patient variability in CD11a saturation and downmodulation. Therefore, since the desired route of administration was SC, this IV dosage was used to estimate an appropriate minimum SC dose of 1 mg/kg/week (based on a 50% bioavailability) that would induce similar changes in PASI, PD measures, and histology. The safety, PK, and PD of a range of SC efalizumab doses (0.5–4.0 mg/kg/week administered for 8–12 weeks) were evaluated initially in two-phase I studies (Gottlieb et al. 2003). To establish whether a higher SC dosage might produce better results, several phase III clinical trials assessed a 2.0 mg/kg/week SC dosage in addition to the 1.0 mg/kg/week dosage. A dose of 1.0 mg/kg/week SC efalizumab was selected as it produced sufficient trough levels in patients to maintain the maximal downmodulation of CD11a expression and binding-site saturation between weekly doses (Joshi et al. 2006). Figure 8.12 depicts the serum efalizumab levels, CD11a expression, and available CD11a-binding sites on T lymphocytes (mean ± SD) after SC administration of 1 mg/kg efalizumab.

SC Administration of Efalizumab

The PK of SC efalizumab was well characterized following multiple SC doses of 1.0 and 2.0 mg/kg/week (Joshi et al. 2006; Mortensen et al. 2005). A phase I study that collected steady-state PK and PD data for 12 weekly SC doses of 1.0 and 2.0 mg/kg in psoriasis patients provided most of the

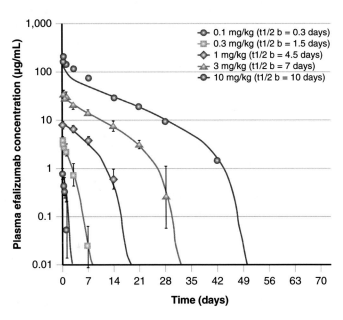

Fig. 8.11 Plasma concentration versus time profile for efalizumab following single IV doses in psoriasis patients (Ng et al. 2005)

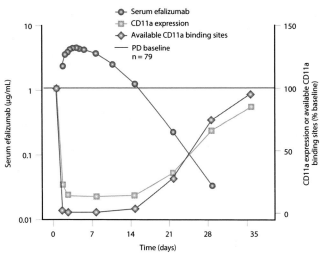

Fig. 8.12 PK/PD profile following efalizumab in humans (1 mg/kg SC) (Joshi et al. 2006)

Fig. 8.13 Serum efalizumab, CD11a expression, and free CD11a binding sites on T lymphocytes, absolute lymphocyte counts, and Psoriasis Area and Severity Index (PASI) score (mean) following 1.0 mg/kg/week SC efalizumab for 12 weeks and 12 weeks post-treatment (Mortensen et al. 2005)

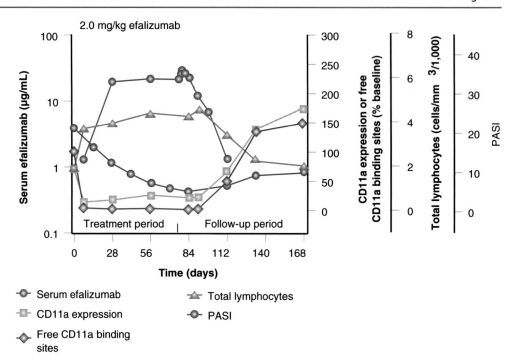

pharmacologic data relevant to the product that was on the market prior to its withdrawal. Although peak serum concentration after the last dose (C_{max}) was observed to be higher for the 2.0 mg/kg/week (30.9 μg/mL) than for the 1.0 mg/kg/week dosage (12.4 μg/mL), no additional changes in PD effects were observed at the higher dosages (Mortensen et al. 2005). Following a dose of 1.0 mg/kg/week, serum efalizumab concentrations were adequate to induce maximal downmodulation of CD11a expression and a reduction in free CD11a-binding sites on T lymphocytes (Fig. 8.13). Steady-state serum efalizumab levels were reached more quickly with the 1.0 mg/kg/week dosage at 4 weeks compared with the 2.0 mg/kg/week dosage at 8 weeks (Mortensen et al. 2005), which is in agreement with the average effective $t_{1/2}$ for SC efalizumab 1.0 mg/kg/week of 5.5 days (Boxenbaum and Battle 1995). The bioavailability was estimated at approximately 50%. Population PK analyses indicated that body weight was the most significant covariate affecting efalizumab SC clearance, thus supporting body weight-based dosing for efalizumab (Sun et al. 2005).

Population Pharmacokinetics of Monoclonal Antibodies

Compared to many small-molecule drugs, mAbs typically exhibit less inter- and intra-subject variability of the standard PK parameters such as volume of distribution and clearance. However, it is possible that certain pathophysiological conditions may result in substantially increased intra- and interpatient variability. Several factors may contribute towards PK

variability, including patient age, body weight, or concomitant disease/drug treatments. Even factors such as diet, lifestyle, ethnicity, and geographic location can differ and may contribute to interindividual variability. These and other "covariates" can have substantial influence on PK parameters. Therefore, good therapeutic practice should always be based on an understanding of both the influence of covariates on PK parameters as well as the PK variability in a given patient population. With this knowledge, dosage adjustments can be made to accommodate differences in PK due to genetic, environmental, physiological, or pathological factors, for instance, in case of compounds with a relatively small therapeutic index. The framework of application of population PK during drug development is summarized in the FDA guidance document entitled "*Guidance for Industry—Population Pharmacokinetics*" (FDA 2022).

For population PK data analysis, there are generally two reliable and practical approaches. One approach is the standard two-stage (STS) method, which estimates parameters from the plasma drug concentration data for an individual subject during the first stage. The estimates from all subjects are then combined to obtain a population mean and variability estimates for the parameters of interest. The method works well when sufficient drug concentration–time data are available for each individual patient; typically, these data are gathered in phase I clinical trials. A second approach, nonlinear mixed effect modeling (NONMEM), attempts to fit the data and partition the differences between theoretical and observed values into random error terms. The influence of fixed effect (i.e., age, sex, and body weight) can be identified through a regression model building process.

The original scope for the NONMEM approach was its applicability even when the amount of time–concentration data obtained from each individual is sparse and conventional compartmental PK analyses are not feasible. This is usually the case during the routine visits in phase III or IV clinical studies. Nowadays, the NONMEM approach is applied far beyond its original scope due to its flexibility and robustness. It has been used to describe data-rich phase I and phase IIa studies or even preclinical data to guide and expedite drug development from early preclinical to clinical studies (Aarons et al. 2001; Chien et al. 2005).

There has been increasing interest in the use of population PK and PD analyses for different antibody products (i.e., antibodies, antibody fragments, or antibody fusion proteins) over the past 15 years (Agoram et al. 2007; Dirks and Meibohm 2010; Gibiansky and Frey 2012; Gibiansky and Gibiansky 2009; Hayashi et al. 2007; Lee et al. 2003; Nestorov et al. 2004; Yim et al. 2005; Zheng et al. 2011; Zhou et al. 2004). One example involving analysis of population plasma concentration data involved a dimeric fusion protein, etanercept. A one-compartment first-order absorption and elimination population PK model with inter-individual and inter-occasion variability on CL, volume of distribution, and absorption rate constant, with covariates of sex and race on apparent CL and body weight on CL and volume of distribution, was developed for etanercept in RA adult patients (Lee et al. 2003). The population PK model for etanercept was further applied to pediatric patients with juvenile RA and established the basis of the 0.8 mg/kg once weekly regimen in pediatric patients with juvenile RA (Yim et al. 2005). Unaltered etanercept PK with concurrent methotrexate in patients with RA has been demonstrated in a phase IIIb study using a population PK modeling approach (Zhou et al. 2004). Thus, no etanercept dose adjustment is needed for patients taking concurrent methotrexate. A simulation exercise of using the final population PK model of subcutaneously administered etanercept in patients with psoriasis indicated that the two different dosing regimens (50 mg once weekly versus 25 mg twice per week) provide a similar steady-state exposure (Nestorov et al. 2004). Therefore, their respective efficacy and safety profiles are likely to be similar as well.

An added feature is the development of a population model involving both PK and PD. Population PK/PD modeling has been used to characterize drug PK and PD with models ranging from simple empirical PK/PD models to advanced mechanistic models by using drug–receptor-binding principles or other physiologically based principles. A mechanism-based population PK- and PD-binding model was developed for a recombinant DNA-derived humanized IgG1 mAb, omalizumab (Hayashi et al. 2007). Clearance and volume of distribution for omalizumab varied with body weight, whereas CL and rate of production

of IgE were predicted accurately by baseline IgE, and overall, these covariates explained much of the interindividual variability. Furthermore, this mechanism-based population PK/PD model enabled the estimation of not only omalizumab disposition but also the binding with its target, IgE, and the rate of production, distribution, and elimination of IgE.

Recently, a platform population PK approach has been used to characterize mAb PK to improve the efficiency of study design, such as optimal dose regimens and PK sampling times. Davda et al. (Davda et al. 2014) determined typical population PK values for four mAbs with linear elimination using model-based meta-analysis, which can be utilized to prospectively optimize FIH study designs. A platform model describing PK properties of vc-MMAE antibody–drug conjugates based on 8 ADCs is reported by Kagedal et al. (Kagedal et al. 2017). The model could be applied to predict PK-profiles of future vc-MMAE ADCs, estimate individual exposure for the subsequent exposure-response analysis, and optimize study design.

Population PK/PD analysis can capture uncertainty and the expected variability in PK/PD data generated in preclinical studies or early phases of clinical development. Understanding the associated PK or PD variability and performing clinical trial simulation by incorporating the uncertainty from the existing PK/PD data allows projecting a plausible range of doses for future clinical studies and final practical uses.

Future Perspective

The success of mAbs and mAb derivatives as new therapeutic agents in several disease areas such as oncology, inflammatory diseases, autoimmune diseases, and transplantation has triggered growing scientific, therapeutic, and business interest in the mAb technology. The market for therapeutic mAbs and mAb derivatives is one of the most dynamic sectors within the pharmaceutical industry. Further growth is expected by developing mAbs toward other surface protein targets, which are not covered yet by marketed mAbs. Particularly, the technological advancement in the area of ADCs, mAb fragments, and bispecifics may overcome some of the limitations of mAbs by providing highly potent drugs selectively to target compartments and to extend the distribution of the active moiety, which are typically not reached by mAbs. ADCs hold great promise for selective drug delivery of potent drugs with unfavorable own selectivity to target cells (e.g., highly potent cytotoxic drugs). Modification of the mAb structure allows adjusting the properties according to therapeutic needs (e.g., adjusting half-life, increasing volume of distribution, and changing clearance pathways). By using modified mAb derivatives, optimized therapeutic agents might become available. For example, this technology has been successfully used for two antibody fragments marketed in inflammatory disease and anti-angiogenesis, abciximab and ranibizumab.

Bispecific antibodies represent another promising new approach to antibody therapy. Technological refinements in antibody engineering have allowed the production of bispecific antibodies that are simultaneously directed toward two distinct target antigens (Holmes 2011). For instance, the CDR consisting of the variable domains (V_L and V_H) at the tip of one arm of an IgG may be asymmetrically designed to bind to a different target than that of the other arm (Fig. 8.1). Symmetrical formats in which each arm can bind two targets are also possible. With advances in antibody engineering, there are more than 60 different kinds of bispecific formats, with molecular weight ranging from 25 to 250 kDa (Spiess et al. 2015).

As of January 2022, there are four bispecific antibodies on the market. In 2009, Removab (Catumaxomab) was the first approved T cell-dependent bispecific antibody that bridges CD3 and EpCAM, but was withdrawn after 4 years owing to commercial reasons, immunogenicity, and toxicity. In December 2014, the FDA approved Blincyto (Blinatumomab) that bridges CD3 on T cells and CD19 on B cells for treatment of relapsed or refractory B-cell acute lymphoblastic leukemia (ALL) and MRD-positive B-cell Precursor ALL. The bispecific antibody Hemlibra (Emicizumab) was approved by FDA in November 2017 that binds to both factor IXa and factor X to prevent bleeding in hemophilia A patients. It has changed the paradigm of hemophilia A treatment from three injections per week to once weekly or biweekly. Rybrevant (Amivantamab) was approved by FDA in May 2021 and binds to the extracellular domains of EGFR and MET to treat locally advanced or metastatic nonsmall cell lung cancer with EGFR exon 20 insertion mutations. In January 2022, FDA approved Kimmtrack (tebentafusp-tebn), an anti-CD3 scFv bispecific that targets peptide-HLA presented gp100, for HLA-A*02:01-positive adult patients with unresectable or metastatic uveal melanoma. There are more than 300 clinical trials testing bispecifics, with a majority of them in the oncology area (Wang et al. 2021).

mAbs have become a key part of the pharmaceutical armamentarium, especially in the oncology and immunology settings and will continue to be a focus area for drug discovery and development. More specifically, the recent approvals of mAbs like pembrolizumab, nivolumab, and atezolizumab in cancer immunotherapy have revolutionized the cancer treatment paradigm. These mAbs, either as monotherapy or in combinations with other cancer immunotherapies, including cancer vaccines, bispecifics, and other modalities, offer tremendous promise for personalized medicine.

Self-Assessment Questions

Questions

1. What are the structural differences among the five immunoglobulin classes?
2. What are key differences in PK/PD among mAbs, ADCs, and small-molecule drugs?
3. Why do IgGs typically show nonlinear PK in the lower plasma (serum) concentration range?
4. What is a surrogate mAb and how can it potentially be used in the drug development process of mAbs?
5. Which other modes of actions apart from ADCC – antibody-dependent cellular cytotoxicity – are known for mAbs? What are the key steps of ADCC?
6. Why do IgGs have a longer in vivo half-life compared with other Igs?
7. What are the development phases for antibody therapeutics? What major activities are involved in each phase?
8. What are the main considerations to assess the drug–drug interaction (DDI) potential for an ADC?

Answers

1. The following structural properties distinguish mAbs:

   The molecular form varies across the five immunoglobulin classes: IgG, IgD, and IgE are monomers; IgM forms a pentamer or hexamer, and IgA exists either as a monomer or dimer. Consequently, the molecular weights of various Igs differ (IgG 150–169 kD, IgA 160–300 kD, IgD 175 kD, IgE 190, IgM 950 kD).

2. Metabolism of mAbs appears to be simpler than for small molecules. In contrast to small-molecule drugs, the typical metabolic enzymes and transporter proteins, such as cytochrome P450, multidrug resistance (MDR) efflux pumps, are not involved in the disposition of mAbs. Therefore, drug–drug interaction studies for those disposition processes are only part of the standard safety assessment for small molecules and not for mAbs.

   mAbs, which have a protein structure, are metabolized by proteases. These enzymes are ubiquitously available in mammalian organisms. In contrast, small-molecule drugs are primarily metabolized in the liver.

   Because of their large molecular weight, intact mAbs are typically not cleared via the renal elimination route. However, renal clearance processes can play a major role in the elimination of small-molecule drugs.

   PK of mAbs usually is dependent on the binding to the pharmacological target protein and shows nonlinear behavior as consequence of its saturation kinetics.

   In general, mAbs have a longer half-life (on the order of days and weeks) than small-molecule drugs (typically on the order of hours).

The distribution of mAbs is very restricted (volume of distribution in the range of 0.1 L/kg). As a consequence, mAbs do have limited access to tissue compartments (e.g., brain) as potential target sites via passive, energy-independent distribution processes only.

The PK of ADCs is more complex with combined characteristics of small-molecule drugs and large-molecule biotherapeutics. There are two alternative clearance pathways for ADCs: proteolytic degradation and deconjugation. Similar to mAbs, ADC clearance through proteolytic degradation is driven primarily by catabolism mediated by target-specific or nonspecific cellular uptake followed by lysosomal degradation. Deconjugation clearance is usually mediated by enzymatic or chemical cleavage (e.g., maleimide exchange) of the linker leading to the release of the cytotoxic drug from the. The released cytotoxic drug is expected to undergo metabolic enzyme and/or transporter-mediated clearance mechanisms consistent with small molecules.

3. At lower concentrations, mAbs generally show nonlinear PK due to receptor-mediated clearance processes, which are characterized by small capacity of the clearance pathway and high affinity to the target protein. Consequently at these low concentrations, mAbs exhibit typically shorter half-life. With increasing doses, these receptors become saturated, and the clearance as well as elimination half-life decreases until it becomes constant. The clearance in the higher concentration range, which is dominated by linear, nontarget-related clearance processes, is therefore also called nonspecific clearance in contrast to the target-related, specific clearance.

4. A surrogate mAb has similar antigen specificity and affinity in experimental animals (e.g., mice and rats) compared to those of the corresponding human antibody in humans. It is quite common that the antigen specificity limits ADME studies of humanized mAbs in rodents. Studies using surrogate antibodies might lead to important information regarding safety, MOA, disposition of the drug, tissue distribution, and receptor pharmacology in the respective animal species, which might be too cumbersome and expensive to be conducted in nonhuman primates. Surrogate mAbs (from mouse or rat) provide a means to gain knowledge of ADME and PD in preclinical rodent models and might facilitate the dose selection for clinical studies.

5. Apart from ADCC, mAbs can exert pharmacological effects by multiple mechanisms that include direct modulation of the target antigen, complement-dependent cytotoxicity (CDC), and apoptosis. The key steps of ADCC are (1) opsonization of the targeted cells, (2) recognition of antibody-coated targeted cells by Fc receptors on the surface of monocytes, macrophages, natural killer cells, and other cells, and (3) destruction of the opsonized targets by phagocytosis of the opsonized targets and/or by toxic substances released after activation of monocytes, macrophages, natural killer cells, and other cells.

6. IgG can bind to neonatal Fc receptor (FcRn) in the endosome, which protects IgG from catabolism via proteolytic degradation. This protection results in a slower clearance and thus longer plasma half-life of IgGs. Consequently, changing the FcRn affinity allows adjustment of the clearance of mAbs (higher affinity—lower clearance), which can be employed to tailor the PK of these molecules.

7. Pre-IND, phases I, II, III, and IV are the major development phases for antibody therapies. Safety pharmacology, toxicokinetics, toxicology, tissue cross-reactivity, local tolerance, PK support for candidate selection, assay support for PK/PD, and PK/PD support for dose/route/regimen are major activities in the pre-IND phase. General toxicity, reproductive toxicity, carcinogenicity, immunogenicity, characterization of dose–concentration–effect relationship, material comparability studies, mechanistic modeling approach, and population PK/predictions are major activities from phase I to phase III. Further studies might be performed as needed after the mAb got market authorization. These studies are called phase IV studies.

8. ADCs have complex molecular structures, combining the characteristics of small-molecule drugs and large-molecule biotherapeutics. The DDI potential for an ADC should be evaluated for the antibody and the small-molecule drug components independently. ADCs can interact with drugs or mAbs via the mechanisms through the antibody components. However, evidence of ADC–drug or ADC–mAb interaction appears to be limited. In comparison, the released payloads are expected to undergo enzyme and/or transporter-mediated clearance mechanisms consistent with small molecules. Thus, DDI of the released payloads may occur through modulation of these clearance pathways. In many cases, the systemic concentration of the released payload might be too low to act as a perpetrator. It might be necessary to evaluate the circulating payload (administered as an ADC) as a victim, which may impact the overall ADC safety profile, as these payloads are highly potent and typically have a very narrow therapeutic index.

References

Aarons L, Karlsson MO, Mentre F, Rombout F, Steimer JL, van Peer A (2001) Role of modelling and simulation in phase I drug development. Eur J Pharm Sci 13(2):115–122

Abramson A, Frederiksen MR, Vegge A, Jensen B, Poulsen M, Mouridsen B et al (2022) Oral delivery of systemic monoclonal antibodies, peptides and small molecules using gastric autoinjectors. Nat Biotechnol 40(1):103–109. https://doi.org/10.1038/s41587-021-01024-0

Agoram BM, Martin SW, van der Graaf PH (2007) The role of mechanism-based pharmacokinetic-pharmacodynamic (PK-PD) modelling in translational research of biologics. Drug Discov Today 12(23–24):1018–1024. https://doi.org/10.1016/j.drudis.2007.10.002

Albrecht H, DeNardo SJ (2006) Recombinant antibodies: from the laboratory to the clinic. Cancer Biother Radiopharm 21(4):285–304. https://doi.org/10.1089/cbr.2006.21.285

Baert F, Noman M, Vermeire S, Van Assche G, D' Haens G, Carbonez A et al (2003) Influence of immunogenicity on the long-term efficacy of infliximab in Crohn's disease. N Engl J Med 348(7):601–608

Bajaj G, Wang X, Agrawal S, Gupta M, Roy A, Feng Y (2017) Model-based population pharmacokinetic analysis of Nivolumab in patients with solid Tumors. CPT Pharmacometrics Syst Pharmacol 6(1):58–66. https://doi.org/10.1002/psp4.12143

Balocco R, De Sousa Guimaraes Koch S, Thorpe R, Weisser K, Malan S (2022) New INN nomenclature for monoclonal antibodies. Lancet 399(10319):24. https://doi.org/10.1016/s0140-6736(21)02732-x

Bauer RJ, Dedrick RL, White ML, Murray MJ, Garovoy MR (1999) Population pharmacokinetics and pharmacodynamics of the anti-CD11a antibody hu1124 in human subjects with psoriasis. J Pharmacokinet Biopharm 27(4):397–420

Baxter LT, Zhu H, Mackensen DG, Jain RK (1994) Physiologically based pharmacokinetic model for specific and nonspecific monoclonal antibodies and fragments in normal tissues and human tumor xenografts in nude mice. Cancer Res 54(6):1517–1528

Baxter LT, Zhu H, Mackensen DG, Butler WF, Jain RK (1995) Biodistribution of monoclonal antibodies: scale-up from mouse to human using a physiologically based pharmacokinetic model. Cancer Res 55(20):4611–4622

Bazin-Redureau MI, Renard CB, Scherrmann JM (1997) Pharmacokinetics of heterologous and homologous immunoglobulin G, F(ab')2 and fab after intravenous administration in the rat. J Pharm Pharmacol 49(3):277–281

Berger MA, Masters GR, Singleton J, Scully MS, Grimm LG, Soltis DA et al (2005) Pharmacokinetics, biodistribution, and radioimmunotherapy with monoclonal antibody 776.1 in a murine model of human ovarian cancer. Cancer Biother Radiopharm 20(6):589–602

Bien-Ly N, Boswell CA, Jeet S, Beach TG, Hoyte K, Luk W et al (2015) Lack of widespread BBB disruption in Alzheimer's disease models: focus on therapeutic antibodies. Neuron 88(2):289–297. https://doi.org/10.1016/j.neuron.2015.09.036

Bittner B, Richter WF, Hourcade-Potelleret F, McIntyre C, Herting F, Zepeda ML et al (2012) Development of a subcutaneous formulation for trastuzumab–nonclinical and clinical bridging approach to the approved intravenous dosing regimen. Arzneimittelforschung 62(9):401–409. https://doi.org/10.1055/s-0032-1321831

Bittner B, Richter W, Schmidt J (2018) Subcutaneous administration of biotherapeutics: an overview of current challenges and opportunities. BioDrugs 32(5):425–440. https://doi.org/10.1007/s40259-018-0295-0

Boswell CA, Brechbiel MW (2007) Development of radioimmunotherapeutic and diagnostic antibodies: an inside-out view. Nucl Med Biol 34(7):757–778. https://doi.org/10.1016/j.nucmedbio.2007.04.001

Boswell CA, Ferl GZ, Mundo EE, Schweiger MG, Marik J, Reich MP et al (2010a) Development and evaluation of a novel method for preclinical measurement of tissue vascular volume. Mol Pharm 7(5):1848–1857. https://doi.org/10.1021/mp100183k

Boswell CA, Tesar DB, Mukhyala K, Theil FP, Fielder PJ, Khawli LA (2010b) Effects of charge on antibody tissue distribution and pharmacokinetics. Bioconjug Chem 21(12):2153–2163. https://doi.org/10.1021/bc100261d

Boswell CA, Bumbaca D, Fielder PJ, Khawli LA (2012) Compartmental tissue distribution of antibody therapeutics: experimental approaches and interpretations. AAPS J 14(3):612–618. https://doi.org/10.1208/s12248-012-9374-1

Boswell CA, Mundo EE, Johnstone B, Ulufatu S, Schweiger MG, Bumbaca D et al (2013) Vascular physiology and protein disposition in a preclinical model of neurodegeneration. Mol Pharm 10(5):1514–1521. https://doi.org/10.1021/mp3004786

Boswell CA, Mundo EE, Ulufatu S, Bumbaca D, Cahaya HS, Majidy N et al (2014) Comparative physiology of mice and rats: radiometric measurement of vascular parameters in rodent tissues. Mol Pharm 11(5):1591–1598. https://doi.org/10.1021/mp400748t

Boxenbaum H, Battle M (1995) Effective half-life in clinical pharmacology. J Clin Pharmacol 35(8):763–766

Brambell FW, Hemmings WA, Morris IG (1964) A theoretical model of gamma-globulin catabolism. Nature 203:1352–1354

Bugelski PJ, Herzyk DJ, Rehm S, Harmsen AG, Gore EV, Williams DM et al (2000) Preclinical development of keliximab, a Primatized anti-CD4 monoclonal antibody, in human CD4 transgenic mice: characterization of the model and safety studies. Hum Exp Toxicol 19(4):230–243. https://doi.org/10.1191/096032700678815783

Bumbaca Yadav D, Sharma VK, Boswell CA, Hotzel I, Tesar D, Shang Y et al (2015) Evaluating the use of antibody variable region (Fv) charge as a risk assessment tool for predicting typical Cynomolgus monkey pharmacokinetics. J Biol Chem 290(50):29732–29741. https://doi.org/10.1074/jbc.M115.692434

Bumbaca D, Wong A, Drake E, Reyes AE 2nd, Lin BC, Stephan JP et al (2011) Highly specific off-target binding identified and eliminated during the humanization of an antibody against FGF receptor 4. MAbs 3(4):376–386

Bumbaca D, Xiang H, Boswell CA, Port RE, Stainton SL, Mundo EE et al (2012) Maximizing tumour exposure to anti-neuropilin-1 antibody requires saturation of non-tumour tissue antigenic sinks in mice. Br J Pharmacol 166(1):368–377. https://doi.org/10.1111/j.1476-5381.2011.01777.x

Bunescu A, Seideman P, Lenkei R, Levin K, Egberg N (2004) Enhanced Fcgamma receptor I, alphaMbeta2 integrin receptor expression by monocytes and neutrophils in rheumatoid arthritis: interaction with platelets. J Rheumatol 31(12):2347–2355

Cartron G, Watier H, Golay J, Solal-Celigny P (2004) From the bench to the bedside: ways to improve rituximab efficacy. Blood 104(9):2635–2642

Castillo AMM, Vu TT, Liva SG, Chen M, Xie Z, Thomas J et al (2021) Murine cancer cachexia models replicate elevated catabolic pembrolizumab clearance in humans. JCSM Rapid Commun 4(2):232–244. https://doi.org/10.1002/rco2.32

CDER (2014) Clinical pharmacology and biopharmaceutical reviews BLA 125522

CDER (2015) Addendum clinical pharmacology review BLA 125509. https://www.accessdata.fda.gov/drugsatfda_docs/nda/2016/125509Orig1s000ClinPharmR.pdf. Accessed 8 May 2018

Chen Y, Samineni D, Mukadam S, Wong H, Shen BQ, Lu D et al (2015) Physiologically based pharmacokinetic modeling as a tool to predict drug interactions for antibody-drug conjugates. Clin Pharmacokinet 54(1):81–93. https://doi.org/10.1007/s40262-014-0182-x

Chen SC, Kagedal M, Gao Y, Wang B, Harle-Yge ML, Girish S et al (2017) Population pharmacokinetics of trastuzumab emtansine in previously treated patients with HER2-positive advanced gastric cancer (AGC). Cancer Chemother Pharmacol 80(6):1147–1159. https://doi.org/10.1007/s00280-017-3443-1

Chien JY, Friedrich S, Heathman MA, de Alwis DP, Sinha V (2005) Pharmacokinetics/pharmacodynamics and the stages of drug development: role of modeling and simulation. AAPS J 7(3):E544–E559

Clarke J, Leach W, Pippig S, Joshi A, Wu B, House R et al (2004) Evaluation of a surrogate antibody for preclinical safety testing of an anti-CD11a monoclonal antibody. Regul Toxicol Pharmacol 40(3):219–226

Coffey GP, Fox JA, Pippig S, Palmieri S, Reitz B, Gonzales M et al (2005) Tissue distribution and receptor-mediated clearance of anti-CD11a antibody in mice. Drug Metab Dispos 33(5):623–629

Cohenuram M, Saif MW (2007) Panitumumab the first fully human monoclonal antibody: from the bench to the clinic. Anti-Cancer Drugs 18(1):7–15

Cornillie F, Shealy D, D'Haens G, Geboes K, Van Assche G, Ceuppens J et al (2001) Infliximab induces potent anti-inflammatory and local immunomodulatory activity but no systemic immune suppression in patients with Crohn's disease. Aliment Pharmacol Ther 15(4):463–473

Dall'Ozzo S, Tartas S, Paintaud G, Cartron G, Colombat P, Bardos P et al (2004) Rituximab-dependent cytotoxicity by natural killer cells: influence of FCGR3A polymorphism on the concentration-effect relationship. Cancer Res 64(13):4664–4669

Danilov SM, Gavrilyuk VD, Franke FE, Pauls K, Harshaw DW, McDonald TD et al (2001) Lung uptake of antibodies to endothelial antigens: key determinants of vascular immunotargeting. Am J Physiol Lung Cell Mol Physiol 280(6):L1335–L1347

Davda JP, Dodds MG, Gibbs MA, Wisdom W, Gibbs J (2014) A model-based meta-analysis of monoclonal antibody pharmacokinetics to guide optimal first-in-human study design. MAbs 6(4):1094–1102. https://doi.org/10.4161/mabs.29095

Davis JD, Bansal A, Hassman D, Akinlade B, Li M, Li Z et al (2018) Evaluation of potential disease-mediated drug-drug interaction in patients with moderate-to-severe atopic dermatitis receiving Dupilumab. Clin Pharmacol Ther 104(6):1146–1154. https://doi.org/10.1002/cpt.1058

Dedrick RL (1973) Animal scale-up. J Pharmacokinet Biopharm 1(5):435–461

den Broeder A, van de Putte L, Rau R, Schattenkirchner M, Van Riel P, Sander O et al (2002) A single dose, placebo controlled study of the fully human anti-tumor necrosis factor-alpha antibody adalimumab (D2E7) in patients with rheumatoid arthritis. J Rheumatol 29(11):2288–2298

Deng R, Iyer S, Theil FP, Mortensen DL, Fielder PJ, Prabhu S (2011) Projecting human pharmacokinetics of therapeutic antibodies from nonclinical data: what have we learned? MAbs 3(1):61–66

Dias C, Abosaleem B, Crispino C, Gao B, Shaywitz A (2015a) Erratum to: tolerability of high-volume subcutaneous injections of a viscous placebo buffer: a randomized, crossover study in healthy subjects. AAPS PharmSciTech 16(6):1500. https://doi.org/10.1208/s12249-015-0324-y

Dias C, Abosaleem B, Crispino C, Gao B, Shaywitz A (2015b) Tolerability of high-volume subcutaneous injections of a viscous placebo buffer: a randomized, crossover study in healthy subjects. AAPS PharmSciTech 16(5):1101–1107. https://doi.org/10.1208/s12249-015-0288-y

Dickinson BL, Badizadegan K, Wu Z, Ahouse JC, Zhu X, Simister NE et al (1999) Bidirectional FcRn-dependent IgG transport in a polarized human intestinal epithelial cell line. J Clin Invest 104(7):903–911

Dirks NL, Meibohm B (2010) Population pharmacokinetics of therapeutic monoclonal antibodies. Clin Pharmacokinet 49(10):633–659. https://doi.org/10.2165/11535960-000000000-00000

Dong JQ, Salinger DH, Endres CJ, Gibbs JP, Hsu CP, Stouch BJ et al (2011) Quantitative prediction of human pharmacokinetics for monoclonal antibodies: retrospective analysis of monkey as a single species for first-in-human prediction. Clin Pharmacokinet 50(2):131–142. https://doi.org/10.2165/11537430-000000000-00000

Dostalek M, Prueksaritanont T, Kelley RF (2017) Pharmacokinetic de-risking tools for selection of monoclonal antibody lead candidates. MAbs 9(5):756–766. https://doi.org/10.1080/19420862.2017.1323160

Druet P, Bariety J, Laliberte F, Bellon B, Belair MF, Paing M (1978) Distribution of heterologous antiperoxidase antibodies and their fragments in the superficial renal cortex of normal Wistar-Munich rat: an ultrastructural study. Lab Investig 39(6):623–631

Duconge J, Castillo R, Crombet T, Alvarez D, Matheu J, Vecino G et al (2004) Integrated pharmacokinetic-pharmacodynamic modeling and allometric scaling for optimizing the dosage regimen of the monoclonal ior EGF/r3 antibody. Eur J Pharm Sci 21(2–3):261–270

Fasanmade AA, Adedokun OJ, Ford J, Hernandez D, Johanns J, Hu C et al (2009) Population pharmacokinetic analysis of infliximab in patients with ulcerative colitis. Eur J Clin Pharmacol 65(12):1211–1228. https://doi.org/10.1007/s00228-009-0718-4

Fasanmade AA, Adedokun OJ, Olson A, Strauss R, Davis HM (2010) Serum albumin concentration: a predictive factor of infliximab pharmacokinetics and clinical response in patients with ulcerative colitis. Int J Clin Pharmacol Ther 48(5):297–308

FDA (2004) Raptiva (Efalizumab) [prescribing information]. https://www.fda.gov/media/75713/download. Accessed 15 June 2022

FDA (2005a) Simulect (Basiliximab) prescribing information. https://www.accessdata.fda.gov/drugsatfda_docs/label/2003/bas-nov010203lb.htm. Accessed 15 June 2022

FDA (2005b) Zenapax (Daclizumab) prescribing information. https://www.accessdata.fda.gov/drugsatfda_docs/label/2002/daclho-f072902lb.pdf. Accessed 15 June 2022

FDA (2006a) Herceptin (Trastuzumab) prescribing information. https://www.accessdata.fda.gov/drugsatfda_docs/label/2010/103792s5250lbl.pdf. Accessed 15 June 2022

FDA (2006b) Xolair (Omalizumab) prescribing information. https://www.accessdata.fda.gov/drugsatfda_docs/label/2016/103976s5225lbl.pdf. Accessed 15 June 2022

FDA (2007) Humira (Adalimumab) prescribing information. https://www.accessdata.fda.gov/drugsatfda_docs/label/2018/125057s406lbl.pdf. Accessed 15 June 2022

FDA (2015) Vectibix (Panitumumab) prescribing information. https://www.accessdata.fda.gov/drugsatfda_docs/label/2009/125147s080lbl.pdf. Accessed 15 June 2022

FDA (2020a) Drug-drug interaction assessment for therapeutic proteins guidance for industry. https://www.fda.gov/regulatory-information/search-fda-guidance-documents/drug-drug-interaction-assessment-therapeutic-proteins-guidance-industry. Accessed 10 July 2022

FDA (2020b) PHESGO prescribing information. https://www.accessdata.fda.gov/drugsatfda_docs/label/2020/761170s000lbl.pdf. Accessed 10 July 2022

FDA (2021) Bispecific antibody development programs guidance for industry. https://www.fda.gov/regulatory-information/search-fda-guidance-documents/bispecific-antibody-development-programs-guidance-industry. Accessed 12 June 2022

FDA (2022) Population pharmacokinetics: guidance for industry. https://www.fda.gov/regulatory-information/search-fda-guidance-documents/population-pharmacokinetics. Accessed 12 June 2022

Fearon KC, Glass DJ, Guttridge DC (2012) Cancer cachexia: mediators, signaling, and metabolic pathways. Cell Metab 16(2):153–166. https://doi.org/10.1016/j.cmet.2012.06.011

Ferl GZ, Wu AM, DiStefano JJ 3rd (2005) A predictive model of therapeutic monoclonal antibody dynamics and regulation by the neonatal fc receptor (FcRn). Ann Biomed Eng 33(11):1640–1652

Frost GI (2007) Recombinant human hyaluronidase (rHuPH20): an enabling platform for subcutaneous drug and fluid administration. Expert Opin Drug Deliv 4(4):427–440. https://doi.org/10.1517/17425247.4.4.427

Garg A, Quartino A, Li J, Jin J, Wada DR, Li H et al (2014) Population pharmacokinetic and covariate analysis of pertuzumab, a HER2-targeted monoclonal antibody, and evaluation of a fixed, non-weight-based dose in patients with a variety of solid tumors. Cancer Chemother Pharmacol 74(4):819–829. https://doi.org/10.1007/s00280-014-2560-3

Genovese MC, Cohen S, Moreland L, Lium D, Robbins S, Newmark R, Bekker P, Study Group (2004) Combination therapy with etanercept and anakinra in the treatment of patients with rheumatoid arthritis who have been treated unsuccessfully with methotrexate. Arthritis Rheum 50(5):1412–1419. https://doi.org/10.1002/art.20221

Ghetie V, Hubbard JG, Kim JK, Tsen MF, Lee Y, Ward ES (1996) Abnormally short serum half-lives of IgG in beta 2-microglobulin-deficient mice. Eur J Immunol 26(3):690–696

Gibiansky L, Frey N (2012) Linking interleukin-6 receptor blockade with tocilizumab and its hematological effects using a modeling approach. J Pharmacokinet Pharmacodyn 39(1):5–16. https://doi.org/10.1007/s10928-011-9227-z

Gibiansky L, Sutjandra L, Doshi S, Zheng J, Sohn W, Peterson MC, Jang GR, Chow AT, Perez-Ruixo JJ (2012) Population pharmacokinetic analysis of denosumab in patients with bone metastases from solid tumours. Clin Pharmacokinet 51 (4):247–60. https://doi.org/10.2165/11598090-000000000-00000

Gibiansky L, Gibiansky E (2009) Target-mediated drug disposition model: relationships with indirect response models and application to population PK-PD analysis. J Pharmacokinet Pharmacodyn 36(4):341–351. https://doi.org/10.1007/s10928-009-9125-9

Gillies SD, Lan Y, Lo KM, Super M, Wesolowski J (1999) Improving the efficacy of antibody-interleukin 2 fusion proteins by reducing their interaction with fc receptors. Cancer Res 59(9):2159–2166

Gillies SD, Lo KM, Burger C, Lan Y, Dahl T, Wong WK (2002) Improved circulating half-life and efficacy of an antibody-interleukin 2 immunocytokine based on reduced intracellular proteolysis. Clin Cancer Res 8(1):210–216

Girish G, Li C (2015) Clinical pharmacology and assay consideration for characterizing pharmacokinetics and understanding efficacy and safety of antibody-drug conjugates. In: Gorovits B, Shord S (eds) Novel methods in bioanalysis and characterization of antibody-drug conjugate. Future Science Ltd, London, pp 36–55

Girish S, Martin SW, Peterson MC (2011) al e AAPS workshop report: strategies to address therapeutic protein–drug interactions during clinical development. APPS J 3:405–416

Goldsby RA, Kindt TJ, Osborine BA, Kuby J (1999) Immunology, 4th edn. W.H. Freeman and Company, New York

Gorovits B, Alley SC, Bilic S, Booth B, Kaur S, Oldfield P et al (2013) Bioanalysis of antibody-drug conjugates: American Association of Pharmaceutical Scientists antibody-drug conjugate working group position paper. Bioanalysis 5(9):997–1006. https://doi.org/10.4155/bio.13.38

Gottlieb AB, Miller B, Lowe N, Shapiro W, Hudson C, Bright R et al (2003) Subcutaneously administered efalizumab (anti-CD11a) improves signs and symptoms of moderate to severe plaque psoriasis. J Cutan Med Surg 7(3):198–207

Han K, Jin J, Maia M, Lowe J, Sersch MA, Allison DE (2014) Lower exposure and faster clearance of bevacizumab in gastric cancer and the impact of patient variables: analysis of individual data from AVAGAST phase III trial. AAPS J 16(5):1056–1063. https://doi.org/10.1208/s12248-014-9631-6

Hayashi N, Tsukamoto Y, Sallas WM, Lowe PJ (2007) A mechanism-based binding model for the population pharmacokinetics and pharmacodynamics of omalizumab. Br J Clin Pharmacol 63(5):548–561. https://doi.org/10.1111/j.1365-2125.2006.02803.x

Hervey PS, Keam SJ (2006) Abatacept. BioDrugs 20(1):53–61. discussion 62

Hinton PR, Xiong JM, Johlfs MG, Tang MT, Keller S, Tsurushita N (2006) An engineered human IgG1 antibody with longer serum half-life. J Immunol 176(1):346–356

Hochhaus G, Brookman L, Fox H, Johnson C, Matthews J, Ren S et al (2003) Pharmacodynamics of omalizumab: implications for optimised dosing strategies and clinical efficacy in the treatment of allergic asthma. Curr Med Res Opin 19(6):491–498. https://doi.org/10.1185/030079903125002171

Holekamp NM, Campochiaro PA, Chang MA, Miller D, Pieramici D, Adamis AP, Brittain C, Evans E, Kaufman D, Maass KF, Patel S, Ranade S, Singh N, Barteselli G, Regillo C, All Archway

Investigators (2022) Archway randomized phase 3 trial of the port delivery system with Ranibizumab for Neovascular age-related macular degeneration. Ophthalmology 129(3):295–307. https://doi.org/10.1016/j.ophtha.2021.09.016

Holmes D (2011) Buy buy bispecific antibodies. Nat Rev Drug Discov 10(11):798–800. https://doi.org/10.1038/nrd3581

Honma W, Gautier A, Paule I, Yamaguchi M, Lowe PJ (2016) Ethnic sensitivity assessment of pharmacokinetics and pharmacodynamics of omalizumab with dosing table expansion. Drug Metab Pharmacokinet 31(3):173–184. https://doi.org/10.1016/j.dmpk.2015.12.003

Hooks MA, Wade CS, Millikan WJ Jr (1991) Muromonab CD-3: a review of its pharmacology, pharmacokinetics, and clinical use in transplantation. Pharmacotherapy 11(1):26–37

Huang SM, Zhao H, Lee JI, Reynolds K, Zhang L, Temple R et al (2010) Therapeutic protein-drug interactions and implications for drug development. Clin Pharmacol Ther 87(4):497–503. https://doi.org/10.1038/clpt.2009.308

ICH (1997a) ICH harmonized tripartite guideline M3: nonclinical safety studies for the conduct of human clinical trials for pharmaceuticals

ICH (1997b) ICH harmonized tripartite guideline S6: preclinical safety evaluation of biotechnology-derived pharmaceuticals

Igawa T, Tsunoda H, Tachibana T, Maeda A, Mimoto F, Moriyama C et al (2010) Reduced elimination of IgG antibodies by engineering the variable region. Protein Eng Des Sel 23(5):385–392. https://doi.org/10.1093/protein/gzq009

Jolling K, Perez Ruixo JJ, Hemeryck A, Vermeulen A, Greway T (2005) Mixed-effects modelling of the interspecies pharmacokinetic scaling of pegylated human erythropoietin. Eur J Pharm Sci 24(5):465–475

Joshi A, Bauer R, Kuebler P, White M, Leddy C, Compton P et al (2006) An overview of the pharmacokinetics and pharmacodynamics of efalizumab: a monoclonal antibody approved for use in psoriasis. J Clin Pharmacol 46(1):10–20

Junghans RP (1997) Finally! The Brambell receptor (FcRB). Mediator of transmission of immunity and protection from catabolism for IgG. Immunol Res 16(1):29–57

Junghans RP, Anderson CL (1996) The protection receptor for IgG catabolism is the beta 2-microglobulin-containing neonatal intestinal transport receptor. Proc Natl Acad Sci U S A 93(11):5512–5516

Kagan L, Abraham AK, Harrold JM, Mager DE (2010) Interspecies scaling of receptor-mediated pharmacokinetics and pharmacodynamics of type I interferons. Pharm Res 27(5):920–932. https://doi.org/10.1007/s11095-010-0098-6

Kagedal M, Gibiansky L, Xu J, Wang X, Samineni D, Chen SC et al (2017) Platform model describing pharmacokinetic properties of vc-MMAE antibody-drug conjugates. J Pharmacokinet Pharmacodyn 44(6):537–548. https://doi.org/10.1007/s10928-017-9544-y

Kairemo KJ, Lappalainen AK, Kaapa E, Laitinen OM, Hyytinen T, Karonen SL et al (2001) In vivo detection of intervertebral disk injury using a radiolabeled monoclonal antibody against keratan sulfate. J Nucl Med 42(3):476–482

Kamath AV (2016) Translational pharmacokinetics and pharmacodynamics of monoclonal antibodies. Drug Discov Today Technol 21–22:75–83. https://doi.org/10.1016/j.ddtec.2016.09.004

Kamath AV, Iyer S (2015) Preclinical pharmacokinetic considerations for the development of antibody drug conjugates. Pharm Res 32(11):3470–3479. https://doi.org/10.1007/s11095-014-1584-z

Kamath AV, Iyer S (2016) Challenges and advances in the assessment of the disposition of antibody-drug conjugates. Biopharm Drug Dispos 37(2):66–74. https://doi.org/10.1002/bdd.1957

Kang YK, Ryu MH, Yoo C, Ryoo BY, Kim HJ, Lee JJ et al (2013) Resumption of imatinib to control metastatic or unresectable gastrointestinal stromal tumours after failure of imatinib and sunitinib (RIGHT): a randomised, placebo-controlled, phase 3 trial. Lancet Oncol 14(12):1175–1182. https://doi.org/10.1016/S1470-2045(13)70453-4

Kaur S, Xu K, Saad OM, Dere RC, Carrasco-Triguero M (2013) Bioanalytical assay strategies for the development of antibody-drug conjugate biotherapeutics. Bioanalysis 5(2):201–226. https://doi.org/10.4155/bio.12.299

Kelley SK, Gelzleichter T, Xie D, Lee WP, Darbonne WC, Qureshi F, Kissler K, Oflazoglu E, Grewal IS (2006) Preclinical pharmacokinetics, pharmacodynamics, and activity of a humanized anti-CD40 antibody (SGN-40) in rodents and non-human primates. Br J Pharmacol 148(8):1116–1123

Kenny JR, Liu MM, Chow AT, Earp JC, Evers R, Slatter JG, Wang DD, Zhang L, Zhou H (2013) Therapeutic protein drug-drug interactions: navigating the knowledge gaps-highlights from the 2012 AAPS NBC roundtable and IQ consortium/FDA workshop. AAPS J 15(4):933–940. https://doi.org/10.1208/s12248-013-9495-1

Keyt BA, Baliga R, Sinclair AM, Carroll SF, Peterson MS (2020) Structure, function, and therapeutic use of IgM antibodies. Antibodies (Basel) 9(4):53. https://doi.org/10.3390/antib9040053

Khawli LA, Goswami S, Hutchinson R, Kwong ZW, Yang J, Wang X, Yao Z, Sreedhara A, Cano T, Tesar D, Nijem I, Allison DE, Wong PY, Kao YH, Quan C, Joshi A, Harris RJ, Motchnik P (2010) Charge variants in IgG1: isolation, characterization, in vitro binding properties and pharmacokinetics in rats. MAbs 2(6):613–624. https://doi.org/10.4161/mabs.2.6.13333

Kiessling P, Lledo-Garcia R, Watanabe S, Langdon G, Tran D, Bari M, Christodoulou L, Jones E, Price G, Smith B, Brennan F, White I, Jolles S (2017) The FcRn inhibitor rozanolixizumab reduces human serum IgG concentration: a randomized phase 1 study. Sci Transl Med 9(414):eaan1208. https://doi.org/10.1126/scitranslmed.aan1208

Kirschbrown WP, Quartino AL, Li H, Mangat R, Wada DR, Garg A, Jin JY, Lum BL (2017) Development of a population pharmacokinetic (PPK) model of intravenous (IV) trastuzumab in patients with a variety of solid tumors to support dosing and treatment recommendations. J Clin Oncol 35(Suppl):2525

Kleiman NS, Raizner AE, Jordan R, Wang AL, Norton D, Mace KF, Joshi A, Coller BS, Weisman HF (1995) Differential inhibition of platelet aggregation induced by adenosine diphosphate or a thrombin receptor-activating peptide in patients treated with bolus chimeric 7E3 fab: implications for inhibition of the internal pool of GPIIb/IIIa receptors. J Am Coll Cardiol 26(7):1665–1671. https://doi.org/10.1016/0735-1097(95)00391-6

Kohler G, Milstein C (1975) Continuous cultures of fused cells secreting antibody of predefined specificity. Nature 256(5517):495–497

Kolar GR, Capra JD (2003) Immunoglobulins: structure and function. In: Paul WE (ed) Fundamental immunology, 5th edn. Lippincott Williams & Wilkins, Philadelphia

Koon HB, Severy P, Hagg DS, Butler K, Hill T, Jones AG, Waldmann TA, Junghans RP (2006) Antileukemic effect of daclizumab in CD25 high-expressing leukemias and impact of tumor burden on antibody dosing. Leuk Res 30(2):190–203

Kovalenko P, DiCioccio AT, Davis JD, Li M, Ardeleanu M, Graham N, Soltys R (2016) Exploratory population PK analysis of Dupilumab, a fully human monoclonal antibody against IL-4Ralpha, in atopic dermatitis patients and Normal volunteers. CPT Pharmacometrics Syst Pharmacol 5(11):617–624. https://doi.org/10.1002/psp4.12136

Kovarik JM, Nashan B, Neuhaus P, Clavien PA, Gerbeau C, Hall ML, Korn A (2001) A population pharmacokinetic screen to identify demographic-clinical covariates of basiliximab in liver transplantation. Clin Pharmacol Ther 69(4):201–209

Krueger JG (2002) The immunologic basis for the treatment of psoriasis with new biologic agents. J Am Acad Dermatol 46(1):1–23. quiz 23-26

Kuus-Reichel K, Grauer LS, Karavodin LM, Knott C, Krusemeier M, Kay NE (1994) Will immunogenicity limit the use, efficacy, and future development of therapeutic monoclonal antibodies? Clin Diagn Lab Immunol 1(4):365–372

Lee H, Kimko HC, Rogge M, Wang D, Nestorov I, Peck CC (2003) Population pharmacokinetic and pharmacodynamic modeling of etanercept using logistic regression analysis. Clin Pharmacol Ther 73(4):348–365

Lee JI, Zhang L, Men AY, Kenna LA, Huang SM (2010) CYP-mediated therapeutic protein-drug interactions: clinical findings, proposed mechanisms and regulatory implications. Clin Pharmacokinet 49(5):295–310. https://doi.org/10.2165/11319980-000000000-00000

Li S, Schmitz KR, Jeffrey PD, Wiltzius JJ, Kussie P, Ferguson KM (2005) Structural basis for inhibition of the epidermal growth factor receptor by cetuximab. Cancer Cell 7(4):301–311. https://doi.org/10.1016/j.ccr.2005.03.003

Li B, Tesar D, Boswell CA, Cahaya HS, Wong A, Zhang J, Meng YG, Eigenbrot C, Pantua H, Diao J, Kapadia SB, Deng R, Kelley RF (2014) Framework selection can influence pharmacokinetics of a humanized therapeutic antibody through differences in molecule charge. MAbs 6(5):1255–1264. https://doi.org/10.4161/mabs.29809

Li H, Yu J, Liu C, Liu J, Subramaniam S, Zhao H, Blumenthal GM, Turner DC, Li C, Ahamadi M, de Greef R, Chatterjee M, Kondic AG, Stone JA, Booth BP, Keegan P, Rahman A, Wang Y (2017) Time dependent pharmacokinetics of pembrolizumab in patients with solid tumor and its correlation with best overall response. J Pharmacokinet Pharmacodyn 44(5):403–414. https://doi.org/10.1007/s10928-017-9528-y

Li C, Zhang C, Deng R, Leipold D, Li D, Latifi B, Gao Y, Zhang C, Li Z, Miles D, Chen SC, Samineni D, Wang B, Agarwal P, Lu D, Prabhu S, Girish S, Kamath AV (2019) Prediction of human pharmacokinetics of antibody-drug conjugates from nonclinical data. Clin Transl Sci 12(5):534–544. https://doi.org/10.1111/cts.12649

Li C, Chen SC, Chen Y, Girish S, Kaagedal M, Lu D, Lu T, Samineni D, Jin JY (2020) Impact of physiologically based pharmacokinetics, population pharmacokinetics and pharmacokinetics/pharmacodynamics in the development of antibody-drug conjugates. J Clin Pharmacol 60(Suppl 1):S105–S119. https://doi.org/10.1002/jcph.1720

Li C, Menon R, Walles M, Singh R, Upreti VV, Brackman D, Lee AJ, Endres CJ, Kumar S, Zhang D, Barletta F, Suri A, Hainzl D, Liao KH, Lalovic B, Beaumont M, Zuo P, Mayer AP, Wei D (2022) Risk-based pharmacokinetic and drug-drug interaction characterization of antibody-drug conjugates in oncology clinical development: an international consortium for innovation and quality in pharmaceutical development perspective. Clin Pharmacol Ther 112(4):754–769. https://doi.org/10.1002/cpt.2448

Lin YS, Nguyen C, Mendoza JL, Escandon E, Fei D, Meng YG, Modi NB (1999) Preclinical pharmacokinetics, interspecies scaling, and tissue distribution of a humanized monoclonal antibody against vascular endothelial growth factor. J Pharmacol Exp Ther 288(1):371–378

Ling J, Zhou H, Jiao Q, Davis HM (2009) Interspecies scaling of therapeutic monoclonal antibodies: initial look. J Clin Pharmacol 49(12):1382–1402. https://doi.org/10.1177/0091270009337134

Liu J, Wang Y, Zhao L (2015) Assessment of exposure-response (E-R) and cse-control (C-C) analyses in oncology using simulation based approach. In: Annual meeting of the American Conference of Pharmacometrics

Lobo ED, Hansen RJ, Balthasar JP (2004) Antibody pharmacokinetics and pharmacodynamics. J Pharm Sci 93(11):2645–2668

Looney RJ, Anolik JH, Campbell D, Felgar RE, Young F, Arend LJ, Sloand JA, Rosenblatt J, Sanz I (2004) B cell depletion as a novel treatment for systemic lupus erythematosus: a phase I/II dose-escalation trial of rituximab. Arthritis Rheum 50(8):2580–2589

Lowe PJ, Georgiou P, Canvin J (2015) Revision of omalizumab dosing table for dosing every 4 instead of 2 weeks for specific ranges of bodyweight and baseline IgE. Regul Toxicol Pharmacol 71(1):68–77. https://doi.org/10.1016/j.yrtph.2014.12.002

Lu D, Burris HA 3rd, Wang B, Dees EC, Cortes J, Joshi A, Gupta M, Yi JH, Chu YW, Shih T, Fang L, Girish S (2012) Drug interaction potential of trastuzumab emtansine (T-DM1) combined with pertuzumab in patients with HER2-positive metastatic breast cancer. Curr Drug Metab 13(7):911–922

Lu D, Girish S, Gao Y, Wang B, Yi JH, Guardino E, Samant M, Cobleigh M, Rimawi M, Conte P, Jin JY (2014) Population pharmacokinetics of trastuzumab emtansine (T-DM1), a HER2-targeted antibody-drug conjugate, in patients with HER2-positive metastatic breast cancer: clinical implications of the effect of covariates. Cancer Chemother Pharmacol 74(2):399–410. https://doi.org/10.1007/s00280-014-2500-2

Ma P, Yang BB, Wang YM, Peterson M, Narayanan A, Sutjandra L, Rodriguez R, Chow A (2009) Population pharmacokinetic analysis of panitumumab in patients with advanced solid tumors. J Clin Pharmacol 49(10):1142–1156. https://doi.org/10.1177/0091270009344989

Mager DE, Mascelli MA, Kleiman NS, Fitzgerald DJ, Abernethy DR (2003) Simultaneous modeling of abciximab plasma concentrations and ex vivo pharmacodynamics in patients undergoing coronary angioplasty. J Pharmacol Exp Ther 307(3):969–976. https://doi.org/10.1124/jpet.103.057299

Mager DE, Woo S, Jusko WJ (2009) Scaling pharmacodynamics from in vitro and preclinical animal studies to humans. Drug Metab Pharmacokinet 24(1):16–24

Mahmood I (2005) Prediction of concentration-time profiles in humans. Pine House Publisher, Rockville

Mahmood I (2009) Pharmacokinetic allometric scaling of antibodies: application to the first-in-human dose estimation. J Pharm Sci 98(10):3850–3861. https://doi.org/10.1002/jps.21682

Mahmood I, Green MD (2005) Pharmacokinetic and pharmacodynamic considerations in the development of therapeutic proteins. Clin Pharmacokinet 44(4):331–347

Maini RN, Breedveld FC, Kalden JR, Smolen JS, Davis D, Macfarlane JD, Antoni C, Leeb B, Elliott MJ, Woody JN, Schaible TF, Feldmann M (1998) Therapeutic efficacy of multiple intravenous infusions of anti-tumor necrosis factor alpha monoclonal antibody combined with low-dose weekly methotrexate in rheumatoid arthritis. Arthritis Rheum 41(9):1552–1563. https://doi.org/10.1002/1529-0131(199809)41:9<1552::AID-ART5>3.0.CO;2-W

Mandikian D, Takahashi N, Lo AA, Li J, Eastham-Anderson J, Slaga D, Ho J, Hristopoulos M, Clark R, Totpal K, Lin K, Joseph SB, Dennis MS, Prabhu S, Junttila TT, Boswell CA (2018) Relative target affinities of T-cell-dependent bispecific antibodies determine biodistribution in a solid tumor mouse model. Mol Cancer Ther 17(4):776–785. https://doi.org/10.1158/1535-7163.MCT-17-0657

Martin-Jimenez T, Riviere JE (2002) Mixed-effects modeling of the interspecies pharmacokinetic scaling of oxytetracycline. J Pharm Sci 91(2):331–341

McClurkan MB, Valentine JL, Arnold L, Owens SM (1993) Disposition of a monoclonal anti-phencyclidine fab fragment of immunoglobulin G in rats. J Pharmacol Exp Ther 266(3):1439–1445

McLaughlin P, Grillo-Lopez AJ, Link BK, Levy R, Czuczman MS, Williams ME, Heyman MR, Bence-Bruckler I, White CA, Cabanillas F, Jain V, Ho AD, Lister J, Wey K, Shen D, Dallaire BK (1998) Rituximab chimeric anti-CD20 monoclonal antibody therapy for relapsed indolent lymphoma: half of patients respond to a four-dose treatment program. J Clin Oncol 16(8):2825–2833

Medesan C, Matesoi D, Radu C, Ghetie V, Ward ES (1997) Delineation of the amino acid residues involved in transcytosis and catabolism of mouse IgG1. J Immunol 158(5):2211–2217

Meibohm B, Derendorf H (2002) Pharmacokinetic/pharmacodynamic studies in drug product development. J Pharm Sci 91(1):18–31

Meijer RT, Koopmans RP, ten Berge IJ, Schellekens PT (2002) Pharmacokinetics of murine anti-human CD3 antibodies in man are determined by the disappearance of target antigen. J Pharmacol Exp Ther 300(1):346–353

Meredith PA, Elliott HL, Donnelly R, Reid JL (1991) Dose-response clarification in early drug development. J Hypertens Suppl 9(6):S356–S357

Morris EC, Rebello P, Thomson KJ, Peggs KS, Kyriakou C, Goldstone AH, Mackinnon S, Hale G (2003) Pharmacokinetics of alemtuzumab used for in vivo and in vitro T-cell depletion in allogeneic transplantations: relevance for early adoptive immunotherapy and infectious complications. Blood 102(1):404–406. https://doi.org/10.1182/blood-2002-09-2687

Mortensen DL, Walicke PA, Wang X, Kwon P, Kuebler P, Gottlieb AB, Krueger JG, Leonardi C, Miller B, Joshi A (2005) Pharmacokinetics and pharmacodynamics of multiple weekly subcutaneous efalizumab doses in patients with plaque psoriasis. J Clin Pharmacol 45(3):286–298

Mould DR, Sweeney KR (2007) The pharmacokinetics and pharmacodynamics of monoclonal antibodies–mechanistic modeling applied to drug development. Curr Opin Drug Discov Devel 10(1):84–96

Mould DR, Davis CB, Minthorn EA, Kwok DC, Elliott MJ, Luggen ME, Totoritis MC (1999) A population pharmacokinetic-pharmacodynamic analysis of single doses of clenoliximab in patients with rheumatoid arthritis. Clin Pharmacol Ther 66(3):246–257

Nakakura EK, McCabe SM, Zheng B, Shorthouse RA, Scheiner TM, Blank G, Jardieu PM, Morris RE (1993) Potent and effective prolongation by anti-LFA-1 monoclonal antibody monotherapy of non-primarily vascularized heart allograft survival in mice without T cell depletion. Transplantation 55(2):412–417

Nestorov I, Zitnik R, Ludden T (2004) Population pharmacokinetic modeling of subcutaneously administered etanercept in patients with psoriasis. J Pharmacokinet Pharmacodyn 31(6):463–490

Ng CM, Joshi A, Dedrick RL, Garovoy MR, Bauer RJ (2005) Pharmacokinetic-pharmacodynamic-efficacy analysis of efalizumab in patients with moderate to severe psoriasis. Pharm Res 22(7):1088–1100

Ng CM, Stefanich E, Anand BS, Fielder PJ, Vaickus L (2006) Pharmacokinetics/pharmacodynamics of nondepleting anti-CD4 monoclonal antibody (TRX1) in healthy human volunteers. Pharm Res 23(1):95–103

Norman DJ, Chatenoud L, Cohen D, Goldman M, Shield CF 3rd (1993) Consensus statement regarding OKT3-induced cytokine-release syndrome and human antimouse antibodies. Transplant Proc 25(2 Suppl 1):89–92

Ober RJ, Radu CG, Ghetie V, Ward ES (2001) Differences in promiscuity for antibody-FcRn interactions across species: implications for therapeutic antibodies. Int Immunol 13(12):1551–1559

Oitate M, Masubuchi N, Ito T, Yabe Y, Karibe T, Aoki T, Murayama N, Kurihara A, Okudaira N, Izumi T (2011) Prediction of human pharmacokinetics of therapeutic monoclonal antibodies from simple allometry of monkey data. Drug Metab Pharmacokinet 26(4):423–430

Page MM, Watts GF (2015) Evolocumab in the treatment of dyslipidemia: pre-clinical and clinical pharmacology. Expert Opin Drug Metab Toxicol 11(9):1505–1515. https://doi.org/10.1517/17425255.2015.1073712

Papp K, Bissonnette R, Krueger JG, Carey W, Gratton D, Gulliver WP, Lui H, Lynde CW, Magee A, Minier D, Ouellet JP, Patel P, Shapiro J, Shear NH, Kramer S, Walicke P, Bauer R, Dedrick RL, Kim SS, White M, Garovoy MR (2001) The treatment of moderate to severe psoriasis with a new anti-CD11a monoclonal antibody. J Am Acad Dermatol 45(5):665–674

Patel D, Bassi R, Hooper A, Prewett M, Hicklin DJ, Kang X (2009) Anti-epidermal growth factor receptor monoclonal antibody cetuximab inhibits EGFR/HER-2 heterodimerization and activation. Int J Oncol 34(1):25–32

Petkova SB, Akilesh S, Sproule TJ, Christianson GJ, Al Khabbaz H, Brown AC, Presta LG, Meng YG, Roopenian DC (2006) Enhanced half-life of genetically engineered human IgG1 antibodies in a humanized FcRn mouse model: potential application in humorally mediated autoimmune disease. Int Immunol 18(12):1759–1769

Prabhu S, Boswell CA, Leipold D, Khawli LA, Li D, Lu D, Theil FP, Joshi A, Lum BL (2011) Antibody delivery of drugs and radionuclides: factors influencing clinical pharmacology. Ther Deliv 2(6):769–791

Presta LG (2002) Engineering antibodies for therapy. Curr Pharm Biotechnol 3(3):237–256

Presta LG, Shields RL, Namenuk AK, Hong K, Meng YG (2002) Engineering therapeutic antibodies for improved function. Biochem Soc Trans 30(4):487–490

Putnam WS, Prabhu S, Zheng Y, Subramanyam M, Wang YM (2010) Pharmacokinetic, pharmacodynamic and immunogenicity comparability assessment strategies for monoclonal antibodies. Trends Biotechnol 28(10):509–516. https://doi.org/10.1016/j.tibtech.2010.07.001

Rafidi H, Estevez A, Ferl GZ, Mandikian D, Stainton S, Sermeno L, Williams SP, Kamath AV, Koerber JT, Boswell CA (2021) Imaging reveals importance of shape and flexibility for glomerular filtration of biologics. Mol Cancer Ther 20(10):2008–2015. https://doi.org/10.1158/1535-7163.MCT-21-0116

Rafidi H, Rajan S, Urban K, Shatz-Binder W, Hui K, Ferl GZ, Kamath AV, Boswell CA (2022) Effect of molecular size on interstitial pharmacokinetics and tissue catabolism of antibodies. MAbs 14(1):2085535. https://doi.org/10.1080/19420862.2022.2085535

Robbie GJ, Criste R, Dall'acqua WF, Jensen K, Patel NK, Losonsky GA, Griffin MP (2013) A novel investigational fc-modified humanized monoclonal antibody, motavizumab-YTE, has an extended half-life in healthy adults. Antimicrob Agents Chemother 57(12):6147–6153. https://doi.org/10.1128/AAC.01285-13

Roopenian DC, Christianson GJ, Sproule TJ, Brown AC, Akilesh S, Jung N, Petkova S, Avanessian L, Choi EY, Shaffer DJ, Eden PA, Anderson CL (2003) The MHC class I-like IgG receptor controls perinatal IgG transport, IgG homeostasis, and fate of IgG-fc-coupled drugs. J Immunol 170(7):3528–3533

Roskos LK, Davis CG, Schwab GM (2004) The clinical pharmacology of therapeutic monoclonal antibodies. Drug Dev Res 61:108–120

Ryman JT, Meibohm B (2017) Pharmacokinetics of monoclonal antibodies. CPT Pharmacometrics Syst Pharmacol 6(9):576–588. https://doi.org/10.1002/psp4.12224

Samineni D, Ding H, Ma F, Shi R, Lu D, Miles D, Mao J, Li C, Jin J, Wright M, Girish S, Chen Y (2020) Physiologically based pharmacokinetic model-informed drug development for Polatuzumab Vedotin: label for drug-drug interactions without dedicated clinical trials. J Clin Pharmacol 60(Suppl 1):S120–S131. https://doi.org/10.1002/jcph.1718

Sampei Z, Igawa T, Soeda T, Okuyama-Nishida Y, Moriyama C, Wakabayashi T, Tanaka E, Muto A, Kojima T, Kitazawa T, Yoshihashi K, Harada A, Funaki M, Haraya K, Tachibana T, Suzuki S, Esaki K, Nabuchi Y, Hattori K (2013) Identification and multidimensional optimization of an asymmetric bispecific IgG antibody mimicking the function of factor VIII cofactor activity. PLoS One 8(2):e57479. https://doi.org/10.1371/journal.pone.0057479

Schmitt C, Kuhn B, Zhang X, Kivitz AJ, Grange S (2011) Disease-drug-drug interaction involving tocilizumab and simvastatin in patients with rheumatoid arthritis. Clin Pharmacol Ther 89(5):735–740. https://doi.org/10.1038/clpt.2011.35

Schrieber SJ, Pfuma-Fletcher E, Wang X, Wang YC, Sagoo S, Madabushi R, Huang SM, Zineh I (2019) Considerations for biologic product drug-drug interactions: a regulatory perspective. Clin Pharmacol Ther 105(6):1332–1334. https://doi.org/10.1002/cpt.1366

Schror K, Weber AA (2003) Comparative pharmacology of GP IIb/IIIa antagonists. J Thromb Thrombolysis 15(2):71–80

Shah DK, Betts AM (2012) Towards a platform PBPK model to characterize the plasma and tissue disposition of monoclonal antibodies in preclinical species and human. J Pharmacokinet Pharmacodyn 39(1):67–86. https://doi.org/10.1007/s10928-011-9232-2

Sheiner LB (1997) Learning versus confirming in clinical drug development. Clin Pharmacol Ther 61(3):275–291

Sheiner L, Wakefield J (1999) Population modelling in drug development. Stat Methods Med Res 8(3):183–193

Shen BQ, Xu K, Liu L, Raab H, Bhakta S, Kenrick M, Parsons-Reponte KL, Tien J, Yu SF, Mai E, Li D, Tibbitts J, Baudys J, Saad OM, Scales SJ, McDonald PJ, Hass PE, Eigenbrot C, Nguyen T, Solis WA, Fuji RN, Flagella KM, Patel D, Spencer SD, Khawli LA, Ebens A, Wong WL, Vandlen R, Kaur S, Sliwkowski MX, Scheller RH, Polakis P, Junutula JR (2012) Conjugation site modulates the in vivo stability and therapeutic activity of antibody-drug conjugates. Nat Biotechnol 30(2):184–189. https://doi.org/10.1038/nbt.2108

Shields RL, Namenuk AK, Hong K, Meng YG, Rae J, Briggs J, Xie D, Lai J, Stadlen A, Li B, Fox JA, Presta LG (2001) High resolution mapping of the binding site on human IgG1 for fc gamma RI, fc gamma RII, fc gamma RIII, and FcRn and design of IgG1 variants with improved binding to the fc gamma R. J Biol Chem 276(9):6591–6604

Sifontis NM, Benedetti E, Vasquez EM (2002) Clinically significant drug interaction between basiliximab and tacrolimus in renal transplant recipients. Transplant Proc 34(5):1730–1732

Simister NE, Mostov KE (1989) Cloning and expression of the neonatal rat intestinal fc receptor, a major histocompatibility complex class I antigen homolog. Cold Spring Harb Symp Quant Biol 54(Pt 1):571–580

Slatter JG, Wienkers LC, Dickmann LC (eds) (2013) Drug interactions of cytokines and anticytokine therapeutic proteins. Drug-drug interactions for therapeutics biologics. Hoboken, NJ, Wiley

Spiekermann GM, Finn PW, Ward ES, Dumont J, Dickinson BL, Blumberg RS, Lencer WI (2002) Receptor-mediated immunoglobulin G transport across mucosal barriers in adult life: functional expression of FcRn in the mammalian lung. J Exp Med 196(3):303–310

Spiess C, Zhai Q, Carter PJ (2015) Alternative molecular formats and therapeutic applications for bispecific antibodies. Mol Immunol 67(2 Pt A):95–106. https://doi.org/10.1016/j.molimm.2015.01.003

Straughn JM Jr, Oliver PG, Zhou T, Wang W, Alvarez RD, Grizzle WE, Buchsbaum DJ (2006) Anti-tumor activity of TRA-8 anti-death receptor 5 (DR5) monoclonal antibody in combination with chemotherapy and radiation therapy in a cervical cancer model. Gynecol Oncol 101(1):46–54. https://doi.org/10.1016/j.ygyno.2005.09.053

Struemper H, Sale M, Patel BR, Ostergaard A, Wierda WG, Hagenbeek A, Coiffier B, Jewell RC (2014) Population pharmacokinetics of ofatumumab in patients with chronic lymphocytic leukemia, follicular lymphoma, and rheumatoid arthritis. J Clin Pharmacol 54(7):818–827. https://doi.org/10.1002/jcph.268

Subramanian GM, Cronin PW, Poley G, Weinstein A, Stoughton SM, Zhong J, Ou Y, Zmuda JF, Osborn BL, Freimuth WW (2005) A phase 1 study of PAmAb, a fully human monoclonal antibody against bacillus anthracis protective antigen, in healthy volunteers. Clin Infect Dis 41(1):12–20

Sun YN, Lu JF, Joshi A, Compton P, Kwon P, Bruno RA (2005) Population pharmacokinetics of efalizumab (humanized monoclonal anti-CD11a antibody) following long-term subcutaneous weekly dosing in psoriasis subjects. J Clin Pharmacol 45(4):468–476

Tabrizi MA, Tseng CM, Roskos LK (2006) Elimination mechanisms of therapeutic monoclonal antibodies. Drug Discov Today 11(1–2):81–88

Tang H, Mayersohn M (2005) Accuracy of allometrically predicted pharmacokinetic parameters in humans: role of species selection. Drug Metab Dispos 33(9):1288–1293

Tang L, Persky AM, Hochhaus G, Meibohm B (2004) Pharmacokinetic aspects of biotechnology products. J Pharm Sci 93(9):2184–2204. https://doi.org/10.1002/jps.20125

Ternant D, Paintaud G (2005) Pharmacokinetics and concentration-effect relationships of therapeutic monoclonal antibodies and fusion proteins. Expert Opin Biol Ther 5(Suppl 1):S37–S47

Tran JQ, Othman AA, Wolstencroft P, Elkins J (2016) Therapeutic protein-drug interaction assessment for daclizumab high-yield process in patients with multiple sclerosis using a cocktail approach. Br J Clin Pharmacol 82(1):160–167. https://doi.org/10.1111/bcp.12936

Trianni (2018) http://trianni.com/technology/mouse/. Accessed 8 May 2018

Turner DC, Kondic AG, Anderson KM, Robinson AG, Garon EB, Riess JW, Jain L, Mayawala K, Kang J, Ebbinghaus SW, Sinha V, de Alwis DP, Stone JA (2018) Pembrolizumab exposure-response assessments challenged by association of cancer cachexia and catabolic clearance. Clin Cancer Res 24(23):5841–5849. https://doi.org/10.1158/1078-0432.CCR-18-0415

Umana P, Jean-Mairet J, Moudry R, Amstutz H, Bailey JE (1999) Engineered glycoforms of an antineuroblastoma IgG1 with optimized antibody-dependent cellular cytotoxic activity. Nat Biotechnol 17(2):176–180

Vaccaro C, Bawdon R, Wanjie S, Ober RJ, Ward ES (2006) Divergent activities of an engineered antibody in murine and human systems have implications for therapeutic antibodies. Proc Natl Acad Sci U S A 103(49):18709–18714

Vaishnaw AK, TenHoor CN (2002) Pharmacokinetics, biologic activity, and tolerability of alefacept by intravenous and intramuscular administration. J Pharmacokinet Pharmacodyn 29(5–6):415–426

Vasquez EM, Pollak R (1997) OKT3 therapy increases cyclosporine blood levels. Clin Transpl 11(1):38–41

Vincenti F, Mendez R, Pescovitz M, Rajagopalan PR, Wilkinson AH, Butt K, Laskow D, Slakey DP, Lorber MI, Garg JP, Garovoy M (2007) A phase I/II randomized open-label multicenter trial of efalizumab, a humanized anti-CD11a, anti-LFA-1 in renal transplantation. Am J Transplant 7(7):1770–1777. https://doi.org/10.1111/j.1600-6143.2007.01845.x

Vugmeyster Y, Guay H, Szklut P, Qian MD, Jin M, Widom A, Spaulding V, Bennett F, Lowe L, Andreyeva T, Lowe D, Lane S, Thom G, Valge-Archer V, Gill D, Young D, Bloom L (2010) In vitro potency, pharmacokinetic profiles, and pharmacological activity of optimized anti-IL-21R antibodies in a mouse model of lupus. MAbs 2(3):335–346

Vugmeyster Y, Szklut P, Wensel D, Ross J, Xu X, Awwad M, Gill D, Tchistiakov L, Warner G (2011) Complex pharmacokinetics of a humanized antibody against human amyloid beta peptide, anti-abeta Ab2, in nonclinical species. Pharm Res 28(7):1696–1706. https://doi.org/10.1007/s11095-011-0405-x

Wang Y (2016) Special considerations for modeling exposure response for biologics. In: American Society for Clinical Pharmacology and Therapeutics Annual Meeting

Wang W, Prueksaritanont T (2010) Prediction of human clearance of therapeutic proteins: simple allometric scaling method revisited. Biopharm Drug Dispos 31(4):253–263. https://doi.org/10.1002/bdd.708

Wang W, Singh S, Zeng DL, King K, Nema S (2007) Antibody structure, instability, and formulation. J Pharm Sci 96(1):1–26

Wang S, Chen K, Lei Q, Ma P, Yuan AQ, Zhao Y, Jiang Y, Fang H, Xing S, Fang Y, Jiang N, Miao H, Zhang M, Sun S, Yu Z, Tao W, Zhu Q, Nie Y, Li N (2021) The state of the art of bispecific antibodies for treating human malignancies. EMBO Mol Med 13(9):e14291. https://doi.org/10.15252/emmm.202114291

Watanabe N, Kuriyama H, Sone H, Neda H, Yamauchi N, Maeda M, Niitsu Y (1988) Continuous internalization of tumor necrosis factor receptors in a human myosarcoma cell line. J Biol Chem 263(21):10262–10266

Weiner LM (2006) Fully human therapeutic monoclonal antibodies. J Immunother 29(1):1–9

Weiskopf K, Weissman IL (2015) Macrophages are critical effectors of antibody therapies for cancer. MAbs 7(2):303–310. https://doi.org/10.1080/19420862.2015.1011450

Weisman MH, Moreland LW, Furst DE, Weinblatt ME, Keystone EC, Paulus HE, Teoh LS, Velagapudi RB, Noertersheuser PA, Granneman GR, Fischkoff SA, Chartash EK (2003) Efficacy, pharmacokinetic, and safety assessment of adalimumab, a fully human anti-tumor necrosis factor-alpha monoclonal antibody, in adults with rheumatoid arthritis receiving concomitant methotrexate: a pilot study. Clin Ther 25(6):1700–1721

Werther WA, Gonzalez TN, O'Connor SJ, McCabe S, Chan B, Hotaling T, Champe M, Fox JA, Jardieu PM, Berman PW, Presta LG (1996) Humanization of an anti-lymphocyte function-associated antigen (LFA)-1 monoclonal antibody and reengineering of the humanized antibody for binding to rhesus LFA-1. J Immunol 157(11):4986–4995

Wiseman GA, White CA, Sparks RB, Erwin WD, Podoloff DA, Lamonica D, Bartlett NL, Parker JA, Dunn WL, Spies SM, Belanger R, Witzig TE, Leigh BR (2001) Biodistribution and dosimetry results from a phase III prospectively randomized controlled trial of Zevalin radioimmunotherapy for low-grade, follicular, or transformed B-cell non-Hodgkin's lymphoma. Crit Rev Oncol Hematol 39(1–2):181–194

Wu B, Sun YN (2014) Pharmacokinetics of peptide-fc fusion proteins. J Pharm Sci 103(1):53–64. https://doi.org/10.1002/jps.23783

Wu H, Pfarr DS, Johnson S, Brewah YA, Woods RM, Patel NK, White WI, Young JF, Kiener PA (2007) Development of motavizumab, an ultra-potent antibody for the prevention of respiratory syncytial virus infection in the upper and lower respiratory tract. J Mol Biol 368(3):652–665. https://doi.org/10.1016/j.jmb.2007.02.024

Wurster U, Haas J (1994) Passage of intravenous immunoglobulin and interaction with the CNS. J Neurol Neurosurg Psychiatry 57:21–25

Yadav DB, Maloney JA, Wildsmith KR, Fuji RN, Meilandt WJ, Solanoy H, Lu Y, Peng K, Wilson B, Chan P, Gadkar K, Kosky A, Goo M, Daugherty A, Couch JA, Keene T, Hayes K, Nikolas LJ, Lane D, Switzer R, Adams E, Watts RJ, Scearce-Levie K, Prabhu S, Shafer L, Thakker DR, Hildebrand K, Atwal JK (2017) Widespread brain distribution and activity following i.c.v. infusion of anti-beta-secretase (BACE1) in nonhuman primates. Br J Pharmacol 174(22):4173–4185. https://doi.org/10.1111/bph.14021

Yang J, Zhao H, Garnett C, Rahman A, Gobburu JV, Pierce W, Schechter G, Summers J, Keegan P, Booth B, Wang Y (2013) The combination of exposure-response and case-control analyses in regulatory decision making. J Clin Pharmacol 53(2):160–166. https://doi.org/10.1177/0091270012445206

Yim DS, Zhou H, Buckwalter M, Nestorov I, Peck CC, Lee H (2005) Population pharmacokinetic analysis and simulation of the time-concentration profile of etanercept in pediatric patients with juvenile rheumatoid arthritis. J Clin Pharmacol 45(3):246–256. https://doi.org/10.1177/0091270004271945

Yip V, Palma E, Tesar DB, Mundo EE, Bumbaca D, Torres EK, Reyes NA, Shen BQ, Fielder PJ, Prabhu S, Khawli LA, Boswell CA (2014) Quantitative cumulative biodistribution of antibodies in mice: effect of modulating binding affinity to the neonatal Fc receptor. MAbs 6(3):689–696. https://doi.org/10.4161/mabs.28254

Younes A, Bartlett NL, Leonard JP, Kennedy DA, Lynch CM, Sievers EL, Forero-Torres A (2010) Brentuximab vedotin (SGN-35) for relapsed CD30-positive lymphomas. N Engl J Med 363(19):1812–1821. https://doi.org/10.1056/NEJMoa1002965

Yu YJ, Atwal JK, Zhang Y, Tong RK, Wildsmith KR, Tan C, Bien-Ly N, Hersom M, Maloney JA, Meilandt WJ, Bumbaca D, Gadkar K, Hoyte K, Luk W, Lu Y, Ernst JA, Scearce-Levie K, Couch JA, Dennis MS, Watts RJ (2014) Therapeutic bispecific antibodies cross the blood-brain barrier in nonhuman primates. Sci Transl Med 6(261):261ra154. https://doi.org/10.1126/scitranslmed.3009835

Zheng Y, Scheerens H, Davis JC Jr, Deng R, Fischer SK, Woods C et al (2011) Translational pharmacokinetics and pharmacodynamics of an FcRn-variant anti-CD4 monoclonal antibody from preclinical model to phase I study. Clin Pharmacol Ther 89(2):283–290. https://doi.org/10.1038/clpt.2010.311

Zhou H (2005) Clinical pharmacokinetics of etanercept: a fully humanized soluble recombinant tumor necrosis factor receptor fusion protein. J Clin Pharmacol 45(5):490–497

Zhou H, Mayer PR, Wajdula J, Fatenejad S (2004) Unaltered etanercept pharmacokinetics with concurrent methotrexate in patients with rheumatoid arthritis. J Clin Pharmacol 44(11):1235–1243. https://doi.org/10.1177/0091270004268049

Zhu Y, Hu C, Lu M, Liao S, Marini JC, Yohrling J et al (2009) Population pharmacokinetic modeling of ustekinumab, a human monoclonal antibody targeting IL-12/23p40, in patients with moderate to severe plaque psoriasis. J Clin Pharmacol 49(2):162–175. https://doi.org/10.1177/0091270008329556

Zhuang Y, de Vries DE, Xu Z, Marciniak SJ Jr, Chen D, Leon F et al (2015) Evaluation of disease-mediated therapeutic protein-drug interactions between an anti-interleukin-6 monoclonal antibody (sirukumab) and cytochrome P450 activities in a phase 1 study in patients with rheumatoid arthritis using a cocktail approach. J Clin Pharmacol 55(12):1386–1394. https://doi.org/10.1002/jcph.561

Zia-Amirhosseini P, Minthorn E, Benincosa LJ, Hart TK, Hottenstein CS, Tobia LA et al (1999) Pharmacokinetics and pharmacodynamics of SB-240563, a humanized monoclonal antibody directed to human interleukin-5, in monkeys. J Pharmacol Exp Ther 291(3):1060–1067

Genomics, Other "OMIC" Technologies, Precision Medicine, and Additional Biotechnology-Related Techniques

9

Robert D. Sindelar

Introduction

Today's regulatory approved medicines look very different than those of the last century. Biologics, especially monoclonal antibodies (mAbs) and recombinant replacement proteins, are becoming the preferred therapeutic entities. The industry pipeline has begun a profound shift in that direction. The products resulting from the techniques and processes of biotechnology continue to grow at an exponential rate, and the expectations are that an even greater percentage of drug development and clinically utilized pharmaceuticals worldwide will be classified as biologics. A recent Pharmaceutical Research and Manufacturers of America report (PhRMA 2021) notes that there are currently over 8000 medicines in clinical development globally and 74% in the pipeline have the potential to be first-in-class treatments. Most pertinent to this textbook, the majority of these medicines in development were impacted directly or indirectly by biotechnologies at one or more points during their lifetime via target identification, and/or lead identification, and/or lead optimization, and/or clinical development and evaluation and/or product production.

Pharmaceutical biotechnology techniques are at the core of most methodologies and approaches utilized today for drug discovery and development of both biologics and small molecule therapeutics. Biotechnology, and in particular modern genomics techniques, enables our understanding of disease etiology and progression, thus pinpointing new therapeutic targets. This facilitates drug lead discovery and lead optimization, followed by clinical development and, hopefully, regulatory approval and clinical use. While recombinant DNA technology and hybridoma techniques are the major methods utilized in pharmaceutical biotechnology through most of its earlier historical timeline, our ever-widening understanding of human cellular function and disease processes, now down to the individual cell, has resulted in a wealth of additional and innovative biotechnologies in order to harvest the information found in the human genome. These technological advances will help explore the association of genomic variation and drug response, enable personalized and precision medicine, enhance pharmaceutical research, and fuel the discovery and development of new and novel biopharmaceuticals. These revolutionary technologies and additional biotechnology-related techniques are improving the very competitive and costly process of drug development of new medicinal agents, diagnostics, and medical devices. Some of the technologies and techniques described in this chapter are both well established and commonly used applications of biotechnology, producing clinically utilized medicines as well as potential therapeutic products now in the developmental pipeline. New techniques are emerging at a rapid and unprecedented pace and their full impact on the future of personalized and precision medicine will turn dreams into realities.

Central to any meaningful discussion of pharmaceutical biotechnology and modern molecular health care are the "OMIC" technologies. The completion almost two decades ago of the Human Genome Project (HGP), one of the great feats of exploration in human history, has provided a wealth of new knowledge that continues to grow exponentially as more genes are sequenced at greater resolution and gene editing techniques have been developed with much greater precision. Researchers are turning increasingly to the task of converting the DNA sequence data into actionable information that will improve, and even revolutionize, drug discovery (see Fig. 9.1), patient-centered pharmaceutical care, and precision medicine. Pharmaceutical scientists are poised to take advantage of a broad range of OMICS technologies, the most relevant ones discussed in this chapter, to create a new and improved drug discovery, development, and clinical translation paradigm. These additional techniques in biotechnology and molecular biology are being rapidly exploited

R. D. Sindelar (✉)
Faculty of Pharmaceutical Sciences, The University of British Columbia (UBC), Vancouver, BC, Canada
e-mail: robert.sindelar@ub.ca

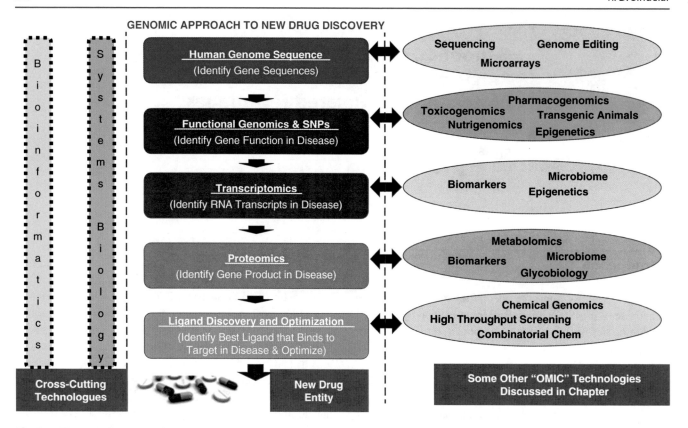

Fig. 9.1 The genomic strategy for new drug discovery

to bring new drugs to market and each topic will be introduced in this chapter.

It is not the intention of this author to detail each and every biotechnology technique exhaustively, since several techniques are mentioned throughout this textbook. Also, numerous printed and web specialized resources already meet that need. Rather, this chapter will illustrate and enumerate various biotechnologies that should be of key interest to pharmacy students, practicing pharmacists, and pharmaceutical scientists because of their effect on many aspects of drug discovery, drug development, and pharmacy practice.

An Introduction to "OMIC" Technologies

Since the discovery of DNA's overall structure in 1953, the world's scientific community has rapidly gained a detailed knowledge of the genetic information encoded by the DNA of a cell or organism so that today we are "personalizing" this information with greater precision. In the 1980s and 1990s, biotechnology techniques produced novel therapeutics and a wealth of information about the mechanisms of various diseases such as cancer at the genetic and molecular level, yet the etiologies of other complex diseases such as obesity, heart disease, and various mental illnesses remain poorly understood. However, today researchers utilizing exciting and ground-breaking "OMIC" technologies and

working closely with clinicians in a "bedside-to-benchtop back to the bedside" paradigm are making serious progress not only toward a molecular-level understanding of the etiology of complex diseases but to clearly identify that there are actually many genetically different diseases called by the single name of cancer, diabetes, depression, etc. Later in this chapter, we will explore the concepts of phenotype. Important here is that most human diseases are manifested through very complex phenotypes that result from genetic, environmental, and other factors. In a large part, the answers were hidden in what was unknown about the human genome. Despite the increasing knowledge of DNA structure and function in the 1990s, the genome, the entire collection of genes and all other functional and nonfunctional DNA sequences in the nucleus of an organism, had yet to be sequenced. DNA may well be the largest, naturally occurring molecule known. Successfully meeting the challenge of sequencing the entire human genome is one of history's great scientific achievements and has opened a path of discovery unprecedented in biology and medicine (Venter et al. 2001; The International Human Genome Sequencing Consortium 2001). While the genetic code for transcription and translation has been known for years, sequencing the human genome provided a blueprint for all human proteins and the sequences of all regulatory elements that govern the developmental interpretation of the genome. The potential significance includes identifying genetic determinants of common and rare diseases, providing a methodology for their diagnosis including clinically useful biomarkers, suggesting interesting new molecular sites for intervention (see Fig. 9.1), and the development of new biotechnologies to bring about their more detailed study and eventual eradication. Unlocking the secrets of the human genome and the technologies to edit the genome is leading to a paradigm shift in medicines research and clinical practice toward better disease understanding and taxonomy, and true personalized precision medicine focused at the molecular level.

Genomics

An organism's complete set of DNA is called its genome. The term genomics is the comprehensive analysis and understanding of DNA structure and function and broadly refers to the analysis of all genes within the genome of an organism. Sequencing the human genome and the genomes of many other organisms has led to an in-depth understanding of not only DNA structure and function but also a fundamental understanding of human biology and disease at the molecular level. While it is a complex and complicated journey from acquiring a DNA sample to sequencing the exact order of bases in a DNA strand, a multitude of technologies and approaches along with exponential enhancements in instrumentation and computation have been employed to sequence genomic DNA faster and less expensively. Many industry analysts predicted a tripling of pharmaceutical R&D productivity due to the sequencing of the human genome, but it is the next generation of whole-genome sequencing technology and the continually reduced cost of sequencing that is driving genomic technology into the clinic (Ha et al. 2018). While the causation of rare diseases was commonly pursued via genetic testing, clinical medicine has historically assessed risk of common diseases in their patients based upon family history. However, the price of whole-genome and whole-exome sequencing (a technique for sequencing all of the protein-coding genes in a genome) has

fallen to the level where these methods are now much more commonly used in clinical medicine for many diseases, both rare and common.

Likewise, the field of genomics is having a fundamental impact on modern drug discovery and development. While validation of viable drug targets identified by genomics has been challenging, great progress has occurred (Dugger et al. 2018). No matter whether it is a better understanding of disease or improved drug discovery, the genomic revolution has been the foundation for an explosion in "OMIC" technologies that find applications in research to address poorly treated and neglected diseases.

Structural Genomics and the Human Genome Project

Genetic analysis initially focused on the area of structural genomics, essentially, the characterization of the macromolecular structure of a genome utilizing computational tools and theoretical frameworks. Structural genomics intersects the techniques of DNA sequencing, cloning, PCR, protein expression, crystallography, and big data analysis. It focuses on the physical aspects of the genome through the construction and analysis of gene sequences and gene maps. To understand the significant new advances developing today in the field of biotechnology, it remains valuable to better understand both structural genomics and also the impact of the Human Genome Project. Proposed in the late 1980s, the publicly funded Human Genome Project (HGP) or Human Genome Initiative (HGI) was officially sanctioned in October 1990 to map the structure and to sequence human DNA (US DOE 2018). For historical context, as described in Table 9.1, HGP structural genomics was envisioned to proceed through increasing levels of genetic resolution: detailed human genetic linkage maps [approximately two megabase pairs (Mb = million base pairs) resolution], complete physical maps (0.1 Mb resolution), and ultimately complete DNA

Table 9.1 The increasing levels of genetic resolution obtained from structural genomic studies of the HGP

Human genome project goals	Base pair resolution
Detailed genetic linkage map	2 Mb
Comments: Poorest resolution depicts relative chromosomal locations of DNA markers, genes, or other markers and the spacing between them on each chromosome	
Complete physical map	0.1 Mb
Comments: Instead of relative distances between markers, maps actual physical distance in base pairs between markers; lower resolution = actual observance of chromosomal banding under microscope; higher resolution is "restriction map" generated in the presence of restriction enzymes	
Complete DNA sequence	1 bp
Comments: The ultimate goal is to determine the base sequence of the genes and markers found in mapping techniques along with the other segments of the entire genome; techniques commonly used include DNA amplification methods such as cloning, PCR, and other techniques described in Chap. 1 along with novel sequencing and bioinformatics techniques	

Mb megabase = one million base pairs, *bp* base pair

sequencing of the approximately three billion base pairs (23 pairs of chromosomes) in a human cell nucleus [one base pair (bp) resolution]. Our level of understanding would sequentially increase as the project progressed. Projected for completion in 2003, the goal of the project was to learn not only what was contained in the genetic code but also how to "mine" the genomic information to cure or help prevent the estimated 4000 genetic diseases afflicting humankind. The project would identify all the genes in the human genome, determine the base pair sequence and store the information in databases, create new tools and improve existing tools for data analysis, and address the ethical, legal, and societal issues (ELSI) that may arise from the project. Earlier than projected, a milestone in genomic science was reached on June 26, 2000, when researchers at the privately funded Celera Genomics and the publicly funded International Human Genome Sequencing Consortium (the international collaboration associated with the HGP) jointly announced that they had completed sequencing 97–99% of the human genome. The journal *Science* rated the mapping of the human genome as its "breakthrough of the year" in its December 22, 2000, issue. The two groups published their results in 2001 (Venter et al. 2001; The International Human Genome Sequencing Consortium 2001).

While both research groups employed the original cloning-based Sanger technique for DNA sequencing (now >35 years old), the genomic DNA sequencing approaches of the HGP and Celera Genomics differed. HGP chopped the human DNA sequence into segments of ever decreasing size. Each DNA segment was further divided or blasted into smaller fragments. Each small fragment was individually sequenced and the sequenced fragments assembled according to their known relative order. The Celera researchers broke the whole-genome into many small fragments at once. Each fragment was sequenced and assembled in order by identifying where they overlapped. Each of the two sequencing approaches required unprecedented computer resources (the field of bioinformatics is described later in this chapter).

Regardless of genome sequencing strategies, the collective results are impressive. More than 27 million high-quality sequence reads provided fivefold coverage of the entire human genome. Genomic studies identified over one million single-nucleotide polymorphisms (SNPs), binary elements of genetic variability (SNPs are described later in this chapter). While original estimates of the number of human genes in the genome varied consistently between 80,000 and 120,000, the genome researchers unveiled a number far short of biologists' predictions, 32,000 (Venter et al. 2001; The International Human Genome Sequencing Consortium 2001). Within months of completion, others suggested that the human genome possesses between 65,000 and 75,000 genes and then much less, approximately 20,000. However,

by 2018, sequencing data from hundreds of human tissue samples were analyzed suggesting that possibly as many as 5000 genes may not have been previously identified including nearly 1200 that carry the code for making proteins. Thus, the exact number of human genes remains an active field of study and likely is at 21,000+ (Willyard 2018). The HGP provided the foundation for rapid growth of sequencing methodologies, instrumentation, sequencing reagents, and bioinformatics approaches that have advanced genome sciences to today's state-of-the-art.

Next-Generation Genome Sequencing (NGS) Including Whole-Genome Sequencing (WGS) and Whole-Exome Sequencing (WES)

The full spectrum of human genetic variation ranges from large chromosomal changes down to the single base pair alterations. The challenge for genomic scientists is to discover the full extent of genomic structural variation, referred to as genotyping, so that the variations and genetic coding may be associated with the encoded trait or traits displayed by the organism (the phenotype). And they wish to do this using as little DNA material as possible, in as short time and for the least cost, all important characteristics of a useful point-of-care clinical technology. The discovery and genotyping of structural variation has been at the core of understanding disease associations as well as identifying possible new drug targets including by searching for disease-susceptibility genes (Sonehara and Okada 2021). In the nearly two decades since the completion of the HGP, sequencing efficiency has increased significantly and the cost of a single whole-genome sequence has decreased from nearly $1 million in 2007 to as low as $400 US depending on the size of the sequence and the reagents utilized. The move toward low-cost, high-throughput sequencing is essential for the successful implementation of genomics into precision medicine and is altering the clinical landscape.

Next-generation genome sequencing methodologies, which differ from the original cloning-based Sanger technique, are massively parallel high-throughput, imaging-based systems with vastly increased speeds and data output. There is no clear definition for *next-generation genome sequencing, also known generally as NGS*, but most are characterized by the direct and parallel sequencing of large numbers of amplified and fragmented DNA without vector-based cloning. The fragmented DNA tends to have sequence reads of 30–400 base pairs.

There is a maturation of single-cell sequencing techniques [single-cell DNA sequencing (scDNA-seq); single-cell RNA sequencing (scRNA-seq); etc.] as well as some single nucleic acid-molecule techniques utilizing commercially available DNA sequencers (Tang et al. 2019). These are techniques effectively creating a "genetic microscope" that can profile genetic, epigenetic, proteomic, and cell-lineage information in individual cells. Utilizing these newer genomic approaches will revolutionize the fields of oncology and immunology where each cell may be distinctly different (Smaglik 2017). The rapid growth of this field of genomic research has been propelled by the development of innovative technologies, novel computational analysis methods, and a growing number of diverse applications across many fields of biology and biomedicine. Research efforts to begin to genetically profile the millions of every single-cell type of the human body may be a formidable task more daunting than the HGP.

Largely used as a research tool until recently, whole-genome sequencing WGS (also known as full genome sequencing, complete genome sequencing, or entire genome sequencing) is a process of determining all of the three billion DNA nucleotides of an individual's DNA sequence, including noncoding sequences at a single time achieving the last base pair resolution goal of the HGP (as listed in Table 9.1) (Schwarze et al. 2018). This entails sequencing all of an individual's chromosomal DNA as well as the DNA contained in the mitochondria. The most comprehensive method for analyzing the genome, the ability to manage the large volume of data generated during analysis, and the rapidly dropping sequencing costs have helped translate this methodology into the clinic (Schwarze et al. 2018). A powerful tool whose development has been driven by research with human material and now initial clinical use, WGS, is equally useful for sequencing any species including microorganisms, livestock, and plants.

The most widely utilized targeted genome sequencing approach is exon sequencing which investigates only the protein-coding regions of the genome; it does not include the nonprotein-coding sequences of the genome. Humans have about 233,785 exons constituting <2% of the human genome, and current research suggests that most known Mendelian and common polygenic disease-related variants are in the exons. Thus, this sequencing approach is a cost-effective alternative to WGS that produces smaller and more manageable data sets for faster analysis. Together, all the exons in a genome are known as the exome and their entire sequencing at one time is called whole-exome sequencing (WES). Note that because DNA variations outside of the exons can affect gene activity and protein production leading to genetic disorders, WES is not always an effective sequencing approach. WGS and WES are valuable methods for researchers and are now being introduced in clinics. Progress is being made on high-resolution 3D genomics. Rapid genome sequencing development coupled with advanced microscopic imaging techniques has enabled us to better understand the 3D spatial organization of the genome (Mohanta et al. 2021). As researchers continue to manipulate genome structure, combining genomics and imaging will deepen our understanding

of the effects of regional proximity of specific gene sequences on biological processes. Recently, 12 leading researchers in the fields of genetics and genomics provided their reflections on the key challenges and opportunities faced by those fields which provides a good look at where we are after almost 20 years (McGuire et al. 2020). Included were the concepts of gene sequencing at the population level, multifactorial phenotyping, and making genomics truly equitable and global including the African genome.

Functional Genomics, Comparative Genomics, and Biobanks

Functional genomics is the subfield of genomics that attempts to answer questions about the function of specific DNA sequences at the levels of transcription and translation, i.e., genes, RNA transcripts, and protein products (Przybyla and Gilbert 2022). Research to relate genomic and intergenic regions of the genome sequence data determined by structural genomics with observed biological function is predicted to fuel new drug discoveries through a better understanding of how the individual components of a biological system work together to produce a particular phenotype. The DNA sequence information itself rarely provides definitive information about the function and regulation of that particular gene. After genome sequencing, a functional genomic approach is the next step in the knowledge chain to identify functional gene products that are potential biotech drug leads and new drug discovery targets (see Fig. 9.1).

To relate functional genomics to therapeutic clinical outcomes, the human genome sequence must reveal the thousands of genetic variations among individuals that will become associated with diseases or symptoms in the patient's lifetime. Sequencing alone is not the solution, simply the end of the beginning of the genomic medicine era. Determining gene functionality in any organism opens the door for linking a disease to specific genes or proteins, which become targets for new drugs, methods to detect organisms (i.e., new diagnostic agents), and/or biomarkers (the presence or change in gene expression profile that correlates with the risk, progression, or susceptibility of a disease). Success with functional genomics will facilitate the ability to observe a clinical problem, take it to the benchtop for structural and functional genomic analysis, and return personalized solutions to the bedside in the form of new therapeutic interventions and medicines.

The face of biology has changed forever with the sequencing of the genomes of numerous organisms. Biotechnologies applied to the sequencing of the human genome are also being utilized to sequence the genomes of comparatively simple organisms as well as other mammals. Often, the proteins encoded by the genomes of more simple organisms and the regulation of those genes closely resemble the proteins and gene regulation in humans. Now that the sequencing of the entire genome is a reality, the chore of sorting through human, pathogen, and other organism diversity factors and correlating them with genomic data to provide real pharmaceutical benefits is an active area of research. Comparative genomics is the field of genomics that studies the relationship of genome structure and function across different biological species or strains and, thus, provides information about the evolutionary processes that act upon a genome. Comparative genomics exploits both similarities and differences in the regulatory regions of genes, as well as RNA and proteins of different organisms to infer how selection has acted upon these elements.

Since model organisms are much easier to maintain in a laboratory setting, researchers have actively pursued "comparative" genomic studies between multiple organisms. Unlocking genomic data for each of these organisms provides valuable insight into the molecular basis of inherited human disease. As an example, *S. cerevisiae*, a yeast, is a good model for studying cancer and is a common organism used in rDNA methodology. It is well known that women who inherit a gene mutation of the *BRCA1* gene have a high risk, perhaps as high as 85%, of developing breast cancer before the age of 50 (Paul and Paul 2014). The first diagnostic product generated from genomic data was the *BRCA1* test for breast cancer predisposition. The gene product of *BRCA1* is a well-characterized protein implicated in both breast and ovarian cancer. Evidence had accumulated suggesting that the Rad9 protein of *S. cerevisiae* was distantly, but significantly, related to the *BRCA1* protein. Thus, *S. cerevisiae* in the lab was well studied during the development of *BRCA1* diagnostics. Similarly, studying *C. elegans*, an unsegmented vermiform, has provided much of our early knowledge of apoptosis, the normal biological process of programmed cell death. Greater than 90% of the proteins identified thus far from a common laboratory animal, the mouse, have structural similarities to known human proteins.

Similarly, mapping the whole of a human cancer cell genome is pinpointing the genes involved in cancer and aids in the understanding of cell changes and treatment of human malignancies utilizing the techniques of both functional and comparative genomics. Cancer as a disease is underlined by distinctive genetic changes. In cancer cells, small changes in the DNA sequence can cause the cell to make a protein that does not allow the cell to function as it should. The genome of a cancer cell can also be used to stratify cancer cells identifying one type of cancer from another or identifying a subtype of cancer within that type, such as HER2+ breast cancer. An unprecedented international collaboration resulted in the Pan-Cancer Analysis of Whole Genomes (PCAWG) of the International Cancer Genome Consortium (ICGC) and The Cancer Genome Atlas (TCGA). Researchers sequenced the tumors of over 2500 patients across 38 different cancer types, as well as the corresponding healthy tissue, with the aim of identifying genome-wide mutations exclusively found in cancer that drive tumor formation (Giunta 2021). Understanding the cancer genome is a quantum step toward improved cancer drug therapy and personalized oncology.

A valuable resource for performing functional and comparative genomics is the "biobank," a collection of biological samples for reference purposes. Repositories of this type also might be referred to as biorepositories or named after the type of tissue depending on the exact type of specimens (i.e., tissue banks or heart tissue banks, etc.). Genomic techniques are fostering the creation of DNA and RNA banks, the collection, storage, and analysis of hundreds of thousands of specimens containing analyzable DNA and/or RNA. All nucleated cells, including cells from blood, hair follicles, buccal swabs, cancer biopsies, and urine specimens, are suitable nucleic acid samples for analysis in the present or at a later date. Note that traditional biopsy is an invasive procedure to obtain a sample of tissue (i.e., cancer cells, heart tissue, liver tissue, etc.) and is not always feasible. New techniques have been developed that allow a less invasive test on a sample of blood called "liquid biopsy." For example, a liquid biopsy can be conducted on a cancer patient to look for cancer cells from a tumor that are circulating in the blood, but also for pieces of DNA or RNA from tumor cells that are in the blood sample (Kwapisz 2017). DNA banks are proving to be valuable tools for genetics research. Many large-scale biobanking efforts are underway at the institutional, national, and international level. When biobanks and their resulting multiple types of data are linked with subject data from medical records, health histories, and patient questionnaires, they serve as invaluable resources in translational medical research. While DNA and RNA banks devoted to cancer research have grown the fastest, there also has been an almost explosive growth in biobanks specializing in research on autism, schizophrenia, Alzheimers, heart disease, diabetes, and many other diseases.

"OMICS"-Enabling Technology: Bioinformatics in the Big Data Era

Recent technological advances in structural genomics, functional genomics, transcriptomics, proteomics, pharmacogenomics, metabolomics, and other "OMIC" techniques have generated an enormous volume of genetic and biochemical data to store and analyze. The continually lowering cost of data generation and analysis is leading to the "big data" era. The term "big data" addresses the challenges of data capture, data storage, data analysis, data search, data mining, data visualization, and data sharing of data sets that are so voluminous and complex that traditional data processing and software are incapable of doing the job. In this case, data mining refers to the bioinformatics approach of "sifting" through volumes of raw data, identifying and extracting relevant information, and developing useful relationships among them. The explosive growth of biological data over the past decade coupled with new big data analysis tools is allowing scientists to perform a wide range of experiments much more rapidly and inexpensively than historically possible. Big data development has focused on the creation and enhancement of predictive analytic methods that extract value from large data sets. New observations are generated that spot trends and correlations that would otherwise be lost in the volume of data being analyzed.

Living in an era of faster computers, bigger and better data storage, and improved methods of data analysis fostered the information superhighway that facilitated the HGP and the "OMIC" revolution. Scientists applied advances in information technology, innovative software algorithms, and massive parallel computing to the ongoing research in biotechnology areas such as genomics to give birth to the fast-growing field of bioinformatics. Bioinformatics, the interdisciplinary field to analyze and interpret biological information with the object of discovering new knowledge and interconnections, is essential to accelerating the rate of biotechnology-related discovery. Traditionally, bioinformatics originally focused on the sequencing of proteins in the early 1950s and then DNA in the 1970s. Now, technological advances in high-throughput profiling of biological systems including next-generation sequencing techniques coupled with the unprecedented ability to data mine and analyze in the big data environment provide scientists with the essential tools required to accelerate the rate of medical discovery that will improve health, well-being, and patient care. With bioinformatics, a researcher can now better exploit the tremendous flood of genomic, transcriptomic, and proteomic data, and more cost-effectively data mine for a drug discovery "needle" in that massive data "haystack." We can now examine the relationship of DNA sequence to structure and function cross-cut with measurements of mRNA expression, transcription factor binding, protein synthesis, metabolite concentrations, and phenotype (to be explained later in this chapter).

Modern drug discovery and the commensurate need to better understand and define disease are utilizing bioinformatics techniques to gather information from multiple sources [such as functional genomic studies, proteomics, phenotyping, patient electronic health records (EHRs; sometimes referred to as electronic medical records, EMRs), and bioassay results including toxicology studies], integrate the data, apply life science developed algorithms, and generate useful target identification and drug lead identification data (Premsrirut 2017). The term "Reverse Informatics" has been used to describe the drug discovery informatics tool that regenerates data in a structured format from primary literature so that it can be easily synchronized with other data sets

and mined to identify new drug targets. As seen in Fig. 9.2, the hierarchy of information collection goes well beyond the biodata contained in the genetic code that is transcribed and translated. Numerous challenges remain including both scientific (technical advances needed to correlate genetic and environmental findings with incidence of disease) and legal and ethical challenges (privacy issues, EHRs, etc.).

The entire encoded human DNA sequence alone requires computer storage of approximately 10^9 bits of information: the equivalent of a thousand 500-page books! GenBank, managed by the National Center for Biotechnology Information, NCBI (NCBI 2022), and the sequence databases of the European Molecular Biology Organization—European Bioinformatics Institute (EMBL-EBI 2022) are two of the many centers worldwide that collaborate on collecting nucleic acid sequences. These databanks (both public and private) store tens of millions of sequences. Once stored, analyzing the volumes of data (i.e., comparing and relating information from various sources) to identify useful and/or predictive characteristics or trends, such as selecting a group of drug targets from all proteins in the human body, presents a Herculean task. This approach has the potential of changing the fundamental way in which basic science is conducted and valid biological conclusions are reached (Sofi et al. 2022).

With OMICS, health care has experienced an explosion in biomedical knowledge fostering dramatic innovations in therapy and expanded capacities to treat challenging medical conditions. Storage and analysis of this wealth of data has

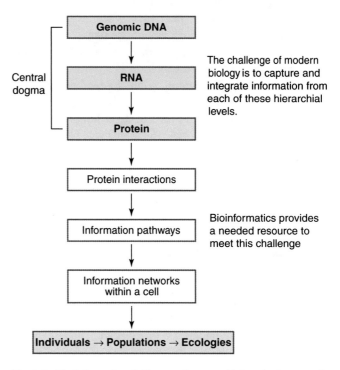

Fig. 9.2 The information challenges of systems biology in the genomic era

been transformed by "cloud computing." The health care system has lagged in adjusting to new discoveries, disseminating data in real time and learning from each and every health care intervention. The concept of a "continuous learning health care system" has been proposed as a paradigm to apply the resources and tools at hand in new science, bioinformatics and big data, and the transformation of the care culture to produce high-quality health care informed by constantly collected and analyzed information and data (IOM 2017). The concept requires the creation of a framework within a health care organization that creates a learning patient-centered continuum of wellness and care ecosystem. Bioinformatics, evidence and great science, knowledge translation, and organizational culture and incentives are defined as the drivers to build this new ecosystem and foster true knowledge-driven innovation. Patients and their families will be active participants in all elements of the process and new knowledge will be captured from each and every intervention and saved in databases as an integral by-product of every care experience. Across the entire health care organization, evidence development and its application should be positioned as natural extensions of exceptional science-driven clinical care to foster health care that learns.

Likewise, the exciting areas of Artificial Intelligence (AI) as applied to health care and delivery, Digital Medicine, the intersection of wearable sensor technologies, mobile computing, the Internet, and the incentives for healthy living, while at an early stage of development related to health care, are rapidly developing (Briganti and Le Moine 2020). Massive data sets connected by Internet of Things (IoT) connected devices (such as smart phones and iPads), cloud-based algorithms, and quantum computing could enable real-time diagnosis and insights that are integrated into our daily lives and shared across multiple health care providers. The continued rapid growth in computer-processing power over the past two decades, the availability of large data sets, and the development of advanced algorithms have driven major improvements in machine learning that have accelerated the examination of AI approaches in health care. By the time the seventh edition of this textbook is published, AI and Digital Medicine likely will have a great impact on pharmaceutical biotechnology and the health care system as a whole. Researchers continue to strive toward "human-level" AI. That is building algorithms that mimic complex actions humans already perform (i.e., interpreting clinical radiography, etc.) but with greater precision and greater insight (sometimes referred to as "deeper learning").

The profession of pharmacy and the pharmaceutical sciences must recognize that optimal patient-centered care will require an effective integration of data and actionable information into new types of patient data systems. The young field of "pharmacy informatics" integrates many of these aspects (Cortess et al. 2019). Pharmacy informatics integrates and uses knowledge, information, technology, and automation in the precision medication optimization process. Patient information including data from genomics, proteomics, pharmacogenomics, metabolomics, and other OMIC technologies can now be integrated with currently collected patient information. The current information includes individual patient characteristics, patient safety, evidence-based medicine, drug information databases, Internet resources, hospital information systems, pharmacy information systems, and drug discovery literature. While it is well beyond the scope of this chapter to explore continuous learning health care systems, AI, Digital Medicine, and Pharmacy Informatics further, the reader can be assured that these fields are becoming important areas for pharmacists and pharmaceutical scientists to be well informed and knowledgeable.

Transcriptomics

Remember that the central dogma of molecular biology is DNA to RNA via the process of transcription and RNA to protein via the process of translation (Fig. 9.2). The transcriptome is the collection of all RNA transcribed elements for a given genome, not only the collection of transcripts that are subsequently translated into proteins (mRNAs). Noncoding transcripts such as noncoding microRNAs (miRNAs) are part of the transcriptome and are described in detail in our chapter on nucleic acid therapies. The term transcriptomics refers to the OMIC technology that examines the complexity of RNA transcripts of an organism under a variety of internal and external conditions reflecting the genes that are being actively expressed at any given time (with the exception of mRNA degradation phenomena such as transcriptional attenuation). Therefore, the transcriptome can vary within individual cells with external environmental conditions, while the genome is roughly fixed for a given cell line (excluding mutations). The variances between distinct cells can have profound functional effects. Single-cell RNA sequencing (scRNA-seq) is a very promising approach to study the transcriptomes of individual cells enabling an unbiased and novel understanding of the cellular composition within specific tissues. This approach has been used extensively in the brain and the central nervous system, thus acting as a bridge between neuroscience, computational biology, and systems biology (Kulkarni et al. 2019). Kidney cells have also received significant scRNA-seq study.

Whole transcriptome analysis can help identify the relationship of sequence to function by exploring genetic networks underlying cellular, physiological, biochemical, and biological systems. This approach is important to modern biomarker discovery. The transcriptomes of stem cells and cancer cells are of particular interest to better understand the processes of cellular differentiation and carcinogenesis. Many OMICS methods are currently being optimized with respect to analysis speed and ease, resolution and accuracy, and the number of genes that can be profiled in a transcriptomics approach. High-throughput techniques based on microarray technology are used to examine the expression level of mRNAs in a given cell population. "Spatial transcriptomics" is the integration of gene transcription data into a spatial coordinate system mapped upon various tissue atlases. This type of approach is critical for the in-depth understanding of individual cell identity and function within the tissue context. The international Human Cell Atlas (HCA 2022) and the US NIH-hosted Human BioMolecular Atlas Program (HuBMAP 2022), both creating comprehensive reference maps of all human cells, include technological components that incorporate spatial transcriptomic mapping as explicit goals.

Proteomics, Structural Proteomics, and Functional Proteomics

Proteins represent the main functional machinery of all cells. Proteomics is the study of an organism's complete complement of proteins. Proteomics seeks to define the function and correlate that with expression profiles of all proteins encoded within an organism's genome or "proteome" (Dupree et al. 2020). Defining protein composition in both the healthy state and in the disease state is a key step in understanding the function of biological systems. The application of functional proteomics in the process of drug discovery has created a field of research referred to as pharmacoproteomics that tries to compare whole protein profiles of healthy persons versus patients with disease. This analysis may point to new and novel targets for drug discovery and precision medicine. While functional genomic research is providing an unprecedented information resource for the study of biochemical pathways at the molecular level, a vast array of the human genes identified in sequencing the human genome are being analyzed to determine if they are functionally important in various disease states (i.e., the druggable genome). These key proteins when identified serve as potential new sites for therapeutic intervention (see Fig. 9.1) (Zhang et al. 2014). The transcription and translation of 21,000+ human genes can produce hundreds of thousands of proteins due to posttranscriptional regulation and posttranslational modification of the protein products (Cf. Chaps. 1 and 2). The variety of posttranslational modifications (PTMs) observed in proteins is responsible for the vast number of isoproteins or proteoforms, which has served as an obstacle in most proteomic experiments. The number, type, and concentration may vary depending on cell or tissue type, disease state, and other factors. The protein's function(s) is dependent on the primary, secondary, and tertiary structure of the protein and the molecules they interact with. Less than 50 years old, the concept of proteomics requires determination of the structural, biochemical, and physiological repertoire of all proteins.

Proteomics is a greater scientific challenge than genomics due to the intricacy of protein expression and the complexity of 3D protein structure (structural proteomics) as it relates to biological activity (functional proteomics). Protein expression, isolation, purification, identification, and characterization are among the key procedures utilized in proteomic research. To perform these procedures, technology platforms such as 2D gel electrophoresis, mass spectrometry, chip-based microarrays (discussed later in this chapter), X-ray crystallography, protein nuclear magnetic resonance (NMR), and phage displays are employed. Mass spectrometry research has developed advanced sample preparation and separation approaches that will facilitate deeper quantitative analysis of the proteome. This has led to the analysis of single-cell proteomes (sc-proteomics). The Human Proteome Project (HPP 2022) is an international project organized by the Human Proteome Organization (HuPO) aims to revolutionize our understanding of the human proteome via a multinational harmonized research effort to map the entire human proteome utilizing currently available and emerging techniques. The project continues today and has shown that protein expression has now been credibly detected for 18,357 (>90%) of the predicted proteins coded in the human genome (Omenn et al. 2021). Successful completion of this multi-year project will enrich and focus our understanding of human biology at the cellular level and lay a foundation for development of new and novel diagnostic, therapeutic, and preventive medical applications.

Pharmaceutical scientists are finding that many of the proteins identified by proteomic research are entirely novel, possessing unknown or little-known functions. This scenario offers not only a unique opportunity to identify previously unknown molecular targets for drug design, but also to develop new biomarkers and ultrasensitive diagnostics to address unmet clinical needs. Often, multiple genes and their protein products are involved in a single disease process. Since few proteins act alone, studying protein interactions will be paramount to a full understanding of functionality (see systems biology later in this chapter). Also, many abnormalities in cell function may result from overexpression of a gene and/or protein, underexpression of a gene and/or protein, a gene mutation causing a malformed protein, and posttranslational modification changes that alter a protein's function. Therefore, the real value of human genome sequence data will only be realized after every protein coded by the approximately 21,000+ genes has a function assigned to it with the completion of the HPP.

"OMICS"-Enabling Technology: Microarrays

The biochips known as microarrays are multiplex lab-on-a-chip assays used in high-throughput screening of biological materials (Schumacher et al. 2015). They have become basic tools of molecular biology. The initial microarrays were

DNA or RNA microarrays that are a collection of hundreds to thousands of immobilized nucleic acid sequences or oligonucleotides in a grid created with specialized equipment that can be simultaneously examined to conduct expression analysis. These gene chips would contain representatives of a particular set of gene sequences, for instance, sequences coding for all human cytochrome P450 isozymes or may contain sequences representing all genes of an organism. Analysis by microarray technology offers a variety of methods to identify genes which might be significant in a specific cellular response mechanism or a particular gene expression pattern that characterizes a particular disease.

Commonly, arrays are prepared on nonporous supports such as glass microscope slides or silicon thin-film cells. DNA microarrays generally contain high-density microspot-

ted cDNA sequences approximately 1 kb in length representing thousands of genes. The field was advanced significantly when technology was developed to synthesize closely spaced oligonucleotides on glass wafers using semiconductory industry photolithographic masking techniques (see Fig. 9.3). Oligonucleotide microarrays contain closely spaced synthetic gene-specific oligonucleotides representing thousands of gene sequences. Microarrays can provide expression analysis for mRNAs. Screening of DNA variation is also possible. Thus, biochips can provide polymorphism detection and genotyping as well as hybridization-based expression monitoring. Microarray analysis has gained increasing significance as a direct result of the genome sequencing studies. Array technology is a logical tool for studying functional genomics since the results obtained may

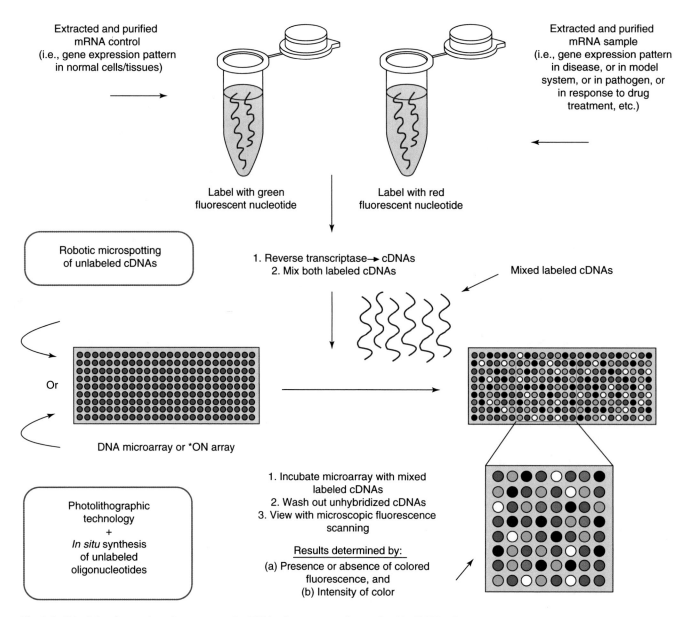

Fig. 9.3 Principle of operation of a representative DNA microarray or oligonucleotide (*ON) microarray

link function to expression. Microarray technology's potential to study key areas of molecular medicine and drug discovery is unlimited at this stage of development. For example, gene expression levels of thousands of mRNA species may be studied simultaneously in normal versus cancer cells, each incubated with potential anticancer drug candidates.

Researchers have developed many types of microarrays beyond the initial DNA or oligonucleotide arrays such as protein, peptide, tissue, cellular, chemical compound, antibody, carbohydrate, phenotype, and microarrays of lysates or serum (Schumacher et al. 2015). The principles are the same, while the immobilized collections differ accordingly.

"OMICS"-Enabled Technology: Brief Introduction to Biomarkers

Biomarkers or biological markers are clinically relevant biological measures used as indicators of normal biological processes, a disease state, predisposition to a disease, disease progression, or pharmaceutical response to a therapeutic intervention (Aronson and Ferner 2017). The significance of biotechnology produced biomarkers has been recognized by the research community, pharma industry, and now the clinic. Detection of or concentration change of a biomarker may indicate a particular disease state (e.g., the presence of a certain antibody may indicate an infection), physiology, or toxicity. A change in expression or state of a protein or other OMIC-related biomarkers may correlate with the risk of or progression of a disease, with the susceptibility of the disease to a given treatment or the drug's safety profile. Biomarkers may be used alone or in combination. Implemented in the form of a medical device, a measured biomarker becomes an in vitro diagnostic tool. While it is well beyond this chapter to provide a detailed discussion of biomarkers, it is important to note that OMIC technologies including OMIC-enabled technologies such as microarrays are being developed as clinical measuring devices for biomarkers. Biomarkers enable focused characterization of patient populations undergoing clinical trials or drug therapy and may accelerate drug development. Modern drug discovery often simultaneously involves biomarker discovery and diagnostic development (Aronson and Ferner 2017). Drug development scientists are hopeful that the development of appropriate biomarkers will facilitate "go" and "no-go" decisions during a preclinical and clinical development processes. Biomarker discovery is closely tied to the other applications of genomics previously described in this chapter. As an indicator of normal biological processes, pathogenic processes, or pharmacological responses to therapeutic intervention, biomarkers may serve as a substitute for a clinical end point and thus be a surrogate end point (Zhao and Brasier 2016). Biomarkers are now available for a wide range of diseases and conditions. A majority are disease-related such as Alzheimer's and Parkinson's disease, cardiac injury, lung injury, acute kidney injury, various cancers, and a host of other diseases and pathological conditions. Others are prognostic or predictive in nature.

A "theranostic" is a rapid diagnostic, possibly a microarray, measuring a clinically significant biomarker, which may identify patients most likely to benefit or be harmed by a new medication. Bundled with a new drug (and likely developed in parallel with that drug), the theranostic's diagnosis of the requisite biomarker (e.g., the overexpression of the HER2 gene product in certain breast cancer patients) influences the physician's therapeutic decisions [i.e., prescribing the drug trastuzumab (Herceptin) for HER2 receptor-positive breast cancer patients]. Thus, the diagnostic and the therapy are distinctly coupled = theranostic. The theranostic predicts clinical success of the drug.

Early-stage cancer detection (premetastasic phase) improves patient response to medical interventions for most cancer types. Unfortunately, many biomarkers shed from early lesions are limited by short circulation times and blood dilution that limit their early detection. To overcome this challenge, "synthetic biomarkers" are being developed. These represent an emerging class of diagnostics that deploy bioengineered sensors inside the body to query early-stage tumors and amplify disease signals to levels that could potentially exceed those of shed biomarkers. This innovative strategy leverages design principles and advances from chemistry, synthetic biology, and cell engineering (Kwong et al. 2021).

Metabolomics and Metabonomics

The human metabolome consists of the large-scale study of the complete set of small molecules that are involved in the energy transmission in the cells by interacting with other biological molecules following metabolic pathways (Lindon and Nicholson 2014). These metabolites may be metabolic intermediates, hormones and other signaling molecules, and secondary metabolites. The techniques and processes for identifying clinically significant biomarkers of human disease and drug safety have fostered the systematic study of the unique chemical fingerprints that specific cellular processes leave behind, specifically their small molecule metabolite profiles. Thus, while genomics and proteomics do not tell the whole story of what might be happening within a cell, metabolic profiling can give an instantaneous snapshot of the physiology of that cell.

Metabolomics is a powerful tool kit for the analysis of phenotype by (1) providing diagnostic patterns via fingerprinting a complex mixture of metabolites in the metabolome, (2) measuring the absolute concentration of targeted metabolites, (3) determining the relative abundance of selected portions of the metabolome, and (4) tracing the biochemical fate of individual metabolites. High-performance liquid chromatography coupled with sophisticated nuclear magnetic resonance (NMR) and mass spectrometry (MS) techniques is used to separate and quantify complex metabolite mixtures found in biological fluids to get a picture of the metabolic continuum of an organism influenced by an internal and external environment (Bingol et al. 2016). The field of "metabonomics" is the holistic study of the metabolic continuum at the equivalent level to the study of genomics and proteomics. However, unlike genomics and proteomics, microarray technology is little used since the molecules assayed in metabonomics are small molecule end products of gene expression and resulting protein function. The term metabolomics has arisen as the metabolic composition of a cell at a specified time, whereas the term metabonomics includes both the static metabolite composition and concentrations and the full-time course fluctuations. Coupling the information being collected in biobanks, large collections of patient's biological samples and medical records, with metabonomic and metabolomic studies will not only detect why a given metabolite level is increasing or decreasing but may reliably predict the onset of disease. Also, the techniques are used in drug safety screening, identification of clinical biomarkers, and systems biology studies (see below).

In oncology, metabolomic studies have been able to interrogate cancer cells for the possible optimal window for therapeutic intervention based upon metabolite concentrations. Scientists have created a number of metabolomics databases that may offer guidance on how metabolomics can be used to study cancer. Recent research studies combine CRISPR techniques (to be discussed later in this chapter) with metabolomic profiling to rapidly elucidate drug mechanisms of action (Yang 2022). The largest such database, the Human Metabolome Database (HMDB; https://hmdb.ca/), contains over 217,920 metabolite entries including both water soluble and lipid soluble metabolites that are considered either abundant in the human metabolome (>1uM) or relatively rare (<1uM) plus greater than 5000 protein sequences linked to the metabolites (Wishart et al. 2022). Each entry contains 130 data fields devoted to chemical, clinical, enzymatic, or biochemical data on the metabolites and many fields are linked to related databases (such as GenBank, DrugBank, PubChem, etc.).

Glycomics or Glycobiology

The scientific field of glycomics, or glycobiology, may be defined most simply as the study of the "glycome" of a biological system. That is, the structure, synthesis, and biological function of all glycans (may be referred to as oligosaccharides or polysaccharides, depending on size) and glycoconjugates in simple and complex systems (Cao et al. 2020). The application of glycomics or glycobiology is sometimes called glycotechnology to distinguish it from biotechnology (referring to glycans rather than proteins and nucleic acids). However, many in the biotech arena consider glycobiology one of the research fields encompassed by the term biotechnology. In the postgenomic era, the intricacies

of protein glycosylation, the mechanisms of genetic control, and the internal and external factors influencing the extent and patterns of glycosylation are important to understanding protein function and proteomics. Like proteins and nucleic acids, glycans are biopolymers. While once referred to as the last frontier of pharmaceutical discovery, recent advances in the biotechnology of discovering, cloning, and harnessing sugar cleaving and synthesizing enzymes have enabled glycobiologists to analyze and manipulate complex carbohydrates more easily (Cao et al. 2020).

Many of the proteins produced by mammalian cells contain attached sugar moieties, making them glycoproteins. The majority of protein-based medicinal agents contain some form of posttranslational modification that can profoundly affect the biological activity of that protein. Patterns of glycosylation significantly affect the biological activity of proteins (Costa et al. 2014). Many of the therapeutically used recombinant DNA-produced proteins are glycosylated including erythropoietin, glucocerebrosidase, and tissue plasminogen activator. Without the appropriate carbohydrates attached, none of these proteins will function therapeutically as the parent glycoprotein does. Glycoforms (variations of the glycosylation pattern of a glycoprotein) of the same protein may differ in physicochemical and biochemical properties. For example, erythropoietin has one O-linked and three N-linked glycosylation sites. The removal of the terminal sugars at each site destroys in vivo activity and removing all sugars results in a more rapid clearance of the molecule and a shorter circulatory half-life (Jiang et al. 2014). Yet, the opposite effect is observed for the deglycosylation of the hematopoietic cytokine granulocyte-macrophage colony-stimulating factor (GM-CSF) (Höglund 1998). In that case, removing the carbohydrate residues increases the specific activity sixfold. The sugars of glycoproteins are known to play a role in the recognition and binding of biomolecules to other molecules in disease states such as asthma, rheumatoid arthritis, cancer, HIV infection, the flu, and other infectious diseases. A useful web resource is the GlyCosmos Portal, a source to access glycoscience data that is integrated with life science data to support glycomics and biopharmaceutical discovery and development (https://glycosmos.org/).

Bacterial hosts for recombinant DNA could produce the animal proteins with identical or nearly identical amino acid sequences. However, early work in bacteria lacked the ability to attach sugar moieties to proteins (a process called glycosylation). Newer biotechnology methodologies helped overcome this issue (cf. Chaps. 1, 2 and 4). Advances in gene editing technologies have provided for the systematic genetic engineering of glycosylation capacities in mammalian cells. This provides new opportunities for studying the glycome and exploiting glycans in biomedicine. Many of the nonglycosylated proteins differ in their biological activity as compared to the native glycoprotein. The production of animal proteins that lacked glycosylation provided an unexpected opportunity to study the functional role of sugar molecules on glycoproteins. Glycoengineering uses chemical, enzymatic, and genetic approaches for the precise exploration and dissection of mammalian species glycomes and their many biological functions (Narimatsu et al. 2021). Genetic engineering of glycosylation in cells also brings studies of the glycome to the single-cell level and may result in future engineered glycoprotein therapeutics. Glycoengineering has become an important component in building designed antibodies and some vaccines.

The complexity of the field can best be illustrated by reviewing the building blocks of glycans, the simple carbohydrates called saccharides or sugars and their derivatives (i.e., amino sugars). Simple carbohydrates can be attached to other types of biological molecules to form glycoconjugates including glycoproteins (predominantly protein), glycolipids, and proteoglycans (about 95% polysaccharide and 5% protein). While carbohydrate chemistry and biology have been active areas of research for centuries, advances in biotechnology have provided techniques and added energy to the study of glycans. Oligosaccharides found conjugated to proteins (glycoproteins) and lipids (glycolipids) display a tremendous structural diversity. The linkages of the monomeric units in proteins and in nucleic acids are generally consistent in all such molecules. Glycans, however, exhibit far greater variability in the linkage between monomeric units than that found in the other biopolymers. As an example, Fig. 9.4 illustrates the common linkage sites to create polymers of glucose. Glucose can be linked at four positions: C-2, C-3, C-4, and C-6 and also can take one of two possible anomeric configurations at C-2 (α and β). It is estimated that for a 10-mer (oligomer of length 10), the number of structurally distinct linear oligomers for each of the biopolymers is for DNA (with four possible bases), 1.04×10^6; for protein (with 20 possible amino acids), 1.28×10^{13}; and for oligosaccharide (with eight monosaccharide types), 1.34×10^{18}.

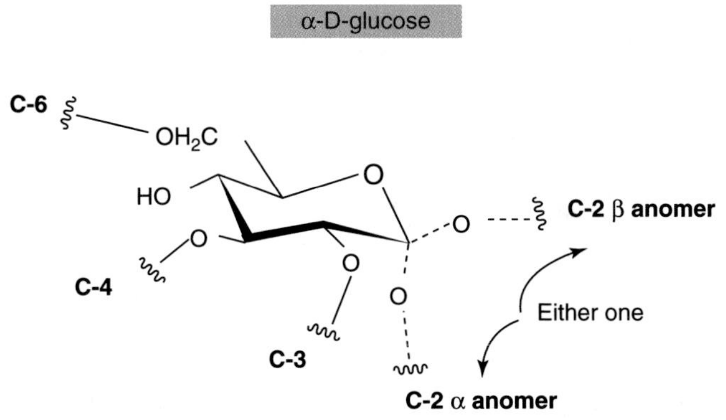

Fig. 9.4 Illustration of the common linkage sites to create biopolymers of glucose. Linkages at four positions: C-2, C-3, C-4, and C-6 and also can take one of two possible anomeric configurations at C-2 (α and β)

Lipidomics

Lipids, the fundamental components of membranes, play multifaceted roles in cell, tissue, and organ physiology. Lipid profiles have been examined extensively in plasma, serum, erythrocytes, blood platelets, fecal matter, urine, as well as biologic tissues including liver, lung, and kidney (Avela and Sirén 2020). The research area of lipidomics is a lipid-targeted metabolomics approach focused on the comprehensive large-scale study of pathways and networks of cellular lipids in biological systems. The metabolome would include the major classes of biological molecules: proteins (and amino acids), nucleic acids, and carbohydrates. The "lipidome" would be a subset of the metabolome that describes the complete lipid profile within a cell, tissue, or whole organism. In lipidomic research, a vast amount of information (structures, functions, interactions, and dynamics) quantitatively describing alterations in the content and composition of different lipid molecular species is accrued after perturbation of a cell, tissue, or organism through changes in its physiological or pathological state. Progress in modern lipidomics and lipid profiling has been greatly accelerated by the development of sensitive analytical techniques such as electrospray ionization (ESI) and matrix-assisted laser desorption/ionization (MALDI). Currently, the isolation and subsequent analysis of lipid mixtures is hampered by extraction and analytical limitations due to characteristics of lipid chemistry.

Abnormal lipid metabolism is implicated in a number of human lifestyle-related diseases. The study of lipidomics is important to a better understanding of many metabolic diseases, as lipids are believed to play a role in obesity, atherosclerosis, stroke, hypertension, diabetes, respiratory disease, and cancer (Zhao et al. 2015). Lipidomics analysis can provide detailed insights into the current state of a patient's metabolism in a way that genomic techniques can not. Lipidomic research may help identify potential biomarkers for establishing preventative or therapeutic interventions against human disease.

Nutrigenetics and Nutrigenomics

The well developed tools and techniques of genomics and bioinformatics have been applied to the investigation of the intricate interaction of mammalian diet and genetic makeup (Keathley et al. 2021). It is critically important to study the effects of genetic variation on dietary responses and the role of nutrients and bioactive food components in gene expression. Nutrigenetics involves the influence of genetic variation on nutrition, whereas nutrigenomics or nutritional genomics has been defined as the influence of nutrition on genome stability, epigenome modifications, transcriptomic alterations, and proteomic changes. This appears to result from gene expression and/or gene variation (e.g., SNP analysis) on a nutrient's absorption, distribution, metabolism, elimination, or biological effects. This includes how nutrients impact on the production and action of specific gene products and how the expressed proteins in turn affect the response to nutrients. Nutrigenomic studies aim to develop predictive means to optimize nutrition, with respect to an individual's genotype. Areas of study include dietary supplements, common foods and beverages, mother's milk, as well as diseases such as cardiovascular disease, obesity, and diabetes. Researchers have noted that this field has suffered from a general lack of consistency to assess the scientific validity across nutrigenetic and nutrigenomic research. With renewed, high-quality research efforts, nutrigenomics could become an important driver to improve an individual's health.

Microbiome

The human microbiome (or sometimes called the human microbiota) is the collection of bacteria, viruses, and fungi and their genomes colonizing the human body. These microorganisms reside primarily in the gut and the rest of the GI tract, but importantly also live on the skin and in the mouth and saliva, in the nostrils, in the eyes, in the genital areas on the body, and in our hair. Led by the tremendous advances in OMICS technology, and a reduction in the costs to perform the analyses, there has recently been tremendous growth and understanding in the collective knowledge of the human microbiome (NIH Human Microbiome Portfolio Analysis Team 2019). As humans, we share our body space with symbiotic, pathogenic, and commensal microorganisms as an ecological community. Sometimes called the "Second Genome" and the Forgotten Organ," the human microbiome contributes more protein-coding genes than the human host. The human microbiome is estimated to contain ten times the number of cells and > 140 times the number of genes that our human bodies contain (please see Fig. 9.5). The set of highly interactive microbial species that constitute the human microbiome is shaped by the environment in which it exists, which includes hosts, and exogenous natural and human factors. Each microbiome host can acquire

Fig. 9.5 The human microbiome emphasizing the gut microbiome. Millions of microbes/cm² on your body. There are more microbes in your intestines than there are human cells in your body

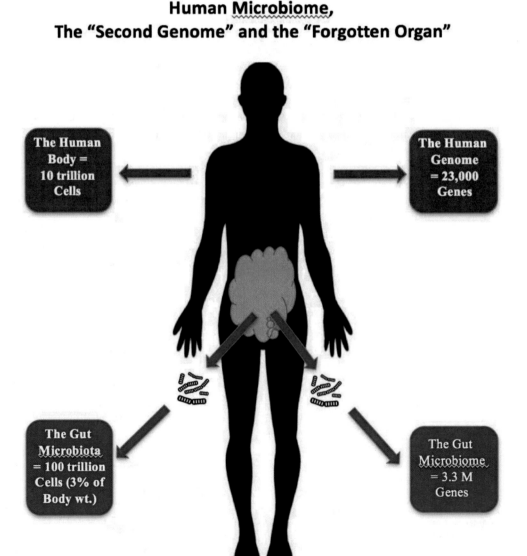

Human Microbiome, The "Second Genome" and the "Forgotten Organ"

The Human Body = 10 trillion Cells

The Human Genome = 23,000 Genes

The Gut Microbiota = 100 trillion Cells (3% of Body wt.)

The Gut Microbiome = 3.3 M Genes

microbes in two ways: (1) inheriting them from the parents; or (2) acquiring them from the environment, including food and interaction with other individuals.

There is little doubt that the microbiome plays a critical role in human health and disease and may influence nearly all aspects of human biology through (a) host–microbe interaction(s) (Chen et al. 2021). Analysis of the functional interactions between the human host and its microbiome under a variety of conditions will provide a better mechanistic understanding of the role of the microbiome in health and disease states and is expected to lead to new diagnostic, prognostic, and treatment interventions. There is strong evidence that diet can influence microbiome composition and function. Diet is known to influence the progression and severity of symptoms of pathogenic states such as inflammatory bowel disease, Crohn's disease, ulcerative colitis, some CNS disorders (such as Parkinson's disease and Alzheimer's), and many other immune-mediated and metabolic diseases. The gut microbiome of an obese twin differs from a lean twin. Research will pave the way for new and novel therapeutic strategies to combat these diseases. There is great promise in the dietary manipulation of the gut microbiome as an approach to treat associated disease, in drug discovery, as well as in maintaining optimal good health.

The Human Microbiome Project (HMP; https://www.hmpdacc.org/) is sponsored by the US NIH and is dedicated to the study of the human "supraorganism" composed of both human and nonhuman cells composing the microbiome. The role of human genome polymorphisms in therapeutic outcomes is well established and is discussed in the next section of this chapter on pharmacogenetics and pharmacogenomics. To date, much less is known about the impact of our microbiome or "second genome" genetic polymorphisms in therapeutic outcomes. The microbiome influences the metabolism of drugs and their metabolites (Zimmermann et al. 2019). Orally administered drugs encounter the gut microbiome prior to reaching many host tissues responsible for metabolism and first-pass metabolism. Thus, the goal of true precision medicine cannot be achieved until we better understand our microbial guests. Unique animal microbiome models (see animal models later in this chapter) are being developed to study microbiome assembly, dynamics, and function and should facilitate the development of potential therapeutic interventions. These include many species frequently used historically in developmental biology and genetics studies including zebrafish, *C. elegans,* numerous small mammals, as well as humanized rats and pigs.

Pharmacogenetics and Pharmacogenomics

It has been noted for decades that patient response to the administration of a drug was highly variable within a diverse patient population. Efficacy as determined in clinical trials is based upon a standard dose range derived from large population studies. Better understanding of the molecular interactions occurring within the pharmacokinetics phase of a drug's action, coupled with new genetics knowledge and then genomic knowledge of the human, has advanced us closer to a rational means to optimize drug therapy. Optimization with respect to the patients' genotype, to ensure maximum efficacy with minimal adverse effects, is the goal. Environment, diet, age, lifestyle, and state of health all can influence a person's response to medicines, but understanding an individual's genetic makeup is thought to be the key to creating personalized drugs with greater efficacy and safety. Approaches such as the related pharmacogenetics and pharmacogenomics promise the advent of "precision medicine," in which drugs and drug combinations are optimized for each individual's unique genetic makeup. This chapter will only serve as an introduction, as entire classes are now offered and many books have been written about pharmacogenetics and pharmacogenomics (for example: Cai et al. 2020; Lam et al. 2021, Scheinfeldt 2022).

Single-Nucleotide Polymorphisms (SNPs)

While comparing the base sequences in the DNA of two individuals reveals them to be approximately 99.5% identical, base differences, or polymorphisms, are scattered throughout the genome. The best-characterized human polymorphisms are single-nucleotide polymorphisms (SNPs) occurring approximately once every 1000 bases in the three billion base pair human genome. The DNA sequence variation is a single nucleotide—A, T, C, or G—in the genome difference between members of a species (or between paired chromosomes in an individual). For example, two sequenced DNA fragments from different individuals, AAGTTCCTA to AAGTTC*T*TA, contain a difference in a single nucleotide. Commonly referred to as "snips," these subtle sequence variations account for most of the genetic differences observed among humans. Thus, they can be utilized to determine inheritance of genes in successive generations.

Research suggests that, in general, humans tolerate SNPs as a probable survival mechanism. This tolerance may result because most SNPs occur in noncoding regions of the genome. Identifying SNPs occurring in gene coding regions (cSNPs) and/or regulatory sequences may hold the key for elucidating complex, polygenic diseases such as cancer, heart disease, and diabetes and understanding the differences in response to drug therapy observed in individual patients. Some cSNPs do not result in amino acid substitutions in their gene's protein product(s) due to the degeneracy of the genetic code. These cSNPs are referred to as synonymous cSNPs. Other cSNPs, known as nonsynonymous, can produce conservative amino acid changes, such as similarity in side chain charge or size or more significant amino acid substitutions.

While SNPs themselves do not cause disease, their presence can help determine the likelihood that an individual may develop a particular disease or malady. SNPs, when associated with epidemiological and pathological data, can be used to track susceptibilities to common diseases such as cancer, heart disease, and diabetes (Biesecker and Spinner 2013; Li et al. 2017). Biomedical researchers have recognized that discovering SNPs linked to diseases will lead potentially to the identification of new drug targets and diagnostic tests. The identification and mapping of hundreds of thousands of SNPs for use in large-scale association studies may turn the SNPs into biomarkers of disease and/or drug response. Genetic factors such as SNPs are believed to likely influence the etiology of diseases such as hypertension, diabetes, and lipidemias directly and via effects on known risk factors. For example, in the chronic metabolic disease type 2 diabetes, a strong association with obesity and its pathogenesis includes defects of both secretion and peripheral actions of insulin. The association between type 2 diabetes and SNPs in three genes was detected in addition to a cluster of new variants on chromosome 10q. However, heritability values range only from 30 to 70% as type 2 diabetes is obviously a heterogeneous disease etiologically and clinically. Thus, SNPs, in the overwhelming majority of cases, will likely not be indicators of disease development by themselves.

Pharmacogenetics Versus Pharmacogenomics

In simplest terms, pharmacogenomics is the whole-genome application of pharmacogenetics, which examines the single-gene interactions with drugs. Tremendous advances in biotechnology are causing a dramatic shift in the way new pharmaceuticals are discovered, developed, and monitored during patient use. Pharmacists will utilize the knowledge gained from genomics and proteomics to tailor drug therapy to meet the needs of their individual patients employing the fields of pharmacogenetics and pharmacogenomics (Papastergiou et al. 2017; Cai et al. 2020; Lam et al. 2021; Scheinfeldt 2022).

Pharmacogenetics is the study of how an individual's genetic differences influence drug action, usage, and dosing. A detailed knowledge of a patient's pharmacogenetics profile in relation to a particular drug therapy may lead to enhanced efficacy and greater safety. Pharmacogenetic analysis may identify the responsive patient population prior to administration, i.e., an example of precision medicine. Of particular interest in the field of pharmacogenetics is our understanding of the genetic influences on drug pharmacokinetic profiles such as genetic variations affecting liver enzymes (i.e., cytochrome P450 group) and drug transporter proteins and the genetic influences on drug pharmacody-

namic profiles such as the variation in receptor protein expression.

In contrast, pharmacogenomics is linked to the whole-genome, not an SNP in a single gene. It is the study of the entire genome of an organism (i.e., human patient), both the expressed and the nonexpressed genes in any given physiologic state. Pharmacogenomics combines traditional pharmaceutical sciences with annotated knowledge of genes, proteins, and single-nucleotide polymorphisms. It might be viewed as a logical convergence of the stepwise advances in genomics with the growing field of pharmacogenetics. Incorrectly, the definitions of pharmacogenetics and pharmacogenomics are often used interchangeably. Whatever the definitions, they share the challenge of clinical translation, moving from bench top research to bedside application for patient care. The overall goal is to match individual patients with the most effective and safest drugs and optimum doses. Since a genomic profile can cause a person to respond to a drug in a way that it was not intended, one strives for the right drug, for the right patient, at the right dose, at the right time. Pharmacogenomics is a big part of what is called "Precision Medicine" (see below).

Genome-Wide Association Studies (GWAS)

The methods of genome-wide association studies (GWAS), also known as whole-genome association studies, are powerful tools to identify genetic loci that affect, for instance, drug response or susceptibility to adverse drug reactions (Relling and Evans 2015; Giacomini et al. 2017; Visscher et al. 2017; McInnes et al. 2021). These studies are an examination of the many genetic variations found in different individuals to determine any association between a variant (genotype) and a biological trait (phenotype). The majority of GWAS typically study associations between SNPs and drug response or SNPs and major disease. Over the past 20 years of intense study, they have emerged as important tools and drivers for modern precision medicine. Challenges have included difficulties identifying the key genetic loci due to two or more genes with small and additive effects on the trait (epistasis), the trait caused by gene mutations at several different chromosomal loci (locus heterogeneity), environmental causes modifying expression of the trait or responsible for the trait, and undetected population structure in the study such as those arising when some study members share a common ancestral heritage. The practical use of this approach and its introduction into the everyday clinical setting remain a challenge, but will undoubtedly be aided by new next-generation sequencing techniques, enhanced bioinformatics capabilities, and running GWAS association studies alongside protein biomarker studies.

On the Path to Precision Medicine: A Brief Introduction
Much of modern medical care decision-making is based upon observations of successful diagnosis and treatment at the larger population level. Health care is undergoing a revolutionary change as actionable data from multiple genomic and other "OMIC" technologies becomes readily available to clinicians at the point-of-care (POC) including doctors' offices, out-patient clinics, and pharmacies. This will foster an emerging new paradigm of patient care, "precision medicine," that will better predict, diagnose, monitor, and treat disease at the level of the specific patient (Akhoon 2021). A fundamental goal of precision medicine is to treat individual patients with significantly greater precision and accuracy via the deep analysis of data (including omics data) from large, diverse populations. This paradigm shift is also being influenced by genome sequencing and other genomic tests becoming much less expensive and even accessible through clinics, public health systems and via direct-to-consumer genomic tests. Medical, pharmacy, and nursing education curricula are evolving rapidly to prepare graduates to practice in a precision medicine practice paradigm. This approach is entirely consistent with the concept of patient-centered care to improve patient outcomes. And it is needed to address the inconsistencies in both diagnosis and treatment across the health care system. Better use of data and technology has the power to transform the quality of health while empowering patients to greater control of their own health and well-being. Precision medicine will enable clinicians to employ the most appropriate course of action for each individual patient managing the extreme complexity of each patient given all of the tools now available in the health care system: OMICS technology data, disease mechanisms, the electronic health record, public health information, big data, etc.

Modern genomics, transcriptomics, proteomics, metabolomics, pharmacogenomics, epigenomics (to be discussed later in this chapter), and other technologies, implemented in the clinic in faster and less expensive instrumentation and methodologies, are now being introduced to identify genetic variants, better inform health care providers about their individual patient, tailor evidence-based medical treatment, and suggest rational approaches toward preventative care. The hopes and realities of precision medicine (sometimes referred to as part of "molecular medicine," pharmacotherapy informed by a patient's individual genomic and proteomic information) are global priorities. As a pharmaceutical biotechnology text, our limited discussion here will focus on precision medicine in a primarily pharmacogenomic and pharmacogenetic context. However, other genomic-type technologies including GWAS, next-generation sequencing, proteomics, and metabolomics will be crucial for the successful implementation of precision medicine. The hope is that "OMIC" science will bring predictability to the optimi-

zation of drug selection and drug dosage to assure safe and effective pharmacotherapy (Fig. 9.6).

Personalized Medicine and P4 Medicine, in a Precision Medicine Context

One of the early rewards expected from the completion of the HGP was to pinpoint specific genes of the 21,000+ discovered that caused common diseases. While most diseases are now recognized as polygenic with more complex systems interactions than single-gene diseases, the data acquired from the HGP has the potential to forever transform health care. Genome-based medicine has been called personalized medicine or molecular medicine and is the next logical step in the evolution of medicine and direct patient care. The in-depth knowledge of an individual's genetics, when coupled with molecular-level understanding of disease mechanisms, and biological systems characteristics of wellness can focus a specific patient's health care in a "personalized" way, i.e., personalized medicine. Often used interchangeably, there is much overlap between the terms personalized medicine (the older term) and precision medicine (the newer term). Since the older term included the word "personalized," there was concern that the approach could be misinterpreted as only involving interventions that are uniquely developed for each individual patient. While this is true to some extent, the context in which it is implemented is very important. Thus, "precision medicine" has evolved as the preferred term to more accurately describe the new paradigm in which interventions will be more effective if they are based upon a deep analysis of broad data sets from large, diverse populations and then focus what is known to tailor the interventions to best address a specific patient, fully understanding the patient's genomic, environmental, and lifestyle factors within the context of these broader data sets (Akhoon 2021).

Leroy Hood of the Institute for Systems Biology coined the term "P4 Medicine." He and his colleagues recognized that medicine was undergoing a revolution as a result of the rapid advances in OMICS technologies and the wealth of information being generated (Sagner et al. 2017). One area of OMICS study, systems biology (described in greater detail later in this chapter), was starting to focus on the incredible complexity of biological systems, both normal and diseased. Recognizing that the current medical framework used to guide health care and manage chronic disease is largely ineffective, they proposed a new way to personalize medicine. A medicine paradigm that is **P**reventative, **P**redictive, **P**ersonalized, and **P**articipatory (i.e., P4) would hold great promise to improve patient health outcomes by harnessing modern biotechnologies and a better understanding of the mechanisms of disease into evidence-based health interven-

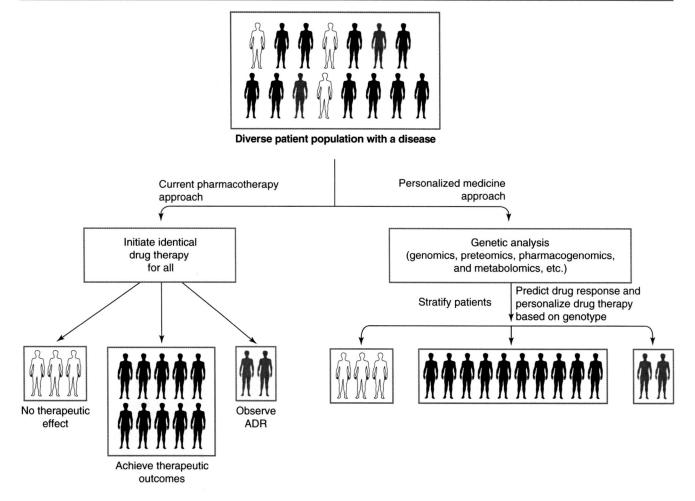

Fig. 9.6 The role of "OMIC" technologies in precision medicine

tions. While many of the key tenets of P4 medicine are providing an important foundation for the implementation of precision medicine, the term P4 medicine is being used less today. To be successful, precision medicine must be conducted in an ecosystem that is a continuously learning health care system (described earlier in this chapter) and links clinicians, medical laboratories, research enterprises, and health information systems. Precision medicine has the potential to profoundly impact the practice of medicine at all levels (ambulatory and primary care, secondary care, tertiary care, and clinical education). Possibly, the best way to consider precision medicine is to consider patients no longer solely "passengers" on an airplane, but now they are "co-pilots."

Implementing Precision Medicine

Optimized precision medicine utilizing pharmacogenomic knowledge as an essential component would not only spot disease before it occurs in a patient or detect a critical variant that will influence treatment, but should increase drug effi-

cacy upon pharmacotherapy and reduce drug toxicity. Also, it would facilitate the drug development process (see Fig. 9.1) including improving clinical development outcomes, reducing overall cost of drug development, and leading to development of new diagnostic tests that impact on therapeutic decisions (Ersek et al. 2018; Primorac et al. 2020). Individualized optimized pharmacotherapy would first require a detailed genetic analysis of a patient, assembling a comprehensive list of SNPs. Patient pharmacogenomic analyses were often conducted using microarray technology based upon clinically validated biomarkers across focused SNP panels (metabolizing enzymes, oncology, etc.) and today has benefited from the development of more widely available, rapid, less costly genomic exome and whole-genome sequencing approaches. The impact of the patient's SNPs on the use of new or existing drugs would thus be predicted and individualized drug therapy would be identified that assures maximal efficacy and minimal toxicity. Precision medicine would also require knowledge of an individual patient's genomic profile to help identify potential drug responders and nonresponders. This might be accomplished by testing for the presence or absence of critical biomarkers that may be associated with prediction of response rates. The US FDA provides an online list of all FDA-approved drugs with pharmacogenomic information in their labels (black boxes). Some, but not all, of the labels include specific actions to be taken based on genetic information. The drug labels contain information on genomic biomarkers that may be predictive of drug exposure and clinical response rate, risk of adverse reactions, genotype-specific dosing, susceptibility to a specific mechanism of drug action, or polymorphic drug target and disposition genes. Rather than reproducing this table in whole or part in this text, the reader may access it in its constantly updated form (FDA 2022).

It is well understood that beyond genomics and proteomics, a patient's behavioral and environmental factors (sometimes referred to as the "exposome") influence clinical outcomes and susceptibility to disease. Emerging fields of nutrigenomics and envirogenomics are studying these additional layers of complexity. Precision medicine will become especially important in cases where the cost of testing is less than either the cost of the drug or the cost of correcting adverse drug reactions caused by the drug. Pharmaceutical care would begin by identifying a patient's susceptibility to a disease, then administering the right drug to the right patient at the right time. For example, the monoclonal antibody trastuzumab (Herceptin) is a personalized breast cancer therapy specifically targeted to the HER2 gene product (25–30% of human breast cancers overexpress the human epidermal growth factor receptor, HER2 protein). Exhibiting reduced side effects as compared to standard chemotherapy due to this protein target specificity, trastuzumab is not prescribed to treat a breast cancer patient unless the patient has first tested positive for HER2 overexpression.

The success of targeted therapy for precision medicine has fostered the concept that the era of the blockbuster drug may be over and will be replaced by the "niche buster" drug, a highly effective medicine individualized for a small group of responding patients identified by genomic and proteomic techniques. (Biospace 2022). Also, while numerous articles predicted that pharmacogenomics would revolutionize medicine, the initial predictions have not been lived up to the hype due to statistical, scientific, and commercial hurdles. With more than 11 million SNP positions believed to be present in the human population, large-scale detection of genetic variation holds the key to successful precision medicine (Bachtiar et al. 2019). Correlation of environmental factors, behavioral factors, genomic and proteomic factors (including pharmacogenomic and metabolomic factors), and phenotypical observables across large populations remains a daunting data-intensive challenge. Yet, pharmacogenetics and pharmacogenomics are having an impact on modern medicine.

Human Genomic Variation Affecting Drug Pharmacokinetics

Genetic variation associated with drug metabolism and drug transport processes resulting from products of gene expression (metabolic enzymes and transport proteins, respectively) play a critical role in determining the concentration of a drug in its active form at the site of its action and also at the site of its possible toxic action(s). Thus, pharmacogenetic and pharmacogenomic analysis of drug metabolism and drug transport is important to a better clinical understanding of and prediction of the effect of genetic variation on drug effectiveness and safety.

It is well recognized that specific drug metabolic phenotypes may cause adverse drug reactions. For instance, some patients lack an enzymatically active form, have a diminished level, or possess a modified version of CYP2D6 (a cytochrome P450 allele) and will metabolize certain classes of pharmaceutical agents differently to other patients expressing the native active enzyme. All pharmacogenetic polymorphisms examined to date differ in frequency among racial and ethnic groups. For example, CYP2D6 enzyme deficiencies may occur in $\leq 2\%$ Asian patients, $\leq 5\%$ black patients, and $\leq 11\%$ white patients (Chaudhry et al. 2014). A diagnostic test to detect CYP2D6 deficiency could be used to identify patients that should not be administered drugs metabolized predominantly by CYP2D6. Table 9.2 provides some selected examples of common drug metabolism polymorphisms and their pharmacokinetic consequences.

With the burgeoning understanding of the genetics of warfarin metabolism, warfarin anticoagulation therapy is becoming a leader in pharmacogenetic analysis for pharmacokinetic prediction (Emery 2017) Adverse drug reactions (ADRs) for warfarin account for 15% of all ADRs in the USA, second only to digoxin. Warfarin dose is adjusted with the goal of achieving an INR (International Normalized Ratio = ratio of patient's prothrombin time as compared to that of a normal control) of 2.0–3.0. The clinical challenge is to limit hemorrhage, the primary ADR, while achieving the optimal degree of protection against thromboembolism. Deviation in the INR has been shown to be the strongest risk factor for bleeding complications. The major routes of metabolism of warfarin are by CYP2C9 and CYP3A4. Some of the compounds, which have been identified to influence positively or negatively warfarin's INR, include cimetidine, clofibrate, propranolol, celecoxib (a competitive inhibition of CYP2C9), fluvoxamine (an inhibitor of several CYP enzymes), various antifungals and antibiotics (e.g., miconazole, fluconazole, erythromycin), omeprazole, alcohol, ginseng, and garlic. Researchers have determined that the majority of individual patient variation observed clinically in

Table 9.2 Some selected examples of common drug metabolism polymorphisms and their pharmacokinetic consequences

Enzyme	Common variant	Potential consequence
CYP1A2	CYP1A2*1F	Increased inducibility
CYP1A2	CYP1A2*1 K	Decreased metabolism
CYP2A6	CYP2A6*2	Decreased metabolism
CYP2B6	CYP2B6*5	No effect
CYP2B6	CYP2B6*6	Increased metabolism
CYP2B6	CYP2B6*7	Increased metabolism
CYP2C8	CYP2C8*2	Decreased metabolism
CYP2C9	CYP2C9*2	Altered affinity
CYP2C9	CYP2C9*3	Decreased metabolism
CYP2D6	CYP2D6*10	Decreased metabolism
CYP2D6	CYP2D6*17	Decreased metabolism
CYP2E1	CYP2E1*2	Decreased metabolism
CYP3A7	CYP3A7*1C	Increased metabolism
Flavin-containing monooxygenase 3	FMO3*2	Decreased metabolism
Flavin-containing monooxygenase 3	FMO3*4	Decreased metabolism

Data from references noted in the Pharmacogenomics section of Chap. 9

response to warfarin therapy is genetic in nature, influenced by the genetic variability of metabolizing enzymes, vitamin K cycle enzymes, and possibly transporter proteins. The CYP2C9 genotype polymorphisms alone explain about 10% of the variability observed in the warfarin maintenance dose. Figure 9.7 shows the proteins involved in warfarin action and indicates the pharmacogenomic variants that more significantly influence warfarin therapy optimal outcome.

Studies at both the basic research and clinical level involve the effect of drug transport proteins on the pharmacokinetic profile of a drug. Some areas of active study of the effect of genetic variation on clinical effectiveness include efflux transporter proteins (for bioavailability, CNS exposure, and tumor resistance) and neurotransmitter uptake transporters (as valid drug targets). Novel transporter proteins are still being identified as a result of the HGP and subsequent proteomic research. More study is needed on the characterization of expression, regulation, and functional properties of known and new transporter proteins to better assess the potential for prediction of altered drug response based on transporter genotypes.

Human Genomic Variation Affecting Drug Pharmacodynamics

Genomic variation such as factors influencing the expression of the target protein directly affects not only the pharmacokinetic profile of drugs, it also strongly influences the pharmacodynamic profile of drugs. Targets include the drug receptor involved in the response as well as the proteins associated with disease risk, pathogenesis, and/or toxicity including infectious disease. There are increasing

numbers of prominent examples of inherited polymorphisms influencing drug pharmacodynamics. To follow on the warfarin example above (see Fig. 9.7), the major component of individual patient variation observed clinically in response to anticoagulant therapy is genetic in nature. However, the CYP2C9 genotype polymorphisms alone only explain about 10% of the variability observed in the warfarin maintenance dose. Warfarin effectiveness is also influenced by the genetic variability of vitamin K cycle enzymes. The drug receptor for warfarin is generally recognized as vitamin K epoxide reductase, the enzyme that recycles vitamin K in the coagulation cascade. Vitamin K epoxide reductase complex1 (VKORC1) has been determined to be highly variant with as much as 50% of the clinical variability observed for warfarin resulting from polymorphisms of this enzyme.

Associations have been implicated between drug response and genetic variations in targets for a variety of drugs including antidepressants (G-protein β3), antipsychotics (dopamine D2, D3, D4; serotonin 5HT2A, 5HT2C), sulfonylureas (sulfonylurea receptor protein), and anesthetics (ryanodine receptor) (Minikel et al. 2020). Likewise, similar associations for efficacy are known such as statins (apolipoprotein E; enhanced survival prolongation with simvastatin) and tacrine (apolipoprotein E; clinical improvement of Alzheimer's symptoms). In addition, similar associations have been studied for drug toxicity and disease polymorphisms including abacavir (major histocompatibility proteins; risk of hypersensitivity), cisapride and terfenadine (HERG, KvLQT1, Mink, MiRP1; increased risk of drug-induced torsade de pointes), and oral contraceptives (prothrombin and factor V; increased deep vein thrombosis).

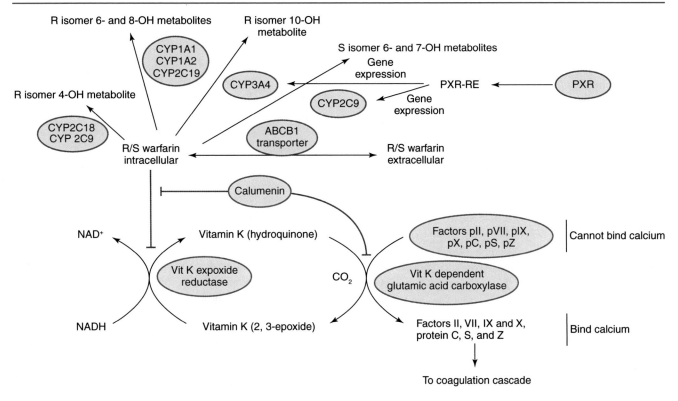

Fig. 9.7 Critical pharmacogenomic variants affecting warfarin drug action and adverse drug reactions

Value of Precision Medicine in Disease

Due to the intimate role of genetics in carcinogenesis, precision medicine is rapidly becoming a success story in oncology based on genetic profiling using proteomic analyses of tumor biopsies (Chakravarty and Solit 2021). As described above, targeted cancer therapies such as trastuzumab (Herceptin) are successful and are viewed as the way of the future. Also, clinically important polymorphisms predict increased toxicity in patients with cancer being treated with chemotherapeutic drugs, for example, 6-mercaptopurine (thiopurine methyltransferase *2, *3A, and *3C variants), 5-fluorouracil (5-FU) (dihydropyrimidine dehydrogenase *2A variant), and irinotecan (UGT1A1*28 allele; FDA-approved Invader UGT1A1 Molecular Assay diagnostic available to screen for the presence of this allele associated with irinotecan toxicity). Likewise, clinically important pharmacogenetics predicts efficacy in oncology patients treated with 5-FU (thymidylate synthase *2 and *3C variants).

A classic clinical application of pharmacogenetics that shows the power of this approach is our understanding of the potentially fatal hematopoietic toxicity that occurs in some patients administered standard doses of the antileukemic agent azathioprine, mercaptopurine, and thioguanine (Chouchana et al. 2014). These drugs are metabolized by the enzyme thiopurine methyltransferase (TPMT) to the inactive S-methylated products. Gene mutations (polymorphisms) may occur in as many as 11% of patients resulting in decreased TPMT-mediated metabolism of the thiopurine drugs. A diagnostic test for TPMT is now available and used clinically. Identified patients with poor TPMT metabolism may need their drug dose lowered 10–15-fold. Mechanisms of multidrug resistance to cancer drugs are influenced by genetic differences. A number of polymorphisms in the MDR-1 gene coding for P-glycoprotein, the transmembrane protein drug efflux pump responsible for multidrug resistance, have been identified. One, known as the T/T genotype and correlated with decreased intestinal expression of P-glycoprotein and increased drug bioavailability, has an allele frequency of 88% in African-American populations, yet only approximately 50% in Caucasian-American populations. Pharmacogenetic and pharmacogenomic analysis of patients is being actively studied in many disease states. However, a detailed discussion goes beyond what this introduction may provide. The reader is encouraged to read further in the pharmacogenetic/pharmacogenomic-related references available, some provided at the end of this chapter.

Challenges in Precision Medicine

There are many keys to success for precision medicine that hinge on continued scientific advancement. While it is great for the advancement of the OMIC sciences, some have questioned how good it is for individual patients, particularly if they are very health challenged, at this stage of its development due to potential exaggerated claims in the public press falling short of the predictive, preventative, and participatory health care paradigm promised. Also, modern medicine may lack proven prevention interventions necessary to address certain genomic traits once discovered (Ward and Ginsburg 2017; Madhavan et al. 2018). There are also economic, societal, and ethical issues that must be addressed to successfully implement genetic testing-based individualized pharmacotherapy. It is fair to state that most drugs will not be effective in all patients all of the time. Thus, the pressure of payers to move from a "payment for product" to a "payment for clinically significant health outcomes" model is reasonable. The use of OMIC health technologies and health informatics approaches to stratify patient populations for drug effectiveness and drug safety is a laudable goal. However, while costs are decreasing, the technologies are challenging to many payers and the resulting drug response predictability requires further clinical validation. Cost-effectiveness and cost-benefit analyses are limited at this date (Chen et al. 2022). Also, the resulting environment created by these technologies in the context of outcome expectations and new drug access/reimbursement models will give rise to a new pharmaceutical business paradigm that is still evolving.

Epigenetics and Epigenomics

DNA is the heritable biomolecule that contains the genetic information resulting in phenotype from parent to offspring. Modern genomics, GWAS and SNP analyses, confirm this and identify genetic variants that may be associated with a different phenotype. However, genome-level information alone does not generally predict the phenotype at an individual level. For instance, researchers and clinicians have known for some time that an individual's response to a drug is affected by the genetic makeup (DNA sequence, genotype) and a set of disease and environmental characteristics working alone or in concert to determine that response. Research has suggested that in addition to DNA sequence, there are a number of other "levels" of information that influence transcription of genomic information. As you are aware, every person's body contains trillions of cells, all of which have essentially the same genome and, therefore, the same genes. Yet some cells are optimized for development into one or more of the 200+ specialized cells that make up our bodies: muscles, bones, brain, etc. For this to transpire from within the same genome, some genes must be turned on or off at different points of cell development in different cell types to affect gene expression, protein production, cell differentiation, growth, and function. The rapidly expanding field of research known as epigenetics (or epigenomics) in some ways can be viewed as a conduit between genotype and phenotype (Prokopuk et al. 2015). Epigenetics literally means "above genetics or over the genetic sequence." It is the factor or factors that influence cell behavior by means other than via a direct effect on the genetic machinery. Epigenetic regulation includes DNA methylation and covalent histone modifications (Fig. 9.8). They result without heritable changes in DNA sequence. Epigenomics is the merged science of genomics and epigenetics (Tost 2020) Functionally, epigenetics acts to regulate gene expression, gene silencing during genomic imprinting, apoptosis, X-chromosome inactivation, and tissue-specific gene activation (such as maintenance of stem cell pluripotency). The epitranscriptome includes all the biochemical modifications of RNA (the transcriptome) within a cell including the epigenetic modifications (such as specific base methylation).

The more we understand epigenetics and epigenomics, the more we are likely to understand those phenotypic traits that are not a result of genetic information alone. This should be facilitated by developments in single-cell epigenomics (Clark et al. 2016; Kelsey et al. 2017). Single-cell epigenomics allows for the study of cellular heterogeneity at different time scales to effectively record a cell's past and predict its future functionality. Epigenetics/epigenomics may also explain low association predictors found in some pharmacogenetic/pharmacogenomic studies. Etiology of disease, such as cancer, likely involves both genetic variants and epigenetic modifications that could result from environmental effects. Age also likely influences epigenetic modifications as studies of identical twins show greater differences in global DNA methylation in older rather than younger sets of twins (Feinberg et al. 2010). Abnormal epigenetic regulation is likely a feature of complex diseases such as diabetes, cancer, and heart disease (Feinberg 2018). Therefore, epigenetic targets are being explored for drug design and development. The first generations of US FDA-approved epigenetics-based drugs are seven agents in three mechanistic classes. The available drugs are three DNA demethylating agents (DNMT) (5-azacytidine, 5-aza-2′-deoxycytidine, and hydralazine) and four histone deacetylase (HDAC) inhibitors (vorinostat, romidepsin, panobinostat, and belinostat). These have been approved mainly for the treatment of blood cancers, in particular myelodysplastic syndromes (MDS) and T-cell lymphoma. Note that hydralazine is an older drug that is approved for the treatment of hypertension.

One of the most studied and best understood molecular mechanisms of epigenetic regulation is methylation of cytosine residues at specific positions in the DNA molecule (Fig. 9.8). Another mechanism of epigenomic control appears to occur at the level of chromatin. In the cell, DNA

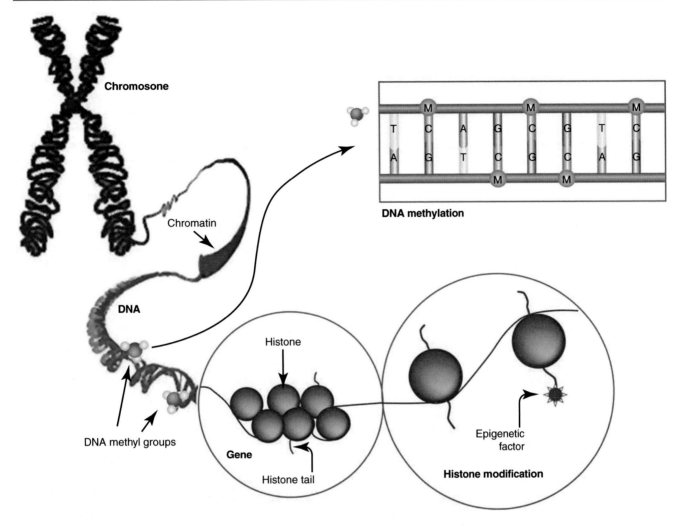

Fig. 9.8 Epigenetic regulation via DNA methylation, histone modifications, and chromatin structure

is wrapped around eight different histone proteins to form chromatin. Packaging of DNA into chromatin can render large regions of the DNA inaccessible and prevent processes such as DNA transcription from occurring. Epigenetic regulation of histone proteins can be by chemical modification including acetylation, methylation, sumoylation, and ubiquitylation. Each can cause structural changes in chromatin affecting DNA accessibility. Nonprotein-coding RNAs, known as ncRNAs, have also been shown to contribute to epigenetic regulation as have mRNAs which can be processed and participate in various interference pathways.

The components involved in human epigenetics serve as promising targets for small molecule drug discovery. Searching for small molecule leads that modulate epigenetic enzymes and processes have the potential to lead to new and novel therapeutic strategies. These can effectively reverse epigenetic abnormalities that are present in and may play a mechanistic role in several human oncogenic, metabolic, inflammatory, cardiovascular, and neurological disorders. As researchers deepen their focus on epigenetic drug targets,

they will also add further new knowledge and understanding to the complex collection of proteins, enzymes, and substrates integral to epigenetic regulation and other processes.

Toxicogenomics

Toxicogenomics, related to pharmacogenomics, combines toxicology, molecular biology, environmental health, and data intense genomic science to elucidate the response of living organisms to stressful environments or xenobiotic agents (including drugs) based upon their genotype (Alexander-Dann et al. 2018). The approach integrates toxicant-specific alterations in gene, protein, and metabolite expression patterns with the response of organisms (phenotypic) and their subcellular components. While toxicogenomics studies how the genome responds to toxic exposures, toxicogenetics studies how an individual's genetic makeup affects his/her response to environmental stresses and toxins such as carcinogens, neurotoxins, and reproductive toxins. Genomic techniques utilized in toxicogenomic studies include gene expression level profiling, SNP analysis of the genetic variation, proteomics, and/or metabolomic methods so that gene expression, protein production, and metabolite production may be studied. The rapid growth in whole-genome and single-cell DNA sequencing capability is rewriting the approach for modern toxicogenomic studies. Toxicogenomics can be very useful in drug discovery and development as new drug candidates can be screened through a combination of gene expression profiling and toxicology to understand gene response, identify general mechanisms of toxicity, and possibly predict drug safety at a better cost (Qin et al. 2016). There have been suggestions that toxicogenomics may decrease the time needed for toxicological investigations of new drug candidates and reduce both cost and animal usage versus conventional toxicity studies.

Toxicogenomic studies attempt to discover associations between the development of drug toxicities and a person's genotype. Clinicians and researchers are attempting to correlate genetic variation in one population to the manifestations of toxicity in other populations to identify and then to predict adverse toxicological effects in clinical trials so that suitable biomarkers for these adverse effects can be developed (Alexander-Dann et al. 2018). Using such methods, toxicogenomics may contribute to precision medicine by testing an individual patient for his or her susceptibility to these adverse effects before prescribing a medication. Patients that would show the marker for an adverse effect would be switched to a different drug. Toxicogenomics may also help identify individual susceptibility to drug dependency and/or addiction. Therefore, toxicogenomics will become increasingly more powerful in predicting toxicity as new biomarkers are identified and validated. Much of the new toxicogenomic technology is developing in the pharmaceutical industry and other corporate laboratories.

Other "OMIC" Technologies

Pharmaceutical scientists and pharmacists may hear about other "OMIC" technologies in which the "OMIC" terms derive from the application of modern genomic techniques to the study of various biological properties and processes. For example, interactomics is the data-intensive broad system study of the interactome, which is the interaction among proteins and other molecules within a cell. Proteogenomics has been used as a broadly encompassing term to describe the merging of genomics, proteomics, small molecules, and informatics. Cellomics has been defined as the study of gene function and the proteins they encode in living cells utilizing light microscopy and especially digital imaging fluorescence microscopy. The field of optogenetics is used by neuroscientists to turn neurons selectively on and off. This combination of genetics and optics utilizes visible light to control well-defined events in cells in living tissues that have been genetically modified to express light-sensitive ion channel, ion pump, or G-Protein-coupled receptor.

"OMICS" Integrating Technologies: Systems Biology and Multiomics

The Human Genome Project (HGP) and the development of bioinformatics technologies have catalyzed fundamental changes in the practice of modern biology and helped unveil a remarkable amount of information about many organisms and their complexity. Biology has become an information science defining all the elements in a complex biological system, simultaneously measuring thousands of data points and placing them in a database for comparative interpretation. As seen in Fig. 9.2, the hierarchy of information collection goes well beyond the biodata contained in the genetic code that is transcribed and translated. The heart of the field called "Systems Biology" involves a generally complex interactive system with insightful views of cells, organisms, and populations. This research area is often described as a noncompetitive or precompetitive technology by the pharmaceutical industry because it is believed to be a foundational technology that must be better developed to be successful at the competitive technology of drug discovery and development. It is the study of the interactions between the components of a biological system and how these interactions give rise to the function and behavior of that system (Tavassoly et al. 2018; Zou and Laubichler 2018). Systems biology is essential for our understanding of how all the individual parts of intricate biological networks are organized and function in living cells. The biological system may involve enzymes and metabolites in a metabolic pathway or other interacting biological molecules affecting a biological process. Molecular biologists have spent the past 60+ years teasing apart cellular pathways down to the molecular level. Characterized by a cycle of theory, computational modeling, and experiments to quantitatively describe cells or cell processes, systems biology is a data-intensive endeavor that results in a conceptual framework for the analysis and understanding of complex biological systems in varying contexts. Statistical mining, data alignment, probabilistic and mathematical modeling, and data visualization into networks are among the mathematical models employed to integrate the data and assemble the systems network. New measurements are stored with existing data, including extensive functional annotations, in molecular databases, and model assembly provides libraries of network models.

As the biological interaction networks are extremely complex, so are graphical representations of these networks. After years of research, a set of guidelines known as the Systems Biology Graphical Notation (SBGN) has been drafted by a community of biochemists, modelers, and computer scientists and is generally accepted to be the standard for graphical representation by all researchers. These standards, very similar to the block diagrams used by electrical engineers, are designed to facilitate the interchange of systems biology information and storage. Due to the complexity of these diagrams depending on the interactions examined and the level of understanding, a figure related to systems biology has not been included in this chapter. However, the reader is referred to the following website authored by the SBGN organization for several excellent examples of complex systems biology-derived protein interaction networks (SBGN 2022). The inability to visualize the complexity of biological systems has in the past impeded the identification and validation of new and novel drug targets. The accepted SBGN standards facilitate the efforts of pharmaceutical scientists to validate new and novel targets for drug design.

Since the major objective of systems biology was to create a model of all the interactions in a system, the experimental techniques utilized attempt to be as complete as possible. These have routinely included genomics, epigenomics, transcriptomics, proteomics, pharmacogenomics, metabolomics, microbiomics, toxicogenomics, and others. As omics technologies became more widely available, faster, and less expensive, and coupled with greatly enhanced data storage and analysis capabilities, the approach known as "Multiomics (Multi-omics)" where the data sets are "multiple omes" has evolved. Combining two or more omics data sets will better power the analysis of the whole system being studied. Thus, today more than ever, systems biology studies are based upon multiomic analysis data (Krassowski et al. 2020; Miao et al. 2021). Multiomics has also been called panomics, integrative omics, or transomics. Pharmaceutical and clinical end points include systems level biomarkers, genetic risk factors, aspects of precision medicine, and drug target identification (Tavassoly et al. 2018; Zou and Laubichler 2018). In the future, applications of systems biology and multiomics to drug discovery promises to have a profound impact on patient-centered medical practice, permitting a comprehensive evaluation of underlying predisposition to disease, disease diagnosis, and disease progression. Also, implementing precision medicine and systems medicine will require new analytical approaches such as advanced systems biology techniques and single-cell multiomics, a technique selected as the science journal *Nature*'s 2019 Method of the Year (Nature 2020) to improve disease prognosis and treatment via a more robust understanding of genotype-to-phenotype relationship in individuals.

"OMICS"-Enabling Technology: Genome Editing
Since the discovery of DNA's overall structure in 1953, the world's scientific community has rapidly gained a detailed knowledge of the genetic information encoded by the DNA of a cell or organism and have been correlating this structure with biological function. The exact base pair sequence of the genome, its genotype that includes the entire collection of genes and all other functional and nonfunctional DNA sequences in the nucleus of an organism, has a direct observable impact on the function of the organism or its phenotype.

In today's biology, genome editing (sometimes called genome engineering) makes specific changes to the DNA of a cell or organism and observes its impact. Genome editing effectively also occurs naturally in organisms including humans via endogenous DNA-repair mechanisms (i.e., HDR, homologous directed repair, and NHEJ, nonhomologous end-joining; natural mechanisms that repair harmful breaks that occur in DNA) and the modifications of DNA that occur as a result of epigenetics. Researchers have wanted to modify the DNA sequence since DNA's discovery to study the correlation of genomic structure to protein structure and to function. One early approach at editing a DNA sequence was site-directed mutagenesis, also called site-specific mutagenesis or oligonucleotide-directed mutagenesis. Site-directed mutagenesis at a single amino acid position in an engineered protein is called a point mutation. Therefore, site-directed mutagenesis techniques can aid in the examination at the molecular level of the relationship of protein 3D -structure (resulting from a transcribed and translated mutated DNA sequence) and the function of the resulting new protein. Since the 1980s, researchers have used the process of HDR to exchange endogenous genomic DNA with exogenous donor DNA with varying success.

Key to genome editing is the ability to selectively and predictively modify the DNA sequence of the target organism and assure the integrity of the resulting edits. Major advances were achieved in the pioneering experiments using yeast meganucleases, a naturally occurring enzyme and engineered versions that can recognize and cut double-stranded DNA sequences of >14 base pairs. There has been explosive development in genome editing in the last decade for DNA targeted gene deletions, integrations, or modifications with the fundamental shift from yeast meganucleases to the latest prokaryotic nucleases used for precise genome manipulation. Genome editing nucleases are engineered enhanced nuclease enzymes or "sequence-specific molecular scissors" based upon native enzymes that specifically cut double-stranded DNA in the target cell. They currently belong to one of three known nuclease categories: <u>z</u>inc-<u>f</u>inger <u>n</u>ucleases (ZFN); <u>t</u>ranscription <u>a</u>ctivator-<u>l</u>ike <u>e</u>ffector <u>n</u>ucleases (TALEN); and <u>c</u>lustered <u>r</u>egularly <u>i</u>nterspaced <u>s</u>hort <u>p</u>alindromic <u>r</u>epeats (CRISPR) and their associated proteins (such as Cas9). Genome editing nuclease-mediated mutagenesis can occur exogenously or endogenously if the nuclease targeting the user's gene of interest is delivered into a parental cell line, either by transfection, electroporation, or viral vector delivery. These topics are discussed in more detail in Chap. 14, Advanced Therapy Medicinal Products.

Applications of genome editing methodologies are extensive and include cell-line optimization (i.e., creation of cell lines that produce higher yields of targeted proteins including antibodies), functional genomics and target validation in drug discovery (i.e., creation of gene knockouts in multiple cell lines, the complete knockout of genes not amenable to RNAi), and cell-based screening (i.e., creation of knock-in cell lines with promoters, fusion tags, or reporters integrated into endogenous genes. Unlocking the secrets of the human genome editing and improving upon specificity of the DNA cuts and repair have many implications including a paradigm shift in medical research and clinical practice toward better disease understanding and taxonomy, and true precision medicine at the molecular level (Dai et al. 2020). These genome editing techniques can introduce unintended genomic alterations cleaving DNA at an off-target site. Thus, genome editing is a very powerful technology that has also far-reaching economic, bioethical, and national security implications. It is expected that genome editing "drugs," enzyme systems that selectively bind, cleave, and enable the direct editing of a specific DNA sequence, based upon these technologies and delivered into a cell in vivo, will become important future therapeutic agents.

The mechanism by which the nuclease genome editing tools including ZFN, TALEN, and CRISPR tools can target and cleave specific double-stranded DNA sequences are generally analogous with differences in binding recognition, associated proteins, specificity, and ease of use. Figure 9.9 provides a very simple generalization of this analogous genome editing mechanism to facilitate understanding. However, there is an important difference that has vaulted CRISPR tools into the forefront of genome editing. Whereas ZFN and TALEN bind to DNA through a direct protein–DNA interaction, requiring the protein to be redesigned for each new target DNA site, the CRISPR-Cas system achieves target specificity through a small RNA that can easily be swapped for other RNAs targeting new sites.

Genome Editing Improvement with Zinc-Finger Nucleases and TALEN

Zinc-finger nucleases (ZFNs) are zinc-containing proteins that occur in several transcription factors. They are used in genome editing as engineered DNA-binding enzymes that enable targeted editing of the genome by creating double-strand breaks in DNA at user-specified locations. Double-strand breaks are important for site-specific mutagenesis as they stimulate the cell's endogenous DNA-repair mechanisms (homologous recombination and nonhomologous end-joining). Meganucleases had a distinct disadvantage because they cut the DNA strand at a specific location (specific sequence of base pairs) making them very selective. Unfortunately, there was little chance of finding or engineering the exact meganuclease requisite to cut a specific DNA sequence. The key discovery was finding a nuclease whose DNA recognition site and cleaving site were separate from each other. ZFN, TALEN, and CRISPR meet this architecture. Thus, all three technologies are nonspecific DNA cutting enzymes which can then be linked to specific DNA sequence recognizing peptides such as zinc fingers, transcription activator-like effectors (TALEs), and proteins such as Cas9. Scientists have identified a large number of zinc fingers that recognize various nucleotide triplets and, with some trial and error, ZFNs are able to recognize their targets with fairly high specificity. ZFN because of their mechanism has potential targetable sites in the average genome of every 500 bp. ZFN were quickly employed as genomic editing tools for the generation of mammalian transgenic animals with possibly the biggest impact in transgenic livestock species (i.e., pigs, cattle, etc.). Gene therapy applications were also explored.

Fig. 9.9 A generalized depiction of the analogous genome editing mechanism for ZFN, TALEN, and CRISPR tools. (*HDR* homologous directed repair, *NHEJ* nonhomologous end-joining; natural mechanisms that repair harmful DNA strand breaks that occur in DNA)

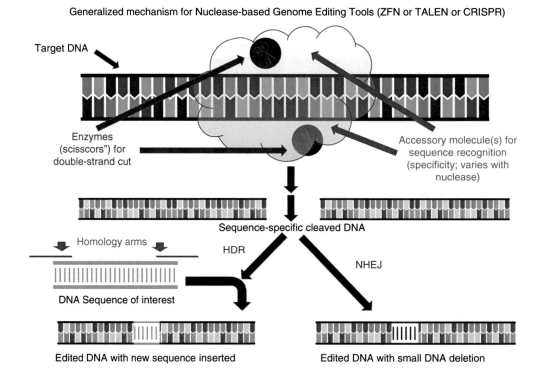

Generalized mechanism for Nuclease-based Genome Editing Tools (ZFN or TALEN or CRISPR)

TALEN (transcription activator-like effector nucleases) are engineered restriction enzymes generated by fusing a specific DNA-binding domain designed to bind any desired targeted DNA sequence, to a nonspecific DNA-cleaving domain that can create double-strand breaks at the target site that can be repaired by error-prone nonhomogenous end-joining (Joung and Sander 2013). The DNA sequence-specific targeting can be engineered by using specific DNA-binding proteins excreted by the plant pathogen Xanthomanos app. These DNA-binding domains are called TAL effectors (TALEs) and consist of repeated domains, each of which contains a highly specific sequence of 34 amino acids. Whereas ZFNs are entirely engineered in the laboratory, TALE proteins exist in nature and many are isolated for use as well as engineered once their precise DNA-binding code was deciphered. They can be effectively designed to bind any desired DNA sequence (with some limitations). TALEN are used in a similar way to the designed zinc-finger nucleases. However, because of their different mechanism, they have potential targetable sites approximately every 35 bp. TALEN was first used to efficiently produce knockout rats and extended to mice. They have been shown to be successful in generating genome-edited mouse strains and transgenic large animal models of human disease. TALEN tools have also been studied for gene therapy applications with a number of preclinical applications including modifying induced pluripotent stem cells in the laboratory (Sebastian and Boch 2021).

Genome Editing with CRISPR Tools

While ZFN and TALEN are powerful genome editing tools, researchers have been looking for more precise genome alterations with high efficiency and ease of laboratory application. Clustered regularly interspaced short palindromic repeats (CRISPR) and CRISPR-associated proteins (Cas) have proven to be the answer (Adli 2018; Doudna 2020; Nidhi et al. 2021). Elements of an adaptive immune system against viruses found in most bacteria, they were recognized early after their discovery to be potential tools for genome editing. CRISPR sequences would be placed on each end of short stretches of DNA that bacteria and other prokaryotes have copied from invading viral phages, preserving a memory of the viruses that have attacked them in the past. These sequences are then transcribed into short RNAs and stored so that they can guide Cas proteins to matching viral sequences when exposed to the same phages in the future. The Cas proteins destroy the matching viral DNA by cutting it.

The initially identified CRISPRs were very simple systems, but were simplified further in the laboratories of Jennifer Doudna and Emmanuelle Charpentier and published in their 2012 milestone paper (Jinek et al. 2012). Literature to date often refers to "CRISPR-Cas9" as the genome editing tool. The CRISPR/Cas9 technology uses a specific RNA sequence called "guide RNA" which binds to another target sequence of DNA referred to as the "target DNA" followed by the cleavage of Cas9 where binding has occurred. The 2020 Nobel Prize in Chemistry honored Emmanuelle Charpentier and Jennifer Doudna for their work on CRISPR-Cas9. The CRISPR/Cas9 system stands out in the field as the most suitable gene editing tool presently as it improves the frequency of precise genome modifications in creating genetically edited animals.

There are numerous CRISPR systems now located in prokaryotes, each associated with a different set of CRISPR-associated proteins. This has become a very actively researched area as improvements, subtle variations are sought and limitations are explored. The most common current CRISPR tool, however, is derived from the CRISPR-Cas system isolated from *Streptococcus pyogenes* (the CRISPR-associated protein is Cas9). An excellent video depicting how CRISPR-Cas9 works can be found on the web at https://www.statnews.com/2018/0404/how-crispr-works--visualized/. In 2015, a second system, called CRISPR-Cpf1, which is even simpler and more specific (Zetsche et al. 2015) has been studied. The CRISPR-associated protein Cpf1 is simpler than systems with Cas as it requires only a single-stranded RNA for base-specific recognition. Also smaller in size than Cas, Cpf1 is thus potentially easier to deliver into cells and tissues. The CRISPR-Cpf1 tool will likely have major implications for genome editing research and medical applications. Research accelerates to increase the specificity of CRISPR systems with engineered RNA secondary structures as new systems are continuing to be discovered and explored (Doudna 2020; Nidhi et al. 2021). Gene editing technologies continue to evolve and improve precise insertions including recent new advances that connect the ability to insert larger segments of DNA into genes by harnessing transposons (known as "jumping genes, "big segments of DNA that can change position within a genome) (Klompe et al. 2019).

Research facilities and drug discovery labs have benefited as genome editing techniques have been reduced to a relatively easy experiment with multiple suppliers of the enzymes, reagents, and whole editing systems. Basic research into the mechanisms of DNA repair, functional genomic studies, and the creation of laboratory animals tailored with very specific gene alterations have proliferated as a result of the discovery of CRISPR tools and the refinement of ZFN and TALEN methodologies. CRISPR systems have provided molecular biology with powerful tools for genome editing both in the laboratory, but potentially also in live animals and humans (Adli 2018; Doudna 2020; Nidhi et al. 2021). Animal models of human genetic-related disorders and diseases are being created. Experiments have been designed for the production of interspecies chimeras. Transgenic animals are now being created with the ability to transcribe and translate human genes inserted into the host species into human proteins to facilitate xenotransplantation (discussed later in this chapter). Gene therapy applications, gene editing "drugs" for injection into the blood stream of patients, and studying gene function in human embryos (human embryo editing) have become possible.

In 2018, a biophysicist surprised the scientific community and unleased global outrage when he announced that the first gene-edited human babies had been created by his laboratory. It must again be stated that gene editing is a very powerful technology that has also far-reaching economic, bioethical, and national security implications. As exciting and promising as CRISPR technologies and other gene editing approaches are, health, scientific, governmental experts, as well as ethicists globally called for appropriate regulation and clinical protocols to limit the risk of ill-considered and unethical human gene editing projects.

Transgenic Animals and Plants and Gene Modified Cells/Tissues in Drug Discovery, Development, and Production

For thousands of years, man has selectively bred animals and plants either to enhance or to create desirable traits in numerous species. The explosive development of recombinant DNA technology, other OMIC technologies, and genome editing tools have made it possible to engineer species possessing particular unique and distinguishing genetic characteristics. As described in Chap. 1, the genetic material of an animal or plant can be manipulated so that extra genes may be inserted (transgenes) or replaced (i.e., human gene homologs coding for related human proteins), or deleted (knockout). New gene editing technologies including CRISPR have enhanced the ability to create novel mutations in numerous species of animals and plants. Theoretically, these approaches enable the introduction of virtually any gene into any organism. A greater understanding of specific gene regulation and expression, and the growing knowledge from functional genomic studies will contribute to important new discoveries made in relevant animal models. Such genetically altered species have found utility in a myriad of research and potential commercial applications including the generation of models of human disease, protein drug production, creation of organs and tissues for xenotransplantation, a host of agricultural uses, and drug discovery.

Transgenic Animals

The term transgenic animal describes an animal in which a foreign DNA segment (a transgene) is incorporated into their genome. Later, the term was extended to also include animals in which their endogenous genomic DNA has had its molecular structure manipulated. While there are some similarities between transgenic technology and gene therapy, it is important to distinguish clearly between them. Technically speaking, the introduction of foreign DNA sequences into a living cell is called gene transfer. Thus, one method to create a transgenic animal involves gene transfer (transgene incorporated into the genome). Gene therapy is also a gene transfer procedure and, in a sense, produces a transgenic human (will be discussed in a later chapter). In transgenic animals, however, the foreign gene is transferred indiscriminately into all cells, including germline cells. The process of gene therapy differs generally from transgenesis since it involves a transfer of the desired gene in such a way that involves only specific somatic and hematopoietic cells, and not germ cells. Thus, unlike in gene therapy, the genetic changes in transgenic organisms are conserved in any offspring according to the general rules of Mendelian inheritance. Please note that the ability to use engineered nuclease tools to edit the genome of an embryo (a germline cell) will conserve the genetic changes in any subsequent somatic and new generations of germline cells. These tools are becoming the predominant methods available to create new transgenic animals. The three production techniques described below may still be utilized to introduce the new DNA into what will become a transgenic animal if the engineered nuclease-based genome editing tools are used to build the mutated DNA to be inserted ex vivo. And the new and improved in vivo genome editing tools will likely become the method of choice in the future. A great resource that is updated frequently is the website "What is Biotechnology" and includes an excellent discussion of transgenic animals (What is Biotechnology? 2022).

The creation of transgenic animals is not new. They have been produced since the 1970s. However, modern biotechnology has greatly improved the methods of inducing the genetic transformation. While the mouse has been the most studied animal species, transgenic technology has been applied to a wide array of small and large mammals, fish (especially zebra fish), poultry, various lower animal forms such as insects, and numerous prokaryotes (Table 9.3). Transgenic animals have already made valuable research contributions to studies involving regulation of gene expression, the function of the immune system, genetic diseases, viral diseases, cardiovascular disease, and the genes responsible for the development of cancer. Transgenic animals have proven to be indispensable in drug lead identification, lead optimization, preclinical drug development, and disease modeling.

Table 9.3 Some key achievements related to the cloning of animals

Year	Cloned animals
1996	Sheep (Dolly)
1998	Cattle (Noto and Kaga)
1998	Goat (Mira)
2000	Mouse (Cumulina)
2000	Pig (a family)
2000	Mouflon (Ombretta, endangered animal)
2001	Cat (copy cat)
2001	Gaur (Noah, Asian wild ox)
2001	Rabbit
2002	Rat (Ralph)
2003	Mule (Idaho Gem)
2003	African wildcat (Ditteaux)
2003	Horse (Prometea)
2003	Deer (Dewey)
2004	Ferrets (Libby and Lilly)
2005	Wolves (Snuwolf and Snuwolffy)
2005	Dog (Snuppy)
2007	Deer
2009	Thoroughbred racing horse
2010	Spanish fighting bull
2011	Braham cattle
2011	Coyote
2015	TALEN gene-edited pigs
2015	CRISPR/Cas9 humanized pigs
2017	Monkey (Zhong Zhong and Hua Hua)
2021	First humanized pig to brain-dead human kidney transplant
2022	First humanized pig to human patient heart transplant

Production of Transgenic Animals by DNA Microinjection and Random Gene Addition

The production of transgenic animals has most commonly involved the microinjection (also called gene transfer) of 100–200 copies of exogenous transgene DNA into the larger, more visible male pronucleus (as compared to the female pronucleus) of a recipient fertilized embryo (see Fig. 9.10) (Bertolini et al. 2016; What is Biotechnology? 2022). The transgene contains both the DNA encoding the desired target amino acid sequence along with regulatory sequences that will mediate the expression of the added gene. The microinjected eggs are then implanted into the reproductive tract of a female and allowed to develop into embryos. The foreign DNA generally becomes randomly inserted at a single site on just one of the host chromosomes (i.e., the founder transgenic animal is heterozygous). Thus, each transgenic founder animal (positive transgene incorporated animals) is a unique species. Interbreeding of founder transgenic animals where the transgene has been incorporated into germ cells may result in the birth of a homozygous progeny provided the transgene incorporation did not induce a mutation of an essential endogenous gene. All cells of the transgenic animal will contain the transgene if DNA insertion occurs prior to the first cell division. However, usually only 20–25% of the offspring contain detectable levels of the transgene. Selection of neonatal animals possessing an incorporated transgene can readily be accomplished either by the direct identification of specific DNA or mRNA sequences or by the observation of gross phenotypic characteristics.

Production of Transgenic Animals by Retroviral Infection

The production of the first genetically altered laboratory mouse embryos was by insertion of a transgene via a modified retroviral vector (More details are provided in a later chapter of this textbook). The nonreplicating viral vector binds to the embryonic host cells, allowing subsequent transfer and insertion of the transgene into the host genome (Bertolini et al. 2016; What is Biotechnology? 2022). Many of the experimental human gene therapy trials have historically employed the same viral vectors. Advantages of this method of transgene production are the ease with which genes can be introduced into embryos at various stages of development, and the characteristic that only a single copy of the transgene is usually integrated into the genome. Disadvantages include possible genetic recombination of the viral vector with other viruses present, the size limitation of the introduced DNA (up to 7 kb of DNA, less than the size of some genes), and the difficulty in preparing certain viral vectors.

Production of Transgenic Animals by Homologous Recombination in Embryonic Stem Cells Following Microinjection of DNA

Transgenic animals can also be produced by the in vitro genetic alteration of pluripotent embryonic stem cells (ES cells) (see Fig. 9.11). ES cell technology is more efficient at creating transgenics than microinjection protocols (Bertolini et al. 2016; Smirnov et al. 2020; What is Biotechnology? 2022). ES cells, a cultured cell line derived from the inner cell mass (blastocyst) of a blastocyte (early preimplantation embryo), are capable of having their genomic DNA modified while retaining their ability to contribute to both somatic and germ cell lineages. The desired gene is incorporated into ES cells by one of several methods such as microinjection. This is followed by introduction of the genetically modified ES

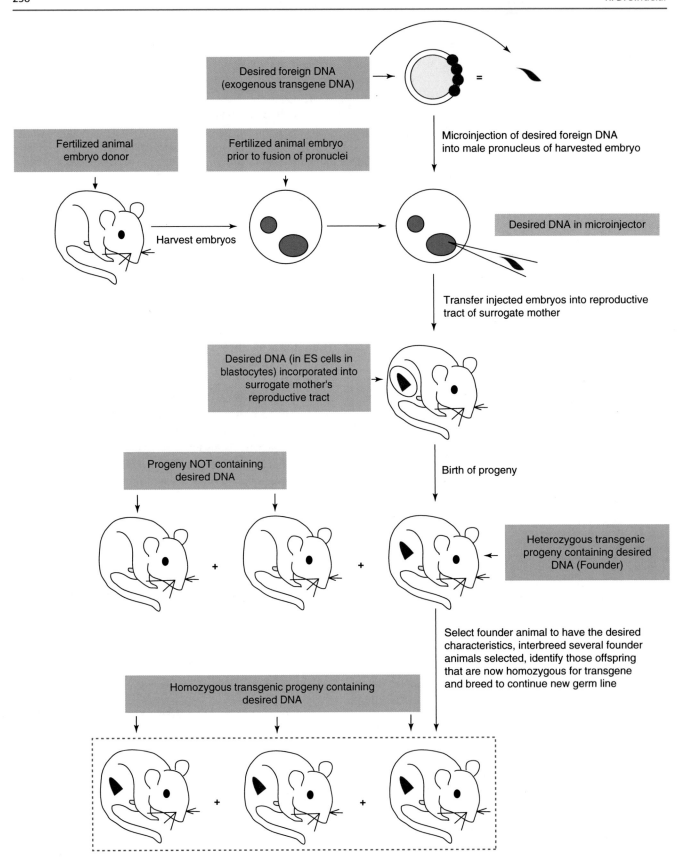

Fig. 9.10 Schematic representation of the production of transgenic animals by DNA microinjection that alters the DNA of all cells of the animal, both somatic and germline

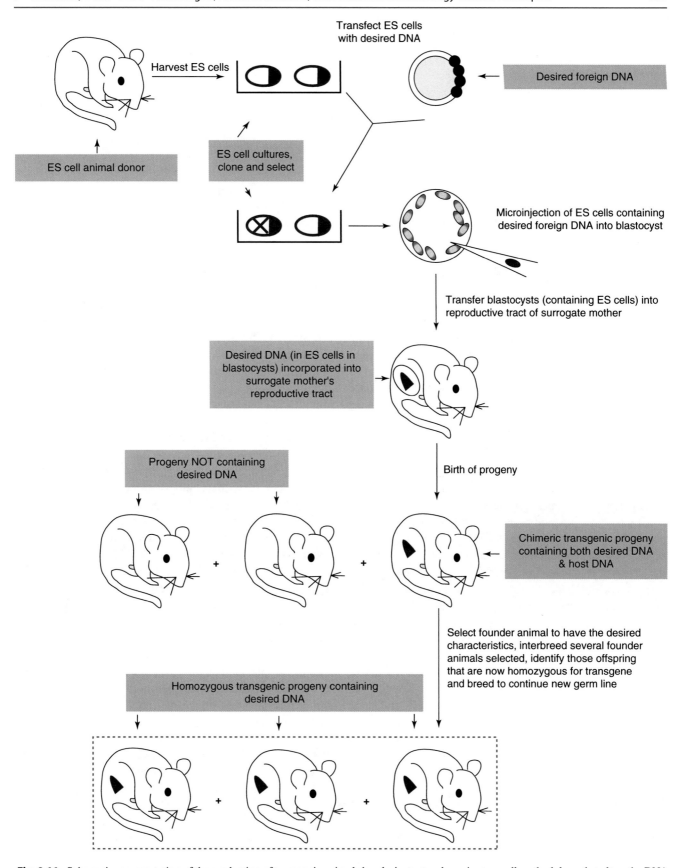

Fig. 9.11 Schematic representation of the production of transgenic animals by pluripotent embryonic stem cell methodology that alters the DNA of all cells of the animal, both somatic and germline

cells into the blastocyst of an early preimplantation embryo, selection, and culturing of targeted ES cells which are transferred subsequently to the reproductive tract of the surrogate host animal. The resulting progeny is screened for evidence that the desired genetic modification is present and selected appropriately. In mice, the process results in approximately 30% of the progeny containing tissue genetically derived from the incorporated ES cells. Interbreeding of selected founder animals can produce species homozygous for the mutation.

While transforming embryonic stem cells is more efficient than the microinjection technique described first, the desired gene must still be inserted into the cultured stem cell's genome to ultimately produce the transgenic animal. The gene insertion could occur in a random or in a targeted process. Nonhomologous recombination, a random process, readily occurs if the desired DNA is introduced into the ES cell genome by a gene recombination process that does not require any sequence homology between genomic DNA and the foreign DNA. While most ES cells fail to insert the foreign DNA, some do. Those that do are selected and injected into the inner cell mass of the animal blastocyst and thus eventually lead to a transgenic species. In still far fewer ES cells, homologous recombination occurs by chance. Segments of DNA base sequence in the vector find homologous sequences in the host genome, and the region between these homologous sequences replaces the matching region in the host DNA. A significant advance in the production of transgenic animals in ES cells is the advent of targeted homologous recombination techniques.

Homologous directed repair, while much rarer in transgenic research than nonhomologous end-joining, can be favored when the researcher carefully designs (engineers) the transferred DNA to have specific sequence homology to the endogenous DNA at the desired integration site and also carefully selects the transfer vector conditions. This targeted homologous recombination at a precise chromosomal position works particularly well in mice and provides an approach that can be used to produce knockout mice (to be discussed later). A modification of the procedure involves the use of hematopoietic bone marrow stem cells rather than pluripotent embryonic stem cells. The use of ES cells results in changes to the whole germ line, while hematopoietic stem cells modified appropriately are expected to repopulate a specific somatic cell line or lines (more similar to gene therapy).

The slow process of conventional breeding has been accelerated using modern genome sequencing techniques to help analyze the genotype–phenotype relationship of new generations produced by both conventional and modern biotech breeding techniques. To generate a herd (or a flock, etc.), alternative approaches would be advantageous. The

technique of nuclear transfer, the replacement of the nuclear genomic DNA of an oocyte (immature egg) or a single-cell fertilized embryo with that from a donor cell, is such an alternative breeding methodology. Animal "cloning" can result from this nuclear transfer technology. Judged a groundbreaking biotech achievement in 1997, creating the sheep Dolly, the first cloned mammal, from a single cell of a 6-year-old ewe was a feat many had thought impossible. Dolly was born after nuclear transfer of the genome from an adult mammary gland cell (See Table 9.3).

Transgenic Plants

A variety of biotechnology genetic engineering techniques have been employed to create a wealth of transgenic plant species as mentioned in Chap. 4: cotton, maize, soybean, potato, petunia, tobacco, papaya, rose, and others (Gerszberg and Hnatuszko-Konka 2022). Agricultural enhancements have resulted by engineering plants to be more herbicide tolerant, insect resistant, fungus resistant, virus resistant, and stress tolerant. Of importance for human health and pharmaceutical biotechnology, gene transfer technology is routinely used to manipulate bulk human protein production in a wide variety of transgenic plant species. It is significant to note that transgenic plants are attractive bulk bioreactors because their posttranslational modification processes often result in plant-derived recombinant human proteins with a similar glycosylation pattern to that found in the corresponding native human protein from a mammalian production system (cf. Chap. 2). Transplantation to the field followed by normal growth, harvest of the biomass, and downstream isolation and protein purification results in a valuable alternative crop for farming. Tobacco fields producing pharmaceutical grade human antibodies (sometimes referred to as "plantibodies") and edible vaccines contained in transgenic potatoes and tomatoes are not futuristic visions, but current research projects in academic and corporate laboratories and test fields. With antibody-based targeted therapeutics becoming increasingly important, the use of transgenic plants continues to expand once research helps to solve problems related to the isolation of the active protein drug and issues concerning cross-fertilization with non-genetically modified organisms (non-GMOs). In 2022, the first plant-based COVID vaccine for adults was approved in Canada. They used *Nicotiana benthamiana*, a close relative of the common tobacco plant, to manufacture a virus subunit protein (introduced into the plant by a gene-engineered bacterium) that mimics the SARS-CoV-2's spike protein (see https://covid-vaccine.canada.ca/covifenz/product-details). The leaves were harvested, the target protein purified and formulated to produce the vaccine [Covifenz®, COVID-19 Vaccine (plant-based virus-like particles [VLP], recombinant, adjuvanted)].

Biopharmaceutical Protein Production in Transgenic Animals and Plants: "Biopharming"

Transgenic animals have been used for a wide range of application including to identify the functions of specific factors in complex homeostatic systems; in toxicology to detect toxicants; in the pharmaceutical industry to produce targeted pharmaceutical proteins, drug production, and product efficacy testing; and to develop animals specially created for use in human xenograph transplantation.

Transgenic farm animals and crop plants continue to be explored as a means to produce large quantities of complex human proteins for the treatment of human disease. Such therapeutic proteins are currently produced in mammalian cell-based reactors, but this production process is expensive (Bertolini et al. 2016; Lamas-Toranzo et al. 2017; Gerszberg and Hnatuszko-Konka 2022) (see Table 9.4). CRISPR and other gene editing technologies are expected to fundamentally accelerate Biopharming successes.

Table 9.4 Some examples of human proteins that have been studied in transgenic animals and plants are provided to show the variety of what may become a viable technology in the future

Species	Protein product	Potential indication(s)
Cow	Collagen	Burns, bone fracture
Cow	Human fertility hormones	Infertility
Cow	Human serum albumin	Surgery, burns, shock, trauma
Cow	Lactoferrin	Bacterial GI infection
Goat	α-1-antiprotease inhibitor	Inherited deficiency
Goat	α-1-antitrypsin	Anti-inflammatory
Goat	Antithrombin III (ATryn)	Associated complications from genetic or acquired deficiency
Goat	Growth hormone	Pituitary dwarfism
Goat	Human fertility hormones	Infertility
Goat	Human serum albumin	Surgery, burns, shock, trauma
Goat	LAtPA2	Venous status ulcers
Goat	Monoclonal antibodies	Colon cancer
Goat	tPA2	Myocardial infarct, pulmonary embolism
Pig	Factor IX	Hemophilia
Pig	Factor VIII	Hemophilia
Pig	Fibrinogen	Burns, surgery
Pig	Human hemoglobin	Blood replacement for transfusion
Pig	Protein C	Deficiency, adjunct to tPA
Rabbit	Insulin-like growth factor	Wound healing
Rabbit	Interleukin-2	Renal cell carcinoma
Rabbit	Protein C	Deficiency, adjunct to tPA
Sheep	α − 1-antitrypsin	Anti-inflammatory
Sheep	Factor VIII	Hemophilia
Sheep	Factor IX	Hemophilia
Sheep	Fibrinogen	Burns, surgery
Sheep	Protein C	Deficiency, adjunct to tPA
Tobacco	IgG	Systemic therapy (rabies virus, hepatitis B virus)
Tobacco	TGF-β2	Ovarian cancer
Tobacco	Vitronectin	Protease
Tobacco	RhinoR	Fusion of human adhesion protein and human IgA for common cold
Tobacco	Ebola virus Mab	Treat Ebola infection
Tobacco	MERS-CoV DPP4-fc	Treat MERS infection
Tobacco-relative	COVID-19 vaccine	Prevent COVID-19 infection
Tomato	Beta-amyloid	Study of Alzheimer's disease
Tomato	Vaccines	Infectious disease
Safflower	Insulin	Diabetes
Carrot	DTP subunit vaccine	Infectious disease
Corn	Meripase	Cystic fibrosis
Cherry	Hep B surface antigen	Hep B vaccine production
Duckweed	Lacteron	Controlled release of α-interferon for hepatitis B and C
Potato	Poultry vaccine	Avian influenza (H5N1)
Cowpea	COVID-19 vaccine	Prevent COVID-19 infection

Data from references noted in this section of Chap. 9
tPA tissue plasminogen activator, *LAtPA* long-acting tissue plasminogen activator, *TGF-β3* tissue growth factor-beta, *DTP* diphtheria and tetanus and pertussis, *Hep B* hepatitis B

Xenotransplantation: Transplantable Transgenic Animal Organs

An innovative use of transgenics and biotechnology-engineered animals is the generation of clinically transplantable animal organs into humans as a potential treatment for end-stage organ disease. The success of human-to-human transplantation of heart, kidney, liver, and other vascularized organs (allotransplantation) created the significant expectation and need for donor organs. Primate-to-human transplantation (xenotransplantation) was successful, but ethical issues and limited number of donor animals were significant barriers. Transplant surgeons recognized early on that organs from the pig were a rational choice for xenotransplantation (due to physiological, anatomical, ethical, and supply reasons) if the serious hyperacute rejection could be overcome. Researchers in academia and industry have pioneered the transgenic engineering of pigs expressing both human complement inhibitory proteins and key human blood group proteins (antigens) including key immunogenic carbohydrates that are detected by the human immune system. These findings begin to pave the way for potential xenograft transplantation of animal components into humans with a lessened chance of acute rejection. However, pigs have a shorter life span (~27 years), and thus their tissues age at a faster rate than humans. Also, there have been concerns because pig organs harbor retroviruses that could be transmitted to humans. However, CRISPR tools have succeeded to inactivate most of the retroviruses found in experimental pig cell lines. In a major advance, in 2020 the US FDA approved a genetic modification of pigs so they do not produce galactose-alpha-1,3-galactose, a highly immuno-reactive sugar not found in humans, but present in numerous other mammalian species. This modification has provided pig organs that have been used for experimental kidney and heart transplants into humans with very limited success (BBC News 2022). With hundreds of thousands of patients on waitlists for heart and kidney transplants, human xenotransplant trials may close.

Knockout Mice (and Knockout Rats)

"Knockout phenotypes" have been extensively used to develop new drugs including the most commonly prescribed medications. While many species including mice, zebra fish, and nematodes have been transformed to lose genetic function for the study of drug discovery and disease modeling, mice have proven to be the most useful. Mice are the laboratory animal species most closely related to humans in which the knockout technique can be easily performed, so they are a favorite subject for knockout experiments and are the gold standard. While a mouse carrying an introduced transgene is called a transgenic mouse, transgenic technologies can also produce a knockout animal. A knockout mouse, also called a gene knockout mouse or a gene-targeted knock-out mouse, is an animal in which an endogenous gene (genomic wild-type allele) has been specifically inactivated or "knocked out" by replacing it with a null allele (National Human Genome Research Institute 2022, 2023). A null allele is a nonfunctional allele of a gene generated by either deletion of the entire gene or mutation of the gene resulting in the synthesis of an inactive protein. Recent advances in intranuclear gene targeting and embryonic stem cell technologies (as described above and in later chapters of this textbook) are expanding the capabilities to produce knockout mice routinely for studying certain human genetic diseases or elucidating the function of a specific gene product.

The procedure for producing knockout mice basically involves a four-step process. A null allele (i.e., knockout allele) is incorporated into one allele of murine embryonic stem (ES) cells. Incorporation is generally quite low; approximately one cell in a million has the required gene replacement. However, the process is designed to impart neomycin and ganciclovir resistance only to those ES cells in which homologous gene integration has resulted. This facilitates the selection and propagation of the correctly engineered ES cells. The resulting ES cells are then injected into early mouse embryos creating chimeric mice (heterozygous for the knockout allele) containing tissues derived from both host cells and ES cells. The chimeric mice are mated to confirm that the null allele is incorporated into the germ line. The confirmed heterozygous chimeric mice are bred to homogeneity-producing progeny that are homozygous knockout mice. Worldwide, there are major mouse knockout programs and collectives that have attempted to pool results and to create a mutation in each of the approximately 20,000 protein-coding genes in the mouse genome using a combination of gene trapping and gene targeting in mouse embryonic stem (ES) cells. These consortia have changed over recent years with most agreeing to work together to achieve this goal in C57BL/6 mouse ES cells and called the International Mouse Phenotyping Consortium (IMPC 2022). This is an exceptional source for background and technical information on knockout mice, links to the many consortia members and labs working in this area of research, and tools to access the 8267 total knockout genes phenotyped thus far. This comprehensive and publicly available resource aids researchers examining the role of each gene in normal physiology and development and sheds light on the pathogenesis of abnormal physiology and diseases. Continuing discoveries will further create better animal models of human monogenic and polygenic diseases such as cancer, diabetes, obesity, cardiovascular disease, and psychiatric and neurodegenerative diseases.

Rats have not routinely been bred as knockouts primarily due to not isolating rat ES cells necessary for the process until much later than mouse ES cells. Since the isolation of rat ES cells plus the development of the genome editing techniques for manipulating human genes now being applied to rats and other animal species, knockout rat stains have become available to the research community, but will not likely replace the important role of knockout ice. The whole field of animal models is rapidly evolving. There is little doubt that CRISPR tools and related techniques will create a boom in genetically modified mouse models. The many C57BL/6 mouse strains now available are revolutionizing drug discovery.

3D Cell Cultures and Organoids

With new biotechnology tools becoming increasingly available especially stem cell methodologies, researchers are continually trying to develop new and improved in vitro technologies to replace some animal models in their study of biology and the drug development process. One approach is to grow in the laboratory miniature versions of human organs that simulate the anatomical, physiological, and mechanical properties of the real thing. Classic 2D cell culture techniques were being enhanced for greater realistic microenvironment simulation by new "3D cell culture" techniques (Mapanao and Voliani 2020). Unlike in 2D cell cultures where the cells are grown as thin layers in petri dishes, the cells in 3D cell cultures are allowed to grow in all three dimensions usually in bioreactors creating cell colonies or spheroid-shaped accumulations of cells. These cultures are derived from one or a few cells from a specific tissue type or organ, or by manipulation of induced pluripotent stem cells (to be discussed in a later chapter in this textbook). When the 3D cell cultures are derived from specific organs, the in vitro 3D-organ cell culture that is created is known as an "organoid" (Nature 2018). *Nature* magazine named the organoid its "Method of the Year 2017." 3D cell culturing has become a commonly used method in the study of cancer and cancer cell growth. Numerous organoids have been grown including brain, liver, lung, stomach, pancreas, ear, thyroid, GI, kidney, and testicular. 3D cell cultures including organoids are transforming medical research and have many potential uses that are being explored. Besides cancer studies, uses include toxicity studies, modeling infectious disease in humans, personalized medicine, preclinical drug development, microbiome studies, and regenerative medicine.

"On-a-Chip" Technology ("Organ-on-Chip")

Recent advances in systems biology, stem cell technology, materials science, 3D cell culture (expansion of the classic 2D cell culture techniques allowing greater cell–cell communication and interaction better mimicking true complex tissue architecture) and microfluidics (the physics and manipulation of extremely small amounts of fluids) have allowed researchers to develop miniature models of human "organs on a chip (OOC)." Basically, an artificial organ, the 3D cell culture grown on a chip is linked to microfluid handling and sensor capabilities that together simulate the ana-

tomical, physiological, and mechanical properties and responses of entire organs and organ systems (Low et al. 2021). The technology opens up great opportunities for next-generation experiments to mimic human organ functionality, microphysiology, and morphology in vitro, replacing traditional animal-based model systems. This technology can also be used as a pharmacokinetic model when new drugs are being developed. The chip support framework is generally a silicon wafer or piece of plastic of only millimeters to a few square centimeters in size.

Lab-on-a-chip devices were being developed to integrate one-or-more laboratory functions onto a single engineered chip that required minute amounts of sample fluid volume and reagents to complete the task. The field developed rapidly in the mid-1990s due to developments in microarray genomics applications, key improvements in micro-electromechanical systems engineering, and the need for portable biological and chemical warfare agent detection systems. Academic laboratories and pharmaceutical companies have embraced the use and further development of this in vitro technology to replace some animal models in the drug development process. Numerous organ systems have been created and tested including lung, liver, heart, artery, and kidney chips that provide an alternative test system to traditional toxicity studies in laboratory animals. The chips are characterized for use as both normal organ tissues as well as diseased organs. Researchers have been experimenting with the integration of multiple organ chips into a single system that may in the future mimic the human body. With further development, "on-a-chip" technology may have a significant impact on high-throughput screening, toxicity testing, drug delivery, and overall drug development.

Synthetic Biology

Modern biotechnology tools have allowed for a number of ways to study very complex biological systems. For example, as described above, systems biology examines complex biological systems as interacting and integrated complex networks. The rapidly developing and disruptive field of study known as "synthetic biology" explores how to build artificial complex biological systems employing many of the same tools and experimental techniques favored by system biologists. While engineering synthetic systems is complex, this approach is increasing in scope with respect to the organisms built, the practical outcomes observed, and the complex integration achieved (Karoui et al. 2019; Hanczyc 2020; Tang et al. 2021). Synthetic biology looks at both the strategic redesign and/or the fabrication of existing biological systems and the design and construction of biological components and systems that do not already exist in nature. Synthetic biology studies often take parts of natural biological systems, characterizing and simplifying them, and then using them as a component of a highly unnatural, engineered, biological system. Synthetic biology studies may provide a more detailed understanding of complex biological systems down to the molecular level. Being able to design and construct a complex system is also one very practical approach to understanding that system under various conditions. The levels a synthetic biologist may work at include the organism, tissue and organ, intercellular, intracellular, biological pathway, and down to the molecular level.

There are many exciting applications for synthetic biology that have been explored or hypothesized across various fields of scientific study including designed and optimized biological pathways, natural product manufacturing, new drug molecule synthesis, and biosensing. From an engineering perspective, synthetic biology could lead to the design and building of engineered biological systems that process information, modify existing chemicals, fabricate new molecules and materials, and maintain and enhance human health and our environment. Because of the obvious societal concerns that synthetic biology experiments raise, the broader science community has engaged in considerable efforts at developing guidelines and regulations and addressing the issues of intellectual property and governance and the ethical, societal, and legal implications. Several bioethics research institutes published reports on ethical concerns and the public perception of synthetic biology.

An early example in 2006 from a research team at the J. Craig Venter Institute constructed and filed a patent application for a synthetic genome of a novel synthetic minimal bacterium named *Mycoplasma laboratorium* (Glass et al. 2007). The team was able to construct an artificial chromosome of 381 genes, and the DNA sequence they have pieced together is based upon the bacterium Mycoplasma genitalium. This is obviously beyond the modification of just a single gene as was the most common approach in early biotechnology approaches. The original bacterium had a fifth of its DNA removed and was able to live successfully with the synthetic chromosome in place. Venter's goal is to make cells that might take carbon dioxide out of the atmosphere and produce methane, used as a feedstock for other fuels.

Synthetic biology-based technologies may transform today's industries including petrochemicals and pharmaceuticals. Many of the first successful applications of synthetic biology have been the bioproducts and services that replace existing products we use every day built upon molecular ingredients previously derived from limited sources such as petroleum (plastics, etc.). Renewable bioacrylics from sugars, green chemicals from agricultural waste, surfactants, biosynthetic rubber, and many more examples are products of this approach. Synthetic biology methods can modify a yeast to produce opioids from sugar (Höhne and Kabisch 2016). New drug delivery platforms, vaccine components, designed proteins, artificial cells, organoids, and novel cell-based platforms are being explored with potential major implications for the pharmaceutical industry and human medicine.

Biotechnology and Drug Discovery

Pharmaceutical scientists have taken advantage of every opportunity or technique available to aid in the long, costly, and unpredictable drug discovery process. In essence, Chap. 9 is an overview of some of the many applications of biotechnology and related techniques useful in target identification and validation, drug discovery or design, lead optimization, and development including preclinical and clinical studies. The techniques described throughout Chap. 9 have changed forever the way drug research is conducted, refining the process that optimizes the useful pharmacological properties of an identified novel molecular lead and minimizes the unwanted properties. The promise of genomics, transcriptomics, proteomics, microarrays, pharmacogenomics/genetics, epigenomics, precision medicine, metabonomics/metabolomics, microbiomics, toxicogenomics, glycomics, systems biology, genome editing, genetically engineered animals, and bioinformatics and big data has radically changed the new drug discovery paradigm. Drug discovery and design are among the most intensively developing modern sciences with its progress accelerated exponentially by biotechnology. Figure 9.12 shows schematically the interaction of three key elements that are essential for modern drug discovery: (1) new targets identified by genomics, proteomics, and related technologies; (2) validation of the identified targets; rapid, sensitive bioassays utilizing high-throughput screening methods; and (3) new molecule creation and optimization employing a host of approaches (Eder and Herrling 2016; Reddy et al. 2016; Doytchinova 2022). The key elements are underpinned at each point by bioinformatics and now big data. The most advanced method for drug discovery is "rational drug design" beginning with the identification of a biological target discovered by or further mechanistically validated by modern biotechnologies. Several of the technologies, methods, and approaches listed in Fig. 9.12 have been described previously in this chapter. Others will be described below.

Modern Small Molecule Drug Screening and Synthesis
Traditionally, drug discovery programs relied heavily upon random screening followed by analog synthesis and lead optimization via structure-activity relationship studies. Discovery of novel, efficacious, and safer small molecule medicinal agents with appropriate "drug-like characteristics is an increasingly costly and complex process." Therefore, any method allowing for a reduction in time and money is

Fig. 9.12 Elements of modern drug discovery: impact of biotechnology

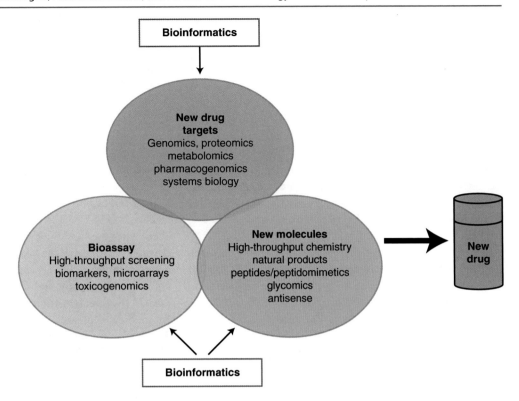

extremely valuable. Advances in biotechnology have contributed to a greater understanding of the cause and progression of disease and have identified new therapeutic targets forming the basis of novel drug screens. New technical discoveries in the fields of proteomics for target discovery and validation and systems biology are expected to facilitate the discovery of new agents with novel mechanisms of action and targets for diseases that were previously difficult or impossible to treat (Finn et al. 2017; Emmerich et al. 2021). Proteins including monoclonal antibodies and RNA molecules have become popular and valuable drug leads and approved pharmaceutical products. Their discovery and development are facilitated by all of the technologies described in this chapter. However, small molecule drugs are still highly desired. Therefore, in an effort to decrease the cost of identifying and optimizing useful, quality drug leads (both small molecule and biologic) against a pharmaceutically important target, researchers have developed newer approaches including high-throughput screening and high-throughput synthesis methods. Applications of biotechnology to in vitro screening include the improved preparation of (1) cloned membrane-bound receptors expressed in cell lines carrying few endogenous receptors; (2) immobilized preparations of receptors, antibodies, and other ligand-binding proteins; and (3) soluble enzymes and extracellular cell-surface expressed protein receptors. In most cases today, biotechnology contributes directly to the understanding, identification, and/or the generation of the drug target being screened (e.g., radioligand binding displacement from a cloned protein receptor).

Previously, libraries of synthetic compounds along with natural products from microbial fermentation, plant extracts, marine organisms, and invertebrates provide a diversity of molecular structures that were screened randomly. Screening can be made more directed if the compounds to be investigated are selected on the basis of structural information about the receptor or natural ligand. The development of sensitive radioligand binding assays and the access to fully automated, robotic screening techniques have accelerated the screening process. High-throughput screening (HTS) provides for the bioassay of thousands of compounds in multiple assays at the same time. The process is automated with robots and utilizes multi-well microtiter plates. Today, companies can conduct 100,000 bioassays a day. In addition, modern drug discovery and lead optimization with DNA microarrays and other biotechnologies allow researchers to track hundreds to thousands of genes.

New and novel methods of "computer-aided drug discovery" (CADD) can significantly accelerate drug lead discovery against new drug targets. In addition, artificial intelligence tools are increasingly being applied to modern drug discovery and development (Schneider et al. 2020). Computationally, "virtual drug screening" is a computer-based approach to predict drug activity by fitting chemical structures to targets. This type of screening is commonly used to rapidly test a library of putative drug leads for their potential to bind and inhibit target receptor or enzyme targets. This CADD approach is routinely used to process virtual libraries containing millions of molecular structures against a variety of drug targets, many discovered using biotechnologies, with known 3D structures. Normally, computational screening is a years-long stage of the drug development process, during which pharmaceutical scientists screen millions of molecular structures in databases for their potential efficacy toward the target of interest. Obviously, it is essential to maximize the number of molecular entities in databases to be evaluated against the target of interest. With the recent explosive growth in structurally diverse chemical libraries beyond a billion molecules, more efficient virtual screening approaches are being developed. These techniques are poised to revolutionize virtual screening technology and can significantly accelerate the effective exploration of ultra-large portions of a chemical space. For example, a new "Deep Docking (DD)" platform has been reported that enables up to 100-fold acceleration of structure-based virtual target screening (Gentile et al. 2020). This exciting new computational screening method results in hundreds- to thousands-fold virtual hit enrichment and, thus, enables the screening of billion molecule-sized chemical libraries without using extraordinary computational resources (Open-source protocol for this method can be found at https://github.com/jamesgleave/DD_protocol).

Traditionally, small drug molecules were synthesized by joining together structural pieces in a set sequence to prepare one product. One of the most powerful tools to optimize drug discovery is automated high-throughput synthesis. When conducted in a combinatorial approach, high-throughput synthesis provides for the simultaneous preparation of hundreds or thousands of related drug candidates. There are two overall approaches to high-throughput synthesis: combinatorial chemistry that randomly mixes various reagents (such as many variations of reagent A with many variations of reagent B to give random mixtures of all products in a reaction vessel) and parallel synthesis that selectively conducts many reactions parallel to each other (such as many variations of reagent A in separate multiple reaction vessels with many variations of reagent B to give many single products in separate vessels) (Please see Fig. 9.13). Assigning the task to automated synthesizing equipment results in the rapid creation of large collections or libraries (as large as 10,000 compounds) of diverse molecules. Ingenious methods have been devised to direct the molecules to be synthesized, to identify the structure of the products, to purify the products via automation, and to isolate compounds. When coupled with high-throughput screening, thousands of compounds can be generated, screened, and evaluated for further drug discovery and development in a matter of weeks.

Chemical Genomics (Chemogenomics)

Having the ability to sequence the human genome coupled with the growth of proteomics now provides researchers with an abundance of potential new drug targets and the need to validate the roles of these newly identified human gene products. In modern drug discovery, chemical genomics (sometimes called chemogenomics or more generally included as a subset of chemical biology) uses chemical probes to help define the complexity of biological systems at the genomics and proteomics level. It involves the screening of large chemical libraries (typically combinatorially derived "druggable" small molecule libraries covering a broad expanse of "diversity space") against all genes or gene products, such as proteins or other targets (i.e., chemical universe screened against target universe) (Finn et al. 2017; Emmerich et al. 2021). In Fig. 9.12, basically chemical genomics would occur when the "new molecules" to be tested in the "bioassay" developed from the "new drug targets." OMIC approaches came from large chemical libraries typically of small molecules synthesized by high-throughput chemistry. As part of the US National Institutes of Health (NIH) Roadmap for Biomedical Research, the National Center for Advanced Translational Medicine (NCATS) Chemical Genomics Center (NCGC) has led an effort to offer public sector biomedical researchers access to libraries of small organic molecules that can be used as chemical probes to study cellular pathways in greater depth (NCATS 2022). It remains difficult to predict which small molecule compounds will be most effective in a given situation. Researchers can maximize the likelihood of a successful match between a chemical compound, its usefulness as a research tool, and its desired therapeutic effect by systematically screening libraries containing thousands of small molecules. Drug candidates are expected from the correlations observed during functional analysis of the molecule–gene product interactions. Genomic profiling by the chemical library may also yield relevant new targets and mechanisms. The target universe will be well-characterized when both the function of the receptor target has been recognized as well as a set of specific molecules that have the ability to bind to the target and modulate it. Chemical genomics has changed the drug discovery paradigm and the approach for the investigation of target pharmacology (see Figs. 9.1 and 9.12). It is expected to be a critical component of drug lead identification and proof of principle determination for selective modulators of complex enzyme systems including proteases, kinases, G-protein-coupled receptors, and nuclear receptors.

Drug Repurposing (Drug Repositioning or Drug Reprofiling)

Even with the tremendous advances in biotechnology including an enhanced understanding of human disease, the discovery and development of a new drug remains slower than expected. In an effort to reduce the average 14-year time frame it takes to translate a new molecule into an approved

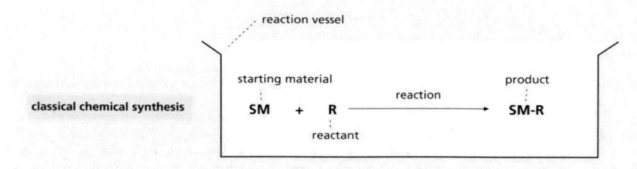

reaction vessel

classical chemical synthesis

starting material

SM   +   R $\xrightarrow{\text{reaction}}$ SM-R

reactant

product

In classical chemical synthesis, a coupling reaction of one starting material (SM) with one reactant (R) would yield just one product, SM-R. One or several reactions may be run simultaneously in separate reaction vessels

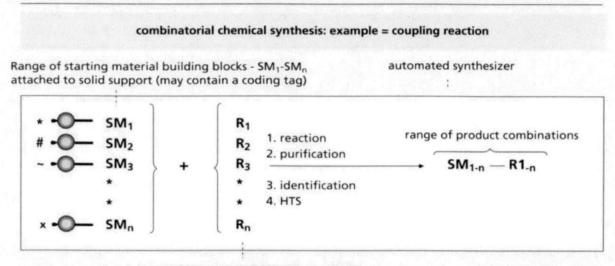

combinatorial chemical synthesis: example = coupling reaction

Range of starting material building blocks - SM_1-SM_n attached to solid support (may contain a coding tag)

automated synthesizer

* -〇— SM_1
\# -〇— SM_2
~ -〇— SM_3
*
*
x -〇— SM_n

+

R_1
R_2
R_3
*
*
R_n

1. reaction
2. purification
3. identification
4. HTS

range of product combinations

$SM_{1\text{-}n}$ — $R1_{\text{-}n}$

range of reactant building blocks - R_1-R_n

In a combinatorial chemical synthesis such as a coupling reaction, a range of starting material building blocks are reacted with a range of reactant building blocks yielding any or all possible product combinations, $SM_{1\text{-}n}$ $R_{1\text{-}n}$. The automated reactions may occur in the same reaction vessel (and coding tag used to separate/identify) or may each occur in small, separate reaction vessels (parallel synthesis)

Fig. 9.13 A schematic representation of a coupling reaction: difference between classical chemical synthesis and combinatorial chemistry

drug, to improve the success rate and to lower the cost of drug discovery and development, drug repurposing has become a useful strategy. Drug repurposing, sometimes called drug repositioning or drug reprofiling, builds upon the existing knowledge of the pharmacology, formulation, toxicity, and human clinical testing of a known drug or drug candidates to quicken the translation from bench to bedside. This approach discovers new indications for approved or significantly developed existing agents thereby decreasing development time and cost (Pushpakom et al. 2019). Drug repurposing has benefited from the rapid growth of proteomics, metabolomics, and other OMIC technologies to help define new pathways and drug targets for the existing

molecules. Repurposing can occur at an early or late stage of drug development. The NCATS also supports research into drug repurposing. A classic example of a successful drug repurposing is the former sedative/hypnotic drug thalidomide used for nausea and to alleviate the symptoms of morning sickness in pregnant women. Removed from the market due to causing significant malformation of the limbs in infants born to mother users of the drug, today thalidomide has been repurposed and approved as a drug to treat certain myelomas and the complications from leprosy. The COVID-19 pandemic has re-emphasized the opportunities for drug repurposing and hastened the repurposing of antiviral drugs and other drug molecules as treatments for the coronavirus-induced disease. Improved access to preclinical compounds and greater access to Phase II–Phase IV clinical data will help realize the full potential of drug repurposing (Pushpakom et al. 2019).

Protein Engineering and Drug Discovery

Structurally complex, proteins are the building blocks of life and are essential for a wide range of cellular activities including enzymatic and structural functions with our cells, tissues, and organs. Proteins can play these roles because they have the ability to achieve complex structural folds with precisely defined structural and dynamic properties. An increasing number of new drugs entering the market today are protein drugs with several hundred introduced since the 1980s. Recombinant DNA (rDNA) technology and a number of other biotechnologies have facilitated the drive to readily modify protein structure and function. Generally known as "protein engineering," this complex process has accelerated exponentially over the last two decade and has become a major catalyst for the design and production of protein drugs. Protein engineering is the selective design and modification of protein structure and function using rDNA technology, often phage technologies, or chemical treatment to produce new proteins and especially enzymes to carry out desired and possibly novel functions for medicine, industry, or agriculture (Lutz and Iamurri 2018; Singh et al. 2018). Structural genomics and functional genomics advances have fueled protein engineering in the postgenomic era. The 3D structures of proteins with known functions and mechanisms facilitate protein engineering approaches to design variants with specific desired properties.

Since protein therapeutics are a rapidly expanding segment of the approved drugs around the world, protein engineering holds the real potential to radically transform our ability to design and create selective customized protein drugs and, thus, significantly improve clinical outcomes. Today, there are generally considered three major approaches of protein engineering research. They are directed evolution, rational protein design, and de novo protein design. The full scope of this exciting area of research, development, and customized protein product production goes well beyond what this introductory textbook may provide. A quick review of the current literature uncovers the tremendous success of protein engineering. Examples include superior or new and novel biological activity, improved enzymatic activity, enhanced binding efficiency and selectivity, broadened substrate specificity, accelerated expression, greater thermostability, novel fusion proteins, raised or reversed stereoselectivity in vivo, etc. The protein engineering product and services market is estimated to be worth $3.02 billion USD by 2027 (Globe News Wire 2022). Thus, only a brief introductory overview of the major strategies for protein engineering follows.

Methods of Protein Engineering

Emerging as one of the most effective protein engineering approaches, directed evolution is an iterative process that expedites the natural evolutionary progression of biomolecules and biological systems. It has an advantage that it requires no prior structural knowledge of the starting state protein of interest. Also, through this approach, it is possible to identify undiscovered protein sequences which have novel functions. Its major disadvantage is that expensive robotic equipment must be used to automate the high-throughput screening needed to analyze the large number of protein variants for the desired properties. The approach basically is similar to that used by breeders of animals and other farm products over the ages. After identifying the starting state protein, the protein-producing gene is subjected to an iterative two-step sequence of gene diversification and modifications via methods to introduce random or region-specific mutagenesis (Step 1). The resulting protein library produced is screened (Step 2), often using phage display methods and proteins with the desired properties selected followed by reproduction and repeat cycles of the process (Molina-Espeja et al. 2016). Francis Arnold, George Smith, and Gregory Winter received the 2018 Nobel Prize in Chemistry for developing tools for protein engineering via the directed evolution approach. Their work and the research that followed has led to a number of commercial products and drugs including, for example, the engineered antibody to TNF-α, called adalimumab (Humira) using Winter's techniques (Service 2018).

Dating back to the 1970s, the most classical method in protein engineering is rational protein design (Lutz and Iamurri 2018; Korendovych 2018). Unlike directed evolution, rational protein design is only an effective approach when the starting state protein 3D structure (generally from X-ray crystallography, cryo-electron microscopy, or NMR analysis), biophysical properties, and its function and mechanism are well understood. Recent advances in computation prediction techniques of protein folding motifs will likely help overcome some of these disadvantages in the future. Nevertheless, rational protein design has the advantage of being relatively inexpensive and technically easy, since the site-directed mutagenesis techniques, peptide synthesis methodology, and artificial gene synthesis approaches utilized are well developed to create specific, targeted DNA insertions, deletions, and substitutions in double-stranded plasmid DNA. In rational protein design, the synthesized target protein is screened to determine if the design variants possess the desired properties.

De novo protein design uses computational protein design algorithms to design customized synthetic proteins by using known 3D structures of natural proteins and their known folding motifs (Lutz and Iamurri 2018; Korendovych and DeGrado 2020). This approach is likely the broadest, most effective tactic for generating functional macromolecules previously unknown to nature. Knowledge of the starting state protein and the desired target properties assists in the computational prediction of new target variant structures. Common computer-assisted protein molecular modeling approaches including ab initio calculations, fragment-based modeling, protein homology modeling, and protein threading are utilized to de novo design the target biomolecule of interest.

The various protein engineering approaches are not mutually exclusive. Often, researchers and biotech engineers apply more than one approach together. Often referred to as semi-rational protein design, or knowledge-based library protein design, this hybrid approach utilizes information on protein sequence, structure, and function similar to rational protein design as well as computational predictive algorithms common to de novo protein design to preselect promising target amino acid residues which are most likely to produce the desired protein function. Mutations of these key amino acid residues create mutant protein libraries that are more likely to have the sought-after function and properties. Newer methods have explored using unnatural amino acids residues to code for novel expanded genetic code-produced protein variants.

Conclusion

Tremendous advances have occurred in biotechnology since Watson and Crick determined the structure of DNA. Improved pharmaceuticals, novel therapeutic agents, unique diagnostic products, and new drug design tools have resulted from the escalating achievements of pharmaceutical biotechnology. While recombinant DNA technology and hybridoma techniques received most of the press in the late 1980s and early1990s, a wealth of additional and innovative biotechnologies and approaches have been, and will continue to be, developed in order to enhance pharmaceutical research and transform our understanding of disease and its precision treatment. Biotechnology has advanced well beyond recombinant proteins and monoclonal antibodies, driven largely by a plethora of new and novel biotech methods. Genomics, transcriptomics, proteomics, microarrays, pharmacogenomics/genetics, epigenomics, personalized and precision medicine, metabonomics/metabolomics, microbiomics, toxicogenomics, glycomics, systems biology, genome editing, genetically engineered animals, synthetic biology, parallel high-throughput screening, virtual screening, automated drug lead

synthesis, and protein engineering are directly influencing the pharmaceutical sciences, medicine, and pharmaceutical care. Advances in the science are accelerating. Application of these, and yet to be discovered biotechnologies, will continue to reshape effective drug therapy as well as improve the competitive, challenging process of drug discovery and development of new medicinal agents and diagnostics. These extremely powerful technologies and their current and future applications present scientists, clinicians including pharmacists, policy makers, and the public far-reaching economic, bioethical, and national security opportunities as well as concerns that must be explored, addressed, and exploited appropriately.

Self-Assessment Questions

Questions
1. What were the increasing levels of genetic resolution of the human genome planned for study as part of the HGP?
2. What is functional genomics?
3. What is proteomics?
4. What are SNPs?
5. What is the difference between pharmacogenetics and pharmacogenomics?
6. Define metabonomics.
7. What is protein engineering?
8. What phase(s) of drug action is (are) affected by genetic variation?
9. Define precision medicine.
10. What is a biomarker?
11. Define systems biology.
12. What is genome editing and what are the four genome editing nucleases?
13. What is the CRISPR-Cas9 tool?
14. Why are engineered animal models valuable to pharmaceutical research?
15. What is a knockout mouse?

Answers
1. HGP structural genomics was envisioned to proceed through increasing levels of genetic resolution: detailed human genetic linkage maps [approx. two megabase pairs (Mb = million base pairs) resolution], complete physical maps (0.1 Mb resolution), and ultimately complete DNA sequencing of the approximately 3.5 billion base pairs (23 pairs of chromosomes) in a human cell nucleus [one base pair (bp) resolution].
2. Functional genomics is an approach to genetic analysis that focuses on genome-wide patterns of gene expression, the mechanisms by which gene expression is coordinated, and the interrelationships of gene expression when a cellular environmental change occurs.

3. The research area called proteomics seeks to define the function and correlate that with expression profiles of all proteins encoded within a genome.

4. While comparing the base sequences in the DNA of two individuals reveals them to be approximately 99.9% identical, base differences, or polymorphisms, are scattered throughout the genome. The best-characterized human polymorphisms are single-nucleotide polymorphisms (SNPs) occurring approximately once every 1000 bases in the 3.5 billion base pair human genome.

5. Pharmacogenetics is the study of how an individual's genetic differences influence drug action, usage, and dosing. A detailed knowledge of a patient's pharmacogenetics in relation to a particular drug therapy may lead to enhanced efficacy and greater safety. While sometimes used interchangeably (especially in pharmacy practice literature), pharmacogenetics and pharmacogenomics are subtly different. Pharmacogenomics introduces the additional element of our present technical ability to pinpoint patient-specific DNA variation using genomic techniques. While overlapping fields of study, pharmacogenomics is a much newer term that correlates an individual patient's DNA variation (SNP level of variation knowledge rather than gene level of variation knowledge) with his or her response to pharmacotherapy.

6. The field of metabonomics is the holistic study of the metabolic continuum at the equivalent level to the study of genomics and proteomics.

7. Protein engineering is the selective design and modification of protein structure and function using rDNA technology or chemical treatment to produce new proteins to carry out desired and possibly new functions for medicine, industry, or agriculture.

8. Genomic variation affects not only the pharmacokinetic profile of drugs (via drug metabolizing enzymes and drug transporter proteins), it also strongly influences the pharmacodynamic profile of drugs via the drug target.

9. The goal of precision medicine is to enable clinicians to employ the most appropriate course of action for each individual patient managing the extreme complexity of each patient given all of the tools now available in the health care system: OMICS technology data, disease mechanisms, the electronic health record, public health information, big data, etc. Sometimes referred to as giving the right drug to the right patient in the right dose at the right time.

10. Biomarkers are clinically relevant substances used as indicators of a biologic state. Detection or concentration change of a biomarker may indicate a particular disease state physiology or toxicity. A change in expression or state of a protein biomarker may correlate with the risk or progression of a disease, with the susceptibility of the disease to a given treatment or the drug's safety profile.

11. Systems biology is the study of the interactions between the components of a biological system, and how these interactions give rise to the function and behavior of that system.

12. Genome editing (sometimes called genome engineering) is the approach in modern biology to make specific changes to the DNA of a cell or organism and observing its impact. Key to genome editing is the ability to selectively and predictively modify the DNA sequence of the target organism and assure the integrity of the resulting edits. The four nucleases are meganucleases, ZFN, TALEN, and CRISPR.

13. **C**lustered **r**egularly **i**nterspaced **s**hort **p**alindromic **r**epeats (CRISPR) and **C**RISPR-**as**sociated proteins (Cas) are elements of an adaptive immune system against viruses found in most bacteria. They are used as sequence-specific genome editing tools. There are numerous CRISPR systems found in prokaryotes, each associated with a different set of CRISPR-associated proteins. The most common current CRISPR tool is derived from the CRISPR-Cas system isolated from *Streptococcus pyogenes* (the CRISPR-associated protein is Cas9).

14. Engineered animal models are proving invaluable since small animal models of disease are often poor mimics of that disease in human patients. Genetic engineering can predispose an animal to a particular disease under scrutiny, and the insertion of human genes into the animal can initiate the development of a more clinically relevant disease condition.

15. A knockout mouse, also called a gene knockout mouse or a gene-targeted knockout mouse, is an animal in which an endogenous gene (genomic wild-type allele) has been specifically inactivated by replacing it with a null allele.

Acknowledgments I wish to acknowledge the tremendous contribution of Dr. Arlene Marie Sindelar, my wife, to some of the graphics found in figures in all six editions of this textbook.

References

Adli M (2018) The CRISPR tool kit for genome editing and beyond. Nat Commun 9:1–13. https://doi.org/10.1038/s41467-018-04252-2

Akhoon N (2021) Precision medicine: a new paradigm in therapeutics. Int J Prev Med 12(12):1–7. https://doi.org/10.4103/ijpvm.IJPVM_375_19

Alexander-Dann B, Pruteanu LL, Oerton E, Sharma N, Berindan-Neagoe I, Módos D, Bender A (2018) Developments in toxicogenomics: understanding and predicting compound-induced toxicity from gene expression data. Mol Omics 14(4):218–236. https://doi.org/10.1039/c8mo00042e

Aronson JK, Ferner RE (2017) Biomarkers-a general review. Curr Protoc Pharmacol 76:9.23.1–9.23.17. https://doi.org/10.1002/cpph.19

Avela HF, Sirén H (2020) Advances in lipidomics. Clin Chim Acta 510:123–141. https://doi.org/10.1016/j.cca.2020.06.049

Bachtiar M, Ooi BNS, Wang J, Jin Y, Tan TW, Chong SS, Lee CGL (2019) Towards precision medicine: interrogating the human genome to identify drug pathways associated with potentially functional, population-differentiated polymorphisms. Pharmacogenomics J 19:516–527. https://doi.org/10.1038/s41397-019-0096-y

BBC News (2022) Man gets genetically-modified pig heart in world-first transplant. https://www.bbc.com/news/world-us-canada-59944889. Accessed 1 Aug 2022

Bertolini LR, Meade H, Lazzarotto CR, Martins LT, Tavares KC, Bertolini M, Murray JD (2016) The transgenic animal platform for biopharmaceutical production. Transgenic Res 25:329–343. https://doi.org/10.1007/s11248-016-9933-9

Biesecker LG, Spinner NB (2013) A genomic view of mosaicism and human disease. Nat Rev Genet 14:307–320. https://doi.org/10.1038/nrg3424

Bingol K, Bruschweller-Li L, Li D, Zhang B, Xie M, Bruschweiler R (2016) Emerging new strategies for successful metabolite identification in metabolomics. Bioanalysis 8:557–573. https://doi.org/10.4155/bio-2015-0004

Biospace (2022) The market for precision medicine. https://www.biospace.com/article/precision-medicine-market-size-to-hit-us-140-69-bn-by-2028/. Accessed 10 Sept 2022

Briganti G, Le Moine O (2020) Artificial intelligence in medicine: today and tomorrow. Front Med 7:1–6. https://doi.org/10.3389/fmed.2020.00027

Cai W, Liu Z, Miao L, Xiang X (2020) Pharmacogenomics in precision medicine: from a perspective of ethnic differences. Springer, Berlin

Cao WQ, Liu MQ, Kong SY, Wu MX, Huang ZZ, Yang PY (2020) Novel methods in glycomics: a 2019 update. Expert Rev Proteomics 17(1):11–25. https://doi.org/10.1080/14789450.2020.1708199

Chakravarty D, Solit DB (2021) Clinical cancer genomic profiling. Nat Rev Genet 22:483–501. https://doi.org/10.1038/s41576-021-00338-8

Chaudhry SR, Muhammad S, Eidens M, Klemm M, Khan D, Efferth T, Weisshaar MP (2014) Pharmacogenetic prediction of individual variability in drug response based on CYP2D6, CYP2C9 and CYP2C19 genetic polymorphisms. Curr Drug Metab 15:711–718. https://doi.org/10.2217/pgs-2017-0194

Chen Y, Zhou J, Wang L (2021) Role and mechanism of gut microbiota in human disease. Front Cell Infect Microbiol 11(3):625913. https://doi.org/10.3389/fcimb.2021.625913

Chen W, Anothaisintawee T, Butani D, Wang Y, Zemlyanska Y, Boon C, Wong N, Virabhak S, Hrishikesh MA, Teerawattananon Y (2022) Assessing the cost-effectiveness of precision medicine: protocol for a systematic review and meta-analysis. BMJ Open 12:e057537. https://doi.org/10.1136/bmjopen-2021-057537

Chouchana L, Narjoz C, Roche D, Golmard JL, Pineau B, Chatellier G, Beaune P, Loriot MA (2014) Interindividual variability in TPMT enzyme activity: 10 years of experience with thiopurine pharmacogenetics and therapeutic drug monitoring. Pharmacogenomics 15:745–757. https://doi.org/10.2217/pgs.14.32

Clark SJ, Lee HJ, Smallwood SA, Kelsey G, Reik W (2016) Single-cell epigenomics: powerful new methods for understanding gene regulation and cell identity. Genome Biol 17:72–82. https://doi.org/10.1186/s13059-016-0944-x

Cortess D, Leung J, Ryl A, Lieu J (2019) Pharmacy informatics: where medication use and technology meet. Can J Hosp Pharm 72(4):320–326

Costa AR, Rodrigues ME, Henriques M, Oliveira R, Azeredo J (2014) Glycosylation: impact, control and improvement during therapeutic protein production. Crit Rev Biotechnol 34:281–299. https://doi.org/10.3109/07388551.2013.793649

Dai X, Blancafort P, Wang P, Sgro A, Thompson EW, Ostrikov KK (2020) Innovative precision gene-editing tools in personalized cancer medicine. Adv Sci 7(12):1902552. https://doi.org/10.1002/advs.201902552

Doudna JA (2020) The promise and challenge of therapeutic genome editing. Nature 578(229):236. https://doi.org/10.1038/s41586-020-1978-5

Doytchinova I (2022) Drug design-past, present, future. Molecules 27(5):1496. https://doi.org/10.3390/molecules27051496

Dugger SA, Platt A, Goldstein DB (2018) Drug development in the era of precision medicine. Nat Rev Drug Discov 17:183–196. https://doi.org/10.1038/nrd.2017.226

Dupree EJ, Jayathirtha M, Yorkey H, Mihasan M, Petre BA, Darie CC (2020) A critical review of bottom-up proteomics: the good, the bad, and the future of this field. Proteomes 8(3):14–40. https://doi.org/10.3390/proteomes8030014

Eder J, Herrling PL (2016) Trends in modern drug discovery. Handb Exp Pharmacol 232:3–20. https://doi.org/10.1007/164_2

EMBL-EBI (2022) European molecular biology organization–European bioinformatics institute. https://wwwemblorg/services--facilities/. Accessed 10 Sept 2022

Emery JD (2017) Pharmacogenomic testing and warfarin: what evidence has the GIFT trial provided? JAMA 318(12):1110–1112. https://doi.org/10.1001/jama.2017.11465

Emmerich CH, Gamboa LM, Hofmann MCJ, Bonin-Andresen M, Arbach O, Schendel P, Gerlach B, Hempel K, Bespalov A, Dirnagl U, Parnham MJ (2021) Improving target assessment in biomedical research: the GOT-IT recommendations. Nat Rev Drug Discov 20:64–81. https://doi.org/10.1038/s41573-020-0087-3

Ersek JL, Black LJ, Thompson MA, Kim ES (2018) Implementing precision medicine programs and clinical trials in the community-based oncology practice: barriers and best practices. Am Soc Clin Oncol Educ Book 38:188–196. https://doi.org/10.1200/EDBK_200633

FDA (2022) Drugs in scientific research. https://www.fda.gov/Drugs/ScienceResearch/ucm572698.htm. Accessed 10 Sept 2022

Feinberg AP (2018) The key role of epigenetics in human disease prevention and mitigation. N Engl J Med 378:1323–1334. https://doi.org/10.1056/NEJMra1402513

Feinberg AP, Irizarry RA, Fradin D, Aryee MJ, Murakami P, Aspelund T, Eiriksdottir G, Harris TB, Launer L, Gudnason V, Fallin MD (2010) Personalized epigenomic signatures that are stable over time and covary with body mass index. Sci Transl Med 2:45–51. https://doi.org/10.1126/scitranslmed.3001262

Finn C, Gaulton A, Kruger FA, Lumbers RT, Shah T, Engmann J (2017) The druggable genome and support for target identification and validation in drug development. Sci Transl Med 9(383):1–15. https://doi.org/10.1126/scitranslmed.aag1166

Gentile F, Agrawal V, Hsing M, Ton A-T, Ban F, Norinder U, Gleave ME, Cherkasov A (2020) Deep docking: a deep learning platform for augmentation of structure based drug discovery. ACS Cent Sci 6(6):939–949. https://doi.org/10.1021/acscentsci.0c00229

Gerszberg A, Hnatuszko-Konka K (2022) Compendium on food crop plants as a platform for pharmaceutical protein production. Int J Mol Sci 23:1–21. https://doi.org/10.3390/ijms23063236

Giacomini KM, Yee SW, Mushiroda T, Weinshilboum RM, Ratain MJ, Kubo M (2017) Genome-wide association studies of drug response and toxicity: an opportunity for genome medicine. Nat Rev Drug Discov 16(1):1. https://doi.org/10.1038/nrd.2016.234

Giunta S (2021) Decoding human cancer with whole genome sequencing: a review of PCAWG project studies published in February 2020. Cancer Metastasis Rev 40:909–924. https://doi.org/10.1007/s10555-021-09969-z

Glass JI, Smith HO, Hutchison III CA, Alperovich NY, Assad-Garcia N (2007) Minimal bacterial genome. United States patent application 20070122826, May 31, 2007

Globe News Wire (2022) The protein engineering product and services market estimates. https://wwwglobenewswirecom/news-release/2022/06/17/2464481/0/en/Protein-Engineering-Market-Size-Worth-USD-3-02-Billion-by-2027-at-7-62-CAGR-Report-by-Market-Research-Future-MRFRhtml. Accessed 10 Sept 2022

Ha VTD, Frizzo-Barker J, Chow-White P (2018) Adopting clinical genomics: a systematic review of genomic literacy among physicians in cancer care. BMC Med Genet 11:18–36. https://doi.org/10.1186/s12920-018-0337-y

Hanczyc MM (2020) Engineering life: a review of synthetic biology. Artif Life 26(2):260–273. https://doi.org/10.1162/artl_a_00318

HCA (2022) The international human cell atlas. https://www.humancellatlas.org/. Accessed 10 Sept 2022

Höglund M (1998) Glycosylated and non-glycosylated recombinant human granulocyte colony-stimulating factor (rhG-CSF)–what is the difference? Med Oncol 15:229–233. https://doi.org/10.1007/BF02787205

Höhne M, Kabisch J (2016) Brewing painkillers: a yeast cell factory for the production of opioids from sugar. Angew Chem Int Ed Engl 55:1248–1125. https://doi.org/10.1002/anie.201510333

HPP (2022) The human proteome project. https://hupo.org/human-proteome-project. Accessed 10 Sept 2022

HuBMAP (2022) Human biomolecular atlas program. https://commonfund.nih.gov/hubmap. Accessed 10 Sept 2022

IMPC (2022) The international mouse phenotyping consortium. http://www.mousephenotype.org/. Accessed 10 Sept 2022

IOM (Institute of Medicine) (2017) Accelerating progress toward continuously learning. The National Academies Press, Washington, DC

Jiang J, Tian F, Cai Y, Qian X, Coatello CE, Ying W (2014) Site-specific qualitative and quantitative analysis of the N- and O-glycoforms in recombinant human erythropoietin. Anal Bioanal Chem 406:6265–6274. https://doi.org/10.1007/s00216-014-8037-8

Jinek M, Chylinski K, Fonfara I, Hauer M, Doudna JA, Charpentier E (2012) A programmable dual-RNA-guided DNA endonuclease in adaptive bacterial immunity. Science 337:816–821. https://doi.org/10.1126/science.1225

Joung JK, Sander JD (2013) TALENs: a widely applicable technology for targeted genome editing. Nat Rev Mol Cell Biol 14:49–55. https://doi.org/10.1038/nrm3486

Karoui ME, Hoyos-Flight M, Fletcher L (2019) Future trends in synthetic biology—a report. Front Bioeng Biotechnol 7(7):175. https://doi.org/10.3389/fbioe.2019.00175

Keathley J, Garneau V, Zavala-Mora D, Heister RR, Gauthier E, Morin-Bernier J, Green R, Vohl M-C (2021) A systematic review and recommendations around frameworks for evaluating scientific validity in nutritional genomics. Front Nutr 8:789215. https://www.frontiersin.org/article/10.3389/fnut.2021.7892158

Kelsey G, Stegle O, Reik W (2017) Single-cell epigenomics: recording the past and predicting the future. Science 358:69–75. https://doi.org/10.1126/science.aan68

Klompe SE, Vo PLH, Halpin-Healy TS, Sternberg SH (2019) Transposon-encoded CRISPR-Cas systems direct RNA-guided DNA integration. Nature 571(7764):219–225. https://doi.org/10.1038/s41586-019-1323-z

Korendovych IV (2018) Rational and Semirational protein design. Methods Mol Biol 1685:15–23. https://doi.org/10.1007/978-1-4939-7366-8_2

Korendovych IV, DeGrado WF (2020) De novo protein design, a retrospective. Q Rev Biophys 53:e3. https://doi.org/10.1017/S0033583519000131

Krassowski M, Vivek D, Sangram SK, Misra BB (2020) State of the field in multi-omics research: from computational needs to data mining and sharing. Front Genet 11:1–27. https://doi.org/10.3389/fgene.2020.610798

Kulkarni A, Anderson AG, Merullo DP, Konopka G (2019) Beyond bulk: a review of single cell transcriptomics methodologies and applications. Curr Opin Biotechnol 58:129–136. https://doi.org/10.1016/j.copbio.2019.03.001

Kwapisz D (2017) The first liquid biopsy test approved. Is it a new era of mutation testing for non-small cell lung cancer? Ann Transl Med 5:46. https://doi.org/10.21037/atm.2017.01.32

Kwong GA, Ghosh S, Gamboa L, Patriotis C, Srivastava S, Bhatia SN (2021) Synthetic biomarkers: a twenty-first century path to early cancer detection. Nat Rev Cancer 21:655–668. https://doi.org/10.1038/s41568-021-00389-3

Lam JT, Gutierrez MA, Shah S (2021) Pharmacogenomics: a primer for clinicians. McGraw Hill, New York

Lamas-Toranzo I, Guerrero-Sánchez J, Miralles-Bover H, Alegre-Cid G, Pericuesta E, Bermejo-Álvarez P (2017) CRISPR is knocking on barn door. Reprod Domest Anim 52(Suppl 4):39–47. https://doi.org/10.1111/rda.13047

Li W, Li M, Pu X, Guo Y (2017) Distinguishing the disease-associated SNPs based on composition frequency analysis. Interdiscip Sci 9:459–467. https://doi.org/10.1007/s12539-017-0248-1

Lindon JC, Nicholson JK (2014) The emergent role of metabolic phenotyping in dynamic patient stratification. Expert Opin Drug Metab Toxicol 10:915–919. https://doi.org/10.1517/17425255.2014.922954

Low LA, Mummery C, Berridge BR, Austin CP, Tagle DA (2021) Organs-on-chips: into the next decade. Nat Rev Drug Discov 20:345–361. https://doi.org/10.1038/s41573-020-0079-3

Lutz S, Iamurri SM (2018) Protein engineering: past, present, and future. Methods Mol Biol 1685:1–12. https://doi.org/10.1007/978-1-4939-7366-8_1

Madhavan S, Subramaniam S, Brown TD, Chen JL (2018) Art and challenges of precision medicine: interpreting and integrating genomic data into clinical practice. Am Soc Clin Oncol Educ Book 38:546–553

Mapanao AK, Voliani V (2020) Three-dimensional tumor models: promoting breakthroughs in nanotheranostics translational research. Appl Mater Today 19:100552. https://doi.org/10.1016/j.apmt.2019.100552

McGuire AL, Gabriel S, Tishkoff SA, Wonkam A, Chakravarti A, Furlong EEM, Treutlein B, Meissner A, Chang HY, Lopez-Bigas N, Segal E, Kim JS (2020) The road ahead in genetics and genomics. Nat Rev Genet 21:581–596. https://doi.org/10.1038/s41576-020-0272-6

McInnes G, Yee SW, Pershad Y, Altman RB (2021) Genomewide association studies in pharmacogenomics. Clin Pharmacol Ther 110:637–648. https://doi.org/10.1002/cpt.2349

Miao Z, Humphreys BD, McMahon AP, Kim J (2021) Multi-omics integration in the age of million single-cell data. Nat Rev Nephrol 17:710–724. https://doi.org/10.1038/s41581-021-00463-x

Minikel EV, Karczewski KJ, Martin HC, Cummings BB, Whiffin N, Rhodes D, Alfoldi J, Trembath RC, van Heel DA, Daly MJ, Genome Aggregation Database Production Team, Genome Aggregation Database Consortium, Schreiber SL, DG MA (2020) Evaluating drug targets through human loss-of-function genetic variation. Nature 581:459–464. https://doi.org/10.1038/s41586-020-2267-z

Mohanta TK, Mishra AK, Al-Harrasi A (2021) The 3D genome: from structure to function. Int J Mol Sci 22(21):11585–11591. https://doi.org/10.3390/ijms222111585

Molina-Espeja P, Vina-Gonzalez J, Gomez-Fernandez BJ, Martin-Diaz J, Garcia-Ruiz E, Miguel A (2016) Beyond the outer limits of nature by directed evolution. Biotechnol Adv 34:754–767. https://doi.org/10.1016/j.biotechadv.2016.03.008

Narimatsu Y, Bull C, Chen YH, Wandall HH, Yang Z, Clausen H (2021) Genetic glycoengineering in mammalian cells. J Biol Chem 296:100448. https://doi.org/10.1016/j.jbc.2021.100448

National Human Genome Institute (2022) Knockout-mice-fact-sheet. https://wwwgenomegov/about-genomics/fact-sheets/Knockout-Mice-Fact-Sheet. Accessed 1 Aug 2022

National Human Genome Research Institute (2023) https://www.genome.gov/about-genomics/fact-sheets/Knockout-Mice-Fact-Sheet. Accessed 13 September 2023

Nature (2018) Method of the year 2017: organoids. Nat Methods 15:1. https://doi.org/10.1038/nmeth.4575

Nature (2020) Method of the year 2019: single-cell multimodal omics. Nat Methods 17:1. https://doi.org/10.1038/s41592-019-0703-5

NCATS (2022) National Center for Advanced Translational Medicine. https://ncats.nih.gov/heal/funding/call-for-proposals/opportunities/heal-target-and-compound-library. Accessed 10 Sept 2022

NCBI (2022) National Center for Biotechnology Information, NCBI, of the National Institutes of Health. http://www.ncbi.nlm.nih.gov. Accessed 10 Sept 2022

Nidhi S, Anand U, Oleksak P, Tripathi P, Lal JA, Thomas G, Kuca K, Tripathi V (2021) Novel CRISPR-Cas systems: an updated review of the current achievements, applications, and future research perspectives. Int J Mol Sci 22(7):1–42. https://doi.org/10.3390/ijms22073327

NIH Human Microbiome Portfolio Analysis Team (2019) A review of 10 years of human microbiome research activities at the US National Institutes of Health, fiscal years 2007-2016. Microbiome 7:31. https://doi.org/10.1186/s40168-019-0620-y

Omenn GS, Lane L, Overall CM, Paik YK, Cristea IM, Corrales FJ, Lindskog C, Weintraub S, Roehrl MHA, Liu S, Bandeira N, Srivastava S, Chen YJ, Aebersold R, Moritz RL, Deutsch EW (2021) Progress identifying and analyzing the human proteome: 2021 metrics from the HUPO human proteome project. J Proteome Res 20(12):5227–5240. https://doi.org/10.1021/acs.jproteome.1c00590

Papastergiou J, Tolios P, Li W, Li J (2017) The innovative Canadian Pharmacogenomic screening initiative in community pharmacy (ICANPIC) study. J Am Pharm Assoc 57:624–629. https://doi.org/10.1016/j.japh.2017.05.006

Paul A, Paul S (2014) The breast cancer susceptibility genes (BRCA) in breast and ovarian cancers. Front Biosci (Landmark Ed) 19:605–618. https://doi.org/10.2741/4230

PhRMA (2021) 2021 industry profile: medicines are transforming the trajectory of many diseases. https://phrma.org/resource-center/Topics/Medicines-in-Development/2021-Industry-Profile-Toolkit-Medicines-are-Transforming-the-Trajectory-of-Many-Diseases Accessed 7 Apr 2022

Premsrirut P (2017) Drug discovery in the age of big data. Drug Discov World 17:8–15

Primorac D, Bach-Rojecky L, Vađunec D, Juginović A, Žunić K, Matišić V, Skelin A, Arsov B, Erceg D, Ivkošić IE, Molnar V, Ćatić J, Mikula I, Boban L, Primorac L, Esquivel B, Donaldson M (2020) Pharmacogenomics at the center of precision medicine: challenges and perspective in an era of big data. Pharmacogenomics 21(2):141–156. https://doi.org/10.2217/pgs-2019-0134

Prokopuk L, Western PS, Stringer JM (2015) Transgenerational epigenetic inheritance: adaptation through the germline epigenome? Epigenomics 7(5):829–846. https://doi.org/10.2217/epi.15.36

Przybyla L, Gilbert LA (2022) A new era in functional genomics screens. Nat Rev Genet 23:9–103. https://doi.org/10.1038/s41576-021-00409-w

Pushpakom S, Iorio F, Eyers P, Escott KJ, Hopper S, Wells A, Doig A, Guilliams T, Latimer J, McNamee C, Norris A, Sanseau P, Cavalla D, Pirmohamed M (2019) Drug repurposing: progress, challenges and recommendations. Nat Rev Drug Discov 18:41–58. https://doi.org/10.1038/nrd.2018.168

Qin C, Tanis KQ, Podtelezhnikov AA, Glaab WE, Sistare FD, DeGeorge JJ (2016) Toxicogenomics in drug development: a match made in heaven? Expert Opin Drug Metab Toxicol 12(8):847–849. https://doi.org/10.1080/17425255.2016.1175437

Reddy AVB, Yusop Z, Jaafar J, Madhavi V, Madhavi G (2016) Advances in drug discovery: impact of genomics and role of analytical instrumentation. Curr Drug Discov Technol 13:211–224. https://doi.org/10.2174/1570163813666160930122643

Relling MV, Evans WE (2015) Pharmacogenomics in the clinic. Nature 526:343–350. https://doi.org/10.1038/nature15817

Sagner M, McNeil A, Puska P, Auffray C, Price ND, Hood L, Lavie CJ, Han Z, Chen Z, Brahmachari SK, McEwen BS, Soares MB, Balling R, Epel E, Arena R (2017) The P4 health spectrum–a predictive, preventive, personalized and participatory continuum for promoting healthspan. Prog Cardiovasc Dis 59:506–521. https://doi.org/10.1016/j.pcad.2016.08.002

SBGN (2022) Examples of complex systems biology-derived protein interaction networks. http://sbgn.github.io/sbgn/examples. Accessed 10 Sept 2022

Scheinfeldt LB (2022) Pharmacogenomics: from basic research to clinical implementation. J Pers Med 11:800

Schneider P, Walters WP, Plowright AT, Sieroka N, Listgarten J, Goodnow RA Jr, Fisher J, Jansen JM, Duca JS, Rush TS, Zentgraf M, Hill JE, Krutoholow E, Kohler M, Blaney J, Funatsu K, Luebkemann C, Schneider G (2020) Rethinking drug design in the artificial intelligence era. Nat Rev Drug Discov 19:353–364. https://doi.org/10.1038/s41573-019-0050-3

Schumacher S, Muekusch S, Seitz H (2015) Up-to-date applications of microarrays and their way to commercialization. Microarrays (Basel) 4:196–213. https://doi.org/10.3390/microarrays4020196

Schwarze K, Buchanan J, Taylor J, Wordsworth S (2018) Are whole-exome and whole-genome sequencing approaches cost-effective? A systematic review of the literature. Gen Med 20(10):1122–1130. https://doi.org/10.1038/gim.2017.247

Sebastian B, Boch J (2021) TALE and TALEN genome editing technologies. Gene Genome Ed 2:1–14. https://doi.org/10.1016/j.ggedit.2021.100007

Service RF (2018) Protein evolution earns chemistry Nobel. Science 362(6411):142. https://doi.org/10.1126/science.362.6411.142

Singh RK, Lee JK, Selvaraj C, Singh R, Li J, Kim SY, Kalia VC (2018) Protein engineering approaches in the post-genomic era. Curr Protein Pept Sci 19(1):5–15. https://doi.org/10.2174/1389203718666161117114243

Smaglik P (2017) The genetic microscope. Nature 545:S25–S27. https://doi.org/10.1038/545S25a

Smirnov A, Fishman V, Yunusova A, Korablev A, Serova I, Skryabin BV, Rozhdestvensky TS, Battulin N (2020) DNA barcoding reveals that injected transgenes are predominantly processed by homologous recombination in mouse zygote. Nucleic Acids Res 48(2):719–735. https://doi.org/10.1093/nar/gkz1085

Sofi MY, Shafi A, Masoodi KZ (2022) Nucleic acid sequence databases. In: Sofi MY, Shafi A, Masoodi KZ (eds) Bioinformatics for everyone. Academic, Cambridge, pp 25–36. https://www.sciencedirect.com/science/article/pii/B9780323911283000161

Sonehara K, Okada Y (2021) Genomics-driven drug discovery based on disease-susceptibility genes. Inflamm Regener 41:8. https://doi.org/10.1186/s41232-021-00158-7

Tang X, Huang Y, Lei J, Luo H (2019) The single-cell sequencing: new developments and medical applications. Cell Biosci 9:53. https://doi.org/10.1186/s13578-019-0314-y

Tang TC, An B, Huang Y, Vasikaran S, Wang Y, Jiang X, Lu TK, Zhong C (2021) Materials design by synthetic biology. Nat Rev Mater 6:332–350. https://doi.org/10.1038/s41578-020-00265-w

Tavassoly I, Goldfarb J, Iyengar R (2018) Systems biology primer: the basic methods and approaches. Essays Biochem 62(4):487–500. https://doi.org/10.1042/EBC20180003

The International Human Genome Sequencing Consortium (2001) Initial sequencing and analysis of the human genome. Nature 409:860–921. https://doi.org/10.1038/35057062

Tost J (2020) 10 years of epigenomics: a journey with the epigenetic community through exciting times. Epigenomics 12(2):81–85. https://doi.org/10.2217/epi-2019-0375

Venter JC et al (2001) The sequence of the human genome. Science 291:1304–1351. https://doi.org/10.1073/pnas.0307971100

Visscher PM, Wray NR, Zhang Q, Sklar P, McCarthy MI, Brown MA, Yang J (2017) 10 years of GWAS discovery: biology, function, and translation. Am J Hum Genet 101:5–22. https://doi.org/10.1016/j.ajhg.2017.06.005

Ward R, Ginsburg GS (2017) Chapter 7–local and global challenges in the clinical implementation of precision medicine. In: Ginsburg GS, Willard HF (eds) Genomic and precision medicine, 3rd edn. Academic, New York, pp 105–117

What is Biotechnology? (2022) Transgenic animals. https://whatisbiotechnology.org/index.php/science/summary/transgenic/transgenic-animals-have-genes-from-other-species-inserted. Accessed 14 May 2022

Willyard C (2018) New human gene tally reignites debate. Nature 558:354–355. https://doi.org/10.1038/d41586-018-05462-w

Wishart DS, Guo A, Oler E, Wang F, Anjum A, Peters H, Dizon R, Sayeeda Z, Tian S, Lee BL, Berjanskii M, Mah R, Yamamoto M, Jovel J, Torres-Calzada C, Hiebert-Giesbrecht M, Lui VW, Varshavi D, Varshavi D, Allen D, Arndt D, Khetarpal N, Sivakumaran A, Harford K, Sanford S, Yee K, Cao X, Budinski Z, Liigand J, Zhang L, Zheng J, Mandal R, Karu N, Dambrova M, Schiöth HB, Greiner R, Gautam V (2022) HMDB 5.0: the human metabolome database for 2022. Nucleic Acids Res 7(50):D622–D631. https://doi.org/10.1093/nar/gkab1062

Yang JH (2022) CRISP(e)R drug discovery. Nat Chem Biol 18:435–436. https://doi.org/10.1038/s41589-022-00979-8

Zetsche B, Gootenberg JS, Abudayyeh OO, Slaymaker IM, Makarova KS, Essletzbichler P, Volz SE, Joung J, van der Oost J, Regev A, Koonin EV, Zhang F (2015) Cpf1 is a single RNA-guided endonuclease of a class 2 CRISPR-Cas system. Cell 163:759–771. https://doi.org/10.1016/j.cell.2015.09.038

Zhang HM, Nan ZR, Hui GQ, Liu XH, Sun Y (2014) Application of genomics and proteomics in drug target discovery. Genet Mol Res 13:198–204. https://doi.org/10.4238/2014.January.10.11

Zhao Y, Brasier AR (2016) Qualification and verification of protein biomarker candidates. Adv Exp Med Biol 919:493–514. https://doi.org/10.1007/978-3-319-41448-5_23

Zhao YY, Cheng XL, Lin RC, Wei F (2015) Lipidomics applications for disease biomarker discovery in mammal models. Biomark Med 9:153–168. https://doi.org/10.2217/bmm.14.81

Zimmermann M, Zimmermann-Kogadeeva M, Wegmann R, Goodman AL (2019) Mapping human microbiome drug metabolism by gut bacteria and their genes. Nature 570:462–467. https://doi.org/10.1038/s41586-019-1291-3

Zou Y, Laubichler MD (2018) From systems to biology: a computational analysis of the research articles on systems biology from 1992 to 2013. PLoS One 13(7):e0200929. https://doi.org/10.1317/journal.pone.0200929

Economic Considerations in Medical Biotechnology

10

Amit S. Patel and Kartick P. Shirur

Introduction

The biotechnology revolution has coincided with another revolution in health care: the emergence of finance and economics as major issues in the use and success of new medical technologies. Health care finance has become a major social issue in nearly every nation, and the evaluation and scrutiny of the pricing and value of new treatments has become an industry unto itself. The most tangible effect of this change is the establishment of the so-called third hurdle for approval of new agents in many nations, after proving safety and efficacy. Beyond the traditional requirements for demonstrating the efficacy and safety of new agents, some nations and many private health care systems now demand data on the economic costs and benefits of new medicines. Although currently required only in a few countries, methods to extend similar prerequisites are being examined by the governments of most developed nations. Many managed care organizations in the USA prefer that an economic dossier be submitted along with the clinical dossier to help inform formulary coverage decisions.

The licensing of new agents in most non-US nations has traditionally been accompanied by a parallel process of price and reimbursement approval, and the development of an economic dossier has emerged as a means of securing the highest possible rates of reimbursement. In recent years, sets of economic guidelines have been developed and adopted by the regulatory authorities of several nations to assist them in their decisions to reimburse new products. As many of the products of biotechnology are used to treat costly disorders and the products themselves are often costly to discover and produce, these new agents have presented new problems to those charged with the financing of medical care delivery. The movement to require an economic rationale for the pricing of new agents brings new challenges to those developing such agents. These requirements also provide firms with new tools to help determine which new technologies will provide the most value to society as well as contribute the greatest financial returns to those developing and marketing the products.

The Value of a New Medical Technology

The task of determining the value of a new agent should fall somewhere within the purview of the marketing function of a firm. Most companies have established health care economic capabilities within the clinical research structure of their organizations. It is essential that the group that addresses the value of a new product does so from the perspective of the market and not of the company or the research team. This is important for two reasons. First, evaluating the product candidate from the perspective of the user, and not from the team that is developing it, can minimize the bias that is inherent in evaluating one's own creations. Second, and most importantly, a market focus will move the evaluation away from the technical and scientifically interesting aspects of the product under evaluation and toward the real utility the product might bring to the medical care marketplace. Although the scientific, or purely clinical, aspects of a new product should never be ignored, when the time comes to measure the economic contribution of a new agent, those developing the new agent must move past these considerations. It is the tangible effects that a new treatment will have on the patient and the health care system that determine its value, not the technology supporting it. The phrase to keep in mind is "value in use."

The importance of a marketing focus when evaluating the economic effects of a new agent, or product candidate, cannot be overstated. Failing to consider the product's value in

A. S. Patel (✉)
Department of Pharmacy, University of Mississippi, Oxford, MS, USA

Department of Pharmacy, Indegene Inc., Oxford, MS, USA
e-mail: amit.patel@indegene.com

K. P. Shirur
Department of Pharmacy, Indegene Inc., Oxford, MS, USA

D. J. A. Crommelin et al. (eds.), *Pharmaceutical Biotechnology*, https://doi.org/10.1007/978-3-031-30023-3_10

use can result in overly optimistic expectations of sales performance and market acceptance. Marketing is often defined as the process of identifying and filling the needs of the market. If this is the case, then the developers of new pharmaceutical technologies must ask two questions: "What does the market need?" and "What does the market want?" Analysis of the pharmaceutical market typically shows that the market needs and wants:

- Lower costs
- Controllable costs
- Predictable cost
- Improved outcomes

Note that this list does not include new therapeutic agents. From the perspective of many payers, authorities, clinicians, and buyers, a new agent, in and of itself, is a challenge. The effort required to evaluate a new agent and prepare recommendations to adopt or reject it takes time away from other efforts. For many in the health care delivery system, a new drug means more work—not that they are opposed to innovation, but newness in and of itself, regardless of the technology behind it, has no intrinsic value. The value of new technologies is in their efficiency and their ability to render results that are not available through other methods or at costs significantly lower than other interventions. Documenting and understanding the economic effects of new technologies on the various health care systems help the firm to allocate its resources more appropriately, accelerate the adoption of new technologies into the health care system, and reap the financial rewards of its innovation.

There are many different aspects of the term "value," depending upon the perspective of the individual or group evaluating a new product and the needs that are met by the product itself. When developing new medical technologies, it is useful to look to the market to determine the aspects of a product that could create and capture the greatest amount of value. Two products that have entered the market provide good examples of the different ways in which value is assessed.

Activase® (tPA, tissue plasminogen activator) from Genentech, one of the first biotechnology entrants in health care, entered the market priced at nearly ten times the price level of streptokinase, its nearest competitor. This product, which is used solely in the hospital setting, significantly increased the cost of medical treatment of patients suffering myocardial infarctions. But the problems associated with streptokinase and the great urgency of need for treatments for acute infarctions were such that many cardiologists believed that any product that proved useful in this area would be worth the added cost. The hospitals, which in the USA are reimbursed on a capitated basis for the bulk of such procedures, were essentially forced to subsidize the use of the agent, as they were unable to pass the added cost of tPA to many of their patients' insurers. The pricing of the product created a significant controversy, but the sales of Activase and its successors have been growing consistently since its launch. The key driver of value for tPA has been, and continues to be, the urgency of the underlying condition. The ability of the product to reduce the rate of immediate mortality is what drives its value. Once the product became a standard of care, incidentally, reimbursement rates were increased to accommodate it, making its economic value positive to hospitals.

An early biotechnology product that delivered a different type of value is the granulocyte-colony stimulating factor Neupogen® from Amgen. The product's primary benefit is in the reduction of serious infections in cancer patients, who often suffer significant decreases in white blood cells [febrile neutropenia (FN)] due to chemotherapy. By bolstering the white blood cell count, Neupogen allows oncologists to use more efficacious doses of cytotoxic oncology agents while decreasing the rate of infection and subsequent hospitalization for cancer patients (estimated costs of US $28,000 per admission). Furthermore, FN reduces cancer patient survival rates through delays, dose reductions, and discontinuations of chemotherapy schedules. Neupogen allows to lower FN incidence and is recommended for the prevention of chemotherapy-induced neutropenia in international guidelines (Cornes et al. 2022). The economic benefits of the product have helped it to gain use rapidly with significantly fewer restrictions than products such as tPA, whose economic value is not as readily apparent.

These two very successful early biotechnology products both provide clear clinical benefits, but their sources of value are quite different. The value of a new product may come from several sources, depending on the needs of clinicians and their perceptions of the situations in which they treat patients. Value can come from the enhancement of the positive aspects of treatment as well. A product that has a higher rate of efficacy than current therapies is the most obvious example of such a case. But any product that provides benefits in an area of critical need, where few or no current treatments are available, will be seen as providing immediate value. This was, and remains, the case for tPA.

Some current treatments bring risk, either because of the uncertainty of their effects on the patient (positive or negative) or because of the effort or cost required to use or understand the treatments. A new product that reduces this risk will be perceived as bringing new value to the market. In such cases, the new product removes or reduces some negative aspects of treatment. Neupogen, by reducing the chance of infection and reducing the average cost of treatment, brought new value to the marketplace in this manner. Any new product under development should be evaluated with these aspects of value in mind. A generalized model of value,

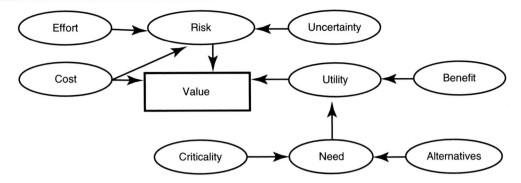

Fig. 10.1 Generalized model of value (Copyright© 2003, Medical Marketing Economics, LLC, Oxford, MS)

presented in Fig. 10.1, can be used to determine the areas of greatest need in the marketplace for a new agent and to provide guidance in product development. By talking with clinicians, patients, and others involved in current treatments and keeping this model in mind, the shortcomings of those current approaches can be evaluated and the sources of new incremental value can be determined.

Understanding the source of the value brought to the market by a new product is crucial to the development of the eventual marketing strategy. Using Fig. 10.1 as a guide, the potential sources of value can be determined for a product candidate and appropriate studies, both clinical and economic, can be designed to measure and demonstrate that value.

An Overview of Economic Analysis for New Technologies

A thorough economic analysis should be used to guide the clinical research protocol to ensure that the end points measured are commercially relevant and useful. The analysis should describe important elements of the market to the firm, helping decision makers to understand the way decisions are made and providing guidance in affecting those decisions. Later, the results of economic analyses should inform and guide marketing and pricing decisions as the product is prepared for launch, as well as help customers to use the product efficiently and effectively.

To prepare a thorough economic analysis, researchers must first have a comprehensive understanding of the flow of patients, services, goods, and money through the various health care systems. This process should begin as soon as the likely indications for a new product have been identified and continue throughout the product's development. The first step is to create basic economic models of the current treatment for the disorder(s) for which the product is likely to be indicated. This step will be used to fine-tune financial assumptions and clinical development process, to assure that the clinical protocols are designed to extract the greatest

clinical and commercial potential from a product. Separate models should be prepared for each indication and level if the product is likely to be used to treat more than one indication and/or several different levels of the same indication (e.g., mild, moderate, and severe).

The purpose of the basic model is to provide a greater understanding of the costs associated with the disorder, and to identify areas and types of cost that provide the greatest potential for the product to generate cost savings. For example, the cost of a disorder that currently requires a significant amount of laboratory testing offers the potential for savings, and thus better pricing, if the new product can reduce or eliminate the need for tests. Similarly, some indications are well treated, but the incidence of side effects is sufficiently high to warrant special attention. When developing a new agent, it is as important to understand the source of the value to be provided as it is to understand the clinical effects of the agent.

Pharmacoeconomics

The field of economic evaluation of medical technologies goes by several names, depending on the discipline of the researchers undertaking the study and the type of technology being measured. For pharmaceutical and biotechnology products, the field has settled on the name of pharmacoeconomics, and an entire discipline has emerged to fill the needs of the area. Contributions to the development of the field have come from several disciplines, including economics, pharmacy administration, and many of the behavioral sciences.

Pharmacoeconomics has been defined as "the description and analysis of the costs of drug therapy to the health care systems and society" (Townsend 1987). Clinical studies assess the efficacy of a biotechnology product; likewise, pharmacoeconomic studies help to evaluate the efficiency of biotechnologically derived drug. In a complete pharmacoeconomic assessment, both the costs and consequences are identified, measured, and compared with other available medical interventions. The increase in the health care expenditure in the USA has resulted in excessive demand for cost-containment measures. Managed care organizations are striving hard to control drug spending and other health care-related costs. Payers have moved from an open formulary system to a more closed formulary system, leading to additional emphasis on pharmacoeconomic type of assessments. Additionally, several states in the USA have passed laws in an attempt to increase transparency of developmental costs for new drugs and cap price increases post launch.

Importance of Pharmacoeconomics

To understand the importance of pharmacoeconomics in the biotechnology industry, it is necessary to understand the differences between the biotechnology products and traditional pharmaceutical products. Szcus and Schneeweiss (2003) have highlighted these differences. They observed that biotechnologically derived products are more expensive than traditional pharmaceutical products and that many biotechnology products are termed "orphan drugs" as they are used in small- or moderate-size patient populations. At times, these products could be the only option to treat the underlying disease condition. Given the high production costs and selling prices of biotechnology products, it is critical for these products to demonstrate adequate cost-effectiveness to justify their high cost. Therefore, pharmacoeconomic analysis is one of the major tools for payers to differentiate between a high-priced traditional pharmaceutical products and costly biotechnology products in certain instances.

Pharmacoeconomic analysis plays a crucial role in disease management. Chang and Nash (1998) outlined the role of pharmacoeconomics in disease management, which includes evaluation and identification of cost-effective medications for the treatment of particular disease conditions. This information can be and is used by payers and hospital personnel to make potential formulary decisions. In such instances, drugs with unfavorable pharmacoeconomic evaluations are unlikely to remain on formularies or will be moved to a restricted status. In addition to formulary decisions, disease management programs often include clinical guidelines that are designed primarily on cost-effectiveness of medications (Joshnson and Nash 1996). When communicated properly, economic analysis can lead physicians to change their prescribing behavior thus decreasing unexplained variation in the treatment of the same disease. Walkom et al. (2006) studied the role of pharmacoeconomics in formulary decision making and found growing importance of pharmacoeconomic evaluations in formulary decision making.

When used appropriately, pharmacoeconomic analysis should help us to answer questions such as:

- What drugs should be included in the outpatient formulary?

- Should these same drugs be included on a hospital formulary?
- What is the best drug for a particular disease in terms of efficacy and cost?
- What is the best drug for a pharmaceutical manufacturer to invest time and money?
- What are the relative cost and benefits of comparable treatment options?

To address the above questions, it becomes necessary for us to understand different costs considered in pharmacoeconomic analysis and the underlying techniques used to perform these pharmacoeconomic evaluations.

Understanding Costs

A comprehensive evaluation of relevant cost and consequences differentiates pharmacoeconomics from traditional cost-containment strategies and drug use evaluations. Costs are defined as the value of the resource consumed by a program or treatment alternative. Health economists use different costs in pharmacoeconomic evaluations, which can be grouped under direct costs, indirect costs, intangible costs, and opportunity costs.

In pharmacoeconomic evaluations, a comparison of two or more treatments extends beyond a simplistic comparison of drug acquisition cost. Including different costs, when appropriate, provides a more accurate estimate of the total economic impact of treatment alternatives and disease management programs in distinguished patients or populations.

Direct Costs

Direct costs are the resources consumed in the prevention or treatment of a disease. The direct costs are further divided into direct medical costs and direct nonmedical costs.

The direct medical costs include expenditures on drugs, medical equipment, laboratory testing, hospital supplies, physician visits, and hospitalization costs. Direct medical costs could be further divided into fixed costs and variable costs. Fixed costs generally represent the overhead costs and are relatively constant. Fixed costs include expenditures on rent, utilities, insurance, accounting, and other administrative activities. These costs are often not included in the pharmacoeconomic evaluations because their use or total cost is unlikely to change as a direct result of a specific intervention. On the other hand, variable costs are an integral part of pharmacoeconomic analysis. Variable costs include drugs, fees for professional services, and supplies. These variable costs increase or decrease depending on the volume.

Direct nonmedical costs are out-of-pocket costs paid by patients (or their caregivers) for nonmedical services which are generally outside health care sector. Direct nonmedical costs included expenditure on transportation to and from the hospital, clinic or physician office, additional trips to emergency rooms, expenses on special diet, family care expenses, and other various forms of out-of-pocket expenses.

Indirect Costs

Indirect costs are those costs that result from morbidity or mortality. Indirect costs assess the overall economic impact of an illness on a patient's life. Typical indirect costs include the loss of earnings due to temporary or permanent disability, loss of income to family member who gave up their job temporarily or permanently to take care of patient, and loss in productivity due to illness. Indirect medical costs are more related to patients and often unknown to or unappreciated by providers and payers.

Intangible Costs

Intangible costs are the most difficult to quantify in monetary terms. These costs represent the nonfinancial outcome of disease and medical care. The examples of intangible costs include pain, suffering, and emotional disturbance due to underlying conditions. Though these costs are identified in an economic analysis, they are not formally calculated. At times, intangible costs are converted into a common unit of outcome measurement such as a quality-adjusted life-year (QALY).

Opportunity Costs

Opportunity costs are often discussed in the economic literature. Opportunity cost is defined as the value of the alternative that was forgone. In simple terms, suppose a person spends $100 to buy a drug to treat a particular disease condition, then the opportunity to use the same $100 to obtain a different medical intervention or treatment for the same disease condition, or for some nonmedical purpose, is lost. This is referred to as an opportunity cost. Although not often included in traditional pharmacoeconomic analysis, opportunity costs are often considered implicitly by patients when cost sharing (e.g., co-pays and coinsurance) is increased in a health benefit plan.

Understanding Pharmacoeconomic Methods

The purpose of this section is to provide an overview of pharmacoeconomic techniques currently used to evaluate drugs or treatment options. Table 10.1 represents the list of pharmacoeconomic methods. Selection of a particular technique depends on the objective of the study and outcome units which are compared. Grauer et al. (2003) stated that "the

fundamental task of economic evaluation is to identify, measure, value and compare the costs and consequences of the alternatives being considered."

Cost of Illness (COI)

Cost of illness analysis is an important pharmacoeconomic tool to examine the economic burden of a particular disease. This technique takes into consideration the direct and indirect costs of a particular disease. A COI analysis thus identifies the overall cost of a particular disease in a defined population. Bootman et al. (1991) argue that COI analysis helps to evaluate the humanistic impact of disease and quantify the resources used in the treatment of disease prior to the discovery of new intervention. This information could be effectively used by pharmacoeconomic researchers to establish a baseline for comparison of new treatment or intervention. COI analysis is not used to compare two alternative treatment options, but to estimate the financial burden of the disease under consideration. Thus, the monetary benefits of prevention and treatment strategies could be measured against the baseline value estimated by cost of illness. In essence, a COI analysis provides the foundation for the measurement of the economic consequences of any treatment for the disorder in question. For example, Segel (2006) points out that a study on the cost-effectiveness of donepezil, published in 1999, relied on a COI study of Alzheimer's disease published a few years earlier. Without the initial COI study, the cost-effectiveness work would have been exponentially more difficult and costly.

Cost-Minimization Analysis (CMA)

Cost-minimization analysis is the simplest pharmacoeconomic evaluation technique. The primary objective of the cost-minimization analysis is to determine the least costly alternative. CMA is used to compare two or more treatment alternatives that are equal in efficacy. An example of CMA would be a comparison of branded product to a generic equivalent or biosimilar. It is assumed that the outcomes associated with the two drugs are equivalent; therefore, costs alone could be compared directly. The cost included in this economic evaluation must extend beyond drug acquisition cost and should include all relevant costs incurred for preparing and administering drugs.

Qiao et al. (2021) performed a CMA to evaluate direct costs of pembrolizumab versus nivolumab plus ipilimumab

Table 10.1 Economic evaluation methodologies

Method	Cost unit	Outcome unit
Cost of illness	Currency	Not assessed
Cost-minimization	Currency	Assumed to be equivalent in comparative groups
Cost–benefit	Currency	Currency
Cost-effectiveness	Currency	Natural units (life-years gained, mg/dl, blood glucose, mm hg blood pressure)
Cost-utility	Currency	Quality-adjusted life-years or other utility

for programmed death ligand 1 (PD-L1)-positive metastatic nonsmall-cell lung cancer (NSCLC) treatment within the first three years following treatment initiation from a US payer perspective. Drug acquisition and administration costs, subsequent treatment costs, costs for adverse events management, and disease and terminal care costs were included to estimate the medical cost. Comparable treatment efficacy and the same unit disease management costs by health state and terminal care costs were assumed across treatments. Cost savings of $54,343, $75,744, and $76,259 per patient were generated with pembrolizumab versus nivolumab plus ipilimumab within one, two, and three years, respectively. Cost savings were mainly attributable to drug acquisition costs, with the rest for drug administration, subsequent treatment, and adverse event management costs. Therefore, it was concluded that pembrolizumab provides higher cost savings versus nivolumab plus ipilimumab as first-line treatment for PD-L1-positive metastatic NSCLC, with comparable levels of effectiveness.

Cost–Benefit Analysis (CBA)

In cost–benefit analysis, costs and benefits are both measured in currency. In a CBA, all the benefits obtained from the program or intervention are converted into some currency value (e.g., US dollars or euros). Likewise, all program costs are identified and assigned a specific currency value. At times, the costs are discounted to their present value. To determine the cost–benefit of a program, the costs are subtracted from the benefit. If the net benefit value is positive, then it can be concluded that the program is worth undertaking from an economic perspective.

The results of the CBA could be expressed either as cost–benefit ratio or a net benefit. For example, a cost associated with a medical program is $1000, and the outcome/benefit resulting from the program is $9000. Therefore, subtracting the cost $1000 from the benefits $9000 will yield net benefit of $8000. When comparing many treatment alternatives, an alternative with the greatest net benefit could be considered as most efficient in terms of use of resources. In CBA, all costs and benefits resulting from the program should be included.

A typical use of CBA is in the decision of whether a national health benefit should include the administration of a specific vaccine. In this case, the cost of vaccinating the population and treating a smaller number of cases of the disease would be compared with the costs that would be incurred if the disease were not to be prevented. At times, however, it is much more difficult to assign a monetary value to benefits. For example, the benefit of a patient's satisfaction with the treatment or improvement in patient's quality of life is very difficult to convert to a monetary sum. This presents a considerable problem. At times, these variables are considered as "intangible benefits," and the decision is left to the

researcher to include in final analysis. Because of this, CBA is seldom used as a pharmacoeconomic method to evaluate a specific treatment, although many who perform different types of analyses often mistakenly refer to their work as "cost–benefit."

Cost-Effectiveness Analysis (CEA)

Cost-effectiveness analysis is a method used to compare treatment alternatives or programs where cost is measured in currency and outcomes/consequences are measured in units of effectiveness or natural units. Therefore, cost-effectiveness analysis helps to establish and promote the most efficient drug therapy for the treatment of particular disease condition. The results of cost-effectiveness analysis are expressed as average cost-effectiveness ratios or as the incremental cost of one alternative over another. CEA is useful in comparing different therapies that have the same outcome units, such as an increase in life expectancy or decrease in blood pressure for hypertension drugs. CEA is a frequently used tool for evaluating different drug therapies to treat a particular disease condition. This type of analysis helps in determining the optimal alternative, which is not always the least costly alternative. CEA has an advantage that it does not require the conversion of health outcomes to monetary units.

CEA is often used to guide formulary management decisions. For example, consider a biotechnology product "X" that provides a 90% efficacy or cure rate for a specific disorder. The total treatment cost for 100 patients with product X is $750,000. Likewise, assume that another biotechnology product "Y" prescribed for the same disorder shows 95% efficacy; however, the treatment cost of 100 patients with product Y is $1000,000. The average cost-effectiveness ratio (ACER) of product X is calculated by dividing the cost $750,000 by the outcome, 90 cures, to yield an ACER of $8333 per cure. Similarly, the ACER for product Y is $10,526 per cure. From this analysis, it is evident that using product Y would cost an additional $2192 per cure, which is the difference between ACERs of product Y and product X.

At times, the incremental cost-effectiveness ratio (ICER) is important in drug selection decisions. From the above example, to calculate the ICER, total cost of product X ($750,000) is subtracted from total cost of product Y ($1000,000). This is then divided by the cures from product X (90), subtracted from the cures resulting for product Y (95). Therefore, the incremental cost for each additional cure with product Y is $250,000 divided by five cures or $50,000 per cure. The incremental cost-effectiveness ratio poses the question of whether one additional cure is worth spending $50,000. The additional cost of cure might be justified by the severity of the disease or condition; this is a decision that is best made with the full knowledge of the economic implications. This provides an example of a situation in which the economic analysis is used to help guide the decision, but not

Table 10.2 Incremental Cost-Effectiveness Ratio (ICER) for two products

Product	Efficacy (%)	No. of patients treated	Total costs ($)	ACER ($)	ICER ($)
Product X	90	100	750,000	8333	50,000
Product Y	95	100	1000,000	10,526	

to make the decision. Table 10.2 represents the ICER for product X and Y.

Good examples of CEAs are the recent comparisons of the use of proprotein convertase subtilisin/kexin type 9 (PCSK9) inhibitors Praluent® (alirocumab) and Repatha® (evolocumab) or ezetimibe with statin therapy. PCSK9 inhibitors were approved in 2015 to lower low-density lipoprotein levels in individuals with heterozygous familial hypercholesterolemia (FH) or atherosclerotic cardiovascular disease (ASCVD). Annual acquisition costs of alirocumab and evolocumab were $14,600 US and $14,100 US, respectively, at launch, which were significantly higher than statins and ezetimibe, for which generic equivalents are available. Kazi et al. (2016) calculated the lifetime major adverse cardiovascular events (MACE), incremental cost per quality-adjusted life-year (QALY), and total cost to the US health care spending over five years. They estimated that adding PCSK9 inhibitors to statin therapy compared to ezetimibe prevented 316,300 MACE at a cost of $503,000 US per QALY in heterozygous FH, and prevented 4.3 million MACE at a cost of $414,000 US per QALY in ASCVD. Use of PCSK9 inhibitors would reduce costs for cardiovascular care by $29 billion US over five years, but add drug costs worth $592 billion US Since PCSK9 inhibitors cost four to five times higher than the generally accepted $100,000 US per QALY threshold, the researchers concluded that annual acquisition costs of PCSK9 inhibitors would have to be reduced to $4536 US in order for them to be cost-effective. Re-analysis of the model with recent data on ASCVD also shows that PCSK9 inhibitors are not cost-effective at 2017 prices (Kazi et al. 2017). Arrieta et al. (2017, b) have made similar conclusions regarding the cost-effectiveness of PCSK9 inhibitors compared to standard statin therapy. A more recent analysis—a systematic review of the cost-effectiveness analysis of PCSK9 inhibitors in cardiovascular diseases concluded that while PCSK9 inhibitors are not cost-effective in general population, their use in FH and statin intolerants may be more cost-effective (Azari et al. 2020).

Cost-Utility Analysis (CUA)

Cost-utility analysis was developed to factor quality of life into economic analysis by comparing the cost of the therapy/intervention with the outcomes measured in quality-adjusted life-years (QALY). The QALYs are calculated by multiplying the length of time in a specific health state by the utility of that health state—the utility of a specific health state is, in

essence, the desirability of life in a specific health state compared with life in perfect health. A utility rating of 0.9 would mean that the health state in question is 90% as desirable as perfect health, while a utility rating of 0.5 would mean that health state is only half a desirable. Death is given a utility score of 0.0. The results of a CUA are expressed in terms of cost per QALY gained as a result of given treatment/intervention. CUA is beneficial when comparing therapies that produce improvements in different or multiple health outcomes. Cost per QALY can be measured and evaluated across several different treatment scenarios, allowing for comparisons of disparate therapies.

Goulart and Ramsey (2011) evaluated cost utility of Avastin® (bevacizumab) and chemotherapy versus chemotherapy alone for the treatment of advanced nonsmall-cell lung cancer (NSCLC). Avastin is currently approved for the treatment of NSCLC in combination with chemotherapy based on two-month median survival proved in clinical trials. Researchers developed a model to determine QALYs and direct medical cost incurred due to treatment with bevacizumab in combination with chemotherapy. The utilities used in calculating QALY were obtained from the literature and costs were obtained from Medicare. The results of the study showed that bevacizumab is not cost-effective when added to chemotherapy. It was found that bevacizumab with chemotherapy increased the mean QALYs by only 0.13 (roughly the equivalent of 1.5 months of perfect health), at an incremental lifetime cost of $72,000 US per patient. The incremental cost-utility ratio (ICUR) was found to be $560,000 US/QALY. The results of these analyses could be potentially used by payers while allocating resources for the treatment of NSCLC care. Table 10.3 represents the base case results of cost-utility analysis.

Sources of Economic Value

The economic value of the product may have elements besides the basic economic efficiency implied by the break-even level just discussed. Quality differences, in terms of reduced side effects, greater efficacy, or other substantive factors, can result in increases in value beyond the break-even point calculated in a simple cost comparison. Should these factors be present, it is crucial to capture their value in the price of the product, but how much value should be captured?

Table 10.3 Base case results of Cost-Utility analysis

Outcomes	CPB	CP	Differences
Effectiveness			
Life expectancy (years)	1.24	1.01	0.23
Progression-free survival (years)	0.72	0.47	0.25
QALYs	0.66	0.53	0.13
Lifetime costs per patients (US$)[a]			
Drug utilization	70,284.75	646.96	69,637.79
Drug administration	4239.87	1495.24	2744.63
Fever and neutropenia	25.32	4.37	20.95
Severe bleeding	19.65	1.33	18.32
Other adverse events	39.06	32.09	6.97
Outpatient visits	1017.90	609.41	408.49
Progressive disease	40,283.71	41,500.96	−1217.25
Total	115,910.26	44,290.36	71,619.90
ICER (US$/life-years gained)			308,981.58
ICUR (US$/QALY gained)			**559,609.48**

CP carboplatin and paclitaxel, *CPB* carboplatin, paclitaxel, and bevacizumab, *ICER* incremental cost-effectiveness ratio, *ICUR* incremental cost-utility ratio, *QALYs* quality-adjusted life-years
[a]Cost in 2010 US dollars

It is important to recognize that a product can provide a significant economic benefit in one indication but none in another. Therefore, it is prudent to perform these studies on all indications considered for a new product. A case in point is that of epoetin alfa (EPO). EPO was initially developed and approved for use in dialysis patients, where its principal benefit is to reduce, or even eliminate, the need for blood transfusion. Studies have shown that EPO doses that drive hematocrit levels to between 33 and 36% result in significantly lower total patient care costs than lower doses of EPO or none at all (Collins et al. 2000). The same product, when used to reduce the need for transfusion in elective surgery, however, has been shown not to be cost-effective (Coyle et al. 1999). Although EPO was shown to reduce the need for transfusion in this study, the cost of the drug far outweighed the savings from reduced transfusions as well as reductions in the transmission and treatment of blood-borne pathogens. Economic efficiency is not automatically transferred from one indication to another.

The lack of economic savings in the surgical indication does not necessarily mean that the product should not be used, only that users must recognize that in this indication use results in substantially higher costs, while in dialysis it actually reduces the total cost of care.

Future US Health Care Changes

Payers within the US health care system have begun to use similar methods of evaluation. Toward this effort, the Institute for Clinical and Economic Review (ICER) was formed in 2006. ICER is an independent nonprofit research institute

that produces reports analyzing the evidence on the effectiveness and value of drugs and other medical services. Since 2007, ICER has published 92 papers examining cost-effectiveness of several different treatments across different therapeutic areas. ICER claims that their analyses help payers, life science companies, patient groups, and others to develop formularies, negotiate drug prices, and explore new ways to apply evidence in a drive toward a health system that can achieve fair prices and fair access.

Several states in the USA have taken steps to control or increase transparency in drug pricing. New York state passed a law in 2017 which enables authorities to determine value-based prices for high-cost drugs, and then negotiate for additional rebates to achieve this price for its Medicaid program (Hwang et al. 2017). Other states have passed bills that require manufacturers to justify price increases above a certain threshold or publish research and development costs for new drugs (Sarpatwari et al. 2016).

Recently, the Inflation Reduction Act (IRA) of 2022 was passed to control Medicare spending on drugs and lower out-of-pocket costs for Medicare enrollees. The IRA will require Medicare to negotiate costs of single-source highest spend drugs: up to ten retail drugs beginning in 2026, which increases gradually to 20 retail and 20 physician-administered drugs by 2029. In addition, the IRA requires manufacturers to pay rebates to Medicare if their drug prices rise faster than overall inflation, places a cap of $2000 per Medicare patient per year on prescription drug spending, and a cap of $35 per month for insulin.

In the pharmaceutical marketing environment of the foreseeable future, it is wise to first consider determining the true medical need for the intervention. Then, if the need is real, to consider surrendering some value to the market—pricing of the product at some point below its full economic value. This is appealing for several reasons:

- The measurement of economics is imprecise and the margin for error can be large.
- If the market is looking for lower costs, filling that need enhances the market potential of the product.
- From a public relations and public policy perspective, launching a new product with the message that it provides savings to the system can also provide positive press and greater awareness.

Biosimilars

Many drugs used to diagnose and treat diseases are biological products that until recently were available from a singular manufacturer. The first biosimilar (a product that is highly similar to the original biological product, described in more detail in Chap. 11 in this textbook) was Zarxio® (filgrastim-sndz), launched by Sandoz in 2015 as a biosimilar to Neupogen® (filgrastim). Subsequently, the biosimilars Inflectra® (infliximab-dyyb), Renflexis™ (infliximab-abda), and Avsola™ (infliximab-axxq) for Remicade® (infliximab) were launched in 2016, 2017, and 2020, respectively. Like generic alternatives launched at a discounted price to their branded counter-parts, Zarxio and Inflectra were launched at a 15% discount to Neupogen and Remicade, respectively. Renflexis was priced at a 35% discount to Remicade and 20% discount to Inflectra. Avsola, the most recent launch in the group, was priced at a 57%, 47%, and 34% discount to Remicade, Inflectra, and Renflexis, respectively.

Biosimilars tend to take much longer than and may be more or equally expensive to develop as their original comparators. However, the current health care system and payers anticipate new biosimilars to be priced less than their original products, an expectation that developers and manufactures of biosimilars need to keep in mind as they pursue R&D of these products. Biosimilars have the potential to reduce overall health care costs spent on biological products compared to their original products; however, their pricing and value and ultimately commercial success remain to be examined. As of January 2022, over 30 biosimilars have been FDA approved, and over 20 of these have been launched in the US market.

In 2023, around ten biosimilars are expected to launch for Humira® (adalimumab), the world's top-selling drug. Many of these biosimilars may not have the FDA designation of interchangeability to Humira. To gain interchangeability status, additional studies are required that show lack of additional risk or reduced efficacy in patients switched back and forth multiple times between the reference product and the biosimilar. Pharmacists can dispense a biosimilar with interchangeable status in place of the originator brand without the need for physician approval. It remains to be seen if adalimumab biosimilars will also be priced at a discount to their reference product, and if product attributes such as interchangeability, concentration, and formulation provide pricing flexibility.

Conclusions

As societies continue to focus on the cost of health care interventions, we must all be concerned about the economic and clinical implications of the products we bring into the system. Delivering value, in the form of improved outcomes, economic savings, or both, is an important part of pharmaceutical science and marketing. Understanding the value that is delivered and the different ways in which it can be measured should be the responsibility of everyone involved with new product development. Further, all of this needs to be done with keeping the changing health care policy and regulatory landscape in mind.

Questions and Answers

1. When conducting pharmacoeconomic evaluations, typically the following costs are calculated or included in the analyses: direct, indirect, intangible, and opportunity costs. Define these cost types and give examples to describe them.

   (a) Direct costs: They are the resources consumed in the prevention or treatment of a disease. Direct costs are further divided into:

      - Direct medical costs: These include expenditures on drugs, medical equipment, laboratory testing, hospital supplies, physician visits, and hospitalization costs. Direct medical costs are further divided into:

         – Variable costs: They are an integral part of pharmacoeconomic analysis. Variable costs increase or decrease depending on the volume, and include drugs, fees for professional services, and supplies.

         – Fixed costs: These include expenditures on rent, utilities, insurance, accounting, and other administrative activities. Fixed costs are *not included* in pharmacoeconomic evaluations because their use or total cost is unlikely to change as a direct result of a specific intervention.

      - Direct nonmedical costs: They are costs are out-of-pocket costs paid by patients (or their caregivers) for nonmedical services, and include expenditure on transportation to and from the hospital, clinic or physician office, additional trips to emergency rooms, expenses on special diet, family care expenses, and other various forms of out-of-pocket expenses.

(b) Indirect costs: They are costs that result from morbidity or mortality. Indirect costs typically include the loss of earnings due to temporary or permanent disability, loss of income to family member who gave up their job temporarily or permanently to take care of patient, and loss in productivity due to illness.

(c) Intangible costs: They represent the nonfinancial outcome of disease and medical care, and are the most difficult to quantify in monetary terms. Examples include pain, suffering, and emotional disturbance due to underlying conditions. Intangible costs are identified but not formally calculated in an economic analysis. At times, they are converted into a common unit of outcome measurement such as a quality-adjusted life-year (QALY).

(d) Opportunity costs: They are defined as the value of the alternative that was forgone due to the purchase of a medical treatment. Opportunity costs are *typically not included* in traditional pharmacoeconomic analysis.

2. List the five pharmacoeconomic techniques used to examine a new medical technology or treatment option?
    (a) Cost of Illness (COI)
    (b) Cost-Minimization Analysis (CMA)
    (c) Cost–Benefit Analysis (CBA)
    (d) Cost-Effectiveness Analysis (CEA)
    (e) Cost-Utility Analysis (CUA)

3. Cost-utility Analysis (CUA) includes outputs shown as QALYs. Define and describe QALY.
    (a) QALYs stand for Quality-Adjusted Life-Years. The QALYs are calculated by multiplying the length of time in a specific health state by the utility of that health state—the utility of a specific health state is, in essence, the desirability of life in a specific health state compared with life in perfect health. The results of a CUA are expressed in terms of cost per QALY gained as a result of given treatment/intervention.

4. Which method is used most often by insurance company payers to make product P&T formulary decisions? Why?
    (a) The cost-effectiveness analysis (CEA) is often used to guide formulary management decisions because:
        • The CEA does not require the conversion of health outcomes to monetary units.
        • The CEA helps in determining the optimal alternative, which may not always the least costly alternative.

References

Arrieta A, Hong JC, Khera R, Virani SS, Krumholz HM, Nasir K (2017a) Updated cost-effectiveness assessments of PCSK9 inhibitors from the perspectives of the health system and private payers: insights derived from the FOURIER trial. JAMA Cardiol 2(12):1369–1374

Arrieta A, Page TF, Veledar E, Nasir K (2017b) Economic evaluation of PCSK9 inhibitors in reducing cardiovascular risk from health system and private payer perspectives. PLoS One 12(1):e0169761

Azari S, Rezapour A, Omidi N, Alipour V, Behzadifar M, Safari H, Tajdini M, Bragazzi NL (2020) Cost-effectiveness analysis of PCSK9 inhibitors in cardiovascular diseases: a systematic review. Heart Fail Rev 25(6):1077–1088

Bootman JL, Townsend JT, McGhan WF (1991) Principles of pharmacoeconomics. Harvey Whitney Books, Cincinnati

Chang K, Nash D (1998) The role of pharmacoeconomic evaluations in disease management. PharmacoEconomics 14(1):11–17

Collins AJ, Li S, Ebben J, Ma JZ, Manning W (2000) Hematocrit levels and associated Medicare expenditures. Am J Kidney Dis 36(2):282–293

Cornes P, Kelton J, Liu R, Zaidi O, Stephens J, Yang J (2022) Real-world cost–effectiveness of primary prophylaxis with G-CSF biosimilars in patients at intermediate/high risk of febrile neutropenia. Future Oncol 18(6):1979–1996

Coyle D, Lee K, Laupacis A, Fergusson D (1999) Economic analysis of erythropoietin in surgery, vol 9. Canadian Coordinating Office for Health Technology Assessment, Ottawa, p 21

Goulart B, Ramsey S (2011) A trial-based assessment of the utility of bevacizumab and chemotherapy versus chemotherapy alone for advanced non-small cell lung cancer. Value Health 14:836–845

Grauer DW, Lee J, Odom TD, Osterhaus JT, Sanchez LA (2003) Pharmacoeconomics and outcomes: applications for patient care, 2nd edn. American College of Clinical Pharmacy, Kansas City

Hwang TJ, Kesselheim AS, Sarpatwari A (2017) Value-based pricing and state reform of prescription drug costs. JAMA 318(7):609–610

Joshnson N, Nash D (eds) (1996) The role of pharmacoeconomics in outcomes management. American Hospital Publishing, Chicago

Kazi DS, Moran AE, Coxson PG, Penko J, Ollendorf DA, Pearson SD, Tice JA, Guzman D, Bibbins-Domingo K (2016) Cost-effectiveness of PCSK9 inhibitor therapy in patients with heterozygous familial hypercholesterolemia or atherosclerotic cardiovascular disease. JAMA 316(7):743–753

Kazi DS, Penko J, Coxson PG, Moran AE, Ollendorf DA, Tice JA, Bibbins-Domingo K (2017) Updated cost-effectiveness analysis of PCSK9 inhibitors based on the results of the FOURIER trial. JAMA 318(8):748–750

Qiao N, Insinga R, Burke T, Lopes G (2021) Cost-minimization analysis of pembrolizumab monotherapy versus nivolumab in combination with ipilimumab as first-line treatment for metastatic pd-L1-positive non-small cell lung cancer: a US payer perspective. PharmacoEconomics 5:765–778

Sarpatwari A, Avorn J, Kesselheim AS (2016) State initiatives to control medication costs — can transparency legislation help? N Engl J Med 374:2301–2304

Segel JE (2006) Cost-of-illness studies—a primer, RTI-UNC Center of excellence in health promotion economics. http://www.rti.org/pubs/coi_primer.pdf

Szcus TD, Schneeweiss S (2003) Pharmacoeconomics and its role in the growth of the biotechnology industry. J Commer Biotechnol 10(2):111–122

Townsend R (1987) Postmarketing drug research and development. Drug Intell Clin Pharm 21(1pt 2):134–136

Walkom E, Robertson J, Newby D, Pillay T (2006) The role of pharmacoeconomics in formulary decision-making: considerations for hospital and managed care pharmacy and therapeutics committees. Formulary 41:374–386

Further Reading

Pharmacoeconomic Methods and Pricing Issues

Bonk RJ (1999) Pharmacoeconomics in perspective. Pharmaceutical Products Press, New York

Drummond MF, O'Brien BJ, Stoddart GL, Torrance GW (1997) Methods for the economic evaluation of health care programmes, 2nd edn. Oxford University Press, Oxford

Kolassa EM (1997) Elements of pharmaceutical pricing. Pharmaceutical Products Press, New York

Pharmacoeconomics of Biotechnology Drugs

Dana WJ, Farthing K (1998) The pharmacoeconomics of high-cost biotechnology products. Pharm Pract Manag Q 18(2):23–31

Hui JW, Yee GC (1998) Pharmacoeconomics of biotechnology drugs (part 2 of 2). J Am Pharm Assoc 38(2):231–233

Hui JW, Yee GC (1998a) Pharmacoeconomics of biotechnology drugs (part 1 of 2). J Am Pharm Assoc 38(1):93–97

Reeder CE (1995) Pharmacoeconomics and health care outcomes management: focus on biotechnology (special supplement). Am J Health Syst Pharm 52(19, S4):S1–S28

Biosimilars: Principles, Regulatory Framework, and Societal Aspects

11

Arnold G. Vulto and Liese Barbier

Introduction

The term "biopharmaceuticals" is used to describe biotechnologically derived drug products. Biopharmaceuticals are protein-based macromolecules and include insulin, human growth hormone, the families of the cytokines and of the monoclonal antibodies, and antibody fragments. A "generic" version of a biopharmaceutical may be introduced after expiration of patents and other exclusivity rights of the originator product. However, the generic paradigm as it has developed for small molecules over the years cannot be used for biopharmaceuticals. In the EU and the US regulatory systems, the term "biosimilar" was coined for copies of brand-name, originator biopharmaceuticals. A biosimilar is a biopharmaceutical product that is highly similar to an already approved biopharmaceutical product (also called the reference product), notwithstanding minor differences in clinically inactive components, and for which there are no clinically meaningful differences between the biosimilar and the reference product in terms of safety, purity, and potency. Different than for small-molecule generic versions, and especially for more complex cases, biosimilar product development generally includes a comparative confirmation of efficacy and safety in an appropriate patient population (Declerck et al. 2017, see further under section Clinical Development).

The aim of this chapter is to provide a comprehensive view on the basic principles underlying biosimilar development, the regulatory framework for the evaluation and the approval of a biosimilar product in both the EU and the US, and societal reflections, needed to understand the current biosimilar landscape. Therefore, this chapter is not about one specific group of biopharmaceuticals; it covers most biopharmaceuticals beyond expiration of exclusivity rights and with a certain regulatory status. Table 11.1 provides definitions of terms relevant to this chapter.

A. G. Vulto (✉)
Hospital Pharmacy, Erasmus University Medical Centre, Rotterdam, The Netherlands

Department of Pharmaceutical and Pharmacological Sciences, KU Leuven, Leuven, Belgium
e-mail: a.vulto@erasmusmc.nl

L. Barbier
Department of Pharmaceutical and Pharmacological Sciences, KU Leuven, Leuven, Belgium
e-mail: liese.barbier@kuleuven.be

Table 11.1 Definitions relevant to this chapter

FDA	US Food and Drug Administration
EMA	European Medicines Agency
Small-molecule drug	Medicinal product prepared by chemical synthesis
Generic product	Nonpatented medicinal product of a small-molecule drug which is therapeutically equivalent
Biopharmaceutical drug	The term "biopharmaceuticals" is used to describe biotechnologically derived drug products. Biopharmaceuticals are protein-based macromolecules and include insulin, human growth hormone, the families of the cytokines and of the monoclonal antibodies, antibody fragments, and nucleotide-based systems such as antisense oligonucleotides, siRNA, mRNA, and DNA preparations for gene delivery[a]
Biosimilar product	A biosimilar is a biopharmaceutical product that is highly similar to an already approved biopharmaceutical product, notwithstanding minor differences in clinically inactive components, and for which there are no clinically meaningful differences between the biosimilar and the approved biopharmaceutical product in terms of safety, purity, and potency EMA definition: "A biosimilar is a biological medicine highly similar to another biological medicine already approved in the EU (the 'reference medicine') in terms of structure, biological activity and efficacy, safety and immunogenicity profile" FDA definition: "A biosimilar is a biological product that is highly similar to and has no clinically meaningful differences from an existing FDA-approved reference product"
Second-generation biopharmaceuticals	A second-generation biopharmaceutical product is derived from an approved biopharmaceutical product which has been deliberately modified to change one or more of the product's characteristics

[a] For the FDA, "biopharmaceuticals" are part of the "biological product" group, which also include viruses, sera, vaccines, and blood products

Understanding Differences in Regulatory Framework for Generics and Biosimilars

In order to have an understanding of the regulatory process involved for a biosimilar, it is essential to appreciate the basic difference between small-molecule drugs and macromolecular biopharmaceuticals (cf. Tables 11.2 and 11.3). Small molecules are chemically synthesized and can be fully characterized. On the other hand, all biopharmaceuticals are – at least in part – produced in a living system such as a microorganism, plant, or animal cells and are difficult to fully characterize. The proteins are typically complex molecules. Differences in a manufacturing process may lead to alteration in the protein structure (Crommelin et al. 2003; Schiestl et al. 2011; Vezer et al. 2016; Kim et al. 2017). Protein structures can differ in at least three ways: in their primary amino acid sequence, in (posttranslational) modification to those amino acid sequences (e.g., glycosylation), and in higher-order structure (folding patterns) and (unwanted) fragments and aggregate formation. Advances in analytical sciences enable protein products to be extensively characterized with respect to their physicochemical and biological properties. Advanced in vitro technologies allow a comprehensive picture of the behavior of the molecule in those aspects that matter, i.e., biological/pharmacological action, predictive for efficacy and safety (Vulto and Jaquez 2016). However, residual uncertainties may require additional testing in humans/patients.

For the approval of a (small-molecule) generic product, it must be pharmaceutically equivalent (same dosage form, strength of active ingredient, route of administration, and labeling as the originator product) and pharmacokinetically equivalent according to ICH guidelines (ICH = International Council for Harmonization of Technical Requirements for Pharmaceuticals for Human Use). Depending on the active substance, it should have the same in vitro dissolution, pharmacokinetic, pharmacodynamic, and clinical outcome profile as the originator product. For the more complex biological products, such a simple assessment of pharmaceutical equivalence and bioequivalence alone is not considered sufficient.

The required dossier for market authorization of a generic focuses on only two aspects. The generic (small-molecule) drug product should be pharmaceutically equivalent and bioequivalent to the brand-name product and is thereby therapeutically equivalent and interchangeable with the brand-name drug product. A generic product has the same active ingredient, and therefore, the safety and efficacy of the active ingredient is already established. The only question is the efficacy of the generic formulation, and this is assured by the bioequivalence study in healthy volunteers or patients. In the case of small molecules, the identity of the active substance is established through a validated chemical synthesis route, full analysis of the active agent, impurity profiling, etc. In the case of biopharmaceuticals, this is, generally speaking, not possible. This copy or follow-on biopharmaceutical product contains the active ingredient which is similar (but not necessarily equal) in characteristics to the reference product. For this reason, generic biopharmaceutical products are referred to as highly similar, *biosimilar* products.

Both for a generic and for a biosimilar product, a complete set of information on Chemistry, Manufacturing, and Controls (CMC) is needed in the dossier submitted for

Table 11.2 Difference between small, low-molecular weight drugs, and biopharmaceuticals

Small, low-molecular weight drugs	Biopharmaceuticals
Low molecular weight	High molecular weight
Relatively simple chemical structure	Complex three-dimensional structure
Chemically synthesized	Produced by living organism
Relatively easy characterization	Difficult to impossible to fully chemically characterize
Synthetically pure	Almost always heterogeneous
Immune responses less frequent	Some molecules prone to eliciting an immune response

Table 11.3 General drug approval requirements for generic product and biosimilar product

Generic (small-molecule) product	Biosimilar product
Pharmaceutical equivalent	Pharmaceutical equivalent
In vitro physicochemical analysis	In vitro physicochemical analysis
	Pharmacological comparability testing
Bioequivalent	Clinically similar
Pharmacokinetics	Clinical comparison: Generally comparative PK/PD study in healthy volunteers. For more
Pharmacodynamics	complex products (such as monoclonal antibodies) also one or more comparative efficacy
Clinical comparison (not standard)	studies in patients may be required

approval to ensure that the drug substance and the drug product are pure, potent, and of high quality. The CMC section should include full analytical characterization, a description of the manufacturing process and test methods, and stability data. The approval process requires manufacturing of the drug product under controlled current good manufacturing practice (cGMP) conditions. The cGMP requirement ensures consistent quality, identity, potency, and purity of the final product between batches.

Of note, biosimilar products are not to be confused with second-generation biopharmaceuticals, e.g., pegylated G-CSF and pegylated interferon alpha, and darbepoetin. Second-generation biopharmaceuticals have improved pharmacological properties/biological activity compared to an already approved biopharmaceutical product which has been deliberately modified (Weise et al. 2011). Second-generation products are marketed with the claim of clinical superiority and require a full new regulatory application.

Principles of Biosimilar Development and the Regulatory Framework

Biosimilar development starts with a comprehensive understanding of the originator product that is the target for "copying." It entails an as detailed as possible analysis of the originator product in chemical, physical, biological, and pharmacological properties. Biological molecules have an intrinsic variability; they are always a mixture of more or less closely related isoforms. The backbone, the amino acid structure (primary and secondary structures), is identical, However, variations may occur in tertiary and quaternary structures. For this reason, the biosimilar developer has to

capture this variability, by analyzing a number of batches (typically 10–20) of the originator product and define the quality attributes of this product including its variability. This essentially sets the quality specifications of the candidate biosimilar (Vulto and Jaquez 2016). The variability of the biosimilar is expected not to be greater than that of the originator product, and all critical quality attributes (i.e., those important for molecule function) must be comparable, or it should be justified why an observed difference is not clinically relevant (EMA 2014a; Weise et al. 2014).

The Similarity Exercise Resulting in Totality-of-Evidence

Biosimilars are licensed on the basis of demonstrating biosimilarity to the originator biological product (i.e., the reference product). This is done via what is called the *similarity exercise* or *biosimilar comparability exercise*. The similarity exercise involves an extensive head-to-head comparison between the biosimilar candidate and the reference product, with the aim of demonstrating, based on the totality of evidence, that the candidate biosimilar is highly similar to its reference product. The term "totality-of-evidence" includes all data and information submitted in the application, including structural and functional characterization, nonclinical evaluation, human pharmacokinetics (PK) and pharmacodynamics (PD) data, clinical immunogenicity data, and comparative clinical study(ies) data, as necessary. Theoretically speaking, the similarity exercise follows a stepwise approach, meaning that each study conducted informs and determines the need, nature, and extent of the subsequent one (EMA 2014a).

The Foundation of Biosimilar Similarity: Physicochemical Characterization

Analytical studies form the foundation of any biosimilar development program. The reference product should be adequately characterized with respect to the critical quality attributes, clinically active components, mechanism of action, and structure–function relationships. In addition, biochemical characterization, and functional characterization should be carried out. Once the reference product is characterized in detail, comparative tests between the reference product and the biosimilar product should be done. The robustness and level of evidence of the physicochemical, biological and nonclinical in vitro data determine the extent and nature of the subsequent clinical studies to be performed (EMA 2014a). The stepwise development of a biosimilar is shown schematically in Fig. 11.1, although in reality many studies are nowadays being run in parallel. At each step, the sponsor should evaluate the level of residual uncertainty about the biosimilarity of the proposed biosimilar product to the reference product and identify the next step to address this. If there is a residual uncertainty about biosimilarity after conducting structural analyses, functional assays, PK and PD studies in healthy volunteers, and a clinical immunogenicity assessment, then additional clinical data in patients may be needed to adequately address

that uncertainty. This can be the case if no validated PD markers are at hand or healthy volunteer studies are not feasible (e.g., with intra-ocular injections). In such case, the aim of the comparative study is to confirm that there are no clinically meaningful differences between the biosimilar and the reference product.

In general, the requirements for approval of biosimilar products are based on the structural complexity and clinical knowledge of and experience with the reference biopharmaceutical product. For example, protein products such as growth hormone have a relatively simple and well-understood chemical structure. In addition, extensive manufacturing and clinical experience is available for these products. On the other hand, recombinant factor VIII is for example a large, highly complex molecule with several isoforms. Because of the varying complexity of biopharmaceuticals, the requirements for the approval process may – within limits – vary and regulators are increasingly open for case-by-case discussions.

Ultimately, the goal of the biosimilar comparability exercise is to exclude any clinically relevant differences between the biosimilar and the reference product. Therefore, studies should be sensitive enough to detect any such differences. Choices in terms of study design, conduct, endpoints, and/or population should be taken into account. This also explains why the approach in biosimilar development differs from

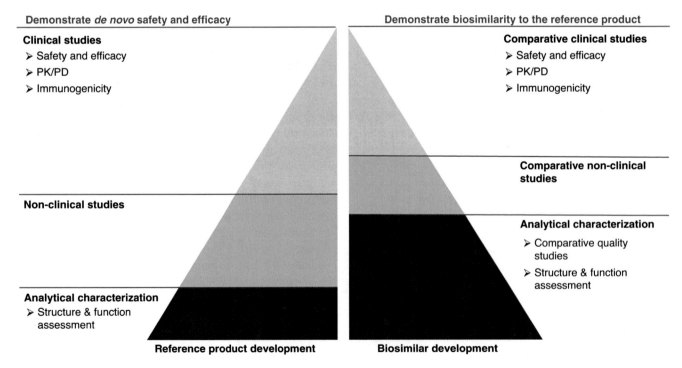

Fig. 11.1 Comparison of reference product versus biosimilar development. Abbreviations: *PK/PD* pharmacokinetic/pharmacodynamic studies. Figure adapted from: European Medicines Agency and European Commission. Biosimilars in the EU—Information guide for healthcare professionals, 2017

that of a new product. While for a new molecule de novo efficacy and safety need to be demonstrated, the aim of biosimilar development is to establish high similarity with the reference product and exclude any clinically meaningful differences (EMA 2014a, 2015; Barbier et al. 2019).

Extrapolation of Indications

Once biosimilarity is demonstrated, a decision has to be made for which indications of the reference product the biosimilar can be registered. For newly introduced, originator products this is straightforward: for each indication efficacy and safety needs to be demonstrated based on appropriately designed clinical studies. For a biosimilar, once a product has been found to be highly similar in terms of structural functional properties and (human) PK, it is unnecessary to repeat studies that have already been performed for each indication of the reference product. Repeating these studies would be costly and time consuming, while not providing new information, and therefore would be deemed unethical. Hence, the principle of extrapolation of indication is essential to biosimilar development. Extrapolation of indication refers to the approval of a biosimilar for use in an indication held by the reference product but not directly studied in a comparative clinical trial for the biosimilar (Weise et al. 2014; Rojas-Chavarro and de Mora 2021). Extrapolation of data is not a new but an already well-established scientific and regulatory principle that has been exercised for many years by regulators e.g., in case of major manufacturing changes of a biological or with the introduction of a new subcutaneous formulation of an existing intravenous used product. Extrapolation of indications for a biosimilar is not granted automatically, but a matter of assessment, and can be considered if biosimilarity to the reference product has been demonstrated with testing in a population that is suitable to detect potentially clinically relevant differences and if there is sufficient scientific justification for extrapolation. This includes a number of factors, elaborated in the reference Weise et al. 2014.

Assessing Safety of Manufacturing Changes Goes Back to 1996

Regulators have accumulated extensive experience in assessing biosimilars. First, manufacturers of biopharmaceuticals (both originators and biosimilars) generally make several manufacturing changes to optimize processes throughout the product lifecycle (Schneider 2013). These changes need to be thoroughly evaluated via the so-called comparability exercise, which involves a pre- and postchange product comparison to ensure that the quality, efficacy, and safety are not adversely impacted (Schneider 2013; Rojas-Chavarro and de Mora 2021). In 1996, the US Food and Drug Administration (FDA) made the first proposals how to perform such a comparative analysis after a manufacturing change (FDA 1996), and it appeared to be a useful basis to start working on formalizing biosimilar assessment. Although this is not the same process, the same scientific principles that underlie the demonstration of pre- and postmanufacturing product similarity also apply to the demonstration of biosimilarity (Declerck and Rezk 2017; Weise et al. 2014). The cornerstone of both being the thorough comparison of physicochemical and functional characteristics using state-of-the-art analytical tools. While comparative clinical testing in patients can be required in case of major changes (e.g., change in cell line), it is rarely requested in the context of a comparability exercise following a manufacturing change. For a demonstration of biosimilarity, generally more data (and in some cases also comparative data in patients) are needed.

The biosimilar regulatory approval pathway is now well-established. Already in 2006, the first biosimilar was evaluated and recommended for approval by the European Medicines Agency (EMA) with the first FDA biosimilar approval following in 2015 (EMA and European Commission 2017; FDA 2022).

Characterization of Biosimilar Products

In the development program of a biosimilar, a comprehensive comparison with the originator's product characteristics (e.g., the primary, secondary, tertiary, and quaternary structures, posttranslational modifications) and functional activity/ies should be considered.

A comprehensive understanding of all steps in the manufacturing process, process controls, and the use of a Quality-by-Design approach will facilitate consistent manufacturing of a high-quality product (Kenett and Kenett 2008).

In the early days of biopharmaceuticals, a detailed characterization and assessment of similarity of the biosimilar and originator's product was not feasible with the arsenal of analytical techniques available at that time. However, modern technologies allow for a more refined and sophisticated comparative characterization with a high predictive value for efficacy and safety (Table 11.4; Chap. 3).

Characterization of the active moiety and impurity/contaminant profiling plays an important role in the development process of a biosimilar drug product. The technological advances in instrumentation significantly improved identification and characterization of biotech products (Chap. 3). It is acknowledged that no one analytical method can fully characterize the biopharmaceutical. A collection of orthogonal analytical methods (Chap. 5) is needed to piece together

Table 11.4 Quality attributes of a monoclonal antibody, how they are measured/quantified, their criticality for molecule properties, and impact on biological activity/clinical function

Quality attributes	Criticality	Impact on biological activity or clinical function	Analytical methods
Primary structure			
Amino acid sequence	+++	Might influence potency	Peptide mapping with LC/MS or LC/MS/MS
N-terminal pyroglutamate	++	Might influence in vivo pharmacokinetics	Edman degradation
C-terminal lysine removal	+	Should not influence in vitro biological activity	
Disulfide bonds	+++	Might influence potency	
Higher-order structures	++	Might influence the receptor or antigen binding, Might influence potency	CD, fluorescence, DSC, FT-IR
Charged variants			
Deamidated form	+	Might influence in vitro biological activity	LC/MS, IEX, CE, IEF, RP-HPLC
Oxidized form	++	Might lead to highly immunogenic aggregates	HPAEC-PAD
Sialylated form	+++	Might influence in vivo clearance	LC/DMB labelling
Mass variants			
Aggregates	+++	Might induce immunogenicity	MS, SEC with UV or MALS, HIC, CGE, AUC, SDS-PAGE
Truncated form	+	Might maintain some biological activity	
Monomer	+	Might maintain some biological activity	
Pegylation	+++	Might influence in vivo clearance	
Oligosaccharides			
Fucose or galactose	+++	Might influence Fc effector functions	HPAEC-PAD
Non-human glycans	+++	Might induce immunogenicity	LC or CE with labelling (e.g., 2AA or 2AB),
High-Man glycans	++	Might induce immunogenicity Might influence in vivo clearance Might influence ADCC	LC/MS, LC/MS/MS
Non-glycosylated form	++	Might influence Fc effector functions	
Biological activities			
ADCC	+++	Might influence mode of action	Cell-based assay, ELISA, SPR
ADCP	++	Might influence mode of action	
CDC	+++	Might influence mode of action	
Apoptosis	++	Might influence mode of action	
Antigen/antibody binding	+++	Might influence mode of action	
FcgR binding	++	Might influence ADCC	
C1q binding	++	Might influence CDC	
FcRn binding	++	Might influence in vivo clearance	
Content	+++	Will affect potency final product	UV spectrometry

(1) O. Kwon, J. Joung, Y. Park, C. W. Kim, S. H. Hong. Considerations of critical quality attributes in the analytical comparability assessment of biosimilar products. Biologicals. 2017
(2) Vulto AG, Jaquez OA. The process defines the product: what really matters in biosimilar design and production? Rheumatology. 2017. doi:10.1093/rheumatology/kex278
AUC Analytical ultracentrifuge, *CD* circular dichroism spectroscopy, CE capillary electrophoresis, *CGE* capillary gel electrophoresis, *DSC* differential scanning calorimetry, *ELISA* enzyme Linked Immunosorbent assay, *FT-IR* Fourier transform infrared spectroscopy, *HI* hydrophobic interaction chromatography, *HPAEC-PAD* high-performance anion exchange chromatography with pulsed amperometric detection, *IEF* isoelectric focusing, *IEX* ion exchange, *LC/DMB* liquid chromatography - 1,2-diamino-4,5-methylenedioxybenzene-2HCl, *LC/MS* liquid chromatography-mass spectrometry, *LC/MS/MS* Liquid chromatography tandem mass spectrometry, *MALS* multi-angle lights scattering, *MS* mass spectrometry, *RP-HPLC* reverse phase high performance liquid chromatography, *SDS-PAGE* sodium dodecyl sulfate–polyacrylamide gel electrophoresis, *SEC* size exclusion chromatography, *SPR* surface plasma resonance

a complete picture of a biotech product. Determining how a small, homogeneous pure protein is folded in absolute terms can be accomplished using crystal X-ray diffraction analysis. This analysis is difficult, expensive, and nonrealistic to perform in the formulated drug product and on a routine basis. Moreover, X-ray diffraction analysis does not pick up low levels of conformational contaminants. The most basic aspect of assessing its identity is to determine its covalent, primary structure using liquid chromatography tandem mass spectrometry (LC/MS/MS), peptide mapping/amino acid sequencing (e.g., via the Edman degradation protocol), and disulfide bond-locating methods. Through circular dichroism, Fourier transform infrared and fluorescence spectroscopy, immunological methods, chromatographic techniques, etc., differences in secondary and higher-order structures can be monitored. Selective analytical methods are used to determine the purity as well as impurities of the biotech product. Methods here include again chromatographic techniques and gel electrophoresis, capillary electrophoresis, isoelectric focusing, static and dynamic light scattering, and ultracentrifugation. Inadequate characterization can result in failure to detect product changes or differences that can impact the safety and efficacy of a product. Therefore, structural factors that impact safety and efficacy of the product and methods to monitor them should be identified in the product development process (cf. Chap. 3). Table 11.4 provides a typical battery of techniques used to characterize different attributes of a monoclonal antibody, the criticality of the attribute, and what the impact can be of the attribute on biological activity or clinical function,

Clinical Development of Biosimilar Products

The main aim of the clinical development program for a biosimilar is to confirm that any differences between the biosimilar candidate and its reference product are not clinically meaningful (EMA 2014a; FDA 2021; WHO 2022). The number and design of the clinical studies conducted in the context of biosimilar development depend, inter alia, on the robustness of the preceding stages of development (analytical in vitro assessment and nonclinical in vivo testing, if required).

The clinical development program generally includes a comparative PK study. If suitable PD markers are available, the PK trial can be combined with PD measurements. The PK(/PD) trial is preferably conducted in healthy volunteers because disease factors may introduce confounding variability. Furthermore, the trial should be designed and powered to demonstrate equivalence to the reference product. The choice of the design of the pharmacokinetic study, i.e., single dose and/or multiple doses, should be justified. Safety and immunogenicity data are normally collected during this trial as well (EMA 2014a). In certain cases, comparative PK/PD studies between the biosimilar product and the reference product using biomarkers may prove sufficient in terms of clinical testing. Notable examples are (peg)filgrastim and insulin biosimilars (EMA 2014a).

In some especially more complex cases such as for monoclonal antibodies (mAb), the comparative PK(/PD) trial is followed by a comparative efficacy study in a relevant patient population to confirm comparable safety and efficacy of the biosimilar candidate with the reference product (EMA 2014a; Wolff-Holz et al. 2019). The rigor of the demonstration of similar structure and function helps determine if a more elaborate or a selective, tailored program of clinical studies in patients can be scientifically justified. In particular in Europe, regulatory requirements for biosimilars are advancing in this regard. The UK Medicines and Healthcare products Regulatory Agency (MHRA) published in 2021 in their biosimilar guideline that, while biosimilar development needs to be evaluated on a case-by-case basis, it is considered that, in most cases, a comparative efficacy trial may not be necessary when supported by a sound scientific justification (MHRA 2021); the MHRA is the UK's standalone medicines and medical devices regulator. EU/EMA regulators acknowledge these advancements in analytical sciences and allow – justified – tailoring of comparative efficacy testing in patients. Posthoc analysis of all licensed biosimilars in Europe did show that not in a single case a comparative efficacy trial in patients played a decisive role in the approval process (Wolff-Holz et al. 2019). Consequently, many see the routine performance of such trials now as redundant and unethical, unless there are unresolved issues that can only be answered by such a trial (Webster et al. 2019; Wolff-Holz et al. 2019; Vulto 2019; Schiestl et al. 2020).

Specifics of the Biosimilar Regulatory Framework in Europe (EMA)

Biosimilar Legal/Regulatory Framework and Guidelines

The regulatory process for the approval of biopharmaceuticals and biosimilars in the EU follows a centralized (i.e., across European member states) route through the EMA. Mandated by the European Commission in 2004, the EMA issued the first guidance documents on the regulatory process for biosimilars in 2005 (EMA 2014a). The basis was laid by the overarching guideline for similar biological medicinal products (CHMP/437/04), first published in 2005 and revised in 2014. Furthermore, EMA published overarching guidelines on biosimilar nonclinical and clinical issues, and quality issues (EMA 2015 and EMA 2014b). From 2006 on, EMA has published a series of product-specific guidance documents on biosimilar medicinal products containing specific biotechnology-derived proteins as active substance, e.g., on recombinant erythropoietin, low-molecular weight heparins (not a protein), interferon beta, somatropin, follicle-stimulating hormone, granulocyte colony-stimulating factor (G-CSF), human insulin, and monoclonal antibodies. These guidelines define key concepts/principles of biotechnology-derived proteins and discuss quality issues, and nonclinical and clinical issues. These guidelines often received updates over the past years (EMA 2022).

EMA Procedures and Protocols

Developers are encouraged by EMA to seek early scientific advice from its experts to align with the expectations of EMA regarding the content of the candidate's dossier. When a sponsor (i.e., applicant) submits a file for assessment by the

EMA, this is evaluated by groups of expert assessors from the different European member states, led by a rapporteur and co-rapporteur. The rapporteurs report to the Committee of Human Medicinal Products (CHMP), which will make a list of questions that need to be addressed by the sponsor. All national regulatory agencies, including Norway and Iceland (non-EU members) represented in the CHMP, participate in this process. After assessments of answers to the raised questions, the CHMP comes to the conclusion whether or not the biosimilar candidate has demonstrated biosimilarity to the reference product and fulfills the criteria for approval. In the case of a positive assessment, a positive recommendation will be issued to the European Commission, which is the body responsible of granting a European-wide marketing authorization (valid in all EU Member States, Norway, Liechtenstein, and Iceland). This is normally done within 67 days of receipt of CHMP's recommendation. The active evaluation time of a biosimilar candidate is the same as for a new molecule, i.e., 210 active assessment days (i.e., exclud-

ing time for clock stops which is time for the sponsor to answer the questions raised by the assessors). The overall assessment process usually takes around 12 months to 15 months when considering the total time from EMA's acceptance of the marketing authorization application until the European Commission granting of the marketing authorization (EMA 2019).

Once a medicine has received an EU-wide marketing authorization, decisions about pricing and reimbursement take place at national and regional level of the individual member states. EMA has no role in these decisions.

Status of Biosimilars in the EU (2023)

In Europe, the first biosimilar was licensed in 2006 (Omnitrope, a biosimilar of somatropin). Since then, the European Commission has approved 84 biosimilars (status: September 2023; overview of EU-approved biosimilars in Table 11.5).

Table 11.5 Overview of centrally approved biosimilars in the EU (evaluation via EMA, and granting of Marketing Authorization by the EC), September 2023

INN	Biosimilar trade name	EC approval date
Adalimumab	Amgevita; Halimatoz; Hefiya; Hulio; Hyrimoz; Imraldi; Amsparity; Hukyndra; Idacio; Libmyris; Yuflyma	March 2017; July 2018; July 2018; September 2018; July 2018; August 2017; February 2020; November 2021; April 2019; November 2021; February 2021
Aflibercept	Yesafili	September 2023
Bevacizumab	Abevmy; Alymsys; Aybintio; Mvasi; Equidacent; Onbevzi; Vegzelma; Zirabev; Oyavas	April 2021; March 2021; August 2020; January 2018; September 2020; January 2021; August 2022; February 2019; March 2021
Eculizumab	Bekemv; Epysqli	April 2023; May 2023
Enoxaparin	Inhixa	September 2016
Epoetin alfa	Abseamed; Binocrit; Epoetin alfa Hexal	August 2007; August 2007; August 2007
Epoetin zeta	Retacrit; Silapo	December 2007; December 2007
Etanercept	Benepali; Erelzi; Nepexto	January 2016; June 2017; May 2020
Filgrastim	Accofil; Filgrastim Hexal; Grastofil; Nivestim; Ratiograstim; Tevagrastim; Zarzio	September 2014; February 2009; October 2013; June 2010; September 2008; September 2008; February 2009
Follitropin alfa	Bemfola; Ovaleap	March 2014; September 2013
Infliximab	Flixabi; Inflectra; Remsima; Zessly	May 2016; September 2013; September 2013; May 2018
Insulin aspart	Insulin aspart Sanofi; Kixelle; Truvelog mix 30	June 2020; February 2021; April 2022
Insulin glargine	Abasaglar; Semglee	September 2014; March 2018
Insulin human (rDNA)	Inpremzia	April 2022
Insulin Lispro	Insulin lispro Sanofi	July 2017
Natalizumab	Tyruko	September 2023
Pegfilgrastim	Cegfila; Fulphila; Grasustek; Nyvepria; Pelgraz; Udenyca; Ziextenzo; Pelmeg; Stimufend	December 2019; November 2018; April 2019; November 2020; September 2018; September 2018; November 2018; November 2018; March 2022
Ranibizumab	Byooviz; Ranivisio; Ximluci	August 2021; August 2022; November 2022
Rituximab	Blitzima; Ritemvia; Rixathon; Riximyo; Truxima; Ruxience	July 2017; July 2017; June 2017; June 2017; February 2017; April 2020
Somatropin	Omnitrope	April 2006
Teriparatide	Livogiva; Sondelbay; Movymia; Terrosa; Kauliv	August 2020; March 2022; January 2017; January 2017; January 2023
Tocilizumab	Tyenne	September 2023
Trastuzumab	Ogivri; Herzuma; Kanjinti; Ontruzant; Trazimera; Zercepac	December 2018; February 2018; May 2018; November 2017; July 2018; July 2020

EC European Commission, *EMA* European Medicines Agency, *INN* international nonproprietary name

Currently, Europe has biosimilars for 24 reference biological products, so often there are multiple biosimilars registered per originator biological and most of them are available on the market, albeit not in all countries. While regulatory evaluation takes place at a central (EU-overarching) level, market entry is organized at the level of the individual member states. Whether to launch a product in a given country is the decision of the marketing authorization holder. Granting of market access by determining price and reimbursement of the product is the decision of the payer in a given country or region. The market share of biosimilars varies strongly per country, per therapeutic area, and per biosimilar product (IQVIA 2021; see further section Societal Aspects).

The EMA is very active in disseminating information on biosimilars, promoting their use as an asset for the European healthcare system (EMA and European Commission 2017; European Commission 2020). Fig. 11.2 shows an overview of the guidelines and documents that are available through the EMA website (Barbier et al. 2022a). Among them is an extensive information guide for healthcare professionals on biosimilars (EMA and European Commission 2017), available in English and all languages of the EU, explaining in detail different aspects of the development and approval process. It also provides definitions on the terms interchangeability and substitution (see more below, in section Switching, Interchangeability, and Substitution in Europe

and the United States). Furthermore, EMA published a patient information Question and Answer guide, also available in English and all official languages of the EU (European Commission 2017) and developed an animated video for patients, explaining key facts on biosimilars and how EMA works to ensure that they are as safe and effective as their reference biological (EMA 2017). In September 2022, EMA published together with the Heads of Medicines Agencies a joint statement on the scientific rationale supporting interchangeability of biosimilar medicines in the EU (EMA and HMA 2022). The statement is accompanied with a Question & Answer document (see also later under section *Switching, Interchangeability, and Substitution in Europe and the United States*).

Naming

The EU law requires that every medicine in the EU has an invented name (trade name/brand name) together with the active substance name (the International Nonproprietary Name (INN)). For the purpose of identifying and tracing biological medicines in the EU, medicines have to be distinguished by their trade name and batch number. The recording of both trade name and batch number ensured that the right medicine can be identified in case any product (and possibly

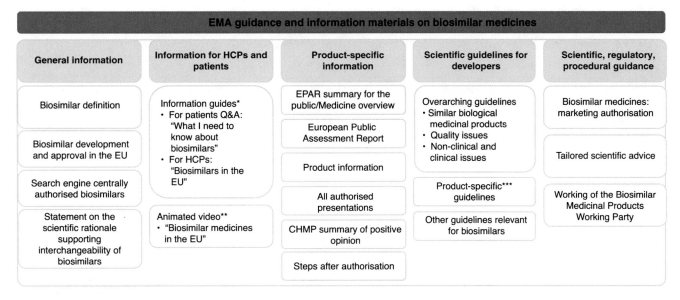

Fig. 11.2 Overview of EMA's information material and guidance documents on biosimilar medicines. Figure adapted from: Barbier et al. Regulatory information and guidance on Biosimilars and their use across Europe: a call for strengthened one voice messaging, 2022, Front Med 9:820755. *Available in 23 official EU languages. **Available in English and other EU languages (Dutch, English, French, German, Italian, Polish, Portuguese, Spanish). ***For recombinant granulocyte colony-stimulating factor, low-molecular-weight heparins, recombinant human insulin and insulin analogs, interferon beta, monoclonal antibodies, recombinant erythropoietins, recombinant follicle-stimulating hormone, somatropin

batch)-specific safety or immunogenicity concern would arise (EMA and European Commission 2017). In summary, reference product and biosimilar and biosimilars of the same reference product are identified by trade name and batch number. This is different as compared to the FDA approach in the United States, where products are distinguished by adding a suffix to the INN (FDA 2017b).

Specifics of the Biosimilar Regulatory Framework in the United States (FDA)

Biosimilar Legal/Regulatory Framework and Guidelines

The Biologics Price Competition and Innovation Act of 2009 (BPCI Act) was enacted as part of the Affordable Care Act on March 23, 2010. The BPCI Act creates an abbreviated licensure pathway for biological products shown to be biosimilar to, or interchangeable with, an FDA-licensed biological referenced product. The objectives of the BPCI Act are conceptually similar to those of the Drug Price Competition and Patent Term Restoration Act of 1984, commonly referred to as the Hatch-Waxman Act that established abbreviated pathways for the approval of drug products under the Federal Food, Drug, and Cosmetic (FDC) Act. Section 351(k) of the Public Health Service (PHS) Act, added by the BPCI Act, sets forth the requirements for an application for a proposed biosimilar

product and an application or a supplement for a proposed interchangeable product. A 351(k) application must contain information demonstrating that the biological product is biosimilar to a reference product based upon the data derived from analytical studies, animal studies, and clinical studies.

The regulatory guidelines for biosimilars are, as they are in Europe, dynamic. In April 2015, the FDA released two final guidance documents on biosimilar product development: (1) Scientific considerations in demonstrating biosimilarity to a reference product and (2) Quality considerations in demonstrating biosimilarity of a therapeutic protein product to a reference product (FDA, 2015). In addition, FDA has published the guidance on clinical pharmacology data to support a demonstration of biosimilarity to a reference product (FDA 2016a), draft guidance on considerations in demonstrating interchangeability with a reference product (FDA 2017a), draft guidance on labeling for biosimilar products (FDA 2016b), and guidance on nonproprietary naming of biological products (FDA 2017b).

Status of Biosimilars in the United States (2023)

In the United States, the first biosimilar was approved in 2015, approximately 10 years after the first biosimilar approval in Europe. As of September 2023, 39 biosimilars have been approved for 11 reference products (overview of US-approved biosimilars in Table 11.6). In contrast to

Table 11.6 Overview of FDA-approved biosimilars in the United States, September 2023

INN	Biosimilar trade name and INN-suffix	FDA approval date
Adalimumab	Amjevita (adalimumab-atto); Cyltezo (adalimumab-adbm)[a]; Hyrimoz (adalimumab-adaz); Hadlima (adalimumab-bwwd); Abrilada (adalimumab-afzb); Hulio (adalimumab-fkjp); Yusimry (adalimumab-aqvh); Idacio (adalimumab-aacf); Yuflyma (adalimumab-aaty)	September 2016; August 2017; October 2018; July 2019; November 2019; July 2020; December 2021; December 2022; May 2023
Bevacizumab	Mvasi (bevacizumab-awwb); Zirabev (bevacizumab-bvzr); Alymsys (bevacizumab-maly); Vegzelma (bevacizumab-adcd)	September 2017; June 2019; April 2022; September 2022
Epoetin-alfa	Retacrit (epoetin alfa-epbx)	May 2018
Etanercept	Erelzi (Etanercept-szzs); Eticovo (etanercept-ykro)	August 2016; April 2019
Filgrastim	Zarxio (Filgrastim-sndz); Nivestym (filgrastim-aafi); Releuko (filgrastim-ayow)	March 2015; July 2018; February 2022
Infliximab	Inflectra (infliximab-dyyb); Renflexis (infliximab-abda); Ixifi (infliximab-qbtx); Avsola (infliximab-axxq)	April 2016; May 2017; December 2017; December 2019
Insulin glargine	Semglee (insulin glargine-yfgn),[a] Rezvoglar (insulin glargine-aglr)[a]	July 2021; December 2021
Natalizumab	Tyruko (natalizumab-sztn)	August 2023
Pegfilgrastim	Fulphila (pegfilgrastim-jmdb); Udenyca (pegfilgrastim-cbqv); Ziextenzo (pegfilgrastim-bmez); Nyvepria (pegfilgrastim-apgf); Fylnetra (pegfilgrastim-pbbk); Stimufend (pegfilgrastim-fpgk)	June 2018; November 2018; November; June 2020; May 2022; September 2022
Ranibizumab	Byooviz (ranibizumab-nuna); Cimerli (ranibizumab-eqrn)[a]	September 2021; August 2022
Rituximab	Truxima (rituximab-abbs); Ruxience (rituximab-pvvr); Riabni (rituximab-arrx)	November 2018; July 2019; December 2020
Tocilizumab	Tofidence (tocilizumab-bavi)	September 2023
Trastuzumab	Ogivri (trastuzumab-dkst); Herzuma (trastuzumab-pkrb); Ontruzant (trastuzumab-dttb); Trazimera (trastuzumab-qyyp); Kanjinti (trastuzumab-anns)	December 2017; December 2018; January 2019; March 2019; June 2019

FDA US Food and Drug Administration, *INN*: international nonproprietary name
[a] Received interchangeability designation by the FDA

Europe, many of these licensed biosimilars are not yet on the market due to a plethora of legal hurdles (see also later in the section Societal Aspects).

As of September 2023, the following biosimilars have received interchangeability designation: adalimumab-adbm, insulin glargine-aglr, insulin glargine-yfgn, and ranibizumab-eqrn (FDA 2023) (see also later in the section *Switching, Interchangeability, and Substitution in Europe and the United States*).

Naming

In the United States, the name for an approved biological product consists of the INN and an FDA-designated suffix. This distinguishing suffix should be devoid of meaning and composed of four lower case letters; it is to be attached with a hyphen to the INN of each originator biological product, related biological product, or biosimilar product, for example: replicamab-cznm and replicamab-jhxf. FDA argued that this naming convention would facilitate pharmacovigilance for biological and biosimilar drug products. Later FDA decided that for the existing innovator products the INN was not required to have a suffix. Some are critical of the effectiveness this naming convention, and arguments have been raised that it may discourage biosimilar prescribing (Kolbe et al. 2021).

Switching, Interchangeability, and Substitution in Europe and the United States

In general, interchangeability refers to the possibility of exchanging one medicine for another with the same clinical effect. In the context of biosimilars, this implies replacing the reference biological product with its biosimilar (or vice versa) or exchanging biosimilars of the same reference product with each other. Such a replacement can occur at two different levels: by the prescribing physician (also termed switching or transitioning) or by the pharmacist (also termed as (automatic) substitution) (EMA and European Commission 2017; Barbier et al. 2021; Rathore et al. 2022).

In Europe, the EMA Heads of Medicines Agencies (HMA) have published in September 2022 a joint position paper on biosimilar interchangeability, confirming that biosimilar medicines approved in the EU are interchangeable with their reference product or with an equivalent biosimilar (EMA and HMA 2022). While interchangeable use of biosimilars (mostly via switching) is already practiced in many EU Member States, this joint position harmonizes the EU regulatory position and provides much needed clarity and guidance on the topic for healthcare professionals, patients, and other stakeholders involved (Barbier et al. 2022a; EMA and HMA 2022).

The policies with regard to switching and substitution are decided at the individual member state level, being not a centralized but national/local responsibility. As such, this falls within the remit of the individual EU member states. The factual decision to prescribe a particular product is with an individual healthcare professional, although this can also be implemented within hospital formularies. While most European countries endorse exchanging between reference biological and biosimilar by the physician (i.e., switching), substitution is currently not allowed or practiced for biological products in most European countries (EMA and European Commission 2017; Barbier and Vulto 2021; Barbier et al. 2021, 2022a; Kurki et al. 2021).

In the United States, a different approach is taken. Here, interchangeability is legally defined in the Biologics Price Competition and Innovation Act of 2009 (BPCIA) and is a matter of regulatory assessment and decision by the FDA. In case a biosimilar receives interchangeability status, it means that it may be substituted at the pharmacy level for the reference biological product without the intervention of the prescribing physician, if also permitted by state laws.

The proposed interchangeable product "can be expected to produce the same clinical result as the reference product in a given patient." To support a demonstration of interchangeability, the sponsor must first show that the proposed product is biosimilar to the reference product. As outlined in FDA guidance to sponsors, the interchangeability designation might require additional clinical data involving two or more alternate switches between a biosimilar and its reference product (FDA 2017a, b). However, the FDA has the statutory right to deviate from this rule.

However, it must be made clear that a designation of interchangeability is unrelated to the biosimilar product's quality, safety, and efficacy, which are established during biosimilar evaluation. It is rather an additional assessment that would enable substitution by the pharmacist at the state level. An interchangeability designation is in essence only relevant for products dispensed in the retail or specialty pharmacy setting in the United States, where automatic substitution could be applicable (Barbier et al. 2021; Park et al. 2022). Regardless of having an interchangeability designation, the prescribing physician can decide to exchange a reference product with a biosimilar or vice versa.

Up until September 2023, four biosimilar products have been designated as interchangeable by the FDA, of which three did not submit any alternating studies/no new clinical studies were required by the FDA to support approval of interchangeability status (FDA 2023).

Societal Aspects of Biosimilars

Biosimilars are in general lower priced versions of at times high-cost originator biological products. Their introduction on the market induces competition, resulting in lowered prices. Beyond price decreases of the reference product, biosimilar competition has also shown to lead to lowering of prices in the broader therapeutic segment (IMS Institute for Healthcare Informatics 2016).

For Europe, the marketing consulting company IQVIA estimates in their 2021 biosimilar report that, based on list prices, biosimilars have reduced the total cost of drug spending by at least 5%. When taking into account (confidential) contract prices, this percentage is estimated to increase to around 10%. It is estimated that over the past 15 years biosimilar competition has resulted in 18 billion euros cumulative savings to healthcare systems. As such, biosimilar market introduction contributes heavily to the sustainability of European healthcare systems. In some countries, biosimilars discounts have been reported to be 50–80% as compared to the original price of the originator product (IQVIA 2018). However, this pattern is not uniform and competition challenges have been reported (IQVIA 2021; Autoriteit Consument & Markt (ACM) 2019; Barbier et al. 2021; Vandenplas et al. 2021).

In the United States, biosimilar market entry and adoption have so far been a moderate success (Cardinal Health 2022a). There are several explanatory factors, including the large number of patents enforced by originator manufacturers, complex litigation processes, and the influence of innovator industry on law-making. In addition, due to the profit-driven nature and complexity of the healthcare system (involving Pharmacy Benefit Managers), patients insufficiently benefit from lower cost biosimilars (Vulto and Barbier 2022; IQVIA 2020a; Cardinal Health 2022b; Van de Wiele et al. 2021).

The advent of biosimilars to the market led to questions on what grounds to choose, either for the originator's product or for the biosimilar product and which one, if multiple biosimilars are available per reference product. The European Journal of Hospital Pharmacists published a document: "Points to consider in the evaluation of biopharmaceuticals," followed by "How to select a biosimilar" in 2013 (Kraemer et al. 2008; Boone et al. 2013). Since then, the biosimilar landscape and the experience with their use have substantially grown, leading to the need for a revision of this work (Barbier et al. 2022b). Besides the competitiveness of the product's price, other criteria should be considered when selecting a best-value biological. Healthcare providers and procurers can use the developed model to make a transparent and documented choice.

The benefits brought by biosimilar introduction are manyfold; (i) biosimilars induce competition, lowering prices of reference biologicals and innovative products; (ii) off-patent biologic and biosimilar competition improves cost-effectiveness of biological therapy which can impact the reimbursement status of the biological or combination therapy, including a biological; (iii) more patients may gain access to biological therapy with the same budget; (iv) patients may access biological therapy earlier and experience ensuing health gains; (v) biosimilars can have differentiating value-added practical characteristics such as an improved injection device compared to the originator product (vice versa is also possible) and improve the quality of care processes; and (vi) off-patent biological and biosimilar competition creates budgetary room for new, innovative medicines and may stimulate developers to innovate (MABEL 2022; Barcina Lacosta et al. 2022; Barbier et al. 2022b).

In concrete terms, biosimilar introduction may not only generate cost savings but also deliver patient benefit. This, for example, by making initially expensive biological treatment cost-effective in an earlier line of treatment. As illustrated in Sweden, the launch of biosimilar filgrastim (a granulocyte colony-stimulating factor) led to reassessment of guidelines on granulocyte colony-stimulating factors, advancing filgrastim to first-line supportive care in cancer, widening, and giving access to patients in an earlier line (IMS Institute for Healthcare Informatics 2016).

The Challenge and the Future

Science-based regulatory policies have been developed and are revised on a continuous basis through advancing analytical techniques and accumulated experience with biosimilar evaluation and use. Biosimilar regulatory guidelines as such have evolved over the years and will continue to do so based on the development of superior analytical techniques to characterize the products, introducing improved manufacturing practices and controls, and on growing clinical and regulatory experience. Rigorous standards of ensuring product safety and efficacy must be maintained and, at the same time, unnecessary and/or unethical duplication trials must be avoided (Wolff-Holz et al. 2019; Kurki et al. 2021; Barbier and Vulto 2021). The development pathway and regulatory requirements of a biosimilar product should depend on the complexity of the molecule. A gradation scheme should be applied in regulatory evaluation rather than using a "one size fits all" model. The advancing knowledge and experience which enables tailoring of data packages, e.g., in terms of comparative efficacy testing in patients, should be adequately reflected through revision of regulatory guidelines. Furthermore, to support biosimilar development efficiency and avoid unnecessary costs, continued harmonization in regulatory requirements between jurisdictions is essential over the upcoming years (Kurki et al. 2022; Niazi 2022b).

Over the next five to ten years, approximately 100 originator biological medicines are expected to lose exclusivity

(Niazi 2022a; Cardinal Health 2022a), representing a significant opportunity for biosimilar development. This wave of products that are losing exclusivity not only includes several multi-billion dollar biologicals, such as pertuzumab and nivolumab, but also products with lower revenue for which development interest may be lower. Also, for orphans biosimilars, specific development challenges such as limited patient numbers are present. This may explain the low number of "orphan biosimilars" currently in clinical development and in regulatory evaluation (IQVIA 2020b). In a recent presentation, IQVIA signaled for a "biosimilar void"; meaning that many originator biologicals (especially those with smaller commercial value) which will be losing exclusivity over the upcoming date are without the development of a corresponding biosimilar to compete with them (GaBI 2023). To stimulate biosimilar development across the board for different types of biologicals (also for smaller patient populations and nonmulti-billion US dollar annual sales), science-based advances and convergence in regulatory requirements are imperative (Niazi 2022a). In a more distant future, biosimilar products for advanced therapy medicinal products (ATMPs) will present a novel frontier.

The regulatory processes for biosimilars are continuously evolving. We learn from new information coming in every day; we evaluate the data, adjust the requirements, and develop new guidelines to make sure that the patient keeps on receiving high-quality, timely, safe, and (cost-)effective biopharmaceuticals.

Self-Assessment Questions

Question 1: Human growth hormone has a molecular weight of around 22 kDa (see Chap. 20) and erythropoietin of 34 kDa (see Chap. 17). Why does the EMA request different clinical protocols for approval of a biosimilar product for these protein drugs?

Question 2: Can a US-approved biosimilar product automatically be interchanged with the brand-name product?

Answer 1: Human growth hormone is a nonglycosylated protein with a well-established primary sequence; erythropoietin is heavily glycosylated with a number of isoforms with more analytical challenges. The EMA biosimilar guidance documents giving more details can be found on the EMA website https://www.ema.europa.eu/en/human-regulatory/research-development/scientific-guidelines/multidisciplinary/multidisciplinary-biosimilar .

Answer 2: Interchanging a biosimilar with its reference product can be either done by the prescriber or the pharmacist. In the first case, when a transition is made between biosimilar and the reference product by the prescriber (this is also termed switching), no FDA interchangeability designation is needed. Switching is a clinical decision that can be made by the prescriber for every approved biosimilar. In the second case, to exchange between a reference product and a biosimilar by the pharmacist (also termed automatic substitution), in the United States, a biosimilar should have interchangeability designation from the FDA. In addition, interchangeability is subject to state legislation.

References

Autoriteit Consument & Markt (ACM) (2019) Sectoronderzoek TNF-alfaremmers. https://www.acm.nl/sites/default/files/documents/2019-09/sectoronderzoek-tnf-alfaremmers.pdf

Barbier L, Simoens S, Soontjens C, Claus B, Vulto AG, Huys I (2021) Off-Patent Biologicals and Biosimilars Tendering in Europe—A Proposal towards More Sustainable Practices. Pharmaceuticals 14: 499

Barbier L, Vulto AG (2021) Interchangeability of biosimilars: overcoming the final hurdles. Drugs 81(16):1897–1903. https://doi.org/10.1007/s40265-021-01629-4

Barbier L, Declerck P, Simoens S, Neven P, Vulto AG, Huys I (2019) The arrival of biosimilar monoclonal antibodies in oncology: clinical studies for trastuzumab biosimilars. Br J Cancer 121(3):199–210. https://doi.org/10.1038/s41416-019-0480-z

Barbier L, Vandenplas Y, Lacosta TB, Vulto AG (2021) Biosimilar interchangeability: regulatory & practical considerations. Biosimilar development. https://www.biosimilardevelopment.com/doc/biosimilar-interchangeability-regulatory-practical-considerations-0001

Barbier L, Mbuaki A, Simoens S, Declerck P, Vulto AG, Huys I (2022a) Regulatory information and guidance on Biosimilars and their use across Europe: a call for strengthened one voice messaging. Front Med 9:820755. https://doi.org/10.3389/fmed.2022.820755

Barbier L, Vandenplas Y, Boone N, Janknegt R, Vulto AG (2022b) How to select a best-value biological medicine? A practical model to support hospital pharmacists. Am J Health Syst Pharm 79:2001. https://doi.org/10.1093/ajhp/zxac235

Barcina Lacosta TB, Vulto AG, Turcu-Stiolica A, Huys I, Simoens S (2022) Qualitative analysis of the design and implementation of benefit-sharing programs for biologics across Europe. BioDrugs 36(2):217–229. https://doi.org/10.1007/s40259-022-00523-z

Boone HN, van der Kuy H, Scott M, Mairs J, Krämer I, Vulto A, Janknegt R (2013) How to select a biosimilar. Eur J Hosp Pharm 20:275–228. https://doi.org/10.1136/ejhpharm-2013-000370

Cardinal Health (2022a) New and upcoming biosimilar launches. https://www.cardinalhealth.com/content/dam/corp/web/documents/Report/cardinal-health-biosimilar-launches.pdf

Cardinal Health (2022b) Biosimilars report: the US journey and path ahead. https://www.cardinalhealth.com/content/dam/corp/web/documents/Report/cardinal-health-2022-biosimilars-report.pdf

Crommelin DJA, Storm G, Verrijk R, de Leede L, Jiskoot W, Hennink WE (2003) Shifting paradigms: biopharmaceuticals versus low molecular weight drugs. Int J Pharm 266:3–16. https://doi.org/10.1016/S0378-5173(03)00376-4

Declerck P, Rezk M (2017) The road from development to approval: evaluating the body of evidence to confirm biosimilarity. Rheumatology 56(suppl_4):iv4–iv13. https://doi.org/10.1093/rheumatology/kex279

Declerck P, Danesi R, Petersel D, Jacobs I (2017) The language of biosimilars: clarification, definitions and regulatory aspects. Drugs 77(6):671–677. https://doi.org/10.1007/s40265-017-0717-1

EMA (2014a) Guideline on similar biological medicinal products (CHMP/437/04 Rev 1). https://www.ema.europa.eu/en/similar-biological-medicinal-products

EMA (2014b) Guideline on similar biological medicinal products containing biotechnology-derived proteins as active substance: quality issues (revision 1)

EMA (2015) Guideline on similar biological medicinal products containing biotechnology-derived proteins as active substance: non-clinical and clinical issues. (EMEA/CHMP/BMWP/42832/2005 Rev1). https://www.ema.europa.eu/en/similar-biological-medicinal-products-containing-biotechnology-derived-proteins-active-substance-non

EMA (2017) Animated video for patients on biosimilars https://www.youtube.com/watch?v=zAt7vd3eiT8. Accessed Aug 2022

EMA (2019) From laboratory to patient: the journey of a medicine assesses by EMA. https://www.ema.europa.eu/en/documents/other/laboratory-patient-journey-centrally-authorised-medicine_en.pdf

EMA (2022) Multidisciplinary: biosimilar http://www.ema.europa.eu/ema/index.jsp?curl=pages/regulation/general/general_content_000408.jsp&mid=WC0b01ac058002958c. Accessed Mar 2022

EMA and European Commission (2017). Biosimilars in the EU–information guide for healthcare professionals. https://www.ema.europa.eu/en/documents/leaflet/biosimilars-eu-information-guide-healthcare-professionals_en.pdf

EMA and HMA (2022) Statement on the scientific rationale supporting interchangeability of biosimilar medicines in the EU https://www.ema.europa.eu/en/documents/public-statement/statement-scientific-rationale-supporting-interchangeability-biosimilar-medicines-eu_en.pdf. Accessed Sept 2022

European Commission (2017) Questions and answers for patients–biosimilar medicines explained. https://ec.europa.eu/docsroom/documents/26643

European Commission (2020) Pharmaceutical strategy for Europe. https://eur-lex.europa.eu/legal-content/EN/TXT/HTML/?uri=CELEX:52020DC0761&from=NL

FDA (1996) Guidance document. Demonstration of comparability of human biological products, including therapeutic biotechnology-derived products. https://www.fda.gov/regulatory-information/search-fda-guidance-documents/demonstration-comparability-human-biological-products-including-therapeutic-biotechnology-derived. Accessed Aug 2022

FDA (2016a) Clinical pharmacology data to support a demonstration of biosimilarity to a reference product. https://www.fda.gov/downloads/Drugs/GuidanceComplianceRegulatoryInformation/Guidances/UCM397017.pdf. Accessed 18 Jan 2018

FDA (2016b) Labeling for biosimilar products guidance for industry. https://www.fda.gov/downloads/Drugs/GuidanceComplianceRegulatoryInformation/Guidances/UCM493439.pdf. Accessed 18 Jan 2018

FDA (2017a) Considerations in demonstrating interchangeability with a reference product guidance for industry. https://www.fda.gov/downloads/Drugs/GuidanceComplianceRegulatoryInformation/Guidances/UCM537135.pdf

FDA (2017b) Nonproprietary naming of biological products: guidance for industry. https://www.fda.gov/downloads/drugs/guidances/ucm459987.pdf. Accessed 18 Jan 2018

FDA (2021) Questions and answers on biosimilar development and the BPCI act guidance for industry. September 2021. https://www.fda.gov/media/119258/download. Accessed Nov 2022

FDA (2022) Biosimilar product information. https://www.fda.gov/drugs/biosimilars/biosimilar-product-information. Accessed 23 Sept 2022

FDA (2023) Biosimilar product information. https://www.fda.gov/drugs/biosimilars/biosimilar-product-information. Accessed 23 Sept 2023

GaBI (2023) Biosimilar development targets limited range of biologicals https://gabionline.net/reports/biosimilar-development-targets-limited-range-of-biologicals

IMS Institute for Healthcare Informatics (2016) The impact of biosimilar competition. https://ec.europa.eu/docsroom/documents/17325/attachments/1/translations/en/renditions/native

IQVIA (2018) The impact of biosimilar competition in Europe https://ec.europa.eu/docsroom/documents/31642/attachments/1/translations/en/renditions/native

IQVIA (2020a) Biosimilars in the United States 2020–2024. https://www.iqvia.com/insights/the-iqvia-institute/reports/biosimilars-in-the-united-states-2020-2024

IQVIA (2020b) The prospects for biosimilars of orphan drugs in Europe. https://www.iqvia.com/insights/the-iqvia-institute/reports/the-prospects-for-biosimilars-of-orphan-drugs-in-europe

IQVIA (2021) The impact of biosimilar competition in Europe. https://www.iqvia.com/library/white-papers/the-impact-of-biosimilar-competition-in-europe-2021

Kenett RS, Kenett DA (2008) Quality by design applications in biosimilar pharmaceutical products. Accred Qual Assur 13(12):681–690. https://doi.org/10.1007/s00769-008-0459-6

Kim S, Song J, Park S, Ham S, Paek K, Kang M, Chae Y, Seo H, Kim H, Flores M (2017) Drifts in ADCC-related quality attributes of Herceptin®: impact on development of a trastuzumab biosimilar. MAbs 9(4):704–714. https://doi.org/10.1080/19420862.2017.1305530

Kolbe AR, Kearsley A, Merchant L, Temkin E, Patel A, Xu J, Jessup A (2021) Physician understanding and willingness to prescribe biosimilars: findings from a US National Survey. BioDrugs 35:363–372

Kraemer I, Tredree R, Vulto A (2008) Points to consider in the evaluation of biopharmaceuticals. EJHP Pract 14:73–76

Kurki P, Barry S, Bourges I, Tsantili P, Wolff-Holz E (2021) Safety, immunogenicity and interchangeability of biosimilar monoclonal antibodies and fusion proteins: a regulatory perspective. Drugs 81(16):1881–1896. https://doi.org/10.1007/s40265-021-01601-2

Kurki P, Kang H, Ekman N, Knezevic I, Weise M, Wolff-Holz E (2022) Regulatory evaluation of biosimilars: refinement of principles based on the scientific evidence and clinical experience. BioDrugs 36(3):359–371. https://doi.org/10.1007/s40259-022-00533-x

MABEL (2022) The MABEL FUND—market analysis of biologics and biosimilars following loss of exclusivity. https://gbiomed.kuleuven.be/english/research/50000715/52577001/mabel/Keyinsights. Accessed 23 Sept 2022

MHRA (2021) Guidance on the licensing of biosimilar products. https://www.gov.uk/government/publications/guidance-on-the-licensing-of-biosimilar-products/guidance-on-the-licensing-of-biosimilar-products. Accessed Aug 2022

Niazi SK (2022a) The coming of age of biosimilars: a personal perspective. Biologics 2:107–127. https://doi.org/10.3390/biologics2020009

Niazi SK (2022b) Biosimilars: harmonizing the approval guidelines. Biologics 2(3):171–195. https://doi.org/10.3390/biologics2030014

Park JP, Jung B, Park HK, Shin D, Jung JA, Ghil J, Han J, Kim KA, Woollett GR (2022) Interchangeability for biologics is a legal distinction in the USA, not a clinical one. BioDrugs 36(4):431–443

Rathore AS, Stevenson JG, Chhabra H, Maharana C (2022) The global landscape on interchangeability of Biosimilars. Expert Opin Biol Ther 22(2):133–148. https://doi.org/10.1080/14712598.2021.1889511

Rojas-Chavarro L, de Mora F (2021) Extrapolation: experience gained from original biologics. Drug Discov Today 26(8):2003–2013. https://doi.org/10.1016/j.drudis.2021.05.006

Schiestl M, Stangler T, Torella C, Cepeljnik T, Toll H, Grau R (2011) Acceptable changes in quality attributes of glycosylated biopharmaceuticals. Nat Biotechnol 29:310–312. https://doi.org/10.1038/nbt.1839

Schiestl M, Ranganna G, Watson K, Jung B, Roth K, Capsius B, Trieb M, Bias P, Maréchal-Jamil J (2020) The path towards a tailored clinical biosimilar development. BioDrugs 34(3):297–306. https://doi.org/10.1007/s40259-020-00422-1

Schneider CK (2013) Biosimilars in rheumatology: the wind of change. Ann Rheum Dis 72(3):315–318. https://doi.org/10.1136/annrheumdis-2012-202941

Van de Wiele VL, Kesselheim AS, Sarpatwari A (2021) Barriers to US biosimilar market growth: lessons from biosimilar patent litigation. Health Aff (Millwood) 40:1198–1205. https://doi.org/10.1377/hlthaff.2020.02484

Vandenplas Y, Simoens S, Van Wilder P, Vulto AG, Huys I (2021) Off-patent biological and biosimilar medicines in Belgium: a market landscape analysis. Front Pharmacol 12:644187. https://doi.org/10.3389/fphar.2021.644187

Vezer B, Buzás Z, Sebeszta M, Zrubka Z (2016) Authorized manufacturing changes for therapeutic monoclonal antibodies (mAbs) in European public Assessment report (EPAR) documents. Curr Med Res Opin 32(5):829–834. https://doi.org/10.1185/03007995.2016.1145579

Vulto AG (2019) Delivering on the promise of biosimilars. BioDrugs 33(6):599–602. https://doi.org/10.1007/s40259-019-00388-9

Vulto AG, Barbier L (2022) When will American patients start benefitting from biosimilars? Mayo Clin Proc 97(6):1044–1047. https://doi.org/10.1016/j.mayocp.2022.04.013

Vulto AG, Jaquez O (2016) The process defines the product: what really matters in biosimilar design and production? Br J Rheumatol 56(4):iv14–iv29. https://doi.org/10.1093/rheumatology/kex278

Webster WJ, Wong WA, Woollet GR (2019) An efficient development paradigm for biosimilars. BioDrugs 33(6):603–611. https://doi.org/10.1007/s40259-019-00371-4

Weise M, Bielsky M-C, De Smet K, Ehmann F, Ekman N, Narayanan G, Heim H-K, Heinonen E, Ho K, Thorpe R, Vleminckx C, Wadhwa M, Schneider CK (2011) Biosimilars-why terminology matters. Nat Biotechnol 29(8):690–693. https://doi.org/10.1038/nbt.1936

Weise M, Kurki P, Wolff-Holz E, Bielsky MC, Schneider CK (2014) Biosimilars: the science of extrapolation. Blood 124(22):3191–3196. https://doi.org/10.1182/blood-2014-06-583617

WHO (2022) Guidelines on evaluation of biosimilars replacement of annex 2 of WHO technical report series, no. 977. https://cdn.who.int/media/docs/default-source/biologicals/bs-documents-(ecbs)/annex-3%2D%2D-who-guidelines-on-evaluation-of-biosimilars_22-apr-2022.pdf

Wolff-Holz E, Tiitso K, Vleminckx C, Weise M (2019) Evolution of the EU biosimilar framework: past and future. BioDrugs 33(6):621–634. https://doi.org/10.1007/s40259-019-00377-y

An Evidence-Based Practice Approach to Evaluating Biotechnologically Derived Medications

12

James P. McCormack

Introduction

Recent advances in pharmaceutical biotechnology have led to the development of many different therapeutic proteins. These technologies have given credence to the legitimate promise that we, at some point, may be able to more closely match some specific patients with the most effective and safe drugs at an individualized dose—personalized/precision medicine.

Background

Despite the clear potential for these technologies, when it comes to treating patients within an evidence-based practice framework, a number of requirements come into the clinical decision-making process with the use of these agents. Clinicians will need to consider all these requirements on a patient-by-patient basis if we are to fully realize the clinical potential of these therapeutic proteins.

These requirements can be separated into evidence and individual treatment decision issues. These are outlined in Table 12.1.

At any clinical encounter where a treatment is to be recommended, the obvious goal would be to "match individual patients with the most effective, and safest drugs and doses" (Sindelar 2013). Tailoring the correct medication and dose to an individual patient has been the goal and the approach of health care providers for millennia.

Over the last decade or so, much of this tailoring concept has been termed personalized medicine or precision medicine. Interestingly, these two terms have recently been co-opted by people interested in the fascinating and potentially very useful area of genomics. Despite this, it is important that clinicians remember that pharmacogenomics is just one of a number of tools or approaches that can be useful as we appropriately attempt to tailor the use of medications in a more personalized way. Obviously, behavioral and environmental factors play an important role in the clinical outcomes associated with treatments and it is of value to remember that

Table 12.1 Requirements for clinical decision-making process

(A) Evidence requirements
Placebo-controlled RCTs of these agents
RCTs comparing these agents directly to presently established therapy
RCTs of direct head-to-head comparisons of monoclonal antibodies that are in the same class/used for the same indication
Cohort data with long-term follow-up to estimate any on-going long-term safety issues
(B) Patient-centered requirements
The promotion of the concept of shared-decision-making
Defining clear, specific, and measurable individual patient outcomes
When possible starting with very low doses and/or continually adjusting doses based on individual response
A discussion of how to address the potential cost issues typically seen with these agents

J. P. McCormack (✉)
Faculty of Pharmaceutical Sciences, The University of British
Columbia (UBC), Vancouver, BC, Canada
e-mail: james.mccormack@ubc.ca

D. J. A. Crommelin et al. (eds.), *Pharmaceutical Biotechnology*, https://doi.org/10.1007/978-3-031-30023-3_12

these and other similar factors will likely never be importantly influenced by genomically targeted medications.

Finally, the present-day personalized medicine discussion typically omits two of the most important aspects of true personalized medicine—first, patient's individual values and preferences, and second, doses of medications need to be individualized based on a clear and objective review of an individual patient's response.

When it comes to these new "omic" technologies, it is helpful to think through the issues outlined in Table 12.1 when trying to figure out where therapeutic proteins fit into the concept of evidence-based practice and true personalized medicine. Using six different examples, many of these issues will be examined. It is useful to breakdown these examples into two specific therapeutic scenarios based on the overall treatment goals of the medication as this impacts the specific issues that require focus. The two therapeutic scenarios are prevention/risk reduction and symptom reduction.

Treatments that Reduce Risk

In prevention, one is taking a treatment to try to reduce the chance in the future of developing symptoms or of having an event such as a heart attack, stroke, hospitalization, or disease exacerbation.

Example 1: Palivizumab for Reducing the Risk of Severe RSV Infection in Children

Bronchiolitis and pneumonia in children are most commonly caused by the Respiratory Syncytial Virus (RSV). Most infants recover from this virus but serious complications can occur, especially in those with underlying medical conditions such as congenital heart disease. For those infants at higher risk of complications, palivizumab (Synagis®), a monoclonal antibody produced by recombinant DNA technology, is used to reduce the risk of developing serious complications. The authors of a 2021 Cochrane review (Garegnani et al. 2021) state "The available evidence suggests that prophylaxis with palivizumab reduces hospitalization due to RSV infection and results in little to no difference in mortality or adverse events" and, to that end, palivizumab prophylaxis was associated with a statistically significant reduction in RSV hospitalizations (Risk Ratio, RR 0.44, 95%; confidence interval, CI 0.30–0.64) and a statistically nonsignificant reduction in all-cause mortality (RR 0.69, 95% CI 0.42–1.15) when compared to placebo. In other words, palivizumab reduced the risk of RSV hospitalizations by just over 50%.

So, to properly use these numbers using an evidence-based practice framework, we need to know the baseline risk of RSV hospitalizations in these infants.

The patients studied were infants with serious medical conditions such as chronic lung disease, congenital heart disease, or those born preterm. In the placebo group, roughly 10% of the infants ended up being hospitalized for RSV over a 5-month period. Giving palivizumab reduces the risk by roughly 50%. Putting this into absolute numbers, the risk went from roughly 10% (in the placebo group) down to 5% (in the palivizumab group). This absolute 5% benefit means for every 20 children who got this agent (typically for 5 months) one benefited, or in other words 95% got no benefit from this treatment. To be fair, even if this treatment prevented all RSV hospitalizations, reducing a 10% risk down to 0% would still only mean one in ten would benefit from treatment. In this example, unfortunately there is no way to predict which of these higher risk infants will benefit and there is also no way to titrate the dose to an effect. So, when it comes to evidence-based practice, we have evidence that this new medication has an effect in the population studied but this benefit needs to be balanced with the fact that 5-monthly injections will cost a total of roughly $10,000 US for a one in 20 chance of benefiting. In addition, cohort studies would be required to determine the long-term effects of immunoprophylaxis on asthma, mortality, and other important clinical outcomes. And finally, at present there are no studies comparing this agent to other potential treatments.

Example 2: Evolucumab for Hyperlipidemia and Reducing the Risk of Heart Disease

Another example of a risk reduction treatment is evolocumab (Repatha®), a monoclonal antibody (proprotein convertase subtilisin/kexin type 9, PCSK9 inhibitor) designed for the treatment of hyperlipidemia and recently approved to reduce the risks of heart attacks, strokes, and coronary revascularization. A Cochrane review from 2020 evaluated three trials that looked at the benefit of evolocumab over placebo over a period of 6–36 months in people with a history of CVD or high LDL-C despite treatment (Schmidt et al. 2020). The main outcome was the risk of a combined CVD endpoint (cardiovascular death, myocardial infarction, stroke, hospitalization for unstable angina, or coronary revascularization). The results were a hazard ratio of 0.84 (0.78–0.91) or in other words a 16% relative benefit. These relative benefits (16%) are numerically less than that seen with statins and/or medications used for blood pressure (25–30%) but there are no head-to-head comparisons between these different treatments so comparisons are at best speculative.

Looking at the absolute numbers, 22.9% in the placebo group ended up with this CVD endpoint and this occurred in 21.3% of the group on evolocumab. This is a 1.6% absolute benefit or 63 people need to be treated to benefit one. As with palivizumab, there can be no dosage titration because only a single dose has been studied and there is no endpoint to which to titrate. The cost of this agent is roughly $6000 US a year for a one in 63 benefit and at present there is not a lot of information on the long-term benefits and harms of this agent. This is clearly an example where patient values and preferences will need to be taken into account via a shared decision-making process.

Example 3: Romosozumab or Alendronate for Fracture Prevention

Fractures increase morbidity and possibly mortality and there are some new biologic treatments that appear to reduce the risk of these fractures. An example of a very useful head-to-head trial in this area is the ARCH study (Saag et al. 2017) where roughly 4100 subjects were randomized to receive either romosozumab (Evenity®) or alendronate in a blinded fashion for 12 months followed by both groups receiving alendronate alone for 12 more months. Over the 24 months, clinical fractures occurred in 9.7% of the romosozumab and 13% in the alendronate group—RR 0.73 (0.61, 0.88). This 27% relative benefit or 3.3% absolute benefit means that 30 people would need to take romosozumab over alendronate for 1 year for one additional person to benefit. Unfortunately, in this trial there was an increase in cardiovascular events in the subjects receiving romosozumab. In the romosozumab group, 2.5% had a serious cardiovascular event compared to 1.9% in the alendronate group. So, for every 167 people taking romosozumab over alendronate, one would have a serious cardiovascular event. There was also a 1.8% absolute increase in injection site reactions in the romosozumab group. In 2022, a systematic review of romosozumab versus placebo reported that romosozumab did not increase the risk of overall serious adverse events but they don't specifically mention cardiovascular events as a safety endpoint (Singh et al. 2022).

So, in this scenario we have a new agent that is more effective at reducing fractures than the gold standard but it may also increase the risk of serious cardiovascular disease. This is where shared decision-making becomes an essential part of the discussion. Does a 3.3% absolute reduction in fractures justify a 0.6% increase in serious cardiovascular events? Only an informed patient can participate in that sort of decision. On top of the benefits and harms discussion, other considerations such as the requirement for injections, the lack of long-term follow-up data, and the cost which is roughly $2000-$2500 US a month, which is considerably more expensive than generic alendronate, need to be incorporated into the shared-decision.

In these three risk reduction examples, the only way clinicians and patients can make decisions about the use of these medications is to have a rough idea of the magnitude of the effect on clinically important endpoints and how these agents compare to either placebo or established therapies. This

information, mixed with the cost and the potential long-term benefits and harms, must be discussed with patients in the realm of a shared-decision.

Treatments that Reduce Symptoms

In contrast to treatments that reduce risk, treatments for symptoms are used with the goal of reducing or eliminating disease-specific symptoms.

Example 4: Guselkumab Used in the Treatment of Plaque Psoriasis

Patients with moderate to severe psoriasis can experience not only physical discomfort but chronic psoriasis can lead to psychological distress secondary to the appearance of lesions over large portions of a person's body. In addition, psoriasis is associated with the development of arthritis, worsened cardiovascular risk factors, and other conditions like inflammatory bowel disease.

The VOYAGE 1 study is an example of a well-designed and informative trial of guselkumab (Tremfya®) for patients with moderate to severe psoriasis (Blauvelt et al. 2017). The VOYAGE 1 trial had three arms—guselkumab (weeks 0–48), adalimumab (weeks 0–48), and a placebo arm (weeks 0–16) after which patients taking placebo crossed over to receive guselkumab from weeks 17–48. This very useful design answers two important questions: Is guselkumab better than placebo and is guselkumab better than an established therapy (adalimumab) from a similar class?

The main endpoint in this trial was a 90% or greater improvement in the Psoriasis Area Severity Index (PASI 90). This score is used to express the overall severity of psoriasis by combining the erythema, induration, and desquamation with the percentage of the affected area.

At week 16, roughly 3% of the placebo subjects had a PASI 90 score, whereas this occurred in 50% in the adalimumab arm and 73% in the guselkumab. The roughly 25% advantage of guselkumab over adalimumab was maintained at week 24 and week 48. In other words, the 50% benefit in the adalimumab group means that for every two people given adalimumab one person will achieve a PASI 90 score. The 25% additional benefit for guselkumab over adalimumab means for every four people who get guselkumab instead of adalimumab one extra person will get a PASI 90 score. When one lowers the benefit threshold to a PASI 75, approximately 90% ended up with improvement in the guselkumab group.

In addition, overall quality of life was improved. A Dermatology Life Quality Index is a score from 0 to 30 with a higher score indicating more severe disease. These subjects started at ~13–14 on this scale. On placebo, this score remained essentially unchanged over the duration of the trial.

However, the score on this scale for subjects on these agents went down by between 9 and 11. For this scale, a minimally important change is considered ~2–3. So not only do these agents clear up psoriatic lesions, this change is also associated with an impressive improvement in a subject's quality of life.

Over this 48-week trial, adverse effect data were collected and there were no greater numbers of people with regard to outcomes such as upper respiratory infections. Adverse events such as malignancies and major adverse coronary events occurred in less than 1% in all groups. While promising, this was only a 48-week study and any impact positively or negatively these agents might have on malignancies or cardiovascular disease may not be seen for years.

So in contrast to the first three examples, in this case we have a treatment that clearly provides clinically important benefits for the vast majority of people for an endpoint that is clear, specific, and measurable (PASI score and quality of life). This allows clinicians and patients the opportunity to figure out if an individual patient is getting a clinically important response from a therapeutic trial. In addition, having an endpoint that is measurable in an individual patient allows clinicians and patients to evaluate different treatments and also to determine the lowest effective dose for an individual patient.

So how could clinicians and patients use this information? From this trial, adalimumab is effective in roughly 50% of subjects so a reasonable approach could be to try adalimumab first as it is less expensive and there is likely more long-term data as adalimumab has been around longer than guselkumab. If an acceptable response is not seen, then one could switch to guselkumab. Regardless of which agent is chosen, once a response has been seen, the next step would be to find the lowest effective dose by either lowering the dose or increasing the interval of the injections and seeing what happens to patient response. This approach would hopefully reduce the cost and potentially reduce the risk of adverse effects given the majority of adverse effects for medications are dose-related. There are also a number of other biologic agents available for plaque psoriasis and a number of them have been compared head-to-head (Sbidian et al. 2022) and the choice of which to use should be based on the balance of efficacy, toxicity and ultimately cost may be the deciding factor.

Example 5: Dupilumab for Uncontrolled Persistent Asthma

Roughly one-quarter of asthmatics have moderate to severe disease with an increased risk for exacerbations, hospitalizations, and an impairment of quality of life. Dupilumab (Dupixent®) is an agent used for eczema but it has also been studied in subjects with moderate to severe asthma. Dupilumab was evaluated in 769 subjects with uncontrolled persistent asthma despite being on inhaled corticosteroids and a long-acting beta-agonist (Wenzel et al. 2016). Subjects received either placebo or one of four different doses of dupilumab for a total of 24 weeks. In addition, in a 2020 systematic review of dupilumab which included three different trials over a period of 24–52 weeks, the overall risk of a severe exacerbation was reduced by roughly 50% (RR 0.51, 95% CI 0.45–0.59) (Agache et al. 2020).

The risk of a severe exacerbation in the Wenzel trial over the 24-week study period was reduced from ~25% in the placebo group down to ~15% in those receiving a dupilumab every 4 weeks and to ~10% of subjects on the twice weekly doses of dupilumab. In other words, one in ten benefitted from monthly injections and one in six benefitted from every 2-week injections.

As with the plaque psoriasis study, these investigators also looked at the impact of this agent on quality of life using an Asthma Quality of Life Questionnaire. The score on this questionnaire ranges from 1 to 7 with higher scores indicating a better quality of life. Subjects started at a score of roughly 4 and the scores increased by 0.9 in the placebo group and by approximately 1.1–1.2 in the dupilumab group. The systematic review by Agache et al. also reported similar findings (Agache et al. 2020). A minimally important change on this scale is considered to be a change of 0.5. Interestingly, in contrast to the previously mentioned psoriasis study, the placebo group in this asthma study experienced a quality of life change (0.9) that would be considered clinically important. However, the difference between the treatment and placebo groups of ~0.2–0.3 (1.1 to 1.2−0.9) on this 1–7 scale would not, on average, be considered clinically important. This makes evaluation of this agent on an individual basis much trickier because of the "benefit" seen in the placebo group and the minimal difference between placebo and dupilumab.

Overall, patients when given this medication will on average experience what they perceive to be a benefit in their quality of life (as an improvement was seen in both the placebo and the active drug group). However, almost all of this improvement is secondary not to the impact of the medication, but likely a combination of regression to the mean, the natural history of the condition, and possibly the placebo effect. For this reason, dose titration to symptom control in this example is not a reasonable approach. Dupilumab did however reduce the risk for severe exacerbations in roughly one in 5–10 people. Finally, cost and the lack of long-term data would need to be given due consideration in the decision-making process.

Example 6: Enzyme Replacement for Fabry Disease

People with the genetic disorder Fabry disease lack the enzyme alpha-galactosidase A. This enzyme is responsible for the breakdown of certain lipid compounds (globotriaosylceramide) and without this enzyme these compounds build-up in blood vessels and affect the function of the eyes, skin, kidney, heart, gastrointestinal system, brain, and nervous system. This build-up can lead to symptoms of pain in the hands and feet, cloudiness in the eye, a decreased ability to sweat, hearing loss, and dark red spots on the skin. In addition, life expectancy is also reduced. Two recombinant enzyme replacement therapies are available, agalsidase alfa (Replagal®) and agalsidase beta (Fabrazyme®).

One of the first trials of enzyme replacement randomized 58 patients to either agalsidase beta 1 mg/kg IV or placebo every 2 weeks for 20 weeks followed by open-label agalsidase beta for 6 months (Eng et al. 2001). The primary endpoint was the percentage of patients in each group who were free of microvascular endothelial deposits of globotriaosylceramide in renal-biopsy specimens. After 20 weeks, in the enzyme replacement arm, 31% patients had deposits but 100% had deposits in the placebo arm. Patients in the enzyme replacement arm also had lower scores on deposits in the heart and skin. Interestingly however, in this trial, overall pain scores and quality of life scores were not different between the two groups. Infusion-related adverse effects were higher in the enzyme group ~50% rigors, ~20% fevers, ~10% headache, ~15% chills than in the placebo group. A roughly 3-year follow-up study suggested that enzyme replacement therapy resulted in continuously decreased plasma globotriaosylceramide levels; however, this follow-up study was not designed to evaluate the impact of enzyme replacement on clinical outcomes (Wilcox et al. 2004).

A 2017 Cochrane review (nine trials of either agalsidase alfa or beta in 351 subjects) concluded "Trials comparing enzyme replacement therapy to placebo show significant improvement with enzyme replacement therapy in regard to microvascular endothelial deposits of globotriaosylceramide and in pain-related quality of life" (El Dib et al. 2017). Pain scores were reduced by ~2 points on a 10-point scale in studies of agalsidase alfa over a period of 6 months. However, the authors also stated "The long-term influence of enzyme replacement therapy on risk of morbidity and mortality related to Anderson-Fabry disease remains to be established."

So, in this example, we have patients with a specific enzyme deficiency that will clearly negatively impact their health over the long-term. We have evidence that enzyme replacement therapy reduces the surrogate marker of endothelial deposits and possibly leads to a reduction in pain but long-term impacts on clinical important outcomes may never truly be known because of the ethics of doing long-term placebo trials in these subjects. These unknowns make funding and medical decisions tricky because these enzyme replacements can cost in excess of $200,000 a year.

Conclusion

Using these examples, it is clear there is no single way to approach clinical decisions around the use of these "omic" technologies. However, even though these novel agents may have unique mechanisms, and in some cases clinically important benefits, they are yet just another treatment option. Given that, they should be incorporated into clinical practice just like any other treatment options which is by using the best available evidence and balancing benefits and harms.

All of the requirements in Table 12.1 come into play as they would with any new medication entering the marketplace and clinical use. The individual decisions revolve very much around what the best available evidence shows and what condition is being treated. A personalized approach in each of these cases is crucial because each patient and response will be very unique. The benefits may be as small as one in 50–100 people benefitting, or in some cases 50–75% of people will derive a clinical benefit. These numbers are very much determined by the baseline risk of a patient and the overall effectiveness of the medication.

Regardless of condition, every decision needs to be informed by the best available evidence, which hopefully includes how these agents compare to not only placebo, but also how they compare to the gold standard treatments and other agents within in the same class.

So as one can see, to effectively use these agents, as with all medications one needs to know what happens clinically in the placebo group, the magnitude of the change in the treatment group and each of these examples brings to light the many different aspects that need to be considered.

For medications like palivizumab, evolucumab, and romosozumab, clinicians need to be able to communicate the magnitude of the risk reduction to patients and to also discuss the adverse effects, the cost, the inconvenience, and the fact that knowledge of the long-term effects is often fairly limited.

For agents like dupilumab and in particular guselkumab where one may be able to evaluate the individual response to a particular medication and dose, individualization and titration to the lowest effective dose becomes a key step in the overall use of these medications.

Finally, for medications like the enzyme replacement therapies for Fabry disease, we may never know if life-long use of these treatments will actually lead to a clinically important improvement in quality of life or a reduction in negative clinical outcomes.

With these new "omic" technologies, there is certainly a possibility that these new agents may be more effective or may in fact be the only effective treatment for a number of difficult to treat or uncommon conditions. However, every one of these new medications must be evaluated using the exact same principles of evidence-based practice presently used for all new and old treatments alike.

Self-Assessment Questions

Questions

1. Is there anything unique about therapeutic proteins and how clinicians need to assess them?
2. If a medication produces a 25% relative reduction in the chance of developing heart disease, does that mean 25% of people who take the medication benefit?
3. For any new medication that has an effect on symptoms, how does one go about figuring out the best dose for an individual patient?
4. When a new medication comes on the market, there is often limited long-term safety data so how should a clinician deal with this problem?
5. Is cost an important issue when it comes to the new "omic" technologies?

Answers

1. Not really. As with all new therapeutic options, they need to be evaluated with appropriately designed RCTs, and then the decision about their value should be based on a combination of evidence and patient-centered requirements.
2. No, any relative benefit needs to be applied to the baseline event rate in the placebo group. For instance, if the baseline risk of a heart attack in the placebo group is 20%, then applying a relative 25% benefit to the 20% baseline risk means that 15% of the treatment group end up getting a heart attack. The absolute risk difference is 20%−15% or an absolute benefit of 5% which translates to having to treat 20 people to benefit one.
3. One can (a) start with the dose used in the studies and then if a benefit is seen a dose titration down can be done to identify the lowest effective dose or (b) start with a dose 1/4 to a 1/8 of that used in the initial clinical trials and then titrate the dose up to the dose that best controls the patients symptoms with a minimum of side effects.
4. In general, unless the new medication provides an important clinical benefit over other existing therapies, it may be best to wait until longer-term adverse effect data is available before incorporating it into practice.
5. Cost is always an important issue when it comes to the selection of any therapeutic treatment. If a new agent doesn't provide a clinically important improvement over older treatments, then cost should be a determining factor in the selection process. If a new agent does provide a clinically relevant improvement in prevention or symptomatic treatment, then the increased cost should be proportional to that increased benefit.

References

Agache I, Song Y, Rocha C, Beltran J, Posso M, Steiner C et al (2020) Efficacy and safety of treatment with dupilumab for severe asthma: a systematic review of the EAACI guidelines-recommendations on the use of biologicals in severe asthma. Allergy 75:1058–1068. https://doi.org/10.1111/all.14268

Blauvelt A, Papp KA, Griffiths CEM et al (2017) Efficacy and safety of guselkumab, an anti-interleukin-23 monoclonal antibody, compared with adalimumab for the continuous treatment of patients with moderate to severe psoriasis: results from the phase III, double-blinded, placebo- and active comparator-controlled VOYAGE 1 trial. J Am Acad Dermatol 76:405–417

El Dib R, Gomaa H, Carvalho RP, Camargo SE, Bazan R, Barretti P, Barreto FC (2017) Enzyme replacement therapy for Anderson-Fabry disease. Cochrane Database Syst Rev 7:CD006663

Eng CM, Guffon N, Wilcox WR, Germain DP, Lee P, Waldek S et al (2001) Safety and efficacy of recombinant human alpha-galactosidase a replacement therapy in Fabry's disease. N Engl J Med 345:9–16

Garegnani L, Styrmisdóttir L, Roson Rodriguez P, Escobar Liquitay CM, Esteban I, Franco JVA (2021) Palivizumab for preventing severe respiratory syncytial virus (RSV) infection in children. Cochrane Database Syst Rev 11:CD013757

Saag KG, Petersen J, Brandi ML, Karaplis AC, Lorentzon M, Thomas T, Maddox J, Fan M, Meisner PD, Grauer A (2017) Romosozumab or alendronate for fracture prevention in women with osteoporosis. N Engl J Med 377:1417–1427

Sbidian E, Chaimani A, Garcia-Doval I, Doney L, Dressler C, Hua C, Hughes C, Naldi L, Afach S, Le Cleach L (2022) Systemic pharmacological treatments for chronic plaque psoriasis: a network meta-analysis. Cochrane Database Syst Rev 12:CD011535

Schmidt AF, Carter JPL, Pearce LS, Wilkins JT, Overington JP, Hingorani AD, Casas J (2020) PCSK9 monoclonal antibodies for the primary and secondary prevention of cardiovascular disease. Cochrane Database Syst Rev 10:CD011748

Sindelar RD (2013) Genomics, other "Omic" technologies, personalized medicine. In: Crommelin DJA, Sindelar RD, Meibohm B (eds) Pharmaceutical biotechnology, 4th edn. Springer, New York, pp 190–221

Singh S, Dutta S, Khasbage S, Kumar T, Sachin J, Sharma J, Varthya SB (2022) A systematic review and meta-analysis of efficacy and safety of Romosozumab in postmenopausal osteoporosis. Osteoporos Int 33(1):1–12. https://doi.org/10.1007/s00198-021-06095-y

Wenzel S, Castro M, Corren J, Maspero J, Wang L, Zhang B, Pirozzi G, Sutherland ER, Evans RR, Joish VN, Eckert L, Graham NM, Stahl N, Yancopoulos GD, Louis-Tisserand M, Teper A (2016) Dupilumab efficacy and safety in adults with uncontrolled persistent asthma despite use of medium-to-high-dose inhaled corticosteroids plus a long-acting beta2 agonist: a randomised double-blind placebo-controlled pivotal phase 2b dose-ranging trial. Lancet 388:31–44

Wilcox WR, Banikazemi M, Guffon N, Waldek S, Lee P, Linthorst GE, Desnick RJ, Germain DP, International Fabry Disease Study Group (2004) Long-term safety and efficacy of enzyme replacement therapy for Fabry disease. Am J Hum Genet 75:65–74

Pharmaceutical Biotechnology: Oligonucleotides, Genes and Cells/Cell Subunits, Vaccines—The Science, Techniques, and Clinical Use

Oligonucleotides and mRNA Therapeutics

Erik Oude Blenke, Raymond M. Schiffelers, and Enrico Mastrobattista

Introduction

In the past decade, several new classes of nucleic acid-based drugs have emerged that complement conventional small molecule and protein-based drugs with their ability to modulate disease pathways at the genetic level. They can be divided into two main classes: short and long RNAs that are also manufactured in a different way. Oligonucleotides are short chains of single-stranded or double-stranded (chemically modified) ribo- or deoxyribonucleotides. Different than messenger RNA (mRNA) therapeutics that are made enzymatically, oligonucleotides are synthetically made, which allows a broader range of chemical modifications to change the pharmacologic and pharmacokinetic behavior of the molecule. Conversely, this method of synthesis limits the length of the sequences that can be made, so oligonucleotides are orders of magnitude shorter than the typical mRNA. Their ability to bind to chromosomal DNA, mRNA, or noncoding RNA (ncRNA) through Watson–Crick and Hoogsteen base pairing offers possibilities for highly specific intervention in gene transcription, mRNA translation, and RNA regulatory pathways for therapeutic applications.

In theory, a specific sequence of 15–17 bases occurs only once in the human genome, which allows specific manipulation of single genes for oligonucleotides with a complementary sequence in this size range. The most widely used and therapeutically most successful way of interference has been the use of complementary oligonucleotides that bind and inhibit or degrade specific mRNAs. Two main mechanisms can be distinguished:

- Antisense inhibition: Based on single-stranded oligonucleotides that bind complementary mRNA and thereby block translation or recruit RNase H for mRNa degradation.
- RNA interference: Based on double-stranded RNA oligonucleotides consisting of a passenger strand and antisense strand that binds complementary mRNA and recruits RNA-induced Silencing Complex (RISC) that degrades the mRNA. These molecules are commonly known as small interfering RNAs, or siRNAs.

Oligonucleotides can be very potent molecules. Yet, interpretation of the mechanism of therapeutic action of a specific oligonucleotide sequence is not straightforward (Moulton 2017). Apart from the desired on-target activity, oligonucleotides are inclined to display sequence-specific, unintended effects. Partial sequence complementarity may affect binding to nucleic acids other than the targeted species (known as off-target effects). This is becoming increasingly complex as new classes of intracellular noncoding nucleic acids continue to be discovered, for example, circular RNAs whose functional role is at present only understood to a limited extent (Yang et al. 2022). This makes it difficult to predict if any of such unintended binding has a negative effect. Immune stimulation by oligonucleotides may also occur and binding to proteins and peptides can alter their activity. For example, the binding of phosphorothioate antisense oligonucleotides (PS-ASOs) inhibits the tenase complex and can increase blood clotting times (Sheehan and Lan 1998). Inside the cell, PS-ASOs can bind to so-called paraspeckle proteins, which induces an intracellular stress response that could lead to apoptosis (Shen et al. 2019).

Recently, the application of mRNA encoding therapeutic proteins has become a reality. Vaccines, protein replacement therapeutics, gene-editing proteins, and enzymes for prodrug activation are just a few of the various approaches currently on the market or in advanced clinical trials. Also, for mRNA, off-target binding to nucleic acid species can be expected. In addition, immune activation, limiting translation efficiency, is an important challenge.

E. Oude Blenke · E. Mastrobattista (✉)
Department of Pharmaceutics, Utrecht Institute for Pharmaceutical Sciences, Utrecht University, Utrecht, The Netherlands
e-mail: e.mastrobattista@uu.nl

R. M. Schiffelers
Laboratory Clinical Chemistry and Haematology, University Medical Center Utrecht, Utrecht, The Netherlands
e-mail: r.schiffelers@umcutrecht.nl

Apart from immune activation and off-target binding, other characteristics of oligonucleotides and mRNA also impede clear-cut application as therapeutics. They are sensitive to degradation by nucleases and their physicochemical characteristics lead to rapid excretion by the kidneys and/or induce uptake by macrophages hindering target tissue accumulation (Geary et al. 2015). In addition, spontaneous passage over cell membranes for these large and charged molecules for intracellular applications is difficult.

Over the years, a number of different modifications have been developed that overcome (part of) these problems (Fig. 13.1) (Hammond et al. 2021). All of the clinically studied oligonucleotides contain one or more of these modified nucleotides. For mRNA, which is generated via enzymatic synthesis, modifications of nucleotides need to be compatible with the in vitro transcription process, which limits the chemical design space. In this chapter, we describe the classes of therapeutic oligonucleotides, categorized according to their mechanism of action.

First, we discuss oligonucleotides that interfere with gene expression and prevent the translation from mRNA. After that, oligonucleotides that change mRNA by

Oligonucleotide Chemistres

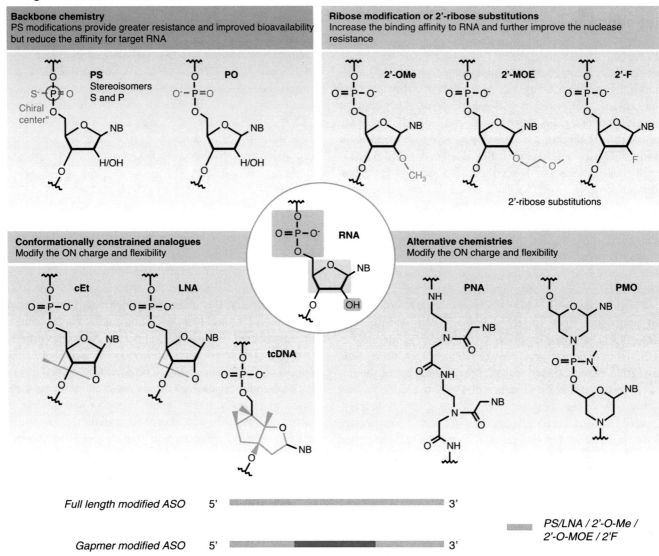

Fig. 13.1 Commonly used oligonucleotide chemistries. Frequently used backbone modifications (purple panel) to replace the naturally occurring phosphodiester bonds (PO) are the phosphorothioates (PS). Modifications to the ribose at the 2′-O position of RNA and 2′-position of DNA include the 2′-O-methyl (2′-OMe), 2′-O-methoxy-ethyl (2′-MOE), and 2′-fluoro (2′-F). The green panels show different conformationally constrained DNA analogues and alternative chemistries. Locked nucleic acid (LNA), constrained 2′-O-ethyl (cEt), and tricyclo-DNA (tcDNA) provide greater binding affinity. Alternative chemistries include changes in the nucleobase, e.g., phosphorodiamidate morpholino oligomers (PMO) and peptide nucleic acid (PNA). Chemical modifications can be distributed over the full length of the oligonucleotide, or restricted to the 5′and 3′ends with unmodified DNA or RNA in the middle (Gapmers). Adapted from Hammond et al. (2021)

skipping unwanted mRNA fragments or introducing base changes are introduced. Finally, we will discuss the use of in vitro transcribed mRNA for vaccination and therapeutic intervention.

The challenges to apply these oligonucleotides and mRNA as therapeutics are faced by essentially all classes: rapid clearance, poor stability, and limited cellular uptake. These issues are addressed for both types of therapeutics, as their different properties and ways of manufacturing require specific solutions for each of them.

Interfering with Gene Expression

Antisense oligonucleotides (ASO), siRNA, miRNA, small activating RNA (saRNA), ADAR recruiting oligonucleotides, and transcription factor decoys are all members of the class of oligonucleotides that can interfere with gene expression, but they function at different stages of the gene expression process (Fig. 13.2). Important to note, ASO- and siRNA-based strategies are by far the most successful approaches in this area and therefore these are discussed in more detail.

Most commonly, the interference in gene expression results in a reduction in protein production. However, improved insight into the transcription and translation process has allowed to develop strategies that actually increase output. Targeted Augmentation of Nuclear Gene Output (TANGO) for example uses ASOs that bind specific regions of pre-mRNA thereby reducing the synthesis of nonproduc-

tive mRNA and increasing the synthesis of productive mRNA, which is especially relevant in haplo-insufficiencies (Wengert et al. 2022). Similarly, small activating RNAs (saRNA) saRNAs recruit endogenous transcriptional complexes to a complementary target gene, leading to increased expression of naturally processed mRNA and upregulation of the target protein expression (Andrikakou et al. 2022).

Antisense Oligonucleotides

The function of oligonucleotides to act as antisense molecules was discovered by Zamecnik and Stephenson in 1978, making it the oldest oligonucleotide-based therapeutic approach (Stephenson and Zamecnik 1978). Many of the difficulties associated with the use of oligonucleotides for medical applications have consequently been encountered for antisense molecules first, explaining why clinical progress has been difficult. Development of siRNA therapies has moved a lot faster, as will be addressed in a later section. But improvements in synthetic chemistry, knowledge on genome, transcriptome and proteome, and new delivery strategies have revived interest in the technology (Shen and Corey 2018). "Classical" antisense oligonucleotides are single-stranded DNA or RNA molecules that generally consist of 13–25 nucleotides. They are complementary to a sense mRNA sequence and can hybridize to it through Watson–Crick base pairing. Three classes of translation inhibiting oligonucleotides can be distinguished based on their mechanism of action (Fig. 13.3):

Fig. 13.2 Overview of different modes of intervention of therapeutic nucleic acids. Gene therapy and gene editing act directly on the genomic DNA and will be discussed in Chap. 14. Other interventions with nucleic acids, including synthetic antisense oligonucleotides (ASO), small interfering RNA (siRNA), adenosine deaminase acting on RNA (ADAR), and mRNA that acts at the level of transcription, RNA splicing, or translation, are discussed in this chapter. Created with BioRender.com

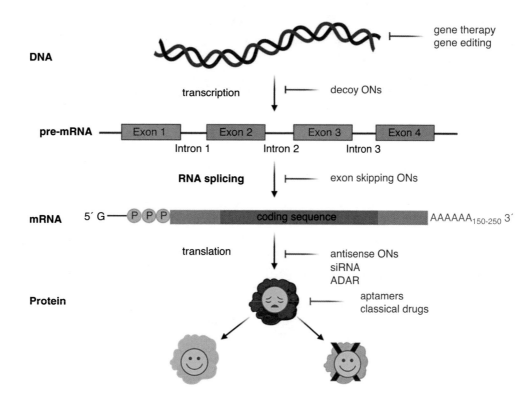

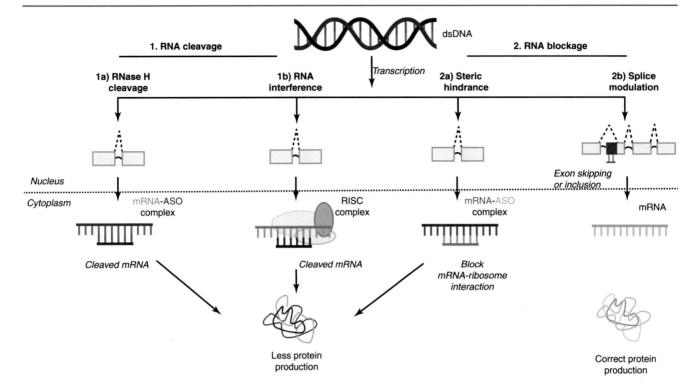

Fig. 13.3 Mechanism of action of oligonucleotides that act at the level of mRNA. Antisense oligonucleotides can inhibit mRNA translation by sterically blocking translation or inducing RNase H-mediated mRNAcleavage. In the ribozyme/DNAzyme approach, the oligonucleotide possesses mRNA degradative properties or functions as a recruiting factor (known as external guide sequences) for endogenous ribozymes (RNase P). Finally, siRNA- and miRNA-based strategies make use of short double-stranded RNA molecules which unwind and bind complementary mRNA in the RNA-induced silencing complex (RISC), which subsequently cleaves the mRNA. Adapted from Dhuri et al. (2020)

- mRNA-blocking oligonucleotides, which physically prevent or inhibit the progression of splicing or translation through binding of complementary mRNA sequences
- mRNA-cleaving oligonucleotides, which induce degradation of mRNA by binding complementary mRNA sequences and recruiting the cytoplasmic nuclease RNase H
- mRNA-cleaving oligonucleotides, which induce degradation of mRNA by recruiting nuclear RNase P via external guide sequences or by nuclease activity of the nucleic acid itself (ribozymes/DNAzymes)

The majority of the clinically studied antisense oligonucleotides act through RNase H. RNase H-mediated knockdown generally reaches >80% downregulation of protein and mRNa expression. In contrast to blocking oligonucleotides, RNase H-recruiting antisense oligonucleotides can inhibit protein expression without a priori restrictions to the region of the mRNA that is targeted. In other words, cleavage of the mRNA will inactivate it, no matter where it is cleaved. Most blocking oligonucleotides, however, need to be targeted to regions within the 5′-untranslated region or AUG initiation codon region to prevent translation, as the ribosome is apparently able to remove bound antisense molecules in the coding region.

The first antisense drug that was introduced to the market in 1998 was fomivirsen (Vitravene®), for the treatment of cytomegalovirus-induced retinitis in AIDS patients. Due to the success of highly active antiretroviral therapy (HAART), the market for this drug has virtually disappeared, and the product has effectively been discontinued. Vitravene® was injected into the vitreous at a dose of 165 μg or 330 μg/eye in 25 μL, once weekly for 3 weeks, followed by 2-week administrations. Reported side effects are related to irritation and inflammation of the eye likely caused by the injection procedure. The phosphate linkages in the backbone are replaced with phosphorothioate linkages (PS) backbone which limits nuclease degradation. Local injection at the pathological site improves target cell accumulation.

Over the past years, four additional antisense products were approved (Table 13.1). Their different chemistries and modifications are summarized in the table. A more detailed description of chemical modifications to enhance stability and affinity will be described later on in this chapter.

Mipomersen is indicated for patients with familial hypercholesterolemia. The drug targets translation of apolipoprotein B-100 in the liver, the main protein constituent in low-density lipoprotein. It is administered by subcutaneous injections of 200 mg once weekly. The nucleotides are linked

Table 13.1 Antisense oligonucleotide therapeutics on the market. ASO chemistry, genetic target, and disease indication are listed

Drug molecule	Genetic target	Disease/condition
Fomivirsen (Vitravene®)	Cytomegalovirus	Cytomegalovirus eye infection
5'-G = C = G = T = T = T = G = C = T = C = T = T = C = T = T = C = T = T = G = C = G-3'		
Mipomersen (Kynamro ®)	Apolipoprotein B-100	Homozygous familial hypercholesterolemia
5'-G* = mC* = mC* = mU* = mC* = A = G = T = mC = T = G = mC = T = T = mC = G* = mC* = A* = mC* = mC*—3'		
Nusinersen (Spinraza®)	Survival motor neuron protein 2	Spinal muscular atrophy
5'-mU* = mC* = A* = mC* = mU* = mU* = mU* = mC* = A* = mU* = A* = A* = mU* = G* = mC* = mU* = G* = G*-3'		
Inotersen	Transthyretin	Familial amyloid polyneuropathy
5'-T* = mC* = T* = T* = G* = G = T = T = A = mC = A = T = G = A = A = A* = T* = mC* = mC* = mC*-3'		
Volanesorsen	Apolipoprotein C-III	Hypertriglyceridemia, familial chylomicronemia syndrome and familial partial lipodystrophy
5'-A* = G* = mC* = T* = T* = mC = T = T = G = T = mC = mC = A = G = mC = T* = T* = T* = A* = T*-3' =thioate, m − 2'-O-methyl, * 2'-O-(2-methoxyethyl)		

by phosphorothioate linkages and the distal ends are 2'-O-methyl modified. It comes with a black box warning for severe liver toxicity (Hair et al. 2013).

Nusinersen is administered at a dose of 12 mg intrathecally per administration. The compound is composed of nucleotides with 2'-O-(2-methoxyethyl) groups and a PS backbone. It targets the splicing of SMN2 and converts part of it to SMN1, the protein that patients with spinal muscular atrophy lack. After four initial loading doses over a period of 30 days, the drug is injected every 4 months for maintenance. Yearly costs to treat a patient are estimated at US$ 750,000 placing it among the most expensive drugs in the world (Aartsma-Rus 2017).

Inotersen is designed to limit translation of mutated transthyretin (TTR) in patients with adult hereditary transthyretin amyloidosis. The compound is also composed of phosphorothioate-linked nucleotides with 2'-O-methyl modifications on the distal ends. In clinical trials, it was administered as 300 mg of inotersen via subcutaneous injection three times on alternate days for the first week, and then once weekly (Keam 2018). Of note, TTR is a very well-characterized genetic target and several newer generation oligonucleotide-based drugs have been developed to replace inotersen; an siRNA against this target is also available (patisiran, Onpattro®).

Volanesorsen has the same chemical design as mipomersen and inotersen, with PS backbone and 2'-MOE flanks. This demonstrates that this type of molecules can be considered a platform technology, where the chemical characteristics remain the same across multiple molecules. Only the genetic sequence is changed based on the target of interest. Volanesorsen targets apolipoprotein C-III translation in the liver and aims to correct the disturbed triglyceride distribution in patients with familial chylomicronemia syndrome and familial partial lipodystrophy. It is administered subcutaneously at a dosage of 300 mg once weekly (Paik and Duggan 2019).

Although sequence specificity is one of the most attractive features for antisense application, there are reports that show that knockdown of related genes with only limited sequence homology can occur, which necessitates careful lead sequence selection.

siRNA/miRNA

MicroRNA (miRNA) and small interfering RNA (siRNA) are double-stranded RNA oligonucleotides of 21–26 base pairs that can cause gene silencing in a process known as RNA interference (RNAi) (Fig. 13.4) (Chakraborty et al. 2017).

Pharmaceutically, the predominant use of RNAi is by employing siRNA at this moment. In a process reminiscent of antisense-mediated degradation of complementary mRNA by recruitment of RNase H, siRNA can degrade complementary mRNa via an enzyme complex known as RNA-induced silencing complex or RISC. The difference between miRNA and siRNA-induced silencing is explained below.

miRNAs are endogenous oligonucleotides produced from transcripts that form stem-loop structures. These are processed in the nucleus into 65–75 nucleotides long pre-miRNA followed by transport to the cytoplasm. Pre-miRNA is further cleaved by an enzyme complex known as Dicer to form miRNA, which is loaded into the RISC (Fig. 13.4).

Synthetic siRNAs are exogenous double-stranded RNAs (dsRNA) of 21–26 nucleotides that are designed to mimic the mature Dicer-cleavage product. siRNAs are similarly loaded into RISC. Subsequently, one strand, known as the passenger strand, leaves the RISC and is degraded. By introducing chemical modifications, the preference for either strand to remain incorporated in the RISC can be influenced to reduce off-target effects. The retained strand remains associated to the RISC. When a homologous mRNA is bound,

Fig. 13.4 Mechanism of action of RNA interference by miRNA and siRNA. RNA is transcribed from specific miRNA genes in the nucleus, which folds into characteristic hairpin loops. These are excised and processed by DROSHA in the nucleus to form pre-miRNA. These pre-miRNA fragments are transported out of the nucleus and into the cytosol. Dicer cleaves the dsRNA, cutting off the loop and separating the RNA strands with the help of Argonaute. Similarly, Dicer cuts invading dsRNA coming from either viruses or synthetic dsRNA which was introduced on purpose. The guide strand is integrated into the RNA-induced silencing complex (RISC). RISC can now regulate gene expression at the mRNA transcript level in two ways. (1) By inducing mRNA degradation. For this route, a near-perfect complementary match is required between the guide strand and the target mRNA as well as a catalytically active Argonaute protein. (2) By interfering with translation. Translational repression only requires a partial sequence match between the guide strand and the target mRNA

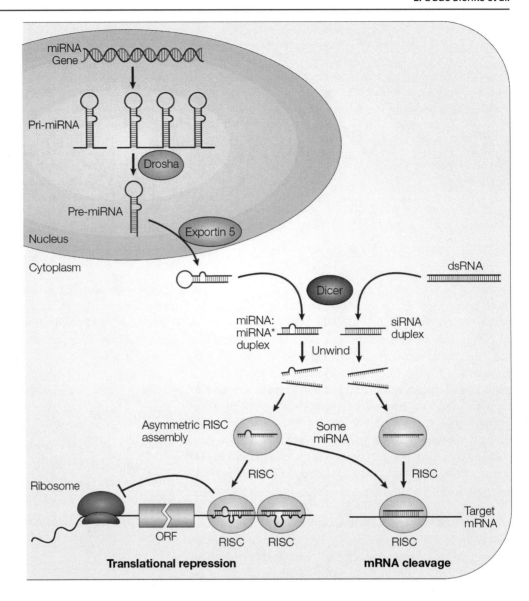

two processes can occur, depending on the level of complementarity between the mRNA and miRNA/siRNA. Complete complementarity induces RISC-mediated cleavage of the mRNA, while lower level complementarity causes translational repression. Endogenous miRNAs generally have a lower level of complementarity between the miRNA strand and mRNA providing broad translational repression of several genes in a genetic program. All currently approved products are siRNAs, underlining that this is currently pharmaceutically the most attractive approach.

A challenge for pharmaceutical development of siRNAs is the fact that the presence of dsRNA in mammalian cells induces an interferon response, which results in nonspecific inhibition of translation and cell death. The longer the dsRNA, the stronger the response. Next to the direct endonucleolytic cleavage of mRNAs via RISC or translational repression, miRNA/siRNA appear also to act at other levels. They have been shown to affect methylation of promoters,

increase degradation of mRNA not mediated by RISC, and enhance protein degradation. These processes can contribute or hamper the desired activity.

Since the discovery of the process, a remarkably rapid progress has been made, and several compounds are currently approved. This rapid progress is partly due to the strong potency of the RNAi technique which seems to silence gene expression far more efficiently than antisense approaches and partly, also, because much has been learned from previous nucleic acid-based clinical trials. Initial clinical studies focused on macular degeneration as local injection of oligonucleotides in this immunoprivileged environment allows relative high doses to be administered with minimal immune reactions.

Alnylam Pharmaceuticals is currently developing a portfolio of siRNA drugs designed to inhibit proteins in the liver. They have two technologies that ensure hepatocyte delivery based on coupling of triantennary N-acetylgalactosamine (GalNAc) to the siRNA or incorporating the siRNA into lipid nanoparticles. With these delivery technology platforms in place, they can now rapidly develop new siRNA therapeutics for a variety of liver diseases. Because oligonucleotides are physicochemically essentially the same molecules, only with different sequences, the delivery technology can be quite generic.

This potential has been demonstrated in many clinical studies. The first drug developed was patisiran (Onpattro®) (Table 13.2). This siRNA targets transthyretin, a protein that is involved in several rare, but severe amyloid diseases. The siRNA is delivered to the hepatocytes by a lipid nanoparticle (cf. Sect. 14.27) These nanoparticles are composed of ionizable lipids designed to be highly potent transfectants. The particle is temporarily stabilized by PEGylated lipids with short fatty acid tails. As these PEGylated lipids dissociate from the particles in vivo, the particles become opsonized in situ with apolipoprotein E, which in turn mediates recognition by the LDL receptor primarily on the hepatocytes. Patisiran was approved in 2018. It is administered through an intravenous infusion every 3 weeks at a dose of 0.3 mg/kg.

This first approval marked an important milestone in the development of siRNA therapeutics, but looking at the pipeline of Alnylam and competing companies, it seems that the siRNA-lipid nanoparticle formulation will be phased out and replaced by the GalNAc conjugates. These conjugates are much more straightforward to produce and easier to administer as they allow subcutaneous injection. All these triantennary GalNAc-targeted formulations consist of alternating 2'-O-(2-methoxyethyl) and 2'-fluoro-modified nucleotides with phosphorothioate linkages on the end.

Givosiran (Givlaari®) is a subcutaneously administered N-acetylgalactosamine-targeted siRNA, designed against aminolevulinic acid synthase 1 mRNA for the treatment of acute hepatic porphyrias. It has, like patisiran, received "Breakthrough Designation" from the FDA during development, illustrating that these new technologies are truly addressing a previously unmet medical need. It was approved in 2019 and is dosed at 2.5 mg/kg every month (Scott 2020).

GalNAc-mediated targeting also forms the basis for a number of subsequently approved products. Approval of the

GalNAc–siRNA conjugate lumasiran (Oxlumo®) followed in 2020. It is indicated for the treatment of primary hyperoxaluria type 1 to lower urinary and plasma oxalate levels in pediatric and adult patients. After loading, the dose for adult patients is set at 3 mg/kg every 3 months (Scott and Keam 2021). Vutrisiran (Amvuttra®) is indicated for the treatment of the polyneuropathy of hereditary transthyretin-mediated amyloidosis in adults. It targets the exact same sequence as patisiran, but is formulated as a GalNAc conjugate instead of a lipid nanoparticle. It was approved in 2022 and is dosed subcutaneously at 25 mg every 3 months. What is clear from these subsequent approvals is that the dose and interval between doses can be reduced for every new product that has been released. This is primarily due to improvements in RNA chemistry that allow for ultra-stable molecules.

Inclisiran (Leqvia®) is the latest example of a heavily modified GalNAc-targeted siRNA that is extremely long-acting. Inclisiran contains 35 ribonucleotides that are chemically modified with 2-O-methyl-ribonucleotide (2'-O-methyl), eight modified with 2'-fluoro-ribonucleotide (2'-fluoro), and one 2'-deoxy-ribonucleotide. In addition, two phosphorothioate groups are added to the 5' end of the sense and both the 5' and 3' ends of the antisense strands. This siRNA targets proprotein convertase subtilisin/kexin type 9 (PCSK9) mRNA. Circulating PCSK9 causes hypercholesterolemia by reducing LDL receptors in hepatocytes. In two phase III studies with each over 150 patients, the drug strongly lowered LDL cholesterol (~50%) in patients that are refractory to other interventions (Ray et al. 2020). The dose of inclisiran in clinical trials is 300 mg given subcutaneously, once every 6 months. This is obviously a huge benefit compared to the conventional statins that are dosed multiple times per day and this patient convenience could be the critical factor for inclisiran to win a (small) part of the huge market for cholesterol lowering drugs. Several other siRNA drug candidates are under late-stage clinical development for lipid-related or cardiovascular disease (see Table 13.2) due to the fact that many metabolic diseases originate in the liver and the attractive proposition to manage cardiovascular disease through infrequent (once every 3–6 months) subcutaneous injections (Goga and Stoffel 2022).

Table 13.2 siRNA therapeutics on the market or in advanced clinical studies

Drug molecule	Genetic target	Disease/condition
Patisiran	Transthyretin	Transthyretin amyloidosis
Givosiran	Aminolevulinic acid synthase 1	Acute hepatic porphyria
Lumasiran	Glycolate oxidase	Primary hyperoxaluria type 1
Inclisiran	PCSK9	Familial hypercholesterolemia
Vutrisiran	Transthyretin	Transthyretin amyloidosis
Fitusiran[a]	Antithrombin	Hemophilia A and B
Olpasiran[a]	Lipoprotein(a)	Atherosclerotic cardiovascular disease
ARO-APOC3[a]	Apolipoprotein C-III	Severe hypertriglyceridemia
ARO-ANG3[a]	Angiopoietin-like protein 3	Familial hypercholesterolemia
Fazirsiran[a]	Mutant alpha-1 antitrypsin (Z-AAT)	Alpha-1 anti-trypsin deficiency

[a]In phase III trials

Besides siRNA, miRNAs may offer some, up to now theoretical, advantages with regard to their natural occurrence and ability to mediate downregulating of pathways rather than specific genes, but clinical activity with miRNAs has been limited. MRX34 reached the phase I clinical trial stage. It is a mimic of endogenous miR-34a, which is a miRNA with tumor-suppressor activity. It was delivered in a lipid nanoparticle. This clinical study was halted in 2016 because of multiple immune-related severe adverse events. Thus, despite recent successes, the delivery challenge of RNA therapeutics is still not entirely solved. Furthermore, the two main delivery technologies we have at hand right now are both only suited for targeting cells in the liver. There are many diseases originating from altered protein expression in the liver, but for diseases involving other tissues, the delivery challenges remain relevant as ever.

Transcript Repair

In addition to transcript removal (or silencing), several mechanisms exist to use oligonucleotides to repair a faulty transcript. This can be beneficial in disease indications where an incorrect version of a protein is expressed from an mRNA, as a result of a mutation in the DNA. Instead of correcting the DNA, the mRNA can be changed to express the correct form of the protein, or one that is more functional than the one that is encoded in the DNA.

Antisense-Induced Exon Skipping

The most widely studied mechanism to change the coding sequence of an mRNA is antisense-induced exon skipping. When a gene is transcribed from the DNA, this initially results in a pre-mRNA, which contains both introns and exons, from which the introns will be removed in a process called RNA splicing. Detrimental effects of certain gene

mutations on protein function can be alleviated by removing the faulty exon during RNA splicing, to excise the mutation that causes a frameshift or translation termination. This so-called exon-skipping technique tries to restore the reading frame by artificially removing one or more exons before or after the deletion or point mutation in the mRNA (Duan et al. 2021). The most popular disease target to work on is Duchenne's muscular dystrophy (DMD) which, using this technique, could be changed into the much milder Becker's dystrophy (Table 13.3).

Exons can be skipped from the mRNA with antisense oligoribonucleotides (Fig. 13.5).

They attach inside the exon to be removed, or at its borders. The oligonucleotides interfere with the splicing machinery so that the targeted exons are no longer included in the mRNA. This results in a shorter form of the dystrophin protein, that is less functional than the wildtype form that is expressed in healthy subjects, but more functional than the mutated form that Duchenne's disease patients are born with. The dystrophin gene is very long, containing 79 exons. Many different mutations exist, so it is impossible to develop a treatment that will help all patients. The most common mutation is in exon 51, but this affects only 13% of total patients. Several exon-skipping oligonucleotides have been in clinical trials, targeting the mutations that are most common and are amenable to repair by splicing out the exon that contains the mutation. Exon 51 has been the main target of interest and several clinical trials have been conducted. Since 2006, the molecule PRO-051/GSK2402968, also known as drisapersen, has been under clinical development by companies Prosensa/GlaxoSmithKline/BioMarin. After initial phase 1 trials, an increase of dystrophin in skeletal muscle biopsies was observed in some patients treated with drisapersen. Average levels rose to 0.93% of the normal amount of dystrophin, as compared to a baseline value of 0.08% in biopsies of untreated Duchenne muscles (Goemans 2016). Subsequently, the drug has been tested in two phase 2 and one phase 3 placebo-controlled trials in more than 300 DMD

Table 13.3 Exon-skipping oligonucleotides for DMD approved or in clinical studies

Molecule	Target
Drisapersen (Kindrisa)[a]	Exon 51
U*=C*=A*=A*=G*=G*=A*=A*=G*=A*=U*=G*=G*=C*=A*=U*=U*=U*=C*=U*	
Eteplirsen (Exondys 51™)	Exon 51
C ≈ T ≈ C ≈ C ≈ A ≈ A ≈ C ≈ A ≈ T ≈ C ≈ A ≈ A ≈ G ≈ G ≈ A ≈ A ≈ G ≈ A ≈ T ≈ G ≈ G ≈ C ≈ A ≈ T ≈ T ≈ T ≈ C ≈ T ≈ A ≈ G	
Golodirsen (Vyondys53™)	Exon 53
G ≈ mU ≈ mU ≈ G ≈ C ≈ C ≈ mU ≈ C ≈ C ≈ G ≈ G ≈ mU ≈ mU ≈ C ≈ mU ≈ G ≈ A ≈ A ≈ G ≈ G ≈ mU ≈ G ≈ mU ≈ mU ≈ C	
Viltolarsen (Viltepso™)	Exon 51
C ≈ C ≈ T ≈ C ≈ C ≈ G ≈ G ≈ T ≈ T ≈ C ≈ T ≈ G ≈ A ≈ A ≈ G ≈ G ≈ T ≈ G ≈ T ≈ T ≈ C	
Casimersen (Amondys45™	Exon 45
C ≈ A ≈ A ≈ mU ≈ G ≈ C ≈ C ≈ A ≈ mU ≈ C ≈ C ≈ mU ≈ G ≈ G ≈ A ≈ G ≈ 5 U ≈ mU ≈ C ≈ C ≈ mU ≈ G	
≈ −phosphoramidate, = −thioate, m-5-methyl,* -2′-O-(2-methoxyethyl)	

[a]Discontinued

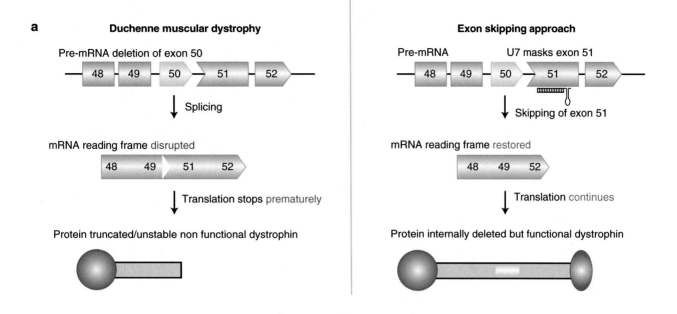

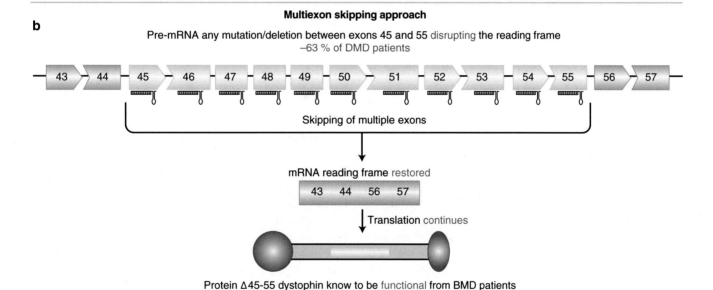

Fig. 13.5 Exon skipping in Duchenne's muscular dystrophy. (**a**) Left panel. In Duchenne's muscular dystrophy, the mutations in the DMD gene encoding dystrophin cause premature termination of translation. This leads to a nonfunctional protein that misses the second attachment point to the cytoskeleton. Right panel. By adding an oligonucleotide that preventing splicing factors from interacting with the pre-mRNA, the affected exon is skipped and the mutated region is not incorporated in the mRNA, leading to translation of a functional (albeit shorter) dystrophin molecule that contains both attachment points to the cytoskeleton. (**b**) The same strategy can also be followed to correct mutations in multiple exons, which is the case in the majority of DMD patients

patients. The primary endpoint in these trials was a 6-min walk test but the trials failed to demonstrate significant clinical benefit. Development of this drug is currently discontinued. This molecule was a 2′O-methyl-modified oligonucleotide with a phosphorothioate backbone.

Another molecule targeting exon 51 is eteplirsen, developed by Sarepta Therapeutics. This molecule is a phosphorodiamidate morpholino oligonucleotide (PMO) and it was anticipated that this different chemistry would be beneficial in exon-skipping applications. In several phase 1 dose-finding studies, a wide range of doses was tested, from 0.5 mg/kg/week to 50 mg/kg/week given as several intramuscular injections. An increase in dystrophin-positive myofibers was reported, but the results were highly variable across biopsies and the results did not appear to be dose-responsive. No improvement on the 6-min walk test was seen either, but 11 subjects were enrolled in an extension study where they received the drug for an additional 3.5 years and had several biopsies taken. In the end, a 0.9% increase in dystrophin levels was reported which eventually led to FDA approval, which caused great controversy within the FDA and the scientific community because many people did not think the drug delivered sufficient benefit to the patients. The drug is currently marketed by Sarepta as Exondys51, but it has never received approval from the European Medicines Agency (Takeda et al. 2021). Nevertheless, this has sped up the development of similar PMO molecules targeting other exons in the dystrophin gene, providing treatment options to additional patient groups affected by DMD. In 2019, golodirsen, marketed as Vyondys53™, was approved. It targets exon 53, which is affected in 8% of the total DMD patients. Similar to eteplirsen, the drug was approved based on the "surrogate endpoint" of improvement of dystrophin levels in biopsies, but not on the 6-min walk test or pulmonary function tests and only by the FDA in the US.

Viltolarsen also targets exon 53, but at a different position in the gene. It is developed by the Japanese company Nippon Shinyaku and received approval in Japan and later in the US (marketed as Viltepso™). Casimersen is the latest PMO exon-skipping drug from Sarepta that received approval. It targets exon 45 that affects another 8% of the total number of DMD patients.

All of these treatments that received approval did so on the basis of improvement of biomarkers rather than functional effects. Although these approvals bring hope to patients and their families, they do not yet provide a cure for this debilitating disease. The clinical evidence that mRNA is correctly spliced and that protein levels are upregulated could mean that initiating treatments earlier could bring a bigger benefit to patients. It is also postulated that this type of therapies can be improved if uptake in the target tissues can be enhanced. For example, by employing different administration routes to target more of the skeletal and heart muscle tissue, by using different types of chemistries, or by conjugating the molecules to increase uptake. Although significant uptake is easier to achieve in different tissues (for example the liver), there are no clinical trials ongoing with exon-skipping oligonucleotides for other applications than treating Duchenne's muscular dystrophy.

ADAR RNA Editing

A newly emerging mechanism to correct (or enhance) mRNA transcripts is the use of oligonucleotides that recruit Adenosine Deaminases Acting on RNA or ADAR enzymes. Similar to RNase-H recruiting oligonucleotides (but unlike CRISPR editing that employs bacterial enzymes), these oligonucleotides recruit endogenous enzymes that play a role in post-transcriptional modification of the mRNA. Three human forms of ADAR have been identified, with ADAR 1 and

ADAR 2 being the best characterized forms (Nishikura 2016). Enzymatic deamidation of an adenosine produces an inosine base, so this process is often called "A-to-I editing." Inosine is read by the ribosome as a guanosine, so this effectively allows to make A-to-G point mutations in an endogenous mRNA and restore a disease-causing mutation. ADARs bind double-stranded RNA (dsRNA), so ADAR recruiting oligonucleotides typically consist of an invariable domain that binds the enzyme and a programmable segment that is complementary to the target sequence. Several different approaches are being explored, with variable lengths of the oligonucleotides and with or without the incorporation of chemically modified bases (Aquino-Jarquin 2020). In preclinical research, ADAR recruiting oligonucleotides have been used to restore α-L-iduronidase catalytic activity in the lysosomal storage disease Hurler syndrome (Qu et al. 2019) and to repair the PiZZ mutation that causes α1-antitrypsin deficiency (Merkle et al. 2019). Clinical development of this type of therapies has not yet been started. The most advanced work that is published utilizes 30 nucleotide long molecules, containing a mix of backbone and sugar modifications, that are conjugated to a GalNAc ligand, to edit a transcript in the liver of cynomolgus monkeys (Monian et al. 2022). Because of the relatively easy delivery of oligonucleotides to the liver, it is likely that first trials in human patients will also employ a GalNAc-conjugated ADAR editing oligonucleotide for correction of a gene in hepatocytes.

Transcription Factor Decoys

Transcription factors are nuclear proteins that usually stimulate and occasionally downregulate gene expression by binding to specific DNA sequences, approximately 6–10 base pairs in length, in promoter, or in enhancer regions of the genes that they influence. The corresponding decoys are oligonucleotides that match the attachment site for the transcription factor, known as consensus sequence, thus luring the transcription factor away from its natural target and thereby altering gene expression (Fig. 13.6) (Mann and Dzau 2000; Hecker and Wagner 2017; Mahjoubin-Tehran et al. 2021).

A target for transcription factor decoys that is under investigation to potentially treat inflammatory diseases such as rheumatoid arthritis, atopic dermatitis, or chronic obstructive pulmonary disease (COPD) is Nuclear factor kappa-light-chain-enhancer of activated B cells κB (NF-κB) (Dreja 2000; Mehta et al. 2021). It is a protein complex involved in cellular responses to stimuli such as stress induced by exposure to free radicals, heavy metals, ultraviolet irradiation, and bacterial and viral antigens. NF-κB decoys have been successfully employed to suppress expression of inflammatory cytokines such as IL-6 and IL-8 in bronchial epithelial cells. Several clinical trials investigating the therapeutic effect of such NF-κB decoys are ongoing (Kunugiza et al.

Fig. 13.6 Mechanism of action of transcription factor decoys. Transcription factor decoys match the consensus attachment site of the factor and thereby prevent it from binding to the DNA, inhibiting the factor's modulating activity on gene expression level

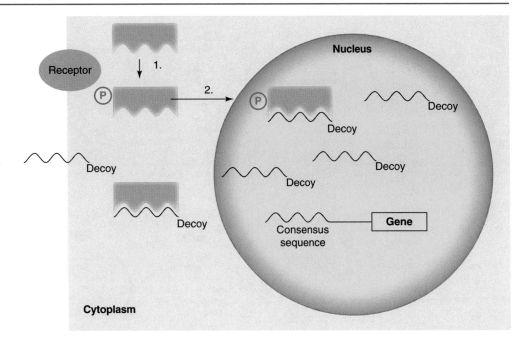

2006; Kato et al. 2021). However, the use of transcription factor decoys has its limitations. The fact that many transcription factors are involved in regulation of a certain gene and that many genes are controlled by a single transcription factor represents important limitations to the decoy approach, especially when decoy action is only desired in the pathological tissue. Clinically, this strategy has been evaluated in patients at risk of postoperative neo-intimal hyperplasia after bypass vein grafting. The oligonucleotide, edifoligide, was delivered to grafts intraoperatively by ex vivo pressure-mediated transfection and was designed to target E2F, a transcription factor that regulates a family of genes involved in smooth muscle cell proliferation. While preclinical studies demonstrated beneficial effects, a series of clinical trials yielded mixed results. Ultimately, no benefit compared to placebo was noted in 5-year graft survival (Lopes et al. 2012). The studies did indicate good safety of this local ex vivo treatment strategy. In another study, the therapeutic effect of decoy ONs was investigated in a mouse model of Marfan syndrome, which is characterized by high expression of matrix metalloproteinases, regulated by the transcription factor activating factor-1 (AP-1) in aortic smooth muscle cells, leading to elastolysis and aortic root aneurysm. AP-1 neutralizing decoy ONs led to reduced MMP activity, elastolysis, and macrophage influx (Arif et al. 2017). The inability to efficiently deliver such decoy ONs into the arterial wall is currently hampering further clinical development.

Current clinical studies focus on topical administration of NF-κB-decoys for atopic dermatitis (AMG0101) and local injection for lumbar disc degeneration (AMG0103). A decoy targeting both STAT6 and NF-κB is aiming to alleviate allergic and autoimmune diseases, such as asthma, rheumatoid arthritis, osteoarthritis, and chronic inflammatory bowel disease (Giridharan and Srinivasan 2018).

Direct Binding to Nonnucleic Acids

For completeness, a number of therapeutic strategies with oligonucleotides are discussed that are not based on base pairing with endogenous nucleic acids. This can be the result from binding to specific immune receptors that recognize nucleic acids at unexpected locations (e.g., extracellular DNA) or specific structural qualities (e.g., CpG motifs). In addition, the ability of nucleic acids to fold into complex three-dimensional structures through internal regions of (partial) complementarity allows them to bind to virtually any molecule with nano- to picomolar affinity (Zhou et al. 2018). This high affinity is supported by data on their extreme specificity. A nucleic acid sequence specifically binding theophylline has a million times higher affinity for theophylline than for caffeine, molecules which differ by only one methyl group (Zimmermann et al. 2000). Clinical success and approvals have, nevertheless, been limited for these classes of nucleic acids.

Aptamers/Riboswitches

Aptamers and riboswitches are single-stranded oligonucleotides of either DNA or RNA, generally about 60 nucleotides long, which fold into well-defined three-dimensional structures (Fig. 13.7). They bind to their target molecule by com-

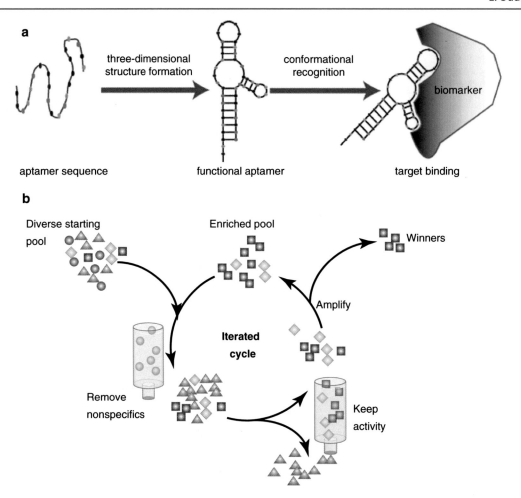

Fig. 13.7 Panel (**a**) Mechanism of action of specific protein binding by aptamers. Panel (**b**) General scheme for SELEX (systematic evolution of ligands by exponential enrichment)-based selection of aptamers. The SELEX process starts by generating a large library of randomized RNA sequences. This library contains up to 10^{15} different nucleic acid molecules that fold into different structures depending on their sequence. The library is incubated with the structure of interest and those RNAs present in the library that bind the protein are separated from those that do not. The obtained RNAs are then amplified by reverse transcriptase-PCR and in vitro transcribed to generate a pool of RNAs that have been enriched for those that bind the target of interest. This selection and amplification process is repeated (usually 8–12 rounds) under increasingly stringent binding conditions to promote Darwinian selection until the RNA ligands with the highest affinity for the target protein are isolated (in the figure: winners). This molecular evolution process can also be performed with DNA, circumventing the need for reverse transcription before PCR and in vitro transcription

plementary shape interactions accompanied by charge and hydrophobic interactions and hydrogen bonds. The target can be small molecules or macromolecules. Whereas ribo-switches can be found in nature in the untranslated regions of some bacterial mRNA to bind specific ligands for gene regulation, aptamers are man-made and isolated by panning large libraries of synthetic oligonucleotides to bind a specific molecule of interest (Fig. 13.7). This selection process has been named SELEX (systematic evolution of ligands by exponential enrichment) (Gao et al. 2016). Usually, the SELEX process is followed by further optimization steps focusing on truncation, chemical modification, multivalent aptamer construction, and mutagenesis.

One aptamer, targeting VEGF, pegaptanib (Macugen®), is marketed for wet age-related macular degeneration (Nimjee et al. 2017) (Table 13.4). The PEGylated aptamer

(for PEGylation see also Chaps. 5 and 22) is injected in the vitreous at a dose of 1.65 mg (0.3 mg of which is aptamer)/eye in 90 μL every 6 weeks. Adverse effects included endophthalmitis and bleeding events in the eye, related to the injection procedure. The compound also appeared to exhibit effects in diabetic retinopathy. To increase stability, several nucleotides are 2′-O-methyl and 2′-O-fluoro modified and the aptamer is conjugated to polyethylene glycol, which stabilizes the aptamer in solution and facilitates clinical delivery.

The pegaptanib product was already approved in 2004, but withdrawn from the European market in 2011. There has not been a lot of clinical activity with aptamers as therapeutics since pegaptanib, but aptamers are still under investigation to be used as targeting moieties to be attached to other types of oligonucleotides.

Table 13.4 Aptamers approved or in advanced clinical studies

Aptamer	Target	Indication
Pegaptanib (Macugen®)	VEGF	Age related macular degeneration
5'-[PEG 40kD]-#CmGmGAA#U#CmAmG#UmGmAmA#UmG#C#U#UmA#UmA#CmA#U#C#CmG-3'T		

= 2' fluoro m = 2' O-methyl

Table 13.5 Immunostimulatory oligonucleotides approved or in advanced clinical studies

CpG 1018 in Heplisav®	Hepatitis B antigen	Hepatitis B vaccine
5'-T = G = A = C = T = G = T = G = A = A = C = G = T = T = C = G = A = G = A = T = G = A-3'(all thioate bonds)		

Stimulating Immune Responses

Specific receptors exist that recognize pathogenic DNA or RNA and subsequently activate a series of genetic programs, for example, the Toll-like receptor (TLR) family, of which TLR3 recognizes double-stranded RNA and TLR9 recognizes CpG DNA fragments, which could indicate the presence of bacterial pathogens. The broad proinflammatory activation that results of this binding, can hamper the application of oligonucleotides as therapeutics. However, it can also be turned around to have applications in antiviral, immune activating, vaccine adjuvant, and antitumor applications. For example, CpG1018 is an approved adjuvant in the marketed Heplisav-B® hepatitis B vaccine (Lee and Lim 2021) (Table 13.5). Several other immune-stimulating oligonucleotides are currently tested in clinical trials for a variety of applications, including cancer.

For both the aptamer-based strategy as well as the immune-stimulating treatment approach, progress over the past decade has been limited, despite the initiation of numerous clinical trials. It is clear that, although the approaches appear deceptively simple, the right balance between target engagement, off-target effects, duration of action, and cost of goods has not been established.

Pharmacokinetics of Oligonucleotide-Based Therapeutics

The drug-like properties of oligonucleotide-based therapeutics are fully dependent on chemical modification of the backbone and sugars. Without any modification or protection, extracellular nucleic acids are extremely short-lived in the body. Because RNA and DNA are normally confined inside the cell (and DNA inside the nucleus) eukaryotes have evolved several mechanisms to clear exogenous nucleic acid-based molecules, as they are normally a sign of invasion by a viral or bacterial pathogen. The main elimination route is nuclease-mediated metabolism and subsequent renal clearance. Extracellular fluids and the bloodstream are full of exo- and endonucleases that cleave bases from the ends, or within the sequence, respectively. Chemical modifications of the backbone make oligonucleotides largely resistant to nucleases but not to renal clearance, due to the size of oligonucleotide therapeutics (5–14 kDa), which lies below the cut-off for glomerular clearance. Also intact oligonucleotides are normally rapidly cleared from the circulation by renal filtration, with plasma elimination half-lives of <10 min. However, oligonucleotides with a modified backbone, especially the phosphorothioate (PS) backbone, bind extensively to plasma proteins. This high plasma protein binding protects oligonucleotides from renal filtration, so that urinary excretion of intact compound is only a minor elimination pathway for highly bound oligonucleotides and plasma elimination half-lives are much longer. This allows oligonucleotides to distribute to peripheral tissues, more or less following the blood flow, with highest accumulation in liver, kidney, bone marrow, skeletal muscle, and skin. Passage over the blood–brain barrier has not been reported. However, direct injection in the cerebrospinal fluid (CSF) results in broad distribution throughout the spinal cord and brain. Plasma protein binding is also enhanced for lipid-modified oligonucleotides, such as cholesterol-oligonucleotide conjugates. In other cases, extended circulation and broad distribution may not be desired, for example if the target cells are the hepatocytes. In this case, oligonucleotides are conjugated to an N-acetylgalactosamine (GalNAc) ligand, to induce rapid uptake during the first passage through the liver (see above).

So, the pharmacokinetics and biodistribution of oligonucleotides are largely driven by the backbone chemistry and conjugation. The modifications on sugars or bases mostly influence the affinity for the target sequence. Within an oligonucleotide class, the pharmacokinetic properties are largely independent of the sequence. Studies with different types of oligonucleotides (in particular phosphorothioate oligonucleotides) have demonstrated that oligonucleotides are rapidly absorbed from injection sites. If the required dose can be dissolved in a small enough volume, a single subcutaneous injection is the preferred—most patient friendly—administration route. Alternatively, intravenous infusion or local administration can be used, e.g., aerosolization for inhaled administration or direct intravitreal injection into the eye. Bioavailability of oligonucleotides can be as high as 90% after intradermal and subcutaneous injections. Oral bioavailability, however, is generally very low due to their large molecular weight, multiple charges at physiological pH, and limited stability in the gastrointestinal tract due to nuclease digestion.

Improving Oligonucleotide Stability

Nuclease resistance of oligonucleotides can be improved by modifying the backbone of the oligonucleotides. Since the early 1960s of the last century, several chemical modifications have been introduced to prevent enzymatic degradation. The mechanism is simple: nucleases cleave the phosphodiester bond between nucleotides, so chemical modifications aim to make that bond look less like the enzyme's natural substrate. The first-generation DNA analogs consisted of the phosphorothioate oligonucleotides, in which one of the nonbridging oxygen atoms in the phosphodiester bond is replaced by sulfur, therefore annotated as PS instead of PO (Fig. 13.1). Apart from making the backbone resistant to nucleases, this substitution also increases plasma protein binding. Because the sulfur atom is larger than the oxygen, the negative charge on each group is distributed over a larger area, thereby making the whole backbone more lipophilic. This enhances plasma protein binding and contributes to the increased circulation half-life seen for phosphorothioate-oligonucleotide (40–60 h). However, every substitution of a PO for a PS lowers the affinity for the target sequence and it was found that at least 10–12 PS bonds per molecule are required to exploit the half-life extending effect. The lower binding affinity of oligonucleotides with a PS backbone can be compensated with additional modifications to the sugars (Crooke et al. 2021).

Another commonly used backbone modification are peptide-nucleic acids (PNA) where the bases of the nucleotides are linked to an amino acid like building block and the backbone is linked with peptide bonds. No products containing such backbones have been approved. Phosphorodiamidate morpholino oligomers (PMO) are another class, where a six-membered morpholine ring replaces the five-membered ribose ring and the backbone is linked by phosphorodiamidate bonds. These bonds are resistant to nucleases and neutral at physiological pH. This lack of charge prohibits loading into RNase H and no compatibility with Ago2/RISC has been described so far either for PMOs. Therefore, supposedly PMOs can only exert an effect via occupancy-only mechanisms such as steric blocking or splice alteration (Crooke et al. 2021). There are four products based on this backbone approved for clinical use, all of them aiming to correct splicing in the dystrophin gene, for patients with Duchenne's muscular dystrophy (DMD) (See Table 13.3). The lack of charge also means there is negligible protein binding and because these molecules need to work in the skeletal muscles to provide any benefits to DMD patients, they must be given in very high doses.

Improving Oligonucleotide Affinity

The majority of single-stranded oligonucleotides on the market and under clinical investigation have a PS backbone to enhance their stability. To compensate for the lower affinity to their target sequence compared to PO oligonucleotides, attention was focused on chemical modification of the sugar groups. In nature, the difference between DNA and RNA is determined by a single hydroxyl group at the 2′-position of the sugar, demonstrating that small changes can have a big impact on how the molecule is processed.

While first-generation oligonucleotides only contained backbone modifications, second-generation PS oligonucleotides all contain modifications at the 2′-position of the ribose unit (Fig. 13.1) that increase their affinity for the target sequence. The mechanism behind this is that the bulkier groups change the conformation of the sugar and the base and thereby "preorganize" the bases, lining them up for easier binding to their target sequences. Of note, the 2′-fluoro substitution is much more hydrophobic and is rarely seen in single-stranded oligonucleotides, but more often in siRNAs. The altered conformation also makes the backbone even more resistant to nucleases and changes the protein binding. For a long time, the 2′-O-methoxyethyl (2′-MOE) oligonucleotides (with a PS backbone) were the preferred choice, as they have greatly improved plasma and tissue half-lives (up to 30 days) and less inflammatory effects.

In the latest generation of oligonucleotides, the conformation of the sugar is "locked" or "constrained" into position by a methylene bridge between the 2′-oxygen of the ribose and the 4′-carbon. Examples of these are locked nucleic acids (LNAs) and constrained ethyl (2′-cEt) groups (Fig. 13.1). These modifications increased the affinity for the target sequence without notably influencing the circulation time or biodistribution compared to 2′-MOE oligonucleotides, allowing a tenfold reduction in dose.

These third-generation oligonucleotides all have superior stability and RNA binding affinities, but fully modified oligonucleotides lack RNase H activation. Therefore, so-called "gapmers" were designed to combine good stability and effective RNase H activation. These chimeric molecules contain flanking regions of 2′-modified oligonucleotides at the 5′ and 3′ termini and a "gap" in the middle containing DNA bases. Commonly used designs contain 3–5 LNA or 2′-cEt bases in the flanks and 8–10 DNA bases in the gap (Crooke et al. 2021). Examples of second-generation oligonucleotides with this "gap" design are mipomersen, nusinersen, and volanesorsen (Table 13.1). They have five 2′-MOE-modified bases on each flank of the molecule (marked with a* in the table) and ten regular DNA bases in the "gap." The 2′-MOE bases increase the affinity of these molecules for their target sequence, but the DNA bases in the middle still allow the RNase H enzyme to bind and cleave the target.

siRNA Chemistry and Modifications

The clinical development of siRNA therapeutics has started later than that of antisense oligonucleotides and has progressed much more rapidly in recent years, but this development has been supported greatly by all the prior work done on chemical modification of oligonucleotides. The obvious difference between siRNAs and single-stranded oligonucleotides is that siRNAs are double stranded. Single-stranded oligonucleotides are amphipathic, with their nucleobases as the hydrophobic surface and their backbone as the hydrophilic surface (with the negative charge spread over a larger area thanks to the PS groups). In siRNA duplexes, the hydrophobic surfaces are facing each other, making the molecule much more hydrophilic and prohibiting any meaningful protein binding to enhance circulation time. The double strands also increase the overall molecular weight of the molecule, making unassisted membrane passage almost impossible. For this reason, siRNAs have been delivered using an LNP or are conjugated to a GalNAc or other types of ligands to aid cellular uptake. More recently, a broad range of lipophilic conjugates has been explored, with the aim to enhance protein binding and biodistribution outside the liver (Tai 2019).

Of the two strands, only the antisense strand is required for the recognition and cleavage of the target mRNA by the RISC. The sense strand or "passenger" strand will be removed inside the cell and therefore conjugations are tolerated without influencing the function of the RISC. The GalNAc conjugate is typically installed on the 3′ end of the passenger strand.

The PS-backbone modification that helped improve the pharmacokinetic properties and stability of antisense oligonucleotides so much is not tolerated in the RISC. Therefore, siRNAs consist entirely of bases with 2′-modified sugars. The original design, called Standard Template Chemistry (STC), consisted of alternating 2′-F bases and 2′-OMe bases. Only two PS bonds were present at the 3′ end of the antisense strand (See Table 13.6). This was the prototype molecule that was tested in clinical trials, but it was found that it was metabolically labile, which required higher doses that were not tolerated well enough. The subsequent design was called "enhanced stabilization chemistry" (ESC) and contained fewer 2′-F modifications and six PS bonds in total. Compared to STC, ESC conjugates achieved up to 200-fold greater exposure, an increased durability of effect thanks to the depot formation as described earlier, and a better safety profile thanks to the lower required dose. The clinically approved conjugates givosiran, lumasiran, and inclisiran all consist of this chemistry and only differ in the sequence they target (Maraganore 2022).

However, for some clinical programs, even the ESC conjugates showed signs of liver toxicity, indicating that these adverse effects could be sequence specific. It was found that these effects are caused by transcriptional repression of sequences that are close, but not identical to the target sequence. To mitigate this, ESC+ chemistry was developed, where the antisense strand is thermally destabilized by the incorporation of a single glycol nucleic acid (GNA) at position 7. This destabilization lowers the affinity for imperfectly matched (off-target) transcripts, without significantly influencing the affinity for the full-length target sequence. This modification allowed the re-deployment of an siRNA sequence targeting hepatitis B virus (ALN-HBV) that was previously found to have an unfavorable safety profile. The identical sequence but with ESC+ chemistry was found to be safe and is reintroduced to clinical development (ALN-HBV2 aka VIR-2218) (Schlegel et al. 2022).

These chemical improvements have turned siRNAs into highly potent drug molecules with an excellent safety profile and conjugation to GalNAc ligands makes them ideally suitable for the therapeutic knockdown of genes in the liver. The

Table 13.6 siRNA chemistry—modifications on the sense and antisense strands of the duplex

Standard Template Chemistry (STC)	
Sense strand	5′-#NmN#NmN#NmN#NmN#NmN#NmN#NmN#NmN#NmN#NmN#N 3′-3xGalNAc
Antisense strand	3′-mN = #N = mN#NmN#NmN#NmN#NmN#NmN#NmN#NmN#NmN#NmN 5′
Enhanced stability chemistry (ESC)	
Sense strand	5′-mN = mN = mNmNmNmN#NmN#N#N#NmNmNmNmNmNmNmNmNmNmN 3′-3xGalNAc
Antisense strand	3′-mN = mN = mNmNmNmNmN#NmN#NmNmNmNmNmN#NmNmNmN = #N = mN 5′
Enhanced stability chemistry+ (ESC+)	
Sense strand	5′-mN = mN = mNmNmNmN#NmN#N#N#NmNmNmNmNmNmNmNmNmNmN 3′-3xGalNAc
Antisense strand	3′-mN = mN = mNmNmNmNmN#NmN#NmNmNmNmNNgN#NmNmNmN = #N = mN 5′
≈ − phosphoramidate, = − thioate, m- 5-methyl, * -′5-O-methyl), # 2′ fluoro, g GNA	

subcutaneous administration route and half-yearly dosing (at least of inclisiran) makes them an attractive option for patients that also ensures great compliance. Ongoing work may improve the potency and durability of siRNAs even further, potentially allowing once-yearly dosing in the future.

Messenger RNA

Introduction

In nature, mRNA forms an intermediate step in the conversion of DNA into protein. This means that by introducing mRNA with a predefined coding sequence into cells, one can produce any protein of interest, making this a powerful tool for therapeutic intervention. This was demonstrated by the enormous success of the mRNA vaccines against coronavirus disease 2019 (COVID-19), in which mRNA was used for the production of the SARS-CoV-2 Spike protein.

In eukaryotic cells, mRNA is produced inside the nucleus by a process called transcription, where an RNA polymerase converts the sequence of a gene into a single-stranded primary transcript, also known as pre-mRNA (Fig. 13.8). The pre-mRNA contains both the introns and exons of the gene but after RNA splicing, the introns are removed leaving only the exons fused together, forming the coding sequence of the transcript. Both the 5′ and 3′ ends of the RNA chain are modified to provide stability (see below).

The mature mRNA is now exported from the nucleus to reach the cytosol, where it can be translated into protein by ribosomes.

For therapeutic use of mRNA, several challenges have to be overcome. Firstly, messenger RNA is generally short-lived, with a half-life varying from minutes to hours in the cytosol. Therefore, the timespan at which therapeutic proteins can be produced is limited. This is compensated by the large number of proteins that can be produced from a single mRNA molecule, with estimates of approx. 100–1000 proteins/mRNA in eukaryotic cells, but mRNA stability remains an issue. Furthermore, even though mRNA is an endogenous molecule in human cells, RNA, including mRNA that enter cells from outside can activate innate immune responses, which can lead to inhibition of protein translation. Thirdly, the sheer size of mRNA makes this a difficult molecule to deliver into the cytosol, requiring delivery systems for efficient delivery of intact mRNA. Finally, long-term storage of mRNA is challenging due to potential degradation and often requires cold-chain storage (Oude Blenke et al. 2022). In this section, we will discuss mRNA as a therapeutic entity. We will provide information about the chemical buildup of mRNA and the chemical modifications that can be introduced, and discuss methods of mRNA manufacturing and purification and ways of delivering those macromolecules into the cytosol of cells. Finally, we provide an overview of application areas where mRNA could be used for therapeutic intervention and highlight a few clinical studies that are currently ongoing.

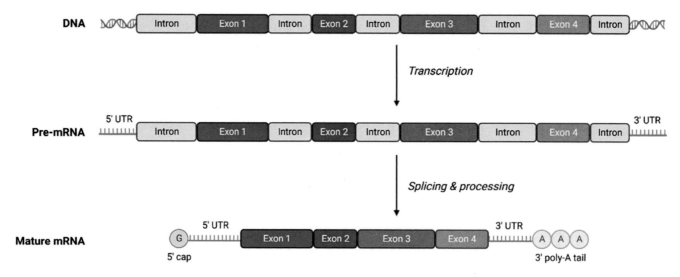

Fig. 13.8 RNA processing in eukaryotes. Binding of transcription factors to the promoter and enhancer regions of a gene leads to initiation of transcription by RNA polymerase II in the nucleus, thereby producing a pre-mRNA strand. The intronic regions are subsequently spliced out by a process called RNA splicing, leaving the exons fused together forming the coding sequence of the transcript. The 5′ end of the mRNA is enzymatically modified with an RNA 7-methylguanosine cap (m7-G) and to the 3′ end a poly-adenine tail is enzymatically added to increase stability of the mRNA in the cytosol. The mature RNA is subsequently transported from the nucleus into the cytosol, where it can be translated into protein by the ribosome complex

mRNA Chemistry

Messenger RNA are delicate macromolecules consisting of a single chain of ribonucleotides. Each ribonucleotide contains a ribose sugar, with one of the four bases (adenine, cytosine, guanine, or uracil) attached to the 1′-position of the ribose pentameric ring. A phosphate group is attached to the 3′-position of the ribose ring and the 5′-position of the next (Fig. 13.9). The presence of a hydroxyl group at the 2′-position of the ribose ring distinguishes RNA from DNA. This group is also responsible for potential instability issues of RNA as it can chemically attack the adjacent phosphodiester bond to cleave the backbone (see below). Because of the flexibility of the mRNA strand and the fact that it is single stranded, intramolecular base pairing can easily occur, which results in mRNA folding. The active folding of secondary structures is implicated in mRNA stability, translatability, localization, and enzyme-mediated RNA editing.

The 5′end and 3′end of the mRNA molecule are modified directly after transcription in the nucleus. At the 5′ end, an RNA 7-methylguanosine cap (m7G cap, also referred to as Cap-0) is enzymatically added. Further modifications include methylation at the 2′-O position of the initiating nucleotide (Cap-1) or first two nucleotides (Cap-2), which prevent activation of innate immune responses via RIG-1 signaling. The presence of a m7G cap is critical for protection from RNases as well as for recognition by the ribosome complex. At the 3′end, around 80–150 adenosine residues are added by a polyadenylate polymerase. This polyadenylation gives protection against degradation by exonucleases but also seem to play a role in translation termination (Fig. 13.10).

Besides end modifications, the four coding bases can be enzymatically modified after transcription to introduce chemical diversity. More than 100 chemical modifications have been detected in eukaryotic mRNAs of which the majority are enzymatically introduced. Apart from the m7G cap incorporated at the 5′end of all mRNAs, N6-methyladenosine (m6A) inosine (I) and pseudouridine (Ψ) are the most common modifications found in eukaryotic mRNA. The importance of these modifications is still under investigation, but seems to play a role in determining the speed and accuracy of protein synthesis by the ribosome (Harcourt et al. 2017). Mature mRNA molecules often contain untranslated regions (UTR) upstream and downstream of the protein coding sequence, which are referred to as the 5′ UTR and 3′UTR. The 5′UTR is critical for ribosome recruitment to the mRNA and secondary structures in this region can play a role in translation efficiency. The 3′UTR often contains regulatory elements to control gene expression at the posttranscriptional level.

mRNA Production and Purification

Therapeutic mRNA is produced by a cell-free in vitro transcription (IVT) reaction, making use of DNA as template (Rosa et al. 2021). The DNA template can consist of a purified plasmid DNA that is linearized with restriction enzymes, or by amplification of the required sequence by PCR, with the latter method being more error-prone as PCR may introduce random mutations within the sequence. IVT relies on the use of an RNA polymerase to catalyze the synthesis of

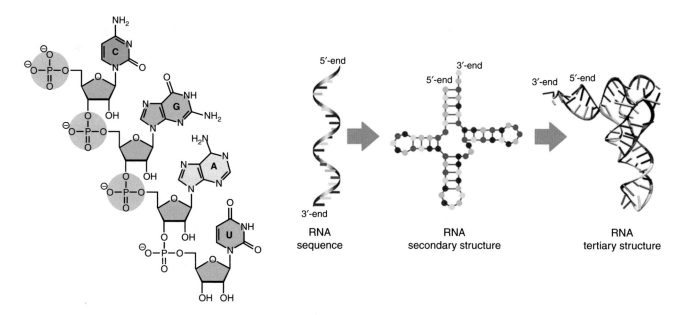

Fig. 13.9 RNA chemistry and structure. RNA is composed of a chain of ribonucleotides with the bases adenine (A), cytosine (C), guanine (G), and uracil (U) attached to the 1′-position of the pentameric ribose ring. The ribonucleotides are connected via phosphodiester bonds (5′-3′). The ribose ring has a hydroxyl group at the 2′-position of the ribose ring. For long strands of RNA, such as mRNA, intrastrand base pairing can lead to secondary and tertiary structure formation. Adapted from Zhao et al. (2021)

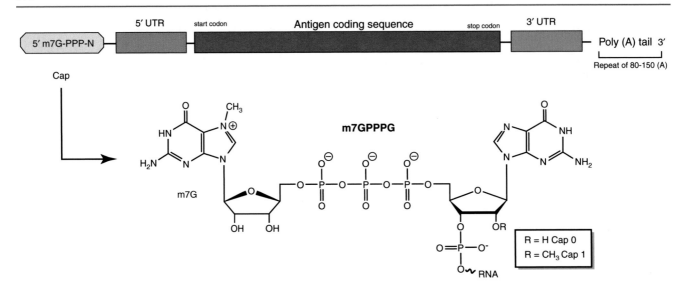

Fig. 13.10 Primary structure of mature mRNA. The mRNA molecule has a m7G cap at the 5'end that is needed for mRNA stability and translation initiation; a 5'-untranslated region (5'UTR) that contains a ribosome binding site; the coding sequence with a start and stop codon; a 3'UTR that often contains sequences for posttranscription regulation of protein expression and a poly(A) tail at the 3'end for stability

the target mRNA from the corresponding DNA template using supplemented nucleotide triphosphates (NTPs) as building blocks. The most frequently used RNA polymerases for this purpose are the bacteriophage T7, SP6, and T3 RNA polymerases, which all require $MgCl_2$ as co-factors. A corresponding promoter sequence (T7, SP6, or T3) should be present in the DNA template to initiate transcription. mRNA capping can be performed during the IVT reaction by substituting part of guanosine triphosphate (GTP) for a cap analog, or the 5'cap can by enzymatically added after the IVT reaction has been completed. This method is to be preferred as the capping efficiency is much higher. The addition of a polyA tail at the 3'end can be obtained in two ways: (1) By encoding the polyA on the DNA template, so that the RNA polymerase will add this during the IVT reaction, or (2) by using a poly(A) polymerase from *E. coli* or yeast to enzymatically add such a poly(A) tail after the IVT reaction.

The IVT reaction is quick and only takes a few hours to produce mRNA. A big advantage of mRNA is that the production process can be standardized as it is largely independent of the mRNA sequence.

In general, milligrams per mL of mRNA can be produced in batch mode. This can be scaled up for industrial scale manufacturing; however, the generation of mRNA by IVT at large scale and under current good manufacturing practice (cGMP) conditions is challenging and costly (Ouranidis et al. 2021). The specialized components of the IVT reaction must be acquired from certified suppliers that guarantee that all the material is animal component-free and GMP grade. Because of the use of enzymes, a continuous mode of production as compared to batch production would have a beneficial effect on the overall manufacturing costs.

Once the mRNA has been produced, it requires purification from the reaction mixture in several successive steps (Fig. 13.11). Components that need to be removed are the template DNA, enzymes for RNA synthesis, capping and polyadenylation, NTPs and unreacted 5'cap analogs, waste products, and aberrant mRNA species such as truncated RNA fragments and double-stranded RNA. The removal of product-related impurities will be needed to prevent low translation efficiency of the mRNA as well as undesired innate immune activation (mostly induced by dsRNA impurities). For purification of mRNA at large scale, a combination of filtration and chromatography techniques is often used. The first step is removal of small-size impurities and concentration of the mRNA product by tangential flow filtration, followed by a chromatography step to isolate the mRNA from large process-related impurities and aberrant RNA. A frequently used method is ion-pair reverse-phase chromatography (IP-RP-HPLC), which makes use of quaternary ammonium compounds (such as triethylammonium acetate) in the mobile phase that bind the negatively charged phosphate groups of RNA and DNA to make them lipophilic. This, in turn, enables interaction with the lipophilic stationary phase of the reverse-phase chromatography column. By gradually changing the polarity of the mobile phase, bound mRNA can be eluted from the column, separating RNA and DNA based on size. Alternative chromatography methods used for mRNA purification include ion exchange chromatography (IEX), affinity chromatography making use of immobilized poly(dT) to capture the polyadenylated mRNA, and multimodal core bead chromatography (Cf. Chaps. 3 and 4). In the latter chromatography method, the mRNA flows through the resin, whereas smaller impurities can enter the functional-

Production

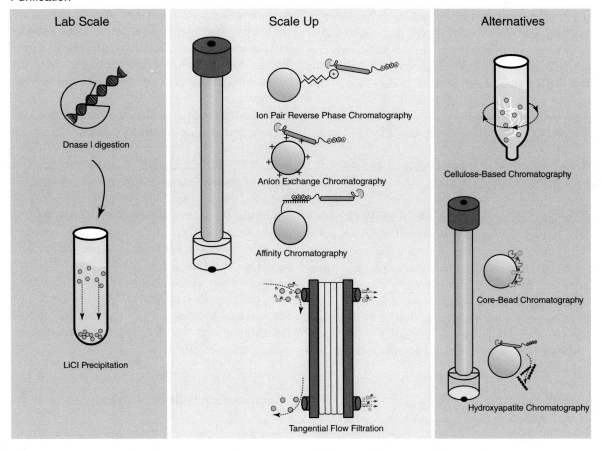

Fig. 13.11 Schematic representation of the production and purification steps of mRNA vaccine manufacturing. mRNA production can be performed in a one-step enzymatic reaction, where a capping analog is used, or in a two-step reaction, where the capping is performed using a capping enzyme from Vaccinia virus. mRNA purification process at lab scale consists of DNase I digestion followed by LiCl precipitation. Purification at a larger scale is obtained using well-established chromatographic strategies coupled with tangential flow filtration. Alternatively, new types of chromatography can be used to complement the standard purification. Figure adapted from Rosa et al. (2021)

ized core of the resin designed to capture IVT reaction components. A new cellulose-based chromatographic technique for purification of mRNA and efficient removal of dsRNA can be used as well. After the first chromatography step, another filtration step will be needed to exchange the buffer into an appropriate storage buffer, followed by sterile filtration and subsequently filling out into sterile storage containers for further use.

mRNA Quality Control and Critical Quality Attributes

The integrity and quality of the mRNA as drug substance are crucial for the effectiveness of the mRNA-based drug product. Even minor degradation events anywhere along the mRNA strand can lead to production of truncated proteins or even complete loss of protein production. Therefore, quality control tests that ensure purity, integrity, and identity of the mRNA are required (Jurga and Barciszewski 2022; Oude Blenke et al. 2022). Product-related impurities are commonly mRNA truncations, cap-less mRNA, and dsRNA, whereas process-related impurities include the presence of endotoxins, residual pDNA, RNA polymerases, and residual bacterial proteins potentially present in the pDNA preparation. Storage conditions and proper formulation for mRNA as well as mRNA formulated as drug product (e.g., LNP-mRNA) are critical to guarantee a good shelf-life (Crommelin et al. 2021; Oude Blenke et al. 2022).

Modifications and Variants of IVT-Produced mRNA to Enhance Protein Yield

The inherent immunogenicity of mRNA hinders its therapeutic use. It might be less of a problem for mRNA-based vaccines, where low levels of antigen production and optimal activation of innate immune responses are required; it is certainly undesired for other therapeutic applications, such as enzyme replacement or antibody therapy, where in general high levels of protein expression and repeated dosing are essential. This has led to various strategies to minimize innate immune responses, enhance mRNA stability, and maximize translation.

mRNA Modifications

Each of the separate parts of an mRNA molecule (5′cap, UTRs, coding sequence, and polyA tail) can be optimized to maximize protein expression. Several synthetic 5′cap analogues have been tested to increase the capping yield and stability of IVT-produced mRNA in comparison with the naturally occurring m^7G cap (Fig. 13.12) (Muttach et al. 2017; Moradian et al. 2022). These synthetic analogs can be enzymatically added after transcription using a Vaccinia

virus capping enzyme or can be co-transcriptionally incorporated by adding the cap analogs directly to the IVT mix. The downside of the latter method is that not all mRNA will be capped as the cap analogs compete with GTP as initiator nucleotide. Another problem is incorporation of the cap in the wrong orientation, which can be prevented by using anti-reverse cap analogs (ARCA) that are methylated at the 3′OH of the N7-methylguanosine ribose and therefore will only incorporate in the right orientation.

The 5′UTR often consist of a sequence derived from the 5′UTR transcript of α-globin or β-globin as this generally gives good translational yields. Optimization of the 5′UTR sequence using bioinformatic tools or machine learning could further optimize translation yields (Castillo-Hair and Seelig 2022).

The incorporation of chemically modified nucleosides, particularly the uridine moieties, can markedly increase protein expression from mRNA. These can be chemical modifications found in nature, of which >100 have been identified or de novo synthesized. Of particular interest are the N1-methylpseudouridines that greatly reduce detection of exogenous mRNA by Toll-like receptors of the innate immune system. Alternatively, mRNA sequences can be codon optimized using algorithms to reduce secondary structure formation in the translated region and reduce the overall uridine content, which can drastically increase protein expression levels even without the use of nucleoside analogs.

Several different strategies have been explored to increase the longevity of mRNA inside the cell with the aim to obtain protein expression over a prolonged period of time (Rohner et al. 2022). Circularization of mRNA (circRNA) has been explored to shield the mRNA from exonuclease degradation. Internal ribosome entry sites (IRES) can be incorporated into this circRNA, so that no expensive 5′capping reagents or polyadenylation is required. Besides prolonged expression, circRNA can also avoid innate immune activation via RIG-I and TLR (Wesselhoeft et al. 2019).

Another approach makes use of the self-replicating capacity of alphaviruses (Fig. 13.13). By co-expressing alphavirus nonstructural proteins nsP1–4 together with the protein of interest, the RNA can be replicated inside the cytosol of transfected cells (Bloom et al. 2021).

Therapeutic Application Areas

The advancements in mRNA technology have spurred the exploration of mRNA for a broad range of therapeutic applications. The five major therapeutic areas in which mRNA therapeutics are currently being explored are cancer immunotherapy, vaccination, protein replacement therapy, in situ production of biopharmaceuticals, and genome editing/genetic reprogramming.

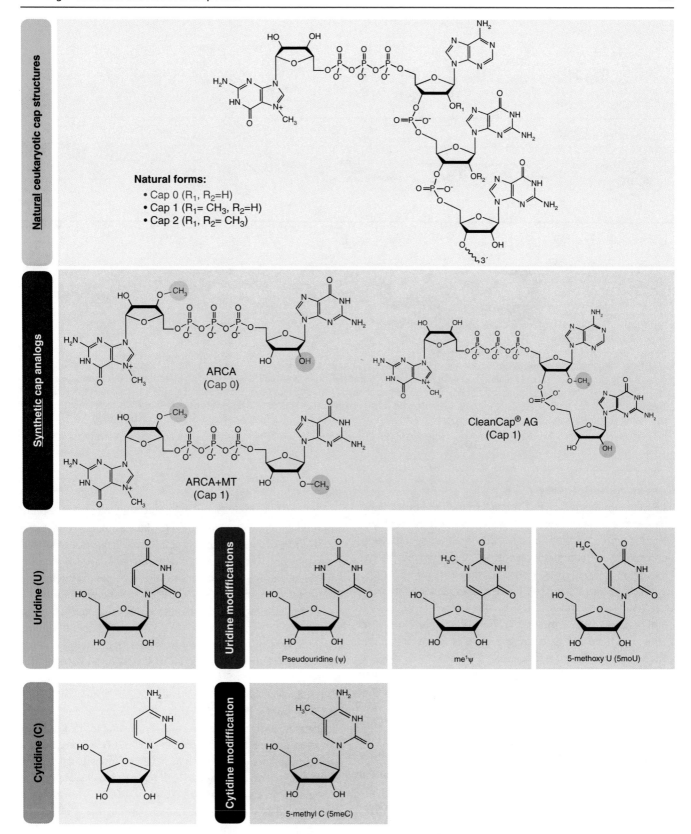

Fig. 13.12 Upper panel: Chemical structures of naturally occurring 5′cap and some frequently used synthetic cap analogs for incorporation into IVT mRNA. Lower panel: Uridine and cytidine nucleoside analogs often used for incorporation in IVT mRNA (ARCA: anti-reverse cap analog, MT: Methyltransferase). Adapted from Moradian et al. (2022)

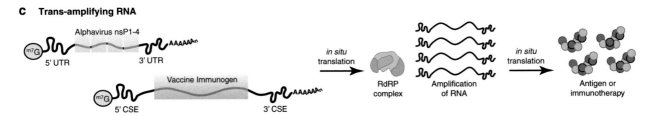

Fig. 13.13 Amplification of mRNA after cellular delivery. Panel (**a**) conventional mRNA is nonreplicating and therefore the yield of protein production by translation is directly related to stability of the mRNA molecule inside the cell. Panel (**b**) self-amplifying RNA contain, besides the coding sequence of the protein of interest, viral conserved sequence elements (CSE) and the coding sequences of the Alphavirus nonstructural proteins nsP1–4. After translation, these viral proteins form an RdRP complex that bind the CSE sequences to induce RNA replication. Panel (**c**) trans-amplifying mRNAs use two different transcripts to achieve a similar amplification effect. Adapted from Bloom et al. (2021)

Cancer Immunotherapy

The main application of IVT mRNA in the field of immunotherapy is as cancer vaccine, in which the mRNA provides a flexible source of tumor-associated or tumor-specific antigens upon transfection into dendritic cells ex vivo or in vivo (Barbier et al. 2022). Many clinical trials are currently ongoing where these IVT mRNA-based cancer vaccines are being tested for safety and efficacy as stand-alone therapy or in combination with immune checkpoint inhibitors (see Table 13.7). mRNA is also very suitable for personalized cancer immunotherapy, where the IVT mRNA is used to encode multiple patient-specific neoantigens identified by sequencing tumor biopsies of individual patients (Sahin et al. 2017). Besides its use as source of tumor antigens, IVT mRNA has also been studied as a means to locally produce cancer-specific and immune checkpoint blocking mAb (Van Hoecke et al. 2021). The IVT mRNA molecule and its mode of delivery can be tweaked to ensure prolonged protein expression at the tumor microenvironment, implying that mAb levels can be sustained locally and over a prolonged period of time. This holds the promise of increased efficacy and decreased toxicity. For example, it could be used for tumor-restricted expression of checkpoint inhibitors to avoid

systemic immune-related adverse events (irAE) associated with such mAb therapies.

IVT mRNA is also being explored as a means to modulate immune-suppressive cell types in the tumor microenvironment (TME). Several immune cells, such as tumor-associated macrophages, fibroblasts, and dendritic cells are key regulators in sustaining an immunosuppressive environment that stimulate tumor growth. IVT mRNA can be used to reprogram these cells to an immune-activating phenotype or to induce local expression of cytokines to activate tumor-resident DCs. For example, TriMix is a cocktail of three mRNA molecules, encoding CD40L, CD70 and a constitutively active form of TLR4, which is used to modulate the TME and activate tumor-resident DCs (Van Lint et al. 2012).

Furthermore, IVT mRNA is being investigated for the transient modulation of immune cells. For example, IVT mRNA encoding antigen-specific T cell receptors (TCR) or chimeric antigen receptors (CAR) have been used to transiently express such receptors on autologous T cells by ex vivo electroporation of the IVT mRNA (Van Hoecke et al. 2021). The transient nature of mRNA reduces the risk of unwanted side effects by the uncontrolled expansion of adoptively transferred immune cells. Clinical trials investigating this approach are ongoing (NCT01897415) (Klampatsa et al. 2021).

Table 13.7 Overview of a selection of clinical trials and approved products with mRNA as active pharmaceutical. The list is not complete and limited to direct in vivo applications of mRNA

Disease	API/encoding sequence	Way of delivery	Administration route	NCT number	Clinical phase
Cancer immunotherapy/vaccines					
Non-small cell lung cancer (NSCLC)	5 mRNA encoding tumor-associated and tumor-specific antigens of NSCLC tumor cells (CV9201)	Protamine-based delivery (RNActive®)	Intradermal	NCT00923312	Phase 1/2
	6 mRNA encoding tumor-associated and tumor-specific antigens of NSCLC tumor cells (CV9202)	Protamine-based delivery (RNActive®)	Intradermal	NCT01915524	Phase 1
Metastatic NSCLC	mRNAs encoding NSCLC-associated antigens (NY-ESO-1, MAGE-C1, MAGE-C2, 5 T4 and MUC-1)	Protamine-based delivery (RNActive®)	Intradermal	NCT03164772	Phase 1/2
Malignant melanoma	mRNA encoding melanoma-associated antigens (Melan-A, mage-A1, mage-A3, Survivin, GP100, and Tyrosinase)	Naked mRNA	Subcutaneous	NCT00204516	Phase 1/2
	mRNA encoding melanoma-associated antigens (Melan-A, mage-A1, mage-A3, Survivin, GP100, and Tyrosinase)	Protamine-based delivery (RNActive®)	Intradermal	NCT00204607	Phase 1/2
	Personalized cancer vaccine targeting 20 tumor-associated antigens (mRNA-4157)	LNP	Intradermal	NCT03897881	Phase 2
	Poly-neoantigen personalized cancer vaccine targeting tumor-associated antigens (IVAC MUTANOME)	Naked mRNA	Intranodal	NCT02035956	Phase 1
	Malignant melanoma-associated antigens (RBL001; RBL002)	Naked mRNA	Intranodal	NCT01684241/	Phase 1
	Malignant melanoma-associated antigens (RBL001.1; RBL002.2; RBL003.1; RBL004.1)	LNP	Intravenous	NCT02410733	Phase 1
Prostate cancer	mRNAs encoding PSA, PSCA, PSMA, STEAP1, PAP, and mucin 1 antigens (CV9104)	Protamine-based delivery (RNActive®)	Intradermal	NCT01817738 NCT01817738	Phase 1/2 Phase 2
	mRNAs encoding PSA, PSCA, PSMA, and STEAP1 antigens (CV9103)	Protamine-based delivery (RNActive®)	Intradermal	NCT00831467 NCT00906243	Phase 1/2
Ovarian cancer	mRNA encoding three ovarian cancer tumor-associated antigens mRNAs	LNP	Intravenous	NCT04163094	Phase 1
Early breast cancer	mRNA encoding TriMix	Naked mRNA	Intratumoral	NCT03788083	Phase 1
Triple-negative breast cancer	Personalized cancer vaccine targeting tumor-associated antigens	LNP	Intravenous	NCT02316457	Phase 1
Solid tumors	Personalized cancer vaccine targeting 20 tumor-associated antigens (mRNA-4157)	LNP	Intramuscular	NCT03313778	Phase 1
Melanoma, Colon cancer, gastrointestinal cancer, genitourinary cancer, hepatocellular cancer	mRNA-based personalized cancer vaccine (NCI-4650)	LNP	Intramuscular	NCT03480152	Phase 1/2
Solid tumors	Personalized cancer vaccine targeting tumor-associated antigens (RO7198457)	LNP	Intravenous	NCT03289962	Phase 1
Relapsed/refractory solid tumor malignancies or lymphoma	mRNA encoding OX40L, IL-23, and IL-36γ (mRNA-2752)	LNP	Intratumoral	NCT03739931	Phase 1
Squamous cell carcinoma, head and neck Neoplasm, cervical neoplasm, penile neoplasms malignant	Human papillomavirus (HPV16) mRNA vaccine/HPV16-derived E6, E7 tumor antigens	Naked mRNA	Intradermal	NCT03418480	Phase 1/2

(continued)

Table 13.7 (continued)

Disease	API/encoding sequence	Way of delivery	Administration route	NCT number	Clinical phase
Infectious disease vaccines					
SARS-CoV-2	mRNA encoding the spike protein of severe acute respiratory syndrome (SARS)-coronavirus 2	LNP	Intramuscular	NCT05477186	Phase 1/2 approved products: Comirnaty (BioNTech/Pfizer) and SpikeVax (Moderna)
Rabies	mRNA encoding rabies virus glycoprotein (RABV-G) (CV7201)	Protamine-based delivery (RNActive®)	Intradermal or intramuscular	NCT02241135	Phase 1
	mRNA encoding rabies virus glycoprotein (RABV-G) (CV7202)	LNP	Intramuscular	NCT03713086	Phase 1
Zika virus	mRNA encoding structural proteins of Zika virus that form VLPs (mRNA-1893)	LNP	Intramuscular	NCT04064905	Phase 1/2
Cytomegalovirus	mRNA-1647 and mRNA-1443/Pentamer complex and full-length membrane-bound glycoprotein B (gB) and pp65 T cell antigen of CMV	LNP	Intradermal	NCT03382405	Phase 1
Tuberculosis	mRNA encoding immunogenic fusion protein (M72) derived from mycobacterium tuberculosis (GSK 692.342)	LNP	Intramuscular	NCT01669096	Phase 2
HIV-1	mRNA encoding TriMix and HTI/APC activation molecules (CD40L + CD70 + caTLR4) and HIV immunogen sequences (gag, pol, Vif and Nef) (iHIVARNA)	Naked mRNA	Intranodal	NCT02413645 NCT02888756	Phase 1/2
Influenza	mRNA encoding H10N8 antigen (AL-506440)	LNP	Intradermal or intramuscular	NCT03076385	Phase 1
	mRNA encoding H7N9 antigen (AL-339851)	LNP	Intradermal or intramuscular	NCT03345043	Phase 1
Protein replacement therapy					
Heart failure	Vascular endothelial growth factor-A (VEGF-A) (AZD8601)	Naked mRNA	Epicardial	NCT03370887	Phase 2
Ulcers associated with type II diabetes	Vascular endothelial growth factor-A (VEGF-A) (AZD8601)	Naked mRNA	Intradermal	NCT02935712	Phase 1
Propionic Acidemia	mRNA encoding the alpha and beta subunits of the mitochondrial enzyme propionyl-CoA Carboxylase (mRNA-3927)	LNP	Intravenous	NCT04159103	Phase 1/2
Isolated Methylmalonic Acidemia	mRNA encoding methylmalonyl-coenzyme A mutase (MUT) (mRNA-3704)	LNP	Intravenous	NCT03810690	Phase 1/2
Ornithine Transcarbamylase deficiency	mRNA encoding ornithine transcarbamylase (MRT5201)	LNP	Intravenous	NCT03767270	Phase 1/2
Cystic fibrosis	mRNA encoding human cystic fibrosis transmembrane regulator protein (CFTR) (MRT5005)	LNP	Intrapulmonary via nebulization	NCT03375047	Phase 1/2
Gene editing					
Hereditary transthyretin (ATTR) amyloidosis with polyneuropathy	mRNA encoding for Cas9 and sgRNA targeting the gene encoding the TTR protein (NTLA-2001)	LNP	Intravenous	NCT0460105	Phase 1
Hereditary angioedema (HAE)	mRNA encoding for Cas9 and sgRNA targeting the *KLKB1* gene encoding Kallikrein B1 (NTLA-2002)	LNP	Intravenous	NCT05120830	Phase 1

Protein Replacement Therapy

Protein replacement therapy relies on the supplementation of proteins that are, often due to an underlying genetic defect, absent, poorly expressed, or nonfunctional. Current strategies rely on systemic delivery of the recombinant version of the missing protein, which often require high doses due to poor tissue or organ targeting and cellular uptake, or frequent administrations in case the protein of interest has a short half-life. Furthermore, substituting membrane proteins is problematic as recombinant versions of such proteins are difficult to formulate and deliver. IVT mRNA could provide a solution to these shortcomings. The use of mRNA provide control over the intracellular targeting of the recombinant protein being produced: by adding a specific signal peptide sequence to the mRNA, the encoded protein can be targeted to remain within the cytosol, shuttled into intracellular organelles, be transported to the cell membrane, or secreted into the extracellular space. Protein-replacement therapies based on mRNA are being developed to treat among others cystic fibrosis, hemophilia A and B, Fabry disease, and phenylketonuria (Gómez-Aguado et al. 2020).

In Situ Production of Biopharmaceuticals

Besides replacing missing or malfunctioning proteins that causes disease, mRNA delivery is also being investigated for the in situ production of biopharmaceuticals, such as cytokines, growth factors, and monoclonal antibodies (Van Hoecke et al. 2021). This would circumvent the often complex production and purification processes and aberrant post-translational modifications associated with biopharmaceuticals and, depending on the mRNA platform being used and the half-life of the biopharmaceutical, might even reduce dosing frequency.

Several groups have investigated the systemic production of mAbs as antiserum to prevent infectious diseases. In a study in humanized mice, it was demonstrated that a single injection of 30 ug of nucleoside-modified mRNA encoding the H and L chain of a broadly neutralizing HIV-1 antibody and delivered to the liver using lipid nanoparticles resulted in antibody serum titers of >150 ug/mL after 24 h with stable levels for approximately 5 days. When compared to bolus injection of the protein mAb at 600 ug/mL, antibody serum titers were twofold higher with the mRNA delivery 24 h after injection. More importantly, humanized mice were fully protected against a challenge with HIV-1, which shows the strength of mRNA (Pardi et al. 2017). Similar passive immunization approaches are being investigated for prevention of infection with Influenza B, Chikungunya virus, Rabies, RSV, and Zika virus (Deal et al. 2021).

Another application is the in situ production of growth factors. For example, mRNA delivered to the liver with LNPs encoding hepatocyte growth factor (HGF) or epidermal-growth factor (EGF) has been investigated to stimulate liver tissue regeneration in nonalcoholic fatty liver disease, with some promising data in mice.

Nonetheless, to successfully replace protein therapeutics by mRNA, several shortcomings still need to be overcome. Firstly, immunogenicity of exogenous mRNA remains an issue. Even though the incorporation of naturally occurring modified nucleosides can partially overcome innate immune activation, thereby enhancing protein expression, they still retain low levels of activation. This could be because the incorporated nucleoside modifications differ from those found in nature in terms of frequency and distribution of the mRNA chain, or due to the presence of impurities in the IVT-produced mRNA that can still trigger innate immune responses. These residual innate immune responses, even though much less compared to unmodified mRNA, can potentially hamper clinical development, especially if high doses or multiple administrations are required.

Secondly, for certain applications the level and duration of mRNA expression will still be a limiting factor. Take for example enzyme replacement therapy, where high doses of >10 mg/kg body weight are often needed for a therapeutic effect. Such high levels of protein expression are currently not possible with administration of mRNA. Further improvements on the stability of mRNA, longevity of mRNA expression, and suppression of innate immune responses are needed before such expression levels can be safely achieved. Alternatively, local delivery of mRNA could lead to local high concentrations of therapeutic proteins.

Thirdly, for effective intracellular delivery of mRNA, delivery systems are needed, which will be further discussed in the section below (cf. section "Improving Cellular Uptake of Both ONs and mRNA").

Genome Editing/Genetic Reprogramming

mRNA can also be used to obtain transient expression of genome editors such as TALEN, Cas9, Base Editors, and Prime Editors (cf. Chap. 14 for further details). The advantage of using mRNA over viral delivery is its transient nature, avoiding the risk of long-term episomal or integrated expression of the nuclease that can cause genotoxic effects. Several studies have demonstrated that mRNA encoding Cas9 can be efficiently co-delivered with a sgRNA using LNPs to form functional Cas9 RNP complexes within target cells for gene knock out. Similarly, co-delivery of Cas9 mRNA, sgRNA, and a single-stranded HDR (homology-directed repair) template can induce genomic correction of singly point mutations. Intellia Therapeutics is currently evaluating the delivery of CRISPR-Cas9 mRNA to the liver using LNPs to treat transthyretin (ATTR) amyloidosis and hereditary angioedema (HAE) in phase 1/2 clinical trials (Gillmore et al. 2021; Huang et al. 2022).

Vaccines

With the approval of two mRNA-based COVID-19 vaccines, mRNA has proven itself as a flexible source of antigen for prophylactic vaccination. mRNA vaccines represent an alternative to conventional vaccine approaches because of their high potency, capacity for rapid development, and potential for low-cost manufacture and safe administration. Several clinical trials are ongoing to test the efficacy of mRNA vaccines against a variety of infectious diseases (Table 13.7). A detailed description of mRNA vaccines can be found in Chap. 15.

Improving Cellular Uptake of Both ONs and mRNA

Besides metabolic elimination by nucleases, the poor cellular uptake of oligonucleotides poses a problem for therapeutic application of (oligo)nucleotides. Compared to conventional drugs, oligonucleotides are relatively large and polyanionic, making passage over cellular membranes virtually impossible (cf. Chap. 5). This is particularly true for mRNAs, siRNAs, and for single-stranded oligonucleotides having morpholino or peptide nucleic acid backbones as these structures poorly bind to cell membranes. However, evidence exists that PS-modified oligonucleotides bound to plasma proteins are taken up by endocytosis. At least the following two distinct uptake mechanisms have been identified: (i) a nonproductive uptake pathway that leads to lysosomal degradation of the plasma protein-bound oligonucleotides and which is the dominant uptake route and (ii) a productive pathway that appears to be clathrin and caveolin independent and which is cell type dependent and accounts for only a minor fraction of the internalized oligonucleotides.

For PS-modified oligonucleotides, the stabilin receptors on liver sinusoidal endothelial cells seem to play an important role in cellular uptake (Miller et al. 2016). Besides direct internalization via these receptors, the involvement of binding to plasma proteins, low-density lipoproteins, or extracellular matrix proteins as an intermediate step in cellular internalization of these PS oligonucleotides cannot be excluded.

Double-stranded RNA also appears to be transported into (nematode) cells by a transmembrane channel: Systemic RNA Interference Deficiency-1, SID-1. It has been suggested that the mammalian SID-1 ortholog, SIDT2, may transport internalized extracellular dsRNA from endocytic compartments into the cytoplasm (Nguyen et al. 2017).

Cellular uptake of oligonucleotides can be improved in vitro and in vivo by physical methods, chemically modifying the oligonucleotides, or by making use of specialized delivery systems. Electroporation of tissue after local injection of oligonucleotides or mRNA results in improved cellular uptake (Cf. Chap. 14).

Due to high-voltage pulses, transient perforations in the cell membrane occur that allow passage of oligonucleotides and even the much larger mRNA into the cytosol. This technique can of course only be applied for delivery of nucleic acids in vitro and in vivo to tissues readily available for electroporation (e.g., skin, skeletal muscle, or superficial tumor tissue).

Grafting oligonucleotides with cationic groups in order to reduce the ionic repulsion between oligonucleotide and the negatively charged cell membrane represents an alternative strategy to enhance cellular uptake. Synthetic guanidinium-containing oligonucleotides showed improved duplex and triplex stability in addition to enhanced cellular uptake. The uptake pattern suggests that these cationic oligonucleotides are internalized by endocytosis, although cytosolic localization could also be observed.

Alternatively, cell-penetrating peptides (CPPs) can be conjugated to oligonucleotides with the purpose to enhance membrane translocation. CPPs are small, cationic peptides derived from protein transduction domains present in a variety of proteins which have strong membrane translocating properties (Lehto et al. 2016; Geng et al. 2022).

Lipid modification of siRNA has also been proven to be beneficial for cellular uptake and subsequent gene silencing (Tai 2019). siRNA against apolipoprotein B-100 mRNA, which was modified by attaching a cholesterol group to the 3′-terminus of the sense strand, showed increased silencing of the gene encoding for apolipoprotein B-100 compared to the unmodified siRNA after intravenous injection into mice. It is suggested that the siRNA is transported by lipoproteins to arrive at the liver.

The most successful strategy in clinically tested oligonucleotides is coupling of triantennary N-acetylgalactosamine (GalNAc) to the therapeutic oligonucleotide. This strategy has been applied to single-stranded ASOs and to siRNAs, and drives efficient uptake by hepatocytes through the asialoglycoprotein receptor

(ASGPR). This strategy has proven to be highly efficient for targeting to the liver, due to the high perfusion and first-pass effect of the liver, the extremely high expression of the ASGPR on hepatocytes, and the relatively low abundance of this receptor on other tissue types.

Similar strategies using other types of ligands can be applied to target different cell types. For example, conjugation to a GLP-1 mimetic peptide enabled uptake in glucagon-like peptide-1 receptor (GLP1R) expressing pancreatic β-cells both in vitro and in vivo (Ämmälä et al. 2018). Conjugation to monoclonal antibodies or antibody fragments further enhances the number of cell types that can be targeted and several of such constructs are being evaluated in clinical trials (Dovgan et al. 2019).

Another strategy involves the use of sophisticated nanoparticle delivery systems to enhance cellular uptake and to target oligonucleotides to specific tissues or cells. Most of the delivery systems for oligonucleotides are based on complexation of oligonucleotides with cationic molecules, of which cationic (ionizable) lipids are nowadays the most common. The field of lipid nanoparticles (LNPs) has taken a huge leap in recent years, first enabling the first-ever siRNA drug to be approved (patisiran) and then forming the basis of the delivery technology for the COVID-19 mRNA vaccines (cf. Chap. 14, Figs. 14.27 and 14.15). Additional improvements to the safety and efficiency of these LNPs are still being made (Hou et al. 2021).

The complexation and encapsulation of oligonucleotide and mRNA drugs in LNPs has a dual function: it protects the nucleic acids from nuclease degradation and, in case of small oligonucleotides renal clearance, it also enhances cellular internalization. By temporarily shielding the LNP complexes with polyethylene glycol, gradual in situ opsonization with apolipoprotein E is promoted (after intravenous administration), leading to hepatocyte uptake via the LDL receptor. This approach is taken for the patisiran product (Akinc et al. 2019) as discussed above.

The approval of the patisiran proved for the first time that these challenging nanomedicine formulations can be developed for clinical applications and that siRNAs can specifically and safely inhibit protein production in the liver. It is clear, however, that still only a fraction of the injected dose arrives at the target site and even more importantly within the target cell cytoplasm.

Recent studies attempted to quantify the cytoplasmic localization of siRNA delivered by lipid nanoparticles. The particles enter cell by clathrin-mediated endocytosis as well as macropinocytosis. Escape of siRNAs from the endosomes into the cytoplasm occurs at a low efficiency of 1–2%; the majority of the cargo is degraded in the lysosomes and as much as 70% of the internalized cargo may be excreted again in a process call endocytic recycling (Gilleron et al. 2013; Sahay et al. 2013). Even so, the progress that has been made

in improving cationic lipid structures has now enabled the clinical feasibility of siRNA therapeutics.

LNPs have also proven to be very suitable for the delivery of mRNA. By adjusting the lipid composition, they can be optimized for systemic (mainly targeting liver and spleen) or local (intramuscular, subcutaneous, intradermal, and intra-pulmonary) delivery of mRNA. Although LNPs have proven to be efficient, clinical data also revealed potential safety issues. Toxicity of mRNA-LNPs can be multifaceted, including direct cellular cytotoxicity at high doses, as well as immune-related toxicity. LNPs have been reported to activate the complement system, which can lead to hypersensitivity reactions known as complement activation-related pseudoallergy (CARPA), even though the occurrence is very rare (Szebeni et al. 2022). The presence of PEG lipids can result in anti-PEG antibody formation, which has a detrimental effect on the circulation times of systemically administered LNPs upon multiple dosing.

Another delivery system suitable for local (intramuscular, intradermal, or intranodal) delivery of mRNA is based on the nucleic acid-binding peptide protamine. Protamine forms a stable complex with mRNA thereby protecting it from nuclease degradation. The protamine-mRNA complex also activates immune cells via TLR7/8 triggering, which is an added benefit for vaccination. Furthermore, protamine can improve storage stability of the mRNA at elevated temperatures. A further improvement to the protamine complexes was developed by CureVac. Their RNActive® technology is based on IVT mRNA, optimized for stability by GC-enrichment of the open reading frame and complexed with protamine (Rauch et al. 2017). This two-component self-adjuvanted vaccine is currently being tested in several clinical trials as prophylactic vaccine against infectious diseases or as cancer immunotherapy (Karam and Daoud 2022).

Chemical modifications of siRNA have also continued to progress and have made the molecules so stable that degradation in the bloodstream is no longer an issue (see section "siRNA Chemistry and Modifications" above). And with GalNAc conjugation to aid uptake in hepatocytes, at least for liver targets oligonucleotide conjugates are preferred over LNP formulations. Patisiran is administered with a 3-week interval, whereas the latest generation of siRNA conjugates can be administered with an interval of three to 6 months. It was recently elucidated that these GalNAc–siRNA conjugates owe their extended pharmacodynamic durability mostly to their enhanced stability in highly acidic subcellular compartments. The GalNAc/ASGPR interaction ensures that the majority of the administered dose is taken up into hepatocytes, forming an intracellular depot inside the lysosomes. The enhanced metabolic stability ensures that the molecules survive even for extended periods of time. And although release or leakage from these lysosomal compartments occurs only slowly and inefficiently, liberated molecules are still functional and will be loaded into newly synthesized Ago2/RISC proteins for months on end, leading to extended target knockdown.

A most intriguing finding is the observation that miRNA (and mRNA) can be transported between cells through endogenous carrier systems. Cellular export of miRNAs via high-density lipoproteins (HDL) was demonstrated to be regulated by neutral sphingomyelinase. HDL-mediated delivery of both exogenous and endogenous miRNAs was shown to inhibit mRNA and was dependent on cellular uptake via scavenger receptor class B type I (Vickers et al. 2011, 2015). In addition, extracellular vesicles, including exosomes and microvesicles, have been shown to contain miRNA and mRNA in their aqueous interior. The functional delivery of mRNA was recently demonstrated (Zomer et al. 2015; You et al. 2023). Mimicking these endogenous delivery systems or hybrid vesicles may enable more efficient functional RNA delivery than those based on synthetic approaches (Cf. Chap. 14).

Perspectives

At present, a handful of oligonucleotide and mRNA based drugs are marketed. With respect to oligonucleotide-based drugs, they all contain chemically modified nucleotides. In addition, they are either delivered locally (directly into the vitreous or CSF) or via the systemic route (to reach the liver). These choices reflect two of the main difficulties in applying oligonucleotides as therapeutics: (1) oligonucleotides are sensitive toward nucleases, and (2) oligonucleotides have difficulties in reaching their target site.

It is not a coincidence that most approved oligonucleotide drugs are for administration into the eye as this organ shows slow clearance rates and is immune privileged. Similarly, intrathecal delivery enables prolonged dosing intervals at relatively low concentrations. Therapeutic applications of oligonucleotides administered via the systemic route are still limited to the liver. By using triantennary N-acetylgalactosamine as targeting ligand conjugated to siR-NAs, efficient liver targeting has been achieved. Current chemical modifications offer nuclease resistance and enhanced cellular uptake that allows a subcutaneously injected oligonucleotide to reach the hepatocyte. This strategy has resulted in the approval of several different siRNA-based products to treat hepatic diseases in a relatively short timeframe, showing that the GalNAc approach can be considered as a plug and play platform for targeting any kind of siRNA to hepatocytes. It is expected that this technology will be further exploited for the development of more siRNA therapeutics for hepatic diseases. Moreover, by using different targeting ligands conjugated to the siRNA, one can even think of reaching other target cells as long as they are readily accessible from the circulation (e.g., blood disorders or vasculature).

The biggest challenge for the development of oligonucleotide therapeutics in the near future is finding ways to target these molecules beyond the liver. Targeted delivery seems especially important in view of the plethora of activities nucleic acids can display. It seems virtually impossible to find nucleic acids that will not interact with any of the other pathways apart from the desired one. Limiting the number of cell types that the nucleic acids can reach and interact with will likely contribute to reduce side effects.

The approval of patisiran, the first siRNA product based on LNPs, was an important hallmark for rapid development of mRNA-based therapeutics as the same LNP delivery system developed for siRNA seemed to be very suitable for delivering mRNA therapeutics as well. With variations in the LNP composition, it is now possible to target hepatocytes or antigen-presenting cells in the liver and spleen. Besides, the LNP platform is also very effective for local (intramuscular and intradermal) vaccination. The success of the COVID-19 mRNA vaccines Comirnaty and SpikeVax has accelerated the development of mRNA vaccines against many other infectious diseases, including Zika virus, Ebola, Influenza, and HIV-1. The coming years will tell us whether mRNA-LNPs will become the method of choice for vaccination against viral infections. Nevertheless, there are still some challenges ahead. The stability issues associated with mRNA-LNP drug products, which require cold-chain storage, might be a limiting factor in global distribution of such vaccines and the relatively high costs of mRNA-LNP production could be a serious constraint for low-income countries where such vaccines are often needed most. Finding ways to make these vaccines stable at room temperature with a decently long shelf-life and reducing the costs of manufacturing are therefore two important incentives for the future.

Besides its use as prophylactic vaccine, mRNA might become an important alternative for protein replacement therapy, especially for applications where the protein is active within the cell. Several clinical studies are ongoing to evaluate the effectiveness of this approach. Protein replacement therapy with mRNA would require repeated administration of high doses of mRNA, which will bring new challenges in terms of toxicity and carrier-mediated adverse reactions. Self-amplifying RNA platforms might be a solution to reduce the frequency of dosing and lower the overall amount of mRNA to be delivered, but these self-amplifying RNA platforms still suffer from innate immune responses against the viral proteins needed for RNA replication.

Finally, another application area in which mRNA will most likely become an important tool is in cancer immunotherapy. mRNA can serve as adjuvants to boost tumor-specific adaptive immune responses by local or systemic expression of cytokines or immune checkpoint inhibitors, and it can serve as a flexible source of tumor antigens. With respect to the latter, the ease by which mRNA can be designed to encode multiple neoantigens makes this the ideal method for personalized cancer vaccines. Ongoing clinical trials will soon teach us how effective this approach will be.

Self-Assessment Questions

Questions

1. Which AS-oligonucleotide modifications are able to recruit RNase H, and which not?
2. Explain the principle of RNAi.
3. What are the major obstacles in therapeutic applications of oligonucleotides?
4. What is the difference between gene correction and gene silencing?
5. Why are the eye and liver popular target organ for oligonucleotide delivery?
6. What are the different requirements for antisense oligonucleotides that are made for exon skipping and those that are made to inhibit translation of a mutated gene?
7. Why is therapeutic mRNA not chemically produced but enzymatically?
8. Which modifications are often applied to therapeutic mRNA?
9. What are the five main areas of therapeutic application of mRNA?
10. Mention three types of delivery systems for oligonucleotides and/or mRNA.

Answers

1. Only charged antisense oligodeoxyribonucleotide phosphodiesters and phosphorothioates elicit efficient RNase H activity. Noncharged oligonucleotides, including, for example, the peptide nucleic acids, morpholino-oligos, and 2-O-alkyloligoribonucleotides, do not recruit RNase H activity and act by physical mRNA blockade.
2. RNAi is a mechanism for RNA-guided regulation of gene expression in which double-stranded ribonucleic acid inhibits the expression of genes with complementary nucleotide sequences. The RNAi pathway is initiated by the enzyme Dicer, which cleaves double-stranded RNA to short double-stranded fragments of 20–25 base pairs. One of the two strands of each fragment, known as the guide strand, is then incorporated into the RNA-induced silencing complex and base pairs with complementary sequences. The most well-studied outcome of this recognition event is a form of posttranscriptional gene silencing; however, the process also affects methylation of promoters, increases degradation of mRNA

(not mediated by RISC), blocks protein translation, and enhances protein degradation.
3. Poor pharmacokinetics, instability, and inability to cross membranes.
4. Gene correction makes use of a homologous recombination process to permanently correct single or multiple point mutations within a region of the gene of interest and therefore acts at the level of DNA. Gene silencing aims at reducing the level of active transcripts of the gene of interest by targeted degradation of the mRNA.
5. The eye is an immunoprivileged organ and allows local injection. The liver can be reached via N-acetylgalactosamine modification of oligonucleotides or by lipid nanoparticle encapsulation.
6. Exons can be skipped from the mRNA with antisense oligoribonucleotides. They attach inside the exon to be removed, or at its borders. The oligonucleotides interfere with the splicing machinery so that the targeted exons are no longer included in the mRNA. Classical antisense oligonucleotides are single-stranded DNA or RNA molecules that generally consist of 13–25 nucleotides. They are complementary to a sense mRNA sequence and can hybridize to it through Watson–Crick base pairing. Translation inhibition can be achieved through physical mRNA blockade or mRNA cleavage by recruitment of RNase H.
7. Complete chemical synthesis of oligonucleotides by solid phase synthesis is currently limited to strands of approximately 150 bases in length. mRNA is often much longer, hence the use of RNA polymerases, capping enzymes, and polyadenylation enzymes to produce mRNA from a DNA template.
8. Therapeutic mRNA often contains a 5'cap structure and a 3'polyadenosine tail for stability. Sometimes, these nucleosides contain modified bases, such as pseudouridines to reduce immunogenicity of the mRNA molecule.
9. Cancer immunotherapy; Protein replacement therapy; Vaccination; In situ production of biopharmaceuticals; Genome editing/genetic reprogramming.
10. Lipid nanoparticles (LNPs), Cell-penetrating peptides (CPPs), protamine complexes, and GalNAc conjugates (only for siRNA).

References

Aartsma-Rus A (2017) FDA approval of nusinersen for spinal muscular atrophy makes 2016 the year of splice modulating oligonucleotides. Nucleic Acid Ther 27:67–69

Akinc A, Maier MA, Manoharan M et al (2019) The Onpattro story and the clinical translation of nanomedicines containing nucleic acid-based drugs. Nat Nanotechnol 14:1084–1087

Ämmälä C, Drury WJ 3rd, Knerr L et al (2018) Targeted delivery of antisense oligonucleotides to pancreatic β-cells. Sci Adv 4:eaat3386

Andrikakou P, Reebye V, Vasconcelos D et al (2022) Enhancing SIRT1 gene expression using small activating RNAs: a novel approach for reversing metabolic syndrome. Nucleic Acid Ther 32:486–496

Aquino-Jarquin G (2020) Novel engineered programmable systems for ADAR-mediated RNA editing. Mol Ther Nucleic Acids 19:1065–1072

Arif R, Zaradzki M, Remes A et al (2017) AP-1 Oligodeoxynucleotides reduce aortic Elastolysis in a murine model of Marfan syndrome. Mol Ther Nucleic Acids 9:69–79

Barbier AJ, Jiang AY, Zhang P et al (2022) The clinical progress of mRNA vaccines and immunotherapies. Nat Biotechnol 40(6):840–854

Bloom K, van den Berg F, Arbuthnot P (2021) Self-amplifying RNA vaccines for infectious diseases. Gene Ther 28:117–129

Castillo-Hair SM, Seelig G (2022) Machine learning for designing next-generation mRNA therapeutics. Acc Chem Res 55:24–34

Chakraborty C, Sharma AR, Sharma G et al (2017) Therapeutic miRNA and siRNA: moving from bench to clinic as next generation medicine. Mol Ther Nucleic Acids 8:132–143

Crommelin DJA, Anchordoquy TJ, Volkin DB et al (2021) Addressing the cold reality of mRNA vaccine stability. J Pharm Sci 110:997–1001

Crooke ST, Baker BF, Crooke RM, Liang X-H (2021) Antisense technology: an overview and prospectus. Nat Rev Drug Discov 20:427–453

Deal CE, Carfi A, Plante OJ (2021) Advancements in mRNA encoded antibodies for passive immunotherapy. Vaccines (Basel) 9:108.

Dhuri K, Bechtold C, Quijano E et al (2020, 2004) Antisense oligonucleotides: an emerging area in drug discovery and development. J Clin Med Res 9.

Dovgan I, Koniev O, Kolodych S, Wagner A (2019) Antibody–oligonucleotide conjugates as therapeutic, imaging, and detection agents. Bioconjug Chem 30:2483–2501

Dreja H (2000) Gene therapy with NF-κB decoy oligodeoxynucleotides in arthritis. Arthritis Res Ther 3:66779

Duan D, Goemans N, Takeda S et al (2021) Duchenne muscular dystrophy. Nat Rev Dis Primers 7:13

Gao S, Zheng X, Jiao B, Wang L (2016) Post-SELEX optimization of aptamers. Anal Bioanal Chem 408:4567–4573

Geary RS, Norris D, Yu R, Bennett CF (2015) Pharmacokinetics, biodistribution and cell uptake of antisense oligonucleotides. Adv Drug Deliv Rev 87:46–51

Geng J, Xia X, Teng L et al (2022) Emerging landscape of cell-penetrating peptide-mediated nucleic acid delivery and their utility in imaging, gene-editing, and RNA-sequencing. J Control Release 341:166–183

Gilleron J, Querbes W, Zeigerer A et al (2013) Image-based analysis of lipid nanoparticle–mediated siRNA delivery, intracellular trafficking and endosomal escape. Nat Biotechnol 31:638

Gillmore JD, Gane E, Taubel J et al (2021) CRISPR-Cas9 in vivo gene editing for transthyretin amyloidosis. N Engl J Med 385:493–502

Giridharan S, Srinivasan M (2018) Mechanisms of NF-κB p65 and strategies for therapeutic manipulation. J Inflamm Res 11:407–419

Goemans NM, Tulinius M, van den Hauwe M, et al (2016) Long-Term Efficacy, Safety, and Pharmacokinetics of Drisapersen in Duchenne Muscular Dystrophy: Results from an Open-Label Extension Study. PLoS One 11:e0161955

Goga A, Stoffel M (2022) Therapeutic RNA-silencing oligonucleotides in metabolic diseases. Nat Rev Drug Discov 21:417–439

Gómez-Aguado I, Rodríguez-Castejón J, Vicente-Pascual M et al (2020) Nanomedicines to deliver mRNA: state of the art and future Perspectives. Nanomaterials (Basel) 10:364.

Hair P, Cameron F, McKeage K (2013) Mipomersen sodium: first global approval. Drugs 73:487–493

Hammond SM, Aartsma-Rus A, Alves S et al (2021) Delivery of oligonucleotide-based therapeutics: challenges and opportunities. EMBO Mol Med 13:e13243

Harcourt EM, Kietrys AM, Kool ET (2017) Chemical and structural effects of base modifications in messenger RNA. Nature 541:339–346

Hecker M, Wagner AH (2017) Transcription factor decoy technology: a therapeutic update. Biochem Pharmacol 144:29–34

Hou X, Zaks T, Langer R, Dong Y (2021) Lipid nanoparticles for mRNA delivery. Nat Rev Mater 6(12):1078–1094

Huang K, Zapata D, Tang Y et al (2022) In vivo delivery of CRISPR-Cas9 genome editing components for therapeutic applications. Biomaterials 291:121876

Jurga S, Barciszewski J (2022) Messenger RNA therapeutics. Springer Nature

Karam M, Daoud G (2022) mRNA vaccines: Past, present, future. Asian J Pharm Sci 17:491–522

Kato K, Akeda K, Miyazaki S et al (2021) NF-kB decoy oligodeoxynucleotide preserves disc height in a rabbit anular-puncture model and reduces pain induction in a rat xenograft-radiculopathy model. Eur Cell Mater 42:90–109

Keam SJ (2018) Inotersen: first global approval. Drugs 78:1371–1376

Klampatsa A, Dimou V, Albelda SM (2021) Mesothelin-targeted CAR-T cell therapy for solid tumors. Expert Opin Biol Ther 21:473–486

Kunugiza Y, Tomita T, Tomita N et al (2006) Inhibitory effect of ribbon-type NF-kappaB decoy oligodeoxynucleotides on osteoclast induction and activity in vitro and in vivo. Arthritis Res Ther 8:R103

Lee G-H, Lim S-G (2021) CpG-Adjuvanted hepatitis B vaccine (HEPLISAV-B®) update. Expert Rev Vaccines 20:487–495

Lehto T, Ezzat K, Wood MJA, El Andaloussi S (2016) Peptides for nucleic acid delivery. Adv Drug Deliv Rev 106:172–182

Lopes RD, Williams JB, Mehta RH et al (2012) Edifoligide and long-term outcomes after coronary artery bypass grafting: PRoject of ex-vivo vein graft ENgineering via transfection IV (PREVENT IV) 5-year results. Am Heart J 164:379–386.e1

Mahjoubin-Tehran M, Rezaei S, Atkin SL et al (2021) Decoys as potential therapeutic tools for diabetes. Drug Discov Today 26:1669–1679

Mann MJ, Dzau VJ (2000) Therapeutic applications of transcription factor decoy oligonucleotides. J Clin Invest 106:1071–1075

Maraganore J (2022) Reflections on Alnylam. Nat Biotechnol, pp 1–10

Mehta M, Paudel KR, Shukla SD et al (2021) Recent trends of NFκB decoy oligodeoxynucleotide-based nanotherapeutics in lung diseases. J Control Release 337:629–644

Merkle T, Merz S, Reautschnig P et al (2019) Precise RNA editing by recruiting endogenous ADARs with antisense oligonucleotides. Nat Biotechnol 37:133–138

Miller CM, Donner AJ, Blank EE et al (2016) Stabilin-1 and Stabilin-2 are specific receptors for the cellular internalization of phosphorothioate-modified antisense oligonucleotides (ASOs) in the liver. Nucleic Acids Res 44:2782–2794

Monian P, Shivalila C, Lu G et al (2022) Endogenous ADAR-mediated RNA editing in non-human primates using stereopure chemically modified oligonucleotides. Nat Biotechnol 40:1093–1102

Moradian H, Roch T, Anthofer L et al (2022) Chemical modification of uridine modulates mRNA-mediated proinflammatory and antiviral response in primary human macrophages. Mol Ther Nucleic Acids 27:854–869

Moulton JD (2017) Making a Morpholino experiment work: controls, favoring specificity, improving efficacy, storage, and dose. Methods Mol Biol 1565:17–29

Muttach F, Muthmann N, Rentmeister A (2017) Synthetic mRNA capping. Beilstein J Org Chem 13:2819–2832

Nguyen TA, Smith BRC, Tate MD et al (2017) SIDT2 transports extracellular dsRNA into the cytoplasm for innate immune recognition. Immunity 47:498–509.e6

Nimjee SM, White RR, Becker RC, Sullenger BA (2017) Aptamers as therapeutics. Annu Rev Pharmacol Toxicol 57:61–79

Nishikura K (2016) A-to-I editing of coding and non-coding RNAs by ADARs. Nat Rev Mol Cell Biol 17:83–96

Oude Blenke E, Örnskov E, Schöneich C et al (2022) The storage and in-use stability of mRNA vaccines and therapeutics: not a cold case. J Pharm Sci 112(2):386–403.

Ouranidis A, Vavilis T, Mandala E et al (2021) mRNA therapeutic modalities design, formulation and manufacturing under pharma 4.0 principles. Biomedicine 10:50.

Paik J, Duggan S (2019) Volanesorsen: First global approval. Drugs 79:1349–1354

Pardi N, Secreto AJ, Shan X et al (2017) Administration of nucleoside-modified mRNA encoding broadly neutralizing antibody protects humanized mice from HIV-1 challenge. Nat Commun 8:14630

Qu L, Yi Z, Zhu S et al (2019) Programmable RNA editing by recruiting endogenous ADAR using engineered RNAs. Nat Biotechnol 37:1059–1069

Rauch S, Lutz J, Kowalczyk A et al (2017) RNActive® technology: generation and testing of stable and immunogenic mRNA vaccines. Methods Mol Biol 1499:89–107

Ray KK, Wright RS, Kallend D et al (2020) Two phase 3 trials of inclisiran in patients with elevated LDL cholesterol. N Engl J Med 382:1507–1519

Rohner E, Yang R, Foo KS et al (2022) Unlocking the promise of mRNA therapeutics. Nat Biotechnol 40:1586–1600

Rosa SS, Prazeres DMF, Azevedo AM, Marques MPC (2021) mRNA vaccines manufacturing: challenges and bottlenecks. Vaccine 39:2190–2200

Sahay G, Querbes W, Alabi C et al (2013) Efficiency of siRNA delivery by lipid nanoparticles is limited by endocytic recycling. Nat Biotechnol 31:653–658

Sahin U, Derhovanessian E, Miller M et al (2017) Personalized RNA mutanome vaccines mobilize poly-specific therapeutic immunity against cancer. Nature 547:222–226

Schlegel MK, Janas MM, Jiang Y et al (2022) From bench to bedside: improving the clinical safety of GalNAc-siRNA conjugates using seed-pairing destabilization. Nucleic Acids Res 50:6656–6670

Scott LJ (2020) Givosiran: First approval. Drugs 80:335–339

Scott LJ, Keam SJ (2021) Lumasiran: first approval. Drugs 81(2):277–282.

Sheehan JP, Lan H-C (1998) Phosphorothioate oligonucleotides inhibit the intrinsic tenase complex. Blood 92:1617–1625

Shen X, Corey DR (2018) Chemistry, mechanism and clinical status of antisense oligonucleotides and duplex RNAs. Nucleic Acids Res 46:1584–1600

Shen W, De Hoyos CL, Migawa MT et al (2019) Chemical modification of PS-ASO therapeutics reduces cellular protein-binding and improves the therapeutic index. Nat Biotechnol 37:640–650

Stephenson ML, Zamecnik PC (1978) Inhibition of Rous sarcoma viral RNA translation by a specific oligodeoxyribonucleotide. Proc Natl Acad Sci U S A 75:285–288

Szebeni J, Storm G, Ljubimova JY et al (2022) Applying lessons learned from nanomedicines to understand rare hypersensitivity reactions to mRNA-based SARS-CoV-2 vaccines. Nat Nanotechnol 17:337–346

Tai W (2019) Chemical modulation of siRNA lipophilicity for efficient delivery. J Control Release 307:98–107

Takeda S, Clemens PR, Hoffman EP (2021) Exon-skipping in Duchenne muscular dystrophy. J Neuromuscul Dis 8:S343–S358

Van Hoecke L, Verbeke R, Dewitte H et al (2021) mRNA in cancer immunotherapy: beyond a source of antigen. Mol Cancer 20:48

Van Lint S, Goyvaerts C, Maenhout S et al (2012) Preclinical evaluation of TriMix and antigen mRNA-based antitumor therapy. Cancer Res 72:1661–1671

Vickers KC, Palmisano BT, Shoucri BM et al (2011) MicroRNAs are transported in plasma and delivered to recipient cells by high-density lipoproteins. Nat Cell Biol 13:423–433

Vickers KC, Palmisano BT, Shoucri BM et al (2015) Erratum: corrigendum: MicroRNAs are transported in plasma and delivered to recipient cells by high-density lipoproteins. Nat Cell Biol 17:104–104

Wengert ER, Wagley PK, Strohm SM et al (2022) Targeted augmentation of nuclear gene output (TANGO) of Scn1a rescues parvalbumin interneuron excitability and reduces seizures in a mouse model of Dravet syndrome. Brain Res 1775:147743

Wesselhoeft RA, Kowalski PS, Parker-Hale FC et al (2019) RNA circularization diminishes immunogenicity and can extend translation duration in vivo. Mol Cell 74:508–520.e4

Yang L, Wilusz JE, Chen L-L (2022) Biogenesis and regulatory roles of circular RNAs. Annu Rev Cell Dev Biol 38:263–289

You Y, Tian Y, Yang Z et al (2023) Intradermally delivered mRNA-encapsulating extracellular vesicles for collagen-replacement therapy. Nat Biomed Eng.

Zhao Q, Zhao Z, Fan X et al (2021) Review of machine learning methods for RNA secondary structure prediction. PLoS Comput Biol 17:e1009291

Zhou G, Latchoumanin O, Hebbard L et al (2018) Aptamers as targeting ligands and therapeutic molecules for overcoming drug resistance in cancers. Adv Drug Deliv Rev 134:107–121

Zimmermann GR, Wick CL, Shields TP et al (2000) Molecular interactions and metal binding in the theophylline-binding core of an RNA aptamer. RNA 6:659–667

Zomer A, Maynard C, Verweij FJ et al (2015) In vivo imaging reveals extracellular vesicle-mediated phenocopying of metastatic behavior. Cell 161:1046–1057

Advanced Therapy Medicinal Products: Clinical, Non-clinical, and Quality Considerations

14

Enrico Mastrobattista, Erik Doevendans, Niek P. van Til,
Vera Kemp, Jeroen de Vrij, and Karin Hoogendoorn

Introduction

The most recent branch of the biotechnology revolution in medicine consists of gene and cell therapy medicinal products and tissue-engineered products. These are collectively called advanced therapies in the US and advanced therapy medicinal products (ATMPs) in the EU. The use of living cells, tissues, or organs in medical practice is not novel. The first successful kidney transplantation in a human took place in 1954, and human hematopoietic stem cell transplantation from a healthy donor to a cancer patient in 1959; the latter is now a routine clinical procedure for bone marrow regeneration. Cell therapies, including stem cell therapies, were not further explored until the early 1990s when the therapeutic relevance of mesenchymal stromal cells (MSCs) was considered for the regeneration of skeletal tissue and later for broader therapeutic use. Furthermore, the development of efficient gene transfer vectors to genetically modify cells

E. Mastrobattista (✉) · E. Doevendans
Pharmaceutics Division, Utrecht Institute for Pharmaceutical Sciences (UIPS), Utrecht University, Utrecht, The Netherlands
e-mail: E.Mastrobattista@uu.nl; E.Doevendans@uu.nl

N. P. van Til
Department of Child Neurology, Amsterdam Leukodystrophy Center, Emma Children's Hospital, Amsterdam University Medical Centers, VU University, and Neuroscience, Cellular & Molecular Mechanisms, Amsterdam, The Netherlands
e-mail: N.P.vanTil@amsterdamumc.nl

V. Kemp
Department of Cell and Chemical Biology, Leiden University Medical Center, Leiden, The Netherlands
e-mail: V.Kemp@LUMC.nl

J. de Vrij
Department of Neurosurgery, Brain Tumor Center, Erasmus MC Cancer Institute, Erasmus University Medical Center, Rotterdam, The Netherlands
e-mail: J.deVrij@ErasmusMC.nl

K. Hoogendoorn
Hospital Pharmacy, Interdivisonal GMP Facility, Leiden University Medical Center, Leiden, The Netherlands

ex vivo or directly in vivo in the 1980s and 1990s further boosted the development of therapies based on genetically corrected or augmented cells. Since the turn of the millennium, there has been a steady increase in the number of clinical studies, with a growing number of target indications. Particularly for the treatment of diseases and tissue/organ defects for which traditional therapies and medicinal products have not always provided positive benefit/risk outcomes, such as multiple sclerosis, Parkinson's disease, cancer, and muscular dystrophy, advanced therapies hold high expectations. The inherent complexity of these products poses unique challenges compared to other therapeutics. The manufacture of gene transfer vectors or "living" materials (i.e., cells and tissues) comes with great challenges in terms of consistency and process and product characterization. Such challenges are analogous, in many ways, to those faced in the past when the first recombinant protein biopharmaceutical products were being developed and regulated. Bringing advanced therapies to market at an acceptable cost, benefit/risk ratio, and quality has proven extremely difficult for certain products.

This chapter discusses the current status and unique aspects of ATMPs. We explain the differences between traditional cell or tissue transplantation versus advanced therapies based on (genetically manipulated) somatic cells or tissue-engineered products. Then, we discuss in detail the various cell technologies and technologies for gene therapy and provide information on the manufacturing of advanced therapies. Finally, the regulatory aspects of ATMPs will be briefly highlighted.

ATMPs: Definitions, Classifications, and Modes of Action

Globally, different names and definitions for gene- and cell-based products are used in different jurisdictions. For example, in the EU, such a product is called an advanced therapy medicinal product (ATMP), whereas the term "advanced therapy" is used in the USA. Minor differences in the definition of (sub-)classes between the two jurisdictions exist, and importantly, in the US, human cells and tissues may also be regulated as devices (similarly to the combined ATMPs in Europe). Furthermore, because of the minor differences in definitions applied in the US and EU, hematopoietic stem cell (HSC) transplantation for the treatment of malignant blood disorders is an example of a product that would be classified as an advanced therapy in the US but would fall outside the scope of ATMPs in the EU (see section "Adult Stem Cells Used as Transplant Product"; Fig. 14.5).

Transplantation or Advanced Therapy?

Advanced therapies, when applied to humans, are considered biological medicinal products, meaning they are typically subject to either or both of the following regulatory regimes: public health legislation and pharmaceutical legislation. However, some clinical interventions for cell- and tissue-based advanced therapies are not considered "medicinal"; these products are subject to public health legislation only. These therapies are often called "cell and tissue transplant products" or "cell and tissue transplantations" and have to meet all of the following criteria (see also Fig. 14.1):

1. A cell or tissue, which is not substantially manipulated. Table 14.1 provides guidance on the definition of substantial and nonsubstantial manipulations.

2. Cells/tissues are used for the same essential function in the donor and recipients (sometimes called "homologous use").
3. It is not combined with a medical device or active implantable medical device.

Cell and tissue transplant products require no clinical trials and no marketing authorization (MA) prior to commercial availability but public health legislations apply.

However, if cells or tissue are substantially manipulated, it fulfills the criteria for an advanced therapy medicinal product (ATMP) and, as such, will be regulated as a medicine, meaning the development must comply with medicines regulations.

ATMPs can be subdivided into somatic cell therapy medicinal products (SCTMP), consisting of somatic cells that have been expanded and/or manipulated ex vivo before being administered back into the patient; gene therapy medicinal products (GTMP), consisting of vectors for transfer of materials that can genetically modify cells ex vivo or in vivo; tissue-engineered products (TEP), involving cells or multiple cell types growing on a scaffold to form a 3D tissue culture; and cells or tissues that are being combined with a medical device (combined ATMP) (Fig. 14.1 and Table 14.2).

Gene or Cell Therapy Medicinal Product?

There is considerable overlap between somatic cell therapy medicinal products (SCTMP) and gene therapy medicinal

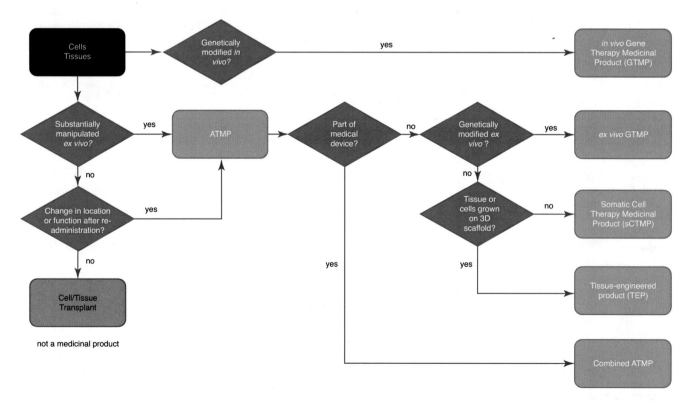

Fig. 14.1 Decision tree for categorizing ATMPs. If cells or tissues are not substantially manipulated and are used for the same essential function in donor and recipient (homologous use), the therapy is not considered an ATMP but categorized as a cell or tissue transplant, for which different regulations apply. If cells are substantially manipulated or genetically modified, these products are considered ATMPs. For definitions of substantial cell manipulation, see Table 14.1

Table 14.1 Substantial and nonsubstantial manipulations of cells or tissues

Substantial manipulation	Non-substantial manipulation
Specific manipulations considered substantial are: 1. Cell expansion (culture; ex vivo) 2. Differentiation and/or activation with growth factors 3. Ex vivo genetic modifications of cells (e.g., with viral vector)	Specific manipulations *not* considered substantial: 1. Cutting 2. Grinding 3. Shaping 4. Centrifugation 5. Soaking in antibiotic or antimicrobial solutions 6. Sterilization 7. Irradiation 8. Cell separation, concentration or purification 9. Filtering 10. Lyophilization 11. Freezing 12. Cryopreservation 13. Vitrification

products (GTMP). If a cell is isolated from the body and substantially manipulated ex vivo before readministration, it is considered a cell therapy medicinal product. Suppose (part of) this substantial manipulation consists of genetic modification; the product of this manipulation is considered as an (ex vivo) gene therapy medicinal product (Fig. 14.1). For example, tabelecleucel (Ebvallo®) consists of allogeneic Epstein-Barr virus (EBV) specific T cells and is registered as a cell therapy medicinal product, whereas Zalmoxis® also consists of allogeneic T cells. Still, these were genetically modified with a retroviral vector to express the human low-affinity nerve growth factor receptor and the herpes simplex virus type 1 (HSV-1) thymidine kinase enzyme. This product was therefore registered as an ex vivo gene therapy medicinal product. Similarly, products based on ex vivo genetic modification of autologous T cells to express a chimeric antigen receptor (CAR-T cells) are categorized as ex vivo gene therapy medicinal products.

Classification

Besides the classification of ATMPs based on regulation as described above, ATMPs can be classified in many other ways, e.g., by:

1. The therapeutic indication they aim to address, e.g., neurological, cardiovascular, or ophthalmological.
2. Whether they comprise cells and/or tissues (see Fig. 14.2):
   (a) derived from and administered to the same human individual (autologous = autogenic), hence the donor = the recipient (patient);
   (b) derived from a human (healthy) donor, who is different from the patient (allogeneic);
   (c) derived from an animal (xenogeneic; see Chap. 9), e.g., porcine islets to treat diabetes mellitus (DM).
3. The potency of the cells, i.e., omnipotent, pluripotent, multipotent, oligopotent, and unipotent (see Table 14.3).
4. The in vivo mode of action (i.e., pharmacological, immunological, metabolic, or regenerative; i.e., regenerate, repair, or replace a human tissue).
5. Their underlying technology, as described in this chapter (Mount et al. 2015):
   (a) somatic cell technologies;
   (b) cell immortalization technologies;
   (c) ex vivo gene modification of cells using viral vector technologies;
   (d) in vivo gene modification of cells using viral vector technologies (see this chapter);
   (e) genome editing technologies;
   (f) cell plasticity technologies;
   (g) three-dimensional technologies;
   (h) combinations of the above technologies.
6. The cell types, e.g., MSCs, dendritic cells (DCs), and T cells.

ATMPs: Possible Mode of Action(s)

The in vivo mode of action(s) (MoA(s)) of an advanced therapy depends on the type of cell/tissue, the ex vivo manipulations performed on the cells/tissue in the manufacturing facility, e.g., genetic modification, the route of administration and the in vivo environment of the cells/tissue:

Table 14.2 EU-ATMP classification definitions according to the EU pharmaceutical legislation, adapted from Smith et al. (2015)

ATMP classification	Definition	Examples
Gene therapy medical product (GTMP)	A GTMP is a biological medicinal product (*excluding vaccines*) that: (a) Contains an active substance which contains or consists of a recombinant nucleic acid used in or administered to human beings with a view to regulating, repairing, replacing, adding or deleting a genetic sequence and; (b) Its therapeutic, prophylactic or diagnostic effect relates *directly* to the recombinant nucleic acid sequence it contains, or to the product of genetic expression of this sequence Gene therapy medicinal products shall not include *vaccines against infectious diseases* (see Chap. 14), which have their own set of vaccine specific guidances	Glybera® (see Chap. 16); Kymriah® (autologous CD19⁺ CAR-T cells); Strimvelis® (genetically modified autologous CD34⁺ cells)
Somatic cell therapy medicinal product (SCTMP)	A SCTMP is a biological medicinal product which fulfils the following two characteristics: (a) Contains or consists of cells or tissues that have been subject to substantial manipulation so that biological characteristics, physiological functions or structural properties relevant for the intended clinical use have been altered or of cells or tissues that are not intended to be used for the same essential function(s) in the recipient and the donor (b) Is presented as having properties for or is used in or administered to human beings with a view to treating, preventing or diagnosing a disease through the pharmacological, immunological or metabolic action of its cells or tissues	Alofisel® (allogeneic MSCs); irradiated plasmacytoid dendritic cell line (allogeneic) loaded with peptides from tumor antigens
Tissue engineered product (TEP)	A TEP is a biological medicinal product that meets the following two characteristics: (a) Contains or consists of engineered cells or tissues, and (b) Is presented as having properties for, or is used in or administered to human beings with a view to regenerating, repairing or replacing a human tissue A TEP may contain cells or tissues of human or animal origin, or both. The cells or tissues may be viable or non-viable. It may also contain additional substances, such as cellular products, biomolecules, biomaterials, chemical substances, scaffolds or matrices. Products containing or consisting exclusively of non-viable human or animal cells and/or tissues, which do not contain pharmacological, immunological or metabolic action, are excluded from this definition. Cells or tissues shall be considered "engineered" if they fulfil at least one of the following conditions: (a) The cells or tissues have been subject to substantial manipulation, so that biological characteristics, physiological functions or structural properties relevant for the intended regeneration, repair or replacement are achieved (b) The cells or tissues are not intended to be used for the same essential function or functions in the recipient as in the donor	Spherox® (autologous chondrocytes); Holoclar® (autologous corneal epithelial cells, which contain stem cells)
Combined ATMP	A combined ATMP fulfills the following conditions: (a) It must incorporate, as an integral part of the product, one or more medical devices or one or more active implantable devices, and (b) Its cellular or tissue part must contain viable cells or tissues, (c) Its cellular or tissue part containing non-viable cells or tissues must be liable to act upon the human body with action that can be considered primary to that of the devices referred to	Allogenic adipose derived regenerative cells encapsulated in hyaluronic acid (TEP + device)[a]; encapsulated allogeneic cells secreting GM-CSF + irradiated autologous tumor cells (GTMP + device)

CD19⁺ (*CAR-T cells*) cluster of differentiation (CD) 19 "chimeric antigen receptor T cells", CAR-T cells, *GM-CSF* granulocyte-macrophage colony-stimulating factor

[a] Hassan et al. (2013)

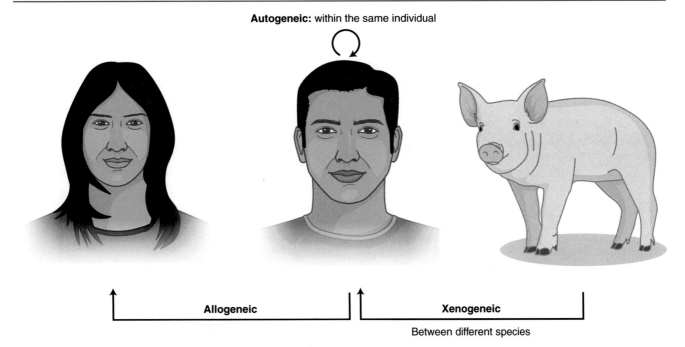

Fig. 14.2 Types of transplants/advanced therapy cell source

Table 14.3 Categorization of stem cells on their potency

Stem cell potency	Explanation and examples
Totipotent (or omnipotent) stem cell	Can differentiate into all embryonic and extraembryonic cell types (i.e., in humans they give rise to the foetus, umbilical cord, and the placenta: morula's cells (0–5 days old embryo)
Pluripotent stem cell	Can differentiate into all three germ cell types (endoderm. mesoderm, or ectoderm lineage) but not the placenta and umbilical cord, and subsequently into all embryonic cell types: ESCs, iPSCs
Multipotent stem cell	Can differentiate into closely related cells, such as all cells in a particular organ: MSCs. other adult (=somatic) stem cells
Oligopotent stem cell	Can differentiate into a restricted closely related group, such as a hematopoietic progenitor cell that can produce a subset of blood cell types, such as B and T cells; vascular stem cell that has the capacity to become both endothelial or smooth muscle cells
Unipotent stem cells (or precursor cell)	Have the property of self-renewal but can only give rise to cells of their own lineage, such as muscle or skin stem cells. This distinguishes these cells from real stem cells as they do not differentiate into other cell phenotypes

1. Pharmacological: cells/tissue release molecules such as cytokines and growth factors upon interaction with their/its environment. An example is the immunoregulatory effect of MSCs. E.g., darvadstrocel (Alofisel®) contains expanded adipose-derived MSCs, which, once activated, impair the proliferation of lymphocytes and reduce the release of pro-inflammatory cytokines at inflammation sites in patients with luminal Crohn's disease. This immunoregulatory activity reduces inflammation and may allow the healing of the tissues around the fistula tract;

2. Regenerative: ex vivo manipulated cells/tissue regenerate, repair, or replace a diseased or damaged human tissue. E.g., to replace damaged ß-cells of a patient suffering from DAM type I, whereby human (h)ESCs are ex vivo differentiated into pre-ß cells, loaded into a device, and administered subcutaneously.

3. Immunological: cells of the immune system are ex vivo activated. E.g., cytotoxic T lymphocytes (CTL) or genetically modified cells (e.g., CAR-T cells) activate the patient's own immune system upon administration, e.g., to treat cancer.

(Dis)similarities with Recombinant Therapeutic Proteins and Other Biologicals

Although ATMPs fall within the group of biologicals, there are substantial differences in the area of chemistry, manufac-

turing, and controls (CMC), nonclinical, clinical, regulatory, and costs/reimbursement compared to recombinant proteins, vaccines, and plasma-derived medicinal products. This is summarized in Table 14.4 and further discussed in this chapter.

Table 14.4 Example of differences between advanced therapies and biopharmaceuticals

Category	Characteristic	Advanced therapies	Other biopharmaceuticals
Non-clinical	Animal models	Often no relevant animal models to predict safety and particularly efficacy in humans available	Relevant animal models to predict safety/efficacy often available
	Safety testing	Tumorigenicity testing may be needed (stem cell derived products)	N.A.
	ADME/pharmacodynamic studies	Often not possible/relevant	Generally performed
Clinical	Disease pathway(s) and mode of action	Often not well understood	Well understood
	First in human trials	Always in patients	Often in healthy volunteers
	PK/PD studies	Often not feasible/relevant	Performed
	Route of administration	Often IV infusion, sometimes local injection, e.g., into tumor, subretinal space of eye; spinal cord: brain: intra-dermal	IV infection or infusion. SC, intradermal. IM
	Patient monitoring	Often long-term follow-up (10–20 years)	Short term follow-up
	Track and traceability	From donor start material (tissue/cell) through manufacturing process to patient and vice versa	From starting material through manufacturing process to patient
Quality/CMC	Product group	Heterogeneous	Less heterogeneous
	Type of formulation	Often a dispersion/suspension of cells	Often a solution (liquid or reconstituted lyophilizate); sometimes emulsion or suspension (vaccines)
	Dose	Mostly number of (viable) cells/kg body weight or cm^2 tissue	Usually milligram range for proteins; microgram range for vaccines: or defined as units activity/mg
	Manufacturing process	Often continuous process, no designated drug substance	Often discontinuous process, designated drug substance and drug product
		Often open and manual process steps: no platform technologies yet. automation in its infancy	Closed and mostly automated process steps: platform technologies
		Often aseptic manufacture, no sterilization possible (no viral removal and/or inactivation steps) due to viability of cell/tissue	Viral removal and/or inactivation steps; sterilization (mostly through ≤0.2 micron filtration)
	Batch definition	Often one batch for one to few patients; off-the-shelf products less common	Off-the-shelf (one batch for multiple patients)
	Safety	Risk for transmission of human viral infections from donor to patient; animal and human derived raw materials and excipients	Risk extremely low due to viral removal/inactivation steps; chemically defined raw materials and excipients
	Product storage and supply	Sometimes 2–8 °C or room temperature—short shelf-life; often vapor phase of liquid nitrogen (at <−150 °C)—longer shelf-life (months–years)	Mostly 2–8 °C and longer shelf-life (years)

Table 14.4 (continued)

Category	Characteristic	Advanced therapies	Other biopharmaceuticals
Regulatory	Landscape	Evolving regulatory landscape	Established regulatory landscape
	Guidances	Specific "advanced therapy" guidances	Guidances for biologicals and vaccines
	Classification	Product classification and product terminology not harmonized globally	Product classification and product terminology mostly harmonized globally
Ethics	Uncontrolled access to non-approved product	Stem cell tourism	Illicit use of biopharmaceuticals
	Acceptability starting material (tissue/cells)	Use of human embryos to manufacture human embryonic stem cell based product not allowed in some countries	N.A.
Reimbursement	Costs	Very high (20,000–1,000,000 Euros) per treatment	Medium–high (500–5000 Euros) per injection

N.A. not applicable, *ADME* absorption, distribution, metabolism, elimination, *PK* pharmacokinetics, *PD* pharmacodynamics, *IV* intravenous, *SC* subcutaneous, *IM* intramuscular, *CMC* chemical, manufacturing, and controls

Part A: Technologies for Cell Therapy and Tissue Engineering

Although ATMPs can be classified by the regulatory regime to be applied (see above), the diversity of this new group of biologicals may be better illustrated by the underlying technology and their potential as therapeutics (Mount et al. 2015). Below, these technologies are briefly discussed with examples of products in clinical development or approved for commercial use.

Somatic Cell Technologies

This technology involves the use of adult stem cells, also known as somatic stem cells. The fundamental property of a stem cell is the capability to multiply, i.e., it has the self-renewal capacity, which is the ability to go through numerous cycles of cell division while maintaining the undifferentiated state and to give rise to a variety of differentiated cells. Stem cells can be characterized by their potency, which is the ability to differentiate into specific cell types. The more cell types it can generate, the higher the potency (Table 14.3). Different stem cells exist (Table 14.5) and can be isolated from embryos, blood cords, tissues, and organs, or they can be derived from differentiated somatic cells (mostly skin fibroblasts) by inducing pluripotency via forced expression of specific transcription factors, the so-called induced pluripotent stem cells (iPSC; see section "iPS Cell Technology").

Adult stem cells are known to be present in many, if not all, individual organs in adults and are generally thought to be "multipotent"; that means that they have the ability to dif-

ferentiate into all cell types within one particular lineage, i.e., they can give rise to the cells found in their organ of origin, but not in other organs (Fig. 14.3). In tissues, also in brain tissue in the subventricular zone and in the dentate gyrus (part of the hippocampus), they exist in an organized environment of supporting cells that define the architecture of the "stem cell niche" (Scadden 2006). A hallmark of adult stem cells is their ability to "self-renew" both in vivo and ex vivo and that they undergo asymmetric cell division.

This means that when they divide, they usually give rise to two different cells, one identical stem cell and the other a partly differentiated progenitor cell (Fig. 14.4). The common pattern in adult tissues is that the resulting progenitor cells are capable of expansion by symmetric division and can subsequently differentiate into the various cell types needed for repair or replenishment of the relevant tissue. Adult stem cells include chondrocytes, HSCs, MSCs, skin stem cells, and immune cells (see Table 14.5). Isolation of adult stem cells from organ-tissues is a challenge because only very small numbers of stem cells reside, and once removed from the body, these cells grow to senescence, a state in which cells stop dividing but do not die. Thus, obtaining large quantities of stem cells is difficult. Also, the separation of stem cells from other (unwanted) cell populations is far from trivial. For some products, master and working cell bank (MCB and WCB) strategies are applied. Still, genetic and phenotypic stability, i.e., certain markers present on the cell surface, must be closely monitored.

Adult Stem Cells Used as Transplant Product

Adult stem cells have been used since the 1950s to treat cancers of blood cells as one of the components of bone marrow

Table 14.5 Origin, characteristics, and uses of "stem" cells

Type of stem cell	Origin	Characteristic potential (see also Table 14.6)	Application
Adult (=somatic) stem cells	Exist in small number in many tissues, often in a well-defined and supportive niche	Multipotent: Give rise to cells of the relevant tissue or local environment	Neural stem cells and limbal stem cells in pre-clinical and clinical development
MSCs (a group of adult stem cells)	A collective term for cells tram mesodermal lineage, sourced from stromal or connective tissue (e g., bone marrow, adipose tissue, and umbilical cord tissue)	Multipotent: A heterogeneous pool of cells. They have a "stem cell-like" character and can differentiate into cells of connective tissues, e.g., chondrocytes, osteoblasts, and adipocytes, but they have also been reported to give rise to many other unrelated cell types	Pre-clinical development & clinical PI-III trials; commercial (Prochymal® and Alofisel®)
Cord blood-derived MSCs (primitive stem cells, somewhere between ESCs and mature adult stem cells)	A specific source of MSCs. Extracted at birth from umbilical cord blood	Multipotent: Yet to be fully determined. Potentially they could be a source of many cell types for individual patients	Private cell banks are established for cryopreservation of cord blood samples; pre-clinical development and clinical phase I/II trials
ESC (no adult stem cells)	Result from ex vivo culture of the inner cell mass of a blastocyst (embryoblast = 5–9 days old embryo)	Pluripotent	Vital source of differentiated cells for different research applications and clinical first in human (FIH) trials ongoing
iPSC (no adult stem cells	Derived by reprogramming of somatic cells (often skin fibroblasts) taken from an adult biopsy	Pluripotent, although methods for full reprogramming are still in development	From autologous source for disease modelling, drug screening including toxicity testing, and FIH trial; pre-clinical development and plans for human leukocyte antigens (HLA)-matched allogeneic iPSCs for FIH trial; research is ongoing with allogeneic iPSCs eliminating HLA-class I expression using genome editing technologies to generate universal cell

transplants (Santos 1983). This procedure involves whole-body irradiation to kill malignant cells in multiple myelomas and leukemia. The patient then receives a bone marrow transplant, not in itself a stem cell product, but the transplant contains a few HSCs which subsequently home to the bone marrow stem cell niches and begin to replenish the blood (Fig. 14.5). Rejection and graft-versus-host disease (GvHD) are still threatening complications of this form of therapy.

Adult Stem Cells for Clinical Application: Immune Cells

Immune cell types currently investigated for their therapeutic value, mostly in the field of cancer, are DCs (see also Chap. 15), tumor-infiltrating lymphocytes (TILs), γδ T cells, regulatory T cells (Tregs), macrophages, and viral reconstitution T cells. Both autologous and allogeneic cells are used as cell sources. These immune cells have a highly specific mode of action and are in different stages of clinical development.

Adult Stem Cells for Clinical Application: MSCs

MSCs, sometimes called multipotent stromal cells or mesenchymal stem cells, have generated considerable interest in cell therapy applications (Bianco et al. 2008). However, the description of the cells, their source, and manufacturing processes are quite heterogeneous. MSCs can, e.g., be isolated from bone marrow, adipose tissue, corneal epithelial cells, and from a gelatinous substance within the umbilical cord (Wharton's jelly) and umbilical cord blood. MSCs differentiate into various phenotypes, including chondrocytes, osteoblasts, and adipocytes. Due to their pleiotropic properties, e.g., growth factors and chemokines producing, anti-apoptotic, angiogenetic, anti-fibrotic, and neuroprotective, have been extensively tested in preclinical models. Hundreds of Phase I-III clinical trials have been performed and are ongoing globally in a wide variety of indications: cardiovascular diseases, GvHD, brain and neurological disorders, muscle, bone, and cartilage diseases, lung and bronchial dis-

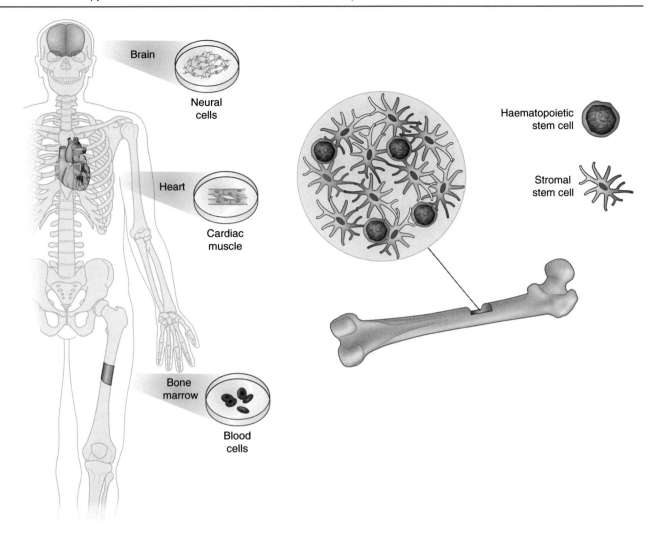

Fig. 14.3 Adult stem cells are present in many tissues in specific stem cell niches, giving rise to a specific group of cells found in the relevant tissue. The examples shown have been studied in detail but adult stem cells, yet to be defined, may be present in many other tissues

eases, wounds and tissue restorations, and immune system diseases (Galderisi et al. 2022), see Fig. 14.6.

MSC can be administered locally, e.g., intralesionally, subcutaneously, or intravascularly. While local administration has been found effective in case of local injury, e.g., to treat bone and joint diseases, heart disease, for the repair of muscle and ligament damage, Crohn's fistulas, and even for the repair of ischemic brain tissue, the systemic infusion is preferable in the case of systemic diseases such as GvHD (Kean et al. 2013). Therefore, both autologous and allogeneic cell sources have been studied. Examples of MSC products that have been approved globally are darvadstrocel (Alofisel®) in the EU for the treatment of Crohn's fistulas,

remestemcel-L (Prochymal®) in Canada and New Zealand, Holoclar and Alofisel in the United States and Temcell HS in Japan for the treatment of pediatric acute graft-versus-host disease (GvHD) (see also Table 14.6).

Cell Immortalization Technologies

Another technology makes use of immortalized cell lines as starting material for the manufacture of cell-based products. An example of such a cell line is the neural stem cell line CTX0E03, derived from the human fetal cortical brain and genetically modified with a retroviral vector encoding the

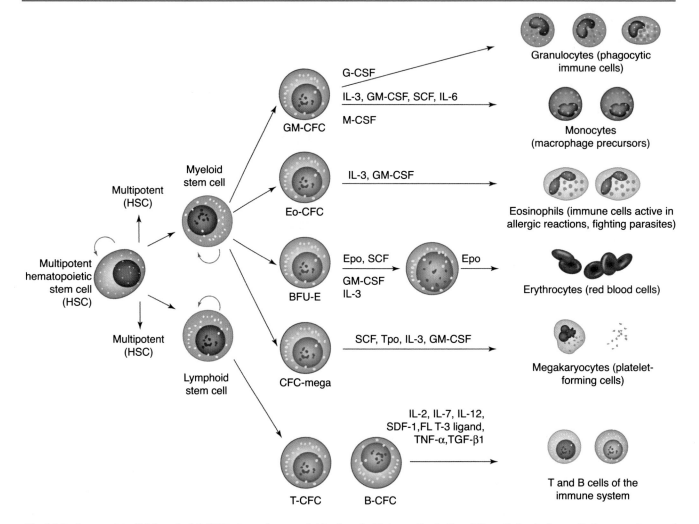

Fig. 14.4 Asymmetric division of adult HSCs, to produce myeloid or lymphoid stem cells, further differentiation to form mitotic progenitors, and subsequently under the control of specific growth factors and cytokines, to form fully differentiated blood cells. The differentiation pathways of the hematopoietic system are better characterized than those of other tissues, but the pattern of differentiation is typical of other tissues. *GM-CSF* = Granulocyte-macrophage colony-stimulating factor, *Eo-CFC* = Eosinophil-leukocyte Colony Forming Cell, *BFU-E* = Bone marrow erythroid progenitor cells, *IL* = interleukin, *SCF* = stem cell factor, *SDF* = stromal cell-derived factor, *TNF* = tumor necrosis factor, *TGF* = transforming growth factor

immortalizing gene, c-mycERTAM (Pollock et al. 2006; Stevanato et al. 2009). Under the conditional regulation by 4-hydroxytamoxifen (4-OHT), this gene enables the large-scale production of the CTX cells using a two-tier cell banking system (MCB and WCB). Clinical testing of the CTX cell-based product is ongoing in a clinical phase II program for stroke. Although cell immortalization technologies have been in development for some time now, this is not a mainstream technology in the pharmaceutical world yet.

Cell Plasticity Technologies

The cell plasticity technology area takes advantage of discoveries that certain cells have the ability to evolve to cell types formerly considered outside their normal differentiation repertoire, i.e., hESCs and iPSCs. This technology has extensive clinical potential due to the high probability of an almost unlimited supply of cells (MCB and WCB approach) and also for the possibility to HLA-match the resulting cell-based product (partly) with the recipient patient. The application of pluripotent stem cells, such as ESCs and iPSCs, goes beyond the administration of cell-based medicinal products and is investigated as a source for tissue engineering and organogenesis (see section "Three-Dimensional Technologies"). In addition, autologous and allogeneic iPSCs are currently extensively used for disease modeling: i.e., patient-specific iPSC-derived cardiomyocytes, cultured in vitro, can be used to identify the genetic basis of cardiac disease, leading to the identification of pharmacogenetic bio-

Fig. 14.5 Schematic representation of bone marrow transplantation, a form of stem cell therapy that was first used over 50 years ago. The transplant contains hematopoietic stem cells from the donor. These cells repopulate niches in the recipient bone marrow

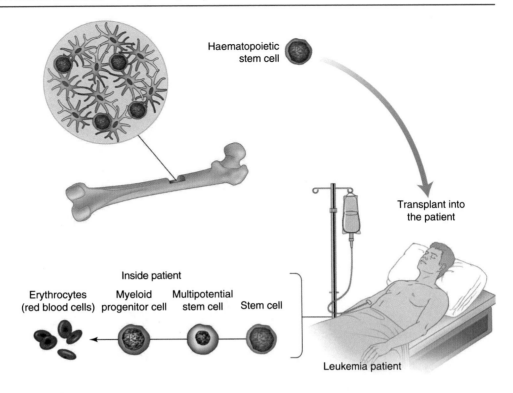

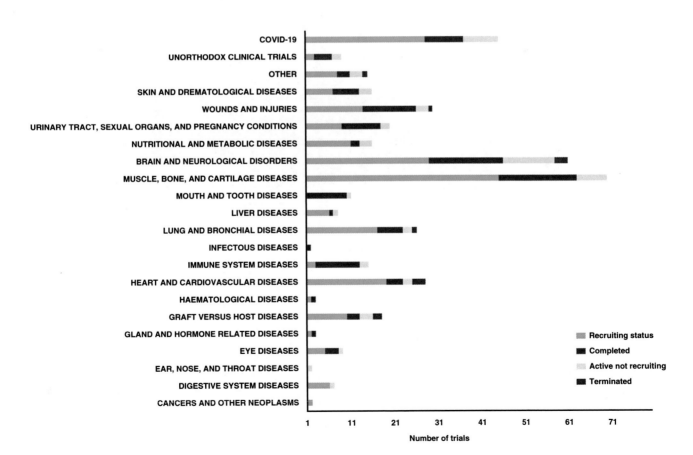

Fig. 14.6 Clinical trials with MSC-derived products and their indication, adapted from Squillaro et al. (2016) (Stem Cell Rev Rep 2022;18: 23–36)

Table 14.6 Available gene and cell therapy medicinal products approved by the EMA and FDA (with the exception of blood cord-derived products; status December 20, 2022)

Brand name	Classification	INN/description	Indication	Company	EMA/FDA approved	Month/year approved
Imlygic™	In vivo GTMP	Talimogene laherparepvec/HSV-1 oncolytic virus containing the cDNA of human GM-CSF	Advanced metastatic melanoma	Amgen	Yes/Yes	December 2015/October 2015
Luxturna™	In vivo GTMP	Voretigene neparvovec/AAV2 vector containing human RPE65 cDNA	Leber congenital amaurosis	Spark Therapeutics	Yes/Yes	November 2018/December 2017
Zolgensma™	In vivo GTMP	Onasemnogene abeparvovec/AAV9 vector containing the SMN1 transgene	Spinal muscular dystrophy (SMA)	AveXis/Novartis	Yes/Yes	May 2020/May 2019
Upstaza™	In vivo GTMP	Eladocagene exuparvovec/AAV2 vector containing the cDNA of human dopa decarboxylase (DDC)	Patients > 18 months with aromatic L-amino acid decarboxylase deficiencies	PTC Tx	Yes/No	July 2022
Roctavian™	In vivo GTMP	Valoctocogene roxaparvovec/AAV5 vector containing the cDNA of the B-domain deleted SQ form of human coagulation factor VIII (hFVIII-SQ)	Adult patients with severe hemophilia A	BioMarin	Yes/No	August 2022
Hemgenix™	In vivo GTMP	Etranacogene dezaparvovec/AAV5 vector containing the cDNA of human coagulation factor IX	Adult patients with hemophilia B	CSL Behring/UniQure	No/Yes	November 2022
Adstiladrin™	In vivo GTMP	Nadofaragene firadenovec-vncg/Nonreplicating AV containing the cDNA of human interferon alfa-2b (rAd-IFNa/Syn3)	BCG-unresponsive high-risk nonmuscle-invasive bladder cancer	Ferring Pharmaceuticals A/S	No/Yes	December 2022
Strimvelis™	Ex vivo GTMP	Autologous CD34+ cells transduced with a retroviral vector containing the ADA cDNA encoding human adenosine deaminase (ADA)	Severe combined immunodeficiency due to adenosine deaminase deficiency (ADA-SCID)	Orchard Tx	Yes/No	May 2016
Zalmoxis™	Ex vivo GTMP	Allogeneic T cells genetically modified with a retroviral vector encoding for a truncated form of the human low-affinity nerve growth factor receptor (ΔLNGFR) and the herpes simplex I virus thymidine kinase (HSV-TK Mut2)	Add-on treatment for patients who have received a haploidentical hematopoietic stem cell transplant	MolMed S.p.A.	Withdrawn/No	August 2016
Kymriah™	Ex vivo GTMP	Tisagenlecleucel/autologous CAR-T cells directed to CD19	Acute Lymphoblastic Leukemia (ALL)	Novartis	Yes/Yes	August 2018/August 2017
Yescarta™	Ex vivo GTMP	Axicabtagene ciloleucel/autologous CAR-T cells directed to CD19	Non-Hodgkin lymphoma: – Large B cell lymphoma – Follicular lymphoma	Kite Pharma	Yes/Yes	August 2018/October 2017
Zynteglo™	Ex vivo GTMP	Betibeglogene autotemcel/Lentiviral vector containing the cDNA of the human HBB gene	Transfusion-dependent beta thalassemia	Bluebird Bio	Yes/Yes	May 2019/August 2022
Libmeldy™	Ex vivo GTMP	Atidarsagene autotemcel/Lentiviral vector containing the human ARSA cDNA encoding arylsulfatase A	Metachromatic Leukodystrophy (MLD)	Orchard Therapeutics	Yes/No	December 2020/
Tecartus™	Ex vivo GTMP	Brexucabtagene autoleucel/autologous CAR-T cells directed to CD19	Mantle Cell Lymphoma Acute Lymphoblastic Leukemia (ALL)	Kite Pharma	Yes/Yes	December 2020/July 2020
Breyanzi™	Ex vivo GTMP	Lisocabtagen maraleucel/autologous CAR-T cells directed to CD19	Relapsed or Refractory Large B-cell lymphoma Non-Hodgkin lymphoma Follicular lymphoma	Bristol Meyers Squibb	Yes/Yes	April 2022/February 2021
Abecma™	Ex vivo GTMP	Idecabtagene vicleucel/autologous CAR-T cells yu76h	Relapsed or Refractory multiple myeloma	Celgene	Yes/Yes	August 2021/March 2021

Product	Type	Description	Indication	Company	Approval	Date
Carvykti™	Ex vivo GTMP	Ciltacabtagene autoleucel/autologous CAR-T cells directed to B cell maturation antigen (BCMA)	Relapsed or Refractory multiple myeloma	Janssen Biotech	Yes/Yes	May 2022/February 2022
Skysona™	Ex vivo GTMP	Elivaldogene autotemcel/autologous CD34+ cells tranduced with retroviral vector containing The cDNA of the human *ABCD1*	Early, active cerebral adrenoleukodystrophy	Bluebird Bio	Withdrawn/Yes	September 2022
Provenge™	SCTMP	Sipuleucel-T/autologous dendritic cells	Metastatic castrate-resistant hormone-refractory prostate cancer	Dendreon Pharma	No/Yes	Withdrawn/April 2010
Holoclar™	SCTMP	Ex vivo expanded autologous human corneal epithelial cells containing stem cells	Patients with corneal damage due to (chemical) burns	Holostem Advanced Therapies	Yes/No	February 2015
Spherox™	SCTMP	Spheroids of human autologous matrix-associated chondrocytes	Symptomatic articular cartilage defects of the femoral condyle and the patella of the knee	CO.DON Ag	Yes/No	July 2017
Alofisel™	SCTMP	Darvadstrocel /Mesenchymal stem cells from fat tissue of adult donors	Complex perianal fistulas in patients with Crohn's disease	Takeda Pharma	Yes/No	March 2018
Ebvallo™	SCTMP	Tabelecleucel/allogeneic, EBV-specific T cells from a matched donor	Second-line treatment for transplant recipients who develop, as a result of immunosuppression treatment, Epstein-Barr virus (EBV) associated posttransplant lymphoproliferative disease (PTLD)	Atara Biotherapeutics	Yes/No	October 2022
GINTUIT™	TEP	Allogeneic cultured keratinocytes and dermal fibroblasts in murine collagen	Topical (nonsubmerged) application to a surgically created vascular wound bed in the treatment of mucogingival conditions in adults (venous leg ulcers)	Organogenesis Inc.	No/Yes	March 2012
MACI™	TEP	Matrix-applied characterized autologous cultured chondrocytes	The repair of single or multiple symptomatic, full-thickness cartilage defects of the knee with or without bone involvement in adults	Vericel	Withdrawn/Yes	December 2016
StrataGraft™	TEP	Allogeneic cultured keratinocytes and dermal fibroblasts in murine collagen	Adults with thermal burns containing intact dermal elements for which surgical intervention is clinically indicated	Mallinckrodt plc	No/Yes	June 2021
Rethymic™	TEP	Human allogeneic processed thymus tissue-agdc	Pediatric patients with Congenital Athymia	Enzyvant	No/Yes	October 2021

markers that support effective and personalized drug therapy and drug discovery including toxicity screening (Sayed et al. 2016).

Embryonic Stem Cells

During the earliest stages of mammalian development, soon after egg and sperm combine, the resulting diploid cells are said to be "totipotent," i.e., they can give rise to both the embryo and placental tissue. At the blastocyst stage of embryogenesis (day 5 in humans), the "inner cell mass" or "embryoblast" is compacted and separated from the surrounding "trophoblast." The latter combines with the maternal endometrium to form the placenta. The inner cell mass can be extracted and grown ex vivo as ESCs, which can give rise to all three germ cell types (mesoderm, endoderm, and ectoderm) and, therefore, potentially any cell type found in the adult (Fig. 14.7).

Mouse ESCs were first isolated in 1981 (Evans and Kaufman 1981; Martin 1981), but it took until 1998 for a similar procedure to be described allowing human ESCs to be grown in culture (Thomson et al. 1998). ESCs can now be grown for many cell divisions, limited only by genetic damage that occurs by mutation after extensive culturing. The pluripotency of ESCs has been demonstrated in mice by injecting cells into a fertilized egg, resulting in the production of chimeric mice, i.e., mice made up of cells derived from both the donor and the injected ESCs, with this technology, transgenic mice for research purposes have been generated. HESCs are currently investigated by a set of cell surface markers (CD markers) and their capacity to differentiate. The criteria for this assessment include the expression of surface markers and transcription factors associated with an undifferentiated state. In addition, extended proliferative capacity, pluripotency, and a euploid karyotype are important characteristics of these cells. Recent advances in human pluripotent stem cell research revealed different subpopulations within stem cell cultures covering a wide spectrum of pluripotent states that hold distinct molecular and functional properties (Goodwin et al. 2020). Moreover, evidence suggests that the epigenetic status of the cells is also a relevant criterion for hESCs. Epigenetic alterations may accumulate when hESCs are cultured in vitro. Therefore, genetic stability over extended periods should be considered as a critical parameter, demonstrating that hESC characteristics do not change over time in terms of karyotype, expression of markers, expression of telomerase and their ability to differentiate into the three germlines (ecto-, meso- and endoderm) (Bar and Benvenisty 2019).

Maintenance and Differentiation of ESCs in Culture

HESCs are grown in the presence of high concentrations of basic fibroblast growth factor-2 (FGF2) and are unresponsive to leukemia inhibitory factor (LIF) (Levenstein et al. 2006). The technical challenge, now that hESCs can be maintained and expanded, is to develop robust methods to control and direct ESC differentiation, so that human cells of any desired phenotype can be obtained (Keller 2005; Murry and Keller 2008) with sufficient purity in terms of the absence of unde-

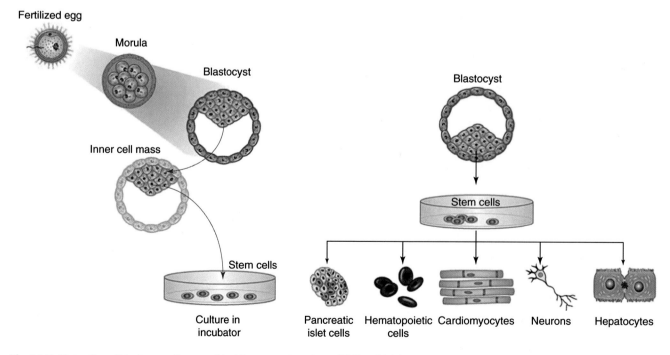

Fig. 14.7 Extraction of the inner cell mass of the blastocyst gives rise to ESCs, which have the capacity to differentiate into all 200+ somatic cell types found in the adult human

sired cells, such as undifferentiated cells, or cells that are capable of de-differentiation into undifferentiated cells or into cells of a different lineage, either of which could cause tumor formation after implantation, both at the site of administration or elsewhere in the body after cell migration. This technology has not fully matured yet. Thus far, attention has focused on the differentiation of human ESCs toward products that could be of obvious use for clinical administration, e.g., midbrain dopaminergic neurons for Parkinson's disease, cardiomyocytes for reinforcement of damaged heart tissue, and pancreatic pre-β-islet cells for implantation in Type I DM.

Since they were isolated for the first time, several first-in-human clinical studies with hESC have been initiated for various indications, including neural diseases (Parkinson's disease, spinal injury), heart disease, cancer, and eye diseases (Table 14.7). In an open-label phase I/II study, hESC-derived retinal pigment epithelial cells (RPE) were given subretinally to patients with Stargardt's disease or patients suffering from macular degeneration. Injected RPE cells showed no sign of hyperproliferation or tumor formation, albeit local immunosuppression was needed to prevent rejection. After 22 months, significant improvement in eyesight was reported for 19 out of 27 patients (Schwartz et al. 2015). Whereas researchers clearly have demonstrated their therapeutic potential, the use of hESC remains controversial. The opinions of scientists, regulators, and public are widely divided, from being very supportive to seeking a regulatory ban on hESC research for ethical/religious reasons. Besides these ethical barriers, some scientific barriers still need to be overcome. Progression towards clinical applications is hampered due to safety concerns, specifically immunogenicity and the unknown potential of undifferentiated escapees for teratomas.

ESC Somatic Cell Nuclear Transfer (Therapeutic Cloning)

An alternative, particularly when an HLA-donor match cannot be found, is to produce ESCs for individual patients, by somatic cell nuclear transfer (SCNT) (Wilmut et al. 2002). This process, also known as "therapeutic cloning," involves the implantation of a cell nucleus from the patient (i.e., genomic DNA extracted from a skin biopsy) into a human egg, which has undergone the removal of its own DNA. The environment in the enucleated egg can reprogram the DNA from the patient, removing epigenetic marks and restoring the DNA to an embryonic state. In addition, the development of an inner cell mass in the egg, after a period of incubation, allows extraction of ESCs that have the patient's exact genotype. These cells could be used subsequently for the production of implants for cell therapy (Fig. 14.8).

Somatic cell nuclear transfer (SCNT), moving nuclear DNA from a donor cell to an enucleated recipient cell to create an exact genetic match of the donor, is an inefficient process. Most eggs that have undergone SCNT cannot completely reprogram the donor DNA, and the surrogate pregnancy is usually unproductive. Moreover, even when the pregnancy comes to term, the cloned offspring are known to carry many epigenetic marks that may compromise normal development.

Given that defects are known to occur after SCNT, the subsequent derivation of cells for clinical uses might also be prone to failure due to defects in ESC differentiation. There is insufficient data available at this stage to judge whether this will be a limitation in practice. However, significant ethical concerns have limited the practice of SCNT. A human egg donor is required, and unless the process becomes more efficient, women who are prepared to donate eggs would need to provide several eggs to produce a single ESC line. There is concern that women could be exploited, particularly women from low economic backgrounds, and as a result, SCNT is not supported by government funding at present in most countries. A restricted number of ESC lines have been produced using spare eggs from in vitro fertilization programs, but SCNT remains a controversial topic and is subject to legal constraints that vary from country to country for mainly biomedical ethical reasons. An alternative source of cells for clinical application is umbilical cord blood stem cells, which are now being banked at childbirth (i.e., in a biobank), at least in private practice, and the first clinical trials have been initiated. Whether cord blood cells can be harnessed to produce all cell phenotypes is not clear at present (see also above "cord blood-derived MSCs"). However, many of the ethical issues surrounding SCNT and uncertainty of cord blood stem cell potency (in vivo activity), may become irrelevant if the promise of iPSCs can be fulfilled.

IPS Cell Technology

Initially, work on pluripotent stem cells (PSCs) was conducted using hESCs; however, the requirement to destroy early-stage embryos in the process of ESC derivation makes their use ethically controversial. In addition, practical considerations hinder their medical applications because any cells or tissues generated from hESCs, by definition, would be allotransplants into the recipient patient (see above).

However, since the discovery of Takahashi and Yamanaka in 2006 that differentiated somatic cells (in particular fibroblasts) can be reprogrammed to produce pluripotent cells by inducing the expression of four transcription factors (Sox2, Oct4, Klf4, and cMyc), the need for human embryo's as a source of pluripotent stem cells has become obsolete (Takahashi and Yamanaka 2006; Takahashi et al. 2007). Over the last decades, induced pluripotent stem cells (iPSC) have exploded, and the technology is now in use in hundreds of

Table 14.7 List of reported and registered hESC-based clinical trials. Adapted from Golchin et al. (2021)

	Indication	Phase	Subject	Start/finish date	Country	Reference
Neural diseases	Spinal cord injury	Phase 1	5	2010–2013	USA	Scott and Magnus (2014)
	Amyotrophic Lateral Sclerosis (ALS)	Phase 1 and Phase 2	21	2018–2020	Israel	NCT03482050
	Parkinson diseases	Phase 1 and Phase 2	50	2017–2020	China	NCT03119636
Heart diseases	Severe heart failure	Phase 1	10	2013–2018	France	NCT02057900
Diabetes	Diabetes type 1	Phase 1 and Phase 2	69	2014–2021	USA	NCT02239354
Reproductive insufficiency	Primary ovarian insufficiency	Phase 1	28	2019–2021	China	NCT03877471
	Infertility	Not applicable	240	2017–2020	China	NCT02713854
	Infertility	–	40	2002–2025	Israel	NCT00353197
Eye diseases	Retinitis pigmentosa	Phase 1	10	2019–2020	China	NCT03944239
	Retinitis Pigmentosa	Phase 1 and Phase 2	12	2019–2021	France	NCT03963154
	Macular degenerative disease	Phase 1 and Phase 2	36	2018–2029	UK	NCT03167203
	Dry Age Related Macular Degeneration Disease (Dry AMD)	Phase 1 and Phase 2	10	2017–2020	China	NCT03046407
	Dry AMD	Phase 1 and Phase 2	16	2015–2023	USA	NCT02590692
	AMD	Phase 1 and Phase 2	10	2018–2020	China	NCT02755428
	Stargardt's Macular Dystrophy (SMD)	–	12	2013–2019	UK	NCT02941991
	Dry AMD	Phase 1	3	2016–2019	South Korea	NCT03305029
	AMD	Phase 1 and Phase 2	12	2012–2020	South Korea	NCT01674829
	Outer retinal degenerations	Phase 1 and Phase 2	18	2015–2019	Brazil	NCT02903576
	SMD	Phase 1 and Phase 2	12	2011–2015	UK	NCT01469832
	SMD	Phase 1 and Phase 2	13	2011–2015	USA	NCT01345006
	Advanced Dry AMD	Phase 1 and Phase 2	13	2011–2015	USA	NCT01344993
	Age-related macular degeneration	Phase 1 and Phase 2	24	2015–2024	Israel	NCT02286089
	SMD	Phase 1	3	2012–2015	South Korea	NCT01625559
	SMD patients	–	13	2012–2019	USA	NCT02445612
	AMD	–	11	2012–2019	USA	NCT02463344
	AMD	Phase 1	2	2015–2019	UK	NCT01691261
	Retinal pigment	–	2	2016–2020	UK	NCT03102138
	Macular degeneration diseases	Phase 1 and Phase 2	15	2015–2019	China	NCT02749734
Immunotherapy	Non-small cell lung cancer	Phase 1	48	2018–2022	UK	NCT03371485
Injury	Meniscus injury	Phase 1	18	2019–2020	China	NCT03839238

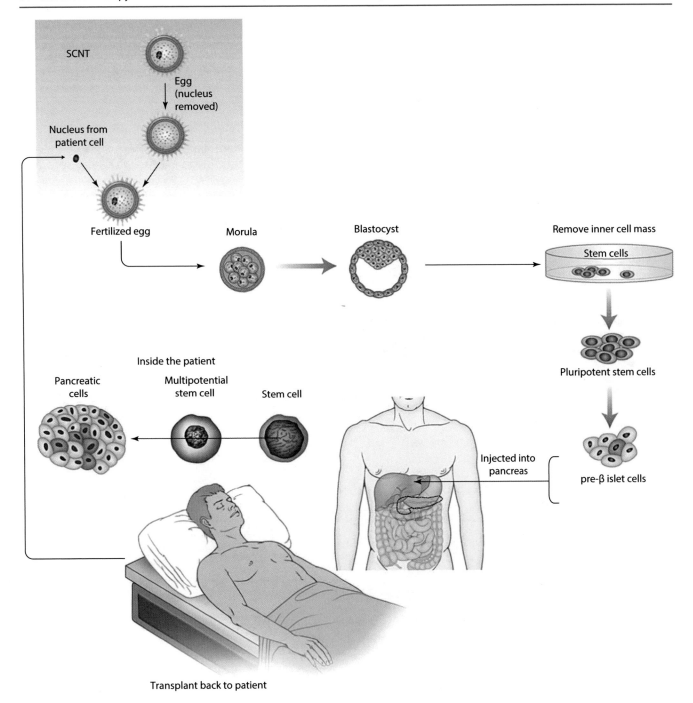

Fig. 14.8 Schematic diagram of the production and clinical use of cell therapies derived using somatic cell nuclear transfer (therapeutic cloning). The example given is for possible treatment of Type I insulin-dependent DM. The final maturation of the pre-β islet cells occurs in the patient's body

stem cell biology laboratories around the world and has been tested for therapeutic applications in clinical trials (Figs. 14.9 and 14.10). The four genes initially identified can be partly substituted by alternatives, and several experiments have shown that integrated lentiviral constructs can be avoided to reduce safety concerns, by using nonviral plasmids (Jia et al. 2010), miRNA and mRNA (Yang et al. 2011; Liu and Verma 2015), protein transduction, and even by substituting some of the factors with small molecules (Yuan et al. 2011).

Considerable effort has been directed at investigating how iPSCs differ from ESCs and whether reprogramming is complete enough to produce truly pluripotent cells. True pluripotency is difficult to demonstrate unequivocally in human iPSCs, so the development of methods to measure the extent

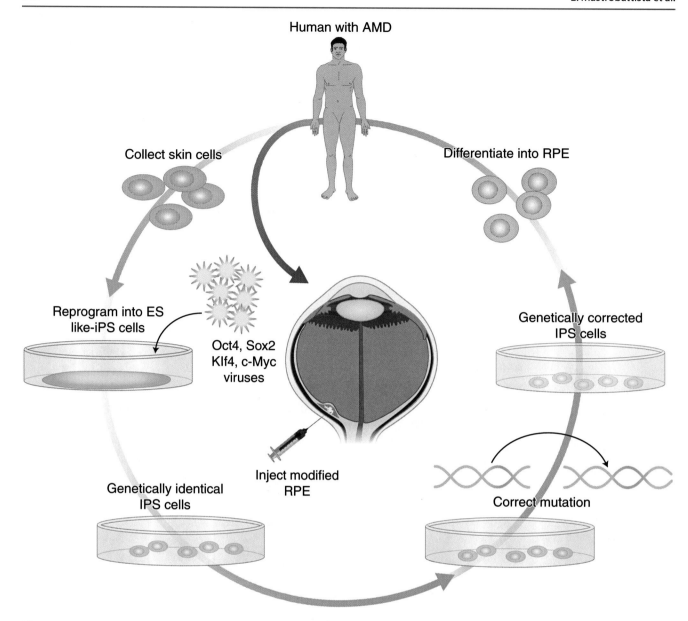

Fig. 14.9 Method used to produce iPS cells, correct a genetic defect responsible for AMD, and implant the corrected stem cells into humans to cure AMD

of reprogramming will be important for practical applications. There are indications that iPSCs can have chromosomal defects and are not fully reprogrammed (Chin et al. 2010). Female human iPSCs appear to maintain the inactivated X chromosome that was present in the skin fibroblasts, although this has not been a problem with mouse iPSCs (Tchieu et al. 2010).

In recent years, progress has been made with improved culture techniques and differentiation protocols, which resulted in safer and clinically relevant iPS cells with lower tumorigenic risk. Various clinical trials with iPSC are being conducted worldwide with Japan and the US being the frontrunners (Fig. 14.10) (Kim et al. 2022). Similar to hESC, most clinical trials focus on retinal degenerative diseases, with Dr. Takahashi at the Riken Center for Developmental Biology in Japan being a pioneer. She led the first team to successfully transplant autologous iPSC-derived RPE cell sheets into a patient with age-related macular degeneration (AMD) in 2014 (Sayed et al. 2016) and is presently conducting studies on the transplantation of iPSC-derived corneal cells. Despite these individual successes, clinical development of iPSC therapies has been slow, possibly related to potential safety issues associated with these cells, including genomic instability and tumorigenicity (Nori et al. 2015). Further clinical research investigating the long-term safety of using iPSC in regenerative medicine is needed.

Fig. 14.10 Distribution of clinical trials involving iPSCs accordingly to different categories. (**a**) Initial classification of 81 clinical trials into observational and interventional studies as addressed by the authors. (**b**) New classification of 81 clinical trials into nontherapeutic and therapeutic studies as defined earlier. (**c**) Classification of nontherapeutic clinical trials according to their use. (**d**) Worldwide distribution of nontherapeutic studies. (**e**) Classification of studies according to the category of targeted disease. Adapted from Kim et al. (2022)

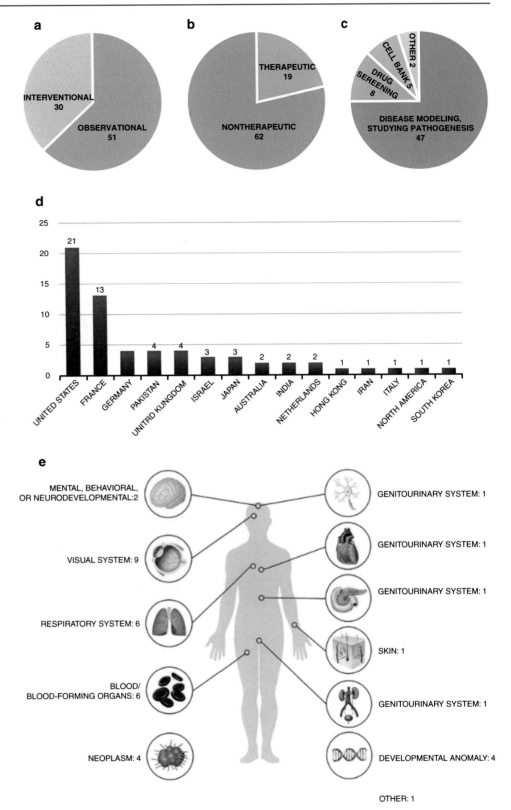

E. Mastrobattista et al.

Transdifferentiation

Transdifferentiation is the process of converting cells from one lineage into another without going through a pluripotent cell stage, as is the case for reprogramming. Forced expression of genes has been used to convert fibroblasts directly into unrelated differentiated cells by skipping the iPSC stage, but also small chemical compounds can be used for this, although less efficient (Hybiak et al. 2020). Upregulation of apoptosis and cancer-related genes occurs in addition to chromatin remodeling during transdifferentiation. An important criterium of successful transdifferentiation is full morphological and molecular differentiation of both initial and final cells. Different cell types have been generated from fibroblasts, including neurons, hepatocytes, endothelial cells, and cardiomyocytes. The technique used is analogous to that used to derive iPSCs, except that genes associated with the desired somatic cell are expressed instead of pluripotency genes. The realization that cellular phenotypes can be transformed in this way has been met with astonishment and is certainly breakthrough technology. It opens the possibility of performing interconversion in vivo, although it does not allow for the expansion of cells in preparation for an implant. However, direct reprogramming of fibroblasts to neural stem cells, as reported in 2012 (Han et al. 2012; Thier et al. 2012),

may be a shortcut to growing neurons. This approach may offer some advantages over the production of neurons by way of iPSCs.

Three-Dimensional Technologies

Another technology, tissue engineering, is combining somatic cell technologies or cell therapy technologies described above, with various types of biocompatible materials to solve structural challenges that are often surgical or immunological in nature. Three-dimensional (3D) technologies, including biomaterial scaffolds, can have many purposes, such as supporting cell viability, induction of cell differentiation, provision of a substrate for cell growth and support of tissue regeneration, provision of the shape, scale, and volume of a desired tissue, provision of growth factors, and encapsulation of cell-based products to protect the product from the host immune system to avoid rejection. This is schematically presented in Figs. 14.11 and 14.12 (Smith and Grande 2015). 3D technologies can be divided into four subtypes of technologies, as shown in Table 14.8. For further reading, see Pina et al. (2019), and Bajaj et al. (2014).

Combinations of the Above Technologies

A combination of the above technologies is currently in preclinical development in the cell therapy area, e.g., the self-

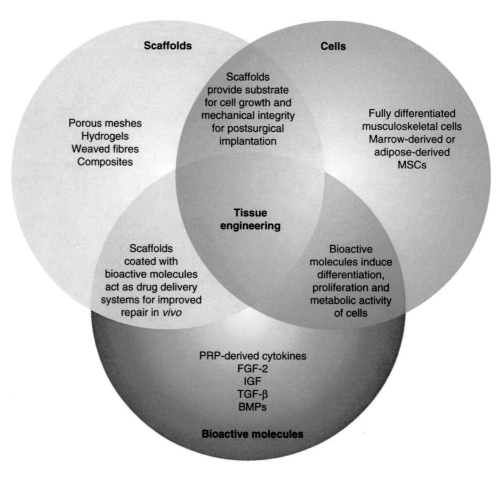

Fig. 14.11 The role of scaffolds in tissue engineering strategies. Scaffolds are an important component of the tissue engineering triad. *BMP* = bone morphogenetic protein, *FGF-2* = fibroblast growth factor 2, *IGF* = insulin-like growth factor, *MSC* = mesenchymal stromal cells, *PRP* = platelet-rich plasma, *TGF-β* = transforming growth factor β. Adapted from Smith and Grande (2015)

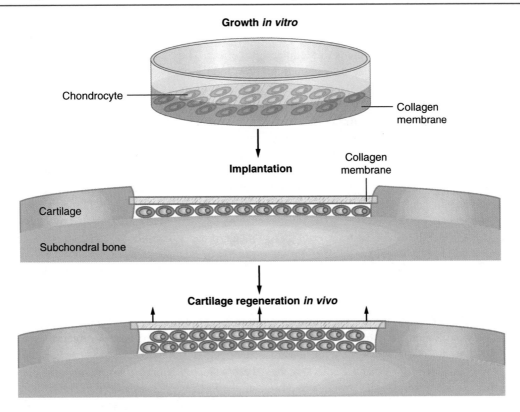

Fig. 14.12 Matrix-assisted chondrocyte implantation (MACI) in cartilage repair. MACI uses chondrocytes that have been seeded into a collagen scaffold and cultured for a period of time prior to surgical implantation, adapted from Smith and Grande (2015)

Table 14.8 3D technologies and examples

Subtype of 3D technologies	Examples of products/organs in pre-clinical or clinical development or commercially used
Simple biomaterials e.g., hyaluronic acid, bone substitutes, alginate-encapsulated islets	Allogenic adipose derived regenerative cells (Keratinocytes) encapsulated in hyaluronic acid to regenerate extracellular matrix-like material to treat corneal blindness; transplantation of pancreatic islets in immune protective alginate capsules to treat DM Type I; MACI® for repair of cartilage defects of the knee (see Fig. 14.13)
3D/shaped scaffolds that provide organ shape and bio-resorbable substrate for cell growth	Bladder; trachea; 3D-printing technologies
Tissue-derived (decellularized) scaffolds that are 3D but with added benefits of native biomechanical strengths and matrix factors	Esophagus; trachea
Smart (second generation) biomaterials that may have thixotropic, thermo-responsive, growth-factor-encapsulating or in situ self-assembly properties	Chitosan and hyaluronic acid are typically used as excipients for thermoset injectable hydrogels encapsulating cells

formation of complex organ buds into organ-like structures, i.e., organoids (Takebe et al. 2015; Brassard and Lutolf 2019).

Part B: Technologies for Gene Therapy

Introduction

Gene therapy aims to treat or cure a disease by inserting, altering, or removing genetic material in affected cells in the human body. The concept of gene therapy arose during the 1960s and early 1970s and was a direct consequence of a series of discoveries made in the preceding years. It started with developing tools to select cells based on a functional trait that could be acquired via the transfer of exogenous DNA. For example, in the mid-1960s, it was demonstrated that cells from patients with Lesch-Nyhan syndrome, which lack the enzyme hypoxanthine guanine phosphoribosyl transferase (HPRT), could be rescued upon uptake of exogenous DNA encoding HPRT, when grown in hypoxanthine-aminopterin-thymidine (HAT) medium which is selective for cells expression HPRT. Despite this proof-of-concept that exogenous DNA can be expressed in mammalian cells, the genetic transformation was inefficient. Methods to facilitate the delivery of exogenous DNA were developed, including the calcium phosphate chemical transfection method. Later, work on polyomaviruses, papovaviruses, and retroviruses provided the tools for more efficient transformation. With the parallel discovery of restriction enzymes in the early 1970s and pioneering work by Paul Berg, Stanley Cohen, and Herbert Boyer, recombinant DNA technology became a fact, enabling the manipulation of DNA, including viral genomes, to insert therapeutic genes that could be efficiently transduced (Cf. Chap. 1). Retroviral vectors were further optimized and became the workhorse for the transfer of therapeutic genes to correct disease phenotypes in vitro. By the mid-1980s, it was demonstrated that T cells from SCID patients lacking the enzyme adenosine deaminase (ADA) could be restored after retroviral vector-mediated delivery of genes encoding ADA. These findings formed the basis of the first human gene therapy clinical trial for ADA-SCID on September 14, 1990, at the National Institute of Health (NIH). The first patient included was Ashanti DeSilva, at the age of four. The medical team isolated the patient's T lymphocytes through apheresis, exposed these cells ex vivo to a genetically engineered live nonvirulent retrovirus carrying the normal ADA gene, and transfused these genetically modified T cells back into the patient's bloodstream. The treatment was successful in the sense that no adverse events were observed, and the low number of T cells that were transduced continued to express the recombinant transgene for over 12 years (Blaese et al. 1995; Muul et al. 2003), but treated patients still relied on enzyme replacement therapy. Now, 30 years later, over 3000 clinical trials with gene therapy were initiated, resulting in 18 gene therapies approved by the Food and Drug Administration (FDA) and European Medicines Agency (EMA).

In this section, we discuss the current state of gene therapy and the gene therapy medicinal products on the market, focusing on in vivo gene transfer. We will discuss how gene therapy can be applied and the various methods of gene transfer, including synthetic and viral vectors. Finally, we highlight the diseases currently subjected to gene therapy and touch upon the regulation of gene therapy products.

Gene Therapy: Definitions and Ways of Application

The definition of gene therapy continues to evolve to keep up with the ongoing technological advances. As a result, older definitions have become obsolete, and others are too broadly defined. Take, for example, the definition as applied by the European Medicines Agency: "gene therapy medicines contain genes that lead to a therapeutic, prophylactic or diagnostic effect. They work by inserting 'recombinant' genes into the body, usually to treat a variety of diseases, including genetic disorders, cancer or long-term diseases." Strictly speaking, any modification applied to the genome that does not involve the insertion of recombinant DNA is not considered gene therapy. With the rise of CRISPR-Cas (clustered regularly interspaced short palindromic repeats) gene editing, base editing, and prime editing (see section "Designer Nucleases for Gene Editing") that can introduce modifications in the human genome without the use of recombinant DNA, this definition falls short. Conversely, the FDA uses a much broader definition: "Human gene therapy seeks to modify or manipulate the expression of a gene or to alter the biological properties of living cells for therapeutic use." FDA generally considers human gene therapy products to include all products that mediate their effects by transcription or translation of transferred genetic material, or by specifically altering host (human) genetic. According to this definition, any intervention that leads to altered gene expression would be considered a gene therapy, including those that do not directly change the genetic makeup of a cell such as gene silencing using RNA interference or the delivery of therapeutic mRNA. To bring some clarity and to draw a clear line as to where gene therapy ends and other nucleic acid therapies start (cf. Chap. 13), we propose the following definition: gene therapy is any intervention that leads to deliberate and long-lasting genomic alteration(s) or episomal expression of recombinant DNA with the aim to treat or cure a disease. This definition excludes interventions that affect gene expression at the level of transcription or translation, such as siRNA therapy, mRNA therapy, and antisense ON therapy, which are often transient and require frequently repeated dosing (see Fig. 14.13).

Gene therapy can be applied in three different ways: gene augmentation or addition, gene correction, and gene knockout.

Gene Augmentation

With **gene augmentation or addition**, an intact copy of a malfunctioning, disease-causing gene is inserted into the genome of patient cells (Fig. 14.14). This approach can only be used if the disease is caused by a "loss-of-function" gene mutation. These are mutations in a gene that causes the encoded protein to (partially) lose its function, which in turn causes disease. Loss-of-function mutations are typically recessive, meaning that both copies of the autosomal gene have defects. The addition of an intact copy of the gene under the control of a strong promoter thereby restores this func-

Fig. 14.13 Overview of different modes of intervention of therapeutic nucleic acids. Gene therapy and gene editing act directly on the genomic DNA and will be discussed in this chapter. Other interventions with nucleic acids, including synthetic oligonucleotides (ONs), small interfering RNA (siRNA), adenosine deaminase acting on RNA (ADAR), and mRNA that act at the level of transcription, RNA-splicing or translation are discussed in Chap. 13. Created with BioRender.com

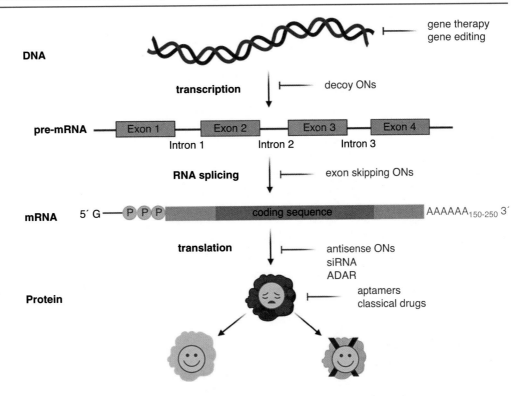

Gene Knock Out

A gene mutation might sometimes lead to a mutant protein that acquires a new, disease-causing function. Such mutations are referred to as "gain-of-function" mutations and are often dominant. Take, for example, the hereditary form of transthyretin amyloidosis (ATTR), which occurs when mutations in the *TTR* gene for transthyretin lead to instability of the tetrameric protein and formation of aggregates and fibrils that damage cells, leading to clinical symptoms. Such a gain-of-function mutation can only be treated if the mutated protein production is halted. This can be done by **gene knock out** in which a targeted disruption of the mutated gene (but not the unaffected allele) is introduced. CRISPR-Cas is a technology often used for creating a targeted gene knock out (see Sect. "CRISPR-Cas9" below).

Gene Correction

Gene correction aims to alter the disease-causing gene mutation. These corrections can be small (point mutations) or require the insertion of large pieces of DNA in case the mutation involves partial deletion of a gene. Gene editing systems such as Zinc-finger nucleases, TALENs (Transcription activator-like effector nucleases), CRISPR-Cas, base editors, or prime editors (see section "Designer Nucleases for Gene Editing" and Cf. Chap. 9) have been developed for this purpose. As such corrections are often very precise and include the use of molecular scissors to cut

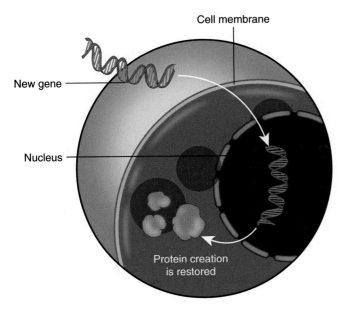

Fig. 14.14 The principles of gene addition to restoring loss-of-function mutations by inserting an intact copy of a gene (randomly) into the genomes of affected cells

tion. Insertion of the therapeutic transgene into the genome can be random or targeted to a specific region within the genome. Both viral and synthetic vectors can be used to deliver such gene constructs into affected cells (see section "Delivery Systems").

open the genomic DNA at precise locations, gene correction is sometimes referred to as gene surgery.

Gene editing tools can be categorized based on those that introduce double-strand cuts, those that only cut one strand of the DNA or those that do not cut at all.

Somatic Versus Germline

Gene therapy can be applied to modify individual cells in the body. This is called somatic gene therapy and only affects the patient being treated. The corrected traits will not be inherited by potential offspring. Conversely, germline gene therapy aims to modify the germ cells (sperm cells and/or egg cells) to correct genetic mutations in the germline. If such modified germ cells are being used for reproduction, this will lead to offspring in which the gene correction is carried by all cells of the body, including the germline cells. Such corrections will therefore be passed down from generation to

generation. Since gene therapy is a relatively new form of therapy, we do not yet know the long-term side effects of this, and as such, germline modifications would be unethical to perform at this stage. Many scientists, therefore, call for a global moratorium on germline gene editing (Lander et al. 2019).

Ex Vivo Versus In Vivo

Ex vivo gene therapy involves the genetic modification of cells outside of the body and their subsequent transplantation back into patients (Fig. 14.15). The advantage of this approach is that there is no patient exposure to the gene transfer vector, which can sometimes be harmful. Furthermore, the correctly modified target cells can be selected, expanded, and, if desired, differentiated before being transferred back to the patient to improve efficacy and safety. The limitation of such an ex vivo approach is that it

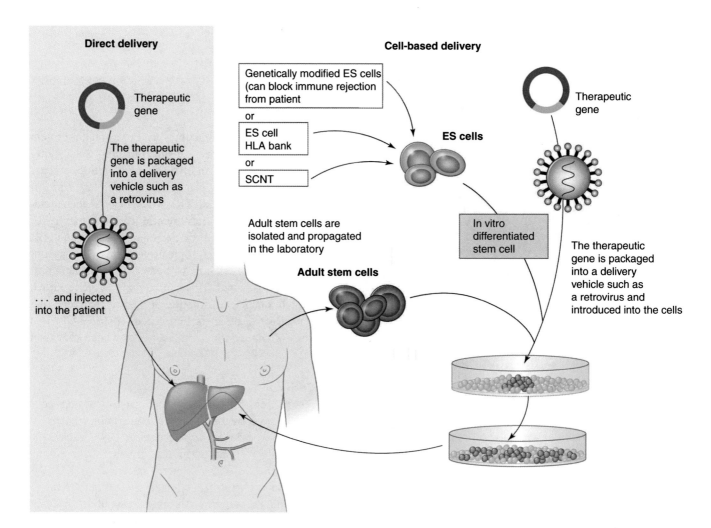

Fig. 14.15 Methods of administration of gene therapy vectors. In vivo gene transfer involves direct administration of the vector in the tissue of interest. Ex vivo gene transfer requires the collection of cellular targets from the patient. The cells are treated in culture with the vector. Cells expressing the therapeutic transgene are harvested and given back to the patient. ES: Embryonic Stem (cell). SCNT: somatic cell nuclear transfer (strategy). From Zwaka 2006; with permission to reprint

can only be applied to cells easily isolated from the human body (e.g., hematopoietic stem cells).

In vivo, gene therapy uses gene transfer vectors locally (e.g., into the eye or muscle) or systemically (via the bloodstream) to reach distant organs or tissues in the patient. Both viral and synthetic vectors have been developed for this purpose, which will be detailed in the section below. The advantage of this approach is that it avoids cumbersome cell isolations and manipulations in the laboratory and can, in principle, be applied by a single injection. The downside is the relatively low gene transfer efficiencies and potential vector-related (immune)toxicities to which the patients are exposed.

Designer Nucleases for Gene Editing

Designer nucleases are engineered nucleases that can be targeted to unique sequences in the human genome to introduce a double-strand DNA cut (Merkert and Martin 2016). As a result of this genomic DNA damage, the endogenous DNA repair system is activated to repair these cuts and ligate the ends of the fragmented DNA. Several different DNA repair pathways can be employed by the cell, which include nonhomologous end-joining (NHEJ), microhomology-mediated end-joining (MMEJ) and homology-directed repair (HDR)(Fig. 14.16).

With NHEJ, the ends of the damaged DNA recruit a protein complex that polishes the DNA: it removes damaged or mismatched nucleotides and randomly fills in nucleotides before the ends are ligated back to each other. As a consequence, NHEJ often leads to small insertions or deletions (indels) at the cut site. If this cut is inside a gene, such indels often lead to a frameshift in the coding sequence, and loss of protein function. It can therefore be used to knock out genes with gain-of-function mutations. MMEJ uses microhomology regions that may be present after end resection of the DNA ends, leading to the base pairing of the single-strand DNA overlaps and removal of the remaining pieces of ssDNA. This leads to deletions at the cut site. Both NHEJ and MMEJ are active throughout the cell cycle and thus active in both dividing and nondividing cells. However, these pathways do not allow precise editing as they tend to lead to errors.

For precise editing, the homology-directed repair is preferred. HDR is mostly active in cells in the G2 and S phases

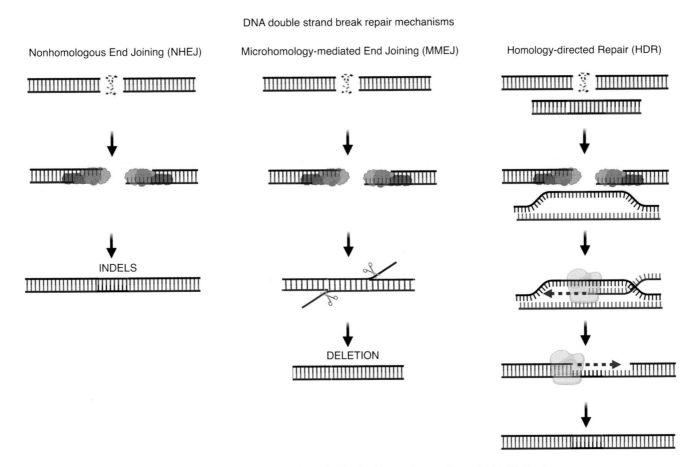

Fig. 14.16 Different pathways for DNA double-strand break repair. For details, see the text. Created with BioRender.com

and is meant to repair broken DNA strands prior to mitosis, making use of the homologous DNA from a sister chromatid to guide the repair. In a similar fashion, DNA double-strand break repair can be guided by providing an exogenous DNA template during repair. This HDR template can be single or double-stranded DNA with on the 5' and 3' ends sequences that are homologous to the sequences surrounding the double-strand cut. The length of such homology arms can vary from only 30 bases up to a few hundred or even thousand bases, depending on the overall size of the DNA to be inserted at the double-strand cut size. The exact working mechanism of HDR is still under debate, but it is generally believed that it involves 5' to 3' resection of the DNA ends at the double-strand cut site, after which the 3' single-strand DNA overlap can invade the homologous DNA duplex of the HDR template to initiate the recombination repair process (Fig. 14.16). Compared to NHEJ and MMEJ, HDR is much less error-prone and often enables precise gene editing. Using single-stranded synthetic oligonucleotides of only 60 bases, specific point mutations can be introduced at a precise location within the genome (Shy et al. 2022). This technology offers great potential to correct debilitating dis-eases caused by a single point mutation in a single gene, such as sickle cell anemia, β-thalassemia, transthyretin amyloidosis (ATTR), or certain genetic eye diseases.

The first designer nucleases to be engineered were the zinc-finger nucleases (ZFNs), followed by transcription activator-like effector nucleases (TALEN), and more recently, the clustered regularly interspaced short palindromic repeats (CRISPR) and its CRISPR associated Cas protein (CRISPR/Cas) (Fig. 14.17) (LaFountaine et al. 2015).

They consist of an engineered restriction endonuclease, often *FokI*, fused to 3-6 zinc-finger proteins, each recognizing a specific three-base pair sequence. *FokI* from the bacterium *Flavobacterium okeanokoites,* only cleaves DNA when it is bound to its recognition sequence GGATG and when it can form dimers. By directing *FokI* monomers to specific sequences on both strands of the DNA, *FokI* monomers can be positioned to form dimers and introduce a double-strand cut in the genomic DNA (Fig. 14.17). Although initially popular, the need for generating complex fusion proteins, which is often time-consuming, and the CRISPR-Cas technology's emergence has made this gene editing technique less popular.

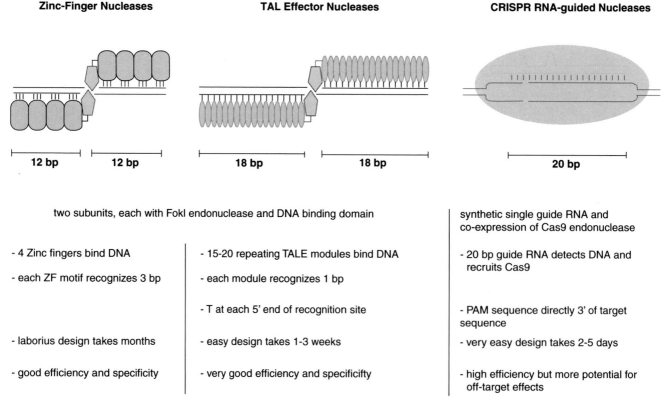

Zinc-Finger Nucleases	**TAL Effector Nucleases**	**CRISPR RNA-guided Nucleases**

| 12 bp | 12 bp | 18 bp | 18 bp | 20 bp |

two subunits, each with FokI endonuclease and DNA binding domain		synthetic single guide RNA and co-expression of Cas9 endonuclease
- 4 Zinc fingers bind DNA	- 15-20 repeating TALE modules bind DNA	- 20 bp guide RNA detects DNA and recruits Cas9
- each ZF motif recognizes 3 bp	- each module recognizes 1 bp	
	- T at each 5' end of recognition site	- PAM sequence directly 3' of target sequence
- laborius design takes months	- easy design takes 1-3 weeks	- very easy design takes 2-5 days
- good efficiency and specificity	- very good efficiency and specificifty	- high efficiency but more potential for off-target effects

Fig. 14.17 Three types of designer nucleases for targeted genome editing. Zinc-finger nucleases and TAL effector nucleases make use of modular DNA-binding proteins to target the FokI nuclease to a specific sequence, whereas CRISPR-Cas uses an associated RNA to target the Cas nuclease to a specific sequence

TALEN

Similar to ZFNs, the transcription activator-like effector nucleases (TALEN) exploit customizable DNA-binding proteins fused to a FokI nuclease. The TALE DNA-binding domains were identified as secreted proteins from the *Xanthomonas spp.* bacteria and consist of highly conversed 33–35 amino acid repeats with two amino acid repeat-variable diresidues (RVD), which dictates individual nucleotide specificity. Assembling repeats into TALE arrays flanked by essential TALE-derived N and C-terminal domains fused to FokI repurposes the system for genome editing. As a result, TALENs are very precise with low off-target editing and are still popular in gene therapy applications. For example, several companies have used this technology for engineering chimeric antigen receptor (CAR) T cells for cancer immunotherapy.

CRISPR-Cas9

CRISPR, an acronym for Clustered Regularly Interspaced Short Palindromic Repeats, and its associated proteins (Cas) are part of the bacterial adaptive immune system to shield invading viruses and which has been repurposed for gene editing in human cells. The Cas9 gene editing system consists of a Cas9 endonuclease interacting with a trans-activating RNA (tracrRNA), which is directed to a target sequence by ~ 20 nt complementary sequences in the CRISPR RNA (crRNA) and flanked by the 3′ protospacer adjacent motif (PAM). The crRNA and tracrRNA can be joined by a tetraloop to form a single guide RNA (sgRNA) (Fig. 14.18). When a sgRNA:Cas9 ribonucleoprotein binds its target sequence in the presence of a flanking PAM sequence, the Cas9 protein will introduce a double-strand break (DSB). If a PAM sequence is lacking, the Cas9 will not cut. Cas9 proteins from different bacteria recognize different PAM sequences. The most frequently used are Cas9 from *Streptococcus pyogenes* (spCas9; PAM 5′-NGG-3′, where N is any nucleobase) and Cas9 from *Staphylococcus aureus* (saCas9; PAM 5′-NNGRRT-3′, where R is a purine), but many different natural and engineered variants of these Cas nucleases exist. For example, Cas12a, previously known as Cpf1 from *Acidoaminococcus sp.*, does not require a tracrRNA but only a crRNA. After binding its target sequence and identifying its PAM (5′-YTN-3′, where Y is a pyrimidine) Cas12a introduces a sticky-end cut, with 4-5 nucleotides overhang. The advantage of CRISPR-Cas for gene editing over the use of ZFN or TALEN is the ease by which the system can be adapted to target a specific sequence in the genome. Whereas for TALEN and ZFN, it is required to engineer a new targeted nuclease for each target sequence, which can take weeks to months as it requires elaborate recombinant protein expression and purification, the target specificity of Cas9 and Cas12a is completely determined by the associated guide RNA, which can be easily synthesized. A downside of CRISPR-Cas gene editing system is its relatively high base mismatch tolerance, meaning that it can introduce DSB at places in the genome with near-identical 20-nt sequences compared to the intended target sequence (as long as a PAM sequence is present). It has been reported that off-target events can occur with as many as 3-5 base pair mismatches. This could introduce unintended mutations in the genome at potentially harmful places. For that reason, Cas9 proteins have been engineered to increase specificity. For example, spCas9-HF1 has very low levels of off-target

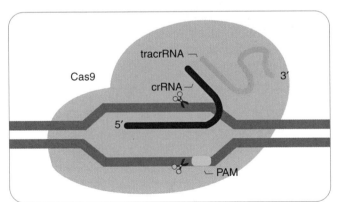

2-part guide RNA

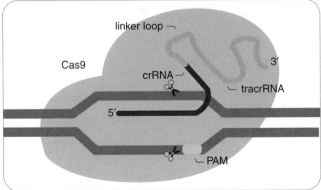

Single guide RNA

Fig. 14.18 The Cas9 gene editing system consists of a Cas9 nuclease with an associated trans-activating RNA (tracrRNA), which is directed to a target sequence by ~ 20 nt complementary sequences in the CRISPR RNA (crRNA) and flanked by the 3′ protospacer adjacent motif (PAM) (left panel). Scissors indicate the location of the double-strand breaks introduced by Cas9. The tracrRNA and crRNA can also be joined by a linker loop to form a single guide RNA (sgRNA; right panel)

effects, albeit at the cost of lower on-target cutting efficiency than wild-type spCas9 (Kleinstiver et al. 2016).

Despite these improved Cas9 nucleases, introducing DSB still imposes a risk during therapeutic gene editing. Besides introducing unintended mutations at off-target sites, generating DSB in genomic DNA can also lead to large deletions and chromosomal translocations. As a result, Cas mutants have been generated that can still be targeted to specific sequences in the genome but are devoid of nuclease activity. These Cas mutants form the basis of an entirely new class of gene editing tools without DSB. The most frequently used will be discussed below.

Base Editors

DNA Base editors allow the introduction of point mutations in the DNA without generating DSB. They consist of two domains: a catalytically "dead" Cas9 enzyme (dCas9) or a nickase (nCas9), fused to a single-stranded DNA modifying enzyme for targeted nucleotide alteration (Fig. 14.19). Two classes of DNA base editors have been described: cytosine base editors (CBE) and adenine base editors (ABE). Cytosine deamination generates uracil, which base pairs as thymidine in DNA and thus converts C:G into A:T base pairs. Conversely, ABEs convert A:T into G:C base pairs. Collectively, all four transition mutations (C→T, T→C,

A→G, and G→A) can be installed. Besides these two major base editors, new variants are being developed, including those that introduce a base transversion (C→G), expanding the scope of disease-causing point mutations that can be edited (Kurt et al. 2021). The distance at which the base editors operate within the protospacer (i.e., the target sequence to which the base editor is bound)

is called the editing window and is dependent on the type of BE being used. For CBE, the editing window spans positions 4-8 of the protospacer, and for ABE positions 4-7 (ABE7.10) or 8-10 (ABE6.3, ABE7.8, or ABE7.9).

Conversion of multiple nucleobases within the editing window is possible and may lead to undesired edits, called bystander edits. Adapted from Antoniou et al. (2021).

Prime Editors

While BEs can, in principle, correct the majority of pathogenic point mutations, they cannot perform all possible single-nucleotide conversions and also cannot mediate targeted insertions or deletions. The solution for this is prime editors (PEs) developed by Anzalone et al. (2019). A PE consists of a reverse transcriptase fused to a nCas9 nickase (Fig. 14.20). In combination with an engineered prime editing guide (pegRNA), this construct is directed to a target sequence in the genome, which nicks the noncoding DNA

Fig. 14.19 Cytosine and Adenine base editors. A. Cytosine base editors (CBE) consist of a nCas9 fused to a cytosine deaminase and a uracil glycosylase inhibitor and convert C:G into T:A base pairs. The editing window within the protospacer is illustrated in green. B. Adenine base editors (ABE) consist of a dCas9 fused to an engineered adenine deaminase from *E. coli* (TadA) and convert A:T into G:C base pairs. The editing window is illustrated in purple

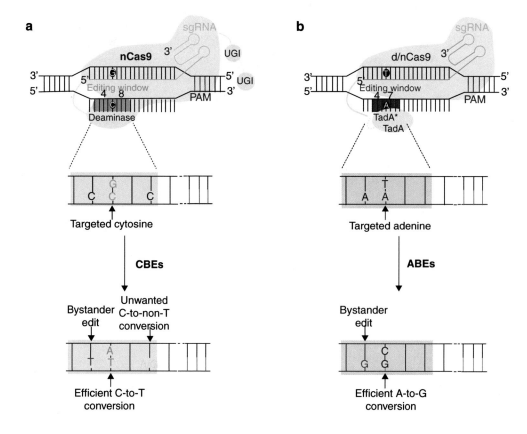

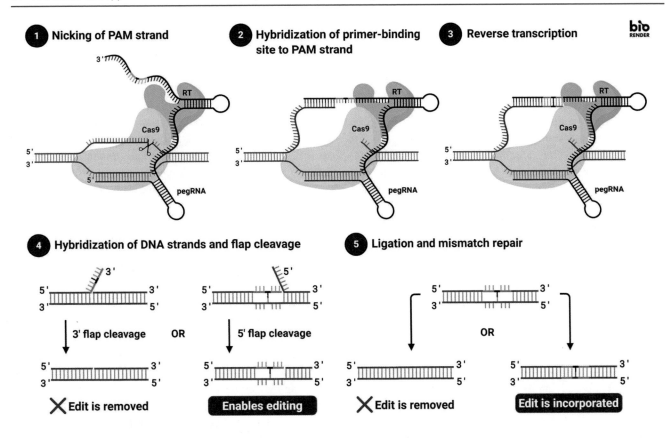

Fig. 14.20 Mechanism of prime editing (PE). Once bound to its cognate targeting sequence, the PE introduces a nick (1), releasing the 3' end of the PAM strand, which can hybridize with the pegRNA extension (2) to initiate reverse transcription (3). The formed branched DNA structure either contains a 3' flap containing the edited sequence or a 5' flap containing the original, unedited sequence, which will be removed by exonucleases, after which the nick is ligated (4). In case of incorporation of the edited strand, a heteroduplex is formed that will be repaired by intrinsic mismatch DNA repair pathways (5)—figure copied from Wikipedia

strand. Hybridization of the 3' end of this DNA strand with the pegRNA initiates reverse transcription using the pegRNA extension as a template. This leads to a branched DNA intermediate with either a 3' flap (containing the edited sequence) or 5' flap (containing the original, unedited sequence) sticking out. Exonucleases remove these flaps, yielding in 50% of the cases a heteroduplex DNA composed of one edited and one unedited strand. The intrinsic mismatch DNA repair pathway can then repair this mismatch with two possible outcomes: the edited strand is copied to the complementary strand, or the unedited strand is restored.

PE is a very flexible gene editing tool that can introduce base transitions, transversions, and deletions (up to 80 bp) or insertions (up to 60 bp). However, it is less suitable for inserting larger pieces of DNA. The downside of PEs is their low efficiency of editing, which might contribute to the edit's various outcomes and instability of the rather large pegRNA. Further research is needed to make this method of gene editing more efficient.

Delivery

For gene therapy to be effective, the transgene construct or gene editing tools must be delivered to diseased cells in the body. For inherited diseases, this would, in principle, be all somatic cells in the body, but this will technically not be feasible. Hence, delivery strategies focus on targeting cells in vivo or ex vivo that are mostly affected by the genetic disease. In the section below, we discuss the various gene delivery vectors that have been developed for gene therapy applications. This includes viral vectors, synthetic vectors, and vectors derived from extracellular vesicles.

Viral Vectors

Viruses have evolved to introduce their genetic material into the host cell efficiently. These properties can create a vehicle to deliver genes for expression into cells. This is called a viral vector. A viral vector differs from the native counterpart in that it cannot replicate and is less pathogenic: genes involved in viral replication and pathogenicity are removed to generate a safe vehicle with enough space to include transgene cassettes. The most commonly used viral vector systems for gene therapy purposes are adeno-associated viral (AAV) and retroviral vectors.

To produce functional viral vector particles, the genes required for the structural viral proteins need to be provided *in trans* with the transfer vector, i.e., a piece of DNA, often plasmid DNA containing the therapeutic gene construct. The genes encoding the viral proteins for replication and packaging of the transfer vector can be placed on plasmids and transfected into producer cells (Fig. 14.21). The transfer vector retains minimal sequences required for stability, replication, or integration. These elements include the terminal sequences called inverted terminal repeats (ITRs) for AAV vectors and long-terminal repeats (LTRs) for gammaretroviral vectors. Additional elements, such as cis-acting elements, e.g., a packaging signal (ψ), may be essential for the efficient incorporation of the transfer vector into the viral particle. Retroviral vectors, besides the packaging signal, may also require truncated leader regions for efficient packaging. Other virus-derived sequences may be included to enhance delivery to the nucleus, such as the central polypurine tract (cPPT) in lentiviral vectors (a subtype of retroviral vectors) (Zennou et al. 2000), or elements to enhance expression of the transgene, such as the commonly incorporated Woodchuck hepatitis posttranslational regulatory element (WPRE). This will be discussed in more detail below. Retroviruses, including gammaretroviral vectors, lentiviral vectors, more recently alpharetroviral vectors, and AAV, are the most extensively studied for ex vivo and in vivo use in monogenic diseases. Their characteristics are summarized in Table 14.9.

Retrovirus

Biology

Retroviruses are membrane-enveloped RNA viruses containing two copies of a positive single-stranded RNA genome (Fig. 14.22). The retroviral vector systems that are commonly used in clinical applications for monogenic diseases are derived from gammaretroviruses and lentiviruses. More recently, alpharetroviral vectors have also been developed. Retroviruses are ~80–145 nm in diameter and have a genome size of about 7–10 kb, composed of a group-specific antigen gene (*gag*), which codes for core and structural proteins of the virus; polymerase (*pol*) gene, which codes for reverse transcriptase, protease, and integrase; and envelope (*env*) gene encoding the retroviral envelope glycoproteins. The long-terminal repeats (LTRs) control the expression of viral genes, hence act as enhancer-promoter. The packaging signal

(ψ) helps efficient incorporation of the viral positive-strand RNA into the virus particle before budding off the cell membrane (Verma 1990).

After viral binding and introducing the viral RNA into the host cell, reverse transcriptase, which has both polymerase and RNase activity, converts the viral RNA to linear double-stranded DNA that integrates into the host genome with the help of the viral integrase. The integrated virus sequence, the provirus, will later undergo transcription and translation to produce viral genomic RNA encoding viral proteins. Virus particles then assemble in the cytoplasm and bud from the host cell to infect other cells.

Suitability of Retroviruses as Vectors for Gene Transfer

To generate replication-deficient retroviral vectors, the sequences encoding the virion proteins (gag, pol, and env) responsible for the viral replication and pathogenicity are removed if redundant or placed in a split packaging system to produce the vector particles. The space that is created by deleting viral genes can be used to insert a transgene cassette. The transgene can be controlled by the native LTRs, which have intrinsic enhancer/promoter activity, or by including exogenous enhancer-promoter sequences. For the production of retroviral particles, the vector containing plasmid is introduced into packaging cell lines, mostly HEK293 cells, to produce the retroviral vectors.

Retroviral vectors have several features for gene transfer applications that are important to consider (Table 14.9). They can accommodate transgene cassettes of 8 kb and integrate them into the host genome. Therefore, retroviral vectors can provide stable, long-term transgene expression in dividing cells with low immunogenic potential, particularly because these vectors are mostly used for ex vivo applications. However, there are several disadvantages to these vectors. Gammaretroviruses and alpharetroviruses cannot transduce nondividing cells, but lentiviral vectors can overcome this hurdle, by transporting the preintegration complex (PIC) through the nuclear pores by an active, energy-dependent process (Bukrinsky et al. 1992), as has been shown for transduction of neurons, hepatocytes and hematopoietic stem cells (HSCs). Current methods of viral vector production generate preparations in which the virus titer is sufficient (1×10^5–1×10^7 active viral vector particles/mL) but can generally only be used for a limited number of patients. Retroviruses are also inactivated by elements of the complement system and rapidly removed from the systemic circulation in response to cellular proteins incorporated in the viral envelope during the budding process. Therefore, there are limited clinical trials with retroviral vectors for direct in vivo gene therapy.

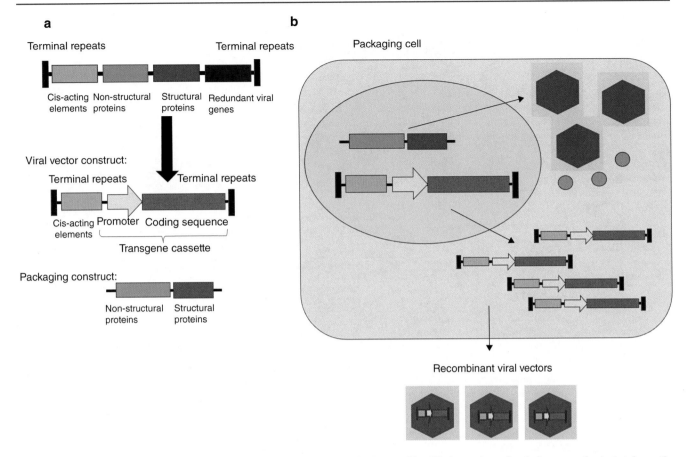

Fig. 14.21 Schematic representation of a virus vector system for gene transfer. (**a**) Simplified overview of a viral genome (top), that forms the basis for the packaging construct (containing gene sequences of structural proteins and nonstructural proteins involved in replication), and the vector transfer construct (containing the transgene cassette). The viral genome contains genes involved in replication, proteins for the virion structure, and the pathogenicity of the virus, which can be removed from the viral vector construct. It is flanked by the terminal sequences (ITRs for AAV or LTRs for retroviral vectors) and cis-acting elements, such as signals for viral packaging. The packaging construct contains only genes that encode nonstructural and structural proteins. The viral vector construct contains the terminal sequences (ITRs or LTRs), required cis-acting elements, and the transgene cassette with promoter and coding sequences (therapeutic gene). The transgene cassette includes a complementary polyadenylation (polyA) termination signal in an AAV vector. In retroviral vectors, this signal is contained in the LTR (**b**). The packaging and vector constructs are brought into the packaging cell by transient transfection, by infection with a helper virus, or by generating stable cell lines applicable to the virus vector system used. The packaging construct expresses replication-related proteins and viral structural proteins. The vector sequences are produced and encapsidated to generate the recombinant viral vector particles

A limitation of retrovirus-based gene therapy is that gammaretrovirus tethers to the transcription start sites and promoters, inserting the genetic cargo semi-randomly into the host genome (Deichmann et al. 2007, 2011; Schwarzwaelder et al. 2007). This can lead to genotoxic events through multiple mechanisms. The vector integration could cause insertional mutagenesis, disrupt and alter the gene expression of a gene close to the integration site. In addition, gammaretroviral vectors have a bias to integrate into proto-oncogenes (Cattoglio et al. 2007), which could cause insertional oncogenesis. Indeed, this serious adverse event was observed in clinical trials, which pushed the field to create safer retroviral vector designs. For instance, modifications have been made to delete part of the U3 region in the 3' LTR to create a self-inactivating (SIN) configuration. After reverse transcription, the 3' U3 region is copied to the 5' LTR creating two LTRs lacking enhancer/promoter activity. This modification enables the use of internal expression cassettes with physio-

Table 14.9 Characteristics of viral vectors for gene transfer[a]

	Gammaretrovirus	Lentivirus	Alpharetrovirus	Adenovirus	Adeno-associated virus
Genetic material	RNA	RNA	RNA	dsDNA	ssDNA
Genome size	7–11 kb	8 kb	9 kb	26–45 kb	4.7 kb
Cloning capacity	8 kb	8 kb	5.8–8.8 kb	7[b]–35[c] kb	<5 kb
Genome forms	Integrated	Integrated	Integrated	Episomal	Stable/episomal
Diameter	100–145 nm	80–120 nm	80–100 nm	80–100 nm	20–12 nm
Tropism	Dividing cells only	Broad, dividing and nondividing cells	Dividing	Broad, dividing and nondividing cells	Broad, not suitable for hematopoietic cells
Virus Protein Expression	No	Yes/no	No	Yes[b]/no[c]	No
Delivery method	Ex vivo	Ex vivo	Ex vivo	In vivo	In vivo
Typical yield (viral particle/ml)	<10^8	<10^7	<10^7	<10^{14}	<10^{11}
Pre-existing immunity	Unlikely	Perhaps, post-entry	Unlikely	Yes	Yes
Immunogenicity	Low	Low	Low	High	Moderate
Potential pathogenicity	Low	High	Low	Low	None
Applications	HSC gene therapy, cellular immunotherapy	HSC gene therapy, cellular immunotherapy	HSC gene therapy; cellular immunotherapy	Oncology	Inherited diseases, postmitotic tissues
Development phase	Clinical stage	Clinical stage	Preclinical stags	Clinical	Clinical
Safely	Insertional mutagenesis	Insertional mutagenesis	Insertional mutagenesis	Potent inflammatory response	Insertional mutagenesis long-term risk not clear. Risk of hepatotoxicity
Physical stability	Poor	Poor	Poor	High	High

[a] Information compiled from references (Edelstein et al. 2004; Weber and Fussenegger 2006)
[b] First-generation, replication-defective adenovirus
[c] Helper-dependent adenovirus

logical or cell type specific promoters reducing the risk of genotoxicity (insertional mutagenesis) and phenotoxicity (ectopic transgene expression) by restricting the expression level to certain cell types or tissues.

Gammaretrovirus

Biology

Gammaretroviruses, such as Moloney murine leukemia virus (MoMLV), were the first retroviral vectors used for clinical application. Gammaretroviruses were originally called oncoretroviruses, which caused tumors in mice. Cell lines derived from HEK293T are commonly used for the production of gammaretroviral vectors. These often generate producer cell lines using ecotropic or amphotropic MLV envelope protein for pseudotyping gammaretroviral vector particles. Pseudotyping means that different envelope glycoproteins are used instead of the wild-type glycoprotein. Amphotropic MLVs are subgroups that infect a broader range of cell types. The structural proteins and transfer vectors are commonly split over three expression plasmids. Vesicular stomatitis virus G (VSVg) glycoprotein can also be used to pseudotype gammaretroviral vectors, stabilizing particles and broadening their tropism. A drawback of using gammaretroviral vectors is that cell division is required for efficient transduction.

Clinical Use of Gammaretrovirus

Retroviral vectors are currently employed mostly for ex vivo gene therapy, such as hematopoietic stem cell (HSC) gene therapy or cellular immunotherapy applications. The gammaretrovirus MoMLV was the first viral vector used in the clinic for treating severe combined immunodeficiency ADA-SCID, a rare inherited disease in which the buildup of toxic deoxyadenosine due to lack of activity of the enzyme adenosine deaminase results in complete lack of T and B lymphocytes (Ferrua and Aiuti 2017). In addition, MoMLV-vector expressing recombinant *ADA* was used for ex vivo genetic modification of autologous peripheral blood lymphocytes (Ferrua and Aiuti 2017).

Other successful clinical trials employing gammaretroviral vectors were performed to treat a rare X-linked SCID (X-SCID) (Kohn and Kohn 2021). MoMLV-vectors expressing *IL2RG* cDNA (also known as common gamma chain) were used to transduce autologous HSCs isolated from

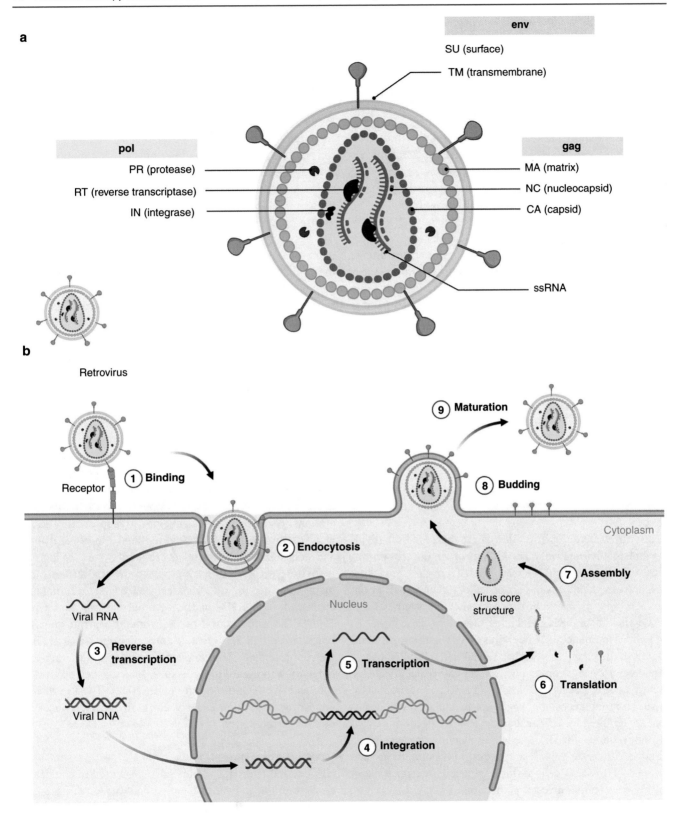

Fig. 14.22 The retrovirus. (**a**) Schematic cross-section of a retrovirus. (**b**) The retrovirus replication cycle. Retroviruses enter cells by receptor-mediated endocytosis (1,2). The viral RNA is reverse-transcribed into double-stranded DNA (3), which is then shuttled into the nucleus, which requires cell division and opening up of the nucleus in most retroviruses. For HIV-1, the viral RNA and core are shuttled into the nucleus, after which reverse transcription occurs (Dharan et al. 2020; Selyutina et al. 2020). Integrase catalyzes the double-stranded viral DNA in the nucleus to integrate into the host genomic DNA as provirus (4). The host cell RNA polymerase transcribes the viral RNA from the integrated provirus in the nucleus (5). Full-length and spliced viral RNA molecules are shuttled out of the nucleus and serve as templates for translation into viral proteins (6) that are cleaved by the protease to form the structural proteins of the virus particle and incorporate two full-length positive-strand RNA copies. The viral cores of the virus particles are assembled (7), and the viral particle buds from the cell membrane (8), after which virus maturation occurs (9). Created with Biorender.com

X-SCID patients without an HLA-identical sibling donor for ex vivo transduction. Subsequently, the genetically modified HSCs were transfused back to patients without preceding cytoreductive chemotherapy to reconstitute lymphocyte and natural killer cell lineages. These initial gene therapy trials enrolled 20 patients (Kohn and Kohn 2021) and showed restoration of immune function in 18 out of 20 patients, but some patients still required immunoglobulin infusions. However, 6 out of 20 patients developed T-cell leukemias due to insertional oncogenesis (Fischer and Hacein-Bey-Abina 2020; Kohn and Kohn 2021). This initiated the construction of self-inactivating gammaretroviral vectors (Kohn and Kohn 2021), which include weaker physiological promoters, such as elongation factor 1 alpha short promoter lacking enhancer sequences. In another trial for Wiskott-Aldrich syndrome, nine out of ten patients developed T-cell leukemia using a similar vector design with transgene transcription driven by the viral LTR (Ferrua and Aiuti 2017). On the other hand, a similar gammaretroviral vector backbone has been used in ADA-SCID, with a much lower genotoxicity profile (Tucci et al. 2022). This emphasizes that vector design and possibly disease background are important factors that determine the outcome in clinical trials.

In addition to safety-enhanced features that have been incorporated into gammaretroviral vectors, other virus-derived vector systems with an enhanced safety profile, such as lentiviral vectors and more recently developed alpharetroviral vector systems, are being exploited.

Lentivirus

Biology

Lentiviruses are retroviruses that can replicate in both dividing and nondividing cells. The biology of lentiviruses resembles that of gammaretroviruses. Apart from the genes *gag*, *pol*, and *env*, lentivirus has six accessory genes, such as *tat*, *rev*, *vpr*, *vpu*, *nef*, and *vif*, which regulate the synthesis and processing of viral RNA and other replicative functions.

Human immunodeficiency virus (HIV) is the most commonly studied lentivirus and has served as the backbone for a viral vector system commonly used for ex vivo applications. Wild-type HIV infects human helper T cells and macrophages. Apart from the genes *gag*, *pol*, and *env*, the accessory genes that are required for HIV replication and pathogenesis in vivo are not required for viral particle production and can all be removed to create sufficient space to incorporate transgene cassettes. HEK293T cells are most frequently used as packaging cells for lentiviral vector production.

Lentiviruses as Vectors for Gene Transfer Vehicles

One of the advantages of using lentiviral vectors is that they can efficiently transduce nondividing cells or terminally differentiated cells such as neurons, macrophages, muscle, and liver cells for in vivo gene therapy, as well as HSCs and immune cells, such as T cells, NK (natural killer) cells and macrophages for ex vivo gene therapy. These cell types are more effectively transduced with lentiviral vectors than using gammaretrovirus-based gene therapy systems. Previous studies have shown that when injected into the rodent brain, liver, muscle, or pancreatic islet cells, lentivirus promoted a sustained gene expression for over 6 months (Miyoshi et al. 1997). Lentiviruses have a typical integration site pattern, which is different from gammaretroviral vectors. The lentiviral LTRs and proteins that bind the integration complex largely determine the integration pattern by tethering the preintegration complex into highly expressed genes (Biffi et al. 2011). These properties reduce the genotoxicity potential of lentiviral vectors compared to gammaretroviral vectors that integrate into transcription start sites (TSS) and have a preference for proto-oncogene integration (Cattoglio et al. 2007; Moiani et al. 2013). The creation of self-inactivating lentiviral vectors has even further reduced the potential of genotoxicity because physiological promoters can drive transgene expression instead of the viral enhancer/promoter (Tucci et al. 2022). The development of the third-generation lentiviral vectors by dividing the plasmid system over four plasmids separating the transfer sequences from gag/pol, the envelope glycoprotein, and the *rev* gene has reduced the risk of generating replication-competent lentivirus (RCL) significantly. The third-generation self-inactivating lentiviral vectors system is the most commonly used as opposed to the earlier generations. As an exogenous envelope glycoprotein, VSVg is also commonly used for pseudotyping to improve tropism and stability. In addition, *rev* is provided in trans to increase lentiviral vector titers by binding to the rev response element sequence in the transfer vector and accumulating unspliced viral RNA in the cytoplasm (Fritz and Green 1996). The introduction of point mutations into the HIV integrase or the LTRs creates lentiviral vectors that are integration defective. This application can provide long-term expression in nondividing tissues, such as skeletal muscle, eye, and liver (Gurumoorthy et al. 2022). The use of integration defective lentiviral vectors obviously reduces the risk for genotoxicity.

Clinical Use of Lentiviral Vectors

HIV-derived lentiviral vectors have been used in clinical trials for multiple indications, including inherited diseases. For example, to treat HIV infection, patient CD4 T cells were genetically modified ex vivo by the lentiviral vector VRX496 (MacGregor 2001). The VRX496 vector contained an antisense sequence targeted to the HIV env gene to interfere with

HIV replication. Although no treatment-related severe adverse events were observed, no statistically significant anti-HIV effects were observed either (Manilla et al. 2005) (cf. the "Cell Therapy" part of this chapter). Lentiviral vectors have been mostly used in clinical trials for primary immune deficiencies (PIDs), such as X-SCID and Wiskott-Aldrich syndrome, and X-linked chronic granulomatous disease (CGD) (Tucci et al. 2022), other blood disorders (sickle cell disease/thalassemia) (Staal et al. 2019) and (neuro)metabolic disorders, such as metachromatic leukodystrophy (MLD), Hurler syndrome or Fabry disease (Chiesa and Bernardo 2022) using genetically modified HSCs. In PIDs and blood disorders, there is an inherent defect in the blood lineages and function development. In (neuro)metabolic disorders, the hematopoietic system may produce recombinant protein for cross-correction and deliver protein in the central nervous system (see also chapter below on monogenic diseases).

Alpharetroviral Vectors

Biology

Alpharetroviral vectors have been more recently developed in addition to gammaretroviral and lentiviral vectors. This system is based on the Rous sarcoma virus (RSV), which was originally isolated from chicken sarcoma cells. RSV contains the viral tyrosine kinase *v-Src* gene, which triggers uncontrolled growth in infected cells (Rubin 2011). The *v-Src* gene increases virulence of RSV. The RSV genome was used to design a split packaging system separating *gag/pro* and *pol* sequences from the transfer vector, with a transfer vector containing self-inactivating configuration and replacing the RSV envelope glycoprotein (Suerth et al. 2014). In this design, the packaging plasmids do not share any homology with the transfer vector, as opposed to the lentiviral vector systems, which do contain sequence overlap between transfer and packaging plasmids. In addition, the alpharetroviral vector leader region is *gag*-sequence free. The integration pattern of alpharetroviral vectors has a more random profile as opposed to gammaretroviral vectors and lentiviral vectors, and aberrant splicing has not been observed (Suerth et al. 2014), which may contribute to safety.

Retroviral Mechanisms of Genotoxicity

Retroviral vectors integrate into the genome and can therefore provide stable transgene expression. However, this could also lead to dysregulation of genes close to the integration site. Three main mechanisms contribute to gene dysregulation and potential genotoxicity. The most common is that the enhancer/promoter elements of the integrated vector could upregulate the expression of genes, potentially proto-oncogenes (Williams et al. 2022). Another mechanism of genotoxicity is related to splice donor and acceptor sites interfering with endogenous gene splicing, generating fusion or truncated transcripts. Finally, genotoxicity may be caused by read-through transcripts from retroviral vectors, and this may induce leukemia. Since the retroviral vector systems most commonly exploit the transcriptional termination signals in the 3' LTR, which are generally poor terminators, transcript read-through may occur. Transcript read-though may be reduced by including elements that improve transcriptional termination (Suerth et al. 2014; David and Doherty 2017).

Genotoxicity risks also depend on cell source selection. For example, lentiviral vector integration sites in postmitotic tissues and dividing cells are different (Bartholomae et al. 2011), and such differences depend on gene expression profiles in the target cells (Biasco et al. 2011).

Adenovirus

Biology

Adenoviruses are nonenveloped (without an outer lipid bilayer), icosahedral, lytic DNA viruses composed of a nucleocapsid and a linear double-stranded genome (Fig. 14.11a). Adenoviruses are capable of infecting both dividing and nondividing cells. More than a hundred (sero) types of adenoviruses have been identified to date. They are grouped into seven subgroups or species (A-G) based on genome size, composition, hemagglutinating properties, and oncogenicity. The adenoviruses serotype 2 and 5 are the most extensively studied and the first to be used as vectors for gene therapy. The adenoviral genome is a linear, nonsegmented dsDNA, between 26 and 45 kb, composed of six early (E1a, E1b, E2a, E2b, E3, and E4) and five late (L1, L2, L3, L4, and L5) genes. The early genes encode proteins necessary for viral replication and prevention of cell death, while the late genes encode proteins for virus assembly, release, and cell death. The genome of adenoviruses is flanked by hairpin-like inverted terminal repeats (ITRs), functioning as self-priming structures that facilitate primase-independent DNA replication (Arrand and Roberts 1979; Shinagawa et al. 1980). A packaging signal sequence promotes viral genome packaging.

Adenovirus infection typically begins with the binding of the fiber knob on the surface of the viral capsid to the CAR and major histocompatibility complex (MHC) class I

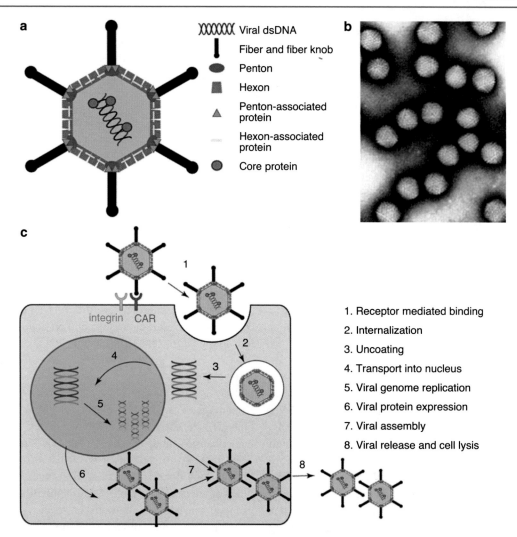

Fig. 14.23 The adenovirus. (**a**) Cross-section of an adenovirus particle. The virus consists of a double-stranded DNA genome encased in a protein capsid. The capsid is primarily made up of hexon proteins, forming 240 trimers. Penton proteins are positioned at each of the 12 vertices of the icosahedral capsid and serve as the base for each fiber protein. Hexon-associated and penton-associated proteins are the glue that holds these proteins together within and across the facets of the capsid. Core proteins bind to penton proteins and serve as a bridge between the virus core and the capsid. (**b**) Electron micrograph of intact adenovirus serotype 5 particles. (**c**) The adenovirus replication cycle. Adenovirus infection begins with the attachment of fiber proteins to cellular receptors such as coxsackie and adenovirus receptor (CAR), and integrins. Through receptor-mediated endocytosis, the virus enters the cytoplasm. In the endosome, capsid proteins are degraded, and viral DNA is released into the cytoplasm and transported to the nucleus for replication. After assembly into new viral particles in the cytoplasm, the host cell is lysed, and the viral progeny is released. In the case of gene therapy, recombinant replication-defective adenoviruses are used to transduce targeted cells. The genome is engineered to accommodate therapeutic transgenes transcribed to mRNA in the nucleus. Messenger RNA is then transported out of the nucleus and into the cytoplasm, where it is translated to therapeutic proteins

(Fig. 14.23). Several alternative entry receptors have been identified, including sialic acid, CD46, and Desmoglein-2 (Gaggar et al. 2003; Nilsson et al. 2011; Wang et al. 2011).

After initial binding, the penton base interacts with integrins on the cell surface to initiate a series of cell signaling processes allowing internalization via receptor-mediated endocytosis (Nemerow and Stewart 1999; Medina-Kauwe 2003). As a result, adenovirus particles enter the nucleus via the nuclear envelope pore complex as early as 30 min after initial cellular contact (Wiethoff and Nemerow 2015). Viral DNA replication and particle assembly in the nucleus starts

8 h after infection and culminates in the release of 10^4–10^5 mature virus particles per cell 30–40 h postinfection by cell lysis (Majhen and Ambriovic-Ristov 2006).

Adenoviruses as Gene Therapy vectors

To construct an adenoviral vector for gene therapy, the E1 and E3 regions of the viral genome are often removed. This both prevents viral replication and creates space to accommodate transgene cassettes. Adenoviruses have a large genome capable of accommodating large transgene cassettes. The adenoviral genome is also easily manipulated to generate a vector with multiple deletions and inserts without affecting its transduction efficiency. Adenoviruses with both E1 and E3 inserts to simultaneously express two therapeutic genes have been reported (Panakanti and Mahato 2009). Moreover, adenoviruses with E1, E3, and E4 deletions and even "gutless" adenovirus (adenoviruses without viral coding regions) have been constructed to drive transgene expression (Armentano et al. 1995; Chen et al. 1997).

Other favorable characteristics of adenoviruses include that the biology of the virus is well understood, that recombinant virus can be generated with high titer and purity, that transgene expression from adenoviruses is rapid and robust, and that adenoviruses can infect a wide range of dividing and nondividing cells. Unfortunately, adenovirus genomes do not integrate into the host genome. While this minimizes the risk of insertional mutagenesis, gene expression is transient, making adenoviruses unsuitable for long-term correction of genetic defects.

The major drawback to the use of recombinant adenoviruses is the ability of the virus to elicit strong innate and adaptive immune responses and the existence of widespread preexisting neutralizing immunity in the population. Innate and adaptive immunity results in the killing of adenovirus-transduced cells and the production of antibodies to adenovirus, resulting in the clearance of the adenoviral vectors from the body (Dai et al. 1995). Preexisting neutralizing antibodies against one or more of the commonly used adenovirus (sero)types immunity in human populations, as a consequence of prior adenovirus infections, significantly reduce the efficacy of these vectors in both preclinical studies and clinical trials (Ertl 2005). To overcome this, researchers have started studying the use of human adenovirus types with low seroprevalence in the human population, such as Ad26. Importantly, these types seem to induce less potent immune responses than the most commonly studied Ad5 (Chen et al. 2010).

The immunogenicity of recombinant adenovirus raises serious safety concerns for its clinical applications. The massive immune responses caused by the administration of adenovirus could lead to multiple organ failures resulting in death. In 1999, a patient died 4 days after injection with an adenoviral vector. This was the first death of a participant in a clinical trial for gene therapy (Stolberg 1999). Another patient experienced a severe immune response syndrome characterized by multiple organ failure and sepsis and died soon after an adenoviral vector dose injection in 2003 (Raper et al. 2003). Preclinical studies also confirmed that the immune response generated by adenoviral vectors must be suppressed before a therapeutic effect can be expected. The transgene expression from adenovirus-transduced cells lasted for about 5–10 days, partially due to the clearance of the transduced cells by the host immune system (Lochmüller et al. 1996). Adenoviruses show an extended duration of

expression when given to nude mice (mice with an "inhibited" immune system) or when an immunosuppressant is administered (Dai et al. 1995). Importantly, in some cases, the strong immunogenicity of adenovirus vectors benefits the therapy, e.g., in cancer or vaccines against pathogens. Less immunogenic adenovirus types such as Ad26 may not be the best choice in these cases. Alternatively, nonhuman primate-derived adenoviruses has been suggested as a source for vector development to avoid preexisting neutralizing immunity while maintaining a strong immune stimulation upon administration (Bots and Hoeben 2020).

A significant effort has been put forth to address the issue of the adenovirus-induced systemic immune response and potential regeneration of replication-competent adenovirus by engineering next-generation adenoviral vectors. First-generation adenoviral vectors were engineered by removing E1 and/or E3 to allow for transgene insertion of up to 4.5–6.5 kb. As the E1 region is vital for adenovirus replication, E1-expressing cell lines have been generated to produce these vectors, such as HEK293. E3 is dispensable for viral propagation in cultured cells.

Second-generation adenovirus vectors lack additional early gene regions (E2a/E2b/E4), enlarging the space for transgene cassettes to 10.5 kb. Like for first-generation adenoviral vectors, the gene deletions are compensated for by producer cell lines expressing the genes. However, titers of second-generation adenoviral vectors are typically lower than for first-generation vectors. Second-generation adenoviral vectors induce notably lower immunogenicity as less viral antigens are being produced, resulting in longer-lasting expression of the encoded transgenes. Nevertheless, the late genes that are still present in the adenovirus vector genome can still trigger undesired immune responses against the vector.

Third-generation adenoviral vectors are called "gutless," "helper-dependent," or "high-capacity" and carry none of the viral sequences except for the ITRs and the packaging signal. These vectors can accommodate ~36 kb of space for transgene cassettes. Production depends on additional helper viruses carrying loxP sites flanking the packaging signal and producer cells that express Cre recombinase. The viral proteins expressed by the genome of the helper viruses allow for replication and packaging. The Cre recombinase ensures the removal of the packaging signal from the helper virus genome to ensure that only the adenovirus genome can be packaged. Third-generation adenoviral vectors have even more reduced immunogenicity, longer-lasting transgene expression in the host cell, and a larger transgene capacity. The main challenge is to eliminate the helper virus from vector batches. Conditionally replicating adenoviral vectors carry tumor-specific gene promoters to make the viruses specifically replicate in tumors. Initially, these were generated by partial deletion of E1B, restricting genome replication to cells that lack p53 such as tumor cells. More recently, condi-

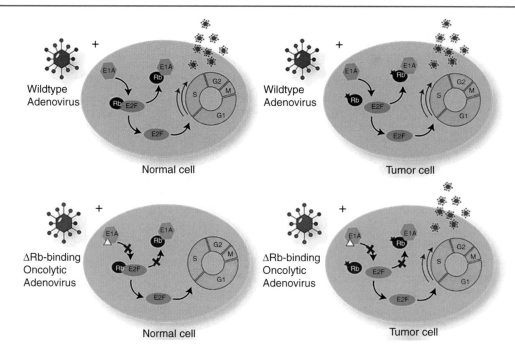

Fig. 14.24 Conditionally replicating adenoviruses. During infection with wild-type adenovirus, the viral E1A protein binds to Rb, releasing E2F. Consequently, the cell cycle progresses, facilitating viral genome replication and virus progeny production. Tumor cells often harbor aberrations in the Rb pathway, resulting in constitutively released E2F. This stimulates cell cycle progression into the S-phase independent of E1A binding to Rb, facilitating virus progeny generation. AdΔRb vectors harbor E1A proteins that are unable to bind to Rb. As a result, Rb remains bound to E2F in normal cells, preventing S-phase and viral genome replication. In tumor cells, however, E2F is already in a released state, so the cell cycle will progress, and virus progeny can be produced. As a consequence, AdΔRb vectors display tumor-selective replication. Courtesy of S.T.F. Bots

tionally replicating adenoviral vectors have been generated by removing a specific stretch of 24 amino acids from the E1A protein. These so-called AdΔ24 or AdΔRb vectors cannot bind to the retinoblastoma (Rb) protein. Rb normally retains E2F, preventing the cells from entering the S-phase and thereby replicating the genome. Cancer cells often have an aberrated Rb pathway and thereby facilitate the S-phase and AdΔRb replication independent from the Rb-binding activities of E1A (Fig. 14.24). These vectors have been widely studied in oncolytic virotherapy, which is more elaborately discussed later in this chapter.

Another hurdle for the use of adenovirus vectors in the human body is the generation of replication-competent adenovirus (RCA). Although the early genes responsible for viral replication and pathogenicity are already removed in the vector construction process, RCA can still be generated by homologous recombination if there is some overlap between sequences in the virus genome and the packaging cell genome. Although adenoviruses typically cause mild respiratory illness, RCA in clinical products could be life-threatening for immune-compromised patients. Several groups have observed the production of RCA from HEK293 cells caused by sequence overlap (Louis et al. 1997). Some new packaging cell lines with less overlap have been reported to overcome such problems. Moreover, there are strict rules

for the purity of vector batches, e.g., the FDA recommends a maximum level of 1 RCA in 3×10^{10} viral particles. RCA assays have been developed to screen vector preparations for the presence of RCA.

Clinical Use of Adenoviral Vectors

Over 500 clinical trials have been initiated using adenovirus vectors (Bulcha et al. 2021). Approximately 50% of all gene therapy clinical trials involving viral vectors use recombinant adenoviruses, making them the most widely applied vector for gene transfer. China was the first to approve a gene therapy, Gendicine, in 2003, for the treatment of head and neck cancer (Pearson et al. 2004). In this adenoviral vector, the E1 gene is replaced by the tumor-suppressor p53. Since its approval, it has been studied in various additional cancers. In 2005, another adenovirus-based gene therapy, Oncorine, was approved in China for the treatment of nasopharyngeal cancer in combination with chemotherapy (Liang 2018). In this vector, E1B is partially deleted. Oncorine is currently also studied in other cancer types. However, both Gendicine and Oncorine have not been approved for clinical use outside of China. A recent addition to the list of marketed adenovirus-based gene therapies is nadofaragene firadenovec (Adstiladrin™). It is a nonreplicating AdV vector expressing recombinant human interferon alfa-2b that is administered into the bladder to treat BCG-unresponsive nonmuscle-invasive bladder cancer. The results of a phase III clinical study showed that 55 out of 103 patients with bladder cancer (with or without a high-grade Ta or T1 tumor) had a complete response within 3 months of the first dose, and this response was maintained in 25 out of 55 patients at 12 months (Boorjian et al. 2021). The FDA approved it in December 2022. Additionally, the immunogenicity of adenovirus vectors has been used in vaccine development (Bulcha et al. 2021). In the last decade, human and chimpanzee Ad vectors have been developed and clinically tested for protection against Ebola, Influenza, or HIV. Especially the Ebola vaccines induced strong and specific cellular and humoral immunity that was long-lasting. Moreover, the global COVID-19 pandemic has prompted the development of additional adenovirus-based vaccines that deliver the SARS-CoV-2 spike protein. This led to the authorization of Vaxzevria (AstraZeneca, full market authorization) and Jcovden (Janssen, conditional market authorization) (see Chap. 15 Vaccines).

Adeno-Associated Virus (AAV)

Biology

The AAV genome is a 4.7 kb linear, single-stranded DNA molecule composed of two open reading frames (ORFs), rep, cap, and two inverted terminal repeats (ITRs) that define the start and end of the viral genome and packaging sequence. The rep genes encode proteins responsible for viral replication, while the cap genes encode structural capsid proteins. ITRs are required for genome replication, packaging, and integration.

The icosahedral AAV capsid is 25 nm in diameter. AAV is deficient in replication, and there are no packaging cells, which can express all the replication-related proteins of the AAV. Therefore, AAV requires coinfection with a helper virus, such as an adenovirus or a herpes simplex virus, to replicate (Fig. 14.25). Thirteen distinct AAV serotypes have been identified, and hundreds of AAV variants have been found in human and nonhuman tissues (Becker et al. 2022; Pupo et al. 2022). The biology of AAV serotype 2 (AAV2) has been the most extensively studied, and this serotype is most often used as a vector for gene transfer. Different serotypes are presumed to recognize different cell receptors and have distinct tissue and cell tropisms.

Suitability of Adeno-Associated Viruses for Gene Transfer

Recombinant AAV vectors have rapidly gained popularity for gene therapy applications within the last decades, due to their lack of pathogenicity and ability to establish long-term gene expression (Table 14.6). The viral genome is simple, making it easy to manipulate. In addition, the virus is resistant to physical and chemical challenges during purification and long-term storage (Croyle et al. 2001; Wright et al.

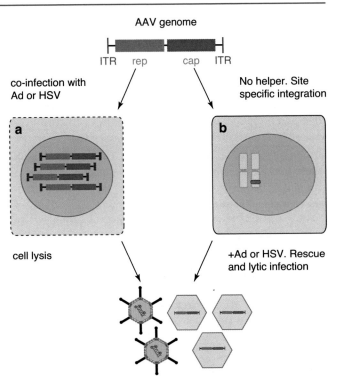

Fig. 14.25 Lifecycle of AAV. AAV can enter cells through receptor-mediated endocytosis. Dependent on pH, it escapes from the endosomes, then trafficking to the nucleus through the nuclear pore complex. Here, its genome is released from the capsid. Once in the nucleus, the virus can follow one of two distinct and interchangeable pathways. (**a**) In the presence of a helper virus (adenovirus or herpes simplex virus), AAV enters a lytic phase. The AAV genome undergoes DNA replication resulting in the amplification of the genome and production of progeny virions. The newly formed AAV viral particles and helper viruses are released from the cell by helper-induced lysis. (**b**) In the absence of a helper virus, it enters a latent phase. During this phase, part of the AAV vectors integrate into host genomic DNA at the preferred site AAVS1 while most AAV vectors persist in an extrachromosomal latent state without integrating into the host genome. The latent AAV genome cannot undergo replication and production of progeny virions in the absence of a helper virus. Similarly, the transgenes carried in the AAV genome are co-expressed without being copied by the host gene expression machinery. ITRs inverted terminal repeats, rep replication. AV Adenovirus; HSV, herpes simplex virus

2003). The ability of the virus to integrate into the human chromosome was an initial concern, but eventually, it turned out that AAV only integrates into a fixed human genomic location called AAVS1, and the integration frequency of recombinant AAV is quite low (Surosky et al. 1997).

The AAV vectors are produced by replacing the rep and cap genes with the transgene. Only one out of 100–1,000 viral particles is infectious. Apart from the production of AAV vectors being laborious, these vectors also have the drawback of limited packaging capacity (4.7 kb) for the transgene. Large genes are, therefore, not suitable for use in a standard AAV vector. To overcome the limited coding capacity, the ITRs of two AAV genomes can anneal to form a head-to-tail structure through trans-splicing between two genomes, almost doubling the capacity of the vector (Yan et al. 2000).

Since recombinant AAV vectors do not contain any viral ORFs, they induce only limited immune responses in humans. Intravenous administration of AAV vectors in mice causes the transient production of pro-inflammatory cytokines and limited infiltration of neutrophils, in contrast to an innate response lasting 24 h or longer induced by aggressive viruses (Zaiss et al. 2002). However, despite the limited innate immunity elicited by AAV vectors, the humoral immunity elicited by AAV is still common. Depending on the serotype, it is estimated that up to 80 % of individuals are positive for AAV antibodies in the human population. The associated neutralizing activity limits the usefulness of AAV in certain applications. To overcome this, capsids have been isolated from nonhuman sources such as nonhuman primates or other vertebrate species. Alternatively, the vector can be retargeted by inserting or removing specific sequences from the AAV that are known to bind to certain receptors, or by directed evolution.

Clinical Use of Adeno-Associated Virus Vectors

The first clinical use of recombinant AAV was to transfer the cDNA of the Cystic Fibrosis Transmembrane Conductance Regulator (CFTR) to the respiratory epithelium for treating cystic fibrosis (Flotte et al. 1996). Since then, hundreds of clinical trials employing recombinant AAV vectors have been initiated worldwide. The interest in AAV for gene therapy has been boosted by the approval of Luxturna and Zolgensma, two AAV-based gene therapies, for retinal dystrophy and spinal muscular atrophy, respectively. The first approval of AAV-based gene therapy was granted by the European Commission to Glybera® (alipogene tiparvovec), which encodes the gene for lipoprotein lipase deficiency for the treatment of patients with familial lipoprotein lipase deficiency (LPLD, synonym: type I hyperlipidemia) (Büning 2013; Salmon et al. 2014). However, this therapy was withdrawn in 2017 due to limited use. Numerous clinical trials using AAV gene therapy have demonstrated its safety and efficacy in neurological, musculoskeletal, hematological, ophthalmological, and metabolic diseases (Kuzmin et al. 2021). Efforts are being made to avoid vector accumulation in the liver and potentially related toxicity and to retarget the vectors to other organs than the liver, eye, nervous system, and muscles (e.g., the heart).

Nonviral Vectors

The inherent problems with recombinant viruses such as limited packaging capacity of transgenes, high production costs, and immunogenicity, a.o. reflected in the generation of neutralizing antibodies, have called for the design of efficient, nonbiological delivery methods for human gene therapy. These nonviral methods can be categorized into physical methods of gene transfection and physicochemical methods that make use of synthetic biological molecules (e.g., lipids, polymers, peptides, or sugars) to encapsulate, complex, or conjugate genetic material (further detailed below).

Like viral vectors, synthetic, nonviral vectors can be used to deliver genetic material for transient, episomal expression, stable transgene integration into the genome, and gene editing.

Transient expression involves the delivery of either plasmid DNA (pDNA) or mRNA. Longevity of expression is dependent on many factors, including the speed at which a cell divides, the stability of the mRNA or pDNA, and the type of nonviral vector that was used. In general, expression fades within 1-2 weeks but can be longer in slowly dividing cells.

Stable expression with synthetic vectors can be achieved by introducing pDNA encoding integration competent genetic elements. These can be of viral origin (Chiang et al. 2020), such as retroviral constructs or derived from engineered transposable elements such as Sleeping Beauty, piggyBac, and Tol2 (Sandoval-Villegas et al. 2021). In both cases, the integrase (for retroviruses) or transposase (for transposable elements) need to be delivered in trans with the DNA inserted, often provided in the form of mRNA. Stable integration has been described for these systems after electroporation or nonviral delivery, but the efficiency drops with the size of the DNA to be integrated.

Synthetic vectors are also very suitable for delivering the necessary components for gene editing. Compared to viral vectors, synthetic vectors offer several advantages for this. First, for some viral vectors, especially the adeno-associated virus, the packaging capacity is limited, which can be a limitation when large or more than 1 gene construct needs to be delivered. Synthetic vectors do not have this intrinsic limitation, as the size of these systems can be readily adjusted to accommodate bulky cargo. Second, unlike viral vectors, synthetic vectors can accommodate a mixed cargo of DNA, RNA, and protein. For example, lipid nanoparticles have been used to co-deliver spCas9 ribonucleoprotein complexes and single-stranded HDR templates to cells in culture, with high editing efficiencies (Walther et al. 2022). Viral vectors would rely on co-expression of the Cas9 nuclease and the sgRNA, each with their own promoter, making it quite bulky, and from a safety point of view, this is not desired. Thirdly, synthetic vectors do not suffer from vector-induced antibody responses, allowing redosing with the same synthetic vector. Despite these advantages, their clinical utility still suffers from low transfection efficiencies, which stems from the nonspecific uptake of the vector by epithelial barriers and extracellular matrix and poor delivery into the therapeutic target cells (Fig. 14.26). New emerging delivery systems and vector-constructing technologies try to address these issues.

Fig. 14.26 Barriers to nonbiological gene delivery. Following systemic administration, the gene medicine (plasmid or siRNA) meets blood nucleases. Then they may traverse the blood vessel barrier and the extracellular matrix compartment before crossing the plasma membrane barrier. Upon entering the cell via receptor-mediated endocytosis, they are trapped in endosomes and need to be released into the cytosol. Endosomal escape is a major rate-limiting step in gene delivery. From Singh et al. (2011); with permission to reprint

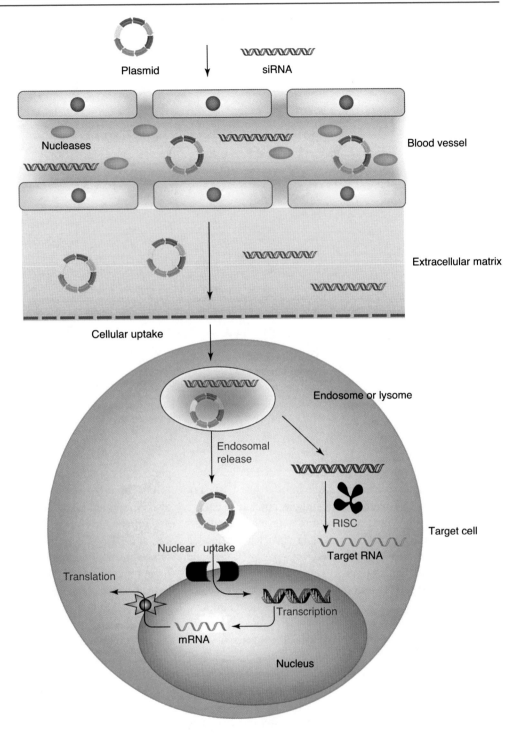

Nonviral delivery methods can be subdivided into physical delivery methods and synthetic and biomimetic vectors for gene transfer.

Physical Methods for Gene Transfer

Physical methods involve the transfer of naked nucleic acids (DNA or RNA) by direct disruption of (target) cell membranes.

The earliest techniques to deliver recombinant DNA to cellular targets include microinjection, particle bombardment, and electroporation (Table 14.10). Microinjection, the direct injection of DNA or RNA into the cytoplasm or nucleus of a single cell, is the simplest and most effective method for the physical delivery of genetic material to cells. This transfects 100 % of the treated cells and minimizes waste of plasmid DNA. However, it requires highly special-

Table 14.10 Summary of nonbiological methods used for gene transfer

	Advantages	Disadvantages
Naked DNA	No special skills needed, easy to produce	Low transduction efficiency, transient gene expression
Physical methods		
Microinjection	Up to 100% transduction efficiency (nuclear injection)	Requires highly specialized skills for delivery
		Limited to ex vivo delivery
Gene gun	Easy to perform	Poor tissue penetration
	Effective immunization with low amount of DNA	
Electroporation	High transduction efficiency	Transient gene expression
		Toxicity, tissue damage
		Highly invasive
Sonoporation	Method well tolerated for other applications	Transient gene expression
		Toxicity not yet established
Laser irradiation	Can achieve 100% transduction efficiency	Special skills and expensive equipment necessary
Magnetofection	Safety of method established in the clinic	Poor efficiency with naked DNA
Chemical methods		
Liposomes	Easy to produce	Protein and tissue binding, transient gene expression
Micelles	Easy to produce and manipulate	Unstable
		Protein and tissue binding
Cationic polymers	High DNA loading	Transient gene expression, toxicity
	Easy to produce and manipulate	
Dendrimers	High DNA loading	Extremely toxic
	High transduction efficiency	
Solid lipid nanoparticles	Low toxicity	NA
	Controlled release and targeting	

NA not applicable

ized equipment and skills. Moreover, microinjection is unsuitable for in vivo or in vitro gene transfer into tissues or organs composed of many cells. Particle bombardment, or gene gun treatment, starts with coating tungsten or gold particles with plasmid DNA. The coated particles are loaded into a gene gun barrel, accelerated with gas pressure, and shot into targeted cells or tissues in a petri dish. Particle bombardment can be used to introduce a variety of DNA vaccines into desirable cells in vitro. However, particle bombardment has a low penetration capacity, making it unsuitable for in vivo gene delivery apart from easily accessible tissue, e.g., the skin. Electroporation generates temporary pores in the plasma membrane to transfer plasmid DNA to the cells by an externally applied high-voltage electrical field. Electroporation increases the gene transfer efficiency by 100–1000 folds compared to naked DNA solutions and has met with great success in laboratory practices and clinical trials (Wells 2004). For example, GMP-compliant electroporation devices have been developed to transfect cells in a closed flow system, thereby preventing potential contamination during transfection. Such closed systems are ideally suited for cell-based therapies in which cells need to be

genetically modified prior to infusion into the patient (see "Cell Therapy" section) (Li et al. 2013). Electroporation can also be applied directly in vivo and has demonstrated safety and efficacy in clinical trials to treat melanoma, prostate cancer, and HIV infection (Daud et al. 2008; Vasan et al. 2011). Other physical methods for gene transfer include sonoporation, photoporation, laser irradiation, magnetofection, and hydroporation (Raes et al. 2021). However, because most of the physical methods induce stress and disruption of cellular structure and function, physical methods are less widely studied compared with the use of viral and synthetic vectors (see below) and are generally restricted to in vitro gene transfer of cultured cells or embryonic stem cells (Table 14.10).

Lipid-Based Vectors

Lipid-based vectors make use of cationic lipids to electrostatically bind negatively charged RNA/DNA to form complexes in the submicron scale (60-400 nm). Since the invention of lipofectamine in 1987, numerous cationic lipids have been synthesized and tested for gene delivery. Most of these cationic lipids are composed of three parts: (i) a hydro-

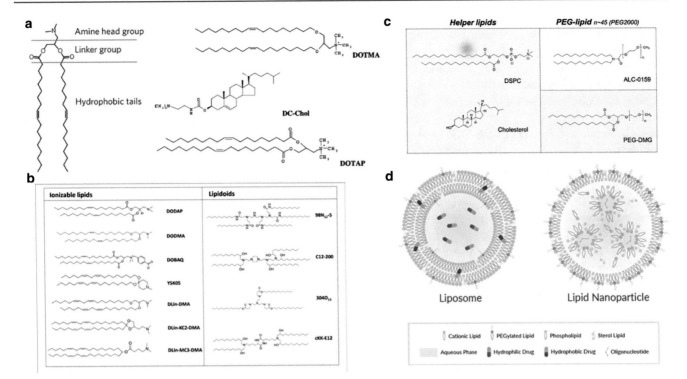

Fig. 14.27 Lipid-based vectors for gene delivery. The key component consists of permanently cationic (**a**) or ionizable lipids (**b**) mixed with neutral helper lipids (often phospholipids) and PEG lipids (**c**). Depending on the exact composition of the lipids and the ratio of nucleic acid cargo to lipid, these vectors can either form liposomes, having an aqueous interior, or LNPs with an interior mainly consisting of electrostatic complexes between RNA/DNA and (ionizable) cationic lipids (**d**)

phobic lipid anchor group; (ii) a linker group, such as an ester, amide, or carbamate; and (iii) a positively charged head group (Fig. 14.27) (Mahato et al. 1997). 2,3-dioleyloxy propyl-1-trimethyl ammonium bromide (DOTMA) and 3-β[N(NV,NV-dimethylaminoethane)-carbamoyl] cholesterol (DC-Chol) are two commonly used cationic lipids with different structures. Cationic lipids are usually mixed with other lipids to change the structural properties as well as toxicity profiles of the formed complexes. Typically, these consist of cholesterol, neutral helper lipids, and lipids with a flexible poly(ethylene glycol) (PEG) polymer attached to their polar headgroup to increase nanoparticle stability and prevent rapid clearance once injected into the circulation.

Depending on the exact composition of the lipids and the ratio of lipids to nucleic acid cargo, the formed structures can either adopt a vesicular nature, containing lipid bilayers enclosing an aqueous core (liposomes), or form solid spherical structures in which the solid core consists of cationic lipid-complexed nucleic acids with some water, which are surrounded by a layer of neutral phospholipids and PEG lipids, the so-called lipid nanoparticles, LNPs.

Initially, permanently charged cationic lipids such as DOTMA were being used for in vivo transfection of pDNA, but direct exposure of cells to these highly cationic nanoparticles caused toxicity along with the activation of the innate immune system. This problem was partially solved by developing ionizable cationic lipids. In the LNP, the amino head groups are only charged at low pH (<6.5). In this way, nucleic acid encapsulation can be performed at low pH to enable optimal complexation, but once the LNPs are formed, and pH is neutralized, the LNPs hardly expose positive charges on their surface that could cause cytotoxicity. Screening of a vast library of such ionizable lipids and lipidoids (i.e., lipid-like materials) has led to the development of highly efficient LNPs for direct in vivo delivery of different therapeutic nucleic acids, including siRNA, mRNA, and pDNA (Sago et al. 2018) (cf. Chap. 5).

The mechanisms by which LNPs deliver their nucleic acid cargo is still a topic of investigation but involves endocytic uptake, followed by the exchange of lipids between the LNP and endosomal membranes, which in sporadic cases, leads to the endosomal escape of the nucleic acid cargo (Fig. 14.26).

To enhance endosomal escape and transport to and through the nuclear membrane, additional functional elements may be attached: for endosomal escape (pH-sensitive fusogenic peptides), for transport in the cytoplasm, and nuclear membrane passage (a nuclear translocation peptide).

LNPs are being explored for gene editing. Hereditary Transthyretin Amyloidosis is a rare genetic disorder that causes the amyloid formation of the TTR protein due to a point mutation in the *TTR* gene. By encapsulating an mRNA encoding spCas9 together with a sgRNA targeting the mouse transthyretin (*Ttr*) gene in the liver, a single intravenous administration of the gene could be effectively knocked out, leading to levels of TTR protein in the serum that were reduced by 97% and which persisted for at least 12 months (Finn et al. 2018). Interim results of a phase I study in patients suffering from hereditary transthyretin amyloidosis showed serum TTR level reduction up to 87%, after a single intravenous dose, with only mild grade 1 adverse effect (Gillmore et al. 2021).

Peptide-Based Vectors

Just like cationic lipids, cationic peptides condense DNA in a similar manner and can be used as gene delivery carriers. Poly(L-lysine) (PLL), a polydisperse, synthetic repeat of the amino acid lysine, was one of the first cationic peptides to deliver genes. However, an increase in the length of PLL leads to increasing cytotoxicity. Besides, PLL shows limited transfection efficiency and needs the addition of endosomolytic agents such as fusogenic peptides to facilitate plasmid release into the cytoplasm. Due to these issues, many researchers have turned to the development of PLL-containing "active" peptides and have met with some success (McKenzie et al. 1999). Such peptides offer many advantages over PLL, such as lower toxicity, precise control of synthesis, and homogeneity of peptide length.

Another class of peptides that have been extensively explored for gene delivery are the cell-penetrating peptides (CPPs). CPPs are short, synthetic peptides that facilitate cellular uptake and endosomal escape of molecules ranging from small chemical compounds, and nucleic acids to entire nanoparticles. CPPs can be categorized as those with a high abundance of cationic amino acids (lysine and arginine) or a sequence with alternating hydrophobic and polar amino acids, creating an amphipathic alpha-helical structure (Cf. Chap. 2). The latter are known for their endosomolytic activity. When complexed with nucleic acids, the nanocomplexes are taken up by endocytosis, where the peptides – because of the low pH – destabilize the endosomal membrane to release the cargo into the cytosol. The stability of the nucleic acid/peptide complexes can be greatly enhanced by the inclusion of a lipid tail to the C- or N-terminus of the CPP.

Despite efficient transfection of a variety of cells in vitro, peptide-based vectors are not yet suitable for systemic in vivo administration because of rapid destabilization, nonspecific plasma protein binding, and uptake by the reticuloendothelial system (Männistö et al. 2002). Another unique challenge for peptide-based gene delivery systems is cytosolic proteasomes, which degrade unneeded or damaged proteins by proteolysis. Co-administration of proteasome inhibitors is the most effective strategy to address this issue.

Polymeric Vectors

Synthetic and naturally occurring cationic polymers constitute another category of nonviral vectors. The advantage of synthetic polymers is that their architecture can be fully tailored: they can be made from biodegradable materials, the molecular architecture can be adapted from linear to branched to star-like dendrimers, and the density of cationic charges can be varied. Several polymers that have been studied for gene delivery are polyethyleneimine (PEI), poly[(2-dimethylamino) ethyl methacrylate (pDMAEMA), polyamidoamine (PAMAM) and biodegradable poly(β-amino ester) polymers (PBAE). Natural polymers, such as chitosan, dextran, gelatin, pullulan, and synthetic analogs, were also explored. Polymeric systems can generally be tailored to specific needs but require PEGylation to prevent undesired aggregation and rapid clearance in vivo. Some polymers, such as PEI, work via the postulated proton-sponge effect: the proton buffering capacity of such polymers at acidic pH causes the buildup of osmotic pressure when they reside in acidifying endocytic compartments, leading to rare endosomal burst events that enable the endosomal escape of the nucleic acid cargo. So far, only a few polymeric vectors for gene therapy have reached clinical stage development. A PEG–PEI–cholesterol lipopolymer is under clinical investigation for immunotherapy of ovarian and colorectal cancers through forced expression of the cytokine interleukin-12 (IL-12) (NCT01489371) (Thaker et al. 2015).

In another study, local delivery of a CRISPR-Cas9 ribonucleoprotein complex with a polymeric vector based on disulfide-crosslinked acrylates into the eye resulted in robust gene editing in the retina of nonhuman primates. It could potentially be used for local treatment of genetic eye diseases (Chen et al. 2019).

Extracellular Vesicles

Extracellular vesicles (EVs) encompass different types of lipid vesicles that are naturally secreted by cells, including exosomes, microvesicles, and apoptotic bodies, and may serve as a new type of delivery system for gene therapies (Yáñez-Mó et al. 2015; Varderidou-Minasian and Lorenowicz 2020) (Figure 14.28). EVs are secreted by almost every cell type and

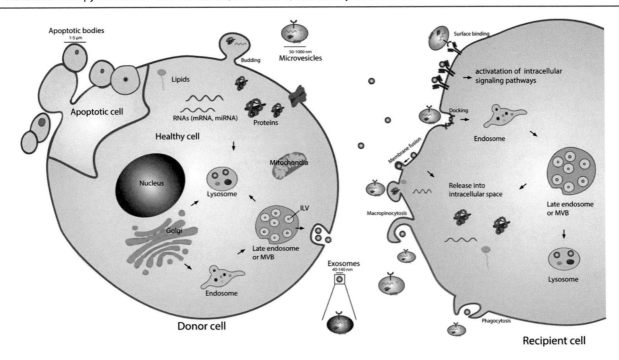

Fig. 14.28 Schematic representation of biogenesis, secretion, and uptake of major EV subpopulations: exosomes, microvesicles, and apoptotic bodies (Source: Varderidou-Minasian et al. 2020)

play pivotal roles in both physiology and pathophysiology. EVs naturally transfer biological payload (DNA, RNA, proteins, lipids) from cell to cell. Also, EVs can overcome cytotoxicity and immunogenicity issues associated with most viral vectors and synthetic nanoparticles, as demonstrated in preclinical models (Corradetti et al. 2021). These characteristics have boosted the development of EV therapeutics, which concerns the (ex vivo) production of EVs that are loaded with therapeutic drugs, proteins, small RNAs, mRNAs, or DNA (Liu and Su 2019). For example, Cas9-gRNA ribonucleoprotein can be loaded in EVs and functionally delivered in recipient cells, which, due to the large size of Cas9, is less feasible with AAV viral vector systems. The first engineered EV therapeutics have found their way into clinical trials, and initial findings on safety and efficacy are promising (e.g., to deliver cytokine profiles locally to tumor cells). Many developments are currently ongoing to fully explore the potential of EVs as a new type of delivery system for gene therapeutics. These not only focus on the engineering aspects to optimize the loading (and unloading) of payload but also involve targeting studies, specificity studies and development of scalable, clinical-grade EV production technologies.

Delivery Systems for CRISPR-Based Gene Editing

CRISPR-Cas9-based genome editing holds great potential to provide an entirely new class of therapeutics. However, to achieve effective therapeutic efficacy, the delivery of CRISPR-Cas9 components to the target cells of the patients is still a major hurdle and a prime topic for research. Several studies suggest an efficient delivery of the ~4 kb Cas9 from

Streptococcus pyogenes into mammalian cells using adenoviral and lentiviral vectors (Eyquem et al. 2017). Nonviral approaches, including cationic polymer-based vectors (Platt et al. 2014), cationic lipid-based vectors (Zuris et al. 2015), and conjugated vectors (Ramakrishna et al. 2014), are also studied as delivery vehicles. For instance, CRISPR-Cas9 was used to target frequently mutated oncogene KRAS alleles in cancer cells and in vivo tumors using lentivirus or AAV expressing Cas9 and single guide RNA (sgRNA) (Kim et al. 2018). Also, more than 97% reduction in serum transthyretin level was achieved in mice when the CRISPR-Cas9 system was delivered using lipid nanoparticles (Finn et al. 2018). Other ex vivo delivery methods such as electroporation and nucleofection, are also extensively applied for the delivery of CRISPR-Cas9 components.

Disease Targets for Gene Therapy

In 2021 3000+ active gene therapy clinical trials were registered worldwide (Fig. 14.29) (John Wiley and Sons LTD 2022). Approximately 65 % of these trials are for cancer treatment. Treatment of monogenetic, cardiovascular, and infectious diseases each takes ~6-12 %, whereas treatment of neurological diseases is close to 2% (Fig. 14.29). Currently, gene therapy trials are primarily performed in the United States (58.3 % of all trials), China (10.1 %), the United Kingdom (7.3 %), and Germany (3.5 %). The geographical distribution of gene therapy clinical trials is summarized in

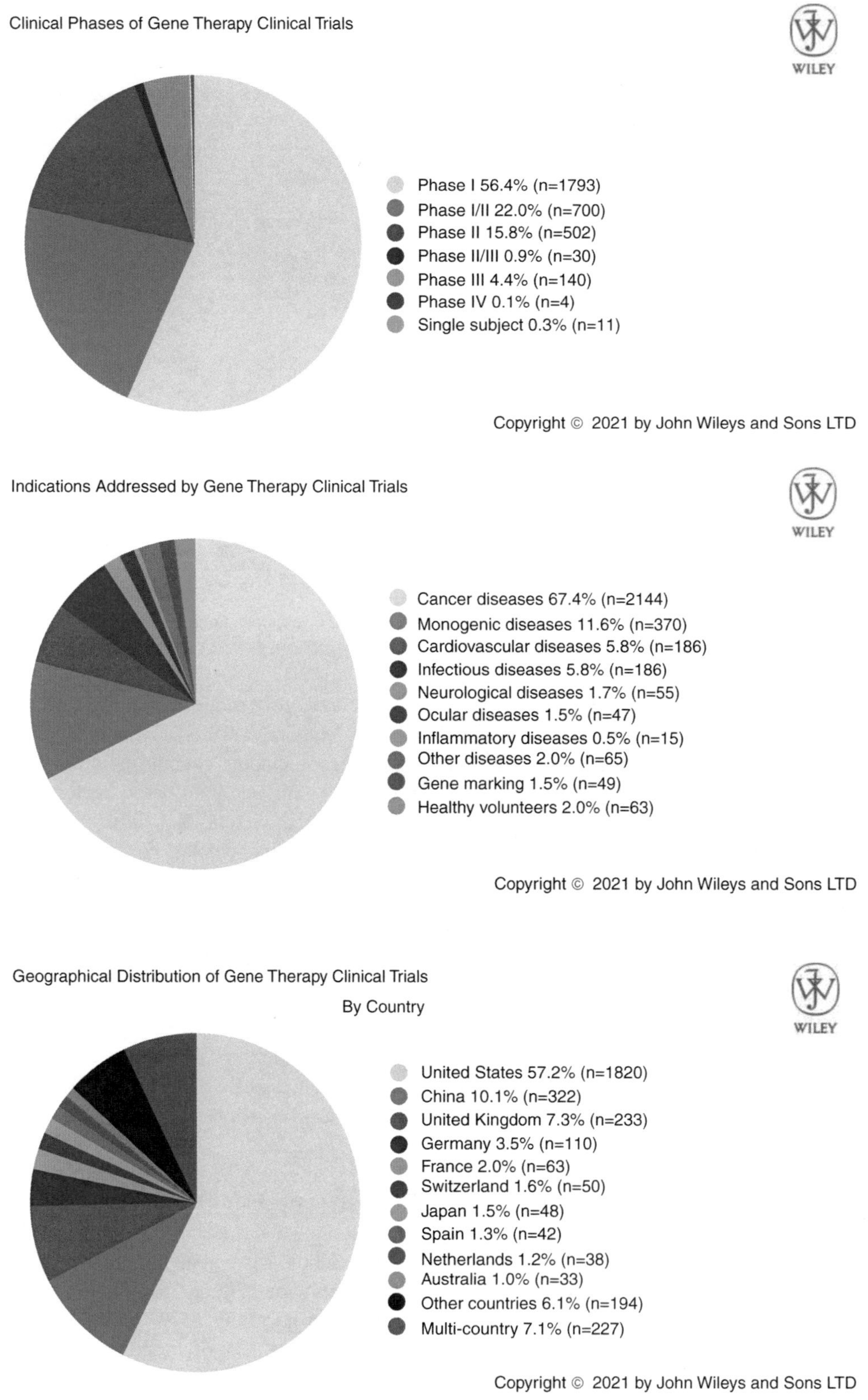

Clinical Phases of Gene Therapy Clinical Trials

- Phase I 56.4% (n=1793)
- Phase I/II 22.0% (n=700)
- Phase II 15.8% (n=502)
- Phase II/III 0.9% (n=30)
- Phase III 4.4% (n=140)
- Phase IV 0.1% (n=4)
- Single subject 0.3% (n=11)

Copyright © 2021 by John Wileys and Sons LTD

Indications Addressed by Gene Therapy Clinical Trials

- Cancer diseases 67.4% (n=2144)
- Monogenic diseases 11.6% (n=370)
- Cardiovascular diseases 5.8% (n=186)
- Infectious diseases 5.8% (n=186)
- Neurological diseases 1.7% (n=55)
- Ocular diseases 1.5% (n=47)
- Inflammatory diseases 0.5% (n=15)
- Other diseases 2.0% (n=65)
- Gene marking 1.5% (n=49)
- Healthy volunteers 2.0% (n=63)

Copyright © 2021 by John Wileys and Sons LTD

Geographical Distribution of Gene Therapy Clinical Trials

By Country

- United States 57.2% (n=1820)
- China 10.1% (n=322)
- United Kingdom 7.3% (n=233)
- Germany 3.5% (n=110)
- France 2.0% (n=63)
- Switzerland 1.6% (n=50)
- Japan 1.5% (n=48)
- Spain 1.3% (n=42)
- Netherlands 1.2% (n=38)
- Australia 1.0% (n=33)
- Other countries 6.1% (n=194)
- Multi-country 7.1% (n=227)

Copyright © 2021 by John Wileys and Sons LTD

Fig. 14.29 Phases of gene therapy in clinical trials (John Wiley and Sons LTD 2022) Disease targets of gene therapy clinical trials (source: Wiley 2021). Other diseases include inflammatory bowel disease, rheumatoid arthritis, chronic renal disease, carpal tunnel syndrome, Alzheimer's disease, diabetic neuropathy, Parkinson's disease, erectile dysfunction, retinitis pigmentosa, and glaucoma International status of gene therapy clinical trials (source: Wiley 2021)

Table 14.11 Conditions for which human gene transfer trials have been approved

Cancer	Other diseases	Cardiovascular disease
Gynecological	Inflammatory bowel disease	Peripheral vascular disease
Breast, ovary, cervix	Rheumatoid arthritis	Intermittent claudication
Nervous system	Chronic renal disease	Critical limb ischemia
Glioblastoma, leptomeningeal carcinomatosis, glioma, astrocytoma, neuroblastoma	Fractures	Myocardial ischemia
Gastrointestinal	Erectile dysfunction	Coronary artery stenosis
Colon, colorectal, liver metastases, posthepatitis liver cancer, pancreas	Anemia of end-stage renal disease	Stable and unstable angina
Genitourinary	Parotid salivary hypofunction	Venous ulcers
Prostate, renal	Type I diabetes	Vascular complications of diabetes
Skin	Detrusor overactivity	Pulmonary hypertension
Melanoma	Graft-versus-host disease	Heart failure
Head and neck		
Nasopharyngeal carcinoma	**Monogenic disorders**	**Infectious disease**
Lung	Cystic fibrosis	HIV/AIDS
Adenocarcinoma, small cell, nonsmall cell	Severe combined immunodeficiency (SCID)	Tetanus
Mesothelioma	Alpha-1 antitrypsin deficiency	Epstein-Barr virus
Hematological	Hemophilia A and B	Cytomegalovirus infection
Leukemia, lymphoma, multiple myeloma	Hurler syndrome	Adenovirus infection
Sarcoma	Hunter syndrome	Japanese encephalitis
Germ cell	Huntington's chorea	Hepatitis C
	Duchenne muscular dystrophy	Hepatitis B
Neurological diseases	Becker muscular dystrophy	Influenza
Alzheimer's disease	Canavan disease	
Carpal tunnel syndrome	Chronic granulomatous disease (CGD)	
Cubital tunnel syndrome	Familial hypercholesterolemia	
Diabetic neuropathy	Gaucher disease	
Epilepsy	Fanconi's anemia	
Multiple sclerosis	Purine nucleoside phosphorylase deficiency	
Myasthenia gravis	Ornithine transcarbamylase deficiency	
Parkinson's disease	Leukocyte adherence deficiency	
Peripheral neuropathy	Gyrate atrophy	
	Fabry disease	
Ocular diseases	Familial amyotrophic lateral sclerosis	
Age-related macular degeneration	Junctional epidermolysis bullosa	
Diabetic macular edema	Wiskott-Aldrich syndrome	
Glaucoma	Lipoprotein lipase deficiency	
Retinitis pigmentosa	Late infantile neuronal ceroid lipofuscinosis	
Superficial corneal opacity	RPE65 mutation (retinal disease)	
	Mucopolysaccharidosis	

Information obtained from reference Ginn et al. (2018)

Fig. 14.29. General indications for gene therapy trials in the clinic are summarized in Table 14.11.

Cancer Gene Therapy

Most of today's gene therapy clinical trials are devoted to treating cancer. Gene therapy has the potential to target and destroy cancer cells in a way that is much more specific and tailored as compared to traditional cancer treatments with chemotherapy and radiation. This property has even become much more relevant in light of the current knowledge on the large intra- and inter-tumor heterogeneity of many cancer types, which advocates the use of therapeutics that attack cancer cells from multiple angles and in a directed fashion. The types of cancer gene therapies being explored vary widely and include oncolytic viruses, ex vivo immune cell modification (e.g., CAR and recombinant T cell receptor T cells), overexpression of pro-apoptotic genes (e.g., p53), introduction of suicide genes, suppression of oncogenes (e.g., using RNA interference), creating synergy toward other treatments (sensitization) and targeted interference with the cancer cell genome using CRISPR-Cas.

Correction of Genetic Abnormalities

In this approach, gene therapy corrects genetic abnormalities contributing to the malignant phenotype by replacing/reintroducing tumor-suppressor genes or downregulating certain oncogenic pathways. Understanding cancer at the molecular level is the starting point for gene correction in cancer therapy. The inactivation or activation of certain genes may contribute to tumor growth. Many cancer gene therapy clinical trials involve overexpression of tumor-suppressor genes such as p53, MDA-7, and ARF (Belete 2021). Mutations in the p53 gene are most commonly seen in a wide spectrum of tumors (Valente et al. 2018). Efficient delivery and expression of the wild-type p53 tumor-suppressor gene prevents the growth of human cancer cells in culture, causes regression of established human tumors in nude mice, or sensitizes existing tumors to the therapeutic effect of conventional chemotherapy and radiotherapy (Valente et al. 2018). The results from clinical trials indicated that the therapeutic effect of the adenoviral vector-based therapy Gendicine, the first gene therapy product, was promising in patients with head and neck squamous cancers (Xia et al. 2020). However, the results were only validated in China. Gendicine has also been used to treat various other cancers, which prolong overall survival when combined with other drugs. Although it does not show any adverse effects on the patients, vector-associated transient fever that lasts for only a few hours cannot be overcome in 50-60 % of the patients (Zhang et al. 2018). Efficient delivery of tumor-suppressor genes deep within tumor tissue is difficult, and restriction of gene expression in malignant tissue is challenging. Possibly this may be improved by the development of novel types of delivery systems. Gene silencing by this approach is also a limited success, especially when a prolonged effect is required. Tumor heterogeneity is another major bottleneck, and it will most likely require co-treatments or simultaneous "all-in-one-vector" delivery of multiple anti-cancer approaches at the same time. Despite these reservations, promising results have been obtained in clinical trials for different cancers, including prostate, lung, pancreatic, and brain tumors.

Immunotherapy

The past decade has witnessed some major breakthroughs in cancer treatment by applying immunotherapies. These consist of different approaches, all having in common that the patient's immune system is being modified or boosted to create additional antitumor responses. One example is the class of the oncolytic viruses (e.g., Imlygic®, talimogene laherparepvec, based on herpex simplex virus-1), which not only kill the tumor cells (in a specific way) but also induce a strong antitumor immune response, as outlined in more detail below. Other more recent developments are the autologous or allogeneic immune cell therapies, most strikingly CAR-T cells (covered in detail elsewhere in this chapter). This concerns the modification of immune cells outside the body (ex vivo) to direct them against tumor targets and subsequent infusion into the patient. Driven by the successes of CAR-T cells for the treatment of different hematological malignancies, many efforts are ongoing to improve the efficacy, e.g., to become active in so-called "cold tumors," that is, tumors with lower presence and influx of immune cells. Methods and technologies are improved to make this therapy more practical, easier, and cheaper. It is also being explored whether in vivo T cell therapy may be viable, which will require the use of vector systems that efficiently and safely deliver the transgenes into the immune cells inside the patient (Xin et al. 2022).

In recent years, a number of viral vectors have been developed that express therapeutic transgenes (Shaw and Suzuki 2019). Many of those represent oncolytic viruses (summarized below). Additionally, several finished and ongoing clinical trials have employed adenoviral vectors expressing tumor-associated antigens such as PSA or MAGEA3, or immunomodulatory molecules like IL-12 and interferon IFN-α/-β. As an example, nonreplicating adenoviral vectors encoding recombinant human interferon α-2b have been approved by the FDA for the treatment of BCG-unresponsive nonmuscle-invasive bladder cancer (Boorjian et al. 2021).

Oncolytic Viruses (Virotherapy)

Oncolytic virotherapy employs viruses that, either by nature or upon engineering, preferentially target tumor cells and induce antitumor immunity (Macedo et al. 2020). Moreover, it can be used to simultaneously deliver therapeutic genes into tumor cells. This represents a relatively novel anticancer approach that has gained interest in recent years. The first oncolytic virus that was approved was Rigvir, an ECHO-7 picornavirus. This virus was authorized in 2004 in Latvia for the treatment of melanoma patients. In 2005, China approved Oncorine, a genetically modified adenovirus (see "Adenovirus" section) for combination treatment with chemotherapy in nasopharyngeal carcinoma. More well-known is Imlygic (also called T-VEC), a modified herpes simplex virus, which was approved in the United States, the European Union, Australia, and Israel in 2015, for use in a subset of advanced melanoma patients (Kaufman et al. 2022). A number of other oncolytic viruses are currently under (pre)clinical investigation, including picorna-, adeno-, reo-, pox-, paramyxo-, rhabdo-, and retroviruses (Macedo et al. 2020). Their clinical antitumor effects as monotherapies, although long-lasting, seem moderate at best. Although combinations with other (chemo- and immuno-) therapies show promising results, alternative (nonviral) approaches with similar effects are often preferred. Many current research efforts are therefore focused on generating oncolytic viruses with enhanced potency. In doing so, viruses have, for example, been engineered with expanded tropisms, enhanced tumor selectivity, and/or increased immunostimulatory properties (Harrington et al. 2019). Additionally, efforts are made to evade circulating immunity in the human population by employing low-prevalent serotypes or nonhuman viruses (Uusi-Kerttula et al. 2015).

Monogenic Diseases

Great successes of gene therapy have been achieved in treating monogenic diseases (see Table 14.6 for a detailed list of approved gene therapy products). The approach has mainly been to apply gene augmentation or addition to restore a loss-of-function mutation, particularly successful for ex vivo applications. Severe combined immunodeficiency diseases (SCID) is a group of diseases in which ex vivo gene therapy has shown a lasting, clinically meaningful therapeutic benefit, e.g., gammaretroviral vector gene therapy for ADA-SCID (approved product Strimvelis™) and X-linked SCID. Gene-based therapy for other blood disorders, such as for β-thalassemia, has also been recently approved (betiglogene autotemcel/Zynteglo™). In inborn errors of metabolism, HSC gene therapy has been used to produce recombinant protein and shown beneficial effects in cerebral adrenoleukodystrophy (CALD, approved product elivaldogene autotemcel/SKYSONA™) and metachromatic leukodystrophy (MLD, approved product atidarsagene autotemce/Libmeldy™)(see Table 14.6) (Eichler et al. 2017; Fumagalli et al. 2022), and other metabolic disorders including Fabry, Gaucher, and mucopolysaccharidosis are underway.

In addition, in vivo gene therapy trials using AAV vectors for the treatment of monogenic diseases, such as spinal muscular atrophy (SMA), have been approved by regulatory agencies (onasemnogene abeparvovec – xioi/Zolgensma®). However, a rare but major risk of this type of in vivo AAV gene therapy is that it can cause acute serious liver injury or acute liver failure (Chand et al. 2021). It was also thought until recently that the safety profile of AAV vector integration is negligible, but recent studies indicate that long-term preclinical studies may be required to thoroughly understand these risks (Nguyen et al. 2021).

Other approaches have used CRISPR/Cas9 editing technology for blood disorders and inborn errors of metabolism. For instance, *BCL11A*, an erythroid-specific enhancer that represses γ-globin expression and fetal hemoglobin in erythroid cells, has been targeted in autologous CD34+ cells (Frangoul et al. 2021; Quintana-Bustamante et al. 2022). In another trial for transthyretin amyloidosis using lipid nanoparticles to deliver CRISPR/Cas9, misfolded transthyretin serum levels were lowered after a single dose and caused durable knockout (Gillmore et al. 2021).

Issues that have prevented gene transfer for monogenic diseases are (a) lack of suitable gene delivery technologies, (b) unfavorable interactions between the host and gene transfer vector, (c) complex biology and pathology of monogenetic diseases and target organs, and (d) lack of relevant measures, i.e., biomarkers, to assess the clinical efficacy and long-term efficacy of gene transfer. Challenges that remain in treating monogenic diseases are to induce gene expression sufficiently to correct or prevent further progression of clinical phenotypes without induction of host immune responses against the vector component or transgene product and minimize the risk of insertional mutagenesis for integrating vectors in target cells. Improvements in vector technology and advancements in the understanding of disease pathology will vastly improve methods for the correction of genetic diseases.

Cardiovascular Diseases

Cardiovascular diseases are the fourth largest group of diseases actively treated by gene therapy clinical trials (John Wiley and Sons LTD 2022). The current understanding of molecular mechanisms of cardiovascular diseases has uncovered many genes that could serve as potential targets for molecular therapies. For example, overexpression of genes involved in vasodilation such as endothelial nitric oxide synthase and heme oxygenase-1 or inhibition of molecules involved in vasoconstriction (angiotensin-converting enzyme, angiotensinogen) have reduced blood pressure in animal models of hypertension (Melo et al. 2005). Most clinical trials for cardiovascular diseases are designed for treating coronary and peripheral ischemia. Overexpression of pro-angiogenic factors such as vascular endothelial growth factor (VEGF), fibroblast growth factor (FGF), and hepatocyte growth factor (HGF) has been effective in myocardial and peripheral ischemia in preclinical studies (Shimamura et al. 2020). Despite the lack of significant benefit in several earlier clinical trials, VEGF gene therapy did show an excellent safety profile and improvement of symptoms in patients following adenovirus or plasmid intramyocardial administration in both pilot studies and long-term follow-ups (Stewart et al. 2006; Reilly et al. 2005). However, limited success was experienced in using gene therapy to treat cardiovascular diseases compared to other areas. The efficacy of gene therapy for cardiovascular disease will most likely be enhanced by strategies that incorporate multiple gene targets with cell-based approaches. Few of the gene therapy approaches for cardiovascular diseases are in phase II or III clinical trials. In 2019, a phase III clinical trial investigating the effect of intramuscular injection of pDNA encoding hepatocyte growth factor for the treatment of critical limb ischemia with ulcerations was successfully finalized, which led to the conditional approval of Collategene in Japan (Ylä-Herttuala 2019).

Infectious Diseases

Genetic vaccines based on DNA or mRNA, which are being discussed in Chap. 15 have shown to be very effective in the prevention of a number of infectious diseases, including COVID-19. For treating chronic infections, gene therapy approaches are being developed as well. Gene therapy for acquired immunodeficiency syndrome (AIDS) is the main application in this category. These interventions share the goal of inducing remission from HIV pathogenesis without the use of antiretroviral therapy (ART). The interest in gene therapy for an HIV cure was inspired by the elimination of the intact virus in Timothy Brown (also known as the Berlin patient) and Adam Casteljo (also known as the London patient), who both received stem-cell transplants from a CCR5-negative donor to treat their underlying malignancies (Gupta et al. 2019). This has spurred the research on genome editing approaches of CCR5 and co-receptor CXCR4 using CRISPR-Cas to confer resistance to CCR5-tropic HIV strains. Other strategies focus on boosting the immune system to reduce or eliminate the HIV pool. Many gene therapy trials for AIDS involve ex vivo transfer of genetic material to autologous T cells using self-inactivating or conditionally replicating viral vectors to improve the immune system of the patients (Manilla et al. 2005; Levine et al. 2006). Other trials employed overexpression of HIV inhibitors such as RevM10 to increase CD4+ T cell survival in HIV-infected individuals (Ranga et al. 1998; Morgan et al. 2005).

Besides HIV, gene therapy approaches are being developed against chronic infections with hepatitis B and hepatitis C virus, herpes simplex virus, malaria, and bacterial infections.

Neurological Diseases

Progress has been made in gene therapy for neurological diseases. For example, the approval of onasemnogene abeporvovec-xioi (Zolgensma®) for spinal muscular atrophy (SMA), which uses an AAV9 serotype with a human survival motor neuron 1 (SMN1) gene, is infused into the circulation but crosses the blood-brain barrier to transduce affected neurons.

Other more commonly investigated neurological diseases for gene therapy are Alzheimer's disease (AD) and Parkinson's disease. For Parkinson's disease, both gene augmentation to restore loss of dopaminergic neurons and restoration of neurotrophic factors have been investigated in clinical trials using AAV vectors and lentiviral vectors (Serva et al. 2022).

AD can be divided into familial and sporadic, with the familial form having mutations in three major genes, amyloid precursor protein (APP), presenilin 1 (PSEN1), or presenilin 2 (PSEN2). Many genetic or environmental factors play a role in sporadic AD, which makes it difficult to apply therapies designed for monogenic disorders. Gene therapy trials have been performed using neurotrophic factors, such as nerve growth factor, to promote neuronal and synaptic repair, but these have been providing mixed results (Lennon et al. 2021).

In the neurogenerative disorder Huntington's disease, many clinical gene therapy trials have been performed aiming at lowering the protein Huntingtin (Htt). Abnormal conformation of Htt results in the toxic gain-of-function. There are approaches using antisense oligonucleotides, but this can only transiently downregulate Htt. AAV5 vectors have been used to deliver miRNA to stably reduce Htt expression long-term (Byun et al. 2022).

Another group of neurological diseases encompasses leukodystrophies, which are a heterogeneous group of genetic disorders affecting the white matter of the central nervous system. Both AAV gene therapies as well as HSC gene ther-

apy approaches have been tested, targeting different cell types in the brain and showing promising results in preclinical and clinical studies (von Jonquieres et al. 2021).

However, delivery of gene therapy technologies to the CNS requires careful selection of vector type and route of administration for optimal biodistribution to the affected cell types requiring correction or modulation of gene expression.

Nonclinical Animal Testing Considerations

A full pre-/nonclinical testing program during drug development, as presented in Chap. 8 for mAbs, may not always be feasible or necessary for advanced therapies due to the nature of these products that consist of a heterogeneous population of human cells or tissues (see also Table 14.2). Generally, the pre-/nonclinical testing package entails studies to provide data on the following:

(i) safety (toxicity, including immunogenicity);
(ii) tolerance (local, systemic);
(iii) biodistribution;
(iv) persistence (duration of exposure);
(v) in vivo proliferation, maturation, and/or differentiation into an unwanted lineage of stem cells (ESCs, iPSCs);
(vi) tumorigenicity;
(vii) reproducibility;
(viii) biological activity (potency) in vivo and/or in vitro; in vivo mechanism of action
(ix) in vitro and in vivo efficacy studies to understand
(x) which cells/cell-subpopulations and cell characteristics have therapeutic value;
(xi) PK/PD to serve dose definition, e.g., number of (viable) cells;
(xii) PK/PD to serve route of administration and schedule;
(xiii) study duration to monitor for toxicity, and;
(xiv) safety of the surgical procedure for local delivery of cells/tissues.

Nonclinical animal safety (toxicology) and efficacy (pharmacology) studies pose significant challenges when applied to advanced therapies, e.g., for the following reasons:

1. Molecular incompatibility and immune rejection in xenogeneic human-animal combinations i.e., human tissues/cells tested in animal models. This is also true for genetically transduced cells, where the genetic modification leads to the expression of human protein(s), e.g., CAR-T cells.
2. Cellular immunotherapy to treat cancer (e.g., TILs) relies on the interaction of the cellular product with the patient's immune system for its effect. The in vivo immunological effect will very likely be different between species.
3. Cells do not undergo ADME in a way conventional medicinal products often do.

Table 14.12 Examples of animal and other pre-clinical models applied for assessment of safety, efficacy, and product potency testing

Animal and other model options	Example	Comment
Immunodeficient or immunosuppressed animal	NOD.SCID-rd1 mouse model of retinitis pigmentosa	See Chap. 9 for details on transgenic animal models
Animal disease model	Diabetic mouse model	Not always possible especially in case of immune based disease
Homologous animal model	AMD mouse model	Copy of human condition regarding pathology, symptoms and prognosis of disease. Use species specific autologous or allogeneic cells instead of human cells and apply the same manufacturing process to produce the animal cell based product; characterize the product to the extent possible; mimic the clinical setting in terms of route of administration, surgical procedure, and dose regime, to the extent possible
Homologous animal model plus use of a vector	ADA-SCID mouse model	See above plus vector encoding the animal homologue for animal cell transduction
Non-invasive whole animal modeling system	Magnetic resonance imaging (MRI) or computed tomography imaging (CTI) techniques	Cell fate/biodistribution studies in animals
Large animal model	Delivery of cells in the sub-retinal space of a pig's eye to train surgeons to safely administer cells in the eye of AMD patients; delivery of stem cells for treating spinal cord injury in pig model	Development of complex surgical procedures which would be technically difficult or impossible in small species
In-vitro assay system	Cell culture system to mimic cell migration upon immune stimulus	Potency test to characterize an advanced therapy

Without nonclinical safety and pharmacology data, it may be difficult to predict the potential safety and efficacy of the cell therapy product in a first-in-human clinical study. Therefore, alternatives should be investigated that could yield evidence of safety and, evidence of efficacy or at least paucity of efficacy, including the use of models explained in Table 14.12.

Relevant Animal Models

Mice are often the species of choice to study advanced therapies. They are relatively inexpensive, reproduce quickly, and can be easily manipulated genetically. However, the ability of mouse experiments to predict the effectiveness of advanced therapies remains controversial. The failure of many mouse models to mimic particular human diseases has compelled investigators to examine animal species that may be more predictive of humans. Larger animals, such as rabbits, dogs, pigs, goats, sheep, and nonhuman primates, are potentially better models than mice. They have a longer life span, which facilitates long-term (e.g., years) studies that are critical for some advanced therapy products with a lifelong effect. In addition, many physiological parameters, e.g., immune system properties that play an important role in the reaction of the host animal to advanced therapies, are much closer to humans than rodents. Large animals also have significant advantages regarding the number and types of cells or amounts of tissues

that can be reproducibly isolated from a single donor animal and ex vivo manipulated in sufficient quantity for analysis and for various nonclinical applications.

In case animal safety data do not provide meaningful information based on which an extrapolation can be made to potential risks posed to humans, those studies may be (partially) waived by regulatory authorities. Study set-up and duration for evaluation of the toxicity and/or biodistribution have to be determined on a case-by-case basis and depend on, e.g.,

- product half-life, which may vary between hours–days and months–years, the latter for cells that engraft in a specific niche in the human body;
- potential alterations of cells over time upon administration;
- dose regime of single or repeat dosing over a period of weeks–months–years;
- chance for migration of the cells in the body to unwanted sites upon administration, e.g., local administration of an adult stem cell in the subretinal space of the eye may be safer than the systemic administration of an ESC/iPSC-derived product;
- type and number of ex vivo cell manipulations performed during manufacture, i.e., in case cells are expanded for multiple passages close to the point where these cells senesce, and animal studies should be performed with cells beyond the cell passage used to manufacture the advanced therapy).

Generally, genotoxicity and specific safety pharmacology studies are not conducted for cell and tissue-based products unless there is a reason for concern, e.g., the use of a novel excipient or novel route of administration for an approved excipient. In addition, reproductive toxicity studies are only required when there is a potential risk for exposure to the reproductive organs. And finally, literature data may be used to support the (lack of) animal data. See Herberts et al. (2011) and Vestergaard et al. (2013) for further reading.

Clinical Testing Considerations

For investigating the safety and efficacy in humans, generally, the same principles apply to advanced therapies as to other medicinal products (see Table 14.4). However, considering the nature and complexity of the products and potential risks and benefits, there are some unique aspects to the clinical programs (Mount et al. 2015):

1. Different set-up of trials compared to most conventional medicinal products:
   (a) First-in-human trials are always in patients and never in healthy volunteers;
   (b) A seamless development path rather than the traditional route of separate formal phase I (safety), phase II (hint of efficacy), and phase III (safety confirmation and efficacy) studies.
2. Traditional PK (ADME)/PD studies may not be feasible.
   (a) Dose (defined as the number of cells/mL; the number of cells/kg body weight) escalation studies may not be feasible as there may not be clear dose-response correlations. A low-, medium-, and high-dose is often selected based on literature data concerning the number of cells that have historically been administered to humans.
   (b) Advanced therapies are frequently administered intravenously and rapidly cleared via the lungs, spleen, and liver (Leibacher and Henschler 2016). Other possible routes are intranodal (DCs to treat rheumatoid arthritis) or local administration via a surgical procedure, e.g., into the eye, brain, spinal cord, or knee.
3. For safety evaluation, the following risks may need to be taken into account, depending on many factors, including the type of product, cell differentiation status upon administration, cell proliferation capacity, cell source being autologous, allogeneic or xenogeneic, the half-life of the cells in the body/lifelong persistence, site and method of administration/implantation, quality of the starting material (derived from a healthy donor or very sick patient), and disease environment(s) which cells may encounter in the patient's body:

   (a) Tumor formation (tumorigenicity), e.g., in case of ESC- and iPSC-derived products which are ex vivo expanded and differentiated;
   (b) Potential adverse reactions at the site of administration, e.g., dimethyl sulfoxide (DMSO) related side effects upon i.v. administration;
   (c) Cells, being subvisible particles, make it difficult to assess subvisible particles potentially present in the product. These foreign particles may damage the tissue upon administration, e.g., in the subretinal space of the eye;
   (d) Inflammatory responses and infections (e.g., side effect of CAR-T cells);
   (e) Implantation procedure for cells or 3-D tissue replacement therapies using a complex surgical procedure, e.g., to administer cells in the subretinal space of the eye, in the spinal cord, or in the brain; 3-D cultured trachea placed in the throat;
   (f) Immuno-mediated side effects (CAR-T cells may cause cytokine release syndrome);
   (g) Immunogenicity, which may depend on:
   - Relative immune privilege of the administration site (e.g., eye);
   - Allelic differences between product and patient cells (e.g., allogeneic dendritic cells);
   - Immune competence of the patient;
   - Need for repeat dosing (more doses may increase the chance of immune rejection of the advanced therapy);
   - Maturation status of the cells (e.g., ESCs).

   Advanced therapies derived from an allogeneic cell source often require immune-suppressant medicines to be administered together with the cell-/tissue-based product. However, some allogeneic cell-/tissue-based products, such as MSCs, have shown relatively low immunogenicity profiles, in part due to the short half-life of the cells in the body. See more details below.
4. Selecting the right patient population for the initial clinical program is challenging as there is a tension between choosing the patients most likely to benefit from an efficacious advanced therapy (e.g., early-stage cancer patients) and limiting the risk to which patients are exposed to the unlicensed therapy (late-stage cancer patients who may not benefit from the therapy at all due to their severe illness).
5. Establishment of surrogate biomarkers for efficacy assessment may be needed to predict long-term clinical outcomes of cells that may persist in the body for years e.g., CAR-T cells which engraft in the peripheral blood and bone marrow and transduced CD34+ cells, which engraft in the bone marrow.

Table 14.13 Example of clinical trials with pluripotent stem cells (hESCs and iPSC), adapted from Trounson and McDonald (2015) and Ilic et al. (2015)

Indication	Active substance	Trial sponsor (country)
AMD	hESC-derived RPEs	Chabiotech (South Korea)
DryAMD; myopic AMO; Stargardt's macular dystrophy	hESC-derived RPEs	Ocata therapeutics (USA)
WetAMD	hESC-derived RPEs	Pfizer (UK)
DryAMD	hESC-derived RPEs	Cell cure neurosciences (Israel)
Type I DM	hESC-derived pancreatic endoderm cell	Viacyte/Johnson& Johnson
Heart failure	hESC-derived CD^{15+} lsl-$^{1+}$ progenitors	Assistance Publique-Hopitaux de Paris (France)
Parkinson's disease	Human parthenogenic-derived neural stem cells	International stem cell Corp. (Australia)
Spinal cord injury	hESC-derived oligodendrocyte precursors	Asterias Biotherapeuticcs (USA)
WetAMD	hESC-derived RPEs	The London project to cure blindness (UK)
WetAMD	iPSC-derived RPEs (autologous)	Aiken institute (Japan)

AMD age-related macular degeneration, *RPEs* retinal pigmented epithelial cells

6. Particularly for genetically modified cells, which may persist in the body for many years or lifelong, long-term (10–20 years) patient follow-up for safety, efficacy, and durability monitoring may be necessary.

Immunological Considerations in Advanced Therapy

The potential application of adult stem cell-based medicinal products derived from allogeneic sources and hESC-based therapies is limited by risks for graft-host rejection issues, as with all therapeutic strategies based on cell, tissue, and organ transplantation, unless the transplant is derived from an autologous source. A way to overcome this challenge is the use of a device to protect the allogeneic cellular product from the host immune system. An example of this strategy is Viacyte's cell-based combination product, where the hESC-derived β-islet progenitors are contained in the Encaptra® cell delivery system, which is placed subcutaneously (see Table 14.13) and later Fig. 14.34). The additional advantage of this system is that cells cannot migrate in the body to unwanted sites, and the device can be taken out in case of, e.g., tumor formation. The disadvantages of such an immune-protective device are fibrosis and the lack of vascularization around the device, which is required for cell viability and insulin production. Certain human body sites have immune privilege, i.e., they tolerate the introduction of nonself-antigens without eliciting an inflammatory immune response. These sites include the eyes, the testicles, the fetus, and certain tumors. There is debate in the cell therapy world regarding the immunogenicity of allogeneic MSCs (Ankrum et al. 2014; Consentius et al. 2015). Clinical trials with standardized immune monitoring programs and a better understanding of the in vivo mode of action of allogeneic MSCs may provide answers.

Administration of drugs to suppress the immune response is standard practice for patients undergoing transplantation, but with immunosuppression come side effects. The hope is that iPSC technology (see above) may overcome rejection problems for which several products are being tested in the clinic (Kim et al. 2022). Another approach is to bank a collection of ESC lines that allows the selection of a matched ABO and HLA haplotype or a close match (Lui et al. 2009). It has been estimated that with a bank of 70–100 ESC lines, a partially matched ESC line that is adequate for each recipient can be chosen. The downside of this approach is that at the time the cell lines are banked, it may not be clear yet for which diseases they will be used in the future, hence what the critical parameters are to characterize the banks, for example, purity of the cells, stability, potency, viral safety, see (Bravery 2015). Preparing cell banks, extensive testing and long-term storage under frozen conditions are very expensive undertakings.

Manufacturing and Testing Considerations

Manufacturing

Cell and tissue-based products are distinct from traditional biopharmaceuticals in that the modified cell/tissue itself is the active ingredient in the medicinal product rather than "simply" the means by which the cells produce an active ingredient (e.g., a recombinant protein; a viral vector). However, many of the production platforms, cell culture media, storage and transport bags, and product excipients and primary containers have been established for traditional cell-based recombinant protein manufacturing processes (cf. Chap. 4) and can be readily applied to these innovative products.

Since the vast majority of advanced therapies contain viable cells/tissue that can be easily destroyed through sterilization procedures and cannot be sterile filtered ($\leq$0.2 μm filter pore size), as cells have a size of 10–30 μm on average and tissues are even bigger, the manufacturing of these products must take place under aseptic conditions. For nonsterile raw and starting materials and excipients, additional steps may need to be taken to ensure subsequent aseptic manufacturing, e.g., heat inactivation, gamma-irradiation or sterile filtration of the material. The facilities, equipment, raw materials, viral vectors, and cells/tissues used must be of suitable quality to allow for good manufacturing practice (GMP) production of the drug product for human application. At every stage of production, materials and the final product should be protected from microbial, viral, and other contamination. The manufacturing of advanced therapies typically requires many or all of the following "manipulation" steps, see Table 14.14.

Control of the Manufacturing Process

As with manufacturing process of biologics, process variables need to be chosen carefully and monitored to allow for adjustments to the process and to ensure a product of high quality is consistently produced. Process variables assessed are, e.g., medium perfusion or exchange rate, feeding regime, biomass, stirring speed, pH, dissolved oxygen (DO), and lactate production. Particularly in the case of open and manual culture steps, this is challenging because any handling of the cells/tissue may impact the quality of the viable material and could potentially contaminate the culture system. Examples of fully closed production systems enabling different manipulation steps in one system are the CliniMACs Prodigy® and the Octane Technology (see Figs. 14.30 and 14.31, respectively).

A fully closed processing system is the CliniMACS Prodigy. This single-use device performs all manufacturing steps (i.e., cell wash, enrichment, activation, genetic modification, expansion, final formulation, and sampling). This contrasts with other manufacturing approaches, which use separate machines for cell culture, cell washing, and other steps in the production chain.

Table 14.14 Typical advanced therapy manipulation steps and equipment used for each step

Manipulation step	Equipment used (examples)
Collection or generation of autologous or allogeneic donor cells; collection of tissue biopsy (i.e., starting material). This step is not considered a GMP manufacturing step and takes place outside the GMP facility at a clinical site	Bone marrow aspiration system, surgical procedure, apheresis/leukopheresis system (Fig. 14.35)
Isolation of specific cell population(s). This is usually where the GMP manufacturing process starts	Knife; fluorescence-activated cell sorting (FACS) (see below): positive/negative selection by e.g., magnetic-activated cell sorting (MACS®) technology (microbeads and column); Elutra®; LOVO spinning membrane filtration device
Cultivation, expansion, and/or (genetic) modification of cells; tissue culture	Cell culture systems (see Chap. 4)
Cell differentiation	Specific raw materials, such as growth factors, are added to the culture medium manually or automatically
Purification of desired cell population(s); purification of tissue	Counter-flow centrifugal elutriation (Ficoll).This technique separates cells by size and density while maintaining cell viability. Cell enrichment kit for the magnetic separation of the desired cells by negative selection. It utilizes antibody magnetic bead complexes. Undesired cells are bound by the antibody and then magnetic beads that, when placed in a magnetic field, leave the desired cells untouched and free in the medium. The same principles and systems can be applied as for isolation of specific cell population(s) (see above)
Cell harvest and cell wash/cell concentration; tissue harvest and wash	Centrifuge; fluidized bed + elutriation-closed system (K-Sep); tangential flow filtration (TFF) technology; spinning-membrane filtration;
Formulation of the harvested cells in excipient mixture; formulation of tissue	Manually; mixing station with disposable bag set-closed system (Invetech)
Filing in the primary container of cell suspension; transfer of tissue to primary container (this is considered the drug product (DP)	Manual vial filling, stopping, and capping (Flexicon pump); manual bag filling and sealing; (semi) automated vial filling (FPC50, Flexicon system)
Labelling of the primary container	Manually; automatically with labelling machine
Short/long term storage of the DP	Refrigerator; controlled rate freezer; freezer, cryopreservation tank
Shipment of the DP to the clinical site	Temperature controlled shipment in cool box, on dry ice, in cryogenic Dewar
Handlings of the DP at the clinical site to allow for administration of the DP to the patient (e.g., thawing, washing, mixing with other ingredient)	Plasmatherm controlled temperature rate dry thawing instrument; centrifuge, mostly manual handlings

The manufacturing process for advanced therapies parallels the processes for *E. coli*/mammalian production cells described in Chap. 4 for therapeutic proteins. But they differ considerably from those processes at a number of critical points. On top of that, the various types of cell therapy products vary widely from each other. Below are examples of manufacturing process flow charts for three different types of advanced therapy medicinal products:

1. Off-the-shelf or nonoff-the-shelf MSC production process, as described below and presented in (Fig. 14.32);
2. Non off-the-shelf CAR-T production process, as this procedure is a prime example of "personalized medicine" (see Chap. 9) the complexity is caught both in the text below and shown in Fig. 14.33;
3. Off-the-shelf human ESC-derived prebeta cell production process, as described below and presented in Fig. 14.34.

Manufacturing of MSC Product

The manufacturing of an off-the-shelf (allogeneic) or nonoff-the-shelf (autologous or allogeneic) cell-based product, e.g., MSC-derived product, is a multi-step process with slight modifications for each specific product (see Fig. 14.32).

- Step 1: Starting material procurement via bone marrow (BM) aspiration (1a) or adipose tissue biopsy (1b) from a healthy donor (allogeneic cell source) or patient (autologous cell source). Other sources of MSCs are not discussed here. The donor (healthy person or patient) is tested for specific human viruses before donating the starting material.
- Step 2: Mononuclear cell separation from BM (2a) using separation techniques; adipose tissue digestion using enzymes, such as collagenase (2b).

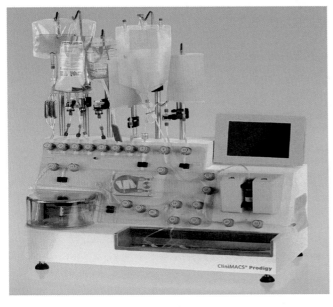

Fig. 14.30 Miltenyi's CliniMACs Prodigy closed processing system for cells grown in suspension (DCs, T cells)

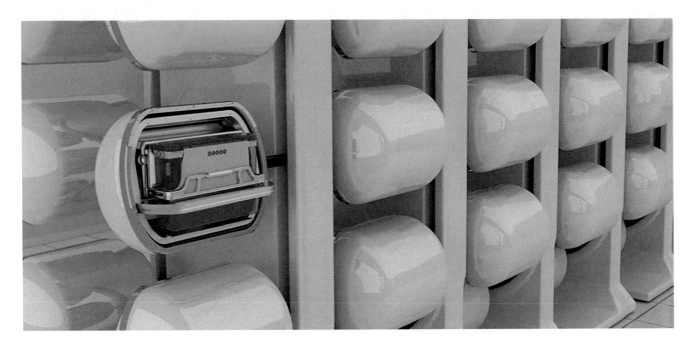

Fig. 14.31 Octane Technology, a fully closed production system for scale-out of autologous or allogeneic tissue- and cell-based products

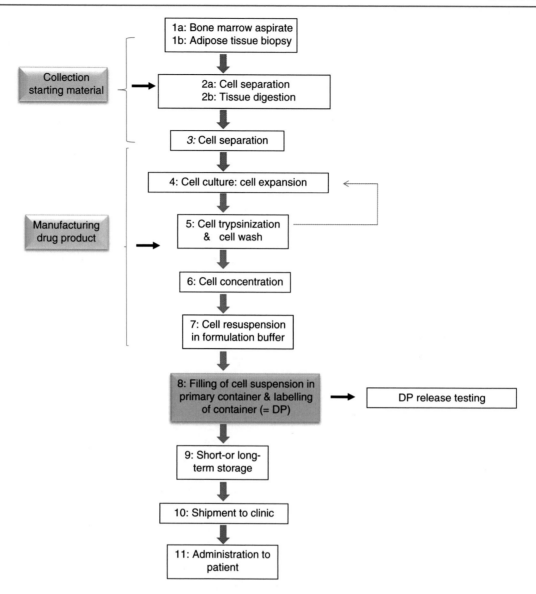

Fig. 14.32 Flow diagram of a manufacturing process for an off-the-shelf (allogeneic) or non-off-the-shelf (autologous or allogeneic) cell-based product based on adherent cells which do expand ex vivo, such as MSCs

- Step 3: Mononuclear cell separation from digested adipose tissue.
- Step 4: MSC expansion: MSCs are adherent cells and can, therefore, either be cultured in a culture flask (2D culture) or on micro-carriers in suspension culture (3D culture). Cells grow and multiply via mitosis and meiosis. By selecting the appropriate surface and culture medium, and culture conditions, unwanted cell populations do not adhere and are separated from the wanted cell populations.
- Step 5: Cell detachment from the surface via trypsinization. Cells are washed to remove dead cells, unwanted cell populations, and trypsin. Steps 4 and 5 are repeated as many times as needed for the targeted dose or to freeze-down a cell bank (MCB/WCB strategy; which is an off-the-shelf product approach).

- Step 6: Cell concentration.
- Step 7: Resuspension of the cells in formulation buffer.
- Step 8: Filling of the cell suspension in the primary container (vial or bag) and labeling of the primary container. This is considered the drug product (DP).
- Step 9: For some products, the cells are immediately shipped by a qualified courier to the side of administration after step 8. In such cases, the hospital should be at a short distance, as the product cells are generally stable for hours to a couple of days at room temperature or at 2–8 °C (short-term storage; nonoff-the-shelf product). To allow for time between product manufacture plus quality control (QC) testing plus the release of the DP and administration, and to allow for easy shipment to distant hospi-

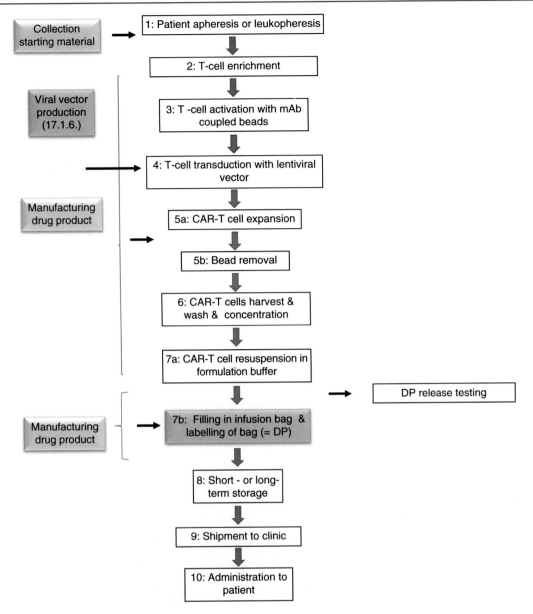

Fig. 14.33 Flow diagram of a CAR-T cell product manufacturing process. At the hospital white blood cells are harvested by leukapheresis (1). The starting material is shipped to the manufacturing facility for enrichment of the wanted T-cell populations (2); T-cell activation (3); transduction (genetic modification) of the T-cells with the lentiviral vector encoding the CAR genetic information (4). Thereafter, transduced cells (CAR-T cells) are ex vivo expanded (5a) and purified via bead removal (5b). Cells are harvested, washed, and concentrated (6); cells are resuspended in formulation buffer (7a) and filled in the primary container (7b), which is labelled. This is considered the drug product. The product is stored (8) and thereafter shipped to the clinic (9). Prior to administration via IV infusion of the CAR-T cells at the hospital (10), the patient is pre-conditioned with chemotherapeutic medicines. Except steps 1 and 10, which take place at the hospital, all other steps take place at a manufacturing facility under GMP conditions. QC testing occurs between steps 1–2 (control of the starting material), in-process (steps 2–7a), and on the final drug product (step 7b)

tals, the product is stored and shipped frozen, often in the vapor phase of liquid nitrogen at < −120 °C (long-term storage).

- Step 10: Shipment of the DP to the clinical site.
- Step 11: Administration to the patient systemically (IV infusion) or locally with/without the use of a surgical procedure.

Manufacturing of CAR-T Product

The manufacturing of genetically modified T cells is a multi-step process with slight modifications for each specific product (Fig. 14.33):

- Step 1: Harvest of blood cells by apheresis (whole blood collection) (Fig. 14.35)) or leukapheresis (collection of

Fig. 14.34 Flow diagram of a hESC-derived combination product manufacturing process to treat DM type I

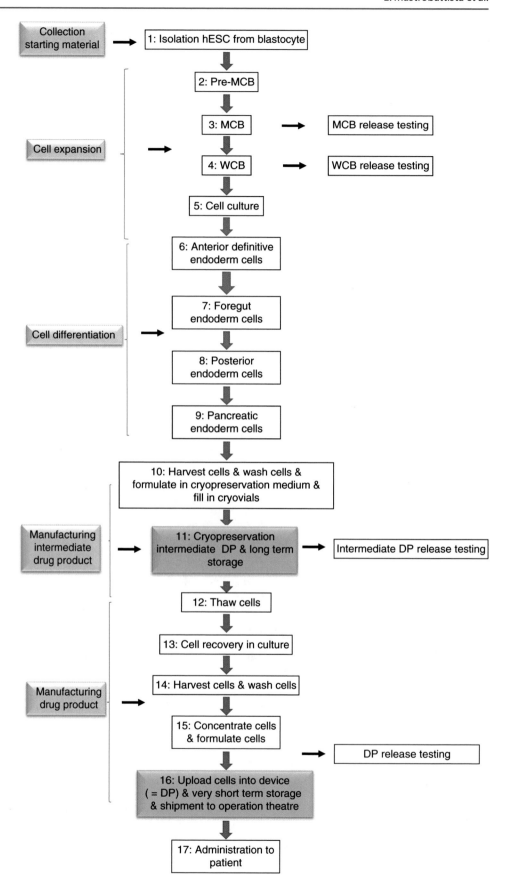

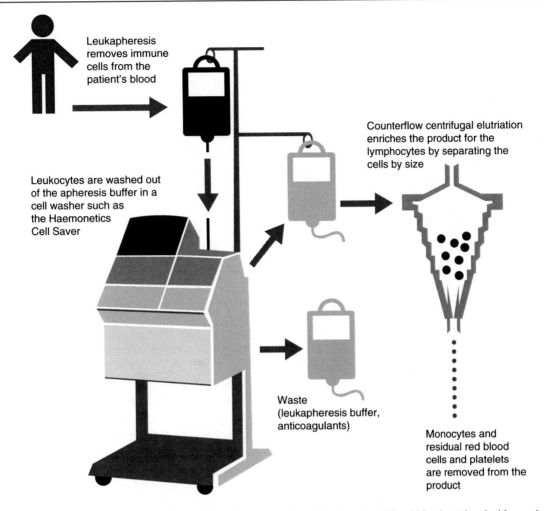

Fig. 14.35 Example of a leukapheresis system, which collects lymphocytes from the donor's peripheral blood, reprinted with permission (Levine et al. 2017)

leukocytes) from the patient (autologous cell source). The so-called "starting material" is shipped either "fresh," i.e., at room temperature or at 2–8 °C, or "frozen" ($\leq$−80 °C) to the GMP manufacturing site. The patient is tested for specific human viruses prior to the donation of the starting material.

- Step 2: From this starting material, lymphocytes can be enriched either by counter-flow centrifugal elutriation or by subset selection according to the cellular phenotype.
- Step 3: The enriched lymphocyte population is placed in culture and stimulated with bead-based artificial antigen-presenting cells, e.g., magnetic beads, coupled with mAbs.
- Step 4: The viral vector is added to transduce the genetic insert (CAR) into the T cells.
- Step 5: The cell culture is expanded in a bioreactor for several days until sufficient numbers of CAR-T cells are obtained for dosing and QC testing. A magnet removes the beads from step 3 as they are considered a process impurity.

- Step 6: The T cells are harvested, washed, and concentrated.
- Step 7: The cells are resuspended in the final product formulation buffer (7a) and filled in the primary container (infusion bag or vial). This is the so-called "DP" (7b). Samples are taken for quality control testing.
- Step 8: For some products, the cells are immediately shipped by a qualified courier to the side of administration after step 7. In such cases, the hospital should be at a short distance, as the product cells are generally stable for hours to a couple of days at room temperature or at 2–8 °C (short-term storage). To allow for time between product manufacture plus QC testing plus the release of the DP and administration, and to allow for easy shipment to distant hospitals, the product is stored and shipped frozen, often in the vapor phase of liquid nitrogen at <−120 °C (long-term storage).
- Step 9: See step 10, manufacturing of MSC product
- Step 10: At the site of administration, the product is either administered directly to the patient or first thawed and

sometimes washed to remove certain excipients such as dimethyl sulfoxide (DMSO) and then administered, often via IV infusion.

The chain of the identity of the entire process, from leukapheresis to infusion and throughout all manufacturing steps and vice versa, i.e., from donor to recipient and from the recipient to donor, is controlled by a computer-based system to ensure the product's identity and product traceability.

Manufacturing of hESC Product

The manufacturing of an hESC-derived combination product (cells in device) to treat DM type I is a multi-step process with expansion and complex differentiation steps, with slight modifications for each specific product (Fig. 14.34):

- Step 1: Isolation of the starting material (hESCs) via the inner cell mass extraction. This procedure can only take place after informed consent from the parent(s) and testing of the mother's blood for specific human viruses. In addition, this step does not occur at a manufacturing facility under GMP but at an accredited tissue establishment, which is often a hospital.
- Step 2: Production of the pre-MCB by hESC culture initiation, cell expansion, cell wash, cell harvest, formulation of the cells in cryogenic medium, fill in a vial, and storage under cryogenic conditions in the vapor phase of liquid nitrogen.
- Step 3: Production of the MCB from a pre-MCB. A pre-MCB vial is thawed, and cells are cultured and expanded as described under "step 2," followed by release testing of the MCB.
- Step 4: Production of a WCB from the MCB (see step 3) and release testing of the WCB.
- Step 5: A WCB vial is thawed, and cells are expanded to obtain the required cell number for cell differentiation. Steps 2 through 5 take a couple of weeks.
- Step 6: Differentiation of undifferentiated hESCs into anterior definitive endoderm cells by adding specific growth factors and other factors to the culture medium. This step takes about 2 days.
- Step 7: Differentiation of anterior definitive endoderm cells into foregut endoderm cells by adding specific growth factors and other factors to the culture medium. This step takes about 3 days.
- Step 8: Differentiation of foregut endoderm cells into posterior foregut cells by adding specific growth factors and other factors to the culture medium. This step takes about 3 days.
- Step 9: Differentiation of posterior foregut cells into pancreatic endoderm cells by adding specific growth factors and other factors to the culture medium. This step takes about 4 days.

- Step 10: Pancreatic endoderm cells are harvested, washed, resuspended in a cryo-preservation medium, and filled in cryovials. The cryovials are labeled. This is considered the "intermediate DP."
- Step 11: The intermediate DP is cryopreserved in the vapor phase of liquid nitrogen at < -120 °C (long-term storage) and extensively QC tested prior to the release of the intermediate DP.
- Step 12: Intermediate DP cryovials are thawed. In case steps 2 through 11 take place at a GMP facility on long distance from the clinical site where the drug product will be administered to the patient, the cryopreserved intermediate DP is shipped frozen to a GMP facility, often the hospital pharmacy, for preparation of the final drug product.
- Step 13: Intermediate DP cells are recovered from the freezing and thawing steps by placing them in culture for another 3–4 days.
- Step 14: The recovered cells are harvested and washed to remove dead cells and culture medium.
- Step 15: Cells are concentrated and formulated in a buffer.
- Step 16: Cells are uploaded into the immune-protective device using a loading device. The pancreatic prebeta cells in the device are considered the DP. Limited QC release testing is performed on the DP.
- Step 17: The device is administered to the patient via a surgical procedure.

Key Factors for a Successful Manufacturing Process

To consistently manufacture advanced therapies at a large-scale, automated manufacturing processes as well as the implementation of functionally closed systems are key success factors for the following reasons: (1) lower the risk of viral and bacterial contamination during manual and open-process steps; (2) decrease costs associated with manual handlings; (3) improve product consistency; (4) shorten production times. Other key factors for success are logistics around the manufacturing, supply chain of the product, and the cost of goods. Particularly animal and human-derived raw materials, for example, growth factors, fetal bovine serum (FBS), antibody-coupled beads, and viral vectors, are very expensive. Considering the high cost and increased risk of validating sterilization cycles of multiple-use bioreactors, these closed-processes for advanced therapies utilize single-use, disposable bioreactors, mimicking current recombinant protein platform approaches (see Chap. 4). Despite some progress made in this field, there remains a requirement for a better understanding of potential manufacturing platforms and how they can be best utilized for advanced therapies, taking the variety of cell and tissue types and clinical applications into account.

Viral Vector Production for Ex Vivo Gene Modification of Cells

Recombinant viral vectors, e.g., retroviruses like lentiviruses (cf. section on Viral Vectors in this chapter and Fig. 14.21), are produced by transfecting packaging cells, cultured with three to four plasmids that encode viral structural proteins, e.g., GAG, POL, Vesicular stomatitis virus (VSV)-G, and REV; the so-called packaging plasmids, and the plasmid encoding the therapeutic gene of interest, e.g., CAR, ADA-SCID; the so-called transfer plasmid. The transfer plasmid encoding the therapeutic gene contains the regulatory sequences that control its expression and a packaging sequence that enables its recognition. Within the packaging cell, e.g., the human embryonic kidney (HEK) 293 cell line, the RNA transcribed from the plasmid encoding the therapeutic gene is recognized by the viral proteins that assemble around it. The recombinant virus is then transported to the plasma membrane of the packaging cell that expresses viral envelope proteins (VSV-G). During budding, the virus acquires the lipid bilayer from the packaging cell surface and incorporates the envelope proteins. The viral vector particles are released from the cells cultured as adherent cells in culture flasks into the cell culture medium. The above-described steps are considered the upstream processing (USP) steps. The virus particles are subsequently harvested from the medium, formulated in a buffer, and filled in the primary container. These production steps are considered the downstream processing (DSP) steps (Morenweiser 2005). DSP steps applied for viral vector production are traditionally used in the biotechnology industry to manufacture recombinant proteins. These are membrane-based (filtration/clarification, concentration/diafiltration using tangential flow filtration, membrane-based chromatography) and chromatography-based (ion-exchange chromatography, affinity chromatography, and size exclusion chromatography) process steps. The combination of these different process steps is variable; in some cases, different purification principles are used for the same purpose. Furthermore, a benzonase/DNase treatment for the degradation of contaminating DNA from the packaging cells is either included in the USP or DSP part of the manufacturing process. Subsequently, the purified virus particles are formulated in a buffer, filled in the primary container, stored frozen, and tested until further use for transduction of the cells to make a genetically modified cell therapy product (Wright 2008). Figure 14.36 provides a schematic overview of the entire viral vector material manufacturing process used to produce a genetically modified cell therapy product. For the production of a viral vector product for in vivo gene therapy (see later in this chapter), the production process is identical.

Excipients

Common excipients used in the formulation of advanced therapies are presented in Table 14.15. Most of these excipients overlap with those used in therapeutic protein products. However, KCl, $MgCl_2$, nucleosides, FBS, and DMSO are not found in therapeutic protein drug products.

Table 14.16 provides an overview of a few commercially available advanced therapies with their formulation and shelf-life

Add to legend under [c]: DMEM = Eagle's minimal essential medium...and then the rest of the text

Primary Container

Generally, two types of containers are used for cell-based products: vials (small volume, low dose) and infusion bags (higher volume and dose), as shown in Fig. 14.37. However, tissue-based products often have a nonstandard container for storage and shipment.

Storage and Shipment

Stability of the starting material (cells or tissue and viral vector) and DP are an important element for the successful production, storage, shipment, and administration of advanced therapies. Starting materials and DPs either have a very short shelf-life of hours–days and are stored and transported at 2–8 °C or at room temperature or have a longer shelf-life (months–years) and are stored and shipped frozen (cryopreserved in the vapor phase of liquid nitrogen at <-120 °C or in a -80 °C or -150 °C freezer).

Manufacturing Model: Scale–Up Versus Scale-Out

Broadly speaking, there are two paradigms in advanced therapy manufacture: off-the-shelf (always allogeneic source of cells/tissue) and patient-specific (commonly autologous source of starting materials, but sometimes allogeneic) DPs. Off-the-shelf products represent a business model akin to current biopharmaceuticals, where one batch can be manufactured to treat multiple patients. This allows for increased economies of scale, which drives down the per-dose cost of the final product. This means that there is a wealth of engineering and process knowledge and technologies that can be leveraged to support the manufacture of off-the-shelf advanced therapies at an increasing scale.

However, scale-up is not just about making the reactor grow the cells bigger. Conventional scale-up bioprocesses typically use cells to produce therapeutic agents (e.g., mAbs),

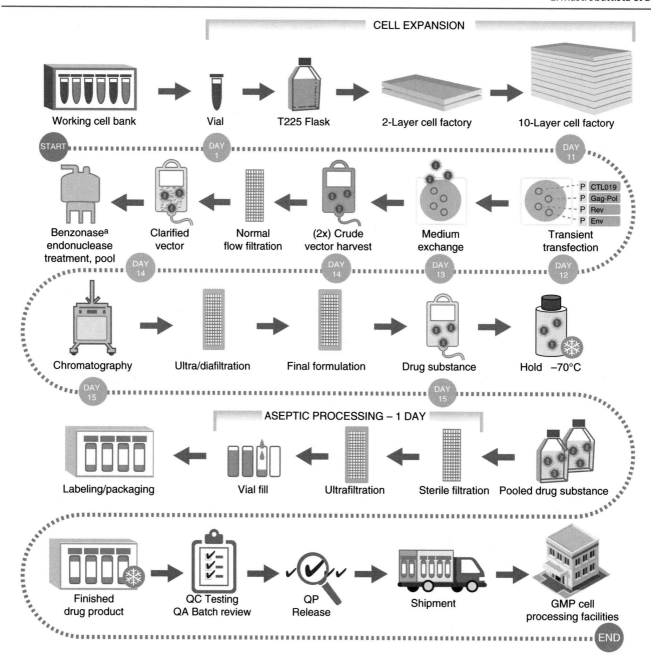

Fig. 14.36 Schematic overview of a lentiviral vector manufacturing process. The produced viral vector is used as starting material for the genetic modification of T-cells in the manufacture of a CAR-T product, reprinted with permission (Levine et al. 2017). A similar production approach is taken for other ex vivo gene therapy as well as in vivo gene therapy products (cf. Chap. 16). QC = quality control, QP = qualified person, QA = quality assurance

which can then be isolated and purified without the need to recover the cell. For the manufacture of advanced therapies, where the cells/tissue culture is the product of interest, retention of cell viability, phenotype, and function to assure quality, is of primary importance in order to preserve product safety and efficacy. As the number of cells increases during expansion, this can become increasingly challenging, as the greater cell numbers lead to an increased chance of inhomogeneity of culture and hence of cellular performance being altered. This means that the desired quality of the cells/tissue

must be maintained through the entire manufacturing process, including the harvest and DSP, storage, shipment, and delivery to the patient. This will require the development of scalable harvesting, DSP, and formulation technologies to cope with the large batch size produced.

Patient-specific advanced therapies offer a new challenge for process scalability, where the manufacturing process must be scaled-out in order to produce one batch for each patient (Fig. 14.38). This introduces the concept of "personalized medicine" (see Chap. 9), where the cost of production

Table 14.15 Examples of excipients used in the formulation of advanced therapy products

Excipients class	Function	Example
Buffer	pH stabilizer	TRIS, histidine, Na-acetate
Salt	Stabilizer	NaCl, KCl, $MgCl_2$
Antioxidant	Prevent oxidation	Methionine
Sugar	Stabilizer, cryoprotectant tonicity modifier	Mannitol, trehalose, sucrose, glucose
Polyol	Collapse temperature modifier	Dextran (low and high molecular weight)
Nucleoside	Stabilizer	Adenosine, guanosine
Protein	Stabilizer, preservative	Fetal bovine serum, human serum albumin
Organic solvent	Stabilizer, cryoprotectant solvent	Glycerol, ethylene glycol, DMSO

Table 14.16 Examples of approved advanced therapies, their formulation, and shelf-lives

Product	Shell-life and storage condition	Composition (active substance)	Excipients/mixtures
Provenge® Suspensbn of cells for IV infusion	18 h at 2–8 °C	$\geq 50 \times 10^6$ autologous CD54$^+$ cells/250 ml activated with PAP-GM-CSP[a]	Lactated Ringer's solution (NaCl, $NaC_3H_5O_3$, KCl, $CaCl_2$)
ChondroCelect® Suspension of cells for implantation	48 h at 15–35 °C	4×10^5 autologous human cartilage cells/ 0.4 ml	DMEM[b]
MACl® Implantation matrix plus cells in solution for implantation	6 days at ≤37 °C and keep out of fridge	0.5×10^5 to 1×10^6 autologous cultured chondrocytes/cm^2 porcine derived type I/III collagen membrane	DMEM, HEPES[c] adjusted for pH with HCl or NaOH and osmality with NaCl
Kymriah® Suspension of calls for IV infusion	9 months at ≤–120 °C in the vapor phase of liquid nitrogen	2×10^5–2.5×10^5 autologous CAR-positive viable T cells	Plasmalyte-A[d], glucose/NaCl, human serum albumin, dextran 40-low molecular weight/glucose, DMSO

[a] Prostatic acid phosphatase granulocyte-macrophage colony-stimulating factor
[b] Calcium Chloride anhydrous, Ferric Nitrate·9H$_2$O, Potassium Chloride, Magnesium Sulphate anhydrous, Sodium Chloride, Sodium Bicarbonate, Potassium Phosphate Monobasic·H$_2$O, D-Glucose, L-Arginine. HCl, L-Cystine·2HCl, L-Glutamine, Glycine, L-Histidine·HCl·H$_2$O, L-Isoleucine. L-Leucine, L-Lysine·HCl, L-Methionine, L-Phenylalanine. L-Serine. L-Threonine, L-Tryptophan, L-Tyrosin·2Na·2H$_2$O, L-Valine, D-Calcium Pantothenate, Choline Chloride, Folic Acid I-Inositol, Niacinamide, Riboflavin·Thiamine·HCl, Pyridoxine·HCl
[c] 4-(2-Hydroxyethyl)piperazine-1-ethanesulfonic acid sodium, *DMEM* Eagle's minimal essential medium
[d] Plasmalyte-A sodium chloride: 5.26 g/l potassium chloride: 0.37 g/l magnesium chloride hexahydrate: 0.30 g/l sodium acetate trihydrate

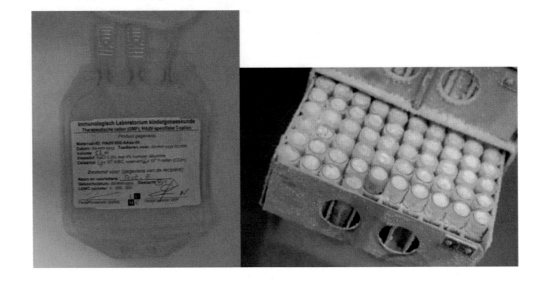

Fig. 14.37 Examples of primary containers for the storage and transport of advanced therapies. Left photo: infusion bag; right photo: cryovials in box to allow for storage in the vapor phase of liquid nitrogen (courtesy of M. de Haan)

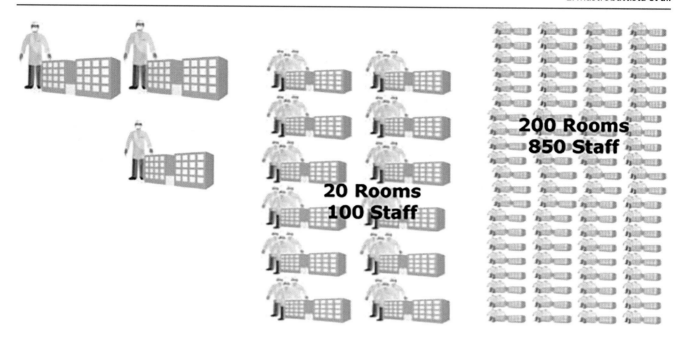

Fig. 14.38 Scale-out of a labor intensive manual process

per batch cannot be reduced by exploiting an increasing economy of scale by simply producing a larger batch. Reducing the cost of these patient-specific cell- and tissue-based products must therefore be achieved by advances in engineering and manufacturing technology, reducing the number of complex, labor-intensive, and open-process steps that are commonplace in the manufacture of these products at research labs. The developments of closed and automated processes as well as process simplification, are key factors for commercial success as this will allow multiple batches to be produced in parallel (scale-out), with reduced burden of oversight by highly-trained scientists. These new processes must be GMP-compliant and closed for sterility.

Testing

As for any DP, cell- and tissue-based therapies are subject to detailed characterization. This involves the assessment of quality attributes, i.e., identity, purity and impurities, viability (Cadena-Herrera et al. 2015), bioactivity (potency; Bravery et al. 2013), safety, quantity, and general attributes, such as appearance, pH, morphology both of the cellular/tissue/vector starting material and the final DP, see Table 14.17. The latter includes QC testing to allow the release of the DP for administration. In addition, at different stages of production, in-process controls are performed to assess the quality and stability of the cells/tissue during manufacture. Finally, a subset of characterization tools is used to assess the stability of the starting material(s) and DP.

However, for a lot of autologous and some allogeneic DPs that are not "off-the-shelf," performing QC tests may be challenging due to the time constraints between manufacture and administration, i.e., the shelf-life of the drug product is hours–days. Moreover, for some autologous products, all the available cell/tissue material is needed for the dose. In such cases, product release may be justified by extensive process validation; in-process control testing and/or QC testing data becoming available after product administration. These approaches require a paradigm shift in the pharma world, where traditional products are only administered after extensive testing and batch release. Adequate QC of starting materials such as cells/ tissue biopsy and viral vectors is crucial as poor-quality starting material will affect the quality of the final product. Autologous or allogeneic cells/tissue can be very heterogeneous due to the inherent donor variability (age, sex, health status, medication), the variable number of cells other than the intended cells, and because the collected cells are not in a synchronized cell cycle. In addition, the origin of the cells, e.g., MSCs of bone marrow, adipose, and cord blood origin, may have a significant impact on the activity and phenotype of the cells after manufacture.

The challenge is that a lot of the techniques used for the characterization of this heterogeneous group of products are not sensitive methods; hence they are not able to pick-up subtle changes to the process and/or to the product.

For further reading oncell- and tissue-based product characterization, see BSI PAS 93:2011. For details on testing (lot release and additional characterization) of viral vectors for

Table 14.17 Examples of techniques applied for the analysis of different quality attributes of cell- and tissue-based therapies

Quality attribute	Explanation	Possible techniques applied
Identity	Distinguish the cellular active substance (s)/tissue from unwanted cell population(s); donor specific test; sometimes a combination of tests	Flow cytometry; karyology, STR, FISH, CGH, microscopy, immunocytochemistry, electrochemiluminescence, protein array, microarray
Active substance purity	Number of viable cells with specific cell surface markers present/absent, unique for the active substance. Closely related to identity	Flow cytometry; ELISA; immunocytochemistry; electrochemiluminescence; protein ligation assay
Cellular (product) impurities	Dead cells (based on total and viable cell numbers); unwanted cell populations. Closely related to identity and purity	Flow cytometry; ELISA; electrochemiluminescence; MS
Process impurities	Depends or process and raw materials used. e.g. antibiotics, cytokines, growth factors, FBS, beads, viral vector starting material	– Cytokines, growth factors, FBS, TryPLESelect: ELISA – Beads: microscopic evaluation; – Antibiotics: LC-MS; – Viral vector: qPCR
Potency/bioactivity	Quantitative measure of relevant biologic function(s) based on the attributes that are linked to relevant in vivo biologic properties; often a combination of assays. Receptors, cellular metabolism, secreted proteins, migration of cells, (lack of) proliferation, differentiation potential, mRNA expression	ELISA; qPCR; Flow cytometry; cell migration in Dunn or Boyden chamber: protein array; LC; MS; animal modal (not quantitative), microarray
Viability and total cell count	Viability is a critical parameter and related to dose, purify and cellular impurities	Colorimetric assay (spectrophotometer), fluorescent assay (including flow cytometry), membrane integrity assay (e.g., trypan blue), microscope. Manual. semi-automated or automated equipment
Dose	Often number of total or viable cells per unit (mL, kg body weight); cm² of tissue	Total call count and viability techniques
Safety	Sterility, endotoxin, mycoplasma, human and animal viruses derived from starting material or raw materials, replication competent viral vector, chromosomal aberrations	Pharmacopoeial tests for sterility, mycoplasma. endotoxin-standard or rapid tests; chromosomal aberrations by karyology FISH, CGH
General attribute	Appearance, pH, osmolality, particles, cell/tissue morphology	Pharmacopoeial tests, microscope for morphology assessment

Some techniques are also used for starting material characterization
Flow cytometry technique is explained below; it can be used for intracellular and cell surface markers
STR short tandem repeat, *FISH* fluorescence in situ hybridization, *CGH* comparative genomic hybridization, *ELISA* enzyme-linked immuno sorbent assay; see Chap. 3 for details on this technique, *MS* mass spectrometry, *LC-MS* liquid chromatography-mass spectrometry; see Chap. 3 for detail on this technique, *qPCR* quantitative polymerase chain reaction; see Chap. 1 for details on PCR

ex vivo and in vivo gene therapy products, see Gombold et al. (2006a, b). Table 14.18 provides an overview of the QC testing panel for an MSC-derived and a CAR-T product.

Flow Cytometry

One of the key technologies in advanced therapy manufacturing is flow cytometry. It can be operated in a QC test environment and in the production of advanced therapies products (see next section). As this technique is not used regularly to characterize therapeutic proteins, it is not discussed in Chap. 3. Therefore, we pay attention to it in this chapter.

Flow cytometry assays may be used to assess cell- and tissue-based product identity, active substance purity, cellular impurity, viability, and potency testing. It is a powerful technique that allows for a specific measurement of cellular components on the cell surface, e.g., CD73, CD90, and CD105, to characterize MSCs, and intracellular components. It is also amenable to the measurement of soluble analyte(s) such as cytokines, released by the cells in the extracellular environment, e.g., upon cell activation.

Flow cytometry is a technology that simultaneously measures and then analyzes multiple physical characteristics of single particles, usually cells, as they flow in a fluid stream through a beam of light (Fig. 14.39). The properties measured include a particle's relative size, relative granularity or internal complexity, and relative fluorescence intensity. These characteristics are determined using an optical-to-electronic coupling system that records how the cell or par-

Table 14.18 Example of QC testing panel for an MSC-derived cell based product and a CAR-T ex vivo gene therapy product

Quality attribute	MSC derived cell based product; allogeneic off-the-shelf (1 batch of multiple vials/bags for multiple patients)	CAR-T ex vivo gene therapy product; autologous (1 batch of 1 infusion bag for 1 patient)
Identity	CD73+, CD90+, CD105+, HLA-DR−, CD3−, CD45- cells by flow cytometry	CAR expression by qPCR
Viability by manual or automated cell count	Number of total cells	Number of total cells
	Number of viable cells	Number of viable cells
	Percentage of viable cells	Percentage of viable cells
Purity by flow cytometry (% of viable cells with a certain CD-marker profile)	Percentage of CD73+, CD90+, CD105+. 7-AAD− cells by flow cytometry	Percentage of viable T cells
		Transduction efficiency by CAR q-PCR
Product = cellular impurities (dead cells and unwanted cell populations) by flow cytometry	Percentages of 7-AAD+ (dead cells), CD3+ (T cells), CD45+ (lymphocytes). CD34+ (HSCs and endothelial cells), CD14+ (monocytes), CD19+ (B cells)	Percentages of red blood cells, granulocytes, dead cells, CD19+ B cells
Process impurities	Residual bovine serum albumin (BSA) by ELISA	Residual antibody conjugated beads (CD3/CD28)
	Residual TryPLESelect by ELISA	BSA by ELISA
	Residual antibiotic by liquid chromatography–mass spectrometry	Residual VSV-G DNA by qPCR-derived from viral vector
Potency	CD marker expression (e.g., adhesion molecules) upon immune activation by flow cytometry	Determination of CAR expression by flow cytometry
		Release of interferon-gamma in response to CD19-expressing target cells
Safety	Sterility	Sterility
	Bacterial endotoxins	Endotoxin
	Mycoplasma	Mycoplasma
	Karyology	PCR-based replication competent lentivirus assay
	Human viral testing; test for the presence of inapparent virus; in-vitro assay for the presence of viral contaminants	N.A.
Dose (calculated)	a–b × 10^6 viable CD73+ CD90+, CD105+, 7-AAD− cells/ml	a–b × 10^6 CD19+ T cells/kg body weight
General attribute	pH	pH
	Osmolality	Osmolality
	Appearance of primary container and content	Appearance of primary container and content
	Content uniformity	N.A.
	Extractable volume from the vial	N.A.

ticle scatters incident laser light and emits fluorescence. A flow cytometer is made up of three main systems: fluidics, optics, and electronics.

- The fluidics system transports single particles (cells) in a stream to the laser beam for interrogation.
- The optics system consists of a light source, mostly lasers, to illuminate the particles in the sample stream and optical filters to direct the resulting light signals to the appropriate detectors. Light scattering or fluorescence emission from auto-fluorescence of the particle or from fluorophores, which are fluorescence labels, e.g., bound to specific antibodies, used to detect the expression of cellular molecules

such as specific proteins or nucleic acids, provides information about the particle's properties. (1) Light that is scattered in the forward direction after interacting with a particle, typically up to 20° offset from the axis of the laser, is collected by a photomultiplier tube or photodiode and is known as the forward scatter (FSC) channel. This FSC measurement can estimate a particle's size, with larger particles refracting more light than smaller particles. (2) Light measured at a 90° angle to the excitation line is called side scatter (SSC). The SSC can provide information about the relative complexity, e.g., granularity and internal structures, of a cell or particle. However, as with forward scatter, this can depend on various factors. Both FSC and SSC are

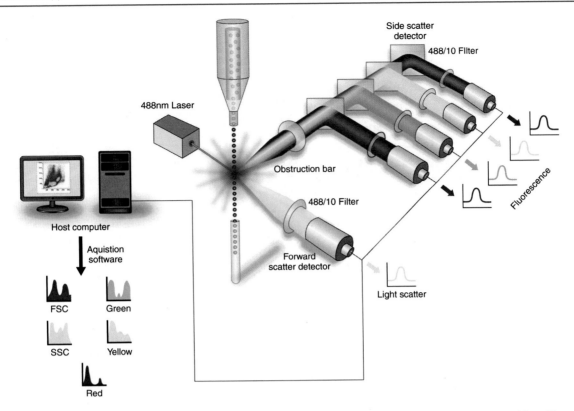

Fig. 14.39 Schematic view of a flow cytometer. Scattered and emitted light signals are converted to electronic pulses, adapted from ThermoFisher Scientific.   http://www.thermofisher.com/nl/en/home/life-science/cell-analysis/cell-analysis-learning-center/molecularprobes-school-of-fluorescence/flow-cytometry-basics/flow-cytometry-fundamentals/how-flow-cytometer-works.html#overview

unique for every particle, and a combination of the two can be used to roughly differentiate cell types in a heterogeneous population such as blood or bone marrow aspirate. However, this scatter information and cell typing depend on the sample type and the quality of sample preparation, so fluorescent labeling is generally required to obtain more detailed information.

– The electronics system converts the detected light signals into electronic signals that the computer can process.

– In the flow cytometer, particles are carried to the laser intercept in a fluid stream. Any suspended particle or cell from 0.2 to 150 μm in size is suitable for analysis. Cells from solid tissue must be desegregated into single cells before analysis. The portion of the fluid stream where particles are located is called the sample core. When particles pass through the laser intercept, they scatter laser light. Any fluorescent molecule present on the particle fluoresces. The scattered and fluorescent light is collected by appropriately positioned lenses. A combination of beam splitters and filters steers the scattered and fluorescent

light to the appropriate detectors. The detectors produce electronic signals proportional to the optical signals striking them. Readouts are collected on each particle or single event. The characteristics or parameters of each event are based on its light scattering and fluorescent properties. The data are collected and stored in the computer. This data can be analyzed to provide information about subpopulations of cells within the sample (see Fig. 14.40).

Fluorescence-Activated Cell Sorting (FACS)

Flow cytometry techniques can also be used to sort specific cell (sub) populations, e.g., to increase product yield and/or reduce the amount of unwanted cell populations, which are considered impurities. A FACS machine provides the ability to separate cells identified by flow cytometry. Droplet-based cell sorters first analyze the particles but also have hardware that can generate droplets and deflect or direct wanted particles into a collection tube. Cell dispersions are often purified based on surface markers such as CD34+ in HSCs or on their

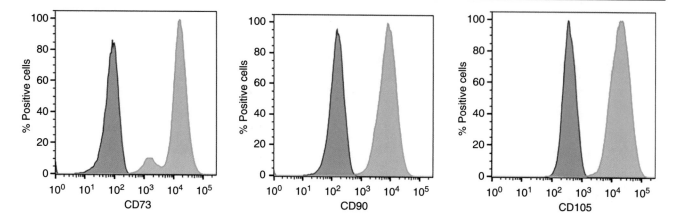

Fig. 14.40 Flow cytometry histograms of MSC product cells. Flow cytometric analysis of MSC product cells against three defined MSC markers (CD73, CD90, and CD105) show that these cells are of mesenchymal cell phenotype. On the X-axis the density of the respective cell surface marker molecule is shown. A single peak is observed for each of the markers tested (blue peak at the right side of each histogram), indicating a single population of cells. The red peak at the left side of each histogram represents the isotype control staining. Courtesy of M. van Pel

viability. Common uses of cell sorting include identifying and isolating cell populations or single cells followed by subsequent downstream applications where DNA, protein, or cellular function is investigated.

Improvements in Testing Strategies Needed

Developing robust, sensitive, rapid, and in-line analytical testing and characterization tools will be required as cell/tissue and viral vector processing platforms continue to evolve. Significant improvements are needed to establish next-generation analytics for (in-process) QC, stability, and additional characterization testing to assess the quality attributes of starting materials, intermediates, and advanced therapy products. Improvements are also to be made in the field of in-line and online testing of cell culture conditions, e.g., pH, morphology, and viability. Reducing the sampling frequency, technical complexity, amount of sample needed, and labor intensiveness of testing are especially critical for a nonoff-the-shelf autologous ex vivo gene therapy product. This contrasts with traditional biophar-

maceuticals, where a single batch of QC tested products may treat hundreds or thousands of patients. Cell processing automation will also be enabled through the development of high throughput in-process and release assays providing results in a very short time frame (minutes–hours). Advanced cell/tissue characterization techniques based on nanofluidics, transcriptomics, and proteomics, and next-generation sequencing techniques may allow a better understanding of what happens to desired cell population(s)/tissue once they are processed and before patient administration (see Chap. 9 for more details on "-omics"), both in the cytosol as well as in the extracellular environment. Examples are changes in intracellular genetic profiles and patterns within the micro-RNA and exosome pools secreted into the culture medium by the cells.

Different advanced therapy technologies are currently at different translation stages and have their particular manufacturing and testing challenges, as summarized in Table 14.19.

Table 14.19 Development stage manufacturing and testing challenges for different advanced therapy technologies, adapted from Mount et al. (2015)

Technologies	Development stage of the field	Current manufacturing technologies	Manufacturing and testing challenges
(a) Somatic cell technologies	Many products in early clinical development phase; few products approved, e.g., Alofisel	Manual process with open handling steps; automated multi-planar flasks and stack systems; micro-carriers in disposable stirred tank systems; hollow fiber growth systems; membrane and contraflow centrifugation systems	Scale-up and control of large scale batches. Recovery of cells from micro-carriers. DSP: Large volume handling, primary container filling at scale using enclosed technologies. Relevant potency assays lacking
(b) Cell immortalization technologies	One product in early clinical development	CompacT Select[a] fully automated and programmable scalable cell culture platform consisting of a robot arm that can access T175 flask or multi-well plate incubator. Standard cell culture activities, such as passage or media change, are conducted and controlled with no manual intervention	Similar to protein manufacturing platform technologies
(c) Ex vivo gene modification of cells using viral vector technologies	Mainly small trials in early and late clinical development phase (gene modified autologous T-cells and HSCs); few products approved, e.g., Strimvelis and Kymriah	Manual processes often not fully enclosed using static bags, gas-permeable pots + lateral movement bioreactors (wave bags) for higher cell yield. Positive or negative cell selection process steps often used. High cell purity becoming possible with sterile cell sorter	Adapting systems to deal with variation in quality and amount of incoming starting material. Lack of product stability pressuring manufacturing and distribution model. Lack of fast QC assays. Low transduction efficiency with non-replicating viral vectors. Enclosed and automated manufacturing systems are becoming available for the entire process (e.g., prodigy)
(d) Cell plasticity technologies	Mainly pre-clinical phase with few ESC and iPSC-derived FIH trials	Current processes are extremely 'manual' and rely on small scale cell culture and harvest technologies. High risk processes with extensive process and product characterization testing to assess product quality, safety, and efficacy	A two-tier banking strategy (MCB/WCB) scale-up process of pluripotent cells prior to differentiation steps needed. Dynamic cell culture systems to expand PSC numbers. Robotic scale-out of current plate-based iPSC technology is also being explored
(e) 3D-technologies	Mainly pre-clinical phase with few FIH trials	A complex manufacturing interplay between (bio)materials, scaffolds, cells, and biological coatings. Incorporates decellularization/recellularization tissue-based products such as trachea, esophagus, and veins	Enclosed bioreactors to control cell and material interface. Improved stability and delivery systems. Robust product quality to ensure large clinical application

[a] Thomas et al. (2009)

Other Aspects of Advanced Therapies

Regulatory Bodies Involved in Regulating Advanced Therapies in Europe

In Europe, the responsibility for regulating transplant products according to the public health legislation lies with the national Competent Authority for tissues and cells in each member state. ATMPs, in contrast, are regulated by pharmaceutical legislation. Hence, marketing approval must be obtained before marketing an ATMP through the centralized procedure, as for any other biological medicinal product. The scientific evaluation of these products is led by a specialized committee within the EMA (the Committee for Advanced Therapies "CAT"). The CAT drafts an opinion for the Committee for Medicinal Products for Human Use (CHMP), which is responsible for providing a second scientific opinion. Based on a positive CHMP opinion, the approval of a marketing authorization application (MAA) is granted by the European Commission. Clinical trials involving ATMPs are regulated and authorized the same manner as

Table 14.20 Regulatory agencies and applicable guidances for advanced therapies in the US and EU

Regulatory agency/institute	Link
EMA	https://www.ema.europa.eu/en/human-regulatory-overview/advanced-therapy-medicinal-products-overview
FDA	https://www.fda.gov/BiologicsBloodVaccines/default.htm
International Conference on Harmonisation (ICH)	https://www.ich.org/
British Standards Institution (BSI)	BSI PAS 83:2012, BSI PAS 84:2012, and BSI PAS 93:2011

other medicinal products, i.e., on a national level by the appropriate national competent authority (NCA).

Regulatory Bodies Involved in Regulating Advanced Therapies in the USA

The situation in the United States is simpler in that the FDA is responsible for both aspects of the legislation: the public health and pharmaceutical legislation. Within the FDA, the responsibility for the regulation of HCT/Ps and human gene therapy products lies with the Center for Biologics Evaluation and Research (CBER), both for clinical trials and marketing authorization. As of 2016, the CBER structure includes the Office of Blood Research and Review (OBRR), the Office of Vaccines Research and Review (OVRR), and the Office of Tissues and Advanced Therapies (OTAT), which was formerly known as the Office of Cellular, Tissues and Gene Therapies (OCTGT). To monitor activity, review data, and anticipate future needs, the FDA operates the Cellular, Tissue, and Gene Therapies Advisory Committee.

Regulatory Guidances

Links to the relevant regulatory bodies involved in advanced therapies in the EU and US, as well as applicable guidances, can be found in Table 14.20.

Stem Cell Tourism

The general interest in advanced therapies worldwide has allowed unregulated practice, particularly of cell-based products, to develop in some countries, i.e., "stem cell tourism." This is a major concern for many stakeholders in the field of ATMPs because treatments are being offered in the absence of a strong safety data package and any proven efficacy. In addition, there is suspicion that the products in use have been manufactured with insufficient attention to GMP, including quality control. Patients must be warned of the dangers of falling prey to unethical operations. An up-to-date source of information on private clinics and stem cell tourism is available at the website of the International Society for Stem Cell Research (www.isscr.org).

Concluding Remarks

Although progress has been made in the area of ATMPs, with about 80 products approved globally and 27 in the USA & EU for commercial use (see examples in Table 14.6) and many products in clinical development, this field was currently struggling with similar problems as the first recombinant proteins 20 years ago. Appropriate manufacturing platforms, supply chain models, healthcare systems, reimbursement models, and regulatory frameworks for these medicinal products need to be established by developers and other key stakeholders, while specific knowledge about quality (production and testing), safety, and efficacy of advanced therapies is steadily growing.

Self-Assessment Questions

Questions

1. What is the difference between embryonic and adult stem cells?
2. How is somatic cell nuclear transfer carried out, and what are the problems with this technique?
3. What are iPSCs, and why are they important?
4. What is the difference between in vivo gene therapy and ex vivo gene therapy?
5. Which disease areas are predominantly investigated clinically with ATMPs?
6. What problems could arise in the use of stem cell-derived products for clinical application?
7. What was the disease target for the first gene therapy clinical trial? What vector was selected for gene transfer?
8. Identify and describe five transcription regulatory elements discussed in the chapter.
9. Several clinical trials involve gene transfer for treating malignant glioma. One approach involves the use of a recombinant retrovirus expressing the HSV-tk transgene. Another involves the use of a recombinant adenovirus expressing the p53 transgene.
   (a) Which of the five current strategies to treat cancer by viral gene therapy does each of these trials employ? Describe the principle behind each strategy.

(b) List two advantages and two disadvantages associated with the vector used in each of these trials.

(c) Outline potential drawbacks to the use of each of these strategies for cancer therapy.

(d) What other approaches could have been selected to prevent the growth and spread of malignant tissue? Explain the principle behind each.

10. What is the purpose of the packaging cell line during the production of recombinant viral vectors for gene transfer? What is the risk associated with using packaging cell lines for vector production?

11. Provide two examples of how gene therapy is used to modulate the immune system to fight infection.

12. Describe one clinical trial for retrovirus-based gene therapy and adenovirus-based gene therapy and identify the most significant adverse effects that have been reported for each trial.

13. What can be incorporated into viral vector design to reduce genotoxic risk?

14. How can preexisting immunity to viral vectors be circumvented?

15. Name two advantages of adenovirus vectors as opposed to AAV vectors and vice versa.

16. How can adenoviruses be engineered to selectively replicate in tumor cells?

Answers

1. Embryonic stem cells are grown ex vivo after extraction of the inner cell mass from a blastocyst. Adult stem cells are found in vivo in many tissues, usually in the specialized environment of a stem cell niche that supports their asymmetric cell division.

2. Somatic cell nuclear transfer (SCNT) involves the injection of a donor genome into an enucleated egg, such that the embryo develops as a clone of the donor genome. This allows the generation of embryonic stem cells using the donor genome and, in principle, allows implantation into the uterus of a recipient female leading to pregnancy. There are ethical problems concerning the supply of fertilized human eggs and technical problems caused by incomplete reprogramming of the donor nucleus.

3. iPSCs are produced by transient expression of pluripotency genes in somatic cells, leading to reprogramming to form pluripotent cells resembling embryonic stem cells. The production of iPSCs allows pluripotent cells to be obtained from a patient without the need for SCNT. iPSCs can be used to derive differentiated cells for producing ATMPs for clinical application or disease modeling purposes.

4. In vivo gene therapy refers to the direct introduction of genetic material into the human body, whereas ex vivo gene therapy refers to the use of cells, which are genetically modified outside the body (i.e., ex vivo) prior to administration of these genetically modified cells into the human body. In the latter case, the genetic material is introduced into the human body using cells as "delivery system". See also "Gene Therapy" section.

5. Various cancers, autoimmune disorders, such as DM type I and Crohn's disease, neurological disorders, such as Parkinson's disease and Alzheimer's disease, myocardial infarction, and macular degeneration.

6. One of the concerns with stem cell-derived ATMPs is the possibility that rare pluripotent or multipotent cells in the product could give rise to tumors after administration to humans, i.e., tumorigenicity risk. Thus, the quality control of medicinal products is of paramount importance. Often, in particular, in the treatment of neurological diseases, it is not clear whether a progenitor, precursor or fully mature cell should be administered. Careful preclinical work is required in each clinical indication to establish the most effective approach. Where the strategy is designed to replace a cell that is lost in a particular disease, the environment into which the cell-based medicinal product is placed may not be supportive of cell survival and integration/persistence. In general, one needs to pay attention to providing a protective environment for the medicinal product

7. The first gene therapy clinical trial was initiated in 1990 for treating adenosine deaminase (ADA) deficiency. In this trial, patients with ADA deficiency were given peripheral blood lymphocytes treated with a retroviral vector expressing the ADA transgene.

8. Promoter is a DNA sequence that enables a gene to be transcribed. The promoter is recognized by RNA polymerase and transcriptional factors. Enhancer is a short DNA sequence that can bind transcription factors or activators to enhance transcription levels of a gene from a distance. Insulators are genetic boundary elements that block the enhancer-promoter interaction or rarely act as a barrier against condensed chromatin proteins. Finally, operators and silencers are usually short DNA sequences close to the promoter with binding affinity to a set of proteins named repressors and inducers.

9. (a) Retrovirus trial

Gene-directed enzyme-prodrug therapy. Cells transduced by the virus express an enzyme capable of converting a prodrug (in this case, ganciclovir) to a cytotoxic metabolite. This conversion cannot occur in cells that do not express the transgene, limiting the cytotoxic effect to transduced cells and their neighbors through the bystander effect.

Adenovirus trial

Correction of genetic mutations that contribute to a malignant phenotype. Cells transduced by the virus express a gene such as p53 that is necessary for controlled cell division and development. This prevents the uncontrolled growth and division associated with malignant disease.

(b) Retrovirus

Advantages—(i) Retroviruses can infect dividing cells which are the therapeutic target in this trial. Despite this fact, the transduction efficiency of this vector in vivo has been low. (ii) Retroviruses can induce long-term gene expression, which should be sufficient to effectively remove malignant tissue.

Disadvantages—(i) Retroviruses have the potential for inducing insertional mutagenesis in normal, healthy cells. (ii) Transgene expression is sometimes limited by the host immune response to cellular components acquired by the virus during large-scale production.

Adenovirus

Advantages—(i) Adenoviruses can infect dividing cells, the therapeutic target in this trial. (ii) Adenoviruses can induce high levels of transgene expression in short periods of time. (iii) Adenoviruses do not have the risk of insertional mutagenesis. (iv) It is relatively easy to produce large amounts of recombinant adenovirus sufficient for clinical use.

Disadvantages—(i) Transgene expression is transient, making readministration necessary for continued effect. (ii) Adenoviral vectors can induce a potent immune response. This limits the success of gene transfer after a second dose of virus and is associated with severe toxicity at certain doses. (iii) Preexisting immunity to adenovirus serotype 5 is common in the general population. This may also limit gene transfer.

(c) Drawbacks to gene-directed enzyme-prodrug therapy.

(i) Efficacy relies on efficient transgene expression and drug bioavailability. (ii) The therapeutic effect may spread to healthy cells through the bystander effect.

Drawbacks to gene correction therapy.

(ii) Gene correction may stop tumor growth but not eliminate it. (ii) Expression is not limited to malignant tissue.

(d) Other approaches for cancer gene therapy

(i) Immunotherapy. A vector expressing pro-inflammatory cytokines, co-stimulatory molecules, or tumor-specific antigens is injected directly into the tumor mass. This facilitates the formation of an antitumor immune response that targets and destroys malignant cells.

(ii) Virotherapy. A replication-competent virus naturally targeting cancers is directly injected into the tumor mass. The virus can induce cell death during replication in malignant tissue by producing cytotoxic proteins and subsequent cell lysis.

10. (i) The primary purpose of the packaging cell line is to provide genetic elements that support virus replication and assembly. These have been eliminated from the vector to prevent it from causing disease in the patient. (ii) The recombinant virus can incorporate elements for replication into its genome through homologous recombination during the production process. The potential for the generation of replication-competent virus (RCV) in this manner does exist for each vector but can vary due to specific features of a given packaging cell line.

11. (i) Gene transfer into autologous immunocytes to increase the immune system of a patient. (ii) Overexpression of protein inhibitors that interfere with virus infection and replication. (iii) Overexpression of known antigenic epitopes of the pathogen by DNA vaccination to stimulate an immune response.

12. (i) One trial employed aerosol administration of a recombinant adenovirus expressing cystic fibrosis transmembrane conductance regulator (CFTR) to treat cystic fibrosis (CF). Another trial employed a recombinant retrovirus expressing recombinant adenosine deaminase (ADA) to transduce autologous T lymphocytes isolated from patients for treating ADA deficiency-induced severe combined immunodeficiency (ADA-SCID). (ii) CF trial. Massive immune response to the recombinant viral vector.

ADA-SCID trial. Lymphoproliferative leukemia is caused by insertional mutagenesis.

13. Self-inactivating configurations can be incorporated to enable the use of physiological and cell type specific promoters, which can reduce potential genotoxic risk. In addition, using viral vector systems with more random integration site profiles and with reduced ability to cause aberrant splicing.

14. Preexisting immunity to viral vectors can be circumvented by using serotypes that have no or low prevalence in the human population, including virus variants of nonhuman origin.

15. Adenovirus vectors can be used to achieve transient transgene expression, and it has a higher packaging capacity than AAV. On the other hand, AAV can be used to get potentially long-term transgene expression, and the risk for strong inducing (too) strong immune responses is lower.

16. Initially, these were generated by partial deletion of E1B, restricting genome replication to cells that lack p53 such as tumor cells. More recently, they have been

generated by removing a specific stretch of 24 amino acids from the E1A protein. These so-called AdΔ24 or AdΔRb vectors are unable to bind to the retinoblastoma (Rb) protein. Rb normally retains E2F, preventing the cells from entering the S-phase and thereby replicating the genome. Cancer cells often have an aberrated Rb pathway, facilitating the S-phase and AdΔRb24 replication independent from the Rb-binding activities of E1A.

References

Ankrum JA, Ong JF, Karp JM (2014) Mesenchymal stem cells: immune evasive, not immune privileged. Nat Biotechnol 32:252–260

Antoniou P, Miccio A, Brusson M (2021) Base and prime editing technologies for blood disorders. Front Genome Ed 3:618406

Anzalone AV, Randolph PB, Davis JR et al (2019) Search-and-replace genome editing without double-strand breaks or donor DNA. Nature. https://doi.org/10.1038/s41586-019-1711-4

Armentano D, Sookdeo CC, Hehir KM et al (1995) Characterization of an adenovirus gene transfer vector containing an E4 deletion. Hum Gene Ther 6:1343–1353

Arrand JR, Roberts RJ (1979) The nucleotide sequences at the termini of adenovirus-2 DNA. J Mol Biol 128:577–594

Bajaj P, Schweller RM, Khademhosseini A et al (2014) 3D biofabrication strategies for tissue engineering and regenerative medicine. Annu Rev Biomed Eng 16:247–276

Bar and Benvenisty. EMBO J. 2019. https://doi.org/10.15252/embj.2018101033

Bartholomae CC, Arens A, Balaggan KS et al (2011) Lentiviral vector integration profiles differ in rodent postmitotic tissues. Mol Ther 19:703–710

Becker J, Fakhiri J, Grimm D (2022) Fantastic AAV gene therapy vectors and how to find them-random diversification, rational design and machine learning. Pathogens 11:756

Belete TM (2021) The Current Status of Gene Therapy for the Treatment of Cancer. Biologics 15:67–77

Bianco P, Robey PG, Simmons PJ (2008) Mesenchymal stem cells: revisiting history, concepts, and assays. Cell Stem Cell 2:313–319

Biasco L, Ambrosi A, Pellin D et al (2011) Integration profile of retroviral vector in gene therapy treated patients is cell-specific according to gene expression and chromatin conformation of target cell. EMBO Mol Med 3:89–101

Biffi A, Bartolomae CC, Cesana D et al (2011) Lentiviral vector common integration sites in preclinical models and a clinical trial reflect a benign integration bias and not oncogenic selection. Blood 117:5332–5339

Blaese RM, Culver KW, Miller AD et al (1995) T lymphocyte-directed gene therapy for ADA- SCID: initial trial results after 4 years. Science 270:475–80

Boorjian SA, Alemozaffar M, Konety BR et al (2021) Intravesical nadofaragene firadenovec gene therapy for BCG-unresponsive non-muscle-invasive bladder cancer: a single-arm, open-label, repeat-dose clinical trial. Lancet Oncol 22:107–117

Bots STF, Hoeben RC (2020) Non-human primate-derived adenoviruses for future use as oncolytic agents? Int J Mol Sci 21:4821

Brassard JA, Lutolf MP (2019) Engineering Stem Cell Self-organization to Build Better Organoids. Cell Stem Cell 24:860–876

Bravery CA (2015) Do human leukocyte antigen-typed cellular therapeutics based on induced pluripotent stem cells make commercial sense? Stem Cells Dev 24:1–10

Bravery CA, Carmen J, Fong T et al (2013) Potency assay development for cellular therapy products: an ISCT review of the requirements and experiences in the industry. Cytotherapy 15:9–19

Bukrinsky MI, Sharova N, Dempsey MP et al (1992) Active nuclear import of human immunodeficiency virus type 1 preintegration complexes. Proc Natl Acad Sci U S A 89:6580–6584

Bulcha JT, Wang Y, Ma H et al (2021) Viral vector platforms within the gene therapy landscape. Signal Transduct Target Ther 6:53

Büning H (2013) Gene therapy enters the pharma market: the short story of a long journey. EMBO Mol Med 5:1–3

Byun S, Lee M, Kim M (2022) Gene therapy for Huntington's disease: The final strategy for a cure? J Mov Disord 15:15–20

Cadena-Herrera D, Esparza-De Lara JE, Ramírez-Ibañez ND et al (2015) Validation of three viable-cell counting methods: Manual, semi-automated, and automated. Biotechnol Rep (Amst) 7:9–16

Cattoglio C, Facchini G, Sartori D et al (2007) Hot spots of retroviral integration in human CD34 hematopoietic cells. Blood 110:1770–1778

Chand D, Mohr F, McMillan H et al (2021) Hepatotoxicity following administration of onasemnogene abeparvovec (AVXS-101) for the treatment of spinal muscular atrophy. J Hepatol 74:560–566

Chen G, Abdeen AA, Wang Y et al (2019) A biodegradable nanocapsule delivers a Cas9 ribonucleoprotein complex for in vivo genome editing. Nat Nanotechnol. https://doi.org/10.1038/s41565-019-0539-2

Chen H, Xiang ZQ, Li Y et al (2010) Adenovirus-based vaccines: comparison of vectors from three species of adenoviridae. J Virol 84:10522–10532

Chen HH, Mack LM, Kelly R et al (1997) Persistence in muscle of an adenoviral vector that lacks all viral genes. Proc Natl Acad Sci U S A 94:1645–1650

Chiang C-Y, Ligunas GD, Chin W-C, Ni C-W (2020) Efficient nonviral stable transgenesis mediated by retroviral integrase. Mol Ther Methods Clin Dev 17:1061–1070

Chiesa R, Bernardo ME (2022) Haematopoietic stem cell gene therapy in inborn errors of metabolism. Br J Haematol 198:227–243

Chin MH, Pellegrini M, Plath K, Lowry WE (2010) Molecular analyses of human induced pluripotent stem cells and embryonic stem cells. Cell Stem Cell 7:263–269

Consentius C, Reinke P, Volk H-D (2015) Immunogenicity of allogeneic mesenchymal stromal cells: what has been seen in vitro and in vivo? Regen Med 10:305–315

Corradetti B, Gonzalez D, Mendes Pinto I, Conlan RS (2021) Editorial: Exosomes as therapeutic systems. Front Cell Dev Biol 9:714743

Croyle MA, Cheng X, Wilson JM (2001) Development of formulations that enhance the physical stability of viral vectors for human gene therapy. Gene Ther 8:1281–1291

Dai Y, Schwarz EM, Gu D et al (1995) Cellular and humoral immune responses to adenoviral vectors containing factor IX gene: tolerization of factor IX and vector antigens allows for long-term expression. Proc Natl Acad Sci U S A 92:1401–1405

Daud AI, DeConti RC, Andrews S et al (2008) Phase I trial of interleukin-12 plasmid electroporation in patients with metastatic melanoma. J Clin Oncol 26:5896–5903

David RM, Doherty AT (2017) Viral vectors: The road to reducing genotoxicity. Toxicol Sci 155:315–325

Deichmann A, Brugman MH, Bartholomae CC et al (2011) Insertion sites in engrafted cells cluster within a limited repertoire of genomic areas after gammaretroviral vector gene therapy. Mol Ther 19:2031–2039

Deichmann A, Hacein-Bey-Abina S, Schmidt M et al (2007) Vector integration is nonrandom and clustered and influences the fate of lymphopoiesis in SCID-X1 gene therapy. J Clin Invest 117:2225–2232

Dharan A, Bachmann N, Talley S et al (2020) Nuclear pore blockade reveals that HIV-1 completes reverse transcription and uncoating in the nucleus. Nat Microbiol 5:1088–1095

Edelstein ML, Abedi MR, Wixon J, Edelstein RM (2004) Gene therapy clinical trials worldwide 1989-2004-an overview. J Gene Med 6:597–602

Eichler F, Duncan C, Musolino PL et al (2017) Hematopoietic stem-cell gene therapy for cerebral adrenoleukodystrophy. N Engl J Med 377:1630–1638

Ertl HCJ (2005) Challenges of immune responses in gene replacement therapy. IDrugs 8:736–738

Evans MJ, Kaufman MH (1981) Establishment in culture of pluripotential cells from mouse embryos. Nature 292:154–156

Eyquem J, Mansilla-Soto J, Giavridis T, et al (2017) Targeting a CAR to the TRAC locus with CRISPR/Cas9 enhances tumour rejection. Nature 543:113–117

Ferrua F, Aiuti A (2017) Twenty-Five Years of Gene Therapy for ADA-SCID: From Bubble Babies to an Approved Drug. Hum Gene Ther 28:972–981

Finn JD, Smith AR, Patel MC et al (2018) A Single Administration of CRISPR/Cas9 Lipid Nanoparticles Achieves Robust and Persistent In Vivo Genome Editing. Cell Rep 22:2227–2235

Fischer A, Hacein-Bey-Abina S (2020) Gene therapy for severe combined immunodeficiencies and beyond. J Exp Med 217. https://doi.org/10.1084/jem.20190607

Flotte T, Carter B, Conrad C et al (1996) A phase I study of an adeno-associated virus-CFTR gene vector in adult CF patients with mild lung disease. Hum Gene Ther 7:1145–1159

Frangoul H, Altshuler D, Cappellini MD et al (2021) CRISPR-Cas9 gene editing for sickle cell disease and β-thalassemia. N Engl J Med 384:252–260

Fritz CC, Green MR (1996) HIV Rev uses a conserved cellular protein export pathway for the nucleocytoplasmic transport of viral RNAs. Curr Biol 6:848–854

Fumagalli F, Calbi V, Natali Sora MG et al (2022) Lentiviral haematopoietic stem-cell gene therapy for early-onset metachromatic leukodystrophy: long-term results from a non-randomised, open-label, phase 1/2 trial and expanded access. Lancet 399:372–383

Gaggar A, Shayakhmetov DM, Lieber A (2003) CD46 is a cellular receptor for group B adenoviruses. Nat Med 9:1408–1412

Galderisi U, Peluso G, Di Bernardo G (2022) Clinical Trials Based on Mesenchymal Stromal Cells are Exponentially Increasing: Where are We in Recent Years? Stem Cell Rev Rep 18:23–36

Gillmore JD, Gane E, Taubel J et al (2021) CRISPR-Cas9 In Vivo Gene Editing for Transthyretin Amyloidosis. N Engl J Med 385:493–502

Ginn SL, Amaya AK, Alexander IE et al (2018) Gene therapy clinical trials worldwide to 2017: An update. J Gene Med 20:e3015

Golchin A, Chatziparasidou A, Ranjbarvan P et al (2021) Embryonic Stem Cells in Clinical Trials: Current Overview of Developments and Challenges. Adv Exp Med Biol 1312:19–37

Gombold J, Peden K, Gavin D et al (2006a) Lot Release and Characterization Testing of Live-Virus-Based Vaccines and Gene Therapy Products, Part 1 Factors Influencing Assay Choices. Bioprocess Int 4:46–54

Gombold J, Peden K, Gavin D et al (2006b) Lot release and characterization testing of live-virus based vaccines and gene therapy products part 2: Case Studies and Discussion. Bioprocess Int 4:56–65

Goodwin J, Laslett AL, Rugg-Gunn PJ (2020) The application of cell surface markers to demarcate distinct human pluripotent states. Exp Cell Res 387:111749

Gupta RK, Abdul-Jawad S, McCoy LE et al (2019) HIV-1 remission following CCR5Δ32/Δ32 haematopoietic stem-cell transplantation. Nature 568:244–248

Gurumoorthy N, Nordin F, Tye GJ et al (2022) Non-Integrating Lentiviral Vectors in Clinical Applications: A Glance Through. Biomedicines 10. https://doi.org/10.3390/biomedicines10010107

Han DW, Tapia N, Hermann A et al (2012) Direct reprogramming of fibroblasts into neural stem cells by defined factors. Cell Stem Cell 10:465–472

Harrington K, Freeman DJ, Kelly B et al (2019) Optimizing oncolytic virotherapy in cancer treatment. Nat Rev Drug Discov 18:689–706

Herberts CA, Kwa MSG, Hermsen HPH (2011) Risk factors in the development of stem cell therapy. J Transl Med 9:29

Hybiak J, Jankowska K, Machaj F et al (2020) Reprogramming and transdifferentiation - two key processes for regenerative medicine. Eur J Pharmacol 882:173202

Ilic D, Devito L, Miere C, Codognotto S (2015) Human embryonic and induced pluripotent stem cells in clinical trials. Br Med Bull 116:19–27

Jia F, Wilson KD, Sun N et al (2010) A nonviral minicircle vector for deriving human iPS cells. Nat Methods 7:197–199

John Wiley and Sons LTD (2022) Gene Therapy Clinical Trials Worldwide. In: The Journal of Gene Medicine Clinical Trial Site. https://a873679.fmphost.com/fmi/webd/GTCT. Accessed 24 Dec 2022

Kaufman HL, Shalhout SZ, Iodice G (2022) Talimogene laherparepvec: Moving from first-in-class to best-in-class. Front Mol Biosci 9:834841

Kean TJ, Lin P, Caplan AI, Dennis JE (2013) MSCs: Delivery Routes and Engraftment, Cell-Targeting Strategies, and Immune Modulation. Stem Cells Int 2013:732742

Keller G (2005) Embryonic stem cell differentiation: emergence of a new era in biology and medicine. Genes Dev 19:1129–1155

Kim JY, Nam Y, Rim YA, Ju JH (2022) Review of the Current Trends in Clinical Trials Involving Induced Pluripotent Stem Cells. Stem Cell Rev Rep 18:142–154

Kim W, Lee S, Kim HS et al (2018) Targeting mutant KRAS with CRISPR-Cas9 controls tumor growth. Genome Res 28:374–382

Kleinstiver BP, Pattanayak V, Prew MS et al (2016) High-fidelity CRISPR–Cas9 nucleases with no detectable genome-wide off-target effects. Nature 529:490–495

Kohn LA, Kohn DB (2021) Gene Therapies for Primary Immune Deficiencies. Front Immunol 12:648951

Kurt IC, Zhou R, Iyer S et al (2021) CRISPR C-to-G base editors for inducing targeted DNA transversions in human cells. Nat Biotechnol 39:41–46

Kuzmin DA, Shutova MV, Johnston NR et al (2021) The clinical landscape for AAV gene therapies. Nat Rev Drug Discov 20:173–174

LaFountaine JS, Fathe K, Smyth HDC (2015) Delivery and therapeutic applications of gene editing technologies ZFNs, TALENs, and CRISPR/Cas9. Int J Pharm 494:180–194

Lander ES, Baylis F, Zhang F et al (2019) Adopt a moratorium on heritable genome editing. Nature Publishing Group UK. https://doi.org/10.1038/d41586-019-00726-5. Accessed 23 Dec 2022

Leibacher J, Henschler R (2016) Biodistribution, migration and homing of systemically applied mesenchymal stem/stromal cells. Stem Cell Res Ther 7:7

Lennon MJ, Rigney G, Raymont V, Sachdev P (2021) Genetic therapies for Alzheimer's disease: A scoping review. J Alzheimers Dis 84:491–504

Levenstein ME, Ludwig TE, Xu R-H et al (2006) Basic fibroblast growth factor support of human embryonic stem cell self-renewal. Stem Cells 24:568–574

Levine BL, Humeau LM, Boyer J et al (2006) Gene transfer in humans using a conditionally replicating lentiviral vector. Proc Natl Acad Sci U S A 103:17372–17377

Levine BL, Miskin J, Wonnacott K, Keir C (2017) Global Manufacturing of CAR T Cell Therapy. Mol Ther Methods Clin Dev 4:92–101

Li L, Allen C, Shivakumar R, Peshwa MV (2013) Large volume flow electroporation of mRNA: clinical scale process. Methods Mol Biol 969:127–138

Liang M (2018) Oncorine, the world first oncolytic virus medicine and its update in China. Curr Cancer Drug Targets 18:171–176

Liu C, Su C (2019) Design strategies and application progress of therapeutic exosomes. Theranostics 9:1015–1028

Liu J, Verma PJ (2015) Synthetic mRNA Reprogramming of Human Fibroblast Cells. Methods Mol Biol 1330:17–28

Lochmüller H, Petrof BJ, Pari G et al (1996) Transient immunosuppression by FK506 permits a sustained high-level dystrophin expression after adenovirus-mediated dystrophin minigene transfer to skeletal muscles of adult dystrophic (mdx) mice. Gene Ther 3:706–716

Louis N, Evelegh C, Graham FL (1997) Cloning and sequencing of the cellular-viral junctions from the human adenovirus type 5 transformed 293 cell line. Virology 233:423–429

Lui KO, Waldmann H, Fairchild PJ (2009) Embryonic stem cells: overcoming the immunological barriers to cell replacement therapy. Curr Stem Cell Res Ther 4:70–80

Macedo N, Miller DM, Haq R, Kaufman HL (2020) Clinical landscape of oncolytic virus research in 2020. J Immunother Cancer 8. https://doi.org/10.1136/jitc-2020-001486

MacGregor RR (2001) Clinical protocol. A phase 1 open-label clinical trial of the safety and tolerability of single escalating doses of autologous CD4 T cells transduced with VRX496 in HIV-positive subjects. Hum Gene Ther 12:2028–2029

Mahato RI, Rolland A, Tomlinson E (1997) Cationic lipid-based gene delivery systems: pharmaceutical perspectives. Pharm Res 14:853–859

Majhen D, Ambriovic-Ristov A (2006) Adenoviral vectors-how to use them in cancer gene therapy? Virus Res 119:121–133

Manilla P, Rebello T, Afable C et al (2005) Regulatory considerations for novel gene therapy products: a review of the process leading to the first clinical lentiviral vector. Hum Gene Ther 16:17–25

Manno CS, Pierce GF, Arruda VR, et al (2006) Successful transduction of liver in hemophilia by AAV-Factor IX and limitations imposed by the host immune response. Nat Med 12:342–347

Martin GR (1981) Isolation of a pluripotent cell line from early mouse embryos cultured in medium conditioned by teratocarcinoma stem cells. Proc Natl Acad Sci U S A 78:7634–7638

Männistö M, Vanderkerken S, Toncheva V, et al (2002) Structure-activity relationships of poly(L-lysines): effects of pegylation and molecular shape on physicochemical and biological properties in gene delivery. J Control Release 83:169–182

McKenzie DL, Collard WT, Rice KG (1999) Comparative gene transfer efficiency of low molecular weight polylysine DNA-condensing peptides. J Pept Res 54:311–318

Medina-Kauwe LK (2003) Endocytosis of adenovirus and adenovirus capsid proteins. Adv Drug Deliv Rev 55:1485–1496

Melo LG, Pachori AS, Gnecchi M, Dzau VJ (2005) Genetic therapies for cardiovascular diseases. Trends Mol Med 11:240–250

Merkert S, Martin U (2016) Targeted genome engineering using designer nucleases: State of the art and practical guidance for application in human pluripotent stem cells. Stem Cell Res 16:377–386

Miyoshi H, Takahashi M, Gage FH, Verma IM (1997) Stable and efficient gene transfer into the retina using an HIVbased lentiviral vector. Proc Natl Acad Sci U S A 94:10319–10323

Moiani A, Miccio A, Rizzi E et al (2013) Deletion of the LTR enhancer/promoter has no impact on the integration profile of MLV vectors in human hematopoietic progenitors. PLoS One 8:e55721

Morenweiser R (2005) Downstream processing of viral vectors and vaccines. Gene Ther 12(Suppl 1):S103–S110

Morgan RA, Walker R, Carter CS et al (2005) Preferential survival of CD4+ T lymphocytes engineered with anti-human immunodeficiency virus (HIV) genes in HIV-infected individuals. Hum Gene Ther 16:1065–1074

Mount NM, Ward SJ, Kefalas P, Hyllner J (2015) Cell-based therapy technology classifications and translational challenges. Philos Trans R Soc Lond B Biol Sci 370:20150017

Muul LM, Tuschong LM, Soenen SL, et al (2003) Persistence and expression of the adenosine deaminase gene for 12 years and immune reaction to gene transfer components: long-term results of the first clinical gene therapy trial. Blood 101:2563–2569

Murry CE, Keller G (2008) Differentiation of embryonic stem cells to clinically relevant populations: lessons from embryonic development. Cell 132:661–680

Nemerow GR, Stewart PL (1999) Role of alpha(v) integrins in adenovirus cell entry and gene delivery. Microbiol Mol Biol Rev 63:725–734

Nguyen GN, Everett JK, Kafle S et al (2021) A long-term study of AAV gene therapy in dogs with hemophilia A identifies clonal expansions of transduced liver cells. Nat Biotechnol 39:47–55

Nilsson EC, Storm RJ, Bauer J et al (2011) The GD1a glycan is a cellular receptor for adenoviruses causing epidemic keratoconjunctivitis. Nat Med 17:105–109

Nori S, Okada Y, Nishimura S et al (2015) Long-term safety issues of iPSC-based cell therapy in a spinal cord injury model: oncogenic transformation with epithelial-mesenchymal transition. Stem Cell Reports 4:360–373

Panakanti R, Mahato RI (2009) Bipartite adenoviral vector encoding hHGF and hIL-1Ra for improved human islet transplantation. Pharm Res 26:587–596

Pearson S, Jia H, Kandachi K (2004) China approves first gene therapy. Nat Biotechnol 22:3–4

Pina S, Ribeiro VP, Marques CF et al (2019) Scaffolding Strategies for Tissue Engineering and Regenerative Medicine Applications. Materials 12. https://doi.org/10.3390/ma12111824

Platt RJ, Chen S, Zhou Y et al (2014) CRISPR-Cas9 knockin mice for genome editing and cancer modeling. Cell 159:440–455

Pollock K, Stroemer P, Patel S et al (2006) A conditionally immortal clonal stem cell line from human cortical neuroepithelium for the treatment of ischemic stroke. Exp Neurol 199:143–155

Pupo A, Fernández A, Low SH et al (2022) AAV vectors: The Rubik's cube of human gene therapy. Mol Ther 30:3515–3541

Quintana-Bustamante O, Fañanas-Baquero S, Dessy-Rodriguez M et al (2022) Gene editing for inherited red blood cell diseases. Front Physiol 13:848261

Raes L, De Smedt SC, Raemdonck K, Braeckmans K (2021) Non-viral transfection technologies for next-generation therapeutic T cell engineering. Biotechnol Adv 49:107760

Ramakrishna S, Kwaku Dad A-B, Beloor J et al (2014) Gene disruption by cell-penetrating peptide-mediated delivery of Cas9 protein and guide RNA. Genome Res 24:1020–1027

Ranga U, Woffendin C, Verma S et al (1998) Enhanced T cell engraftment after retroviral delivery of an antiviral gene in HIV-infected individuals. Proc Natl Acad Sci U S A 95:1201–1206

Raper SE, Chirmule N, Lee FS et al (2003) Fatal systemic inflammatory response syndrome in a ornithine transcarbamylase deficient patient following adenoviral gene transfer. Mol Genet Metab 80:148–158

Reilly JP, Grise MA, Fortuin FD, et al (2005) Long-term (2-year) clinical events following transthoracic intramyocardial gene transfer of VEGF-2 in no-option patients. J Interv Cardiol 18:27–31

Rubin H (2011) The early history of tumor virology: Rous, RIF, and RAV. Proc Natl Acad Sci U S A 108:14389–14396

Sago CD, Lokugamage MP, Paunovska K et al (2018) High-throughput in vivo screen of functional mRNA delivery identifies nanoparticles for endothelial cell gene editing. Proc Natl Acad Sci U S A 115:E9944–E9952

Salmon F, Grosios K, Petry H (2014) Safety profile of recombinant adeno-associated viral vectors: focus on alipogene tiparvovec (Glybera®). Expert Rev Clin Pharmacol 7:53–65

Sandoval-Villegas N, Nurieva W, Amberger M, Ivics Z (2021) Contemporary Transposon Tools: A Review and Guide through Mechanisms and Applications of Sleeping Beauty, piggyBac and Tol2 for Genome Engineering. Int J Mol Sci 22. https://doi.org/10.3390/ijms22105084

Santos GW (1983) History of bone marrow transplantation. Clin Haematol 12:611–639

Sayed N, Liu C, Wu JC (2016) Translation of Human-Induced Pluripotent Stem Cells: From Clinical Trial in a Dish to Precision Medicine. J Am Coll Cardiol 67:2161–2176

Scadden DT (2006) The stem-cell niche as an entity of action. Nature 441:1075–1079

Schwartz SD, Regillo CD, Lam BL et al (2015) Human embryonic stem cell-derived retinal pigment epithelium in patients with age-related macular degeneration and Stargardt's macular dystrophy: follow-up of two open-label phase 1/2 studies. Lancet 385:509–516

Schwarzwaelder K, Howe SJ, Schmidt M et al (2007) Gammaretrovirus-mediated correction of SCID-X1 is associated with skewed vector integration site distribution in vivo. J Clin Invest 117:2241–2249

Selyutina A, Persaud M, Lee K et al (2020) Nuclear Import of the HIV-1 Core Precedes Reverse Transcription and Uncoating. Cell Rep 32:108201

Serva SN, Bernstein J, Thompson JA et al (2022) An update on advanced therapies for Parkinson's disease: From gene therapy to neuromodulation. Front Surg 9:863921

Shaw AR, Suzuki M (2019) Immunology of Adenoviral vectors in cancer therapy. Mol Ther Methods Clin Dev 15:418–429

Shimamura M, Nakagami H, Sanada F, Morishita R (2020) Progress of Gene Therapy in Cardiovascular Disease. Hypertension 76:1038–1044

Shinagawa M, Padmanabhan RV, Padmanabhan R (1980) The nucleotide sequence of the right-hand terminal SmaI-K fragment of adenovirus type 2 DNA. Gene 9:99–114

Shy BR, Vykunta VS, Ha A et al (2022) High-yield genome engineering in primary cells using a hybrid ssDNA repair template and small-molecule cocktails. Nat Biotechnol. https://doi.org/10.1038/s41587-022-01418-8

Singh S, Narang AS, Mahato RI (2011) Subcellular fate and off-target effects of siRNA, shRNA, and miRNA. Pharm Res 28:2996–3015

Smith BD, Grande DA (2015) The current state of scaffolds for musculoskeletal regenerative applications. Nat Rev Rheumatol 11:213–222

Staal FJT, Aiuti A, Cavazzana M (2019) Autologous Stem-Cell-Based Gene Therapy for Inherited Disorders: State of the Art and Perspectives. Front Pediatr 7:443

Stewart DJ, Hilton JD, Arnold JMO, et al (2006) Angiogenic gene therapy in patients with nonrevascularizable ischemic heart disease: a phase 2 randomized, controlled trial of AdVEGF(121) versus maximum medical treatment. Gene Ther 13:1503–1511

Stevanato L, Corteling RL, Stroemer P et al (2009) c-MycERTAM transgene silencing in a genetically modified human neural stem cell line implanted into MCAo rodent brain. BMC Neurosci 10:86

Stolberg SG (1999) The biotech death of Jesse Gelsinger. N Y Times Mag 136–140:149–150

Suerth JD, Labenski V, Schambach A (2014) Alpharetroviral vectors: from a cancer-causing agent to a useful tool for human gene therapy. Viruses 6:4811–4838

Surosky RT, Urabe M, Godwin SG et al (1997) Adeno-associated virus Rep proteins target DNA sequences to a unique locus in the human genome. J Virol 71:7951–7959

Takahashi K, Tanabe K, Ohnuki M et al (2007) Induction of pluripotent stem cells from adult human fibroblasts by defined factors. Cell 131:861–872

Takahashi K, Yamanaka S (2006) Induction of pluripotent stem cells from mouse embryonic and adult fibroblast cultures by defined factors. Cell 126:663–676

Takebe T, Enomura M, Yoshizawa E et al (2015) Vascularized and Complex Organ Buds from Diverse Tissues via Mesenchymal Cell-Driven Condensation. Cell Stem Cell 16:556–565

Tchieu J, Kuoy E, Chin MH et al (2010) Female human iPSCs retain an inactive X chromosome. Cell Stem Cell 7:329–342

Thaker PH, Brady WE, Bradley WH et al (2015) Phase I study of intraperitoneal IL-12 plasmid formulated with PEG-PEI-cholesterol lipopolymer administered in combination with pegylated liposomal doxorubicin in recurrent or persistent epithelial ovarian, Fallopian tube, or primary peritoneal cancer patients: An NRG/GOG study. J Clin Orthod 33:5541–5541

Thier M, Wörsdörfer P, Lakes YB et al (2012) Direct conversion of fibroblasts into stably expandable neural stem cells. Cell Stem Cell 10:473–479

Thomson JA, Itskovitz-Eldor J, Shapiro SS et al (1998) Embryonic stem cell lines derived from human blastocysts. Science 282:1145–1147

Trounson A, McDonald C (2015) Stem cell therapies in clinical trials: progress and challenges. Cell Stem Cell 17:11–22

Tucci F, Galimberti S, Naldini L et al (2022) A systematic review and meta-analysis of gene therapy with hematopoietic stem and progenitor cells for monogenic disorders. Nat Commun 13:1315

Uusi-Kerttula H, Hulin-Curtis S, Davies J, Parker AL (2015) Oncolytic Adenovirus: Strategies and insights for vector design and immuno-oncolytic applications. Viruses 7:6009–6042

Valente JFA, Queiroz JA, Sousa F (2018) p53 as the Focus of Gene Therapy: Past, Present and Future. Curr Drug Targets 19:1801–1817

Varderidou-Minasian S, Lorenowicz MJ (2020) Mesenchymal stromal/stem cell-derived extracellular vesicles in tissue repair: challenges and opportunities. Theranostics 10:5979–5997

Vasan S, Hurley A, Schlesinger SJ et al (2011) In vivo electroporation enhances the immunogenicity of an HIV-1 DNA vaccine candidate in healthy volunteers. PLoS One 6:e19252

Vestergaard HT, D'Apote L, Schneider CK, Herberts C (2013) The evolution of nonclinical regulatory science: advanced therapy medicinal products as a paradigm. Mol Ther 21:1644–1648

Verma IM (By Inder M. Verma on November 1 1990) Gene Therapy. Scientific American.

von Jonquieres G, Rae CD, Housley GD (2021) Emerging concepts in vector development for glial gene therapy: Implications for leukodystrophies. Front Cell Neurosci 15:661857

Walther J, Wilbie D, Tissingh VSJ et al (2022) Impact of Formulation Conditions on Lipid Nanoparticle Characteristics and Functional Delivery of CRISPR RNP for Gene Knock-Out and Correction. Pharmaceutics 14:213

Wang H, Li Z-Y, Liu Y et al (2011) Desmoglein 2 is a receptor for adenovirus serotypes 3, 7, 11 and 14. Nat Med 17:96–104

Weber W, Fussenegger M (2006) Pharmacologic transgene control systems for gene therapy. J Gene Med 8:535–556

Wells DJ (2004) Gene therapy progress and prospects: electroporation and other physical methods. Gene Ther 11:1363–1369

Wiethoff CM, Nemerow GR (2015) Adenovirus membrane penetration: Tickling the tail of a sleeping dragon. Virology 479–480:591–599

Williams DA, Bledsoe JR, Duncan CN, et al (2022) Myelodysplastic syndromes after eli-cel gene therapy for cerebral adrenoleukodystrophy (CALD). In: MOLECULAR THERAPY. CELL PRESS 50 HAMPSHIRE ST, FLOOR 5, CAMBRIDGE, MA 02139 USA, pp 6–6

Wilmut I, Beaujean N, de Sousa PA et al (2002) Somatic cell nuclear transfer. Nature 419:583–586

Wright JF (2008) Manufacturing and characterizing AAV-based vectors for use in clinical studies. Gene Ther 15:840–848

Wright JF, Qu G, Tang C, Sommer JM (2003) Recombinant adeno-associated virus: formulation challenges and strategies for a gene therapy vector. Curr Opin Drug Discov Devel 6:174–178

Xia Y, Li X, Sun W (2020) Applications of Recombinant Adenovirus-p53 Gene Therapy for Cancers in the Clinic in China. Curr Gene Ther 20:127–141

Xin T, Cheng L, Zhou C et al (2022) In-Vivo Induced CAR-T Cell for the Potential Breakthrough to Overcome the Barriers of Current CAR-T Cell Therapy. Front Oncol 12:809754

Yan Z, Zhang Y, Duan D, Engelhardt JF (2000) Trans-splicing vectors expand the utility of adeno-associated virus for gene therapy. Proc Natl Acad Sci U S A 97:6716–6721

Yáñez-Mó M, Siljander PR-M, Andreu Z et al (2015) Biological properties of extracellular vesicles and their physiological functions. J Extracell Vesicles 4:27066

Yang C-S, Li Z, Rana TM (2011) microRNAs modulate iPS cell generation. RNA 17:1451–1460

Ylä-Herttuala S (2019) Gene Therapy of Critical Limb Ischemia Enters Clinical Use. Mol. Ther. 27:2053

Yuan X, Wan H, Zhao X et al (2011) Brief report: combined chemical treatment enables Oct4-induced reprogramming from mouse embryonic fibroblasts. Stem Cells 29:549–553

Zaiss A-K, Liu Q, Bowen GP et al (2002) Differential activation of innate immune responses by adenovirus and adeno-associated virus vectors. J Virol 76:4580–4590

Zennou V, Petit C, Guetard D et al (2000) HIV-1 genome nuclear import is mediated by a central DNA flap. Cell 101:173–185

Zhang W-W, Li L, Li D et al (2018) The first approved gene therapy product for cancer Ad-p53 (Gendicine): 12 years in the clinic. Hum Gene Ther 29:160–179

Zuris JA, Thompson DB, Shu Y et al (2015) Cationic lipid-mediated delivery of proteins enables efficient protein-based genome editing in vitro and in vivo. Nat Biotechnol 33:73–80

Further Reading

BSI PAS 83:2012 (2012) Developing human cells for clinical applications in the European Union and the United States of America – guide. Publicly Available Specification PAS83:2012, The British Standards Institution, ISBN 978 0 580 71052 0

BSI PAS 84:2012 (2012) Cell therapy and regenerative medicine – glossary. Publicly Available Specification PAS84:2012, The British Standards Institution, ISBN 978 0 580 74904 9

BSI PAS 93:2011 (2011) Characterization of human cells for clinical applications – guide. Publicly Available Specification PAS93:2011, The British Standards Institution, ISBN 978 0 580 69850 7

Galli MC et al (2015) Regulatory aspects of gene therapy and cell therapy products : a global perspective. Springer, Cham [Switzerland]

Herzog RW, Popplewell L (2020) Gene therapy : a nefvw therapeutic direction? S. Karger, Basel

National Institutes of Health (2006) Regenerative medicine. National Institutes of Health, Bethesda, DC

Narang A, Mahato RI (2010) Targeted delivery of small and macromolecular drugs. CRC Press, Boca Raton, FL

Nóbrega C et al (2020) A handbook of gene and cell therapy. Springer, Cham

Schleef M et al (2001) Plasmids for therapy and vaccination. Wiley-VCH, New York, NY

Schleef M (2005) DNA pharmaceuticals: formulation and delivery in gene therapy. DNA vaccination and immunotherapy. John Wiley & Sons, Hoboken

Vaccines

Wim Jiskoot, Gideon F. A. Kersten, Enrico Mastrobattista, and Bram Slütter

Introduction

Since vaccination was documented by Edward Jenner in 1798, it has become the most successful means of preventing infectious diseases, saving millions of lives every year. Rapid development of many safe and efficacious anti-SARS-CoV-2 vaccines, including some based on novel technology has again demonstrated that vaccines are crucial for infectious disease prevention. Application of vaccines is not limited to the prevention of infectious diseases. Licensed vaccines and vaccines in the pipeline include, among others, therapeutic vaccines against allergies, cancer, and neurodegenerative diseases such as Alzheimer's disease. In this chapter, a vaccine is defined as a drug product containing immunogenic material (pathogen, allergen, tumor) with the aim of inducing or suppressing antigen-specific immune responses.

Modern biotechnology has an enormous impact on current vaccine development. The elucidation of the molecular structures of pathogens and the tremendous progress made in immunology as well as developments in proteomics and bio-informatics have led to the identification of protective antigens and ways to deliver them. Together with technological advances, this has caused a move from empirical vaccine development to more rational approaches to develop effective and safe vaccines. In addition, modern methodologies may provide simpler and cheaper production processes for selected vaccine components.

Although vaccines resemble other biopharmaceuticals such as therapeutic proteins in some aspects, there are several important differences (Table 15.1). Unique features of vaccines include the low dose and frequency of administration, and the widely different vaccine categories (Table 15.2). Also, the target group is not only patients but basically every human being on the planet, with the emphasis on very young, healthy children. These differences have a huge impact on the requirements for vaccine admission on the market and release of vaccine batches, putting safety requirements on par with efficacy.

In the following section, immunological principles that are important for vaccine design are summarized. Subsequently, vaccine categories will be discussed, including current developments, as well as lessons learned from the COVID-19 pandemic and our preparedness for future emerging infectious diseases. It is not our intent to provide a comprehensive review. Rather, we will explain current approaches to vaccine development and illustrate these approaches with representative examples. Routes of administration will be discussed in a separate section. In the last section, pharmaceutical aspects of vaccines, including issues related with production, formulation, characterization and storage, are dealt with.

Sadly, Wim Jiskoot passed away by the time of publication of this sixth edition.

W. Jiskoot (Deceased) · B. Slütter
Division of Biotherapeutics, Leiden Academic Centre for Drug Research (LACDR), Leiden University, Leiden, The Netherlands
e-mail: d.j.a.crommelin@uu.nl; b.a.slutter@lacdr.leidenuniv.nl

G. F. A. Kersten (✉)
Division of Biotherapeutics, Leiden Academic Centre for Drug Research (LACDR), Leiden University, Leiden, The Netherlands

Coriolis Pharma, Martinsried, Germany
e-mail: gideon.kersten@coriolis-pharma.com

E. Mastrobattista
Department of Pharmaceutics, Utrecht University, Utrecht, The Netherlands
e-mail: E.Mastrobattista@uu.nl

Table 15.1 Exemplary differences between vaccines and most other biopharmaceuticals

Characteristic	Vaccines	Other biopharmaceuticals
Dose	Low (microgram range)	High (usually milligram range)
Frequency	Low (months—decades)	High (days—weeks)
Product group	Heterogeneous	Less heterogeneous
Characteristics	Sometimes ill-defined	Mostly well-defined
Indication	Mostly prophylactic	Therapeutic
Target group	Every human being	Patients

Table 15.2 Vaccine categories based on antigen source

Vaccine (sub) type	Example(s)	Antigen	Disease	Prophylactic/ therapeutic	Status of development[a]
Live					
Attenuated viruses	Poliovirus (Sabin)	Whole attenuated virus	Polio	P	2
	Rotavirus		Diarrhea	P	2
	Measles virus		Measles	P	2
	Mumps virus		Mumps	P	2
	Varicella zoster		Chickenpox	P	2
	Yellow fever virus		Yellow fever	P	2
Attenuated bacteria	BCG	Whole attenuated bacteria	Tuberculosis	P	2
	Salmonella typhi		Typhoid fever	P	2
Vectored	SARS-CoV-2	Spike protein	COVID-19	P	2
	Ebola antigens using adenoviral or vesicular stomatitis virus vectors	cAd3-EBO Z, VSV ZEBOV	Ebola	P	2
Human cells	Autologous dendritic cells (Provenge)	Prostate acid phosphatase (PAP)	Prostate cancer	T	2
Inactivated					
Whole virus	Poliovirus (2nd generation)	Formaldehyde-inactivated poliovirus	Polio	P	2
	Rabies virus	Propriolactone-inactivated Pitman-Moore L503 rabies virus strain	Rabies	P	2
	Hepatitis A virus	Formaldehyde-inactivated Hepatitis A virus strain CR 326F antigen	Hepatitis	P	2
Whole bacteria	*Bordetella pertussis*	Heat or formaldehyde inactivated bacteria	Whooping cough	P	2
	Vibrio cholerae		Cholera	P	2
Human cells	Melacine	Melanoma cell extract	Melanoma	T	2
Subunits					
Proteins	Mosquirix	CSP-HBsAg fusion protein	Malaria	P	2
	Hepatitis B virus	HBsAg VLP	Hepatitis B	P	2
	Gardasil	HPV-6/11/16/18 VP1 VLPs	Cervical cancer	P	2
	Clostridium tetani	Tetanus toxoid	Tetanus	P	2
	Corynebacterium diphtheriae	Diphtheria toxoid	Diphtheria	P	2
Peptides	ISA101 synthetic long peptides (SLP)	HPV16 E6 and E7 T cell epitopes	Cervical cancer	T	1
Polysaccharides	*Haemophilis influenzae* type B (Hib)	Capsular polysaccharide	Invasive Hib disease	P	2
	Neisseria meningitidis	Capsular polysaccharides	Meningitis	P	2
	Streptococcus pneumoniae	Capsular polysaccharides, either free or conjugated to CRM197 diphtheria toxoid	Pneumonia, meningitis	P	2
Nucleic acids					
DNA	Influenza vaccine	H5N1, H1N1	Flu	P	1
	Human papilloma virus vaccine	HPV 16/18 fusion consensus antigens	Cervical cancer	T	1
	HIV vaccine	HIV-1 gag, env and pol	AIDS	T	1
RNA	SARS-CoV-2	Spike protein	COVID-19	P	2
	Self-adjuvanted mRNA	PSA, PSCA, PSMA, STEAP1	Prostate cancer	T	1

[a] (1) Clinical evaluation, (2) Marketed product Abbreviations: *HIV* human immunodeficiency virus, *HPV* human papillomavirus, *cAd3 EBO-Z* chimp adenovirus serotype 3-vectored Zaire ebolavirus antigens; *VSV ZEBOV* vesicular stomatitis virus-vectored Zaire ebolavirus antigens, *CSP-HBsAg* circumsporozoite protein fused to Hepatitis virus B surface antigen, *VLP* virus-like particle, *CRM197* genetically detoxified form of diphtheria toxin, *H5N1, H1N1* influenza virus A subtypes, with differences in the surface antigens hemagglutinin (H) and Neuraminidase (N), *HIV-1 gag, env and pol* the 3 major proteins of HIV-1, with gag being a polyprotein that is processed to matrix and core proteins, env a protein residing in the lipid envelope and pol the reverse transcriptase, *PSA* prostate-specific antigen, *PSCA* prostate stem cell antigen, *PSMA* prostate-specific membrane antigen, *STEAP1* prostate-specific metalloreductase

Immunological Principles

Introduction

As a reaction to infection, the human immune system launches a series of immunological responses with the goal of eliminating the pathogen. Innate immune cells will be the first to respond and will attempt to clear the pathogen through phagocytosis and/or lysis. As pathogens have developed strategies to evade the innate immune response, all vertebrates are capable of eliciting a highly specific response by virtue of their adaptive immune system. The adaptive immune system can generate humoral immunity and cell-mediated immunity (see Fig. 15.1 and Table 15.3). Antibodies, produced by B-cells, are the typical representatives of humoral immunity. An antibody belongs to one of four different immunoglobulin classes (IgM, IgG, IgA, or IgE) (cf. Chap. 8). Antibodies are able to prevent infection or disease through several mechanisms:

1. Binding of antibodies covers the pathogen with Fc (constant fragment), the "rear end" of immunoglobulins. Phagocytic cells, such as macrophages, express surface receptors for Fc. This allows targeting of the opsonized (antibody-coated) antigen to these cells, followed by enhanced phagocytosis.
2. Immune complexes (i.e., complexes of antibodies bound to target antigens) can activate complement, a system of proteins that then becomes cytolytic to bacteria, enveloped viruses, or infected cells.
3. Phagocytic cells may express receptors for complement factors associated with immune complexes. Binding of these activated complement factors enhances phagocytosis.
4. Viruses can be neutralized by antibodies through binding at or near receptor binding sites on the virus surface. This may prevent binding to and entry into the host cell.

Antibodies are effective against many, but not all infectious microorganisms. They may have limited value when pathogens occupy intracellular niches (such as intracellular bacteria and parasites), which are not easily reached by antibodies. In this case, cell-mediated immunity is required to clear the infected cells. T-cells are the major representative of cell-mediated immunity and can clear infections by the following mechanisms:

1. Cytotoxic T-lymphocytes (CTLs, also called cytotoxic T-cells) react with target cells and kill them by release of cytolytic proteins like perforin.
2. T-helper cells (Th1-type, see below) activate macrophages, allowing them to kill intracellular pathogens.

In contrast to the innate response, the adaptive immune response is very specific to the invading pathogen (Fig. 15.2). The adaptive immune system comprises B-cells and T-cells with a wide range of specificities, owing to the unique compositions of their B-cell receptor (BCR) and T-cell receptor (TCR). During an infection the innate immune system instructs those B- and T-cells that have BCRs and TCRs specific for the invading pathogen to proliferate and gain effector functions. When the infection is cleared, most of these B- and T-cells are obsolete and many will die by apoptosis. Antibodies produced by B-cells, however, can persist in the circulation for an extended period of time. Moreover, some of the B- and T-cells resist apoptosis and can maintain themselves for many years as memory B- and T-cells. In contrast to their naive counterparts, these memory cells are rapidly activated and clonally expanded when they re-encounter the same pathogen or booster vaccine on a later occasion. Therefore, unlike the primary response, the response after repeated infection or vaccination is very fast and usually sufficiently strong to prevent reoccurrence of the disease (Fig. 15.2a).

Vaccination exploits the formation of this immunological memory by the adaptive immune system. The principle of vaccination is mimicking an infection in such a way that the natural specific defense mechanism of the host against the pathogen will be activated and immunological memory is established, but the host will remain free of the disease that normally results from a natural infection (Fig. 15.2b). This is effectuated by administration of immunogenic components that consist of, are derived from, or are related to the pathogen. The immune response is highly specific: it discriminates not only between pathogen species but often also between different strains within one species (e.g., strains of meningococci, polioviruses, coronaviruses, influenza viruses). Albeit sometimes a hurdle for vaccine developers, this high specificity of the immune system allows an almost perfect balance between responsiveness to foreign antigens and tolerance to self-antigens.

In the next paragraphs we will discuss the immunological principles leading to effector and memory responses.

Generation of an Immune Response and Immunological Memory

The generation of an immune response by vaccination follows several distinct steps that should ultimately lead to a potent effector response and/or long-lasting memory. After administration of the vaccine, the first step is uptake by professional antigen-presenting cells (APCs) at the site of application. APCs are able to shuttle the vaccine components to secondary lymphoid organs and present the antigens to T-

a Activation of T-helper Cells

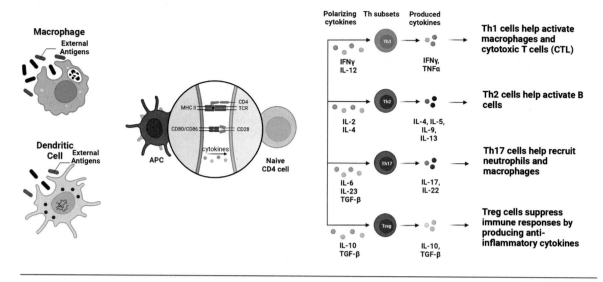

b Steps in B-cell Differentiation

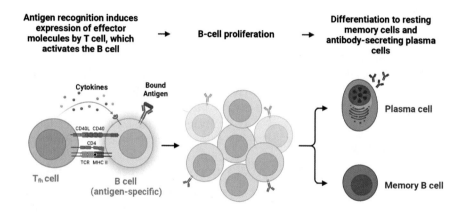

c Activation of Cytotoxic T lymphocytes

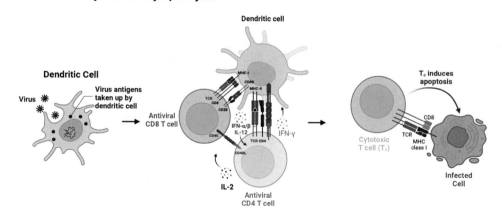

Fig. 15.1 Schematic representation of antigen-dependent immune responses. (**a**) Activation of T-helper cells (Th-cells). An antigen-presenting cell (APC), e.g., a dendritic cell, phagocytoses exogenous antigens (bacteria or soluble antigens) and degrades them partially. Antigen fragments are presented by MHC class II molecules to a CD4-positive Th-cell; the MHC-antigen complex on the APC is recognized by the T-cell receptor (TCR) and CD4 molecules on the Th-cell. The APC-Th-cell interaction leads to activation of the Th-cell. The activated Th-cell produces cytokines, resulting in the activation of macrophages (Th1 help), B cells (Th2 help; panel **b**), or cytotoxic T cells (panel **c**). (**b**) Antibody production. The presence of antigen and Th2-type cytokines causes proliferation and differentiation of B cells. Only B cells specific for the antigen become activated. The B cells, now called plasma cells, produce and secrete large amounts of antibody. Some B cells differentiate into memory cells. (**c**) Activation of cytotoxic T lymphocytes (CTLs). CTLs recognize nonself antigens expressed by MHC class I molecules on the surface of virally infected cells or tumor cells. Cytolytic proteins are produced by the CTL upon interaction with the target cell. Created by BioRender.com

Table 15.3 Important immune products protecting against infectious diseases

Immune response	Immune product	Accessory factors	Infectious agents
Humoral	IgG	Complement, neutrophils	Bacteria and viruses
	IgA	Alternative complement pathway	Microorganisms causing respiratory and enteric infections
	IgM	Complement, macrophages	(Encapsulated) bacteria
	IgE	Mast cells	Extracellular parasites
Cell mediated	CTL	Cytolytic proteins	Viruses, mycobacteria, intracellular parasites
	Th1	Macrophages	Mycobacteria, treponema (syphilis), fungi

Fig. 15.2 Principle of adaptive immune responses following infection and vaccination. (**a**) Schematic representation of adaptive immune responses upon primary and secondary infection. Upon primary infection T- and B-cell responses take time to develop, allowing pathogens to proliferate and cause disease. Upon secondary infection, circulating antibodies and memory T-cells quickly respond, preventing proliferation and dissemination of the pathogen. (**b**) Application of a vaccine inducing an adaptive immune response like a natural infection, but without associated disease

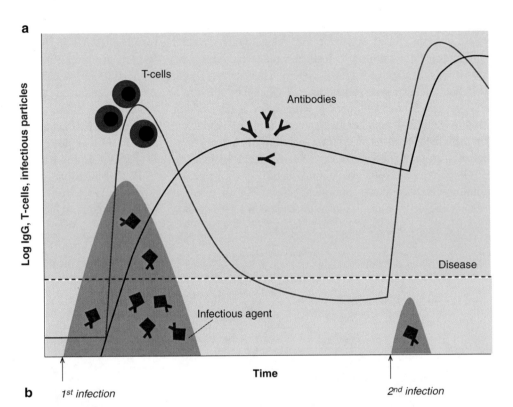

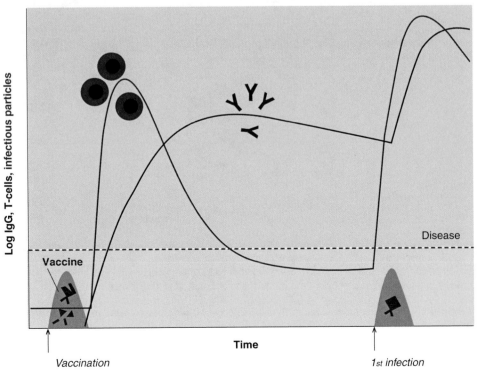

and B-lymphocytes, which—under the right conditions—results in activation of these lymphocytes. This simplified process is illustrated in Fig. 15.3. Below, we describe in more detail the successive steps leading to an immune response, in particular the steps relevant for the design of vaccines.

Activation of the Innate Immune System

Every immune reaction against a pathogen or a vaccine starts with activation of the innate immune system. Although the innate response itself does not lead to pathogen specific immunological memory, it is instrumental in activating and educating the adaptive immune system. There is increasing evidence that exposure to pathogens or vaccines leads to a kind of innate immune memory which acts against pathogens in general. This is called trained immunity. Evidence for the existence of trained innate immunity is the observation that immunization with potent live vaccines (measles, BCG) can cause some protection against pathogens not immunologically related to these vaccines. The vaccination causes durable changes in innate immune cells, for instance

by changing genomic methylation, thereby affecting gene expression or by metabolic reprogramming of cells (Netea et al. 2020). Important constituents of the innate immune system are APCs like macrophages and dendritic cells (DCs), which reside in tissues. By continuously endocytosing extracellular material, they sample their environment for potential harmful materials. To distinguish harmful from innocuous substances, APCs are equipped with pattern recognition receptors (PRRs) that allow detection of conserved microbial and viral structures, called pathogen-associated molecular patterns (PAMPs) (Kawai and Akira 2009). Examples of PAMPs are viral RNA and bacterial cell wall constituents, such as lipopolysaccharide (LPS) and flagellin (Table 15.4). As pathogens occupy different cellular niches, PRRs can be found either on the cell surface and endosomes (often for bacterial PAMPs) or in the cytoplasm (often for viral PAMPs). Examples of PRRs are toll- like receptors (TLRs), C-type lectins and RIG-I-like receptors (Table 15.4).

PRR activation induces a maturation program, which switches APCs from an antigen sampling to an antigen presentation mode, which is critical for their role as intermedi-

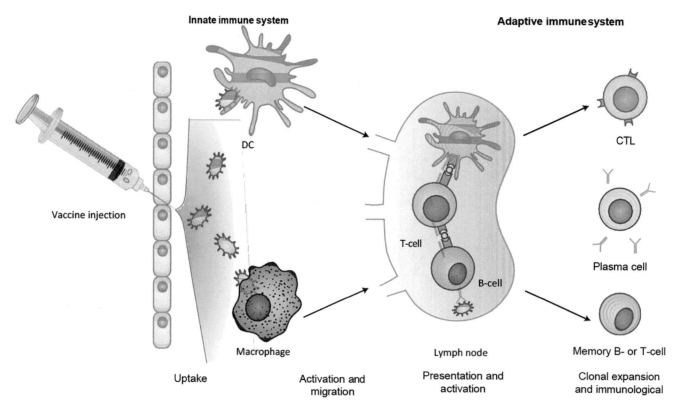

Fig. 15.3 Overview of the steps leading to immunity after administration of a vaccine. Upon subcutaneous or intramuscular administration, the vaccine components are taken up by phagocytic cells such as macrophages and dendritic cells (DCs) that reside in the peripheral tissue and express pattern recognition receptors (PRRs) that recognize pathogen-associated molecular patterns (PAMPs). Professional antigen-presenting cells (APCs) that have taken up antigens become activated and start migrating toward nearby lymph nodes. Inside the lymph nodes, the antigen processed by the APCs is presented to lymphocytes, which, when recognizing the antigen and receiving the appropriate co-stimulatory signals, become activated. These antigen-specific B- and T-cells clonally expand to produce multiple progenitors recognizing the same antigen. In addition, memory B- and T-cells are formed that provide long-term (sometimes lifelong) protection against infection with the pathogen

Table 15.4 Examples of pattern recognition receptors (PRRs), their ligands (PAMPs) and ligand source[a]

PRR	PRR location	PAMP[b]	Source
TLR1–TLR2	Cell surface	Triacyl lipopeptides	Bacteria
TLR2–TLR6	Cell surface	Diacyl lipopeptides	Bacteria
		Zymosan	Fungus
TLR3	Endosome	dsRNA	Virus
TLR4	Cell surface	LPS	Bacteria
TLR5	Cell surface	Flagellin	Bacteria
TLR7	Endosome	Single stranded (ss) RNA	RNA viruses
TLR8	Endosome	ssRNA	RNA viruses
LR9	Endosome	CpG DNA	Bacteria
TLR10[c]	Cell surface	Triacyl lipopeptides(?), some viral proteins	Bacteria, viruses
TLR11[d]	Endosome	Flagellin, profilin	Bacteria, parasites
TLR12[d]	Endosome	Profilin	Parasites
TLR13[d]	Endosome	Ribosomal RNA	Bacteria
RIG-I	Cytosol	ssRNA and short double stranded RNA	Viruses
MDA5	Cytosol	Long dsRNA	Viruses
LGP2 (helicase)	Cytosol	RNA	Viruses
NOD1/NLRC1	Cytosol	iE-DAP	Bacteria
NOD2/NLRC2	Cytosol	MDP	Bacteria
		ATP	Bacteria/ host
		Uric acid, CPPD, amyloid-β	Host
NALP1/ NLRP1	Cytosol	Anthrax lethal toxin	Bacteria
IPAF/NLRC4	Cytosol	Flagellin	Bacteria
NAIP5	Cytosol	Flagellin	Bacteria
Dectin-1	Cell surface	β-Glucan	Fungi

[a]Adapted from Kawai and Akira, Int. Immunol. 2009
[b]Several of the PAMPs listed here are used as vaccine adjuvants (see section "Formulation" and Table 15.5)
[c]TLR10 is anti-inflammatory. Mice lack the TLR10 gene
[d]Humans lack genes coding for TLR11, TLR12, and TLR13

ates for lymphocyte activation. PRR activation induces expression of MHC class I (MHCI) and MHC class II (MHCII) molecules, increasing the APCs' capacity to present antigen to T-cells. Moreover, APCs will gain expression of chemokine receptors (e.g., CCR7) that allow them to migrate to secondary lymphoid tissue. Finally, PRR stimulation induces upregulation of co-stimulatory molecules and pro-inflammatory cytokines, which provide import activation signals to T-cells during antigen presentation.

PRR activation is an essential step in the vaccination process and therefore important to consider when designing a vaccine. Live attenuated or inactivated vaccines naturally contain PAMPs to activate PRRs, however subunit vaccines may lack these PAMPs and may require addition of adjuvants (see section "Formulation").

Antigen Presentation

The peripheral lymphoid organs are the primary meeting place between cells of the innate immune system (APCs) and cells of the adaptive immune system (T-cells and B-cells). Whereas APCs are distributed throughout peripheral tissues, T- and B-cells are primarily located in secondary lymphoid organs, such as lymph nodes, spleen and Peyer's patches. An important reason for this is that, although the human body harbors a large number of lymphocytes (ca. 10^{12}), only few

T- and B-cells will have a TCR or BCR that is specific for the antigen of interest. By concentrating lymphocytes in secondary lymphoid organs and having APCs presenting antigen there, the chance of antigen specific T- and B-cells encountering their cognate antigen is increased. Upon interaction with APCs, antigen specific T-cells and B-cells will be activated, provided that they acquire the appropriate signals.

The first of these signals is antigen presentation, which allows selection of antigen specific B- and T-cells. B-cells and T-cells recognize antigens in different ways. B-cells must recognize antigens in their native form as their BCR (containing immunoglobulin D) contains the same variable part of the antibodies that the B-cells will excrete after developing into a plasma cell. The antibodies need to bind to the pathogen that also expresses intact antigen and therefore the primary interaction of the B-cell must be with native antigen. Therefore, B-cell antigens do not require major processing. In fact, B-cells can take up antigens that are small enough to drain to lymph nodes without the help of APCs. To shuttle larger antigens to lymphoid tissue in their native form, APCs express receptors that allow presentation of intact antigens to B-cells (Batista and Harwood 2009).

Some antigens are able to directly stimulate antibody production by B-cells without T-cell involvement. These thymus-independent antigens include certain linear antigens that are not readily degraded in the body and have a repeating determinant, such as bacterial polysaccharides. The epitope repeats enable cross-linking of BCR which leads to B-cell activation. Thymus-independent antigens do not induce immunological memory and are therefore less interesting from a vaccination standpoint. It is possible, however, to render these antigens thymus dependent by chemically coupling them to a protein carrier (see sections "B-cell and T-cell Activation" and "Polysaccharide Vaccines").

T-cells are unable to directly interact with antigen, but depend on the APCs to process antigens into peptide fragments (T-cell epitopes) and present them to the T-cells in the context of MHCI (to CD8+ T-cells; CD cluster of differentiation) or MHCII molecules (CD4+ T-cells) on the APC surface. Whether antigens are presented on MHCI or MHCII molecules is dependent on the intracellular location of the antigen processing. Exogenous antigens, acquired by endocytosis, can undergo limited proteolysis in the endosome and associate with MHCII molecules (Fig. 15.1a). MHCII molecules loaded with protein fragments return to the surface and can interact with CD4+ T-helper cells. Endogenous antigens, such as viral or mutated proteins produced by the host cell, are generated by proteasomal processing in the cytosol. The resulting peptides can associate with MHCI in the endoplasmic reticulum and can interact with (CD8+) cytotoxic T-cells (Fig. 15.1c).

These different antigen presentation pathways have consequences for vaccine design. As T-cells only recognize processed antigen fragments, T-cell responses rely on linear continuous epitopes, which are linear peptide sequences (usually consisting of up to ten amino acid residues) of the protein (see Fig. 15.1a). In contrast, B-cell epitopes can be conformational and even discontinuous comprising amino acid residues sometimes far apart in the primary sequence, which are brought together through the unique folding of the protein (see Fig. 15.1b). Antibody recognition of B-cell epitopes, whether continuous or discontinuous, is usually dependent on the conformation (=three-dimensional structure) of the antigen. For vaccines aimed to induce high levels of neutralizing antibodies (for instance, diphtheria and tetanus vaccines), one should take great care that the antigen remains in its native form. Vaccines that should induce CTL responses (e.g., some virus vaccines, cancer vaccines) will not necessarily require the antigen to be in its native form, as the antigen will have to be degraded before presentation anyway. A major challenge for these types of vaccines, however, is that MHCI presentation requires antigen to enter the cytosol rather than the endosomal compartment of an APC (see Fig. 15.1c). Professional APCs, especially DCs, have the capacity to transfer exogenously acquired antigens from the endosomal compartment into the MHCI processing pathway. The process is referred to as cross-presentation.

B-Cell and T-Cell Activation

Next to TCR stimulation through peptide loaded MHCI or MHCII molecules, the second signal T-cells require is co-stimulation via interaction of accessory and co-stimulatory molecules on the APCs (Fig. 15.4). This cell–cell interaction is essential for proper stimulation of lymphocytes, and without those accessory signals, antigen-specific T-cells will not proliferate and may become anergic (i.e., acquire a state of unresponsiveness). As co-stimulatory molecules, such as CD80/86, CD40, and ICAM-1, are upregulated on APCs after PRR stimulation, this signal functions as an additional safety check to prevent unwanted immune responses against self-antigens.

T-cells receiving TCR stimulation and co-stimulation will become activated, clonally expand and generate multiple progenitors all recognizing the same antigen. In contrast, T-cells can also receive co-inhibitory signals from the APC (e.g., PD1, CTLA4 activation). These signals reduce T-cell activation and provide a negative control mechanism against uncontrolled or unwanted T-cell responses.

Before and during clonal expansion, T-cells receive cytokine signals that influence their fate (signal 3). Cytokines can promote T-cell proliferation and also affect their effector function (see Fig. 15.4). For instance, interleukin 12 (IL-12) and type I interferon (IFN) are cytokines that are essential for the development of CTLs. Lack of these cytokines results in reduced proliferation of CTLs and a reduced capacity to kill target cells.

Especially for the CD4+ T-helper cells the cytokine signal during priming is crucial, as T-helper cells can have various effector functions. For instance, in the presence of cytokines, such as IL-12 and type I IFN, CD4+ T-cells develop into T-helper 1 (Th1) cells. These cells produce cytokines, such as IFN-γ and tumor necrosis factor alfa (TNF-α), which potentiate the effector function of phagocytes and increase inflammation. Therefore, induction of memory Th1 cells is a major goal for vaccines that aim to protect against intracellular pathogens. T-helper 2 (Th2) cells develop under influence of IL-4 signaling. These Th2 cells produce another set of cytokines that prevent Th1 differentiation and support B-cell proliferation and differentiation. Th2 cells have therefore been associated with increased humoral responses. However, as the cytokines produced by Th2 cells have been linked to IgE production by B-cells, reducing the number of memory Th2 cells has become an important focus in the design of vaccines aiming to reduce allergic responses.

Next to Th1 and Th2 cells, various other T-helper subsets have been identified, each having unique functional properties. Th17 cells develop when CD4+ T-cells receive transforming growth factor beta (TGF-β), IL-6 and IL-23 signals, and produce IL-17 and IL-22. These cytokines support the defense of mucosal surfaces, but have also been linked to

Fig. 15.4 The 3 signals of T-helper cell activation. (1) Antigen presentation. Peptides derived from a vaccine are loaded on MHCII molecules by the APC and presented to the T-cell receptor (TCR) on T-cells. (2) Co-stimulation. Activated APCs express co- stimulatory molecules, such as CD80/86 which support T-cell activation through inter-action with CD28 on T-cells. (3) Cytokines. APCs can produce different cytokines depending on the type of PAMP that has activated the APC. These cytokines provide a third signal to the T-cell by engaging their cognate receptors on the T-cell surface. Whereas IL-12 (red) signaling leads to Th1 polarization of the CD4$^+$ T-cell, IL-4 (yellow) signaling induces Th2 polarization and IL-6/IL23 (blue/purple) signaling pro- vides a pathway toward Th17 CD4$^+$ T-cells

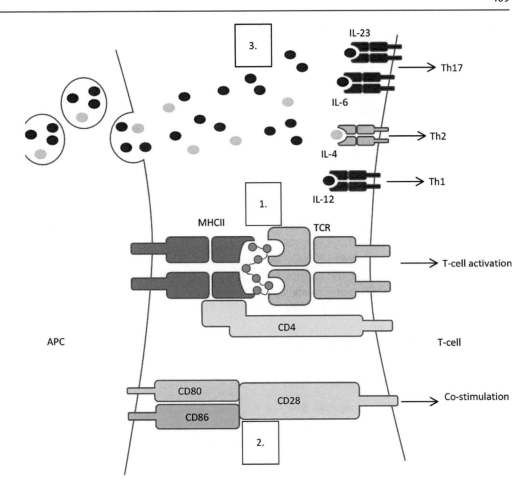

inflammatory disease, such as inflammatory bowel disease and psoriasis. Regulatory T-cells (Tregs) are subsets of CD4+ T-cells that play an important role in limiting inflammation through secretion of anti-inflammatory cytokines, such as IL-10 and TGF-β. Induction of Tregs may be of interest for vaccines that aim to reduce inflammation in autoimmune diseases.

One particular subset of Th-cells is devoted to providing help to B-cells, the so-called T-follicular helper cells (Tfhs). Under influence of IL-6 and IL-21, Tfhs upregulate molecules, such as C-X-C chemokine receptor type 5 (CXCR5), allowing them to migrate into B-cell zones. There, Tfhs can interact with B-cells that present cognate antigen on MHCII molecules. Only B-cells that receive co-stimulatory signals from Tfhs will be able to generate high-affinity IgG antibodies or mature into memory B-cells. This can have consequences for vaccine design, as vaccines that are aimed to generate B-cell memory need to contain both B-cell and T-cell epitopes in one entity. For instance, polysaccharides derived from *Haemophilus influenzae* type b, *Neisseria meningitidis*, or *Streptococcus pneumoniae* are targets for neutralizing antibodies, but require conjugation to a protein to allow T-cell help and development of B-cell memory.

B-cells differentiate into antibody-producing plasma cells. The effectivity of antibodies depends on their specificity, binding affinity, and concentration. After initial selection and activation by intact antigen and T-helper cells, antigen-specific B-cells undergo a complex process of fine tuning and further selection via recombination events, somatic hypermutation, and class switch to optimal isotypes. This leads ideally to high affinity antibodies with optimal effector functions such as ability to activate complement.

After clearance of the infection or vaccine the immune response enters a contraction phase. Many B- and T-effector cells will enter apoptosis and die but a small fraction of the antigen specific B- and T-cells may differentiate into memory cells. Memory cells are long-lived and reactivate upon repeated contact with antigen they recognize. Memory T-cells can differentiate into different subsets with specific roles in response to pathogens. Central memory T-cells possess a high proliferative capacity and populate secondary lymphoid organs. This allows them to quickly expand upon re-exposure to their cognate antigens. In contrast, effector memory T-cells do not rapidly expand upon re-exposure, but maintain effector functions such as cytokine production and cytolysis. They circulate through the body including inflamed tissues, which allows them to rapidly respond to infections. An important branch of effector memory T-cells are resident memory T-cells, which do not circulate but permanently reside in the tissue of original infection or vaccination. Due to their effector function and strategic location, resident memory T-cells can immediately respond to infection, either neutralizing the first infected host cells or rapidly recruit immune cells to the side of infection (Schenkel et al. 2014).

Vaccine Categories

Vaccines can be classified based on whether they are aimed to prevent (prophylactic) or cure (therapeutic) a disease, the type of disease to treat (infectious diseases, allergy, autoimmune disease, cancer, etc.), or the antigen source used for vaccination (e.g., whole pathogens, subunits, peptides, or nucleic acids), as illustrated in Fig. 15.5. Below, we first discuss vaccine categories based on antigen source. Next, current developments in therapeutic vaccines against cancer and other diseases are highlighted.

Classification Based on Antigen Source

Traditional vaccines originate from viruses or bacteria and can be divided in vaccines consisting of live attenuated pathogens and nonliving (inactivated) pathogens. In case the antigens that can convey immunity are known, specific subunits derived from the pathogen, such as proteins or polysaccharides, can be formulated into a vaccine. Nowadays, such subunit vaccines can also be made recombinantly (in case of proteins), or by chemical conjugation to a carrier protein (in case of polysaccharides) to enhance the immune response to the antigenic components. Moreover, with our current knowledge of immune recognition, both B- and T-cell epitopes can be identified and synthetically made, especially the latter because T-cell epitopes are linear peptide sequences.

Finally, nucleic acids form a separate class of antigen source, in which the DNA or RNA encoding the antigen(s) of interest is transfected into host cells to enable endogenous production and presentation of protein antigens. An overview of the various categories of vaccines and examples thereof is given in Table 15.2 and will be detailed in the sections below.

Live-Attenuated Vaccines

Before the introduction of recombinant DNA (rDNA) technology, live vaccines were made by the attenuation of virulent microorganisms by serial passage in experimental animals or in vitro and selection of mutant strains with reduced virulence or toxicity. Examples are vaccine strains for current vaccines such as oral polio vaccine, measles–mumps–rubella (MMR) combination vaccine, yellow fever vaccine, and tuberculosis vaccine consisting of bacille Calmette-Guérin (BCG). An alternative approach is chemical mutagenesis. For instance, by treating *Salmonella typhi* with nitrosoguanidine, a mutant strain lacking some enzymes that are responsible for the virulence was isolated (Germanier and Fuer 1975).

Live attenuated vaccines have the advantage that after administration they may replicate in the host, similar to their pathogenic counterparts. This confronts the host with a larger and more sustained dose of antigen and PAMPs, which means that few and low doses are required. In general, the vaccines give long-lasting humoral and cell-mediated immunity.

Live attenuated vaccines also have drawbacks. Live viral vaccines bear the risk of reverting to a virulent form, although this is unlikely when the attenuated seed strain contains several mutations. Nevertheless, for diseases such as viral hepatitis, AIDS, and cancers, this drawback makes the use of traditional live vaccines virtually unthinkable. Furthermore, it is important to recognize that immunization of immune-deficient children or immunocompromised adults with live organisms can lead to serious complications. For instance, a

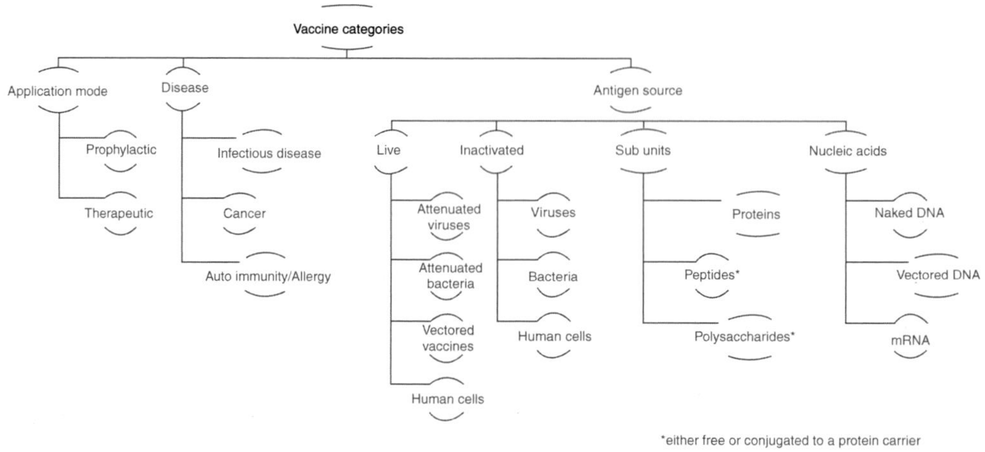

Fig. 15.5 Vaccine categories based on type of treatment, type of disease and antigen source

*either free or conjugated to a protein carrier

child with T-cell deficiency may become overwhelmed with BCG and die. Similarly, patients using certain immunosuppressive drugs (e.g., cyclosporin, methotrexate) should not be vaccinated with live attenuated vaccines.

Genetically Attenuated Live Vaccines

Emerging insights in molecular pathogenesis of many infectious diseases make it possible to attenuate micro-organisms more efficiently nowadays. By making multiple deletions, the risk of reversion to a virulent state during production or after administration can be virtually eliminated. A prerequisite for attenuation by genetic engineering is that the factors responsible for virulence and the life cycle of the pathogen are known in detail. It is also obvious that the protective antigens or epitopes must be known: attenuation must not result in reduced immunogenicity.

An example of an improved live vaccine obtained by homologous genetic engineering is the oral cholera vaccine Vaxchora. An effective cholera vaccine should induce a local, humoral response in order to prevent colonization of the small intestine. Initial trials with Vibrio cholerae cholera toxin (CT) mutants caused mild diarrhea, which was thought to be caused by the expression of accessory toxins. A natural mutant was isolated that was negative for these toxins. Next, CT was detoxified by rDNA technology. The resulting vaccine strain, called CVD 103, is well tolerated and challenge experiments with adult volunteers showed protection (Levine et al. 2017; Garcia et al. 2005).

Genetically attenuated live vaccines have the general drawbacks mentioned in the paragraph about classically attenuated live vaccines. For these reasons, it is not surpris-

ing that homologous engineering is mainly restricted to pathogens that are used as starting materials for the production of subunit vaccines (see the section "Subunit Vaccines" below).

Live Vectored Vaccines

A way to improve the safety or efficacy of vaccines is to use live, avirulent, or attenuated bacteria or viruses as a carrier to express protective antigens from a pathogen (see Table 15.2 for examples). Live vectored vaccines are created by recombinant technology, wherein one or more genes of the vector organism are replaced by one or more protective genes from the pathogen. Administration of such live vectored vaccines results in expression of the antigen-encoding genes either by the vaccinated individual's own cells or by the vector organism itself (e.g., in case of a bacterial vector).

The first licensed vector vaccine, in 2019, was an Ebola vaccine with Vesicular Stomatitis Virus (VSV) as vector. The VSV vector is attenuated by deletion of its glycoprotein gene but it remains replication competent. Instead of VSV glycoprotein it expresses the Ebola glycoprotein as antigen of interest.

For safety reasons also replication-deficient vectors have been developed (Fig. 15.6). These vectors are made incapable of replication by deleting genes. Being replication incompetent poses problems for the manufacturing of such a vaccine because in normal host cells the virus will not replicate. Therefore, producer cell lines have to be designed providing the missing gene product. After vaccination, the viral vector will be able to infect cells once and enter one replication cycle. The progeny virus will however miss the deleted

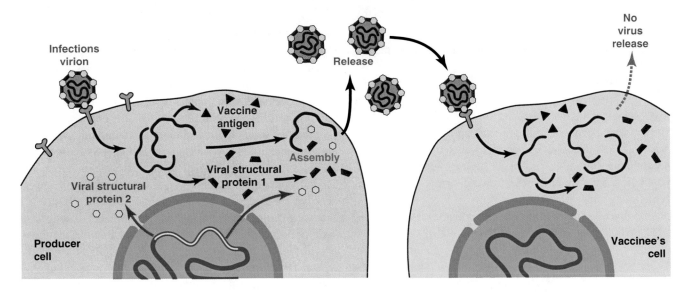

Fig. 15.6 Replication deficient viral vaccine vector. The vaccine strain is genetically modified by deleting at least one gene essential for the viral life cycle and adding the gene coding for the protein antigen of interest. The vaccine is produced on a producer cell line having the missing viral gene(s). Progeny virus cultured on these cells is phenotypically identical to wild type virus but cannot replicate in normal host cells. After vaccination the vector is able to infect host cells but these cells will not form functional virus since one essential protein is missing. Other proteins will be produced, including the antigen. Adapted from Pinschewer (2017)

gene product and the viral lifecycle is interrupted, adding to the safety profile of the vaccine.

Adenoviruses are also being used as vaccine vectors. Among the first approved SARS-CoV-2 vaccines were replication-deficient adenoviral vectors, carrying the gene coding for the S protein of SARS-CoV-2. Adenoviruses have several characteristics that make them suitable as vaccine vectors: (1) they can infect a broad range of both dividing and non-dividing mammalian cells, which expands possibilities to select production cell lines; (2) transgene expression is generally high and can be further increased by using heterologous promoter sequences; (3) adenovirus vectors are mostly replication deficient and do not integrate their genomes into the chromosomes of host cells, making these vectors very safe to use; and (4) upon parenteral administration, adenovirus vectors induce strong immunity and evoke both humoral and cellular responses against the expressed antigen. The use of adenoviral SARS-CoV-2 vaccines is associated with a slightly increased risk of serious blood clotting problems. In the USA, the frequency is 4 cases per million doses for the J&J/Janssen vaccine (Center for Disease Control and Prevention US, CDC 2019), which is considered acceptable by regulatory authorities since the benefits of vaccination outweigh the risks by far.

Another important viral vector is vaccinia virus. Advantages of vaccinia virus as vector include (1) its proven safety in humans as a smallpox vaccine, (2) the possibility for multiple antigen expression, (3) the ease of production, (4) its relative heat resistance, and (5) its various possible administration routes. A multitude of live recombinant vaccinia vaccines with viral and tumor antigens have been constructed, several of which have been tested in the clinic (Buchbinder et al. 2017; Payne et al. 2017; Guo et al. 2019). The products of genes coding for viral envelope proteins could be correctly processed and inserted into the plasma membrane of infected cells.

A potential limitation of the use of live vectored vaccines is the prevalence of preexisting immunity against the vector itself, which could neutralize the vaccine before the immune system can be primed. Such pre-existing immunity has been described for adenoviral vectors, for which the prevalence of neutralizing antibodies can be as high as 90% of the total population. The use of human or nonhuman (e.g., chimpanzee) strains with no or low prevalence of preexisting immunity as live vectors is therefore recommended (Ahi et al. 2011; Wong et al. 2018). Another potential problem is the induction of vector immunity after repeated immunization. Using heterologous prime-boost vectors may minimize risks. Developers of SARS-CoV-2 vector vaccines indeed used these strategies: Oxford/AstraZeneca used a chimpanzee adenovirus, J&J/Janssen vaccine is given only once and the Russian Sputnik vaccine consists of two different adenoviral vectors for primer and booster vaccination.

Inactivated Vaccines

An early approach for preparing vaccines is the inactivation of whole bacteria or viruses. A number of chemical reagents (e.g., formaldehyde, glutaraldehyde, β-propiolactone) and heat are commonly used for inactivation. Examples of inactivated vaccines are whole-cell pertussis, cholera, typhoid fever, and some polio vaccines. Inactivation may result in the loss of relevant epitopes due to covalent changes or partial unfolding of antigens. Also, since these vaccines do not replicate in vivo, often a higher dose is needed to induce protection, as compared to live attenuated vaccines. This may increase the price.

Subunit Vaccines

Given the complexity and batch-to-batch variability of vaccines consisting of inactivated whole pathogens, the use of well-defined antigenic subunits of pathogens is desired. Such subunits can be antigens (proteins or polysaccharides) directly purified from the pathogen, recombinantly produced protein antigens, or synthetic peptides.

Diphtheria Toxoid and Tetanus Toxoid Vaccines

Some bacteria such as *Corynebacterium diphtheriae* and *Clostridium tetani* form toxins. Antibody-mediated immunity to the toxins is the main protection mechanism against the adverse effects of infections with these bacteria. Both toxins are proteins and are inactivated with formaldehyde for inclusion in vaccines. The immunogenicity of such toxoids is relatively low and is improved by the adsorption of the toxoids to colloidal aluminum salts. This combination of an antigen and an adjuvant is still used in combination vaccines.

Polysaccharide Vaccines

Bacterial capsular polysaccharides consist of pathogen-specific multiple repeating carbohydrate epitopes, which are isolated from cultures of the pathogenic species. Plain capsular polysaccharides are thymus-independent antigens that are poorly immunogenic in infants and show poor immunological memory when applied in older children and adults. The immunogenicity of polysaccharides is highly increased when they are chemically coupled to carrier proteins containing Th-cell epitopes. This coupling makes them T-cell dependent, which is due to the participation of Th-cells that are activated during the response to the carrier. Examples of such polysaccharide conjugate vaccines include meningococcal type C, pneumococcal, and *Haemophilus influenzae* type b (Hib) polysaccharide vaccines that are included in many national immunization programs.

Acellular Pertussis Vaccines

The relatively frequent occurrence of side effects of whole cell pertussis vaccine was the main reason to develop subunit pertussis vaccines. The development of such acellular pertussis vaccines in the 1980s exemplifies how a better insight into factors that are important for pathogenesis and immunogenicity can lead to improved vaccines: it was conceived that a subunit vaccine consisting of a limited number of purified immunogenic components and devoid of (toxic) bacterial LPS would significantly reduce undesired effects. Current licensed acellular pertussis vaccines contain one to four protein antigens. Although these vaccines are effective and safe, they cannot prevent regular epidemics of whooping cough in many Western countries. Short-lived immunity and vaccine-induced selection of circulating strains resisting the primed immune system may contribute to this. Therefore, attempts are made to improve vaccination schemes and to develop new pertussis vaccines.

Recombinant Subunit Vaccines

To improve the yield, facilitate the production, and/or improve the safety of protein-based vaccines, protein antigens are nowadays often produced recombinantly, i.e., expressed by host cells that are safe to handle and/or allow high expression levels.

Heterologous hosts used for the expression of protein antigens include yeasts, bacteria, insect cells, plant cells, and mammalian cell lines. Hepatitis B surface antigen (HBsAg), which previously was obtained from the plasma of infected individuals, has been expressed in baker's yeast, *Saccharomyces cerevisiae* (Vanlandschoot et al. 2002), and in mammalian cells, such as Chinese hamster ovary cells (Raz et al. 2001), by transforming the host cell with a plasmid containing the HBsAg-encoding gene. Both expression systems yield 22-nm HBsAg particles (also called virus-like particles or VLPs) that are structurally identical to the native virus. Advantages are safety, consistent quality, and high yields. The yeast-derived vaccine has become available worldwide and appears to be as safe and efficacious as the classical plasma-derived vaccine.

The two human papillomavirus (HPV) vaccines currently on the market are produced as recombinant proteins which, like HBsAg, assemble spontaneously into virus-like particles. Antigens for Gardasil, available as a quadrivalent or nonvalent HPV vaccine, are produced in yeast, whereas antigens for the bivalent vaccine Cervarix are produced in insect cells.

Recombinant Peptide Vaccines

After identification of a protective epitope, it is possible to incorporate the corresponding peptide sequence through genetic fusion into a carrier protein, such as HBsAg, hepatitis B core antigen, and β-galactosidase (Francis and Larche 2005). The peptide-encoding DNA sequence is synthesized and inserted into the carrier protein gene. An example of the recombinant peptide approach is a malaria vaccine based on a 16-fold repeat of the Asn-Ala-Asn-Pro sequence of a *Plasmodium falciparum* surface antigen. The gene encoding this peptide was fused with the HBsAg gene, and the fusion product was expressed by yeast cells (Vreden et al. 1991). Clinical trials with this candidate malaria vaccine demonstrated moderate efficacy in children and infants in Africa (RTS,S Clinical Trials Partnership 2015). Despite its efficacy of about only 30% (reduction of deadly malaria), this vaccine was approved by WHO in 2021 for childhood immunization because it still is cost-effective in areas with moderate to high malaria transmission and will save thousands of lives when used in these areas.

Genetic fusion of peptides with proteins offers the possibility to produce protective epitopes of toxic antigens derived from pathogenic species as part of nontoxic proteins expressed by harmless species. Furthermore, a uniform product is obtained in comparison with the variability of chemical conjugates (see the section "Synthetic Peptide Vaccines," below).

Synthetic Peptide Vaccines

In principle, a vaccine could consist of only the relevant epitopes instead of intact pathogens or proteins. Peptide epitopes are small enough to be produced synthetically, and a peptide-based vaccine would be much better defined than traditional vaccines, making the concept of peptide vaccines attractive. However, it turned out to be difficult to develop these vaccines, and today there are no licensed peptide-based vaccines available yet. Nevertheless, important progress has been made, and some synthetic peptide vaccines have now entered the clinic, e.g., for immunotherapy of cancer (Melief and van der Burg 2008; van Poelgeest et al. 2013; Chen et al. 2020). To understand the complexity of peptide vaccines, one has to distinguish the different types of epitopes.

B-Cell Epitope-Based Peptide Vaccines

Epitopes recognized by antibodies or B-cells are very often conformation dependent (see above, section "Immunological Principles"). For this reason, it is difficult to identify them accurately and to synthesize them in the correct conformation. Manipulation of the antigen, such as digestion or the cloning of parts of the gene, will often affect B-cell epitope integrity. Attempts are made to predict B-cell epitopes using bioinformatics tools but this is in its infancy. Once the epitope is identified, synthesizing it as a functional peptide has proven to be difficult as well. The peptides often need to be conformationally restrained. This can be achieved by cyclization of the peptide (Oomen et al. 2005) or by the use of scaffolds to synthesize complex peptide structures (Timmerman et al. 2007).

T-Cell Epitope-Based Peptide Vaccines

Regarding conformation, T-cell epitopes are less demanding because they are presented naturally as processed peptides by APCs to T-cells. As a result, T-cell epitopes are linear. Here, we discern CD8+ epitopes (8–10 amino acid residues; MHC class I restricted) and CD4+ epitopes (>12 amino acid residues; MHC class II restricted). The main requirement is that they fit into binding grooves of MHC molecules with high enough affinity. Studies with peptide-based cancer vaccines have shown that these should contain both CD8+ and CD4+ epitopes in order to elicit a protective immune response. Furthermore, minimal peptides that can be externally loaded on MHC molecules of cells have been shown to induce less robust responses than longer peptides that require intracellular processing after uptake by DCs. Another point to consider is the variable repertoire of MHC molecules in a population, implying that a T-cell epitope-based peptide vaccine should contain several T-cell epitopes in order to be effective in the majority of the vaccinated population. Following these concepts, clinical trials with overlapping long peptide vaccines have shown promising results in the immune-therapy of patients with HPV-induced malignancies (Melief and van der Burg 2008; Farmer et al. 2021).

Nucleic Acid Vaccines

Immunization with nucleic acid vaccines involves the administration of genetic material, plasmid DNA or messenger RNA (mRNA), encoding the desired antigen. The encoded antigen is then expressed by the host cells and after which an immune response against the expressed antigen is raised. Nucleic acid vaccines offer the safety of subunit vaccines and the advantages of live recombinant vaccines. They can induce strong CTL responses against the encoded antigen as well as antibody responses.

mRNA Vaccines

mRNA vaccines (Fig. 15.7) made a spectacular entry as the first SARS-CoV-2 vaccines within a year of the emergence of SARS-CoV2. This was only possible, because the development of the science and technology began decades ago (Dolgin 2021). Initial problems with stability, delivery and toxicity were solved (See Chap. 13 Oligonucleotide & mRNA Therapies) and current products have an excellent safety profile, transient, non-integrative protein expression, and enhanced immunogenicity as compared to plasmid DNA vaccines. Initially, mRNA-based vaccines coped with stability problems, poor expression levels, and strong inflamma-

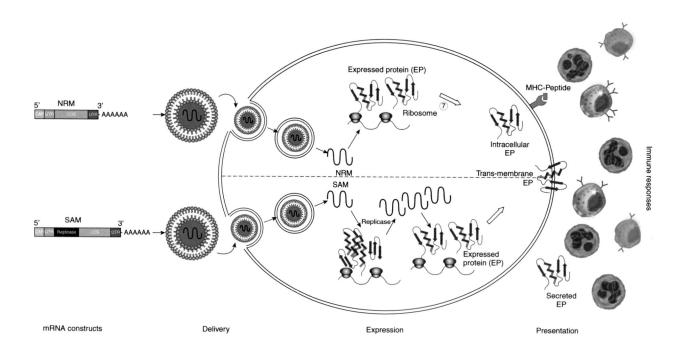

Fig. 15.7 In situ antigen production by mRNA. The mRNA is delivered to the cytosol via endosomal uptake and escape. The mRNA is directly translated into protein. The gene products are either protein antigen in the case of non-replicating RNA (NRM) or proteins enabling self-amplification of RNA in the case of self-amplifying constructs (SAM). The in situ produced antigen can attach to the cell membrane or excreted to become detectable by B-cells. Part of the antigen will be processed by the host cell and fragments will be presented by MHC molecules to T-cells. Adapted from Jackson et al. (2020)

tory adverse effects. To enhance immunogenicity, reduce inflammatory responses, and prolong protein expression, mRNAs were either chemically modified (both backbone and nucleoside modifications), sequence optimized, and/or formulated in nanocarriers, in particular lipid nanoparticles (LNP)(Chap. 5). These modifications resulted in slower degradation, an improved safety profile, and enhanced immune activation primarily through TLR-7 signaling. Optimized mRNA vaccines have been shown to elicit strong and balanced Th1/Th2 immune responses in animal models. The unprecedented success of the SARS-CoV-2 mRNA vaccines probably has cleared the way for other vaccines, both against infectious diseases and cancer vaccines (e.g., Kubler et al. 2015; Sebastian et al. 2014). The mRNA concept is a true platform approach. This means that standardized concepts of product and production process can be used again and again and a new vaccine needs only minimal adaptations because only the mRNA sequence is different. That does not mean that all attempts to develop new vaccines will be successful. Some antigens produced in situ by the mRNA may be intrinsically non-immunogenic, unstable or a different type of innate immune activation may be needed. Despite being a platform technology optimization cycles will remain needed.

For some antigens, a higher dose may be needed. mRNA is transient, leading to potentially too short antigen expression times, unfavorable for proper immune activation. This can be circumvented by making use of self-amplifying RNAs based on the alphavirus replication machinery. Four alphavirus genes responsible for RNA replication are co-expressed with the gene of interest encoding the desired antigen (Fig. 15.7). Transfection of this single RNA construct into cells leads to prolonged and 10–50-fold enhanced antigen expression in mice. Recent clinical data, however, show that such enhanced antigen expression was not observed in humans, probably due to cytoplasmic molecular pattern receptors (RIG-1, MDA-5) that have restricted RNA self-amplification (Pollock et al. 2022).

DNA Vaccines

DNA-based vaccines were until recently considered as the way to go for genetic vaccination, mainly because its stability. Also, bacterial plasmids are ideal for activating innate immunity as Toll-like receptor-9 (TLR-9) expressed on many phagocytic cells can recognize unmethylated bacterial DNA (see section "Adjuvants"). The main disadvantage of nucleic acid immunization is the poor immunogenicity in man. Therefore, they often require, like subunit vaccines, adjuvants, or delivery systems to boost the immune response against the DNA-encoded antigen(s). The DNA must be delivered into the cell nucleus which is more complicated than cytoplasmic delivery that is sufficient for mRNA. Nevertheless, DNA has proven to be very effective when used in combination with protein antigens in heterolo-

gous DNA-prime/protein-boost strategies. Examples of DNA vaccines that have been tested in clinical trials comprise plasmids encoding SARS-CoV-2 S protein, HIV-1 antigens and malaria antigens.

Delivery of Nucleic Acid Vaccines

Since nucleic acids do not easily enter cells but require intracellular delivery in their intact form for their activity, therapeutic application of these biomacromolecules requires sophisticated delivery methods or systems. More information of nucleic acid delivery systems can be found in Chaps. 5 and 14.

For vaccination purposes, naked nucleic acids (i.e., without a delivery system) can be administered to animals and humans via intramuscular injection but this is not very effective. Delivery of nucleic acids with lipidic or polymeric nanocarriers can increase both the cellular uptake and immune activation. COVID-19 mRNA vaccines contain lipid nanoparticles as delivery system. Nanocarriers protect the nucleic acids from premature degradation and enhance their cellular uptake by professional APCs. Instead of synthetic nanocarriers, viruses can be used as vectors as well (see also section Live vectored vaccines, above). Besides viruses, bacteria that replicate inside cells can also be used to deliver plasmid DNA into host cells for the expression of pathogen-derived antigens. Attenuated strains of *Shigella flexneri* and *Listeria monocytogenes* have been used for this purpose.

The use of naked DNA for vaccination requires high doses, e.g., in the milligram range, most likely because of its poor delivery, and has so far shown poor immunogenicity in human trials. Physical methods of DNA delivery can be used as well. These include ballistic approaches using a gene gun to inject DNA-coated gold nanoparticles into the epidermis, jet-injectors, electroporation (Kraynyak et al. 2022), and DNA tattooing (Samuels et al. 2017).

Therapeutic Vaccines

Most classical vaccine applications are prophylactic: they prevent an infectious disease from developing. Besides prophylactic applications, vaccines may be used to treat already established diseases, such as infectious diseases, cancer, and inflammatory disorders. Although the development of therapeutic vaccines is still in its infancy, especially in the field of cancer vaccines the insights and developments are rapidly progressing and some examples will be highlighted here.

Cancer Vaccines

Cancer is a collection of diseases characterized by uncontrolled cell division with the potential to invade and spread to other parts of the body. These characteristics are caused by gene mutations that are inherited or were accumulated during life by environmental factors. Such mutations may also lead to subtle changes in the antigenic repertoire of tumor cells as compared to healthy cells. This provides a basis for the development of therapeutic cancer vaccines aimed at inducing specific cellular immune responses and to a lesser extent humoral immune responses to pre-established cancer (Melief et al. 2015; van der Burg et al. 2016). A distinction can be made between so-called tumor- associated antigens that are present in normal tissues, but over expressed in tumors, and neoantigens, which are newly formed tumor-specific antigens caused by somatic DNA mutations.

Tumor-Associated Antigen Vaccines

Initially, clinical trials with cancer vaccines focused on the use of a single tumor-associated antigen (e.g., melanoma-associated antigen-1, prostate-specific antigen, mucin-1, carcinoembryonic antigen), mixtures of ill-defined antigens from whole tumor cell lysates, or whole tumor cells. The latter can be autologous tumor cells directly isolated from the patients or allogeneic tumor cells that have been genetically modified to express cytokines (e.g., GM-CSF) or other immune stimulating molecules. An advantage of using whole tumor cells is the presence of a wide array of tumor-specific antigens that could potentially lead to tumor-specific immune responses. A disadvantage is that ill-defined tumor cell lysates will mostly express self-antigens. Breaking immunological tolerance against these self-antigens can result in transient or persistent autoimmune reactions. Therefore, tumor vaccines containing well defined antigens are preferred. It has been shown in clinical trials that carefully selected antigens can break tolerance against conserved tumor antigens and induce potent T-cell responses, as was shown with an mRNA-based vaccine (Sahin et al. 2020).

Neoantigen Vaccines

Neoantigens are attractive for use in cancer vaccines, as they are foreign protein sequences that are absent in healthy tissue. However, since most neoantigens are unique to an individual's tumor, neoantigen vaccination requires a personalized approach, in which the vaccine composition is adjusted to the patient's needs. This is a labor-intensive and costly procedure which must be performed fast because the patient is waiting for treatment. Various neoantigen vaccination platforms have entered the clinic for the treatment of various cancers. Synthetic long peptide (SLP) vaccines consist of sets of peptides containing both Th and CTL neoepitopes that need to be processed by professional APCs and cross presented on MHC class I in order to elicit antigen-specific cellular responses. An advantage of SLPs over synthetic peptide epitopes that can directly bind MHC class I molecules is that the need for antigen processing prevents T-cell anergy. Since the length of peptides that can be synthe-

sized has its technical limitations, multiple SLPs need to be manufactured separately and combined to cover the breadth of neoantigens identified per individual. SLP vaccines have been successfully applied as therapeutic vaccines to treat cervical cancer as well as melanoma (Ott et al. 2017). Neoantigens can also be delivered as nucleic acids (both DNA and mRNA). An advantage of this approach is the intrinsic adjuvant properties of bacterially derived plasmid DNA and mRNA and the ease at which multiple epitopes can be combined in a single construct. In addition, endogenous expression of antigen leads to efficient MHC class I presentation and subsequent CD8+ T-cell induction.

Both SLP- and nucleic acid-based approaches can also be used for application in an ex vivo setting, in which patient-derived DCs are loaded with the antigen source and stimulated with cytokines before being administered to the patient (see also Chap. 14). Overall, the results with neoantigen vaccination look promising with reported partial and complete cancer regressions in several trials.

Combining vaccination with other cancer therapy may further increase the effectiveness of cancer treatments. An example is immune checkpoint therapy and vaccination. Tumor cells are able to tolerize tumor-specific T-cells by expressing membrane proteins that bind to T-cells and have a tolerizing effect. This can be circumvented by checkpoint inhibitors; e.g., monoclonal antibodies that prevent this tumor cell—T-cell interaction.

Exploiting Innate Immunity

Apart from antigen-specific vaccination, inducing an immune response against tumors can also be successful via innate immune activation. The notion that viral infections in cancer patients sometimes induced tumor regression led to the use of viruses with oncolytic potential. These viruses are able to kill tumor cells by preferentially infecting and lysing tumor cells, releasing tumor-specific antigens in the process or by activating the innate immune system which in turn results in non-specific (e.g., NK cells) as well as specific (e.g., T-cells) anti-tumor activity (Bommareddy et al. 2018). The potential of oncolytic viruses can further be increased by adding genes to the viral genome coding for cytokines (cf. Chap. 14). T-Vec (Imlygic) is an approved herpes simplex virus producing GM-CSF indicated for melanoma. Some bacteria also have anti-tumor activity by boosting innate responses. BCG, the vaccine initially developed against tuberculosis is also indicated for some types of bladder cancer (Ji et al. 2022).

Other Therapeutic Vaccine Applications

Besides prevention of infectious diseases or treatment of cancer, vaccines are also being developed for other therapeutic applications. These include treatment of Alzheimer's disease, induction of tolerance against food components and prevention of drug abuse. Most of these vaccines are still in an experimental phase. A few of these developments will be highlighted below.

Tolerogenic Vaccines to Treat Allergy or Autoimmune Diseases

Vaccines can be designed to induce immunological tolerance via the generation of regulatory T-cells (Tregs) with the aim to durably suppress undesired immune responses. For example, patients with autoimmune diseases in which the immune system attacks self-antigens and causes irreversible damage of tissues and cells would benefit from a vaccine that could specifically induce tolerance to the self-antigens. For multiple sclerosis, the main self-antigens are known and several vaccination approaches have been followed to induce tolerance. These range from injection of T-cell epitopes derived from self-antigens to vaccination with tolerogenic nanoparticles containing self-antigens and immunosuppressive drugs to mRNA vaccines (Hunter et al. 2014; Northrup et al. 2016; Krienke et al. 2021). Induction of Treg requires a different strategy than induction of effector T-cells and B-cells, described in the immunological principles paragraph. Rather than stimulation of PRR and induction of inflammatory signal 3 cytokines, the absence of co-stimulatory molecules and induction of anti-inflammatory signal 3 cytokines such as TGF-beta and IL-10 have been suggested as the proper strategy to induce Treg (PMID: 35392079). Similarly, the administration of low doses of antigens in the absence of immune stimulating molecules that lead to inflammation, also called allergy-specific immunotherapy, to desensitize against food (e.g., shrimp, peanut, cow's milk) or other (e.g., birch pollen, house dust mite) allergies are applied (Shamji and Durham 2017; Berings et al. 2017). The resulting immunological tolerance may result from induction of Treg but also involves the process of exhaustion and induction of antibodies with the IgG4 subtype (Lighaam and Rispens 2016).

Route of Administration

Introduction

The immunological response to a vaccine is dependent on the route of administration. Most current vaccines are administered intramuscularly or subcutaneously. Parenteral immunization (here defined as administration via those routes where a conventional hypodermic needle is used) usually induces systemic immunity but has disadvantages compared to other routes, e.g., needle phobia, infections caused by needlestick injuries and needle re-use, required vaccine sterility and injection skills. Moreover, parenterally administered vaccines generally do not result in effective immune responses at mucosal surfaces. As mucosal surfaces are a common port of entry for many pathogens, induction of a mucosal secretory IgA response may prevent the attachment and entry of pathogens into the host. For example, antibodies against cholera need to be in the gut lumen to inhibit adherence to and colonization of the intestinal wall. Therefore, mucosal immunization (e.g., oral, intranasal, or intravaginal) may be preferred, because it may induce both mucosal and systemic immunity. For instance, orally administered live attenuated *Salmonella typhi* vaccine not only invades the mucosal lining of the gut but also infects cells of the phagocytic system throughout the body, thereby stimulating the production of both secretory and systemic antibodies. Additional advantages of mucosal immunization are the ease of administration and the avoidance of systemic side effects (Czerkinsky and Holmgren 2012; Holmgren and Czerkinsky 2005).

The Oral Route of Administration

From a receiver perspective, oral delivery of vaccines would be preferable in many cases, because it is vaccinee friendly. Up to now, however, only a limited number of oral vaccines (e.g., oral polio, cholera, typhoid fever, and rotavirus vaccines) have made it to the market. Most of these vaccines are based on attenuated versions of pathogens for which the route of administration is the same as the natural route of infection. The gut is relatively immune tolerant to prevent immune responses against food antigens. Therefore, a relatively high dose of antigen is required to induce significant responses. A replicating vaccine provides this more easily than an inactivated vaccine. In addition, oral bio-availability is usually very low because of (1) degradation of protein antigens in the gastrointestinal (GI) tract and (2) poor permeability of the wall of the GI tract in case of a passive transport process.

Still, for the category of oral vaccines, the above-mentioned hurdles of degradation and permeation are not necessarily prohibitive. For oral immunization, only a (small) fraction of the antigen has to reach its target site to elicit an immune response. The target cells are lymphocytes and antigen-presenting accessory cells located in Peyer's patches (Fig. 15.8). The B-lymphocyte population includes cells that produce secretory IgA antibodies.

These Peyer's patches are macroscopically identifiable follicular structures located in the wall of the gastrointestinal tract. Peyer's patches are overlaid with microfold (M) cells that separate the luminal contents from the lymphocytes. These M cells have little lysosomal degradation capacity, are specialized in the uptake of particulate matter, and allow for antigen sampling and delivery to underlying APCs. Moreover, the density of mucus-producing goblet cells is lower in Peyer's patches than in surrounding parts of the GI tract.

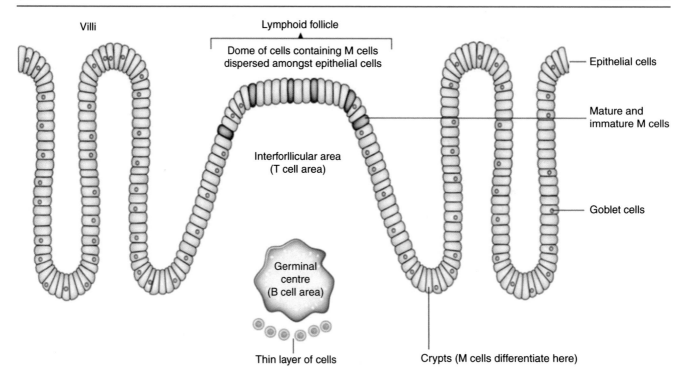

Fig. 15.8 Schematic diagram of the structure of intestinal Peyer's patches. M cells within the follicle-associated epithelium are enlarged for emphasis (from O'Hagan 1990)

This reduces mucus production and facilitates access to the M cell surface for luminal contents (Delves and Roitt 2011), which is of particular importance for the uptake of nano- and microparticle-based vaccines. Consequently, attempts to improve antigen delivery via the Peyer's patches and to enhance the immune response are made by using microspheres, liposomes, or modified live vectors, such as attenuated bacteria and viruses (Vela Ramirez et al. 2017). The latter have the additional advantage of replication-induced dose increase.

Other Routes of Administration

Apart from the oral route, the nose, lungs, rectum, oral cavity, and skin have been selected as potential sites of noninvasive vaccine administration. Most vaccines administered via these routes are still under development. Today, only a nasal influenza vaccine is licensed (FluMist, branded as Fluenz in Europe), and in China, Iran, Russia, and India, nasal COVID-19 vaccines have been approved that are based on adenovirus-vectored vaccines (Waltz 2022).

Besides mucosal vaccines, a number of intradermal vaccine delivery systems have been developed. These include needle-free jet injection of vaccines in liquid form and intradermal delivery with microneedles. Intradermal jet injectors are in clinical development, e.g., for inactivated polio vaccine. The classical liquid jet injectors deliver small volumes (microliter range) of liquid vaccine formulation with a high velocity. Depending on fluid velocity and nozzle design, the vaccine is deposited intradermally or dispersed deeper, i.e., subcutaneously of intramuscularly. Current versions use pre-filled disposable delivery units for single use to avoid contamination. Jet injectors for subcutaneous/intramuscular as well as intradermal delivery have been licensed combined with influenza vaccine or polio vaccine (Pharmajet.com).

Another attractive, potentially pain free approach for intradermal vaccine delivery is the use of microneedles or microneedle arrays with small individual needles in the 100–1000 μm range. There are multiple microneedle types and formats, such as solid microneedle arrays on which the vaccine components are coated, hollow microneedles through which a liquid vaccine formulation can be delivered via a micropump or syringe, and dissolvable microneedles containing the antigen/adjuvant embedded in, e.g., a sugar or polymer matrix, which dissolves rapidly after application (Mitragotri 2005; Kis et al. 2012; van der Maaden et al. 2012). Microneedle applications are in early clinical development (Rouphael et al. 2021) and show competitive potential with respect to dose, ease of administration and thermostability. This would allow easier logistics and administration by sending microneedle patches in the mail followed by self-administration. Examples are shown in Fig. 15.9.

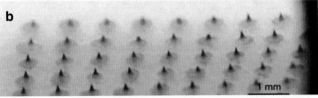

Fig. 15.9 (a) Hollow silicon microneedles, 300 μm in length, fabricated using a combination of wet and dry etch micromachining technologies (blue) and a 26-gauge syringe needle (brown) for comparison (Courtesy: Joe O'Brien & Conor O'Mahony, Tyndall National Institute). (b) Array of dissolvable microneedles, 280 μm in length at a density of 144 needles per cm², composed of sugars and polymers, fabricated in PDMS molds of master silicon microneedle arrays using a proprietary method (UK Patent Application Number 1107642.9) (Courtesy: Anne Moore, Anto Vrdoljak, School of Pharmacy, University College Cork)

Pharmaceutical Aspects

Production

Except for synthetic peptides, the antigenic components of vaccines are derived from microorganisms or animal cells. For optimal expression of the required vaccine component(s), these microorganisms or animal cells can be genetically modified. Animal cells are used for the cultivation of viruses and for the production of some subunit vaccine components and have the advantage that the vaccine components are released into the culture medium. However, some viruses cause cell lysis and consequently the culture medium will contain high concentrations of host cell proteins and host cell DNA, requiring extensive purification steps.

Three stages can be discerned in the manufacture of cell-derived vaccines: (1) cultivation or upstream processing, (2) purification or downstream processing, and (3) formulation. For the first two stages, the reader is referred to Chap. 4 and formulation of biopharmaceuticals is addressed in Chap. 5. The following section deals with formulation aspects specifically related to vaccines.

Formulation

Adjuvants: Immune Potentiators and Delivery Systems

The formulation of the vaccine is one of the major determinants that influence the type of immune response that is elicited, as it determines the type of co-stimulatory molecules and cytokines that are expressed by APCs. Through their various PRRs, APCs are more or less capable of "sensing" the type of vaccine that is encountered. This determines the set of co-stimulatory signals and proinflammatory cytokines that APCs will generate when presenting the antigen to T-cells in the peripheral lymphoid organs (Pulendran and Ahmed 2011). For instance, pathogens or vaccines containing lipoproteins or peptidoglycans will trigger DCs via TLR-2, which predominantly generates a Th2 response, whereas stimulation of DCs through TLR-3, or TLR-9 is known to yield robust Th1 and CTL responses. Therefore, vaccines should be formulated in such a way that the appropriate T-cell response will be triggered. This can be done by presenting the antigen in its native format, as is the case for the live-attenuated vaccines, or by formulating the native antigen with adjuvants that stimulate the desired response. Besides immune stimulatory molecules, a vehicle to deliver antigen to antigen-presenting cells and B-cells may be crucial, especially for highly purified subunit antigens. Immune stimulatory molecules and delivery systems are called adjuvants. Adjuvants are defined as any material that can increase or modulate the immune response against an antigen by activating the innate immune system. Adjuvants can stimulate the immune system by several, not mutually exclusive mechanisms (Guy 2007): (1) a depot effect leading to slow antigen release and prolonged antigen presentation, (2) attraction and stimulation of APCs by some local tissue damage and binding to PRRs present on or in APCs, and (3) delivery of the antigen to regional lymph nodes by improved antigen uptake, transport, and presentation by APCs.

Colloidal aluminum salts (hydroxide, phosphate) are widely used adjuvants in many classical vaccine formulations. In recent years other adjuvants have been introduced in marketed vaccines such as vaccines against Varicella zoster, HPV, Hepatitis B, malaria, corona, and influenza. These adjuvants include monophosphoryl lipid A, QS-21 saponin, CpG oligonucleotides and particulate adjuvants (delivery systems) like oil-in-water emulsions and LNP. A considerable part of the success of SARS-CoV-2 mRNA vaccines is due to sophisticated LNP used to protect and deliver mRNA from the endosomes to the cytoplasm where it can be translated into protein (Verbeke et al. 2021) (Fig. 15.7). An important component of the LNPs are the ionizable lipids, which are responsible for complexing the mRNA in the core of the nanoparticle, but also for initiating endosomal escape once the nanoparticles have been taken up by cells. Even viral vectors can be considered as adjuvants as they deliver the antigen and activate the innate immune system. To optimize and balance the immune response sometimes combinations of adjuvants are used. Varicella zoster vaccine contains monophosphoryl lipid A as well as QS-21 saponin. Table 15.5 shows some examples of well-known adjuvants.

Combination Vaccines

Since oral immunization is not possible for most available vaccines (see the section "Route of Administration" above), the strategy to mix individual vaccines in order to limit the number of injections has been common practice since many decades. Currently, vaccines are available containing up to six nonrelated antigens: diphtheria-tetanus-pertussis-hepatitis B-polio-*Haemophilus influenzae* type b vaccine. Another example is MMR vaccine, alone or in combination with varicella vaccine. Sometimes a vaccine contains antigens from several subtypes of a particular pathogen. Pneumococcal conjugate vaccine 20 (PCV20) is an example. This vaccine contains polysaccharides from 20 pneumococcal strains, conjugated to an immunologically non-related (a harmless form of diphtheria toxin) carrier protein to provide T-helper cell recognition and, as a result, induce immunological memory.

Combining vaccine components sometimes results in pharmaceutical as well as immunological problems. For instance, formaldehyde-containing components may chemically react with other components; an unstable antigen may need freeze drying, whereas other antigens should not be frozen. Components that are not compatible can be mixed prior to injection, if there is no short-term incompatibility. To this end, dual-chamber cartridges (cf. Chap. 5) have been developed. From an immunological point of view, the immunization schedules of the individual components of combination vaccines should match. Pertussis vaccine, for instance, should be given as early in life as possible, since whooping cough is most dangerous in very young children, whereas diphtheria toxoid can be given some time later. Even when this condition of matching immunization schedules is met and the components are pharmaceutically compatible, the success of a combination vaccine is not warranted. Vaccine components in combination vaccines may exhibit a different behavior in vivo compared to separate administration of the components. For instance, enhancement (Paradiso et al. 1993) as well as suppression (Mallet et al. 2004) of humoral immune responses has been reported.

Characterization

Modern vaccines have to meet similar standards as other biotechnological pharmaceuticals and can be characterized with a combination of appropriate biochemical, physicochemical, and immunochemical techniques (cf. Chaps. 3 and 5). The use of state-of-the art analytical techniques for the design and release of new vaccines is gaining importance. Currently,

Table 15.5 Examples of adjuvants used in vaccine formulations[a]

Immune potentiators	Examples	Characteristics
Bacterial origin		
Triacyl lipopeptides	Pam3Cys; Pam3CSK4	TLR1–TLR2 agonists
Diacyl lipopeptides	Pam2Cys; MALP2	TLR2–TLR6 agonists
LPS analogs	MPL; RC-529	Endotoxins; TLR4 agonists
Cell wall components	Peptidoglycan; muramyl peptides	TLR2–TLR4 agonists
Flagellin		TLR5 agonist
CpG DNA		TLR9 agonist
Toxins	Cholera toxin B subunit; heat labile enterotoxin subunit B	
Viral origin		
Double stranded RNA	Poly(I:C); poly(rA:rU)	TLR3 agonists
Guanoside analogs	Imiquimod; resiquimod	TLR7–TLR8 agonists
Other origin		
Plant-derived	QuilA; QS21	Triterpene glycosides; crucial components of Matrix-M adjuvant
Mineral	Aluminum hydroxide; aluminum phosphate	Colloidal suspensions; antigen adsorption is crucial
Synthetic lipids	Avridine; DDA	Used as liposome components
Delivery systems	*Examples*	*Characteristics*
Oil-in-water emulsions	AF03; MF59	
Water-in-oil emulsions	Montanide ISA 51; Montanide ISA 720	
Particulate carriers	Liposomes; virosomes; LNPs; polymeric nano- and microparticles; bacterial ghosts	Antigen association with carrier is crucial
Combination adjuvants	*Examples*	*Characteristics*
Miscellaneous	AS01; AS02; CAF01; Montanide ISA 51 plus GM-CSF	

[a]Adapted from Amorij et al. (2012)

animal experiments are needed for quality control of some vaccines but in vitro analytical techniques may eventually (partly) substitute tests in vivo. During the development of the production process of a vaccine component, a combination of suitable assays can be defined. These assays can subsequently be applied during its routine production.

Column chromatographic (HPLC) and electrophoretic techniques, such as gel electrophoresis and capillary electrophoresis, provide information about the purity, molecular weight, and other physicochemical properties of antigens. Physicochemical assays comprise mass spectrometry and spectroscopy, including circular dichroism and fluorescence spectroscopy. Information is obtained mainly about the molecular weight and the conformation of the antigen(s). Immunochemical assays, such as enzyme-linked immunoassays, are powerful methods for the quantification of the antigen(s). By using monoclonal antibodies (preferably with the same specificity as those of protective human antibodies) information can be obtained about the conformation and accessibility of the epitope to which the antibodies are directed. Moreover, the use of biosensors makes it possible to measure antigen-antibody interactions momentarily, allowing accurate determination of binding kinetics and affinity constants. Furthermore, since practically all vaccines are particulate in nature, it is sensible to use state-of-the-art particle sizing and counting methods to characterize them (Roesch et al. 2022).

Storage

Depending on their specific characteristics, vaccines are stored as solution or as a freeze-dried formulation, usually at 2–8 °C. The present generation of mRNA COVID-19 vaccines needs subzero storage temperatures (−20 or around −80 °C). The shelf life of vaccines depends on the composition and physicochemical characteristics of the vaccine formulation and on the storage conditions and typically is in the order of several years. The quality of the primary container can influence the long-term stability of vaccines, e.g., through adsorption or pH changes resulting from contact with the vial wall or vial stopper. The use of pH indicators or temperature- or time-sensitive labels ("vial vaccine monitors," which change color when too long exposed to too high temperatures) can avoid unintentional administration of an inappropriately stored vaccine.

Concluding Remarks

Despite the tremendous success of the classical vaccines, there are still many infectious diseases and other diseases (e.g., cancer) against which no effective vaccine exists.

Emerging diseases, often caused by viruses, will remain a continuous threat. The COVID-19 pandemic has shown that, contrary to many expectations upfront-even by experts it is possible to develop potent and safe vaccines much faster than anticipated. This was evident for relatively new technologies (viral vectors, mRNA), but some COVID-19 vaccines based on classical concepts were approved not much later. Although modern vaccines—like other biopharmaceuticals—are expensive, calculations may indicate cost-effectiveness (the costs of a medical intervention relative to savings in terms of health gain) for vaccination against many of these diseases. In addition, the growing microbial resistance to existing antibiotics increases the need to develop vaccines against common bacterial infections. It is expected that novel vaccines against several of these diseases continue to become available, and in these cases, the preferred type of vaccine will be chosen from one of the different options described in this chapter.

Self-Assessment Questions

Questions

1. Imagine three vaccine types against the same viral disease: (1) formaldehyde inactivated virus, (2) genetically attenuated live virus and (3) highly purified viral protein.
   (a) One vaccine is supplemented with an adjuvant. What is an adjuvant?
   (b) Which of the three vaccines should contain added adjuvant and why?
   (c) Vaccines 1 and 2 have almost the same antigen composition. Despite this, one vaccine can be given in a considerably lower dose than the other one to induce the same level of protection. Which one and why?
   (d) Which vaccine is able to induce cellular cytotoxic T-cell responses and why?
2. How do antibodies prevent infection or disease?
3. How does the immune system prevent unwanted T-cell responses against self-antigens and how does this affect vaccine design?
4. What is the definition of a subunit vaccine? Give three different types of subunit vaccines.
5. Development and production of mRNA vaccines and vector vaccines can be considered as platform technology. What does this mean?
6. Mention at least three advantages of mucosal vaccination. What are M cells and why are they important in mucosal vaccination?
7. Mention two or more examples of currently available combination vaccines. Which pharmaceutical and immunological conditions have to be fulfilled when formulating combination vaccines?

Answers

1.
   (a) An adjuvant is a vaccine component improving qualitatively and/or quantitatively the immune response against an antigen. Adjuvants act on the innate immune system.
   (b) Vaccine 3, because it only contains pure antigen and lacks an innate immune stimulus. Vaccines 1 and 2 consist of complete viruses which in general contain innate immune potentiators, such as double stranded RNA.
   (c) Vaccine 2 is a live vaccine. Therefore, it can replicate to some extent after administration, increasing the effective dose and extending the contact time with the immune system.
   (d) Vaccine 2, because it infects cells. Infected cells produce progeny virus. This endogenous antigen source is partially processed and presented in MHC class 1 molecules to Th-cells. This results in induction of CD8 T-cells.

2. Antibodies are able to neutralize pathogens by at least four mechanisms:
   (a) Fc-mediated phagocytosis.
   (b) Complement activation resulting in cytolytic activity.
   (c) Complement-mediated phagocytosis.
   (d) Competitive binding on sites that are crucial for the biological activity of the antigen.

3. Besides antigen presentation through MHCI or MHCII molecules, T-cells require a second signal from an APC before they will proliferate. This second signal supplied by the APC is referred to as co-stimulation and only occurs when the APC has sensed danger, by detecting PAMPs. Therefore, an effective vaccine needs to contain both an antigen and a PAMP (often in the form of an adjuvant).

4. Subunit vaccines are vaccines that contain one or more individual components of a pathogen, e.g., proteins, oligosaccharides or peptide epitopes. These can be either isolated from the pathogen (in case of oligosaccharides, toxins or other protein antigens), recombinantly produced (in case of protein antigens) or synthesized (in case of peptide epitopes).

5. Platform technology means that:
   (a) the product (vaccines against different pathogens) is very similar e.g., mRNA formulated in lipid nanoparticles or a replication deficient adenovirus expressing the antigen of choice.
   (b) the production process and the analytics are more or less standardized.

6. Advantages of mucosal vaccination over vaccination by injection are that it:
   (a) avoids infections caused by needlestick injuries and needle re-use,
   (b) is easier to perform and more vaccinee friendly,
   (c) can induce mucosal immunity.

   M cells are cells present in mucosal surfaces (such as the nasal cavity and the Peyer's patches in the gastrointestinal tract). M cells have little lysosomal degradation capacity, are specialized in the uptake of particulate matter, such as nano- and microparticulate vaccines. They can sample particulate antigens and deliver them to under-lying APCs. The density of mucus-producing goblet cells is low in Peyer's patches and M cells do not produce mucus, which facilitates the access of (particulate) antigens to the M cell surface.

7. Examples of combination vaccines include.

diphtheria-tetanus-pertussis(−polio) vaccines and measles-mumps-rubella(−varicella) vaccines. Prerequisites for combining vaccine components are:

   (e) Pharmaceutical compatibility of vaccine components and additives.
   (f) Compatibility of immunization schedules.
   (g) No interference between immune responses to individual components.

References

Ahi YS, Bangari DS, Mittal SK (2011) Adenoviral vector immunity: its implications and circumvention strategies. Curr Gene Ther 11(4):307–320

Amorij JP, Kersten GF, Saluja V, Tonnis WF, Hinrichs WL, Slütter B, Bal SM, Bouwstra JA, Huckriede A, Jiskoot W (2012) Towards tailored vaccine delivery: needs, challenges and perspectives. J Control Release 161(2):363–376

Batista FD, Harwood NE (2009) The who, how and where of antigen presentation to B cells. Nat Rev Immunol 9(1):15–27

Berings M, Karaaslan C, Altunbulakli C, Gevaert P, Akdis M, Bachert C, Akdis CA (2017) Advances and highlights in allergen immunotherapy: on the way to sustained clinical and immunologic tolerance. J Allergy Clin Immunol 140(5):1250–1267

Bommareddy PK, Shettigar M, Kaufman HL (2018) Integrating oncolytic viruses in combination cancer immunotherapy. Nature Rev 18:498–513

Buchbinder SP, Grunenberg NA, Sanchez BJ, Seaton KE, Ferrari G, Moody MA, Frahm N, Montefiori DC, Hay CM, Goepfert PA, Baden LR, Robinson HL, Yu X, Gilbert PB, MJ ME, Huang Y, Tomaras GD, Group HIVVTNS (2017) Immunogenicity of a novel clade B HIV-1 vaccine combination: results of phase 1 randomized placebo controlled trial of an HIV-1 GM-CSF-expressing DNA prime with a modified vaccinia Ankara vaccine boost in healthy HIV-1 uninfected adults. PLoS One 12(7):e0179597

CDC (2019). https://www.cdc.gov/coronavirus/2019-ncov/vaccines/safety/adverse-events.html. Accessed 24 Aug 2022

Chen X, Yang J, Wang L, Liu B (2020) Personalized neoantigen vaccination with synthetic long peptides: recent advances and future perspectives. Theranostics 10:6011–6023

Czerkinsky C, Holmgren J (2012) Mucosal delivery routes for optimal immunization: targeting immunity to the right tissues. Curr Top Microbiol Immunol 354:1–18

Delves PJ, Roitt IM (2011) Roitt's essential immunology, 12th edn. Wiley, Chichester

Dolgin E (2021) The tangled history of mRNA vaccines. Nature 597:318–324

Farmer E, Cheng MA, Hung C-F, Wu T-C (2021) Vaccination strategies for the control and treatment of HPV infection and HPV-associated cancer. Recent Results Cancer Res 217:157–195

Francis JN, Larche M (2005) Peptide-based vaccination: where do we stand? Curr Opin Allergy Clin Immunol 5(6):537–543

Garcia L, Jidy MD, Garcia H, Rodriguez BL, Fernandez R, Ano G, Cedre B, Valmaseda T, Suzarte E, Ramirez M, Pino Y, Campos J, Menendez J, Valera R, Gonzalez D, Gonzalez I, Perez O, Serrano T, Lastre M, Miralles F, Del Campo J, Maestre JL, Perez JL, Talavera A, Perez A, Marrero K, Ledon T, Fando R (2005) The vaccine candidate vibrio cholerae 638 is protective against cholera in healthy volunteers. Infect Immun 73(5):3018–3024

Germanier R, Fuer E (1975) Isolation and characterization of gal E mutant ty 21a of salmonella typhi: a candi- date strain for a live, oral typhoid vaccine. J Infect Dis 131(5):553–558

Guo ZS, Lu B, Guo Z, Giehl E, Feist M, Dai E, Liu W, Storkus WJ, He Y, Liu Z, Bartlett DL (2019) Vaccinia virus-mediated cancer immunotherapy: cancer vaccines and oncolytics. J ImmunoTherapy Cancer 7:6. https://doi.org/10.1186/s40425-018-0495-7

Guy B (2007) The perfect mix: recent progress in adjuvant research. Nat Rev Microbiol 5(7):505–517

Holmgren J, Czerkinsky C (2005) Mucosal immunity and vaccines. Nat Med 11(4 Suppl):S45–S53

Hunter Z, McCarthy DP, Yap WT, Harp CT, Getts DR, Shea LD, Miller SD (2014) A biodegradable nanoparticle platform for the induction of antigen-specific immune tolerance for treatment of autoimmune disease. ACS Nano 8(3):2148–2160

Jackson NAC, Kerster KE, Casimiro D, Gurunathan DRF (2020) The promise of mRNA vaccines: a biotech and industrial perspective. NPJ Vaccines 5:11. https://doi.org/10.1038/s41541-020-0159-8

Ji N, Long M, Garcia-Vilanova A, Ault R, Moliva JI, Yusoof KA, Mukherjee N, Curiel TJ, Dixon H, Torrelles JB, Svatek RS (2022) Selective delipidation of Mycobacterium bovis BCG retains antitumor efficacy against non-muscle invasive bladder cancer. Cancer Immunol Immunother 72(1):125–136. https://doi.org/10.1007/s00262-022-03236-y

Kawai T, Akira S (2009) The roles of TLRs, RLRs and NLRs in pathogen recognition. Int Immunol 21(4):317–337

Kis EE, Winter G, Myschik J (2012) Devices for intradermal vaccination. Vaccine 30(3):523–538

Kraynyak KA, Blackwood E, Agnes J, Tebas P, Giffear M et al (2022) SARS-CoV-2 DNA vaccine INO-4800 induces durable immune responses capable of being boosted in a phase 1 open-label trial. J Infect Dis 225:1923–1932

Krienke C, Kolb L, Diken E, Streuber M, Kirchhoff S, Bukur T, Akilli-Öztürk Ö, Kranz LM, Berger H, Petschenka J, Diken M, Kreiter S, Yogev N, Waisman A, Karikó K, Türeci Ö, Sahin U (2021) A non-inflammatory mRNA vaccine for treatment of experimental autoimmune encephalomyelitis. Science 371:145–153

Kubler H, Scheel B, Gnad-Vogt U, Miller K, Schultze-Seemann W, Vom Dorp F, Parmiani G, Hampel C, Wedel S, Trojan L, Jocham D, Maurer T, Rippin G, Fotin-Mleczek M, von der Mulbe F, Probst J, Hoerr I, Kallen KJ, Lander T, Stenzl A (2015) Self-adjuvanted mRNA vaccination in advanced prostate cancer patients: a first-in-man phase I/IIa study. J Immunother Cancer 3:26

Levine MM, Chen WH, Kaper JB, Lock M, Danzig L, Gurwith M (2017) PaxVax CVD 103-HgR single-dose live oral cholera vaccine. Expert Rev Vaccines 16(3):197–213

Lighaam LC, Rispens T (2016) The immunobiology of immunoglobulin G4. Semin Liver Dis 3:200–215

Mallet E, Belohradsky BH, Lagos R, Gothefors L, Camier P, Carriere JP, Kanra G, Hoffenbach A, Langue J, Undreiner F, Roussel F, Reinert P, Flodmark CE, Stojanov S, Liese J, Levine MM, Munoz A, Schodel F, Hessel L, Hexavalent Vaccine Trial Study G (2004) A liquid hexavalent combined vaccine against diphtheria, tetanus, pertussis, poliomyelitis, Haemophilus influenzae type B and hepatitis B: review of immunogenicity and safety. Vaccine 22(11–12):1343–1357

Melief CJ, van der Burg SH (2008) Immunotherapy of established (pre)malignant disease by synthetic long peptide vaccines. Nat Rev Cancer 8(5):351–360

Melief CJ, van Hall T, Arens R, Ossendorp F, van der Burg SH (2015) Therapeutic cancer vaccines. J Clin Invest 125(9):3401–3412

Mitragotri S (2005) Immunization without needles. Nat Rev Immunol 5(12):905–916

Netea MG, Domínguez-Andrés J, Barreiro LB, Chavakis T, Divangahi M, Fuchs E, Joosten LAB, van der Meer JWM, Mhlanga MM, Mulder WJM, Riksen NP, Schlitzer A, Schultze JL, Stabell Benn C, Sun JC, Xavier RJ, Latz E (2020) Defining trained immunity and its role in health and disease. Nat Rev Immunol 20:375–388

Northrup L, Christopher MA, Sullivan BP, Berkland C (2016) Combining antigen and immunomodulators: emerging trends in antigen-specific immunotherapy for auto-immunity. Adv Drug Deliv Rev 98:86–98

O'Hagan DT (1990) Intestinal translocation of particulates — implications for drug and antigen delivery. Adv Drug Deliv Rev 5(3):265–285

Oomen CJ, Hoogerhout P, Kuipers B, Vidarsson G, van Alphen L, Gros P (2005) Crystal structure of an anti- meningococcal subtype P1.4 PorA antibody pro- vides basis for peptide-vaccine design. J Mol Biol 351(5):1070–1080

Ott PA, Hu Z, Keskin DB, Shukla SA, Sun J, Bozym DJ, Zhang W, Luoma A, Giobbie-Hurder A, Peter L, Chen C, Olive O, Carter TA, Li S, Lieb DJ, Eisenhaure T, Gjini E, Stevens J, Lane WJ, Javeri I, Nellaiappan K, Salazar AM, Daley H, Seaman M, Buchbinder EI, Yoon CH, Harden M, Lennon N, Gabriel S, Rodig SJ, Barouch DH, Aster JC, Getz G, Wucherpfennig K, Neuberg D, Ritz J, Lander ES, Fritsch EF, Hacohen N, Wu CJ (2017) An immunogenic personal neoantigen vaccine for patients with melanoma. Nature 547(7662):217–221

Paradiso PR, Hogerman DA, Madore DV, Keyserling H, King J, Reisinger KS, Blatter MM, Rothstein E, Bernstein HH, Hackell J (1993) Safety and immunogenicity of a combined diphtheria, tetanus, pertussis and Haemophilus influenzae type b vaccine in young infants. Pediatrics 92(6):827–832

Payne RO, Silk SE, Elias SC, Miura K, Diouf A, Galaway F, de Graaf H, Brendish NJ, Poulton ID, Griffiths OJ, Edwards NJ, Jin J, Labbe GM, Alanine DG, Siani L, Di Marco S, Roberts R, Green N, Berrie E, Ishizuka AS, Nielsen CM, Bardelli M, Partey FD, Ofori MF, Barfod L, Wambua J, Murungi LM, Osier FH, Biswas S, McCarthy JS, Minassian AM, Ashfield R, Viebig NK, Nugent FL, Douglas AD, Vekemans J, Wright GJ, Faust SN, Hill AV, Long CA, Lawrie AM, Draper SJ (2017) Human vaccination against RH5 induces neutralizing antimalarial antibodies that inhibit RH5 invasion complex interactions. JCI Insight 2(21):96381

Pharmajet.com/regulatory-clearances (2022). Accessed 29 Aug 2022

Pinschewer DD (2017) Virally vectored vaccine delivery: medical needs, mechanisms, advantages and challenges. Swiss Med Wkly 147:w14465

Pollock KMCHM, Szubert AJ, Libri V, Boffito M, Owen D, Bern H, McFarlane LR, O'Hara J, Lemm N, McKay P, Rampling T, YTM Y, Milinkovic A, Kingsley C, Cole T, Fagerbrink S, Aban M, Tanaka M, Mehdipour S, Robbins A, Budd W, Faust S, Hassanin H, Cosgrove

CA, Winston A, Fidler S, Dunn D, McCormack S, Shattocka RJ, on behalf of the COVAC1 study Group (2022) Safety and immunogenicity of a self-amplifying RNA vaccine against COVID-19: COVAC1, a phase I, dose-ranging trial. EClinicalMedicine 44:101262

Pulendran B, Ahmed R (2011) Immunological mechanisms of vaccination. Nat Immunol 12(6):509–517

Raz R, Koren R, Bass D (2001) Safety and immunogenicity of a new mammalian cell-derived recombinant hepatitis B vaccine containing pre-S1 and pre-S2 antigens in adults. Isr Med Assoc J 3(5):328–332

Roesch A, Zolls S, Stadler D, Helbig C, Wuchner K, Kersten G, Hawe A, Jiskoot W, Menzen T (2022) Particles in biopharmaceutical formulations, part 2: an update on analytical techniques and applications for therapeutic proteins, viruses, vaccines and cells. J Pharm Sci 111:933–950

Rouphael NG, Lai L, Tandon S, McCullough MP, Kong Y, Kabbani S, Natrajan MS, Xu Y, Zhu Y, Wang D, O'Shea J, Sherman A, Yu T, Henry S, McAllister D, Stadlbauer D, Khurana S, Golding H, Krammer F, Mulligan MJ, Prausnitz MR (2021) Immunologic mechanisms of seasonal influenza vaccination administered by microneedle patch from a randomized phase I trial. NPJ Vaccines 6:89

RTS,S Clinical Trials Partnership (2015) Efficacy and safety of RTS,S/AS01 malaria vaccine with or without a booster dose in infants and children in Africa: final results of a phase 3, individually randomised, controlled trial. Lancet 386(9988):31–45

Sahin U, Oehm P, Derhovanessian E, Jabulowsky RA, Vormehr M, Gold M, Maurus D, Schwarck-Kokarakis D, Kuhn AN, Omokoko T, Kranz TM, Diken M, Kreiter S, Haas H, Attig S, Rae R, Cuk K, Kemmer-Brück A, Breitkreuz A, Tolliver C, Caspar J, Quinkhardt J, Hebich L, Stein M, Hohberger A, Vogler I, Liebig I, Renken S, Sikorski J, Leierer M, Müller V, Mitzel-Rink H, Miederer M, Huber C, Grabbe S, Utikal J, Pinter A, Kaufmann R, Hassel JC, Loquai C, Türeci Ö (2020) An RNA vaccine drives immunity in checkpoint-inhibitor-treated melanoma. Nature 585:107–112

Samuels S, Marijne Heeren A, Zijlmans H, Welters MJP, van den Berg JH, Philips D, Kvistborg P, Ehsan I, Scholl SME, Nuijen B, Schumacher TNM, van Beurden M, Jordanova ES, Haanen J, van der Burg SH, Kenter GG (2017) HPV16 E7 DNA tattooing: safety, immunogenicity, and clinical response in patients with HPV-positive vulvar intraepithelial neoplasia. Cancer Immunol Immunother 66(9):1163–1173

Schenkel JM, Fraser KA, Beura LK, Pauken KE, Vezys V, Masopust D (2014) T cell memory. Resident memory CD8 T cells trigger protective innate and adaptive immune responses. Science 346(6205):98–101

Sebastian M, Papachristofilou A, Weiss C, Fruh M, Cathomas R, Hilbe W, Wehler T, Rippin G, Koch SD, Scheel B, Fotin-Mleczek M, Heidenreich R, Kallen KJ, Gnad-Vogt U, Zippelius A (2014) Phase Ib study evaluating a self-adjuvanted mRNA cancer vaccine (RNActive(R)) combined with local radiation as con- solidation and maintenance treatment for patients with stage IV non-small cell lung cancer. BMC Cancer 14:748

Shamji MH, Durham SR (2017) Mechanisms of allergen immunotherapy for inhaled allergens and predictive biomarkers. J Allergy Clin Immunol 140(6):1485–1498

Timmerman P, Puijk WC, Meloen RH (2007) Functional reconstruction and synthetic mimicry of a conformational epitope using CLIPS technology. J Mol Recognit 20(5):283–299

van der Burg SH, Arens R, Ossendorp F, van Hall T, Melief CJ (2016) Vaccines for established cancer: overcoming the challenges posed by immune evasion. Nat Rev Cancer 16(4):219–233

van der Maaden K, Jiskoot W, Bouwstra J (2012) Microneedle technologies for (trans)dermal drug and vaccine delivery. J Control Release 161(2):645–655

van Poelgeest MI, Welters MJ, van Esch EM, Stynenbosch LF, Kerpershoek G, van Persijn van Meerten EL, van den Hende M, Lowik MJ, Berends-van der Meer DM, Fathers LM, Valentijn AR, Oostendorp J, Fleuren GJ, Melief CJ, Kenter GG, van der Burg SH (2013) HPV16 synthetic long peptide (HPV16-SLP) vaccination therapy of patients with advanced or recurrent HPV16-induced gynecological carcinoma, a phase II trial. J Transl Med 11:88

Vanlandschoot P, Roobrouck A, Van Houtte F, Leroux-Roels G (2002) Recombinant HBsAg, an apoptotic-like lipo-protein, interferes with the LPS-induced activation of ERK-1/2 and JNK-1/2 in monocytes. Biochem Biophys Res Commun 297(3):486–491

Vela Ramirez JE, Sharpe LA, Peppas NA (2017) Current state and challenges in developing oral vaccines. Adv Drug Deliv Rev 114:116–131

Verbeke R, Lentacker I, De Smedt S, Dewitte H (2021) The dawn of mRNA vaccines: the COVID-19 case. J Control Release 333:511–520

Vreden SG, Verhave JP, Oettinger T, Sauerwein RW, Meuwissen JH (1991) Phase I clinical trial of a recombinant malaria vaccine consisting of the circumsporozoite repeat region of plasmodium falciparum coupled to hepatitis B surface antigen. Am J Trop Med Hyg 45(5):533–538

Waltz E (2022) China and India approve nasal Civid vaccines. Nature 609:450

Wong G, Mendoza EJ, Plummer FA, Gao GF, Kobinger GP, Qiu X (2018) From bench to almost bedside: the long road to a licensed Ebola virus vaccine. Expert Opin Biol Ther 18(2):159–173

Further Reading

Abbas AK, Lichtman AH, Pillai S (2019) Basic immunology: functions and disorders of the immune system, 6th edn. Elsevier/Saunders, Philadelphia

Delves PJ, Martin SJ, Burton DR, Roitt IM (2017) Roitt's essential immunology, 13th edn. Wiley, Hoboken

Loffler P (2021) Review: Vaccine myth-buster — Cleaning up with prejudices and dangerous misinformation. Front Immunol 12:663280. https://doi.org/10.3389/fimmu.2021.663280

Orenstein WA, Offit PA, Edwards KM, Plotkin SA (2023) Plotkin's vaccines, 8th edn. Elsevier, Cambridge

Pollard AJ, Bijker EM (2021) A guide to vaccinology: from basic principles to new developments. Nat Reviews 22:83–100

Pharmaceutical Biotechnology: The Protein Products of Biotechnology and Their Clinical Use— Endogenous Proteins and Their Variations

Insulin

16

Chad D. Paavola, Michael R. De Felippis, David P. Allen,
Ashish Garg, James L. Sabatowski, Rattan Juneja,
and D. Bruce Baldwin

Introduction

Insulin was discovered over 100 years ago by Banting and Best in 1921 (Bliss 2007a). Soon afterwards, manufacturing processes were developed to extract insulin from porcine and bovine pancreata. From 1921 to 1980, innovation focused on increasing the purity of the insulin and providing different formulations for enhanced glucose control by altering time action (Brange 1987a, b; Owens et al. 2022). Purification was improved by optimizing extraction and processing conditions along with implementing chromatographic processes [size exclusion, ion exchange, and reversed phase (Kroef et al. 1989)]. These improvements reduced the levels of both general protein impurities and undesired insulin-related proteins such as proinsulin and insulin polymers. Formulation development primarily focused on improving chemical stability by moving from acidic to neutral formulations and modifying the time-action profile through the use of various levels of zinc and protamine. The evolution of recombinant DNA technology led to the widespread availability of human insulin, which has eliminated issues with sourcing constraints while providing the patient with an exogenous source of native insulin. Combining the improved purification meth-

odologies and recombinant DNA (rDNA) technology, manufacturers of insulin are now able to provide the purest human insulin ever made available, >98% pure. Further advances in rDNA technology, coupled with a detailed understanding of the molecular properties of insulin and knowledge of its endogenous secretion profile, enabled the development of insulin analogs with improved pharmacology relative to previous human insulin products.

Chemical Description

Insulin, a 51-amino acid protein, is a hormone that is composed of two polypeptide chains that are connected by two inter-chain disulfide bonds (Fig. 16.1) (Baker et al. 1988). The A-chain is composed of 21 amino acid residues, and the B-chain is composed of 30 amino acid residues. The inter-chain disulfide linkages occur between A^7–B^7 and A^{20}–B^{19}, respectively. A third intra-chain disulfide bond is located in the A-chain, between residues A^6 and A^{11}. Endogenous insulin is synthesized in the β-cells of the pancreas as a single-chain proinsulin precursor. This precursor contains a leader sequence, the B-chain, a connecting peptide (C peptide), and an A-chain. The proinsulin precursor is converted to insulin by enzymatic cleavage. Current manufacturing strategies mimic this enzymatic approach for generating the two-chain molecule and most analogs.

Bovine and porcine insulin preparations were the first commercially available insulin therapies. However, difficulties obtaining sufficient supplies of bovine or porcine pancreata and concerns over transmissible spongiform encephalopathies associated with the use of animal-derived materials were major reasons for the eventual discontinuation of these products. All major manufacturers of insulin have moved to production based on recombinant DNA technology.

Recombinant DNA technology enabled the production of human insulin from microorganisms, such as bacteria or yeast. In most cases, a precursor proinsulin is produced by

C. D. Paavola
Lilly Research Laboratories, Biotechnology Discovery Research,
Eli Lilly and Company, Indianapolis, IN, USA

M. R. De Felippis · D. P. Allen · A. Garg
Lilly Research Laboratories, Bioproduct Research and
Development, Eli Lilly and Company, Indianapolis, IN, USA

J. L. Sabatowski
Lilly Global Quality, Indianapolis, IN, USA

R. Juneja
Diabetes Global Medical Affairs, Eli Lilly and Company,
Indianapolis, IN, USA

D. B. Baldwin (✉)
Lilly Research Laboratories, Technical Services and
Manufacturing Sciences, Eli Lilly and Company,
Indianapolis, IN, USA
e-mail: baldwin_bruce@lilly.com

© The Author(s), under exclusive license to Springer Nature Switzerland AG 2024
D. J. A. Crommelin et al. (eds.), *Pharmaceutical Biotechnology*, https://doi.org/10.1007/978-3-031-30023-3_16

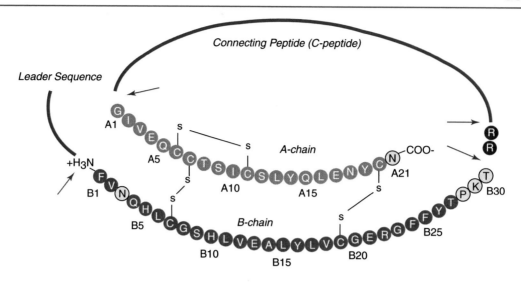

Fig. 16.1 Primary sequence of insulin. The A-chain is displayed in tan. The B-chain is displayed in teal. The yellow-highlighted amino acids represent sites of sequence alterations denoted in Table 16.1. The arrows indicate sites of enzymatic processing that remove the leader sequence and connecting peptide from the expressed proinsulin. The sequences of the leader and connecting peptide are not shown, with the exception of the two arginine residues (red) that are retained in an analog to alter their time action (e.g., insulin glargine)

Table 16.1 Amino acid substitutions in insulin analogs compared to human insulin

Species	A²¹	B³	B²⁸	B²⁹	B³⁰	B³¹	B³²
Human insulin (Humulin®, Novolin®, Afrezza®, Myxredlin®)	Asn	Asn	Pro	Lys	Thr	–	–
Insulin lispro (Humalog®, Liprolog®, Admelog®)	Asn	Asn	Lys	Pro	Thr	–	–
Insulin aspart (NovoRapid®, NovoLog®)	Asn	Asn	Asp	Lys	Thr	–	–
Insulin glulisine (Apidra®)	Asn	Lys	Pro	Glu	Thr	–	–
Insulin glargine (Lantus®, Basaglar®, Toujeo®, Semglee®, Rezvoglar®)	Gly	Asn	Pro	Lys	Thr	Arg	Arg
Insulin detemir (Levemir®)	Asn	Asn	Pro	Lys-(N-tetradecanoyl)	–	–	–
Insulin degludec (Tresiba®)	Asn	Asn	Pro	Lys-(Nᵋ-hexadecandioyl-γ-Glu)	–	–	–

the introduction of an exogenous gene into the microorganism. This proinsulin contains a leader sequence on the amino terminus of the B-chain and a connecting peptide between the A- and B-chain. The connecting peptide enables proper folding of the proinsulin in the cells or during in vitro processing. The leader sequence and connecting peptide can vary in their amino acid composition and size, dependent on the expression system being used. In most cases, the leader sequence and connecting peptide are removed by enzymatic cleavage during the purification process. After processing and final purification, the two-chain insulin is delivered with sufficient purity.

The establishment of these biotechnological processes for generating insulin, coupled with an understanding of its chemical and physical properties, enabled the engineering of new analogs of insulin. These new analogs (listed with human insulin in Table 16.1) contain various mutations and/or chemical modifications of the human insulin protein that

enable differing in vivo time action profiles and stability within their commercial drug product formulations.

Some insulin analogs are now available from more than one manufacturer. Products such as Basaglar®, Semglee®, Rezvoglar®, and Admelog® in Table 16.1 came to market in the United States by different regulatory pathways. Since March 2020, in the United States, insulin products have been regulated as biologics. As such, new manufacturers of an already approved insulin analog are eligible to seek approval of their product as a biosimilar or interchangeable-biosimilar to the originator product (White and Goldman 2019). In general, worldwide regulatory agencies require the use of comparative analytical assessments as well as preclinical and clinical evaluations to demonstrate that the proposed product is highly similar to the originator product (Heinemann et al. 2015). Unbranded biological products are also a feature of the insulin market in the United States due to complexities of reimbursement.

The net charge on human insulin is produced from the ionization potential of four glutamic acid residues, four tyrosine residues, two histidine residues, a lysine residue, and an arginine residue, in conjunction with two α-carboxyl and two α-amino groups. Human insulin has an isoelectric point (pI) of 5.3 in the denatured state; thus, the insulin molecule is negatively charged at neutral pH (Kaarsholm et al. 1990). This net negative charge state of insulin has been used in formulation development, as will be discussed later.

In addition to the net charge on insulin, another important intrinsic property of the molecule is its ability to readily associate into dimers and higher-order associated states (Figs. 16.2 and 16.3) (Pekar and Frank 1972). The driving force for dimerization appears to be the formation of favorable hydrophobic interactions at the carboxy-terminus of the B-chain (Ciszak et al. 1995). Insulin can associate into discrete hexameric complexes in the presence of various divalent metal ions, such as zinc at 0.33 equivalents per monomer (Goldman and Carpenter 1974), where each zinc ion (a total of two) is coordinated by a His^{B10} residue from three adjacent monomers. Physiologically, insulin is stored as a zinc-containing hexamer in the β-cells of the pancreas. As will be discussed later, the ability to form discrete hexamers in the presence of zinc has been used to develop therapeutically useful formulations of insulin.

Most commercial insulin formulations contain phenolic excipients (e.g., phenol and m-cresol) as antimicrobial agents. As represented in Figs. 16.2 and 16.3d, these phenolic species also bind to specific sites on insulin hexamers, causing a conformational change that increases the chemical stability of insulin in commercial preparations (Brange and Langkjaer 1992). X-ray crystallographic studies have identified the location of six phenolic ligand binding sites on the insulin hexamer and the nature of the conformational change induced by the binding of these ligands (Derewenda et al. 1989). The phenolic ligands occupy a binding pocket between monomers of adjacent dimers by hydrogen bonds with the carbonyl oxygen of Cys^{A6} and the amide proton of Cys^{A11} as well as numerous van der Waals contacts. The binding of these ligands stabilizes a conformational change that occurs at the amino-terminus of the B-chain in each insulin monomer, shifting the conformational equilibrium of residues B1 to B8 from an extended structure (T-state) to an α-helical structure (R-state). This conformational change is referred to as the T↔R transition (Brader and Dunn 1991) and is illustrated in Fig. 16.3c, d.

In addition to the presence of zinc and phenolic preservatives, modern insulin formulations may contain a tonicity agent (e.g., glycerol or NaCl), a buffer (e.g., sodium phosphate or Tris listed as trometamol or tromethamine), and surfactants (e.g., polysorbate 20 or polysorbate 80). The tonicity agent is used to minimize subcutaneous tissue damage and pain upon injection. The buffer is present to minimize pH drift in some pH-sensitive formulations. The surfactant is present to increase the physical stability of the formulation. Afrezza® is a dry powder inhalation product, so it differs from the solution formulations described above. This pulmonary insulin formulation contains fumaryl diketopiperazine to create technospheres® on which human insulin is coated (Pfützner et al. 2004).

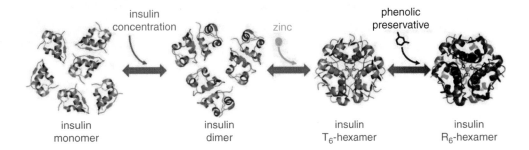

Fig. 16.2 Schematic representation of insulin association in the presence and absence of zinc and phenolic preservatives, e.g., phenol or m-cresol

insulin concentration

zinc

phenolic preservative

insulin monomer

insulin dimer

insulin T$_6$-hexamer

insulin R$_6$-hexamer

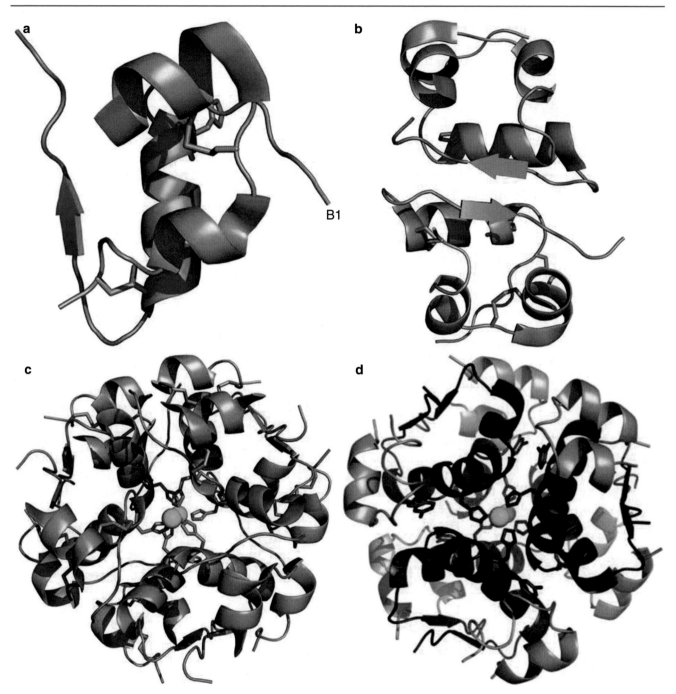

Fig. 16.3 (**a**) A cartoon representation of the secondary and tertiary structures of an insulin monomer, with the B1–B8 region in an extended conformation (T-state). The A-chain is colored tan and the B-chain is colored teal. The α-helices are depicted as coils and the β-sheet formed during dimerization is depicted as arrow. (**b**) A cartoon representation of the secondary and tertiary structures of an insulin dimer, with the B1–B8 regions in an extended conformation (T-state). The A-chains are colored tan, and the B-chains are colored teal. (**c**) A cartoon representation of the secondary and tertiary structures of a zinc-induced T_6-state insulin hexamer, with the B1–B8 regions in an extended conformation (T-state). The A-chains are colored tan, the B-chains are colored teal, and zinc is colored green. (**d**) A cartoon representation of the secondary and tertiary structures of zinc- and phenolic preservative-induced R_6-state insulin hexamer, wherein the B1–B8 regions are locked in the α-helical conformation (R-state). The A-chains are colored tan, the B-chains are colored red, zinc is colored green, and preservative is colored blue

Pharmacology and Formulations

Normal insulin secretion in a person without diabetes falls into two categories: (1) insulin that is secreted in response to a meal and (2) the background or basal insulin that is continually secreted between meals and during the nighttime hours (Fig. 16.4). The pancreatic response to a meal typically results in peak serum insulin levels of 60–80 µU/mL (2.08–2.77 ng/mL or 360–480 pM), whereas basal serum insulin levels fall within the 5–15 µU/mL (0.17–0.52 ng/mL or 30–90 pM) range (Galloway and Chance 1994). Because of these vastly different insulin demands, considerable effort has been expended to develop insulin formulations with pharmacokinetic (PK) and pharmacodynamic (PD) properties suitable for mealtime and basal requirements. Development has continued on insulin analogs and insulin analog formulations for further improved PK and PD properties, including both ultra-rapid and ultra-long profiles.

Regular and Rapid-Acting Soluble Preparations

Initial soluble insulin formulations were prepared under acidic conditions and were chemically unstable. In these early formulations, considerable deamidation was identified at Asn[A21], significant potency loss was observed during prolonged storage under acidic conditions, and time action was impaired due to the pH transition across the pI of the insulin molecule, which resulted in protein precipitation at the site of injection. Efforts to improve the chemical stability of these soluble formulations led to the development of neutral, zinc-stabilized solutions.

The insulin in these neutral pH, regular formulations is chemically stabilized by the addition of zinc (~0.4% (w/w) relative to the insulin concentration) and phenolic preservatives. As mentioned above, the addition of zinc leads to the formation of discrete T_6 hexameric structures (containing 2 Zn^{+2} atoms per hexamer) that can bind six molecules of phenolic preservatives, e.g., m-cresol to form R_6 hexamers (Figs. 16.2 and 16.3c). This in turn decreases the availability of residues involved in deamidation and high-molecular-weight polymer formation (Brange et al. 1992a, b).

The pharmacodynamic profile of these soluble formulations (marketed in the US under the designation type R) is listed in Table 16.2. The neutral pH, regular formulations show peak insulin activity between 2 and 3 h with a maximum duration of 5 to 8 h. As with other formulations, the variations in time action can be attributed to factors such as dose, site of injection, temperature, and the patient's physical activity. Despite the soluble state of insulin in these formulations, a delay in activity is still observed. This delay has been attributed to the time required for the hexamer to dissociate into the dimeric and/or monomeric constituents prior to absorption from the interstitium. This dissociation requires time-dependent diffusion of the preservative, zinc, and insulin from the site of injection, effectively diluting the protein and shifting the equilibrium from hexamers to dimers and monomers (Fig. 16.5a) (Brange et al. 1990). Studies explor-

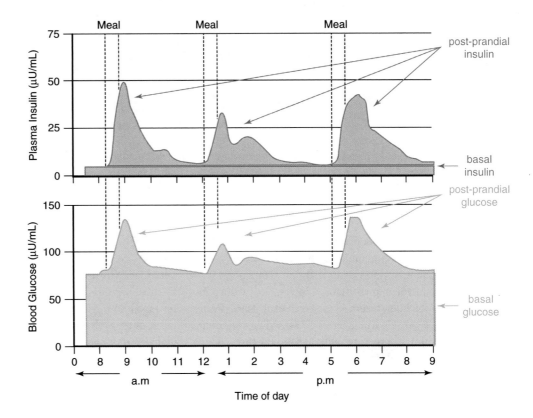

Fig. 16.4 A schematic representation of glucose and insulin profiles during the day in nondiabetic individuals Adapted and reprinted from (Schade et al. 1983)

Table 16.2 A list of human insulin formulations

Type[a]	Description	Appearance	Components	Action (h)[b]		
				Onset	Peak	Duration
R[c]	Regular soluble insulin human injection	Clear solution	Metal ion: zinc (10–40 µg/mL) Buffer: none Preservative: m-cresol (2.5 mg/mL) Tonicity agent: glycerin (16 mg/mL) pH: 7.25–7.6 Concentration: U100	0.5–1	2–4	5–8
Pulmonary	Insulin human inhalation powder	Dry powder	Metal ion: none Buffer: none Preservative: none Tonicity agent: none pH: N/A Carrier particles: fumaryl diketopiperazine and polysorbate 80 Dosages: 4 U and 8 U	0.1–0.2	0.5–0.9	2.5–3.0
N	NPH insulin human isophane suspension	Turbid or cloudy suspension	Metal ion: zinc (21–40 µg/mL) Buffer: dibasic sodium phosphate (3.78 mg/mL) Preservatives: m-cresol (1.6 mg/mL), phenol (0.73 mg/mL) Tonicity agent: glycerin (16 mg/mL) Modifying protein: protamine (~0.35 mg/mL) pH: 7.0–7.5 Concentration: U100	1–2	2–8	14–24
70/30	70% insulin human isophane suspension, 30% regular insulin human injection	Turbid or cloudy suspension	Metal ion: zinc (21–35 µg/mL) Buffer: dibasic sodium phosphate (3.78 mg/mL) Preservatives: m-cresol (1.6 mg/mL), phenol (0.73 mg/mL) Tonicity agent: glycerin (16 mg/mL) Modifying protein: protamine (~0.241 mg/mL) pH: 7.0–7.8 Concentration: U100	0.5	2–4	14–24
R[c]	Regular soluble insulin human injection	Clear solution	Metal ion: zinc (~85 µg/mL) Buffer: none Preservative: m-cresol (2.5 mg/mL) Tonicity agent: glycerin (16 mg/mL) pH: 7.0–7.8 Concentration: U500	0.5–0.75	3.5–8.5	6–11.5
IV	Regular soluble insulin human in sodium chloride injection	Clear solution	Metal ion: none Buffer: dibasic sodium phosphate (1.12 mg/mL) Tonicity agent: sodium chloride (9 mg/mL) pH: 6.5–7.2 Concentration: U1	0.35	N/A[d]	1.5

[a]US designation

[b]The time-action profiles of Lilly insulins are the average onset, peak action, and duration of action that are taken from a composite of studies. The onset, peak, and duration of insulin action depend on numerous factors, such as dose, injection site, presence of insulin antibodies, and physical activity. The action times listed represent the generally accepted values in the medical community

[c]Another notable designation is S (Britain). Other soluble formulations have been designed for pump use and include Velosulin® and HOE 21PH®

[d]Intravenous infusions do not generally exhibit a pronounced peak in activity

ing the relationship of molecular weight and cumulative dose recovery of various compounds in the popliteal lymph following subcutaneous injection suggest that lymphatic transport may account for approximately 20% of the absorption of insulin from the interstitium (Supersaxo et al. 1990; Porter and Charman 2000; Charman et al. 2001). The remaining balance of insulin is predominantly absorbed through the microvascular endothelium.

Insulin analogs with monomerizing mutations were designed to weaken the dimerization interface and achieve a faster response to prandial glucose increases. The pharmacological properties of these formulations are listed in Table 16.3 (insulin prandial analog formulations), Table 16.4 (insulin mixture analog formulations), and Table 16.5 (insulin basal analog formulations). The development of monomeric analogs of insulin for the treatment of insulin-dependent diabetes mellitus has focused on shifting the self-association properties of insulin to favor the monomeric species and consequently minimizing the delay in time action (Brange et al. 1988, 1990; Brems et al. 1992). One such monomeric ana-

Fig. 16.5 A schematic representation of regular or rapid-acting insulin dissociation and absorption after subcutaneous administration; (**a**) regular human insulin, (**b**) insulin lispro or insulin aspart, and (**c**) insulin glulisine

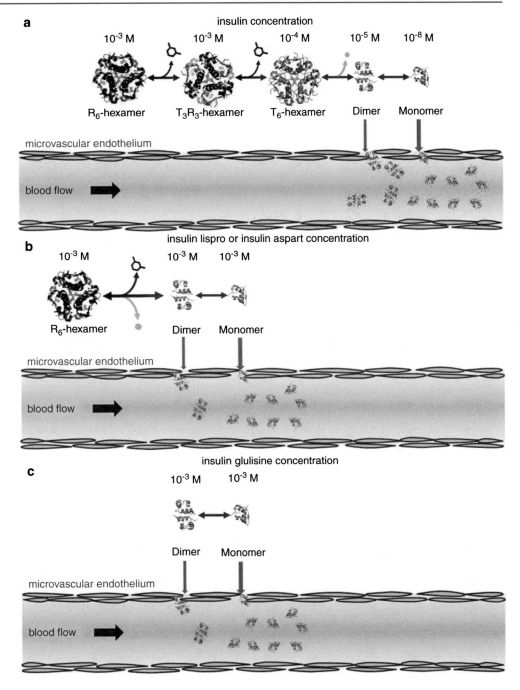

log, Lys^{B28}Pro^{B29}-human insulin (insulin lispro; CAS Number 133107-64-9; Humalog®, Liprolog®, Admelog®), has demonstrated a more rapid time-action profile, with a peak activity of approximately 1 h (Howey et al. 1994). The sequence inversion of amino acids at positions B28 and B29 yields an analog with reduced self-association behavior compared to human insulin (Fig. 16.1; Table 16.1); yet, insulin lispro can be stabilized in a preservative- and zinc-dependent hexameric complex that provides the necessary chemical and physical stability required for insulin preparations. Despite the hexameric complexation of this analog, insulin lispro retains its rapid time action. Based on the crystal structure of

the insulin lispro hexameric complex (Ciszak et al. 1995) and the self-association behavior in solution (Bakaysa et al. 1996), it is hypothesized that the reduced dimerization properties of the analog, coupled with the zinc- and preservative-dependent hexamerization requirements, yield a hexameric complex that readily dissociates into monomers after rapid diffusion of the phenolic preservative into the subcutaneous tissue at the site of injection (Fig. 16.5b). Consequently, the time-dependent diffusion, dilution, and dissociation of the zinc hexamers is not necessary for the analog to generate monomers/dimers, which are required for absorption across the microvascular endothelium.

Table 16.3 A list of insulin prandial (fastacting) analog formulations

Type[a]	Description	Appearance	Components	Action (h)[b]		
				Onset	Peak	Duration
Humalog®, Admelog®, Liprolog®	Rapid-acting soluble insulin lispro for injection	Aqueous, clear, and colorless solution	Metal ion: zinc (19.7 μg/mL U100; 46 μg/mL U200) Buffer: dibasic sodium phosphate (1.88 mg/mL U100) tromethamine (5 mg/mL U200) Preservatives: m-cresol (3.15 mg/mL), phenol (trace) Tonicity agent: glycerin (16 mg/mL) pH: 7.0–7.8 Concentration: Humalog® U100 and U200: Admelog® U100	0.25–0.5	~2.5[c]	3–4[c]
NovoLog®	Rapid-acting soluble insulin aspart for injection	Aqueous, clear, and colorless solution	Metal ion: zinc (19.6 μg/mL) Buffer: disodium hydrogen phosphate dihydrate (1.25 mg/mL) Preservative: m-cresol (1.72 mg/mL), phenol (1.50 mg/mL) Tonicity agents: glycerin (16 mg/mL), sodium chloride (0.58 mg/mL) pH: 7.2–7.6 Concentration: U100	0.25[c]	1–3[c]	3–5[c]
Fiasp®	Rapid-acting soluble insulin aspart for injection	Aqueous, clear, and colorless solution	Metal ion: zinc Buffer: disodium hydrogen phosphate dihydrate Preservative: m-cresol, phenol Tonicity agents: glycerol Other: arginine hydrochloride, niacinamide (vitamin B3), pH: 7.1 Concentration: U100	0.26–0.33[c]	1.5–2.2[c]	5–7[c]
Apidra®	Rapid-acting soluble insulin glulisine for injection	Aqueous, clear, and colorless solution	Metal ion: none Buffer: tromethamine (6 mg/mL) Preservative: m-cresol (3.15 mg/mL) Tonicity agent: sodium chloride (5 mg/mL) Stabilizing agent: polysorbate 20, (0.01 mg/mL in vials) pH: ~7.3 Concentration: U100	0.2–0.5[c]	1.5[c]	~5.3[c]
Lyumjev®	Rapid-acting soluble insulin lispro for injection	Aqueous, clear, and colorless solution	Metal ion: zinc (39 μg/mL U100; 52 μg/mL U200) Buffer: sodium citrate Dihydrate (4.4 mg/mL) Preservatives: m-cresol (3.15 mg/mL), phenol (trace) Tonicity agent: glycerin (12 mg/mL) Other: Treprostinil (1 μg/mL in both U100 and U200), magnesium chloride hexahydrate (1 mg/mL in both U100 and U200) pH: 7.0–7.8 Concentration: U100 and U200	0.28[c]	2–3[c]	4–7[c]

[a]US designation
[b]The time-action profiles of Lilly insulins are the average onset, peak action, and duration of action taken from a composite of studies. The onset, peak, and duration of insulin action depend on numerous factors, such as dose, injection site, presence of insulin antibodies, and physical activity. The action times listed represent the generally accepted values in the medical community
[c]Micromedex [database online]. IBM Watson Health; 2022 Updated periodically

It is important to highlight that the properties engineered into insulin lispro not only provide the patient with a more convenient therapy but also improve control of postprandial hyperglycemia and reduce the frequency of non-severe hypoglycemic events (Anderson et al. 1997; Holleman et al. 1997).

Since the introduction of insulin lispro, two additional rapid-acting insulin analogs have been introduced to the market. The amino acid modifications made to the human insulin sequence to produce these analogs are depicted in Table 16.1. Like insulin lispro, both analogs are supplied as neutral pH solutions containing phenolic preservative. The

Table 16.4 A list of insulin mixture (fast-acting/long-acting) analog formulations

Type[a]	Description	Appearance	Components	Action (h)[b]		
				Onset	Peak	Duration
Humalog® Mix75/25™	75% insulin lispro protamine suspension and 25% insulin lispro for injection	Turbid or cloudy suspension	Metal ion: zinc (25 µg/mL) Buffer: dibasic sodium phosphate (3.78 mg/mL) Preservative: m-cresol (1.76 mg/mL), phenol (0.715 mg/mL) Tonicity agent: glycerin (16 mg/mL) Modifying protein: protamine sulfate (0.28 mg/mL) pH: 7.0–7.8 Concentration: U100	0.25–0.5	2–3	14–24[c]
Humalog® Mix50/50™	50% insulin lispro protamine suspension and 50% insulin lispro for injection	Turbid or cloudy suspension	Metal ion: zinc (30.5 µg/mL) Buffer: dibasic sodium phosphate (3.78 mg/mL) Preservative: m-cresol (2.20 mg/mL), phenol (0.89 mg/mL) Tonicity agent: glycerin (16 mg/mL) Modifying protein: protamine sulfate (0.19 mg/mL) pH: 7.0–7.8 Concentration: U100	0.25–0.5	2–3	14–20[d]
Novolog® mix 70/30	70% insulin aspart protamine suspension and 30% insulin aspart for injection	Turbid or cloudy suspension	Metal ion: zinc (19.6 µg/mL) Buffer: dibasic sodium phosphate (1.25 mg/mL) Preservatives: m-cresol (1.72 mg/mL), phenol (1.50 mg/mL) Tonicity agents: sodium chloride (0.58 mg/mL), mannitol (36.4 mg/mL) Modifying protein: protamine (0.33 mg/mL) pH: 7.2–7.44 Concentration: U100	0.17–0.33[d]	1.8–3.6[d]	13[d]
Ryzodeg® 70/30	70% long-acting soluble insulin degludec and 30% rapid-acting insulin aspart for injection	Aqueous, clear, and colorless solution	Metal ion: zinc (27.4 µg/mL) Buffer: none Preservatives: m-cresol (1.72 mg/mL), phenol (1.50 mg/mL) Tonicity agents: glycerol (19 mg/mL), sodium chloride (0.58 mg/mL) Modifying protein: none pH: 7.4		2.3	24 after attaining steady state after 3–4 days)[c]

[a]US designation

[b]The time-action profiles of Lilly insulins are the average onset, peak action, and duration of action taken from a composite of studies. The onset, peak, and duration of insulin action depend on numerous factors, such as dose, injection site, presence of insulin antibodies, and physical activity. The action times listed represent the generally accepted values in the medical community

[c] Prescribing Information insert

[d]Micromedex [database online]. IBM Watson Health; 2022 Updated periodically

design strategy for Asp[B28]-human insulin (insulin aspart, CAS – Chemical Abstract Service—Number 116094–23-6; NovoRapid® or NovoLog®) (Brange et al. 1988, 1990) involves the replacement of Pro[B28] with a negatively charged aspartic acid residue. Like Lys[B28]Pro[B29]-human insulin, Asp[B28]-human insulin has a more rapid time action following subcutaneous injection (Fig. 16.5b) (Heinemann et al. 1997). This rapid action is also achieved through a reduction in the self-association behavior compared to human insulin (Brange et al. 1990; Whittingham et al. 1998). The other rapid-acting analog, Lys[B3]-Glu[B29]-human insulin (insulin glulisine; CAS

Number 160337-95-1, Apidra®), involves a substitution of the lysine residue at position 29 of the B-chain with a negatively charged glutamic acid. Additionally, this analog replaces the Asn[B3] with a positively charged lysine. Scientific reports describing the impact of these changes on the molecular properties of this analog are lacking. However, the glutamic acid substitution occurs at a position known to be involved in dimer formation (Brange et al. 1990) and may result in the disruption of key interactions at the monomer-monomer interface. The Asn residue at position 3 of the B-chain plays no direct role in insulin self-association

Table 16.5 A list of insulin basal (long-acting) analog formulations

Type[a]	Description	Appearance	Components	Action (h)[b]		
				Onset	Peak	Duration
Lantus®, Basaglar®, Semglee®, Rezvoglar®	Long-acting soluble insulin glargine for injection	Aqueous, clear, and colorless solution	Metal ion: Zinc (30 µg/mL) Buffer: None Preservative: m-cresol (2.7 mg/mL) Tonicity agent: Glycerin (20 mg 8%/mL) Modifying protein: None Stabilizing agent: Polysorbate 20 (20 µg/mL – Vials only) pH: ~ 4 Concentration: U100		Constant release with no pronounced peak[c]	24[c]
Toujeo®	Long-acting soluble insulin glargine for injection	Aqueous, clear, and colorless solution	Metal ion: Zinc (90 µg/mL) Buffer: None Preservative: m-cresol (2.7 mg/mL) Tonicity agent: Glycerin (20 mg 85%/mL) Modifying protein: None pH: ~ 4 Concentration: U300	0–6[c]	Steady state after 5 days[c]	24[c]
Levemir®	Long-acting soluble insulin detemir for injection	Aqueous, clear, and colorless solution	Metal ion: Zinc (65.4 µg/mL) Buffer: Dibasic sodium phosphate (0.89 mg/mL) Preservatives: m-cresol (2.06 mg/mL), phenol (1.8 mg/mL) Tonicity agents: Sodium chloride (1.17 mg/mL) Modifying protein: None pH: 7.4 Concentration: U100		3–14[c]	5.7–23.2[c]
Tresiba®	Long-acting soluble insulin degludec for injection	Aqueous, clear, and colorless solution	Metal ion: Zinc (32.7 µg/mL U100; 71.9 µg/mL U200) Buffer: None Preservatives: m-cresol (1.72 mg/mL), phenol (1.50 mg/mL) Tonicity agents: None Modifying protein: None pH: 7.6 Concentration: U100 and U200		9[c]	24 at steady state[c]

[a]US designation

[b]The time-action profiles of Lilly insulins are the average onset, peak action, and duration of action taken from a composite of studies. The onset, peak, and duration of insulin action depend on numerous factors, such as dose, injection site, presence of insulin antibodies, and physical activity. The action times listed represent the generally accepted values in the medical community

[c]Micromedex [database online]. IBM Watson Health; 2022 Updated periodically

(Brange et al. 1990), but it is flanked by two amino acids involved in the assembly of the insulin hexamer. Despite the limited physicochemical information on insulin glulisine, studies conducted in persons with either type 1 (T1DM) or type 2 diabetes (T2DM) (Dailey et al. 2004; Dreyer et al. 2005) confirm that the analog displays similar pharmacological properties as insulin lispro. Interestingly, insulin glulisine is not formulated in the presence of zinc, as are the other rapid-acting analogs. Instead, insulin glulisine is, in vials, formulated in the presence of a stabilizing agent (polysorbate 20) (Table 16.3). The surfactant in the vial formulation presumably minimizes aggregation. Apidra®, which is a non-hexameric formulation of insulin glulisine, exhibits greater exposure at the same dose relative to Humalog® (insulin lispro) and Novolog® (insulin aspart) (Heise et al. 2007; Arnolds et al. 2010). Because of that difference, PK metrics that depend on exposure, like time to onset and area under the curve from 0–30 min, appear to suggest faster absorption, while parameters that do not depend on overall exposure, like time to maximum insulin concentration and time to early half maximum insulin concentration (T_{max} and $T_{early\ \frac{1}{2}\ max}$), are consistent with similar absorption rate across all three products. The lack of meaningful differences in T_{max} and $T_{early\ \frac{1}{2}\ max}$ between Apidra®, Humalog®, and Novolog® suggests breakdown of hexamers from the two latter formulations must be rapid relative to the rate-limiting step, subcutaneous absorption (Fig. 16.5c) (Home 2012). In any event, the significance of any differences in time action between these products has yet to be demonstrated clinically (Dreyer et al. 2005).

The aforementioned commercially available rapid-acting insulin analogs are approved for use in external infusion pumps. Buffer, and surfactant in the case of Apidra®, is included in these formulations to minimize the physical aggregation of insulin that can lead to clogging of the infusion sets. In early pump systems, gas-permeable infusion tubing was used with the external pumps. Consequently, a buffer was added to the formulation in order to minimize pH changes due to dissolved carbon dioxide. With continued improvements in external pumping systems, tubing composed of materials having greater resistance to carbon dioxide diffusion has been introduced and the potential for pH-induced precipitation of insulin is greatly reduced.

Ultra-Rapid Insulins

Products with increasingly fast time action represent the most recent development in prandial insulins. The first product in the ultra-rapid space is a reformulation of insulin aspart marketed as Fiasp® (insulin aspart). The formulation differs from NovoLog® primarily in the addition of nicotinamide and arginine. In a study with T1DM participants, the onset of action across a range of doses was reduced relative to NovoLog® by 5–6 min, and glucose infusion over the first 30 min was 1.5 to two-fold greater than for NovoLog®. No change in the duration of action was reported (Heise and Mathieu 2017). A second ultra-rapid insulin product is marketed as Lyumjev®(insulin lispro, also known as Ultra Rapid Lispro or URLi). This reformulation of insulin lispro incorporates treprostinil and citrate to increase blood flow through the injection site and to increase vascular permeability, respectively (Leohr et al. 2020). In a study with T1DM participants, the onset of action was 10.8 min faster than Humalog® and glucose infusion over the first 30 min was 2.8-fold greater. The duration of action was reported to be 43.6 min shorter than Humalog ® (Linnebjerg et al. 2020). Lyumjev®, Fiasp®, Humalog®, and NovoRapid® have been evaluated in a head-to-head clinical study. Lyumjev exhibited the fastest absorption and resulted in the smallest glucose excursion following a test meal (Heise et al. 2020).

Basal Insulin Suspension Preparations

The normal human pancreas secretes approximately 1 unit of insulin (0.035 mg) per hour to maintain basal glycemic control (Waldhäusl et al. 1979). Adequate basal insulin levels are a critical component of diabetes therapy because they regulate hepatic glucose output, which is essential for proper maintenance of glucose homeostasis during the diurnal cycling of the body. Consequently, basal insulin formulations must provide a very different PK profile than "mealtime" insulin formulations.

This detailed understanding of specific insulin requirements was unknown at the time the original NPH suspension product was introduced. This first commercially available basal insulin preparation is known as Neutral Protamine Hagedorn (NPH) and is named after its inventor H. C. Hagedorn (Hagedorn et al. 1936). This preparation is a neutral microcrystalline suspension that is prepared by the cocrystallization of insulin with protamine. Originally produced using animal-derived insulins, the currently marked products are exclusively manufactured using human insulin. The cocrystallizing agent, protamine, consists of a closely related group of very basic peptides that are isolated from fish sperm. Protamine is heterogeneous in composition; however, four primary components have been identified, and each shows a high degree of sequence homology (Hoffmann et al. 1990). In general, protamine is ~30 amino acids in length and has an amino acid composition that is primarily composed of arginine, 65–70%. Using crystallization conditions later identified by Krayenbuhl and Rosenberg (1946), oblong tetragonal NPH insulin crystals with volumes between 1 and 20 μm^3 can be consistently prepared from protamine and insulin (Deckert 1980). These formulations, by design, have very minimal levels of soluble insulin in solution. The condition at which no measurable protamine or insulin exists in solution after crystallization is referred to as the isophane point of insulin. Crystal dissolution is presumed to be the rate-limiting step in the absorption of NPH insulin. Consequently, the time action of the formulation is prolonged by further delaying the dissociation of the hexamer into dimers and monomers (Fig. 16.6a).

NPH has an onset of action from 1 to 2 h, peak activity from 6 to 12 h, and duration of activity from 18 to 24 h (Table 16.2). As with other formulations, the variations in time action are due to factors such as dose, site of injection, temperature, and the patient's physical activity. In T2DM patients, NPH can be used as either once-daily or twice-daily therapy; however, in T1DM patients, NPH is predominately used as a twice-daily therapy.

NPH can be readily mixed with regular insulin either extemporaneously by the patient or as obtained from the manufacturer in a premixed formulation (Table 16.2). Premixed insulin, e.g., Humulin® 70/30 or Humulin® 50/50 (NPH/regular), has been shown to provide the patient with

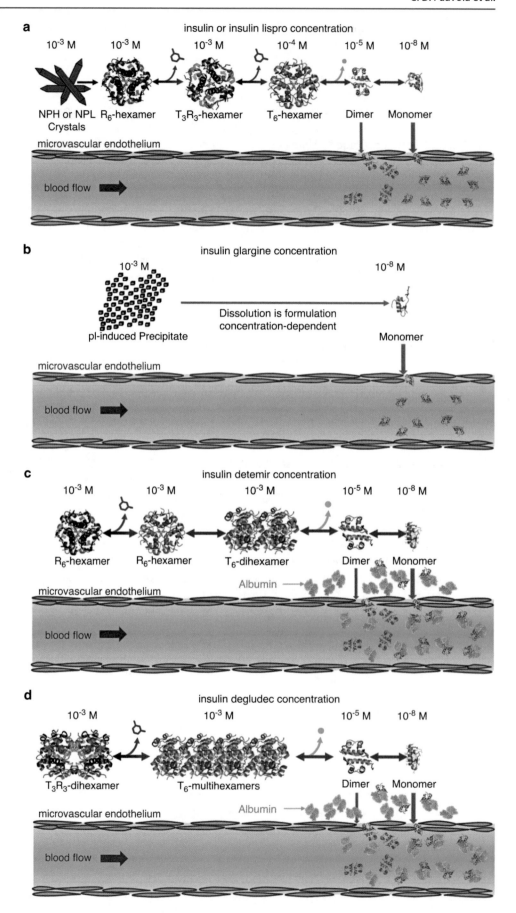

improved dose accuracy and consequently improved glycemic control (Bell et al. 1991). In these preparations, a portion of the soluble regular insulin will reversibly adsorb to the surface of the NPH crystals through an electrostatically mediated interaction under formulation conditions (Dodd et al. 1995); however, this adsorption is reversible under physiological conditions and consequently has no clinical significance (Hamaguchi et al. 1990; Davis et al. 1991). Due, in part, to the reversibility of the adsorption process, NPH/regular mixtures are uniquely stable and have a 3-year shelf life.

The rapid-acting insulin analog, insulin lispro, can be extemporaneously mixed with NPH; however, such mixtures must be injected immediately after the two insulins are combined due to the potential for exchange between the soluble and suspension components upon long-term storage. Exchange refers to the release of human insulin from the NPH crystals into the solution phase and concomitant loss of the analog into the crystalline phase. The presence of human insulin in solution could diminish the rapid time-action effect of the rapid-acting analog. One way to overcome the problem of exchange is to prepare mixtures containing the same insulin species in both the suspension and the solution phases, analogous to human insulin regular/NPH preparations. However, this approach requires an NPH-like preparation of the rapid-acting analog.

An NPH-like suspension of insulin lispro, referred to as neutral protamine lispro (NPL), and its physicochemical properties relative to human insulin NPH have been described (De Felippis et al. 1998). In order to prepare the appropriate crystalline form of the analog, significant modifications to the NPH crystallization procedure are required. The differences between the crystallization conditions have been proposed to result from the reduced self-association properties of insulin lispro.

Pharmacological studies reported for NPL (Janssen et al. 1997; De Felippis et al. 1998), indicate that the PK and PD properties of this analog suspension are analogous to human insulin NPH (Table 16.2). Clinical trials of insulin lispro protamine suspension alone in T2DM and in combination with insulin lispro in T1DM have been reported (Fogelfeld et al. 2010; Chacra et al. 2010; Strojek et al. 2010). In T2DM patients, the PK/PD profile of NPL can support a once-daily therapy regimen (Hompesch et al. 2009).

Homogeneous, biphasic, premixture preparations containing NPL and rapid-acting insulin lispro, Humalog® Mix 75/25 and Humalog® Mix 50/50, are marketed (Table 16.4). As with insulin lispro, premixed preparations containing insulin aspart combined with a protamine-containing microcrystalline suspension of insulin aspart (Balschmidt 1996) are commercially available (Novolog® 70/30). Clinical data on insulin lispro mixtures and those composed of insulin aspart have been reported in the literature (Weyer et al. 1997;

Heise et al. 1998). The pharmacological properties of the rapid-acting analogs are preserved in these stable mixtures (Table 16.3).

Immunogenicity issues with protamine have been documented in a small percentage of diabetic patients (Kurtz et al. 1983; Nell and Thomas 1988). Individuals who show sensitivity to the protamine in NPH formulations (or premixed formulations of insulins lispro and aspart) are routinely switched to other long-acting insulin formulations (e.g., Lantus®, Toujeo®, Levemir®, Tresiba®, Basaglar®, Semglee®, or Rezvoglar®) to control their basal glucose levels.

Basal Insulin Solution Preparations

While the NPH and NPH-like analog preparations are still widely used in diabetes treatment regimens, research and development efforts continuing into present time remain focused on strategies to design improved basal insulin products. Advancements in basal insulin therapy have concentrated on developing solution formulations that exhibit flatter and peakless PK profiles and a longer duration of action. The result of this work has produced three commercially available basal insulin analog products that address hypoglycemia risk and decrease the number of injections needed on a daily basis to achieve glycemic control.

Lantus® (insulin glargine) and Levemir® (insulin detemir) were the first of these new basal insulin analog preparations to be approved for diabetes treatment (Table 16.1; Fig. 16.1). Lantus® derives its protracted time-action profile from the slow and relatively constant dissolution of solid particles that form as a result of a pH shift of the acidic formulation to neutral pH in the subcutaneous tissue. This slow dissolution precedes the dissociation of insulin into absorbable units, and thus the rate of absorption (units per hour) into the bloodstream is significantly decreased in comparison to that of prandial or bolus (mealtime) formulations. Levemir®, on the other hand, achieves its protracted effect by a combination of structural interactions and physiological binding events (Havelund et al. 2004).

The active ingredient in Lantus®, Toujeo®, Basaglar®, Semglee®, and Rezvoglar® is the insulin analog, insulin glargine (Gly^{A21}, Arg^{B31}, Arg^{B32}-human insulin; CAS Number 160337-95-1), whose amino acid sequence modifications are highlighted in Table 16.1 and Fig. 16.1. This analog differs from human insulin in that the amino acid asparagine is replaced with glycine at position Asn^{A21} and two arginine residues have been added to the carboxy terminus of the B-chain. The additional arginine residues shift the isoelectric point from ~5.7 to ~6.9, thereby producing an insulin analog that is soluble at acidic pH values but is less soluble at the neutral pH environment within subcutaneous tissue. As a result of this property, Lantus® is a solution formulation of insulin glargine prepared at pH 4.0. The introduction of glycine at position Asn^{A21} yields a protein with acceptable chemical stability since the asparagine side chain of human insulin is susceptible to acid-mediated degradation (deamidation and increased high-molecular-weight proteins) and reduced potency. Thus, the changes to the molecular sequence of insulin have been made to improve chemical stability and to modulate absorption from the subcutaneous tissue, resulting in an analog that has approximately the same potency as human insulin. The Lantus® formulation is a clear solution that incorporates zinc and m-cresol (anti-microbial preservative). Consequently, Lantus® does not need to be resuspended prior to dosing like the protamine-containing insulin suspension products. Immediately following injection into the subcutaneous tissue, the insulin glargine precipitates due to the shift to neutral pH conditions. The slowly dissolving precipitate results in a relatively constant rate of absorption over 24 h with no pronounced peak (Fig. 16.6b; Table 16.5). This profile allows once-daily dosing and can satisfy basal insulin requirements of persons with diabetes. As with all insulin preparations, the time course of Lantus® may vary in different individuals or at different times in the same individual, and the rate of absorption is dependent on blood supply, temperature, and the patient's physical activity. Lantus® should not be diluted or mixed with any other solution or insulin, as will be discussed below.

Insulin glargine products from other manufacturers, Basaglar® (Abasaglar® in Europe), Semglee® and Rezvoglar®, are approved for use in patients with type 1 or 2 diabetes. These products have comparable physiochemical and pharmacological properties to the originator product, Lantus®.

Insulin detemir (LysB^{29}(N-tetradecanoyl)des(B30)human insulin; CAS Number 169148–63-4) is the active ingredient in Levemir®, and the analog strategy employs acylation of insulin with a fatty acid moiety as a means to achieve a protracted pharmacological effect. As shown in Table 16.1 and Fig. 16.1, the B30 threonine residue of human insulin is eliminated in insulin detemir, and a 14-carbon, myristoyl fatty acid is covalently attached to the ε-amino group of Lys^{B29}. The analog forms a zinc hexamer at neutral pH in a preserved solution. Clinical studies have reported that insulin detemir displays lower PK and PD variability than NPH and/or insulin glargine (Hermansen et al. 2001; Vague et al. 2003; Heise et al. 2004; Porcellati et al. 2011). An approximate description of the PD profile of Levemir® is listed in Table 16.5. This analog appears to display a slower onset of action than NPH without a pronounced peak (Heinemann et al. 1999). The binding of the tetradecanoyl-acylated insulin to albumin was originally proposed as the underlying mechanism behind the observed prolonged effect for insulin detemir analog; however, investigations on insulin detemir have determined that the mechanism is more complex (Havelund et al. 2004). It has been proposed that subcutaneous absorption is initially delayed as a result of hexamer sta-

bility and dihexamerization (Fig. 16.6c). Such interactions between hexamers are likely a consequence of the symmetrical arrangement of fatty acid moieties around the outside of the hexamers as shown by X-ray crystallographic studies (Whittingham et al. 2004). These associated forms further bind to albumin within the injection site depot, and further prolongation may result due to this binding.

Although Lantus® and Levemir® improved basal insulin therapy, these insulin formulations may not achieve full 24 h coverage in all patients. Moreover, the desire to eliminate or minimize nocturnal hypoglycemia has driven the exploration of improved basal insulin therapies. Two commercial products have emerged from research efforts in this area. Tresiba® builds upon the design strategy of Levemir® by incorporating an acylated insulin analog as its active ingredient. The second product, Toujeo®, represents a departure from a molecular engineering approach and utilizes changes in precipitation behavior at higher formulation concentrations of insulin glargine.

The active ingredient in Tresiba® is insulin degludec (Lys^{B29}(Nε-hexadecandioyl-γ-Glu) des(B30) human insulin; CAS Number 844439–96-9). As shown in Table 16.1, it differs from human insulin in that the amino acid threonine in position B30 has been omitted and the ε-amino group of Lys^{B29} is covalently modified with a γ-L-glutamic acid linker containing a bound C16 fatty acid with a terminal carboxylic acid. Tresiba® is a clear, colorless solution that contains either insulin degludec 100 units/mL (U100) or 200 units/mL (U200). In the presence of the phenolic preservatives and zinc, insulin degludec forms a soluble and stable dihexamer described as a closed configuration that results from an interaction between one of the fatty diacid side chains of one hexamer and the zinc atom of the adjacent hexamer. After injection, as the bound preservative molecules diffuse away, the dihexamers adopt an open configuration and interact with each other to form multihexamer units facilitated by the interaction of fatty diacid chains and zinc atoms between adjacent hexamers (Fig. 16.6d). Subsequent diffusion of zinc from each end of the chain results in terminal hexamers breaking apart into dimers, which then dissociate into monomers. This process enables a gradual release of monomers that are absorbed into the systemic circulation. The complex self-association and dissociation characteristics result in insulin degludec exhibiting the longest duration of action at the injection depot compared to all other basal insulins (Jonassen et al. 2012; Steensgaard et al. 2013).

Tresiba® U100 and U200 have shown to be bioequivalent and exhibit a mean half-life of 25 h and at least a 42-h duration of action at a steady state using once-daily administration. Tresiba® achieves a steady state in 3–5 days with serum insulin concentration increasing slowly until a very flat profile is achieved. This results in a reduced risk of nocturnal hypoglycemia versus Lantus® for both T1DM and T2DM patients. Because of its slow accumulation, long duration of action and flat, and predictable PD profile, Tresiba® can be dosed any time of the day versus Lantus® which has to be dosed at the same time everyday (Heise and Mathieu 2017).

Toujeo® is a higher concentration preparation of insulin glargine formulated at 300 U/mL and is thus three times more concentrated than the 100 U/mL Lantus® product. The amino acid composition of the active ingredient, insulin glargine, is the same in Toujeo® and Lantus® and maintains the same physical and chemical properties. However, the higher concentration allows a dose to be administered in a third of the injection volume of Lantus®. It is believed that this lower volume produces a subcutaneous precipitate with reduced surface area, resulting in more prolonged and constant release of insulin into the bloodstream (Fig. 16.6b) (Ritzel et al. 2015). PK and PD studies have shown that Toujeo® exhibits a flatter profile and more prolonged glucose-lowering activity over ≥24 h compared to Lantus®. Toujeo® achived steady state in 3–4 days and exhibits a half-life of 19 h versus Lantus®, which achieves steady state in 2–4 days with a half-life of 12 h. In addition, lower within day and day-to-day glycemic variability is also observed at clinically relevant basal insulin doses. It is important to note that Lantus® U100 and Toujeo® U300 are *not bioequivalent*, since the time actions are different and 12% more U300 is needed for a similar glycemic effect (Ritzel et al. 2015). This effect possibly results from increased enzymatic inactivation of the drug by tissue peptidases in the subcutaneous tissue due to longer residence time of Toujeo®. Toujeo® offers comparable glycemic control to Lantus® in both T1DM and T2DM but the main clinically relevant benefit of Toujeo® is that it provides lower nocturnal hypoglycemia risk (Lau et al. 2017).

The commercialization of soluble basal insulin analogs has been a tremendous advancement in the treatment of patients with type 2 diabetes. However, the once-daily administration regime still poses issues with compliance, adherence, and convenience for some patients resulting in delayed initiation of treatment and consequently increasing the risk of worsening diabetic late-stage complications. Minimizing the burden of injection has been a key driver in current research and development efforts to design insulin analogs having a duration longer than 24 h. While clinical investigations are still in progress and no product has obtained a marketing authorization, two next-generation basal insulin analogs have emerged that are demonstrating the intriguing possibility of once-weekly administration.

Insulin icodec (Glu^{A14}, His^{B16}, His^{B25}, Lys^{B29}(Nε-icosanedioyl-γGlu-[2-(2-{2-[2-(2-aminoethoxy)ethoxy] acetylamino}ethoxy)ethoxy]acetyl)) des(B30) human insulin; CAS Number 118837943-2) represents an iteration on the covalent acylation approach and is based on re-engineering of the ultra-long oral basal insulin OI338 with a plasma half-life of 70 h in humans. Albumin binding was further increased by changing the fatty diacid conjugated to the epsilon amino group of B29 lysine from a 1,18-octadecanedioic acid (C18) to a 1,20-icosanedioic acid (C20) and a Tyr^{B16} → His substitution was made to reduce insulin receptor affinity. Other amino acid modifications include Tyr^{A14} → Glu, Phe^{B25} → His, and removal of the Thr residue at position B30 (Kjeldsen et al. 2021). In a phase 1 clinical study insulin icodec had a half-life of 196 h. Three phase 2 clinical trials have been completed and insulin icodec was shown to be comparable to daily administration of insulin glargine U100 in controlling blood glucose in a 26 week phase-2 study (Rosenstock et al. 2020).

Insulin efsitora alfa (Glu^{B16}, His^{B25}, Gly^{B27}, Gly^{B28}, Gly^{B29}, Gly^{B30}-Gly-Ser-Gly-Gly-Gly-Gly-Thr^{A10}, Asp^{A14}, Gly^{A21}, human insulin fusion with Gly-Gly-Gly-Gly-Gly-Gln-Gly-Gly-Gly-Gly-Gln-Gly-Gly-Gly-Gly-Gln-Gly-Gly-Gly-Gly-Gly-(IgG2 Fc); CAS Number 2131038–11-2) represents a completely novel molecular design strategy for an insulin analog intended for once-weekly administration. The analog is an Fc-fusion protein that is expressed from CHO cells as a disulfide-linked homodimer with a molecular weight of 64.1 kDa. Each monomer of the homodimer is comprised of a single-chain variant of insulin with the B-chain linked to the A-chain by a short linker (Linker 1) and inter-domain linker (Linker 2) connecting the A-chain of insulin to the constant domain derived from an antibody of isotype IgG2 (Fc). The single-chain variant of insulin includes the amino acid modifications Tyr^{B16} → Glu, Phe^{B25} → His, Thr^{B27} → Gly, Pro^{B28} → Gly, Lys^{B29} → Gly, Thr^{B30} → Gly, Ile^{A10} → Thr, Tyr^{A14} → Asp, and Asn^{A14} → Gly compared to human insulin. The first linker, which connects insulin chain A with chain B, is comprised of a seven amino acid, flexible linker, Gly-Gly-Ser-Gly-Gly-Gly-Gly, while the second linker, which connects the insulin with the Fc, contains the sequence Gly-Gly-Gly-Gly-Gly-Gln-Gly-Gly-Gly-Gly-Gln-Gly-Gly--Gly-Gly-Gln-Gly-Gly-Gly-Gly-Gly. Studies using streptozotocin-treated diabetic rats demonstrated a significant decrease in blood glucose compared to vehicle-treated animals 24 h post-injection that persisted through 336 h with a t1/2 of ~120 h following subcutaneous administration. The results from this preclinical study support once-weekly dosing in humans (Moyers et al. 2022).

The results of a Phase 2 safety and efficacy study comparing insulin efsitora alfa to insulin degludec over 32 weeks in patients with T2DM previously treated with oral antidiabetic drugs and a basal insulin have been reported. Weekly administration of insulin efsitora alfa was noninferior to insulin degludec for glycemic control as measured by the change in HbA1c after 32 weeks with a lower rate of documented and nocturnal hypoglycemia ≤70 mg/dL and less weight gain. No safety signals were detected (Frias et al. 2021). Clinical development for insulin efsitora alfa remains in progress.

Basal Insulin Preparations co-Formulated with Rapid-Acting Insulin Analog or GLP-1R Agonists

Following the successful commercialization of Tresiba® and Lantus®, premixture preparations incorporating their respective basal insulin analog active ingredients with either a rapid-acting insulin analog or a human incretin analog of glucagon-like peptide-1 (GLP-1) were introduced. A stable premixture of insulin degludec and insulin aspart is marketed as Ryzodeg® 70/30. The rationale for combining a basal insulin with rapid-acting insulin in a single product is similar to the biphasic solution/suspension products discussed earlier. However, Ryzodeg® is a stable solution composed of 70% insulin degludec and 30% insulin aspart (Havelund et al. 2015). Clinical studies have shown that Ryzodeg® 70/30 can be effectively used in patients with T1DM in combination with Novolog® (Hirsch et al. 2017) and patients with T2DM (Park et al. 2017). However, it should be noted that the Ryzodeg® label states that the administration of insulin formulation should be subcutaneously once or twice daily with any main meal(s), thus, requiring administration of a rapid- or short-acting insulin at other meals, if needed.

Two basal insulin analog/GLP-1 incretin premixture preparations are commercially available for use in patients with T2DM, Xultophy® and Soliqua®. Both the basal insulin analog and incretin components lower blood glucose, but do so in different ways. GLP-1 facilitates a glucose-dependent increase in insulin secretion by β-cells and decreased glucagon secretion by α-cells with a lower risk of hypoglycemia. Additionally, GLP-1 can reduce body weight, but may not cause sufficient insulin secretion to achieve desired glycemic control. Basal insulin therapy increases circulating insulin in a non-glucose-dependent manner, improves β-cell function, and plays a role in regulation and production of glucose, but may cause weight gain and carries the risk of hypoglycemia. Therefore, a combination of the two therapies compliments their respective therapeutic benefits while compensating for undesired effects (Rodbard et al. 2016; Gough et al. 2016).

A premixture preparation of 100 units/mL insulin degludec and 3.6 mg/mL of the GLP-1 analog liraglutide is marketed as Xultophy®. This product is supplied as a prefilled, single-patient multi-use disposable pen. The formulation contains 19.7 mg/mL glycerol, 5.70 mg/mL phenol, and 55 µg/mL zinc, and has a pH of approximately 8.15. The product is considered an alternative to basal–bolus therapy in the treatment of Type 2 Diabetes (Rodbard et al. 2016; Hughes 2016).

The other commercial premixture preparation marketed is Soliqua® 100/33, which combines insulin glargine with the GLP-1 analog lixisenatide for the treatment of patients with T2DM (Goldman and Trujillo 2017). This product is supplied as a prefilled, single-patient multi-use disposable pen. The solution formulation contains 100 units insulin glargine, 33 µg/mL lixisenatide, 3 mg/mL methionine, 2.7 mg/mL metacresol, 20 mg/mL glycerol, and 30 µg/mL of zinc. The pH of the formulation is not clearly stated in the prescribing information.

As newer insulin analogs continue to emerge, interest remains in formulating premixtures with GLP-1 incretins. IcoSema, a combination of once-weekly basal insulin icodec 350U and semaglutide 1 mg, is currently in Phase 3 clinical development. This fixed ratio combination is recommended for once-weekly administration, irrespective of meals, and is anticipated to have superior HbA1c reduction compared to each component administered alone, weight benefits compared to basal insulin, and reduced risk of hypoglycemia (Kalra et al. 2021). It is important to note that these purported outcomes for IcoSema require clinical validation and final approval of product labeling by regulatory authorities.

Concentrated Human Insulin Formulations

Most insulin and insulin analog formulations are manufactured at a concentration of 100 insulin units/mL, referred to as U100. Manufacturers have however steadily introduced more concentrated insulin formulations (U200, U300, and U500). Most concentrated insulins have been developed to facilitate the administration of larger quantities of insulin in lower volumes. These lower volumes may be associated with lesser pain from injections which could improve patient compliance and enable clinicians to tailor insulin therapy for the needs of their patients; particularly those with obesity and/or severe insulin resistance. As discussed above, many human insulin and insulin analog preparations are now marketed in dedicated devices at concentrations >U100 (Johnson et al. 2017; Ovalle et al. 2018). The need to have dedicated devices for the concentrated insulins is particularly important to prevent dosing errors. In addition, it cannot be assumed that all concentrated insulins have PK and PD properties equivalent to their U100 counterparts. Humalog® U200 and Tresiba® U200 have been shown to be bioequivalent to their U100 counterparts (Korsatko et al. 2013; de la Peña et al. 2016). Conversely, Humulin® R U500 and Toujeo® (U300) are not bioequivalent to their U100 counterparts (de la Peña et al. 2011; Becker et al. 2015). Thus, safety concerns can be addressed with patient education and the use of precise and accurate, dedicated insulin delivery devices. However, Humulin® R U500 human insulin formulations warrant separate discussion, as it is used in patients that exhibit insulin resistance and possess a different time action profile. This higher concentration insulin provides a preferential treatment option for those needing higher doses.

Early pharmacological studies demonstrated reduced absorption associated with increasing concentrations of insulin (Binder 1969; Binder et al. 1984). Galloway showed no statistically significant differences in PK serum insulin levels with increasing concentrations of pork regular insulin (at 0.25 U/kg) from U40 to U500 (Galloway et al. 1981); however, time to peak glucose responses were mildly delayed, and peak effect was variably reduced as concentration increased. The first PK/PD study of human U500 vs. U100 regular insulin in healthy obese subjects was published (de la Peña et al. 2011). Overall insulin exposure and effect were similar at both 50 U and 100 U doses (0.5 and 1.0 U/kg) with both formulations. However, the two formulations were not bioequivalent: peak insulin concentration (C_{max}) and effect (R_{max}) were significantly reduced for U500 vs. U100 for both doses. Time to peak concentration (t_{max}) and time to maximal effect (tR_{max}) were significantly longer for U500 vs. U100 only at the U100 dose. Duration of action (tR_{last}) was prolonged for U500 at both doses vs. U100 (50 U: 19.7 vs. 18.3 h; 100 U: 21.5 vs. 18.3 h; $p < 0.05$ for both). The onset of action (t_{onset}) was within 20 min for both formulations and supports the clinical use of human U500 regular 30 min before meals to leverage the prandial effect. Basal insulin between-meal requirements are expected to be covered by the long "tail" of action of the U500 formulation (de la Peña et al. 2011); further work is needed to assess if overnight coverage in T2D patients is possible.

A randomized controlled trial comparing twice-daily and thrice-daily U500 in insulin-resistant T2DM demonstrated that both dosing regimes were efficacious and safe, although higher hypoglycemia rates were observed in higher-dose groups (Wysham et al. 2016). Moreover, a series of other studies have demonstrated reductions in HbA1c (glycated hemoglobin) of 1.0–1.7% over 3–98 months of use (Lane et al. 2009; Boldo and Comi 2012; Ziesmer et al. 2012). Paradoxically, insulin dose generally did not statistically increase after conversion to human U500 regular insulin, although one large case series did report an increase in the total daily dose by 0.44 U/kg (Boldo and Comi 2012). Weight gain with treatment was variable, up to 4.2–6.8 kg (Lane et al. 2009; Boldo and Comi 2012). Reports of severe hypoglycemia have been infrequent, although an increase in non-severe hypoglycemia was reported in one large study (Boldo and Comi 2012). Most studies have used twice-daily or thrice-daily regimens (Lane et al. 2009; Boldo and Comi 2012; Ziesmer et al. 2012). A simplified dosing algorithm has also been published (Segal et al. 2010).

In the past, safety concerns with concentrated insulin therapy in diabetes patients, besides hypoglycemia and weight gain, mainly relate to the risk of dose confusion due to the lack of a dedicated injection device for U500 insulin. In November of 2016, Becton, Dickinson and Company in collaboration with Eli Lilly and Company launched the first insulin syringe specifically developed for patients prescribed Humulin® R U500, thus eliminating complex dosing conversion tables and formulas required when having to use the standard U100 insulin syringes. Also early in 2016 Eli Lilly and Company launched a Humulin® R U500 single-patient multiple-use disposable pen. For product differentiation, the Humulin® R U500 single-patient multiple-use disposable pen has a distinctive aqua color. The U500 insulin vial and labeling of vial and box are distinctive from U100 insulins, with black and white-colored and green-colored lettering, aqua-colored labeling and a larger box size (20 mL containing 10,000 U) with the green coloring matching the secondary packaging and needle shield color for the U500 syringe. The aqua-color labeling was selected to align with the distinctive aqua-colored use for the Humulin® R U500 single-patient multiple-use disposable pen.

Pharmaceutical Concerns

Chemical Stability of Insulin Formulations

The purity of insulin formulations is typically assessed by high-performance liquid chromatography using reversed-phase and size exclusion separation modes (USP 2021). Insulin has two primary routes of chemical degradation upon storage and use: hydrolytic transformation of amide to acid groups and formation of covalent dimers and higher-order polymers. Primarily the pH, the storage temperature, and the components of the specific formulation influence the rate of formation of these degradation products. In an acidic solution, the main degradation reaction is the transformation of asparagine (Asn) at the terminal 21 position of the A-chain to aspartic acid. This reaction is relatively facile at low pH but is extremely slow at neutral pH (Brange et al. 1992b). This was the primary degradation route in early soluble (acidic) insulin formulations. However, the development of neutral solutions and suspensions has diminished the importance of this degradation route. Stability studies of neutral solutions indicate that the amount of A21 desamido insulin does not change upon storage. Thus, the relatively small amounts of this bioactive material present in the formulation arise either from the source of insulin or from pharmaceutical process operations.

The deamidation of the Asn^{B3} of the B-chain is the primary degradation mechanism at neutral pH. The reaction proceeds through the formation of a cyclic imide that results in two products, aspartic acid (Asp) and iso-aspartic acid (iso-Asp) (Brennan and Clarke 1995). This reaction occurs relatively slowly in neutral solution (approximately 1/12 the rate of A21 desamido formation in acid solution). The relative amounts of these products are influenced by the flexibility of the B-chain, with approximate ratios of Asp:iso-Asp of 1:2 and 2:1 for solution and crystalline formulations, respectively. As noted earlier, the use of phenolic preservatives provides a stabilizing effect on the insulin hexamer that reduces the formation of the cyclic imide, as evidenced by reduced deamidation. The rate of formation also depends on temperature; typical rates of formation are approximately 2% per year at 5 °C. Studies have shown B3 deamidated insulin to be essentially fully potent (Brange 1987a).

High-molecular-weight protein (HMWP) products form at both refrigerated and room temperature storage conditions. Accelerated conditions (37 °C for 1 year) were used to enrich and then characterize the HMWP species from neutral pH insulin formulations (Hjorth et al. 2015). The primary species identified were covalent dimers of insulin linked at the Asn^{A21} and Lys^{B29} positions originating from the attack of the Lys^{B29} on the anhydride intermediate of Asn^{A21}. These purified covalent dimers were then further characterized for biological activity, conformational structure including self-association and fibrillation properties (Hjorth et al. 2016). The covalent dimer showed little to no biological activity, and the covalent dimer tertiary structure was identical to the non-covalent dimer such that in the presence of Zn it participated in the hexamer assembly. In studies aimed at evaluating physical attributes like self-association and fibrillation, the non-covalent dimer reduced the propensity for self-association and delayed the onset of fibril formation. There is evidence that insulin–protamine heterodimers also form in NPH suspensions (Brange et al. 1992a). At higher temperatures, the probability of forming higher-order insulin oligomers increases. The rate of formation of HMWPs is less than that of hydrolytic reactions; typical rates are less than 0.5% per year for soluble neutral regular insulin formulations at 5 °C. The rate of formation can be affected by the strength of the insulin formulation or by the addition of glycerol as a tonicity agent. The latter increases the rate of HMWP formation presumably by introducing impurities such as glyceraldehyde. HMWP formation is believed to also occur as a result of a reaction between the Phe^{B1} terminal amino group and the Asn^{A21} terminal carboxy group of a second insulin molecule via a cyclic anhydride (or succinimide, based on unpublished results of the authors) intermediate (Darrington and Anderson 1995). Reaction with the intermediate may also occur via the amino-terminus of the A-chain. Disulfide exchange leading to polymer formation is also possible at basic pH; however, the rate for these reactions is very slow under neutral pH formulation conditions. The quality of excipients such as glycerol is also critical because small amounts of aldehyde and other glycerol-related chemical impurities can accelerate the formation of HMWPs. The biopotency of HMWPs is significantly less (1/12–1/5 of insulin) than monomeric species (Brange 1987c).

Visible (400 to 800 nm) and ultraviolet (UV) light (200 to 400 nm) may degrade protein biopharmaceuticals. There is more absorbance of ultraviolet light with minimal absorption of lower energy visible light; thus, exposure to UV light is expected to degrade insulin to a greater degree. Photostability testing, controlling for temperature, has shown that insulin and analogs are photosensitive. HMWPs are the most sensitive indicator of photo instability. Light may be absorbed directly by the insulin amino acids or indirectly by the formulation components (e.g., metacresol or protamine) of formulated insulin drug products. HWMPs is thought to be the result of photo-oxidation of susceptible amino acids (tyrosine, phenylalanine, and cysteine). A reaction involving tyrosine and cysteine on the A chain may lead to dimer formation. Phe^{B1} and Asn^{A21} are also possible dimerization sites. Light absorption may also break internal disulfide bonds permitting crosslinking via free sulfhydryls with other insulin molecules. Photodegradation typically takes place directly through the absorption of UV light or through electron transfer from tryptophan to tyrosine. Controls should be practiced for minimizing exposures to light to minimize undesirable chemical degradation. For formulated drug products, the patient information leaflets supplied with these products should be consulted for the most current information regarding protection from light exposure.

Only limited chemical stability data has been published in the scientific literature for the insulin analog formulations containing insulin lispro, insulin aspart, insulin glulisine, insulin glargine, insulin detemir, or insulin degludec; however, it is reasonable to presume that similar chemical degradation pathways are present to varying extents in these compounds. Nevertheless, since some analogs are formulated under acidic conditions, e.g., Lantus®, Toujeo®, Basaglar®, Semglee® and Rezvoglar®, or have been modified with hydrophobic moieties, e.g., Levemir® or Tresiba®, it is reasonable to presume that alternate chemical degradation pathways may be possible. It should be noted that the amino acid substitution of glycine for asparagine at position 21 of the insulin glargine A-chain effectively eliminates the potential for deamidation that would occur under the acidic pH conditions used in the Lantus®, Toujeo®, Basaglar®, Semglee®, and Rezvoglar® formulations. This glycine substitution may also result in lower covalent dimer formation.

Physical Stability of Insulin Formulations

The physical stability of insulin formulations is mediated by noncovalent aggregation of insulin. Hydrophobic forces typically drive the aggregation, although electrostatics plays a subtle but important role. Aggregation typically leads to a loss in potency of the formulation, and therefore conditions promoting this type of physical degradation (i.e., extreme mechanical agitation or exposure to air-liquid interfaces often in combination with elevated temperatures) should be avoided for all insulin products. A particularly severe type of nonreversible aggregation results in the formation of insulin fibrils. The mechanism of insulin fibrillation is widely believed to result from destabilization of hexamers (i.e., the predominant self-associated form of most insulin solution preparations), causing an increase in the population of monomers that can partially unfold and initiate the aggregation process (Jansen et al. 2005). Physical attributes of insulin formulations are readily assessed by visual observation for macroscopic characteristics as well as by instrumental methods such as light and differential phase contrast microscopy. Insulin fibrillation can be confirmed using atomic force microscopy (Jansen et al. 2005). Various particle-sizing techniques also may be used to characterize physical degradation phenomena. Fluorescence spectroscopy using specific dyes has proven useful in monitoring the time course of insulin fibrillation process (Nielsen et al. 2001).

In general, insulin solutions have good physical stability. Physical changes in soluble formulations may be manifested as color or clarity change or, in extreme situations, increases in solution viscosity, a phenomenon referred to as gelation, or the formation of a precipitate that could be an indication of fibrillation. Insulin suspensions, such as NPH, are the most susceptible to changes in physical stability. Such physical instability typically occurs as a result of both elevated temperature and mechanical stress to the insulin preparation. The increase in temperature favors hydrophobic interactions, while mechanical agitation serves to provide mixing and stress across interfacial boundaries. Nucleation and higher-order forms of aggregation in suspensions can lead to conditions described as visible clumping of the insulin microcrystalline particles or adherence of the aggregates to the inner wall of the glass storage container. The latter phenomenon is referred to as frosting. In severe cases, resuspension may be nearly impossible because of caking of the suspension in the vial. Temperatures above ambient (>25 °C) can accelerate the aggregation process, especially those at or above body temperature (37 °C). Normal mechanical mixing of suspensions to achieve dispersion of the microcrystalline insulin particles prior to administration is not deleterious to physical stability. However, vigorous shaking or mixing should be avoided. Consequently, this latter constraint has, in part, led to the observation that patients do not place enough effort into resuspension. Thus, proper emphasis must be placed on training the patient in the resuspension of crystalline, amorphous, and premixed suspension formulations of insulin and insulin analogs. The necessity of rigorous resuspension may be the first sign of aggregation and should prompt a careful examination of the formulation to verify its suitability for use.

As with the chemical stability data, published information regarding the physical stability of the newer insulin analog formulations containing insulin lispro, insulin aspart, insulin glulisine, insulin glargine, or insulin detemir is limited. However, it is reasonable to assume that similar controls are practiced for preventing exposures to extreme agitation and thermal excursions to minimize undesirable physical transformations such as precipitation, aggregation, gelation, or fibrillation. The patient should be advised of the resuspension technique for specific insoluble insulin and insulin analog formulations, which is detailed in the package insert.

Clinical and Practice Aspects

Vial Presentations

Insulin is commonly available in 10-mL vials, and a strength of U100 (100 U/mL) is standard. The introduction of U100 insulins (Humalog®, Humulin® N, R, and 70/30) in 5-mL vials (filled to 3 mL: 300 U) has met a need for smaller volumes and less waste in hospital usage. The result is a combined economic impact of reduced acquisition cost and purchased volume (Edmondson et al. 2014; Eby et al. 2014).

It is essential to obtain the proper strength and formulation of insulin in order to maintain glycemic control (Hirsch et al. 2020). In addition, the brand/method of manufacture is important. Any change in insulin should be made cautiously and only under medical supervision. Common formulations, such as regular and NPH, are listed in Table 16.2, and the newer insulin analog formulations are listed in Tables 16.3, 16.4, and 16.5. Mixtures of rapid- or fast-acting with intermediate-acting insulin formulations are popular choices for glycemic control, particularly for people with type 2 diabetes (Table 16.4). The ratio is defined as the ratio of protamine-containing fraction/rapid- or fast-acting fraction, e.g., Humalog® Mix 75/25, where 75% of a dose is available as insulin lispro protamine suspension and 25% as insulin lispro for injection. With regard to NPH regular mixtures, caution must be used in the nomenclature because it may vary depending on the country of sale and the governing regulatory body. In the USA, for example, the predominant species is listed first as in N/R 70/30, but in Europe the same formulation is described as R/N 30/70 (Soluble/Isophane) where the base ("normal") ingredient is listed first. Currently, an effort is being made to standardize worldwide to the European nomenclature. The predominant human insulin mixture formulations sold globally are N/R 70/30; however, the number of suppliers may vary from country to country.

Injection Devices

Insulin syringes should be purchased to match the strength of the insulin that is to be administered (e.g., for U100 strength use 30-, 50-, or 100-unit syringes designated for U100 and for U500 strength use the syringe designated for U500). The gauge of needles available for insulin administration has been reduced to very fine gauges (30–32 G) in order to minimize pain during injection. In addition to finer and ultra-thin gauge needles, the length of needles has shortened to a minimum of 4 mm, in part, to prevent unintended IM injection, but also to reduce the perception of pain during injection. Recently, studies have shown that skin thickness is rarely

>3 mm and that needles of 4–5 mm consistently deliver insulin into the subcutaneous adipose tissue (Gibney et al. 2010). The use of a new needle for each dose maintains the sharp point, siliconization, and sterility of the needle, enabling a sterile and relatively pain-free injection.

The first pen injector used a 1.5-mL cartridge of U100 insulin (NovoPen® by Novo Nordisk in 1985). A needle was attached to the end of the pen, and the proper dose was selected and then injected by the patient. The cartridge was replaced when the content was exhausted, typically 3–7 days. Currently, 3.0-mL cartridges in U100 strength for regular, NPH, and the range of R/N mixtures, as well as the various rapid- and long-acting insulin analogs, have become the market standard, particularly in disposable pen devices with pre-filled insulin reservoirs. The advantages of the pen devices are primarily better compliance for the patient through a variety of factors including more accurate and reproducible dose control, easier transport of the drug, more discrete dose administration, timelier dose administration, and greater convenience.

Advances in engineering and manufacturing robustness are enabling pen devices with ½ unit dosing accuracy, which means they can accurately deliver doses as small as 5 μL. To meet the need for larger doses, some pens now have a maximum dose size of 800 μL, which translates to 80 units/injection from a U100 formulation. Pre-filled, disposable pens also enable safe injection of higher concentrations. U300 and U500 long-acting pens are now available for patients requiring large doses, as has been previously discussed.

In addition, several manufacturers now offer connected delivery systems, which electronically track the patient's dosing, and send this information to an app or directly to the cloud. This data is then used by either the patient or a health care provider (HCP) to help titrate the dosing regimen for improved blood sugar control. These dosing counters can either be integrated into an insulin pen or pump or can be installed onto a pen by the patient. Improved continuous glucose meters, together with connected insulin delivery devices and phone-based apps, can provide the patient with the data required to better control their blood sugar levels.

Continuous Subcutaneous Insulin Infusion: External Pumps

As previously mentioned, solution formulations of human insulin specifically designed for continuous subcutaneous insulin infusion (CSII) are commercially available. CSII systems were traditionally used by a small population of patients with diabetes but have become more popular with the recent introduction of rapid-acting insulin analogs. Currently, all three rapid-acting insulin analog formulations and two ultra-rapid formulations have received regulatory approval for this mode of delivery. Specific in vitro data demonstrating physicochemical stability for CSII has been reported for Humalog®

(De Felippis et al. 2006; Sharrow et al. 2012), Novolog®, and Apidra® (Senstius et al. 2007a, 2007b). Early pump devices contained glass or plastic reservoirs that must be hand filled from vial presentations by the patient. Some newer pumps have been specifically designed to accept the same glass 3-mL cartridges used in pen injector systems. Due to concerns over the impact of elevated temperature exposure and mechanical stress on the integrity of the insulin molecule along with the potential increased risk of microbial contamination, the patient information leaflets for the rapid-acting insulin analog products specify time intervals for changing the CSII infusion set and infusion site. The time period that insulin analog products can remain in a pump reservoir varies with 7, 6, or 2 days listed for Humalog®, Novolog®, and Apidra®, respectively, due to formulation stability. As always, the patient information leaflets supplied with these products should be consulted for the most current information related to in-use periods.

A number of systems have been developed or are under development to automatically dose insulin in response to continuous glucose monitoring data. The first step in this direction is a relatively simple system to turn off basal insulin delivery when glucose drops below a predetermined threshold (low glucose suspend), as is implemented in the Medtronic 530G pump. The Medtronic 640G and t:slim X2 pumps incorporate an algorithm to predict future glucose levels and prospectively reduce or stop insulin delivery (predictive low glucose suspend). This was followed by semi-automated insulin delivery systems (also known as hybrid closed-loop systems) that offer varying levels of sophistication driven by closed-loop algorithms embedded in the pump or in a mobile app that is designed to automate insulin delivery based on data from a continuous glucose monitor (CGM). This technology is often referred to as Automated Insulin Delivery (AID) although carbohydrates consumed as well as exercise still need to be entered into the system for appropriate insulin dosing(American Diabetes Association Professional Practice Committee 2022). So far, no study has been able to demonstrate superior glycemic control with ultra-rapid insulins when used with current AID algorithms when compared with rapid-acting analogs. Whether the PK of ultra-rapid insulins would require changes in the algorithms of the AID system in order to capitalize on their rapid onset of action to treat post-prandial glucose is unknown at this time although studies are ongoing (Lee et al. 2013).

Numerous systems are also under development to more completely automate insulin delivery decisions. Some of these include delivery of both insulin at high glucose and glucagon at low glucose, while others depend only on modulation of insulin dosing. The latter, which are broadly described as "closed-loop insulin pumps," are being advanced by several companies, including Medtronic, which has launched the first pump widely recognized as closed-loop in the 670G. The 670G has been followed by Tandem Control IQ, Omnipod 5, and the Medtronic 780G systems.

Clinical studies are currently underway to better understand the extent to which systems employing both glucagon and insulin may or may not deliver better outcomes than those that deliver only insulin. The diabetes community, impatient with the pace of commercial development and regulatory approval, has also developed open-source closed-loop insulin delivery systems based on commodity electronics and medical devices. The best known of these is the Do It Yourself (DIY) community like Loop (loopkit.github.io) (Lum et al. 2021) and the Open Artificial Pancreas System (https://openaps.org/) (Lewis and Leibrand 2016). While assembling the OpenAPS system requires a high degree of technical sophistication, users report a lower daily burden in managing their diabetes, and their experience is potentially valuable in speeding the development of commercial systems.

Noninvasive Delivery

Since the discovery of insulin, there has been a strong desire to overcome the need for injection-based therapy. Progress has been made in the form of needle-free injector systems (Robertson et al. 2000), but these devices have not gained widespread acceptance presumably because administration is not entirely pain-free, device costs are high, and other factors make it less desirable than traditional injection. Extensive research efforts have also focused on noninvasive routes of administration with attempts made to demonstrate the feasibility of transdermal, nasal, buccal, ocular, pulmonary, oral, and even rectal delivery of insulin (Heinemann et al. 2005; Heise 2021). Unfortunately, most attempts failed to progress beyond the proof of concept stage because low bioavailability, dose–response variability, and other adverse factors seriously called into question commercial viability. This situation has changed to some extent for pulmonary delivery of insulin.

As noted above, Afrezza® is an approved pulmonary insulin product for the treatment of post-prandial hyperglycemia. The formulation uses the dry powder Technosphere® technology wherein, particles composed of fumaryl diketopiperazine and polysorbate 80 are coated with human insulin. The delivery system utilizes single-use plastic cartridges containing 4 U, 8 U, or 12 U of the formulated insulin powder. The powder is aerosolized and delivered to the lung via oral inhalation using an inhaler.

Two oral insulin products are in late-stage development: IN-105 and ORMD-0801. IN-105 is modified with a short-chain polyethylene glycol that is claimed to improve resistance to proteolysis and promote absorption and is intended to be administered at mealtime (Khedkar et al. 2010). ORMD-0801 is formulated with excipients that are intended to inhibit proteolysis in the small intestine and improve translocation across the gut epithelial lining. Clinical studies have administered the drug once daily (Eldor et al. 2021). Novo Nordisk also developed an oral product for basal insulin delivery, insulin-338, but it was discontinued by 2019 due to high doses that were determined to be commercially nonviable (Halberg et al. 2019). Low bioavailability and the resulting challenges with commercial scaling and product cost, which are common to most noninvasive delivery approaches, suggest a need for disruptive innovation to achieve success in this area.

Storage

Insulin solution formulations should be stored in a cool place that avoids sunlight. Vials or cartridges that are not in active use should be stored under refrigerated (2–8 °C) conditions. Vials or cartridges in active use may be stored at ambient temperature. Vials in active use may also be stored at refrigerated conditions. Some cartridge devices may restrict storage during active use to ambient temperature. The in-use period for insulin formulations ranges from 28 to 56 days depending upon the product and its chemical, physical, and microbiological stability during use. High temperatures, such as those found in non-air-conditioned vehicles in the summer or other non-climate-controlled conditions, should be avoided due to the potential for chemical and/or physical changes to the formulation properties. Insulin formulations should not be frozen; if this occurs, the product should be disposed of immediately, since either the formulation or the container-closure integrity may be compromised. Insulin formulations should never be purchased or used past the expiration date on the package. Further information on storage and use of specific insulin products are contained in their respective patient information leaflets.

Usage

Resuspension

Insulin suspensions (e.g., NPH, NPL, premixtures) should be resuspended by gentle back-and-forth mixing and rolling of the vial between the palms to obtain a uniform, milky suspension. The patient should be advised of the resuspension technique for specific insoluble insulin and insulin analog formulations, which is detailed in the package insert. The homogeneity of suspensions is critical to obtaining an accurate dose. Any suspension that fails to provide a homogeneous dispersion of particles should not be used. Insulin formulations contained in cartridges in pen injectors may be suspended in a similar manner; however, the smaller size of the container and shape of the pen injector may require slight modification of the resuspension method to ensure complete resuspension. A bead (glass or metal) is typically added to cartridges to aid in the resuspension of suspension formulations.

Dosing

Dose withdrawal should immediately follow the resuspension of any insulin suspension. The patient should be instructed by their doctor, pharmacist, or nurse educator in proper procedures for dose administration. Of particular importance are procedures for disinfecting the container top and injection site. The patient is also advised to use a new needle and syringe for each injection. Reuse of these components, even after cleaning, may lead to contamination of the insulin formulation by microorganisms or by other materials, such as cleaning agents.

Extemporaneous Mixing

As discussed above in the section on "Intermediate-Acting Insulin," regular insulin can be mixed in the syringe with NPH and is stable enough to be stored for extended periods of time. Regular insulin analogs may also be diluted with a sterile diluent as well. For example, Humalog®, may be diluted 1:10 or 1:2 using a sterile vial and diluent and stored for extended periods of time. Further information use of specific insulin products are contained in their respective patient information leaflets.

With regard to the extemporaneous mixing of the newer insulin analogs, caution must be used. Lantus®, due to its acidic pH, should not be mixed with other fast- or rapid-acting insulin formulations which are formulated at neutral pH. If Lantus® is mixed with other insulin formulations, the solution may become cloudy due to isoelectric point (pI) precipitation of both the insulin glargine and the fast- or rapid-acting insulin resulting from pH changes. Consequently, the PK/PD profile, e.g., onset of action and time to peak effect, of Lantus® and/or the mixed insulin may be altered in an unpredictable manner. With regard to rapid-acting insulin analogs, extemporaneous mixing with human insulin NPH formulations is acceptable if used immediately. Under no circumstances should these formulations be stored as mixtures, as human insulin and insulin analog exchange can occur between solution and the crystalline matter, thereby potentially altering time-action profiles of the solution insulin analog. With regard to Levemir®, the human prescription drug label states that the product should not be diluted or mixed with any other insulin or solution to avoid altered and unpredictable changes in PK or PD profile (e.g., onset of action, time to peak effect).

Intravenous Infusion

Under medical supervision in a clinical setting, regular insulins may be administered intravenously. Insulin suspensions are not to be administered intravenously. Typical concentrations may be from 0.05 unit/mL to 1 unit/mL using 0.9% sodium chloride with compatible (e.g. polypropylene) infusion bags and tubing sets. A regular human insulin product prediluted for intravenous infusion is also available under the trade name Myxredlin®. There is no advantage to using rapid-acting insulins in preparing infusions because the rate of absorption is no longer a factor; however, it may be allowed for in the respective patient information leaflet.

Further information on storage and use of specific insulin products are contained in their respective patient information leaflets.

Self-Assessment Questions

Questions
1. Which insulin analog formulations cannot be mixed and stored? Why?
2. What are the primary chemical and physical stability issues with human insulin formulations?
3. What are some benefits and concerns around higher concentration insulin formulations?
4. Why are current marketed insulins made using recombinant microbial expression?
5. Why would someone want a longer-acting analog vs. a more rapid-acting analog?
6. What are some amino acid substitutions in insulin analogs, and why were they introduced?

Answers
1. Lantus®, Toujeo®, Basaglar®, Semglee® and Rezvoglar® are long-acting insulin formulations which are formulated at pH 4.0, which should not be mixed with rapid- or fast-acting insulin formulated under neutral pH. If Lantus® is mixed with other insulin formulations, the solution may become cloudy due to isoelectric point precipitation of both the insulin glargine and the fast- or rapid-acting insulin resulting from pH changes. Consequently, the onset of action and time to peak effect, of Lantus® and/or the mixed insulin may be altered in an unpredictable manner.
2. The two primary modes of chemical degradation are deamidation and HMWP formation. These routes of chemical degradation occur in all formulations. However, they are generally slower in suspension formulations. Physical instability is most often observed in insulin suspension formulations and pump formulations. In suspension formulations, particle agglomeration can occur resulting in the visible clumping of the crystalline and/or amorphous insulin. The soluble insulin in pump formulations can also precipitate or aggregate.
3. The main benefit of higher insulin concentration is the ability for insulin resistant patients to deliver larger doses of insulin with lower injection volumes, potentially reducing patient discomfort and number of injections, thus enhancing patient compliance. Major concerns with higher concentration insulin formulations relate to the potential for dose confusion that could lead to hypoglyce-

mic events. The coordination of insulin packaging with specialized syringes is an attempt to reduce this risk.

4. Increases in both the total patient population and potential reactions to animal-sourced insulins pushed a conversion to insulins sourced from recombinant microbial systems. The ability to manipulate the primary sequence of insulin then enabled engineering efforts that resulted in the introduction of analogs designed to improve patient therapy.

5. Glucose homeostasis is attained by delivering insulin at a constant low baseline level and then an increased level in response to meals. The basal level is maintained with the longer-acting analogs, where minimization in the differences in the concentration of the insulin over time is desired. The mealtime or prandial insulin requirement is covered by the rapid-acting analogs, where a rapid onset and rapid clearance are desired in response to carbohydrate absorbed from a meal.

6. In the design of the first insulin analogs, amino acid substitutions in the carboxy-terminal region of the B-chain were introduced to disrupt the dimerization domain of insulin, to enable a more rapid uptake upon subcutaneous injection (see insulins lispro, aspart, and glulisine). Subsequently, amino acid substitutions have been inserted to address formulation stability (see insulins glulisine (B3K) and glargine (A21G)). In addition, glargine also retains the arginine residues from the C-peptide on the carboxy-terminus of the B-chain, which shifts its isoelectric point to facilitate precipitation on injection (see Table 16.1).

Acknowledgements The authors thank Tony Schaff, Ingrid Hensley, Ryan Murray and Francisco Valenzuela for their invaluable input.

References

American Diabetes Association Professional Practice Committee (2022) 7. Diabetes technology: standards of medical Care in Diabetes—2022. Diabetes Care 45:S97–S112. https://doi.org/10.2337/dc22-S007

Anderson JH, Brunelle RL, Keohane P et al (1997) Mealtime treatment with insulin analog improves postprandial hyperglycemia and hypoglycemia in patients with non-insulin-dependent diabetes mellitus. Arch Intern Med 157:1249–1255. https://doi.org/10.1001/ARCHINTE.1997.00440320157015

Arnolds S, Rave K, Hövelmann U et al (2010) Insulin glulisine has a faster onset of action compared with insulin aspart in healthy volunteers. Exp Clin Endocrinol Diabetes 118:662–664. https://doi.org/10.1055/S-0030-1252067/ID/12

Bakaysa DL, Radziuk J, Havel HA et al (1996) Physicochemical basis for the rapid time-action of LysB28ProB29-insulin: dissociation of a protein-ligand complex. Protein Sci 5:2521–2531. https://doi.org/10.1002/PRO.5560051215

Baker EN, Blundell TL, Cutfield JF et al (1988) The structure of 2Zn pig insulin crystals at 1.5 a resolution. Philos Trans R Soc Lond Ser B Biol Sci 319:369–456. https://doi.org/10.1098/rstb.1988.0058

Balschmidt P (1996) AspB28 insulin crystals. US Patent 5,547,930 (20 Aug 1996)

Becker RHA, Dahmen R, Bergmann K et al (2015) New insulin glargine 300 units·mL−1 provides a more even activity profile and prolonged glycemic control at steady state compared with insulin glargine 100 units·mL−1. Diabetes Care 38:637–643. https://doi.org/10.2337/DC14-0006

Bell DSH, Clements RS, Perentesis G et al (1991) Dosage accuracy of self-mixed vs premixed insulin. Arch Intern Med 151:2265–2269. https://doi.org/10.1001/ARCHINTE.1991.00400110111022

Binder C (1969) Absorption of injected insulin: a clinical-pharmacological study. Acta Pharmacol Toxicol (Copenh) 27:1–83. https://doi.org/10.1111/j.1600-0773.1969.tb03069.x

Binder C, Lauritzen T, Faber O, Pramming S (1984) Insulin pharmacokinetics. Diabetes Care 7:188–199. https://doi.org/10.2337/DIACARE.7.2.188

Bliss M (2007) The discovery of insulin, 25th anniv edn. University of Toronto Press, Toronto

Boldo A, Comi RJ (2012) Clinical experience with U500 insulin: risks and benefits. Endocr Pract 18:56–61. https://doi.org/10.4158/EP11163.OR

Brader ML, Dunn MF (1991) Insulin hexamers: new conformations and applications. Trends Biochem Sci 16:341–345. https://doi.org/10.1016/0968-0004(91)90140-Q

Brange J (1987a) Insulin preparations. In: Galenics of insulin. Springer, Berlin Heidelberg, pp 58–60

Brange J (1987b) Production of bovine and porcine insulin. In: Galenics of insulin. Springer, Berlin, Heidelberg, pp 1–6

Brange J (1987c) Insulin preparations. In: Galenics of insulin. Springer, Berlin Heidelberg, pp 17–39

Brange J, Havelund S, Hougaard P (1992a) Chemical stability of insulin. 2. Formation of higher molecular weight transformation products during storage of pharmaceutical preparations. Pharm Res 9(6):727–734. https://doi.org/10.1023/A:1015887001987

Brange J, Langkjaer L (1992) Chemical stability of insulin. 3. Influence of excipients, formulation, and pH. Acta Pharm Nord 4:149–158

Brange J, Langkjaer L, Havelund S, Vølund A (1992b) Chemical stability of insulin. 1. Hydrolytic degradation during storage of pharmaceutical preparations. Pharm Res An Off J Am Assoc Pharm Sci 9:715–726. https://doi.org/10.1023/A:1015835017916

Brange J, Owens DR, Kang S, Vølund A (1990) Monomeric insulins and their experimental and clinical implications. Diabetes Care 13:923–954. https://doi.org/10.2337/DIACARE.13.9.923

Brange J, Ribel U, Hansen JF et al (1988) Monomeric insulins obtained by protein engineering and their medical implications. Nature 333:679–682. https://doi.org/10.1038/333679a0

Brems DN, Alter LA, Beckage MJ et al (1992) Altering the association properties of insulin by amino acid replacement. Protein Eng Des Sel 5:527–533. https://doi.org/10.1093/PROTEIN/5.6.527

Brennan TV, Clarke S (1995) Deamidation and isoasparate formation in model synthetic peptides. In: Aswad DW (ed) Deamidation and isoaspartate formation in peptides and proteins. CRC Press, Boca Raton, pp 65–90

Chacra AR, Kipnes M, Ilag LL et al (2010) Comparison of insulin lispro protamine suspension and insulin detemir in basal-bolus therapy in patients with type 1 diabetes. Diabet Med 27:563–569. https://doi.org/10.1111/J.1464-5491.2010.02986.X

Charman SA, McLennan DN, Edwards GA (2001) Porter CJH (2001) lymphatic absorption is a significant contributor to the subcutaneous bioavailability of insulin in a sheep model. Pharm Res 1811(18):1620–1626. https://doi.org/10.1023/A:1013046918190

Ciszak E, Beals JM, Frank BH et al (1995) Role of C-terminal B-chain residues in insulin assembly: the structure of hexameric LysB28ProB29-human insulin. Structure 3:615–622. https://doi.org/10.1016/S0969-2126(01)00195-2

Dailey G, Rosenstock J, Moses RG, Ways K (2004) Insulin Glulisine provides improved glycemic control in patients with type 2 diabetes. Diabetes Care 27:2363–2368. https://doi.org/10.2337/DIACARE.27.10.2363

Darrington RT, Anderson BD (1995) Effects of insulin concentration and self-association on the partitioning of its A-21 cyclic anhydride intermediate to Desamido insulin and covalent dimer. Pharm Res 127(12):1077–1084. https://doi.org/10.1023/A:1016231019677

Davis SN, Thompson CJ, Brown MD et al (1991) A comparison of the pharmacokinetics and metabolic effects of human regular and NPH insulin mixtures. Diabetes Res Clin Pract 13:107–117. https://doi.org/10.1016/0168-8227(91)90041-B

de la Peña A, Riddle M, Morrow LA et al (2011) Pharmacokinetics and pharmacodynamics of high-dose human regular U-500 insulin versus human regular U-100 insulin in healthy obese subjects. Diabetes Care 34:2496–2501. https://doi.org/10.2337/DC11-0721

de la Peña A, Seger M, Soon D et al (2016) Bioequivalence and comparative pharmacodynamics of insulin lispro 200 U/mL relative to insulin lispro (Humalog®) 100 U/mL. Clin Pharmacol Drug Dev 5:69–75. https://doi.org/10.1002/CPDD.221

Deckert T (1980) Intermediate-acting insulin preparations: NPH and Lente. Diabetes Care 3:623–626. https://doi.org/10.2337/DIACARE.3.5.623

De Felippis MR, Bakaysa DL, Bell MA et al (1998) preparation and characterization of a cocrystalline suspension of [LysB28,ProB29]-human insulin analogue. J Pharm Sci 87:170–176. https://doi.org/10.1021/JS970285M

De Felippis MR, Bell MA, Heyob JA, Storms SM (2006) In vitro stability of insulin lispro in continuous subcutaneous insulin infusion. Diabetes Technol Ther 8:358–368. https://doi.org/10.1089/dia.2006.8.358

Derewenda U, Derewenda Z, Dodson EJ et al (1989) Phenol stabilizes more helix in a new symmetrical zinc insulin hexamer. Nature 3386216(338):594–596. https://doi.org/10.1038/338594a0

Dodd SW, Havel HA, Kovach PM et al (1995) (1995) reversible adsorption of soluble Hexameric insulin onto the surface of insulin crystals Cocrystallized with protamine: an electrostatic interaction. Pharm Res 121(12):60–68. https://doi.org/10.1023/A:1016231019793

Dreyer M, Prager R, Robinson A et al (2005) Efficacy and safety of insulin glulisine in patients with type 1 diabetes. Horm Metab Res 37:702–707. https://doi.org/10.1055/S-2005-870584/ID/9

Eby E, Smolen L, Pitts A et al (2014) Economic impact of converting from pen and 10-mL vial to 3-mL vial for insulin delivery in a hospital setting. Hosp Pharm 49:1033–1038. https://doi.org/10.1310/HJP4911-1033

Edmondson G, Criswell J, Krueger L, Eby EL (2014) Economic impact of converting from 10-mL insulin vials to 3-mL vials and pens in a hospital setting. Am J Heal Pharm 71:1485–1489. https://doi.org/10.2146/AJHP130515

Eldor R, Neutel J, Homer K, Kidron M (2021) Efficacy and safety of 28-day treatment with oral insulin (ORMD-0801) in patients with type 2 diabetes: a randomized, placebo-controlled trial. Diabetes Obes Metab 23:2529–2538. https://doi.org/10.1111/dom.14499

Fogelfeld L, Dharmalingam M, Robling K et al (2010) A randomized, treat-to-target trial comparing insulin lispro protamine suspension and insulin detemir in insulin-naive patients with type 2 diabetes. Diabet Med 27:181–188. https://doi.org/10.1111/J.1464-5491.2009.02899.X

Frias JP, Chien J, Zhang Q et al (2021) Once weekly basal insulin fc (BIF) is safe and efficacious in patients with type 2 diabetes mellitus (T2DM) previously treated with basal insulin. J Endocr Soc 5:A448–A449. https://doi.org/10.1210/jendso/bvab048.916

Galloway JA, Chance RE (1994) Improving insulin therapy: achievements and challenges. Horm Metab Res 26:591–598. https://doi.org/10.1055/S-2007-1001766/BIB

Galloway JA, Spradlin CT, Nelson RL et al (1981) Factors influencing the absorption, serum insulin concentration, and blood glucose responses after injections of regular insulin and various insulin mixtures. Diabetes Care 4:366–376. https://doi.org/10.2337/DIACARE.4.3.366

Gibney MA, Arce CH, Byron KJ, Hirsch LJ (2010) Skin and subcutaneous adipose layer thickness in adults with diabetes at sites used for insulin injections: implications for needle length recommendations. Curr Med Res Opin 26:1519–1530. https://doi.org/10.1185/03007995.2010.481203/SUPPL_FILE/ICMO_A_481203_SM0001.PDF

Goldman J, Carpenter FH (1974) Zinc binding, circular dichroism, and equilibrium sedimentation studies on insulin (bovine) and several of its derivatives. Biochemistry 13:4566–4574. https://doi.org/10.1021/bi00719a015

Goldman J, Trujillo JM (2017) iGlarLixi: a fixed-ratio combination of insulin glargine 100 U/mL and Lixisenatide for the treatment of type 2 diabetes. Ann Pharmacother 51:990–999. https://doi.org/10.1177/1060028017717281

Gough SCL, Jain R, Woo VC (2016) Insulin degludec/liraglutide (IDegLira) for the treatment of type 2 diabetes. Expert Rev Endocrinol Metab 11:7–19. https://doi.org/10.1586/17446651.2016.1113129

Hagedorn HC, Jensen BN, Krarup NB, Wodstrup I (1936) Protamine Insulinate. J Am Med Assoc 106:177–180. https://doi.org/10.1001/JAMA.1936.02770030007002

Halberg IB, Lyby K, Wassermann K et al (2019) Efficacy and safety of oral basal insulin versus subcutaneous insulin glargine in type 2 diabetes: a randomised, double-blind, phase 2 trial. Lancet Diabetes Endocrinol 7:179–188. https://doi.org/10.1016/S2213-8587(18)30372-3

Hamaguchi T, Hashimoto Y, Miyata T et al (1990) Effect of mixing short-and intermediate (NPH insulin or Zn insulin suspension)-acting human insulin on plasma free insulin levels and action profiles. J Japan Diabetes Soc 33:223–229. https://doi.org/10.11213/tonyobyo1958.33.223

Havelund S, Plum A, Ribel U et al (2004) (2004) the mechanism of protraction of insulin Detemir, a long-acting, Acylated analog of human insulin. Pharm Res 218(21):1498–1504. https://doi.org/10.1023/B:PHAM.0000036926.54824.37

Havelund S, Ribel U, Hubálek F et al (2015) Investigation of the physicochemical properties that enable co-formulation of basal insulin degludec with fast-acting insulin aspart. Pharm Res 32:2250–2258. https://doi.org/10.1007/S11095-014-1614-X/FIGURES/4

Heinemann L, Khatami H, McKinnon R, Home P (2015) An overview of current regulatory requirements for approval of biosimilar insulins. Diabetes Technol Ther 17:510–526. https://doi.org/10.1089/dia.2014.0362

Heinemann L, Pfutzner A, Heise T (2005) Alternative routes of administration as an approach to improve insulin therapy: update on dermal, Oral, nasal and pulmonary insulin delivery. Curr Pharm Des 7:1327–1351. https://doi.org/10.2174/1381612013397384

Heinemann L, Sinha K, Weyer C et al (1999) Time-action profile of the soluble, fatty acid acylated, long-acting insulin analogue NN304. Diabet Med 16:332–338. https://doi.org/10.1046/J.1464-5491.1999.00081.X

Heinemann L, Weyer C, Rave K et al (1997) Comparison of the time-action profiles of U40- and U100-regular human insulin and the rapid-acting insulin analogue B28 asp. Exp Clin Endocrinol Diabetes 105:140–144. https://doi.org/10.1055/S-0029-1211742/BIB

Heise T (2021) The future of insulin therapy. Diabetes Res Clin Pract 175:108820. https://doi.org/10.1016/J.DIABRES.2021.108820

Heise T, Linnebjerg H, Coutant D et al (2020) Ultra rapid lispro lowers postprandial glucose and more closely matches normal physiological glucose response compared to other rapid insulin analogues:

a phase 1 randomized, crossover study. Diabetes Obes Metab 22:1789–1798. https://doi.org/10.1111/DOM.14094

Heise T, Mathieu C (2017) Impact of the mode of protraction of basal insulin therapies on their pharmacokinetic and pharmacodynamic properties and resulting clinical outcomes. Diabetes Obes Metab 19:3–12. https://doi.org/10.1111/DOM.12782

Heise T, Nosek L, Roønn BB et al (2004) Lower within-subject variability of insulin Detemir in comparison to NPH insulin and insulin glargine in people with type 1 diabetes. Diabetes 53:1614–1620. https://doi.org/10.2337/DIABETES.53.6.1614

Heise T, Nosek L, Spitzer H et al (2007) Insulin glulisine: a faster onset of action compared with insulin lispro. Diabetes Obes Metab 9:746–753. https://doi.org/10.1111/J.1463-1326.2007.00746.X

Heise T, Weyer C, Serwas A et al (1998) Time-action profiles of novel premixed preparations of insulin Lispro and NPL insulin. Diabetes Care 21:800–803. https://doi.org/10.2337/DIACARE.21.5.800

Hermansen K, Madsbad S, Perrild H et al (2001) Comparison of the soluble basal insulin analog insulin Detemir with NPH InsulinA randomized open crossover trial in type 1 diabetic subjects on basal-bolus therapy. Diabetes Care 24:296–301. https://doi.org/10.2337/DIACARE.24.2.296

Hirsch IB, Franek E, Mersebach H et al (2017) Safety and efficacy of insulin degludec/insulin aspart with bolus mealtime insulin aspart compared with standard basal–bolus treatment in people with type 1 diabetes: 1–year results from a randomized clinical trial (BOOST® T1). Diabet Med 34:167–173. https://doi.org/10.1111/DME.13068

Hirsch IB, Juneja R, Beals JM et al (2020) The evolution of insulin and how it informs therapy and treatment choices. Endocr Rev 41:733–755. https://doi.org/10.1210/ENDREV/BNAA015

Hjorth CF, Hubálek F, Andersson J et al (2015) Purification and identification of high molecular weight products formed during storage of neutral formulation of human insulin. Pharm Res 32:2072–2085. https://doi.org/10.1007/S11095-014-1600-3/FIGURES/22

Hjorth CF, Norrman M, Wahlund PO et al (2016) Structure, aggregation, and activity of a covalent insulin dimer formed during storage of neutral formulation of human insulin. J Pharm Sci 105:1376–1386. https://doi.org/10.1016/J.XPHS.2016.01.003

Hoffmann JA, Chance RE, Johnson MG (1990) Purification and analysis of the major components of chum salmon protamine contained in insulin formulations using high-performance liquid chromatography. Protein Expr Purif 1:127–133. https://doi.org/10.1016/1046-5928(90)90005-J

Holleman F, Schmitt H, Rottiers R et al (1997) Reduced frequency of severe hypoglycemia and coma in well-controlled IDDM patients treated with insulin Lispro. Diabetes Care 20:1827–1832. https://doi.org/10.2337/DIACARE.20.12.1827

Home PD (2012) The pharmacokinetics and pharmacodynamics of rapid-acting insulin analogues and their clinical consequences. Diabetes Obes Metab 14:780–788. https://doi.org/10.1111/j.1463-1326.2012.01580.x

Hompesch M, Ocheltree SM, Wondmagegnehu ET et al (2009) Pharmacokinetics and pharmacodynamics of insulin lispro protamine suspension compared with insulin glargine and insulin detemir in type 2 diabetes. Curr Med Res Opin 25:2679–2687. https://doi.org/10.1185/03007990903223739

Howey DC, Bowsher RR, Brunelle RL, Woodworth JR (1994) [Lys(B28), pro(B29)]-human insulin: a rapidly absorbed analogue of human insulin. Diabetes 43:396–402. https://doi.org/10.2337/DIAB.43.3.396

Hughes E (2016) IDegLira: redefining insulin optimisation using a single injection in patients with type 2 diabetes. Prim Care Diabetes 10:202–209. https://doi.org/10.1016/J.PCD.2015.12.005

Jansen R, Dzwolak W, Winter R (2005) Amyloidogenic self-assembly of insulin aggregates probed by high resolution atomic force microscopy. Biophys J 88:1344–1353. https://doi.org/10.1529/BIOPHYSJ.104.048843

Janssen MMJ, Casteleijn S, Devillé W et al (1997) Nighttime insulin kinetics and glycemic control in type 1 diabetes patients following Administration of an Intermediate-Acting Lispro Preparation. Diabetes Care 20:1870–1873. https://doi.org/10.2337/DIACARE.20.12.1870

Johnson JL, Downes JM, Obi CK, Asante NB (2017) Novel concentrated insulin delivery devices: developments for safe and simple dose conversions. J Diabetes Sci Technol 11:618–622. https://doi.org/10.1177/1932296816680830

Jonassen I, Havelund S, Hoeg-Jensen T et al (2012) Design of the novel protraction mechanism of insulin degludec, an ultra-long-acting basal insulin. Pharm Res 29:2104–2114. https://doi.org/10.1007/S11095-012-0739-Z/FIGURES/5

Kaarsholm NC, Havelund S, Hougaard P (1990) Ionization behavior of native and mutant insulins: pK perturbation of B13-Glu in aggregated species. Arch Biochem Biophys 283:496–502. https://doi.org/10.1016/0003-9861(90)90673-M

Kalra S, Bhattacharya S, Kapoor N (2021) Contemporary classification of glucagon-like peptide 1 receptor agonists (GLP1RAs). Diabetes Ther 12:2133–2147. https://doi.org/10.1007/S13300-021-01113-Y/TABLES/4

Khedkar A, Iyer H, Anand A et al (2010) A dose range finding study of novel oral insulin (IN-105) under fed conditions in type 2 diabetes mellitus subjects. Diabetes Obes Metab 12:659–664. https://doi.org/10.1111/J.1463-1326.2010.01213.X

Kjeldsen TB, Hubálek F, Hjørringgaard CU et al (2021) Molecular engineering of insulin Icodec, the first Acylated insulin analog for once-weekly Administration in Humans. J Med Chem 64:8942. https://doi.org/10.1021/ACS.JMEDCHEM.1C00257/SUPPL_FILE/JM1C00257_SI_001.PDF

Korsatko S, Deller S, Koehler G et al (2013) A comparison of the steady-state pharmacokinetic and Pharmacodynamic profiles of 100 and 200 U/mL formulations of ultra-long-acting insulin Degludec. Clin Drug Investig 337(33):515–521. https://doi.org/10.1007/S40261-013-0096-7

Krayenbuhl C, Rosenberg TH (1946) Crystalline protamine insulin. Rep Steno Mem Hosp Nord Insul 1:60–73

Kroef EP, Owens RA, Campbell EL et al (1989) Production scale purification of biosynthetic human insulin by reversed-phase high-performance liquid chromatography. J Chromatogr A 461:45–61. https://doi.org/10.1016/S0021-9673(00)94274-2

Kurtz AB, Gray RS, Markanday S (1983) Nabarro JDN (1983) circulating IgG antibody to protamine in patients treated with protamine-insulins. Diabetologia 254(25):322–324. https://doi.org/10.1007/BF00253194

Lane WS, Cochran EK, Jackson JA et al (2009) High-dose insulin therapy: is it time for U-500 insulin? Endocr Pract 15:71–79. https://doi.org/10.4158/EP.15.1.71

Lau IT, Lee KF, So WY et al (2017) Insulin glargine 300 U/mL for basal insulin therapy in type 1 and type 2 diabetes mellitus. Diabetes, Metab Syndr Obes Targets Ther 10:273. https://doi.org/10.2147/DMSO.S131358

Lee JJ, Dassau E, Zisser H et al (2013) The impact of insulin pharmacokinetics and pharmacodynamics on the closed-loop artificial pancreas. Proc IEEE Conf Decis Control 127–132. https://doi.org/10.1109/CDC.2013.6759870

Leohr J, Dellva MA, LaBell E et al (2020) Pharmacokinetic and Glucodynamic responses of ultra rapid Lispro vs Lispro across a clinically relevant range of subcutaneous doses in healthy subjects. Clin Ther 42:1762–1777.e4. https://doi.org/10.1016/J.CLINTHERA.2020.07.005

Lewis D, Leibrand S (2016) Real-world use of open source artificial pancreas systems. J Diabetes Sci Technol 10:1411. https://doi.org/10.1177/1932296816665635

Linnebjerg H, Zhang Q, LaBell E et al (2020) Pharmacokinetics and Glucodynamics of ultra rapid Lispro (URLi) versus Humalog®

(Lispro) in younger adults and elderly patients with type 1 diabetes mellitus: a randomised controlled trial. Clin Pharmacokinet 59:1589–1599. https://doi.org/10.1007/S40262-020-00903-0/TABLES/3

Lum JW, Bailey RJ, Barnes-Lomen V et al (2021) A real-world prospective study of the safety and effectiveness of the loop open source automated insulin delivery system. Diabetes Technol Ther 23:367–375. https://doi.org/10.1089/DIA.2020.0535/SUPPL_FILE/SUPP_TABLE12.DOCX

Moyers JS, Hansen RJ, Day JW et al (2022) Preclinical characterization of LY3209590, a novel weekly basal insulin fc-fusion protein. J Pharmacol Exp Ther. JPET-AR-2022-001105 382(3):346–355. https://doi.org/10.1124/jpet.122.001105

Nell LJ, Thomas JW (1988) Frequency and specificity of protamine antibodies in diabetic and control subjects. Diabetes 37:172–176. https://doi.org/10.2337/DIAB.37.2.172

Nielsen L, Khurana R, Coats A et al (2001) Effect of environmental factors on the kinetics of insulin fibril formation: elucidation of the molecular mechanism†. Biochemistry 40:6036–6046. https://doi.org/10.1021/BI002555C

Ovalle F, Segal AR, Anderson JE et al (2018) Understanding concentrated insulins: a systematic review of randomized controlled trials. Curr Med Res Opin 34:1029–1043

Owens DR, Monnier L, Ceriello A, Bolli GB (2022) Insulin centennial: milestones influencing the development of insulin preparations since 1922. Diabetes Obes Metab 24:27–42. https://doi.org/10.1111/DOM.14587

Park SW, Bebakar WMW, Hernandez PG et al (2017) Insulin degludec/insulin aspart once daily in type 2 diabetes: a comparison of simple or stepwise titration algorithms (BOOST®: SIMPLE USE). Diabet Med 34:174–179. https://doi.org/10.1111/DME.13069

Pekar AH, Frank BH (1972) Conformation of proinsulin. A comparison of insulin and proinsulin self-Association at Neutral pH. Biochemistry 11:4013–4016. https://doi.org/10.1021/BI00772A001/ASSET/BI00772A001.FP.PNG_V03

Pfützner A, Mann AE, Steiner SS (2004) Technosphere™/Insulin—a new approach for effective delivery of human insulin via the pulmonary route. Diabetes Technol Ther. https://home.liebertpub.com/dia 4:589–594. https://doi.org/10.1089/152091502320798204

Porcellati F, Bolli GB, Fanelli CG (2011) Pharmacokinetics and pharmacodynamics of basal insulins. Diabetes Technol Ther 13:S-15–S-24. https://doi.org/10.1089/dia.2011.0038

Porter CJH, Charman SA (2000) Lymphatic transport of proteins after subcutaneous administration. J Pharm Sci 89:297–310. https://doi.org/10.1002/(sici)1520-6017(200003)89:3<297::aid-jps2>3.0.co;2-p

Ritzel R, Roussel R, Bolli GB et al (2015) Patient-level meta-analysis of the EDITION 1, 2 and 3 studies: glycaemic control and hypoglycaemia with new insulin glargine 300 U/ml versus glargine 100 U/ml in people with type 2 diabetes. Diabetes Obes Metab 17:859–867. https://doi.org/10.1111/DOM.12485

Robertson KE, Glazer NB, Campbell RK (2000) The latest developments in insulin injection devices. Diabetes Educ 26:135–152. https://doi.org/10.1177/014572170002600114

Rodbard HW, Buse JB, Woo V et al (2016) Benefits of combination of insulin degludec and liraglutide are independent of baseline glycated haemoglobin level and duration of type 2 diabetes. Diabetes Obes Metab 18:40–48. https://doi.org/10.1111/DOM.12574

Rosenstock J, Bajaj HS, Janež A et al (2020) Once-weekly insulin for type 2 diabetes without previous insulin treatment. N Engl J Med 383:2107–2116. https://doi.org/10.1056/NEJMoa2022474

Schade DS, Santiago JV, Skyler JS, Rizza RA (1983) Intensive insulin therapy. Medical Examination Pub. Co, Princeton, NJ

Segal AR, Brunner JE, Burch FT, Jackson JA (2010) Use of concentrated insulin human regular (U-500) for patients with diabetes. Am J Heal Pharm 67:1526–1535. https://doi.org/10.2146/AJHP090554

Senstius J, Harboe E, Westermann H (2007a) In vitro stability of insulin aspart in simulated continuous subcutaneous insulin infusion using a MiniMed® 508 insulin pump. Diabetes Technol Ther 9:75–79. https://doi.org/10.1089/dia.2006.0041

Senstius J, Poulsen C, Hvass A (2007b) Comparison of in vitro stability for insulin aspart and insulin glulisine during simulated use in insulin pumps. Diabetes Technol Ther 9:517–521. https://doi.org/10.1089/dia.2007.0233

Sharrow SD, Glass LC, Dobbins MA (2012) 14-day in vitro chemical stability of insulin lispro in the MiniMed paradigm pump. Diabetes Technol Ther 14:264–270. https://doi.org/10.1089/DIA.2011.0125/ASSET/IMAGES/LARGE/FIGURE5.JPEG

Steensgaard DB, Schluckebier G, Strauss HM et al (2013) Ligand-controlled assembly of hexamers, dihexamers, and linear multihexamer structures by the engineered acylated insulin degludec. Biochemistry 52:295–309. https://doi.org/10.1021/bi3008609

Strojek K, Shi C, Carey MA, Jacober SJ (2010) Addition of insulin lispro protamine suspension or insulin glargine to oral type 2 diabetes regimens: a randomized trial. Diabetes Obes Metab 12:916–922. https://doi.org/10.1111/J.1463-1326.2010.01257.X

Supersaxo A, Hein WR, Steffen H (1990) Effect of molecular weight on the lymphatic absorption of water-soluble compounds following subcutaneous administration. Pharm Res 7:167–169. https://doi.org/10.1023/A:1015880819328

USP (2021) Insulin Human. USP-NF. USP, Rockville, MD

Vague P, Selam JL, Skeie S et al (2003) Insulin Detemir is associated with more predictable glycemic control and reduced risk of hypoglycemia than NPH insulin in patients with type 1 diabetes on a basal-bolus regimen with Premeal insulin Aspart. Diabetes Care 26:590–596. https://doi.org/10.2337/DIACARE.26.3.590

Waldhäusl W, Bratusch-Marrain P, Gasic S et al (1979) Insulin production rate following glucose ingestion estimated by splanchnic C-peptide output in normal man. Diabetologia 17:221–227. https://doi.org/10.1007/BF01235858

Weyer C, Heise T, Heinemann L (1997) Insulin Aspart in a 30/70 premixed formulation: Pharmacodynamic properties of a rapid-acting insulin analog in stable mixture. Diabetes Care 20:1612–1614. https://doi.org/10.2337/DIACARE.20.10.1612

White J, Goldman J (2019) Biosimilar and follow-on insulin: the ins, outs, and interchangeability. J Pharm Technol 35:25–35. https://doi.org/10.1177/8755122518802268/FORMAT/EPUB

Whittingham JL, Edwards DJ, Antson AA et al (1998) Interactions of phenol and m-cresol in the insulin hexamer, and their effect on the association properties of B28 pro → asp insulin analogues. Biochemistry 37:11516–11523. https://doi.org/10.1021/bi980807s

Whittingham JL, Jonassen I, Havelund S et al (2004) Crystallographic and solution studies of N-lithocholyl insulin: a new generation of prolonged-acting human insulins. Biochemistry 43:5987–5995. https://doi.org/10.1021/bi036163s

Wysham C, Hood RC, Warren ML et al (2016) Effect of Total daily dose on efficacy, dosing, and safety of 2 dose titration regimens of human regular U-500 insulin in severely insulin-resistant patients with type 2 diabetes. Endocr Pract 22:653–665. https://doi.org/10.4158/EP15959.OR

Ziesmer AE, Kelly KC, Guerra PA et al (2012) U500 regular insulin use in insulin-resistant type 2 diabetic veteran patients. Endocr Pract 18:34–38. https://doi.org/10.4158/EP11043.OR

Further Reading

American Diabetes Association (2019) Practical insulin: a handbook for prescribing providers, 5th edn. American Diabetes Association, Arlington

Beals JM (2015) Chapter 3: Insulin analogs—"improving the therapy of diabetes". In: Fischer J, Rotella DP (eds) Successful drug discovery, vol 1. Wiley, Hoboken, NJ

Bliss M (2007) The discovery of insulin, 25th anniv edn. University of Toronto Press, Toronto

Burant C (ed) (2012) Medical management of type 2 diabetes, 7th edn. American Diabetes Association, Arlington

Cooper T, Ainsburg A (2010) Breakthrough: Elizabeth Hughes, the discovery of insulin, and the making of a medical miracle. St. Martin's Press, New York

Wolfsdorf JI (2012) Intensive diabetes management, 5th edn. American Diabetes Association, Arlington

Zaykov AN, Mayer JP, DiMarchi RD (2016) Pursuit of a perfect insulin. Nat Rev Drug Discov 15(6):425–439

Hematopoietic Growth Factors

Juan Jose Pérez-Ruixo and Wojciech Krzyzanski

Introduction

Hematopoiesis is an intricate, well-regulated, and homeostatic multistep process that allows immature precursor cells in the bone marrow to proliferate, differentiate, mature, and become functional blood cells that transport oxygen and carbon dioxide; contribute to host immunity; and facilitate blood clotting. In the early 1900s, scientists recognized the presence of circulating factors that regulate hematopoiesis. It took approximately 50 years to develop in vitro cell culture systems in order to definitively prove that the growth and survival of early blood cells require the presence of specific circulating factors, called hematopoietic growth factors (HGF). The presence of many HGF with different targets at extremely small amounts in blood, bone marrow, and urine confounded the search for a single HGF with a specific activity. Scientific progress was slow until it became possible to purify sufficient quantities to evaluate the characteristics and biologic potential of the isolated materials. The introduction of recombinant DNA technology triggered a flurry of studies and an information explosion, which confirmed hematopoiesis is mediated by a series of HGF that acts individually and in various combinations involving complex feedback mechanisms. Today, many HGF have been isolated; some have been studied extensively, and a few have been manufactured for clinical use.

Different mature blood cells have been identified, each derived from primitive hematopoietic stem cells in the bone marrow. The most primitive pool of pluripotent stem cells comprises approximately 0.1% of the nucleated cells of the bone marrow, and 5% of these cells may be actively cycling at a given time. The stem cell pool maintains itself, seemingly without extensive depletion, by asymmetrical cell division. When a stem cell divides, one daughter cell remains in the stem cell pool and the other becomes a committed colony-forming unit (CFU). The CFU proliferates at a greater rate than the other stem cells and is more limited in self-renewal than pluripotent hematopoietic stem cells. Proliferation and differentiation are regulated by different mechanisms that necessarily involve HGF, which eventually convert the dividing cells into a population of terminally differentiated functional cells committed to the myeloid or the lymphoid pathway. Functional hematopoietic-derived blood cells from the myeloid pathway are red blood cells (erythrocytes), granulocytes (neutrophils, eosinophils, and basophils), monocytes and macrophages, tissue mast cells, and platelets (thrombocytes). Cells committed to the lymphoid pathway give rise to B- or T-lymphocytes and plasma cells.

Most HGF are glycosylated single-chain polypeptides encoded by a specific gene. Production of a recombinant HGF protein is accomplished by first identifying and isolating the particular HGF gene coding region, inserting the HGF DNA into a plasmid, and then expressing the recombinant growth factor protein in a biologic system (e.g., bacteria, yeast, or mammalian cells). The carbohydrate content of HGF varies by the particular protein and production method, which affects not only the molecular weight of the glycoprotein but potentially the specific biologic activity and the circulating half-life as well. These are the main reasons why the recombinant copies of HGF proteins cannot be exactly identical to the original HGF protein; however, they might become biosimilars of the original HGF protein. Extensive reviews of the characteristics of biosimilar products have been recently published (Schellekens et al. 2016; Kidanewold et al. 2021). In addition, a summary of the HGF and their activities is provided in Table 17.1.

This chapter will focus on reviewing the molecular structure, mechanism of action, pharmacokinetics and pharmacodynamics, clinical indications, and adverse events of HGF proteins stimulating erythropoiesis, granulopoiesis, and

J. J. Pérez-Ruixo (✉)
Department of Clinical Pharmacology and Pharmacometrics, Johnson & Johnson Innovative Medicine, Beerse, Belgium
e-mail: jjperezr@its.jnj.com

W. Krzyzanski
Department of Pharmaceutical Sciences, State University of New York, Buffalo, NY, USA

Table 17.1 Hematopoietic growth factors and their activities

Factor	Molecular weight (kDa)	Target cells	Actions
Erythropoietin (EPO)	34–39	Erythroid progenitors	Increase red blood cell counts
Granulocyte colony-stimulating factor (G-CSF)	18	Granulocyte progenitors and mature neutrophils	Increase neutrophil counts
Granulocyte-macrophage colony-stimulating factor (GM-CSF)	14–35	Granulocyte-macrophage progenitors and eosinophil progenitors	Increase neutrophil, eosinophil, and monocyte counts
Stem cell factor (SCF)		Granulocyte-erythroid progenitors, lymphoid progenitors, and natural killer cells	Increase pluripotent stem cells and progenitor cells for all other cell types
Thrombopoietin (TPO)	35	Stem cells, megakaryocytes, and erythroid progenitors	Increase platelet counts

thrombopoiesis. The common pharmacokinetic and pharmacodynamic features across the HGF are presented in detail for erythropoietin-stimulating agents (ESA), granulocyte colony-stimulating factors (G-CSF), and thrombopoietin receptor agonists (TRA) and then briefly discussed for other HGF. In this context, the existence of flip-flop pharmacokinetics justifying the efficiency of the subcutaneous administration relative to the intravenous administration, as well as the concentration-dependent disposition mediated by its binding to the target receptor and the time-dependent pharmacokinetics, consequence of its pharmacodynamic action extending the receptor pool over time, are common features of the recombinant proteins targeting the receptors for erythropoietin, G-CSF and thrombopoietin.

Erythropoiesis-Stimulating Agents

Erythropoietin (EPO) is a 30.4 kDa glycoprotein hormone secreted by the kidneys in response to tissue hypoxia, which stimulates red blood cell (RBC) production. EPO requires glycosylation to regulate erythrocyte production by activating the EPO receptor (EPOR) and stimulating the proliferation and differentiation of erythrocytic progenitors in the bone marrow, which leads to reticulocytosis, erythrocytosis, and the increase of hemoglobin concentration in the blood. The gene that encodes EPO is located on chromosome 7. The cloning of the *EPO* gene in the early 1980s allowed for the development of recombinant erythropoietins and analogs (erythropoiesis-stimulating agents [ESAs]), offering an alternative to transfusion as a method of raising hemoglobin levels that has successfully been used for over 30 years to treat anemia in millions of anemic patients.

Epoetin alfa (Epogen®), the first commercial form of recombinant human erythropoietin (rHuEPO) marketed in the USA, EU, Japan, and China, and epoetin beta (Recormon®, NeoRecormon®), marketed outside of the USA, are both expressed in Chinese hamster ovary cells. Both have the same 165 amino acid sequence, which is identical to human EPO, and contain two disulfide bonds and three N-linked and one O-linked sialic acid-containing carbohydrate chains (Halstenson et al. 1991) and lead to the same biological effects as endogenous EPO (Egrie et al. 1986). No important differences in clinical efficacy are apparent between epoetin alfa and beta (Jelkmann 2000). Darbepoetin alfa (Aranesp®) is a hyperglycosylated erythropoietin analog with five amino acid changes and two additional N-linked carbohydrate chains, which has the same mechanism of action as EPO (Elliott et al. 2004a). However, darbepoetin alfa has a threefold increased serum half-life (Macdougall et al. 1999; Elliott et al. 2003; Sinclair and Elliott 2005) and increased in vivo potency (Egrie et al. 2003), allowing for more convenient modes of administration, including extended dosing intervals (Vansteenkiste et al. 2002; Nissenson et al. 2002) up to monthly dosing as described in the US label. It is marketed in more than 50 countries and is indicated to treat the anemia of patients with chronic kidney disease and chemotherapy-induced anemia in cancer patients.

A large methoxy polyethylene glycol (PEG) polymer chain was integrated into the epoetin beta molecule via amide bonds between either the N-terminal amino group or the ε-amino group of lysine by means of a succinimidyl butanoic acid linker (Macdougall 2005). The resulting pegylated epoetin beta molecule has been marketed as Mircera® to treat the anemia of patients with chronic kidney disease (CKD), but its clinical development as treatment for chemotherapy-induced anemia was stopped (Gascon et al. 2010). The pegylated epoetin beta stimulates erythropoiesis by binding to EPOR; however, the EPOR binding affinity is reduced (Jarsch et al. 2008). This biologic disadvantage is counterbalanced with an extended half-life in humans, which allows for extended dosing intervals in CKD patients (Chanu et al. 2010; Sulowicz et al. 2007; Klinger et al. 2007; Kessler et al. 2010), similar to the dosing interval of darbepoetin.

Five rHuEPO biosimilars manufactured by two companies have been approved in the EU. Abseamed®, Binocrit®, and Epoetin alfa HEXAL® all produced by Rentschler Biotechnologie GmbH but marketed by three different companies are epoetin alfa biosimilars of the reference product

Eprex®. Comparable safety and efficacy between these three biosimilars and Eprex® was demonstrated in randomized controlled trials in hemodialysis patients with renal anemia. Although the EMA regulatory guidelines for rHuEPO biosimilars recommend that comparable efficacy and safety are demonstrated with two randomized trials in the nephrology setting, these biosimilars were approved based on a single nephrology trial. Two additional biosimilar versions of Eprex®, Retacrit®, and Silapo® are manufactured by Norbitec GmbH, under the international nonproprietary name (INN) of epoetin zeta. The comparability of epoetin zeta to Eprex® was demonstrated in two randomized clinical trials, a correction phase study and a maintenance phase study, involving hemodialysis patients with renal anemia. In the correction phase study, the comparability between epoetin zeta and Eprex® over the evaluation period was demonstrated for mean hemoglobin levels, but not for mean dose. Similar results were reported in the maintenance phase study, suggesting a possible difference in the bioactivity of epoetin zeta and Eprex®. Data from studies in cancer patients receiving chemotherapy and treated with epoetin alfa biosimilars and epoetin zeta were also submitted for approval, but these studies were not adequately powered to demonstrate therapeutic equivalence to the reference product in this patient population. However, epoetin alfa biosimilars and epoetin zeta were approved in the EU for indications in renal anemia, chemotherapy-induced anemia, and for pre-donation of blood prior to surgery for autologous transfusion (Schellekens and Moors 2010). Retacrit® was approved in May 2018 in the US for the treatment of anemia caused by chronic kidney disease, chemotherapy or use of zidovudine in patients with HIV infection (Table 17.2).

Regulation of Erythropoietin
The primary site of EPO synthesis in adults is the peritubular cells of the kidney (Jelkmann 2000; Jelkmann 1992). The liver is a secondary site of EPO production, with synthesis occurring in both hepatocytes and fibroblastoid interstitial cells (Spivak 1998). No preformed stores of EPO exist, and serum EPO concentrations are maintained at a constant concentration by homeostatic turnover, which consists of the basal production and elimination of the hormone (Fisher 2003). Within a healthy individual, the serum EPO concentration tends to be controlled tightly; however, large interindividual variability is evident from the normal range, 5–35 IU/L (Fisher 2003). Maintenance of normal serum concentrations of endogenous EPO requires the synthesis of about 2–3 IU/kg/day, or approx. 1000–1500 IU/week for a 70-kg man. Sex differences and regular-to-moderate athletic training do not appear to affect endogenous EPO serum concentrations. The blood flow in the kidney has a circadian rhythm in normal individuals; therefore, the endogenous production of EPO has diurnal variations with the highest levels in the evening and at night (Wide et al. 1989).

The overexpression of EPO occurs in a number of adaptive and pathologic conditions. In response to acute hypoxic stress, such as severe blood loss or severe anemia, the EPO production rate can increase 100- to 1000-fold. Numerous studies have shown an exponential increase in serum EPO, with increasing degrees of anemia, although the maximal bone marrow response to such stimulation is only a four- to six-fold increase in the RBC production rate (Jelkmann 2000). Overproduction of EPO with accompanying erythrocytosis may be an adaptive response to conditions that produce chronic tissue hypoxia, such as living at high altitude, chronic respiratory diseases, cyanotic heart disease, sleep apnea, smoking, localized renal hypoxia, radiotherapy, or hemoglobinopathies with increased oxygen affinity. Paraneoplastic production of EPO from some tumors and cysts can also result in high serum concentrations of EPO. Following bone marrow ablation, aplastic anemia, or anemia in patients with hypoplastic marrows, serum EPO levels are disproportionately increased relative to slightly decreased hemoglobin levels. Conversely, individuals with hyperactive marrow owing to hemolytic anemia had disproportionately low serum EPO levels and rapid EPO serum disappearance.

In chronic kidney disease, up to 60% of patients have hemoglobin concentrations below 11 g/dL before beginning dialysis (Jungers et al. 2002). Multiple mechanisms contribute to the low hemoglobin levels (Fisher 2003), but the most important is the inability of the diseased kidneys to produce an appropriate EPO response for the given degree of anemia

Table 17.2 ESAs approved for treatment in the United States (US), European Union (EU), and other countries (ROW)

Active ingredient	Drug name	Region
Epoetin alfa	Epogen/procrit	US, ROW
	Binocrit, hexal, abseamed	EU
	Eprex, repoitin, epiao	ROW
Epoetin alfa-EPBX	Retacrit	US
Epoetin beta	Recormon, neorecormon	EU
Epoetin zeta	Retacrit, silapo	EU
Epoetin theta	Biopoin	EU
Darbepoetin alfa	Aranesp	US, EU, ROW
Methoxy polyethylene glycol-epoetin beta	Mircera	US, EU, ROW

or an inability to meet the increased RBC demands of uremic patients (Adamson and Eschbach 1990). In addition, the uremic state itself appears to blunt the bone marrow response to EPO (Fisher 2003). It is of interest that serum EPO concentrations in chronically anemic dialysis patients increase to some extent in response to acute hypoxic stress (from either acute bleeding or systemic hypoxemia), suggesting that kidney failure does not result in a complete inability to produce EPO (Kato et al. 1994).

In cancer patients, anemia is of multifactorial etiology (Fisher 2003), and there are three distinct types of anemia: cancer-related anemia (nontreatment related), anemia related to myelosuppressive chemotherapy, and anemia related to other causes such as bleeding, nutritional deficiency, or iron deficiency, among others. As with other anemias of chronic disease, including those associated with chronic infection and inflammatory disorders, the anemia of cancer is characterized by a decreased production of endogenous EPO (Miller et al. 1990), cytokine-induced suppression of bone marrow function, disordered iron absorption and metabolism (Bron et al. 2001), and decreased erythrocyte survival. In the anemia related to chemotherapy treatment, the amount of endogenous EPO transiently increases up to six-fold within the 48 h after the administration of chemotherapy and returns to baseline within a week (Glaspy et al. 2005). After myeloablative chemotherapy, severe thrombocytopenia and bleeding might contribute to a significant loss of RBC. Finally, the anemias associated with infant prematurity, pregnancy, allogeneic bone marrow transplantation, and HIV infection are often characterized by inappropriately low EPO concentrations (Spivak 1998).

Pharmacokinetics
Absorption

After subcutaneous (s.c.) dosing of rHuEPO, its absorption is slow, leading to peak serum concentrations at 5–30 h and a longer terminal half-life (24–79 h) than that obtained after intravenous (i.v.) administration (McMahon et al. 1990). These results indicate the presence of flip-flop pharmacokinetics, where the rate of absorption is slower than the rate of elimination. Thus, the absorption process is the rate-limiting process for its disposition, and the observed terminal half-life after s.c. dosing reflects the absorption rate rather than elimination rate.

Following s.c. administration, protein therapeutics, including the marketed recombinant HGF proteins, typically enter the systemic circulation via the blood capillaries or the lymphatic system (Porter and Charman 2000; McLennan et al. 2006). The lymphatic system is considered to be the primary route of absorption from the s.c. injection site for protein therapeutics greater than 16 kD, due to the restricted vascular access afforded by the continuous endothelial layer of blood capillaries (Supersaxo et al. 1990). In both healthy subjects and cancer patients, the fraction of dose absorbed via the lymphatics is about 80–90% and increases at doses higher than 300 IU/kg (Olsson-Gisleskog et al. 2007; Ait-Oudhia et al. 2011; Krzyzanski et al. 2005; Ramakrishnan et al. 2004). The s.c. absorption rates of rHuEPO vary according to the administration site, with a more rapid and extensive absorption when injected into the thigh compared with the abdomen or arm (Jensen et al. 1994). This relatively small difference is most likely reflecting regional differences in blood and lymph flow and not considered to be clinically relevant as the pharmacodynamic profile (i.e., reticulocytes time course) did not evidence any difference across the site of administration. Small differences in the absorption due to the administration site have been observed for other HGF, such as G-CSF and romiplostim, but they are also of limited clinical relevance.

The s.c. absorption of darbepoetin alfa in humans is also slow, with peak concentrations reached at 34–58 h post-dose, followed by a generally monophasic decline. Similarly to rHuEPO, darbepoetin alfa also displays flip-flop pharmacokinetics, with a longer half-life after s.c. dosing than after i.v. dosing (Agoram et al. 2007). The mean terminal half-life of darbepoetin alfa, 73 h, is associated with large variability between patients, consistent with the variability observed for other ESAs (Glaspy et al. 2005). The mean absorption time of darbepoetin alpha is 56 h, substantially longer than the mean absorption time reported for rHuEPO (Olsson-Gisleskog et al. 2007).

The reported 20–30% reduction in the darbepoetin alfa absorption rate per decade of age (Agoram et al. 2007) is consistent with the estimated effect of age on the rHuEPO absorption rate in healthy subjects (Olsson-Gisleskog et al. 2007) and cancer patients (Ait-Oudhia et al. 2011) and reflects the longer terminal half-life and the larger exposure to both drugs in older patients. It has been hypothesized (Agoram et al. 2007) that the age-dependent reduction in lymphatic flow rate could be the physiological reason behind this relationship, as it has also been reported for monoclonal antibodies administered by s.c. route (Sutjandra et al. 2011; Kakkar et al. 2011). The data available also suggest that the pharmacokinetic profile of rHuEPO and darbepoetin alfa after s.c. administration is similar in adults and children; however, s.c. absorption in children may be more rapid than in adults for both drugs (Heatherington 2003).

Bioavailability

Initial bioavailability estimates for rHuEPO after s.c. administration range from about 15 to 40% and are similar for epoetin alfa and beta (Deicher and Horl 2004). When the pharmacokinetics of s.c. rHuEPO and darbepoetin alfa were studied over a wider dose range in healthy volunteers and after accounting for the rHuEPO nonlinear clearance, exposure was found to increase more than proportional with dose (Olsson-Gisleskog et al. 2007; Agoram et al. 2007; Cheung et al. 1998, 2001). The s.c. bioavailability of darbepoetin alfa increases from 57 to 69% when the 200 µg dose is increased up to 400 µg, while the s.c. bioavailability of rHuEPO increases from 54 to 65% when the 40 kIU dose is increased up to 80 kIU. The apparent increase in s.c. bioavailability with dose of ESA might indicate saturable presystemic processes. Nevertheless, despite the apparent low bioavailability, s.c. administration of ESA has been reported to produce equivalent or better efficacy relative to i.v. administration, although there is a wide range of interpatient variability (Kaufman et al. 1998). The flip-flop kinetics together with the absolute bioavailability following s.c. dosing results in a substantial increase in the efficiency of the ESA s.c. administration relative to i.v. administration that have been also reported for the G-CSF agonists (filgrastim, lenograstim and pegfilgrastim) as well as the c-Mpl agonist (romiplostim).

Distribution

During i.v. infusion, serum rHuEPO and darbepoetin alfa concentrations rise rapidly and then decline in a bi-exponential manner (Olsson-Gisleskog et al. 2007; Doshi et al. 2010). The peak serum rHuEPO and darbepoetin alfa concentrations correlate linearly with dose. A rHuEPO dose of 50 IU/kg produces concentrations of about 1000 mIU/mL 15 min after the end of the infusion, while a darbepoetin alfa dose of 0.75 µg/kg generates serum concentrations of about 10–20 ng/mL after the end of the infusion (Doshi et al. 2010). As expected from its large molecular weight, the volume of distribution of rHuEPO is similar to the plasma volume (40–60 mL/kg), suggesting confinement of rHuEPO within the plasma circulation (McMahon et al. 1990; Olsson-Gisleskog et al. 2007). The data available also suggest that the volume of distribution, normalized by body weight, in adults and children is similar after i.v. administration of rHuEPO and darbepoetin alfa.

Elimination

Despite the long clinical experience with ESAs, the mechanism(s) of their clearance have not been fully elucidated, and there is a paucity of information regarding which organ(s) and tissue(s) are important in the metabolism of these drugs. Two ESA clearance pathways have been suggested to explain ESA elimination: (1) a capacity-limited clearance pathway utilizing EPO receptor-mediated endocytosis by erythroid progenitor cells and (2) an EPOR-independent linear clearance, although still not well understood, reflecting other mechanism(s). In vivo studies demonstrate that the kidney, liver, and lymph exert a negligible effect on in vivo EPOR-independent clearance. Clearly, our understanding of the nature of the EPOR-independent clearance pathways is incomplete. However, it is important to recognize that renal excretion and hepatic metabolism of ESAs plays a minor role in their elimination and altered renal or hepatic function does not warrant dose adjustments.

Notably, the presence of two clearance pathways also determines the elimination of the G-CSF agonists (filgrastim and lenograstim – see below) as well as the c-Mpl agonist (romiplostim).

An investigation of the trafficking and degradation of rHuEPO by EPOR-expressing cells (BsF3) in cell culture found that rHuEPO was subjected to EPO receptor-mediated endocytosis followed by degradation in lysosomes (Gross and Lodish 2006). The rHuEPO receptor-binding, dissociation, and trafficking properties affected the relative rate of rHuEPO cellular uptake and intracellular degradation (Walrafen et al. 2005). About 57% of surface-bound rHuEPO was internalized ($k_{in} = 0.06$ min^{-1}) and, after internalization, 60% of the ligand was recycled intact to the cell surface, while 40% was degraded. In spite of the in vitro results suggesting the role of EPOR on ESA clearance, the in vivo evidence is indirect and mostly arises from chemotherapy studies in patients treated with rHuEPO and darbepoetin alfa (Chapel et al. 2001). Chemotherapy-based approaches may also affect EPOR-independent clearance mechanisms due to the destruction of macrophages or neutrophils. The reduction in the number of these cells may explain, or at least contribute to, the decrease in ESA clearance observed after chemotherapy treatment.

Studies investigating the pharmacokinetics of rHuEPO analogs with different EPOR binding activity suggested that EPOR-independent pathway plays a major role in the ESA clearance since decreasing the number of receptors with chemotherapy or, blocking the EPOR pathway with analogs without binding activity, were unable to completely shut down the elimination of rHuEPO. In addition, since pegylation has been shown to mainly affect the EPOR-independent clearance pathway, EPOR-mediated clearance may not be the dominant route of ESA elimination (Agoram et al. 2009).

It has been shown that carbohydrate side chains of EPO are necessary for the persistence and in vivo biologic activity of the molecule, but not for in vitro receptor binding or stimulation of proliferation. Indeed rHuEPO molecules with increased sialic acid content have less affinity for the EPOR (Sinclair and Elliott 2005; Elliott et al. 2004b). Darbepoetin alfa is a hyperglycosylated analog of rHuEPO, with three- to five-fold lower affinity for the EPOR compared to rHuEPO (Gross and Lodish 2006; Elliott et al. 2004b), but has 3- to four-fold longer serum half-life and greater in vivo activity than rHuEPO (Egrie et al. 2003). Surface-bound darbepoetin alfa was internalized at the same rate as rHuEPO, and after internalization, 60% of each ligand was re-secreted intact and 40% degraded (Gross and Lodish 2006). While in vitro experiments suggested that relative to rHuEPO, darbepoetin may have reduced clearance in vivo because of reduced EPOR-mediated endocytosis and degradation, darbepoetin alfa has other biophysical characteristics, such as increased molecular size and decreased isoelectric point, suggesting that the reduced clearance might be better explained by other mechanisms. In this context, studies investigating the pharmacokinetics of rHuEPO analogs with different EPOR binding activity suggest that hyperglycosylation mainly impacts the EPOR-independent clearance pathway, which also supports the hypothesis that EPOR-mediated clearance may not play a dominant role in ESA elimination (Agoram et al. 2009).

A population pharmacokinetic meta-analysis of rHuEPO in 533 healthy subjects enrolled in 16 clinical studies, where a wide range of i.v. and s.c. rHuEPO doses were administered, has helped in quantifying the two separate elimination pathways and understanding the influence of demographic characteristics and other covariates on the pharmacokinetic parameters of rHuEPO (Olsson-Gisleskog et al. 2007). At low concentrations, including the endogenous EPO concentrations observed at baseline or in ESA-untreated states, the nonlinear clearance operates at full capacity, giving a total clearance of about 0.9 L/h. As concentrations increase, the nonlinearity of pharmacokinetics becomes more important and, at the concentration of 394 IU/L, the clearance is 0.6 L/h. When the concentration is above 3546 IU/L, the nonlinear clearance of rHuEPO was fully (>90%) saturated, and the total clearance decreased to almost one-third, being mainly represented by the linear component. At concentrations higher than 3546 IU/L, rHuEPO pharmacokinetics is approximately linear. The concentration-dependent clearance appears to be independent of the type of rHuEPO (epoetin alfa vs. epoetin beta) or population (healthy subjects or patients with chronic renal failure).

A further indication of the possible involvement of EPOR binding in the disposition of rHuEPO can be found when investigating the rHuEPO pharmacokinetics after multiple dosing. A rHuEPO time-dependent clearance, with a 10–30% increase after several weeks of treatments with no subsequent changes (McMahon et al. 1990; Cheung et al. 1998; Yan et al. 2012) has been attributed to the limited number of EPOR located on the finite, but expandable, number of bone marrow erythroid progenitors. The pharmacodynamic action of rHuEPO increases Burst-Forming Unit-Erythroid (BFU-E) and Colony-Forming Unit-Erythroid (CFU-E) cell expansion and, consequently, the number of EPOR, which in turn results in an increase in rHuEPO clearance, a decrease in the apparent volume of distribution and a reduction in terminal half-life. The term pharmacodynamic-mediated drug disposition (PDMDD) has been coined to describe these types of TMDD models where pharmacodynamics affects the size of the target pool and influences the drug clearance, as has been described for ESAs. Consequently, the pharmacokinetics of rHuEPO is considered nonlinear because it is concentration dependent and nonstationary (time-dependent) (Yan et al. 2012).

The rHuEPO pharmacokinetic models for healthy subjects can be applied to patients with anemia due to renal insufficiency; however, they may have limited predictive value when applied to patients receiving chemotherapy. The consequences of the chemotherapy effect on the pharmacokinetics of rHuEPO in oncology patients are derived from the reduced number of EPOR available to clear rHuEPO in progenitor cells and the reduction of non-EPOR-mediated clearance (Olsson-Gisleskog et al. 2007). In cancer patients treated with chemotherapy, a correlation between the decline in the absolute reticulocyte count and the decrease in the clearance of rHuEPO over time has been observed (Ait-Oudhia et al. 2011). As a consequence, the rHuEPO elimination process becomes slower than the absorption process, and the flip-flop phenomenon observed in healthy subjects disappears when rHuEPO is given s.c. to cancer patients receiving chemotherapy (Olsson-Gisleskog et al. 2007; Ait-Oudhia et al. 2011). Furthermore, this phenomenon has clinical implications with respect to the synchronicity of ESA and chemotherapy administration, suggesting asynchronous dosing might be superior (Glaspy et al. 2005).

Pharmacodynamics

After rHuEPO is administered, it binds to the EPOR on the surface of the BFU-E, CFU-E, and proerythroblast and activates the signal transduction pathways previously described. CFU-E cells have the highest EPOR density (1000 receptors per cell) and are the most sensitive to EPO. Experimental data suggest that approximately only 5–10% of EPOR must be continuously occupied with rHuEPO in order to prevent apoptosis and stimulate proliferation and differentiation of erythroid precursors. Then, CFU-Es will differentiate into normoblasts (including proerythroblast, basophilic erythroblast, polychromatophilic erythroblast, and orthochromatic erythroblast) and, upon normoblast denucleation, reticulocytes will be formed and reside in the bone marrow for 1 day before they are released into the bloodstream, where they circulate for about 1 day before maturing to erythrocytes. In healthy adults, the RBC life span is about 120 days and shows a relatively narrow distribution. The RBC life span is similar in cancer patients but markedly reduced in patients with chronic kidney disease, and it has been estimated to be about 60–65 days in dialysis patients and slightly longer in nondialysis patients (82 days), with a moderate interindividual variability (Uehlinger et al. 1992; Chanu et al. 2010).

Previous studies have demonstrated that highly glycosylated rHuEPO has increased in vivo biological activity and serum half-life, but decreased receptor binding affinity (Egrie and Browne 2001). Given these relationships, a comparison of clearance among different ESAs has to be interpreted in conjunction with EPOR binding affinity and/or in vivo activity. Darbepoetin alfa stimulates erythropoiesis by the same mechanisms as those previously discussed for endogenous EPO and rHuEPO. In vitro, the affinity of darbe-poetin alfa for the EPOR is one-third to one-fifth of the rHuEPO affinity (Gross and Lodish 2006); however, the increase in mean residence time of darbepoetin alfa results in a prolonged period above an erythropoietic threshold that more than compensates for the reduced receptor affinity, yielding an increased in vivo activity (Elliott et al. 2003; Egrie et al. 2003).

Besides this information, there is limited clinical data quantifying ESA potency in humans, which makes it diffi-cult to establish the net balance of in vivo activity associated with simultaneous modifications of drug clearance and potency. Furthermore, in most of the clinical PK/PD rela-tionships published, ESA pharmacodynamic actions have been linked to serum concentration, under the assumption that the pharmacodynamic effect is proportional to the amount of drug–receptor complex. The PK/PD models developed for rHuEPO and darbepoetin alfa suggest that the reduction in darbepoetin alfa binding affinity to human EPOR relative to rHuEPO is translated into a less than two-fold reduction in vivo potency (EC_{50} = 82 IU/L, assuming ng/mL is equivalent to 200 IU/L, vs. EC_{50} = 58 IU/L). Additionally, darbepoetin alfa clearance is at least one-third of rHuEPO clearance, while the maximum stimulatory effect of erythropoiesis (E_{max}) is similar for both ESAs (Krzyzanski et al. 2005). The darbepoetin alfa reduction in clearance is more pronounced than its reduction of in vivo potency; therefore, the longer half-life associated with darbepoetin alfa predominates over its reduction of in vivo potency, which explains the overall improvement in dose efficiency of darbepoetin alfa relative to rHuEPO.

Different mechanisms have been proposed to explain the pharmacodynamic tolerance of the rHuEPO effect. Besides the increase in rHuEPO clearance over time due to the increase in the number of EPOR, an oxygen-mediated feed-back mechanism, erythroid precursor pool depletion, and iron-restricted erythropoiesis have also been proposed as tol-erance mechanisms (Krzyzanski et al. 2005; Ramakrishnan et al. 2004). The oxygen feedback mechanism is regulated through an oxygen-sensing system: a high hemoglobin level leads to an increased oxygen level and eventually inhibits the production of endogenous EPO. Erythroid progenitor cells are EPO dependent; they cannot survive without EPO. On the other hand, extensive rHuEPO treatment results in ane-mia due to the depletion of the erythroid precursor pool (Piron et al. 2001). This anemia is not due to low endogenous EPO levels but rather exhaustion of erythroid progenitors (Krzyzanski et al. 2005; Perez-Ruixo et al. 2009). Furthermore, iron-restricted erythropoiesis occurs in the presence of absolute iron deficiency, functional iron defi-ciency, and/or iron sequestration. Absolute iron deficiency is a common nutritional deficiency in women's health, pediat-rics, and the elderly. Functional iron deficiency occurs in

patients with significant EPO-mediated erythropoiesis or therapy with ESAs, even when storage iron is present. Iron sequestration, mediated by hepcidin, is an underappreciated but common cause of iron-restricted erythropoiesis in patients with chronic inflammatory disease. It has been shown that iron supplementation improves the hematopoietic response of ESAs used for chemotherapy-induced anemia. In multiple-dosing regimens, even though the endogenous EPO production might be suppressed, the total concentration of EPO is still high, and tolerance may occur due to precur-sor pool depletion and/or iron-restricted erythropoiesis. However, the oxygen feedback mechanism might be present, especially at the end of dosing intervals in regimens that extend longer than four rHuEPO half-lives.

Indications for Cancer Patients and Potential Adverse Events

Unless otherwise indicated, the information pertaining to ESA indications in cancer patients provided in this section is derived from the product prescribing information package inserts as well as the National Comprehensive Cancer Network for cancer- and chemotherapy-induced anemia. ESAs are indicated for the treatment of anemia due to the effects of concomitantly administered chemotherapy for a duration of ≥2 months in patients with metastatic, nonmy-eloid malignancies. However, ESA treatment is not indicated for patients receiving hormonal agents, biologics, or radio-therapy, unless they are receiving concomitant myelosup-pressive chemotherapy. Notably, ESA therapy should not be used to treat anemia associated with malignancy or anemia of cancer in patients with either solid or nonmyeloid hemato-logical malignancies who are not receiving concurrent che-motherapy (Rizzo et al. 2008). Furthermore, ESA treatment is not indicated for patients receiving myelosuppressive ther-apy when the anticipated outcome is cure, due to the absence of studies that adequately characterize the impact of ESA therapy on progression-free survival and overall survival. ESA therapy is also not indicated for the treatment of anemia in cancer patients due to other factors such as absolute or functional iron deficiency, folate deficiencies, hemolysis, or gastrointestinal bleeding. ESA use in cancer patients has not been demonstrated in controlled clinical trials to improve symptoms of anemia, quality of life, fatigue, or patient well-being.

Depending on the clinical situation and the severity of anemia, red blood cell transfusion could be an alternative option to ESA therapy (Rizzo et al. 2008). Otherwise, a s.c. rHuEPO dose of 150 IU/kg three times a week (TIW) or 40 kIU weekly (QW) is recommended to increase hemoglobin and decrease transfusions in patients with chemotherapy-associated anemia when the hemoglobin concentration is approaching or has fallen below, 10 g/dL. Alternatively, s.c. rHuEPO dose of 80 kIU biweekly (Q2W) or 120 kIU every

3 weeks (Q3W) can be used as initial dosing because these two dosage schedules have not been found to have any differences in efficacy with respect to the approved TIW and QW dosing schedules. The dose of ESA therapy should be titrated for each patient to achieve and maintain the lowest hemoglobin level sufficient to avoid blood transfusion. Therefore, the TIW s.c. dose of rHuEPO should be increased to 300 IU/kg if no reduction in transfusion requirements or rise in hemoglobin after 8 weeks of treatment has been observed. Similarly, the QW dose should increase to 60 kIU if no increase in hemoglobin by at least 1 g/dL after 4 weeks of treatment is observed. In addition, if hemoglobin exceeds 11 g/dL, but not 12 g/dL, the dose should be reduced by 25%. However, if hemoglobin exceeds 12 g/dL, therapy should be held until hemoglobin falls below 11 g/dL and then restarted at a 25% dose reduction. The pediatric dosing guidance is based on an initial i.v. dose of 600 IU/kg QW (maximum 40 kIU). If there is no increase in hemoglobin by at least 1 g/dL after 4 weeks of treatment (in the absence of RBC transfusion), the rHuEPO dose should be increased to 900 IU/kg (maximum 60 kIU) in order to maintain the lowest hemoglobin level sufficient to avoid RBC transfusion.

The recommended initial s.c. dose of darbepoetin alfa is 2.25 µg/kg QW or 500 µg once every 3 weeks (Q3W). The initial darbepoetin alfa s.c. dose of 2.25 µg/kg QW should be increased to 4.5 µg/kg QW if hemoglobin increase is less than 1 g/dL after 6 weeks of treatment. In addition, if hemoglobin increases by more than 1 g/dL in any 2-week period or when the hemoglobin reaches a level that avoids transfusion, the dose should be reduced by 40%. If hemoglobin exceeds a level needed to avoid transfusion, therapy should be held until hemoglobin approaches a level where transfusions may be required, then restarted at a 40% dose reduction. A s.c. darbepoetin alfa dose of 100 µg QW, 200 µg Q2W, or 300 µg Q3W can be used as alternative initial dosing since differences in efficacy have not been found. These initial dose levels should be increased to 150–200 µg QW, 300 µg Q2W, or 500 µg Q3W, respectively. At this time the safety and efficacy of darbepoetin alfa in children receiving chemotherapy has not been established.

Although no specific serum rHuEPO level has been established which predicts the patients unlikely to respond to epoetin alfa therapy, treatment is not recommended for patients with grossly elevated serum rHuEPO levels (e.g., greater than 200 mUnits/mL). The hemoglobin should be monitored on a weekly basis in patients receiving ESA therapy until hemoglobin becomes stable and then at regular intervals thereafter.

Patients with multiple myeloma, especially those with renal failure, may benefit from adjunctive ESA therapy to treat anemia. Endogenous EPO levels should be monitored in order to assist in planning ESA therapy. No high-quality, published studies support the exclusive use of epoetin or darbe-

poetin in anemic myeloma, non-Hodgkin's lymphoma, or chronic lymphocytic leukemia in the absence of chemotherapy. Treatment with chemotherapy and/or corticosteroids should be initiated first. If a rise in hemoglobin does not result, treatment with epoetin or darbepoetin may begin in patients with particular caution exercised with chemotherapeutic agents and disease states where the risk of thromboembolism is increased. Blood transfusion is also an option (Rizzo et al. 2008). The current standard of care for symptomatic anemia in patients with myelodysplastic syndrome (MDS) is supportive care with RBC transfusion. Patients with serum EPO levels less than or equal to 500 IU/L, normal cytogenetics, and less than 15% marrow-ringed sideroblasts may respond to relatively high doses of rHuEPO (40–60 kIU s.c.TIW) or darbepoetin alfa (150–300 µg QW s.c.). Evidence supports the use of epoetin or darbepoetin in patients with anemia associated with low-risk myelodysplasia (Rizzo et al. 2008). Supportive care with RBC transfusion is the standard of care for symptomatic anemia in patients with hematologic malignancies (non-Hodgkin's lymphoma, chronic lymphocytic leukemia). There is insufficient data to recommend ESA therapy for patients responding to treatment with good prognosis and persistent transfusion-dependent anemia.

Iron supplementation improves the hematopoietic response of ESAs used for chemotherapy-induced anemia. A recent meta-analysis of randomized, controlled trials, comparing parenteral or oral iron or no iron, when added to ESAs in anemic cancer patients, evidenced that overall parenteral iron reduces the risk of transfusions by 23% and increases the chance of hematopoietic response by 29% when compared with ESAs alone. On the contrary, oral iron does not increase hematopoietic response or transfusion rate. The significance of these results is that the proportion of nonresponders to ESAs treated with parenteral iron will have strongly improved quality of life and cost ameliorated (Petrelli et al. 2012).

Several studies have reported a possible decreased survival rate in cancer patients receiving ESA for the correction of anemia. Analyses of eight randomized studies in patients with cancer found a decrease in overall survival and/or locoregional disease control associated with ESA therapy for the correction of anemia with an off-label target hemoglobin level greater than 12 g/dL. These results were confirmed in three recent meta-analyses (Bennett et al. 2008; Bohlius et al. 2009; Tonelli et al. 2009) and refuted in other two meta-analyses (Ludwig et al. 2009; Glaspy et al. 2010). There are also observational data and data from randomized studies that show no increase in mortality with ESA use according to prescribing label, specifically in patients receiving chemotherapy. In addition, an increased risk for thromboembolic events has been reported with ESA therapy in cancer patients. Besides the intrinsic risk associated with the malignancy itself, the chemotherapy, and other concomitant factors, results from several meta-analyses established a significant association between the increased risk for thrombotic events and ESA use, with relative risk point estimates ranging from 1.48 to 1.69 (Bennett et al. 2008; Tonelli et al. 2009; Ludwig et al. 2009; Glaspy et al. 2010). The increased risk for mortality and thrombotic events in cancer patients receiving ESA therapy is specified in the black box warning included in the FDA label. Seizures and antibody-associated pure red cell aplasia (PRCA) have occurred in chronic renal failure patients receiving ESA therapy. While it is unclear whether cancer patients receiving ESA therapy are at risk of seizures and/or PRCA, ESA treatments should be closely monitored.

Myeloid Hematopoietic Growth Factors

Granulocyte Colony-Stimulating Factor (G-CSF)

The chemical properties of the myeloid hematopoietic growth factors, G-CSF (cf. Chap. 2) and granulocyte-macrophage colony-stimulating factor (GM-CSF), have been characterized (Table 17.3) and extensively reviewed (Armitage 1998). The gene that encodes G-CSF is located on chromosome 17; the mature G-CSF polypeptide has 174 amino acids and is produced in monocytes, fibroblasts, endothelial cells, and bone marrow stromal cells. Filgrastim, a non-glycosylated r-metHuG-CSF, is marketed by several companies under various trade names throughout the world, and several filgrastim biosimilars have also been approved. Lenograstim, a glycosylated rHuG-CSF, is not marketed in the United States but is marketed in other countries under several trade names. Pegfilgrastim, a sustained-duration form of filgrastim to which a 20 kDa polyethylene glycol molecule is covalently bound to the N-terminal methionine residue, is marketed as Neulasta® in the European Union, the United States, and other countries, and several pegfilgrastim biosimilars are in development. Although not all indications are approved in every country, filgrastim, lenograstim, and pegfilgrastim are indicated for the prevention and treatment of chemotherapy-induced febrile neutropenia in cancer patients receiving chemotherapy, mobilization of stem cells for transplantation in oncology patients, and support of induction/consolidation chemotherapy for AML and hematopoiesis after bone marrow transplantation, among others (Aapro et al. 2011). Filgrastim and pegfilgrastim are the first FDA-approved medical countermeasures to increase survival in patients exposed to myelosuppressive doses of radiation (Harrold et al. 2020).

Table 17.4 summarizes the biosimilars manufactured that have been approved in the EU, the US, and/or rest of the world (ROW) based on comparable safety and efficacy between

Table 17.3 Characteristics of the marketed myeloid growth factors, rhG-CSF, and rhGM-CSF

	G-CSF	GM-CSF
Nonproprietary name	Filgrastim, lenograstim, and pegfilgrastim	Molgramostim and sargramostim
Chromosome location	17	4
Amino acids	174[a]	127 or 128[b]
Glycosylation	O-linked (lenograstim)	N-linked (sargramostim)
Pegylation	Pegfilgrastim	None
Source of gene	Bladder carcinoma cell line (filgrastim, pegfilgrastim) and squamous carcinoma cell line (lenograstim)	Human monocyte cell line (molgramostim) and mouse T-lymphoma cell line (sargramostim)
Expression system	E. coli (bacteria): Filgrastim and pegfilgrastim	E. coli (bacteria): Molgramostim
	Chinese hamster ovary cell line (mammalian): Lenograstim	Saccharomyces cerevisiae (yeast): Sargramostim

[a]Native G-CSF has two forms, one with 177, which is less active than the other form with 174 amino acids; filgrastim has an N-terminal methionine
[b]Molgramostim has 128 amino acids; sargramostim 127

Table 17.4 Filgrastim, pegfilgrastim, and their biosimilars approved for treatment in the United States (US), European Union (EU) and other countries (ROW)

Active ingredient	Drug name	Country
Filgrastim	Neupogen	US, EU, ROW
	Nivestym/nivestim	US, EU, ROW
	Zarxio/zarzio	US, EU
	Grasofil, tevagrastim	EU, ROW
	Reluko	US
	Accofil, cegfila, hexal, ratiograstim	EU
Pegfilgrastim	Neulasta	US, EU
	Ziextenzo, fulphila, nyvepria	US, EU, ROW
	Fylentra, stimufend, udenyca	US
	Grasustek, pelmeg, stimufend	EU
	Pelgraz/lapelga/neupeg/neutropeg	EU, ROW

these biosimilars and the reference product (filgrastim or peg-filgrastim) in randomized controlled trials in healthy subjects or patients with cancer receiving chemotherapy.

Pharmacokinetics

Absorption and bioavailability

Absorption of filgrastim and pegfilgrastim after s.c. administration, similarly to other protein drugs, is controlled by a faster transport through the endothelial wall of blood capillaries and slower delivery to the systemic circulation via the lymphatic system. The serum concentration after a s.c. administration rapidly increases within less than an hour to reach a shallower peak at about 4–5 h and continue to decline in a bi-phasic manner (Wang et al. 2001; Wiczling et al. 2009). The onset rate decreases with the drug dose. A $t_{max} = 4.3$ h has been reported for lenograstim (Hayashi et al. 1999). The time course of pegfilgrastim serum concentrations after a s.c. dose administration follows the same pattern as for filgrastim, but with slower rates. The t_{max} values are dose dependent and range from 8 h (30 μg/kg) to 24 h (300 μg/kg) (Roskos et al. 2006). The mean filgrastim absorption time for the lymphatic route is 2.5–5 h with about 60% of the s.c. dose transported this way (Wang et al. 2001; Wiczling et al. 2009). The mean absorption time of the remaining 40% of the s.c. through the endothelial wall is about 3.3–3.8 h. The mean absorption time for lenograstim is 2.5 h (Hayashi et al. 1999), and for filgrastim 44 h (Roskos et al. 2006). The slower filgrastim absorption rate is attributed to its larger molecular weight, as for proteins that are larger than 16 kDa, the lymphatic pathway is a predominant route of absorption (Supersaxo et al. 1990). The absolute bioavailability for filgrastim is about 60–70% (Wang et al. 2001; Wiczling et al. 2009). After s.c. administration, both filgrastim and pegfilgrastim exhibits a flip-flop phenomenon, justifying efficiency of the s.c. administration relative to the i.v. dosing, as well as nonlinear and nonstationary pharmacokinetics due to pharmacodynamic-mediated drug disposition. These findings were quantitatively characterized in a population pharmacokinetic and pharmacodynamic meta-analysis using data from 10 phase I–III clinical studies, conducted in 110 healthy adults, and 618 adult and 52 pediatric patients on chemotherapy, following administration of a wide range of i.v. and s.c. doses of filgrastim or pegfilgrastim (Melhem et al. 2018).

Distribution

The binding of filgrastim and pegfilgrastim to G-CSFR expressed on neutrophils that circulate in the blood and extravascular tissues obscures assessment of the volume of distribution based on the serum concentration of unbound drug. There is no clear distributional phase in available PK data following i.v. administration. The estimate of filgrastim volume distribution based on a TMDD PK model (Wiczling et al. 2009; Melhem et al. 2018) is less than the plasma vol-

ume of adult humans, but slightly larger for pegfilgrastim (Melhem et al. 2018).

Elimination

Filgrastim is predominantly cleared by renal elimination and G-CSFR mediated endocytosis and degradation. Isotopic studies reported 50% of the administered dose was excreted the urine (1998). PK data in patients with hepatic impairment implies that filgrastim is minimally cleared by the liver (Lau et al. 1996). The filgrastim half-life is dose dependent. For an i.v. dose 34.5 μg/kg in healthy subjects $t_{1/2} = 3.8$ h (FDA 2022) and for 3.45 μg/kg $t_{1/2} = 2.7$ h (Roskos et al. 1998). Pegylation was designed to reduce pegfilgrastim renal clearance. The half-life of pegfilgrastim in healthy subjects following s.c. administration does not depend on dose and ranges between 46 and 62 h (Roskos et al. 2006).

Both filgrastim and pegfilgrastim are subjects to G-CSFR mediated clearance. Similarly to ESAs, upon G-CSFR binding and endocytosis, the drug is degraded in the lysosome or recycled to the cell membrane (Sarkar and Lauffenburger 2003). Since G-CSFRs are expressed mostly on the bone marrow precursor cells for neutrophils and circulating neutrophils, the G-CSFR mediated clearance is a saturable process limited by the pool of neutrophils and density of G-CSFR they express. There are approximately 50–500 receptors per cell (Nicola and Metclaf 1985). The in vitro equilibrium dissociation rate constant K_D for filgrastim is about 70–140 pM (Avalos et al. 1990). The pegfilgrastim affinity for the G-CSFR is only marginally lower than that of the parent protein (EMA 2009). TMDD models of filgrastim report K_D values in the range 16–77 pM (Wiczling et al. 2009; Krzyzanski et al. 2010). Simultaneous fits of a TMDD model to PK data revealed that K_D values are lower for filgrastim than pegfilgrastim (24 pM and 96 pM, respectively) (Melhem et al. 2018) indicating that in vivo affinity of filgrastim for G-CSFR is stronger than for pegfilgrastim. The reported values of the Michaelis-Menten constant K_m for the healthy subjects are 28.9 ng/mL for filgrastim (Wang et al. 2001) and 5.9 ng/mL for pegfilgrastim (Roskos et al. 2006). The constant K_m sets reference values for drug serum concentrations above which the receptor mediated clearance becomes saturated. Following multiple dosing, the filgrastim clearance increases due to an increased neutrophil count. The ratios of the first dose AUC_{0-24} to the tenth dose AUC_{0-24} were 2.5, 3.3, 5.9, and 7.2 following s.c. dosing of 75, 150, 300, and 600 μg/day, respectively (Roskos et al. 1998).

Pharmacodynamics

Filgrastim and pegfilgrastim increase the proliferation and differentiation of neutrophils from committed progenitor cells, induce maturation, and enhance the survival and function of mature neutrophils, resulting in dose-dependent increases in neutrophils counts. The mobilization of banded cells and segmented neutrophils from the bone marrow to the blood is enhanced. Small increases of monocyte and lymphocyte counts are observed. Eosinophilia and basophilia have not been seen (Roskos et al. 1998). The absolute neutrophil count (ANC) after a single i.v. dose 375 µg of filgrastim results in a rapid increase from a baseline (4.1 10^3cells/µL) to reach a peak 20 10^3cells/µL at about 10 h followed by a slower decline to the baseline at more than 70 h (Wang et al. 2001). An increase in dose increases both the ANC peak and peak time. An identical dose given s.c. results in a slightly higher peak and later peak time of 12–15 h. The differences in ANC responses to the same i.v. and s.c. dose increase with dose. This indicates higher filgrastim efficacy of the s.c. administration relative to the i.v.. This phenomenon is attributed to the flip-flop kinetics of filgrastim that results in its longer half-life and increased exposure for the s.c. administration, despite a fractional bioavailability. The ANC peak time after a single filgrastim dose correlates with the mean lifespan of neutrophils in the circulation of about 10 h (McKenzie 1996). A similar correlation holds for other hematopoietic growth factors (Krzyzanski et al. 1999). The ANC responses after a single dose of lenograstim are very similar to those for filgrastim (Hayashi et al. 1999). The ANC responses to a 30 µk/kg s.c. dose of pegfilgrastim result in a peak of 30.4 10^3cells/µL, a peak time of 2.5 days, and duration of the response of 5.8 days (Molineux et al. 1999). Similarly to filgrastim, all of these parameters increase with dose.

The daily i.v. doses of filgrastim result in increasing peaks after each administration until a plateau is reached at about 2–4 days for doses less than 11.5 µk/kg/day and 6–8 days for higher doses (Roskos et al. 1998). The plateau is dose dependent. The steady state ANC peak response increases threefold for dose 1.15 µk/kg/day and six-fold for dose 11.5 µk/kg/day. A similar ANC response and dose-response relationship can be observed for daily s.c. doses of filgrastim. After termination of the treatment, the ANC returns to the baseline within 24 h. The increased ANC for repeated doses increases the pool of G-CSFR, that in turns increases the filgrastim plasma clearance, the hallmark of pharmacodynamic-mediated drug disposition. Consequently, both drugs exhibit non-stationary PK for repeated dosing. Despite the lower pegfilgrastim in vivo affinity for the G-CSF receptor than filgrastim, the net stimulatory effects of pegfilgrastim are significantly greater than those of filgrastim. The longer half-life of pegfilgrastim, relative to filgrastim, counterbalances the lower receptor affinity.

Pharmacokinetics and Pharmacodynamics in Cancer Patients with Neutropenia

A primary indication of filgrastim and pegfilgrastim is to decrease infections in cancer patients with febrile neutropenia caused by myelosuppressive chemotherapy. Chemotherapy-induced neutropenia decreases filgrastim and pegfilgrastim G-CSFR-mediated clearance. Patients with metastatic breast cancer treated with melphalan received a continuous s.c. infusion of filgrastim 10 µg/kg/day for 5 days before melphalan or 20 µg/kg/day for 10 days after chemotherapy. Serum concentrations of filgrastim before chemotherapy decreased while after chemotherapy serum filgrastim accumulated during the time of the ANC nadir (Layton et al. 1989). In a clinical study of patients with non-small-cell lung cancer a single s.c. dose 100 µg/kg pegfilgrastim was administered 14 days before and 24 h after chemotherapy. The before and after chemotherapy AUC values were 5640 and 7150 µg/kg·h (Johnston et al. 2000). The high serum concentration of pegfilgrastim was sustained until ANC recovery began, after which serum pegfilgrastim concentration rapidly declined. Since neutropenia reduces G-CSFR-mediated clearance, filgrastim pharmacokinetics during the nadir ANC becomes linear. C_{max} and AUC were proportional to dose for breast cancer patients receiving i.v. single filgrastim doses 11.5, 34.5, or 69 µg/kg (Roskos et al. 1998). The half-life ranged from 1.8–5.9 h. Linear pharmacokinetics has also been reported for high doses of filgrastim, presumably due to saturation of receptor-mediated clearance by high filgrastim concentrations. Data from a pivotal study confirmed that a once-per-chemotherapy-cycle injection of pegfilgrastim at 6 mg was as safe and effective as 11 daily injections of filgrastim at 5 µg/kg in reducing neutropenia and its complications in patients with breast cancer receiving four cycles of doxorubicin/docetaxel chemotherapy (Green et al. 2003). Because of the highly efficient regulation of pegfilgrastim clearance via neutrophils and neutrophil precursors, a single fixed dose of pegfilgrastim can be given once per chemotherapy cycle in conjunction with a variety of myelosuppressive chemotherapy regimens (Yang and Kido 2011). Extensive clinical reviews on the myeloid growth factors have been published elsewhere (Keating 2011; Crawford et al. 2009).

Granulocyte-Macrophage Colony-Stimulating Factor (GM-CSF) and Stem Cell Factor (SCF)

The granulocyte-macrophage colony-stimulating factor (GM-CSF) is a polypeptide of 128 amino acids encoded by a gene located on chromosome 4, secreted by macrophages, T cells, mast cells, NK cells, endothelial cells, and fibroblasts. Molgramostim (marketed in the EU) and sargramostim (marketed in the USA) are two versions of rHuGM-CSF rarely used today. rHuGM-CSF is indicated for neutropenia associated with bone marrow transplantation and antiviral therapy for AIDS-related cytomegalovirus. rHuGM-CSF is also indicated for failed bone marrow transplantation or delayed engraftment and for use in mobilization and after transplantation of autologous PBPCs.

Similarly to G-CSF and GM-CSF, stem cell factor (SCF), encoded on chromosome 12, is a membrane-bound polypeptide of 248 amino acids that proteolytically release a soluble SCF containing 165 amino acids. SCF is an early-acting hematopoietic growth factor that stimulates the proliferation of primitive hematopoietic and non-hematopoietic cells. In vitro, SCF alone has minimal colony-stimulating activity on hematopoietic progenitor cells; however, it synergistically increases colony-forming or stimulatory activity of other HGF. Unlike most hematopoietic growth factors, SCF circulates in relatively high concentrations in normal human plasma. Ancestim® is a non-glycosylated version of the soluble r-metHuSCF marketed in Canada, Australia, and New Zealand and is rarely used in combination with G-CSF to increase the mobilization of peripheral blood progenitor cells (PBPC) for harvesting and support of autologous transplantation after myeloablative chemotherapy in patients with cancer. Comprehensive reviews of r-metHuSCF have been recently published (Langley 2004).

Megakaryocyte Hematopoietic Growth Factors

Megakaryocytopoiesis is a continuous developmental process of platelet production regulated by a complex network of HGF. In this process, hematopoietic stem cells undergo proliferation, differentiation, and maturation, generating megakaryocytes and platelets. Platelet production is controlled by signaling through the hematopoietic c-Mpl receptor. The ligand for this receptor, thrombopoietin (TPO) is the primary regulator of megakaryocyte development and subsequent platelet formation. TPO is an HGF encoded on chromosome 3 and produced in the liver and bone marrow stroma. Depending on the source, the mature polypeptide has between 305 and 355 amino acids, which may undergo cleavage to a smaller polypeptide that retains biologic activity. Upon binding to the c-Mpl receptor, TPO triggers several cellular signal transduction processes, which involve the following pathways: JAK-STAT and TYK2 tyrosine kinase, mitogen-activated protein kinase (MAPK), phosphatidylinositol 3-kinase (PI3K), and nuclear factor kappa B (NF-κB).

Early recombinant forms of TPO, rHu-TPO and the pegylated megakaryocyte growth and development factor (Peg-MGDF), showed promising results in clinical trials. However, later studies failed to meet their clinical endpoints, because the recombinant proteins generated antibodies that cross-reacted with c-Mpl ligands and resulted in paradoxical thrombocytopenia (Li et al. 2001). Further clinical development of these molecules was therefore suspended. An extensive compilation of the biology of rHu-TPO and Peg-MGDF has been published (Kuter et al. 1997).

Romiplostim (Nplate®), previously known as AMG 531, is a novel biological thrombopoiesis receptor agonist that was developed to overcome the problem of cross-reacting autoantibodies by use of a peptide sequence with no homology to endogenous TPO to activate the c-Mpl receptor. Structurally, romiplostim is a 59 kDa fusion protein that con-

sists of two identical subunits, each containing a human IgG$_1$ Fc domain covalently linked at the C-terminus to a peptide consisting of two c-Mpl binding domains. The four copies of the TPO mimetic peptide stimulate megakaryocytopoiesis by binding the TPO receptor, yet because they bear no sequence homology with TPO, there is a reduced potential for the generation of anti-TPO antibodies. In vitro, romiplostim competes with TPO for binding to the c-Mpl receptor on normal platelets and Mpl-transfected cells (BaF3-Mpl cells). Upon binding to the c-Mpl receptor on megakaryocyte precursors in bone marrow, romiplostim activates the Janus kinase/signal transducers and activators of transcription (JAK-STAT) and other pathways in the same way as endogenous TPO (Broudy and Lin 2004), leading to sustained improvement of platelet counts with continued treatment. When co-cultured with murine bone marrow cells, romiplostim promotes the growth of CFU-megakaryocytes and promotes the proliferation as well as the maturation of megakaryocytes (Bussel et al. 2021). Much less is known about the effects mediated by the Fc region of the molecule and any immunomodulatory effects that may occur by specifically binding with Fc receptors and thereby affecting various immune responses. Interaction with the Fcγ receptors may allow romiplostim to modify the maintenance of humoral tolerance, cell maturation, antigen presentation, and Treg expansion. Finally, romiplostim may be capable of activating regulatory T cells through two epitopes of the Fc region termed *Tregitopes* (Bussel et al. 2021).

During preclinical development, romiplostim led to robust platelet responses in mice, rats, rabbits, and monkeys. The pharmacokinetics and pharmacodynamics of romiplostim in animals, healthy subjects, and patients with immune thrombocytopenia purpura (ITP) have been extensively investigated during clinical development (Wang et al. 2010, 2011; Yan and Krzyzanski 2013; Perez-Ruixo et al. 2012). In brief, among healthy volunteers, platelet counts increased 1 to 3 days after i.v. administration and 4 to 9 days after s.c. administration, peaking on days 12 to 16. Similar to erythropoietin stimulating agents and G-CSF analogs, romiplostim exhibits a nonlinear pharmacokinetics and drug increased more than dose proportionally after i.v. administration. Based on noncompartmental analysis, as the romiplostim i.v. dose increased from 0.3 to 10 μg/kg, the mean central volume of distribution decreased from 122 to 48.2 mL/kg, and the mean clearance decreased from 754 to 6.69 mL/h/kg. Similar to TPO, the disposition of romiplostim presumably involves TPOR on platelets and other cells in the thrombopoiesis lineage, such as megakaryocytes; binding of romiplostim to c-Mpl receptor also triggers subsequent internalization and degradation. Because the number of c-Mpl receptors on the cell surface is limited, binding to the c-Mpl receptor is saturated at high romiplostim concentrations, resulting in nonlinear volume of distribution and clearance, which translates into a nonlinear and nonstationary pharmacokinetics due to pharmacodynamic-mediated drug disposition (Wang et al. 2010).

Romiplostim absolute bioavailability after s.c. administration was estimated to be approximately 50% based on pharmacokinetic/pharmacodynamic modeling (Wang et al. 2010). Most serum concentrations of romiplostim after s.c. administration (dose range 0.1–2 µg/kg) were not measurable, but the observed thrombocytosis precluded increasing the s.c. doses higher than 2 µg/kg in healthy subjects. At this s.c. dose level, absorption of romiplostim from the subcutaneous site appeared to be slow, with the peak concentrations observed between 24 and 36 h after dosing. The romiplostim s.c. absorption was slower than its elimination after i.v. dosing, which suggests the presence of a flip-flop phenomenon and justifies the higher efficiency of the s.c. route relative to the i.v. dosing (Wang et al. 2010).

Romiplostim administered as a single i.v. (0.3–10.0 µg/kg) or s.c. dose (0.1–2.0 µg/kg) in healthy subjects produced dose-dependent and prolonged increases in circulating platelet counts. Platelet counts rose above baseline (average 200–260 × 10^9/L) 1–3 days after i.v. administration and 4–9 days after s.c. administration, peaked on days 12–16, and returned to baseline by day 28. This platelet time course was similar to those obtained from previous studies of PEG-rHuMGDF and rHuTPO. The delay in platelet response after romiplostim administration may reflect enhanced maturation of megakaryocytes in the bone marrow or platelet release into circulation, or both; the subsequent rise in platelet counts may be related to an expansion of the megakaryocytes. After reaching the peak of platelet count, the decrease in platelet counts is determined by the rate of platelet production (megakaryocytopoiesis and thrombopoiesis) as well as the life span of the platelet (8–11 days). Similar platelet responses were observed after i.v. and s.c. administration of romiplostim at the same dose level (1 µg/kg), although serum concentrations of romiplostim were markedly lower (barely measurable or not measurable) after s.c. administration (Wang et al. 2004). This pharmacodynamic response independent of the route of administration has also been observed in animals receiving romiplostim (Krzyzanski et al. 2013) or with other hematopoietic growth factors, suggesting that platelet response is not driven by the magnitude of concentrations achieved, but by the length of time that romiplostim concentrations remained above a threshold. Based on the study data, the threshold for romiplostim to elicit the pharmacologic effect appears to be low and indicative of the high potency of romiplostim in humans. Similar to rHu-EPO, clinical data suggest that approximately only 20–30% of c-Mpl receptors must be occupied with thrombopoietin receptor agonist in order to have 50% of the maximal effect in stimulating the proliferation and differentiation of precursors cells (Wang et al. 2010; Samtani et al. 2009).

Romiplostim was initially approved in the USA and the EU to increase the number of platelets in order to decrease the risk of bleeding in adults who have idiopathic thrombocytopenic purpura (ITP), an ongoing condition that may cause easy bruising or bleeding due to an abnormally low number of platelets in the blood, and have had an insufficient response to corticosteroids, immunoglobulins, or splenectomy (Bussel et al. 2006). An extensive review of the use of romiplostim in ITP patients has been published (Keating 2012). Romiplostim is also used to increase the number of platelets in order to decrease the risk of bleeding in children at least 1 year of age who have had ITP for at least 6 months. Romiplostim should only be used in adults and children 1 year of age or older who cannot be treated or have not been helped by other treatments, including other medications or surgery to remove the spleen. Romiplostim is also indicated for hematopoietic syndrome of acute radiation to increase survival in adults and in pediatric patients (including term neonates) acutely exposed to myelosuppressive doses of radiation (>2 Gy) (Bussel et al. 2021).

Romiplostim is used to increase the number of platelets enough to lower the risk of bleeding, but it is not used to increase the number of platelets to a normal level. Romiplostim should not be used to treat people who have low platelet levels caused by myelodysplastic syndrome (a group of conditions in which the bone marrow produces blood cells that are misshapen and does not produce enough healthy blood cells) or any other conditions that cause low platelet levels other than ITP. At this time, romiplostim or other protein-based c-Mpl ligands are not approved for clinical use in cancer patients; however, clinical trial data in oncology patients have been recently reported (Kantarjian et al. 2010a, b; Bussel et al. 2021).

Self-Assessment Questions

Questions

1. What do hematopoietic factors do?
2. What are the major lineages or types of mature blood cells?
3. Generally, chemically describe the hematopoietic growth factors.
4. How do hematopoietic growth factors function?
5. What are the in vivo actions of rhG-CSF and rhGM-CSF in patients with advanced cancer?
6. What is the physiologic role of EPO?
7. What are the currently commercially available hematopoietic growth factors?
8. What are the indications for rhG-CSF?
9. What are the indications for rhEPO?
10. What is the indication for romiplostim?
11. What are the relevant and common pharmacokinetic and pharmacodynamic properties of the HGF?

Answers

1. Hematopoietic growth factors regulate both hematopoiesis and the functional activity of blood cells (including proliferation, differentiation, and maturation). Some hematopoietic growth factors mobilize progenitor cells to move from the bone marrow to the peripheral blood.

2. The myeloid pathway gives rise to red blood cells (erythrocytes), platelets, monocytes/macrophages, and granulocytes (neutrophils, eosinophils, and basophils). The lymphoid pathway gives rise to lymphocytes.

3. They are glycoproteins, which can be distinguished by their amino acid sequence and glycosylation (carbohydrate linkages). Hematopoietic growth factors have folding patterns that are dictated by physical interactions and covalent cysteine-cysteine disulfide bridges. Correct folding is necessary for biologic activity. Most hematopoietic growth factors are single-chain polypeptides weighing approximately 14–35 kDa. The carbohydrate content varies depending on the growth factor and production method, which in turn affects the molecular weight but not necessarily the biologic activity.

4. Hematopoietic growth factors act by binding to specific cell surface receptors. The resultant complex sends a signal to the cell to express genes, which in turn induce cellular proliferation, differentiation, or activation. A hematopoietic growth factor may also act indirectly if the cell expresses a gene that causes the production of a different hematopoietic growth factor or another cytokine, which in turn binds to and stimulates a different cell.

5. Both growth factors cause a transient leucopenia that is followed by a dose-dependent increase in the number of circulating mature and immature neutrophils. Both growth factors enhance the in vitro function of neutrophils obtained from treated patients. rhGM-CSF, but not rhG-CSF, also increases the number of circulating monocytes/macrophages and eosinophils, as well as in vitro monocyte cytotoxicity and cytokine production.

6. EPO maintains a normal red blood cell count by causing committed erythroid progenitor cells to proliferate and differentiate into normoblasts. EPO also shifts marrow reticulocytes into circulation.

7. Five hematopoietic growth factors are commercially available, rhG-CSF (filgrastim, lenograstim, pegfilgrastim), rhGM-CSF (molgramostim and sargramostim), rhEPO (epoetin alfa, epoetin beta, darbepoetin alfa), rhSCF (ancestim), and rhIL-11 (oprelvekin).

8. Approval for marketing varies by country and not all countries have all labeled uses. rhG-CSF is indicated for neutropenia associated with myelosuppressive cancer chemotherapy, bone marrow transplantation, and severe chronic neutropenia; rhG-CSF is also indicated to mobilize peripheral blood progenitor cells (PBPC) for PBPC transplantation; and rhG-CSF is indicated for the reversal of clinically significant neutropenia and subsequent maintenance or adequate neutrophil counts in patients with HIV infection during treatment with antiviral and/or other myelosuppressive medications.

9. rhEPO is indicated to treat anemia associated with chronic renal failure, zidovudine-induced anemia in HIV-infected patients, and chemotherapy-induced anemia. rhEPO is also indicated to reduce allogeneic blood transfusions and hasten erythroid recovery in surgery patients.

10. Romiplostim is indicated for the treatment of thrombocytopenia in patients with chronic immune thrombocytopenia (ITP) who have had an insufficient response to corticosteroids, immunoglobulins, or splenectomy.

11. There are two main characteristics that are common to erythropoietin stimulating agents, G-CSF analogs and thrombopoietin receptor agonist. The first one is the presence of the flip-flop pharmacokinetics that justifies the efficiency of the s.c. administration relative to the i.v. dosing. The second is the nonlinear (concentration-dependent) and nonstationary pharmacokinetics due to pharmacodynamic-mediated drug disposition, which justify the dose approved since they achieve the level of receptor coverage to achieve clinically relevant endpoints.

Acknowledgements Parts of this chapter are updated versions of previously published portions of several other chapters and/or review manuscripts, that include:

1. Heatherington AC (Heatherington 2003) Clinical pharmacokinetic properties of rHuEPO: a review. In: Molineux G, Foote MA, Elliott S (eds) Erythropoietins and erythropoiesis: molecular, cellular, preclinical, and clinical biology. Birkhauser, Basel, pp. 87–112.

2. Elliot S, Heatherington AC, Foote MA (2004) Erythropoietic factors: clinical pharmacology and pharmacokinetics. In: Morstyn G, Foote MA, and Lieschke GJ (eds) Hematopoietic growths factors in oncology. Humana Press, Totowa, pp. 97–123.

3. Foote AN (2008) Hematopoietic growth factors. In: Crommelin DJA, Sindelar RD, Meibohm B (eds) Pharmaceutical biotechnology. Fundamentals and applications, third edn. Informa Healthcare USA, New York, pp. 225–242.

4. Doshi S, Perez-Ruixo JJ, Jang GR, Chow A, Elliot S (2008) Pharmacocinétique de les agents stimulant l'érythropoïèse. In: Rossert J, Casadevall N, Gisselbrecht C (eds) Les agents stimulant l'érythropoïèse. Paris, France.

5. Doshi S, Perez-Ruixo JJ, Jang GR, Chow AT (2009) Pharmacokinetics of erythropoiesis-stimulating agents. In: Molineux G, Foote MA, Elliott S (eds) Erythropoietins and erythropoiesis: molecular, cellular, preclinical, and clinical biology. second edn. Birkhäuser Verlag AG, Basel, pp. 195–224.
6. Doshi S, Krzyzanski W, Yue S, Elliott S, Chow A, Pérez-Ruixo JJ. Clinical pharmacokinetics and pharmacodynamics of erythropoiesis-stimulating agents. Clin Pharmacokinet. 2013 Dec;52(12):1063-83.

References

Aapro MS, Bohlius J, Cameron DA et al (2011) 2010 update of EORTC guidelines for the use of granulocyte-colony stimulating factor to reduce the incidence of chemotherapy-induced febrile neutropenia in adult patients with lymphoproliferative disorders and solid tumours. Eur J Cancer 47:8–32

Adamson JW, Eschbach JW (1990) Treatment of the anemia of chronic renal failure with recombinant human erythropoietin. Annu Rev Med 41:349–360

Agoram B, Sutjandra L, Sullivan JT (2007) Population pharmacokinetics of darbepoetin alfa in healthy subjects. Br J Clin Pharmacol 63:41–52

Agoram B, Aoki K, Doshi S et al (2009) Investigation of the effects of altered receptor binding activity on the clearance of erythropoiesis-stimulating proteins: nonerythropoietin receptor-mediated pathways may play a major role. J Pharm Sci 98:2198–2211

Ait-Oudhia S, Vermeulen A, Krzyzanski W (2011) Non-linear mixed effect modeling of the time-variant disposition of erythropoietin in anemic cancer patients. Biopharm Drug Dispos 32:1–15

Armitage JO (1998) Emerging applications of recombinant human granulocyte colony-stimulating factor. Blood 92:4491–4508

Avalos BR, Gasson JC, Hedvat C, Quan SG, Baldwin GC, Weisbart RH, Williams RE, Golde DW, DiPersio JF (1990) Human granulocyte colony-stimulating factor: biologic activities and receptor characterization on hematopoietic cells and small cell lung cancer cell lines. Blood 75:851–857

Bennett CL, Silver SM, Djulbegovic B et al (2008) Venous thromboembolism and mortality associated with recombinant erythropoietin and darbepoetin administration for the treatment of cancer-associated anemia. JAMA 299:914–924

Bohlius J, Schmidlin K, Brillant C et al (2009) Recombinant human erythropoiesis-stimulating agents and mortality in patients with cancer: a meta-analysis of randomised trials. Lancet 373:1532–1542

Bron D, Meuleman N, Mascaux C (2001) Biological basis of anemia. Semin Oncol 28:1–6

Broudy VC, Lin NL (2004) AMG531 stimulates megakaryopoiesis in vitro by binding to Mpl. Cytokine 25:52–60

Bussel JB, Kuter DJ, George JN et al (2006) AMG 531, a thrombopoiesis-stimulating protein, for chronic ITP. N Engl J Med 355:1672–1681

Bussel JB, Soff G, Balduzzi A, Cooper N, Lawrence T, Semple JW (2021) A review of romiplostim mechanism of action and clinical applicability. Drug Des Devel Ther 26(15):2243–2268

Chanu P, Gieschke R, Charoin JE, Pannier A, Reigner B (2010) Population pharmacokinetic – pharmacodynamic model for a C.E.R.a. in both ESA-naive and ESA-treated chronic kidney disease patients with renal anemia. J Clin Pharmacol 50:507–520

Chapel S, Veng-Pedersen P, Hohl RJ, Schmidt RL, McGuire EM, Widness JA (2001) Changes in erythropoietin pharmacokinetics following busulfan-induced bone marrow ablation in sheep: evidence for bone marrow as a major erythropoietin elimination pathway. J Pharmacol Exp Ther 298:820–824

Cheung WK, Goon BL, Guilfoyle MC, Wacholtz MC (1998) Pharmacokinetics and pharmacodynamics of recombinant human erythropoietin after single and multiple subcutaneous doses to healthy subjects. Clin Pharmacol Ther 64:412–423

Cheung WK, Minton N, Gunawardena K (2001) Pharmacokinetics and pharmacodynamics of epoetin alfa once weekly and three times weekly. Eur J Clin Pharmacol 57:411–418

Crawford J, Armitage J, Balducci L et al (2009) Myeloid growth factors. J Natl Compr Cancer Netw 7:64–83

Deicher R, Horl WH (2004) Differentiating factors between erythropoiesis-stimulating agents: a guide to selection for anemia of chronic kidney disease. Drugs 64:499–509

Doshi S, Chow A, Pérez Ruixo JJ (2010) Exposure-response modeling of darbepoetin alfa in anemic patients with chronic kidney disease not receiving dialysis. J Clin Pharmacol 50(9 Suppl):75S–90S

Egrie JC, Browne JK (2001) Development and characterization of novel erythropoiesis stimulating protein (NESP). Br J Cancer 84(Suppl 1):3–10

Egrie JC, Strickland TW, Lane J et al (1986) Characterization and biological effects of recombinant human erythropoietin. Immunobiology 72:213–224

Egrie JC, Dwyer E, Browne JK, Hitz A, Lykos MA (2003) Darbepoetin alfa has a longer circulating half-life and greater in vivo potency than recombinant human erythropoietin. Exp Hematol 31:290–299

Elliott S, Lorenzini T, Asher S et al (2003) Enhancement of therapeutic protein in vivo activities through glycoengineering. Nat Biotechnol 21:414–421

Elliott S, Heatherington AC, Foote MA (2004a) Erythropoietic factors. In: Morstyn G, Foote MA, Lieschke GJ (eds) Hematopoietic growth factors in oncology: basic science and clinical therapeutics. Humana Press Inc, Totowa, pp 97–123

Elliott S, Egrie J, Browne J et al (2004b) Control of rHuEPO biological activity: the role of carbohydrate. Exp Hematol 32:1146–1155

EMA (2009) Neulasta: EPAR Scientific discussion. https://www.ema.europa.eu/en/documents/scientific-discussion/neulasta-epar-scientific-discussion_en.pdf

FDA's labeling resources for human prescription drugs (2022) Neupogen (filgrastim). Highlights of prescribing information. https://www.accessdata.fda.gov/spl/data/b77e2e68-e746-4922-8b98-7cfeb5f710a6/b77e2e68-e746-4922-8b98-7cfeb5f710a6.xml

Fisher JW (2003) Erythropoietin: physiology and pharmacology update. Exp Biol Med 228:1–14

Gascon P, Pirker R, Del Mastro L, Durrwell L (2010) Effects of CERA (continuous erythropoietin receptor activator) in patients with advanced non-small-cell lung cancer (NSCLC) receiving chemotherapy: results of a phase II study. Ann Oncol 21:2029–2039

Glaspy J, Henry D, Patel R et al (2005) The effects of chemotherapy on endogenous erythropoietin levels and the pharmacokinetics and erythropoietic response of darbepoetin alfa: a randomised clinical trial of synchronous versus asynchronous dosing of darbepoetin alfa. Eur J Cancer 41:1140–1149

Glaspy J, Crawford J, Vansteenkiste J et al (2010) Erythropoiesis stimulating agents in oncology: a study-level meta-analysis of survival and other safety outcomes. Br J Cancer 102:301–315

Green MD, Koelbl H, Baselga J et al (2003) A randomized double-blind multicenter phase III study of fixed-dose single-administration pegfilgrastim versus daily filgrastim in patients receiving myelosuppressive chemotherapy. Ann Oncol 14:29–35

Gross AW, Lodish HF (2006) Cellular trafficking and degradation of erythropoietin and novel erythropoiesis stimulating protein (NESP). J Biol Chem 281:2024–2032

Halstenson CE, Macres M, Katz SA et al (1991) Comparative pharmacokinetics and pharmacodynamics of epoetin alfa and epoetin beta. Clin Pharmacol Ther 50:702–712

Harrold H, Gisleskog PO, Perez-Ruixo JJ, Delor I, Chow A, Jacqmin P, Melhem M (2020) Prediction of survival benefit of filgrastim in adult and pediatric patients with acute radiation syndrome. Clin Transl Sci 13:807–817

Hayashi N, Kinoshita H, Yukawa E, Higuchi S (1999) Pharmacokinetic and pharmacodynamic analysis of subcutaneous recombinant human granulocyte colony stimulating factor (lenograstim) administration. J Clin Pharmacol 39:583–592

Heatherington AC (2003) Clinical pharmacokinetic properties of rHuEPO: a review. In: Molineux G, Foote MA, Elliott S (eds) Erythropoietins and erythropoiesis: molecular, cellular, preclinical, and clinical biology. Birkhauser, Basel, pp 87–112

Jarsch M, Brandt M, Lanzendörfer M, Haselbeck A (2008) Comparative erythropoietin receptor binding kinetics of C.E.R.a. and epoetin-beta determined by surface plasmon resonance and competition binding assay. Pharmacology 81:63–69

Jelkmann W (1992) Erythropoietin: structure, control of production, and function. Physiol Rev 72:449–489

Jelkmann W (2000) Use of recombinant human erythropoietin as an antianemic and performance enhancing drug. Curr Pharm Biotechnol 1:11–31

Jensen JD, Jensen LW, Madsen JK (1994) The pharmacokinetics of recombinant human erythropoietin after subcutaneous injection at different sites. Eur J Clin Pharmacol 46:333–337

Johnston E, Crawford J, Blackwell S et al (2000) Randomized, dose escalation study of SD/01 compared with daily filgrastim in patients receiving chemotherapy. J Clin Oncol 18:2522–2528

Jungers PY, Robino C, Choukroun G, Nguyen-Khoa T, Massy ZA, Jungers P (2002) Incidence of anaemia, and use of Epoetin therapy in pre-dialysis patients: a prospective study in 403 patients. Nephrol Dial Transplant 17:1621–1627

Kakkar T, Sung C, Gibiansky L et al (2011) Population PK and IgE pharmacodynamic analysis of a fully human monoclonal antibody against IL4 receptor. Pharm Res 28:2530–2542

Kantarjian H, Fenaux P, Sekeres MA et al (2010a) Safety and efficacy of romiplostim in patients with lower-risk myelodysplastic syndrome and thrombocytopenia. J Clin Oncol 28:437–444

Kantarjian HM, Giles FJ, Greenberg PL et al (2010b) Phase 2 study of romiplostim in patients with low- or intermediate-risk myelodysplastic syndrome receiving azacitidine therapy. Blood 116:3163–3170

Kato A, Hishida A, Kumagai H, Furuya R, Nakajima T, Honda N (1994) Erythropoietin production in patients with chronic renal failure. Ren Fail 16:645–651

Kaufman JS, Reda DJ, Fye CL et al (1998) Subcutaneous compared with intravenous epoetin in patients receiving hemodialysis. Department of Veterans Affairs Cooperative Study Group on erythropoietin in hemodialysis patients. N Engl J Med 339:578–583

Keating GM (2011) Lenograstim: a review of its use in chemotherapy-induced neutropenia, for acceleration of neutrophil recovery following haematopoietic stem cell transplantation and in peripheral blood stem cell mobilization. Drugs 71:679–707

Keating GM (2012) Romiplostim. Drugs 72:415–435

Kessler M, Martinez-Castelao A, Siamopoulos KC, Villa G, Spinowitz B, Dougherty FC, Beyer U (2010) C.E.R.A. once every 4 weeks in patients with chronic kidney disease not on dialysis: the ARCTOS extension study. Hemodial Int 14:233–239

Kidanewold A, Woldu B, Enawgaw B (2021) Role of erythropoiesis stimulating agents in the treatment of anemia: a literature review. Clin Lab 67(4)

Klinger M, Arias M, Vargemezis V, Besarab A, Sulowicz W, Gerntholtz T, Ciechanowski K, Dougherty FC, Beyer U (2007) Efficacy of intravenous methoxy polyethylene glycol-epoetin beta administered

every 2 weeks compared with epoetin administered 3 times weekly in patients treated by hemodialysis or peritoneal dialysis: a randomized trial. Am J Kidney Dis 50:989–1000

Krzyzanski W, Ramakrishnan R, Jusko WJ (1999) Basic models for agents that alter production of natural cells. J Pharmacokin Biopharm 27:467–489

Krzyzanski W, Jusko WJ, Wacholtz MC, Minton N, Cheung WK (2005) Pharmacokinetic and pharmacodynamic modeling of recombinant human erythropoietin after multiple subcutaneous doses in healthy subjects. Eur J Pharm Sci 26:295–306

Krzyzanski W, Wiczling P, Lowe P, Pigeolet E, Fink M, Berghout A, Balser S (2010) Population modeling of filgrastim PK-PD in healthy adults following intravenous and subcutaneous administrations. J Clin Pharmacol 50:101S–112S

Krzyzanski W, Sutjandra L, Perez-Ruixo JJ, Sloey B, Chow AT, Wang YM (2013) Pharmacokinetic and pharmacodynamic modeling of romiplostim in animals. Pharm Res 30:655–669

Kuter DJ, Hunt P, Sheridan W, Zucker-Franklin D (eds) (1997) Thrombopoiesis and thrombopoetin. Humana Press Inc, Totowa, p 412

Langley KE (2004) Stem cell factor and its receptor, c-kit. In: Morstyn G, Foote MA, Lieschke GJ (eds) Hematopoietic growth factors in oncology. Humana Press Inc, Totowa, pp 153–184

Lau D, Pilz D, Schwab G (1996) Phase I pharmacokinetic and pharmacodynamic studies of G-CSF (filgrastim) in patients with renal or liver impairment compared to healthy volunteers. Br J Haematol 93(Suppl. 2):277

Layton JE, Hockman H, Sheridan WP, Morstyn G (1989) Evidence for a novel in vivo control mechanism of granulopoiesis: mature cell-related control of a regulatory growth factor. Blood 74:1303–1307

Li J, Yang C, Xia Y et al (2001) Thrombocytopenia caused by the development of antibodies to thrombopoietin. Blood 98:3241–3248

Ludwig H, Crawford J, Osterborg A et al (2009) Pooled analysis of individual patient-level data from all randomized, double-blind, placebo controlled trials of darbepoetin alfa in the treatment of patients with chemotherapy-induced anemia. J Clin Oncol 27:2838–2847

Macdougall IC (2005) CERA (continuous erythropoietin receptor activator): a new erythropoiesis-stimulating agent for the treatment of anemia. Curr Hematol Rep 4:436–440

Macdougall IC, Gray SJ, Elston O et al (1999) Pharmacokinetics of novel erythropoiesis stimulating protein compared with epoetin alfa in dialysis patients. J Am Soc Nephrol 10:2392–2395

McKenzie SB (1996) Texbook of hematology, 2nd edn. Williams & Wilkins, Baltimore

McLennan DN, Porter CJ, Edwards GA, Heatherington AC, Martin SW, Charman SA (2006) The absorption of darbepoetin alfa occurs predominantly via the lymphatics following subcutaneous administration to sheep. Pharm Res 23:2060–2066

McMahon FG, Vargas R, Ryan M et al (1990) Pharmacokinetics and effects of recombinant human erythropoietin after intravenous and subcutaneous injections in healthy volunteers. Blood 76:1718–1722

Melhem M, Delor I, Pérez-Ruixo JJ et al (2018) Pharmacokinetic-pharmacodynamic modelling of neutrophil response to G-CSF in healthy subjects and patients with chemotherapy-induced neutropenia. Br J Clin Pharmacol 84:911–925

Miller CB, Jones RJ, Piantadosi S, Abeloff MD, Spivak JL (1990) Decreased erythropoietin response in patients with the anemia of cancer. N Engl J Med 322:1689–1692

Molineux G, Kinstler O, Briddell B et al (1999) A new form of Filgrastim with sustained duration in vivo and enhanced ability to mobilize PBPC in both mice and humans. Exp Hematol 27:1724–1734

Nicola MA, Metclaf D (1985) Binding of 125I-labeled granulocyte colony-stimulating factor to normal murine hemopoietic cells. J Cell Physiol 124:313–321

Nissenson AR, Swan SK, Lindberg JS et al (2002) Randomized, controlled trial of darbepoetin alfa for the treatment of anemia in hemodialysis patients. Am J Kidney Dis 40:110–118

Olsson-Gisleskog P, Jacqmin P, Perez-Ruixo JJ (2007) Population pharmacokinetics meta-analysis of recombinant human erythropoietin in healthy subjects. Clin Pharmacokinet 46:159–173

Perez-Ruixo JJ, Krzyzanski W, Bouman-Thio E et al (2009) Pharmacokinetics and pharmacodynamics of the erythropoietin mimetibody construct CNTO 528 in healthy subjects. Clin Pharmacokinet 48:601–613

Perez-Ruixo JJ, Green B, Doshi S, Wang YM, Mould DR (2012) Romiplostim dose-response in patients with immune thrombocytopenia. J Clin Pharmacol 52:1540–1551

Petrelli F, Borgonovo K, Cabiddu M, Lonati V, Barni S (2012) Addition of iron to erythropoiesis-stimulating agents in cancer patients: a meta-analysis of randomized trials. J Cancer Res Clin Oncol 138:179–187

Piron M, Loo M, Gothot A, Tassin F, Fillet G, Beguin Y (2001) Cessation of intensive treatment with recombinant human erythropoietin is followed by secondary anemia. Blood 97:442–448

Porter CJ, Charman SA (2000) Lymphatic transport of proteins after subcutaneous administration. J Pharm Sci 89:297–310

Ramakrishnan R, Cheung WK, Wacholtz MC, Minton N, Jusko WJ (2004) Pharmacokinetic and pharmacodynamic modeling of recombinant human erythropoietin after single and multiple doses in healthy volunteers. J Clin Pharmacol 44:991–1002

Rizzo JD, Somerfield MR, Hagerty KL et al (2008) Use of epoetin and darbepoetin in patients with cancer: 2007 American Society of Clinical Oncology/American Society of Hematology clinical practice guideline update. J Clin Oncol 26:132–149

Roskos LK, Cheung EN, Vincent M, Foote M, Morstyn G (1998) Pharmacology of filgrastim (r-meHuG-CSF). In: Morstyn G, Dexter TM, Foote M (eds) Filgrastim (r-meHuG-CSF) in clinical practice, 2nd edn. Marcel Dekker, New York

Roskos LK, Lum P, Lockbaum P, Schwab G, Yang BB (2006) Pharmacokinetic/pharmacodynamic modeling of pegfilgrastim in healthy subjects. J Clin Pharmacol 46:747–757

Samtani MN, Perez-Ruixo JJ, Brown KH et al (2009) Pharmacokinetic and pharmacodynamic modeling of pegylated thrombopoietin mimetic peptide (PEG-TPOm) after single intravenous dose administration in healthy subjects. J Clin Pharmacol 49:336–350

Sarkar CA, Lauffenburger DA (2003) Cell-level pharmacokinetic model of granulocyte colony-stimulating factor: implications for ligand lifetime and potency in vivo. Mol Pharmacol 63:147–158

Schellekens H, Moors E (2010) Clinical comparability and European biosimilar regulations. Nat Biotechnol 28:28–31

Schellekens H, Smolen JS, Dicato M, Rifkin RM (2016) Safety and efficacy of biosimilars in oncology. Lancet Oncol 17:e502–e509

Sinclair AM, Elliott S (2005) Glycoengineering: the effect of glycosylation on the properties of therapeutic proteins. J Pharm Sci 94:1626–1635

Spivak JL (1998) The biology and clinical applications of recombinant erythropoietin. Semin Oncol 25(3 suppl 7):7–11

Sulowicz W, Locatelli F, Ryckelynck JP, Balla J, Csiky B, Harris K, Ehrhard P, Beyer U (2007) Once-monthly subcutaneous C.E.R.a. maintains stable hemoglobin control in patients with chronic kidney disease on dialysis and converted directly from epoetin one to three times weekly. Clin J Am Soc Nephrol 2:637–646

Supersaxo A, Hein WR, Steffen H (1990) Effect of molecular weight on the lymphatic absorption of water-soluble compounds following subcutaneous administration. Pharm Res 7:167–169

Sutjandra L, Rodriguez RD, Doshi S et al (2011) Population pharmacokinetic meta-analysis of denosumab in healthy subjects and postmenopausal women with osteopenia or osteoporosis. Clin Pharmacokinet 50:793–807

Tonelli M, Hemmelgarn B, Reiman T et al (2009) Benefits and harms of erythropoiesis-stimulating agents for anemia related to cancer: a meta-analysis. CMAJ 180:E62–E71

Uehlinger DE, Goth FA, Sheiner LB (1992) A pharmacodynamic model of erythropoietin therapy for uremic anemia. Clin Pharmacol Ther 51:76–89

Vansteenkiste J, Pirker R, Massuti B et al (2002) Double-blind, placebo-controlled, randomized phase III trial of darbepoetin alfa in lung cancer patients receiving chemotherapy. J Natl Cancer Inst 94:1211–1220

Walrafen P, Verdier F, Kadri Z, Chrétien S, Lacombe C, Mayeux P (2005) Both proteasomes and lysosomes degrade the activated erythropoietin receptor. Blood 105:600–608

Wang B, Ludden TM, Cheung EN EN, Schwab GG, Roskos LK (2001) Population pharmacokinetic–pharmacodynamic modeling of filgrastim (r-metHuG-CSF) in healthy volunteers. J Pharmacokinet Pharmacodyn 28:321–342

Wang B, Nichol JL, Sullivan JT (2004) Pharmacodynamics and pharmacokinetics of AMG 531, a novel thrombopoietin receptor ligand. Clin Pharmacol Ther 76:628–638

Wang YM, Krzyzanski W, Doshi S, Xiao JJ, Pérez-Ruixo JJ, Chow AT (2010) Pharmacodynamics-mediated drug disposition (PDMDD) and precursor pool lifespan model for single dose of romiplostim in healthy subjects. AAPS J 12:729–740

Wang YM, Sloey B, Wong T, Khandelwal P, Melara R, Sun YN (2011) Investigation of the pharmacokinetics of romiplostim in rodents with a focus on the clearance mechanism. Pharm Res 28:1931–1938

Wiczling P, Lowe P, Pigeolet E, Ludicke F, Balser S, Krzyzanski W (2009) Population pharmacokinetic modelling of filgrastim in healthy adults following intravenous and subcutaneous administrations. Clin Pharmacokinet 48:817–826

Wide L, Bengtsson C, Birgegard G (1989) Circadian rhythm of erythropoietin in human serum. Br J Haematol 72:85–90

Yan X, Krzyzanski W (2013) Quantitative assessment of minimal effective concentration of erythropoiesis-stimulating agents. CPT Pharmacometrics Syst Pharmacol 2:e62. https://doi.org/10.1038/psp.2013.39

Yan X, Lowe PJ, Fink M, Berghout A, Balser S, Krzyzanski W (2012) Population pharmacokinetic and pharmacodynamic model-based comparability assessment of a recombinant human epoetin alfa and the biosimilar HX575. J Clin Pharmacol 52:1624–1644

Yang BB, Kido A (2011) Pharmacokinetics and pharmacodynamics of pegfilgrastim. Clin Pharmacokinet 50:295–306

Recombinant Coagulation Factors and Thrombolytic Agents

Koen Mertens and Alexander B. Meijer

Introduction

Blood coagulation and fibrinolysis are two sides of the same coin: the hemostatic system. Hemostasis involves the interplay of the coagulation cascade with activated blood platelets and the vessel wall. This results in the formation of a thrombus, which comprises activated platelets and cross-linked fibrin. Thrombus formation is counterbalanced by coagulation inhibitors on the one hand and the fibrinolytic system on the other hand. Upon vascular injury, coagulation *(thrombus formation)* serves to achieve local bleeding arrest, while fibrinolysis *(thrombus dissolution)* prevents obstruction of blood flow elsewhere in the circulation. Disruption of the balance between these mechanisms can cause malfunction of the hemostatic system and becomes manifest as bleeding or thrombosis. Bleeding disorders are usually due to a defect or deficiency in one of the constituents of the coagulation cascade and can be treated by substitution therapy to compensate for the missing component. Thrombotic disorders comprise two distinct forms: venous and arterial thrombosis. Venous thrombosis results from excessive coagulation and can ultimately lead to pulmonary embolism, while arterial thrombosis is associated with atherosclerosis, stroke, and acute coronary syndrome, the major indication for thrombolytic therapy. This chapter focuses on the recombinant biopharmaceuticals that are currently available for the treatment of bleeding and thrombosis.

K. Mertens (✉)
Department of Pharmaceutical Sciences, Utrecht Institute for Pharmaceutical Sciences, Utrecht University, Utrecht, The Netherlands
e-mail: k.mertens@uu.nl

A. B. Meijer
Department of Molecular Hematology, Sanquin Research, Amsterdam, The Netherlands
e-mail: s.meijer@sanquin.nl

The Essentials of Blood Coagulation and Fibrinolysis

Formation and Dissolution of the Hemostatic Plug

While the network of hemostasis and thrombosis is generally perceived as being a complex issue in physiology, understanding the essentials thereof will greatly facilitate appreciation of the biopharmaceuticals used in this area, and in particular, of the pharmacodynamics thereof. As schematically depicted in Fig. 18.1, the system can be divided into three distinct sections. The left part (panel a) comprises the coagulation cascade. This is an ordered conversion of proenzymes into active enzymes that leads to the activation of prothrombin into thrombin. Once the initial amounts of thrombin are formed, it catalyzes a variety of enzymatic conversions. These include a self-amplifying loop by feedback activation of several other components upstream in the cascade and self-dampening of its own formation by activating an enzyme that specifically inactivates essential cofactors in the cascade. The central part (panel b) summarizes how thrombin drives the formation of the hemostatic plug, initially by activating platelets and converting fibrinogen into fibrin polymers, and subsequently by activating factor XIII, which then cross-links fibrin into a stable network. The right section (panel c) represents the fibrinolytic system, in which activators like tissue-type plasminogen activator (t-PA) convert plasminogen into the enzyme plasmin that degrades cross-linked fibrin into soluble fragments, resulting in lysis of the hemostatic plug.

The Network of Coagulation Factors

Why do we need so many coagulation factors? The answer lies in the various regulatory steps, which allow the mechanism to act as a biological amplifier that rapidly responds to injury, while at the same time the process should remain restricted to the site of injury. The apparent complexity of the coagulation scheme in Fig. 15.1 (panel a) merits some further explanation. Apart from a few exceptions, the proteins in

D. J. A. Crommelin et al. (eds.), *Pharmaceutical Biotechnology*, https://doi.org/10.1007/978-3-031-30023-3_18

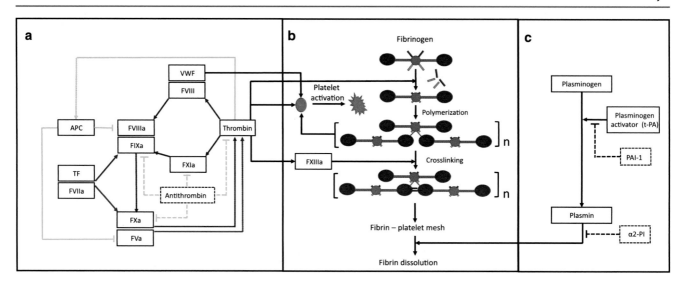

Fig. 18.1 Schematic representation of hemostasis. Panel (**a**) the coagulation cascade; Panel (**b**): the hemostatic plug; Panel (**c**): the fibrinolytic system. The various phases in the coagulation cascade comprise initiation (*blue*), propagation (*red*), and termination (*yellow*). The constituents thereof are further described in the text. Roman numerals refer to individual coagulation factors. Von Willebrand factor is abbreviated as *VWF*, *APC* activated protein C, *TF* tissue factor, *PAI-1* plasminogen activator inhibitor-1, *α2-PI* α2-plasmin inhibitor

this system ('factors') are indicated by roman numerals, with the suffix 'a' referring to their activated form. This nomenclature dates from 1959 and has been implemented to resolve confusion by the coexistence of multiple names for the same component. For some factors, however, the old names have remained in use, such as fibrinogen (factor I), prothrombin (factor II), and Tissue Factor (factor III). The terms factor IV and factor VI are obsolete, as they have never been assigned to a specific protein. Of the other coagulation factors, numbers V and VIII are cofactors, while the remaining numbers VII and IX to XII are proenzymes of serine proteases that catalyze the sequential steps in the cascade (Furie and Furie 1988).

The concept of an ordered sequence of reactions as a cascade or waterfall has first been described in 1964 (Davie and Ratnoff 1964; Macfarlane 1964) and has remained essentially unaltered since then. What did change, however, is the insight into the various feedback effects of thrombin upstream in the cascade. This introduced a distinction between successive stages in thrombin formation, which are often referred to as *initiation*, *propagation*, and *termination* (Mann et al. 2009). In each of these phases, different sections of the cascade predominate. These can be summarized as follows:

1. In the *initiation phase*, indicated by *blue arrows* in Fig. 18.1a, vascular injury leads to the exposure of tissue factor (TF). The cascade then runs from the FVIIa/TF to FXa/FVa complex to the conversion of prothrombin into thrombin. The initially formed thrombin starts activating platelets and cleaving fibrinogen into fibrin (Fig. 18.1b), and at the same time, initiates the self-amplifying propagation loop.

2. This *propagation phase*, indicated by *red arrows,* involves the activation of factor XI (FXI) and the cofactors factor VIII (FVIII) and factor V (FV). The cascade then runs from the FXIa (or FVIIa) to FIXa/FVIIIa complex and further from the FXa/FVa complex to generate large amounts of thrombin. The main difference between the initiation and the propagation phase is that the latter comprises an additional amplifying step, which involves FVIII and FIX.

3. Finally, in the *termination phase*, indicated by the *yellow arrows*, thrombin binds to an endothelial receptor (thrombomodulin) and activates Protein C. Once activated, this inactivates the cofactors FVIIIa and FVa and thereby causes thrombin formation to shut-down. The activated coagulation enzymes are neutralized by antithrombin (*yellow dashed lines*).

Local Control Mechanisms

How do the various mechanisms shown in Fig. 18.1 remain localized to the site of injury? One reason is that assembly of enzyme–cofactor complexes in the coagulation cascade requires membranes containing acidic phospholipids, which are exposed on perturbed cells such as activated platelets, but not on resting cells. Moreover, any excess of activated factors is neutralized by protease inhibitors that occur in plasma. These include tissue factor pathway inhibitor, which inhibits at the level of factor VIIa, and antithrombin, an inhibitor of most of the other enzymes from the cascade. Similar considerations apply to the fibrinolytic system (Fig. 18.1c). First, most t-PA is stored in vascular cells and released upon perturbation, while levels of t-PA in circulation are low. Second, plasmin needs its substrate, fibrin, to express full fibrinolytic activity. Finally, the circulating inhibitors plasminogen activator inhibitor-1 (PAI-1) and α2-plasmin inhibitor (α2-PI) neutralize the excess of their target proteases (indicated by dashed lines in Fig. 18.1c).

Hemostatic Disorders

There is little or no redundancy within the intricate protein network involved in the formation and dissolution of the hemostatic plug. Therefore, deficiency or dysfunction of a single component may have a severe impact. The consequences thereof can include both bleeding and thrombosis. Deficiency of most of the coagulation factors results in a bleeding tendency, while a shortage of coagulation inhibitors such as antithrombin or protein C predisposes to thrombosis. Likewise, thrombosis may also result from a defect in the fibrinolytic pathway, while bleeding has been associated with a deficiency in α2-plasmin inhibitor. The incidence of deficiencies in hemostatic proteins is summarized in Table 18.1 (adapted from Bishop and Lawson 2004 and Palla et al. 2015). In affected patients, bleeding can be treated by administering the missing plasma protein. This may be of human origin, as a concentrate derived from human plasma, or a biopharmaceutical from recombinant sources. Table 18.1 lists the availability of such protein therapeutics. It is evident that, with a few exceptions, recombinant products have been clustering around the field of hemophilia, which due to its relatively high incidence, represents the most significant market.

Hemophilia and Other Bleeding Disorders

Hemophilia is the best-known bleeding disorder, as it occurred in several European royal families, including the Romanov dynasty in Russia. These descendants of the British Queen Victoria had a defect in the gene encoding in factor IX and thus suffered from hemophilia B (Rogaev et al. 2009). The more frequent disorder, however, is hemophilia A, or factor VIII deficiency. Hemophilia A and B are clinically indistinguishable, which is to be expected because factor VIII and factor IX participate in the same step in the coagulation cascade: the activation of factor X. Lack of either the cofactor or the enzyme moiety in the FVIIIa/FIXa complex disrupts the biological amplifier that drives the propagation phase in thrombin generation (see Fig. 18.1). Hemophilia differs from the other inherited bleeding disorders in that the genes encoding factor VIII and factor IX are located on the X-chromosome. The inheritance pattern is X-linked recessive, and mutation in the single X-chromosome in males is sufficient to cause the deficiency. In females, however, the same mutation in one X-chromosome results in heterozygosity and in factor levels which, although reduced, still are sufficient to support normal hemostasis.

The other coagulation factor deficiencies do not display this sex-linked pattern, and a complete deficiency thereof will need a defect in both chromosomes (homozygous or compound heterozygous). For most of these deficiencies, often referred to as the Rare Bleeding Disorders, substitution therapy is available, albeit to a limited extent, in the form of concentrates from human blood (see Table 18.1).

The most common bleeding disorder is von Willebrand's disease. This is caused by a deficiency or dysfunction of the von Willebrand factor, a large multimeric adhesion protein that mediates the interaction between activated platelets and the perturbed vessel wall at the site of injury. As such, the von Willebrand factor is an essential constituent of the hemostatic plug (see Fig. 18.1b). Besides its role as an adhesion protein, the von Willebrand factor serves as carrier of factor VIII in the circulation and protects factor VIII from degradation and premature clearance. Within von Willebrand's disease, a variety of subtypes can be distinguished, with partial quantitative deficiency and different molecular defects (for

Table 18.1 Deficiency of hemostatic proteins: incidence and options for substitution therapy

Deficiency	Incidence	Bleeding phenotype	Availability of products for substitution therapy	
			Human plasma-derived	Recombinant
Fibrinogen (factor I)	1 in 1 million	None to severe	Few	None
Prothrombin (factor II)	1 in 2 million	Mild to moderate	Common	None
Factor V	1 in 1 million	Moderate	One (under evaluation)	None
Factor VII	1 in 500,000	Mild to severe	Few	Factor VIIa only
Factor VIII (hemophilia A)	1 in 5000 males	Moderate to severe	Common	Several (see Table 18.2)
Factor IX (hemophilia B)	1 in 25,000 males	Moderate to severe	Common	Several (see Table 18.3)
Factor X	1 in 1 million	Mild to severe	Common	None
Factor XI	1 in 1 million	Mild to moderate	Few	None
Factor XIII	1 in 2 million	Moderate to severe	Few	One
von Willebrand factor	~1 in 1000	Mild to severe	Common	One
α2-plasmin inhibitor	very rare	Mild to severe	None	None
Antithrombin	~1 in 5000	Thrombotic phenotype	Common	One

review, see Castaman and Linari 2017). Only the most severe deficiency (type III, 1–2% of all cases) requires substitution therapy with von Willebrand factor-containing therapeutics.

Thrombosis

While protein substitution therapy is a straightforward approach to stopping or preventing bleeding, the management of thrombosis is more complex. Partial deficiencies of antithrombin or protein C are risk factors that predispose to venous thrombosis, but usually multiple, often acquired risk factors, are involved. The mechanisms involved in thrombosis are beyond the scope of the present chapter and have been reviewed elsewhere (Mackman 2008). Treatment involves dampening the coagulation cascade or inhibiting platelet activation, usually by small molecules rather than protein therapeutics. Protein substitution therapy remains limited to some specific cases of antithrombin deficiency.

As for arterial thrombosis, this is associated with atherosclerosis. After the rupture of an atherosclerotic plaque, a platelet-rich thrombus may lead to arterial occlusion, which then may result in ischemia, myocardial infarction, or stroke (Lusis 2000; Mackman 2008). In such situations, lysis of the occluding thrombus is the immediate target. This can be accomplished by infusing large amounts of plasminogen activators. The medical need for such thrombolytic therapy is high. This explains why tissue-type plasminogen activator, after human insulin and human growth hormone, was among the first recombinant protein therapeutics that reached the market.

Recombinant Factor VIII

In the early 1980s, the factor VIII cDNA has been cloned by two independent groups, one collaborating with Genentech (Vehar et al. 1984), and another with Genetics Institute (Toole et al. 1984). This provided access to the biotechnological development of recombinant factor VIII. Factor VIII was among the largest proteins characterized at that time. The factor VIII gene spans 180 kb and comprises 26 exons that together encode a polypeptide of 2351 amino acids. This includes a signal peptide of 19 amino acids and a mature protein of 2332 amino acids. As reviewed in detail elsewhere (Lenting et al. 1998; Fay 2004), factor VIII displays the ordered domain structure A1-*a1*-A2-*a2*-B-*a3*-A3-C1-C2 (see Fig. 18.2a). The triplicated A-domains share homology with the copper-binding protein ceruloplasmin, while the two C-domains are structurally related to lipid-binding proteins such as lactadherin. The A-domains are bordered by short spacers (*a1*, *a2*, and *a3*) that are also known as acidic regions. These spacers contain sulfated tyrosine residues that are needed for full biological activity.

The central B-domain carries the vast majority of the glycosylation sites in factor VIII. Due to processing at the junction B-*a3*, mature factor VIII is secreted as a heterodimer of the heavy chain (A1-*a1*-A2-*a2*-B) and light chain (*a3*-A3-C1-C2). In the circulation, the B-domain is subject to cleavage at multiple sites, resulting in size heterogeneity without apparent consequence for biological activity (Andersson et al. 1986). Thus, about one-third of the factor VIII protein seems dispensable for biological activity. Indeed, recombinant expression of a factor VIII variant lacking a major part of the B-domain proved functionally normal (Toole et al. 1986). For B-domain deleted factor VIII, several crystal structures are available. These reveal a compact cluster of the three A-domains, supported by a tandem of the two C-domains, in a side-by-side orientation (Fig. 18.2b). This structure represents the inactive factor VIII, and conversion into factor VIIIa involves cleavage at the *a1*-A2, *a2*-B, and *a3*-A3 junctions. The resulting heterotrimer is the molecular form that assembles with factor IXa to activate factor X in the coagulation cascade.

Factor VIII Products

Due to the size and complexity of the factor VIII protein, mammalian cells need to be used as an expression system. Commonly used cells are CHO (Chinese hamster ovary), BHK (baby hamster kidney), and HEK (human embryonic kidney) or engineered derivatives thereof. Post-translational modifications (see Fig. 18.2a) include glysosylation and the sulfation of specific tyrosine residues in the *a1*, *a2*, and *a3* segments. One particular challenge is its low expression, as factor VIII tends to accumulate intracellularly, which is cytotoxic for the host cells. Another challenge is the instability of factor VIII once secreted. Over the past 40 years, these issues have been resolved by a limited number of manufacturers only. An overview of the various products is given in Table 18.2.

Full-Length Factor VIII

The first generation of recombinant factor VIII was intended to provide exact copies of the natural protein from human plasma, which had been established as being effective in the treatment of hemophilia A since the 1960s. The first products, Kogenate (developed at Genentech and Bayer) and Recombinate (developed at Genetics Institute and Baxter) were licensed in 1992. In order to enhance secretion and stability, Recombinate was produced by co-expression with the von Willebrand factor, which was removed during downstream processing. The simultaneous expression of these two exceptionally large polypeptides was a remarkable achievement in the early years of biopharmaceuticals.

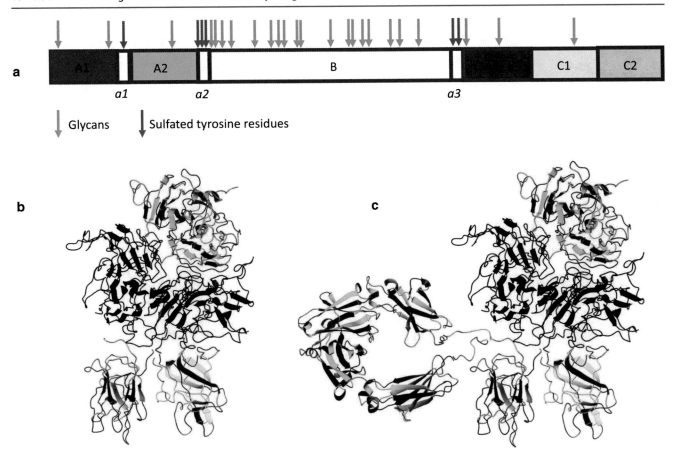

Fig. 18.2 Structure of factor VIII. Panel (**a**) displays the domain structure of full-length factor VIII. Panel (**b**) represents the 3D-structure of B-domain-deleted factor VIII (derived from pdb 2R7E), and panel (**c**) is the same structure with the Fc fragment of IgG1 (from pdb 1HZA) fused to the C-terminus of the C2-domain

Both Kogenate and Recombinate were produced using the human albumin-containing culture medium and employed an albumin-containing formulation for the final product. As such, these initial products were actually plasma-derived, except that the active ingredient was produced through recombinant technology. In 2000, Kogenate has been replaced by Kogenate-FS, with an albumin-free formulation. In 2016, this was followed by Kovaltry, which employs co-expression with heat shock protein-70 to facilitate factor VIII expression in protein-free medium (Table 18.2). As for Recombinate, a protein-free equivalent (Advate) became available in 2003. A PEGylated variant thereof, Adynovate, was licensed in 2015 (see below, section "Extended Half-Life Factor VIII").

B-Domain Deleted (BDD) Factor VIII

The finding that the central B-domain is dispensable for in vitro activity (Toole et al. 1986) formed the basis for the design of a variety of deletion mutants. An advantage of B-domain deletion is that this greatly improves factor VIII expression. The first BDD-factor VIII was ReFacto

(developed at Pharmacia/Pfizer), which was followed in 2009 by the protein-free equivalent ReFacto-AF/Xyntha. Other BDD-factor VIII products have slightly different deletions (see Table 18.2). All of them have conserved a small N-terminal part of the B-domain, in order to maintain the thrombin cleavage site at the *a2*-B junction (position 740–741). Moreover, most constructs have retained the processing site at the B-*a3* junction (position 1648–1649) and consequently are, like wild-type factor VIII, heterodimeric molecules (NovoEight, Nuwiq). An exception is Afstyla, which is a single-chain molecule because the deletion extends into the *a3* segment. Despite the differences between the individual constructs, these BDD-factor VIII products, once activated, are fully indistinguishable from wild-type factor VIIIa, because the B-domain remnant and the *a3*-section are released upon activation. Other BDD factor VIII products, in particular those that have been engineered to extend half-life (see Table 18.2), may still contain the introduced modification within their activated form (see following section).

Table 18.2 Recombinant factor VIII products

Product type	INN name	(Investigational) product name(s)	First approval (USA/EU)	Half-life (mean, h)	Production technology	Reference
Full-length FVIII	octocog alfa	Recombinate	1992	12	CHO cells, co-expression with VWF, albumin in culture and formulation	
		Advate	2003	10-14	as Recombinate, but without protein in culture, albumin-free formulation	
		Kogenate Helixate	1992	14-17	BHK cells, albumin in culture and formulation	
		Kogenate-FS	2000	14-17	as Kogenate, but with albumin-free formulation	
		Kovaltry, Iblias (BAY 81-8973)	2016	13	BHK, co-expression with HSP-70, protein-free culture and formulation,	Maas Enriquez et al. (2016)
	rurioctocog alfa pegol	Adynovate (BAX 855)	2015	14-16	as Advate, 20kDa pegylation to extend half-life	Turecek et al. (2012)
B-Domain Deleted FVIII	moroctocog alfa	ReFacto	1999	14	CHO cells, deletion of amino acids 744-1637, albumin in culture, not in formulation	
		ReFacto-AF Xyntha	2008	13	as ReFacto, albumin-free culture and formulation	
	turoctogoc alfa	NovoEight (N8)	2013	11	CHO cells, deletion of amino acids 751-1637, albumin-free culture and formulation	Thim et al. (2010)
	simoctocog alfa	Nuwiq	2014	17	HEK cells, B-domain replaced by 16 amino acid linker, protein-free culture and formulation	Sandberg et al. (2012)
	lonoctocog alfa	Afstyla (rVIII-SingleChain)	2016	14	CHO cells, deletion of amino acids 765-1652, single-chain molecule, albumin-free	Schmidbauer et al. (2015)
	efmoroctocog alfa	Eloctate, Elocta (rFVIIIFc)	2014	20	HEK cells, deletion of amino acids 744-1637, dimeric IgG1-Fc fused to FVIII C2-domain	Leksa et al. (2017)
	damoctocog alfa pegol	Jivi (BAY 94-9027)	2018	~19	HEK-hybrid cells, deletion of amino acids 743-1636, Lys-1804 substituted by Cys for site-specific 40kDa pegylation in the A3-domain	Mei et al. (2010)
	turoctocog alfa pegol	Esperoct (N8-GP)	2019	~19	as NovoEight, with site-specific glycopegylation in residual B-domain segment	Stennicke et al. (2013)
	efanesoctocog alfa	Altuviiio (BIVV001)	2023	~41	dimeric IgG1-Fc fusion as in Eloctate, with added fusion with FVIII-binding region of VWF and two half-life extending (XTEN) insertions	Lissitchkov et al. (2022)
FVIII mimic	Emicizumab	Hemlibra (ACE910)	2018	4-5 weeks	Recombinant humanized bispecific antibody against FIXa and FX for subcutaneous administration	Uchida et al. (2016)

Extended Half-Life Factor VIII

One limitation of factor VIII replacement therapy is the relatively short half-life, which may vary between 10 and 17 h (see Table 18.2). Half-life extension would provide an obvious therapeutic advancement. Over the past few years, several products have been developed to pursue this goal. Thus, the original paradigm that recombinant factor VIII should be identical to the natural protein has started shifting toward engineering to improve on nature.

Adynovate is a PEG (polyethylene glycol)-ylated full-length factor VIII, based on Advate. It contains on average 2 moles of branched 20 kDa PEG per factor VIII molecule.

Approximately 60% of the PEG chains are located in the B-domain and thus are lost upon activation. The remaining modifications are randomly dispersed over the rest of the protein and, by virtue of the low extent of PEGylation, do not interfere with any known functional interaction (Turecek et al. 2012).

Eloctate is a fusion between BDD-factor VIII and the Fc fragment of IgG. This serves to target factor VIII to the FcRn receptor recycling pathway that endows IgG and albumin with their long half-life (see Chap. 8). The fusion of the Fc fragment to the factor VIII C-terminus yields a derivative in which the activated form is non-natural in that it still carries the Fc moiety. Surprisingly, this C-terminal extension conserves factor VIII's potency, presumably because the Fc portion is loosely attached and, besides the linker sequence, has no specific interaction with the factor VIII surface (Leksa et al. 2017). On the other hand, the fusion protein displays an aberrant response in some activity assays used for therapy monitoring (see section "Clinical Considerations").

Another product in this category is Esperoct. This is a PEGylated version of NovoEight (N8-GP, see Table 18.2). This is modified with a branched 40 kDa PEG using enzymatic glycoconjugation, which is targeted to the single O-linked glycan in the short B-domain remnant. Because the PEG moiety is released upon activation, the active form is the natural, unmodified factor VIIIa. A different strategy has been used for the recently approved Jivi (BAY 94–9027, see Table 18.2), developed by Bayer. This employs cysteine-targeted PEGylation with a branched 60 kDa PEG. This cysteine is introduced by mutagenesis in the factor VIII A3 domain. The PEG moiety remains conserved after activation, and thus the active species in this product is non-natural. This has some impact on activity assays used for monitoring (see section "Clinical Considerations").

The benefit of these half-life extensions seems limited (Table 18.2). For the current four products, cross-over studies reported a 1.5- to 1.7-fold half-life prolongation over an unmodified comparator (Peyvandi et al. 2013). So far, this seems to be the maximum achievable prolongation by such modification strategies. This limitation is because factor VIII, whether or not modified to extend its half-life, circulates in complex with the recipient's endogenous von Willebrand factor. Therefore, the clearance of the von Willebrand factor remains the major driver of factor VIII's half-life (Pipe et al. 2016). To overcome this limitation, a novel engineered factor VIII variant (BIVV001, Efanesoctocog alfa) has been designed in which factor VIII is fused with a fragment of the von Willebrand factor comprising the factor VIII-binding region. This fusion construct further comprises a dimeric Fc fragment and two unstructured, so-called 'XTEN' insertions, together resulting in a 3 to four-fold half-life extension (see Table 18.2).

Factor VIII Mimics

A remarkably innovative approach is to design proteins that are totally unrelated to factor VIII but share its biological activity. This is based on the notion that factor VIIIa assembles with FIXa, and facilitates the proper alignment with factor X that is needed for full activity in the propagation phase of coagulation (see Fig. 18.1). This has been accomplished by creating a humanized bispecific monoclonal antibody, wherein one IgG-arm binds to factor IXa, and the other to factor X. This compound, initially called ACE910, now Emicizumab, was developed at Chugai Pharmaceutical as a treatment for hemophilia A patients with inhibitors against

FVIII but finally has been licensed for use in hemophiliacs both with and without inhibitors (Hemlibra, Table 18.2). A major advantage of an antibody over factor VIII is that it has a much longer half-life and allows for subcutaneous administration. (see Chap. 8 for details on monoclonal antibodies). Another difference is that factor VIII circulates in an inactive form, while the antibody is not under such proteolytic control. This carries the inherent risk of causing thrombotic events. For Hemlibra, safe dosage regimens for prophylaxis have been established between 1.5 mg/kg once weekly and 6 mg/kg every 4 weeks. As such, Hemlibra provides a valuable innovation in the repertoire of factor VIII-based therapeutics.

A novel factor VIII mimetic antibody, designated Mim8, has been developed at Novo Nordisk (Østergaard et al. 2021). One difference with Hemlibra is that the factor IXa-binding arm has been specifically optimized to also enhance the proteolytic activity of factor IXa. Preclinical data suggest that Mim8 is significantly more potent than Hemlibra. Whether or not this translates into a clinical benefit is currently under investigation.

Pharmacology

The term factor VIII, or Anti-Hemophilic Factor in the early US literature, has been introduced before the protein had been identified. Factor VIII was defined by its function as the activity that corrects the clotting defect of hemophilic plasma. This functional definition has remained the basis for the quantification of factor VIII since then, and factor VIII concentrations are expressed in units, where 1 unit represents the amount of factor VIII activity in 1 mL of normal human plasma. Besides in units/mL, factor VIII levels in plasma are often also expressed as % of normal. Thus, the assessment of product potency, dosing, and pharmacokinetics is based on bioassays, with their inherent variability.

Pharmacokinetics

The pharmacokinetics of factor VIII is largely dependent on the plasma level of the von Willebrand factor, which stabilizes factor VIII and protects against premature clearance. As reviewed in detail elsewhere (Björkman and Berntorp 2001), systemic clearance in patients with normal von Willebrand factor levels is typically 3 mL/h/kg, with an apparent distribution volume that slightly exceeds the patient's plasma volume, and an average elimination half-life of approximately 14 h. While assessment of pharmacokinetics is mandatory for all individual factor VIII products, there is quite some variability between the data reported (see Iorio 2017 and Table 18.2). Apart from variation between patients, technical issues contribute to this variation, such as the sampling time points, and the dependence on bioassays for the determination of the administered dose and plasma levels post-infusion. Given this variability, the pharmacokinetics of the various recombinant factor VIII products can be considered as being essentially similar, with the exception of those products that have been engineered to extend half-life.

Pharmacodynamics

Hemophilia A is a congenital deficiency or dysfunction of factor VIII which, dependent on the residual level of factor VIII activity, is categorized as mild (6–40%), moderate (1–5%), or severe (<1% of normal) (for review, see Franchini and Mannucci 2013). Administration of factor VIII serves as a substitution therapy, with the objective of restoring the defect in thrombin generation. Normalization of factor VIII levels in the circulation requires administration by the intravenous route. The therapeutic range is rather narrow, because supra-normal factor VIII levels increase the risk of thrombosis, while the risk of bleeding tends to increase when levels drop below 40%. A usual strategy is to maintain a dosing schedule that ensures that factor VIII trough levels remain at least at 1–4% of normal to avoid the bleeding risk of the severe form of hemophilia A.

For all licensed factor VIII products, efficacy has been established for both on-demand and prophylactic treatment. Dosing is generally based on the empirical finding that 1 unit factor VIII per kg body weight raises the plasma factor VIII activity by 2%, which implies the simple formula (Björkman and Berntorp 2001):

$$\text{dose}(\text{units} / \text{kg}) = 50 \times (\text{required rise in units} / \text{mL factor VIII})$$

On-demand treatment involves factor VIII administration once bleeding occurs. Dependent on the type of bleeding, this requires administration of 20–100 units/kg, which can be repeated every 8–24 h if needed (Franchini and Mannucci 2013). In developed countries, the general treatment has the objective not to treat but rather to prevent bleeding episodes. This can be short-term prophylaxis to provide protection during surgical or invasive procedures or long-term prophylaxis to prevent spontaneous bleeding. The latter involves regular treatment at a dose of 20–40 units/kg, which needs to be repeated 2–3 times per week in order to maintain the factor VIII level above 1% of normal. For extended half-life products, a recommended regimen is 50 units/kg every 4–5 days (Cafuir and Kempton 2017).

For all factor VIII products, the main side effect is the formation of antibodies against factor VIII (inhibitors) once replacement therapy is started. This involves 30–40% of all severe patients. In the majority of cases, these antibodies are transient and can be eliminated by immuno-tolerance therapy with high doses of factor VIII. In a significant proportion of the patient population, however, the antibodies remain persistent and thereby preclude further treatment with factor VIII (Franchini and Mannucci 2013). In those cases, bleeding can be controlled using factor VIII-bypassing agents, including recombinant factor VIIa (see below) and bispecific antibody factor VIII mimics.

Pharmaceutical Considerations

While the factor VIII concentration in human plasma is approximately 0.1 µg/mL, the potency of factor VIII products is expressed in International Units (IU). This is an activity unit relative to a reference preparation that is established from time to time by the World Health Organization. Products are available in fillings ranging between 250 and 2000 (or 3000) IU in order to provide convenient dosing for children and adults using a single vial. Current products are freeze-dried and need to be reconstituted before infusion. Most products are stabilized in a sucrose-containing formulation and have a shelf life of 24–36 months upon storage at 2–8 °C. For most products, data are available that establish stability at room temperature for 6–12 months, thus facilitating patients to travel while maintaining their prophylactic treatment.

As for impurities, most products may contain low amounts of residual proteins from the host expression system. With the exception of Nuwiq, which is expressed in a human cell line, these products may cause side effects in patients with hypersensitivity against rodent proteins. This adverse event, however, remains rare. To further enhance safety, the downstream processing of most recombinant factor VIII products includes virus-eliminating steps that have been previously established for plasma-derived factor VIII products, such as immuno-affinity chromatography, solvent/detergent treatment, ultrafiltration, and combinations thereof (Franchini and Mannucci 2013).

Clinical Considerations

The sole indication for recombinant factor VIII is congenital hemophilia A that is not complicated by factor VIII inhibitors. Similarly, acquired hemophilia A, which is due to the spontaneous formation of anti-factor VIII antibodies should be treated using alternative therapies such as factor VIII bypassing agents.

For all recombinant factor VIII products, appropriate treatment protocols have been established as part of the licensing procedure. Nevertheless, there is an increasing trend toward personalized prophylactic treatment based on the pharmacokinetic profile of individual patient/product combinations. The advantage of this policy is that it allows the precise prediction of trough values and adaptation of dosing schedules accordingly. As for the extended half-life products, these tend to be used not to increase the dosing intervals but rather to maintain higher trough levels and thus to offer better protection against spontaneous bleeding (Iorio 2017). A crucial requirement for this approach to be allowable is the availability of robust methods for monitoring post-infusion factor VIII levels. While this is feasible for full-length products, this remains a problematic issue for many of the engineered factor VIII products, including some products with B-domain deletions, the single-chain factor VIII product, some of the PEGylated products, and the Fc-fusion protein. While some manufacturers recommend the use of specific assay reagents, or reagent-dependent conversion factors, it seems evident that this approach lacks robustness, and as such is complicating the personalized use of these products (Kitchen et al. 2017).

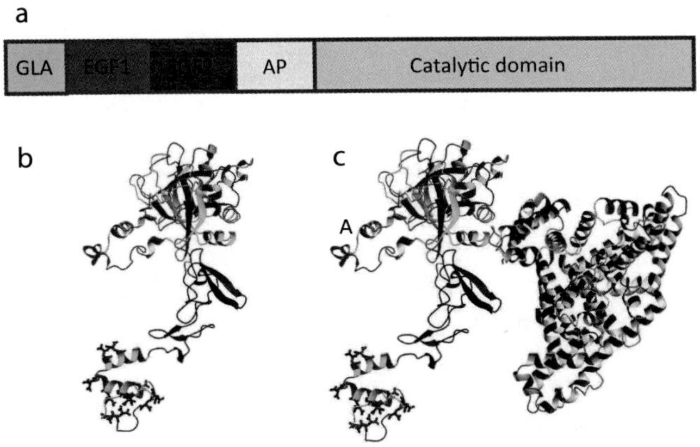

Recombinant Factor IX

While the cDNA encoding factor IX has been cloned in 1982 (Choo et al. 1982; Kurachi and Davie 1982), it took until 1997 before the first recombinant factor IX was licensed. Expressing the protein proved challenging in that factor IX requires excessive post-translational modification for full biological activity. These include 12 glutamine residues in the N-terminal section of the mature protein, which need to be modified into γ-carboxyglutamic acid (conversion of Glu into Gla) by vitamin K-dependent carboxylase. The modification involves interaction between carboxylase and the 46 amino acid long propeptide of factor IX, which is cleaved off by a furin-like processing enzyme prior to secretion (Jorgensen et al. 1987; Wasley et al. 1993). This results in a mature protein of 416 amino acids with the Gla-rich segment (the Gla-domain) at its N-terminus. Factor IX further comprises two epidermal growth factor-like (EGF) domains, an activation peptide (AP), and the catalytic domain (see Fig. 18.3a). Cleavage at both sides of the activation peptide yields a two-chain protein, consisting of a heavy chain (the catalytic domain) and a light chain (the Gla-EGF1-EGF2 section). Structural data indicate that factor IX is a stretched molecule, with the catalytic domain on one end and the Gla-domain, with its Gla-residues protruding, at the opposite end (Perera et al. 2001, see Fig. 18.3c).

Why are these Gla residues so important? Like in other vitamin K-dependent coagulation factors, such as factor VII, factor X, protein C, and prothrombin, they provide high-affinity binding sites for Ca^{2+}-ions and thereby mediate the interaction with negatively charged lipid membranes. Thus, the appropriate assembly of coagulation factors at the site of injury is fully dependent on the presence of these clustered

Fig. 18.3 Structure of factor IX. Panel (**a**) displays the domain structure. The colors of the individual domains correspond to those in panel (**b**), which provides a 3D structure of factor IX (Perera et al. 2001). In the Gla-domain, the γ-carboxylated residues are indicated in purple. Panel (**c**) provides the same structure with human albumin (from pdb 1AO6) fused to the C-terminus of the catalytic domain. *EGF* epidermal growth factor, *AP* activation peptide

Ca^{2+}-binding sites (Furie and Furie 1988). Therefore, full carboxylation is a major requirement in the production of recombinant factor IX.

Factor IX Products

The complexity of the post-translational modifications remains a major challenge in the production of recombinant factor IX. Mammalian cells need to be used as an expression system, because these contain the necessary carboxylase enzyme complex. However, the capacity of host cells to perform this post-translational modification tends to be limiting, leading to under-carboxylated forms that lack full activity. Biological activity further requires the removal of the propeptide, which usually implies the need for co-expression with a furin-like processing enzyme (Harrison et al. 1998). Most other post-translation modifications are located in the activation peptide and include two N-linked glycosylation sites, tyrosine sulfation, and serine phosphorylation. With respect to these modifications, recombinant factor IX remains non-identical to its plasma-derived counterpart (Peters et al. 2010; Monroe et al. 2016).

Wild-Type Factor IX
Normal human plasma factor IX exists in two allelic forms, with alanine or threonine in position 148 in the activation peptide. This dimorphism, however, is without any known functional implication. Some recombinant factor IX products provide the Ala-148 variant, while others have Thr in position 148 (see Table 18.3). The first recombinant factor

Table 18.3 Recombinant factor IX products

Product type	INN name	(Investigational) product name(s)	First approval (USA/EU)	Half-life (mean, h)	Production technology	Reference
Wild-type FIX	nonacog alfa	BeneFIX	1997	17–36	CHO cells, co-expression with furin, Ala-148 polymorphic variant	Harrison et al. (1998)
	nonacog gamma	Rixubis (Bax326)	2013	27	CHO cells, co-expression with furin, Ala-148 polymorphic variant	Dietrich et al. (2013)
	trenonacog alfa	IXinity (IB1001)	2015	24	CHO cells, Thr-148 polymorphic variant	Monroe et al. (2016)
Engineered FIX	eftrenonacog alfa	Alprolix (FIX-Fc)	2014	82	HEK cells, Thr-148 variant, Fc fused to FIX C-terminus, co-expression with Fc alone and with processing enzyme PC5	Peters et al. (2010)
	albutrepenonacog alfa	Idelvion (F9-fusion protein)	2016	92	CHO cells, Thr-148 variant, albumin fused to FIX C-terminus by cleavable linker	Metzner et al. (2009)
	nonacog beta pegol	Rebinyn, Refixia (N9-GP)	2017	96	CHO cells, Ala-148 variant, glycoPEGylation in activation peptide region	Østergaard et al. (2011)

IX was BeneFIX, developed at Genetics Institute. More than 15 years later, two other products have been licensed, Rixubis and IXinity, developed at Baxter and Inspiration Biopharmaceuticals, respectively. All three products are produced in CHO cells under serum-free conditions and undergo extensive downstream processing to remove impurities, including at least one virus eliminating step. For IXinity, the purification process includes one polishing step to specifically remove CHO-derived proteins (Monroe et al. 2016). Despite the differences in production and purification methods, the final products of these wild-type recombinant factor IX products are similar in terms of efficacy and pharmacokinetics.

Extended Half-Life Factor IX

Like factor VIII, also factor IX replacement therapy is limited by its relatively short half-life of about 24 h (see Table 18.3). For factor IX, engineering has proven more rewarding than for factor VIII. Currently, three products have been licensed that offer 3–four fold extended half-life (see Table 18.3 and references therein).

Alprolix has been developed by Biogen Idec and is a fusion between factor IX and the Fc fragment of IgG. The targeting of factor IX to the FcRn receptor prolongs the half-life to approximately 82 h. The fusion of the Fc fragment to the factor IX C-terminus yields a derivative in which the activated form still carries the Fc-moiety. The downside thereof is that the activity of this fusion protein is considerably lower than that of 'wild-type' Factor IX.

Idelvion has been developed by CSL Behring and is a fusion between factor IX and human albumin (Fig. 18.3c). The presence of the relatively large albumin at the C-terminus of the catalytic domain is incompatible with biological activity. This has been partially resolved by using a linker sequence that comprises the sequence of one of the two natural cleavage sites in factor IX (Metzner et al. 2009). Thus, proteolytic activation of this fusion protein involves three cleavages, the extra one serving to release the albumin moi-

ety. Nevertheless, this fusion protein displays reduced in vitro activity, possibly because albumin release is relatively slow or incomplete. Its reported half-life is ~92 h (Iorio 2017).

Rebinyne/Refixia has been developed by Novo Nordisk. In this product, recombinant factor IX is modified with a branched 40 kDa PEG. This modification is targeted to the native N-glycans in the activation peptide by enzymatic glycoconjugation. Because the PEG moiety is released with the activation peptide upon activation, the active form is the natural, unmodified factor IXa. The in vitro activity of this factor IX derivative seems normal, although this remains dependent of assay reagents used.

Several other engineered factor IX products are under development, including dalcinonacog alfa (ISU304, Catalyst Biosciences). This factor IX variant carries three point substitutions in the catalytic domain in order to enhance biological activity. Dalcinonacog alfa is currently being evaluated for subcutaneous use (Mahlangu et al. 2021).

Pharmacology

Like factor VIII, the term factor IX has been used before the protein had been identified. Factor IX was defined by its function to correct the clotting defect of plasma of patients suffering from hemophilia B. Biological activity has remained the basis for the quantification of factor IX, and concentrations are expressed in units, where 1 unit represents the amount of factor IX activity in 1 mL of normal human plasma. Plasma factor IX levels can also be expressed as % of normal. Thus, assessment of product potency, dosing, and pharmacokinetics are based on bioassays, and not on factor IX protein concentrations. Dosing is generally based on the empirical finding that 1 unit factor IX per kg body weight raises the plasma factor IX activity by 1% (or 0.01 unit/mL), according to the formula (Björkman and Berntorp 2001):

$$dose(units / kg) = 100 \times$$
$$(required\ rise\ in\ units / mL\ factor\ IX)$$

Pharmacokinetics

The pharmacokinetics of factor IX have been reviewed in detail elsewhere (Björkman and Berntorp 2001; Iorio 2017). Systemic clearance is around 5 mL/h/kg for plasma-derived factor IX, and somewhat higher for recombinant wild-type factor IX. The distribution volume considerably exceeds the plasma volume, presumably because factor IX rapidly binds to the vascular surface. Elimination generally follows a bi-exponential pattern, with a terminal half-life of 20–34 h. Assessment of pharmacokinetics is mandatory for all individual factor IX products and seems more consistent than that of factor VIII (see Iorio 2017 and Table 18.3). Similar to factor VIII, factor IX pharmacokinetic data remain dependent on technical issues such as sampling time points, and the bioassays for the determination of the dose and post-infusion plasma levels. Despite this variability, it is evident that the engineered factor IX products do display a substantial half-life extension.

Pharmacodynamics

Hemophilia B is a congenital deficiency or dysfunction of factor IX which, dependent on the residual level of factor IX activity, is categorized as mild, moderate, or severe (Björkman and Berntorp 2001). Like factor VIII, normalization of factor IX levels in the circulation requires intravenous administration. The therapeutic range is narrow and similar to that of factor VIII.

For all licensed factor IX products, efficacy has been established for both on-demand and prophylactic treatment. On-demand treatment generally involves the administration of 20–100 units/kg, depending on the bleeding type. Long-term prophylaxis involves regular treatment at a dose of 20–40 units/kg, with intervals of 3–4 days. For extended half-life products, various regimens have been assessed, varying from 10 units/kg every 7 days for one product to 75 units/kg every 14 days or 100 units/kg every 10 days for other products (Young and Mahlangu 2016). Thus, no general recommendation can be given so far. For the time being, dosing should strictly adhere to the product-specific recommendations for the individual products given in the Summary of Product Characteristics as issued by regulatory authorities such as EMA.

Unlike factor VIII, an immune response against factor IX is rare. Hypersensitivity and allergic reactions have been reported, and factor IX inhibitory antibodies may occur in a very small proportion of patients (1% or less). Antibodies against hamster proteins have been reported for one particu-

lar product (IXinity), but this has been resolved by introducing an additional step to further reduce these impurities (Monroe et al. 2016).

Pharmaceutical Considerations

The factor IX concentration in human plasma is approximately 4 μg/mL. However, like factor VIII, activity units are used for factor IX. The potency of factor IX products is expressed in International Units (IU) relative to an international standard established by the World Health Organization. Most products are available in fillings ranging between 250 and 3000 IU in order to provide appropriate, body-weight guided, dosing for children and adults. Current products are freeze dried and need to be reconstituted before infusion. Most products are stabilized in a sucrose-containing formulation and have a shelf life of 24–36 months upon storage at 2–8 °C. For most products, data are available that establish stability at room temperature for 6–24 months.

As for impurities, most products may contain low amounts of residual proteins from the host expression system. With the exception of Alprolix, which is expressed in a human cell line, these products may cause side effects in patients which hypersensitivity against rodent proteins. To enhance safety further, the downstream processing of most recombinant factor IX products includes one or more virus-eliminating steps, usually including nanofiltration.

Clinical Considerations

The sole indication for recombinant factor IX is congenital hemophilia B. For all factor IX products, appropriate treatment protocols have been established as part of the licensing procedure. Like for hemophilia A, there is an increasing interest in personalized prophylactic treatment based on the pharmacokinetic profile of individual patient/product combinations (Iorio 2017). This remains challenging for the extended half-life products, because of the lack of uniform, robust assays for monitoring these engineered factor IX variants (Kitchen et al. 2017).

An additional complication is that the Fc- and albumin fusion proteins display reduced activity, indicating that patients are treated with factor IX that is partially inactive. This might seem irrelevant because dosing is based on activity (Peters et al. 2010). On the other hand, this introduces an extra, assay-dependent variable in personalized, pharmacokinetics-based dosing. One important implication is that patients cannot be switched from one extended half-life product to another while maintaining the same dosing schedule.

Other Hemostatic Proteins

While a variety of recombinant factor VIII and IX products is available from multiple biotechnology companies, other hemostatic proteins have received limited attention. For some proteins, no more than one or two products have been licensed so far. These are reviewed in the present section.

Recombinant Factor VIIa

Factor VIIa is the enzyme that triggers the initiation phase of thrombin generation and acts directly upstream of factor X in the cascade. Based on theoretical considerations, it has been proposed that the presence of high concentrations of factor VIIa should have the potential of activating factor X to a sufficient extent to bypass the need for factor VIII or factor IX in the coagulation cascade (see Fig. 18.1). Once this concept had been verified (Hedner and Kisiel 1983), the use of factor VIIa became an option for treatment of bleeding in patients with inhibitory antibodies against factor VIII or IX, and a few other bleeding disorders (see Clinical considerations).

The factor VII concentration in human plasma is low (0.5 µg/mL) and as such is insufficient to provide a source for large amounts of factor VIIa. This made factor VIIa an attractive target for production by recombinant technology. Factor VII is a glycoprotein of approximately 57 kDa, which has the same domain structure as factor IX (Fig. 18.3). It shares the same requirement for γ-carboxylation in its Gla-domain at the N-terminus of the mature protein. Recombinant factor VII has been developed at Novo Nordisk and is produced in BHK cells under serum-free conditions. During downstream processing, single-chain factor VII is converted into two-chain factor VIIa by autoactivation (Hedner 2006). Recombinant factor VIIa (INN name eptacog alfa) has been licensed since 1996 under the product name NovoSeven.

A nearly identical product (eptacog beta, investigational name LR769) has been developed at LFB Biotechnologies and is licensed in the USA and EU since 2020 and 2022 under the names Sevenfact and Cevenfacta, respectively. It differs from most other recombinant products in that it is harvested from the milk of transgenic rabbits expressing human factor VII. Downstream processing results in fully active factor VIIa which carries all relevant post-translational modifications (Chevreux et al. 2017). Another biosimilar is produced by AryoGen Biopharma. This product (AryoSeven) has only been licensed in Iran so far. Marzeptacog alfa is an engineered factor VIIa, with four amino acid substitutions to enhance stability and activity, and is under investigation for subcutaneous administration (Mahlangu et al. 2021). Another engineered variant (vatrep-tacog alfa) has been designed to enhance biological activity, but this variant proved to be immunogenic due to the amino acid substitutions made, and phase III trials have been terminated (Mahlangu et al. 2015). Although the recent advent of transgenic factor VIIa holds promise for the future, NovoSeven is the sole recombinant factor VIIa product that is widely used so far.

Pharmacology

Factor VIIa normally activates factor X when bound to tissue factor at the site of vascular injury. NovoSeven dosing, however, is in large excess over physiological concentrations. Under such conditions, factor VIIa is believed to activate factor X on the surface of activated platelets in a tissue factor-independent manner (Hedner 2006). Factor VIIa potency and dosing is based on bioassays but, unlike factor VIII and IX, is expressed in terms of protein concentration, and not in units. Like other coagulation factors, Factor VIIa therapy requires intravenous administration. The pharmacokinetics of recombinant factor VIIa has been assessed in various patient groups. In non-bleeding patients, the overall mean clearance is approximately 30 mL/h/kg, and the half-life is around 2–3 h (Björkman and Berntorp 2001). For hemophilia A or B patients with inhibitory antibodies, the recommended dosage is 90 µg/kg every 2 h until hemostasis is achieved. Dosage is slightly different for other indications (see section "Clinical Considerations").

Pharmaceutical Considerations

NovoSeven is available in fillings of 1–8 mg per vial. The product is freeze-dried and needs to be reconstituted in a specific histidine-containing diluent before use. The product is stable for 3 years at room temperature. As for impurities, downstream processing includes a virus inactivating solvent/detergent step.

Clinical Considerations

Apart from treatment of hemophilia A and B with inhibitory antibodies, recombinant factor VIIa is also indicated for treatment of acquired hemophilia due to antibodies in non-hemophilic patients. Another indication is congenital factor VII deficiency. The recommended dosage then is 15–30 µg/kg every 4–6 h. NovoSeven is also indicated for the platelet-based bleeding disorder Glanzmann Thrombasthenia. The recommended dosage then is 90 µg/kg every 2–6 h, until hemostasis has been achieved (Hedner and Ezban 2008). The most common and serious adverse reactions are thrombotic events. As for Sevenfact/Cevenfacta, the only indication is hemophilia with inhibitory antibodies. Due to its transgenic origin, this product is not recommended for patients with known hypersensitivity to rabbits.

Recombinant von Willebrand Factor

Von Willebrand factor is one of the largest proteins in circulation. Monomeric von Willebrand factor is synthesized as a preproprotein, with a signal peptide of 22 amino acids, a propeptide that is unusually large (741 amino acids) and a mature subunit of 2015 amino acids (for review, see Castaman and Linari 2017). Monomers form tail-to-tail dimers via a cysteine-rich region at the C-terminus. These dimers can multimerize by the formation of cysteine bridges at the N-terminus of the mature subunit in a process that requires the presence of the propeptide. The molecular size of the von Willebrand factor thus can vary between a dimer of 500 kDa and multimers of up to 20,000 kDa. The primary function of the von Willebrand factor is to promote platelet aggregation and adhesion, which is essential for appropriate thrombus formation (Fig. 18.1b). The secondary function is to carry factor VIII in the circulation and thereby to prevent its premature clearance (Pipe et al. 2016; Castaman and Linari 2017).

It seems evident that the size of this protein and the complexity of its processing into dimers and multimers make the production of a recombinant von Willebrand factor a challenge. Nevertheless, this has already been achieved in the 1980s at Genetics Institute by the co-expression of von Willebrand factor and factor VIII. While the factor VIII products resulting from co-expression (Recombinate and Advate, see Table 18.2) were depleted of the von Willebrand factor, this byproduct has subsequently been developed into a separate product. Downstream processing includes in vitro processing with recombinant processing enzyme (furin) to remove the propeptide and expose optimal factor VIII binding (Turecek et al. 2009). This recombinant product (INN name vonicog alfa) has first been licensed as Vonvendi in the US in 2015 and is called Veyvondi in the EU. It is the first recombinant von Willebrand factor available as an alternative to the current plasma-derived counterparts (Franchini and Mannucci 2013).

Pharmacology and Pharmaceutical Considerations

Recombinant von Willebrand factor serves to control bleeding episodes in patients with severe type-III deficiency. The deficiency of von Willebrand factor in these patients implies not only a defective platelet aggregation and adhesion but also abnormally fast clearance of their endogenous factor VIII. Therefore, factor VIII needs to be supplemented too, usually to a level of 35–50% of normal, in order to achieve hemostasis (Mannucci 2004).

The concentration of recombinant von Willebrand factor is expressed in terms of biological activity by its effect on platelet aggregation in the presence of ristocetin (so-called 'ristocetin cofactor activity'). Potency is expressed in International Units (IU) and is based on the International Standard for von Willebrand factor concentrate. Vonvendi is available as lyophilized powder for reconstitution in fillings of 650 and 1300 IU and needs to be administered intravenously. Its shelf-life is 36 months at 3–5 °C or 12 months at room temperature. Being derived from CHO cells under serum- and protein-free conditions, the most relevant impurities may represent residual hamster protein, which might cause hypersensitivity reactions in some recipients. The pharmacokinetics have been studied extensively (Gill et al. 2015), both for recombinant von Willebrand factor alone and in combination with recombinant factor VIII (see under Clinical considerations). The overall clearance was 3 mL/h/kg and half-life approximately 22 h. In the phase III study, the initial dosage was 40–80 IU/kg, followed by 40–60 kg every 8–24 h if clinically required.

Clinical Considerations

The vast experience with the treatment of severe von Willebrand's disease is based on the use of plasma-derived products, in particular concentrates that contain both von Willebrand factor and factor VIII. The recommended regimen is 30–50 IU/kg von Willebrand factor while keeping the trough factor VIII level > 30–50% of normal (Mannucci 2004). In Vonvendi, the sole active component is the recombinant von Willebrand factor. To control bleeding, the first dosage should therefore be combined with recombinant factor VIII. Because the recombinant von Willebrand factor sustains the stability of the patient's endogenous factor VIII, co-administration with factor VIII for subsequent infusions might prove unnecessary (Gill et al. 2015). For the moment, the burden of co-administration and dual monitoring seems to be prohibitive for general use in less specialized centers. Ongoing post-marketing trials serve to clarify further the exact place of recombinant von Willebrand factor in the management of this complex bleeding disorder (Franchini and Mannucci 2016).

Recombinant Factor XIII

Factor XIII is the pro-enzyme of a transglutaminase that, after activation by thrombin, reinforces fibrin polymers by the formation of γ-glutamyl-ε-lysyl amide cross-links. A congenital deficiency of factor XIII is associated with severe bleeding due to impaired thrombus stability (see Fig. 18.1b). Factor XIII is a hetero-tetramer comprising two catalytic A-subunits of 82 kDa and two carrier B-subunits of 73 kDa, which are linked by noncovalent interactions (Komaromi et al. 2011). The A- and B-subunits are encoded by different genes, and in plasma, the B-chain circulates in excess over the A-chain. An intracellular form of factor XIII is present in platelets, macrophages, and other cells and is a homodimer of two A-subunits only. Most patients with factor XIII deficiency have a defect in the A-subunit, which makes this a suitable target for recombinant factor XIII substitution therapy (Inbal et al. 2012).

Pharmacology and Pharmaceutical Considerations

Recombinant factor XIII A-subunit has been developed at ZymoGenetics and Novo Nordisk and is expressed in *Saccharomyces cerevisiae* under protein-free conditions. The purified yeast-derived protein is a dimer of two non-glycosylated factor XIII A-subunits (Lovejoy et al. 2006), which upon infusion spontaneously associates with B-subunits to form the factor XIII heterotetramer in the circulation. The product (INN name catridecacog) has been licensed in 2012 and is available under the product name Tretten in the US and as NovoThirteen elsewhere.

The potency of recombinant factor XIII is expressed in International Units (IU) and is based on the International Standard for factor XIII concentrate. NovoThirteen is available as lyophilized powder for reconstitution in vials of 2500 IU, to be reconstituted in physiological saline for intravenous administration. The shelf-life is 24 months at 2–8 °C. Pharmacokinetic studies have shown that the half-life varies between 6 and 9 days (Lovejoy et al. 2006). Prophylactic treatment has proven efficacious and safe using a regimen of 35 IU/kg once monthly (Inbal et al. 2012).

Clinical Considerations

Recombinant factor XIII is the first recombinant product for a rare bleeding disorder. So far, plasma-derived, hetero-tetrameric factor XIII has been used for the treatment of congenital factor XIII deficiency. Due to the lack of post-translation glycosylation, the yeast protein is non-identical to the natural A-subunit homodimer. Nevertheless, this has not resulted in any detectable immunogenicity so far. For a majority of patients, therefore, recombinant A-subunit dimer proves a viable alternative for plasma-derived factor XIII.

Recombinant Antithrombin

Antithrombin belongs to the class of serine protease inhibitors ('serpins'), and as such might be considered as not being a coagulation factor in the strict sense. However, antithrombin is the major inhibitor of factor IXa, factor Xa, factor Xia, and thrombin and thereby plays a regulatory role at multiple levels in the coagulation cascade. Antithrombin is a single-chain glycoprotein and has a molecular mass of 58 kDa. Its inhibitory potential is greatly enhanced by binding to hepa-rin or other glycosaminoglycans. This drives antithrombin, like other serpins, into a conformation that favors the interaction with its target proteases (Huntington 2003). A partial deficiency of antithrombin reduces the downregulation of the coagulation cascade and thereby enhances the risk of thrombosis. In antithrombin deficiency, prophylaxis usually consists of standard antithrombotic therapy using heparin or small molecules that inhibit coagulation (Mackman 2008). However, the inherent bleeding risk thereof makes such therapy undesirable for use in high-risk situations such as surgery or delivery. Suppletion of antithrombin then provides an alternative treatment.

Recombinant antithrombin is licensed in the EU and US since 2006 and 2009, respectively. The product has been developed by Genzyme Transgenics (GTC Biotherapeutics) and is available as ATryn (INN name antithrombin alfa). ATryn is particularly remarkable from a biotechnological point of view, because it is harvested from the milk of transgenic goats (Edmunds et al. 1998). This technology, which is introduced in Chap. 9, has proven feasible for the expression of a wide range of therapeutic proteins in milk or other body fluids (Lubon et al. 1996, see also Table 9.4). During further development, however, transgenic technology had to deal with numerous issues, including stability of transgenic inheritance, appropriate post-translational modification and stability of heterologous proteins in milk, and regulatory affairs (Lubon et al. 1996). Upon its approval by EMA and FDA, ATryn became the first biopharmaceutical that took all hurdles of production in transgenic animals. Since then, a few more transgenic biopharmaceuticals have been licensed, including recombinant factor VIIa (see above).

Pharmacology and Pharmaceutical Considerations

The potency of ATryn is expressed in International Units, which relate to the International Standard for antithrombin in concentrate as established by WHO. The unitage is based on in vitro inhibitory activity in the presence of heparin. ATryn is available as lyophilized powder for reconstitution and infusion, in fillings of 1750 IU per vial. When stored at 2–8 °C, the shelf-life is 4 years. The transgenic antithrombin in ATryn is identical to plasma-derived antithrombin, with the exception of its glycosylation, which is less complex and comprises fewer sialic acid moieties (Edmunds et al. 1998). Rigorous pharmacokinetic studies have not been reported, but the estimated mean elimination half-life is 10 h, which is six-fold shorter than that of the fully glycosylated antithrombin from human plasma. Therefore, dosing recommendations for plasma-derived and transgenic antithrombin are different. Dosing aims to maintain plasma levels between 80 and 120% of normal. For ATryn, the recommended dosing is (Tiede et al. 2008):

$$\text{Loading dose}\left(\text{IU}/\text{kg}\right)=\left(\frac{100-\text{pretreatment}}{\text{antithrombin level in}\%}\right)/2.28$$

$$\text{Maintenance}\left(\text{IU}/\text{kg}/\text{hr}\right)=\left(\frac{100-\text{pretreatment}}{\text{antithrombin level in}\%}\right)/10.22$$

The loading dose can be administered as a bolus, while maintenance requires continuous infusion.

Clinical Considerations

ATryn is indicated for the prophylaxis of venous thrombosis in surgery in adult patients with congenital antithrombin deficiency. In the absence of sufficient data, ATryn initially has not been recommended for use during pregnancy. Subsequent studies have generated additional support for using ATryn in both perioperative and peripartum settings (Paidas et al. 2014). One concern could be the potential immunogenicity of the transgenic protein due to the glycosylation differences between transgenic and native human antithrombin. However, no antibodies have been reported, neither to the transgenic protein nor to any other goat protein (Paidas et al. 2014).

Recombinant Thrombolytic Agents

The need for thrombolytic agents dates from the 1970s, when myocardial infarction was recognized as being caused by the rupture of an atherosclerotic plaque in a coronary artery, followed by the formation of an occluding thrombus (Lusis 2000). Recanalization of the occluded vessel requires thrombus dissolution, which can be achieved by activating the fibrinolytic system (see Fig. 18.1c). The first thrombolytic agent used was streptokinase, a bacterial plasminogen activator isolated from *Streptococcus haemolyticus*. This agent has been widely used in the treatment of myocardial infarction and as such has developed into a worldwide blockbuster. However, the use of a bacterial protein implies the risk of immunogenicity, and this may hamper repeated dosing. The search for a human plasminogen activator has resulted in the identification of a tissue-type plasminogen activator (t-PA). Its low concentration in tissues and in the blood is prohibitive for obtaining natural t-PA in substantial amounts. This made t-PA an obvious candidate for production by recombinant technology (for review, see Collen and Lijnen 2005).

Recombinant Tissue-Type Plasminogen Activator (t-PA)

The primary structure of human t-PA has been established since the cloning and expression of its cDNA (Pennica et al. 1983). The mature protein comprises an N-terminal region that is called the finger domain, followed by a single EGF-like domain, two so-called kringle domains, and the catalytic domain (see Fig. 18.4a). There are four sites for N-linked glycosylation, one in kringle-1, two in kringle-2, and one in the protease domain. Human t-PA is a single-chain protein with a molecular mass of about 70 kDa. It is converted into a two-chain form by plasmin, by cleavage at the junction between the kringle-2 and the catalytic domain. While the two-chain form represents the fully activated enzyme, the single-chain form displays very similar fibrinolytic activity. Both forms of t-PA share high-affinity interaction with fibrin and inhibition by plasminogen activator inhibitor-1 (PAI-1) (Rijken et al. 1982).

Wild-Type Recombinant t-PA (Alteplase)

Wild-type t-PA has initially been expressed in *Escherichia coli* (Pennica et al. 1983). However, the numerous (17) internal disulfide bridges and multiple glycosylation sites make production in mammalian cells more efficient. Recombinant t-PA (INN name alteplase) has been developed at Genentech and is produced in CHO cells. Alteplase displays the fibrin-specificity of natural t-PA, which is mediated by the finger and kringle-2 domain. The biological half-life is only 5–6 min. Half-life is limited by the rapid inhibition by PAI-1 and by the interaction of the finger and/or EGF-like domains with clearance receptors (Collen and Lijnen 2005). Alteplase has first been licensed in 1987 and is available under multiple product names, including Activase (Genentech) and Actilyse (Boehringer Ingelheim).

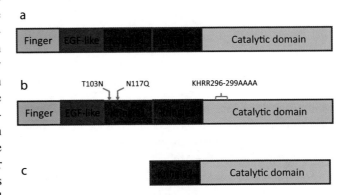

Fig. 18.4 Structure of recombinant tissue-type plasminogen activator. Panel (**a**) represents the domain structure of full-length t-PA (alteplase). Panel (**b**) displays the variant tenecteplase, and panel (**c**) the truncated variant reteplase. Further details are given in the text. EGF: epidermal growth factor

Engineered Recombinant t-PA

Tenecteplase is the INN name of a t-PA variant that has been engineered at Genentech, with the aim to prolong half-life, while maintaining the fibrin specificity of natural t-PA. This has been accomplished by a few substitutions in the full-length protein (see Fig. 18.4b). The Thr103 to Asn (T103N) substitution creates a new N-linked glycosylation site, and the Asn117 to Gln (N117Q) mutation eliminates a high-mannose glycosylation site in the kringle-1 domain. In addition, the amino acids Lys295-His-Arg-Arg299 have been replaced by four Ala residues. The modified glycosylation in the kringle-1 domain serves to reduce clearance, while the KHRR296-299AAAA substitutions confer increased resistance against inhibition by PAI-1 (Keyt et al. 1994). Tenecteplase is produced in CHO cells and has first been licensed in the US in 2000. It is available under the product names TNKase and Metalyse.

Reteplase is the INN name for a truncated variant of t-PA which consists of only the kringle-2 and the catalytic domain (see Fig. 18.4c). This variant was developed at Boehringer Mannheim, with the objective of eliminating the domains that drive clearance, while conserving the kringle-2 domain for fibrin-specificity. This major truncation allows for production in *Escherichia coli*. Reteplase is a single-chain, non-glycosylated protein with a molecular mass of 39 kDa. Its downstream processing includes extraction from inclusion bodies and in vitro renaturation and refolding (Moser et al. 1998; Simpson et al. 2006). Reteplase was first licensed in 1996 and is available under the product names Rapsilysin and Retevase.

A variety of other t-PA variants has been designed, including lanoteplase. Lanoteplase has been developed at Genetics Institute. It lacks the finger and EFG-like domains, as well as the glycosylation in position Asn117 (Collen and Lijnen 2005). Like other, more recently engineered t-PA variants, lanoteplase has not been licensed and therefore remains beyond the scope of this chapter.

Pharmacology

Multiple methods are being used for the quantification of t-PA. First, potency is expressed in terms of International Units (IU), which reflects biological activity against the International Standard as established by WHO. Alternatively, t-PA concentration can be defined by protein content and expressed in mg. Which method is used for potency labeling varies between products. For reteplase, potency is expressed in units (U) using a product-specific reference. For patient monitoring or for pharmacokinetic studies, t-PA concentrations are usually based on immunological assays.

Pharmacokinetics

Assessment of the pharmacokinetics of t-PA is complex. It is reflecting a combination of 'regular' plasma elimination of free t-PA, the adsorption of t-PA to insoluble fibrin, the neutralization by PAI-1, and the elimination of t-PA/PAI-1 complexes. The reader is referred to the Summary of Product Characteristics of the individual products for details (EMA). As a general parameter, most studies just use the biological half-life as estimated from levels of circulating t-PA. In these terms, the half-life of alteplase is 5–6 min, whereas the half-life of tenecteplase and reteplase is 17–20 min and 14–18 min, respectively (Collen and Lijnen 2005).

Pharmacodynamics

The recombinant plasminogen activators differ from streptokinase in that they are fibrin-specific. Fibrin has a high affinity for both t-PA and plasminogen and serves as a surface for local plasminogen activation. In contrast to free plasmin, fibrin-bound plasmin is relatively insensitive to inactivation by α2-plasmin inhibitors. These interactions make fibrin the central regulator of its own degradation (Collen and Lijnen 2005). Normally, this subtle mechanism is triggered adequately by the small amounts of t-PA in plasma (5–10 ng/mL).

Numerous clinical studies have addressed recombinant t-PA in acute myocardial infarction, usually in comparison with a standard streptokinase regimen. Due to the difference in pharmacokinetics, dosage regimens are product-specific. For alteplase, a typical regimen is a total of 100 mg, of which 60 mg during the first hour, and 20 mg over the second hour, and 20 mg over the third hour. An alternative is 100 mg divided over a bolus of 15 mg, followed by 0.75 mg/kg over 30 min, and 0.5 mg/kg over 60 min (Collen and Lijnen 2004).

In contrast, tenecteplase usually is administered as a single intravenous bolus of 30–50 mg (6000–10,000 IU), dependent on body weight (Guillermin et al. 2016). For reteplase, a recommended regimen is a double bolus (10 U + 10 U, 30 min apart), independent of body weight (Simpson et al. 2006). Apparently, the engineering to extend half-life of tenecteplase and reteplase indeed resulted in the intended dosing advantage.

In thrombolytic therapy, t-PA levels are achieved that are in large excess (up to 500-fold) over the physiological concentration. This carries the inherent risk that t-PA also acts in a fibrin-independent manner, and degrades fibrinogen (fibrinogenolysis). This may lead to systemic fibrinogen shortage and concomitant bleeding risk. Bleeding is the most common complication of thrombolysis, also because it is usually combined with anticoagulant therapy to prevent re-occlusion. This may vary between superficial bleeding at the site of injection and more severe episodes, including intracranial bleeding.

Pharmaceutical Considerations

Alteplase, tenecteplase and reteplase are lyophilized products that need to be reconstituted for intravenous infusion. Shelf-life is at least 2 years at temperatures not exceeding 25 °C. Once reconstituted, alteplase can be further diluted with a physiological saline solution to facilitate continuous infusion. Usually, thrombolytic therapy is combined with continuous infusion of heparin. This should not be co-administered through the same cannula, because of solubility issues, in particular for reteplase.

As for impurities, alteplase and tenecteplase are produced in CHO cells and may include residual hamster protein. Reteplase is produced in *Escherichia coli* and is subjected to a validated process of denaturation and refolding to produce the active fibrinolytic enzyme. Reteplase has a lower affinity for fibrin than alteplase and tenecteplase. It also displays lower in vitro fibrinolytic activity than alteplase (Simpson et al. 2006). Because of its different unitage used for potency labeling, the content of active agent in reteplase cannot be directly related to that of alteplase or tenecteplase.

Clinical Considerations

The primary indication for alteplase, tenecteplase, and reteplase is acute myocardial infarction. Alteplase has further been licensed for use in acute pulmonary embolism and acute ischaemic stroke. For tenecteplase and reteplase, studies are addressing the extension into these indications too. For the treatment of stroke, it is essential to start therapy only after the prior exclusion of intracranial bleeding by appropriate imaging techniques.

In the majority of clinical trials, the recombinant agents have been compared with standard streptokinase therapy, often in combination with heparin or other anticoagulants, to prevent re-occlusion (Collins et al. 1997). Overall, all agents seem to offer adequate thrombolytic therapy of acute myocardial infarction, with comparable survival rates and bleeding complications. In the absence of a direct comparison between recombinant thrombolytic agents, it remains difficult to conclude that one agent would be preferable over another. A meta-analysis suggests that tenecteplase and alteplase are equally effective in acute coronary syndrome, with a slightly lower bleeding risk for tenecteplase (Guillermin et al. 2016). It should be noted, however, that clinical studies are complex and require the inclusion of large numbers of patients to reach satisfying conclusions for all individual indications. Therefore, the current thrombolytic agents continue to be the subject of clinical studies to further explore their therapeutic potential in life-threatening thrombotic complications.

Concluding Remarks

In the early 1980s, the advent of biotechnology provided the perspective of unlimited access to recombinant coagulation factors and thrombolytic agents. This was particularly promising for low-abundance proteins such as t-PA, and factors VII and VIII, for which the medical need was by far exceeding the availability of pharmaceutical production from human sources. Now, four decades later, one may wonder whether biotechnology actually has fulfilled its initial promise in this field. It remains difficult to give a fully unequivocal answer at this point.

On the positive side, there are some major achievements. The most prominent example is t-PA. Its cDNA was cloned in 1983, and recombinant t-PA became available as a life-saving product already 4 years later. It is further remarkable that for recombinant factor VIII, being a much more complex protein than t-PA, the gap between cloning and licensing could be closed in only 8 years. Another success is activated factor VII. Recombinant technology has been instrumental in obtaining this trace protein in sufficient amounts to meet the clinical need. More recently, we have seen examples of how protein engineering may generate biopharmaceuticals that display more favorable pharmacokinetics than their natural counterparts.

At the same time, however, some limitations remain apparent. For instance, recombinant t-PA now is the generally established agent for thrombolytic therapy in the US and EU, but not in territories where t-PA is not affordable, and streptokinase is still being used. Another example is factor VIII. Since the introduction of recombinant factor VIII supplies has been a limiting factor, and prices have remained prohibitive for many countries. While factor VIII is included in WHO's List of Essential Medicines, the majority (~75%) of the world's patients still lack appropriate access to hemophilia care, despite intense efforts to maximize the production of conventional factor VIII from human plasma. Apparently, recombinant factor VIII did not meet the full medical need. In this regard, one may question why so many new recombinant factor VIII products have been recently developed without taking the affordability issue into account. Moreover, with the exception of factor XIII deficiency, the rare bleeding disorders have remained dependent on blood-derived products, with their inherent limited supply. The initial expectation was that plasma-derived coagulation factors may soon become obsolete. It is evident that this optimism has been premature.

The next few decades offer excellent perspectives on fulfilling the initial promise of biotechnology. It may become possible to replace costly mammalian cell expression with cheaper technology. For relatively simple proteins such as t-PA, expression in transgenic plants has proven feasible.

Production of transgenic animals has, after many hurdles, resulted in the first tangible products on the market. Moreover, for most of the hemostatic proteins, the dominant patents have expired, thus opening the way towards more affordable biosimilars (cf. Chap. 11). Finally, protein engineering continues to generate therapeutics that display improved biological activity at a much lower dosage. It remains an attractive challenge to a new generation of researchers to accomplish these goals in the near future.

Self-Assessment Questions

Questions

1. *Proteolytic Processing*
   Furin-like enzymes play a role in the post-translational modification of a several hemostatic proteins.
   (a) Give three examples and describe the relevance of processing in these proteins.
   (b) How is this modification accomplished in biotechnological production?

2. *Transgenic animal bioreactors*
   A variety of transgenic animals has been used to express hemostatic proteins in their milk, including rabbits, goats, sheep, pigs and cows. Recombinant antithrombin (ATryn) is produced in goats.
   (a) ATryn differs from plasma-derived antithrombin. What is the major difference? How could this be caused?
   (b) Would you consider using goats as bioreactor for producing transgenic factor VIII or IX, or would you prefer an alternative? Why?

3. *Extended half-life factor IX*
   The factor IX-albumin fusion protein is one of the available products with extended half-life.
   (a) What is the mechanism underlying the half-life prolongation?
   (b) Are there particular disadvantages or limitations of this particular fusion strategy?
   (c) What are the implications for patients switching from normal factor IX to this fusion product?

4. *Thrombolytic agents*
   (a) Describe the mechanism by which fibrinolysis is under local control.
   (b) What is the main adverse effect of thrombolysis, and how is this caused.
   (c) Reteplase differs from other thrombolytic agents in that it is produced in E. coli. What are the potential advantages or disadvantages of this approach?

Answers

1. *Proteolytic processing*
   (a) Three examples are: (1) factor IX, wherein the propeptide that is needed for appropriate γ-carboxylation, needs to be cleaved off in order to generate the Gla-

domain as the natural N-terminus of the mature factor IX, (2) von Willebrand factor, wherein the propeptide is important for the formation of multimers, but is removed prior to secretion in order to facilitate its function as carrier of factor VIII in the circulation, and (3) factor VIII, which is processed into a two chain heterodimer as in plasma-derived, natural factor VIII; the role of this cleavage is unclear, and one recombinant product is deliberately constructed to be single-chain (see Table 18.2).
   (b) If cleavage by endogenous processing proteases is limiting (as for instance in CHO cells), the usual approach is to co-express furin (or a related protease) in the same cells, or to perform in vitro maturation using a recombinant processing protease (as in recombinant von Willebrand factor).

2. *Transgenic animal bioreactors*
   (a) The major difference between ATryn and plasma-derived antithrombin is the glycosylation, with less complicated glycans that carry fewer end-standing sialic acid moieties. This suggests that glycosylation in goat mammary gland cells is incomplete. Apparently, this translates into a shorter half-life.
   (b) Half-life of factor VIII or IX is a key issue because these proteins are predominantly used for long-term prophylaxis. The use of transgenic goats would imply a risk of undesirably short half-life. Factor VIII and IX have been expressed in pigs, and glycosylation patterns seem less aberrant than in goats. Establishing correct glycosylation is a crucial element in assessing the potential benefit of transgenic technology for proteins that are to be used in non-acute situations.

3. *Extended half-life factor IX*
   (a) The factor IX-albumin fusion binds to the FcRn receptor via its albumin moiety. Due to the binding to this recycling receptor, the fusion protein escapes endosomal degradation after endocytosis and recycles back to the circulation.
   (b) The disadvantage is that factor IX is inactive as long as albumin remains fused to its catalytic domain. This has been solved by creating a cleavable linker. However, partial cleavage of the linker at the site of injury will reduce biological activity.
   (c) For the patient, dosing intervals will be significantly longer. Alternatively, trough levels of factor IX will be higher, which adds protection against spontaneous bleeding. However, because monitoring is more complex than for other factor IX products, it is more difficult to develop an individualized dosing regimen.

4. *Thrombolytic agents*
   (a) Fibrinolysis is controlled by fibrin, which binds t-PA and plasminogen (see text), and thereby limits plasmin formation to the site of the occluding thrombus.

(b) The main adverse effect of thrombolysis is bleeding. This is largely due to insufficient fibrin-specificity, leading to the degradation and consumption of fibrinogen (fibrinogenolysis).

(c) The advantage of production in *E. coli* is the lower production cost. The disadvantage is the need for denaturation and refolding. Reteplase lacks a few domains of normal t-PA, including the fibrin-binding finger domain. Therefore, its affinity for fibrin is lower, which theoretically reduces its fibrin-specificity.

References

Andersson LO, Forsman N, Huang K, Larsen K, Lundin A, Pavlu B, Sandberg H, Sewerin K, Smart J (1986) Isolation and characterization of human factor VIII: molecular forms in commercial factor VIII concentrate, cryoprecipitate and plasma. Proc Natl Acad Sci 83:2979–2983

Björkman S, Berntorp E (2001) Pharmacokinetics of coagulation factors: clinical relevance for patients with haemophilia. Clin Pharmacokinet 40:815–832

Cafuir LA, Kempton CL (2017) Current and emerging factor VIII replacement products for hemophilia a. Ther Adv Hematol 8:303–313

Castaman G, Linari S (2017) Diagnosis and treatment of von Willebrand disease and rare bleeding disorders. J Clin Med 6:E4

Chevreux G, Tilly N, Leblanc Y, Ramon C, Faid V, Martin M, Dhainaut F, Bihoreau N (2017) Biochemical characterization of LR769, a new recombinant factor VIIa bypassing agent produced in the milk of transgenic rabbits. Haemophilia 23:e324–e334

Choo KH, Gould KG, Rees DL, Brownlee GG (1982) Molecular cloning of the gene for human anti-haemophilic factor IX. Nature 299:178–180

Collen D, Lijnen HR (2004) Tissue-type plasminogen activator: a historical perspective and personal account. J Thromb Haemost 2:541–546

Collen D, Lijnen HR (2005) Thrombolytic agents. Thromb Haemost 93:627–630

Collins R, Peto R, Baigent C, Sleight P (1997) Aspirin, heparin, and fibrinolytic therapy in suspected acute myocardial infarction. N Engl J Med 336:847–860

Davie EW, Ratnoff OD (1964) Waterfall sequence for intrinsic blood clotting. Science 145:1310–1312

Dietrich B, Schiviz A, Hoellriegl W, Horling F, Benamara K, Rottensteiner H, Turecek PL, Schwarz HP, Scheiflinger F, Muchitsch EM (2013) Preclinical safety and efficacy of a new recombinant FIX drug product for treatment of hemophilia B. Int J Hematol 98:525–532

Edmunds T, Van Patten SM, Pollock J, Hanson E, Bernasconi R, Higgins E, Manvalan P, Ziomek C, Meade H, McPherson JM, Cole ES (1998) Transgenically produced human antithrombin: structural and functional comparison to human plasma-derived antithrombin. Blood 91:4561–4571

Fay PJ (2004) Activation of factor VIII and mechanisms of cofactor action. Blood Rev 18:1–15

Franchini M, Mannucci PM (2013) Hemophilia a in the third millennium. Blood Rev 27:179–184

Franchini M, Mannucci PM (2016) Von Willebrand factor (Vonvendi®): the first recombinant product licensed for the treatment of von Willebrand disease. Expert Rev Hematol 9:825–830

Furie B, Furie BC (1988) The molecular basis of blood coagulation. Cell 53:505–518

Gill JC, Castaman G, Windyga J, Kouides P, Ragni M, Leebeek FWG, Obermann-Slupetzky O, Chapman M, Fritsch S, Pavlova BG, Presch I, Ewenstein B (2015) Hemostatic efficacy, safety, and pharmacokinetics of a recombinant von Willebrand factor in severe von Willebrand disease. Blood 126:2038–2046

Guillermin A, Yan DJ, Perrier A, Marti C (2016) Safety and efficacy of tenecteplase versus alteplase in acute coronary syndrome: a systematic review and meta-analysis of randomized trials. Arch Med Sci 12:1181–1187

Harrison S, Adamson S, Bonam D, Brodeur S, Charlebois T, Clancy B, Costigan R, Drapeau D, Hamilton M, Hanley K, Kelley B, Knight A, Leonard M, McCarthy M, Oakes P, Sterl K, Switzer M, Walsh R, Foster W (1998) The manufacturing process for recombinant factor IX. Semin Hematol 35(Suppl 2):4–10

Hedner U (2006) Mechanism of action, development and clinical experience of recombinant FVIIa. J Biotechnol 124:747–757

Hedner U, Ezban M (2008) Tissue factor and factor VIIa as therapeutic targets in disorders of hemostasis. Annu Rev Med 59:29–41

Hedner U, Kisiel W (1983) Use of human factor VIIa in the treatment of two hemophilia a patients with high-titer inhibitors. J Clin Invest 71:1836–1841

Huntington JA (2003) Mechanisms of glycosaminoglycan activation of the serpins in hemostasis. J Thromb Haemost 1:1535–1549

Inbal A, Oldenburg J, Carcao M, Rosholm A, Tehranchi R, Nugent D (2012) Recombinant factor XIII: a safe and novel treatment for congenital factor XIII deficiency. Blood 119:5111–5117

Iorio A (2017) Using pharmacokinetics to individualize hemophilia therapy. Hematology 2017. Am Soc Hematol Educ Program 2017:595–604

Jorgensen MJ, Cantor AB, Furie BC, Brown CL, Shoemaker CB, Furie B (1987) Recognition site directing vitamin K-dependent γ-carboxylation resides on the propeptide of factor IX. Cell 48:185–191

Keyt BA, Paoni NF, Refino CJ, Berleau L, Nguyen H, Chow A, Lai J, Pena L, Pater C, Ogez J, Etscheverry T, Botstein D, Bennett WF (1994) A faster-acting and more potent form of tissue plasminogen activator. Proc Natl Acad Sci U S A 91:3670–3674

Kitchen S, Tiefenbacher S, Gosselin R (2017) Factor activity assays for monitoring extended half-life FVIII and factor IX replacement therapies. Semin Thromb Hemost 43:331–337

Komaromi I, Bagoly Z, Muszbek L (2011) Factor XIII: novel structural and functional aspects. J Thromb Haemost 9:9–20

Kurachi K, Davie EW (1982) Isolation and characterization of a cDNA coding for human factor IX. Proc Natl Acad Sci U S A 79:6461–6464

Leksa NC, Chiu PL, Bou-Assaf GM, Quan C, Liu Z, Goodman AB, Chambers MG, Tsutakawa SE, Hammel M, Peters RT, Waltz T, Kulman JD (2017) The structural basis for the functional comparability of factor VIII and the long-acting variant factor VIII fc fusion protein. J Thromb Haemost 15:1167–1179

Lenting PJ, van Mourik JA, Mertens K (1998) The life cycle of coagulation factor VIII in view of its structure and function. Blood 92:3983–3996

Lissitchkov T, Willemze A, Katragadda S, Rice K, Poloskey S, Benson C (2022) Efanesoctocog alfa for hemophilia a: results from a phase 1 repeat-dose study. Blood Adv 6:1089–1094

Lovejoy AE, Reynolds TC, Visich JE, Butine MD, Young G, Belvedere MA, Blain RC, Pederson SM, Ishak LM, Nugent DJ (2006) Safety and pharmacokinetics of recombinant factor XIII-A₂ administration in patients with congenital factor XIII deficiency. Blood 108:57–62

Lusis AJ (2000) Atherosclerosis. Nature 407:233–241

Maas Enriquez M, Thrift J, Garger S, Katterle Y (2016) Bay 81-8973, a full-length recombinant factor VIII: human heat shock protein 70 improves the manufacturing process without affecting clinical safety. Protein Expr Purif 127:111–115

Macfarlane RG (1964) An enzyme cascade in the blood clotting mechanism, and its function as a biochemical amplifier. Nature 202:498–499

Mahlangu JN, Weldingh KN, Lentz SR, Kaicker S, Karim FA, Matsushita T, Recht M, Tomczak W, Windyga J, Ehrenfort S, Knobe K (2015) Changes in the amino acid sequence of the recombinant human factor VIIa analog, vatreptacog alfa, are associated with clinical immunogenicity. J Thromb Haemost 13:1989–1998

Mahlangu JN, Levy H, Kosinova MV, Khachatryan H, Korczowski B, Makhaldiani L, Iosava G, Lee M, Del Greco F (2021) Subcutaneous engineered factor VIIa marzeptacog alfa (activated) in hemophilia with inhibitors: phase 2 trial of pharmacokinetics, pharmacodynamics, efficacy, and safety. Res Pract Thromb Haemost 5:e12576

Mann KG, Orfeo T, Butenas S, Undas A, Brummel-Ziedins K (2009) Blood coagulation dynamics in haemostasis. Hamostaseologie 29:7–16

Mannucci PM (2004) Treatment of von Willebrand's disease. N Engl J Med 531:683–694

Mei B, Pan C, Jiang H, Tjandra H, Strauss J, Chen Y, Liu T, Zhang X, Severs J, Newgren J, Chen J, Gu J-M, Subramanyam B, Fournel MA, Pierce GF (2010) Rational design of a fully active, long-acting PEGylated factor VIII for hemophilia a treatment. Blood 116:270–279

Metzner HJ, Weimer T, Kronthaler U, Lang W, Schulte S (2009) Genetic fusion to albumin improves the pharmacokinetic properties of factor IX. Thromb Haemost 102:634–644

Monroe DM, Jenny RJ, Van Cott KE, Buhay S, Saward LL (2016) Characterization of IXINITY (trenonacog alfa), a recombinant factor IX with primary sequence coresponding to the threonine-148 polymorph. Adv Hematol 2016:7678901

Moser M, Kohler B, Schmittner M, Bode C (1998) Recombinant plasminogen activators: a comparative review of the clinical pharmacology and therapeutic use of alteplase and reteplase. BioDrugs 9:455–463

Østergaard H, Bjelke JR, Hansen L, Petersen LC, Pedersen AA, Elm T, Møller F, Hermit MB, Holm PK, Krogh TN, Petersen LM, Ezban M, Sørensen BB, Andersen MD, Agersø H, Ahmandian H, Balling KW, Christiansen MLS, Knobe K, Nichols TC, Bjørn SE, Tranholm M (2011) Prolonged half-life and preserved enzymatic properties of factor IX selectively PEGylated on native N-glycans in the activation peptide. Blood 118:2333–2341

Østergaard H, Lund J, Greisen PJ, Kjellev S, Henriksen A, Lorenzen N, Johansson E, Røder G, Rasch MG, Johnsen LB, Egebjerg T, Lund S, Rahbek-Nielsen H, Gandhi PS, Lamberth K, Loftager M, Andersen LM, Bonde AC, Stavenuiter F, Madsen DE, Li X, Holm TL, Ley CD, Thygesen P, Zhu H, Zhou R, Thorn K, Yang Z, Hermit MB, Bjelke JR, Hansen BG, Hilden I (2021) A factor VIIIa-mimetic bispecific antibody, Mim8, ameliorates bleeding upon severe vascular challenge in hemophilia a mice. Blood 138:1258–1268

Paidas MJ, Forsyth C, Quéré I, Rodger M, Frieling JTM, Tait RC (2014) Perioperative and peripartum prevention of venous thromboembolism in patients with hereditary antithrombin deficiency using recombinant antithrombin therapy. Blood Coagul Fibrinolysis 25:444–450

Palla R, Peyvandi F, Shapiro A (2015) Rare bleeding disorders: diagnosis and treatment. Blood 125:2052–2061

Pennica D, Holmes WE, Kohr WJ, Harkins RN, Vehar GA, Ward CA, Bennett WF, Yelverton E, Seeburg PH, Heyneker HL, Goeddel DV, Collen D (1983) Cloning and expression of human tissue-type plasminogen activator cDNA in E. coli. Nature 301:214–221

Perera L, Darden T, Pedersen LG (2001) Modeling human zymogen factor IX. Thromb Haemost 85:596–603

Peters RT, Low SC, Kamphaus GD, Dumont JA, Amari JV, Lu Q, Zarbis-Papastoitsis G, Reidy TJ, Merricks EP, Nichols TC, Bitonti AJ (2010) Prolonged activity of factor IX as a monomeric fc fusion protein. Blood 115:2057–2064

Pipe SW, Montgomery RR, Pratt KP, Lenting PJ, Lillicrap D (2016) Life in the shadow of a dominant partner: the FVIII-VWF association and its clinical implications for hemophilia a. Blood 128:2007–2016

Rijken DC, Hoylaerts M, Collen D (1982) Fibrinolytic properties of one-chain and two-chain human extrinsic (tissue-type) plasminogen activator. J Biol Chem 257:2920–2925

Rogaev EI, Grigorenko AP, Faskhutdinova G, Kittler ELW, Moliaka YK (2009) Science 326:817

Sandberg H, Kannicht C, Stenlund P, Dadaian M, Oswaldsson U, Cordula C, Walter O (2012) Functional characteristics of the novel, human-derived recombinant FVIII protein product, human-cl rhFVIII. Thromb Res 130:808–817

Schmidbauer S, Witzel R, Robbel L, Sebastian P, Grammel N, Metzner HJ, Schulte S (2015) Physicochemical characterisation of rVIII-SingleChain, a novel recombinant single-chain factor VIII. Thromb Res 136:388–395

Simpson D, Siddiqui MAA, Scott LJ, Hilleman DE (2006) Reteplase: a review of its use in the management of thrombotic occlusive disorders. Am J Cardiovasc Drugs 6:265–285

Stennicke HR, Kjalke M, Karpf DM, Baling KW, Johansen PB, Elm T, Øvlisen K, Möller F, Holmberg HL, Gudme CN, Persson E, Hilden I, Pelzer H, Rahbeck-Nielsen H, Jespersgaard C, Bogsnes A, Pedersen AA, Kristensen AK, Peschke B, Kappers W, Rode F, Thim L, Tranholm M, Ezban M, Olsen EHN, Bjørn SE (2013) A novel B-domain O-glycoPEGylated FVIII (N8-GP) demonstrates full efficacy and prolonged effect in hemophilic mice models. Blood 121:2108–2116

Thim L, Vandahl B, Karlsson J, Klausen NK, Pedersen J, Krogh TN, Kjalke M, Petersen JM, Johnsen LB, Bolt G, Nørby PL, Steenstrup TD (2010) Purification and characterization of a new recombinant factor VIII (N8). Haemophilia 16:349–359

Tiede A, Tait RC, Shaffer DW, Baudo F, Boneu B, Dempfle CE, Horrelou MH, Klamroth R, Lazarchick J, Mumford AD, Schulman S, Shiach C, Bonfiglio LJ, Frieling JTM, Conard J, von Depka M (2008) Antithrombin alfa in hereditary antithrombin deficient patients: a phase 3 study of prophylactic intravenous administration in high risk situations. Thromb Haemost 99:616–622

Toole JJ, Knopf JL, Wozney JM, Sultzman LA, Bueker JL, Pittman DD, Kaufman RJ, Brown E, Shoemaker C, Orr EC, Amphlett GW, Foster WB, Coe ML, Knutson GJ, Fass DN, Hewick RM (1984) Molecular cloning of a cDNA encoding human antihaemophilic factor. Nature 312:342–347

Toole JJ, Pittman DD, Orr EC, Murtha P, Wasley LC, Kaufman RJ (1986) A large region (approximately equal to 95 kDa) of human factor VIII is dispensable for in vitro procoagulant activity. Proc Natl Acad Sci U S A 83:5939–5942

Turecek PL, Mitterer A, Matthiessen HP, Gritsch H, Varadi K, Siekmann J, Schnecker K, Plaimauer B, Kaliwoda M, Purtscher M, Woehrer W, Mundt W, Muchitsch E-M, Suiter T, Ewenstein BM, Ehrlich HJ, Schwarz HP (2009) Development of a plasma- and albumin-free recombinant von Willebrand factor. Hamostaseologie 29(Suppl 1):S32–S38

Turecek PL, Bossard M, Graniger M, Gritsch H, Höllriegl W, Kaliwoda M, Matthiessen P, Mitterer A, Muchitsch E-M, Purtscher M, Rottensteiner H, Schiviz A, Schrenk G, Siekmann J, Varadi K, Riley T, Ehrlich HJ, Schwarz HP, Scheiflinger F (2012) BAX 855, a PEGylated rFVIII product with prolonged half-life: development, functional and structural characterisation. Hamostaseologie 32(Suppl 1):S29–S38

Uchida N, Sambe T, Yoneyama K, Fukazawa N, Kawanishi T, Kobayashi S, Shima M (2016) A first-in-human phase 1 study of ACE910, a novel factor VIII-mimetic bispecific antibody, in healthy subjects. Blood 127:1633–1641

Vehar GA, Keyt B, Eaton D, Rodriguez H, O'Brien DP, Rotblatt F, Oppermann H, Keck R, Wood WI, Harkins RN, Tuddenham EGD, Lawn RM, Capon DJ (1984) Structure of human factor VIII. Nature 312:337–342

Wasley LC, Rehemtulla A, Bristol JA, Kaufman RJ (1993) PACE/furin can process the vitamin K-dependent pro-factor IX precursor within the secretory pathway. J Biol Chem 268:8458–8465

Young G, Mahlangu JN (2016) Extended half-life clotting factor concentrates: results from published clinical trials. Haemophilia 22(Suppl 5):25–30

Further Reading

Bishop P, Lawson J (2004) Recombinant biologics for treatment of bleeding disorders. Nat Rev Drug Discov 3:684–694

Lubon H, Paleyanda RK, Velander WH, Drohan WN (1996) Blood proteins from transgenic animal bioreactors. Transfus Med Rev 10:131–143

Mackman N (2008) Triggers, targets and treatments for thrombosis. Nature 451:914–918

Peyvandi F, Garagiola I, Seregni S (2013) Future of coagulation factor replacement therapy. J Thromb Haemost 11(Suppl 1):84–98

Follicle-Stimulating Hormone

19

Tom Sam, Marc Bastiaansen, and Keith Gordon

Introduction

About 15% of all couples experience infertility at some time during their reproductive lives. Nowadays, infertility can be treated by the use of assisted reproductive technologies (ART), such as in vitro fertilization (IVF) and intracytoplasmic sperm injection (ICSI). A common element of these programs is the treatment with follicle-stimulating hormone (FSH) to increase the number of oocytes retrievable for the IVF or ICSI procedure, a procedure called controlled ovarian stimulation (with a goal of multifollicular development). Patients suffering from female infertility because of chronic anovulation may also be treated with lower doses of FSH, with the aim of achieving mono-follicular development.

Natural FSH is produced and secreted by the anterior lobe of the pituitary, a gland that sits in the sella turcica just below the hypothalamus at the base of the brain. The main target for FSH is the FSH receptor on the surface of the granulosa cells that surround the oocyte. FSH acts synergistically with estrogens and luteinizing hormone (LH) to stimulate proliferation of these granulosa cells, which leads to follicular growth. As the primary function of FSH in the female is the regulation of follicle growth and development, this process explains why deficient endogenous production of FSH may cause infertility. In males, FSH plays a pivotal role in spermatogenesis.

The first clinical use of exogenous gonadotropin to induce ovulation in women was achieved in the 1950s with pituitary extracts (Gemzell et al. 1958). However, the use of this cadaveric source material was stopped because of supply problems, and moreover, it became apparent that it could transmit diseases such as Creutzfeldt–Jakob disease (Lunenfeld 2002). Fortunately, gonadotropins extracted from pituitary glands were soon superseded by human menopausal gonadotropin (HMG) products extracted from the urine of post-menopausal women in the late 1950s (Lunenfeld 2002).

Up until 1995, FSH preparations for infertility treatment were derived from urine from (post)menopausal women. As over 100,000 liters of urine was required for a single batch, many thousands of donors are needed. Hence, the source of urinary FSH is heterogeneous and the sourcing cumbersome. Moreover, in addition to FSH, these urinary preparations contain impurities including pharmaceutically active proteins such as LH and many uncharacterized proteins and other substances (Van de Weijer et al. 2003). Recombinant DNA technology allows the reproducible manufacturing of FSH preparations of high purity and specific activity, devoid of urinary contaminants. Recombinant FSH was originally produced using a Chinese hamster ovary (CHO) cell line, transfected with the genes encoding for the two human FSH subunits (Van Wezenbeek et al. 1990; Howles 1996). The isolation procedures render a product of high purity (at least 99%), devoid of LH activity and very similar to natural FSH. A recombinant FSH derived from a cell line of human fetal retinal origin (PER.C6) (follitropin delta) was introduced (Olson et al. 2014), which is dosed in micrograms rather than international units (IUs).

Currently, there are several clinically approved recombinant FSH-containing drug products on various markets. The most widely used products are Gonal-F®, manufactured by Merck Serono S.A. (EMA 2010), and Puregon®, with the brand name of Follistim® in the USA and Japan, manufactured by Organon (EMA 2006). Regulatory authorities have issued distinct international non-proprietary names (INN) for the three corresponding recombinant FSH drug substances, i.e., follitropin alfa (Gonal-F®), follitropin beta

This chapter is dedicated in honor to a lifetime of scientific contributions of Tom Sam (deceased June 2020); NV Organon, Oss, The Netherlands.

T. Sam · M. Bastiaansen (✉)
NV Organon, Oss, The Netherlands
e-mail: marc.bastiaansen@organon.com

K. Gordon
Organon, Inc., Jersey City, New Jersey, USA
e-mail: keith.gordon@organon.com

(Puregon®/Follistim®), and follitropin delta (Rekovelle®) (EMA 2016). In addition, a few other recombinant FSH preparations, "biosimilar" to follitropin alfa, were developed and are available under the following names: Bemfola® and Ovaleap®) (Rettenbacher et al. 2015).

FSH Is a Glycoprotein Hormone

Follicle-stimulating hormone belongs to a family of structurally related glycoproteins that also includes luteinizing hormone (LH), chorionic gonadotropin (CG), collectively called the gonadotropins, and thyroid-stimulating hormone (TSH, also named thyrotropin). These hormones belong to the "cystine-knot protein family." FSH is a heterodimeric protein consisting of two non-covalently associated glycoprotein subunits, denoted α and β. The α-subunit contains five intra-subunit disulfide bonds and is identical for all these glycoproteins, and it is the β-subunit with six intra-subunit disulfide bonds that provides each hormone with its specific biological function.

The glycoprotein subunits of FSH consist of two polypeptide backbones with carbohydrate side chains attached to the two asparagine (Asn) amino acid residues on each subunit. The oligosaccharides are attached to Asn-52 and Asn-78 on the α-subunit (92 amino acids) and to Asn-7 and Asn-24 on the β-subunit (111 amino acids). The glycoprotein FSH has a molecular mass of approximately 35 kDa. For the FSH preparation to be biologically active, the two subunits must be correctly assembled into their three-dimensional heterodimeric protein structure and post-translationally modified (Fig. 19.1).

Assembly and glycosylation are intracellular processes that take place in the endoplasmic reticulum and in the Golgi apparatus. This glycosylation process leads to the formation of a population of hormone isoforms differing in their carbohydrate side-chain composition. The carbohydrate side chains of FSH are essential for its biological activity since they (1) influence FSH receptor binding, (2) play an important role in the signal transduction into the FSH target cell, and (3) affect the plasma residence time of the hormone. The glycosylation process is subtly different between the different manufacturing processes, resulting in slightly different isoform profiles (see below).

Recombinant FSH contains approximately one-third carbohydrate on a mass-per-mass basis. The carbohydrate side chains are composed of mannose, fucose, N-acetylglucosamine, galactose, and sialic acid. Structure analysis by proton nuclear magnetic resonance (^{1}H-NMR) spectroscopy on oligosaccharides enzymatically cleaved from follitropin beta reveals minor differences with natural FSH. For instance, the bisecting GlcNAc residues are lacking in the recombinant molecule, simply because the FSH-producing CHO cells do not possess the enzymes to incorporate these residues. Furthermore, the carbohydrate side chains of recombinant FSH exclusively contain α2–3 linked sialic acid, whereas in the natural hormone, α1–6 linked sialic acid occurs as well. However, all carbohydrate side chains identified in recombinant FSH are moieties normally found in other natural human glycoproteins. The amino acid sequences of the α-subunit and the β-subunit of the recombinant FSH derived from human fetal retina (follitropin delta) are identical to the sequence of natural human FSH and CHO-derived FSH products in clinical use, while the sialic acid content is higher (Olson et al. 2014).

Fig. 19.1 A three-dimensional model of FSH. The ribbons represent the polypeptide backbones of the α-subunit (green ribbon) and the β-subunit (blue ribbon). The carbohydrate side chains (violet and pink space filled globules) cover large areas of the surface of the polypeptide subunits

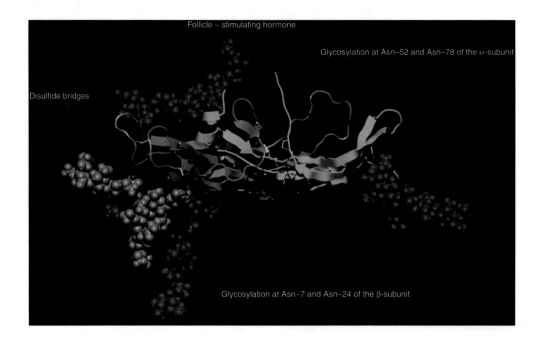

Follicle – stimulating hormone

Glycosylation at Asn–52 and Asn–78 of the α-subunit

Disulfide bridges

Glycosylation at Asn–7 and Asn–24 of the β-subunit

Whereas FSH only contains N-linked carbohydrates, human chorionic gonadotropin (hCG) also carries four O-linked (at serine or threonine residues) carbohydrates, all located at the carboxy-terminal peptide (CTP) of its beta subunit. This glycosylated CTP is the major difference with the beta subunit of LH and is demonstrated to be responsible for the much longer plasma residence time of hCG compared to natural LH (Matzuk et al. 1990).

Production of Recombinant FSH

The genes coding for the human FSH α-subunit and β-subunit were inserted in cloning vectors (plasmids) to enable efficient transfer into recipient cells. These vectors also contained promoters that could direct transcription of foreign genes in recipient cells. CHO cells were selected as recipient cells since they were easily transfected with foreign DNA and are capable of synthesizing complex glycoproteins as present in the urine FSH. Attempts have been made to produce FSH via yeast to increase yield, e.g., bovine, porcine, ovine, and primate recombinant FSH via *Komagataella pastoris (K. pastoris)*. However, none of these attempts has resulted in a marketed recombinant *human* FSH product (Gifre et al. 2017). This is due to the complex human gonadotropin structure and glycosylation pattern that cannot be mimicked sufficiently by these yeasts. As a result, commercial recombinant gonadotropins are still produced using the Chinese hamster ovary (CHO) cell line or other mammalian cells.

These CHO cells can be grown in cell cultures on a large scale. To construct an FSH-producing cell line, N.V. Organon, the manufacturer of Puregon®/Follistim®, used one single vector containing the coding sequences for both subunit genes (Olijve et al. 1996). Merck Serono S.A., the manufacturer of Gonal-F®, used two separate vectors, one for each subunit gene (Howles 1996). Following transfection, a genetically stable transformant producing biologically active recombinant FSH was isolated. For the CHO cell line used for manufacturing Puregon®/Follistim®, it was shown that approximately 150–450 gene copies were present.

To establish a master cell bank (MCB), identical homogeneous cell preparations of the selected clone are stored in individual vials and cryopreserved until needed. Subsequently, a working cell bank (WCB) is established by the expansion of cells derived from a single vial of the MCB, and aliquots are put into vials and cryopreserved as well. Each time a production run is started, cells from one or more vials of the WCB are cultured (cf. Chap. 14).

Both recombinant FSH products are isolated from cell culture supernatants. These supernatants are collected from a perfusion-type bioreactor containing recombinant FSH-producing CHO cells grown on microcarriers. This is because the CHO cell lines used are anchorage-dependent cells, which implies that a proper surface must be provided for cell growth. The reactor is perfused with a growth-promoting medium during a period that may continue for up to 3 months (see also Chap. 4). The downstream purification processes for the isolation of the two recombinant FSH products are different. For Puregon®/Follistim®, a series of chromatographic steps, including anion and cation exchange chromatography, hydrophobic chromatography, and size-exclusion chromatography, are used. Recombinant FSH in Gonal-F® is obtained by a similar process of five chromatographic steps but also includes an immunoaffinity step using a murine FSH-specific monoclonal antibody. In both production processes, each purification step is rigorously controlled in order to ensure the batch-to-batch consistency of the purified product.

Description of Recombinant FSH

Structural Characteristics

Like urinary sourced (natural) FSH, the recombinant versions exist in several distinct molecular forms (isohormones), with identical polypeptide backbones but with differences in oligosaccharide structure, in particular in the degree of terminal sialylation. These isohormones can be separated by chromatofocusing or isoelectric focusing on the basis of their different isoelectric points (pI) as has been demonstrated for follitropin beta (De Leeuw et al. 1996; Fig. 19.2).

The typical pattern for FSH indicates an isohormone distribution between pI values of 6 and 4. To obtain structural information at the subunit level, the two subunits were separated by reversed phase high performance liquid chromatography (RP-HPLC) and treated to release the N-linked carbohydrate side chains. Fractions with low pI values (acidic fractions; high sialic acid/galactose ratio) displayed a high content of tri- and tetrasialo-oligosaccharides and a low

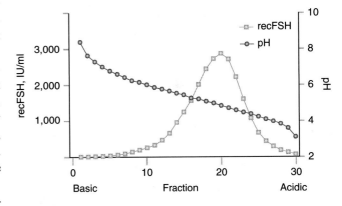

Fig. 19.2 Isohormone profile of recombinant follicle-stimulating hormone (follitropin beta) after preparative free flow focusing (De Leeuw et al. 1996). The FSH concentration was determined by a two-site immunoassay that is capable of quantifying the various isohormones equally well

content of neutral and monosialo-oligosaccharides. For fractions with a high pI (basic fractions) value, the reverse was found.

One of the tools for further characterization is the immunoassay. Due to the specific recognition characteristics of the antibodies used, this assay determines FSH-specific structural features and provides a relative measure for the quantity of FSH, as it is not sensitive to the differences in glycosylation.

Biological Properties of Recombinant FSH Isohormones

An FSH preparation can be biologically characterized with several essentially different assays, each having its own specific merits (Mannaerts et al. 1992). The receptor binding assay provides information on the proper conformation for interaction with the FSH receptor. Receptor binding studies with calf testis membranes have shown that FSH isoform activity in follitropin beta decreases when going from high to low pI isoforms. The in vitro bioassay measures the capability of FSH to transduce signals into target cells (the intrinsic bioactivity). The in vitro bioactivity, assessed in the rat Sertoli cell bioassay, also decreases when going from high to low pI isoforms. The in vivo bioassay provides the overall bioactivity of an FSH preparation. It is determined by the number of molecules, the plasma residence time, the receptor binding activity, and the signal transduction. Interestingly, in contrast to the receptor binding and in vitro bioassays, the in vivo biological activity determined in rats shows an approximately 20-fold increase between isoforms with a pI value of 5.49, as compared to those with a pI value of 4.27. These results indicate that the basic isohormones exhibit the highest receptor binding and signal transduction activity, whereas the acidic isohormones are the more active forms under in vivo conditions. This notion also warrants pharmacokinetic studies to further characterize the biological properties of FSH preparations; see below.

Pharmacokinetic Behavior of Recombinant FSH Isohormones

A typical batch of recombinant follicle-stimulating hormone (rFSH) displays a distribution of FSH immunoreactivity from pI 6–4 (see Fig. 19.2). Fractions or pools of fractions were used for the biological characterization (De Leeuw et al. 1996). The pharmacokinetic behavior of follitropin beta and its isohormones was investigated in beagle dogs that were given an intramuscular bolus injection of a number of FSH isohormone fractions, each with a specific pI value (De Leeuw et al. 1996). With a decrease in pI value from 5.49

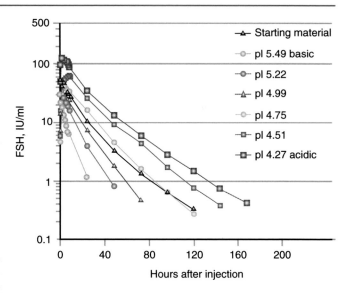

Fig. 19.3 Kinetic behavior of FSH isoforms after a single intramuscular injection (20 IU/kg) in beagle dogs (De Leeuw et al. 1996)

(basic) to 4.27 (acidic), the area under the curve (AUC) increased and the clearance decreased, each more than tenfold. A more than twofold difference in elimination half-life between the most acidic and the most basic FSH isohormone fraction was calculated. The absorption rate of the two most acidic isoforms was higher than the absorption rates of all other isoforms. The AUC and the clearance for the follitropin beta preparation, being a mixture of all isohormone fractions, corresponded with the center of the isohormone profile (Fig. 19.3). In contrast, the elimination of the follitropin beta preparation occurred at a rate similar to that of the most acidic fractions, indicating that the elimination rate is largely determined by the removal of the most acidic isoforms from the plasma.

Thus, for follitropin beta isohormone fractions, a clear correlation exists between pI value and pharmacokinetic behavior. Increasing acidity leads to an increase in the extent of absorption and elimination half-life and to a decrease in clearance (Fig. 19.3).

Pharmaceutical Formulations

Recombinant FSH preparations distinguish themselves from the earlier urinary FSH preparations by their high purity (at least 99%). However, pure proteins are relatively unstable and are therefore often lyophilized, unless some specific stabilizing measures can be taken. FSH preparations are available in different strengths and presentation forms, both as freeze-dried products (powder, cake) and as solution for injection. Follitropin alfa was originally formulated with sucrose (bulking agent, lyoprotectant), sodium dihydrogen phosphate/disodium hydrogen phosphate, phosphoric acid, and sodium hydroxide (for pH adjustment). In 2002, L-methionine (antioxidant) and polysorbate 20 (to prevent adsorption losses) were added to the single-dose formulation.

Follitropin beta was originally formulated with sucrose (lyoprotectant), sodium citrate (stabilizer), polysorbate 20 (to prevent adsorption losses), and hydrochloride/sodium hydroxide (for pH adjustment). This product was initially marketed as a freeze-dried presentation form; however, a solution for injection with several strengths of follitropin beta was developed subsequently. To stabilize the solutions against oxidation, 0.25 mg of L-methionine had to be added. Furthermore, the solution in the cartridge contains benzyl alcohol as preservative. For follitropin alfa, a multidose solution for injection in a pre-filled pen became available in 2004. This solution contains poloxamer 188 instead of polysorbate 20, and m-cresol has been added as preservative.

Pen injectors for follitropin beta solution for injection have also been developed with multidose cartridges containing solution for injection, giving the patient improved convenience. Studies showed that drug delivery was more precise, efficient (18% less loss), and better tolerated using ready-to-use pens than syringe injections (Platteau et al. 2003).

The solutions for injection for follitropin beta should be stored in the refrigerator for a maximum of 3 years with the container kept in the outer carton to protect the solution from light. The patient can keep the solutions at room temperature for a maximum of 3 months. The multidose solution of follitropin beta has a shelf-life of 3 years and can be stored for 1 month at room temperature.

Clinical Aspects

Recombinant FSH products on the market have been approved for two female indications. The first indication is anovulation (including polycystic ovarian disease) in women who are unresponsive to clomiphene citrate (an estrogen receptor modulator). The second indication is controlled ovarian hyperstimulation to induce the development of multiple follicles in medically assisted reproduction programs, such as in vitro fertilization (IVF) and intracytoplasmic sperm injection (ICSI). In addition, recombinant FSH may be used in men with congenital or acquired hypogonadotropic hypogonadism to stimulate spermatogenesis.

The clinical treatment scheme for recombinant FSH products (Gonal-F®, Puregon®, and Rekovelle®) depends on the exact pharmacokinetics and pharmacodynamics of these products in humans. A typical clinical treatment scheme for the FSH product Puregon® (follitropin beta) is given below as an illustration. For the treatment of anovulatory patients (aiming at mono-follicular growth), it is recommended to start follitropin beta treatment with 50 IU per day for

7–14 days and gradually increase dosing with steps of 25–50 IU if no sufficient response is seen. This gradual dose-increasing schedule is followed in order to minimize multi-follicular development and to avoid the induction of ovarian hyperstimulation syndrome (a serious condition of unwanted hyperstimulation). ART cycles are typically programmed or scheduled (often with combined oral contraceptives (COCs)) to accommodate the clinic's timelines and the woman's commitments. When the stimulation is to begin the COCs are stopped and a few days later daily injections of FSH and/or LH containing products are given and the stimulation cycle has begun. These injections are generally self-administered at home. With varying frequencies, the woman returns to the clinic for monitoring. This usually consists of a transvaginal ultrasound assessment of follicle size and the drawing of some blood for hormone determination (typically estradiol, progesterone, and, to a lesser extent, LH and FSH). On stimulation day 5, she typically starts to take daily injections of a gonadotropin-releasing hormone (GnRH) antagonist to prevent the occurrence of a premature LH surge, which could jeopardize the oocyte collection. This continues (both FSH and GnRH antagonist injections) until >2–3 follicles reach a diameter of 17–18 mm at which point a single dose of hCG (10,000 IU of urinary CG or 250 µg rhCG) is given to induce final oocyte maturation and to facilitate collection of the oocytes via transvaginal oocyte aspiration. If the woman happens to be a high responder (>15–18 large follicles), she will likely receive a GnRH agonist trigger to induce an endogenous LH/FSH surge rather than the hCG, since this greatly reduces the risk of ovarian hyperstimulation syndrome (OHSS). Oocyte pickup is performed between 34 and 36 h after the hCG or GnRH agonist injection. Once collected, oocytes can be used for

- Immediate use as part of an ART cycle (with or without embryo cryopreservation)
- Future use in women undergoing medical therapy that may negatively impact ovarian function
- Future use in women who may want to conceive in the future

As mentioned, similar treatment regimens are recommended for Gonal-F and Rekovelle. Although for Rekovelle initial dosing recommendations are based on an algorithm that takes into consideration her body weight and serum anti-Mullerian hormone levels (EMA 2016).

After subcutaneous administration, follitropin beta has an elimination half-life of approximately 33 h (Voortman et al. 1999). Steady-state levels of follitropin beta are therefore obtained after four to five daily doses reaching therapeutically effective plasma concentrations of FSH. Follitropin beta is administered via the subcutaneous route with good local tolerance. Bioavailability is approximately 77%. In the patient population, the formation of antibodies against recombinant FSH or CHO-cell-derived proteins was rare. The CHO-cell-derived proteins present after the cell culture process are removed during the FSH purification process to very low levels to ensure patient safety.

A Newly Developed FSH Analog

The need for daily injections of FSH is a burden for the women treated in an ART regimen. Therefore, several different approaches have been undertaken to arrive at FSH preparations that need fewer injections, such as slow-release formulations, addition of N-linked carbohydrates, and other chemical modifications including PEGylation (Fauser et al. 2009). An elegant approach pioneered by Irving Boime and collaborators (Fares et al. 1992; LaPolt et al. 1992) is based on the longer in vivo half-life of hCG compared to LH. The half-life of hCG is longer than that of LH because hCG has more amino acid and carbohydrate residues at the carboxyl terminal. The longer half-life of the FSH analog was created by clipping off the end of the rFSH gene and constructing the end of the hCG gene on it. The new gene was then transfected into Chinese hamster ovary cells, which then produce the resulting chimeric protein FSH-C-terminal peptide (FSH-CTP), also known as Org 36286 and by its INN, corifollitropin alfa. In this FSH analog, the beta subunit is extended by a single CTP (28 amino acids). Thorough biochemical analysis demonstrated the expected amino acid sequence of the alpha subunit and the extended beta subunit but revealed two additional O-linked glycosylation sites in corifollitropin alfa (compared to the four to five sites reported in hCG (Fig. 19.4)).

Nonclinical evaluation demonstrated that the receptor binding and transactivation profile of this new molecular entity was specific and comparable to that of recFSH without intrinsic TSH-receptor or LH-receptor activation. However, the in vivo half-life was increased 1.5 to twofold in the species tested, and a two- to fourfold increase of bioactivity was found across all in vivo pharmacodynamic parameters tested (Verbost et al. 2011). These observations were corroborated by a very extensive data set obtained in a broad panel of clinical trials (phases I, II, and III), including the largest comparator-controlled trial of its kind in fertility (the comparator being recFSH) (Devroey et al. 2009; Fauser et al. 2010). A single subcutaneous dose of corifollitropin alfa (Elonva®) can be used to initiate and sustain multifollicular growth for 7 days, while the efficacy and safety of this novel biopharmaceutical were similar to that of daily injections with recombinant FSH. Whereas normally more than 7 days of FSH treatment has to be given after the first injection, in

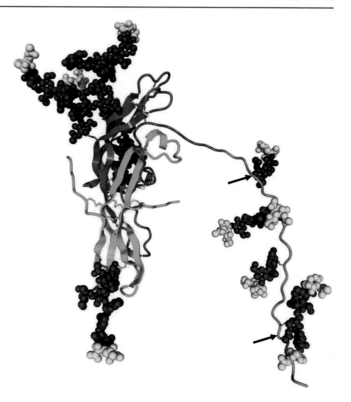

Fig. 19.4 A three-dimensional model of corifollitropin alfa. The ribbons represent the polypeptide backbones of the α-subunit (green ribbon) and the β-subunit (blue ribbon). The carbohydrate side chains (pink and yellow space filled globules) represent N-linked and O-linked carbohydrates. The sialic acid carbohydrates are depicted in yellow. The arrows indicate additional O-linked carbohydrate sites. Courtesy of M.L.C.E. Kouwijzer and R. Bosch

about one-third of the women treated with FSH-CTP, no additional FSH treatment was needed. For the others, stimulation could be continued with daily FSH injections until the final oocyte maturation had been reached (Balen et al. 2004).

Dedicated clinical research revealed no clinically relevant immunogenicity against the FSH analog (Norman et al. 2011), despite being a fusion protein. Hence, by virtue of its ~2-fold increased in vivo half-life, corifollitropin alfa has demonstrated to provide a valuable alternative for FSH by acting as a sustained follicle stimulant. Elonva® is approved (EU) for controlled ovarian stimulation in combination with a GnRH antagonist for the development of multiple follicles in women participating in an ART program.

FSH provides a great example of the evolution of biopharmaceuticals, starting from the natural form (pituitary then urine derived), via close imitations thereof (recombinant FSH), toward further improved biopharmaceuticals (FSH analogs, corifollitropin alfa being the first form that made it to the market). Such developments in pharmaceutical biotechnology are clearly to the benefit of the patients in need for effective, safe, and convenient treatment options.

References

Balen AH, Mulders AG, Fauser BC, Schoot BC, Renier MA, Devroey P et al (2004) Pharmacodynamics of a single low dose of long-acting recombinant follicle- stimulating hormone (FSH-carboxy terminal peptide, corifollitropin alfa) in women with World Health Organization group II anovulatory infertility. J Clin Endocrinol Metab 89(12):6297–6304

De Leeuw R, Mulders J, Voortman G, Rombout F, Damm J, Kloosterboer L (1996) Structure-function relationship of recombinant follicle stimulating hormone (Puregon®). Mol Hum Reprod 2:361–369

Devroey P, Boostanfar R, Koper NP, Ijzerman P, Mannaerts BMJL, Fauser BC, for the ENGAGE Investigators (2009) A double-blind, non-inferiority RCT comparing corifollitropin alfa and recombinant FSH during the first seven days of ovarian stimulation using a GnRH antagonist protocol. Hum Reprod 24:3063–3072

EMA (2006) European Public Assessment Report Puregon (Follitropin beta) https://www.ema.europa.eu/en/documents/product-information/puregon-epar-product-information_en.pdf. Accessed 16 Mar 2022

EMA (2010) European Public Assessment Report Gonal-F (Follitropin alfa) https://www.ema.europa.eu/en/documents/product-information/gonal-f-epar-product-information_en.pdf. Accessed 16 Mar 2022

EMA (2016) European Public Assessment Report Rekovelle (Follitropin delta). https://www.ema.europa.eu/en/documents/product-information/rekovelle-epar-product-information_en.pdf. Accessed 16 Mar 2022

Fares FA, Suganuma N, Nishimori K et al (1992) Design of a long-acting follitropin agonist by fusing the C-terminal sequence of chorionic gonadotropin B subunit to the follitropin B subunit. PNAS 89:4304–4308

Fauser BCJM, Alper MM, Ledger W, Schoolcraft WB, Zandvliet A, Mannaerts BMJL (2010) Pharmacokinetics and follicular dynamics of corifollitropin alfa versus recombinant FSH during controlled ovarian stimulation for in vitro fertilisation. Reprod Biomed Online 21:593–601

Fauser BCJM, Mannaerts BMJL, Devroey P, Leader A, Boime I, Baird DT (2009) Advances in recombinant DNA technology: corifollitropin alfa, a hybrid molecule with sustained follicle-stimulating activity and reduced injection frequency. Hum Reprod Update 15:309–321

Gemzell CA, Diczfalusy E, Tillinger G (1958) Clinical effect of human pituitary follicle-stimulating hormone (FSH). J Clin Endocrinol Metab 18:1333–1348

Gifre L, Aris A, Bach A, Garcia-Fruitos E (2017) Microb Cell Factories 16:40

Howles CM (1996) Genetic engineering of human FSH (Gonal-F®). Hum Reprod Update 2:172–191

LaPolt PS, Nishimori K, Fares FA, Perlas E, Boime I, Hsueh AJW (1992) Enhanced stimulation of follicle maturation and ovulatory potential by long-acting follicle-stimulating hormone agonists with extended carboxy-terminal peptides. Endocrinology 131:2514–2520

Lunenfeld B (2002) Development of gonadotrophins for clinical use. RBM Online 4(suppl 1):11–17

Mannaerts BMJL, De Leeuw R, Geelen J, Van Ravenstein A, Van Wezenbeek P, Schuurs A, Kloosterboer L (1992) Comparative in vitro and in vivo studies on the biological properties of recom-binant human follicle stimulating hormone. Endocrinology 129:2623–2630

Matzuk MM, Hsueh AJ, Lapolt P, Tsafriri A, Keene JL, Boime I (1990) The biological role of the carboxy-terminal extension of human chorionic gonadotropin (corrected) beta-subunit. Endocrinology 126:376–383

Norman RJ, Zegers-Hochschild F, Salle BS, Elbers J, Heijnen E, Marintcheva-Petrova M, Mannaerts B (2011) Repeated ovarian stimulation with corifollitropin alfa in patients in a GnRH antagonist protocol: no concern for immunogenicity. Hum Reprod 26(8):2200–2208

Olijve W, de Boer W, Mulders JWM, van Wezenbeek PMGF (1996) Molecular biology and biochemistry of human recombinant follicle stimulating hormone (Puregon®). Mol Hum Reprod 2:371–382

Olson H, Standström R, Grundemar L (2014) Different pharmacokinetic and pharmacodynamic properties of recombinant follicle-stimulating hormone (rFSH) derived from a human cell line compared with rFSH from a non-human cell line. J Clin Pharmacol 54:1299–1307

Platteau P, Laurent E, Albano C, Osmanagaoglu K, Vernaeve V, Tournaye H et al (2003) An open, randomized single-Centre study to compare the efficacy and convenience of follitropin beta administered by a pen device with follitropin alpha administered by a conventional syringe in women undergoing ovarian stimulation for IVF/ICSI. Hum Reprod 18(6):1200–1204

Rettenbacher M, Andersen AN, Garcia-Velasco JA, Sator M, Barri P, Lindenberg S, van der Veen K, Khalaf Y, Bentin-Ley U, Obruca A, Tews G, Schenk M, Stowitzki T, Narvekar N, Sator K, Inthurn B (2015) A multi-center phase 3 study comparing efficacy and safety of Bemfola versus Gonal-F in women undergoing ovarian stimulation for IVF. Reprod Biomed Online 30:504–513

Van de Weijer BHM, Mulders JWM, Bods ES, Verhaert PDEM, van den Hooven HW (2003) Compositional analyses of human menopausal gonadotrophin preparation extracted from urine (menotropin). Identification of some of its major impurities. RBM Online 7(5):547–557

Van Wezenbeek P, Draaier J, Van Meel F, Olijve W (1990) Recombinant follicle stimulating hormone. I. Construction, selection and characterization of a cell line. In: Crommelin DJA, Schellekens H (eds) From clone to clinic, developments in biotherapy, vol 1. Kluwer, Dordrecht, pp 245–251

Verbost P, Sloot WN, Rose UM, de Leeuw R, Hanssen RGJM, Verheijden GFM (2011) Pharmacologic profiling of corifollitropin alfa, the first developed sustained follicle stimulant. Eur J Pharmacol 651:227–233

Voortman G, van de Post J, Schoemaker RC, van Gerwen J (1999) Bioequivalence of subcutaneous injections of recombinant human follicle stimulating hormone (Puregon®) by pen-injector and syringe. Hum Reprod 14:1698–1702

Further Reading

Dias JA (2021) New human follitropin preparations: How glycan structural differences may affect biochemical and biological function and clinical effect. Front Endocrinol 12:Article 636038

Seyhan A, Ata B (2011) The role of corifollitropin alfa in controlled ovarian stimulation for IVF in combination with GnRH antagonist. Int J Women's Health 3:243–255

Human Growth Hormone

20

Le N. Dao, Barbara Lippe, Michael Laird,
and Daan J. A. Crommelin

Introduction

Human growth hormone (hGH) is a protein hormone essential for normal growth and development in humans. hGH affects many aspects of human metabolism, including lipolysis, the stimulation of protein synthesis, and the inhibition of glucose metabolism. Human growth hormone was first isolated and identified in the late 1950s from extracts of pituitary glands obtained from cadavers and from patients undergoing hypophysectomy. The first clinical use of these pituitary-extracted hGHs for stimulation of growth in hypopituitary children occurred in 1957 and 1958 (Raben 1958). From 1958 to 1985, the primary material used for clinical studies was pituitary-derived growth hormone (pit-hGH). Human growth hormone was first cloned in 1979 (Goeddel et al. 1979; Martial et al. 1979). The first use in humans of recombinant human growth hormone (rhGH) was reported in the literature in 1982 (Hintz et al. 1982). The introduction of rhGH coincided with reports of a number of cases of Creutzfeldt-Jakob disease, a fatal degenerative neurological disorder, in patients receiving pituitary-derived hGH. Concern over possible contamination of the pituitary-derived hGH preparations by the prion responsible for Creutzfeldt-Jakob

Updated by Daan J.A. Crommelin for the sixth edition

L. N. Dao
Clinical Pharmacology, Genentech Inc.,
South San Francisco, CA, USA

B. Lippe
Endocrine Care, Genentech Inc., South San Francisco, CA, USA

M. Laird
Department of Bioprocess Development, Division of Medical
Affairs, Genentech Inc., South San Francisco, CA, USA

D. J. A. Crommelin (✉)
Department of Pharmaceutics, Utrecht Institute for Pharmaceutical
Sciences, Utrecht University, Utrecht, The Netherlands
e-mail: d.j.a.crommelin@uu.nl

disease led to the removal of pit-hGH products from the market in the USA in 1985 followed by the Food and Drug Administration (FDA) approval of rhGH later in the year. The initial rhGH preparations were produced in bacteria (*Escherichia coli*) but, unlike endogenous hGH, contained an N-terminal methionine group (met-rhGH). Natural sequence recombinant hGH products have subsequently been produced in bacteria, yeast, and mammalian cells.

hGH Structure and Isohormones

The major, circulating form of hGH is a non-glycosylated, 22 kDa protein composed of 191 amino acid residues linked by disulfide bridges in two peptide loops (Fig. 20.1). The three-dimensional structure of hGH includes four antiparallel alpha-helical regions (Fig. 20.2) and three mini-helices. Helix 4 and helix 1 have been determined to contain the primary sites for binding to the growth hormone receptor. In addition, two of the three mini-helices located within the connecting link between helix 1 and helix 2 have been shown to play an important role in the binding of growth hormone to its receptor (Root et al. 2002; Wells et al. 1993). Endogenous growth hormone contains a variety of other isoforms including a 20 kDa monomer, disulfide-linked dimers, oligomers, proteolytic fragments, and other modified forms (Boguszewski 2003; Lewis et al. 2000). The 20 kDa monomer, dimers, oligomers, and other modified forms occur as a result of different gene products, different splicing of hGH mRNA, and posttranslational modifications. These isoforms are generally expressed at lower amounts compared to the 22 kDa protein (Baumann 2009).

There are two hGH genes in humans: the "normal" hGH-N gene and the "variant" hGH-V gene. The hGH-N gene is expressed in the pituitary gland. The hGH-V gene is expressed in the placenta and is responsible for the production of several variant forms of hGH found in pregnant women. Non-glycosylated and glycosylated isoforms of hGH-V have been identified (Ray et al. 1989; Baumann 1991).

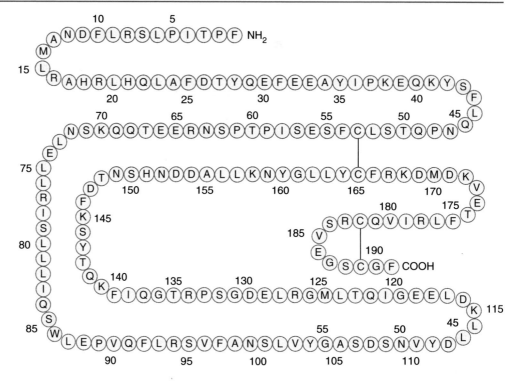

Fig. 20.1 Primary structure of recombinant human growth hormone

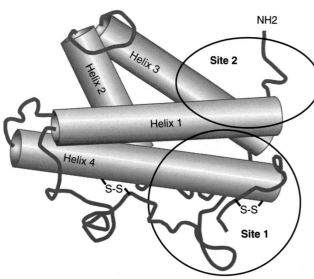

Fig. 20.2 Schematic 3D structure of hGH showing four antiparallel α-helices and receptor binding sites 1 and 2. Approximate positions of the two disulfide bridges (*S-S*) are also indicated. (Modified from Wells et al. (1993))

Pharmacology

Growth Hormone Secretion and Regulation

Growth hormone is secreted in a pulsatile manner from somatotrophs in the anterior pituitary. Multiple feedback loops are present in normal regulation of hGH secretion (Casanueva 1992; Giustina and Veldhuis 1998) (Fig. 20.3). Growth hormone release from the pituitary is regulated by a "short loop" of two coupled hypothalamic peptides – a stimulatory peptide, growth hormone-releasing hormone (GHRH), and an inhibitory peptide, somatostatin (also known as somatotropin release-inhibiting factor (SRIF)). GHRH and somatostatin are, in turn, regulated by neuronal input to the hypothalamus and the GH secretagogue, ghrelin (Kojima et al. 2001). There is possibly also an "ultrashort loop" in which hGH release is feedback regulated by growth hormone receptors present on the somatotrophs of the pituitary themselves. Growth hormone secretion is also regulated by a "long loop" of indirect peripheral signals including negative feedback via insulin-like growth factor 1 (IGF-1) and positive feedback via ghrelin. Growth hormone-induced peripheral IGF-1 inhibits somatotroph release of hGH and stimulates somatostatin release.

Growth hormone secretion changes during human development, with the highest production rates observed during gestation and puberty (Giustina and Veldhuis 1998; Brook and Hindmarsh 1992). Growth hormone production declines approximately 10–15% each decade from age 20 to 70. Endogenous hGH secretion also varies with sex, age, nutritional status, obesity, physical activity, and in a variety of disease states. Endogenous hGH is secreted in periodic bursts over a 24 h period with great variability in burst frequency, amplitude, and duration. There is little detectable hGH released from the pituitary between bursts. The highest endogenous hGH serum concentrations of 10–30 ng/mL usually occur at night when the secretory bursts are largest and most frequent.

Fig. 20.3 Schematic representation of hGH regulation and biologic actions in man. "Short loop" regulation of hGH secretion occurs between the hypothalamus and pituitary. GHRH stimulates GH release. Somatostatin inhibits GH release. "Long loop" regulation of hGH secretion occurs through peripheral feedback signals, primarily negative feedback from insulin-like growth factor 1 (IGF-1). hGH acts directly on muscle, bone, and adipose tissue. Other anabolic actions are generally mediated through IGF-1

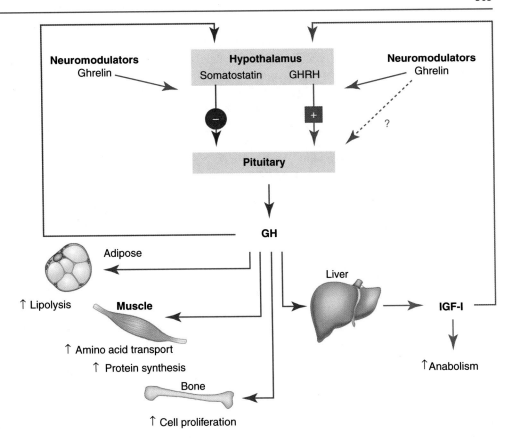

Growth Hormone Biologic Actions

hGH has well-defined growth-promoting and metabolic actions. hGH stimulates the growth of cartilage and bone directly, through hGH receptors in those tissues, and indirectly, via local increases in IGF-1 (Isaksson et al. 2000; Bouillon 1991). Metabolic actions, which may be directly controlled by hGH, include the elevation of circulating glucose levels (diabetogenic effect) and acute increases in circulating concentrations of free fatty acids (lipolytic effect). Other hGH anabolic and metabolic actions believed to be mediated through increases in local or systemic IGF-1 concentrations include the following: increases in net muscle protein synthesis (anabolic effect), skeletal muscle growth, chondroblast and osteoblast proliferation, and linear growth; modulation of reproduction in both males and females; maintenance, control, and modulation of lymphocyte functions; increases in glomerular filtration rate and renal plasma flow rate (osmoregulation); influences on the release and metabolism of insulin, glucagon, and thyroid hormones (T3, T4); and possible direct effects on pituitary function and neural tissue development (Casanueva 1992; Strobl and Thomas 1994; Le Roith et al. 1991).

hGH Receptor and Binding Proteins

The hGH receptor (GHR) is a member of the hematopoietic cytokine receptor family. It has an extracellular domain con-sisting of 246 amino acids, a single 24-amino-acid transmembrane domain, and a 350-amino-acid intracellular domain (Fisker 2006). The extracellular domain has at least six potential N-glycosylation sites and is usually extensively glycosylated. GHRs are found in most tissues in humans. However, the greatest concentration of receptors in humans and other mammals occurs in the liver (Mertani et al. 1995).

As much as 40–45% of monomeric hGH circulating in plasma is bound to one of two growth hormone-binding proteins (GHBP) (Fisker 2006). Binding proteins decrease the clearance of hGH from the circulation (Baumann 1991) and may also serve to dampen the biological effects of hGH by competing with cell receptors for circulating free hGH. The major form of GHBP in humans is a high-affinity ($K_a = 10^{-9}$ to 10^{-8} M), low-capacity form that preferentially binds the 22 kDa form of hGH (Baumann 1991; Herington et al. 1986). Another low-affinity ($K_a = 10^{-5}$ M), high-capacity GHBP is also present that binds the 20 kDa form with equal or slightly greater affinity than the 22 kDa form. In humans, the high-affinity GHBP is identical to the extracellular domain of the hGH receptor and arises by proteolytic cleavage of hGH receptors by a process called ecto-domain shedding. Since the high-affinity binding protein is derived from hGH receptors, circulating levels of GHBP generally reflect hGH receptor status in many tissues (Fisker 2006; Hansen 2002).

Molecular Endocrinology and Signal Transduction

X-ray crystallographic studies and functional studies of the extracellular domain of the hGH receptor suggest that two receptor molecules form a dimer with a single growth hormone molecule by sequentially binding to site 1 on helix 4 of hGH and then to site 2 on helix 3 (Fig. 20.4) (Wells et al. 1993). Signal transduction may occur by activation/phosphorylation of JAK-2 tyrosine kinase followed by activation/phosphorylation of multiple signaling cascades (Herrington and Carter-Su 2001; Piwien-Pilipuk et al. 2002; Brooks and Waters 2010).

Dosing Schedules and Routes

The dosing levels and routes for exogenously administered growth hormone were first established for pit-hGH in growth hormone deficient (GHD) patients (Laursen 2004; Jorgensen 1991). The initial pit-hGH regimen, three times weekly by intramuscular (IM) injection, was based on a number of factors including patient compliance and limited availability of hGH derived from cadaver pituitaries and the assumption that intramuscular injections would be less immunogenic. Subsequent clinical evaluations found a very strong patient preference for subcutaneous (SC) administration and data

showing no increased immunogenicity. Furthermore, increased growth rates were observed with daily SC injections compared to the three times weekly injection schedules (MacGillivray et al. 1996). The abdomen, deltoid area, and thigh are commonly used subcutaneous injection sites. Current dosing schedules usually dictate daily SC injections, often self-administered with a variety of injection devices (Table 20.1).

Several different approaches were developed to prolong the plasma half-life of hGH (Pampanini et al. 2022). In 2020, the FDA approved somapacitan-beco, a weekly subcutaneously dosed hGH derivative for use in adults. A similar path as with GLP-1RA albumin binding derivatives (Chap. 5) was followed. An albumin binding lipid moiety is attached to the hGH amino acid backbone terminus. The interaction with albumin in the blood compartment prolongs the plasma half-life (cf. next paragraph).

Another long-circulating hGH analog is based on attaching polyethylene glycol (PEG; 40 kDa) to the protein, a strategy earlier developed for other therapeutic proteins with a short plasma half-life (Table 5.10). The analog lonapegsomatropin-tcgd was added to the hGH arsenal for pediatric treatment in 2021. The hGH is slowly

Fig. 20.4 Growth hormone secreted isoforms, binding proteins, and receptor interactions. Both 22 and 20 kDa forms are secreted by the pituitary. Pituitary hGH is stored bound to zinc (Zn^{2+}), which is released upon secretion from the pituitary. Secreted hGH is free or bound to either the low- or high-affinity GHBP in plasma. Receptor activation involves dimerization of two receptor molecules with one molecule of hGH. (Modified from Wells et al. (1993))

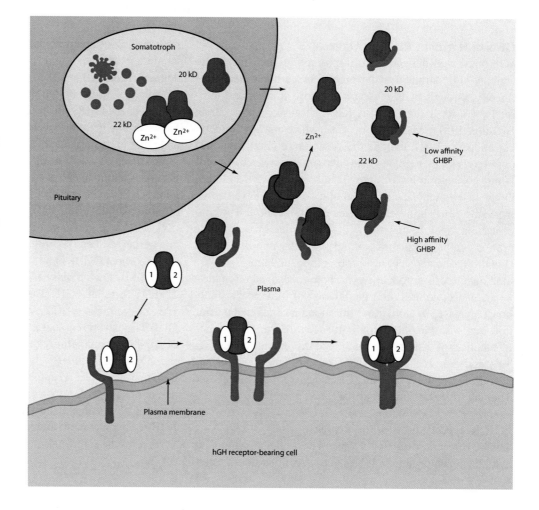

Table 20.1 Recombinant hGH products

Source	Brand name	Product	Container	Injection device	Manufacturer
Recombinant protein produced in bacteria (*E. coli*)	Genotropin®	Lyophilized powder	Two-chamber cartridge	Genotropin Pen®	Pfizer
		Multiple-dose cartridge: 5 and 12 mg			
		Single-dose MiniQuick cartridge: 0.2–2 mg		Genotropin MiniQuick®	
	Norditropin®	Liquid 5 mg/1.5 mL, 10 mg/1.5 mL, 15 mg/1.5 mL, 30 mg/3 mL	Prefilled, preloaded pen device	Norditropin FlexPro®	Novo Nordisk
	Nutropin AQ®	Liquid 5 mg/2 mL, 10 mg/2 mL, 20 mg/2 mL	Prefilled, preloaded pen device	Nutropin AQ® NuSpin	Genentech, Inc.
	Humatrope®	Lyophilized powder Vial: 5 mg/ 5 mL Cartridge: 6 mg/3 mL, 12 mg/3 mL, 24 mg/3 mL	Vial and cartridge	Single-use syringe HumatroPen™	Eli Lilly & Co.
	Zomacton®	Lyophilized powder 5 mg/5 mL, 10 mg/1 mL	Vial	Single-use syringe Needle-free device	Ferring
	Omnitrope®	Lyophilized powder Vial: 5.8 mg/1.14 mL	Vial	Single-use syringe	Sandoz
		Liquid Cartridge: 5 mg/1.5 mL, 10 mg/1.5 mL	Cartridge	Omnitrope® pen	
	Sogroya® Somapacitan-beco	Liquid 5 mg/1.5 mL, 10 mg/1.5 mL	Prefilled pen	Sogroya® pen	Novo Nordisk
	Skytrofa® Lonapegsomatropin-tcgd	Lyophilized powder Per cartridge 3–13.3 mg	Prefilled cartridge	Skytrofa® autoinjector	Ascendis Pharma, Inc.
	Ngenla® Somatrogon	Liquid 20 or 50 mg/mL	Prefilled pen		Pfizer (only EMA approved in 2022)
Recombinant protein produced in mammalian cells (C127 mouse cell line derived)	Serostim® (AIDS wasting)	Lyophilized powder Single-dose vial: 5 mg, 6 mg; diluent: sterile water for injection Multidose vial: 4 mg; diluent: bacteriostatic water for injection	Vial	Syringe Needle-free device: SeroJet™	Serono
	Saizen® (growth inadequacy)	Lyophilized powder Vial: 5 mg, 8.8 mg; diluent: bacteriostatic water for injection	Vial	Single-use syringe	
		Saizenprep® single-use reconstitution mixing device with Saizen 8.8 mg vial and bacteriostatic water for injection cartridge		One.Click® pen Cool.Click® Easypod® drug delivery device	
	Zorbtive® (short bowel syndrome)	Lyophilized powder Multidose vial: 8.8 mg; diluent: bacteriostatic water for injection	Vial	Single-use syringe	

Note: This table represents rhGH products currently available in the USA and is not meant to be an exhaustive list of globally available marketed hGH products; all products are checked via https://www.accessdata.fda.gov/scripts/cder/daf/ (FDA 2022)

released by autocleavage of the linker between hGH and PEG. A third hGH derivative (EMA 2022) with a week dosing interval is somatrogon. Here, one C-terminal peptide (CTP) from the beta chain of human chorionic gonadotropin (hCG) is coupled to the N-terminus of hGH and two copies of CTP (in tandem) to the C-terminus. The presence of the glycosylated CTP domains prolongs the half-life of somatrogon.

Pharmacokinetics and Metabolism

The earliest pharmacokinetic studies were conducted with pituitary-derived hGH (pit-hGH). The pharmacokinetic profiles of pit-hGH, met-rhGH, and rhGH have been compared (Hansen 2002; Laursen 2004) and shown to be very similar. The pharmacokinetics of hGH have been studied in normal, healthy children and adults and a variety of patient populations (Hansen 2002; Laursen 2004; Jorgensen 1991).

Exogenously administered pit-hGH, met-rhGH, and rhGH are rapidly cleared following intravenous (IV) injection with terminal half-lives of approximately 15–20 min (Hansen 2002; Laursen 2004). Distribution volumes usually approximate the plasma volume. hGH clearance in normal subjects ranges from 2.2 to 3.0 mL/kg/min. hGH clearance decreases with increasing serum GH concentrations, most likely due to saturation of hGH receptors at concentrations >10–15 µg/L (Hansen 2002). Comparative analyses of total hGH clearance have not shown consistent population differences based on age, sex, or body composition. However, hGH clearance is controlled by a complex interaction between free hGH, GHBP-bound hGH, and GH receptor status (Hansen 2002). Individual subject variations in GHBP or GH receptor levels may result in substantial differences in hGH clearance.

Human growth hormone is slowly, but relatively completely, absorbed after either IM or SC injection. Time to peak concentration ranges from 2 to 4 h following IM bolus administration and 4 to 6 h following SC bolus administration (Laursen 2004; Jorgensen 1991). Subcutaneously administered rhGH is approximately 50–80% bioavailable (Laursen 2004). The rate of absorption of hGH is slightly faster after injection in the abdomen compared with the thigh (Laursen 2004), but the extent of absorption is comparable. Elimination half-lives following extravascular administration (2–5 h) are usually longer than the IV terminal half-lives indicating absorption rate-limited kinetics. Plasma half-life information of the three approved hGH analogs (status March 2022) demonstrates their prolonged pharmacokinetic profile. The hGH analog somapacitan-beco strongly interacts with plasma albumin and is slowly released from its carrier molecule. This leads to an elimination half-life of 2–3 days (FDA 2020). With lonapegsomatropin-tcgd, the apparent half-life of hGH is approximately 25 h (FDA 2021). For somatrogon, the plasma half-life is 28 h (EMA 2022).

hGH pharmacokinetics in the presence of growth hormone deficiency, diabetes, obesity, and critical illness or diseases of the thyroid, liver, and kidney have been evaluated. Results suggest disposition is not significantly altered compared with normal subjects except in severe liver or kidney dysfunction (Hansen 2002; Haffner et al. 1994; Owens et al. 1973; Cameron et al. 1972). The reduction in clearance observed in severe liver (30%) or kidney dysfunction (40–75%) is consistent with the role of the liver and kidney as major organs of hGH elimination.

Both the kidney and the liver have been shown to be important in the clearance of hGH in humans (Hansen 2002). The relative contribution of each organ has not been rigorously quantitated in humans, but the preponderance of studies in laboratory animals and in isolated perfused organ systems suggests a dominant role for the kidney at pharmacologic levels of hGH. Receptor-mediated uptake of hGH by the liver is the major extrarenal clearance mechanism (Harvey 1995).

Protein Manufacture, Formulation, and Stability

Commercially available hGH preparations are summarized in Table 20.1. All recombinant growth hormones (analogs) except Serostim®/Saizen®/Zorbtive® are produced in bacteria (E. coli). Serostim®, Saizen®, and Zorbtive® are produced in mammalian cells (C127 mouse cells). Growth hormone produced in the cytoplasm of E. coli may contain an N-terminal methionine residue. Natural sequence rhGH is produced either by enzymatic cleavage of the methionine residue during the purification process or by secreting the rhGH into the periplasmic space where the signal peptide is removed by the cell and the native N-terminus of rhGH (no methionine) is revealed. rhGH can be produced in the periplasm in a soluble, properly folded form (Chang et al. 1989) or as refractile/inclusion bodies that require the insoluble rhGH to be extracted, denatured, and refolded (Shin et al. 1998) (cf. Chap. 4). The rhGH is released from the cells by osmotic shock (periplasm) or mechanical lysis (cytoplasm and periplasm), and the protein is recovered and purified. rhGH synthesized in mammalian cells is transported across the endoplasmic reticulum and secreted directly into the culture medium from which it is recovered and purified (Catzel et al. 2003). Historically, the potency of hGH products was expressed in International Units per mg (IU/mg). However, the use of IU dosages is no longer necessary due to the high level of purity and consistent potency of recombinant hGH products.

All current rhGH products are available as lyophilized or liquid preparations. Lyophilized formulations typically include 5 or 10 mg of protein in a glycine and mannitol or sucrose-containing phosphate buffer excipient. The materials are usually reconstituted with sterile water for injection for single use or with bacteriostatic water or bacteriostatic saline for multiple injection use. Liquid formulations of rhGH (Norditropin®, Nutropin AQ®, Omnitrope®) contain excipients such as mannitol or sodium chloride, histidine or citrate buffer, poloxamer 188 or polysorbate 20, and phenol or benzyl alcohol. Product stability has been very good with shelf lives of approximately 2 years at 2–8 °C. The lyophilized rhGH preparation of Omnitrope® was approved for marketing as the first "biosimilar" rhGH product in the EU and the USA in 2006. Two strengths of Omnitrope® (5 and 10 mg cartridges) were approved for marketing in 2008 for use in pen devices (Omnitrope® pen).

Clinical Usage

Clinical usage of rhGH has been reviewed for pediatric and adult indications (Franklin and Geffner 2011). Investigations of clinical usage of hGH have focused, generally, on two major areas of hGH biologic action: (1) linear growth promotion and (2) modulation of metabolism. Growth-promoting indications in children that have been approved for the market include growth hormone deficiency, idiopathic short stature, growth failure associated with chronic renal insufficiency, growth failure in children born small for gestational age, and short stature in Prader-Willi syndrome, Turner syndrome, Noonan syndrome, and short stature homeobox-containing gene deficiency on the X chromosome (SHOX). Modulation of metabolism is the primary biologic action in long-term replacement therapy in adults with GH deficiency of either childhood or adult onset or for GH supplementation in AIDS wasting or cachexia and in short bowel syndrome. Contraindications to rhGH use include use in patients with active malignancy, active proliferative or severe nonproliferative retinopathy, acute critical illness, children with Prader-Willi syndrome (PWS) who are severely obese or have severe respiratory impairment, children with closed epiphyses, and hypersensitivity to somatropin or excipients.

Growth Hormone Deficiency (GHD)

The major indication for therapeutic use of hGH is the long-term replacement treatment for children with classic growth hormone deficiency in whom growth failure is due to a lack of adequate endogenous hGH secretion. Children with GHD fall into a variety of etiologic categories including genetic defects in the hypothalamic-pituitary axis, developmental anomalies of the brain with or without identifiable syndromes, acquired events such as central nervous system (CNS) lesions (craniopharyngioma most commonly) or from the treatment of CNS tumors (medulloblastoma, glioblastoma, etc.), trauma, or other events requiring CNS irradiation. When diagnosed in otherwise healthy children, it is called idiopathic. In patients suspected of having GHD, the diagnosis is usually defined based on an inadequate response to two hGH provocation tests implying a functional deficiency in the production or secretion of hGH from the pituitary gland. In patients with documented organic causes, especially if panhypopituitarism is present, two stimulation tests may not be required. Usual doses range from 0.24 to 0.30 mg/kg/week administered as daily SC injections in prepubertal children. Doses up to 0.7 mg/kg/week have been approved for GHD adolescent subjects to improve final height based on a clinical trial (Mauras et al. 2000) showing hGH treatment results in increased growth velocity and enhancement in final adult height. For most GHD children, the growth response is greatest in the first year of treatment and correlates positively with hGH dose, degree of short stature, and frequency of injections and negatively with chronological age at onset of treatment. hGH therapy in children is usually continued until growth has been completed, as evidenced by epiphyseal fusion. rhGH treatment of idiopathic GH-deficient children has a positive overall safety profile documented in long-term clinical registries (Bell et al. 2010; Darendeliler et al. 2007). However, in those with organic causes, the safety profile may be dependent on the underlying medical condition and its prior treatment. In childhood cancer survivors, an increased risk of a second neoplasm has been reported in patients treated with somatropin after their first neoplasm (Sklar et al. 2002), and this information is communicated under warnings and precautions in the product labels.

Idiopathic Short Stature (ISS)

Idiopathic short stature (ISS) comprises a heterogeneous group of growth failure states for which the proximate cause remains unknown. Postulated defects include impaired spontaneous hGH secretion, hGH resistance due to low levels of hGH receptors, or other defects in either secreted hGH, hGH receptors or post-receptor events, or other genetic defects yet to be elucidated. Nevertheless, studies have documented that many patients in the "idiopathic" group respond to exogenous growth hormone treatment with acceleration of growth and improvement in final height. As a consequence of several long-term multicenter trials to final height, rhGH was approved for the treatment of ISS in the USA (Hintz et al. 1999; Leschek et al. 2004). The risk-benefit assessment of improved growth in what is generally considered to be a population of "healthy" children continues to be debated, and while short-term safety has been documented (Bell et al. 2010), long-term safety is still a question that is being addressed by a multi-country study in Europe looking at adult patients treated with rhGH as children. Based on preliminary data and published reports (Carel et al. 2012; Savendahl et al. 2012), the FDA and the European Medicines Agency (EMA) preliminary assessments are that there should be no change in prescribing information with emphasis on adherence to approved dosages. Final assessments are expected to be available in 2013.

Turner Syndrome (TS)

Turner syndrome is a disease of females caused by partial or total loss of one sex chromosome and is characterized by decreased intrauterine and postnatal growth, short final adult height, incomplete development of the ovaries and secondary sexual characteristics, and other physical abnormalities. Although serum levels of hGH and IGF-1 are not consistently low in this population, hGH treatment (0.375 mg/kg/week), alone or in combination with oxandrolone, significantly improves growth rate and final adult height in this patient group, and its use in Turner syndrome is now an accepted indication (Rosenfeld et al. 1998; Kappelgaard and Laursen 2011).

Prader-Willi Syndrome (PWS)

Prader-Willi syndrome is a genetic disease usually caused by the functional deletion of a gene on the paternal allele of chromosome 15. Clinical manifestations in childhood include lack of satiety, obesity, hypotonia, short stature, hypogonadism, and behavior abnormalities. PWS children have especially high rates of morbidity due to obesity-related illnesses. Growth hormone treatment (0.24–0.48 mg/kg/week) has been shown to improve height and, perhaps more importantly, improve body composition, physical strength, and agility in PWS children (Allen and Carrel 2004). However, growth hormone treatment is contraindicated in the face of severe obesity or severe respiratory impairment since sudden deaths in this group of patients, when treated with growth hormone, have been reported.

Small for Gestational Age (SGA)

Growth hormone is approved for use in long-term treatment of growth failure in children born small for gestational age who fail to manifest catch-up growth. Children born at birth weights and birth lengths more than two standard deviations below the mean are considered small for gestational age. Children who fail to catch up by age 2–3 are at risk for growing into adults with substantial height deficits (Rappaport 2004). Growth hormone treatment at doses of 0.24–0.48 mg/kg/week can induce catch-up growth, with the potential to normalize height at an earlier age and the potential to improve adult height.

Chronic Renal Insufficiency (CRI)/Chronic Kidney Disease (CKD)

Children with renal insufficiency, which is termed chronic kidney disease (CKD), grow slowly, possibly related to defects in metabolism, nutrition, metabolic bone disease, and/or defects in the IGF-1/hGH axis. Basal serum hGH concentrations may be normal or high, and IGF-1 responses to hGH stimulation is usually normal. However, there are reported abnormalities in the IGF-binding protein levels in CKD patients suggesting possible problems with GH/IGF-1 action. Growth hormone therapy (0.35 mg/kg/week) in children with chronic renal insufficiency results in significant increases in height velocity (Greenbaum et al. 2004). Increases are best during the first year of treatment for younger children with stable renal disease. Responses are less for children on dialysis. Growth hormone has not been approved for children posttransplant.

Noonan Syndrome

First called male Turner syndrome due to similar phenotypic characteristics, Noonan syndrome, an autosomal dominant disorder, was later recognized as a separate condition. It occurs in both males and females. Its key features are short stature (although some patients will achieve normal adult height), congenital heart disease, most commonly pulmonic stenosis or hypertrophic cardiomyopathy, a short and often webbed neck, ptosis, and chest/sternum deformities. While the GH/IGF axis is intact in Noonan syndrome and the mechanisms through which the mutations cause short stature are unknown, clinical trials with rhGH demonstrated significant increases in growth rate and modest increases in final height (Osio et al. 2005), resulting in FDA approval of this indication in 2007. Pediatric patients with short stature associated with Noonan syndrome are given up to an rhGH dose of 0.066 mg/kg/day. Safety concerns in treatment with rhGH have been raised with respect to progression of cardiomyopathy, although data to date do not support this clinical concern.

Short Stature Homeobox-Containing Gene (SHOX)

The SHOX gene is located on the pseudoautosomal region of the X chromosome and the homologous distal region on the Y chromosome. Healthy males and females express two active copies of the SHOX gene, one from each of the sex chromosomes. Females with TS missing an X chromosome or part of an X chromosome have one copy of the SHOX gene. A significant percentage of the growth failure in TS females is secondary to this gene loss. The variably expressed SHOX gene in long bones tends to be in the mesomelic segments. Mutations resulting in haploinsufficiency of SHOX are also responsible for the short stature in some patients with a pseudoautosomal dominant condition of mesomelic dyschondrosteosis called Leri-Weill syndrome. In addition, sporadic mutations are also responsible for short stature in a small percentage of patients who would otherwise be characterized as idiopathic short stature (Rao et al. 1997). A multicenter study of a heterogeneous group of patients with SHOX haploinsufficiency demonstrated a clinically significant effect of rhGH on growth in children with SHOX mutations compared to the untreated control group. In addition, the efficacy of rhGH treatment was similar to that seen in a comparable group of girls with TS, leading to the approval of the SHOX indication (Blum et al. 2007). The FDA-approved rhGH dose for SHOX deficiency is 0.35 mg/kg/week. To date, no specific safety signals attributable to rhGH treatment have emerged in these children.

Growth Hormone Deficient Adults

Early limitations in hGH supply severely limited treatment of adults with GHD. With the increased supply of recombinant rhGH products, replacement therapy for adults was evaluated and, ultimately, approved as a clinical indication. Growth hormone has been approved for two growth hormone deficient adult populations: (a) adults with childhood-onset GHD and (b) adults with adult-onset GHD usually due to pituitary tumors, CNS irradiation, or head trauma. Increases in bone density have been observed in some bone types, although treatment duration greater than 1 year may be necessary to see significant effects. hGH treatment consistently elevates both serum IGF-1 and insulin levels. Women have also been shown to require higher doses to normalize IGF-1 levels than men, especially women taking oral estrogens.

Clinical Malnutrition and Wasting Syndromes

Growth hormone is approved for the treatment of short bowel syndrome (SBS) in adults, a congenital or acquired condition in which less than 200 cm of small intestine is present. Short bowel syndrome patients have severe fluid and nutrient malabsorption and are often dependent upon intravenous parenteral nutrition (IPN). Administration of growth hormone, 0.1 mg/kg/day to a maximum of 8 mg for 4 weeks, alone or in combination with glutamine, reduces the volume and frequency of required IPN (Keating and Wellington 2004). Growth hormone is indicated for use in adult patients who are also receiving specialized nutritional support. Usage for periods >4 weeks, or in children, has not been investigated. Usage of growth hormone for SBS remains controversial due to potential risks associated with IGF-1-related fibrosis and cancer (Theiss et al. 2004).

Growth hormone is also approved for use in wasting associated with AIDS. Growth hormone treatment (~0.1 mg/kg daily, max. 6 mg/day), when used with controlled diets, increases body weight and nitrogen retention. rhGH treatment is also under investigation for HIV-associated lipodystrophy, a syndrome of fat redistribution and metabolic complications resulting from the highly active antiretroviral therapy commonly used in HIV infection (Burgess and Wanke 2005).

Other Conditions Under Investigation

Growth hormone levels and IGF decline with age, prompting the initiation of multiple clinical trials for use in adults over age 60 (Di Somma et al. 2011). However, clear long-term efficacy in muscle strength or improvements in activities of daily life have not been sufficiently demonstrated to gain regulatory approval for this indication. The use of hGH therapy to ameliorate the negative nitrogen balance seen in patients following surgery, injury, or infections has been investigated in a number of studies (Takala et al. 1999; Jeevanandam et al. 1995; Ponting et al. 1988; Voerman et al. 1995). However, due to the increased mortality found in a study of severe critical illness (Takala et al. 1999) and the subsequent contraindication for use in acute critical illness,

very few registration trials examining the use of hGH for these conditions have been initiated. Studies of hGH effects in burns have shown significant effectiveness in acceleration of healing in skin graft donor sites and improvements in growth in burned children (Herndon and Tompkins 2004). Growth hormone has been shown to significantly reduce multiple disease symptoms and improve well-being and growth in children and adults with Crohn's disease, a chronic inflammatory disorder of the bowel (Theiss et al. 2004; Slonim et al. 2000; Denson et al. 2010). Growth hormone has also shown benefit in cardiovascular recovery and function in congestive heart failure (Colao et al. 2004). Recent studies indicate that growth hormone treatment improves growth, pulmonary function, and clinical status in children with cystic fibrosis (Stalvey et al. 2012).

Safety Concerns

hGH has been widely used for many years and has been proven to have a positive safety profile for most pediatric indications (Growth Hormone Research Society 2001). However, sudden death in some patients with PWS and severe obesity associated with rhGH treatment resulted in a contraindication to its use in severely obese or respiratory compromised PWS children (Eiholzer 2005). Adverse events have been reported in a small number of children and include benign intracranial hypertension, glucose intolerance, and the rare development of anti-hGH antibodies. In most cases, the formation of anti-hGH antibodies following rhGH treatment has not been positively correlated with a loss in efficacy.

Growth hormone therapy is also not associated with increased risk of primary malignancies or tumor recurrence (Growth Hormone Research Society 2001; Sklar et al. 2002). However, an increase in secondary malignancies in childhood cancer survivors, especially those treated with CNS irradiation, has been described (Sklar et al. 2002).

Growth hormone inhibits 11β-hydroxysteroid dehydrogenase type 1 (11β-HSD1) activity in adipose/hepatic tissue and may impact the metabolism of cortisol and cortisone (Gelding et al. 1998). Treatment with rhGH could potentially unmask undiagnosed central adrenal insufficiency (secondary hypoadrenalism) or increase the requirement for maintenance or stress doses of replacement corticosteroid in those already diagnosed with adrenal insufficiency.

Growth hormone has caused significant, dose-limiting fluid retention in adult populations resulting in increased body weight, swollen joints and arthralgias, and carpal tunnel syndrome (Carroll and van den Berghe 2001). Symptoms were usually transient and resolved upon reduction of hGH dosage or upon discontinuation of the hGH treatment. Growth hormone administration has been associated with increased mortality in clinical trials in critically ill, intensive-care patients with acute catabolism (Takala et al. 1999) and is, therefore, contraindicated for use in critically ill patients.

Growth hormone's anabolic and lipolytic effects have made it attractive as a performance enhancement drug among athletes. Illicit hGH usage has been anecdotally reported for the last 20 years. Detection of rhGH abuse proximate to the time of testing is now possible due to the development of assays that rely on detecting changed ratios of exogenous rhGH (22 kDa only) and endogenous hGH (22 kDa, 20 kDa, and other forms). Screening for proximate rhGH abuse, based on the new ratio assays, was included in the 2006 Olympic Games for the first time (McHugh et al. 2005).

Concluding Remarks

The abundant supply of rhGH, made possible by recombinant DNA technology, has allowed enormous advances to be made in understanding the basic structure, function, and physiology of hGH over the past 30+ years. As a result of those advances, recombinant hGH has been developed into a safe and efficacious therapy for a variety of growth and metabolic disorders in children and adults. Continuing basic research in GH and IGF-1 biology, genomics, and GH-related diseases and continuing clinical investigation into additional uses in pediatric growth disorders or disorders of metabolism may yield as yet new indications for treatment.

Self-Assessment Questions

Questions

1. One molecule of hGH is required to sequentially bind to two receptor molecules for receptor activation. What consequences might the requirement for sequential dimerization have on observed dose-response relationships?

2. Growth hormone is known or presumed to act directly upon which tissues?

3. You are investigating the use of hGH as an adjunct therapy for malnutrition/wasting in a clinical population that also has severe liver disease. What effects would you expect the liver disease to have on the observed plasma levels of hGH after dosing and on possible efficacy (improvement in nitrogen retention, prevention of hypoglycemia, etc.)?

Answers

1. Sequential dimerization will potentially result in a "bell-shaped" dose-response curve; that is, response is stimulated at low concentrations and inhibited at high concentrations. The inhibition of responses at high concentrations is due to blocking of dimerization caused by the excess hGH saturating all the available receptors. Inhibition of in vitro hGH binding is observed at high hGH (mM) concentrations. Reductions in biological responses (total IGF-1 increase and weight gain) have also been seen with increasing hGH doses in animal studies. However, inhibitory effects of high concentrations of hGH are not seen in treatment of human patients since hGH dose levels are maintained within normal physiological ranges and never approach inhibitory levels.

2. Growth hormone is known to act directly on both bone and cartilage and possibly also on muscle and adipose tissue. Growth hormone effects on other tissues appear to be mediated through the IGF-1 axis or other effectors.

3. Severe liver disease may reduce the clearance of the exogenously administered hGH, and observed plasma levels may be higher and persist longer compared to patients without liver disease. However, the increased drug exposure may not result in increased anabolic effects. The desired anabolic effects require the production/release of IGF-1 from the liver. Both IGF-1 production and the number of hGH receptors may be reduced due to the liver disease. To understand the results (or lack of results) from the treatment, it is important to monitor effect parameters (i.e., IGF-1 and possibly IGF-1 binding protein levels, liver function enzymes) in addition to hGH levels.

References

Allen DB, Carrel AL (2004) Growth hormone therapy for Prader-Willi syndrome: a critical appraisal. J Pediatr Endocrinol Metab 17:1297–1306

Baumann G (1991) Growth hormone heterogeneity: genes, isohormones, variants and binding proteins. Endocr Rev 12:424–449

Baumann GP (2009) Growth hormone isoforms. Growth Hormon IGF Res 19:333–340

Bell J, Parker KL, Swinford RD, Hoffman AR, Maneatis T, Lippe B (2010) Long-term safety of recombinant human growth hormone in children. J Clin Endocrinol Metab 95:167–177

Blum WF, Crowe BJ, Quigley CA, Jung H, Cao D, Ross JL, Braun L, Rappold G, Shox Study Group (2007) Growth hormone is effective in treatment of short stature associated with short stature homeobox-containing gene deficiency: two-year results of a randomized, controlled, multicenter trial. J Clin Endocrinol Metab 92:219–228

Boguszewski CL (2003) Molecular heterogeneity of human GH: from basic research to clinical implications. J Endocrinol Investig 26:274–288

Bouillon R (1991) Growth hormone and bone. Horm Res 36(Suppl 1):49–55

Brook CGD, Hindmarsh PC (1992) The somatotropic axis in puberty. Endocrinol Metab Clin N Am 21:767–782

Brooks AJ, Waters MJ (2010) The growth hormone receptor: mechanism of activation and clinical implications. Nat Rev Endocrinol 6(9):515–525

Burgess E, Wanke C (2005) Use of recombinant human growth hormone in HIV-associated lipodystrophy. Curr Opin Infect Dis 18:17–24

Cameron DP, Burger HG, Catt KJ et al (1972) Metabolic clearance of human growth hormone in patients with hepatic and renal failure, and in the isolated perfused pig liver. Metabolism 21:895–904

Carel J-C, Ecosse E, Landier F, Meguellati-Hakkas D, Kaguelidou F, Rey G, Coste J (2012) Long-term mortality after recombinant growth hormone treatment for isolated growth hormone deficiency or childhood short stature: preliminary report of the French SAGhE study. J Clin Endocrinol Metab 97:416–425. https://doi.org/10.1210/jc.2011-1995

Carroll PV, van den Berghe G (2001) Safety aspects of pharmacological GH therapy in adults. Growth Hormon IGF Res 11:166–172

Casanueva F (1992) Physiology of growth hormone secretion and action. Endocrinol Metab Clin N Am 21:483–517

Catzel D, Lalevski H, Marquis CP, Gray PP, Van Dyk D, Mahler SM (2003) Purification of recombinant human growth hormone from CHO cell culture supernatant by Gradiflow preparative electrophoresis technology. Protein Expr Purif 32(1):126–234

Chang JY, Pai RC, Bennett WF, Bochner BR (1989) Periplasmic secretion of human growth hormone by Escherichia coli. Biochem Soc Trans 17(2):335–337

Colao A, Vitale G, Pivonello R et al (2004) The heart: an end-organ of GH action. Eur J Endocrinol 151:S93–S101

Growth Hormone Research Society (2001) Critical evaluation of the safety of recombinant human growth hormone administration: statement from the Growth Hormone Research Society. J Clin Endocrinol Metab 86(5):1868–1870

Darendeliler F, Karagiannis G, Wilton P (2007) Headache, idiopathic intracranial hypertension and slipped capital femoral epiphysis during growth hormone treatment: a safety update from KIGS. Horm Res 68(Suppl 5):41–47

Denson LA, Kim MO, Bezold R et al (2010) A randomized controlled trial of growth hormone in active pediatric Crohn disease. J Pediatr Gastroenterol Nutr 51:130–139

Di Somma C, Brunelli V, Savanelli MC et al (2011) Somatopause: state of the art. Minerva Endocrinol 36:243–255

Eiholzer U (2005) Deaths in children with Prader-Willi syndrome: a contribution to the debate about the safety of growth hormone treatment in children with PWS. Horm Res 63(1):33–39

EMA (2022) Ngenla (somatrogon) https://www.ema.europa.eu/en/documents/product-information/ngenla-epar-product-information_en.pdf. Accessed 4 March 2022

FDA (2020) SOGROYA® (somapacitan-beco) injection, for subcutaneous use. https://www.accessdata.fda.gov/drugsatfda_docs/label/2021/761156s001s002lbl.pdf. accessed 3 March 2022

FDA (2021) SKYTROFA™ (lonapegsomatropin-tcgd) for injection, for subcutaneous use. https://www.accessdata.fda.gov/drugsatfda_docs/label/2021/761177lbl.pdf. Accessed 3 March 2022

Fisker S (2006) Physiology and pathophysiology of growth hormone binding protein: methodological and clinical aspects. Growth Hormon IGF Res 16:1–28

Franklin SL, Geffner ME (2011) Growth hormone: the expansion of available products and indications. Endocrinol Metab Clin N Am 38:587–611

Gelding SV, Taylor NF, Wood PJ, Noonan K, Weaver JU, Wood DF, Monson JP (1998) The effect of growth hormone replacement therapy on cortisol-cortisone interconversion in hypopituitary adults: evidence for growth hormone modulation of extrarenal 11 beta-hydroxysteroid dehydrogenase activity. Clin Endocrinol 48(2):153–162

Giustina A, Veldhuis JD (1998) Pathophysiology of the neuroregulation of growth hormone secretion in experimental animals and in the human. Endocr Rev 19(6):717–797

Goeddel DV, Heyreker HL, Hozumi T et al (1979) Direct expression in Escherichia coli of a DNA sequence coding for human growth hormone. Nature 281:544–548

Greenbaum LA, Del Rio M, Bamgbola F et al (2004) Rationale for growth hormone therapy in children with chronic kidney disease. Adv Chronic Kidney Dis 11(4):377–386

Haffner D, Schaefer F, Girard J et al (1994) Metabolic clearance of recombinant human growth hormone in health and chronic renal failure. J Clin Invest 93:1163–1171

Hansen TK (2002) Pharmacokinetics and acute lipolytic actions of growth hormone: impact of age, body composition, binding proteins and other hormones. Growth Hormon IGF Res 12:342–358

Harvey S (1995) Growth hormone metabolism. In: Harvey S, Scanes CG, Daughaday WH (eds) Growth hormone. CRC Press, Inc., Boca Raton, FL, pp 285–301

Herington AC, Ymer S, Stevenson J (1986) Identification and characterization of specific binding proteins for growth hormone in normal human sera. J Clin Invest 77:1817–1823

Herndon DN, Tompkins RG (2004) Support of the metabolic response to burn injury. Lancet 363:1895–1902

Herrington J, Carter-Su C (2001) Signaling pathways activated by the growth hormone receptor. Trends Endocrinol Metab 12(6):252–257

Hintz RL, Rosenfeld RG, Wilson DM et al (1982) Biosynthetic methionyl human growth hormone is biologically active in adult man. Lancet 1:1276–1279

Hintz RL, Attie KM, Baptista J, Roche A (1999) Effect of growth hormone treatment on adult height of children with idiopathic short stature. N Engl J Med 340:502–507

Isaksson OG, Ohlsson C, Bengtsson B et al (2000) GH and bone-experimental and clinical studies. Endocr J 47(Suppl):S9–S16

Jeevanandam M, Ali MR, Holaday NJ et al (1995) Adjuvant recombinant human hormone normalizes plasma amino acids in parenterally fed trauma patients. J Parenter Enter Nutr 19:137–144

Jorgensen JOL (1991) Human growth hormone replacement therapy: pharmacological and clinical aspects. Endocr Rev 12:189–207

Kappelgaard AM, Laursen T (2011) The benefits of growth hormone therapy in patients with turner syndrome, Noonan syndrome, and children born small for gestational age. Growth Hormon IGF Res 21(6):305–313

Keating GM, Wellington K (2004) Somatropin (Zorbtive™) in short bowel syndrome. Drugs 64(12):1375–1381

Kojima M, Hosoda H, Matsuo H et al (2001) Ghrelin: discovery of the natural endogenous ligand for the growth hormone secretagogue receptor. Trends Endocrinol Metab 12(3):118–126

Laursen T (2004) Clinical pharmacological aspects of growth hormone administration. Growth Hormon IGF Res 14:16–44

Le Roith D, Adamo M, Werner H, Roberts CT Jr (1991) Insulin-like growth factors and their receptors as growth regulators in normal physiology and pathologic states. Trends Endocrinol Metab 2:134–139

Leschek EW, Ross SR, Yanovski JA, Troendle JF, Quigley CA, Chipman JJ, Crowe BJ et al (2004) Effect of growth hormone treatment on adult height in peripubertal children with idiopathic short stature: a randomized, double blind, placebo-controlled trial. J Clin Endocrinol Metab 89:3140–3148

Lewis UJ, Sinhda YN, Lewis GP (2000) Structure and properties of members of the hGH family: a review. Endocr J 47:S1–S8

MacGillivray MH, Baptista J, Johanson A (1996) Outcome of a four-year randomized study of daily versus three times weekly somatropin treatment in prepubertal naïve growth hormone deficient children. J Clin Endocrinol Metab 81:1806–1809

Martial JA, Hallewell RA, Baxter JD (1979) Human growth hormone: complementary DNA cloning and expression in bacteria. Science 205:602–607

Mauras N, Attie KM, Reiter EO et al (2000) High dose recombinant human growth hormone (GH) treatment of GH-deficient patients in puberty increases near-final height: a randomized, multicenter trial. J Clin Endocrinol Metab 85:3653–3660

McHugh CM, Park RT, Sonksen PH et al (2005) Challenges in detecting the abuse of growth hormone in sport. Clin Chem 51(9):1587–1593

Mertani HC, Delehaye-Zervas MC, Martini JF et al (1995) Localization of growth hormone receptor messenger RNA in human tissues. Endocrine 3:135–142

Osio D, Dahlgren J, Wikland KA, Westphal O (2005) Improved final height with long-term growth hormone treatment in Noonan syndrome. Acta Paediatr 94(9):1232–1237

Owens D, Srivastava MC, Tompkins CV et al (1973) Studies on the metabolic clearance rate, apparent distribution space and plasma half-disappearance time of unlabelled human growth hormone in normal subjects and in patients with liver disease, renal disease, thyroid disease and diabetes mellitus. Eur J Clin Investig 3:284–294

Pampanini V, Deodati A, Inzaghi E, Cianfarani S (2022) Long-acting growth hormone preparations and their use in children with growth hormone deficiency. Horm Res Paediatr. https://doi.org/10.1159/000523791

Piwien-Pilipuk G, Huo JS, Schwartz J (2002) Growth hormone signal transduction. J Pediatr Endocrinol Metab 15:771–786

Ponting GA, Halliday D, Teale JD et al (1988) Postoperative positive nitrogen balance with intravenous hyponutrition and growth hormone. Lancet 1:438–440

Raben MS (1958) Treatment of a pituitary dwarf with human growth hormone. J Clin Endocrinol Metab 18:901–903

Rao E, Weiss B, Fukami M et al (1997) Pseudoautosomal deletions encompassing a novel homeobox gene cause growth failure in idiopathic short stature and turner syndrome. Nat Genet 16:54–63

Rappaport R (2004) Growth and growth hormone in children born small for gestational age. Growth Hormon IGF Res 14:S3–S6

Ray J, Jones BK, Liebhaber SA, Cooke NE (1989) Glycosylated human growth hormone variant. Endocrinology 125(1):566–568

Root AW, Root MJ et al (2002) Clinical pharmacology of human growth hormone and its secretagogues. Curr Drug Targets Immune Endocr Metabol Disord 2:27–52

Rosenfeld RG, Attie KM, Frane J et al (1998) Growth hormone therapy of Turner's syndrome: beneficial effect on adult height. J Pediatr 132:319–324

Savendahl L, Maes M, Albersson K, Borgstrom B, Carel J-C, Henrad S, Speybroeck N, Thomas M, Xandwijken G, Hokken-Koelega A (2012) Long-term mortality and causes of death in isolated GHD, ISS and SGA patients treated with recombinant growth hormone during childhood in Belgium, The Netherlands, and Sweden: preliminary report of a 3 countries participating in the EU SAGhE study. J Clin Endocrinol Metab 97:E213–E217. https://doi.org/10.1210/jc.2011-2882

Shin NK, Kim DY, Shin CS, Hong MS, Lee J, Shin HC (1998) High-level production of human growth hormone in Escherichia coli by a simple recombinant process. J Biotechnol 62(2):143–151

Sklar CA, Mertens AC, Mitby P et al (2002) Risk of disease recurrence and second neoplasms in survivors of children cancer treated with growth hormone: a report from the childhood cancer survivor study. J Clin Endocrinol Metab 87(7):3136–3141

Slonim AE, Bulone L, Damore MB et al (2000) A preliminary study of growth hormone therapy for Crohn's disease. N Engl J Med 342:1633–1637

Stalvey MS, Anbar RD, Konstan MVV, Jacobs JR, Bakker B, Lippe B, Geller DE (2012) A multi-center controlled trial of growth hormone treatment in children with cystic fibrosis. Pediatr Pulmonol 47:252–263. https://doi.org/10.1002/ppul.21546

Strobl JS, Thomas MJ (1994) Human growth hormone. Pharm Rev 46:1–34

Takala J, Ruokonen E, Webster NR et al (1999) Increased mortality associated with growth hormone treatment in critically ill adults. N Engl J Med 341(11):785–792

Theiss AL, Fruchtman S, Lund PK (2004) Growth factors in inflammatory bowel disease. The actions and interactions of growth hormone and insulin-like growth factor-I. Inflamm Bowel Dis 10(6):871–880

Voerman BJ, van Schijndel RJMS, Goreneveld ABJ et al (1995) Effects of human growth hormone in critically ill non-septic patients: results from a prospective, randomized, placebo-controlled trial. Crit Care Med 23:665–673

Wells JA, Cunningham BC, Fuh G et al (1993) The molecular basis for growth hormone-receptor interactions. Recent Prog Horm Res 48:253–275

Further Reading

Boguszewski CL (2003) Molecular heterogeneity of human GH: from basic research to clinical implications. J Endocrinol Investig 26:274–288

Brooks AJ, Waters MJ (2010) The growth hormone receptor: mechanism of activation and clinical implications. Nat Rev. Endocrinol 6(9):515–525

Fisker S (2006) Physiology and pathophysiology of growth hormone binding protein: methodological and clinical aspects. Growth Hormon IGF Res 16:1–28

Giustina A, Veldhuis JD (1998) Pathophysiology of the neuroregulation of growth hormone secretion in experimental animals and in the human. Endocr Rev 19(6):717–797

Harvey S, Scanes CG, Daughaday WH (eds) (1995) Growth hormone. CRC Press, Inc., Boca Raton, FL

Harris M, Hofman PL, Cutfield WS (2004) Growth hormone treatment in children. Pediatr Drugs 6(2):93–106

Laursen T (2004) Clinical pharmacological aspects of growth hormone administration. Growth Hormon IGF Res 14:16–44

Simpson H, Savine R, Sonksen P et al (2002) Growth hormone replacement therapy for adults: into the new millennium. Growth Hormon IGF Res 12:1–33

Recombinant Human Deoxyribonuclease I

Robert A. Lazarus and Jeffrey S. Wagener

Introduction

Human deoxyribonuclease I (DNase I) is an endonuclease that catalyzes the hydrolysis of extracellular DNA and is just one of the numerous types of nucleases found in nature (Horton 2008; Yang 2011). It is the most extensively studied member in the family of DNase I-like nucleases (Keyel 2017; Lazarus 2002; Baranovskii et al. 2004; Shiokawa and Tanuma 2001); the homologous bovine DNase I has received even greater attention historically (Laskowski Sr 1971; Moore 1981; Chen and Liao 2006). Mammalian DNases have been broadly divided into several families initially based upon their products, pH optima, and divalent metal ion requirements. These include the neutral DNase I family (EC 3.1.21.1), the acidic DNase II family (EC 3.1.22.1), as well as apoptotic nucleases such as DFF40/CAD and endonuclease G (Keyel 2017; Lazarus 2002; Evans and Aguilera 2003; Widlak and Garrard 2005). The human DNase I gene resides on chromosome 16p13.3 and contains ten exons and nine introns, which span 15 kb of genomic DNA (Kominato et al. 2006). DNase I is synthesized as a precursor and contains a 22-residue signal sequence that is cleaved upon secretion, resulting in the 260-residue mature enzyme. It is secreted by the pancreas and parotid glands, consistent with its proposed primary role of digesting nucleic acids in the gastrointestinal tract. However, it is also present in blood and urine as well as other tissues, suggesting additional functions.

Jeffrey S. Wagener was deceased at the time of publication.

This chapter is dedicated in honor of a lifetime of scientific and clinical contributions by Jeffrey S. Wagener.

R. A. Lazarus (✉)
Department of Early Discovery Biochemistry, Genentech Inc., South San Francisco, CA, USA
e-mail: lazarus.bob@gene.com

J. S. Wagener (Deceased)
Department of Pediatrics, University of Colorado School of Medicine, Aurora, CO, USA

Recombinant human DNase I (rhDNase I, rhDNase, Pulmozyme®, dornase alfa) has been developed clinically where it is aerosolized into the airways for treatment of pulmonary disease in patients with cystic fibrosis (CF) (Suri 2005; Wagener and Kupfer 2012). Cystic fibrosis is an autosomal recessive disease caused by mutations in the CF transmembrane conductance regulator (CFTR) gene (Kerem et al. 1989; Riordan et al. 1989). Mutations of this gene result in both abnormal quantity and function of an apical membrane protein responsible for chloride ion transfer. The CFTR protein is a member of the adenosine triphosphate (ATP)-binding cassette transporter superfamily (member ABCC7) and, in addition to transporting chloride, has many other functions including the regulation of epithelial sodium channels, ATP-release mechanisms, anion exchangers, sodium bicarbonate transporters, and aquaporin water channels found in airways, intestine, pancreas, sweat duct, and other fluid-transporting tissues (Guggino and Stanton 2006). Clinical manifestations of the disease include chronic obstructive airway disease, increased sweat electrolyte excretion, male infertility due to obstruction of the vas deferens, and exocrine pancreatic insufficiency.

In the airways, abnormal CFTR results in altered secretions and mucociliary clearance, leading to a cycle of obstruction, chronic bacterial infection, and neutrophil-dominated inflammation. This bacterial infection and neutrophil-dominated airway inflammation begin early in the patient's life, and while initially it helps to control infection, the degree of inflammation remains excessive to the degree of infection. Poorly regulated neutrophil-dominated inflammation damages the airways over time due to the release of potent oxidants and proteases. Additionally, necrosis of neutrophils leads to the accumulation of extracellular DNA and actin, increasing the viscosity of mucus and creating further obstruction, and a downward spiral of lung damage, loss of lung function, and ultimately premature death (Fig. 21.1).

rhDNase I has also been studied in a variety of other diseases where extracellular DNA has been postulated to play a pathological role, including prolonged mechanical ventilation due to persistent airway obstruction (Riethmueller et al. 2006), ventila-

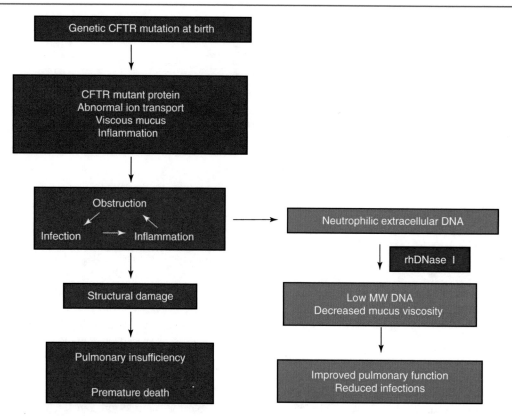

Fig. 21.1 Cystic fibrosis and rhDNase I. CFTR genetic mutation at birth leads to either reduced or improperly folded CFTR protein, which results in altered ion transport, viscous mucus, and inflammation in the airways. Eventually, this leads to obstruction of the airways, bacterial infection, and further inflammation. After neutrophils arrive to fight the infection, they die and release cellular contents, one of which is DNA. Persistent obstruction, infection, and inflammation lead to structural damage and eventually pulmonary insufficiency and premature death. rhDNase I is aerosolized into the airways where it degrades DNA to lower molecular weight fragments, thus reducing CF mucus viscosity and allowing expectoration, which improves lung function and reduces bacterial infections

tor-associated pneumonia in infants (Scala et al. 2017), atelectasis (Hendriks et al. 2005), chronic sinusitis (Cimmino et al. 2005), primary ciliary dyskinesia (Desai et al. 1995; ten Berge et al. 1999; El Abiad et al. 2007), other non-CF lung diseases in children (Boogaard et al. 2007), and empyema (Simpson et al. 2003; Rahman et al. 2011). rhDNase I has also been studied to examine its effectiveness in the presence or absence of antibiotics against biofilm-producing strains of *Staphylococcus aureus* and *Staphylococcus epidermidis* (Kaplan et al. 2012). rhDNase I exhibits potent antibiofilm and antimicrobial-sensitizing activities at clinically achievable concentrations.

Finally, the use of rhDNase I has been investigated in systemic lupus erythematosus (SLE) where degradation or prevention of immune complexes containing anti-DNA antigens may have therapeutic benefit (Lachmann 2003; Davis Jr et al. 1999).

Historical Perspective and Rationale

Macromolecules that contribute to the physical properties of lung secretions include mucus glycoproteins, filamentous actin, and DNA. Experiments in the 1950s and 1960s revealed that DNA is present in very high concentrations (3–14 mg/mL) only in infected lung secretions (Matthews et al. 1963). This

implied that the DNA that contributes to the high viscoelastic nature of CF sputum is derived from neutrophils responding to chronic infections (Potter et al. 1969). These DNA-rich secretions also bind aminoglycoside antibiotics commonly used for treatment of pulmonary infections and thus may reduce their efficacy (Ramphal et al. 1988; Bataillon et al. 1992).

Early in vitro studies in which lung secretions were incubated for several hours with partially purified bovine pancreatic DNase I showed a large reduction in viscosity (Armstrong and White 1950; Chernick et al. 1961). Based on these observations, bovine pancreatic DNase I (Dornavac or pancreatic dornase) was approved in the United States for human use in 1958. Numerous uncontrolled clinical studies in patients with pneumonia and one study in patients with CF suggested that bovine pancreatic DNase I was effective in reducing the viscosity of lung secretions (Lieberman 1968). However, severe adverse reactions occurred occasionally, perhaps due to allergic reactions to a foreign protein or from contaminating proteases, since up to 2% of trypsin and chymotrypsin were present in the final product (Raskin 1968; Lieberman 1962). Both bovine DNase I products were eventually withdrawn from the market.

In the late 1980s, human deoxyribonuclease I was cloned from a human pancreatic cDNA library, sequenced and expressed recombinantly using mammalian cell culture in Chinese hamster ovary (CHO) cells to reevaluate the potential of DNase I as a therapeutic for cystic fibrosis (Shak et al. 1990). In vitro incubation of purulent sputum from CF patients with catalytic concentrations of rhDNase I reduced its viscoelasticity (Shak et al. 1990). The reduction in viscoelasticity was directly related to both rhDNase I concentration and reduction in the size of the DNA in the samples. Therefore, reduction of high molecular weight DNA into smaller fragments by treatment with aerosolized rhDNase I was proposed as a mechanism to reduce the mucus viscosity and improve mucus clearability from obstructed airways in patients. It was hoped that improved clearance of the purulent mucus would enhance pulmonary function and reduce recurrent exacerbations of respiratory symptoms requiring parenteral antibiotics. This proved to be the case, and rhDNase I was approved by the Food and Drug Administration in 1993. Since that time, the clinical use of rhDNase I has continued to increase with over 67% of CF patients currently receiving chronic therapy (Konstan et al. 2010).

Protein Chemistry, Enzymology, and Structure

The protein chemistry of human DNases including DNase I has been reviewed (Lazarus 2002; Baranovskii et al. 2004). Recombinant human DNase I is a monomeric, 260-amino-acid glycoprotein (Fig. 21.2) produced by mammalian CHO cells (Shak et al. 1990). The protein has four cysteines, which are oxidized into two disulfides between Cys101–Cys104

and Cys173–Cys209 as well as two potential N-linked glycosylation sites at Asn18 and Asn106 (Fig. 21.2). rhDNase I is glycosylated at both sites and migrates as a broad band on polyacrylamide gel electrophoresis gels with an approximate molecular weight of 37 kDa, which is significantly higher than the predicted molecular mass from the amino acid sequence of 29.3 kDa. rhDNase I is an acidic protein and has a calculated pI of 4.58. The primary amino acid sequence is identical to that of the native human enzyme purified from urine.

DNase I cleaves double-stranded DNA and, to a much lesser degree, single-stranded DNA, nonspecifically by nicking phosphodiester linkages in one of the strands between the 3′-oxygen atom and the phosphorus to yield 3′-hydroxyl and 5′-phosphoryl oligonucleotides with inversion of configuration at the phosphorus. rhDNase I enzymatic activity is dependent upon the presence of divalent metal ions for structure, as there are two tightly bound Ca^{2+} atoms and catalysis, which requires either Mg^{2+} or Mn^{2+} (Pan and Lazarus 1999). The active site includes two histidine residues (His134 and His252) and two acidic residues (Glu78 and Asp212), all of which are critical for the general acid–base catalysis of phosphodiester bonds since alanine substitution of any of these results in a total loss of activity (Ulmer et al. 1996). Other residues involved in the coordination of divalent metal ions at the active site and DNA contact residues have also been identified by mutational analysis and include Asn7, Glu39, Asp168, Asn170, and Asp251 (Pan et al. 1998; Parsiegla et al. 2012). The two Ca^{2+} binding sites require acidic or polar residues for coordination of Ca^{2+}; for site 1, these include Asp201 and Thr203, and for site 2, these include Asp99, Asp107, and Glu112 (Pan and Lazarus 1999). DNase I is a relatively stable enzyme and shows optimal activity at

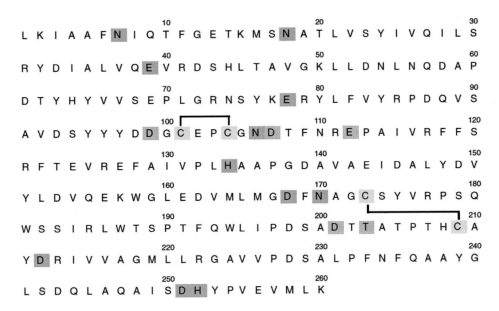

Fig. 21.2 Primary amino acid sequence of rhDNase I. Active site residues are highlighted in *blue*, cysteine residues that form disulfide bonds are shown in *yellow*, N-linked glycosylation sites are highlighted in *pink*, and residues involved in Ca^{2+} coordination are shown in *beige*. The 22-residue signal sequence that is cleaved prior to secretion is not shown

pH 5.5–7.5. It is inactivated by heat and is potently inhibited by EDTA and G-actin. Surprisingly, DNase I is also inhibited by NaCl and has only ca. 30% of the maximal activity in physiological saline.

rhDNase I belongs to the DNase I-like structural superfamily according to the structural classification of proteins (SCOP) database and is also related to the endonuclease–exonuclease–phosphatase family (Andreeva et al. 2008; Dlakic 2000; Wang et al. 2010). The X-ray crystal structure of rhDNase I was initially solved at 2.2 Å resolution and superimposes with the biochemically more widely studied bovine DNase I, which shares 78% sequence identity, with a root mean square deviation for main chain atoms of 0.56 Å (Wolf et al. 1995). DNase I is a compact α/β protein having a core of two tightly packed six-stranded β-sheets surrounded by eight α-helices and several loop regions (Fig. 21.3). Bovine DNase I has also been crystallized in complex with G-actin (Kabsch et al. 1990) as well as with several short oligonucleotides, revealing key features of DNA recognition in the minor groove and catalytic hydrolysis (Suck 1994).

The structure of rhDNase I containing a divalent Mg^{2+} and a phosphate in the active site has been solved at 1.95 Å

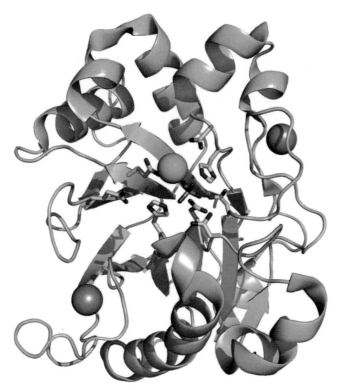

Fig. 21.3 Cartoon representation of rhDNase I depicting the active site of the enzyme. The active site residues are shown as *beige sticks* where oxygen and nitrogen atoms are in *red* and *blue*, respectively. The Mg^{2+} ion is shown as a *cyan sphere*, and Ca^{2+} ions are shown as *gray* spheres. The phosphate ion is shown in sticks with the phosphorus atom in *orange*. The figure was made using PyMOL (www.pymol.org) using accession code 4AWN for rhDNase I (Parsiegla et al. 2012)

resolution (Parsiegla et al. 2012). The combined structural and mutagenesis data suggest a Mg^{2+}-assisted pentavalent phosphate transition state during catalysis of rhDNase I, where Asp168 may play a key role as a catalytic general base. Asn170 is also in close proximity to both the attacking water molecule and the phosphoryl oxygen. His134 and His252 appear to act as general acids in stabilizing the pentavalent transition state. There is also a critical catalytic role for rhDNase I Asn7, a residue that is highly conserved among mammalian DNase I enzymes and members of the DNase I-like superfamily. The Mg^{2+} cation resides at the computationally predicted site IVb (Gueroult et al. 2010) and interacts with Asn7, Glu39, and Asp251 via a complex set of water-mediated hydrogen bond interactions. A comprehensive analysis of the rhDNase I with members of the DNase I-like structural superfamily (Andreeva et al. 2008; Dlakic 2000) such as the apurinic/apyrimidinic endonucleases from humans (APE1) (Mol et al. 2000) and *Neisseria meningitidis* (NApe) (Carpenter et al. 2007), sphingomyelin phosphodiesterase (SMase) from *Bacillus cereus* (Ago et al. 2006), or the C-terminal domain of human CNOT6L nuclease (Wang et al. 2010) solved in complex with cations or DNA has revealed new insights into the catalytic mechanism of DNA hydrolysis.

Several variants of rhDNase I with greatly improved enzymatic properties have been engineered by site-directed mutagenesis (Pan and Lazarus 1997; Pan et al. 1998). The methods for production of the variants and the assays to characterize them have been reviewed recently (Pan et al. 2001; Sinicropi and Lazarus 2001). The rationale for improving activity was to increase binding affinity to DNA by introducing positively charged residues (Arg or Lys) on rhDNase I loops at the DNA binding interface to form a salt bridge with phosphates on the DNA backbone. These so-called "hyperactive" rhDNase I variants are substantially more active than wild-type rhDNase I and are no longer inhibited by physiological saline. The greater catalytic activity of the hyperactive variants is due to a change in the catalytic mechanism from a "single nicking" activity in the case of wild-type rhDNase I to a "processive nicking" activity in the hyperactive rhDNase I variants (Pan and Lazarus 1997), where gaps rather than nicks result in a higher frequency of double-stranded cleavages.

It is interesting to note that significantly greater activity can result from just a few mutations on the surface that are not important for structural integrity. For whatever reason, DNase I is not as efficient an enzyme as it could be for degrading DNA into small fragments. Furthermore, the inhibition by G-actin can be eliminated by a single amino acid substitution (see below). Thus, DNase I is under some degree of regulation in vivo. One can only speculate that nature may have wanted to avoid an enzyme with too much DNA degrading activity that could result in undesired mutations in the genome.

Emerging Biology

Neutrophil Extracellular Traps

Neutrophils are important effector cells that play a key role in innate immunity. They have well-documented "first-line defense" roles in phagocytosis of foreign organisms such as bacteria and fungi. NETosis, the formation and release of neutrophil extracellular traps (NETs), is another defense mechanism that has emerged more recently (Brinkmann et al. 2004). Neutrophils can degranulate, releasing NETs, which are structures of DNA filaments coated with toxic histones, proteases, oxidative enzymes, and other proteins that can immobilize and neutralize bacteria in the extracellular environment. NETosis can lead to a cell death process that is distinct from both apoptosis and necrosis, but it can also be independent of cell death (Jorch and Kubes 2017; Honda and Kubes 2018). NETs and NETosis have been broadly studied in biology, and their role in a wide variety of diseases is of great interest (Jorch and Kubes 2017; Brinkmann 2018). These have included autoimmune, inflammatory, thrombotic, cancer, and other diseases such as SLE, vasculitis, acute pancreatitis, rheumatoid arthritis, type 1 diabetes mellitus and wound healing, gout, inflammatory bowel disease, vascular occlusion, sepsis, acute respiratory distress syndrome (ARDS), metastasis, and others (Cools-Lartigue et al. 2014; Martinod and Wagner 2014; Merza et al. 2015; Wong et al. 2015; Lood et al. 2016; Park et al. 2016; Gupta and Kaplan 2016; Jiménez-Alcázar et al. 2017; Apel et al. 2018; Honda and Kubes 2018). It is particularly significant that DNase I or rhDNase I was almost always used in the biological studies on the abovementioned disease areas to show the effect of NET degradation, most often resulting in beneficial effects. Notably, serum DNase I had been shown to be essential for degradation of NETs (Hakkim et al. 2010); however, other DNase I-like nucleases like DNase1L3 can also degrade NETs (Jiménez-Alcázar et al. 2017).

While it is clear that NETs and NETosis play an important role in biology, that role is quite complex. In many cases, they likely play a beneficial role; however, in others, they may play a pathological role. Like many areas of biology and disease, it is likely the right balance as well as their locality that makes the difference. Furthermore, it is not always clear if they are the cause or result of a given pathology; there is evidence supporting both in thrombosis (Jiménez-Alcázar et al. 2017). There is much evidence showing that diseases involve a multitude of pathways. For example, there are strong connections between coagulation, inflammation, and cancer pathways. It is not surprising that NETs also have connections with these pathways. NETosis is a factor in tumor progression as well as cancer-associated thrombosis (Demers and Wagner 2014). Both neutrophils and circulating extracellular DNA have been suggested to play an important role in cancer (Hawes et al. 2015).

With respect to rhDNase I, it is especially noteworthy that NETs play a significant role in CF (Gray et al. 2015; Law and Gray 2017), since this is the major use of rhDNase I clinically. While rhDNase I has a rheological effect in reducing the viscoelasticity in CF sputum, it also has biological effects related to releasing various enzymes and proteins from the NETs upon hydrolysis of the DNA. In the CF lung, NETs may play less of a role as an antibacterial and more of a role in fostering inflammation. rhDNase I likely exerts its anti-inflammatory function due to degradation of NETs.

Pharmacology

In Vitro Activity in CF Sputum

In vitro, rhDNase I hydrolyzes the DNA in sputum of CF patients and reduces sputum viscoelasticity (Shak et al. 1990). Effects of rhDNase I were initially examined using a relatively crude "pourability" assay. Pourability was assessed qualitatively by inverting a tube of sputum and observing the movement of sputum after a tap on the side of the tube. Catalytic amounts of rhDNase I (50 μg/mL) greatly reduced the viscosity of the sputum, rapidly transforming it from a viscous gel to a flowing liquid. More than 50% of the sputum moved down the tube within 15 min of incubation, and all the sputum moved freely down the tube within 30 min. The qualitative results of the pourability assay were confirmed by quantitative measurement of viscosity using a Brookfield Cone–Plate viscometer (Fig. 21.4). The reduction of viscosity by rhDNase I is rhDNase I concentration dependent and is associated with reduction in size of sputum DNA as measured by agarose gel electrophoresis (Fig. 21.5).

Additional in vitro studies of CF mucus samples treated with rhDNase I demonstrated a dose-dependent improvement in cough transport and mucociliary transport of CF mucus using a frog palate model and a reduction in adhesiveness as measured by mucus contact angle (Zahm et al. 1995). The improvements in mucus transport properties and adhesiveness were associated with a decrease in mucus viscosity and mucus surface tension, suggesting rhDNase I treatment may improve the clearance of mucus from airways. The in vitro viscoelastic properties of rhDNase I have also been studied in combination with normal saline, 3% hypertonic saline, or nacystelyn, the L-lysine salt of N-acetyl cysteine (King et al. 1997; Dasgupta and King 1996). The major impact of rhDNase I on CF sputum is to decrease spinnability, which is the thread forming ability of mucus under the influence of low amplitude stretching. CF sputum spinnability decreases 25% after 30 min incubation with rhDNase I (King et al. 1997). rhDNase I in normal saline and saline alone both increased the cough clearability index. With the combination of rhDNase I and 3% hypertonic saline, there was minimal effect on spinnability; however, mucus rigidity

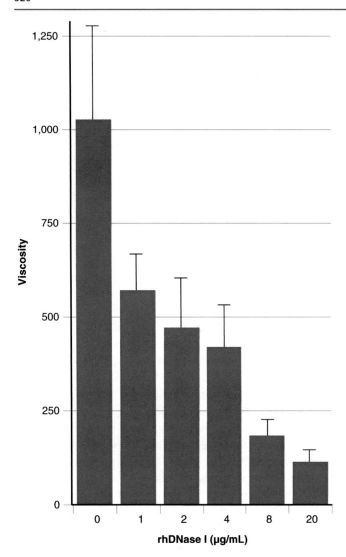

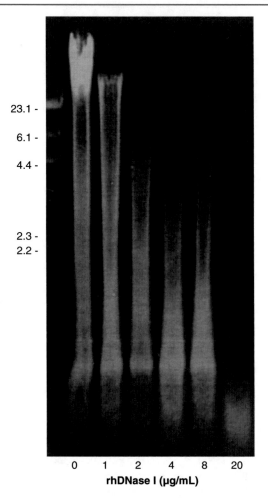

Fig. 21.4 In vitro reduction in viscosity (in centipoise) of cystic fibrosis sputum by cone–plate viscometry. Cystic fibrosis sputum was incubated with various concentrations of rhDNase I for 15 min at 37 °C

Fig. 21.5 In vitro reduction in sputum DNA size as measured by agarose gel electrophoresis. Cystic fibrosis sputum was incubated with increasing concentrations (0–20 µg/mL) of rhDNase I for 150 min at 37 °C. Molecular weight standards for DNA in kb are indicated

and cough clearability improved greater than with either agent alone. The predicted mucociliary clearance did not significantly increase with 3% saline either alone or in combination with rhDNase I. Combining rhDNase I with nacystelyn has an additive benefit on spinnability but no effect on mucous rigidity or cough clearability (Dasgupta and King 1996). These effects of rhDNase I can be variable in vivo and do not necessarily correlate with the level of DNA in sputum. For example, sputum from CF patients that clinically responded to rhDNase I contains significantly higher levels of magnesium ions compared with sputum from patients who do not have a clear response (Sanders et al. 2006). Although this response is consistent with the requirement for divalent cations and their mode of action on DNase I (Campbell and Jackson 1980), the mechanism of increased rhDNase I activity by magnesium ions has been attributed to altering the polymerization state of actin such that equilib-

rium favors increased F-actin and decreased G-actin (see below).

The mechanism of action of rhDNase I to reduce CF sputum viscosity has been ascribed to DNA hydrolysis (Shak et al. 1990). However, an alternative mechanism involving depolymerization of filamentous actin (F-actin) has been suggested since F-actin contributes to the viscoelastic properties of CF sputum and the actin-depolymerizing protein gelsolin also reduces sputum viscoelasticity (Vasconcellos et al. 1994). F-actin is in equilibrium with its monomeric form (G-actin), which binds to rhDNase I with high affinity and is also a potent inhibitor of DNase I activity (Lazarides and Lindberg 1974). DNase I is known to depolymerize F-actin by binding to G-actin with high affinity, shifting the equilibrium in favor of rhDNase I/G-actin complexes (Hitchcock et al. 1976). To elucidate the mechanism of rhDNase I in CF sputum, the activity of two types of rhDNase I variants was compared in CF sputum (Ulmer et al. 1996). Active site variants were engineered that were unable to catalyze DNA hydrolysis but retained wild-type G-actin binding. Actin-resistant variants that no longer bound G-actin but retained wild-type DNA hydrolytic activity were also characterized. The active site variants did not degrade DNA in CF sputum and did not decrease sputum viscoelasticity (Fig. 21.6). Since the active site variants retained the ability to bind G-actin, these results argue against depolymerization of F-actin as the mechanism of action. In contrast, the actin-resistant variants were more potent than wild-type DNase I in their ability to degrade DNA and reduce sputum viscoelasticity (Fig. 21.6). The increased potency of the actin-resistant variants indicated that G-actin was a significant inhibitor of wild-type DNase I in CF sputum and confirmed that hydrolysis of DNA was the mechanism by which rhDNase I decreases sputum viscoelasticity. The mechanism for reduction of sputum viscosity by gelsolin was subsequently determined to result from an unexpected second binding site on actin that competes with DNase I, thus relieving the inhibition by G-actin (Davoodian et al. 1997). Additional in vitro studies characterizing the relative potency of actin-resistant and hyperactive rhDNase I variants in serum and CF sputum have been reported (Pan et al. 1998).

In Vivo Activity in CF Sputum

In vivo confirmation of the proposed mechanism of action for rhDNase I has been obtained from direct characterization of apparent DNA size (Fig. 21.7) and measurements of enzymatic and immunoreactive (ELISA) activity of rhDNase I (Fig. 21.8) in sputum from cystic fibrosis patients (Sinicropi

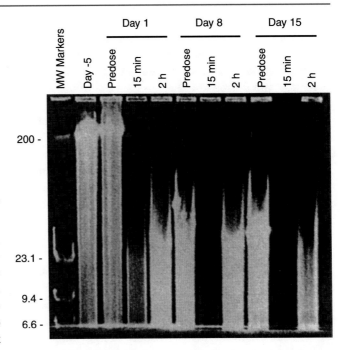

Fig. 21.7 Sustained reduction in DNA length in sputum recovered from a CF patient treated with 2.5 mg of rhDNase I BID for up to 15 days. Samples were analyzed by pulsed-field gel electrophoresis on an agarose gel. Molecular weight standards for DNA in kb are indicated

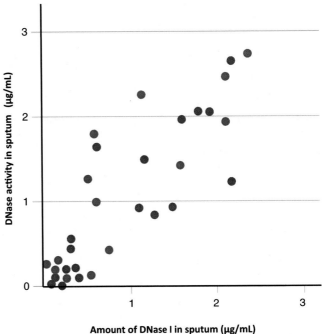

Fig. 21.8 Immunoreactive concentrations and enzymatic activity of rhDNase I in sputum following aerosol administration of either 10 mg (●) or 20 mg (●) of rhDNase I to patients with cystic fibrosis. Each data point is a separate sample measured in duplicate

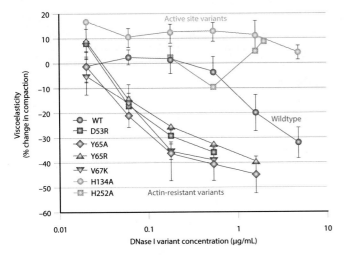

Fig. 21.6 Mechanism of action in CF mucus for rhDNase I. The change in viscoelasticity in CF mucus as a function of DNase concentration was determined for wild-type rhDNase I, two active site variants that no longer catalyze DNA hydrolysis, and four variants that are no longer inhibited by G-actin (Ulmer et al. 1996)

et al. 1994). Sputum samples were obtained 1–6 h post-dose from adult cystic fibrosis patients after inhalation of 5–20 mg of rhDNase I. rhDNase I therapy produced a sustained reduction in DNA size in recovered sputum (Fig. 21.7), in good agreement with the in vitro data.

Inhalation of the therapeutic dose of rhDNase I produced sputum levels of rhDNase I that have been shown to be effective in vitro (Fig. 21.8) (Shak 1995). Similarly, delivery to CF patients as young as 3 months can produce bronchoalveolar lavage fluid levels similar to that of older patients (Wagener et al. 1998). The recovered rhDNase I was also enzymatically active. Enzymatic activity was directly correlated with rhDNase I concentrations in the sputum. Viscoelasticity was reduced in the recovered sputum as well. Furthermore, results from scintigraphic studies in using twice daily 2.5 mg of rhDNase I in CF patients suggested possible reductions in pulmonary obstruction and increased rates of mucociliary sputum clearance from the inner zone of the lung compared to controls (Laube et al. 1996). This finding was not confirmed in a crossover design study using once-daily dosing, suggesting that improvement of mucociliary clearance may require higher doses (Robinson et al. 2000). Epidemiologic evaluation of patients changing from once-daily dosing to twice daily and vice versa has also shown improvement in lung function and fewer pulmonary exacerbations on higher doses, although the clinical indication for a dose change is not clear (VanDevanter et al. 2018).

Pharmacokinetics and Metabolism

Nonclinical pharmacokinetic data in rats and monkeys suggest minimal systemic absorption of rhDNase I following aerosol inhalation of clinically equivalent doses. rhDNase I is cleared from the systemic circulation without any accumulation in tissues following acute exposure (Green 1994). Additionally, nonclinical metabolism studies suggest that the low rhDNase I concentrations present in serum following inhalation will be bound to binding proteins (Green 1994; Mohler et al. 1993). The low concentrations of endogenous DNase I normally present in serum and the low concentrations of rhDNase I in serum following inhalation are inactive due to the ionic composition and presence of binding proteins in serum (Prince et al. 1998).

When 2.5 mg of rhDNase I was administered twice daily by inhalation to 18 CF patients, mean sputum concentrations of 2 µg/mL DNase I were measurable within 15 min after the first dose on day 1 (Fig. 21.9). Mean sputum concentrations declined to an average of 0.6 µg/mL 2 h following inhalation. The peak rhDNase I concentration measured 2 h after inhalation on days 8 and 15 increased to 3.0 and 3.6 µg/mL, respectively. Sputum rhDNase I concentrations measured 6 h after inhalation on days 8 and 15 were similar to day 1. Pre-dose trough concentrations of 0.3–0.4 µg/mL rhDNase I measured on day 8 and day 15 (sample taken approximately 12 h after

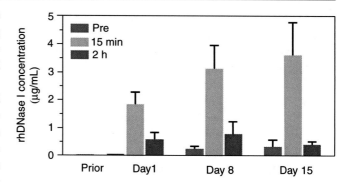

Fig. 21.9 rhDNase I concentration in sputum following administration of 2.5 mg of rhDNase I twice daily by inhalation to CF patients. The blue, orange, and purple bars represent concentrations at pre-dose, 15 min post-dose, and 2 h post-dose, respectively. The mean ± SD is shown with $N = 18$

the previous dose) were, however, higher than day 1, suggesting possible modest accumulation of rhDNase I with repeated dosing. Inhalation of up to 10 mg three times daily of rhDNase I by four CF patients for six consecutive days did not result in significant elevation of serum concentrations of DNase above normal endogenous levels (Aitken et al. 1992; Hubbard et al. 1992). After administration of up to 2.5 mg of rhDNase I twice daily for 6 months to 321 CF patients, no accumulation of serum DNase was noted (assay limit of detection is approximately 0.5 ng DNase/mL serum).

Protein Manufacturing and Formulation

rhDNase I is expressed in mammalian cell culture and purified to homogeneity using a variety of chromatographic steps. The development of the formulation of rhDNase I is especially important in that a suitable formulation is required to take into account protein stability, aerosolization properties, tonicity, and the sealed container for storage (Shire 1996). rhDNase I (Pulmozyme®, dornase alfa) is manufactured by Genentech, Inc. and formulated as a sterile, clear, and colorless aqueous solution containing 1.0 mg/mL dornase alfa, 0.15 mg/mL calcium chloride dihydrate, and 8.77 mg/mL sodium chloride. The solution contains no preservative and has a nominal pH of 6.3. Pulmozyme® is administered by the inhalation of an aerosol mist produced by a compressed air-driven nebulizer system. Pulmozyme® is supplied as single-use ampoules, which deliver 2.5 mL of solution to the nebulizer.

The choice of formulation components was determined by a need to provide 1–2 years of stability and to meet additional requirements unique to aerosol delivery (Shire 1996). A simple colorimetric assay for rhDNase I activity was used to evaluate the stability of rhDNase I in various formulations (Sinicropi et al. 1994). In order to avoid adverse pulmonary reactions, such as cough or bronchoconstriction, aerosols for local pulmonary delivery should be formulated as isotonic solutions with minimal or no buffer components and should maintain pH > 5.0. rhDNase I has an additional requirement for calcium to be present for optimal enzymatic activity. Limiting formulation components raised concerns about pH control, since protein stability and solubility can be highly pH dependent. Fortunately, the protein itself provided sufficient buffering capacity at 1 mg/mL to maintain pH stability over the storage life of the product.

Drug Delivery

The droplet or particle size of an aerosol is a critical factor in defining the site of deposition of the drug in the patient's airways (Gonda 1990). A distribution of particle or droplet size of 1–6 μm is optimal for the uniform deposition of rhDNase I in the airways (Cipolla et al. 1994). Jet nebulizers are the simplest method of producing aerosols in the desired respirable range. However, recirculation of protein solutions under high shear rates in the nebulizer bowl can present risks to the integrity of the protein molecule. rhDNase I survived recirculation and high shear rates during the nebulization process with no apparent degradation in protein quality or enzymatic activity (Cipolla et al. 1994). Ultrasonic nebulizers produce greater heat than jet nebulizers, and protein breakdown prevents their use with rhDNase I. Significant advances in nebulizer technology have occurred since the original approval of rhDNase I. Newer nebulizers using a vibrating, perforated membrane do not produce protein breakdown and provide more rapid and efficient delivery of particles in the respirable range (Scherer et al. 2011).

Approved jet nebulizers produce aerosol droplets in the respirable range (1–6 μm) with a mass median aerodynamic diameter (MMAD) of 4–6 μm. The delivery of rhDNase I with a device that produces smaller droplets leads to more peripheral deposition in the smaller airways and thereby improves efficacy (Geller et al. 1998). Results obtained in 749 CF patients with mild disease confirmed that patients randomized to the SideStream nebulizer powered by the Mobil Aire Compressor (MMAD = 2.1 μm) tended to have greater improvement in pulmonary function than patients using the Hudson T nebulizer with Pulmo-Aide Compressor (MMAD = 4.9 μm). These results indicate that the efficacy of rhDNase I is dependent, in part, on the physical properties of the aerosol produced by the delivery system. Nebulizers with vibrating mesh technology (Pari eFlow®) produce similarly small particles, suggesting these may result in further improved efficacy of rhDNase I. Furthermore, "smart" nebulizers are now available that coach the patient on taking a proper breath to improve delivery to the lower airways. Delivery of rhDNase I improves and lung function improves more when these nebulizers are used (Bakker et al. 2011). A randomized trial of efficacy and safety of dornase alfa delivered by the eRapid® nebulizer in CF patients showed comparable efficacy and safety, shorter nebulization times, and higher patient preference when compared to standard jet nebulizer systems such as the LC Plus (Sawicki et al. 2015).

Clinical Use

Indication and Clinical Dosage
rhDNase I (Pulmozyme®, dornase alfa) is currently approved for use in CF patients, in conjunction with standard therapies, to reduce the frequency of respiratory infections requiring parenteral antibiotics and to improve pulmonary function (Fig. 21.1). The recommended dose for use in most CF patients is one 2.5 mg dose inhaled daily.

Cystic Fibrosis
rhDNase I was evaluated in a large, randomized, placebo-controlled trial of clinically stable CF patients, 5 years of age or older, with baseline forced vital capacity (FVC) > 40% of the predicted value (based on sex, age, and height) (Fuchs et al. 1994). All patients received additional standard therapies for CF. Patients were treated with placebo or 2.5 mg of rhDNase I once or twice a day for 6 months. When compared to placebo, both once-daily and twice-daily doses of rhDNase I resulted in a 28–37% reduction in respiratory tract infections requiring the use of parenteral antibiotics (Fig. 21.10). Within 8 days of the start of treatment with rhDNase I, mean forced expiratory volume in 1 s (FEV_1) increased 7.9% in patients treated once a day and 9.0% in those treated twice a day compared to the baseline values. The mean FEV_1 observed during long-term therapy increased 5.8% from baseline at the 2.5 mg daily dose level and 5.6% from baseline at the 2.5 mg twice-daily dose level (Fig. 21.11). The

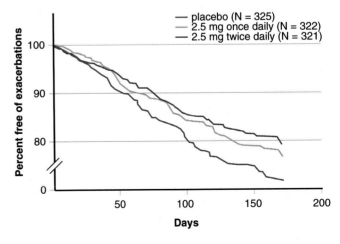

Fig. 21.10 Proportion of patients free of exacerbations of respiratory symptoms requiring parenteral antibiotic therapy from a 24-week study

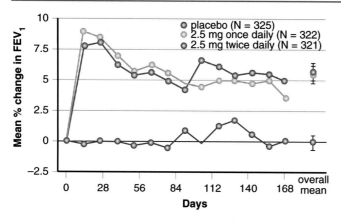

Fig. 21.11 Mean percent change in FEV₁ from baseline through a 24-week study

risk of respiratory tract infection was reduced even in patients whose pulmonary function (FEV₁) did not improve. This finding may be due to improved clearance of mucus from the small airways in the lung, which will have little effect on FEV₁ (Shak 1995). Supporting this concept is the finding that rhDNase I improves the lung clearance index in 6–18-year-old CF patients with normal lung function (Amin et al. 2011). Alternatively, the use of rhDNase I may be altering the neutrophilic inflammatory response that occurs early in the course of CF lung disease, similar to the use of other anti-inflammatory therapies (Konstan and Ratjen 2012). The administration of rhDNase I also lessened shortness of breath, increased the general perception of well-being, and reduced the severity of other cystic fibrosis-related symptoms. Based on these findings, the US Cystic Fibrosis Foundation, in their chronic therapy pulmonary guidelines, strongly recommends the use of rhDNase I in patients 6 years old and older with moderate to severe lung disease and recommends its use in patients with mild lung disease (Flume et al. 2007).

The safety and deposition, but not efficacy, of rhDNase I have been studied in CF patients <5 years old (3 months to 5 years) since therapy may provide clinical benefit for young CF patients with mild disease (Wagener et al. 1998). After 2 weeks of daily administration of 2.5 mg of rhDNase I, levels of rhDNase I deposited in the lower airways were similar for children <5 years compared to a group of 5–10-year-olds. Moreover, rhDNase I was well tolerated in the younger age group with an adverse event frequency similar to that in the older age group. To further understand how rhDNase I might alter the progression of lung disease, children with mild CF-related lung disease were treated for 2 years in a randomized controlled trial (Quan et al. 2001). Children with a mean age of 8.4 years and an FVC ≥ 85% predicted were treated once daily with either placebo or 2.5 mg of rhDNase I. After 96 weeks, lung function was significantly better in the treated

group compared with placebo, particularly for tests measuring function of smaller airways. Respiratory exacerbations were also reduced in the treated group.

Clinical trials have indicated that rhDNase I therapy can be continued or initiated during an acute respiratory exacerbation (Wilmott et al. 1996). rhDNase I, however, does not produce a pulmonary function benefit when used short term in the most severely ill CF patients (FVC < 40% predicted) (Shah et al. 1995; McCoy et al. 1996). Short-term dose ranging studies demonstrated that doses in excess of 2.5 mg twice daily did not provide further significant improvement in FEV₁ (Aitken et al. 1992; Hubbard et al. 1992; Ramsey et al. 1993). Patients who have received the drug on a cyclical regimen (i.e., administration of rhDNase I 10 mg twice daily for 14 days, followed by a 14-day washout period) showed rapid improvement in FEV₁ with the initiation of rhDNase I and a return to baseline following withdrawal (Eisenberg et al. 1997). rhDNase I use improves quality of life as measured by the validated cystic fibrosis questionnaire-revised (CFQ-R) questionnaire (Rozov et al. 2010).

Concomitant therapy with rhDNase I and other standard CF therapies often show additive effects. The intermittent administration of aerosolized tobramycin was approved for use in CF patients with or without concomitant use of rhDNase I (Ramsey et al. 1999). Aerosolized tobramycin was well tolerated, enhanced pulmonary function, and decreased the density of *Pseudomonas aeruginosa* in sputum. In combination with rhDNase I, a larger treatment effect was noted but did not reach statistical significance. No differences in safety profile were observed following aerosolized tobramycin in patients that did or did not use rhDNase I. Chronic use of azithromycin has also been studied in CF patients chronically infected with *P. aeruginosa* (Saiman et al. 2005). Similar improvement in lung function and reduction in respiratory exacerbations was seen in patients receiving rhDNase I as those who did not, suggesting an additive but not synergistic benefit of the two therapies when used together. The combination of hypertonic saline therapy with chronic use of rhDNase I has similar additive benefits (Elkins et al. 2006), in agreement with the previously mentioned in vitro studies. Finally, combining ivacaftor, which potentiates chloride transport in CF patients with the G551D gene mutation, with chronic rhDNase I use produces an additive benefit in both improved lung function and reduced respiratory exacerbations (Ramsey et al. 2011). Notably, there was no evidence of a change in adverse events related to combination therapy in any of these studies.

Following Food and Drug Administration (FDA) approval of rhDNase I in 1993, a large epidemiologic study of CF patients was initiated (the Epidemiologic Study of Cystic Fibrosis or ESCF), which continued until 2005 (Morgan et al. 1999). This study was designed to evaluate practice patterns in CF patients and has included data from over 24,000 patients. Recent analysis of ESCF data showed that chronic use of rhDNase I is associated with a decreased rate of decline in lung function overtime (Konstan et al. 2011). This reduced rate of decline in lung function is similar to findings with oral ibuprofen (Konstan et al. 1995, 2007) and inhaled corticosteroids (Ren et al. 2008), suggesting that there may be a long-term anti-inflammatory benefit with the use of rhDNase I. This potential anti-inflammatory effect is supported by a randomized trial in 105 CF patients with mild lung disease (FEV_1 > 80% predicted) (Paul et al. 2004). Based on an initial bronchoscopy and alveolar lavage, patients were divided into two groups: those without airway inflammation ($n = 20$) and those with airway inflammation. The patients with inflammation were then randomized to treatment with ($n = 46$) or without rhDNase I ($n = 39$). Follow-up bronchoscopy and lavage were performed at 18 and 36 months. In the patients treated with rhDNase I, there was no change in inflammation as measured by elastase and IL-8 levels and neutrophil number. Patients not treated with rhDNase I and patients who did not have inflammation at baseline all had worsening neutrophilic inflammation on follow-up. Although this study was not designed to evaluate the rate of lung function decline, patients treated with rhDNase I dropped FEV_1 by 1.99% predicted per year compared to a 3.26% predicted drop per year in untreated subjects. Finally, rhDNase I is associated with a 15% reduction in the odds of subsequent year mortality in patients with CF (Sawicki et al. 2012).

Non-cystic Fibrosis Respiratory Disease

Although originally considered beneficial for the treatment of non-CF-related bronchiectasis (Wills et al. 1996), rhDNase I had no effect on pulmonary function or the frequency of respiratory exacerbations in a randomized controlled trial (O'Donnell et al. 1998). In another randomized controlled trial of rhDNase I, young children had shorter periods of ventilatory support following cardiac surgery when rhDNase I was instilled twice daily into the endotracheal tube (Riethmueller et al. 2006). Complicating atelectasis was less frequent in the treated group, consistent with numerous case reports suggesting that rhDNase I decreases, and can be used to treat, atelectasis when directly instilled into the airway (Hendriks et al. 2005). This effectiveness in treating atelectasis seems particularly true for newborns with lung disease requiring mechanical ventilation (MacKinnon et al. 2011; Dilmen et al. 2011; Fedakar et al. 2012; Altunhan et al. 2012). Limited benefit has been seen in children with asthma (Puterman and Weinberg 1997), although no benefit has been seen in adults (Silverman et al. 2012), consistent with the lack of neutrophil-dominated inflammation in asthma.

Finally, while there is increased free DNA in the secretions of infants with respiratory syncytial virus-caused bronchiolitis, early suggestions of benefit (Nasr et al. 2001) have not translated into reduced hospitalization or the need for supplemental oxygen (Boogaard et al. 2007).

Empyema and Parapneumonic Effusion

In principle, rhDNase I may be useful for treating any condition where high levels of extracellular DNA and associated viscoelastic properties are pathological. Pulmonary empyema involves the collection of purulent material in the pleural space, and the use of rhDNase I instilled into the pleural space has been reviewed (Simpson et al. 2003; Corcoran et al. 2015). In one large, multicenter clinical trial for the treatment of empyema in adults, twice-daily intrapleural administration over 3 days was evaluated in four groups: 5 mg rhDNase, 10 mg tissue plasminogen activator (tPA), a combination of both, and a double placebo. Patients receiving the combination therapy had improved fluid drainage and a reduced frequency of surgical referral (Rahman et al. 2011). Additional studies using similar doses have demonstrated similar clinical benefits in children, although determining when and if to perform surgery remains unclear (Piccolo et al. 2015; Gilbert and Gorden 2017). Importantly, it appears that using either tPA or rhDNase 1 alone does not produce the benefit of combined therapy.

Other Medical Conditions

rhDNase I has also been instilled into the nasal sinuses after surgery for chronic infections (Cimmino et al. 2005; Raynor et al. 2000). While daily use over 28 days of nasal nebulized rhDNase I in patients with CF did not produce a significant change in the sinuses on magnetic resonance imaging (MRI), there was a significant clinical improvement as measured by a quality-of-life questionnaire (Mainz et al. 2011; Mainz et al. 2014).

Children with otitis media develop chronic neutrophil inflammation and an associated increase in NETs. rhDNase I has been proposed for the management of chronic otitis disease in patients not responding well to antibiotics, although no significant clinical trials have been conducted (Thornton et al. 2013).

The effect of rhDNase I has been studied in a small group of advanced head and neck cancer patients with thick, tenacious upper airway secretions while receiving chemoradiation therapy (Mittal et al. 2013). Preliminary results did not show a significant difference versus placebo in quality-of-life measures but did show improvement in secondary endpoints of oropharyngeal secretions and DNA concentrations.

Safety

The administration of rhDNase I has not been associated with an increase in any major adverse events. Most adverse events were not more common with rhDNase I than with placebo treatment and probably reflect complications related to the underlying lung disease. Most events associated with dosing were mild, transient in nature, and did not require alterations in dosing. Observed symptoms included hoarseness, pharyngitis, laryngitis, rash, chest pain, and conjunctivitis. Within all the studies, a small percentage (average 2–4%) of patients treated with rhDNase I developed serum antibodies to rhDNase I. None of these patients developed anaphylaxis, and the clinical significance of serum antibodies to rhDNase I is unknown. rhDNase I has also been associated with a slight increased risk of allergic bronchopulmonary aspergillosis in CF patients, although this most likely represents the chronic use of a wet nebulizer and not a complication of rhDNase I (Jubin et al. 2010).

Summary

DNase I, a secreted human enzyme whose normal function is thought to be for digestion of extracellular DNA, has been developed as a safe and effective adjunctive agent in the treatment of pulmonary disease in cystic fibrosis patients. rhDNase I reduces the viscoelasticity and improves the transport properties of viscous mucus both in vitro and in vivo. Inhalation of aerosolized rhDNase I reduces the risk of infections requiring antibiotics and improves pulmonary function and the well-being of CF patients with mild to moderate disease. Studies also suggest that rhDNase I has benefit in infants and young children with CF and in patients with early disease who may have "normal" lung function. Additional studies may assess the usefulness of rhDNase I in early-stage CF pulmonary disease and other diseases where extracellular DNA and NETs may play a pathological role.

Self-Assessment Questions

Questions

1. Mutations of the CF gene result in abnormalities of the CF transmembrane conductance regulator protein. How do abnormalities of this protein lead to eventual lung damage?
2. How does rhDNase I result in improved pulmonary function in patients with cystic fibrosis?
3. rhDNase I is strongly recommended by the US Cystic Fibrosis Foundation for use in CF patients with moderate and severe lung disease. What other CF patients may benefit from using this therapy?
4. In addition to improving lung function in CF patients, what benefits have been demonstrated in clinical trials of rhDNase I?
5. What other medical conditions may benefit from treatment with rhDNase I?
6. Why should ultrasonic nebulizers not be used to administer rhDNase I? What other types of nebulizers might be effective for delivering rhDNase I?

Answers

1. Abnormal CFTR protein results in abnormal airway surface liquid, which leads to a vicious cycle of airway obstruction, infection, and inflammation. The chronic, excessive neutrophilic inflammation in the airway results in the release of neutrophil elastase, oxidants, and extracellular DNA. These substances result in worsening obstruction and progressive airway damage.
2. rhDNase I cleaves extracellular DNA in the airway of CF patients, resulting in improved airway clearance of secretions. Additionally, rhDNase I may have anti-inflammatory properties, resulting in a slower progression of lung damage and reduced rate of decline in lung function. This anti-inflammatory benefit is more likely present in patients with earlier, less severe lung disease.
3. rhDNase I is also recommended in CF patients over age 6 with mild lung disease. It has also been demonstrated as safe in younger patients, although efficacy has not been studied.
4. rhDNase I use decreases the frequency of respiratory exacerbations. rhDNase I also lessened shortness of breath, increased the general perception of well-being, and reduced the severity of other cystic fibrosis-related symptoms.
5. Although only approved by the FDA for treatment of lung disease in patients with CF, controlled trials have also shown efficacy for treating empyema (in combination with tPA) and sinus disease in patients with CF.
6. Ultrasonic nebulizers generate heat, which breaks down the protein in rhDNase I. Vibrating permeable membrane nebulizers and "smart" nebulizers do not damage the drug and may improve the effectiveness of rhDNase.

References

Ago H, Oda M, Takahashi M, Tsuge H, Ochi S, Katunuma N, Miyano M, Sakurai J (2006) Structural basis of the sphingomyelin phosphodiesterase activity in neutral sphingomyelinase from *Bacillus cereus*. J Biol Chem 281:16157–16167

Aitken ML, Burke W, McDonald G, Shak S, Montgomery AB, Smith A (1992) Recombinant human DNase inhalation in normal subjects and patients with cystic fibrosis. A phase 1 study. JAMA 267:1947–1951

Altunhan H, Annagür A, Pekcan S, Ors R, Koç H (2012) Comparing the efficacy of nebulizer recombinant human DNase and hypertonic saline as monotherapy and combined treatment in the treatment of persistent atelectasis in mechanically ventilated newborns. Pediatr Int 54:131–136

Amin R, Subbarao P, Lou W, Jabar A, Balkovec S, Jensen R, Kerrigan S, Gustafsson P, Ratjen F (2011) The effect of dornase alfa on ventilation in homogeneity in patients with cystic fibrosis. Eur Respir J 37:806–812

Andreeva A, Howorth D, Chandonia JM, Brenner SE, Hubbard TJ, Chothia C, Murzin AG (2008) Data growth and its impact on the SCOP database: new developments. Nucleic Acids Res 36:D419–D425

Apel F, Zychlinsky A, Kenny EF (2018) The role of neutrophil extracellular traps in rheumatic diseases. Nat Rev Rheumatol. 14:467–475

Armstrong JB, White JC (1950) Liquefaction of viscous purulent exudates by deoxyribonuclease. Lancet 2:739–742

Bakker EM, Volpi S, Salonini E, van der Wiel-Kooij EC, Sintnicolaas CJJCM, Hop WCJ, Assael BM, Merkus PJFM, Tiddens HAWM (2011) Improved treatment response to dornase alfa in cystic fibrosis patients using controlled inhalation. Eur Respir J 38:1328–1335

Baranovskii AG, Buneva VN, Nevinsky GA (2004) Human deoxyribonucleases. Biochemistry (Mosc) 69:587–601

Bataillon V, Lhermitte M, Lafitte JJ, Pommery J, Roussel P (1992) The binding of amikacin to macromolecules from the sputum of patients suffering from respiratory diseases. J Antimicrob Chemother 29:499–508

Boogaard R, de Jongste JC, Merkus PJ (2007) Pharmacotherapy of impaired mucociliary clearance in non-CF pediatric lung disease. A review of the literature. Pediatr Pulmonol 42:989–1001

Boogaard R, Hulsmann AR, van Veen L, Vaessen-Verberne AA, Yap YN, Sprij AJ, Brinkhorst G, Sibbles B, Hendriks T, Feith SW, Lincke CR, Brandsma AE, Brand PL, Hop WC, de Hoog M, Merkus PJ (2007) Recombinant human deoxyribonuclease in infants with respiratory syncytial virus bronchiolitis. Chest 131:788–795

Brinkmann V (2018) Neutrophil extracellular traps in the second decade. J Innate Immun 10:414–421

Brinkmann V, Reichard U, Goosmann C, Fauler B, Uhlemann Y, Weiss DS, Weinrauch Y, Zychlinsky A (2004) Neutrophil extracellular traps kill bacteria. Science 303:1532–1535

Campbell VW, Jackson DA (1980) The effect of divalent cations on the mode of action of DNase I. the initial reaction products produced from covalently closed circular DNA. J Biol Chem 255:3726–3735

Carpenter EP, Corbett A, Thomson H, Adacha J, Jensen K, Bergeron J, Kasampalidis I, Exley R, Winterbotham M, Tang C, Baldwin GS, Freemont P (2007) AP endonuclease paralogues with distinct activities in DNA repair and bacterial pathogenesis. EMBO J 26:1363–1372

Chen WJ, Liao TH (2006) Structure and function of bovine pancreatic deoxyribonuclease I. Protein Pept Lett 13:447–453

Chernick WS, Barbero GJ, Eichel HJ (1961) In vitro evaluation of effect of enzymes on tracheobronchial secretions from patients with cystic fibrosis. Pediatrics 27:589–596

Cimmino M, Nardone M, Cavaliere M, Plantulli A, Sepe A, Esposito V, Mazzarella G, Raia V (2005) Dornase alfa as postoperative therapy in cystic fibrosis sinonasal disease. Arch Otolaryngol Head Neck Surg 131:1097–1101

Cipolla D, Gonda I, Shire SJ (1994) Characterization of aerosols of human recombinant deoxyribonuclease I (rhDNase) generated by jet nebulizers. Pharm Res 11:491–498

Cools-Lartigue J, Spicer J, Najmeh S, Ferri L (2014) Neutrophil extracellular traps in cancer progression. Cell Mol Life Sci 71:4179–4194

Corcoran JP, Wrightson JM, Belcher E, DeCamp MM, Feller-Kopman D, Rahman NM (2015) Pleural infection: past, present, and future directions. Lancet Respir Med 3:563–577

Dasgupta B, King M (1996) Reduction in viscoelasticity in cystic fibrosis sputum in vitro using combined treatment with nacystelyn and rhDNase. Pediatr Pulmonol 22:161–166

Davis JC Jr, Manzi S, Yarboro C, Rairie J, McInnes I, Averthelyi D, Sinicropi D, Hale VG, Balow J, Austin H, Boumpas DT, Klippel JH (1999) Recombinant human DNase I (rhDNase) in patients with lupus nephritis. Lupus 8:68–76

Davoodian K, Ritchings BW, Ramphal R, Bubb MR (1997) Gelsolin activates DNase I in vitro and cystic fibrosis sputum. Biochemistry 36:9637–9641

Demers M, Wagner DD (2014) NETosis: a new factor in tumor progression and cancer-associated thrombosis. Semin Thromb Hemost 40:277–283

Desai M, Weller PH, Spencer DA (1995) Clinical benefit from nebulized human recombinant DNase in Kartagener's syndrome. Pediatr Pulmonol 20:307–308

Dilmen U, Karagol BS, Oguz SS (2011) Nebulized hypertonic saline and recombinant human DNase in the treatment of pulmonary atelectasis in newborns. Pediatr Int 53:328–331

Dlakic M (2000) Functionally unrelated signalling proteins contain a fold similar to Mg^{2+}-dependent endonucleases. Trends Biochem Sci 25:272–273

Eisenberg JD, Aitken ML, Dorkin HL, Harwood IR, Ramsey BW, Schidlow DV, Wilmott RW, Wohl ME, Fuchs HJ, Christiansen DH, Smith AL (1997) Safety of repeated intermittent courses of aerosolized recombinant human deoxyribonuclease in patients with cystic fibrosis. J Pediatr 131:118–124

El Abiad NM, Clifton S, Nasr SZ (2007) Long-term use of nebulized human recombinant DNase I in two siblings with primary ciliary dyskinesia. Respir Med 101:2224–2226

Elkins MR, Robinson M, Rose BR, Harbour C, Moriarty CP, Marks GB, Belousova EG, Xuan W, Bye PT (2006) A controlled trial of long-term inhaled hypertonic saline in patients with cystic fibrosis. N Engl J Med 354:229–240

Evans CJ, Aguilera RJ (2003) DNase II: genes, enzymes and function. Gene 322:1–15

Fedakar A, Aydogdu C, Fedakar A, Ugurlucan M, Bolu S, Iskender M (2012) Safety of recombinant human deoxyribonuclease as a rescue treatment for persistent atelectasis in newborns. Ann Saudi Med 32:131–136

Flume PA, O'Sullivan BP, Robinson KA, Goss CH, Mogayzel PJ Jr, Willey-Courand DB, Bujan J, Finder J, Lester M, Quittell L, Rosenblatt R, Vender RL, Hazle L, Sabadosa K, Marshall B (2007) Cystic fibrosis pulmonary guidelines: chronic medications for maintenance of lung health. Am J Respir Crit Care Med 176:957–969

Fuchs HJ, Borowitz DS, Christiansen DH, Morris EM, Nash ML, Ramsey BW, Rosenstein BJ, Smith AL, Wohl ME (1994) Effect of aerosolized recombinant human DNase on exacerbations of respiratory symptoms and on pulmonary function in patients with cystic fibrosis. N Engl J Med 331:637–642

Geller DE, Eigen H, Fiel SB, Clark A, Lamarre AP, Johnson CA, Konstan MW (1998) Effect of smaller droplet size of dornase alfa on lung function in mild cystic fibrosis. Pediatr Pulmonol 25:83–87

Gilbert CR, Gorden JA (2017) Use of intrapleural tissue plasminogen activator and deoxyribonuclease in pleural space infections: an update on alternative regimens. Curr Opin Pulm Med 23:371–375

Gonda I (1990) Aerosols for delivery of therapeutic and diagnostic agents to the respiratory tract. Crit Rev Ther Drug Carrier Syst 6:273–313

Gray RD, McCullagh BN, McCray PB (2015) NETs and CF lung disease: current status and future prospects. Antibiotics 4:62–75

Green JD (1994) Pharmaco-toxicological expert report Pulmozyme rhDNase Genentech, Inc. Hum Exp Toxicol 13:S1–S42

Gueroult M, Picot D, Abi-Ghanem J, Hartmann B, Baaden M (2010) How cations can assist DNase I in DNA binding and hydrolysis. PLoS Comput Biol 6:e1001000

Guggino WB, Stanton BA (2006) New insights into cystic fibrosis: molecular switches that regulate CFTR. Nat Rev Mol Cell Biol 7:426–436

Gupta S, Kaplan MJ (2016) The role of neutrophils and NETosis in autoimmune and renal diseases. Nat Rev Nephrol 12:402–413

Hakkim A, Fürnrohr BG, Amann K, Laube B, Abed UA, Brinkmann V, Herrmann M, Voll RE, Zychlinsky A (2010) Impairment of neutrophil extracellular trap degradation is associated with lupus nephritis. Proc Natl Acad Sci USA 107:9813–9818

Hawes MC, Wen F, Elquza E (2015) Extracellular DNA: a bridge to cancer. Cancer Res 75:4260–4264

Hendriks T, de Hoog M, Lequin MH, Devos AS, Merkus PJ (2005) DNase and atelectasis in non-cystic fibrosis pediatric patients. Crit Care 9:R351–R356

Hitchcock SE, Carisson L, Lindberg U (1976) Depolymerization of F-actin by deoxyribonuclease I. Cell 7:531–542

Honda M, Kubes P (2018) Neutrophils and neutrophil extracellular traps in the liver and gastrointestinal system. Nat Rev Gastroenterol Hepatol 15:206–221

Horton NC (2008) DNA nucleases. In: Rice PA, Correll CC (eds) Protein-nucleic acid interactions: structural biology. Royal Society of Chemistry Publishing, Cambridge, MA, pp 333–363

Hubbard RC, McElvaney NG, Birrer P, Shak S, Robinson WW, Jolley C, Wu M, Chernick MS, Crystal RG (1992) A preliminary study of aerosolized recombinant human deoxyribonuclease I in the treatment of cystic fibrosis. N Engl J Med 326:812–815

Jiménez-Alcázar M, Kim N, Fuchs TA (2017) Circulating extracellular DNA: cause or consequence of thrombosis? Semin Thromb Hemost 243:553–561

Jiménez-Alcázar M, Rangaswamy C, Panda R, Bitterling J, Simsek YJ, Long AT, Bilyy R, Krenn V, Renné C, Renné T, Kluge S, Panzer U, Mizuta R, Mannherz HG, Kitamura D, Herrmann M, Napirei M, Fuchs TA (2017) Host DNases prevent vascular occlusion by neutrophil extracellular traps. Science 358:1202–1206

Jorch SK, Kubes P (2017) An emerging role for neutrophil extracellular traps in noninfectious disease. Nature Med 23:279–287

Jubin V, Ranque S, Le Bel NS, Sarles J, Dubus J-C (2010) Risk factors for Aspergillus colonization and allergic bronchopulmonary aspergillosis in children with cystic fibrosis. Pediatr Pulmonol 45:764–771

Kabsch W, Mannherz HG, Suck D, Pai EF, Holmes KC (1990) Atomic structure of the actin: DNase I complex. Nature 347:37–44

Kaplan JB, LoVetri K, Cardona ST, Madhyastha S, Sadovskaya I, Jabbouri S, Izano EA (2012) Recombinant human DNase I decreases biofilm and increases antimicrobial susceptibility in Staphylococci. J Antibiot (Tokyo) 65:73–77

Kerem B, Rommens JM, Buchanan JA, Markiewicz D, Cox TK, Chakravarti A, Buchwald M, Tsui LC (1989) Identification of the cystic fibrosis gene: genetic analysis. Science 245:1073–1080

Keyel PA (2017) Dnases in health and disease. Dev Biol 429:1–11

King M, Dasgupta B, Tomkiewicz RP, Brown NE (1997) Rheology of cystic fibrosis sputum after in vitro treatment with hypertonic saline alone and in combination with recombinant human deoxyribonuclease I. Am J Respir Crit Care Med 156:173–177

Kominato Y, Ueki M, Iida R, Kawai Y, Nakajima T, Makita C, Itoi M, Tajima Y, Kishi K, Yasuda T (2006) Characterization of human deoxyribonuclease I gene (DNASE1) promoters reveals the utilization of two transcription-starting exons and the involvement of Sp1 in its transcriptional regulation. FEBS J 273:3094–3105

Konstan MW, Byard PJ, Hoppel CL, Davis PB (1995) Effect of high-dose ibuprofen in patients with cystic fibrosis. N Engl J Med 332:848–854

Konstan MW, Ratjen F (2012) Effect of dornase alfa on inflammation and lung function: potential role in the early treatment of cystic fibrosis. J Cyst Fibros 11:78–83

Konstan MW, Schluchter MD, Xue W, Davis PB (2007) Clinical use of ibuprofen is associated with slower FEV$_1$ decline in children with cystic fibrosis. Am J Respir Crit Care Med 176:1084–1089

Konstan MW, VanDevanter DR, Rasouliyan L, Pasta DJ, Yegin A, Morgan WJ, Wagener JS (2010) Trends in the use of routine therapies in cystic fibrosis: 1995–2005. Pediatr Pulmonol 45:1167–1172

Konstan MW, Wagener JS, Pasta DJ, Millar SJ, Jacobs JR, Yegin A, Morgan WJ (2011) Clinical use of dornase alfa is associated with a slower rate of FEV$_1$ decline in cystic fibrosis. Pediatr Pulmonol 46:545–553

Lachmann PJ (2003) Lupus and desoxyribonuclease. Lupus 12:202–206

Laskowski M Sr (1971) Deoxyribonuclease I. In: Boyer PD (ed) The enzymes, vol 4, 3rd edn. Academic, New York, pp 289–311

Laube BL, Auci RM, Shields DE, Christiansen DH, Lucas MK, Fuchs HJ, Rosenstein BJ (1996) Effect of rhDNase on airflow obstruction and mucociliary clearance in cystic fibrosis. Am J Respir Crit Care Med 153:752–760

Law SM, Gray RD (2017) Neutrophil extracellular traps and the dysfunctional innate immune response of cystic fibrosis lung disease: a review. J Inflamm 14:29

Lazarides E, Lindberg U (1974) Actin is the naturally occurring inhibitor of deoxyribonuclease I. Proc Natl Acad Sci U S A 71:4742–4746

Lazarus RA (2002) Human deoxyribonucleases. In: Creighton TE (ed) Wiley encyclopedia of molecular medicine. Wiley, New York, pp 1025–1028

Lieberman J (1962) Enzymatic dissolution of pulmonary secretions. An in vitro study of sputum from patients with cystic fibrosis of pancreas. Am J Dis Child 104:342–348

Lieberman J (1968) Dornase aerosol effect on sputum viscosity in cases of cystic fibrosis. JAMA 205:312–313

Lood C, Blanco LP, Purmalek MM, Carmona-Rivera C, De Ravin SS, Smith CK, Malech HL, Ledbetter JA, Elkon KB, Kaplan MJ (2016) Neutrophil extracellular traps enriched in oxidized mitochondrial DNA are interferogenic and contribute to lupus-like disease. Nat Med 22:146–153

MacKinnon R, Wheeler KI, Sokol J (2011) Endotracheal DNase for atelectasis in ventilated neonates. J Perinatol 31:799–801

Mainz JG, Schien C, Schiller I, Schädlich K, Koitschev A, Koitschev C, Riethmüller J, Graepler-Mainka U, Wiedemann B, Beck JF (2014) Sinonasal inhalation of dornase alfa administered by vibrating aerosol to cystic fibrosis patients: a double-blind placebo-controlled cross-over trial. J Cyst Fibros 13:461–470

Mainz JG, Schiller I, Ritschel C, Mentzel H-J, Riethmuller J, Koitschev A, Schneider G, Beck JF, Wiedemann B (2011) Sinonasal inhalation of dornase alfa in CF: a double-blind placebo-controlled cross-over pilot trial. Auris Nasus Larynx 38:220–227

Martinod K, Wagner DD (2014) Thrombosis: tangled up in NETs. Blood 18:2768–2776

Matthews LW, Specter S, Lemm J, Potter JL (1963) The over-all chemical composition of pulmonary secretions from patients with cystic fibrosis, bronchiectasis and laryngectomy. Am Rev Respir Dis 88:119–204

McCoy K, Hamilton S, Johnson C (1996) Effects of 12-week administration of dornase alfa in patients with advanced cystic fibrosis lung disease. Chest 110:889–895

Merza M, Hartman H, Rahman M, Hwaiz R, Zhang E, Renström E, Luo L, Mörgelin M, Regner S, Thorlacius H (2015) Neutrophil extracellular traps induce trypsin activation, inflammation, and tissue damage in mice with severe acute pancreatitis. Gastroenterology 149:1920–1931

Mittal BB, Wang E, Sejpal S, Agulnik M, Mittal A, Harris K (2013) Effect of recombinant human deoxyribonuclease on oropharyngeal secretions in patients with head-and-neck cancers treated with radiochemotherapy. Int J Radiation Oncol Biol Phys 87:282–289

Mohler M, Cook J, Lewis D, Moore J, Sinicropi D, Championsmith A, Ferraiolo B, Mordenti J (1993) Altered pharmacokinetics of recombinant human deoxyribonuclease in rats due to the presence of a binding protein. Drug Metab Dispos 21:71–75

Mol CD, Izumi T, Mitra S, Tainer JA (2000) DNA-bound structures and mutants reveal abasic DNA binding by APE1 and DNA repair coordination. Nature 403:451–456

Moore S (1981) Pancreatic DNase. In: Boyer PD (ed) The enzymes, vol 14, 3rd edn. Academic, New York, pp 281–296

Morgan WJ, Butler SM, Johnson CA, Colin AA, FitzSimmons SC, Geller DE, Konstan MW, Light MJ, Rabin HR, Regelmann WE, Schidlow DV, Stokes DC, Wohl ME, Kaplowitz H, Wyatt MM, Stryker S (1999) Epidemiologic study of cystic fibrosis: design and implementation of a prospective, multicenter, observational study of patients with cystic fibrosis in the U.S. and Canada. Pediatr Pulmonol 28:231–241

Nasr SZ, Strouse PJ, Soskolne E, Maxvold NJ, Garver KA, Rubin BK, Moler FW (2001) Efficacy of recombinant human deoxyribonuclease I in the hospital management of respiratory syncytial virus bronchiolitis. Chest 120:203–208

O'Donnell AE, Barker AF, Ilowite JS, Fick RB (1998) Treatment of idiopathic bronchiectasis with aerosolized recombinant human DNase I. rhDNase study group. Chest 113:1329–1334

Pan CQ, Dodge TH, Baker DL, Prince WS, Sinicropi DV, Lazarus RA (1998) Improved potency of hyperactive and actin-resistant human DNase I variants for treatment of cystic fibrosis and systemic lupus erythematosus. J Biol Chem 273:18374–18381

Pan CQ, Lazarus RA (1997) Engineering hyperactive variants of human deoxyribonuclease I by altering its functional mechanism. Biochemistry 36:6624–6632

Pan CQ, Lazarus RA (1999) Ca^{2+}–dependent activity of human DNase I and its hyperactive variants. Protein Sci 8:1780–1788

Pan CQ, Sinicropi DV, Lazarus RA (2001) Engineered properties and assays for human DNase I mutants. Methods Mol Biol 160:309–321

Pan CQ, Ulmer JS, Herzka A, Lazarus RA (1998) Mutational analysis of human DNase I at the DNA binding interface: implications for DNA recognition, catalysis, and metal ion dependence. Protein Sci 7:628–636

Park J, Wysocki RW, Amoozgar Z, Maiorino L, Fein MR, Jorns J, Schott AF, Kinugasa-Katayama Y, Lee Y, Won NH, Nakasone ES, Hearn SA, Küttner V, Qiu J, Almeida AS, Perurena N, Kessenbrock K, Goldberg MS, Egeblad M (2016) Cancer cells induce metastasis-supporting neutrophil extracellular DNA traps. Sci Transl Med 8:361ra138

Parsiegla G, Noguere C, Santell L, Lazarus RA, Bourne Y (2012) The structure of human DNase I bound to magnesium and phosphate ions points to a catalytic mechanism common to members of the DNase I-like superfamily. Biochemistry 51:10250–10258

Paul K, Rietschel E, Ballmann M, Griese M, Worlitzsch D, Shute J, Chen C, Schink T, Döring G, van Koningsbruggen S, Wahn U, Ratjen F (2004) Effect of treatment with dornase alpha on airway inflammation in patients with cystic fibrosis. Am J Respir Crit Care Med 169:719–725

Piccolo F, Popowicz N, Wong D, Lee YC (2015) Intrapleural tissue plasminogen activator and deoxyribonuclease therapy for pleural infection. J Thorac Dis 7:999–1008

Potter JL, Specter S, Matthews LW, Lemm J (1969) Studies on pulmonary secretions. 3. The nucleic acids in whole pulmonary secretions from patients with cystic fibrosis bronchiectasis and laryngectomy. Am Rev Respir Dis 99:909–915

Prince WS, Baker DL, Dodge AH, Ahmed AE, Chestnut RW, Sinicropi DV (1998) Pharmacodynamics of recombinant human DNase I in serum. Clin Exp Immunol 113:289–296

Puterman AS, Weinberg EG (1997) rhDNase in acute asthma. Pediatr Pulmonol 23:316–317

Quan JM, Tiddens HA, Sy JP, McKenzie SG, Montgomery MD, Robinson PJ, Wohl ME, Konstan MW (2001) A two-year randomized, placebo-controlled trial of dornase alfa in young patients with cystic fibrosis with mild lung function abnormalities. J Pediatr 139:813–820

Rahman NM, Maskell NA, West A, Teoh R, Arnold A, Mackinlay C, Peckham D, Davies CW, Ali N, Kinnear W, Bentley A, Kahan BC, Wrightson JM, Davies HE, Hooper CE, Lee YC, Hedley EL, Crosthwaite N, Choo L, Helm EJ, Gleeson FV, Nunn AJ, Davies RJ (2011) Intrapleural use of tissue plasminogen activator and DNase in pleural infection. N Engl J Med 365:518–526

Ramphal R, Lhermitte M, Filliat M, Roussel P (1988) The binding of anti-pseudomonal antibiotics to macromolecules from cystic fibrosis sputum. J Antimicrob Chemother 22:483–490

Ramsey BW, Astley SJ, Aitken ML, Burke W, Colin AA, Dorkin HL, Eisenberg JD, Gibson RL, Harwood IR, Schidlow DV, Wilmott RW, Wohl ME, Meyerson LJ, Shak S, Fuchs H, Smith AL (1993) Efficacy and safety of short-term administration of aerosolized recombinant human deoxyribonuclease in patients with cystic fibrosis. Am Rev Respir Dis 148:145–151

Ramsey BW, Davies J, McElvaney NG, Tullis E, Bell SC, Dřevínek P, Griese M, McKone EF, Wainwright CE, Konstan MW, Moss R, Ratjen F, Sermet-Gaudelus I, Rowe SM, Dong Q, Rodriguez S, Yen K, Ordoñez C, Elborn JS, VX08-770-102 Study Group (2011) A CFTR potentiator in patients with cystic fibrosis and the G551D mutation. New Engl J Med 365:1663–1672

Ramsey BW, Pepe MS, Quan JM, Otto KL, Montgomery AB, Williams-Warren J, Vasiljev KM, Borowitz D, Bowman CM, Marshall BC, Marshall S, Smith AL (1999) Intermittent administration of inhaled tobramycin in patients with cystic fibrosis. N Engl J Med 340:23–30

Raskin P (1968) Bronchospasm after inhalation of pancreatic dornase. Am Rev Respir Dis 98:697–698

Raynor EM, Butler A, Guill M, Bent JP 3rd (2000) Nasally inhaled dornase alfa in the postoperative management of chronic sinusitis due to cystic fibrosis. Arch Otolaryngol Head Neck Surg 126:581–583

Ren CL, Pasta DJ, Rasouliyan L, Wagener JS, Konstan MW, Morgan WJ (2008) Relationship between inhaled corticosteroid therapy and rate of lung function decline in children with cystic fibrosis. J Pediatr 153:746–751

Riethmueller J, Borth-Bruhns T, Kumpf M, Vonthein R, Wiskirchen J, Stern M, Hofbeck M, Baden W (2006) Recombinant human deoxyribonuclease shortens ventilation time in young, mechanically ventilated children. Pediatr Pulmonol 41:61–66

Riordan JR, Rommens JM, Kerem B, Alon N, Rozmahel R, Grzelczak Z, Zielenski J, Lok S, Plavsic N, Chou JL, Drumm ML, Iannuzzi MC, Collins FS, Tsui LC (1989) Identification of the cystic fibrosis gene: cloning and characterization of complementary DNA. Science 245:1066–1073

Robinson M, Hemming AL, Moriarty C, Eberl S, Bye PT (2000) Effect of a short course of rhDNase on cough and mucociliary clearance in patients with cystic fibrosis. Pediatr Pulmonol 30:16–24

Rozov T, de Oliveira VZ, Santana MA, Adde FV, Mendes RH, Paschoal IA, Reis FJC, Higa LYS, de Castro Toledo AC Jr, Pahl M (2010) Dornase alfa improves the health-related quality of life among Brazilian patients with cystic fibrosis – a one-year prospective study. Pediatr Pulmonol 45:874–882

Saiman L, Mayer-Hamblett N, Campbell P, Marshall BC (2005) Heterogeneity of treatment response to azithromycin in patients with cystic fibrosis. Am J Respir Crit Care Med 172:1008–1012

Sanders NN, Franckx H, De Boeck K, Haustraete J, De Smedt SC, Demeester J (2006) Role of magnesium in the failure of rhDNase therapy in patients with cystic fibrosis. Thorax 61:962–968

Sawicki GS, Chou W, Raimundo K, Trzaskoma B, Konstan MW (2015) Randomized trial of efficacy and safety of dornase alfa delivered by eRapid nebulizer in cystic fibrosis patients. J Cyst Fibros 14:777–783

Sawicki GS, Signorovitch JE, Zhang J, Latremouille-Viau D, von Wartburg M, Wu EQ, Shi L (2012) Reduced mortality in cystic fibrosis patients treated with tobramycin inhalation solution. Pediatr Pulmonol 47:44–52

Scala M, Hoy D, Bautista M, Palafoutas JJ, Abubakar K (2017) Pilot study of dornase alfa (Pulmozyme) therapy for acquired ventilator-associated infection in preterm infants. Pediatr Pulmonol 52:787–791

Scherer T, Geller DE, Owyang L, Tservistas M, Keller M, Boden N, Kesser KC, Shire SJ (2011) A technical feasibility study of dornase alfa delivery with eFlow vibrating membrane nebulizers: aerosol characteristics and physicochemical stability. J Pharm Sci 100:98–109

Shah PI, Bush A, Canny GJ, Colin AA, Fuchs HJ, Geddes DM, Johnson CA, Light MC, Scott SF, Tullis DE, De Vault A, Wohl ME, Hodson ME (1995) Recombinant human DNase I in cystic fibrosis patients with severe pulmonary disease: a short-term, double-blind study followed by six months open-label treatment. Eur Respir J 8:954–958

Shak S (1995) Aerosolized recombinant human DNase I for the treatment of cystic fibrosis. Chest 107:65S–70S

Shak S, Capon DJ, Hellmiss R, Marsters SA, Baker CL (1990) Recombinant human DNase I reduces the viscosity of cystic fibrosis sputum. Proc Natl Acad Sci USA 87:9188–9192

Shiokawa D, Tanuma S (2001) Characterization of human DNase I family endonucleases and activation of DNase gamma during apoptosis. Biochemistry 40:143–152

Shire SJ (1996) Stability characterization and formulation development of recombinant human deoxyribonuclease I [Pulmozyme, (dornase alfa)]. In: Pearlman R, Wang YJ (eds) Pharmaceutical biotechnology: formulation, characterization and stability of protein drugs, vol 9. Plenum Press, New York, pp 393–426

Silverman RA, Foley F, Dalipi R, Kline M, Lesser M (2012) The use of rhDNAse in severely ill, non-intubated adult asthmatics refractory to bronchodilators: a pilot study. Respir Med 106:1096–1102

Simpson G, Roomes D, Reeves B (2003) Successful treatment of empyema thoracis with human recombinant deoxyribonuclease. Thorax 58:365–366

Sinicropi D, Baker DL, Prince WS, Shiffer K, Shak S (1994) Colorimetric determination of DNase I activity with a DNA-methyl green substrate. Anal Biochem 222:351–358

Sinicropi DV, Lazarus RA (2001) Assays for human DNase I activity in biological matrices. Methods Mol Biol 160:325–333

Sinicropi DV, Prince WS, Lofgren JA, Williams M, Lucas M, DeVault A (1994) Sputum pharmacodynamics and pharmacokinetics of recombinant human DNase I in cystic fibrosis. Am J Respir Crit Care Med 149:A671

Suck D (1994) DNA recognition by DNase I. J Mol Recognit 7:65–70

Suri R (2005) The use of human deoxyribonuclease (rhDNase) in the management of cystic fibrosis. BioDrugs 19:135–144

ten Berge M, Brinkhorst G, Kroon AA, de Jongste JC (1999) DNase treatment in primary ciliary dyskinesia – assessment by nocturnal pulse oximetry. Pediatr Pulmonol 27:59–61

Thornton RB, Wiertsema SP, Kirkham LS, Rigby PJ, Vijayasekaran S, Coates HL, Richmond PC (2013) Neutrophil extracellular traps and bacterial biofilms in middle ear effusion of children with recurrent acute otitis media – a potential treatment target. PLoS One 8:e53837

Ulmer JS, Herzka A, Toy KJ, Baker DL, Dodge AH, Sinicropi D, Shak S, Lazarus RA (1996) Engineering actin-resistant human DNase I for treatment of cystic fibrosis. Proc Natl Acad Sci USA 93:8225–8229

VanDevanter DR, Craib ML, Pasta DJ, Millar SJ, Morgan WJ, Konstan MW (2018) Cystic fibrosis clinical characteristics associated with dornase alfa treatment regimen change. Pediatr Pulmonol 53:43–49

Vasconcellos CA, Allen PG, Wohl ME, Drazen JM, Janmey PA, Stossel TP (1994) Reduction in viscosity of cystic fibrosis sputum in vitro by gelsolin. Science 263:969–971

Wagener JS, Kupfer O (2012) Dornase alfa (Pulmozyme). Curr Opin Pulm Med 18:609–614

Wagener JS, Rock MJ, McCubbin MM, Hamilton SD, Johnson CA, Ahrens RC (1998) Aerosol delivery and safety of recombinant human deoxyribonuclease in young children with cystic fibrosis: a bronchoscopic study. J Pediatr 133:486–491

Wang H, Morita M, Yang X, Suzuki T, Yang W, Wang J, Ito K, Wang Q, Zhao C, Bartlam M, Yamamoto T, Rao Z (2010) Crystal structure of the human CNOT6L nuclease domain reveals strict poly(a) substrate specificity. EMBO J 29:2566–2576

Widlak P, Garrard WT (2005) Discovery, regulation, and action of the major apoptotic nucleases DFF40/CAD and endonuclease G. J Cell Biochem 94:1078–1087

Wills PJ, Wodehouse T, Corkery K, Mallon K, Wilson R, Cole PJ (1996) Short-term recombinant human DNase in bronchiectasis. Effect on clinical state and in vitro sputum transportability. Am J Respir Crit Care Med 154:413–417

Wilmott RW, Amin RS, Colin AA, DeVault A, Dozor AJ, Eigen H, Johnson C, Lester LA, McCoy K, McKean LP, Moss R, Nash ML, Jue CP, Regelmann W, Stokes DC, Fuchs HJ (1996) Aerosolized recombinant human DNase in hospitalized cystic fibrosis patients with acute pulmonary exacerbations. Am J Respir Crit Care Med 153:1914–1917

Wolf E, Frenz J, Suck D (1995) Structure of human pancreatic DNase I at 2.2 Å resolution. Protein Eng 8:79

Wong SL, Demers M, Martinod K, Gallant M, Wang Y, Goldfine AB, Kahn CR, Wagner DD (2015) Diabetes primes neutrophils to undergo NETosis, which impairs wound healing. Nat Med 21:815–819

Yang W (2011) Nucleases: diversity of structure, function and mechanism. Q Rev Biophys 44:1–93

Zahm JM, Girod de Bentzmann S, Deneuville E, Perrot-Minnot C, Dabadie A, Pennaforte F, Roussey M, Shak S, Puchelle E (1995) Dose-dependent in vitro effect of recombinant human DNase on rheological and transport properties of cystic fibrosis respiratory mucus. Eur Respir J 8:381–386

Interferons and Interleukins

22

Adapted for the Sixth Edition
by Daan J. A. Crommelin

Jean-Charles Ryff, Sidney Pestka,
and Daan J. A. Crommelin

Introduction

Endogenous interferons and interleukins, member of the cytokine family, are major protagonists of our defense system; they act in a concerted way within the immune system to defend against, contain, or eliminate toxic or invasive agents. Because of their toxic effects, several containment mechanisms such as short half-life, downregulation, and neutralization factors ensure that their action is strictly localized. Various diseases are accompanied by acute systemic or chronic-relapsing symptoms of inflammation. These are a sign of containment failure, i.e., dysregulation of interleukin production. During the past few years, our understanding of the cellular and molecular mechanisms of immune regulation in allergy, chronic inflammatory and autoimmune diseases, tumor development, and chronic infections has grown at a rapid pace. A better insight into the functions, reciprocal regulation, and antagonism of interleukins, interferons, and other cytokines offers opportunities for novel treatment approaches. Because of their toxicity, systemically administered interferons and interleukins have a number of unwanted effects. This is the reason why only a handful of interferons and even fewer interleukins were approved by regulatory authorities. The main effort in research for their therapeutic use is therefore directed toward mitigating their toxic effects

---

Sidney Pestka was deceased at the time of publication.

J.-C. Ryff
BRAIN, Biotech Research and Innovation Network,
Basel, Switzerland

S. Pestka (Deceased)
Department of Molecular Genetics, Microbiology, and
Immunology, Robert Wood Johnson Medical School, University of
Medicine and Dentistry of New Jersey, Piscataway, NJ, USA

D. J. A. Crommelin (✉)
Department of Pharmaceutics, Utrecht Institute for Pharmaceutical
Sciences (UIPS), Utrecht University, Utrecht, The Netherlands
e-mail: d.j.a.crommelin@uu.nl

and improving their pharmacokinetic properties. Protein engineering of interferons or interleukins has led to improved pharmacological properties with subsequent reduction of toxicity. Other research efforts have resulted in engineered fusion protein molecules with target specificity as well as an improved pharmacologic profile.

The therapeutic success of cytokine biologicals may be limited. However, therapy targeted at modifying the activity of members of the cytokine family e.g., by using mAbs against interleukin (IL)-1beta, IL-4, IL-5, IL-6, IL-12/23, IL-17A, or IL-6 or their receptors (cf. Chap. 24), is highly successful. Enough reason to pay attention to the general structures, overall mechanisms of action, activity, and safety issues of interferons and interleukins.

In 1957, Alick Isaacs and Jean Lindenmann described a substance that was produced by virus-infected cell cultures and "interfered" with infection by other viruses; it was called interferon. Over the following decades, it was found that "interferon" comprises a family of related proteins with several additional properties. Starting in the 1960s, various "factors" produced primarily by white blood cell (WBC) as well as other cell supernatants were described, which acted in various ways on other WBCs or somatic cells. They were usually given a descriptive name either associated with their cell of origin or their activity on other cells, resulting in a myriad of names. The application of molecular biological techniques allowed us to determine that some of these "cytokines" had multiple activities and that different cytokines had similar overlapping activities. A classification system based on genetic structure and protein characterization is being used. The interactive networks and cascades of cytokines, i.e., interferons (IFN), interleukins (IL), growth factors (GF), chemokines (CK), their receptors (r or R), and signaling pathways, are highly complex and will be further explored in this chapter.

Cytokine is a term coined in 1974 by Stanley Cohen in an attempt to develop a more systematic approach to the numerous regulatory proteins secreted by hematopoietic and non-hematopoietic cells. Cytokines play a critical role in modulating the innate and adaptive immune systems. They are multifunctional peptides that are now known to be produced by normal and neoplastic cells, as well as by cells of the immune system. These local messengers and signaling molecules are involved in the development of the immune system, cell growth and differentiation, repair mechanisms, and the inflammatory cascade and are categorized in four protein families:

(a) **Interferons**: Proteins produced by eukaryotic cells in response to viral infections, tumors, and other biological inducers. They promote an antiviral state in other, neighboring cells and also help to regulate the immune response. They exhibit a variety of activities and represent a wide family of proteins.

(b) **Interleukins**: A group of cytokines mainly secreted by leukocytes and primarily affecting growth and differentiation of hematopoietic and immune cells. They are also produced by other normal and malignant cells and are of central importance in the regulation of hematopoiesis, immunity, inflammation, tissue remodeling, and embryonic development. Interleukins can be excreted by T-helper cells type 1 (Th1; pro-inflammatory), e.g., IL-2, IL-12, and IL-18, or T-helper cells type 2 (Th2; anti-inflammatory) stimulating, e.g., IL-4, IL-10, and IL-13. A third category of T-helper cells 17 (Th17) has been described, which is associated with autoimmunity and excretes IL-17A, IL-17F, and IL-22.

Thus, all interleukins are cytokines; however, not all cytokines are interleukins.

(c) **Growth factors**: Proteins that activate cellular proliferation and/or differentiation. Many growth factors stimulate cellular division in numerous different cell types; others are specific to a particular cell type. They also promote proliferation of connective tissue, glial, and smooth muscle cells, enhance normal wound healing, and promote proliferation and differentiation of erythrocytes (erythropoietin). Hematopoietic growth factors are reviewed in Chap. 17. Some ILs have a functional overlap with growth factors, e.g., IL-2, IL-3, and IL-11.

(d) **Chemokines** (**chemo**tactic cyto**kines**): A large family of structurally related low molecular weight proteins with potent leukocyte activation and/or chemotactic activity. "CXC" (or α) and "C-C" (or β) chemokine subsets are based on the presence or absence of an amino acid between the first two of four conserved cysteines. A third subset, "C," has only two cysteines, and to date, only one member, IL-16, has been identified. The fourth subgroup, the C-X3-C chemokine, has three amino acid residues between the first two cysteines (Hughes and Nibbs 2018).

(e) **Others**, such as tumor necrosis factors (TNF) lymphotoxin alpha (LT-α) and LT-β and transforming growth factor (TGF)-α and TGF-β.

All cytokines including interferons and interleukins act by binding to specific transmembrane receptors. In general, these receptors have two main components: a low-affinity ligand-binding domain that ensures ligand specificity and a high-affinity effector domain activating target gene promoters via an intracellular signaling pathway. Because cytokines can bind to their receptors only where these are expressed on the cell membrane, a functional tissue or cell specificity is ensured.

Cytokine signaling is tightly controlled within the cell through the action of multiple different negative regulators. Members of the suppressors of cytokine signaling (SOCS) family specifically interfere with cytokine signaling by several different mechanisms including direct binding and inhibition of Janus kinase (JAK) proteins, competition with JAK-activated signal transducer and activator of transcription (STAT) for binding sites on the cytokine receptor (see below), and activation of proteasomal degradation of signaling components (Croker et al. 2008; Durham et al. 2019).

Some basic pharmacological terms relevant to cytokines are (cf. Chap. 5, Table 5.8):

- **Autocrine**: If the cytokine acts on the cell that secretes it
- **Paracrine**: If the action is restricted to the immediate vicinity of a cytokine's secretion
- **Endocrine**: If the cytokine diffuses or is otherwise transported to distant regions of the body to affect different tissues

They can act on many targets, can act in concert, or can antagonize one another:

- **Synergy**: Action together to induce a different response than either can induce alone.
- **Antagonism**: Cytokines can counteract one another.
- **Pleiotropy**: Action in a similar way on more than one "target" cell.
- **Redundancy**: More than one cytokine triggers identical or similar responses in a given "target" cell.
- **Pathway activation**: Triggered sequential induction or "cascade."

Interferons: Nomenclature and Functions

Interferons are a family of naturally occurring proteins and glycoproteins with molecular weights of 16,776 to 22,093 Da produced and secreted by cells in response to viral infections and to synthetic or biological inducers. By interacting with their specific heterodimeric receptors on the surface of cells, the interferons initiate a broad and varied array of signals that induce cellular antiviral states, modulate inflammatory responses, inhibit or stimulate cell growth, produce or inhibit apoptosis, and modulate many components of the immune system. Structurally, they are part of the helical cytokine family (Fig. 22.1). Major research efforts have been undertaken to understand the signaling mechanisms through which these cytokines induce their effects. As a general example, Fig. 22.2 illustrates the JAK-STAT (Janus kinase, originally "just another kinase"—signal transducer and activator of transcription) pathway, the best characterized IFN signaling pathway. However, coordination and cooperation of multiple

Fig. 22.1 Class 1 helical cytokines. Class 1 helical cytokines fold into a bundle of four tightly packed α-helices. On the basis of their helix length, class 1 helical cytokines are characterized as (**a**) long chain, such as IL-6, or (**b**) short chain, such as IL-4. From Huising et al. (2006) with permission

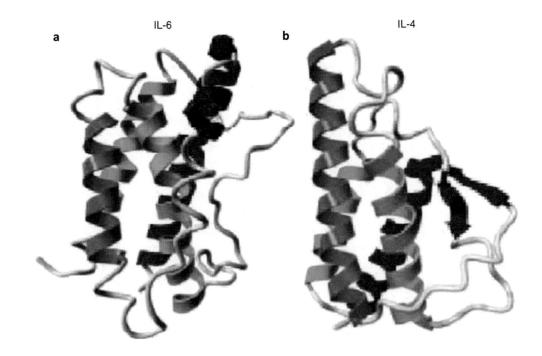

Fig. 22.2 Generic JAK-STAT signaling pathway mediated by most cytokine receptors

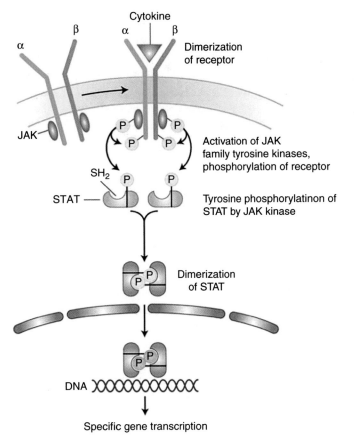

distinct signaling cascades, including the mitogen-activated protein kinase p38 cascade and the phosphatidylinositol 3-kinase cascade, are required for the generation of responses to interferons (Platanias 2005). Many of the symptoms of acute viral infections are the consequence of the high systemic IFNα response induced by the infecting viruses particularly during the viremic phase.

Human type I interferons comprise 13 different IFNα isoforms or subtypes with varying specificities, e.g., affinities to different cell types, and downstream activities. Although there are 13 human IFNα proteins, two of them (IFNα1 and IFNα13) are identical proteins so that the total number of type I IFNs often is listed as 12. There is also one subtype each for IFNβ (beta), IFNε (epsilon), IFNκ (kappa), and IFNω (omega). Their ability to establish an "antiviral state" is the distinctive fundamental property of type I IFNs. They are produced by most cells. However, certain types seem to be more selectively expressed, e.g., IFNκ by keratinocytes.

Type II IFN consists of a single representative: IFNγ (gamma). IFNγ or immune interferon plays an essential role in cell-mediated immune responses. It is produced by natural killer (NK) cells, dendritic cells, cytotoxic T cells, and progenitor Th0 cells and Th1 cells. IFNα2, IFNβ, and IFNγ are the most extensively studied to date. Overviews of the IFN-like cytokines can be found in reviews by Pestka et al. (2004) and Alexander et al. (2017).

Interleukins: Nomenclature and Functions

The first interleukins were identified in the 1970s. Initially, it was believed that interleukins were made chiefly by leukocytes to act primarily on other leukocytes. For this reason, they were named interleukins, meaning "between leukocytes." It was later realized that they are also produced by and interact with a host of cells not involved in immunity and are involved in many other physiological functions. The role that interleukins play in the body is much greater than was initially understood. They are, however, primarily a collection of immune cell growth, differentiation, and maturation factors. Collectively, they orchestrate a precise and efficient immune response to toxins and pathogens, including cancer cells, recognized as foreign. As is the case for IFNs, ILs bind to related specific cell surface receptors that activate similar intracellular signaling cascades. Many interleukins, mainly those with pro-inflammatory function, are intrinsically toxic either directly or indirectly, i.e., through induction of toxic gene products. Therefore, the human body has an elaborate system of checks and balances that, under (patho)physiological conditions, regulates the magnitude and duration of an immune response. ILs are produced upon appropriate stimulation and have a short circulation time. Their production is regulated by positive and negative feedback loops.

Furthermore, their effect is mostly localized, and in some cases, soluble receptors or neutralizing antibodies limit their dissemination. Specific receptor antagonists can also control their activity.

Overviews of the 40 interleukins identified today and their structure and function can be found in reviews by, e.g., Alexander et al. (2017), Vaillant and Qurie (2022), and data banks such as UniProt (UniProt 2021) and RCSB (Protein Data Bank 2021).

Under pathophysiological conditions, the sequential concentrations of agonistic and antagonistic interleukins establish a delicate balance in driving pro- and anti-inflammatory phases. This process can be disturbed by various pathogenic agents or mechanisms:

- Infectious agents or toxins
- Allergens
- Malignant tumors
- Genetic variants

These pathogenic agents or mechanisms can result in a self-limited or protracted disequilibrium. Symptoms of disease are the consequence of an adequate immune response at the end of which the steady state is reestablished. A brisk inflammatory response is the sign of a healthy immune reaction. In some instances, an inadequate response can manifest itself as relapsing remitting progressive disease, e.g., rheumatoid arthritis, asthma, psoriasis, chronic inflammatory bowel disease, multiple sclerosis, chronic hepatitis, or chronic insulitis leading to diabetes mellitus. All have in common that they need a genetic predisposition and an environmental trigger factor to become active and are, at best, only partially understood. In many cases, these diseases are caused by either insufficient production or overproduction of key interleukins. Thus, in principle, once the diagnosis is made, these interleukins can be therapeutically supplemented or suppressed to restore proper balance.

In spite of years of growing insights into their behavior and the role they play in many diseases, only three interleukin-based products were approved by regulatory authorities today, i.e., interleukin-2, IL-11, and anakinra (IL-1Ra, a human interleukin-1 receptor antagonist) (Table 22.2). Details are provided in the "Therapeutic Use of Recombinant Interleukins" section.

Therapeutic Use of Recombinant Interferons

IFNα Therapeutics

Together with recombinant human insulin and growth hormone, recombinant IFNα was one of the first recombinant DNA-derived approved pharmaceuticals that used the then new biotech technologies to produce large amounts of a well-defined, purified protein for large-scale therapeutic use. These techniques improved over time (Sidney 1986; Castro et al. 2021). Starting in the early 1980s, a number of cytokines produced by recombinant gene technology were developed. Table 22.1 summarizes the recombinant IFNs approved for therapeutic use.

Interferon alpha-2 (a modified generic name for IFNα2) was developed independently by Hoffmann-La Roche Ltd. (interferon alpha-2a; Roferon®A) and Schering-Plough Corporation (interferon alpha-2b; Intron®A). Infergen (interferon alphacon-1) from Amgen followed later. This was a synthetic "consensus" interferon consisting of 166 amino acids and not occurring in nature. The supply of all three abovementioned products was discontinued. Finally, interferon alpha-n3, Alferon N®, from Hemispherx Biopharma was approved in 1998; this product is still on the market (Table 22.1).

The adverse event profile for these IFNα products is the same. It is generally more or less well tolerated depending on the dose regimen used and consists primarily of the "influenza-like symptoms" named as such because they mimic the symptoms of early influenza. This, of course, should come as no surprise as these symptoms are similar to peaks of endogenous interferon stimulated by the influenza virus infection.

The half-life of the interferon alpha products in blood is short, typically 2–5 h. Given the principle that the toxicity of a given medication may be defined by its peak concentration and by the time it is above a toxic threshold concentration and the efficacy by the time the substance is above the minimal therapeutic level, it would be desirable to obtain a therapeutic regimen that minimizes fluctuations in the range below the toxic and above the therapeutic threshold concentration. A constant therapeutic drug concentration would be an ideal goal. The first step toward that goal, as a proof of concept, was to model a long-acting interferon using an insulin pump to inject patients with chronic hepatitis C with interferon α-2a at predetermined rates per hour for 28 days. A similar study was performed in patients with renal cell carcinoma. These studies indicated that interferon α-2a at a constant dose was indeed better tolerated while showing activity when administered by continuous subcutaneous (SC) infusion with a syringe pump (Carreño et al. 1992). The next step therefore was to develop a new longer acting molecule by attaching several polyethylene glycol (PEG) chains to the native interferon molecule (see the "PEGylation" section).

Pharmaceutical Formulations and Dosing for Interferon Alpha Therapeutics

Alferon N, the only non-modified-interferon alpha product left on the market, has a narrow indication: the following dosing scheme is used for the treatment of refractory or recurring external condylomata acuminata: 0.05 mL (250,000 international units) per wart, injected intralesionally two times a week for up to 8 weeks (Interferon-Alpha-n3 2023).

Table 22.1 Interferons approved as biopharmaceuticals approved in the United States and Europe

Recombinant interferons	Company	First indication	First approval
Interferon-α			
IFN-α2a produced in *E. coli;* Roferon A®	Hoffmann-La Roche (Basel, Switzerland)	Hairy cell leukemia, hepatitis C, discontinued	1986 (EU and USA)
IFN-α2b produced in *E. coli;* Intron® A; Viraferon®; Alfatronol®	Schering-Plough (Kenilworth, NJ, USA)	Hairy cell leukemia, hepatitis C, discontinued	1986 (USA and EU)
IFN-αcon1, synthetic type I IFN produced in *E. coli;* Infergen®	Amgen (Thousand Oaks, CA), Yamanouchi Europe (Leiderdorp, The Netherlands, EU)	Chronic hepatitis C, discontinued	1999 (EU), 2001 (USA)
IFNα-n3; Alferon N®	Hemispherx Biopharma (USA)	Genital warts	1989 (USA)
Interferon-β			
IFN-β1a produced in CHO cells; Rebif®	Serono (Geneva, Switzerland)	Relapsing/remitting multiple sclerosis	1998 (EU) 2002 (USA)
IFN-β1a produced in CHO cells; Avonex®	Biogen (Cambridge, MA, USA)	Relapsing/remitting multiple sclerosis	1997 (EU), 1996 (USA)
IFN-β1b Cys17 Ser substitution; produced in *E. coli;* Betaferon®	Schering AG (Berlin, Germany)	Relapsing/remitting multiple sclerosis	1995 (EU)
IFN-β1b, Cys17 Ser substitution; produced in *E. coli;* Betaseron®	Berlex Labs/Chiron (Richmond/Emeryville, CA, USA)	Relapsing/remitting multiple sclerosis	1993 (USA)
Interferon-γ			
Actimmune® (IFN-γ1b; produced in *E. coli*)	Genentech (San Francisco, CA, USA), InterMune (Palo Alto, CA, USA), Horizon Pharma Inc.	Chronic granulomatous disease	1999 (USA)

IFNβ Therapeutics

Three IFNβ products (Table 22.1) are marketed worldwide for the treatment of multiple sclerosis: the first was Berlex' Betaseron®, marketed by Schering AG as Betaferon® in Europe. It is interferon β-1b with 165 amino acids and an approximate molecular weight of 18,500 Da, with a cysteine-17-serine substitution. It is produced in *Escherichia. coli*, which was then the standard method. It is non-glycosylated, as without further engineering glycosylation is not possible in the *E. coli* system (see Chap. 4 "Production and Purification of Recombinant Proteins") (EMA Betaferon 2006). Independently, Biogen and Serono developed a glycosylated IFNβ-1a produced in Chinese hamster ovary cells. Thus, not only is the amino acid sequence of these IFNβs identical to that of natural fibroblast-derived human interferon beta, they are also glycosylated, each containing a single N-linked complex carbohydrate moiety. The two products are marketed as Avonex® and Rebif®, respectively. All three products are indicated for the treatment of multiple sclerosis.

Glycosylating proteins fundamentally alters their pharmacokinetic and pharmacodynamic properties. The non-glycosylated interferon β-1b (IFNβ$_{ser17}$) has the expected short circulation time: time to peak concentration (C_{max}) between 1 and 8 h with a mean peak serum interferon concentration of 40 IU/mL after a single SC injection of 0.5 mg (16 million international units [MIU]). Bioavailability is about 50%. Patients receiving single intravenous (IV) doses up to 2.0 mg (64 MIU) show an increase in serum concentrations, which is dose proportional. Mean terminal elimination half-life values ranged from 8.0 min to 4.3 h. Thrice weekly IV dosing for 2 weeks resulted in no accumulation of IFNβ-1b in sera of patients. Pharmacokinetic parameters after single and multiple IV doses were comparable. Following every other day SC administration of 0.25 mg (8 MIU) IFNβ-1b in healthy volunteers, biologic response marker levels (neopterin, β2-microglobulin, myxovirus resistance protein 1 [MxA protein], and IL-10) increased significantly above baseline for 6–12 h after the first dose. Biologic response marker levels peaked between 40 and 124 h and remained elevated above baseline throughout the 7-day (168-h) study.

Glycosylated IFNβ-1a such as Rebif®, on the other hand, is slower to reach C_{max}, with a median of 16 h, and the serum elimination half-life is 69 ± 37 h (mean ± SD). In healthy volunteers, a single SC injection of 60 μg (~18 MIU) of interferon β-1a resulted in a C_{max} of 5.1 ± 1.7 IU/mL. Following every other day SC injections in healthy volunteers, an increase in area under the curve (AUC) of approximately 240% was observed, suggesting that accumulation of IFNβ-1a occurs after repeated administration. Biological response markers (e.g., 2′, 5′-oligoadenylate synthetase [2′,5′-OAS], neopterin, and β$_2$-microglobulin) are induced by IFNβ-1a following a single SC administration of 60 μg.

Intracellular 2′,5′-OAS peaked between 12 and 24 h, and β$_2$-microglobulin and neopterin serum concentrations showed a maximum at approximately 24–48 h. All three markers remained elevated for up to 4 days. Administration of 22 μg (6 MIU) IFNβ-1a three times per week inhibited mitogen-induced release of pro-inflammatory cytokines (IFNγ, IL-1, IL-6, TNFα, and TNFβ) by peripheral blood mononuclear cells that, on average, was near double that observed with IFNβ-1a administered once per week at either a 22 (6 MIU) or 66 μg (12 MIU) dose (FDA Interferon Beta-1a Revised 2003).

Pharmaceutical Formulations and Dosing for Interferon Beta Therapeutics

Betaseron®/Betaferon® is formulated as a sterile powder with a 0.54% sodium chloride solution as diluent. Reconstituted, it presents as 0.25 mg (8 MIU of antiviral activity) per mL. The recommended dose is 0.25 mg injected SC every other day.

Avonex® is formulated as a lyophilized powder for intramuscular (IM) injection. After reconstitution with the supplied diluent (sterile water for injection), each vial contains 30 μg of IFNβ-1a, 15 mg human serum albumin (HSA), 5.8 mg sodium chloride, 5.7 mg dibasic sodium phosphate, and 1.2 mg monobasic sodium phosphate in 1.0 mL at a pH of approximately 7.3, or as a prefilled syringe with a sterile solution for IM injection containing 0.5 mL with 30 μg of interferon β-1a, 0.79 mg sodium acetate trihydrate, 0.25 mg glacial acetic acid, 15.8 mg arginine hydrochloride, and 0.025 mg polysorbate 20 in water for injection at a pH of approximately 4.8. The recommended dosage is 30 μg injected IM once a week.

Rebif® is supplied in pre-filled 0.5 mL syringes: each 0.5 mL contains either 22 μg (6 MIU) or 44 μg (12 MIU) of IFNβ-1a, 2 or 4 mg HSA, 27.3 mg mannitol, 0.4 mg sodium acetate, and water for injection. The recommended dosage is 22 μg (6 MIU) given three times per week by SC injection. This dose is effective in the majority of patients to delay progression of the disease. Patients with a higher degree of disability EDSS (Kurtzke Expanded Disability Status Scale) of 4 or higher may require a dose of 44 μg (12 MIU) three times per week.

The adverse event profile for the three IFNβs is similar to IFNα. It is generally reasonably well tolerated and subjectively again consists primarily of the "influenza-like symptoms." For a detailed reporting of all adverse events, the reader is referred to the product information for each biopharmaceutical.

IFNγ Therapeutics

Actimmune® (recombinant interferon γ-1b; immune IFN) is a single-chain polypeptide containing 140 amino acids. It is produced by genetically engineered *E. coli* containing the

DNA encoding the human protein. It is a highly purified sterile solution consisting of non-covalent dimers of two identical 16,465 Da monomers. Actimmune® is slowly absorbed; after IM injection of 100 µg/m², a C_{max} of 1.5 ng/mL is reached in approximately 4 h, and after SC injection, a C_{max} of 0.6 ng/mL is reached in 7 h. The apparent fraction of dose absorbed is >89%. The mean half-life after IV administration was 38 min and after IM and SC dosing with 100 µg/m², 2.9 and 5.9 h, respectively. Multiple-dose SC pharmacokinetics showed no accumulation of Actimmune® after 12 consecutive daily injections of 100 µg/m² (FDA Recombinant Interferon Gamma-1b 1990).

Pharmaceutical Formulations and Dosing for Interferon Gamma Therapeutics

Actimmune® is a solution filled in a single-dose vial for SC injection. Each 0.5 mL contains 100 µg (2 million IU) of IFNγ-1b, formulated in 20 mg mannitol, 0.36 mg sodium succinate, 0.05 mg polysorbate 20, and sterile water for injection. The dosage for the treatment of patients with chronic granulomatous disease or severe, malignant osteopetrosis is 50 µg/m² (1 million IU/m²) for patients with a body surface area greater than 0.5 m² and 1.5 mcg/kg/dose for patients with a body surface area equal to or less than 0.5 m².

The adverse event profile of IFNγ is similar to IFNα; it is generally well tolerated and subjectively consists primarily of the "influenza-like symptoms." For a detailed reporting of all adverse events, the reader is referred to the Actimmune® product information.

Therapeutic Use of Recombinant Interleukins

In general, the approach to the development of interleukins as a therapeutic modality is even more challenging than for IFNs. Most interleukins are embedded in a complex regulatory network, and so far, the therapeutic use of interleukins has been somewhat disappointing. This was largely due to our lack of understanding of the role of these molecules and of the best way to use them; they are less well studied than IFNs. IL-2, for example, was initially developed by oncologists in the days when "go in fast, hit them hard, and get out"

was the prevalent strategy. Terms such as "maximal tolerated dose" actually defined the dose at which a given drug was in most cases no longer tolerated. Thus, IL-2 got an undeserved bad reputation. Interleukins currently approved as biologicals worldwide are listed in Table 22.2.

Aldesleukin

Proleukin® (aldesleukin), a non-glycosylated human recombinant interleukin-2 product, is a protein with a molecular weight of approximately 15 kDa. The chemical name is desalanyl-1, serine-125 human interleukin-2. It is produced by recombinant DNA technology using a genetically engineered *E. coli* containing an analog of the human interleukin-2 gene. The modified human IL-2 gene encodes a modified human IL-2 differing from the native form: the molecule has no N-terminal alanine; the codon for this amino acid was deleted during the genetic engineering procedure; moreover, serine was substituted for cysteine at amino acid position 125. Aldesleukin exists as biologically active, non-covalently bound microaggregates with an average size of 27 recombinant interleukin-2 molecules. The pharmacokinetic profile of aldesleukin is characterized by high plasma concentrations following a short IV infusion, rapid distribution into the extravascular space, and elimination from the body by metabolism in the kidneys with little or no bioactive protein excreted in the urine. Studies of aldesleukin administered IV indicate that upon completion of infusion, approximately 30% of the administered dose is detectable in plasma. Observed serum levels are dose proportional. The distribution and elimination half-life after a 5-min IV infusion are 13 and 85 min, respectively. In humans and animals, aldesleukin is cleared from the circulation by both glomerular filtration and peritubular extraction in the kidney. The rapid clearance of aldesleukin has led to dosage schedules characterized by frequent, short infusions. The adverse event profile of IL-2 is similar to that seen for IFNs and many ILs; it is generally reasonably well tolerated and subjectively consists primarily of the "influenza-like symptoms." For a detailed reporting of all adverse events, rarely severe, and pharmacological properties, the reader is referred to the product information for Proleukin® (FDA Aldesleukin revised 2012).

Table 22.2 Interleukins approved as biopharmaceuticals worldwide

Recombinant interleukins	Company	First indication	First approval
Proleukin® (aldesleukin; IL-2, lacking N-terminal alanine, C125 S substitution, produced in *E. coli*)	Chiron Therapeutics (Emeryville, CA), now Novartis Vaccines and Diagnostics, Inc.	RCC (renal cell carcinoma)	1992 (EU and USA)
Neumega® (oprelvekin; IL-11, lacking N-terminal proline produced in *E. coli*)	Genetics Institute (Cambridge, MA), now Pfizer Inc.	Prevention of chemotherapy-induced thrombocytopenia	1997 (USA)
Kineret® (anakinra; IL-1 receptor antagonist, produced in *E. coli*)	Amgen (Thousand Oaks, CA), now Swedish Orphan Biovitrum AB	RA (rheumatoid arthritis)	2001 (USA)

Pharmaceutical Formulations and Dosing of Aldesleukin

Proleukin® is supplied as a sterile, lyophilized cake in single-use vials intended for IV injection. After reconstitution with 1.2 mL sterile water for injection, each mL contains 18 million IU (1.1 mg) aldesleukin, 50 mg mannitol, and 0.18 mg sodium dodecyl sulfate, without preservatives, buffered with approximately 0.17 mg monobasic and 0.89 mg dibasic sodium phosphate to a pH of 7.5. It is indicated for the treatment of adults with metastatic renal cell carcinoma or metastatic melanoma. Each treatment course consists of two 5-day treatment cycles: 600,000 IU/kg (0.037 mg/kg) is administered every 8 h by a 15-min IV infusion for a maximum of 14 doses. Following 9 days of rest, the schedule is repeated for another 14 doses, or a maximum of 28 doses per course, as tolerated (Table 22.2).

Oprelvekin

Neumega® (oprelvekin), a non-glycosylated form of IL-11, is produced in *E. coli* by recombinant DNA technology and consists of a 177-amino-acid sequence and has a molecular mass of approximately 19 kDa. It differs from the 178-amino-acid primary sequence of native IL-11 in lacking the amino-terminal proline residue. It is used as a thrombopoietic growth factor that directly stimulates the proliferation of hematopoietic stem cells and megakaryocyte progenitor cells and induces megakaryocyte maturation resulting in increased platelet production. Pharmacokinetics show a rapid clearance from the serum and distribution to highly perfused organs. The kidneys are the primary route of elimination, and little intact product can be found in the urine. After injection, the C_{max} of 17.4 ± 5.4 ng/mL is reached after 3.2 ± 2.4 h (T_{max}) with a half-life of 6.9 ± 1.7 h. The absolute bioavailability is >80%. There is no accumulation after multiple doses. Patients with severely impaired renal function show a marked decrease in clearance to 40% of that seen in subjects with normal renal function.

Pharmaceutical Formulations and Dosing of Oprelvekin

Neumega® is supplied as single-use vials containing 5 mg of oprelvekin (specific activity approximately 8×10^6 U/mg) as a sterile lyophilized powder with 23 mg of glycine, 1.6 mg of dibasic sodium phosphate heptahydrate, and 0.55 mg of monobasic sodium phosphate monohydrate. When reconstituted with 1 mL of sterile water for injection, the solution has a pH of 7.0. It is indicated for the prevention of severe thrombocytopenia following myelosuppressive chemotherapy. The recommended dose is 50 μg/kg given once daily by SC injection after a chemotherapy cycle in courses of 10–21 days. Platelet counts should be monitored to assess the optimal course of therapy. Treatment beyond 21 days is not recommended. Oprelvekin is generally well tolerated. Reported adverse events, mainly as a consequence of fluid retention, include edema, tachycardia/palpitations, dyspnea, and oral moniliasis. For a detailed reporting of all adverse events, rarely severe, the reader is referred to the product information for Neumega® (FDA oprelvekin 1997).

Anakinra

Kineret® (anakinra) is a recombinant, non-glycosylated form of the human interleukin-1 receptor antagonist (IL-1Ra) produced using an *E. coli* bacterial expression system. It consists of 153 amino acids, has a molecular weight of 17.3 kDa, and differs from native human IL-1Ra in that it has a single methionine residue added to its amino terminus. The absolute bioavailability of Kineret® after a 70 mg SC bolus injection is 95%. C_{max} occurs 3–7 h after SC administration at clinically relevant doses (1–2 mg/kg), and the half-life ranges from 4 to 6 h. There is no accumulation of Kineret® after daily SC doses for up to 24 weeks. The mean plasma clearance with mild and moderate (creatinine clearance 50–80 mL/min and 30–49 mL/min) renal insufficiency was reduced by 16% and 50%, respectively. In severe renal insufficiency and end-stage renal disease (creatinine clearance <30 mL/min), mean plasma clearance declined by 70% and 75%, respectively. Less than 2.5% of the administered dose is removed by hemodialysis or continuous peritoneal dialysis. A dose schedule change should be considered for subjects with severe renal insufficiency or end-stage renal disease.

Pharmaceutical Formulations and Dosing of Anakinra

Kineret® is supplied in single-use prefilled glass syringes with 27-gauge needles as a sterile, clear, preservative-free solution for daily SC administration. Each prefilled glass syringe contains 0.67 mL (100 mg) of anakinra in a solution (pH 6.5) containing 1.29 mg sodium citrate, 5.48 mg sodium chloride, 0.12 mg disodium EDTA, and 0.70 mg polysorbate 80 in water for injection. It is indicated for the reduction of signs and symptoms and slowing of the progression of structural damage in moderately to severely active rheumatoid arthritis and can be used alone or in combination with disease-modifying anti-rheumatic drugs (DMARD) other than TNF-blocking agents (see Chap. 24 Antibody-Based Biotherapeutics in Inflammatory Diseases). The recommended dose for the treatment of patients with rheumatoid arthritis is 100 mg/day. Patients with severe renal insufficiency or end-stage renal disease should receive 100 mg every other day. Twenty years after its introduction (2001), Anakinra was approved to treat COVID-19 patients with pneumonia on oxygen support. Anakinra is generally well tolerated; the most common adverse reactions are injection-site reactions; the most serious adverse reactions are neutropenia, particularly when used in combination with TNF-blocking agents, and serious infections. For a detailed reporting of all adverse events, rarely severe, the reader is referred to the product information for anakinra (FDA Anakinra 2001).

Engineering IFNs and ILs: A Continuing Story

PEGylation (Table 22.3)

Since 1977, it has been known that polyethylene glycol (PEG) conjugated proteins are frequently more effective than their native parent molecule. Our understanding of PEG chemistry and how it affects the behavior of a biopharmaceutical has increased with the number of PEGylated proteins developed as therapeutic agents. More extensive reporting on protein-PEG combinations can be found in Chap. 5, Table 5.10. PEG is hydrophilic and assumed to be inert, nontoxic, and non-immunogenic, although the Food and Drug Administration (FDA) has issued a document entitled "Potential Safety Issues" on the basis of the Adverse Event Reporting System (AERS) (FDA AERS 2011). In its most common form either linear or branched PEG terminated with hydroxyl groups that can be activated for coupling to the desired target protein is used. Its general structure is

Bifunctional linear PEG (diol)

For polypeptide modification, one hydroxyl group is usually inactivated by conversion to monomethoxy or mPEG, which is monofunctional; that is, only one hydroxyl group is activated during the PEGylation process, thus avoiding the formation of interprotein (oligomerization) or intraprotein bridges:

Monofunctional linear mPEG

To couple PEG to molecules such as polypeptides, polysaccharides, polynucleotides, or small organic molecules, it is necessary to chemically activate it. This is done by preparing a PEG derivative with a functional group chosen according to the desired profile for the final product. In addition to the linear PEGs, branched structures have proven useful for peptide and protein modifications:

Branched PEG

Branched PEG or PEG2 has a number of advantages over linear structures:

- Attached to proteins, they "act" much larger than a linear mPEG of the same molecular weight.
- Two PEG chains are added per attachment site, reducing the chance of protein inactivation.
- They are more effective in protecting proteins from proteolysis, reducing antigenicity and immunogenicity.

Depending on the desired use for the PEG-modified molecule, different PEGylation strategies can be chosen, for example:

- Multiple shorter-chain PEGylation if the biological activity should be preserved
- A weak PEG-protein bond if a slow-release effect is desired
- A branched chain with high molecular weight and a strong bond if prolonged circulation and receptor saturation are the goal

The development of rhIFNα from the native, unmodified molecule to the PEGylated form with the desired pharmacological profile is an example of how the understanding of PEG chemistry progressed with experience (Zeuzem et al. 2003). Increasing the length of the PEG chain resulted in

Table 22.3 PEGylated interferons approved in the United States and Europe

PEGylated recombinant interferons	Company	First indication	First approval
Pegasys® (PEGylated IFNα-2a produced in *E. coli*)	Hoffmann-La Roche (Basel, Switzerland)	Chronic hepatitis B and C	2002 (EU and USA)
PegIntron® (PEGylated IFNα-2b produced in *E. coli*)	Schering-Plough (Kenilworth, NJ, USA)	Chronic hepatitis C, discontinued in EU	2000 (EU), 2001 (USA)

progressively longer circulating half-life due to protracted resorption and lower clearance, ultimately resulting in a near constant serum concentration over an entire week summarized in Fig. 22.3.

The first PEGylated interferon, IFN alpha-2a, used a linear, 5 kDa mPEG with a weak urethane PEG-IFN alpha-2a link. Clinical trials conducted with this compound were unsuccessful because the blood circulation half-life for the conjugate (Fig. 22.3b) was only slightly improved relative to that of the native protein (Fig. 22.3a). Development of the product was therefore halted at phase II clinical trials. The second compound was developed by Schering-Plough, Kenilworth, NJ, in collaboration with Enzon Pharmaceuticals Inc., Bridgewater, NJ. It consists of a longer (12 kDa), linear PEG with a urethane linkage to IFN alpha-2b. The chosen strategy was to combine the advantages of high specific activity with slower serum clearance resulting in PegIntron® with markedly improved pharmacological properties allowing once-a-week administration (Fig. 22.3c) (Glue et al. 2000). PegIntron®is approved for the treatment of chronic hepatitis C (FDA Pegintron 2001). In 2021, it was withdrawn from the market in the EU (Table 22.3).

The development of the third PEGylated interferon, IFN alpha-2a, took a different approach. The strategic goal was to achieve lasting and constant serum concentrations over an entire week. In a collaboration of Roche with Shearwater Polymers in Huntsville, AL, now Nektar, San Carlos, CA, IFNα-2a was linked by a stable amide bond to four different PEG chains of various sizes, structures, and site-attachment numbers. The resulting products were tested for antiviral activity and a variety of pharmacokinetic parameters including half-life, absorption rate, and mean residence time:

- 20 kDa linear mono-PEG-IFN alpha-2a
- 40 kDa linear di-PEG-IFN alpha-2a
- 20 kDa branched mono-PEG-IFN alpha-2a
- 40 kDa branched mono-PEG-IFN alpha-2a

The 40 kDa, branched PEGylated molecule (later named Pegasys®) exhibited sustained blood levels, decreased systemic clearance, and an approximately tenfold increase in serum half-life over regular interferon. The biological activity was similarly prolonged resulting in an optimal pharmacological profile (Fig. 22.3d). It was therefore chosen for further clinical development leading to its approval worldwide for the treatment of chronic hepatitis B and C (Reddy et al. 2002; FDA Pegasys 2002) (Table 22.3).

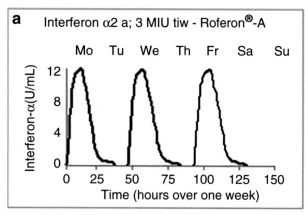

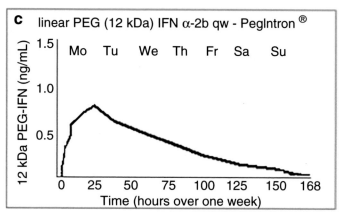

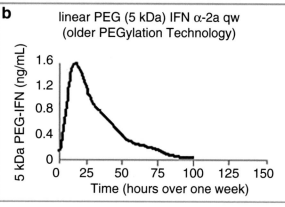

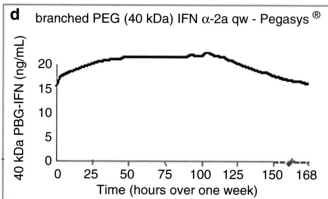

Fig. 22.3 Pharmacokinetic profiles for IFN-alpha2, after repeated dosing (panel **a**) and after one dose of different PEG-IFN-alpha2 molecules (linear or branched PEG with different molecular weights) (panels **b**, **c** and **d**)

Further Cytokine Engineering

Based on the understanding of the function and limitations of a given therapeutic protein product (TPP), rational protein engineering allows the creation of a new product with improved and expanded activities. Having shown a degree of activity in the treatment of certain cancers, IL-2 is a good example to illustrate this line of thought. Systemic IL-2 (aldesleukin) treatment has shown significant clinical benefit in a minority of renal cell and melanoma patients, with long-term survival in some cases. However, a number of limiting factors have been identified. Its pharmacological properties, short half-life, and its adverse effects, mainly vascular leak syndrome (VLS) with different organ manifestations—a pathophysiological manifestation of acute inflammation—make it difficult to handle. Acute inflammation is a process typical of vascularized tissues whereby interstitial fluid and white blood cells accumulate at the site of injury. Thus, flooding the body with exogenously administered IL-2 can induce a dose-dependent "vascular leak syndrome" in any vascularized organ. So far, we have been "playing the piano with boxing gloves—now is the time to take off our boxing gloves." What is needed is a possibility to specifically target those cells we wish to impact and only those.

IL-2 has dual properties: one, the ability to expand and activate innate and adaptive effector cells, which is the basis of its anticancer activity, and two, to coordinate an immuno-suppressive microenvironment by recruiting regulatory T cells (Tregs) and myeloid-derived suppressor cells (MDSCs) as a regulatory mechanism that prevents excessive immune responses and autoimmunity. Unfortunately, the expansion of immune-suppressive Treg cells as well as other immune dysregulation limits or impedes IL-2's anticancer activity (Setrerrahmane and Xu 2017).

While PEGylating IL-2 may have resolved the issue of short half-life and ensuing peak toxicity, we are still repeatedly flooding the whole organism with a TPP of known toxicity. With a better understanding of the factors limiting its mechanism of action as well as the structure-function relationship of proteins, rational design and engineering strategies allow adaptation of its beneficial or deleterious (toxic) activity or the creation of new activities.

IL-2 is reengineered by creating a recombinant fusion protein composed of a genetically engineered human monoclonal antibody directed against carcinoembryonic antigen (CEA), i.e., cergutuzumab, linked to an engineered, variant form of interleukin-2 (IL-2v): amunaleukin. Upon administration of cergutuzumab amunaleukin (Fig. 22.4), the antibody moiety recognizes and binds to CEA, thereby specifically targeting IL-2v to CEA-expressing tumor tissue. Subsequently, the IL-2v moiety stimulates a local immune response, which activates both natural killer cells and cytotoxic T cells and eventually leads to tumor cell killing. CEA

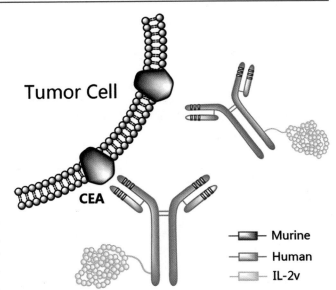

Fig. 22.4 Graphic model of the cergutuzumab amunaleukin fusion protein

is a cell surface protein that is expressed on a wide variety of cancer cells. The mutations found in IL-2v inhibit its binding to the IL-2 receptor-alpha (IL-2Rα), which prevents the activation of Treg. However, it can still bind to and induce signaling through the IL-2Rβγ, which allows the preferential expansion of NK cells and CD8-positive T cells. The Fc domain of cergutuzumab has been modified to prevent Fc-gamma binding and downstream cell activation. This idea of modifying the IL-2 structure to increase its therapeutic index is still in the clinical test phase (Mullard 2021).

Outlook and Conclusions

There is a very precise and organized order in the intricate function of the immune system to make it work effectively, and we are well on our way to map it. But a lot of hard work still lies ahead. The fundamental approach to cytokine or cytokine antagonist therapy with biopharmaceuticals is to identify diseases caused by insufficient or excessive cytokine production. In the first case, e.g., with certain chronic viral diseases or cancers, appropriate cytokines are used pharmacologically to boost the immune response. Examples include IFNα with antiviral as well as immunomodulatory properties in chronic viral hepatitis, or IL-2 or IL-12 in renal cell cancer and malignant melanoma. For chronic inflammatory or atopic diseases caused by unchecked overproduction of interleukins, two options are available: one can either inhibit the interleukin or its receptor(s), e.g., with human(ized) monoclonal antibodies such as infliximab, reslizumab, mepolizumab, siltuximab, ixekizumab, secukinumab, adalimumab, and ustekinumab, or etanercept (see Chap. 24 Monoclonal antibodies in anti-inflammatory therapy) for various indications, or the IL-1R antagonist (Kineret®; see above) in rheumatoid arthritis. Another option would be to downregulate excessively produced interleukin using its antagonistic cytokine, e.g., PEGylated IL-12 in asthma or IL-10 in psoriasis. To date, it appears that the first option is more successful than downregulation by anti-inflammatory cytokines, which has so far not resulted in an approved product.

Self-Assessment Questions

Questions

Decide whether each of the statements below is true or false. If you believe a statement is false, explain why.

1. Interferons are defined:
   (a) By the cell type that produces them
   (b) By their anti-inflammatory properties
   (c) By their antiviral activity
   (d) By their protein structure
   (e) By their genetic structure
2. Human interferon alpha:
   (a) Is produced selectively by leukocytes
   (b) Is a virucidal substance
   (c) Triggers antiviral effects in cells expressing appropriate receptors
   (d) Acts on the immune system to booster specific antiviral response
3. Interleukins are characterized by:
   (a) Their action on target cells
   (b) Their protein structure
   (c) Their genetic structure
   (d) Pro- or anti-inflammatory effect
4. The following interleukins are generally considered to be "pro-inflammatory," i.e., induce and/or be part of a Th1 response:
   (a) The IL-1 family, IL-2
   (b) IL-3
   (c) IL-4, IL-10, IL-13
5. Interleukins are:
   (a) Secreted specifically by leukocytes to act on other leukocytes
   (b) Bound to a specific receptor complex to exert their effect
   (c) A family of proteins that co-regulate the immune response
   (d) Non-toxic products of the body in response to pathogens and other potentially harmful agents
   (e) Long-acting immune modulators

6. Interferons and interleukins can be toxic; several (patho) physiological containment mechanisms exist to counteract excessive production:
   (a) Soluble receptors
   (b) Binding to cell surface receptors
   (c) Neutralizing antibodies
   (d) Negative feedback mechanisms
   (e) Naturally occurring IL receptor antagonists
7. The following interleukins are approved for therapeutic use:
   (a) IL-1
   (b) IL-2
   (c) IL-10
   (d) IL-11
       Where appropriate, specify some of the indications they are used for.
8. Protein PEGylation:
   (a) Prolongs circulation half-life of the PEGylated protein
   (b) Decreases antigenicity of the PEGylated protein
   (c) Protects the protein from proteolysis
   (d) Is difficult due to the toxicity of polyethylene glycol
   (e) Improves the therapeutic efficacy of the PEGylated protein
9. The following PEGylated IFNs and ILs have been approved for therapeutic use:
   (a) Interferon α2
   (b) Interferon β
   (c) Interleukin-1
   (d) Interleukin-2
   (e) Interleukin-12

Answers

1. Interferons are defined:
   (a) False. Although IFNα used to be called "leukocyte interferon" and IFNβ "fibroblast interferon" because they were initially produced from buffy coats (leukocytes) infected with Sendai virus and human diploid fibroblasts stimulated with poly(I)-poly(C) or Newcastle disease virus (NDV), respectively, interferons and their units (IU) are defined by their antiviral activity.
   (b) False. While they can act as immune modulators and on occasion have anti-inflammatory properties (e.g., IFNβ, for the treatment of multiple sclerosis), they will more often induce a Th1 or pro-inflammatory response. IFNγ is one of the classical pro-inflammatory markers.
   (c) True.

(d) and (e) False. The full protein and genetic sequences of the different interferons and their subtypes were only defined long after the initial crude IFN mixtures had been tested in the clinic initially against viral diseases and subsequently against cancers.

Today, however, the protein and genetic sequences are necessary to specify an interferon and its purity during the production by biotechnological techniques. Also new interferons or interleukins will be accepted as such by the HUGO Gene Nomenclature Committee (HGNC) based on their function and a previously unknown genetic sequence.

2. Human interferon alpha:
   (a) False. Interferon alpha is produced by many cell types, including T cells and B cells, macrophages, fibroblasts, endothelial cells, and osteoblasts among others.
   (b) False. By interacting with their specific heterodimeric receptors on the surface of cells, the interferons initiate a broad and varied array of signals that induce antiviral state.
   (c) True.
   (d) True.
3. Interleukins are characterized by:
   (a) and (e) False. ILs are characterized by their protein and gene structures registered in the HGNC database (and similar centralized databases) (HGNC). Their names and symbols must be approved by the HGNC.
   (b) True.
   (c) True.
   (d) False. While some ILs can be classified as pro- or anti-inflammatory, this is not what basically defines them.
4. The following interleukins are generally considered to be "pro-inflammatory," i.e., induce and/or be part of a Th1 response:
   (a) True.
   (b) False. IL-3 is a multicolony-stimulating, hematopoietic growth factor that stimulates the generation of hematopoietic progenitors of every lineage.
   (c) False. These three interleukins all play a role in the differentiation and activation of basophils and eosinophils leading to a Th2 response.
5. Interleukins are:
   (a) False. Interleukins are mainly secreted by leukocytes and primarily affecting growth and differentiation of hematopoietic and immune cells. They are also produced by other normal and malignant cells and are of central importance in the regulation of hematopoiesis, immunity, inflammation, tissue remodeling, and embryonic development.

(b) True.
(c) True.
(d) False. Many interleukins, primarily those with pro-inflammatory function, are intrinsically toxic either directly or indirectly, i.e., through induction of toxic gene products.
(e) False. Interleukins usually have a short circulation time, and their production is regulated by positive and negative feedback loops.
6. Interferons and interleukins can be toxic; several (patho) physiological containment mechanisms exist to counteract excessive production.
   (a) True.
   (b) False. Binding to cell surface receptor is a physiological process and has negligible effect on "circulating" interferons or interleukins.
   (c) to (e) are true.
7. The following interleukins are approved for therapeutic use:
   (a) True. An IL-1 analog/antagonist (Kineret®) is indicated for the treatment of rheumatoid arthritis.
   (b) True. IL-2 (Proleukin®) is indicated for the treatment of adults with metastatic renal cell carcinoma or metastatic melanoma.
   (c) False. Clinical development of IL-10 (Tenovil™) as an anti-inflammatory drug for several indications such as psoriasis, Crohn's disease, and rheumatoid arthritis was discontinued in phase III due to insufficient efficacy to warrant further development.
   (d) True. IL-11 (Neumega®) is indicated for the prevention of severe thrombocytopenia following myelosuppressive chemotherapy.
8. Protein PEGylation:
   (a) True.
   (b) True.
   (c) True.
   (d) False. PEG is inert, nontoxic, non-immunogenic, and, in its most common form, either linear or branched terminated with hydroxyl groups that can be activated to couple to the desired target protein. But there is a caveat (FDA AERS 2011).
   (e) True.
9. The following PEGylated IFNs and ILs have been approved for therapeutic use:
   (a) True: For chronic hepatitis C and B. Limited clinical trials have also been conducted in renal cell carcinoma, malignant melanoma, and non-Hodgkin lymphoma.
   (b) (c), (d), and (e) are false, although early clinical trials have been conducted with PEGylated IL-2 in renal cell carcinoma (RCC) and malignant melanoma and pharmacokinetic studies with PEGylated IFNβ in animal models.

Bibliography

Alexander SP, Fabbro D, Kelly E et al (2017) The concise guide to pharmacology 2017/18: catalytic receptors. Br J Pharmacol 174:225–271. https://doi.org/10.1111/bph.13876/full

Carreño V, Tapia L, Ryff JC et al (1992) Treatment of chronic hepatitis C by continuous subcutaneous infusion of interferon-alpha. J Med Virol 37:215–219. https://doi.org/10.1002/jmv.1890370312

Castro LS, Lobo GS, Pereira P et al (2021) Interferon-based biopharmaceuticals: overview on the production, purification, and formulation. Vaccines (Basel) 9:328. https://doi.org/10.3390/vaccines9040328

Croker BA, Kiu H, Nicholson SE (2008) SOCS regulation of the JAK/STAT signalling pathway. Semin Cell Dev Biol 19:414–422

Durham GA, Williams JJL, Nasim MT, Palmer TM (2019) Targeting SOCS proteins to control JAK-STAT signalling in disease. Trends Pharmacol Sci 40:298–308. https://doi.org/10.1016/j.tips.2019.03.001

EMA Betaferon (2006) Betaferon SMPC revised version. https://www.ema.europa.eu/en/documents/product-information/betaferon-epar-product-information_en.pdf. Accessed 24 Jan 2022

FDA AERS (2011) Potential Signals of Serious Risks/New Safety Information Identified by the Adverse Event Reporting System (AERS). https://www.fda.gov/drugs/questions-and-answers-fdas-adverse-event-reporting-system-faers/october-december-2011-potential-signals-serious-risksnew-safety-information-identified-adverse-event. Accessed 24 Jan 2022

FDA Aldesleukin revised (2012) Aldesleukin. https://www.accessdata.fda.gov/drugsatfda_docs/label/2012/103293s5130lbl.pdf. Accessed 22 Jan 2022

FDA Anakinra (2001) Anakinra smpc. https://www.accessdata.fda.gov/drugsatfda_docs/label/2012/103950s5136lbl.pdf. Accessed 22 Jan 2022

FDA oprelvekin (1997) Oprelvekin. https://www.accessdata.fda.gov/drugsatfda_docs/label/2009/103694s1008lbl.pdf. Accessed 24 Jan 2022

FDA Pegasys (2002) Pegasys. https://www.accessdata.fda.gov/drugsatfda_docs/label/2011/103964s5204lbl.pdf. Accessed 24 Jan 2022

FDA Pegintron (2001) Pegintron. https://www.accessdata.fda.gov/drugsatfda_docs/label/2013/103949s5263lbl.pdf. Accessed 24 Jan 2022

FDA Recombinant Interferon Gamma-1b (1990) Recombinant interferon gamma-1b. https://www.accessdata.fda.gov/drugsatfda_docs/label/2015/103836s5182lbl.pdf. Accessed 23 Jan 2022

FDA Revised (2003) Rebif revised 2003. In: revised 2003. https://www.accessdata.fda.gov/drugsatfda_docs/label/2003/ifnbser050203LB.pdf. Accessed 24 Jan 2022

Glue P, Fang JWS, Rouzier-Panis R et al (2000) Pegylated interferon-α2b: pharmacokinetics, pharmacodynamics, safety, and preliminary efficacy data. Clin Pharmacol Ther 68:556–567. https://doi.org/10.1067/mcp.2000.110973

HGNC (n.d.) The resource for approved human gene nomenclature. https://www.genenames.org/. Accessed 24 Jan 2022

Hughes CE, Nibbs RJB (2018) A guide to chemokines and their receptors. FEBS J 285:2944. https://doi.org/10.1111/febs.14466

Huising MO, Kruiswijk CP, Flik G (2006) REVIEW phylogeny and evolution of class-I helical cytokines. J Endocrinol 189:1–25. https://doi.org/10.1677/joe.1.06591

Interferon-Alfa-n3 (2023) Alferon N. https://www.drugs.com/dosage/interferon-alfa-n3.htm. Accessed 26 Jan 2022

Mullard A (2021) Restoring IL-2 to its cancer immunotherapy glory. Nat Rev. Drug Discov 20:163–165

Pestka S, Krause CD, Walter MR, Petska S (2004) Interferons, interferon-like cytokines, and their receptors. Immunol Rev 202:8–32

Platanias LC (2005) Mechanisms of type-I- and type-II-interferon-mediated signalling. Nat Rev. Immunol 5:375–386

Protein Data Bank (2021) RCSB PDB. In: RCSB PDB. https://www.rcsb.org/structure/4HSA. Accessed 22 Jan 2022

Reddy KR, Modi MW, Pedder S (2002) Use of peginterferon alfa-2a (40 KD) (Pegasys) for the treatment of hepatitis C. Adv Drug Deliv Rev 54:571–586

Setrerrahmane S, Xu H (2017) Tumor-related interleukins: old validated targets for new anti-cancer drug development. Molecul Cancer 16(1):1–7. https://doi.org/10.1186/s12943-017-0721-9

Sidney P (1986) Interferon from 1981-1986. Methods Enzymol 119:3–14

UniProt (2021) UniProt Knowledgebase. https://www.uniprot.org/uniprot/. Accessed 22 Jan 2022

Vaillant AAJ, Qurie A (2022) Interleukin. StatPearls Publishing LLC. https://www.ncbi.nlm.nih.gov/books/NBK499840/#__NBK499840_dtls__. Accessed 4 Nov 2022

Zeuzem S, Welsch C, Herrmann E (2003) Pharmacokinetics of peginterferons. Semin Liver Dis 23(Suppl 1):23–28. https://doi.org/10.1055/s-2003-41631

Pharmaceutical Biotechnology: The Protein Products of Biotechnology and Their Clinical Use— Monoclonal Antibodies and Their Variations

Antibody-Based Biotherapeutics in Cancer

23

Jürgen Barth

Introduction

More than 80 years ago, in the 1940 ies the first chemotherapeutic drugs were used to support surgery and/or radiation therapy against cancer. In the following decades, chemotherapy developed to a mainstay of therapy rather than being supporting in character. In the last two decades, monoclonal antibodies (mAbs) have faced a growing interest as antitumor therapeutics. In November 1997, rituximab was the first-in-class therapeutic mAb approved by the Food and Drug Administration (FDA). In June 1998 followed the approval in the European Union. At that time, even the manufacturer believed the approval as monotherapy for relapsed/refractory, CD20-positive, low-grade, or follicular B-cell non-Hodgkin's lymphoma would be a niche indication. The following rapid development of many different mAbs in multiple different cancer indications shows how incorrect this assessment was. Nowadays, a large number of mAbs have been approved for use in numerous cancer indications with many more in different stages of clinical development.

Classification of Monoclonal Antibodies

Structure, functionality, classes, and possible modifications of mAbs and their mode of action are described in Chap. 8. mAbs can be roughly divided into unconjugated antibodies and conjugated antibodies. The latter ones can be subdivided into toxin-coupled antibodies, the so-called chemo-immunoconjugates now designated as antibody–drug conjugates (ADC), and radionuclide-coupled antibodies, the radio immune conjugates (RIC). These "naked," uncoupled toxins given alone would be too toxic for the patient. The antibody works as a tugboat, tugging the toxin to accumulate selectively nearby or in the tumor. A RIC can be considered "inner radiation therapy." Figure 23.1 gives a rough systematic overview about the different mAbs used in cancer therapy.

J. Barth (✉)
Justus-Liebig-Universität, Medizinische Klinik IV, Universitätsklinik, Giessen, Germany
e-mail: Juergen.Barth@innere.med.uni-giessen.de

Fig. 23.1 Schematic
classification of different
mAbs used in cancer therapy;
for details, see text

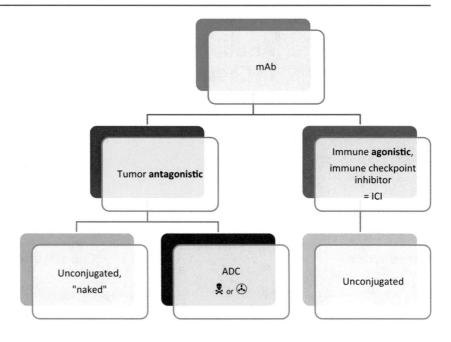

Pharmacological Targets of Oncologic Antibodies

mAbs can be directed against surface structures of malignant cells. These structures comprise receptors or proteins, such as the epidermal growth factor receptor (EGFR) or the cluster of differentiation (CD) antigens. Besides that, some mAbs intercept soluble (growth) factors in the blood and peripheral tissue fluids, which could lead to a tumor survival benefit. One example in oncology is the vascular endothelial growth factor (VEGF), which is intercepted by bevacizumab or aflibercept ("*VEGF-Trap*"). The most recent and rapidly growing class of mAbs are immune agonistic antibodies, also called immune checkpoint inhibitors (ICIs), currently directed against the cytotoxic T-lymphocyte antigen 4 (CTLA4) and the programmed cell death receptor1 (PD 1) or against its ligand (PD-L1), lymphocyte activation gene-3 (LAG-3) blocking antibodies, and T-cell immunoreceptor

with Ig and ITIM domains (TIGIT), which is an immunoreceptor tyrosine-based inhibitory motif. The main difference in the mode of action compared to all the other classes of mAbs and even chemotherapeutics is that not the tumor itself is attacked, but the immune system of the patient is held "under fire" to re-recognize the tumor. Table 23.1 gives an overview over the present mAbs with an oncology indication, ordered by the target structure.

All mentioned targets can also be found on nonmalignant tissues, which explains the observed toxicity. At present time, none of the approved mAbs is directed against a tumor-specific molecular driver, as they are known, for instance, in Philadelphia chromosome-positive CML (the fusion protein BCR/Abl), in anaplastic lymphoma kinase (ALK)-positive non-small-cell lung cancer (NSCLC), in EGFR-mutated NSCLC (mutated EGFR), or in the FLT3-positive AML. Her2 (= ErbB2), for example, can also be found in normal tissues, but in Her2-positive breast cancers, it is overexpressed.

Table 23.1 Therapeutic mAbs used in cancer therapy clustered by their targets (in alphabetical/numerical order)

INN	Target	Type	Molecular weight (kDa)
Anti-CD antibodies			
Belantamab mafodotin	Anti-BCMA (= CD269) (+ MMAF)	Humanized	146
Tafasitamab	Anti-CD19	Humanized	150
Loncastuximab tesirine	Anti-CD19 + toxin	Humanized	151
Blinatumomab	Anti-CD19 x anti-CD3 (T and B cells; bispecific)	Murine	54
Obinutuzumab	Anti-CD20 (B lymphocytes)	Humanized	146
Ofatumumab	Anti-CD20 (B lymphocytes)	Human	149
Rituximab	Anti-CD20 (B lymphocytes)	Chimeric	145
I^{131} tositumomab tiuxetan	Anti-CD20 (B lymphocytes) + radionuclide	Murine	150
Y^{90} ibritumomab tiuxetan	Anti-CD20 (B lymphocytes) + radionuclide	Murine	153
Mosunetuzumab	Anti-CD20 x anti-CD3 (bispecific)	Humanized	146
Inotuzumab ozogamicin	Anti-CD22 + toxin	Humanized	160
Brentuximab vedotin	Anti-CD30 + toxin (HD, sALCL)	Chimeric	149–151
Gemtuzumab ozogamicin	Anti-CD33 + toxin (myeloid cells)	Humanized	151–153
Daratumumab	Anti-CD38	Human	148
Isatuximab	Anti-CD38	Chimeric	145
Alemtuzumab	Anti-CD52 (lymphocytes)	Humanized	145
Polatuzumab vedotin	Anti-CD79b + toxin	Humanized	150
Elotuzumab	Anti-SLAMF7 (= anti-CD319)	Humanized	148
Anti-EGFR antibodies			
Amivantamab	Anti-EGFR x anti-MET (bispecific)	Human	148
Cetuximab	Anti-EGFR (ErbB1)	Chimeric	145
Margetuximab	Anti-HER2/neu (ErbB2)	Chimeric	145
Necitumumab	Anti-EGFR	Human	144
Panitumumab	Anti-EGFR (ErbB1)	Human	144
Pertuzumab	Anti-HER2/neu (ErbB2)	Humanized	148
Trastuzumab	Anti-HER2/neu (ErbB2)	Humanized	146
Trastuzumab deruxtecan	Anti-HER2/neu (ErbB2) + toxin	Humanized	153
Trastuzumab emtansine (T-DM1)	Anti-HER2/neu (ErbB2) + toxin	Humanized	153
Anti-angiogenic antibodies			
Bevacizumab	Anti-VEGF ("ligand capture")	Humanized	149
Aflibercept	Anti-VEGF ("VEGF-trap")	Human	97
Ramucirumab	Anti-VEGF-R2	Human	147
Immune agonistic antibodies, immune checkpoint inhibitors (ICIs)			
Ipilimumab	Anti-CTLA4 (T cells)	Human	148
Tremelimumab	Anti-CTLA4	Human	149
Cemiplimab	Anti-PD-1	Human	143
Dostarlimab	Anti-PD-1	Humanized	143
Nivolumab	Anti-PD-1	Human	146
Pembrolizumab	Anti-PD-1	Humanized	149
Atezolizumab	Anti-PD-L1 (expressed on tumor tissue)	Humanized	145
Avelumab	Anti-PD-L1 (expressed on tumor tissue)	Human	147
Durvalumab	Anti-PD-L1 (expressed on tumor tissue)	Human	146
Relatlimab	Anti-Lag-3 (expressed on activated T cells, NK cells, and dendritic cells)	Human	145
Other antibodies			
Denosumab	Anti-RANKL	Human	146
Dinutuximab	Anti-GD2	Chimeric	147–150
Enfortumab vedotin	Anti-nectin 4 + toxin	Human	152
Mirvetuximab soravtansine	Anti-folate-receptor alpha (FRα)	Chimeric	150
Mogamulizumab	Anti-CCR4	Humanized	149
Sacituzumab govitecan	Trop-2 (trophoblast cell-surface antigen-2) + toxin	Humanized	160
Tisotumab vedotin	Tissue factor (TF)	Human	153

Pharmacokinetics of mAbs: General Remarks

Unlike small, defined molecules, the pharmacokinetic properties of mAbs differ markedly. Distribution into tissue is slow (internalization) because of the molecular size of mAbs. Volumes of distribution are generally low. Metabolism is similar to catabolic degradation of endogenous and dietetic peptides and proteins. The half-life is usually relatively long, which allows for long dosing intervals during maintenance therapy. Rituximab, for instance, is given every 2 months in the first line setting and every 3 months in the relapsed or refractory situation. Possible factors influencing the pharmacokinetics include the amount of the target antigen, which corresponds to the total tumor mass, immune reactions to the antibody (anti-drug antibodies), and patient demographics such as gender or age (see section on rituximab). Population pharmacokinetic analyses have been applied in assessing covariates in the disposition of mAbs. Both linear and nonlinear eliminations have been reported for mAbs, which is probably caused by target-mediated disposition (overview (Ryman and Meibohm 2017)). If a tumor responds to a mAb therapy, the amount of the remaining target antigen diminishes, and the half-life can be prolonged. However, mAb dosing is based on different pharmacokinetic models, leading to approved dosing strategies of body surface area-based (rituximab), body weight-based (trastuzumab, pertuzumab), or flat dosing (atezolizumab, obinutuzumab).

Anti-CD Antibodies

The cluster of differentiation (also known as cluster of designation or classification determinant and often abbreviated as CD) denotes groups of immune phenotyping surface properties for the identification and investigation of cell surface molecules providing targets on cells in order to classify these cells according to their biochemical or functional characteristics. CD molecules are membrane-bound proteins, often glycosylated, in part cell specific. In terms of physiology, CD molecules can act in numerous ways, often acting as receptors important to cell functioning and/or survival. Others have enzymatic activity. A signaling cascade is usually initiated or modulated, altering the behavior of the cell. Additionally to the mechanisms of ADCC and CDC (Chap. 8), mAbs can alter or even block signal transduction pathways, leading to cell death via apoptosis (overview (Ludwig et al. 2003)).

Anti-CD20 Antibodies

In the 1980s, CD20 was identified as a B-cell marker by Stashenko and coworkers (Stashenko et al. 1980). CD20 is expressed on early pre-B cells, it remains through B-cell development and is then lost from plasma cells (Tedder and Engel 1994). It is not found on hematopoietic stem cells. Therefore, anti-CD20 antibodies are also designated as anti-lymphocytic antibodies. The antigen is neither shed nor internalized. CD20 is expressed on the majority of B-cell lymphomas (Stashenko et al. 1980; Nadler et al. 1981; Tedder and Engel 1994). It is a non-glycosylated member of the membrane-spanning 4-A (MS4A) family (Ishibashi et al. 2001). The antigen consists of three hydrophobic regions forming a tetraspan transmembrane molecule with a single extracellular loop and intracellular N- and C-terminal regions (Tedder et al. 1988). CD20 has been shown to be present on the cell surface as a homo-multimer, in tetramer complexes (Polyak and Deans 2002; Polyak et al. 2008). Despite decades of research, no natural ligand for CD20 could be detected (Cragg et al. 2005). However, subsequent data revealed that CD20 is resident in lipid raft domains of the plasma membrane where it probably functions as a store-operated calcium (SOC) channel following ligation of the B-cell receptor (BCR) with the antigen, implying that it plays a role in regulating cytoplasmic calcium levels after antigen engagement (Bubien et al. 1993; Kanzaki et al. 1995; Li et al. 2003).

Pharmacology. Anti-CD20 antibodies can be classified into type I and type II antibodies with different pharmacodynamics (Cragg and Glennie 2004). Type I (rituximab-like) mAbs induce CD20 to redistribute into large detergent-resistant microdomains (lipid rafts), whereas type II (tositumomab-like) mAbs do not (Deans et al. 1998). These lipid microenvironments on the cell surface—known as lipid rafts—also take part in the process of signal transduction. Lipid rafts containing a given set of proteins can change their size and composition in response to intra- or extracellular stimuli. This favors specific protein–protein interactions, resulting in the activation of signaling cascades. For details, see Simons and Toomre (2000). Rituximab and ofatumumab are type I, whereas obinutuzumab is the first glyco-engineered type II mAb. Type I antibodies display a remarkable ability to activate complement and elicit complement-dependent cytotoxicity (CDC, Chap. 8). This is a result of enhanced recruitment of the complement system component C1q. This ability, however, appears to be directly linked to their (CD20-bound) translocation into lipid rafts, which cluster the antibody Fc regions, thus enabling improved C1q binding (Cragg et al. 2003). Type II mAbs do not show these characteristics: They do not change CD20 distribution after binding; no concomitant clustering is observed, and they are relatively ineffective in CDC. Interestingly, they evoke far more homotypic adhesion and direct killing of target cells in a caspase-independent manner (Chan et al. 2003; Ivanov et al. 2008). In summary, type I mAbs have to cross-link the homo-tetrameric CD20 antigen for a pharmacodynamic effect. In diseases like CLL, with a lower density of CD20 antigens compared with other B-cell lymphomas, the tetramers have to be bridged with more than one molecule of rituximab. That is the explanation why rituximab in CLL is given at a higher dose of 500 mg/m^2 in subsequent cycles after the usual starting dose of 375 mg/m^2. For illustration, see Fig. 23.2.

Type II mAbs bind within a CD20 tetramer, without lipid raft redistribution. They induce a "closed conformation" (Beers et al. 2010; Niederfellner et al. 2011) (Fig. 23.3). The antigens don't have to be bridged. The type II mAb-induced cell death is dependent on homotypic adhesion, requires cholesterol, and is energy dependent, involving the relocaliza-

tion of mitochondria to the vicinity of the cell–cell contact (Beers et al. 2010).

The pharmacodynamic mechanism of action and the resulting effects of rituximab are illustrated in Fig. 23.4 (from Stolz and Schuler (2009)).

Rituximab

Rituximab is used as a monotherapy as well as in combination with chemotherapy. It has become a mainstay in the treatment of B-cell malignancies such as aggressive and indolent non-Hodgkin's lymphomas (NHLs) and chronic lymphocytic leukemia (CLL), as well as other B-cell-

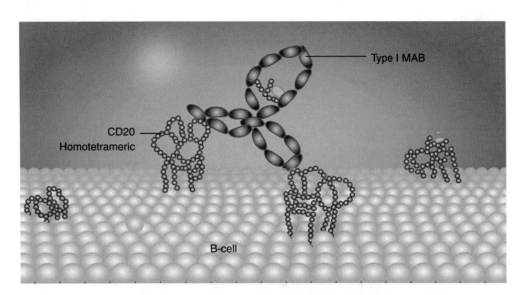

Fig. 23.2 Anti-CD20 mAbs can be divided into two distinct subtypes, termed types I and II. Type I mAbs induce CD20 to redistribute into large detergent-resistant microdomains (rafts), whereas type II mAbs do not. This differential ability of anti-CD20 mAbs to redistribute CD20 in the cell membrane impacts on many of the binding properties and effector functions that control the therapeutic success of anti-CD20 mAbs. Type I and II mAbs have the ability to evoke different effects: type I mAbs engage ADCC and cause CD20 modulation but do not elicit direct cell death, whereas type II mAbs mediate direct cell death and engage ADCC but do not promote CD20 modulation (Beers et al. 2010). In diseases with a low CD20 density, more than one rituximab molecule must bridge the CD20 antigens

Fig. 23.3 Type II mAbs (tositumomab-like) bind within a CD20 tetramer, and the antigens do not have to be bridged. No redistribution into lipid rafts occurs (Beers et al. 2010, Niederfellner et al. 2011)

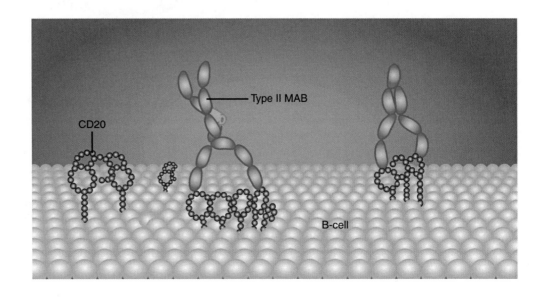

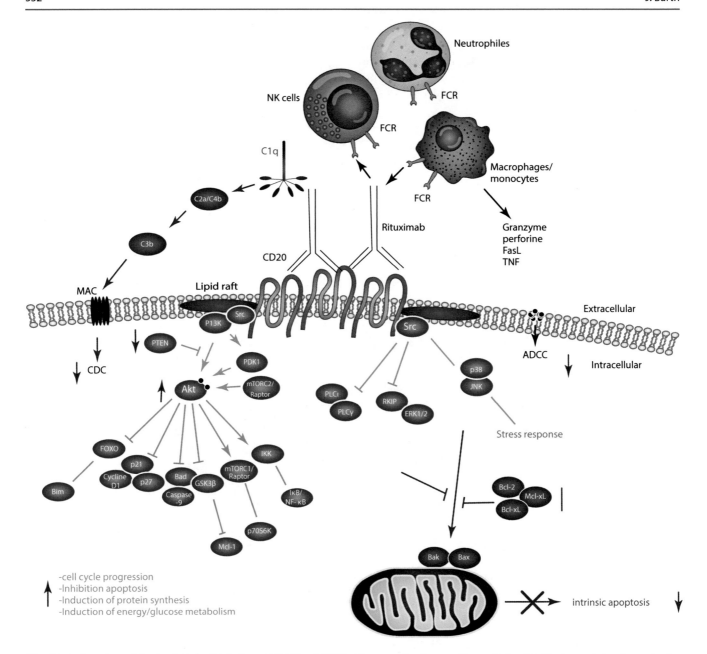

Fig. 23.4 Overview of rituximab-inducible indirect (ADCC und CDC) effector mechanisms and intracellular, CD20-triggered signal transduction cascades after CD20 cross-linking. FCR is Fc receptor (Stolz and Schuler 2009)

triggered diseases. It is also widely utilized in an off-label fashion for numerous clinical conditions like immune thrombocytopenia (ITP). Histology subtype has been described as the main tumor parameter that influences rituximab efficacy. In an early trial, when rituximab was given for four doses as monotherapy, small lymphocytic lymphoma histology negatively influenced the objective response rate compared with other low-grade lymphomas (McLaughlin et al. 1998). Several studies have described a low objective response rate in patients with CLL treated with the standard dose of 375 mg/m² rituximab monotherapy used in other NHL

(Nguyen et al. 1999; Huhn et al. 2001; O'Brien et al. 2001). As mentioned above, the level of CD20 expression (density) may differ depending on tumor histology (Manches et al. 2003). This was initially used to explain the variation in rituximab efficacy among different histological subtypes. Despite conflicting in vitro data, the dosing in CLL was adjusted to 500 mg/m² for subsequent doses after an initial dosing of the well-known 375 mg/m² finally. The reduced dose for the first infusion is necessary because of a usually high tumor load in need for therapy in CLL blast crisis. Responding to rituximab therapy bears a high risk of tumor lysis and cytokine release syndrome as well as other infusion-related reactions (IRR); Rituximab Safety, section below. For induction therapy as well as for maintenance therapy (used in follicular lymphoma (FL)), a minimum serum concentration has to be achieved for tumor response (Berinstein et al. 1998; Gordan et al. 2005).

Responses to therapeutic mAbs have also been reported to correlate with specific polymorphisms in the Fcγ receptor (FcγR), especially when ADCC is a major part in the mechanism of action (Mellor et al. 2013). These polymorphisms are associated with differential affinity of the receptors to mAbs. At present, these functional FcγR polymorphisms are not suitable to be used as pharmacogenetic biomarkers that could be used to better target the use of mAbs in cancer patients. The available studies do not describe a consistent effect of FcγR genotype on the clinical antitumor activity of therapeutic IgG1 mAbs. Inconsistencies observed in studies are likely related to differences in tumor type, the cytotoxic agents used in combination, the clinical settings, like metastatic vs. adjuvant situation, the clinical benefit parameters measured, the therapeutic antibody used, as well as the magnitude and even the direction of the effect. However, even patients with an "unfavorable" FcγR genotype do benefit from an antibody-containing regimen (overview (Mellor et al. 2013)).

To minimize IRR, the infusion rate is increased slowly from the initial infusion rate of 50 mg/h every 30 min by doubling the subsequent infusion rate if no reactions occur up to 400 mg/h. Shortly before approval of rituximab biosimilars, a fixed dose, subcutaneous (s.c.) formulation was launched in the EU. Rituximab is concentrated up to 120 mg/mL in recombinant hyaluronidase, a penetration enhancer ("spreading factor"). The dose for follicular lymphoma is 1400 mg (12 mL) and for CLL 1600 mg (16 mL). The first dose has still to be given by the i.v. route due to tolerability concerns. The subsequent s.c. doses can be given over 6 min—a substantial advantage from the patient's point of view.

As shown by the SABRINA study, the pharmacokinetic profile of s.c. rituximab was non-inferior to i.v. rituximab. I.v. and s.c. rituximab had similar efficacy and safety profiles, and no new safety concerns were noted. S.c. administration does not compromise the anti-lymphoma activity of rituximab when given with chemotherapy.

Rituximab Safety

From today's point of view, dose finding and scheduling of rituximab administration, the first approved therapeutic anticancer antibody, were more empirical than rational or biologic target based. As no dose-limiting toxicity or clear dose–response relationship was found during a phase I study with single doses of 10, 50, 100, 250, or 500 mg/m^2 (Maloney et al. 1994), the 375 mg/m^2 dose level with a 4-weekly dosing scheme was chosen for phase II studies (Maloney et al. 1997; McLaughlin et al. 1998) without an apparent reasoning. A dose escalation trial was conducted in CLL, and even at 2250 mg/m^2 (monotherapy), the maximum tolerated dose was not reached (O'Brien et al. 2001). Rituximab has significant activity in patients with CLL at the higher dose levels. However, the dose–response curve is not steep. Myelosuppression and infections are uncommon.

Rituximab is generally well tolerated in patients with both malignant and nonmalignant diseases, including children (Quartier et al. 2001) and even pregnant women (Herold et al. 2001). Severe adverse events are rare but possible. The most common adverse events are infusion-/foreign protein-related and occur most frequently during or shortly after the first infusion. However, extremely rare, hypersensitivity reactions can occur even after years of maintenance therapy, leading to permanent discontinuation of its use. This syndrome consists of chills, fever, headache, rhinitis, pruritus, vasodilation, asthenia, and angioedema. A cytokine release syndrome is possible. Less often reported are hypotension, rash, bronchospasm, pain at tumor sites, and rash or severe skin toxicity inclusive Stevens–Johnson syndrome and toxic epidermal necrolysis (= Lyell syndrome). Routine premedication consists of corticosteroids, if they are not already a component of the chemotherapy- or antiemetic scheme. The same drug classes are used for intervention in case of hypersensitivity reactions (HSR) despite correct premedication.

Rituximab induces a rapid depletion of CD20$^+$ B cells in the peripheral blood. B cells remain at low levels for at least 2–6 months with recovery to pretreatment values occurring within 12 months (Maloney et al. 1997). Other hematological toxicities comprise temporary reduction of platelets or neutrophils and occasionally reduced immunoglobulin levels, which also can be caused by the underlying disease and its bone marrow involvement. Despite the fact that the drug is generally well tolerated, a small number of patients experience unexpected and severe toxicities. It is speculated that these patients have a rituximab serum level dramatically above the average. However, so far, this could not be proven in clinical routine.

Also, gender and age seem to influence rituximab kinetics and clinical outcome. The German High-Grade Non-Hodgkin Lymphoma Study Group first detected in elderly women with diffuse large B-cell lymphoma (DLBCL) a slower elimination of rituximab compared with men, which translated in a better outcome for these women (Pfreundschuh et al. 2014). Consequently, the Study Group conducted a phase II trial increasing the number of rituximab infusions to achieve high rituximab levels early during treatment. In this DENSE-R-CHOP-14 trial, 100 elderly patients with aggressive CD20$^+$ B-cell lymphoma received sic cycles of biweekly CHOP-14 combined with 12 × rituximab (375 mg/m^2) on days 0, 1, 4, 8, 15, 22, 29, 43, 57, 71, 85, and 99 (CHOP-14 treatment regimen: cyclophosphamide 600 mg/m² day 1; doxorubicin (= hydroxydaunorubicin) 50 mg/m² day 1; vincristine (**O**ncovin®) 1.4 mg/m² max. 2 mg day 1; predniso(lo)ne 100 mg days 1–5 every 14 days).

This intensification of rituximab administration achieved higher rituximab serum levels and resulted in higher complete remission and event-free survival rates in elderly patients with poor-prognosis DLBCL (Pfreundschuh et al. 2007). The negative impact of male gender in DLBCL was confirmed by a meta-analysis (Yildirim et al. 2015). While body weight contributes to a faster elimination in males according to Muller et al. (2012), others found a significant longer 5-year OS in a high BMI group (>22.55 kg/m²) when compared to that of the low BMI group (Weiss et al. 2017).

With the increasing use of Rituximab, a growing number of rare adverse effects have been recognized. Late-onset neutropenia (LON) is, according to the National Cancer Institute Common Toxicity Criteria, defined as grade 3–4 neutropenia, in which absolute neutrophil count is less than $1.0 \times 10^3/L$, occurring 4 weeks after the last rituximab administration. LON has been reported in 5–27% of rituximab-treated lymphoma patients. Similar figures apply for autoimmune patients, but those appear to have more infections during the neutropenic period (Tesfa et al. 2011; Tesfa and Palmblad 2011).

The mechanisms of LON after rituximab treatment still have not been elucidated. One hypothesis is that LON may result from hematopoietic lineage competition due to an excessive B-cell recovery in the bone marrow by promotion of B-cell lymphopoiesis over granulopoiesis (Anolik et al. 2003). Others see a correlation between a high-affinity *FCGR3 158 V* allele and LON in lymphoma patients (Weng et al. 2010). A clinically relevant question is if it is safe to re-treat patients with rituximab who previously developed LON. Recurrence of LON episodes upon reexposed patients has been reported. Nevertheless, a close clinical follow-up or complete blood count monitoring is considered adequate in most LON cases (Tesfa and Palmblad 2011).

Following rituximab administration, reactivation of hepatitis B, in some cases resulting in fulminant hepatitis, hepatic failure, and death, has been reported. This led to testing of active replicating hepatitis as well as to screening of all patients for hepatitis B viral infection by measuring HbsAg and anti-HBc before initiating treatment with rituximab as a therapeutic precondition. Monitoring and/or reactivation of antiviral prophylaxis should be considered and is advised by the FDA's Full Prescribing Information (FPI) and the European Medicines Agency's Summary of Product Characteristics (SmPC). HBV reactivation can occur up to 24 months following completion of rituximab therapy.

The immune-suppressing antibody with or without chemotherapy can lead to reactivation of latent JC polyoma virus, leading to potentially fatal progressive multifocal leukoencephalopathy (PML). Patients presenting with onset of neurologic manifestations should consult a neurologist with this suspected diagnosis. For natalizumab, an anti-alpha4-integrin antibody used against multiple sclerosis, which induced PML in some patients, removal of the drug via plasma exchange was an important part of the therapy, as it reconstituted the immune system. In a case report, the first successful removal of rituximab with plasma exchange therapy (rituximab apheresis) and accompanying complement-dependent cytotoxicity (CDC) test assay for monitoring has been described. Neurologic symptoms of the patient improved within the first 2 weeks (Burchardt et al. 2012).

Ofatumumab

Ofatumumab is a type I human immunoglobulin (Ig) G1k antibody. Ofatumumab binds with greater avidity than rituximab to another epitope of CD20, which encompasses the small extracellular loop (residues 74–80) and the N-terminal region of the second large extracellular loop (Fig. 23.5). Several crystal structure-based analyses of the Fab fragment of the antibody suggest that ofatumumab can bind closer and tighter to the cell membrane than rituximab, leading to more effective CDC. Induction of CDC appears to be greater than with rituximab and occurs even at a lower density of CD20 on the cell surface (Cheson 2010). After binding to CD20 on malignant B cells, ofatumumab induces clustering of CD20 into lipid rafts, similar to other type I antibodies as described above. In contrast to rituximab, ofatumumab does not induce apoptosis of B-cell lines (Cheson 2010).

Initially licensed for double refractory CLL, that is, those resistant or refractory to fludarabine and alemtuzumab, ofatumumab can be used in previously untreated as well as relapsed patients after conventional chemotherapy. As already mentioned, CLL in need for therapy (blast crisis) bears a risk of rapid cell breakdown in the sense of a tumor lysis syndrome. Therefore, a reduced initial flat dose of 300 mg is given. The subsequent flat dose is 1000 mg or 2000 mg, depending on the clinical situation (for details, see FPI and SmPC).

Ofatumumab Safety

Qualitatively, the toxicity profile of ofatumumab is similar to that of rituximab. This includes hepatitis B virus reactivation and the potential for PML. No additional toxicity or safety signals have been observed so far.

Obinutuzumab

Obinutuzumab is the first glycoengineered, type II humanized anti-CD20 mAb. It is posttranslational defucosylated, resulting in the absence of a fucose sugar residue from IgG oligosaccharides in the Fc region of the mAb molecule (Tobinai et al. 2017). Obinutuzumab binds within the CD20 tetramer, as depicted in Fig. 23.3. CDC, probably, does not

Fig. 23.5 Distinct CD20-binding epitopes of rituximab, ofatumumab, and obinutuzumab (designated as GA101). Ofatumumab binds to the so-called small loop. (From Klein et al. (2013)).

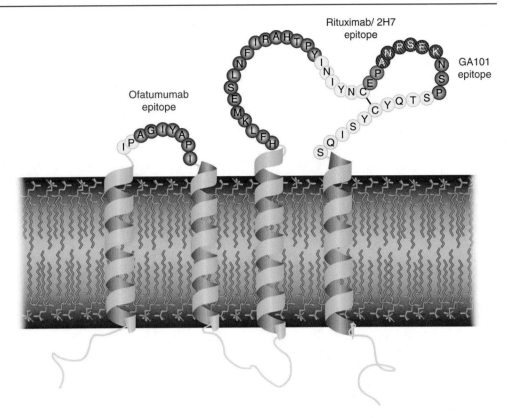

contribute to the overall activity of obinutuzumab. The limited capacity of obinutuzumab to fix complement via its Fc portion may further enhance its capability to bind to FcγRIII and mediate ADCC (summarized in Tobinai et al. (2017)). Obinutuzumab was significantly more effective than rituximab in depleting B cells in whole blood samples from healthy donors (n = 10) and from an individual with CLL (Mossner et al. 2010). In CLL, the dosing of obinutuzumab is 100 mg on day 1, followed by 900 mg on day 2 in cycle 1. If tolerated and no IRR occur, the second part of the split dose can be given on day 1. In subsequent cycles, there is no dose splitting on day 1. As mentioned before, starting therapy of CLL may lead to the tumor lysis syndrome. Therefore, the split dose is a safety measure and not necessary in other diseases, e.g., indolent NHL. According to the FPI and SmPC, no dose splitting is advised for follicular lymphoma.

In contrast to rituximab, obinutuzumab is ramped up in cycle 1 with a flat dose of 1000 mg on days 1, 8, and 15. Consequently, patients in head-to-head comparisons of obinutuzumab with rituximab received 3000 mg obinutuzumab in cycle 1, whereas rituximab patients received 375 mg/m². This was often seen as a dosing imbalance. However, the manufacturer did not study a second experimental arm with the same flat dose for rituximab, which would have been considered a real head-to-head study. Different dosing regimens were tested for obinutuzumab. For achieving the desired serum concentration, 1600 mg on days 1 and 8 in cycle 1, followed by 800 mg on day 1 in subsequent cycles, was determined. 1600 mg might need very long infusion times due to possible IRR, especially in CLL. Results in the GAUGUIN study (Morschhauser et al. 2013) and pharmacokinetic modeling showed that obinutuzumab 1000 mg per cycle with additional 1000 mg doses on days 8 and 15 of cycle 1 can achieve exposures similar to the 1600/800 mg regimen. This simplified flat-dose 1000 mg schedule was adopted for subsequent phase II and III investigations (Cartron et al. 2016).

Obinutuzumab Safety

The safety profile of obinutuzumab is similar to the other anti-CD20 antibodies rituximab and ofatumumab.

Y[90] ibritumomab and I[131] tositumomab tiuxetan are murine RICs. Yttrium-90 is a beta-emitter, while iodine-131 is a beta-emitter and an emitter of gamma radiation. Their activity is mainly achieved through their radioisotopes rather than their intrinsic antibody activity (Fig. 23.6). The radiation can penetrate through several cell layers, which is termed "cross fire" or "bystander" effect (Fig. 23.7). Adjacent cells not marked by the antibody are also affected. Although active, these antibodies never gained a broad acceptance in the medical community. As of February 2014, the production of tositumomab and I[131] tositumomab has been discontinued by the manufacturer and is no longer available (NCI, 2014). Y[90] ibritumomab tiuxetan is still approved in the EU but is rarely used.

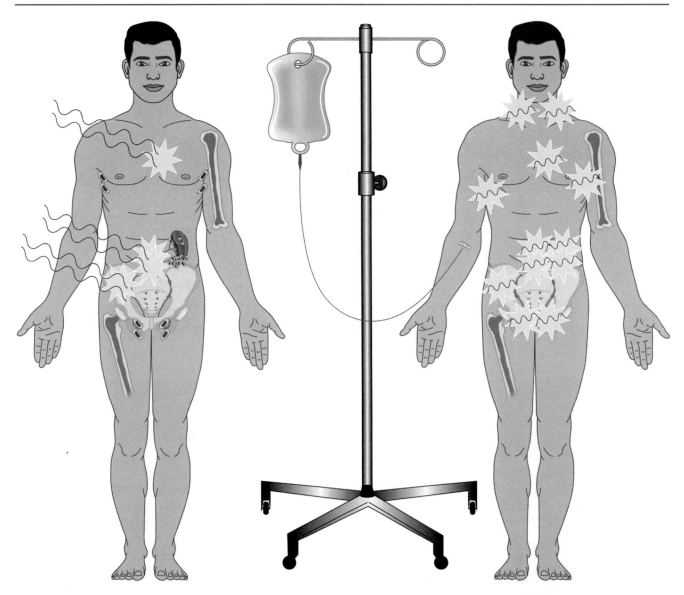

Fig. 23.6 Comparison of conventional radiotherapy (left) with an external radiation source. Radiation from inside (right) by RICs

Due to the historical development, up to this point, anti-CD20 mAbs (unconjugated as well as conjugated RICs) were presented together. For the remainder of the chapter, however, ADCs will be discussed separately, even if the therapeutic target of an ADC is the same as that of an unconjugated mAb (e.g., anti-HER2). The mode of action of the payload will be addressed in a separate section but will be cross-referenced in the corresponding target antigen sections.

Other Anti-CD Antibodies: Unconjugated

Alemtuzumab was another anti-lymphocytic antibody, directed against the CD52 antigen. The CD52 antigen is mainly found on mature B and T lymphocytes but can also be found on monocytes/macrophages, NK cells, eosinophils, and epithelial cells of the male reproductive tract. Alemtuzumab was used for the treatment of double refractory B-CLL (refractory to an alkylator plus fludarabine). This hematological approval was retracted in 2012 to favor the development as an immunosuppressant against multiple sclerosis, for which it is currently approved.

Anti-myeloma Antibodies

Elotuzumab

Elotuzumab is an immunostimulatory monoclonal antibody that recognizes signaling lymphocytic activation molecule

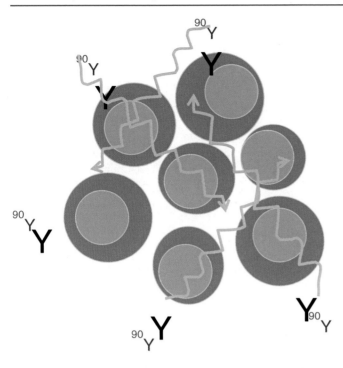

Fig. 23.7 "Cross fire" or "bystander" effect of radiation, penetrating more cell layers than marked by the antibody

family member 7 (SLAMF7; CD319), a glycoprotein highly expressed on myeloma and natural killer cells but not on normal tissues (Hsi et al. 2008). Elotuzumab causes myeloma cell death via a dual mechanism of action (Collins et al. 2013) (Fig. 23.8):

1. *Direct activation*: Binding to SLAMF7 directly activates natural killer cells (Collins et al. 2013) but not myeloma cells (Guo et al. 2015).
2. *Tagging for recognition*: Elotuzumab activates natural killer cells via CD16, enabling selective killing of myeloma cells via antibody-dependent cellular cytotoxicity (ADCC) with minimal effects on normal tissue (Collins et al. 2013).

Elotuzumab is not active as a monotherapy (Radhakrishnan et al. 2017), so it is combined with an immune-modulating drug, lenalidomide or pomalidomide, and dexamethasone. This combination lowers the killing threshold for myeloma cells. Further combinations, for example, with proteasome inhibitors, are currently being tested.

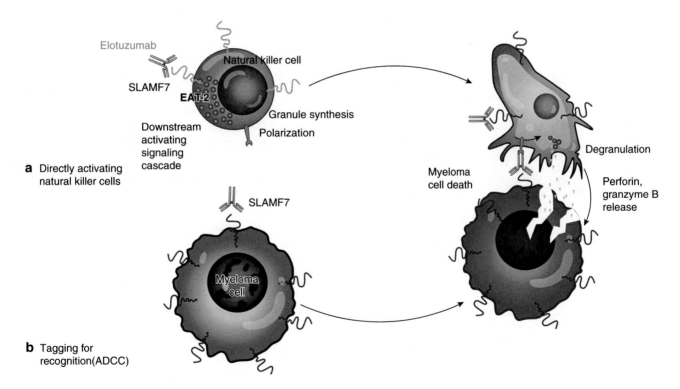

Fig. 23.8 Dual mechanism of action of elotuzumab. (**a**) Direct activation. (**b**) Tagging for recognition (Collins et al. 2013)

Elotuzumab Safety

Elotuzumab needs an intensive premedication. Despite correct pretreatment with antipyretics, antihistamines, and multiple doses of oral and i.v. corticosteroids, IRRs occur during elotuzumab administration, which makes a close patient monitoring mandatory. A specified infusion rate may be increased in a stepwise fashion as described in the FPI/SmPC. Several interruptions of the administration might be necessary before completion. According to the SmPC, patients experiencing an IRR shall be supervised in regard to their vital parameters every 30 min for 2 h after completing the elotuzumab infusion (EuropeanMedicinesAgency 2022b).

Anti-CD38 Antibodies

Daratumumab

Daratumumab is directed against the CD38 antigen, found on T cells (precursors, activated), B cells (precursors, activated), myeloid cells (monocytes, macrophages, dendritic cells), NK cells, erythrocytes, and platelets. However, current results show that CD38 is not lymphocyte specific. It is ubiquitously expressed in virtually all tissues. It is present not only on cell surfaces but also in various intracellular organelles, including the nucleus (Lee 2006).

CD38 has multiple functions. It acts as a receptor in close contact with the B-cell receptor complex and CXCR4. In engagement with CD31 or hyaluronic acid, it activates NF-kappaB, ZAP-70, and ERK1/2 pathways. It also works as an ectoenzyme. CD38 interacts in a pH-dependent manner (Fig. 23.9) with NAD^+ and $NADP^+$, which are converted to cADPR, ADPR, and NAADP, all intracellular Ca^{2+}-mobilizing agents (Malavasi et al. 2011). The ectoenzyme activity of CD38 is independent of its receptor functions (Deaglio et al. 2007).

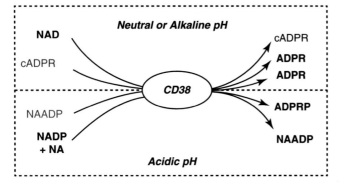

Fig. 23.9 Multiple pH-dependent enzymatic reactions catalyzed by CD38. cADPR and NAADP are two different Ca^{2+} messengers (Lee 2006) in the sense of "new" second messengers. Abbreviations used: *cADPR* cyclic ADP-ribose, *ADPR* ADP-ribose, *NAADP* nicotinic acid adenine dinucleotide phosphate, *ADPRP* ADP-ribose-2′-phosphate

Because of marked quantitative differences in expression levels of CD38 between normal cells and leukemic cells, combined with its role in cell signaling, suggests that CD38 is an attractive target for immunotherapy treatment of multiple myeloma (MM) (Malavasi et al. 2008). It is highly and uniformly expressed on myeloma cells (Lin et al. 2004; Santonocito et al. 2004), whereas only a relatively low expression was detected on normal lymphoid and myeloid cells and in some tissues of non-hematopoietic origin (Deaglio et al. 2001). In summary, daratumumab binding to CD38 elicits a signaling cascade and immune effector function engagement, leading to (de Weers et al. 2011):

- Complement-dependent cytotoxicity (CDC).
- Antibody-dependent cell-mediated cytotoxicity (ADCC).
- Antibody-dependent cell-mediated phagocytosis (ADCP), a rather new/unknown mode of action (Overdijk et al. 2015).
- Induction of apoptosis.
- Modulation of cellular enzymatic activities associated with calcium mobilization and signaling.
- Combination of these activities leads to elimination of plasma cells from the bone marrow in MM patients.

Daratumumab also kills myeloid-derived suppressor cells Tregs or negatively regulating T cells.

Isatuximab

Isatuximab is the second approved anti-CD38 antibody with qualitatively the same mode of action (MOA). There are, however, several differences: First, both mAbs bind to CD38 but on different epitopes. Isatuximab and daratumumab recognize 23 and 27 amino acids of human CD38, respectively (Fig. 23.10) (Martin et al. 2019).

A graphical overview of the mode of action of CD38 mAbs is presented in Fig. 23.11.

In vitro measurable differences, without any evidence regarding their clinical relevance, are listed in Table 23.2 (Martin et al. 2019).

Another difference between these two mAbs is that daratumumab can induce programmed cell death through cross-linking (bridging) several CD38 molecules (compare type I and type II anti-CD20 antibodies in "Rituximab" section), whereas isatuximab exerts a strong pro-apoptotic activity independent of cross-linking. Isatuximab and its F(ab')2 fragments also induce apoptosis of myeloma cells highly expressing CD38 through the activation of caspases 3 and 7, lysosome-dependent cell death, and by upregulation of reactive oxygen species (Aits and Jaattela 2013). It has also been demonstrated that isatuximab-mediated killing of myeloma cells is enhanced by compounds like pomalidomide (Morandi et al. 2018) (see also "Anti-CD20 Antibodies" section).

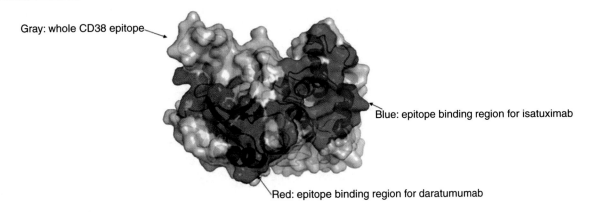

Fig. 23.10 Different CD38 epitope binding sites for daratumumab and isatuximab (Martin et al. 2019)

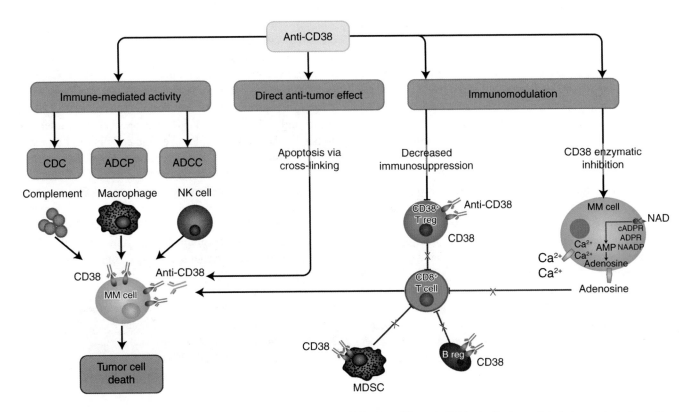

Fig. 23.11 Modes of action of anti-CD38 antibodies (de Weers et al. 2011; Overdijk et al. 2015). Antibody-mediated phagocytosis is a rather newly recognized pharmacodynamic process. For details, see text

Table 23.2 In vivo measurable distinctions between daratumumab and isatuximab. This is explicitly not a comparison in the sense of "*is better than*" (Martin et al. 2019)

Daratumumab	Isatuximab
Limited inhibition of cADPR synthesis with no dose response	Almost complete inhibition of CD38 cyclase activity
Limited NAD ⇨ cADPR conversion with no dose response	Dose-dependent NAD ⇨ cADPR conversion
	Additionally induced inhibition of CD38 enzymatic activity

Daratumumab and Isatuximab Safety

Both antibodies need a premedication. Despite correct pretreatment with antipyretics, antihistamines, and corticosteroids, IRRs occur during daratumumab and isatuximab administration. Especially for daratumumab, IRRs can occur hours after completing the infusion. The range of IRR onset was up to over 70 h. Administration of an intermediate- or long-acting corticosteroid orally for 2 days starting the day after the administration of daratumumab is recommended for the monotherapy setting. In combination regimens, an inter-mediate- or long-acting corticosteroid beginning the day after the administration of a daratumumab infusion shall be given. If a background regimen-specific corticosteroid (e.g., dexamethasone, prednisone) is administered the day after the daratumumab infusion, additional corticosteroids may not be needed (JanssenBiotech 2022a, b). While some cancer centers administer montelukast prophylactically in off-label use, others change to s.c. administration (see below).

For patients with a history of chronic obstructive pulmonary disease, prescribing short- and long-acting bronchodilators and inhaled corticosteroids should be considered (JanssenBiotech 2022a, b). A suddenly stuffy nose, cough, throat irritation, allergic rhinitis, and hoarseness during the infusion can be regarded as early signs of an upcoming IRR.

During clinical trials, the median duration of first intravenous (i.v.) infusion of daratumumab with 16 mg/kg was 7.0 h. As longer infusion times to minimize the risk of IRR were impractical, the idea came up to split the very first dose of daratumumab from 16 mg/kg day 1 to 8 mg/kg days 1 and 2 for week 1 in cycle 1. This was justifiable by the PK, investigated through computer simulations for various recommended dosing regimens and in a multi-cohort, phase 1b study. A transient difference in concentration on day 1 in cycle 1 is not expected to have any impact on overall clinical outcomes (Xu et al. 2018). The safety of this modification was evaluated in a retrospective observational study at community-based oncology clinics. These findings provided real-world evidence on the infusion time and safety of the first infusion of daratumumab (Rifkin et al. 2018). This split dosing was approved in the EU (EuropeanMedicinesAgency 2022a).

Due to necessary i.v. infusion interruptions caused by these IRR, a s.c. formulation as described for rituximab was developed to resolve the problem. Dissolved in recombinant hyaluronidase, the fixed dose of 1800 mg daratumumab (15 mL) achieves the same trough level on day 1 of cycle 3, compared to the i.v. dosage form. The IRR rate could be reduced from approximately >30% to 11%. Delayed systemic administration-related reactions could be reduced to 1% (JanssenBiotech 2022a, b). Of note, body weight seems to have an influence on the trough levels and mAb exposition.

Although delayed IRRs were not observed during the ICARIA study with isatuximab, the magnitude of IRRs was similar, with grade 3/4 reactions below 3%. Because CD38 is also expressed on erythrocytes, the indirect Coombs test used prior to blood transfusions is false positive, up to 6 months after the last infusion of anti-CD38 mAbs. Blood typing at baseline is therefore highly recommended before the first daratumumab/isatuximab infusion. Meanwhile, daratumumab and isatuximab in blood samples can be destroyed by incubation with dithiothreitol (DTT) prior to the blood compatibility testing. DTT cracks the structure stabilizing disulfide bridges of the antibody.

All three antibodies elotuzumab, daratumumab, and isatuximab also interfere with serum protein electrophoresis or immune fixation assays (IFA), leading to false-positive results in patients with IgGκ myeloma protein, making the assessment of the initial response difficult. Meanwhile, specific assays were developed to circumvent these problems. The DIRA assay contains murine anti-daratumumab antibodies, which bind daratumumab to keep it out of the M-protein region of the immune fixation assay (van de Donk et al. 2016). For isatuximab, a so-called Hydrashift assay was developed (Finn et al. 2020).

Belantamab Mafodotin

Belantamab mafodotin is the first anti-BCMA-directed monoclonal antibody, a humanized, afucosylated IgG1 anti-**B**-**c**ell **m**aturation **a**ntigen mAb. BCMA is a transmembrane glycoprotein in the tumor necrosis factor receptor (TNFR) superfamily 17 (TNFRSF17). It is expressed at significantly higher levels in all multiple myeloma cells but not on other normal tissues, except normal plasma cells. Its level in patient serum is further correlated with disease status and prognosis. **B**-cell **a**ctivating **f**actor (BAFF) and **a** **p**roliferation-**i**nducing **l**igand (APRIL; CD256), together with other growth factors, bind to BCMA. BAFF and APRIL are important for the development and function of B cells.

Altogether, there are four mechanisms of action for belantamab mafodotin:

1. ADC toxicity (see "Vedotins and Mafodotins in the Auristatins section" in the ADC section).
2. ADCC, via FcγRIIIa receptors.
3. Immunogenic cell death (ICD)—by cytokines.
4. Inhibition of the BCMA-receptor pathway, in the sense of a receptor antagonism.

During ICD, calreticulin is expressed pre-apoptotic on the cell surface, thereby generating an "eat me" signal for dendritic cells (DC). Autophagy facilitates the release of ATP from dying cells, constituting a "find me" signal for DCs and their precursors as well as a proinflammatory stimulus. Additionally, ICD is associated with the post-apoptotic release of the non-histone chromatin-binding protein high-mobility group box 1 (HMGB1). HMGB1 binds to the toll-like receptor 4 on DCs and thus stimulates their antigen-presenting functions (Martins et al. 2014). Belantamab mafodotin induces the ICD (Bauzon et al. 2019) and possesses the abovementioned effector mechanisms (Fig. 23.12).

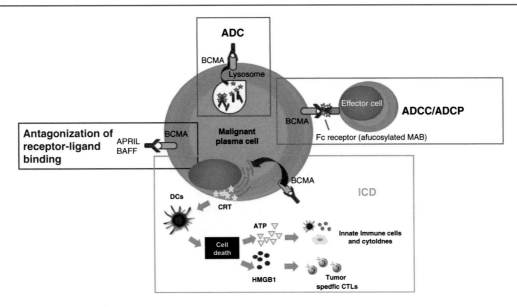

Fig. 23.12 Mechanism of action of belantamab mafodotin in a clockwise direction. 12 o'clock: belantamab mafodotin binds to cell surface BCMA and is rapidly internalized. Inside the myeloma cell, the payload is released, disrupting the microtubule network, leading to cell cycle arrest and apoptosis. 3 o'clock: the known effector mechanisms ADCC and ADCP. 6 o'clock: ICD; for details, see text. 9 o'clock: classical antagonizing ligand binding to the BCMA receptor, resulting in signal transduction inhibition. Abbreviations used: *ADC* antibody–drug conjugate, *ADCC/ADCP* antibody-dependent cell-mediated cytotoxicity/phagocytosis, *APRIL* a proliferation-inducing ligand, *BAFF* B-cell activating factor, *BCMA* B-cell maturation antigen, *CRT* calreticulin, an "eat me" signal for phagocytes, *CTLs* cytotoxic T cells, *DC* dendritic cells, *HMGB1* high-mobility group box 1 protein, *ICD* immunogenic cell death

Belantamab Mafodotin Safety

Likewise other mAbs, IRRs are possible under belantamab mafodotin. However, a premedication is not mandatory. If grade 2 or higher IRRs occur during administration, the infusion rate has to be reduced or the infusion has to be stopped. Supportive care is recommended according to the local institutional rules. The infusion can be resumed with a reduced infusion rate by at least 50% once the symptoms have resolved. If grade 2 or higher IRRs occur, premedication, not otherwise specified, is recommended for subsequent infusions. Not surprisingly, like all other anti-myeloma therapeutics, belantamab mafodotin causes a certain hematotoxicity. New in this context is an eye disorder, designated as keratopathy or corneal adverse reaction, with or without changes in visual acuity. The underlying pathology—an unspecific, reversible off-target mechanism—is described by Farooq et al. (Farooq et al. 2020) and is recommended for the interested reader. Ophthalmic exams are required at baseline, prior to each dose, and promptly for worsening symptoms (GlaxoSmothKline 2020).

Bispecific Antibodies

Blinatumomab

Blinatumomab is a first-in-class bispecific antibody, directed against CD3 on T cells and CD19 on B cells. It is also designated as a BiTE, a bispecific T-cell engager. This antibody construct is made of two single-chain variable fragments (scFv) of anti-CD3/CD19, enabling a patient's T cells to recognize malignant B cells (Fig. 23.13). CD3 is part of the T-cell receptor. CD19, originally a B-cell marker, is expressed on the majority of B-cell malignancies in normal to high levels. By linking these two cell types, T cells are activated to exert cytotoxic activity on the target cell. The usual clinical use is the Philadelphia chromosome negative B (precursor) ALL.

Blinatumomab has a molecular weight of 54.1 kDa, approximately one-third of the size of a traditional monoclonal antibody (Wu et al. 2015). It is eliminated rather quickly by the kidneys with a half-life of 1.2 h. Therefore, blinatumomab has to be continuously infused over 1 month.

Blinatumomab Safety

Blinatumomab can cause cytokine release as well as fulminant neurological toxicities even with fatal outcome. Some of them were caused by prescription, calculation, or handling errors. When changing the infusion lines, for instance, the used lines must not be flushed. This would result in an erroneous bolus application with severe toxicities or death. To avoid all this and to make blinatumomab therapy safer, the manufacturer developed risk minimization information brochures for physicians, other healthcare professionals, and patients. The preparation protocol is rather complicated, and a high-precision infusion pump is necessary.

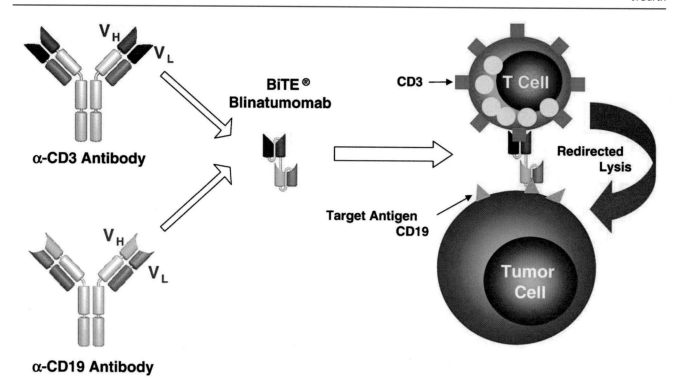

Fig. 23.13 Two fused scFv of anti-CD3/anti-CD19 antibodies result in the blinatumomab construct. One arm of blinatumomab binds to CD3; the other binds to CD19. This engages unstimulated T cells, which destroys the CD19[+] cells

Mosunetuzumab

Mosunetuzumab is a full-length, humanized T-cell-recruiting bispecific antibody targeting CD20-expressing B cells (anti-CD20 x anti-CD3). The antibody contains a $N^{297}G$ amino acid substitution in the Fc region. This substitution results in a non-glycosylated heavy chain that has minimal binding to Fcγ receptors, which reduces Fc effector functions (minimization of cytokine release probability; see below). Similar to blinatumomab, the mechanism of action of mosunetuzumab involves engaging T cells via CD3 with CD20[+]-expressing cells, leading to T-cell activation and T-cell-mediated cytolysis of the CD20[+]-expressing cells. The initial approval for mosunetuzumab is the double relapse/refractory follicular lymphoma after at least two previous systemic therapies.

Mosunetuzumab Safety

A cytokine release syndrome (CRS), like that associated with CAR-T cells, might occur with bispecific mAbs as a consequence of T-cell activation. This reaction can be life threatening. Therefore, patients have to be pretreated with corticosteroids, antihistamines, and antipyretics at a minimum up to cycle 2. An intervention for the treatment of an emerging CRS is, besides corticosteroids, tocilizumab (approved indication for the treatment of CAR-T-cell-induced CRS).

Amivantamab

Amivantamab is a fully human anti-EGFR x anti-MET bispecific mAb. It was designed to engage these two mentioned distinct driver pathways in non-small-cell lung cancer (NSCLC) (Moores et al. 2016; Vijayaraghavan et al. 2020; Yun et al. 2020). By binding to each receptor's extracellular domain, amivantamab can inhibit ligand binding, promote receptor–antibody complex endocytosis and degradation, and induce antibody-dependent cellular cytotoxicity by natural killer cells and trogocytosis by macrophages. Trogocytosis (derived from the ancient Greek *trogo*, meaning "gnaw") is a process whereby lymphocytes conjugated to antigen-presenting cells extract surface molecules from these cells and express them on their own surface. In contrast to phagocytosis, both cells survive this contact.

NSCLC with epidermal growth factor receptor (EGFR) exon 20 insertion mutations exhibits inherent resistance to approved tyrosine kinase inhibitors. Amivantamab with its immune cell-directing activity binds to each receptor's extracellular domain, bypassing resistance at the tyrosine kinase inhibitor binding site.

Dosing of amivantamab appears a little bit exceptional. Patients with a body weight below 80 kg at baseline receive 1050 mg, and those with equal or greater are dosed with 1400 mg. This two-tiered weight-based dosing was estab-

Table 23.3 Initial and subsequent dosing of amivantamab depending on baseline body weight

Week	Dose 1050 mg (<80 kg)	Dose 1400 mg (≥80 kg)
1, day 1	350 mg	350 mg
1, day 2	700 mg	1050 mg
2, day 1	1050 mg	1400 mg
Subsequent weeks, day 1	1050 mg	1400 mg

lished using population pharmacokinetic analysis. This dosing reduces pharmacokinetic variability and exposure differences. Of note, this dosage is based on baseline body weight and only changed in case of adverse events, as specified by the FPI/SmPC. Amivantamab exhibits linear pharmacokinetics at 350-1,750 mg and nonlinear pharmacokinetics below 350 mg. A dose reduction below 350 mg therefore makes no sense (compare dosing week 1, day 1, Table 23.3). The dose of 1050 mg, however, provides saturation of circulating serum EGFR and MET targets (Park et al. 2021). To reduce the IRR risk, the initial infusion must be split to be administered in two consecutive days (Table 23.3).

Amivantamab Safety

An IRR prophylaxis has to be administered. Antihistamine and antipyretic prophylaxis is required at all doses. Glucocorticoid is also required at the initial dose (week 1, days 1 and 2), then optional for subsequent doses in the absence of IRR.

The antibody can cause a drug-induced interstitial lung disease (ILD), sometimes called nonbacterial, nonviral pneumonitis. Recognized too late, an ILD can be fatal within a very short time period. Patients have to be monitored for new or worsening symptoms indicative of ILD/pneumonitis.

Ocular toxicity, including keratitis, dry eye symptoms, conjunctival redness, blurred vision, visual impairment, ocular itching, and uveitis, can be associated with amivantamab (keratitis in 0.7%, uveitis in 0.3% of the safety population). All reported events were grades 1–2. Patients presenting with eye symptoms shall be promptly referred to an ophthalmologist.

Other toxicities are consistent with on-target inhibition of the EGFR and MET pathways. EGFR inhibition causes dermatologic adverse reactions as described in the anti-EGFR antibodies section, including the long-eyelash syndrome. MET inhibition causes hypoalbuminemia, typically during the first 8 weeks of treatment, and peripheral edema.

Anti-Growth Factor Receptor Antibodies: Anti-EGFR

Anti-EGFR Strategies

The cell membranous receptors of the epidermal growth factor family consist of four related, functionally different members with tyrosine kinase activity, the homologues EGFR1 to EGFR4 (also called HER1 to HER4 or ErbB1 to ErbB4). The inactive monomers homo- (HER1 with HER1) or hetero-dimerize (HER1 with HER2 or HER3 or HER4) after external ligand binding. Conformational change leads to auto-phosphorylation and subsequent signal transduction for diverse cellular functions including proliferation, cell survival—comprising inhibition of apoptosis, adhesion, and DNA damage repair. In tumors, this pro-survival signal transduction is activated permanently, without binding of a physiological, external ligand such as transforming growth factor-alpha (TGFA), heparin-binding EGF-like growth factor, betacellulin, amphiregulin, epiregulin, epigen, and neuregulins (Rubin and Yarden 2001; Schneider and Wolf 2009; Singh et al. 2016). Diseases with malignancy-associated overexpression of EGFR1 (HER1) are, for example, colorectal carcinomas (CRC), NSCLC, pancreas carcinomas, and squamous cell cancers of the head and neck. Her2 is overexpressed by certain breast and gastric cancers. Over-expression is also known to occur in ovarian (Tuefferd et al. 2007; English et al. 2013; Teplinsky and Muggia 2014), stomach, and aggressive forms of uterine cancer, such as uterine serous endometrial carcinoma (Santin et al. 2008; Buza et al. 2014). HER2 is over-expressed in 30% of salivary duct carcinomas (Chiosea et al. 2015), however, without any therapeutic consequences up to now. Overexpression must be demonstrated by immunohistochemistry (IHC) or fluorescent in situ hybridization (FISH) as a prerequisite for the use of anti-EGFR mAbs.

Cetuximab

Cetuximab is a chimeric IgG1 antibody composed of the Fv region of a murine anti-EGFR antibody with human IgG heavy-chain and kappa light-chain constant regions. It binds to the extracellular domain of EGFR1 with an affinity five- to tenfold greater than endogenous ligands, resulting in inhibition of EGFR signaling. Moreover, it also exerts the cytotoxic immune effector mechanisms described in Chap. 8. It is indicated in the treatment of metastatic colorectal cancer as monotherapy and combination therapy. Cetuximab also binds to a number of EGFR antigen-negative tissues; therefore a saturating loading dose of 400 mg/m² has to be given as a first dose and again, if the weekly interval is exceeded. The weekly follow-up dose is 250 mg/m². Many commonly used chemotherapy regimens for CRC, including irinotecan monotherapy or in combination with infusional fluorouracil, like FolFOx or FolFIri, are administered every 2 weeks (FolFIri treatment regimen (= folinic acid, fluorouracil, irinotecan): irinotecan 180 mg/m² day 1; calcium folinate 400 mg/m² day 1; fluorouracil (5-FU) 400 mg/m² day 1 undiluted bolus injection over 5 min; fluorouracil (5-FU) 2400 mg/m² days 1–2 as 48 h as prolonged infusion; FolFOx treatment regimen (= folinic acid, fluorouracil, oxaliplatin): oxaliplatin 100 mg/m² day 1; calcium folinate 400 mg/m² day 1; fluorouracil (5-FU) 400 mg/m² day 1 undiluted bolus injection over 5 min; fluorouracil (5-FU) 2400 mg/m² day 1–2 as 48 h as prolonged infusion).

The ability to synchronize the administration of cetuximab and concomitant chemotherapy is more convenient for the patients. Dosing of 500 mg/m² every 2 weeks exhibited predictable pharmacokinetics, which were similar to those of the approved weekly dosing regimen. No differences between the weekly and two-weekly regimen on pharmacodynamics were observed. Efficacy and safety were also similar (Tabernero et al. 2008). Cetuximab is also used in the treatment of certain squamous cell cancers of the head and neck in combination with radiation or in combination with a platin-based chemotherapy.

Panitumumab

Panitumumab is the fully human version of an anti-EGFR1 mAb and shares the same target as cetuximab with slightly different binding affinity and specificity. In contrast to cetuximab, no loading dose is necessary—it is continuously dosed with 6 mg/kg q2w.

Both antibodies are only effective in pan-RAS mutation-negative tumors (wild type). K- and N-RAS proteins are G proteins (GTPases) downstream of EGFR and components of EGFR signaling, propagating EGFR signaling events. Activating mutations result in constitutive activation without necessity of ligand binding to the EGFR (Fig. 23.14). In other words, an anti-EGFR antibody therapy for patients

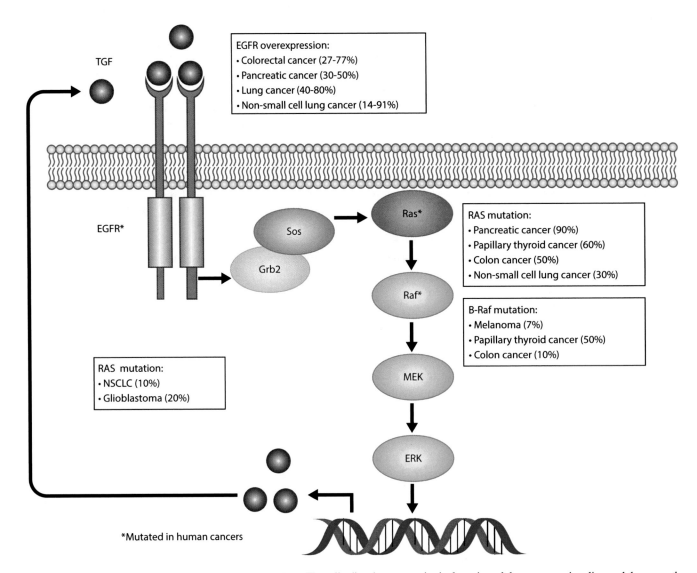

Fig. 23.14 Mutated RAS carcinomas are resistant to anti-EGFR antibodies due to constitutively activated downstream signaling and thus growth promotion

with activating RAS mutations has no therapeutic benefit but exposes the patient to the risk of anti-EGFR mAb-related toxicity.

Pan-RAS tissue testing is mandatory according to clinical guidelines (Van Cutsem et al. 2014) before initiating a cetuximab or panitumumab therapy. The benefit of adding cetuximab to RAS wild type was shown by the pooled analysis of the OPUS and CRYSTAL studies (Bokemeyer et al. 2012).

Approximately up to 10% of CRC tumors also carry a BRAF mutation. RAS mutations and BRAF mutations are usually mutually exclusive (De Roock et al. 2010), so double testing is not necessary. A BRAF mutation is a strong negative prognostic biomarker, and patients with a BRAF mutant CRC have a very poor prognosis (Van Cutsem et al. 2011). However, patients with a KRAS wild type but a BRAF mutation benefit from a cetuximab-containing chemotherapy (Van Cutsem et al. 2011).

Side Matters: Location of the CRC. It has been observed for years that patients with right-sided colorectal cancer had worse outcomes than those with left-sided disease (Brule et al. 2015). Meanwhile, right- and left-sided tumors are regarded as molecular distinct (Tejpar et al. 2016). This translates into therapeutic consequences. Whereas treatment of patients with right-sided tumors should be started with an anti-angiogenic component (bevacizumab), patients with left-sided tumors benefit from an anti-EGFR treatment (Venook et al. 2016) (Fig. 23.15).

Necitumumab

Necitumumab is an anti-EGFR recombinant human monoclonal antibody of the IgG1κ isotype, currently used against NSCLC in combination with cisplatin (75–80 mg/m^2 on day 1) and gemcitabine 1200–1250 mg/m^2 on days 1 and 8). It was approved with a statistically significant difference in median overall survival (OS) of 1.6 months, compared to cisplatin and gemcitabine without antibody (11.5 vs 9.9 months). There is a fixed dosing for necitumumab of 800 mg q3w.

Anti-EGFR mAb Safety

Common anti-EGFR therapy associated adverse events include acneiform rash, diarrhea, hypomagnesemia, hypocalcemia, and infusion reactions. Most common are papulopustular rash of the upper trunk and face skin (60–90%), dry and itchy skin (12–16%), and resulting microbial infections (38–70%), conditioned by open pustules as portal of entry for germs. Less frequently, pruritus, hair modifications, and paronychial inflammation occur (Holcmann and Sibilia 2015). By now, the mechanisms underlying skin disorders induced by EGFR inhibitors are well understood (Holcmann and Sibilia 2015). Alterations in chemokine and cytokine production in keratinocytes may result in attraction of inflammatory cells. Disturbed keratinocyte differentiation impairs proper formation of tight junctions and physiological barrier function. The barrier defect and reduced expression of antimicrobial peptides result in bacterial infections.

Fig. 23.15 Median survival dependent on the tumor location and first-line regimen in colorectal cancer (CRC (based on Venook et al. (2016)). KRAS-WT = KRAS wild type, not mutated. *HR* hazard ratio

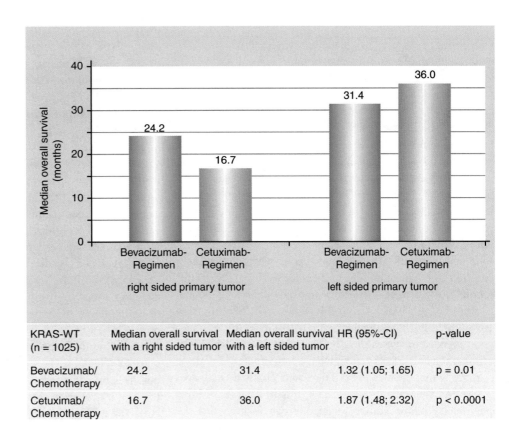

KRAS-WT (n = 1025)	Median overall survival with a right sided tumor	Median overall survival with a left sided tumor	HR (95%-CI)	p-value
Bevacizumab/ Chemotherapy	24.2	31.4	1.32 (1.05; 1.65)	p = 0.01
Cetuximab/ Chemotherapy	16.7	36.0	1.87 (1.48; 2.32)	p < 0.0001

Prophylactic—rather than reactive—management of skin reactions for all patients receiving EGFR inhibitors is recommended. Appropriate prophylaxis can effectively reduce the severity of skin reactions and also a stigmatization as a cancer patient, discernible by their skin eruptions. Skin care has the potential to directly benefit patients and their quality of life. Several recommendations and guidelines have been published (assorted samples: Gutzmer et al. (2011), Lacouture et al. (2011), Potthoff et al. (2011), Hofheinz et al. (2016)). Skin changes during therapy proceed in three phases. Skin care has to be adapted to these phases:

- Phase I: acneiform skin changes.
- Phase II: desiccation phase.
- Phase III: dry, sensitive skin.

During phase III, the skin is very sensitive to sunlight. An unnecessary sun exposure should be avoided, long-sleeved outer clothing should be worn, and the use of sun blockers should be considered. Furthermore, micro-traumatization should be avoided. That is:

- No mechanical manipulation of alleged spots.
- No hot hair drying.
- No use of curlers, especially no tight wrapped curlers.
- No long and hot showering or bathing.
- No vigorous rubbing with towels (hard to comply in case of additional, tantalizing itching).
- No occlusion, for instance, with rubber gloves.
- No too tight footwear.
- Shaving, electric or blade shaving, always causes (unavoidable) micro-traumatization.

Skin cracking that is hardly healing and possibly painful can be sealed with instant glue (like cyanoacrylate glue). This can keep these lesions from worsening or becoming infected. Superglue is also supposed to have some local anesthetic properties (Lacouture et al. 2011).

Electrolyte disturbances (Ca^{2+}, Mg^{2+}) were initially "simply reported" in the SmPC/FPI. However, especially Mg^{2+} loss occurs frequently. The EGFR is also found in the distal part of the collecting duct and other parts of the kidneys but with a rather high expression in the ascending part of Henle's loop and in the distal convoluted tubule, where active reabsorption of magnesium ions takes place (~70%). The EGFR regulates the protein TRPM6 (= transient receptor potential cation channel, under family M member 6) (Schlingmann et al. 2007). By blocking this pump with anti-EGFR antibodies, magnesium losses develop (Izzedine et al. 2010). The elimination half-life of cetuximab is approximately 3–7 days. Assuming that the half-life is 3 days and 5 half-lives are necessary to quantitatively eliminate a drug to have no residual pharmacodynamic activity, then one can assume that the EGFR-dependent magnesium pump is more or less perma-

nently inhibited during the weekly dosing schedules. The same applies for panitumumab with its two-weekly schedule and a half-life of 7.5 days. As a consequence, cumulative hypomagnesaemia of grade 3/4 can develop. This is termed *magnesium wasting syndrome* (MWS). The (randomly) observed frequency of MWS ranges between 27% (Fakih et al. 2006) and 36% (Schrag et al. 2005), although in a prospective cohort study, 97% of the patients were affected (Tejpar et al. 2007). Besides the known symptoms of magnesium loss such as excitability and cardiac effects, a severe fatigue syndrome can be observed. To detect a MWS early, magnesium levels have to be monitored from baseline (as comparative value), because a loss can occur up to 8 weeks after completing therapy. A hypomagnesemia grade 1/2 can be treated by oral substitution. A grade 3/4 hypomagnesemia must be treated by parenteral substitution. Fakih et al. recommend 6–10 g (!) magnesium sulfate daily or twice weekly (Fakih 2008). This magnitude of necessary substitution was confirmed for cetuximab- (Schrag et al. 2005) as well as for panitumumab-induced MWS (Cheng et al. 2009). Such an amount of magnesium sulfate has to be given as a protracted infusion, because of a quick compensatory elimination from short infusions by the kidneys. Ambulatory pumps such as those for infusional 5-FU can be used. In contrast to anti-EGFR mAbs, no magnesium deficiency—not to mention MWS—has been reported with the use of small molecular inhibitors EGFR tyrosine kinase inhibitors such as erlotinib or gefitinib, except in context with diarrhea. During a phase I study with erlotinib combined with the multikinase inhibitor sorafenib, phosphate deficiencies were observed (Izzedine et al. 2010). MWS seems to be a class effect of anti-EGFR mAbs. Detected relatively late after approval of cetuximab and panitumumab, a frequency for hypomagnesemia of 83% is stated in the FPI/SmPC of necitumumab. Meanwhile, electrolyte monitoring (potassium, calcium, magnesium) is recommended to prevent cardiopulmonary arrest.

Hypersensitivity Risks by Tick Bite: An Unexpected Connection with Cetuximab's Structure. As with other mAbs, hypersensitivity prophylaxis by premedication with an antihistamine is recommended. European product specifications also recommend the use of a corticosteroid premedication. However, severe anaphylaxis during the first exposure to cetuximab has been observed. These reactions to cetuximab developed rapidly, and symptoms often peaked during or within 20 min following the first infusion of the antibody and occasionally proved fatal. A lot of patients also reported to be intolerant to mammalian (red) meat such as beef and pork. These first events were associated with a specific geographical region: a group of southern US states. It was not until 2006 that the true severity of the reactions became obvious. Tick bites can induce an immunological reaction against the oligosaccharide galactose-alpha-1,3-galactose (alpha-gal). The enzyme beta-galactosyl alpha 1,3 galactosyltransferase, needed for the formation of alpha-gal, is inactivated in humans and higher mammals due to an evolutionary process. As a result, immunocompetent individuals may form IgG isotype antibodies to alpha-gal, which makes alpha-gal an immunogenic carbohydrate. The distribution of these antibodies first became clear from the states in which reactions to cetuximab were occurring, i.e., Virginia, North Carolina, Tennessee, Arkansas, Oklahoma, and Missouri (Chung et al. 2008). Subsequently, it has become clear that the syndrome of delayed anaphylaxis to red meat is most common in these same states (Commins et al. 2009). In fact, it was the similarity between the region for reactions to cetuximab and the maximum incidence of Rocky Mountain spotted fever that suggested that tick bites might be relevant to these reactions. The alpha-gal component is also a partial structure Fab fragment of cetuximab's heavy chain. Each cetuximab molecule contains two alpha-gal epitopes that can cross-link the high-affinity receptor for IgE on mast cells, leading to mast cell activation and release of hypersensitivity mediators (Saleh et al. 2012) (Fig. 23.16). IgE binding to alpha-gal was later linked to allergic reactions to red meat in America and Europe.

Despite the development of an ELISA test for anti-cetuximab IgE for the identification of patients with high risk (Mariotte et al. 2011), no recommendations for pretherapeutic testing exist up to now. The phenomenon, however, is mentioned in the SmPC. For further reading, consult Chung et al. (2008), Commins et al. (2009), Commins et al. (2011), Saleh et al. (2012), Berg et al. (2014), and Steinke et al. (2015).

Anti-Growth Factor Receptor Antibodies: Anti-EGFR2 (= Anti HER2/neu)

Trastuzumab

Trastuzumab is mainly indicated for HER2/*neu* overexpressing breast cancers in different clinical situations, as monotherapy as well as in combination therapy. It is also approved for HER2/*neu* overexpressing gastric cancers in combination with chemotherapy. Overexpression has to be verified by IHC or FISH testing. Dosing depends on the interval. For the weekly regimen, a 4 mg/kg loading dose is necessary, followed by 2 mg/kg weekly. The loading dose for the 3-weekly interval is 8 mg/kg followed by 6 mg/kg. Like rituximab, a fixed dose preparation of s.c. trastuzumab together with recombinant hyaluronidase, 2000 U/mL, is available (trastuzumab and hyaluronidase-oysk in the United States). 600 mg is given q3w over 2–5 min in the thigh. No loading dose is necessary. However, this time-saving procedure compared to i.v. infusions is counteracted by the EMA's demand to supervise the patients for 6 h (!) after the first and for 2 h after subsequent injections.

Pertuzumab

Pertuzumab is another anti-HER2 antibody. Trastuzumab binds to the HER2 subdomain IV and disrupts ligand-*in*dependent HER2 signaling (Fig. 23.17). Pertuzumab binds to subdomain II and blocks ligand-dependent HER2 heterodimerization with HER1, HER3, and HER4. Therefore, pertuzumab has been classified as the first HDI = HER2 dimerization inhibitor. The combination of pertuzumab and trastuzumab significantly augmented antitumor activity in HER2-overexpressing xenograft models, which was the rationale for the development of this double antibody-based anti-Her2 strategy. Improved anticancer activity was proven in patients treated with pertuzumab in combination with trastuzumab compared to either drug alone (Baselga et al. 2012; Cortes et al. 2012).

Pertuzumab is given as a combination therapy with trastuzumab on a 3-weekly basis (with docetaxel). The initial loading dose is 840 mg followed by 240 mg q3w (trastuzumab: 8 mg/kg loading dose followed by 6 mg/kg q3W). Since 2020, a fixed combination of pertuzumab, trastuzumab, and hyaluronidase for s.c. injection with 1,200 mg pertuzumab, 600 mg trastuzumab, and 30,000 units hyaluronidase for initial dosing and 600 mg pertuzumab, 600 mg trastuzumab, and 20,000 units hyaluronidase for maintenance dosing is available.

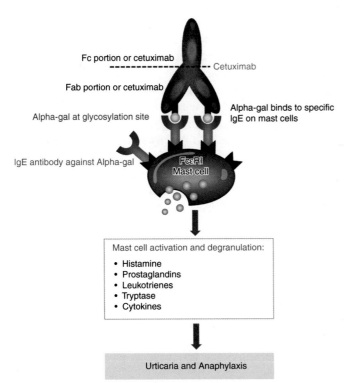

Fig. 23.16 Infusion reactions with cetuximab are linked to the presence of IgE antibodies directed against the alpha-gal component of the Fab fragment of the cetuximab heavy chain

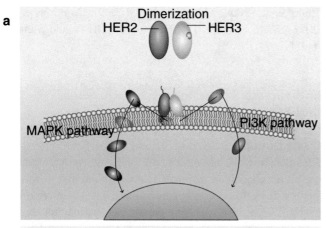

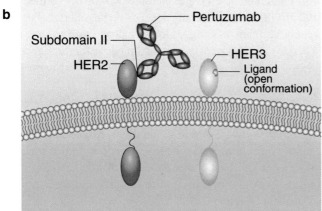

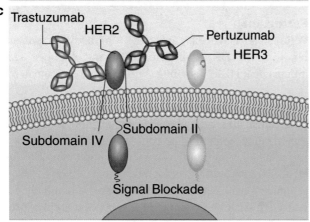

Fig. 23.17 (**a**) HER2/3 heterodimerization with subsequently the strongest mitogenic signaling and activation of two key pathways that regulate cell survival and growth. (**b**) Pertuzumab binds to subdomain II and blocks ligand-dependent HER2 heterodimerization with HER1, HER3, and HER4. (**c**) Trastuzumab binds to subdomain IV and disrupts ligand-independent HER2 signaling. Trastuzumab in combination with pertuzumab provides a more comprehensive blockade of HER2-driven signaling pathways

Ado-trastuzumab/Trastuzumab Emtansine (T-DM1)

Ado-trastuzumab/trastuzumab emtansine (T-DM1) is the third anti-HER2 antibody, an ADC coupled with a toxin (Fig. 23.24; see "ADC" section of this chapter). Dosing is based on body weight with 3.6 mg/kg q3w.

Trastuzumab Deruxtecan

Trastuzumab deruxtecan (fam-trastuzumab deruxtecan-nxki in the United States) is a HER2-directed antibody with a topoisomerase I inhibitor as payload (see "ADC" section of this chapter).

Margetuximab

The latest anti-Her2 family member is margetuximab. Margetuximab is a chimeric, Fc-engineered, anti-Her2 IgG1 mAb that shares epitope specificity and Fc-independent anti-proliferative effects with trastuzumab. Fc engineering of margetuximab alters 5 amino acids from wild-type IgG1 to increase affinity for activating Fcγ receptor FcγRIIIa and to decrease affinity for inhibitory FcγRIIb. Margetuximab was approved on the basis of an international, randomized, open-label, phase 3 study (SOPHIA). This was a direct comparison with trastuzumab, each combined with single-agent chemotherapy, in pretreated patients with Her2+ advanced breast cancer. Patients must have had progressive disease after two or more lines of prior Her2-targeted therapy, including pertuzumab, and 1–3 lines of nonhormonal metastatic breast cancer therapy. Allowed combination chemotherapy was capecitabine, eribulin, gemcitabine, or vinorelbine, based on the investigator's choice. The primary endpoint was PFS. The centrally assessed PFS of margetuximab plus chemotherapy was prolonged for 0.9 months in median, compared with trastuzumab plus CTX (5.8 months vs. 4.9 months with $p = 0.03$) (Rugo et al. 2021).

Anti-HER2 mAb Safety

The Her2 receptor has essential roles in embryonal and fetal development and tissue protection, i.e., via anti-apoptosis. Particularly neuronal and non-neuronal tissues, including cardiac myocytes, require Her2 for normal development (Zhao et al. 1998; Negro et al. 2004). By blocking these protective functions, the corresponding toxicity, that is, cardiotoxicity, is observed. This applies to all three anti-HER2 mAbs. All of them carry a warning message in their FPI/SmPC for cardiomyopathy and/or the development of a reduced left ventricular ejection function. That means all anti-HER2 antibodies have the potential to cause ventricular dysfunction and congestive heart failure. There is a higher risk for patients who receive anthracyclines, taxanes, or cyclophosphamide in combination and/or during previous therapy. An evaluation of left ventricular function in all patients prior to and during treatment with these mAbs is mandatory. Post approval surveys revealed a potential for lung toxicity, including interstitial lung disease. Some (rare) cases developed respiratory distress syndrome, including some with fatal outcome. Severe IRRs, including hypersensitivity and anaphylaxis, have also been reported with trastuzumab and hyaluronidase-oysk (s.c.). In the HannaH and SafeHER trials, 9% and 4.2% of patients experienced grade 1–4 hypersensitivity and anaphylaxis, respectively, whereas grade 3–4 hypersensitivity and anaphylactic reactions occurred in 1% and <1% of the patients (GenentechInc 2019). Exacerbation of chemotherapy-induced neutropenia can occur with trastuzumab and can also occur when combining pertuzumab and trastuzumab. For T-DM1, hepatotoxicity has been reported, predominantly as asymptomatic elevations of transaminases, and neurotoxicity. This is most probably attributable to MMAE. The most common adverse reactions are infusion reactions (usually mild to moderate), which rarely require discontinuation of therapy. A routine prophylaxis with antihistamines, antipyretics, and/or corticosteroids is not recommended. Skin toxicity like rash, pruritus, and dry skin is described. However, an intense prophylaxis therapy as required for the anti-EGFR mAbs is apparently not necessary. For trastuzumab deruxtecan, there is an explicit warning for the development of an ILD (DaiichiSankyo 2022). Of all ≥ grade 3 toxicities, an ILD was observed in 3% of the patients. Of these, 2.6% had a fatal outcome. Patients should be informed about a potential lung toxicity, monitored for signs and symptoms of ILD/pneumonitis, and should be advised to immediately report cough, dyspnea, fever, and/or any new or worsening respiratory symptoms.

Anti-Angiogenic Antibodies

A small (undetectable) tumor is nourished by passive diffusion. When it has reached dimensions of 1–2 mm across, diffusion is not sufficient any longer. This leads to evolutionary pressure for the tumor and, in the sense of gain of function, it can secrete autocrine pro-angiogenic growth factors such as vascular endothelial growth factor (VEGF). The angiogenic switch has been turned on. The subsequent formation of new blood vessels from preexisting vessels (angiogenesis) is crucial for tumor development, growth, and metastasis (Ohta et al. 1996). VEGF, with its different subtypes (VEGF-A to VEGF-E), is a secretory, pro-angiogenic cytokine and one of the major regulators of the process of neovascularization (Ohta et al. 1996). An important regulating process in angiogenesis is the interaction of the VEGF subtypes with their receptors. Figure 23.18 shows the members of the VEGF family with their receptors and the resulting effects.

Anti-angiogenic therapy leads to

- Regression of existing microvasculature, known as antivascularization.
- Normalization of the surviving vasculature offering optimal chemotherapy delivery within the tumor, thus enhancing antitumor properties.
- Anti-angiogenic effect leading to the inhibition of new vessel growth, offering improved response rates and tumor death with the potential of improving patient relevant outcomes (progression free survival, PFS; time to progression, TTP; overall survival, OS).

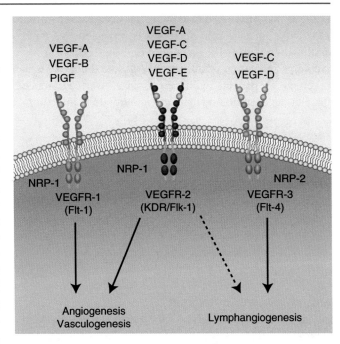

Fig. 23.18 VEGF family and their corresponding receptors. PlGF was discovered in human placenta and shares a 50% homology to VEGF-A. *VEGF* vascular endothelial growth factor, *PlGF* placenta growth factor, *VEGF-R* vascular endothelial growth factor receptor, *NRP* neuropilin

Up to now, there are two principles of anti-angiogenic therapy:

- Intercepting the soluble growth factors in the peripheral blood before they can interact with their receptors (bevacizumab, aflibercept).
- Blocking the receptor (a) by a mAb (ramucirumab) from the extracellular space and (b) by a small-molecule kinase inhibitor from the intracellular space (e.g., axitinib, cabozantinib, nintedanib, pazopanib, regorafenib, sorafenib, sunitinib—not further discussed here).

Bevacizumab

Bevacizumab is directed against VEGF-A and its isoforms. VEGF is intercepted in the peripheral blood and tumor microenvironment and thus neutralized before it can exert pro-angiogenic effects. Bevacizumab is used against several solid tumors like CRC (initial approval), NSCLC, and breast and renal cancer. Dosing depends on the schedule interval and the underlying disease. 5–10 mg/kg q2w and 7.5–15 mg/kg q3w are given.

Ziv-aflibercept/Aflibercept

Ziv-aflibercept/aflibercept (in EU) is a recombinant fusion protein consisting of the VEGF-binding extracellular domain of human VEGF receptors 1 and 2, fused to the Fc part of human IgG1. It binds to all VEGF-A isoforms with a 100-fold higher affinity than bevacizumab. It also binds to

VEGF-C and VEGF-D and to PlGF. It is designated as VEGF trap. The clinical use in oncology is restricted to CRC in combination with FolFIri at a dose of 4 mg/kg q2w. A non-oncologic indication of aflibercept (of note the INN) is the treatment of patients with neovascular (wet) age-related macular degeneration by ophthalmic intravitreal injection.

Ramucirumab

Ramucirumab is a recombinant human IgG1 monoclonal antibody that specifically binds to vascular endothelial growth factor receptor 2. Its clinical use comprises NSCLC and gastric and colorectal cancer as single agent (gastric or gastroesophageal junction adenocarcinomas) or in combination with selected chemotherapy regimens. The dosage for gastric cancer/CRC is 8 mg/kg qw2, whereas for NSCLC, it is 10 mg/kg q3w.

Anti-angiogenic mAb Safety

Initially, it was assumed that VEGF-targeted therapies would be toxicity free. However, clinical trials and experience revealed several adverse events associated with anti-angiogenic agents, mAbs, as well as small kinase inhibitors, which can be summarized as a class effect (Hutson et al. 2008; Chen and Cleck 2009). These treatment-emergent adverse events comprise:

- Hypertension.
- Impaired wound healing.
- Hemorrhage, including severe courses.
- Proteinuria.
- Nephrotic syndrome or thrombotic microangiopathy.
- Gastrointestinal perforation.
- Fistula formation.
- Arterial thromboembolic events.
- Grade 4 venous thromboembolic events (including pulmonary embolism).
- Posterior reversible encephalopathy syndrome (PRES), also known as reversible posterior leukoencephalopathy syndrome (RPLS), colloquially referred to as "cortical blindness."

Pre-therapeutic hypertension has to be adjusted and monitored through therapy. Sometimes, "aggressive" interventions for blood pressure control have to be undertaken, or therapy has to be discontinued. Blood pressure is mainly driven by the VEGF-R2, e.g., by NO and prostacyclin I_2 release from endothelial cells. By blocking VEGF-R2 and intercepting its ligands, the synthesis of vasodilators is suppressed, which in turn increases the peripheral resistance and in consequence the blood pressure (Verheul and Pinedo 2007). Elevated peripheral vascular resistance can also contribute to a reduced left ventricular ejection function and thus be the cause for (congestive) heart insufficiency.

Wound healing depends on neoangiogenesis. Therefore, it is easy to understand that wound healing is impaired by anti-angiogenic therapy. The mentioned mAbs may be used at the earliest 4 weeks after surgery and after complete wound healing. They also must be interrupted for 4 weeks before a planned surgery. Gastrointestinal perforations, especially along surgical sutures, have occurred.

PRES or RPLS is a rare (<1%) but severe side effect of anti-angiogenic or blood pressure-elevating drugs (examples of the latter: proteasome inhibitors). RPLS is a brain–capillary leak syndrome related to hypertension, fluid retention, and the cytotoxic effects of immunosuppressive agents on the vascular endothelium. It seems that severe hypertensive encephalopathy leads to RPLS and vasogenic edema of the posterior cerebral white matter, induced by endothelial dysfunction and a disrupted blood–brain barrier (Verheul and Pinedo 2007; Widakowich et al. 2007). This rare syndrome regained attention after the approval of bevacizumab, but for most of the anti-angiogenic kinase inhibitors, this side effect has been described, as well as for proteasome inhibitors such as carfilzomib, which also elevates blood pressure. Patients who experience a RPLS must never be exposed again to the causative drug.

Toxin-Conjugated Antibodies: Antibody–Drug Conjugates (ADCs)

Antibody–drug conjugates (ADCs) are chemically linked combinations of mAbs and small-molecule drugs with anti-tumor activity. As mentioned above, the unconjugated drug alone is too toxic for the patient. Therefore, common features of toxin-coupled antibodies are:

- The ADC has to be stable in circulation in vivo.
- The antigen has to be predominantly tumor specific.
- After binding to the tumor antigen, the ADC has to be internalized.
- When internalized, the ADC has to be degradable inside the cell to release the lethal cargo.

The linker between the mAb and toxin can be distinguished into enzymatically cleavable linkers, such as the dipeptide linkers used in brentuximab vedotin, or enzymatically uncleavable linkers, such as the thioether linker in ado-trastuzumab emtansine (trastuzumab emtansine in the EU, syn. T-DM1). Linker technologies are depicted in short in Table 23.4.

The linker itself contributes to the properties of the ADC including PK and ADME (overview (Han and Zhao 2014)).

Not applying to the ADC characterization are the fusion toxins moxetumomab pasudotox and tagraxofusp. The former is the murine immunoglobulin variable domain of CD22,

Table 23.4 Linker technology and designated release mechanisms, according to Pettinato (2021)

ADC linker technology	Release mechanism
Disulfide	To be cleaved through disulfide exchange with an intracellular thiol, such as glutathione
Hydrazone	For serum stability and degradation in acidic compartments within the cytoplasm
Peptide	To be enzymatically hydrolyzed by lysosomal proteases such as cathepsin B
Thioether	Nonreducible, designed for intracellular proteolytic degradation

Table 23.5 Currently approved ADCs, in alphabetical order of the INN

INN (in alphabetical order)	Target	Linker
Belantamab mafodotin	Anti-BCMA❶ + toxin	Protease resistant
Brentuximab vedotin	Anti-CD30 + toxin	Protease cleavable
Enfortumab vedotin	Anti-nectin 4 + toxin	Protease cleavable
Gemtuzumab ozogamicin	Anti-CD33 + toxin	Protease cleavable
Inotuzumab ozogamicin	Anti-CD22 + toxin	Protease cleavable
Loncastuximab tesirine	Anti-CD19 + toxin	Protease cleavable
Mirvetuximab soravtansine	Folate receptor alpha (FRα)	Protease cleavable
Polatuzumab vedotin	Anti-CD79b + toxin	Protease cleavable
Sacituzumab govitecan	Anti-Trop-2❷	Protease cleavable
Tisotumab vedotin	Anti-tissue factor + toxin	Protease cleavable
Trastuzumab deruxtecan (DXd)	Anti-HER2/neu (ErbB2) + toxin	Protease cleavable
Trastuzumab emtansine (T-DM1)	Anti-HER2/neu (ErbB2) + toxin	Protease resistant

❶ B-cell maturation antigen
❷ Trophoblast cell-surface antigen-2

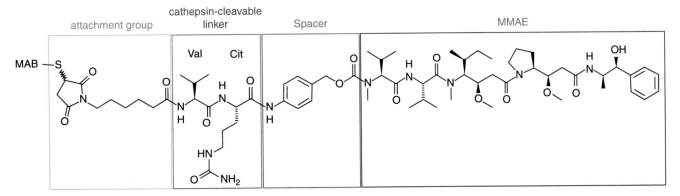

Fig. 23.19 The suffix in the naming of an ADC is comprised of the payload itself, a linker, and sometimes a spacer and/or attachment group. Exemplifying the ADC of brentuximab vedotin with MMAE and the attached mAb-linker conjugate via a spacer (para-aminobenzyl carbamate), MMAE is attached to the cathepsin cleavable amino acid VC linker, followed by an attachment group, consisting of maleimide and caproic acid

genetically fused to a truncated form of *pseudomonas* exotoxin, PE38. The latter is composed of recombinant human interleukin-3 and truncated diphtheria toxin fusion protein, not even a mAb. These fusion proteins are not further discussed in this chapter. Table 23.5 lists the currently used ADCs.

Naming of the Conjugate

The name of an ADC consists of a prefix and a suffix. The prefix denominates the mAb component. The suffix consists of a linker (see above) plus the pure payload, i.e., the toxin itself. Sometimes the suffix terms the payload plus a linker plus a necessary spacer and/or attachment group for steric reasons, as exemplarily shown in Fig. 23.19.

There are no general guidelines for the designation of ADC suffixes. At the time of writing, for example, in all approved vedotin ADCs (brentuximab vedotin, enfortumab vedotin, polatuzumab vedotin, tisotumab vedotin), the vedotin part consists of maleimidocaproyl-valine-citrulline-p-aminobenzyloxycarbonyl (VC) linker plus MMAE.

Characterization of the Toxins: Mechanisms of Action

Ozogamicins and Tesirines

The payload of the ozogamicins is a cytotoxic enediyne-antibiotic *N*-acetyl derivative from *Micromonospora echinospora* ssp. *Calichenensis*, termed calicheamicin γ_1^I. The calicheamicin cleaves double-stranded DNA via a radical mechanism. Like bleomycin, it belongs to the so-called free radical-based DNA-cleaving natural products. Calicheamicin binds with sequence specificity to TCCT-, TCTC-, and TTTT <-- in the minor grove of the DNA (minor groove-binding alkylating agent).

Gemtuzumab Ozogamicin

Gemtuzumab ozogamicin is an anti-CD33-directed mAb, coupled with calicheamicin γ_1^I (gamma-one-iodine, Fig. 23.20).

Gemtuzumab ozogamicin originally received accelerated approval in May 2000 as a stand-alone treatment for older patients with CD33-positive AML who had experienced a relapse. It was voluntarily withdrawn from the market after subsequent confirmatory trials failed to verify clinical benefit and demonstrated safety concerns, including a high number of early deaths. Gemtuzumab ozogamicin was reapproved in September 2017 with a modified dosing regimen. This approval includes a lower recommended dose, a different schedule in combination with chemotherapy or on its own, and a new patient population (FDA 2017).

Gemtuzumab Ozogamicin Safety. Worrisome adverse events are end organ damages such as veno-occlusive disease or the sinusoidal obstruction syndrome, which can lead to fatalities. Other, in part severe, treatment-emergent adverse events are hemorrhage, mucositis, nausea, vomiting, constipation, headache, increased liver enzymes (AST and ALT), rash, fever, and infection.

Inotuzumab Ozogamicin

Inotuzumab ozogamicin carries the same toxin as gemtuzumab in gemtuzumab ozogamicin, but the mAb component of the ADC is directed against the CD22 antigen on (precursor) B-ALL cells. Safety and treatment emergent adverse events are similar to those of gemtuzumab ozogamicin.

Another group of minor groove-binding alkylating agents are the payloads of the tesirines—a pyrrolobenzodiazepine (PDB) dimer. PDBs are a class of natural products produced by various *actinomycetes*. The first pyrrolobenzodiazepine antitumor antibiotic, anthramycin, synthesized by *Streptomyces refuineus*, was discovered in 1965 (Fig. 23.21). PDBs are cardiotoxic, which has precluded their continued clinical use as unmodified molecules.

PDBs act as sequence-selective DNA alkylating compounds. PBD dimers bind in the minor groove of DNA, where they form covalent amine cross-links between the N2 of guanine and the C11 position of the PBD. The resulting PBD–DNA adducts cause replication forks to stall cell proliferation and tumor cells to arrest at the G2-M boundary, ultimately resulting in apoptosis at low nanomolar to pico-

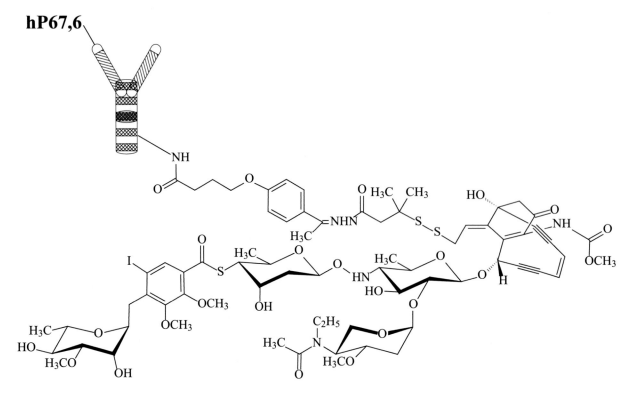

Fig. 23.20 The complex chemical structure of *N*-acetyl-calicheamicin γ_1^I coupled to an anti-CD33 mAb (symbolized in the upper left; © J. Barth)

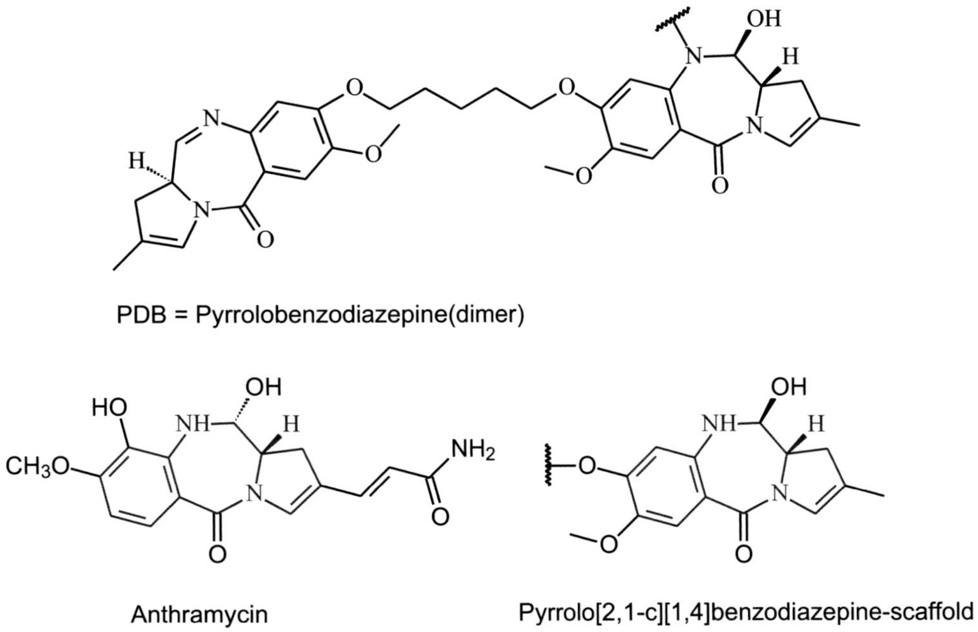

Fig. 23.21 Above: the PDB dimer active "warhead" used in the ADC loncastuximab tesirine. Below: the naturally occurring precursor anthramycin and the PDB scaffold (formulas © by J. Barth)

PDB = Pyrrolobenzodiazepine(dimer)

Anthramycin

Pyrrolo[2,1-c][1,4]benzodiazepine-scaffold

molar concentrations (EditorialReviewTeam 2022). A dramatic increase in cytotoxicity and sequence selectivity has been achieved by linking two PBD units to form PBD dimers as cross-linking agents on opposite DNA strands (e.g., interstrand cross-links). Pyrrolobenzodiazepines typically demonstrate IC_{50} values in the low to mid picomolar range in a variety of cell types in vitro.

Loncastuximab Tesirine

Loncastuximab tesirine is a CD19-directed ADC indicated from third line onward, i.e., after two or more lines of systemic therapy, of adult patients with relapsed or refractory large B-cell lymphoma. This includes diffuse large B-cell lymphoma (DLBCL), not otherwise specified (NOS) DLBCL arising from low-grade lymphoma, and high-grade B-cell lymphoma. NOS means without *Myc* and *BCL2* and/or *BCL6* rearrangement, thus no "double-" or "triple-hit" lymphoma. Loncastuximab tesirine is composed of a humanized IgG1κ mAb, conjugated to SG3199 PDB dimer. SG3199 is the most cytotoxic payload employed in a marketed ADC to date. This PBD dimer acts as a minor groove-binding alkylating agent as described above, linked through a protease-cleavable valine–alanine linker to the anti-CD19 mAb. Baseline dosing is 0.15 mg/kg q3w for two cycles. From cycle 3 onward, the dose is bisected to 0.075 mg/kg q3w. For patients with a body mass index (BMI) ≥35 kg/m², an adjusted body weight (ABW) has to be used. The ABW calculates as follows: ABW [kg] = 35 kg/m²× [height in meters]. The monoclonal antibody portion of loncastuximab tesirine is expected to be metabolized into small peptides by catabolic protein pathways. The small-molecule cytotoxin SG3199 is metabolized by CYP3A4/5 in vitro.

Loncastuximab Tesirine Safety

A premedication with dexamethasone 4 mg orally or intravenously twice daily for 3 days beginning the day before loncastuximab tesirine infusion should be given. If dexamethasone administration does not begin the day before the planned infusion, dexamethasone should begin at least 2 h prior to administration of loncastuximab tesirine. Typical for alkylating agents, the PDB dimer is myelotoxic. Patients should be monitored for the development of pleural and pericardial effusions, ascites, and peripheral and general edema. Serious cutaneous reactions occurred in patients treated with loncastuximab tesirine. Grade 3 cutaneous reactions occurred in 4% and included photosensitivity reaction, rash (including exfoliative and maculo-papular), and erythema. Monitoring and dermatologic consultation should be considered.

Vedotins and Mafodotins: Auristatins

Auristatins are dolastatin 10-based analogues. Clinical trials of dolastatin 10 did not progress because of its nonspecific toxicity. The cytotoxic compounds of vedotins and mafodotins are synthetic derivatives of dolastatin 10, which are pentapeptides found in the wedge sea hare, a shell-less mollusk, *Dolabella auricularia*. They are designated as monomethyl auristatin E (MMAE) and monomethyl auristatin F (MMAF)—mnemonic: vEdotin = MMAE and maFodotin = MMAF (Fig. 23.22). The difference between these two molecules is a phenylalanine present at the C-terminus of MMAF. The charged carboxylic acid terminus limits its passive diffusion into surrounding cells. That means that MMAF has hardly any bystander effect cell killing, whereas MMAE can enter (and leave) cells via passive diffusion (Doronina et al. 2006). Monomethyl auristatins are 100–1,000 times

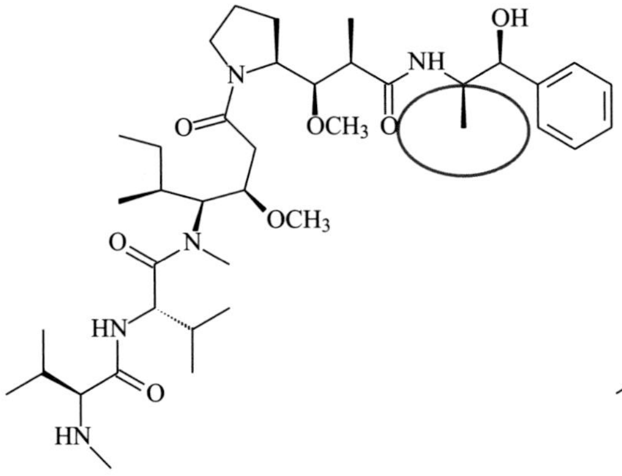

MMAE (cis Konformer) + linker = Vedotin

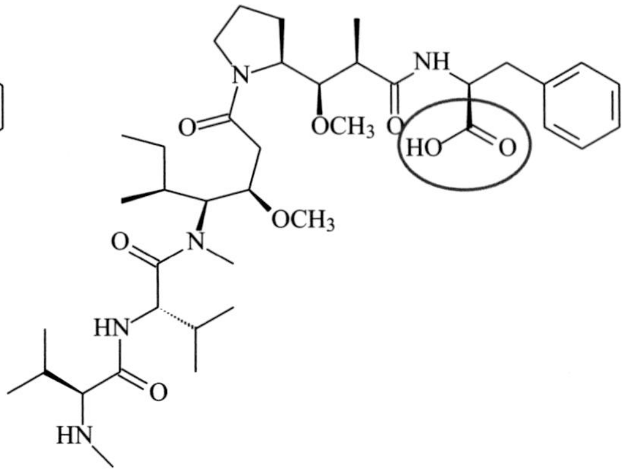

MMAF (cis Konformer) + linker = Vedotin

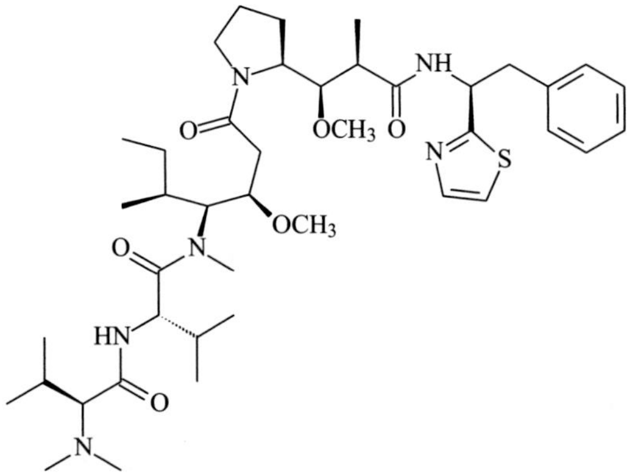

Dolastatin 10

Fig. 23.22 Above: the synthetic dolastatin derivatives MMAE and MMAF, the latter with an ionizable, C-terminal phenylalanine moiety (for comparison, notice the pink circles). This prevents passive diffusion out of the target cell in contrast to its uncharged counterpart, MMAE. Below: the naturally occurring parent substance dolastatin 10 (formulas © by J. Barth)

more toxic than doxorubicin (Nilsson et al. 2010). These synthetic agents interact with the vinca alkaloid binding site on α-tubulin and block its polymerization and prevent the formation of the mitotic apparatus (Bouchard et al. 2014), thus inducing cell cycle arrest in G2/M phase, causing cells to undergo apoptosis (Chen et al. 2017).

MMAE is primarily metabolized by CYP3A4 in vitro, whereas MMAF is mainly hydrolyzed and dehydrated to a cyclized isomeric form. MMAE and MMAF are hemato-, neuro-, and hepatotoxic in general and show some skin and ocular toxicity (see below). Toxicity of the gastrointestinal tract can also occur (Mecklenburg 2018).

Brentuximab Vedotin

Brentuximab vedotin is directed against the CD30 antigen, expressed on classical Hodgkin's lymphoma Reed–Sternberg and anaplastic large cell lymphoma cells and, in part, mycosis fungoides. It is expressed in embryonal carcinomas but not in seminomas and is thus a useful marker in distinguishing between these germ cell tumors. The dosing is 1.8 mg/kg with a cutoff weight of max. 100 kg. In combination with AVD (= doxorubicin, vindesine, dacarbazine), the dosing is 1.2 mg/kg.

Brentuximab Vedotin Safety

The drug has a myelotoxic potential with possible grade 3/4 neutropenia, thrombocytopenia, and anemia. Neutropenia can result in severe infections and/or opportunistic infections. In the presence of severe renal or hepatic impairment, a higher frequency of ≥ grade 3 toxicities and deaths was observed. Besides severe but rare dermatologic reactions including Stevens–Johnson syndrome and toxic epidermal necrolysis, a cumulative, peripheral sensory neurotoxicity has been observed. This may require a delay, change in dose, or discontinuation of treatment.

Brentuximab and ado-trastuzumab (see below) each carry a toxin that, in case of extravasation, could cause skin necrosis like the vinca alkaloids. As mentioned, the antibodies are conjugated with a cleavable or an uncleavable linker to the toxin. Whether cleavable linkers will be lysed in the extravasation area is unclear at the current time. ADCs with an uncleavable liker will be degraded like proteins, and after that, the toxin is released. If and how this may happen to a clinically relevant degree in an extravasation area is also unclear at the current time. Limited clinical experience so far suggests that extravasated ADCs may only cause slight skin reactions.

Enfortumab Vedotin

Enfortumab vedotin is a fully humanized, IgG1 anti-nectin-4 mAb, conjugated to the microtubule inhibitor MMAE through a protease-cleavable linker. Nectin-4 is a type I transmembrane glycoprotein belonging to the Ig superfamily of proteins. Nectins are primarily involved in the formation and maintenance of adherence and tight junctions [(Takai et al. 2008). Physiologically, nectin-4 is mainly expressed by embryo and fetal tissues, while its expression is low in normal adult tissues (Nishiwada et al. 2015). Pathologically, it has been found to be overexpressed by several cancers and demonstrated to promote tumor growth and proliferation (Takano et al. 2009, Derycke et al. 2010, Athanassiadou et al. 2011). Enfortumab vedotin is indicated for the treatment of adult patients with locally advanced or metastatic urothelial cancer who have previously received a checkpoint inhibitor (anti-PD-1 or anti-PD-L1) and a platinum-containing chemotherapy in the neoadjuvant/adjuvant, locally advanced or metastatic setting. The dose is 1.25 mg/kg up to a maximum dose of 125 mg, accordingly to 100 kg body weight.

Enfortumab Vedotin Safety

Hyperglycemia occurred in patients treated with enfortumab vedotin, including death, and diabetic ketoacidosis. Diabetic ketoacidosis may occur in patients with and without preexisting diabetes mellitus. Blood glucose level monitoring is recommended, especially in patients with, or at risk for, diabetes mellitus or hyperglycemia. Patients should also be monitored for peripheral neuropathy. This neuropathy is predominantly sensory in nature. Prophylactic artificial tears should be considered to minimize ocular disorders like dry eyes. However, vision changes may occur. Ophthalmic topical steroids after an ophthalmic exam might be necessary. Skin reactions can occur during enfortumab vedotin treatment (maculopapular rash, pruritus). Grade 3–4 skin reactions occurred in 10% of patients and included symmetrical drug-related intertriginous and flexural exanthema, bullous dermatitis, exfoliative dermatitis, and hand–foot syndrome. For severe reactions, the treatment has to be interrupted until improvement or resolution.

Polatuzumab Vedotin

Polatuzumab vedotin is an ADC with an anti-CD79b mAb designed for the treatment of DLBCL. The current approval is first-line therapy in combination with CHP. CHP is the cyclophosphamide, doxorubicin, vincristine, and predni-sone/prednisolone (CHOP) regimen, where the vinca alkaloid vincristine (O = Oncovin) is replaced by polatuzumab vedotin. For the relapsed/refractory situation, polatuzumab vedotin is combined with bendamustine–rituximab. This means that polatuzumab vedotin is approved in combination with another oncologic mAb, the anti-CD20 antibody rituximab. Anti-CD20 monoclonal antibodies are part of almost any regimen for CD20+ B-cell lymphomas. The molecular rationale for combining these two mAbs in DLBCL was elucidated by Kawasaki et al. (2022) in vitro.

In short, both AKT signaling and ERK signaling have been reported to regulate CD20 expression. Immunoblotting analysis identified that AKT and ERK phosphorylation was increased after polatuzumab treatment. However, it was noticed that the anti-CD79b antibody increased the phosphorylation of AKT but inhibited the phosphorylation of ERK. In contrast, MMAE potentiated phosphorylation of ERK but slightly attenuated the phosphorylation of AKT. Taken together, these results suggested that polatuzumab vedotin activates both AKT signaling and ERK signaling and shows a significant effect on the upregulation of CD20 expression levels with an increased sensitivity to rituximab, in terms of complement mediated cytotoxicity and antibody-dependent cellular cytotoxicity. These effects occur as a mechanism of action independent of the payload. Conversely, a decreased activity of polatuzumab vedotin can be deduced if the DLBCL cells display a low density of the CD79b antigen and potentially in a monotherapy setting of polatuzumab vedotin. A sequence-specific administration (polatuzumab vedotin always before rituximab) is especially in the steady-state situation due to the long half-life of polatuzumab vedotin not necessary (adults with non-Hodgkin's lymphoma 22 days [range 6.1–52]). Dosing is based on body weight with 1.8 mg/kg. Due to limited clinical experience in patients treated with 1.8 mg/kg polatuzumab vedotin at a total dose >240 mg, it is recommended not to exceed the dose 240 mg/cycle according to EuropeanMedicinesAgency (2022d). This recommendation cannot be found in the US FPI. 240 mg corresponds to a body weight of 133.3 kg.

Polatuzumab Vedotin Safety

The payload MMAE is a mitotic spindle poison. Therefore, peripheral neuropathies are typical to the spectrum of side effects. Serious cases of hepatotoxicity, including elevations of transaminases and/or bilirubin, have occurred in patients treated with polatuzumab vedotin. Grade 3 and 4 transaminase elevations develop rarely (1.9% and 1.9%, respectively).

Tisotumab Vedotin

One might be surprised that a fully humanized anti-tissue factor mAb such as tisotumab vedotin is conjugated with the MMAE payload. However, tissue factor (TF) contributes to tumor growth, angiogenesis, metastasis, and thrombosis in

Fig. 23.23 Interplay of TF tumor growth, angiogenesis, metastasis, and thrombosis in cancer patients. TF expression by tumor cells may increase the growth of tumors by increasing cell survival and/or increasing angiogenesis as well as enhancing metastasis by activating coagulation and platelets. Microparticles released by TF-positive tumor cells and host cells into the blood may trigger venous thromboembolism (Kasthuri et al. 2009)

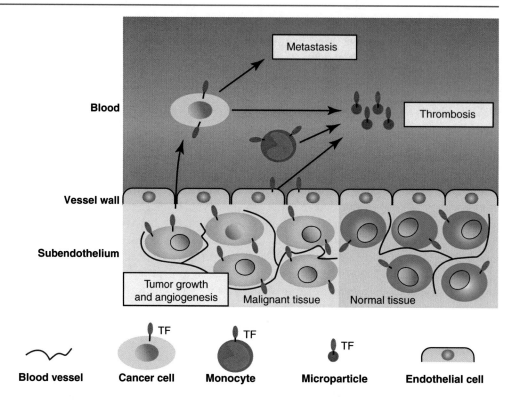

patients with cancer. TF expression by tumor cells may increase the growth of tumors by increasing cell survival and/or increasing angiogenesis. TF expression by tumor cells in the blood enhances metastasis by activating coagulation and platelets. It is well known that coagulation in cancer patients is shifted to the procoagulatory side. It is ascertained that release of TF-positive microparticles by tumor cells and host cells into the blood may trigger venous thromboembolism (Fig. 23.23) (Kasthuri et al. 2009).

Preclinical evidence about tisotumab vedotin showed that in contrast to other TF-targeting mAbs, it does not alter coagulation parameters while maintaining high cytotoxic power (Breij et al. 2014). The approved indication is recurrent or metastatic cervical cancer with disease progression on or after chemotherapy. Dosing is 2 mg/kg with a maximum dose of 200 mg, given as an intravenous infusion every 3 weeks.

Tisotumab Vedotin Safety

Because of the broad distribution of TF in the human body, one might expect a broad spectrum of toxicity. Tisotumab vedotin shares the same potential of peripheral neuropathy as all the other MMAE/MMAF conjugates. By targeting a component of the coagulation cascade, there is an elevated risk for bleeding. Hemorrhage occurred in 62% of patients with cervical cancer treated with tisotumab vedotin across clinical trials. Grade 3 hemorrhage occurred in 5% of patients. Severe, life-threatening, or fatal pneumonitis/ILD can occur in patients treated with antibody–drug conjugates containing vedotin including tisotumab vedotin. Patients have to be monitored concerning neuropathies and pulmonary symp-

toms. Several ocular adverse reactions occurred in 60% of patients with cervical cancer treated with tisotumab vedotin, making a widespread eye care necessary. Besides an ophthalmic exam including visual acuity and slit lamp exam at baseline, prior to each dose, and as clinically indicated, the patient needs to use corticosteroid eye drops for 72 h after each infusion, vasoconstrictor eye drops immediately prior to each infusion, cooling eye pads during the infusion of tisotumab vedotin, and lubricating eye drops for the duration of therapy and for 30 days after the last dose of tisotumab vedotin. The patient also needs to avoid wearing contact lenses unless advised by their eye care provider for the entire duration of therapy (SeagenInc. 2021).

Maytansinoids (DMs)

Maytansinoids (DMs) are thiol derivatives of maytansine (Fig. 23.24). Maytansines bind to tubulin at the rhizoxin/vinblastine binding site and thus inhibit microtubule formation. They are believed to have a high affinity for tubulin located at the ends of microtubules. Their suppression of microtubule dynamics causes cells to arrest in the G2/M phase of the cell cycle, ultimately resulting in apoptosis (Bouchard et al. 2014).

Trastuzumab Emtansine

Trastuzumab emtansine is an anti-Her2 mAb with DM1 as payload (Fig. 23.25). T-DM1 was the first-in-class ADC approved for the treatment of solid tumors. The ADC treatment of solid tumors fell short due to numerous biological barriers in the tumor microenvironment (e.g., poor vascularization, diffusion through dense stroma, overcoming tumor

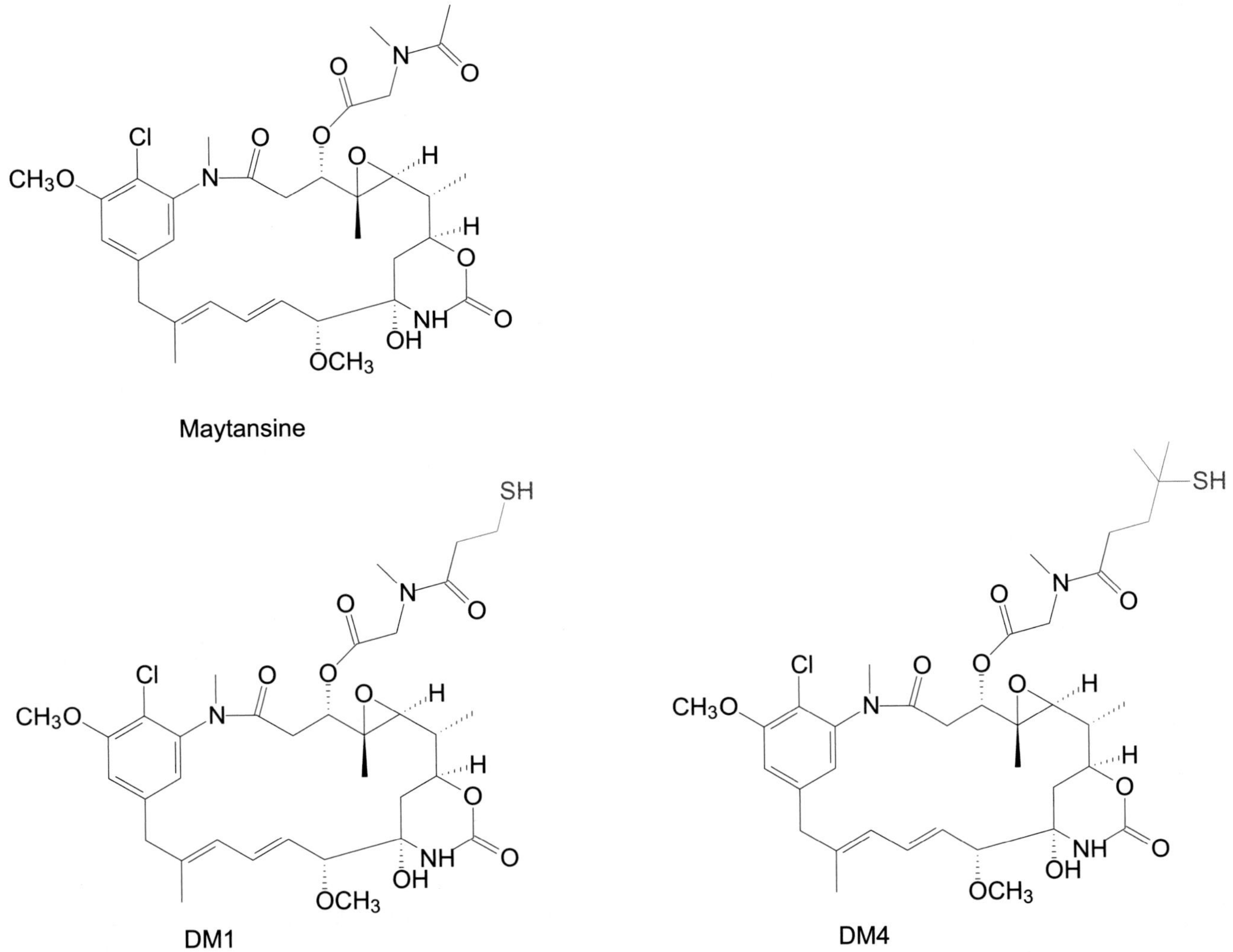

Maytansine

DM1

DM4

Fig. 23.24 The maytansine parent compound and the thiol derivatives DM1 and DM4. DM1 is known from trastuzumab emtansine (syn. T-DM1). The DM4 compound is under investigation as ravtansine

Fig. 23.25 The antibody–toxin construct in T-DM1

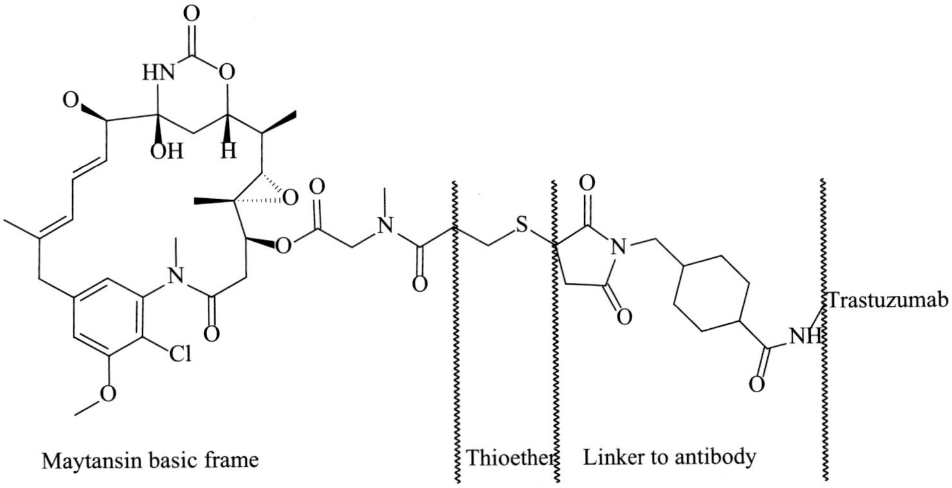

Maytansin basic frame   Thioether Linker to antibody

interstitial fluid pressure). These circumstances limited drug penetration. Unlike hematologic malignancies, the concept of lineage-specific antigen expression (e.g., anti-CD20, anti-CD33) is not applicable to solid tumors. The solid tumor antigens are expressed mainly "tumor associated" rather than "tumor specific" (Criscitiello et al. 2021). This means that the antigens are expressed on tumor cells but also on normal cells, weakly or limited to a given tissue type. This implies that ADCs in these indications have on-target/off-tumor toxicity dependent on the expression of the specific target by normal cells.

Camptothecins

Deruxtecan (DXd)

The toxin component of trastuzumab deruxtecan is an exatecan derivative, DX-8951 (syn. DXd). It is a structural analogue of camptothecin, a topoisomerase I (TOP I) inhibitor (see Fig. 23.27), with strong antineoplastic activity. DXd is 10 times more effective than the active metabolite of irinotecan SN-38 (EuropeanMedicinesAgency 2022c). The TOP I inhibitory concentration (IC_{50}) is 2.78 µM for SN38 compared to 0.31 µM for DXd, i.e., ~9-fold lower.

Trastuzumab Deruxtecan

The recommended dosage of trastuzumab deruxtecan (T-DXd) is 5.4 mg/kg given as an intravenous infusion once every 3 weeks (21-day cycle) for unresectable Her2⁺ breast cancer after two or more prior regimens, until disease progression or unacceptable toxicity. Observed with high interest is the apparent potential of T-DXd against brain metastases. This is more or less unexpected because mAbs are believed not to be able to cross the blood–brain barrier to a significant degree. However, a subanalysis of the DESTINY-Breast03 study comparing T-DXd vs. T-DM1 demonstrated superior progression-free survival with T-DXd vs. T-DM1 (Hurvitz et al. 2023) (Fig. 23.26).

Govitecan

Govitecan is a hydrolyzable antibody linker, bearing the active metabolite of irinotecan, that is, SN38, a TOP I inhibitor. Figure 23.27 compares the camptothecin payloads and the unconjugated used TOP I inhibitors.

Sacituzumab Govitecan

Sacituzumab govitecan consists of a fully humanized IgG1 anti-Trop-2 antibody conjugated to the payload SN-38, a topoisomerase I inhibitor. Trop-2 (trophoblast antigen 2) is a transmembrane glycoprotein coded by the gene *TACSTD2*, which primarily acts as intracellular calcium signal transducer (Shvartsur and Bonavida 2015). Trop2 is expressed in several normal tissues, including the skin, uterus, bladder, esophagus, oral mucosa, nasopharynx, and lungs. It is overexpressed in many epithelial tumors, like breast, urothelial, lung, and gynecological and gastrointestinal carcinomas (Stepan et al. 2011). Its overexpression directly promotes tumor growth by inducing several different oncogenic pathways and has been associated to poor prognosis. The ADC gained FDA approval as a treatment for triple-negative breast cancer (TNBC) after at least two prior therapies for metastatic disease with the recommended dose of 10 mg/kg. Other entities than TNBC are under investigation (NSCLC and urothelial carcinoma).

Sacituzumab Govitecan Safety

Just like irinotecan, sacituzumab govitecan can cause severe diarrhea, which can occur early (cholinergic early symptom) or even delayed (cAMP mediated). For the cholinergic early

Brain Metastases at Baseline

	T-DXd	T-DM1
Median PFS [months], (95% CI)	15.0 (12.5-22.2)	3.0 (2.8-5.8)
12-mo PFS rate, [%] (95% CI)	72.0 (55.0-83.5)	20.9 (8.7-36.6)
HR (95% CI)	0.25 (0.13-0.45)	

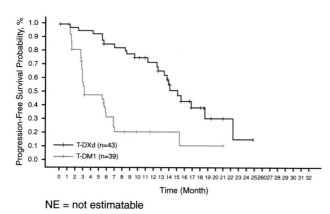

NE = not estimatable

No Brain Metastases at Baseline

	T-DXd	T-DM1
Median PFS, [months] (95% CI)	NE (22.2-NE)	7.1 (5.6-9.7)
12-mo PFS rate, [%] (95% CI)	76.5 (70.0-81.8)	36.4 (29.4-43.4)
HR (95% CI)	0.30 (0.22-0.40)	

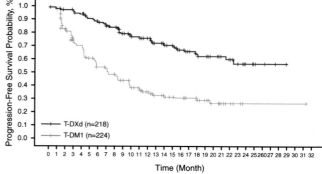

Fig. 23.26 PFS curves of a direct comparison of T-DXd vs. T-DM1 in patients with HER2⁺ metastatic breast cancer, with and without brain metastasis at baseline; for details, see Hurvitz et al. (2023)

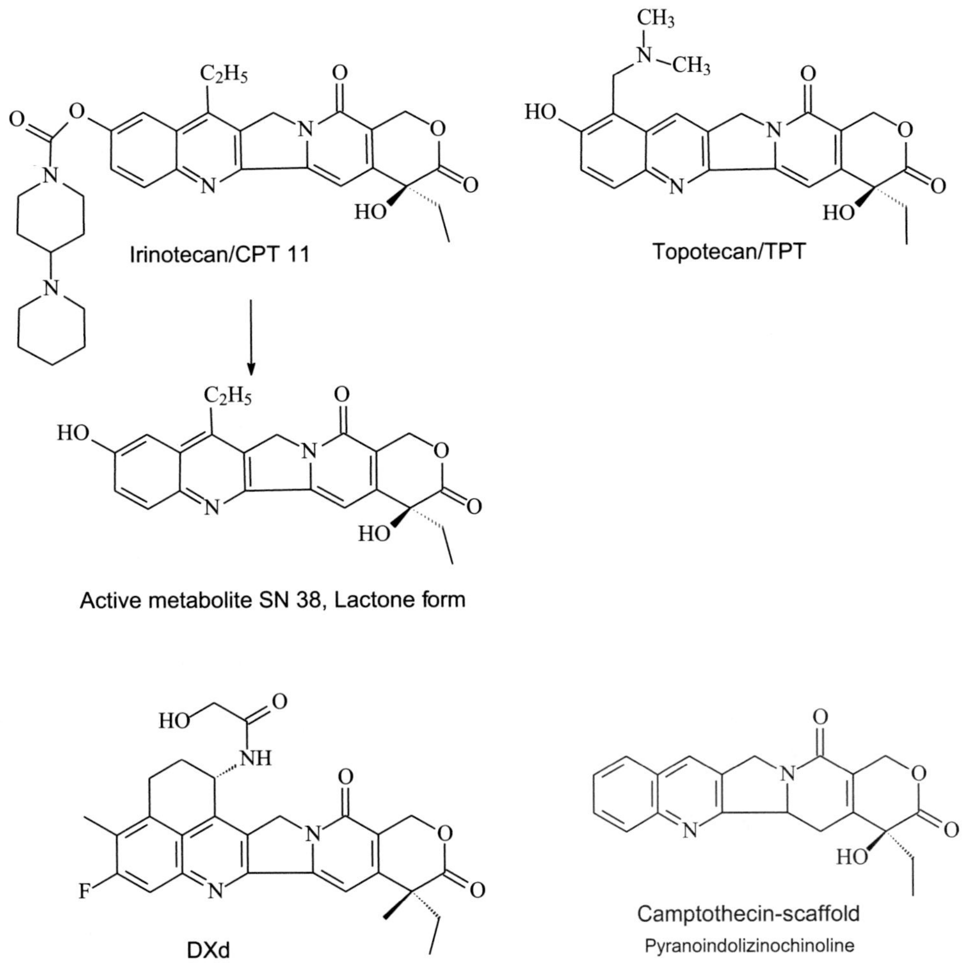

Fig. 23.27 Comparison of the camptothecin parent compound camptothecin (in blue, right below) with the unconjugated TOP I inhibitors topotecan and irinotecan, its active metabolite SN-38, the payload of sacituzumab govitecan and DXd, and the payload of trastuzumab deruxtecan (formula © by J. Barth)

Irinotecan/CPT 11

Topotecan/TPT

Active metabolite SN 38, Lactone form

DXd

Camptothecin-scaffold
Pyranoindolizinochinoline

onset, an intervention with atropine is recommended (plus fluids and electrolytes as needed). Antiemetic prevention is necessary and the use of G-CSF as secondary prophylaxis should be considered. Especially patients with a reduced UGT1A1 activity (homozygous for the uridine diphosphate-glucuronosyltransferase 1A1 (UGT1A1*28 allele) are at increased risk for (febrile) neutropenia following initiation of sacituzumab govitecan.

Immune Oncology

Immune Agonistic Antibodies

The group of immune agonistic antibodies are also designated as immune checkpoint inhibitors (ICIs), checkpoint inhibitors (CI), or immune oncologics (IO), expressions that can all be considered synonymous. In principle, the human immune system can prevent tumorigenesis. Tumor cells, however, have the capability to escape the immune system (Vesely et al. 2011). (Re-)activating the immune system to eliminate cancer cells to produce clinically relevant responses

has been a long-standing "dream of mankind." T cells have an important role to play in fighting cancers. To avoid collateral damage of host tissues and organs, T-cell action is balanced by inhibitory signals and molecules—the so-called immune checkpoints. They are critical for maintaining self-tolerance, on the one hand, and modulating the duration and amplitude of immune responses, on the other.

Normally, after T-cell activation, cytotoxic t-lymphocyte antigen 4 (CTLA4; CD152) is upregulated on the plasma membrane. Its functions is to downregulate T-cell activity through a variety of mechanisms, including preventing co-stimulation by outcompeting CD28 for its ligand, B7, and also by inducing T-cell cycle arrest. Through these mechanisms and others, CTLA4 has an essential role in maintaining normal immunologic homeostasis. CTLA4 downmodulates the amplitude of T-cell activation. Blockade of CTLA4 by mAbs such as ipilimumab keeps T cells stimulated (Fig. 23.28).

While CTLA4 primarily regulates the amplitude of the early stages of T-cell activation, the major role of PD-1 is to limit the T-cell activity in peripheral tissues. Despite its name, programmed cell death protein-1- (PD-1) does not

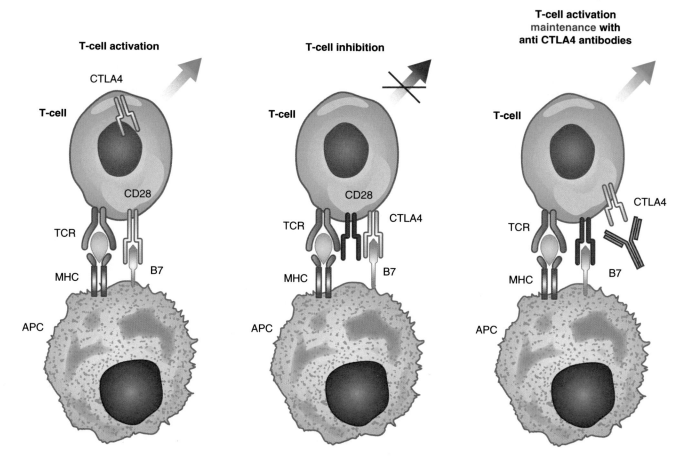

Fig. 23.28 Simplified mechanism of action of anti-CTLA4 antibodies. *Left*: T-cell activation by antigen presenting to the T-cell receptor with the co-stimulatory signal B7 to CD28. *Middle*: Under physiological conditions, the activated T cells are downregulated after 48–72 h by the displacement of costimulatory CD28 with CTLA4. The B7–CTLA4 axis slows down the T cells up to complete inhibition and maintains self-tolerance in the periphery. *Right*: By blocking CTLA4 with mAbs, the T cell remains activated, attacking tumor tissues. *APC* antigen-presenting cell, *CTLA4* cytotoxic T-lymphocyte antigen 4, *MHC* major histocompatibility complex, *TCR* T-cell receptor

induce cell death directly. When engaged with one of its ligands, PD-L1 or PD-L2, PD-1 inhibits kinases that are involved in T-cell activation. In summary, this pathway is a "stop signal" for T cells. Tumors use these inhibitory pathways to evade an immune attack through overexpression of PD-L1. As a result, this blocks the generation of an immune response to the tumor (cf. Figs. 23.28 and 23.29).

Long-term exposure to antigens in the presence of inflammatory cytokines induces a distinct phenotype in T cells, characterized by loss of effector functions, sustained expression of inhibitory receptors, poor proliferative capacity, and decreased cytotoxic functions. This progressive loss of T-cell effector functions is commonly seen during chronic viral infections and in cancer. It is termed "T-cell exhaustion." T cells detect the tumor, accumulate in the tumor microenvironment, however, and are silenced by inhibitory molecules expressed on the tumor surface such as PD-L1/2. PD-1/PD-L1 inhibitors pharmacologically prevent the PD-1/PD-L1 interaction, thus facilitating a positive immune response to kill the tumor. For further immune oncology

information, the reader is referred to Pardoll (2012), Ferris (2013), Intlekofer and Thompson (2013), Li et al. (2016), Suzuki et al. (2016), Alsaab et al. (2017), and Rotte et al. (2018).

Other known negative or inhibitory checkpoint molecules include:

- Lymphocyte activation gene-3 (AG-3) suppresses an immune response by action on Tregs as well as direct effects on CD8+ T cells.
- T-cell Immunoglobulin domain and mucin domain-3 (TIM-3) is expressed on activated human CD4+ T cells and regulates Th1 and Th17 cytokines. TIM-3 acts as a negative regulator of Th1/Tc1 function by triggering cell death upon interaction with its ligand, galectin-9.
- Indoleamine 2,3-dioxygenase (IDO) is a tryptophan catabolic enzyme with immune-inhibitory properties. IDO is known to suppress T and NK cells, generate and activate Tregs and myeloid-derived suppressor cells, and promote tumor angiogenesis.

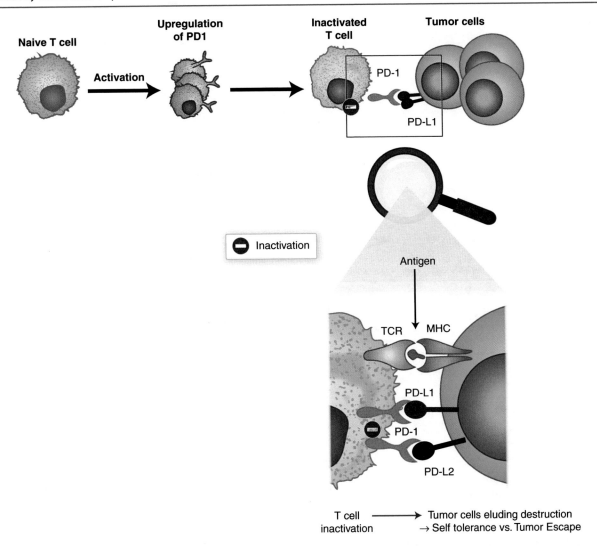

Fig. 23.29 Naive T cells are activated by the presentation of tumor antigens. During this priming phase, PD-1 is upregulated as a physiological reaction to protect unaffected host tissue. The corresponding ligand, PD-L1, expressed on tumor cells, downregulates T cells, pretending to be "friendly," healthy tissue. The magnifier shows the details of the T-cell receptor (TCR) interaction with tumor antigen-presenting major histocompatibility complex (MHC). Inactivating programmed cell death receptor 1 (PD-1) is stimulated by the corresponding ligands (PD-L1/2), resulting in T-cell exhaustion. *MHC* major histocompatibility complex, *PD-1* programmed cell death receptor1, *PD-L1/2* PD-1/2-ligand, *TCR* T-cell receptor

Immune oncology mAbs do not show a clear dose–response relationship. They are either dosed by body weight or with a fixed dose. For the future, it may be possible that antibodies currently dosed by body weight may be switched to fixed dosing.

Due to their mechanism of action, infiltrating the tumor tissue followed by tumor cell killing, an initial increase of tumor lesions was observed. From the formal point of view, these patients were RECIST progressive (RECIST = Response Evaluation Criteria In Solid Tumors). However, it would have been a mistake to stop treatment too early, as biopsies confirmed inflammatory cell infiltrates with an apparent enlargement of the tumor. This phenomenon is termed pseudoprogression or "tumor flare" (short overview (Chiou and Burotto 2015)). Thus,

to ascertain a clinical benefit, it can take 12 weeks or more after the first infusion and may include the emergence of (temporary) new lesions. To avoid misclassification according WHO in immune oncology, immune-related response criteria (irRC) were developed in 2009 (Wolchok et al. 2009).

Anti-CTLA4 Antibodies

Ipilimumab *is* the first anti-CTLA4, recombinant human monoclonal antibody. It is used as monotherapy for the treatment of melanoma in different clinical situations. 3 mg/kg q3w for 4 doses is used for unresectable or metastatic melanoma. In a certain adjuvant setting (cutaneous melanoma

Table 23.6 Approved anti-PD-1 and anti-PD-L1 antibodies, in alphabetical order per column

Anti-PD-1 antibodies	Anti-PD-L1 antibodies
Cemiplimab	Atezolizumab
Dostarlimab	Avelumab
Nivolumab	Durvalumab
Pembrolizumab	

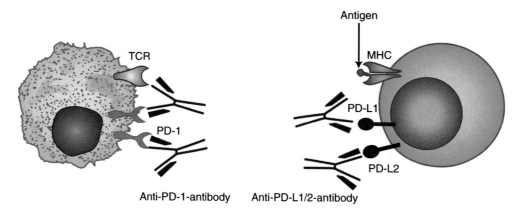

Fig. 23.30 Anti-PD-1 and/or anti-PD-L antibodies prevent the inactivation of immune competent T cells

with pathologic involvement of regional lymph nodes of more than 1 mm that have undergone complete resection, including total lymphadenectomy), 10 mg/kg q3w for 4 doses followed by 10 mg/kg q12w up to 3 years is recommended. Figure 23.28 illustrates the simplified mechanism of action. Tremelimumab is the second approved anti-CTLA4 antibody (only in combination with durvalumab).

Anti-PD-1 and Anti-PD-L1 Antibodies

While development of new anti-PD-1/anti-PD-L1 antibodies is rapidly progressing and will likely provide new agents, the following mAbs were on the market at the time of the creation of this manuscript (Table 23.6).

The anti-PD-1/anti-PD-L1 antibodies are used against different solid tumors. Depending on the underlying disease and the clinical situation (first-line treatment or relapsed tumor), PD-L1 testing is sometimes mandatory. The relevance and usefulness of PD-L1 as a predicting biomarker are still under debate. Figure 23.29 illustrates the pathological mechanisms and Fig. 23.30 the pharmacodynamic intervention with mAbs.

LAG-3 Antibodies

LAG-3 (CD223) stands for "lymphocyte-activation gene 3," a transmembrane protein of the immunoglobulin family. It is a co-inhibitory receptor to suppress T-cell activation and cytokine secretion. This ensures a state of immune homeostasis (Andrews et al. 2017). LAG-3 exerts differential inhibitory effects on various types of lymphocytes. The precise molecular mechanisms of LAG-3 inhibitory signaling and interaction with other immune checkpoints are mostly unclear. However, LAG-3 shows a striking synergy with PD-1 (Long et al. 2018). In particular, combination therapy of the anti-LAG-3 mAb relatlimab plus anti-PD-1 mAb nivolumab has shown impressive clinical efficacy in melanoma patients and was approved as a fixed combination consisting of 240 mg nivolumab and 80 mg relatlimab in a single-dose vial. Patients 12 years of age or older receive a total dose of 480 mg nivolumab and 160 mg relatlimab.

Immune Oncology mAb Safety: Playing with Fire?

Due to their mechanism of action, immune checkpoint inhibitors in the form of mAbs against CTLA4, PD-1, PD-L1, and LAG-3 have a unique spectrum of toxicity that differs from the typical adverse events seen with chemotherapeutic agents. By inhibiting checkpoint molecules, the immune system is activated or even over-activated or, from the viewpoint of the normal tissue, deregulated (non-recognizing "self"). These new kinds of autoimmune-like symptoms are based on the loss of self-tolerance and termed immune-related adverse events (irAEs). For CTLA4-blocking antibodies, toxicities seem to be dose related, because the rate of grade 3–4 drug-related serious irAEs increased from 5% to 18% when the dose was increased from 3 to 10 mg/kg, whereas it was 0% at a dose of 0.3 mg/kg. In contrast, the toxicities related to PD-1 blockade are similar at doses ranging from 0.3 to 10 mg/kg, exemplary observed with nivolumab. The observed toxicities depend on:

- The patient population/the underlying disease
- The dose (especially for CTLA4 antibodies)
- The schedule

That means that the same dose or schedule in different diseases might result in different toxicity profiles. Combinational checkpoint blockade can result in an increase of grade 3/4 treatment emergent adverse events (TEAE) up to 53% (Wolchok et al. 2013). The toxicity pattern is based on induced autoimmunity and comprises:

- Skin irAEs, including Stevens–Johnson syndrome and toxic epidermal necrolysis (both being rare).
- Gastrointestinal irAEs—autoimmune colitis.
- Autoimmune hepatitis.
- Autoimmune pancreatitis.
- Autoimmune thyroiditis.
- Autoimmune hypophysitis.
- Neurological irAEs.
- Autoimmune pneumonitis.
- Ocular irAEs (rare), like autoimmune uveitis and autoimmune episcleritis.

In December 2016, the Federal Institute for Drugs and Medical Devices, Germany, informed the medical professional circles about suspected cases of pancytopenia/agranulocytosis. Onset was between 12 and 274 days after start of therapy. Three instances were fatal. Another rare irAE was myocarditis (nine cases), with two of them having an additional myositis and rhabdomyolysis. Four cases had a fatal outcome. Immune-mediated myocarditis and hematologic toxicity were underrated or even unexpected at that time. This opinion is not understandable from the present point of view. A large retrospective analysis of the World Health Organization pharmacovigilance database (VigiLyze) comprising more than 16,000,000 adverse drug reactions, reported from 2009 through January 2018 in VigiLyze, was evaluated. 613 fatalities due to irAEs in patients treated with immune checkpoint inhibitors were detected. Patients were exposed to anti-CTLA4 (ipilimumab or tremelimumab), anti-PD-1 (nivolumab, pembrolizumab), or anti-PD-L1 (atezolizumab, avelumab, durvalumab). The spectrum differed widely between regimens. In a total of 193 anti-CTLA4 deaths, most were usually from colitis (135 [70%]). Anti-PD-1/PD-L1–related fatalities were often from pneumonitis (333 [35%]), hepatitis (115 [22%]), and neurotoxic effects (50 [15%]). Combination PD-1/CTLA4 deaths were frequently from colitis (32 [37%]) and myocarditis (22 [25%]). Myocarditis had the highest fatality rate (52 [39.7%] of 131 reported cases), whereas endocrine events and colitis had only 2–5% reported fatalities; 10% to 17% of other organ–system toxic effects reported had fatal outcomes (Wang et al. 2018). Hematological immune-related adverse events are meanwhile termed as hem-irAEs. Case reports or case series of the following hem-irAEs are published:

- Aplastic anemia/bone marrow failure.
- Hemophilia A.
- Acute thrombosis.
- Large granular lymphocytosis.
- Hemophagocytic lymphohistiocytosis/macrophage activation syndrome.
- Eosinophilia.
- Hematological cytopenias affecting one or more hematological cell lines. Literature reports include cases of ir-neutropenia.
- Autoimmune hemolytic anemia.
- Ir-thrombocytopenia (ir-TCP).
- Pancytopenia.

Awareness about potential myocarditis should be kept in mind. Recommendations for diagnostics can be found in Spallarossa et al. (2019) and Guo et al. (2020). However, in principle, every tissue and organ can be affected (Fig. 23.31).

Toxicity/irAE Management

Physicians and pharmacists should know about and be able to recognize irAE as such and the recommended interventions (Weber et al. 2012; Haanen et al. 2015; Weber et al. 2015; Davies et al. 2017).

In severe diarrhea with associated signs of colitis, the usual countermeasures (fluid + electrolytes ⇒ loperamide ⇒ budesonide) are insufficient. Intravenous or oral steroid therapy has to be initiated. For patients in whom intravenous steroids followed by high-dose oral steroid therapy do not lead to initial resolution of symptoms within 48–72 h, treatment with infliximab at 5 mg/kg can be used as an "emergency brake." Once relief of symptoms is achieved with infliximab, it should be discontinued. A prolonged steroid taper over 45–60 days should be instituted. There may be a waxing and waning of the GI adverse effects. As steroids are tapered, there can be a recrudescence of symptoms, mandating a retapering of steroids starting at a higher dose, a more prolonged taper, and the (re-)use of infliximab. Prophylactic use of budesonide cannot be recommended, based on a phase II study (Wolchok et al. 2010).

Autoimmune hepatitis is likewise treated with (high-dose) corticosteroids. If serum transaminase levels do not decrease within 48 h after initiation of systemic steroids, oral mycophenolate mofetil 500 mg q12h should be considered. Infliximab has to be avoided, because of its potential own

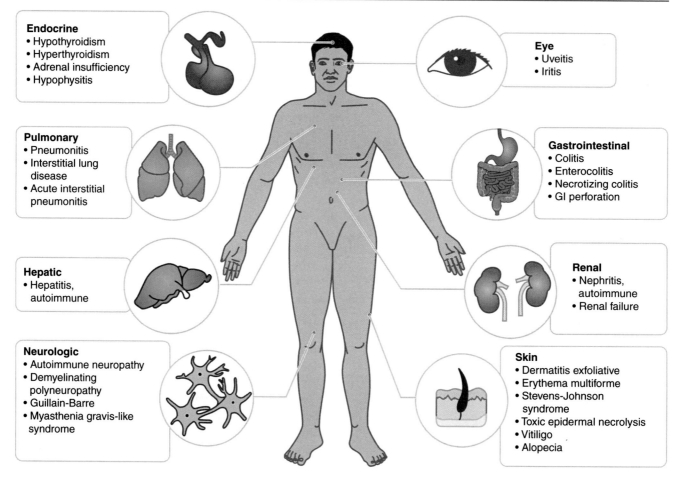

Fig. 23.31 Affected tissues and organs and symptoms by autoimmune-related adverse reactions. Listing is not intended to be exhaustive

hepatotoxicity. As described above, resurgence of symptoms and steroid tapering is necessary.

Difficult in diagnosis can be the auto-hypophysis with symptoms of headache, nausea, vertigo, behavior change, visual disturbances such as diplopia, and weakness. In this context, new occurrence of brain metastases has to be excluded. Baseline measurement of pituitary, thyroid, adrenal, and gonadal status, i.e., serum morning cortisol, adrenocorticotropic hormone [ACTH], free triiodothyronine [T3], free thyroxine [T4], thyroid-stimulating hormone [TSH], testosterone in males and follicle-stimulating hormone, luteinizing hormone, and prolactin in females, is recommended.

Immune-related pancreatitis generally manifests as an asymptomatic increase of amylase and lipase. Some patients experience more or less unspecific symptoms like fevers, malaise, nausea and vomiting, and/or abdominal pain. As described for other irAEs, a steroid taper is indicated, but often, this has minimal *immediate* effects. The symptoms of an autoimmune pancreatitis resolve slowly.

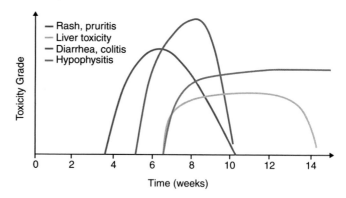

Fig. 23.32 irAEs exhibit a characteristic pattern in the timing of their occurrence. The frequency is not considered. (From Weber et al. (2012))

The irAEs caused by checkpoint blockade exhibit a characteristic pattern in the timing of their occurrence, shown in Fig. 23.32 (Weber et al. 2012).

After 2–3 weeks of initiation of a checkpoint blocker therapy, skin-related irAEs can be expected. Gastrointestinal side effects emerge after approximately 5–6 weeks. Hepatic irAEs occur after 6–7 weeks and endocrinological irAEs (autoimmune hypothyroidis or hypophysitis) after an average of 9 weeks. Long-term effects of immune oncology therapy are currently still largely unknown.

Self-Assessment Questions

Questions

1. Against what kind of molecular targets can mAbs be directed? Give examples.
2. By which factors can the pharmacokinetics of mAbs be influenced?
3. What does ADC stand for?
4. Why is rituximab in CLL dosed with 375 mg/m² initially but escalated to 500 mg/m²in subsequent cycles?
5. Describe the differences in clinical effects between type I and type II anti-lymphoma antibodies.
6. How many subtypes of the epidermal growth factor receptors exist, and how are they designated?
7. Which diagnostic actions are mandatory before using anti-EGFR/anti-Her2 antibodies?
8. Does it make therapeutically a difference if a colorectal carcinoma is right sided or left sided?
9. Does it make sense to combine different mAbs?
10. What is the main difference in the mechanism of action between "classical" antibody treatment of cancers and immune agonistic antibodies?
11. Describe the toxicity profile of checkpoint inhibitors.

Answers

1. mAbs can be directed against:
   - Surface antigens like CD antigens (CD20, CD30)
   - Receptors (EGFR)
   - Growth factors (VEGF)
   - Activating immune checkpoints on T cells (PD-1)
   - Suppressing immune checkpoints on tumor cells (PD-L1)
2. Tumor burden that corresponds to the amount of target antigen:
   - Neutralizing anti-mAb-antibodies
   - Gender
   - Age
3. ADC is the abbreviation for antibody–drug conjugate, where an antibody is chemically linked to a toxin with antitumor activity. The toxins are too toxic to be given in unconjugated form.
4. In CLL, the CD20 antigen density is lower than in other CD20-positive lymphomas. The reduced dose for the first infusion is necessary because of a usually high tumor load in need for therapy in CLL blast crisis.

Responding to rituximab therapy bears a high risk of tumor lysis and cytokine release syndrome as well as other infusion-related reactions.

5. Type I antibodies cause predominant CDC, while type II antibodies ADCC.
6. Four subtypes, EGFR1–4 or ErbB1–4 or Her1–4
7. Verification of EGFR or Her2 overexpression, in case of Her2 with IHC or FISH. Pan-RAS wild-type testing for cetuximab and panitumumab.
8. Depending on the localization, these tumors are regarded as biologically different. Patients with a right-sided tumor benefit from an anti-angiogenic drug containing first-line therapy, whereas left-sided tumors benefit from an anti-EGFR strategy if they are not RAS mutated.
9. Yes, for instance, trastuzumab plus pertuzumab, polatuzumab vedotin plus rituximab, and nivolumab plus relatlimab.
10. The so-called checkpoint-inhibiting antibodies do not attack the tumor itself; they restore their immunogenicity and/or (re-)activate immune-competent cells.
11. The toxicity pattern consists of loss of self-tolerance, resulting in autoimmune reactions (immune-related adverse events) such as:
    - Skin irAEs, including Stevens–Johnson syndrome and toxic epidermal necrolysis (both being rare)
    - Gastrointestinal irAEs—autoimmune colitis
    - Autoimmune hepatitis
    - Autoimmune pancreatitis
    - Autoimmune thyroiditis
    - Autoimmune hypophysitis
    - Neurological irAEs
    - Autoimmune pneumonitis
    - Ocular irAEs (rare), such as autoimmune uveitis and autoimmune episcleritis

References

Aits S, Jaattela M (2013) Lysosomal cell death at a glance. J Cell Sci 126(Pt 9):1905–1912
Alsaab HO et al (2017) PD-1 and PD-L1 checkpoint signaling inhibition for cancer immunotherapy: mechanism, combinations, and clinical outcome. Front Pharmacol 8:561
Andrews LP, Marciscano AE, Drake CG, Vignali DA (2017) LAG3 (CD223) as a cancer immunotherapy target. Immunol Rev 276(1):80–96
Anolik JH et al (2003) The relationship of FcgammaRIIIa genotype to degree of B cell depletion by rituximab in the treatment of systemic lupus erythematosus. Arthritis Rheum 48(2):455–459
Athanassiadou AM et al (2011) The significance of Survivin and Nectin-4 expression in the prognosis of breast carcinoma. Folia Histochem Cytobiol 49(1):26–33
Baselga J et al (2012) Pertuzumab plus trastuzumab plus docetaxel for metastatic breast cancer. N Engl J Med 366(2):109–119

Bauzon M et al (2019) Maytansine-bearing antibody-drug conjugates induce in vitro hallmarks of immunogenic cell death selectively in antigen-positive target cells. Onco Targets Ther 8(4):e1565859

Beers SA et al (2010) CD20 as a target for therapeutic type I and II monoclonal antibodies. Semin Hematol 47(2):107–114

Berg EA, Platts-Mills TA, Commins SP (2014) Drug allergens and food—the cetuximab and galactose-alpha-1,3-galactose story. Ann Allergy Asthma Immunol 112(2):97–101

Berinstein NL et al (1998) Association of serum rituximab (IDEC-C2B8) concentration and anti-tumor response in the treatment of recurrent low-grade or follicular non-Hodgkin's lymphoma. Ann Oncol 9(9):995–1001

Bokemeyer C et al (2012) Addition of cetuximab to chemotherapy as first-line treatment for KRAS wild-type metastatic colorectal cancer: pooled analysis of the CRYSTAL and OPUS randomised clinical trials. Eur J Cancer 48(10):1466–1475

Bouchard H, Viskov C, Garcia-Echeverria C (2014) Antibody-drug conjugates-a new wave of cancer drugs. Bioorg Med Chem Lett 24(23):5357–5363

Breij EC et al (2014) An antibody-drug conjugate that targets tissue factor exhibits potent therapeutic activity against a broad range of solid tumors. Cancer Res 74(4):1214–1226

Brule SY et al (2015) Location of colon cancer (right-sided versus left-sided) as a prognostic factor and a predictor of benefit from cetuximab in NCIC CO.17. Eur J Cancer 51(11):1405–1414

Bubien JK et al (1993) Transfection of the CD20 cell surface molecule into ectopic cell types generates a Ca2+ conductance found constitutively in B lymphocytes. J Cell Biol 121(5):1121–1132

Burchardt A, Wienzek-Lischka S, Schoelz C, Hackstein H, Rummel M (2012) Plasma exchange (PE) therapy (rituximab apheresis) for rituximab (R) induced progressive multifocal leukoencephalopathy (PML) in hematologic disorders. Onkologie 133

Buza N, Roque DM, Santin AD (2014) HER2/neu in endometrial cancer: a promising therapeutic target with diagnostic challenges. Arch Pathol Lab Med 138(3):343–350

Cartron G et al (2016) Rationale for optimal obinutuzumab/GA101 dosing regimen in B-cell non-Hodgkin lymphoma. Haematologica 101(2):226–234

Chan HT et al (2003) CD20-induced lymphoma cell death is independent of both caspases and its redistribution into triton X-100 insoluble membrane rafts. Cancer Res 63(17):5480–5489

Chen HX, Cleck JN (2009) Adverse effects of anticancer agents that target the VEGF pathway. Nat Rev Clin Oncol 6(8):465–477

Chen H et al (2017) Tubulin inhibitor-based antibody-drug conjugates for cancer therapy. Molecules 22(8):1281

Cheng H, Gammon D, Dutton TM, Piperdi B (2009) Panitumumab-related Hypomagnesemia in patients with colorectal cancer. Hosp Pharm 44:234–238

Cheson BD (2010) Ofatumumab, a novel anti-CD20 monoclonal antibody for the treatment of B-cell malignancies. J Clin Oncol 28(21):3525–3530

Chiosea SI et al (2015) Molecular characterization of apocrine salivary duct carcinoma. Am J Surg Pathol 39(6):744–752

Chiou VL, Burotto M (2015) Pseudoprogression and immune-related response in solid tumors. J Clin Oncol 33(31):3541–3543

Chung CH et al (2008) Cetuximab-induced anaphylaxis and IgE specific for galactose-alpha-1,3-galactose. N Engl J Med 358(11):1109–1117

Collins SM et al (2013) Elotuzumab directly enhances NK cell cytotoxicity against myeloma via CS1 ligation: evidence for augmented NK cell function complementing ADCC. Cancer Immunol Immunother 62(12):1841–1849

Commins SP et al (2009) Delayed anaphylaxis, angioedema, or urticaria after consumption of red meat in patients with IgE antibodies specific for galactose-alpha-1,3-galactose. J Allergy Clin Immunol 123(2):426–433

Commins SP et al (2011) The relevance of tick bites to the production of IgE antibodies to the mammalian oligosaccharide galactose-alpha-1,3-galactose. J Allergy Clin Immunol 127(5):1286–1293 e1286

Cortes J et al (2012) Pertuzumab monotherapy after trastuzumab-based treatment and subsequent reintroduction of trastuzumab: activity and tolerability in patients with advanced human epidermal growth factor receptor 2-positive breast cancer. J Clin Oncol 30(14):1594–1600

Cragg MS, Glennie MJ (2004) Antibody specificity controls in vivo effector mechanisms of anti-CD20 reagents. Blood 103(7):2738–2743

Cragg MS, Walshe CA, Ivanov AO, Glennie MJ (2005) The biology of CD20 and its potential as a target for mAb therapy. Curr Dir Autoimmun 8:140–174

Cragg MS et al (2003) Complement-mediated lysis by anti-CD20 mAb correlates with segregation into lipid rafts. Blood 101(3):1045–1052

Criscitiello C, Morganti S, Curigliano G (2021) Antibody-drug conjugates in solid tumors: a look into novel targets. J Hematol Oncol 14(1):20

DaiichiSankyo (2022) Enhertu Prescribing Information. https://daiichisankyo.us/prescribing-information-portlet/getPIContent?productName=Enhertu&inline=true

Davies A et al (2017) Efficacy and safety of subcutaneous rituximab versus intravenous rituximab for first-line treatment of follicular lymphoma (SABRINA): a randomised, open-label, phase 3 trial. Lancet Haematol 4(6):e272–e282

De Roock W et al (2010) Effects of KRAS, BRAF, NRAS, and PIK3CA mutations on the efficacy of cetuximab plus chemotherapy in chemotherapy-refractory metastatic colorectal cancer: a retrospective consortium analysis. Lancet Oncol 11(8):753–762

de Weers M et al (2011) Daratumumab, a novel therapeutic human CD38 monoclonal antibody, induces killing of multiple myeloma and other hematological tumors. J Immunol 186(3):1840–1848

Deaglio S, Mehta K, Malavasi F (2001) Human CD38: a (r)evolutionary story of enzymes and receptors. Leuk Res 25(1):1–12

Deaglio S et al (2007) CD38/CD19: a lipid raft-dependent signaling complex in human B cells. Blood 109(12):5390–5398

Deans JP, Robbins SM, Polyak MJ, Savage JA (1998) Rapid redistribution of CD20 to a low density detergent-insoluble membrane compartment. J Biol Chem 273(1):344–348

Derycke MS et al (2010) Nectin 4 overexpression in ovarian cancer tissues and serum: potential role as a serum biomarker. Am J Clin Pathol 134(5):835–845

Doronina SO et al (2006) Enhanced activity of monomethylauristatin F through monoclonal antibody delivery: effects of linker technology on efficacy and toxicity. Bioconjug Chem 17(1):114–124

EditorialReviewTeam (2022) Pyrrolobenzodiazepine (PBD). J Antibody Drug Conjugates

English DP, Roque DM, Santin AD (2013) HER2 expression beyond breast cancer: therapeutic implications for gynecologic malignancies. Mol Diagn Ther 17(2):85–99

EuropeanMedicinesAgency (2022a) Darzalex - European Public Assessment Report: Product Information. https://www.ema.europa.eu/en/documents/product-information/darzalex-epar-product-information_en.pdf

EuropeanMedicinesAgency (2022b) Empliciti - European Public Assessment Report: Product Information. https://www.ema.europa.eu/en/documents/product-information/empliciti-epar-product-information_en.pdf

EuropeanMedicinesAgency (2022c) Enhertu - European Public Assessment Report: Product Information. https://www.ema.europa.eu/en/documents/product-information/enhertu-epar-product-information_en.pdf

EuropeanMedicinesAgency (2022d) Polivy - European Public Assessment Report: Product Information. https://www.ema.europa.eu/en/documents/product-information/polivy-epar-product-information_en.pdf

Fakih M (2008) Management of anti-EGFR-targeting monoclonal antibody-induced hypomagnesemia. Oncology (Williston Park) 22(1):74–76

Fakih MG, Wilding G, Lombardo J (2006) Cetuximab-induced hypomagnesemia in patients with colorectal cancer. Clin Colorectal Cancer 6(2):152–156

Farooq AV et al (2020) Corneal epithelial findings in patients with multiple myeloma treated with antibody-drug conjugate belantamab mafodotin in the pivotal, randomized, DREAMM-2 study. Ophthalmol Ther 9(4):889–911

FDA (2017) FDA approves Mylotarg for treatment of acute myeloid leukemia. https://www.fda.gov/NewsEvents/Newsroom/PressAnnouncements/ucm574507.htm

Ferris R (2013) PD-1 targeting in cancer immunotherapy. Cancer 119(23):E1–E3

Finn GMS, Chu R, de Velde H, Menad S, Melki MT, Nouadje G (2020) Development of a Hydrashift 2/4 Isatuximab assay to mitigate interference with monoclonal protein detection on immunofixation electrophoresis in vitro diagnostic tests in multiple myeloma. Blood 136:15

GenentechInc (2019) Herceptin Hylecta Prescribing Information. https://www.gene.com/download/pdf/herceptin_hylecta_prescribing.pdf

GlaxoSmithKline (2020) Blenrep Prescribing Information. https://gskpro.com/content/dam/global/hcpportal/en_US/Prescribing_Information/Blenrep/pdf/BLENREP-PI-MG.PDF

Gordan LN et al (2005) Phase II trial of individualized rituximab dosing for patients with CD20-positive lymphoproliferative disorders. J Clin Oncol 23(6):1096–1102

Guo H et al (2015) Immune cell inhibition by SLAMF7 is mediated by a mechanism requiring src kinases, CD45, and SHIP-1 that is defective in multiple myeloma cells. Mol Cell Biol 35(1):41–51

Guo CW et al (2020) A closer look at immune-mediated myocarditis in the era of combined checkpoint blockade and targeted therapies. Eur J Cancer 124:15–24

Gutzmer R et al (2011) Management of cutaneous side effects of EGFR inhibitors: recommendations from a German expert panel for the primary treating physician. J Dtsch Dermatol Ges 9(3):195–203

Haanen JB, Thienen H, Blank CU (2015) Toxicity patterns with immunomodulating antibodies and their combinations. Semin Oncol 42(3):423–428

Han TH, Zhao B (2014) Absorption, distribution, metabolism, and excretion considerations for the development of antibody-drug conjugates. Drug Metab Dispos 42(11):1914–1920

Herold M, Schnohr S, Bittrich H (2001) Efficacy and safety of a combined rituximab chemotherapy during pregnancy. J Clin Oncol 19(14):3439

Hofheinz RD et al (2016) Recommendations for the prophylactic management of skin reactions induced by epidermal growth factor receptor inhibitors in patients with solid tumors. Oncologist 21(12):1483–1491

Holcmann M, Sibilia M (2015) Mechanisms underlying skin disorders induced by EGFR inhibitors. Mol Cell Oncol 2(4):e1004969

Hsi ED et al (2008) CS1, a potential new therapeutic antibody target for the treatment of multiple myeloma. Clin Cancer Res 14(9):2775–2784

Huhn D et al (2001) Rituximab therapy of patients with B-cell chronic lymphocytic leukemia. Blood 98(5):1326–1331

Hurvitz SA et al (2023) Trastuzumab deruxtecan versus trastuzumab emtansine in patients with HER2-positive metastatic breast cancer: updated results from DESTINY-Breast03, a randomised, open-label, phase 3 trial. Lancet 401(10371):105–117

Hutson TE, Figlin RA, Kuhn JG, Motzer RJ (2008) Targeted therapies for metastatic renal cell carcinoma: an overview of toxicity and dosing strategies. Oncologist 13(10):1084–1096

Intlekofer AM, Thompson CB (2013) At the bench: preclinical rationale for CTLA-4 and PD-1 blockade as cancer immunotherapy. J Leukoc Biol 94(1):25–39

Ishibashi K, Suzuki M, Sasaki S, Imai M (2001) Identification of a new multigene four-transmembrane family (MS4A) related to CD20, HTm4 and beta subunit of the high-affinity IgE receptor. Gene 264(1):87–93

Ivanov A, Krysov S, Cragg MS, Illidge T (2008) Radiation therapy with tositumomab (B1) anti-CD20 monoclonal antibody initiates extracellular signal-regulated kinase/mitogen-activated protein kinase-dependent cell death that overcomes resistance to apoptosis. Clin Cancer Res 14(15):4925–4934

Izzedine H et al (2010) Electrolyte disorders related to EGFR-targeting drugs. Crit Rev Oncol Hematol 73(3):213–219

JanssenBiotech (2022a) Darzalex Faspro Prescribing Information. https://www.janssenlabels.com/package-insert/product-monograph/prescribing-information/DARZALEX+Faspro-pi.pdf

JanssenBiotech (2022b) Darzalex Prescribing Infromation. https://www.janssenlabels.com/package-insert/product-monograph/prescribing-information/DARZALEX-pi.pdf

Kanzaki M, Shibata H, Mogami H, Kojima I (1995) Expression of calcium-permeable cation channel CD20 accelerates progression through the G1 phase in Balb/c 3T3 cells. J Biol Chem 270(22):13099–13104

Kasthuri RS, Taubman MB, Mackman N (2009) Role of tissue factor in cancer. J Clin Oncol 27(29):4834–4838

Kawasaki N, Nishito Y, Yoshimura Y, Yoshiura S (2022) The molecular rationale for the combination of polatuzumab vedotin plus rituximab in diffuse large B-cell lymphoma. Br J Haematol 199(2):245–255

Klein C et al (2013) Epitope interactions of monoclonal antibodies targeting CD20 and their relationship to functional properties. MAbs 5(1):22–33

Lacouture ME et al (2011) Clinical practice guidelines for the prevention and treatment of EGFR inhibitor-associated dermatologic toxicities. Support Care Cancer 19(8):1079–1095

Lee HC (2006) Structure and enzymatic functions of human CD38. Mol Med 12(11-12):317–323

Li H, Ayer LM, Lytton J, Deans JP (2003) Store-operated cation entry mediated by CD20 in membrane rafts. J Biol Chem 278(43):42427–42434

Li Y et al (2016) A mini-review for cancer immunotherapy: molecular understanding of PD-1/PD-L1 pathway and translational blockade of immune checkpoints. Int J Mol Sci 17(7):1151

Lin P, Owens R, Tricot G, Wilson CS (2004) Flow cytometric immunophenotypic analysis of 306 cases of multiple myeloma. Am J Clin Pathol 121(4):482–488

Long L et al (2018) The promising immune checkpoint LAG-3: from tumor microenvironment to cancer immunotherapy. Genes Cancer 9(5-6):176–189

Ludwig DL et al (2003) Monoclonal antibody therapeutics and apoptosis. Oncogene 22(56):9097–9106

Malavasi F et al (2008) Evolution and function of the ADP ribosyl cyclase/CD38 gene family in physiology and pathology. Physiol Rev 88(3):841–886

Malavasi F et al (2011) CD38 and chronic lymphocytic leukemia: a decade later. Blood 118(13):3470–3478

Maloney DG et al (1994) Phase I clinical trial using escalating single-dose infusion of chimeric anti-CD20 monoclonal antibody (IDEC-C2B8) in patients with recurrent B-cell lymphoma. Blood 84(8):2457–2466

Maloney DG et al (1997) IDEC-C2B8 (rituximab) anti-CD20 monoclonal antibody therapy in patients with relapsed low-grade non-Hodgkin's lymphoma. Blood 90(6):2188–2195

Manches O et al (2003) In vitro mechanisms of action of rituximab on primary non-Hodgkin lymphomas. Blood 101(3):949–954

Mariotte D et al (2011) Anti-cetuximab IgE ELISA for identification of patients at a high risk of cetuximab-induced anaphylaxis. MAbs 3(4):396–401

Martin TG et al (2019) Therapeutic opportunities with pharmacological inhibition of CD38 with Isatuximab. Cell 8(12):1522

Martins I et al (2014) Molecular mechanisms of ATP secretion during immunogenic cell death. Cell Death Differ 21(1):79–91

McLaughlin P et al (1998) Rituximab chimeric anti-CD20 monoclonal antibody therapy for relapsed indolent lymphoma: half of patients respond to a four-dose treatment program. J Clin Oncol 16(8):2825–2833

Mecklenburg L (2018) A brief introduction to antibody-drug conjugates for toxicologic pathologists. Toxicol Pathol 46(7):746–752

Mellor JD et al (2013) A critical review of the role of fc gamma receptor polymorphisms in the response to monoclonal antibodies in cancer. J Hematol Oncol 6:1

Moores SL et al (2016) A novel bispecific antibody targeting EGFR and cMet is effective against EGFR inhibitor-resistant lung tumors. Cancer Res 76(13):3942–3953

Morandi F et al (2018) CD38: a target for immunotherapeutic approaches in multiple myeloma. Front Immunol 9:2722

Morschhauser FA et al (2013) Obinutuzumab (GA101) monotherapy in relapsed/refractory diffuse large b-cell lymphoma or mantle-cell lymphoma: results from the phase II GAUGUIN study. J Clin Oncol 31(23):2912–2919

Mossner E et al (2010) Increasing the efficacy of CD20 antibody therapy through the engineering of a new type II anti-CD20 antibody with enhanced direct and immune effector cell-mediated B-cell cytotoxicity. Blood 115(22):4393–4402

Muller C et al (2012) The role of sex and weight on rituximab clearance and serum elimination half-life in elderly patients with DLBCL. Blood 119(14):3276–3284

Nadler LM et al (1981) A unique cell surface antigen identifying lymphoid malignancies of B cell origin. J Clin Invest 67(1):134–140

NCI. N, Cancer, Institute (2014) FDA approval for tositumomab and iodine I 131 tositumomab. Internet. https://www.cancer.gov/about-cancer/treatment/drugs/fda-tositumomab-I131iodine-tositumomab

Negro A, Brar BK, Lee KF (2004) Essential roles of Her2/erbB2 in cardiac development and function. Recent Prog Horm Res 59:1–12

Nguyen DT et al (1999) IDEC-C2B8 anti-CD20 (rituximab) immunotherapy in patients with low-grade non-Hodgkin's lymphoma and lymphoproliferative disorders: evaluation of response on 48 patients. Eur J Haematol 62(2):76–82

Niederfellner G et al (2011) Epitope characterization and crystal structure of GA101 provide insights into the molecular basis for type I/II distinction of CD20 antibodies. Blood 118(2):358–367

Nilsson R et al (2010) Toxicity-reducing potential of extracorporeal affinity adsorption treatment in combination with the auristatin-conjugated monoclonal antibody BR96 in a syngeneic rat tumor model. Cancer 116(4 Suppl):1033–1042

Nishiwada S et al (2015) Nectin-4 expression contributes to tumor proliferation, angiogenesis and patient prognosis in human pancreatic cancer. J Exp Clin Cancer Res 34(1):30

O'Brien SM et al (2001) Rituximab dose-escalation trial in chronic lymphocytic leukemia. J Clin Oncol 19(8):2165–2170

Ohta Y et al (1996) Significance of vascular endothelial growth factor messenger RNA expression in primary lung cancer. Clin Cancer Res 2(8):1411–1416

Overdijk MB et al (2015) Antibody-mediated phagocytosis contributes to the anti-tumor activity of the therapeutic antibody daratumumab in lymphoma and multiple myeloma. MAbs 7(2):311–321

Pardoll DM (2012) The blockade of immune checkpoints in cancer immunotherapy. Nat Rev Cancer 12(4):252–264

Park K et al (2021) Amivantamab in EGFR exon 20 insertion-mutated non-small-cell lung cancer progressing on platinum chemotherapy: initial results from the CHRYSALIS phase I study. J Clin Oncol 39(30):3391–3402

Pettinato MC (2021) Introduction to antibody-drug conjugates. Antibodies (Basel) 10(4):42

Pfreundschuh M, Zeynalova S, Poeschel V, Haenel M, Schmitz N, Hensel M, Reiser M, Loeffler M, Schubert J (2007) Dose-dense rituximab improves outcome of elderly patients with poor-prognosis diffuse large B-cell lymphoma (DLBCL): Results of the DENSE-R-CHOP-14 Trial of the German High-Grade Non-Hodgkin Lymphoma Study Group (DSHNHL). Blood 110(11):789

Pfreundschuh M et al (2014) Suboptimal dosing of rituximab in male and female patients with DLBCL. Blood 123(5):640–646

Polyak MJ, Deans JP (2002) Alanine-170 and proline-172 are critical determinants for extracellular CD20 epitopes; heterogeneity in the fine specificity of CD20 monoclonal antibodies is defined by additional requirements imposed by both amino acid sequence and quaternary structure. Blood 99(9):3256–3262

Polyak MJ, Li H, Shariat N, Deans JP (2008) CD20 homo-oligomers physically associate with the B cell antigen receptor. Dissociation upon receptor engagement and recruitment of phosphoproteins and calmodulin-binding proteins. J Biol Chem 283(27):18545–18552

Potthoff K et al (2011) Interdisciplinary management of EGFR-inhibitor-induced skin reactions: a German expert opinion. Ann Oncol 22(3):524–535

Quartier P et al (2001) Treatment of childhood autoimmune haemolytic anaemia with rituximab. Lancet 358(9292):1511–1513

Radhakrishnan SV et al (2017) Elotuzumab as a novel anti-myeloma immunotherapy. Hum Vaccin Immunother 13(8):1751–1757

Rifkin RMSD, Aguilar KM, Baidoo B, Maiese EM (2018) Safety of split first dosing vs standard dosing administration of daratumumab among multiple myeloma patients treated in a US community oncology setting: a real-world observational study. Blood 132:4846

Rotte A, Jin JY, Lemaire V (2018) Mechanistic overview of immune checkpoints to support the rational design of their combinations in cancer immunotherapy. Ann Oncol 29(1):71–83

Rubin I, Yarden Y (2001) The basic biology of HER2. Ann Oncol 12(Suppl 1):S3–S8

Rugo HS et al (2021) Efficacy of margetuximab vs trastuzumab in patients with pretreated ERBB2-positive advanced breast cancer: a phase 3 randomized clinical trial. JAMA Oncol 7(4):573–584

Ryman JT, Meibohm B (2017) Pharmacokinetics of monoclonal antibodies. CPT Pharmacometrics Syst Pharmacol 6(9):576–588

Saleh H et al (2012) Anaphylactic reactions to oligosaccharides in red meat: a syndrome in evolution. Clin Mol Allergy 10(1):5

Santin AD et al (2008) Trastuzumab treatment in patients with advanced or recurrent endometrial carcinoma overexpressing HER2/neu. Int J Gynaecol Obstet 102(2):128–131

Santonocito AM et al (2004) Flow cytometric detection of aneuploid CD38(++) plasmacells and CD19(+) B-lymphocytes in bone marrow, peripheral blood and PBSC harvest in multiple myeloma patients. Leuk Res 28(5):469–477

Schlingmann KP, et al. (2007) TRPM6 and TRPM7--Gatekeepers of human magnesium metabolism. Biochim Biophys Acta 1772(8):813-821

Schneider MR, Wolf E (2009) The epidermal growth factor receptor ligands at a glance. J Cell Physiol 218(3):460–466

Schrag D, Chung KY, Flombaum C, Saltz L (2005) Cetuximab therapy and symptomatic hypomagnesemia. J Natl Cancer Inst 97(16):1221–1224

SeagenInc (2021) Tivdak Prescribing Information. https://seagendocs.com/Tivdak_Full_Ltr_Master.pdf

Shvartsur A, Bonavida B (2015) Trop2 and its overexpression in cancers: regulation and clinical/therapeutic implications. Genes Cancer 6(3-4):84–105

Simons K, Toomre D (2000) Lipid rafts and signal transduction. Nat Rev Mol Cell Biol 1(1):31–39

Singh B, Carpenter G, Coffey RJ (2016) EGF receptor ligands: recent advances. F1000Res 5

Spallarossa P et al (2019) Identification and management of immune checkpoint inhibitor-related myocarditis: use troponin wisely. J Clin Oncol 37(25):2201–2205

Stashenko P, Nadler LM, Hardy R, Schlossman SF (1980) Characterization of a human B lymphocyte-specific antigen. J Immunol 125(4):1678–1685

Steinke JW, Platts-Mills TA, Commins SP (2015) The alpha-gal story: lessons learned from connecting the dots. J Allergy Clin Immunol 135(3):589–596. quiz 597

Stepan LP et al (2011) Expression of Trop2 cell surface glycoprotein in normal and tumor tissues: potential implications as a cancer therapeutic target. J Histochem Cytochem 59(7):701–710

Stolz C, Schuler M (2009) Molecular mechanisms of resistance to rituximab and pharmacologic strategies for its circumvention. Leuk Lymphoma 50(6):873–885

Suzuki S, Ishida T, Yoshikawa K, Ueda R (2016) Current status of immunotherapy. Jpn J Clin Oncol 46(3):191–203

Tabernero J, Pfeiffer P, Cervantes A (2008) Administration of cetuximab every 2 weeks in the treatment of metastatic colorectal cancer: an effective, more convenient alternative to weekly administration? Oncologist 13(2):113–119

Takai Y, Miyoshi J, Ikeda W, Ogita H (2008) Nectins and nectin-like molecules: roles in contact inhibition of cell movement and proliferation. Nat Rev Mol Cell Biol 9(8):603–615

Takano A et al (2009) Identification of nectin-4 oncoprotein as a diagnostic and therapeutic target for lung cancer. Cancer Res 69(16):6694–6703

Tedder TF, Engel P (1994) CD20: a regulator of cell-cycle progression of B lymphocytes. Immunol Today 15(9):450–454

Tedder TF, Streuli M, Schlossman SF, Saito H (1988) Isolation and structure of a cDNA encoding the B1 (CD20) cell-surface antigen of human B lymphocytes. Proc Natl Acad Sci U S A 85(1):208–212

Tejpar S et al (2007) Magnesium wasting associated with epidermal-growth-factor receptor-targeting antibodies in colorectal cancer: a prospective study. Lancet Oncol 8(5):387–394

Tejpar S et al (2016) Prognostic and predictive relevance of primary tumor location in patients with RAS wild-type metastatic colorectal cancer: retrospective analyses of the CRYSTAL and FIRE-3 trials. JAMA Oncol 3(2):194–201

Teplinsky E, Muggia F (2014) Targeting HER2 in ovarian and uterine cancers: challenges and future directions. Gynecol Oncol 135(2):364–370

Tesfa D, Palmblad J (2011) Late-onset neutropenia following rituximab therapy: incidence, clinical features and possible mechanisms. Expert Rev Hematol 4(6):619–625

Tesfa D et al (2011) Late-onset neutropenia following rituximab therapy in rheumatic diseases: association with B lymphocyte depletion and infections. Arthritis Rheum 63(8):2209–2214

Tobinai K, Klein C, Oya N, Fingerle-Rowson G (2017) A review of Obinutuzumab (GA101), a novel type II anti-CD20 monoclonal antibody, for the treatment of patients with B-cell malignancies. Adv Ther 34(2):324–356

Tuefferd M et al (2007) HER2 status in ovarian carcinomas: a multi-center GINECO study of 320 patients. PLoS One 2(11):e1138

Van Cutsem E et al (2011) Cetuximab plus irinotecan, fluorouracil, and leucovorin as first-line treatment for metastatic colorectal cancer: updated analysis of overall survival according to tumor KRAS and BRAF mutation status. J Clin Oncol 29(15):2011–2019

Van Cutsem E et al (2014) Metastatic colorectal cancer: ESMO clinical practice guidelines for diagnosis, treatment and follow-up. Ann Oncol 25 Suppl 3:iii1-9

van de Donk NW et al (2016) Clinical efficacy and management of monoclonal antibodies targeting CD38 and SLAMF7 in multiple myeloma. Blood 127(6):681–695

Venook A, Niedzwiecki D, Innocenti F, Fruth B, Greene B, O'Neil BH, Shaw JE, Atkins JN, Horvath LE, Polite BN, Meyerhardt JA, O'Reilly EM, Goldberg RM, Hochster HS, Blanke CD, Schilsky RL, Mayer RJ, Bertagnolli MM, Lenz H-J (2016) Impact of primary (1°) tumor location on overall survival (OS) and progression-free survival (PFS) in patients (pts) with metastatic colorectal cancer (mCRC): analysis of CALGB/SWOG 80405 (Alliance). J Clin Oncol 34(Suppl):abstr 3504

Verheul HM, Pinedo HM (2007) Possible molecular mechanisms involved in the toxicity of angiogenesis inhibition. Nat Rev Cancer 7(6):475–485

Vesely MD, Kershaw MH, Schreiber RD, Smyth MJ (2011) Natural innate and adaptive immunity to cancer. Annu Rev Immunol 29:235–271

Vijayaraghavan S et al (2020) Amivantamab (JNJ-61186372), an fc enhanced EGFR/cMet bispecific antibody, induces receptor down-modulation and antitumor activity by monocyte/macrophage trogocytosis. Mol Cancer Ther 19(10):2044–2056

Wang DY et al (2018) Fatal toxic effects associated with immune checkpoint inhibitors: a systematic review and meta-analysis. JAMA Oncol 4(12):1721–1728

Weber JS, Kahler KC, Hauschild A (2012) Management of immune-related adverse events and kinetics of response with ipilimumab. J Clin Oncol 30(21):2691–2697

Weber JS, Yang JC, Atkins MB, Disis ML (2015) Toxicities of immunotherapy for the practitioner. J Clin Oncol 33(18):2092–2099

Weiss L et al (2017) Influence of body mass index on survival in indolent and mantle cell lymphomas: analysis of the StiL NHL1 trial. Ann Hematol 96(7):1155–1162

Weng WK, Negrin RS, Lavori P, Horning SJ (2010) Immunoglobulin G Fc receptor FcgammaRIIIa 158 V/F polymorphism correlates with rituximab-induced neutropenia after autologous transplantation in patients with non-Hodgkin's lymphoma. J Clin Oncol 28(2):279–284

Widakowich C et al (2007) Review: side effects of approved molecular targeted therapies in solid cancers. Oncologist 12(12):1443–1455

Wolchok JD et al (2009) Guidelines for the evaluation of immune therapy activity in solid tumors: immune-related response criteria. Clin Cancer Res 15(23):7412–7420

Wolchok JD et al (2010) Ipilimumab monotherapy in patients with pre-treated advanced melanoma: a randomised, double-blind, multicentre, phase 2, dose-ranging study. Lancet Oncol 11(2):155–164

Wolchok JD et al (2013) Nivolumab plus ipilimumab in advanced melanoma. N Engl J Med 369(2):122–133

Wu J, Fu J, Zhang M, Liu D (2015) Blinatumomab: a bispecific T cell engager (BiTE) antibody against CD19/CD3 for refractory acute lymphoid leukemia. J Hematol Oncol 8:104

Xu SMP, Usmani SZ, Lonial S, Jakubowiak A, Oriol A, Krishnan A, Bladé J, Luo M, Sun YN, Zhang L, Deraedt W, Qi M, Ukropec J, Clemens PL (2018) Split first dose Administration of Daratumumab for the treatment of patients with multiple myeloma (MM): clinical pharmacology and population pharmacokinetic (PK) analyses. Blood 132:1970

Yildirim M, Kaya V, Demirpence O, Paydas S (2015) The role of gender in patients with diffuse large B cell lymphoma treated with rituximab-containing regimens: a meta-analysis. Arch Med Sci 11(4):708–714

Yun J et al (2020) Antitumor activity of Amivantamab (JNJ-61186372), an EGFR-MET bispecific antibody, in diverse models of EGFR exon 20 insertion-driven NSCLC. Cancer Discov 10(8):1194–1209

Zhao YY et al (1998) Neuregulins promote survival and growth of cardiac myocytes. Persistence of ErbB2 and ErbB4 expression in neonatal and adult ventricular myocytes. J Biol Chem 273(17):10261–10269

Antibody-Based Biotherapeutics in Inflammatory Diseases

Yan Xu, Jia Chen, and Honghui Zhou

Introduction

Inflammatory diseases encompass a broad and diverse spectrum of serious chronic disorders, many of which have significant need for safe and effective pharmacotherapies. The conventional drugs used to treat immune-mediated inflammatory diseases include nonsteroidal anti-inflammatory drugs (NSAIDs), corticosteroids, sulfasalazine, 5-aminosalicylates, methotrexate, azathioprine, and 6-mercaptopurine which have exhibited limited efficacy with significant side effects. The initial rationale and promise of antibody-based biotherapeutics, such as monoclonal antibodies (mAbs), was focused on oncology and organ transplantation (Ehrlich 1891; Gura 2002). Over the last two decades, there has been significant success in developing a number of antibody-based biotherapeutics as a very effective and relatively safe treatment for several inflammatory diseases, and this area of research and development is rapidly expanding. In 2021, 3 of the top-20 selling drugs were mAbs for the treatment of chronic inflammatory conditions (Dunleavy 2021).

Antibody-based biotherapeutics are a subclass of protein therapeutics. These are large molecular weight glycoproteins designed and produced through recombinant DNA technology and require production in eukaryotic cells using bioreactor technology. These modalities have provided many efficacious therapeutic options for patients and are providing significant insights into the underlying complex pathological pathways of these disorders, which, in turn, are identifying new targets for treatment of these diseases. A significant translational insight derived from the clinical development programs of antibody-based pharmacotherapy is the dysregulation of common proinflammatory mediators, such as tumor necrosis factor alpha (TNFα), across diverse rheumatologic, dermatologic, and gastroenterological pathologies. In addition, the observation of patient subsets that are refractory to a particular therapy indicates that dysregulation of different mediators (targets) may be the primary driver of the underlying disease and require a different treatment. Antibody-based biotherapeutics embody structural, biochemical, and pharmacologic properties distinct from other biologic or chemically synthesized molecular drugs. In general, they exhibit relatively long half-lives (~2–3 weeks) at therapeutic doses, and have high affinity and target specificity with minimum off-target effects, which usually translate into potent and sustained pharmacodynamic (PD) effects.

Currently approved antibody-based therapies for autoimmune/inflammatory disorders include chimeric, humanized, human mAbs and fusion proteins. The mechanism of action of these agents includes either neutralizing a soluble ligand(s) such as cytokines or binding to receptors to block signaling through ligands, or acting as direct agonist or antagonist. The examples of neutralizing soluble ligands include TNFα (infliximab, golimumab, adalimumab, etanercept, certolizumab), interleukin (IL)-12/IL-23 (ustekinumab), IL-13 (tralokinumab), IL-23 (guselkumab, risankizumab, tildrakizumab), IL-17A (secukinumab and ixekizumab), IL-5 (mepolizumab and reslizumab), IL-1β (canakinumab and rilonacept), soluble immunoglobulin E (IgE, omalizumab), B-lymphocyte stimulator (BLyS, belimumab), and thymic stromal lymphopoietin (TSLP, tezepelumab). The examples of binding to receptors include IL-36 receptor (spesolimab), IL-4 receptor (dupilumab), IL-5 receptor (benralizumab), IL-6 receptor (tocilizumab, sarilumab, satralizumab) IL-17 receptor A (brodalumab) and type I interferon receptor (anifrolumab). The examples of direct agonist or antagonists include anti-CD80/CD86 agents to inhibit lymphocyte acti-

Y. Xu
Clinical Pharmacology and Pharmacometrics, Biogen, Cambridge, MA, USA
e-mail: yan.xu1@biogen.com

J. Chen
Clinical Pharmacology, Simcere Pharmaceuticals, Shanghai, China
e-mail: chenjia@simcere.com

H. Zhou (✉)
Global Clinical Pharmacology and Pharmacometrics, Jazz Pharmaceuticals, Philadelphia, PA, USA
e-mail: honghui.zhou@jazzpharma.com

D. J. A. Crommelin et al. (eds.), *Pharmaceutical Biotechnology*, https://doi.org/10.1007/978-3-031-30023-3_24

vation (abatacept), CD20 directed cytolytic agents (rituximab and ocrelizumab), anti-CD19 cytolytic agent (inebilizumab), and anti-integrin agents to inhibit lymphocyte migration (natalizumab and vedolizumab). Table 24.1 summarizes the 33 antibody-based biotherapeutics that are currently approved and on the market for the treatment of immune-mediated inflammatory diseases.

One challenge in the long-term treatment with antibody-based biotherapeutics of immune-mediated disorders is to avoid the side effects due to the potent and sustained suppression of the immune system. The expectation for these newer therapies is that they can be used earlier in the course of disease to not only maintain control over episodic disease flares but also prevent the less reversible organ damage posed by long-term uncontrolled chronic inflammation or even reversal of disease such as joint damage caused by rheumatoid arthritis (Taylor et al. 2004).

The primary focus of this chapter is on describing the pharmacologic properties of approved biologic therapies

for major classes of inflammatory diseases, such as arthritides, systemic lupus erythematosus (SLE), psoriasis, atopic dermatitis (AD), inflammatory bowel disease (IBD), asthma, and a number of less common inflammatory disorders (including CAR T cell-induced cytokine release syndrome [CRS], chronic idiopathic urticaria [CIU], cryopyrin-associated periodic syndrome [CAPS], chronic rhinosinusitis with nasal polyps [CRSwNP], deficiency of the interleukin-1 receptor antagonist [DIRA], eosinophilic granulomatosis with polyangiitis [EGPA], generalized pustular psoriasis [GPP] flare, giant cell arteritis [GCA], hidradenitis suppurativa [HS], lupus nephritis [LN], neuromyelitis optica spectrum disorder [NMOSD], multiple sclerosis [MS], recurrent pericarditis [RR], and uveitis [UV]). Within each of these disease categories, biologic agents will be introduced according to their mechanisms of action, and listed alphabetically when having a same mechanism of action. Notably information described in this chapter is based on the original "innovator" products (as of November 2022).

Table 24.1 Use of antibody-based biotherapeutics in immune-mediated inflammatory diseases

Product	Molecular weight	Modality	Target	Indication[a]	Mechanism of action	Recommended dose regimen[a]
Abatacept (Orencia®)	92 kDa	Soluble fusion protein that consists of extracellular domain of human CTLA-4 linked to the modified Fc portion of human IgG1	CTLA-4	RA, PsA, and pJIA	Abatacept is a selective costimulation modulator that inhibits T lymphocyte activation by binding to CD80 and CD86, thereby blocking interaction with CD28. This interaction provides a costimulatory signal necessary for full activation of T lymphocytes	*IV infusion for RA and PsA*: weight-based dosing (~10 mg/kg, i.e., 500, 750, and 1000 mg for patients weighing <60 kg, 60–100 kg, and >100 kg, respectively) administered as IV infusion at Weeks 0, 2, and 4 and q4w thereafter *SC injection for RA*: 125 mg SC qw with or without an initial IV loading dose (according to body weight categories described above) *SC injection for PsA*: 125 mg SC qw (without the need of an IV loading dose) *IV infusion for pJIA (6 years and older)*: For pediatrics weighing <75 kg: 10 mg/kg at Weeks 0, 2, and 4 and q4w by IV infusion; for pediatric patients weighing ≥75 kg: following the IV dose regimen for adult RA (not to exceed 1000 mg/injection) *SC injection for pJIA (2 years and older)*: SC qw at a body-weight-tiered dosing of 50, 87.5 and 125 mg for pJIA patients weighing 10 to <25 kg, 25 to <50 kg and ≥50 kg, respectively

Table 24.1 (continued)

Product	Molecular weight	Modality	Target	Indication[a]	Mechanism of action	Recommended dose regimen[a]
Adalimumab (Humira®)	148 kDa	Human IgG1 mAb	TNFα	RA, PsA, AS, pJIA, PsO, CD, UC, pediatric CD, pediatric UC, HS, and UV	Adalimumab binds specifically to TNFα and blocks its interaction with the p55 and p75 cell surface TNF receptors. Adalimumab also lyses surface TNF expressing cells in vitro in the presence of complement. Adalimumab does not bind or inactivate lymphotoxin (TNF-β)	*RA, PsA, and AS:* 40 mg SC q2w. Some patients with RA not receiving methotrexate may benefit from increasing the dosage to 40 mg qw or 80 mg q2w *pJIA and pediatric UV (2 years and older):* SC q2w at a dose level of 10, 20, and 40 mg for pediatric patients weighting 10 to <15 kg, 15 to <30 kg and ≥30 kg, respectively *PsO and UV:* 80 mg initial SC dose followed by 40 mg SC q2w *CD and UC:* 160 mg initial dose, 80 mg dose 2 weeks later, followed by 40 mg SC q2w *Pediatric CD (6 years and older):* For pediatric patients weighing 17 to <40 kg: 80 mg initial dose, 40 mg dose 2 weeks later, followed by 20 mg SC q2w. Pediatric patients weighing ≥40 kg: Following the adult dose regimen *Pediatric UC (5 years and older):* For pediatric patients weighing 20 to <40 kg: 80 mg initial dose, 40 mg dose on days 8 and 15, followed by 40 mg SC q2w or 20 mg qw from day 29. Pediatric patients weighing ≥40 kg: 160 mg initial dose, 80 mg dose on days 8 and 15, followed by 80 mg SC q2w or 40 mg qw from day 29 *HS (adults and adolescents 12 years and older):* For adults and adolescents weighing 60 kg and above, 160 mg initial dose, 80 mg dose 2 weeks later, followed by 40 mg SC qw or 80 mg q2w from Day 29; for adolescents weighing 30 kg to <60 kg, 80 mg initial dose, followed by 40 mg SC q2w from Day 8
Anifrolumab-fnia (Saphnelo®)	148 kDa	Human IgG1κ mAb	Type I interferon receptor	SLE	Anifrolumab-fnia binds to subunit 1 of IFNAR with high specificity and affinity. This binding inhibits type I IFN signaling, thereby blocking the biologic activity of type I IFNs. Anifrolumab-fnia also induces the internalization of IFNAR1, thereby reducing the levels of cell surface IFNAR1 available for receptor assembly. Blockade of receptor mediated type I IFN signaling inhibits IFN responsive gene expression as well as downstream inflammatory and immunological processes. Inhibition of type I IFN blocks plasma cell differentiation and normalizes peripheral T-cell subsets	*SLE:* 300 mg IV over a 30-min for q4w

(continued)

Table 24.1 (continued)

Product	Molecular weight	Modality	Target	Indication[a]	Mechanism of action	Recommended dose regimen[a]
Belimumab (Benlysta®)	147 kDa	Human IgG1λ mAb	BLyS	SLE	Belimumab is a BLyS-specific inhibitor that blocks the binding of soluble BLyS, a B-cell survival factor, to its receptors on B cells. Belimumab does not bind B cells directly, but by binding BLyS Belimumab inhibits the survival of B cells, including autoreactive B cells, and reduces the differentiation of B cells into immunoglobulin-producing plasma cells	*IV infusion for SLE and LN (aged 5 years and older)*: 10 mg/kg q2w for the first three doses, followed by 10 mg/kg q4w thereafter *SC injection for adults with SLE*: 200 mg qw *SC injection for adults with LN*: 400 mg qw for 4 doses, then 200 mg qw thereafter
Benralizumab (Fasenra®)	150 kDa	Human IgG1κ mAb	IL-5 receptor	Asthma	Benralizumab binds to the α subunit of the human IL-5 receptor, which expressed on the surface of eosinophils and basophils The absence of fucose in the Fc domain of benralizumab facilitates binding to FcγRIII receptors on immune effectors cells, such as natural killer cells, leading to apoptosis of eosinophils and basophils through antibody-dependent cell-mediated cytotoxicity (ADCC)	*Asthma (12 years of age and older)*: 30 mg SC at Weeks 0, 4, and 8, followed by 30 mg q8w
Brodalumab (Siliq®)	144 kDa	Human IgG2κ mAb	IL-17 receptor A	PsO	Brodalumab selectively binds to human IL-17 receptor A and inhibits its interactions with cytokines IL-17A, IL-17F, IL-17C, IL-17A/F heterodimer and IL-25 IL-17 receptor A is a protein expressed on the cell surface and is a required component of receptor complexes utilized by multiple IL-17 family cytokines. Blocking IL17 receptor A inhibits IL-17 cytokine-induced responses including the release of pro-inflammatory cytokines and chemokines	*PsO*: 210 mg SC at Weeks 0, 1, and 2 followed by 210 mg q2w

Table 24.1 (continued)

Product	Molecular weight	Modality	Target	Indication[a]	Mechanism of action	Recommended dose regimen[a]
Canakinumab (Ilaris®)	145 kDa	Human IgG1κ mAb	IL-1β	Still's disease (AOSD and sJIA), CAPS, TRAPS, HIDS/ MKD, FMF	Canakinumab binds to human IL-1β and neutralizes its activity by blocking its interaction with IL-1 receptors, but it does not bind IL-1α or IL-1 receptor antagonist	*AOSD and sJIA (2 years and older):* 4 mg/kg (with maximum of 300 mg) SC q4w for patients weighing ≥7.5 kg *CAPS (4 years and older):* 150 mg SC q8w for patients weighing >40 kg and 2 mg/kg SC q8w for patients weighing ≥15 kg and ≤40 kg. For children 15–40 kg with an inadequate response, the dose can be increased to 3 mg/kg SC q8w *TRAPS, HIDS/MKD, FMF:* 2 mg/kg SC q4w for patients weighing ≤40 kg and the dose can be increased to 4 mg/kg q4w if the clinical response is not adequate; 150 mg SC q4w for patients weighing >40 kg and the dose can be increased to 300 mg q4w if the clinical response is not adequate
Certolizumab pegol (Cimzia®)	91 kDa	Humanized antibody Fab' fragment, conjugated to polyethylene glycol	TNFα	RA, PsA, AS, CD, PsO, and nr-axSpA	Certolizumab pegol binds to human TNFα and selectively neutralizes TNFα (but not TNFβ)	*RA and PsA:* 400 mg SC initially and at Weeks 2 and 4, followed by 200 mg SC q2w. For maintenance dosing, 400 mg SC q4w can also be considered *AS and nr-axSpA:* 400 mg SC initially and at Weeks 2 and 4, followed by 200 mg SC q2w or 400 mg SC q4w *PsO:* 400 mg SC q2w. For some patients weighing ≤90 kg, a dose of 400 mg SC initially and at Weeks 2 and 4, followed by 200 mg SC q2w *CD:* 400 mg SC initially and at Weeks 2 and 4. For patients achieving clinical response followed by 400 mg SC q4w

(continued)

Table 24.1 (continued)

Product	Molecular weight	Modality	Target	Indication[a]	Mechanism of action	Recommended dose regimen[a]
Dupilumab (Dupixent®)	147 kDa	IgG4	IL-4 receptor subunit α	AD, asthma, and CRSwNP	Dupilumab inhibits IL-4 and IL-13 signaling by binding to the IL-4 receptor subunit α shared by the IL-4 and IL-13 receptor complexes Blocking IL-4 receptor subunit α with dupilumab inhibits IL-4 and IL-13 cytokine-induced responses, including the release of proinflammatory cytokines, chemokines, and IgE	*Adult AD*: 600 mg (two 300 mg injections) SC initially, followed by 300 mg SC q2w *Pediatric AD (6–17 years of age)*: 600 mg as initial loading dose followed by 300 mg SC q4w for patients weighing ≥15 kg and <30 kg; 400 mg as initial loading dose followed by 200 mg SC q2w for patients weighing ≥30 kg and <60 kg; 600 mg as initial loading dose followed by 300 mg SC q2w for patients weighing ≥60 kg; *Pediatric AD (6-month to 5 years of age)*: 200 mg SC q4w for patients weighing ≥5 kg and <15 kg; 300 mg SC q4w for patients weighing ≥15 kg and <30 kg; *Asthma (12 years of age and older)*: 400 mg as initial loading dose followed by 200 mg SC q2w or 600 mg as initial loading dose follow by 300 mg SC q2w *Dosage for patients with oral corticosteroid-dependent asthma or with co-morbid moderate-to-severe AD or adults with comorbid CRSwNP*: 600 mg as initial loading dose follow by 300 mg SC q2w *Asthma (6–11 years of age)*: 100 mg SC q2w or 300 mg SC q4w for patients weighing ≥15 kg and <30 kg; 200 mg SC q2w for patients weighing ≥30 kg *CRSwNP*: 300 mg SC q2w
Etanercept (Enbrel®)	150 kDa	Dimeric fusion protein linked to the Fc portion of human IgG1	TNFα	RA, PsA, AS, pJIA, and PsO	Etanercept is a dimeric soluble form of the p75 TNF receptor that can bind TNF molecules. Etanercept inhibits binding of TNF-α and TNF-β to cell surface TNF receptors, rendering TNF biologically inactive	*RA, PsA, and AS*: 50 mg SC qw with or without methotrexate *Adult PsO*: 50 mg twice weekly for 3 months, followed by 50 mg SC qw *pJIA (2 years and older) and pediatric PsO (4 years and older)*: 0.8 mg/kg SC qw, with a maximum of 50 mg per week

Table 24.1 (continued)

Product	Molecular weight	Modality	Target	Indication[a]	Mechanism of action	Recommended dose regimen[a]
Golimumab (Simponi®, Simponi Aria®)	150–151 kDa	Human IgG1κ mAb	TNFα	RA, PsA, AS, UC, and pJIA; nr-axSpA (EU only, SmPC)	Golimumab binds to both the soluble and transmembrane bioactive forms of human TNFα. This interaction prevents the binding of TNFα to its receptors, thereby inhibiting the biological activity of TNFα	*IV infusion for RA, PsA, and AS:* 2 mg/kg IV infusion over 30 min at Weeks 0 and 4, then q8w thereafter *IV injection for pediatric pJIA and PsA (2 years of age and older):* 80 mg/m² IV infusion over 30 min at Weeks 0 and 4, then q8w thereafter *SC injection for RA, PsA AS, and nr-axSpA:* 50 mg SC monthly. Consider 100 mg SC monthly if no adequate clinical response after 3 or 4 doses *SC injection for UC:* 200 mg SC at Week 0, followed by 100 mg SC at Week 2 and then 100 mg SC q4w *SC injection for pJIA (≥40 kg):* 50 mg SC once a month
Guselkumab (Tremfya®)	143.6 kDa	Human IgG1λ mAb	IL-23	PsO, PsA	Guselkumab selectively binds to the p19 subunit of IL-23 and inhibits its interaction with the IL-23 receptor Guselkumab inhibits the release of proinflammatory cytokines and chemokines	*PsO and PsA:* 100 mg SC at Weeks 0 and 4, followed by 100 mg SC q8w thereafter
Inebilizumab-cdon (Uplizna®)	149 kDa	Humanized afucosylated IgG1 mAb	CD19	NMOSD	The precise mechanism by which inebilizumab-cdon exerts its therapeutic effects in NMOSD is unknown but is presumed to involve binding to CD19, a cell surface antigen present on pre-B and mature B lymphocytes. Following cell surface binding to B lymphocytes, inebilizumab-cdon results in antibody-dependent cellular cytolysis	*NMOSD in adult patients who are anti-aquaporin-4 (AQP4) antibody positive:* 300 mg SC at Weeks 0, 2, followed by a maintenance dosage starting 6 months from the first infusion of 300 mg q6m
Infliximab (Remicade®)	149 kDa	Chimeric IgG1κ mAb	TNFα	RA, PsA, AS, PsO, CD, UC, pediatric CD, pediatric UC	Infliximab neutralizes the biological activity of TNFα by binding to the soluble and transmembrane forms of TNFα and inhibits binding of TNFα with its receptors. Infliximab does not neutralize TNFβ	*RA:* 3 mg/kg IV at Weeks 0, 2, and 6, followed by 3 mg/kg q8w thereafter. Some patients may benefit from increasing the dose up to 10 mg/kg or treating as often as q4w *PsA and PsO:* 5 mg/kg IV at Weeks 0, 2, and 6, then q8w thereafter *AS:* 5 mg/kg IV at Weeks 0, 2, and 6, then q6w thereafter *CD and UC:* 5 mg/kg IV at Weeks 0, 2, and 6, then q8w thereafter. Some CD patients who initially respond to treatment may benefit from increasing the dose to 10 mg/kg IV if they later lose their response *Pediatric CD and UC (6 years of age and older):* 5 mg/kg IV at Weeks 0, 2, and 6, then q8w thereafter

(continued)

Table 24.1 (continued)

Product	Molecular weight	Modality	Target	Indication[a]	Mechanism of action	Recommended dose regimen[a]
Ixekizumab (Taltz®)	146 kDa	Human IgG4 mAb	IL-17A	PsA, PsO, AS, and nr-axSpA	Ixekizumab selectively binds with IL-17A and inhibits its interaction with the IL-17 receptor. IL-17A is a naturally occurring cytokine that is involved in normal inflammatory and immune responses. Ixekizumab inhibits the release of proinflammatory cytokines and chemokines	*PsO*: 160 mg SC at Week 0, followed by 80 mg SC at Weeks 2, 4, 6, 8, 10, and 12, then 80 mg SC q4w thereafter *Pediatric PsO (6 years and older)*: 160 mg SC at Week 0, followed by 80 mg q4w for patients weighing >50 kg; 80 mg SC at Week 0, followed by 40 mg q4w for patients weighing between 25 and 50 kg; 40 mg SC at Week 0, followed by 20 mg q4w for patients weighing <25 kg *PsA and AS*: 160 mg SC at Week 0, followed by 80 mg SC q4w thereafter. For PsA patients with coexistent moderate-to-severe PsO, the dose regimen for PsO can be used *Nr-axSpA:* 80 mg SC q4w
Mepolizumab (Nucala®)	149 kDa	Humanized IgG1 mAb	IL-5	Asthma, EGPA, CRSwNP	Mepolizumab binds to IL-5, inhibiting the bioactivity of IL-5 by blocking its binding to the alpha chain of the IL-5 receptor complex expressed on the eosinophil cell surface. IL-5 is the major cytokine responsible for the growth and differentiation, recruitment, activation, and survival of eosinophils, which plays a role in the pathogenesis of asthma and EGPA	*Severe asthma aged 12 years and older and CRSwNP aged 18 years and older*: 100 mg SC injections q4w *Severe asthma aged 6–11 years old*: 40 mg SC injections q4w *EGPA*: 300 mg administrated as 3 separate 100 mg SC injections q4w
Natalizumab (Tysabri®)	149 kDa	Human IgG4κ mAb	Integrin	CD and MS	Natalizumab binds to the α4-subunit of α4β1 and α4β7 integrins expressed on the surface of all leukocytes except neutrophils, and inhibits the α4-mediated adhesion of leukocytes to their counter-receptor(s)	*CD and MS*: 300 mg q4w by IV infusion over 1 h
Ocrelizumab (Ocrevus®)	145 kDa	Human glycosylated IgG1 mAb	CD20	MS	Ocrelizumab directed against CD20-expressing B-cells. CD20 is a cell surface antigen present on pre-B and mature B lymphocytes. Following cell surface binding to B lymphocytes, ocrelizumab results in antibody-dependent cell-mediated cytotoxicity (ADCC) and complement-mediated lysis	*MS*: 300 mg IV infusion initially, followed 2 weeks later by a second 300 mg IV infusion, and then 600 mg IV infusion every 6 months Pre-medicate with methylprednisolone (or an equivalent corticosteroid) and an antihistamine (e.g., diphenhydramine) prior to each infusion

Table 24.1 (continued)

Product	Molecular weight	Modality	Target	Indication[a]	Mechanism of action	Recommended dose regimen[a]
Omalizumab (Xolair®)	149 kDa	Human IgG1κ mAb	IgE	Asthma, CIU	Omalizumab binds to IgE and lowers free IgE levels. Subsequently, IgE receptors (FcεRI) on cells down-regulate. Reduction in surface-bound IgE on FcεRI-bearing cells limits the degree of release of mediators of the allergic response	*Asthma (6 years of age and older)*: 75–375 mg SC q2w or q4w, with dose and dosing frequency determined by pretreatment serum IgE levels and body weight *CIU (12 years of age and older)*: 150 or 300 mg SC q4w
Reslizumab (Cinqair®)	147 kDa	Human IgG4κ mAb	IL-5	Asthma	Reslizumab binds to IL-5, inhibiting the bioactivity of IL-5 by blocking its binding to the α chain of the IL-5 receptor complex expressed on the eosinophil cell surface. IL-5 is the major cytokine responsible for the growth and differentiation, recruitment, activation, and survival of eosinophils, which plays a role in the pathogenesis of asthma	*Asthma*: 3 mg/kg q4w by IV infusion over 20–50 min
Rilonacept (Arcalyst®)	251 kDa	Dimeric fusion protein linked to the Fc portion of human IgG1	IL-1β	CAPS, RP, DIRA	Rilonacept blocks IL-1β signaling by acting as a soluble decoy receptor that binds IL-1β and prevents its interaction with cell surface receptors. Rilonacept also binds IL-1α and IL-1 receptor antagonist with reduced affinity	*Adult CAPS and RP*: 320 mg SC as a loading dose, followed by a maintenance dose regime of 160 mg SC qw *Pediatric CAPS and RP (12–17 years of age)*: 4.4 mg/kg (up to 320 mg) SC as a loading dose, followed by a maintenance dose regime of 2.2 mg/kg (up to 160 mg) SC qw *Adult and pediatric DIRA patients weighing >10 kg*: 4.4 mg/kg (up to 320 mg) SC qw
Risankizumab-rzaa (Skyrizi®)	149 kDa	Human IgG1 mAb	IL-23	PsO, PsA, CD	Risankizumab-rzaa selectively binds to the p19 subunit of human IL-23 cytokine and inhibits its interaction with the IL-23 receptor. Risankizumab-rzaa inhibits the release of pro-inflammatory cytokines and chemokines	*PsO and PsA*: 150 mg SC at Weeks 0, 4, then q12w thereafter *CD*: 600 mg by IV infusion over at least 1 h at Weeks 0, 4, and 8. The recommended maintenance dosage is 360 mg SC at Week 12, then q8w thereafter
Rituximab (Rituxan®)	145 kDa	Chimeric IgG1κ mAb	CD20	RA	Rituximab targets the CD20 antigen expressed on the surface of pre-B and mature B-lymphocytes. Upon binding to CD20, rituximab mediates B-cell lysis	*RA*: Rituximab in combination with methotrexate is two 1000 mg IV infusions separated by 2 weeks (one course) every 24 weeks or based on clinical evaluation, but no more frequent than every 16 weeks. Methylprednisolone 100 mg IV or equivalent glucocorticoid is recommended 30 min prior to each infusion

(continued)

Table 24.1 (continued)

Product	Molecular weight	Modality	Target	Indication[a]	Mechanism of action	Recommended dose regimen[a]
Sarilumab (Kevzara®)	150 kDa	Human IgG1 mAb	IL-6 receptor	RA	Sarilumab binds to both soluble and membrane-bound IL-6 receptors, and has been shown to inhibit IL-6-mediated signaling through these receptors IL-6 is a pleiotropic pro-inflammatory cytokine produced by a variety of cell types including T-and B-cells, lymphocytes, monocytes, and fibroblasts. IL-6 is also produced by synovial and endothelial cells leading to local production of IL-6 in joints affected by inflammatory processes	*RA*: 200 mg SC q2w
Satralizumab-mwge (Ensprynd®)	143 kDa	Human IgG2 mAb	IL-6 receptor	NMOSD	The precise mechanism by which satralizumab-mwge exerts therapeutic effects in NMOSD is unknown but is presumed to involve inhibition of IL-6-mediated signaling through binding to soluble and membrane-bound IL-6 receptors	*NMOSD in adult patients who are anti-aquaporin-4 (AQP4) antibody positive*: 120 mg SC at Weeks 0, 2, and 4, followed by a maintenance dosage of 120 mg q4w
Secukinumab (Cosentyx®)	151 kDa	Human IgG1κ mAb	IL-17A	PsA, pediatric PsA, AS, PsO, pediatric PsO, nr-axSpA and ERA	Secukinumab binds to IL-17A and inhibits its interaction with the IL-17 receptor IL-17A is a naturally occurring cytokine that is involved in normal inflammatory and immune responses. Secukinumab inhibits the release of proinflammatory cytokines and chemokines	*PsO*: 300 mg SC at Weeks 0, 1, 2, 3, and 4 followed by 300 mg SC q4w. For some patients, a dose of 150 mg may be acceptable *PsA. AS and nr-axSpA*: For PsA patients with coexistent moderate-to-severe PsO: same dose for PsO. For AS, nr-axSpA and other PsA patients: 150 mg SC q4w with or without of a loading dosage (i.e., 150 mg SC at Weeks 0, 1, 2, 3, and 4). If a patient continues to have active PsA, consider a dosage of 300 mg. *ERA, pediatric PsO (6 years and older) and pediatric PsA (2 years and older)*: 75 mg SC for patients weighing <50 kg and 150 mg SC for patients weighing ≥50 kg at Weeks 0, 1, 2, 3, and 4, then q4w thereafter

Table 24.1   (continued)

Product	Molecular weight	Modality	Target	Indication[a]	Mechanism of action	Recommended dose regimen[a]
Spesolimab-sbzo (Spevigo®)	146 kDa	Human IgG1 mAb	IL-36 receptor	GPP flare	Spesolimab-sbzo inhibits IL-36 signaling by specifically binding to the IL-36R. Binding of spesolimab-sbzo to IL-36R prevents the subsequentactivation of IL-36R by cognate ligands (IL-36 α, β and γ) and downstream activation of pro-inflammatory andpro-fibrotic pathways. The precise mechanism linking reduced IL-36R activity and the treatment of flares ofGPP is unclear	*GPP flare*: a single dose of 900 mg IV over 90 min. If flare symptoms persist, may administer an additional 900 mg IV dose 1 week after the initial dose
Tezepelumab-ekko (Tezspire®)	147 kDa	Human IgG2λ mAb	TSLP	Asthma	Blocking TSLP with tezepelumab-ekko reduces biomarkers and cytokines associated with inflammation including blood eosinophils, airway submucosal eosinophils, IgE, FeNO, IL-5, and IL-13; however, the mechanism of tezepelumab-ekko action in asthma has not been definitively established	*Asthma*: For the adult patients and pediatric patients aged 12 years and older, 210 mg SC q4w
Tildrakizumab (Ilumya®)	147 kDa	Human IgG1κ mAb	IL-23	PsO	Tildrakizumab selectively binds to the p19 subunit of human IL-23 cytokine and inhibits its interaction with the IL-23 receptor. Tildrakizumab inhibits the release of pro-inflammatory cytokines and chemokines	*PsO:* 100 mg at Weeks 0, 4, then q12w thereafter

(continued)

Table 24.1 (continued)

Product	Molecular weight	Modality	Target	Indication[a]	Mechanism of action	Recommended dose regimen[a]
Tocilizumab (Actemra®)	148 kDa	Human IgG1κ mAb	IL-6 receptor	RA, pJIA, sJIA, GCA, CRS, and SSc-ILD	Tocilizumab binds to both soluble and membrane-bound IL-6 receptors, and has been shown to inhibit IL-6-mediated signaling through these receptors IL-6 is a pleiotropic pro-inflammatory cytokine produced by a variety of cell types including T-and B-cells, lymphocytes, monocytes, and fibroblasts. IL-6 is also produced by synovial and endothelial cells leading to local production of IL-6 in joints affected by inflammatory processes	*RA*: For IV use: 4 mg/kg IV q4w, followed by an increase to 8 mg/kg IV q4w based on clinical response For SC use: 162 mg SC q2w, followed by an increase to qw based on clinical response for patients <100 kg; the recommended dose for patients ≥100 kg is 162 mg SC qw *pJIA (2 years of age and older)*: 10 mg/kg or 8 mg/kg IV q4w for patients weighing <30 kg or ≥30 kg, respectively, or 162 mg SC Q3W or 162 mg SC Q2W for patients weighing <30 kg or ≥30 kg, respectively, alone or in combination with methotrexate *sJIA (2 years of age and older)*: 12 mg/kg or 8 mg/kg IV q2w for patients weighing <30 kg or ≥30 kg, respectively, or 162 mg SC Q3W or 162 mg SC Q2W for patients weighing <30 kg or ≥30 kg, respectively, alone or in combination with methotrexate *GCA*: 6 mg/kg IV q4w or 162 mg SC qw, in combination with a tapering course of glucocorticoids. Dose regimen of 162 mg SC q2w in combination with a tapering course of glucocorticoids may be prescribed based on clinical considerations. Tocilizumab can be used alone following discontinuation of glucocorticoids *CRS (2 years of age and older)*: 12 mg/kg or 8 mg/kg IV for patients weighing <30 kg or ≥30 kg, respectively, alone or in combination with corticosteroids. If no clinical improvement after the first dose, up to 3 additional doses of tocilizumab may be administered. The interval between consecutive doses should be at least 8 h. Doses exceeding 800 mg per infusion are not recommended *SSc-ILD*: 162 mg SC qw
Tralokinumab-ldrm (Adbry®)	147 kDa	Human IgG4 mAb	IL-13	AD	Tralokinumab-ldrm specifically binds to human IL-13 and inhibits its interaction with the IL-13Rα1 and IL-13Rα2 Tralokinumab-ldrm inhibits the bioactivity of IL-13 by blocking IL-13 interaction with IL-13Rα1/IL-4Rα receptor complex and inhibits IL-13-induced responses including the release of proinflammatory cytokines, chemokines and IgE	*AD*: Initial dose of 600 mg, followed by 300 mg SC q2w. A dosage of 300 mg q4w may be considered for patients below 100 kg who achieve clear or almost clear skin after 16 weeks of treatment

Table 24.1 (continued)

Product	Molecular weight	Modality	Target	Indication[a]	Mechanism of action	Recommended dose regimen[a]
Ustekinumab (Stelara®)	148–149 kDa	Human IgG1κ mAb	IL12/IL23	PsO, PsA, pediatric PsO, pediatric PsA, CD, and UC	Ustekinumab binds to the p40 protein subunit used by both the IL-12 and IL-23 cytokines. IL-12 and IL-23 are naturally occurring cytokines that are involved in inflammatory and immune responses, such as natural killer cell activation and CD4+ T-cell differentiation and activation. IL-12 and IL-23 have been implicated as important contributors to chronic inflammations	*PsO*: 45 mg SC initially and 4 weeks later, followed by 45 mg SC q12w for patients weighing ≤100 kg; 90 mg SC initially and 4 weeks later, followed by 90 mg SC q12w for patients weighing >100 kg *PsA*: 45 mg SC initially and 4 weeks later, followed by 45 mg SC q12w. For patients with co-existent moderate-to-severe PsO weighing >100 kg:90 mg *Pediatric PsO and pediatric PsA (6–17 years of age)*: SC weight-based-dosing (0.75 mg/kg, 45 mg and 90 mg for weight range of <60 kg, 60–100 kg and ≥100 kg, respectively) at the initial dose, 4 weeks later, then q12w thereafter *CD and UC*: a single IV infusion using weight-based-dosing (~ 6 mg/kg, i.e., 260 mg, 390 mg, and 520 mg for weight range of ≤55 kg, 55–85 kg and >85 kg, respectively). The recommended maintenance dose of ustekinumab is 90 mg SC 8 weeks after the initial IV dose, then q8w thereafter
Vedolizumab (Entyvio®)	147 kDa	Humanized IgG1 mAb	Integrin	CD and UC	Vedolizumab binds to the α4β7 integrin and blocks the interaction of α4β7 integrin with MAdCAM-1 and inhibits the migration of memory T-lymphocytes across the endothelium into inflamed gastrointestinal parenchymal tissue. Vedolizumab does not bind to or inhibit function of the α4β1 and αEβ7 integrins and does not antagonize the interaction of α4 integrins with VCAM-1	*CD and UC*: 300 mg at Weeks 0, 2, and 6 by IV infusion over approximately 30 min, then q8w thereafter. Following the first two IV doses administered at Week 0 and Week 2 in UC, vedolizumab may be switched to SC injection of 108 mg q2w at Week 6.

Abbreviations: AS ankylosing spondylitis, *AD* atopic dermatitis, *AOSD* adult-onset Still's disease, *BLyS* B-lymphocyte stimulator, *CAPS* cryopyrin-associated periodic syndromes, *CD* Crohn's disease, *CD2* T-lymphocyte antigen CD2, *CD20/80/86* B-lymphocyte antigen CD20, CD80, CD86, *CIU* chronic idiopathic urticaria, *CRS* cytokine release syndrome, *CRSwNP* chronic rhinosinusitis with nasal polyps, *CTLA-4* cytotoxic T-lymphocyte-associated antigen 4, *DIRA* deficiency of the interleukin-1 receptor antagonist, *EGPA* eosinophilic granulomatosis with polyangiitis, *ERA* enthesitis-related arthritis, *FeNO* fractional exhaled nitric oxide, *FMF* familial mediterranean fever, *GCA* giant cell arteritis, *HS* hidradenitis suppurativa, *GPP* generalized pustular psoriasis, *HIDS/MKD* hyperimmunoglobulin D syndrome/mevalonate kinase deficiency, *IgE* immunoglobulin E, *IL* interleukin, *IV* intravenous, *kDa* kilodalton, *LN* lupus nephritis, *mAb* monoclonal antibody, *MAdCAM-1* mucosal addressin cell adhesion molecule-1, *MW* molecular weight, *MS* multiple sclerosis, *NLR-3* nucleotide-binding domain, leucine-rich family (NLR), pyrin domain-containing 3, *NMOSD* neuromyelitis optica spectrum disorder, nr-*axSpA* non-radiographic axial spondyloarthritis, *pJIA* polyarticular juvenile idiopathic arthritis, *PsA* psoriatic arthritis, *PsO* plaque psoriasis, *q12w* every 12 weeks, *q2w* every 2 weeks, *q4w* every 4 weeks, *q8w* every 8 weeks, *qw* every week, *RA* rheumatoid arthritis, *RP* recurrent pericarditis, *SC* subcutaneous, *SLE* systemic lupus erythematosus, *sJIA* systemic juvenile idiopathic arthritis, *SSc-ILD* systemic sclerosis-associated interstitial lung disease, *TNFα* tumor necrosis factor alpha, *TRAPS* tumor necrosis factor receptor-associated periodic syndrome, *UC* ulcerative colitis, *UV* uveitis, *VCAM-1* vascular cell adhesion molecule-1

[a] From US prescription information (USPI), unless otherwise indicated

Arthritides

Arthritides are a class of chronic autoimmune inflammatory conditions of unknown etiology, characterized by pain and stiffness of the affected joints and tissues (Davis and Mease 2008; McInnes and Schett 2011). Arthritides consist of a variety of clinical diseases, such as rheumatoid arthritis (RA), psoriatic arthritis (PsA), and ankylosing spondylitis (AS), involving skeletal joints. Juvenile idiopathic arthritis (JIA) is a chronic inflammatory arthropathy with an age of onset of <16 years, including polyarticular JIA (pJIA) and systemic JIA (sJIA). Although these arthritic diseases may have different clinical manifestations, they are considered to have a similar underlying etiology. RA is the most common of the autoimmune arthritides, affecting at least 1% of the general population in the USA. If not properly treated, the chronic inflammation can result in progressive and irreversible joint destruction. Although early intervention with corticosteroids and conventional non-biologic disease-modifying antirheumatic drugs (DMARDs) has been proven to attenuate inflammation, these therapies are not very effective in slowing down the progression of joint damage in large subsets of patients. In addition, these conventional therapies often have significant off-target side effects (O'Dell 2004).

Several biologic agents have been approved for patients with RA, PsA, AS, or JIA (Table 24.1). These therapeutic proteins provide valuable treatment options for patients, particularly for those who experience significant side effects and/or have inadequate clinical efficacy with conventional DMARDs. These biotherapeutics significantly improve the signs and symptoms of the disease, effectively inhibit (and sometimes even reverse) the progression of joint damage, and greatly improve physical functions and quality of life. Anti-TNFα agents (adalimumab, certolizumab, etanercept, golimumab, and infliximab) are considered the gold standard biologic therapy for RA; however, the availability of biologic agents with different mechanisms of action such as anti-IL-6 receptor agents (sarilumab and tocilizumab) provides alternative options when patients do not achieve adequate response to anti-TNFα agents. Other non-TNFα targeting biologic agents demonstrating effectiveness for the treatment of arthritides include agents neutralizing soluble cytokines such as IL-1β (canakinumab), IL-17A (ixekizumab and secukinumab), and IL-12/IL-23 (ustekinumab), and acting as direct agonist or antagonists to inhibit T-cell activation (abatacept) or depletion of B lymphocytes (rituximab). Compared to conventional DMARDs, the most attractive attribute of antibody-based therapeutic proteins is their binding to the target with high specificity, consequently producing greater efficacy and fewer off-target adverse effects. In addition, antibody-based therapeutic proteins usually have long half-lives (up to 2–3 weeks), which allow for infrequent dosing which is desirable for the patients with chronic diseases.

Anti-TNFα Biologic Agents

TNFα is a naturally occurring cytokine that is involved in normal inflammatory and immune responses. Elevated levels of TNFα are found in the synovial fluid of patients with RA, JIA, PsA, and AS and play an important role in both the pathologic inflammation and the joint destruction that are hallmarks of these diseases.

The approved anti-TNFα agents such as etanercept, infliximab, adalimumab, golimumab, and certolizumab pegol are presently one of the most successful classes of therapeutic proteins with many approved clinical indications. Etanercept, a dimeric fusion protein consisting of the p75 human TNFα receptor linked to the Fc portion of human IgG1, was the first anti-TNFα biologic agent approved for an arthritide indication. Infliximab is a chimeric IgG1 mAb containing ~25% mouse sequence and ~75% human sequence. Certolizumab pegol is a Fab antibody fragment linked to polyethylene glycol that enhances solubility and prolongs elimination half-life. Adalimumab and golimumab are two human IgG1 mAbs, which were created using phage display libraries and the expression of human immunoglobulin genes by transgenic mice, respectively. Human antibodies were developed to minimize immunogenicity; however, patients treated with either adalimumab or golimumab still develop antidrug antibodies. Future efforts are still required to generate therapeutic mAbs that not only have the human sequence as the primary structure, but also have secondary and tertiary structures like natural human immunoglobulins. Although no head-to-head comparative trials are currently available, these anti-TNFα agents appear to have similar efficacy for the treatment of adult RA patients alone or as an add-on to methotrexate (Salliot et al. 2011). These anti-TNFα agents have different elimination half-lives and offer a variety of dosing options. Infliximab is administered intravenously, while the other four anti-TNFα agents can be administered subcutaneously. Golimumab can also be administrated either intravenously or subcutaneously. Etanercept has the shortest half-life (~ 4 days) and needs to be dosed once or twice a week. Infliximab is administered intravenously every 4–8 weeks, while adalimumab, certolizumab, and golimumab are administered subcutaneously every 2 weeks, every 2–4 weeks, or monthly, respectively (Tracey et al. 2008).

Although these antibody-based drugs offer targeted therapy with high specificity, there are adverse effects associated with them that need to be closely monitored (Bongartz et al. 2006; Brown et al. 2002; Ellerin et al. 2003). Certain adverse events such as infections are the result of inhibition of the protective functions of the targeted cytokines and related immune cells. Serious infections due to bacterial, mycobacterial, invasive fungal, viral, protozoal, or other opportunistic pathogens have been reported in patients with RA, PsA, AS, or JIA who received TNFα blockers and other immunosuppressant therapeutic proteins. Patients should be closely monitored for the development of signs and symptoms of infection during and after treatment with these therapeutic proteins, including the development of tuberculosis in patients who tested negative for latent tuberculosis infection prior to the initiation of therapy. Malignancy, albeit rare, has been another important concern when using these immunosuppressant therapeutic proteins. In controlled clinical trials of TNFα blockers, more cases of lymphoma and leukemia have been observed among patients receiving anti-TNFα treatment compared to patients in the control groups; however, there are confounders when assessing the risk of malignancy associated with the use of these therapeutic proteins in patients with chronic inflammatory diseases. Patients with chronic inflammatory diseases, particularly patients with highly active disease and/or chronic exposure to immunosuppressant therapies, may be at higher risk (up to several folds) than the general population for the development of lymphoma and leukemia, even in the absence of TNF-blocking therapy (Smedby et al. 2008). Other notable adverse events associated with these complex proteins include demyelinating disorders, liver enzyme elevation, autoimmune diseases (such as lupus), immunogenicity (formation of antibodies to the therapeutic protein), infusion/injection site reactions, and other hypersensitivity reactions. Overall, a large number of clinical trials have demonstrated that the benefits outweigh the risks for anti-TNF biologic agents in patients with various arthritides.

Adalimumab

Adalimumab (Humira®) is a recombinant human IgG1 mAb, which was created using phage display technology resulting in a human antibody. Adalimumab binds specifically to TNF-α and blocks its interaction with the p55 and p75 cell surface TNF receptors. Adalimumab does not bind or inactivate lymphotoxin (Humira®, US prescribing information 2022).

The efficacy and safety of adalimumab have been assessed in various adult RA populations (DMARD [including methotrexate]-inadequate responders and methotrexate-naive patients), polyarticular JIA, PsA, and AS. In a randomized, double-blind, controlled Phase III study in patients with active RA despite methotrexate therapy, 6-month treatment with subcutaneous adalimumab in combination with methotrexate at the recommended dose regimen (40 mg every 2 weeks) resulted in 63, 39, and 21% of RA patients achieving ACR20, ACR50, and ACR70 (20, 50, and 70% improvement in the American College of Rheumatology (ACR) response criterion), respectively, while the control group (methotrexate plus placebo) had only 30, 10, and 3% of patients with ACR20, ACR50, and ACR70 responses, respectively.

Serious infection and malignancy Black-Box warnings were placed on the adalimumab label (Humira®, US prescribing information 2022), similar to the labels of other TNF antagonists. Patients treated with adalimumab are at increased risk for developing serious infections involving various organ systems and sites that may lead to hospitalization or death. Adalimumab should not be started during an active infection. If an infection develops, monitor carefully, and stop adalimumab if infection becomes serious. Lymphoma and other malignancies, some fatal, have been reported in children and adolescent patients treated with TNF blockers including adalimumab. Risks and benefits of TNF-blocker treatment including adalimumab should be considered prior to initiating therapy in patients with a known malignancy other than a successfully treated non-melanoma skin cancer (NMSC) or when continuing a TNF blocker in patients who develop a malignancy.

The concomitant use of a TNF blocker and abatacept (lymphocyte activation inhibitor) or anakinra (IL-1 receptor antagonist) was associated with a higher risk of serious infections in patients with RA. Therefore, the concomitant use of adalimumab and these biologic products is not recommended in the treatment of patients with RA (Humira®, US prescribing information 2022).

For the treatment of arthritides, adalimumab is indicated for the treatment of adult patients with moderately to severely active RA, active PsA, or active AS. Adalimumab is also indicated for reducing signs and symptoms in pediatric patients 2 years of age and older with moderately to severely active polyarticular JIA (Humira®, US prescribing information 2022).

Certolizumab Pegol

Certolizumab pegol (Cimzia®) is a recombinant, humanized antibody Fab fragment that is conjugated to an approximately 40 kDa polyethylene glycol. Certolizumab pegol binds to human TNFα with high affinity and selectively neutralizes TNFα activity, but does not neutralize lymphotoxin (Cimzia®, US prescribing information 2019).

The efficacy and safety of certolizumab pegol have been assessed in adult RA patients who had active disease despite methotrexate therapy or who had failed at least one conventional DMARD other than methotrexate. Assessments in adult patients with active PsA, active AS, and active non-radiographic axial spondyloarthritis (nr-axSpA) have also been conducted. In a randomized, double-blind, controlled Phase III study in patients with active RA despite methotrexate therapy, 6-month treatment with subcutaneous certolizumab pegol at the recommended dose regimen (400 mg initially and at Weeks 2 and 4, following by 200 mg every 2 weeks) in combination with methotrexate resulted in 59, 37, and 21% of RA patients achieving ACR20, ACR50, and ACR70, respectively, while the control group (methotrexate plus placebo) had only 14, 8, and 3% of patients with ACR20, ACR50, and ACR70 responses, respectively.

Serious infection and malignancy Black-Box warnings were placed on the certolizumab pegol label (Cimzia®, US prescribing information 2019), similar to the labels of other TNF antagonists.

Of the rheumatic diseases, certolizumab pegol is indicated for the treatment of adult patients with moderately to severely active RA, active PsA, active AS, and active nr-axSpA. Certolizumab pegol is not indicated for use in pediatric patients (Cimzia®, US prescribing information 2019).

Etanercept

Etanercept (Enbrel®) is the first anti-TNF biologic agent approved for arthritide indication. Etanercept is a dimeric fusion protein consisting of the extracellular ligand-binding portion of the human 75 kDa (p75) TNF receptor linked to the Fc portion of human IgG1. Etanercept inhibits binding of TNF-α and TNF-β (lymphotoxin alpha) to cell surface TNF receptors, rendering TNF biologically inactive (Enbrel®, US prescribing information 2022).

The efficacy and safety of etanercept have been assessed in various adult RA populations (DMARD [including methotrexate]-inadequate responders and methotrexate-naive patients), active polyarticular JIA, active PsA, and active AS. In a randomized, double-blind, controlled Phase III study in patients with active RA despite methotrexate therapy, 6-month treatment with subcutaneous etanercept in combination with methotrexate at the recommended dose regimen (50 mg weekly) resulted in 71, 39, and 15% of RA patients achieving ACR20, ACR50, and ACR70, respectively, while the control group (methotrexate plus placebo) had only 27, 3, and 0% of patients exhibiting ACR20, ACR50, and ACR70 responses, respectively.

Serious infection and malignancy Black-Box warnings were placed on the etanercept label (Enbrel®, US prescribing information 2022), similar to the labels of other TNF antagonists.

Use of etanercept with anakinra (IL-1 receptor antagonist) or abatacept (T-lymphocyte activation inhibitor) is not recommended. Concurrent administration of etanercept with anakinra or abatacept resulted in increased incidences of serious adverse events, including infections, and did not demonstrate increased clinical benefit. In a 24-week study in patients with active RA on background methotrexate, the ACR50 response rate was 31% for patients treated with the combination of anakinra and etanercept and 41% for patients treated with etanercept alone, indicating no added clinical benefit of the combination over etanercept alone. A higher rate of serious infections was observed in RA patients with concurrent anakinra and etanercept therapy (7%) than in patients treated with etanercept alone (0%). Therefore, use of anakinra in combination with TNF blocking agents is not recommended (Enbrel®, US prescribing information 2022). In controlled clinical trials in patients with active RA, patients receiving concomitant intravenous abatacept and etanercept therapy experienced more infections (63%) and serious infections (4.4%) compared to patients treated with only etanercept (43% and 0.8%, respectively). In addition, these trials failed to demonstrate an enhancement of efficacy with concomitant administration of abatacept with etanercept (Orencia®, US prescribing information 2021). As a result, abatacept should not be given concomitantly with TNF antagonists. The use of etanercept in patients receiving concurrent cyclophosphamide therapy is also not recommended.

For the treatment of arthritides, etanercept is indicated for the treatment of adult patients with moderately to severely active RA, active PsA, or active AS. Etanercept is also indicated for reducing signs and symptoms of polyarticular JIA in pediatric patients 2 years of age and older (Enbrel®, US prescribing information 2022).

Golimumab

Golimumab (Simponi®[subcutaneous], Simponi Aria®[intravenous]) is a human IgG1κ mAb which was created using genetically engineered mice immunized with human TNF. Golimumab binds to both the soluble and transmembrane bioactive forms of human TNFα and therefore inhibits the biologic activity of TNFα. There is no evidence of golimumab binding to other TNF superfamily ligands; golimumab does not bind or neutralize human lymphotoxin. Golimumab does not lyse human monocytes expressing transmembrane TNF in the presence of complement or effector cells (Simponi®, Simponi Aria®, US prescribing information 2021).

The efficacy and safety of golimumab have been assessed in various adult RA populations (methotrexate-inadequate responders, methotrexate-naive patients, and patients with previous use of other anti-TNFα agents), active PsA, active AS, active nr-axSpA, and pediatric patients with active polyarticular JIA and active PsA. In a randomized, double-blind, controlled Phase III study in patients with active RA despite methotrexate therapy, 6-month treatment with subcutaneous golimumab in combination with methotrexate at the recommended golimumab dose regimen (50 mg every 4 weeks) resulted in 60, 37, and 20% of RA patients achieving ACR20, ACR50, and ACR70, respectively, while the control group (methotrexate plus placebo) had only 28, 14, and 5% of patients exhibiting ACR20, ACR50, and ACR70 responses, respectively. Recently, another Phase III study demonstrated similar efficacy and safety for treating adult RA patients with intravenous golimumab with a longer dosing interval (every 8 weeks) as an alternative to the previously established subcutaneous route of administration (every 4 weeks).

Serious infection and malignancy Black-Box warnings were placed on the golimumab label, similar to the labels of other TNF antagonists. The concomitant use of golimumab with biologics approved to treat RA, PsA, or AS is not recommended because of the possibility of an increased risk of infection (Simponi®, Simponi Aria®, US prescribing information 2021).

Within arthritides, golimumab is indicated for the treatment of adult patients with moderately to severely active RA, or active AS (Simponi®, Simponi Aria®, US prescribing information 2021). For active PsA and pJIA, golimumab can be used in patients 2 years of age and older. Golimumab is also indicated for the treatment of adult patients with nr-axSpA (Simponi®, SmPC 2021).

Infliximab

Infliximab (Remicade®) is a chimeric IgG1κ mAb that is composed of human constant and murine variable regions. Infliximab neutralizes the biologic activity of TNFα by binding with high affinity to the soluble and transmembrane forms of TNFα and inhibits binding of TNFα with its receptors. Infliximab does not neutralize TNFβ (lymphotoxin-α) (Remicade®, US prescribing information 2021).

The efficacy and safety of infliximab have been assessed in various adult RA populations (methotrexate-inadequate responders and methotrexate-naive patients), active PsA, and active AS. In a randomized, double-blind, controlled Phase III study in patients with active RA despite methotrexate therapy, 30-week treatment with intravenous infliximab (3 mg/kg at Weeks 0, 2, and 6, following by 3 mg/kg every 8 weeks thereafter) in combination with methotrexate resulted in 50, 27, and 8% of RA patients achieving ACR20, ACR50, and ACR70, respectively, while the control group (methotrexate plus placebo) had only 20, 5, and 0% of patients exhibiting ACR20, ACR50, and ACR70 responses, respectively.

Serious infection and malignancy Black-Box warnings were placed on the infliximab label (Remicade®, US prescribing information 2021), similar to the labels of other TNF antagonists.

There is insufficient information regarding the concomitant use of infliximab with other biological therapeutics used to treat the same conditions as infliximab. The concomitant use of infliximab with these biologics is not recommended because of the possibility of an increased risk of infection. An increased risk of serious infections has been seen in clinical RA trials of other TNF blockers used in combination with anakinra or abatacept, with no added benefit (described earlier in this chapter); therefore, use of infliximab with abatacept or anakinra is not recommended. The use of tocilizumab (IL-6 receptor antagonist) in combination with TNF antagonists, including infliximab, should also be avoided because of the possibility of increased immunosuppression and

increased risk of infection (Remicade®, US prescribing information 2021).

Within arthritides, infliximab is indicated for reducing signs and symptoms, inhibiting the progression of structural damage, and improving physical function in patients with moderately to severe active RA. Infliximab is also indicated for reducing signs and symptoms in patients with active AS and in patients with active PsA, inhibiting the progression of structural damage, and improving physical function in patients with PsA (Remicade®, US prescribing information 2021).

Anti-IL-1β Biologic Agents

As of November 2022, canakinumab (Ilaris®) is the only anti-IL-β biologic agent approved for arthritides-related disorders, i.e., systemic JIA (sJIA). Systemic JIA (sJIA) is a severe autoinflammatory disease, driven by innate immunity by means of proinflammatory cytokines such as IL-1β.

Canakinumab

Canakinumab (Ilaris®) is a recombinant human IgG1κ anti-IL-1β mAb. Canakinumab binds to human IL-1β and neutralizes its activity by blocking its interaction with IL-1 receptors, but it does not bind IL-1α or IL-1 receptor antagonist (IL-1Ra) (Ilaris®, US prescribing information 2020).

The efficacy and safety of canakinumab have been assessed in two Phase III studies (Study 1 and Study 2) in sJIA patients aged 2 to less than 20 years. Study 1 was a randomized, double-blind, placebo-controlled, single-dose 4-week study in sJIA patients who were randomized to receive a single subcutaneous dose of 4 mg/kg canakinumab or placebo. At Day 29, 81, 79, and 67% of sJIA patients who received a single subcutaneous dose of 4 mg/kg canakinumab achieved (Pediatric American College of Rheumatology Criteria) PEDACR30, PEDACR50, and PEDACR70 responses (30, 50, and 70% improvement in an adapted Pediatric American College of Rheumatology response criterion), respectively, while the control group had only 10, 5, and 2% of patients with PEDACR30, PEDACR50, and PEDACR70 responses, respectively. Study 2 was a two-part study to assess the flare prevention by canakinumab in patients with active sJIA, with an open-label, single-arm active treatment period (Part I) followed by a randomized, double-blind, placebo-controlled, event-driven withdrawal design (Part II). The probability of experiencing a flare (defined by worsening of greater than or equal to 30% in at least 3 of the 6 core Pediatric ACR response variables combined with improvement of greater than or equal to 30% in no more than 1 of the 6 variables, or reappearance of fever not due to infection for at least 2 consecutive days) over time in Part II was statistically lower for the canakinumab group than for the placebo group. This corresponded to a 64% relative reduction in the risk of flare for patients in the canakinumab group as compared to those in the placebo group.

The efficacy of canakinumab in adult-onset Still's disease (AOSD) is based on the pharmacokinetic exposure and extrapolation of the established efficacy of canakinumab in sJIA patients. Efficacy of canakinumab was also assessed in a randomized, double-blind, placebo-controlled study that enrolled 36 patients (22–70 years old) diagnosed with AOSD. The efficacy data were generally consistent with the results of a pooled efficacy analysis of sJIA patients.

An increased incidence of serious infections and an increased risk of neutropenia have been associated with administration of another IL-1 blocker in combination with TNF inhibitors (anakinra in combination with etanercept) in another patient population (adult RA). Use of canakinumab with TNF inhibitors may also result in similar toxicities and is not recommended because this may increase the risk of serious infections (Ilaris®, US prescribing information 2020).

Of the rheumatic diseases, canakinumab is indicated for the treatment of active Still's disease, including AOSD and sJIA in patients aged 2 years and older (Ilaris®, US prescribing information 2020).

Anti-IL-17A Biologic Agents

Two anti-IL-17A biologic agents have been approved for arthritides-related disorders, ixekizumab (Taltz®) and secukinumab (Cosentyx®). Ixekizumab is a humanized IgG4 mAb while secukinumab (Cosentyx®) is a human IgG1κ mAb. Each of these two agents selectively binds with the IL-17A cytokine and inhibits its interaction with the IL-17 receptor. IL-17A is a naturally occurring cytokine that is involved in normal inflammatory and immune responses. Ixekizumab or secukinumab inhibits the release of proinflammatory cytokines and chemokines.

Ixekizumab

Ixekizumab (Taltz®) is a humanized anti-IL17A IgG4 mAb produced by recombinant DNA technology in a recombinant mammalian cell line (Taltz®, US prescribing information 2022).

The efficacy and safety of ixekizumab have been assessed in various adult populations including active PsA, active AS, and active nr-axSpA. For the treatment of active PsA, the efficacy and safety of ixekizumab have been assessed in two randomized, double-blind, controlled Phase III studies in patients with active PsA despite non-steroidal anti-inflammatory drug (NSAIDs), corticosteroid, or non-biologic conventional DMARD therapy. In the randomized, double-blind, controlled Phase III study in biologic-naive

PsA patients (43% patients with concomitant methotrexate use), 6-month treatment with subcutaneous ixekizumab at the recommended dose regimen (160 mg at Week 0, followed by 80 mg every 4 weeks thereafter) resulted in 58, 40, and 23% of RA patients achieving ACR20, ACR50, and ACR70, respectively, while the control group had only 30, 15, and 6% of patients with ACR20, ACR50, and ACR70 responses, respectively. In the other Phase III study in anti-TNFα experienced PsA patients, greater responses compared to placebo were seen regardless of prior anti-TNF exposure.

For the treatment of active AS, the safety and efficacy of ixekizumab have been assessed in two randomized, double-blind, placebo-controlled studies (AS1 and AS2) in adult patients with active ankylosing spondylitis. In both studies, patients treated with ixekizumab 80 mg every 4 weeks demonstrated greater improvements in ASAS40 and ASAS20 responses (64% and 48% in AS1 and 48% and 25% in AS2, respectively) compared to placebo (40% and 18% in AS1 and 30% and 13% in AS2, respectively) at Week 16. Responses were seen regardless of concomitant therapies. In AS2, responses were seen regardless of prior TNF-inhibitor exposure.

For the treatment of active nr-axSpA with objective signs of inflammation, the safety and efficacy of ixekizumab have been assessed in a randomized, double-blind, 52-week placebo-controlled study (nr-axSpA1) in adult patients with active axial spondyloarthritis for at least 3 months. At both Weeks 16 and 52, a greater proportion of patients treated with ixekizumab 80 mg every 4 weeks had ASAS40 response (35.4% and 30.2%, respectively) compared to patients treated with placebo (19.0% and 13.3%, respectively).

Of the rheumatic diseases, ixekizumab is indicative for the treatment of adult patients with moderately to severely active PsA, active AS, and active nr-axSpA with objective signs of inflammation (Taltz®, US prescribing information 2022).

Secukinumab

Secukinumab (Cosentyx®) is a recombinant human anti-IL-17A IgG1/κ mAb expressed in a recombinant Chinese hamster ovary cell line (Cosentyx®, US prescribing information 2021).

The efficacy and safety of secukinumab have been assessed in adult patients with active PsA, active AS, and active nr-axSpA with objective signs of inflammation. The efficacy and safety of secukinumab have also been assessed in pediatric patients with active PsA (≥2 years old) and enthesitis-related arthritis (ERA, (≥4 years old). In a randomized, double-blind, controlled Phase III study in patients with active PsA despite NSAIDs, corticosteroid, or DMARD therapy, 6-month treatment with subcutaneous secukinumab at the recommended dose regimen (150 mg every 4 weeks) resulted in 51, 35, and 21% of RA patients (55% of these patients had concomitant methotrexate use) achieving ACR20, ACR50, and ACR70, respectively, while the control group had only 15, 7, and 1% of patients with ACR20, ACR50, and ACR70 responses, respectively.

Of the rheumatic diseases, secukinumab is indicated for the treatment of active PsA in patients 2 years of age and older, active ERA, and adult patients with AS and with active nr-axSpA with objective signs of inflammation (Cosentyx®, US prescribing information 2021).

Anti-IL-12/IL-23 Biologic Agents

As of November 2022, ustekinumab (Stelara®) is the only anti-IL-12/IL-23 biologic agent approved for arthritides-related disorders, i.e., PsA. IL-12 and IL-23 are naturally occurring cytokines that are involved in inflammatory and immune responses, such as natural killer cell activation and CD4+ T-cell differentiation and activation.

Ustekinumab
Ustekinumab (Stelara®) is a human IgG1κ mAb that binds with high affinity and specificity to the p40 protein subunit used by both the IL-12 and IL-23 cytokines. In in vitro models, ustekinumab was shown to disrupt IL-12- and IL-23-mediated signaling and cytokine cascades by disrupting the interaction of these cytokines with a shared cell surface receptor chain, IL-12β1 (Stelara®, US prescribing information 2022).

The safety and efficacy of ustekinumab have been assessed in adult patients with active PsA (≥5 swollen joints and ≥5 tender joints) despite NSAIDs or DMARD therapy. In a randomized, double-blind, controlled Phase III study in TNF blocker naive PsA patients (approximately 50% of patients continued on stable doses of methotrexate), 6-month treatment in patients received subcutaneous ustekinumab treatment at the recommended dose regimen (45 mg at Weeks 0 and 4 followed by every 12 weeks thereafter) resulted in 42, 25, 12, and 57% of PsA patients achieving ACR20, ACR50, ACR70, and PASI75 (at least a 75% reduction in PASI [Psoriasis Activity and Severity Index] score from baseline) respectively, while the control group (placebo) had only 23, 9, 2, and 11% of patients with ACR20, ACR20, ACR50, ACR70, and PASI75, respectively.

Use of ustekinumab in the age of 6–17 is supported by evidence from adequate and well-controlled studies of ustekinumab in adults with psoriasis and PsA, pharmacokinetic data from adult patients with psoriasis, adult patients with PsA, and pediatric patients with psoriasis, and safety data from two clinical studies in 44 pediatric patients 6–11 years old with psoriasis and 110 pediatric patients

12–17 years old with psoriasis. The observed pre-dose (trough) concentrations are generally comparable between adult patients with psoriasis, adult patients with PsA, and pediatric patients with psoriasis, and the PK (pharmacokinetic) exposure is expected to be comparable between adult and pediatric patients with PsA.

Of the rheumatic diseases, ustekinumab is indicated for the treatment of patients aged 6 years or older with active PsA (Stelara®, US prescribing information 2022).

Anti-IL-23 Biologic Agents

Guselkumab (Tremfya®) and risankizumab-rzaa (Skyrizi®) are the two anti-IL-23 biologic agents approved for arthritides-related disorders, i.e., PsA. The IL-23 pathway is suggested contributing to the chronic inflammation underlying the pathophysiology of PsA.

Guselkumab
Guselkumab (Tremfya®) is a human IgG1λ mAb, which is produced in a mammalian cell line using recombinant DNA technology. It selectively binds to the p19 subunit of IL-23 and inhibits its interaction with the IL-23 receptor (Tremfya®, US prescribing information 2022). IL-23 is a naturally occurring cytokine that is involved in inflammatory and immune responses.

The safety and efficacy of guselkumab have been evaluated in two multicenter, randomized, double-blind, placebo-controlled Phase III trials in adult patients with active PsA who had inadequate response to standard therapies. In both trials (PSA-1 and PSA-2), subjects treated with the recommended guselkumab subcutaneous dose regimen (100 mg at Week 0, Week 4, and every 8 weeks thereafter) demonstrated a greater clinical response compared to placebo at Week 24. Compared to the placebo group, the difference of response rates in ACR20/ACR50/ACR70 at Week 24 for the guselkumab treatment groups in PSA-1 and PSA-2 were 30/21/6% and 31/17/15%, respectively. Similar responses were seen regardless of prior anti-TNFα exposure in PSA1, and in both trials similar responses were seen regardless of concomitant conventional DMARD use, previous treatment with conventional DMARDs, gender, and body weight.

Of the rheumatic diseases, guselkumab is indicated for the treatment of adult patients with active PsA (Tremfya®, US prescribing information 2022).

Risankizumab-rzaa
Risankizumab-rzaa, an IL-23 antagonist, is a humanized immunoglobulin G1 (IgG1) monoclonal antibody. Risankizumab-rzaa is produced by recombinant DNA technology in Chinese hamster ovary cells and selectively binds to the p19 unit of human IL-23 cytokine and inhibits its inter-

action with the IL-23 receptor (Skyrizi®, US prescribing information 2022).

The safety and efficacy of risankizumab-rzaa for PsA have been assessed in two randomized, double-blind, placebo-controlled studies (PsA-1 and PsA-2) in adult patients with active PsA. In PsA-1, all subjects had a previous inadequate response or intolerance to non-biologic DMARD therapy and were biologic naive. In PsA-2, 53.5% of subjects had a previous inadequate response or intolerance to non-biologic DMARD therapy, and 46.5% of subjects had a previous inadequate response or intolerance to biologic therapy. Six-month treatment in PsA patients who received subcutaneous risankizumab-rzaa treatment at the recommended dose regimen (150 mg at Weeks 0, 4, and 16 followed by every 12 weeks after Week 28) achieved at Week 24 ACR20, ACR50, and ACR70 responses of 57.3, 33.4, and 15.3% in PsA1 and 51.3, 26.3, and 12.0%, respectively, while the control group (placebo) had only ACR20, ACR50, and ACR70 responses with 33.5, 11.3, and 4.7% in PsA1 and 26.5, 9.3, and 5.9% in PsA2, respectively.

Of the rheumatic diseases, risankizumab-rzaa is indicated for the treatment of patients with active psoriatic arthritis in adults (Skyrizi®, US prescribing information 2022).

Anti-IL-6 Receptor Biologic Agents

Two anti-IL-6 receptor biologic agents have been approved for arthritides-related disorders, sarilumab (Kevzara®) and tocilizumab (Actemra®). Sarilumab is a human IgG1 mAb while tocilizumab is a humanized IgG1/κ mAb. Each of these two agents selectively binds to both soluble and membrane-bound IL-6 receptors (sIL-6R and mIL-6R), and has been shown to inhibit IL-6-mediated signaling through these receptors. IL-6 is a pleiotropic pro-inflammatory cytokine produced by a variety of cell types including lymphocytes, monocytes, and fibroblasts. IL-6 has been shown to be involved in diverse physiological processes such as T-cell activation, induction of immunoglobulin secretion, initiation of hepatic acute phase protein synthesis, and stimulation of hematopoietic precursor cell proliferation and differentiation. IL-6 is also produced by synovial and endothelial cells leading to local production of IL-6 in joints affected by inflammatory processes such as RA (Actemra®, US prescribing information 2021).

Sarilumab
Sarilumab (Kevzara®) is a human recombinant anti-IL-6 receptor mAb of the IgG1 subclass, which is produced by recombinant DNA technology in Chinese hamster ovary cell suspension culture (Kevzara®, US prescribing information 2018).

The efficacy and safety of sarilumab have been assessed in patients with moderately to severely active RA who had inadequate clinical response to methotrexate and who had an inadequate clinical response or were intolerant to one or more TNFα antagonists. In a randomized, double-blind, controlled Phase III study in RA patients with inadequate clinical response to methotrexate, 6-month treatment with subcutaneous sarilumab at the recommended dose regimen (200 mg every 2 weeks) with concomitant methotrexate use resulted in 66, 46, and 25% of RA patients achieving ACR20, ACR50, and ACR70, respectively, while the control group had only 33, 17, and 7% of patients with ACR20, ACR50, and ACR70 responses, respectively.

Serious infection Black-Box warning was placed on the sarilumab label (Kevzara®, US prescribing information 2018). Serious infections leading to hospitalization or death including bacterial, viral, invasive fungal, and other opportunistic infections have occurred in patients receiving sarilumab. In the 52-week placebo-controlled population, the rate of serious infections in the 200 mg and 150 mg sarilumab with concomitant DMARD group was 4.3 and 3.0 events per 100 patient-years, respectively, compared to 3.1 events per 100 patient-years in the placebo plus DMARD group. In the long-term safety population, the overall rate of serious infections was consistent with rates in the controlled periods of the studies. The most frequently observed serious infections included pneumonia and cellulitis. Cases of opportunistic infection have been reported. Therefore, signs and symptoms of infection should be closely monitored during treatment with sarilumab. Sarilumab use should be avoided during an active infection.

Sarilumab was approved for the treatment of adult patients with moderately to severely active RA who have had an inadequate response or intolerance to one or more DMARDs (Kevzara®, US prescribing information 2018).

Tocilizumab
Tocilizumab (Actemra®), a recombinant humanized anti-IL-6 receptor mAb of IgG 1κ, was the first drug approved in this class (Actemra®, US prescribing information 2021).

The efficacy and safety of tocilizumab have been assessed in various adult RA populations (DMARD [including methotrexate]-inadequate responders, and methotrexate-naive patients, and patients with an inadequate clinical response or intolerant to one or more TNF antagonist therapies), and pediatrics with polyarticular JIA and systemic JIA. In a randomized, double-blind, controlled Phase III study in RA patients, treatment with intravenous tocilizumab (4 or 8 mg/kg) in combination with methotrexate resulted in 51–56%, 25–32%, and 11–13% of RA patients achieving ACR20, ACR50, and ACR70, respectively, at Week 24, while the control group (methotrexate plus placebo) had only 27, 10, and 2% of patients exhibiting ACR20, ACR50, and

ACR70 responses, respectively. Recently, another Phase III study demonstrated similar efficacy and safety for treating RA patients using a subcutaneous route of administration as an alternative to the previously established intravenous one.

Serious infection Black-Box warning was placed on the tocilizumab label (Actemra®, US prescribing information 2021). Serious infections leading to hospitalization or death including tuberculosis (TB), bacterial, viral, invasive fungal, and other opportunistic infections have occurred in patients receiving tocilizumab. In the 24-week, controlled clinical studies, the rate of serious infections in the tocilizumab intravenously administered 8 mg/kg monotherapy group was 3.6 per 100 patient-years compared to 1.5 per 100 patient-years in the methotrexate group. The rate of serious infections in the 8 mg/kg intravenously administered tocilizumab plus DMARD group was 5.3 events per 100 patient-years, respectively, compared to 3.9 events per 100 patient-years in the placebo plus DMARD group. Therefore, tocilizumab should not be administered during an active infection, including localized infections. If a serious infection develops, tocilizumab therapy needs to be interrupted until the infection is controlled.

Of the rheumatic diseases, tocilizumab is indicated as a second-line biologic therapy for the treatment of adult patients with moderately to severely active RA who have had an inadequate response to one or more TNF antagonist therapies. It is also indicated for the treatment of pediatric patients 2 years of age and older with active pJIA and sJIA. For RA, pJIA, and sJIA, tocilizumab may be used alone or in combination with methotrexate (Actemra®, US prescribing information 2021).

Anti-CD80/CD86 to Inhibit Lymphocyte Activation

Abatacept

Abatacept (Orencia®) is a soluble fusion protein that consists of the extracellular domain of human cytotoxic T-lymphocyte-associated antigen 4 (CTLA-4) linked to the Fc portion of human IgG1. Abatacept inhibits T-lymphocyte activation by binding to CD80 and CD86 on antigen presenting cells (APC), thereby blocking interaction with CD28 on T-cells which is required for their activation (Orencia®, US prescribing information 2021). Activated T lymphocytes are implicated in the pathogenesis of RA.

The efficacy and safety of abatacept have been assessed in various adult RA populations (DMARD [including methotrexate]-inadequate responders, anti-TNFα-inadequate responders, and methotrexate-naive patients), active pJIA, and active PsA. In a randomized, double-blind, controlled Phase III study in patients with active RA despite methotrexate therapy, 6-month treatment with intravenous abatacept in combination with methotrexate at the recommended dose regimen (500, 750, and 1000 mg for patients weighing <60 kg, 60–100 kg, and >100 kg, respectively) resulted in 68, 40, and 20% of RA patients achieving ACR20, ACR50, and ACR70, respectively, while the control group (methotrexate plus placebo) had only 40, 17, and 7% of patients achieving ACR20, ACR50, and ACR70 responses, respectively. Recently, another Phase III study demonstrated similar efficacy and safety for treating RA patients using a subcutaneous route of administration as an alternative to the previously established intravenous one.

As described earlier in this chapter, abatacept should not be given concomitantly with TNF antagonists due to an increased risk of serious adverse events, including infections, whereas no clear signal in clinical benefit when compared to anti-TNF therapy alone.

Abatacept is indicated for the treatment of adult patients with moderately to severely active RA and PsA. It is also indicated for reducing signs and symptoms in pediatric patients 2 years of age and older with moderately to severely active pJIA (Orencia®, US prescribing information 2021).

Anti-CD20 Cytolytic Agents

Rituximab

Rituximab (Rituxan®) is a genetically engineered chimeric murine/human IgG1κ mAb that binds specifically to the antigen CD20 on pre-B and mature B lymphocytes. In the pathogenesis of RA, B cells may be involved in the autoimmune/inflammatory process through production of rheumatoid factor (RF) and other autoantibodies, antigen presentation, T-cell activation, and/or proinflammatory cytokine production. Rituximab binds to the CD20 antigen on B lymphocytes, and the Fc domain recruits immune effector functions to mediate B-cell lysis, resulting in B-cell depletion of circulating and tissue-based B cells (Rituxan®, US prescribing information 2021).

The efficacy and safety of rituximab have been assessed in adult patients with moderately to severely active RA who had a prior inadequate response to at least one anti-TNFα agent. In a randomized, double-blind, controlled Phase III study in RA patients, treatment with one course of intravenous rituximab (2 doses of 1000 mg rituximab separated by 2 weeks) in combination with methotrexate resulted in 51, 27, and 12% of RA patients achieving ACR20, ACR50, and ACR70, respectively, at Week 24, while the control group (methotrexate plus placebo) had only 18, 5, and 1% of patients exhibiting ACR20, ACR50, and ACR70 responses, respectively.

Black-Box warnings of fatal infusion reactions, severe mucocutaneous reactions, hepatitis B virus (HBV) reactivation, and progressive multifocal leukoencephalopathy (PML) were placed on the rituximab label (Rituxan®, US prescribing information 2021). Rituximab can cause severe, including fatal, infusion reactions. Severe reactions typically occurred during the first infusion with time to onset of 30–120 min. Therefore, rituximab should only be administered by a healthcare professional with appropriate medical support to manage severe infusion reactions that can be fatal if they occur. The incidence of serious infections was 2% in the rituximab-treated patients with RA and 1% in the placebo group in the pre-marketing trials. Mucocutaneous reactions, some with fatal outcome, can occur in patients treated with rituximab. These reactions include paraneoplastic pemphigus, Stevens-Johnson syndrome, lichenoid dermatitis, vesiculobullous dermatitis, and toxic epidermal necrolysis. The onset of these reactions has been variable and includes reports with onset on the first day of rituximab exposure. HBV reactivation, in some cases resulting in fulminant hepatitis, hepatic failure, and death, can occur in patients treated with drugs classified as CD20-directed cytolytic antibodies, including rituximab. John Cunningham virus infection resulting in PML and death can occur in rituximab-treated patients with autoimmune diseases. Most cases of PML were diagnosed within 12 months of their last infusion of rituximab. Rituximab should be discontinued in patients who experienced/developed these reactions, and rituximab is not recommended for use in patients with severe, active infections.

Within arthritides, rituximab is indicated as a second-line biologic therapy for the treatment of adult patients with moderately to severely active RA who have had an inadequate response to one or more TNF antagonist therapies. The estimated median terminal half-life of rituximab is 18 days in RA patients; however, the pharmacodynamic effect on B cells lasts much longer than the drug level. B-cell recovery begins at approximately 6 months, and median B-cell levels return to normal by 12 months following completion of treatment with rituximab. Consequently, treatment courses of rituximab should be administered every 24 weeks or based on clinical evaluation, but no more frequent than every 16 weeks (Rituxan®, US prescribing information 2021).

Systemic Lupus Erythematosus

Systemic lupus erythematosus (SLE) is an autoimmune disease characterized by the involvement of multiple organ systems, a clinical pattern of unpredictable exacerbations and remissions, and the presence of autoantibodies. The immunopathogenic characteristic of this disease is polyclonal B-cell activation that leads to hyperglobulinemia, autoantibody production, and immune complex formation, which in turn leads to inflammation and damage that can affect multiple organ systems. Generalized SLE symptoms may include fever, fatigue, rash, oral ulceration, hair loss, and arthralgias. US prevalence estimates for various types of lupus, including SLE, vary greatly, with estimates as high as 100 per 100,000 persons affected. Approximately 80–90% of patients with SLE are women.

The primary causes of SLE remain unclear. Current therapies tend to be generally immunosuppressive, often providing suboptimal control over disease manifestation and long-term outcomes due to ineffectiveness or side effects. These SLE therapies have targeted nonspecific sites for inflammatory reduction (e.g., NSAIDs and antimalarials) and immune system suppression (e.g., corticosteroids, azathioprine, cyclophosphamide, methotrexate, and mycophenolate). Belimumab (Benlysta®) and anifrolumab-fnia (Saphnelo®) are the two currently approved biologic therapies for SLE.

Anti- BLyS Biologic Agents

Belimumab

Belimumab (Benlysta®) is a human IgG1λ mAb specific for B-lymphocyte stimulator (BLyS) produced by recombinant DNA technology in a murine cell (NS0) expression system. Belimumab blocks the binding of soluble BLyS, a B-cell survival factor, to its receptors on B cells. Belimumab does not bind B cells directly, but by binding BLyS, it inhibits the survival of B cells, including autoreactive B cells, and reduces the differentiation of B cells into immunoglobulin-producing plasma cells (Benlysta®, US prescribing information 2022).

The safety and effectiveness of belimumab were evaluated in four randomized, double-blind, placebo-controlled studies in patients with SLE according to criteria from the American College of Rheumatology, with intravenous belimumab administration in Trials 1–3 and subcutaneous belimumab administration in Trial 4. Patients were on a stable SLE standard of care treatment regimen comprising any of the following (alone or in combination): corticosteroids, antimalarials, NSAIDs, and immunosuppressives.

The first Phase III study (Trial 1) evaluated intravenous doses of 1, 4, and 10 mg/kg belimumab plus standard of care compared to placebo plus standard of care over 52 weeks in patients with SLE. The patients were administered belimumab intravenously on Days 0, 14, 28, and then every 28 days for 48 weeks. The co-primary endpoints were percent changes in SELENA-SLEDAI score (a sum of all marked SLE-related descriptors) at Week 24 and time to flare over 52 weeks. Exploratory analysis of this study identified a subgroup of patients (72%), who were autoantibody positive, in whom belimumab appeared to offer benefit. The results of this study informed the design of the next two Phase III studies (Trial 2 and Trial 3) and led to the selection of a target population and indication that was limited to autoantibody-positive SLE patients. Trial 2 and Trial 3 were both randomized, double-blind, placebo-controlled trials in patients with SLE. The studies were similar in design except Trial 2 had a 52-week duration and Trial 3 had a 76-week duration. Both studies compared intravenous belimumab 1 and 10 mg/kg plus standard of care to placebo plus standard of care. Eligible patients had active SLE disease, defined as a SELENA-SLEDAI score ≥ 6, and positive autoantibody test results at screening. The patients were administered belimumab intravenously over a 1-h period on Days 0, 14, 28, and then every 28 days for 48 weeks in Trial 3 and for 72 weeks in Trial 2. The primary efficacy determination was based on a composite endpoint (SLE Responder Index or SRI) response at Week 52 compared to baseline. The proportion of SLE patients achieving a SRI response in Trial 2 and Trial 3 was significantly higher in the belimumab 10 mg/kg group than in the placebo group (odds ratio of 1.5–1.8 after 52 weeks of treatment). Trial 4 is another randomized, double-blind, placebo-controlled Phase III study in autoantibody-positive SLE patients, who received subcutaneous administrated belimumab (200 mg weekly dosing plus standard of care) or placebo plus standard of care. Statistically significant efficacy was achieved in the subcutaneous belimumab group versus the placebo group, with odds ratio of 1.7 after 52 weeks of treatment.

In Phase III trials, reports suggest that overall rates for adverse events, infections, treatment discontinuations due to adverse events, and fatalities were not significantly different between the belimumab- and the placebo-treated patients; however, serious and severe infusion-related reactions were reported more often in belimumab-treated patients.

The safety and efficacy of belimumab in pediatric SLE patients were evaluated in an international, randomized, double-blind, placebo-controlled, 52-week, pharmacokinetic (PK), efficacy and safety study. The proportion of SLE patients achieving a SRI-4 response in this trial was significantly higher in the belimumab 10 mg/kg plus standard therapy group than in the placebo plus standard therapy group (odds ratio of 1.49 after 52 weeks of treatment). In the trial, the probability of experiencing a severe SLE flare, as measured by the modified SELENA-SLEDAI Flare Index, excluding severe flares triggered only by an increase of the SELENA-SLEDAI score to >12, was calculated. The proportion of pediatric patients reporting at least one severe flare during the study was numerically lower in pediatric patients receiving belimumab plus standard therapy (17%) compared with those receiving placebo plus standard therapy (35%). Pediatric patients receiving belimumab 10 mg/kg plus standard therapy had a 64% lower risk of experiencing a severe flare during the 52 weeks of observation, relative to the placebo plus standard therapy group. Of the pediatric patients experiencing a severe flare, the median time to the first severe flare was 150 days in pediatric patients receiving belimumab plus standard therapy compared with 113 days in pediatric patients receiving placebo plus standard therapy. Overall PK in pediatric patients with SLE was consistent with the adult population with SLE.

Belimumab is indicated for the treatment of patients aged 5 years and older with active autoantibody-positive SLE who are receiving standard therapy. The efficacy of belimumab has not been evaluated in patients with severe active central nervous system lupus. Belimumab has not been studied in combination with other biologics. Use of belimumab is not recommended in these situations (Benlysta®, US prescribing information 2022).

Type I Interferon Receptor Antagonists

Anifrolumab-fnia

Anifrolumab-fnia (Saphnelo®) is a human IgG1κ monoclonal antibody that binds to subunit 1 of the type I interferon receptor (IFNAR) with high specificity and affinity. This binding inhibits type I IFN signaling, thereby blocking the biologic activity of type I IFNs. Type I IFNs play a role in the pathogenesis of SLE. Approximately 60–80% of adult patients with active SLE express elevated levels of type I IFN inducible genes. Anifrolumab-fnia also induces the internalization of IFNAR1, thereby reducing the levels of cell surface IFNAR1 available for receptor assembly. Blockade of receptor-mediated type I IFN signaling inhibits IFN responsive gene expression as well as downstream inflammatory and immunological processes. Inhibition of type I IFN blocks plasma cell differentiation and normalizes peripheral T-cell subsets.

The safety and efficacy of anifrolumab-fnia were evaluated in three 52-week treatment period, multicenter, randomized, double-blind, placebo-controlled studies (Trial 1, Trial 2, and Trial 3). Efficacy of anifrolumab-fnia was established based on assessment of clinical response using the composite endpoints, the British Isles Lupus Assessment Group-based Composite Lupus Assessment (BICLA) and the SLE Responder Index (SRI-4). The reduction in disease activity seen in the BICLA and SRI-4 was related primarily to improvement in the mucocutaneous and musculoskeletal organ systems. Compared to the placebo group, the difference of response rates in BICLA at Week 52 for anifrolumab treatment group in Trial 1, Trial 2, and Trial 3 was 28.8%, 17.0%, and 16.3%, respectively. Compared to the placebo group, the difference of response rates in SRI-4 at Week 52 for anifrolumab treatment group in Trial 1, Trial 2, and Trial 3 was 24.0%, 6.0%, and 18.2%, respectively. Flare rate was reduced in patients receiving anifrolumab compared to patients who received placebo although the difference was not statistically significant.

Anifrolumab-fnia is indicated for the treatment of adult patients with moderate-to-severe SLE, who are receiving standard therapy (Saphnelo®, US prescribing information 2022).

Psoriasis

Psoriasis is the most common chronic, immune-mediated skin disorder, affecting approximately 2% of the world's population (Nestle et al. 2009). Thickened epidermal layers resulting from excessive keratinocyte cell proliferation characterize psoriasis. The majority of sufferers are afflicted with psoriasis for most of their lives. Symptoms typically present between the ages of 15 and 35, with the majority of individuals diagnosed before the age of 40. Plaque psoriasis is the most common form, affecting approximately 85–90% of psoriatic individuals with the condition. The disease manifests as raised, well-demarcated, erythematous, and frequently pruritic and painful plaques with silvery scales (Christophers 2001; Griffiths and Barker 2007). Approximately 25% of individuals with psoriasis develop moderate-to-severe disease with widely disseminated lesions. In clinical development and in managing patient care, the Psoriasis Activity and Severity Index (PASI) is commonly used as an instrument to measure and evaluate patient care and treatment effects of anti-psoriasis therapies (Feldman and Krueger 2005).

Prior to the availability of biologic agents in psoriasis, multiple therapeutic options existed for the treatment of the disease; however, a significant unmet need remained for a safe, highly effective, convenient systemic therapy for patients with moderate-to-severe forms of the disease (Papp et al. 2011). Psoralen plus ultraviolet A light therapy, while effective, is inconvenient and is associated with an increased risk of skin malignancies and photodamage. Significant safety concerns and organ toxicity are associated with chronic administration of conventional systemic agents such as methotrexate, cyclosporine, and acitretin, thus limiting their use in long-term psoriasis management.

Four anti-TNFα biologic agents are approved for use in psoriasis: etanercept, adalimumab, certolizumab pegol, and infliximab. Two anti-IL-17A mAbs are approved for treatment of psoriasis, ixekizumab and secukinumab. Other approved biologic agents for the treatment of psoriasis include ustekinumab (anti-IL-12/IL-23), guselkumab (anti-IL-23), risankizumab (anti-IL-23), tildrakizumab (anti-IL-23), and brodalumab (anti-IL-17 receptor A) (Table 24.1). Efalizumab (Raptiva®) is a humanized IgG1 mAb directed against CD11 and inhibits leukocyte function. Efalizumab was approved in 2003 for the treatment of moderate-to-severe psoriasis but was voluntarily withdrawn in 2009 due to reports of PML, an opportunistic viral infection of the brain that usually leads to death or severe disability. Alefacept (Amevive®) is a CD2-directed human leukocyte function antigen-3 (LFA-3)/Fc fusion protein. It interferes with lymphocyte activation via CD2 and causes a reduction in subsets of CD2+ T lymphocytes and circulating total CD4+ and CD8+ T-lymphocyte counts. Alefacept was approved in 2003 for the treatment of moderate-to-severe psoriasis; however, because of the availability of better tolerated and more effective biologics for psoriasis, alefacept was withdrawn from use by its sponsor in 2011.

Anti-TNFα Biologic Agents

TNFα antagonists (adalimumab, etanercept, infliximab, and certolizumab) block the binding of TNFα to its receptor, interrupting the subsequent signaling and inflammatory pathways driven by TNFα. This activity suppresses inflammation and the increased activation of T cells that are characteristics of psoriasis (Humira®, US prescribing information 2022; Enbrel®, US prescribing information 2022; Cimzia® US prescribing information, 2019).

An evidence-based comparison from clinical trials of three TNFα antagonists (adalimumab, etanercept, and infliximab) has indicated better efficacy of infliximab and adalimumab than etanercept in treating psoriasis (Langley 2012).

A meta-analysis comparing three TNFα antagonists and traditional systemic therapy (e.g., cyclosporine) used in the treatment of moderate-to-severe psoriasis demonstrated high efficacy and tolerability of TNFα antagonists (Langley 2012). Due to the mode of action, there is a concern that patients receiving TNFα antagonists may become more susceptible to infection; however, a meta-analysis of trial data for these TNFα antagonists showed that serious infection rates were not much higher than those in placebo-treated patients.

Another meta-analysis was completed from 20 trials employing anti-psoriatic biologics. Based on an indirect comparison and a placebo PASI50 (50% improvement in PASI score from baseline) response of 13%, infliximab had the highest predicted mean probability of response of 93, 80, and 54% for PASI50, PASI75, and PASI90 (50%, 75%, and 90% improvement in PASI score from baseline), respectively, followed by ustekinumab 90 mg at 90, 74, and 46%, respectively, and then ustekinumab 45 mg, adalimumab, etanercept, and efalizumab (Reich et al. 2012).

A more recent network meta-analysis (Gomez-Garccia et al. 2017) based on 27 randomized controlled trials was conducted to assess the short-term efficacy and safety of biologic therapies for psoriasis. From the available evidence (direct and indirect comparison) collected from six biologics (infliximab, adalimumab, etanercept, secukinumab, and ustekinumab), infliximab and secukinumab were shown to be the most effective short-term treatments (as ranked by PASI75 and PASI90 responses), but were the biologics most likely to produce at least one adverse event or an infectious adverse event, respectively. Ustekinumab demonstrated the best efficacy–safety profile among the six biologics (it had the third most efficacious treatment effect, and was the only agent that did not show increased risk of adverse events compared with placebo). These results are consistent with results from another recent meta-analysis (Jabbar-Lopez et al. 2017) where ustekinumab was shown to fall into the category of high efficacy and tolerability for the treatment of (together with adalimumab and secukinumab).

Adalimumab, etanercept, certolizumab pegol, and infliximab have been indicative for the treatment of plaque psoriasis in adult patients. In addition, etanercept is approved for pediatric plaque psoriasis (age 4 years and older) based on the beneficial clinical evidence from a randomized, double-blind, placebo-controlled study in children 4–17 years of age with moderate-to-severe plaque psoriasis.

Anti-IL-17A Biologic Agents

Two anti-IL-17A biologic agents have been approved for psoriasis, ixekizumab (Taltz®), and secukinumab (Cosentyx®).

Ixekizumab

An overview of ixekizumab, a humanized anti-IL-17A mAb, has been provided earlier in this chapter (Taltz®, US prescribing information 2022). In addition to the use of this antibody for the treatment of PsA, ixekizumab has also been evaluated in patients with moderate-to-severe plaque psoriasis. Elevated levels of IL-17A are found in psoriatic plaques. Treatment with IL-17A antagonists may reduce epidermal neutrophils and IL-17A levels in psoriatic plaques.

The safety and effectiveness of ixekizumab have been evaluated in three randomized, double-blind, placebo-controlled Phase III trials (Trials 1, 2, and 3) in adult patients with moderate-to-severe psoriasis. In the two active comparator trials (Trials 2 and 3), patients who randomized to the etanercept arm received subcutaneous etanercept 50 mg twice weekly. Three-month treatment with subcutaneous ixekizumab at the recommended dose regimen (160 mg at Week 0, followed by 80 mg at Weeks 2, 4, 6, 8 10, and 12, then 80 mg every 4 weeks thereafter) resulted in 87–90%, 68–71%, and 35–40% of psoriatic patients achieving PASI75, PASI90, and PASI100, respectively, while the etanercept group had 41%, 18%, and 4% of patients achieving PASI75, PASI90, and PASI100, respectively; and the placebo control group had only 2–7%, 1–3%, and 0–1% of patients exhibiting PASI75, PASI90, and PASI100 responses, respectively. Among clinical responders at Week 12, the percentage of patients who maintained this response (i.e., static Physician Global Assessment [sPGA] score 0 or 1) at Week 60 (48 weeks following re-randomization) in Trials 1 and 2

was higher for patients treated with ixekizumab (75%) compared to those treated with placebo (7%).

The safety and effectiveness of ixekizumab have also been established in pediatric patients aged 6 years and older with moderate-to-severe plaque psoriasis. A randomized, double-blind, multicenter, placebo-controlled trial enrolled 171 pediatric patients 6 to less than 18 years of age, with moderate-to-severe plaque psoriasis who were candidates for phototherapy or systemic therapy or were inadequately controlled on topical therapy. Subjects in the ixekizumab were dosed stratified by weight: <25 kg, 40 mg at Week 0 followed by 20 mg every 4 weeks (Q4W); 25 kg to 50 kg, 80 mg at Week 0 followed by 40 mg Q4W; >50 kg, 160 mg at Week 0 followed by 80 mg Q4W. Three-month treatment with subcutaneous ixekizumab resulted in 89%, 78%, and 50% of psoriatic patients achieving PASI75, PASI90, and PASI100, respectively, while the placebo control group had only 25%, 5%, and 2% of patients exhibiting PASI75, PASI90, and PASI100 responses, respectively.

Ixekizumab is indicated for the treatment of patients aged 6 years or older with moderate-to-severe plaque psoriasis who are candidates for systemic therapy or phototherapy (Taltz®, US prescribing information 2022).

Secukinumab

An overview of secukinumab has been provided earlier in this chapter (Cosentyx®, US prescribing information 2021). In addition to the use of this human anti-IL17A mAb for the treatment of PsA and AS, secukinumab has also been evaluated in patients with plaque psoriasis.

Four randomized, double-blind, placebo-controlled Phase III trials (Trials 1, 2, 3, and 4) have been conducted for secukinumab in patients with moderate-to-severe psoriasis. In the active comparator trial (Trial 2), patients who randomized to the etanercept arm received subcutaneous etanercept 50 mg twice weekly. Three-month treatment with secukinumab of 300 mg subcutaneously administered at Weeks 0, 1, 2, 3, and 4, and every 4 weeks thereafter resulted in 75–87% of psoriatic patients achieving PASI75, while the etanercept group had 44% of patients achieving PASI75; and the placebo control group had only 0–5% of patients exhibiting PASI75 response (Langley et al. 2014). With continued treatment over 52 weeks, PASI75 responders maintained their responses in 81–84% patients treated with secukinumab (300 mg every 4 weeks).

A head-to-head randomized controlled trial was conducted comparing the efficacy and safety between secukinumab and ustekinumab in patients with moderate-to-severe psoriasis (Blauvelt et al. 2017). Secukinumab was administered 300 mg subcutaneously at baseline and at Weeks 1, 2, and 3, and then every 4 weeks from Week 4 onward. Ustekinumab was dosed subcutaneously at 45 mg in patients ≤100 kg and at 90 mg in those >100 kg, and given at baseline, at Week 4, and then every 12 weeks. Secukinumab has demonstrated superior efficacy to ustekinumab at Weeks 4 and 16 in patients with plaque psoriasis. This superior efficacy is sustained over 52 weeks in the proportion of patients achieving PASI90 (76% vs. 61%) and PASI100 (46% vs. 36%). In addition, patients received secukinumab had greater improvement in health-related quality of life and comparable safety.

A 52-week, multicenter randomized, double-blind, placebo and active-controlled trial enrolled 162 pediatric subjects 6 years of age and older, with severe plaque psoriasis (as defined by a PASI score ≥20, an IGA modified 2011 score of 4, and involving ≥10% of the body surface area) who were candidates for systemic therapy. Subjects were randomized to receive placebo, secukinumab, or a biologic active control. In the secukinumab groups, subjects with body weight < 25 kg received 75 mg, subjects with body weight 25 to <50 kg received either 75 mg or 150 mg (2 times the recommended dose), and subjects with body weight ≥50 kg received either 150 mg or 300 mg (2 times the recommended dose). The co-primary endpoints were the proportion of subjects who achieved a reduction in PASI score of at least 75% (PASI75) from baseline to Week 12 and the proportion of subjects who achieved an IGA modified 2011 score of "clear" or "almost clear" (0 or 1) with at least a 2-point improvement from baseline to Week 12. The % patients with PASI75 responses was 70% in the combined secukinumab group versus 15% in the placebo group at Week 12; the % patients with IGA of "clear" or "almost clear" was 56% in the combined secukinumab group versus 5% in the placebo group at Week 12.

Secukinumab is indicated for the treatment of patients 6 years and older with moderate-to-severe plaque psoriasis who are candidates for systemic therapy or phototherapy (Cosentyx®, US prescribing information 2021).

Anti-IL-17 Receptor a Biologic Agent

As of November 2022, brodalumab (Siliq®) is the only anti-IL-17 receptor A (IL-17RA) biologic agent approved for psoriasis. IL-17RA is a protein expressed on the cell surface and is a required component of receptor complexes utilized by multiple IL-17 family cytokines. Blocking IL-17RA inhibits IL-17 cytokine-induced responses including the release of pro-inflammatory cytokines and chemokines.

Brodalumab

Brodalumab (Siliq®) is a human IgG2κ mAb expressed in a Chinese hamster ovary cell line. It selectively binds to human IL-17 receptor A (IL-17RA) and inhibits its interactions with cytokines IL-17A, IL-17F, IL-17C, IL-17A/F heterodimer, and IL-25 (Siliq®, US prescribing information 2020).

The safety and effectiveness of brodalumab have been evaluated in three multicenter, randomized, double-blind, controlled Phase III trials (Trials 1, 2, and 3) in patients with moderate-to-severe plaque psoriasis. In these trials, patients were randomized to placebo or subcutaneous brodalumab treatment of 210 mg at Weeks 0, 1, and 2, followed by 210 mg every 2 weeks thereafter. In the two active comparator trials (Trials 2 and 3), patients randomized to the subcutaneous ustekinumab group received a 45-mg dose if their weight was ≤100 kg and a 90-mg dose if their weight was >100 kg at Weeks 0, 4, and 16, followed by the same dose every 12 weeks. Treatment with brodalumab resulted in 83–86% and 37–42% of psoriatic patients achieving PASI75 and PASI100, respectively, at Week 12, while the ustekinumab group had 69–70% and 19–22% of patients achieving PASI75 and PASI100, respectively; and the placebo control group had only 3–8% and 0–1% of patients exhibiting PASI75 and PASI100 responses, respectively. Maintenance of the treatment effect of brodalumab was demonstrated. Among PASI100 responders at Week 12, 72% of the patients who continued on brodalumab 210 mg every 2 weeks maintained the response at Week 52.

Suicidal ideation and behavior Black-Box warnings were placed on the brodalumab label (Siliq®, US prescribing information 2020). Suicidal ideation and behavior, including four completed suicides, occurred in patients treated with brodalumab in the psoriasis pre-marketing trials. There were no completed suicides in the 12-week placebo-controlled portion of the trials. A causal association between treatment with brodalumab and increased risk of suicidal ideation and behavior has not been established. Because of the observed suicidal ideation and behavior in subjects treated with brodalumab, in the USA, brodalumab is only available through a restricted program under REMS (Risk Evaluation and Mitigation Strategy). Prescribers should weigh the potential risks and benefits before using brodalumab in patients with a history of depression or suicidality. Patients with new or worsening suicidal thoughts and behavior should be referred to a mental health professional, as appropriate.

Brodalumab is indicated for the treatment of moderate-to-severe plaque psoriasis in adult patients who are candidates for systemic therapy or phototherapy and have failed to respond or have lost response to other systemic therapies (Siliq®, US prescribing information 2020).

Anti-IL-12/IL-23 Biologic Agents

Ustekinumab

An overview of ustekinumab has been provided earlier in this chapter (Stelara®, US prescribing information 2022). In addition to its use for the treatment of PsA, the efficacy and safety of ustekinumab have been assessed in adult and adolescent patients (12 years or older) with moderate-to-severe psoriasis.

The safety and efficacy of ustekinumab in treatment of psoriasis were assessed in two Phase III trials (PHOENIX 1 and PHOENIX 2). The results from these two trials demonstrated that ustekinumab was effective in ameliorating psoriatic plaques, pruritus, and nail psoriasis (Leonardi et al. 2008; Papp et al. 2008). Within 12 weeks of initiating subcutaneous ustekinumab treatment (45 or 90 mg/kg at Weeks 0 and 4), more than two-thirds of patients exhibited ≥75% reduction in PASI (PASI75 response). Maximum efficacy was achieved at approximately 24 weeks after initiation of therapy, with approximately 75% of ustekinumab-treated patients achieving a PASI75 response. Clinical response to ustekinumab was associated with serum ustekinumab concentrations that were somewhat correlated with patient body weight. While efficacy of the 45 and 90 mg doses of ustekinumab was similar in patients weighing ≤100 kg, the 90-mg dose was more effective than the 45-mg dose in patients weighing >100 kg, who represented approximately one-third of the combined PHOENIX 1 and PHOENIX 2 population. Thus, to optimize efficacy in all patients while minimizing unnecessary drug exposure, fixed dose administration of ustekinumab based on body weight is indicated, i.e., for patients weighing >100 kg, the recommended dose is 90 mg initially and 4 weeks later, followed by 90 mg every 12 weeks; for patients weighing ≤100 kg, the recommended dose is 45 mg initially and 4 weeks later, followed by 45 mg every 12 weeks.

Results of the Phase III psoriasis clinical trials indicated that ustekinumab was generally well tolerated. Rates and types of adverse events, serious adverse events, adverse events leading to treatment discontinuation, and laboratory abnormalities were generally comparable among patients receiving subcutaneous administration of placebo, ustekinumab 45 or 90 mg during the 12-week placebo-controlled phases of PHOENIX 1 and PHOENIX 2. Dose–response relationships for safety events were not apparent. Rates of serious infections and malignancies in PHOENIX 1 and PHOENIX 2 were low and comparable across treatment groups during the placebo-controlled phases, no apparent increase in the frequency of these adverse events was observed through 18 months of treatment, and no mycobacterial or salmonella infections were reported (Leonardi et al. 2008; Papp et al. 2008).

A head-to-head controlled trial was conducted comparing the efficacy and safety of a TNFα antagonist, etanercept, and ustekinumab in patients with moderate-to-severe psoriasis (Griffiths et al. 2010). Ustekinumab was administered subcutaneously at either 45 or 90 mg at Weeks 0 and 4, and etanercept was administered subcutaneously 50 mg twice weekly for 12 weeks. There was at least 75% improvement in the PASI (PASI75) at Week 12 in 67.5% of patients who received 45 mg of ustekinumab and 73.8% of patients who received 90 mg of ustekinumab, as compared to 56.8% of those who received etanercept ($p = 0.01$ and $p < 0.001$, respectively). The efficacy of ustekinumab at 45 or 90 mg was superior to that of etanercept over a 12-week period while the safety profiles were similar.

Ustekinumab is indicated for the treatment of adult patients with moderate-to-severe plaque psoriasis who are candidates for systemic therapy or phototherapy. Ustekinumab is also indicative for pediatric patients (6–17 years old) with moderate-to-severe plaque psoriasis (Stelara®, US prescribing information 2022).

Anti-IL-23 Biologic Agent

Three anti-IL-23 biologic agents have been approved for psoriasis, guselkumab (Tremfya®), risankizumab-rzaa (Skyrizi®), and tildrakizumab (Ilumya®). The IL-23 pathway is suggested contributing to the chronic inflammation underlying the pathophysiology of psoriasis.

Guselkumab and risankizumab-rzaa have also been approved for PsA, and the overview of these two mAbs was provided earlier in this chapter.

Guselkumab

In addition to its use for the treatment of PsA, the safety and efficacy of guselkumab have been evaluated in three multi-center, randomized, double-blind Phase III trials (VOYAGE 1, VOYAGE 2, and NAVIGATE) in adult patients with moderate-to-severe plaque psoriasis. In VOYAGE 1 and VOYAGE 2, patients were randomized to either guselkumab (100 mg SC at Weeks 0 and 4 and every 8 weeks thereafter), placebo, or US licensed adalimumab (80 mg SC at Week 0 and 40 mg at Week 1, followed by 40 mg every other week thereafter). Sixteen-weeks treatment with subcutaneous guselkumab resulted in 70–73% of psoriatic patients achieving PASI90, while the placebo control group had only 2–3% patients exhibiting PASI90 response. An analysis based on clinical data from all the North America sites (i.e., the USA and Canada) demonstrated the superiority of guselkumab to US licensed adalimumab. The PASI90 response rates at Weeks 16, 24, and 48 were 64–73%, 71–80%, and 73%, respectively, for guselkumab and 41–42%, 44–51%, and 46%, respectively, for adalimumab. NAVIGATE evaluated the efficacy of guselkumab in patients who had not achieved an adequate response at Week 16 after initial treatment with ustekinumab (dosed 45 mg or 90 mg according to the subject's baseline weight [≤ and >100 kg, respectively] at Week 0 and Week 4). These patients were randomized to either continue with ustekinumab treatment every 12 weeks or switch to guselkumab. Twelve weeks after randomization, a greater proportion of patients on guselkumab treatment compared to ustekinumab (31% vs. 14%) achieved an Investigator's Global Assessment (IGA) score of 0 or 1 with a ≥2 grade improvement.

Guselkumab is indicated for the treatment of adult patients with moderate-to-severe plaque psoriasis who are candidates for systemic therapy or phototherapy (Tremfya®, US prescribing information 2022).

Risankizumab-rzaa

The safety and efficacy of risankizumab-rzaa for the patients aged 18 years and older with moderate-to-severe plaque psoriasis was assessed in four multicenter, randomized, double-blind studies (PsO-1, PsO-2, PsO-3, and PsO-4). In PsO-1 and PsO-2, patients were randomized to the risankizumab-rzaa 150 mg treatment group, placebo group, and biologic active control group. Subjects received treatment at Weeks 0, 4, and every 12 weeks thereafter. At Week 16 subjects receiving risankizumab-rzaa achieved sPGA 0, PASI 90, and PASI 100 of 37%, 75%, and 36%, respectively, in PsO-1 and 51%, 75%, and 51%, respectively, in PsO-2 while the placebo group only got the sPGA 0, PASI 90, and PASI 100 of 2%, 5%, and 0% in PsO-1 and 3%, 2%, and 2% in PsO-2, respectively. In PsO-3 at Week 16, risankizumab-rzaa was superior to placebo on the co-primary endpoints of sPGA 0 or 1 (84% risankizumab-rzaa and 7% placebo) and PASI 90 (73% risankizumab-rzaa and 2% placebo). The respective response rates for risankizumab-rzaa and placebo at Week 16 were: sPGA 0 (46% risankizumab-rzaa and 1% placebo); PASI 100 (47% risankizuman-rzaa and 1% placebo); and PASI 75 (89% risankizumab-rzaa and 8% placebo).

Risankizumab-rzaa is indicated for the treatment of moderate-to-severe plaque psoriasis in adults who are candidates for systemic therapy or phototherapy (Skyrizi®, US prescribing information 2022).

Tildrakizumab

Tildrakizumab (Ilumya®) is a humanized IgG1 monoclonal antibody that selectively binds to the p19 subunit of IL-23 and inhibits its interaction with the IL-23 receptor. IL-23 is a naturally occurring cytokine that is involved in inflammatory and immune responses. Tildrakizumab inhibits the release of proinflammatory cytokines and chemokines (Ilumya®, US prescribing information 2020).

The safety and efficacy of Tildrakizumab was assessed in two multicenter, randomized, double-blind, placebo-controlled trials. In both trials, subjects were randomized to either placebo or tildrakizumab (100 mg at Week 0, Week 4, and every 12 weeks thereafter up to 64 weeks). The co-primary endpoints PGA of 0 or 1 and PASI 75 were 58% and 64% in Trial 2, and 55% and 61% in Trial 3, respectively, while the primary endpoints for the placebo group were 7% and 6% in Trial 2, and 4% and 6% in Trial 3, respectively.

Tildrakizumab is indicated for the treatment of adults with moderate-to-severe plaque psoriasis who are candidates for systemic therapy or phototherapy (Ilumya®, US prescribing information 2020).

Atopic Dermatitis

Atopic dermatitis (AD) is a common chronic pruritic inflammatory skin disease with the prevalence rates worldwide up to 2–5% for adults and up to 10–20% for children. It is the most common form of eczema and is characterized by rashes on the skin that can include symptoms such as intense, persistent itching and skin dryness, cracking, redness, crusting, and oozing. The pathogenesis of AD is multifactorial and is not completely understood. The abnormalities in the skin of AD patients include an activation of the immune system (especially the type 2 T helper [Th2]-pathway) without an adequate stimulus, epithelial barrier dysfunction (often associated with a mutation of filaggrin gene), and an imbalance in the skin microbiota composition. Treatment of patients suffering from mild or moderate AD includes the use of emollients and topical glucocorticoids or topical calcineurin inhibitors. Patients with chronic and severe AD where topical therapy is usually insufficient require the use of systemic immunosuppressive drugs, which is often limited due to toxicity and severe adverse effects. There is an immense unmet need for new therapy concepts for moderate-to-severe AD, which impacts the quality of life of patients and has become a global health problem.

With a better understanding of the pathophysiology of the disease, multiple new therapeutic proteins have been developed for the treatment of AD (Boguniewicz et al. 2017). These biologic agents exhibit several novel mechanisms of action including anti-IL-31 receptor A (nemolizumab) (Kabashima and Irie 2021), anti-IL-4 receptor α (dupilumab) (Kraft and Worm 2017), and anti-IL-13 (tralokinumab [Wollenberg et al. 2021; Silverberg et al. 2020]) and lebrikizumab [NCT02340234, NCT02465606, ClinicalTrials.gov, accessed 22 Jan 2018]). Of these agents, dupilumab (Dupixent®) and tralokinumab (Adbry® for the USA and Adtralza® for outside the USA) were recently approved by the FDA or EMA for the treatment of adults with inadequately controlled moderate-to-severe AD. Nemolizumab (Mitchga® Syringes) was approved in Japan on 28 March 2022 for use in adults and children over the age of 13 years for the treatment of itch associated with AD (only when existing treatment is insufficiently effective).

Anti-IL4 Receptor α Agent

Dupilumab

Dupilumab (Dupixent®) is a human mAb of the IgG4 subclass, which is produced by recombinant DNA technology in Chinese hamster ovary cell suspension culture. Dupilumab inhibits IL-4 and IL-13 signaling by specifically binding to the IL-4 receptor α (IL-4Rα) subunit shared by the IL-4 and IL-13 receptor complexes. Blocking IL-4Rα with dupilumab inhibits IL-4 and IL-13 cytokine-induced responses, including the release of proinflammatory cytokines, chemokines, and IgE (Dupixent®, US prescribing information 2022).

The efficacy and safety of dupilumab have been evaluated in three randomized, double-blind, controlled Phase III trials in adult patients with moderate-to-severe AD who were not adequately controlled by topical medication(s). Disease severity was defined by an Investigator's Global Assessment (IGA) score ≥3 in the overall assessment of AD lesions on a severity scale of 0–4, and an Eczema Area and Severity Index (EASI) score ≥16 on a scale of 0–72. In the two dupilumab monotherapy Phase III trials, 12-week treatment with subcutaneous dupilumab at the recommended dose regimen (600 mg initially followed by 300 mg every 2 weeks) resulted in 36–38% and 44–51% of AD patients achieving IGA 0/1 (IGA "clear" or "almost clear" with a reduction of ≥2 points on a 0–4 IGA scale) and EASI 75 (improvement of at least 75% in EASI score from baseline) responses, respectively, while the control group had only 9–10% and 12–15% of patients with IGA 0/1 and EASI 75 responses, respectively. Clinical efficacy was also shown in the concomitant Phase III trial, where 12-week treatment with dupilumab at the recommended subcutaneous dose regimen (600 mg initially followed by 300 mg every 2 weeks) in combination with topical corticosteroids resulted in 39 and 69% of AD patients achieving IGA 0/1 and EASI 75 responses, respectively, while the control group (placebo plus topical corticosteroids) had only 12 and 23% of patients with IGA 0/1 and EASI 75 responses, respectively.

Dupilumab is indicated for the treatment of patients aged 6 months and older with moderate-to-severe atopic dermatitis (AD) whose disease is not adequately controlled with topical prescription therapies or when those therapies are not advisable. Dupilumab can be used with or without topical corticosteroids (Dupixent®, US prescribing information 2022).

Anti-IL13 Biologic Agent

Tralokinumab

Tralokinumab-ldrm (Adbry®, US prescribing information 2022) is a human IgG4 monoclonal antibody that specifi-

cally binds to human interleukin13 (IL-13) and inhibits its interaction with the IL-13 receptor α1 and α2 subunits (IL-13Rα1 and IL13Rα2). IL-13 is a naturally occurring cytokine of the type 2 immune response. Tralokinumab-ldrm inhibits the bioactivity of IL-13 by blocking IL-13 interaction with IL-13Rα1/IL-4Rα receptor complex. Tralokinumab-ldrm inhibits IL-13-induced responses including the release of proinflammatory cytokines, chemokines, and IgE (Adbry®, US prescribing information 2022).

The efficacy and safety of tralokinumab as monotherapy and with concomitant topical corticosteroids have been evaluated in three pivotal randomized, double-blind, placebo-controlled studies (ECZTRA 1, ECZTRA 2, and ECZTRA 3) in adult patients with moderate-to-severe atopic dermatitis. In ECZTRA 1 and ECZTRA 2, from baseline to Week 16, a significantly greater proportion of patients randomized and dosed to tralokinumab achieved IGA 0 or 1, EASI-75, and/or an improvement of ≥4 points on the Worst Daily Pruritus NRS compared to placebo. In both monotherapy studies (ECZTRA 1 and ECZTRA 2), tralokinumab treatment (300 mg SC every 2 weeks) reduced itch, as measured by the percent change from baseline in Worst Daily Pruritus NRS, already at Week 1 compared to placebo. The reduction in itch was observed in parallel with improvements in objective signs and symptoms of atopic dermatitis and quality of life.

To evaluate maintenance of response, subjects from ECZTRA 1 and ECZTRA 2 treated with tralokinumab 300 mg administered subcutaneously once every 2 weeks (Q2W) for 16 weeks who achieved IGA 0 or 1 or EASI-75 at Week 16 were re-randomized to an additional 36-week treatment of (1) 300 mg tralokinumab every 2 weeks (Q2W) or (2) alternating tralokinumab 300 mg and placebo Q2W (tralokinumab Q4W) or (3) placebo Q2W, for a cumulative 52-week study treatment. Of the subjects randomized to tralokinumab, who did not achieve IGA 0 or 1 or EASI-75 at Week 16 and were transferred to open-label tralokinumab 300 mg Q2W + optional topical corticosteroids (TCS), 20.8% in ECZTRA 1 and 19.3% in ECZTRA 2 achieved IGA 0 or 1 at Week 52, and 46.1% in ECZTRA 1 and 39.3% in ECZTRA 2 achieved EASI-75 at Week 52. The clinical response was mainly driven by continued tralokinumab treatment rather than optional topical corticosteroid treatment.

In ECZTRA 3, a significantly greater proportion of patients randomized to tralokinumab 300 mg Q2W + TCS achieved IGA 0 or 1, EASI-75, and/or an improvement of ≥4 points on the Worst Daily Pruritus NRS at week 16 compared to placebo + TCS. For the maintenance response evaluation in ECZTRA3, among all the subjects who achieved either IGA 0 or 1 or EASI-75 at Week 16, the mean percentage improvement in EASI score from baseline was 93.5% at Week 32 when maintained on tralokinumab 300 mg

Q2W + TCS and 91.5% at Week 32 for subjects on tralokinumab 300 mg Q4W + TCS. Of the subjects randomized to tralokinumab 300 mg Q2W + TCS who did not achieve IGA 0 or 1 or EASI-75 at Week 16, 30.5% achieved IGA 0/1 and 55.8% achieved EASI-75 at Week 32 when treated continuously with tralokinumab 300 mg Q2W + TCS for additional 16 weeks.

Tralokinumab-ldrm is indicated for the treatment of moderate-to-severe atopic dermatitis in adult patients who are candidates for systemic therapy (Adbry®, US prescribing information 2022).

Inflammatory Bowel Disease

Inflammatory bowel disease (IBD) refers to a group of chronic inflammatory diseases of the gastrointestinal tract that mainly comprise two well-defined clinical entities, Crohn's disease (CD) and ulcerative colitis (UC). A study in 2012 (Molodecky et al. 2012) found that prevalence of UC in North America between 1980 and 2008 was 198.1–298.5 cases per 100,000 persons and prevalence of CD was 135.7–318.5 cases/100,000 persons. CD and UC are two idiopathic IBD. They have some similarities and unique differences (Baumgart and Sandborn 2007).

CD is a relapsing, transmural inflammatory disease of the gastrointestinal mucosa that can affect the entire gastrointestinal tract from the mouth to the anus, while UC is a relapsing, nontransmural inflammatory disease that is restricted to the colon. CD may involve all layers of the intestine, and there can be normal healthy bowel between patches of diseased bowel, while UC does not affect all layers of the bowel but only affects the top layers of the colon in an even and continuous distribution. In CD, pain is commonly experienced in the lower right abdomen, while in UC, in the lower left part of the abdomen. In CD, the colon wall may be thickened and may have a rocky appearance, while in UC the colon wall is thinner and shows continuous inflammation. In clinical practice, disease activity of CD is typically described as mild to moderate (ambulatory patients able to tolerate oral alimentation without manifestations of dehydration, toxicity, abdominal tenderness, painful mass, obstruction, or >10% weight loss), moderate-to-severe disease (failure to respond to treatment for mild disease, more prominent symptoms of fever, weight loss, abdominal pain or tenderness, intermittent nausea and vomiting without obstruction, or significant anemia), and severe-to-fulminant disease (persisting symptoms on corticosteroids, high fevers, persistent vomiting, evidence of intestinal obstruction, rebound tenderness, cachexia, or evidence of an abscess). While disease activity of UC is typically described as mild (up to four bloody stools daily and no systemic toxicity), moderate (four to six blood stools daily and minimal toxicity), or severe (more than six stools daily

and signs of toxicity, such as fever, tachycardia, anemia, and raised erythrocyte sedimentation rate), patients with fulminant UC usually have more than ten bloody stools daily, continuous bleeding, anemia requiring blood transfusion, abdominal tenderness, and colonic dilation on plain abdominal radiographs (Baumgart and Sandborn 2007).

Conventional pharmacologic treatments for IBD include aminosalicylates, corticosteroids, immunomodulators (azathioprine, 6-mecaptopurine, methotrexate, cyclosporine), and antibiotics (metronidazole, ciprofloxacin, clarithromycin). The aim of traditional therapy is to induce and maintain remission in patients. Treatment guidelines generally recommend initiating treatment with first-line agents such as sulfasalazine and systemic corticosteroids, followed by immunomodulators. These conventional pharmacologic therapies are often effective in patients with IBD, particularly in those with mildly to moderately active disease; however, a significant proportion of patients have severely active disease that is often refractory to these conventional therapies. Furthermore, these small molecule drugs have limitations in the treatment of IBD. Corticosteroids have many side effects and are not suitable for long-term maintenance therapy. Corticosteroids are also ineffective for healing bowel ulcerations (Modigliani et al. 1990). Immunomodulators promote mucosal healing, but the onset of action is slow. The use of anti-TNFα agents can overcome the shortcomings of the conventional treatment options and provide greater improvement for severe or refractory IBD. Anti-TNFα therapy can rapidly improve signs and symptoms (i.e., induce and maintain clinical response and clinical remission), promote mucosal healing, eliminate corticosteroid use, and has the potential to alter the natural history of IBD. Historically, therapeutic proteins have been used as rescue therapy for patients with IBD refractory to conventional therapies. Recently, evidence has emerged that early use of anti-TNFα therapy in patients at high risk may induce a greater response and prevent irreversible damage to the intestine (D'Haens et al. 2008). There are also concerns with respect to increased risks of infections and malignancy associated with the use of anti-TNFα agents in patients with IBD (Hoentjen and van Bodegraven 2009). The timing of initiating therapy with therapeutic proteins and the identification of the subset of patients who can achieve maximal benefit from treatment using therapeutic proteins remain active areas of debate, and further clinical research is required to provide evidence-based guidelines. Anti-integrin and anti-IL-12/IL-23 agents are non-TNF biologic therapies developed recently for the treatment of IBD; they provide alternative in case of treatment failures to conventional drugs (such as glucocorticoids and immunomodulators) and/or anti-TNF-α therapies.

There are eight biologic agents approved for the treatment of CD and/or UC; those are four anti-TNFα agents (infliximab, adalimumab, certolizumab pegol, and golimumab), one anti-IL-12/IL-23 agent (ustekinumab), one anti-IL-23 agent (risankizumab), and two anti-integrin agents (natalizumab and vedolizumab) (Table 26.1).

Anti-TNFα Biologic Agents

Four anti-TNFα agents (infliximab, adalimumab, certolizumab pegol, and golimumab) have been approved for the treatment of CD and/or UC. Not all anti-TNFα agents have been shown to be effective for IBD. For example, infliximab, the first-in-class anti-TNFα biologic agent approved for treating CD, has been shown to be highly effective in the treatment of CD, but etanercept was shown to be ineffective for this disease. A mechanism postulated to explain the differential effects of infliximab and etanercept for CD was that infliximab could bind membrane-associated TNFα and induce apoptosis of activated T cells and macrophages, but etanercept only binds to soluble TNFα (Van den Brande et al. 2003). However, this theory is questioned by later data showing induction of apoptosis by etanercept and clinical efficacy of certolizumab pegol, a non-apoptotic anti-TNFα agent (Chaudhary et al. 2006; Sandborn et al. 2006). As of November 2022, infliximab and adalimumab are the only antibody-based therapeutic proteins approved for pediatric patients with CD or UC.

Adalimumab

An overview of adalimumab, a human anti-TNFα mAb, has been provided earlier in this chapter (Humira®, US prescribing information 2022). In addition to the use of this therapeutic protein for the treatment of arthritides and psoriasis, the efficacy and safety of adalimumab have been assessed in adult and pediatric patients with active CD and active UC.

The efficacy and safety of adalimumab in the treatment of adult CD have been evaluated in patients with moderately to severely active CD (Crohn's Disease Activity Index [CDAI] ≥220 and ≤450) in three randomized, double-blind, controlled Phase III studies (CD-1I, CD-II, and CD-III). In a Phase III Study (CD-I) in CD patients who were naive to TNF blocker, treatment with subcutaneous adalimumab at the recommended induction dose regimen (160 and 80 mg at Weeks 0 and 2, respectively) resulted in 58 and 36% of patients achieving clinical response (defined as a decrease in CDAI of at least 70 points) and clinical remission (defined as CDAI <150), respectively, at Week 4, while the control group (placebo) had 34 and 12% of patients with clinical response and clinical remission, respectively. A greater percentage of the patients treated with adalimumab also achieved induction of clinical response and remission versus placebo at Week 4 in CD patients who had lost response to or were intolerant

to infliximab in another Phase III trial (CD-II). The adalimumab group had 52 and 21% of patients who were in clinical response and clinical remission, respectively, at Week 4, while the placebo group had 34 and 7% of patients in clinical response and clinical remission, respectively. In the third Phase III trial (CD-III), maintenance of clinical remission was evaluated. Among clinical responders at Week 4, further maintenance treatment with 40 mg adalimumab administrated subcutaneously every other week demonstrated greater efficacy than the placebo maintenance group. The adalimumab maintenance group had 43 and 36% of patients who were in clinical response and clinical remission, respectively, at Week 56, while the placebo maintenance group had 18 and 12% of patients in clinical response and clinical remission, respectively.

Two randomized, double-blind, placebo-controlled Phase III studies (UC-I and UC-II) have been conducted for adalimumab in patients with moderately to severely active UC (Mayo score 6–12 on a 12-point scale, with an endoscopy subscore of 2–3 on a scale of 0–3) despite concurrent or prior treatment with immunosuppressants such as corticosteroids, azathioprine, or 6-mercaptopurine. All patients in Study UC-I were TNF blocker naive, and 40% patients enrolled in Study UC-II had previously used another TNF-blocker. In both UC-I and UC-II studies, a greater percentage of the patients treated with subcutaneous adalimumab at the recommended subcutaneous dose regimen (160 and 80 mg at Weeks 0 and 2, respectively) compared to patients treated with placebo (16.5–18.5% vs. ~9%) achieved induction of clinical remission (defined as Mayo score ≤2 with no individual subscores >1). In Study UC-II, a greater percentage of the patients treated with maintenance adalimumab of 40 mg every other week compared to patients treated with placebo (8.5% vs. 4.1%) achieved sustained clinical remission (defined as clinical remission at both Weeks 8 and 52). The subgroup of patients with prior TNF-blocker use in Study UC-II achieved induction of clinical remission at 9% in the adalimumab group versus 7% in the placebo group, and sustained clinical remission at 5% in the adalimumab group versus 1% in the placebo group.

In addition to adult CD and UC, safety and efficacy of adalimumab for pediatric patients 6 years of age or older with CD and 5 years of age or older with UC have also been established. Use of adalimumab in this age group is supported by evidence from adult CD trials with additional data in pediatric CD patients (6–17 years of age) and in pediatric UC patients (5–17 years of age) from respective randomized, double-blind Phase III pediatric studies.

Among IBD indications, adalimumab is indicated for treatment of CD by reducing signs and symptoms and inducing and maintaining clinical remission in adult patients with moderately to severely active CD who have had an inadequate response to conventional therapy, and by reducing signs and symptoms and inducing clinical remission in these patients if they have also lost response to or are intolerant to infliximab. Adalimumab is also indicated for treatment of UC by inducing and sustaining clinical remission in adult patients with moderately to severely active ulcerative colitis who have had an inadequate response to immunosuppressants such as corticosteroids, azathioprine, or 6-mercaptopurine. Adalimumab is also indicated for treatment of pediatric CD in patients 6 years of age and older or for treatment of pediatric UC in patients 5 years of age and older who have had an inadequate response to corticosteroids or immunomodulators (Humira®, US prescribing information 2022).

Certolizumab Pegol

An overview of certolizumab pegol, a TNF blocker, has been provided earlier in this chapter (Cimzia®, US prescribing information 2019). In addition to its use for the treatment of RA and PsA, the efficacy and safety of certolizumab pegol have also been assessed in adult patients with moderately to severely active CD.

In a randomized, double-blind, controlled Phase III study in patients with active CD (CDAI $\geq$220 and $\leq$450), treatment with subcutaneous certolizumab pegol at the recommended induction dose regimen (400 mg at Weeks 0, 2, and 4) resulted in 35 and 22% of patients achieving clinical response (defined as a decrease in CDAI $\geq$100) and clinical remission (defined as CDAI $\leq$150), respectively, at Week 6, while the control group (placebo) had 27 and 17% of patients with clinical response and clinical remission, respectively. Among clinical responders at Week 6, further maintenance treatment with 400 mg certolizumab pegol administered subcutaneously every 4 weeks demonstrated greater efficacy than the placebo maintenance group. The certolizumab pegol maintenance group had 63 and 48% of patients who were in clinical response and clinical remission, respectively, at Week 26, while the placebo maintenance group had 36 and 29% of patients in clinical response and clinical remission, respectively.

Among IBD indications, Certolizumab pegol is indicated for reducing signs and symptoms of CD and maintaining clinical response in adult patients with moderately to severely active disease who have had an inadequate response to conventional therapy (Cimzia®, US prescribing information 2019).

Golimumab

An overview of golimumab, an anti-TNFα human mAb, has been provided earlier in this chapter (Simponi®, Simponi Aria®, US prescribing information 2021). In addition to its use for the treatment of RA, PsA, and AS, the efficacy and safety of golimumab have been assessed in adult patients with moderately to severely active UC.

In a randomized, double-blind, controlled Phase III study in patients with active UC (Mayo score of 6–12), treatment with subcutaneous golimumab at the recommended induction dose regimen (200 mg at Week 0, followed by 100 mg at Week 2) resulted in 51 and 18% of patients achieving clinical response (defined as a decrease from baseline in the Mayo score of $\geq$30% and $\geq$3 points, accompanied by a decrease in the rectal bleeding subscore of $\geq$1 or a rectal bleeding subscore of 0 or 1) and clinical remission (defined as a Mayo score $\leq$2 points, with no individual subscore >1; improvement of endoscopic appearance of the mucosa was defined as a Mayo endoscopy subscore of 0 [normal or inactive disease] or 1 [erythema, decreased vascular pattern, mild friability]), respectively, at Week 6, while the control group (placebo) had 30 and 6% of patients with clinical response and clinical remission, respectively. Among clinical responders at Week 6, further maintenance treatment with 100 mg subcutaneous golimumab every 4 weeks demonstrated greater efficacy than the placebo maintenance group. The golimumab maintenance group had 50 and 28% of patients who were in clinical response at Week 54 and clinical remission at both Week 30 and Week 54, respectively, while the placebo maintenance group had 31 and 16% of patients in clinical response and clinical remission, respectively.

Among IBD indications, golimumab is indicated for adult patients with moderately to severely active UC with an inadequate response or intolerant to prior treatment or requiring continuous steroid therapy who have had an inadequate response to conventional therapy, for inducing and maintaining clinical response, improving endoscopic appearance of the mucosa during induction, inducing clinical remission, and achieving and sustaining clinical remission in induction responders (Simponi®, US prescribing information 2021).

Infliximab

An overview of infliximab, an anti-TNFα chimeric mAb, has been provided earlier in this chapter (Remicade®, US prescribing information 2021). In addition to its use for the treatment of arthritides and psoriasis, the efficacy and safety of infliximab have also been assessed in adult and pediatric patients with moderately to severely active CD or UC.

In a randomized, double-blind, controlled Phase III study in patients with moderately to severely active CD (CDAI $\geq$220 and $\leq$400), treatment with an initial intravenous dose of 5 mg/kg infliximab at Week 0 resulted in 57% of patients achieving clinical response (defined as a decrease in CDAI $\geq$70) at Week 2. All patients who received 5 mg/kg infliximab at Week 0 were then randomized to placebo or infliximab maintenance groups (5 or 10 mg/kg at Weeks 2 and 6, followed by every 8 weeks). Maintenance treatment with infliximab demonstrated greater efficacy than placebo maintenance treatment. The 5 and 10 mg/kg infliximab maintenance groups had 25 and 34% of patients who were in clinical remission and discontinued corticosteroid use, respectively, at Week 54, while the placebo maintenance group had 11% of patients in clinical remission with corticosteroid discontinuation.

In another randomized, double-blind, controlled Phase III study in patients with moderately to severely active UC (Mayo score of 6–12, Endoscopy subscore ≥ 2), patients were randomized at Week 0 to receive either placebo or infliximab at Weeks 0, 2, and 6 and every 8 weeks thereafter. At Week 8, a greater proportion of patients in the 5 mg/kg infliximab treatment group were in clinical response (defined as a decrease in Mayo score by $\geq 30\%$ and ≥ 3 points;) and clinical remission (defined as a Mayo score ≤ 2 points with no individual subscore >1) compared to the placebo treatment group (69% vs. 37% for clinical response; 39% vs. 15% for clinical remission). The clinical efficacy was maintained over time. At Week 54, the 5 mg/kg infliximab maintenance group had 45 and 35% of patients who were in clinical response and clinical remission, respectively, while the placebo maintenance group had 20 and 17% of patients in clinical response and clinical remission, respectively.

For both CD and UC, maintenance therapy with infliximab every 8 weeks significantly reduced disease-related hospitalizations and surgeries. A reduction in corticosteroid use and improvements in quality of life were observed. In addition, the safety and efficacy of infliximab for pediatric patients 6 years of age or older with CD or UC have also been established.

Among IBD indications, infliximab is indicated for treatment of CD by reducing signs and symptoms and inducing and maintaining clinical remission in adult patients with moderately to severely active disease who have had an inadequate response to conventional therapy and by reducing the number of draining enterocutaneous and rectovaginal fistulas and maintaining fistula closure in adult patients with fistulizing disease. Infliximab is also indicated for treatment of UC by reducing signs and symptoms, inducing and maintaining clinical remission and mucosal healing, and eliminating corticosteroid use in adult patients with moderately to severely active disease who have had an inadequate response to conventional therapy. Additionally, infliximab is indicated for treatment of pediatric patients 6 years of age or older with CD or UC who have had an inadequate response to conventional therapy (Remicade®, US prescribing information 2021).

Anti-IL-12/IL-23 Biologic Agents

Ustekinumab

An overview of ustekinumab, a human anti-IL-12/IL-23 mAb, has been provided earlier in this chapter (Stelara®, US prescribing information 2022). In addition to its use for the treatment of PsA and psoriasis, the safety and efficacy of ustekinumab, an IL-12/IL-23 inhibitor, in treatment of CD has been recently established (Stelara®, US prescribing information 2022). IL-12 and IL-23 have been implicated as important contributors to the chronic inflammation that is a hallmark of CD and UC.

The efficacy and safety of ustekinumab have been assessed in three randomized, double-blind, placebo-controlled Phase III studies in adult patients with moderately to severely active CD (CDAI ≥ 220 and ≤ 450). These were two 8-week intravenous induction studies (CD-1 and CD-2) followed by a 44-week subcutaneous randomized withdrawal maintenance study (CD-3) representing 52 weeks of therapy. In both induction studies (CD-1 and CD-2), patients were randomized to receive a single intravenous administration of ustekinumab at either approximately 6 mg/kg (recommended induction dose), placebo, or 130 mg (a lower dose than recommended). Patients who had failed or were intolerant to prior treatment with a TNF blocker were enrolled for Study CD-1, while patients who had never received a TNF blocker or previously received but had not failed a TNF blocker were enrolled for Study CD-2. Concomitant stable dosages of aminosalicylates, corticosteroids, and immunomodulators were allowed in both studies. In Study CD-1, the ustekinumab group had 34 and 21% of patients who were in clinical response (defined as reduction in CDAI score by at least 100 points or being in clinical remission) at Week 6 and clinical remission (defined as CDAI score < 150) at Week 8, respectively, while the placebo group had 21 and 7% of patients in clinical response at Week 6 and clinical remission at Week 8, respectively. In Study CD-2, the ustekinumab group had 56 and 40% of patients who were in clinical response at Week 6 and clinical remission at Week 8, respectively, while the placebo group had 29 and 20% of patients in clinical response at Week 6 and clinical remission at Week 8, respectively. The maintenance study (CD-3) evaluated patients who achieved clinical response at Week 8 of induction in studies CD-1 or CD-2. At Week 44 of the maintenance treatment (i.e., Week 52 from the initiation of induction therapy), 47% of patients received 90 mg subcutaneous ustekinumab every 8 weeks maintenance treatment demonstrated corticosteroid-free and in clinical remission, compared to 30% of patients in the placebo group.

For the treatment of UC, the safety and efficacy of ustekinumab was assessed in two randomized, double-blind, placebo-controlled clinical studies [UC-1 and UC-2] in adult patients with moderately to severely active UC who had an inadequate response to or failed to tolerate biologic (i.e., TNF blocker and/or vedolizumab), corticosteroids, and/or 6-MP or AZA therapy. In UC-1, a significantly greater proportion of patients treated with ustekinumab (at the recommended dose of approximately 6 mg/kg dose) were in clinical remission (19% versus 7%) and response (58% versus 31%) and achieved endoscopic improvement (25% versus 13%) and histologic-endoscopic mucosal improvement (17% versus 8%) compared to placebo. In UC-2, the maintenance effect at Week 44 with recommended dosage of 90 mg every 8 weeks was assessed. A significantly greater proportion of patients treated with ustekinumab were in primary clinical endpoint at Week 44 compared to placebo.

Among IBD indications, ustekinumab is indicated for treatment of adult patients with moderately to severely active CD and UC who have failed or were intolerant to treatment with immunomodulators or corticosteroids, but never failed a TNF blocker or who failed or were intolerant to treatment with one or more TNF blockers (Stelara®, US prescribing information 2022).

Anti-IL-23 Biologic Agent

As of November 2022, risankizumab-rzaa (Skyrizi®) is the only anti-IL-23 biologic agent approved for the treatment of CD. No anti-IL-23 biologic agent has been approved for UC yet, though several Phase III trials are on-going, including those for guselkumab, risankizumab, and mirikizumab (another anti-IL-23 mAb) (ClinicalTrials.gov).

Risankizumab-rzaa

An overview of risankizumab-rzaa has been provided earlier in this chapter. In addition to the use of this therapeutic protein for the treatment of PsO and PsA, risankizumab-rzaa has also been evaluated in patients with IBD (Skyrizi®, US prescribing information 2022).

For the indication of CD, the induction treatment effect of risankizumab was assessed in two randomized, double-blind, parallel group, placebo-controlled pivotal trials, CD-1 and CD-2, and maintenance treatment effect was assessed in another pivotal trial, CD-3. In CD-1 and CD-2, the patients received the treatment of risankizumab-rzaa 1200 mg, or placebo as an intravenous infusion at Weeks 0, 4, and 8. The co-primary endpoints were clinical remission and endoscopic response at Week 12. Secondary endpoints included clinical response and endoscopic remission. Compared to placebo group, the treatment difference of clinical remission, endoscopic response, clinical response, and endoscopic remission were 21%, 28%, 23%, and 15%, respectively, in CD-1 and 22%, 18%, 29%, and 15%, respectively, in CD-2. In CD-3, subjects were randomized to receive a maintenance regimen of risankizumab-rzaa 360 mg or placebo at Week 12 and every 8 weeks thereafter for up to an additional 52 weeks. Compared to placebo group, the treatment difference of clinical remission, endoscopic response at week 52 was 14% and 31%, respectively. Endoscopic remission was observed at Week 52 in 41% of subjects treated with the risankizumab-rzaa maintenance regimen and 13% of subjects treated with placebo. This endpoint was not statistically significant under the prespecified multiple testing procedure.

Risankizumab-rzaa is indicated for the treatment of moderately to severely active CD in adults (Skyrizi®, US prescribing information 2022).

Anti-Integrin Biologic Agent

Natalizumab and vedolizumab are both anti-integrin agents approved for treatment of CD (vedolizumab is also indicated for the treatment of UC). Natalizumab has a more restricted use with a requirement for patient registration due to its potential risk of PML. It has been hypothesized that preventing α4β1/α4β7 integrin binding to vascular cell adhesion molecule-1 (VCAM-1) may result in decreased immune surveillance within the central nervous system (CNS), in turn increasing the risk of developing PML. Unlike natalizumab, vedolizumab specifically targets α4β7 and does not inhibit binding at VCAM-1 (Soler et al. 2009). Overall, vedolizumab seems to be safe with respect to the risk of PML, but continuous and careful monitoring of patients is needed to explore its full safety profile.

Natalizumab

Natalizumab (Tysabri®) is a recombinant humanized IgG4κ mAb that is produced in murine myeloma cells. Natalizumab binds to the α4-subunit of α4β1 and α4β7 integrins expressed on the surface of all leukocytes, and inhibits the α4-mediated adhesion of leukocytes to their counter-receptor(s). Disruption of these molecular interactions prevents transmigration of leukocytes across the endothelium into inflamed parenchymal tissue (Tysabri®, US prescribing information 2021).

The efficacy and safety of natalizumab have been assessed in adult patients with moderately to severely active CD (CDAI ≥220 and ≤450). In a randomized, double-blind, controlled Phase III study in patients with active CD, treatment with intravenous natalizumab at the recommended induction dose regimen (300 mg every 4 weeks) resulted in 56% of patients achieving clinical response (defined as a decrease in CDAI ≥70 from baseline) at Week 10, while the control group (placebo) had 49% of patients achieving clinical response ($p = 0.067$). Among clinical responders at both Weeks 10 and 12, maintenance treatment with 300 mg natalizumab every 4 weeks demonstrated greater efficacy than that observed in the placebo maintenance group. The natalizumab maintenance group had 54 and 40% of patients who were in clinical response and clinical remission (defined as CDAI score < 150) at month 15, respectively, while the placebo maintenance group had 20 and 15% of patients in clinical response and clinical remission, respectively.

Natalizumab was first approved in November 2004 and was suspended soon after (February 2005) because of the occurrence of three cases of PML, an opportunistic viral infection of the brain that usually leads to death or severe disability. Two of the three PML cases were reported in multiple sclerosis patients (another indication of natalizumab) and one in a patient with CD. In June 2006, the FDA and European Medicine Agency (EMA) granted approval for the reintroduction of natalizumab under a specific risk management plan designed to redefine the safety profile of natalizumab (with Black-Box Warning in label).

Among IBD indications, natalizumab is indicated for the induction and maintenance of clinical response and remission in adult patients with moderately to severely active CD with evidence of inflammation who have had an inadequate response to, or are unable to tolerate, conventional CD therapies and TNF-α inhibitors. Natalizumab should not be used in combination with immunosuppressants or inhibitors of TNFα in CD (Tysabri®, US prescribing information 2021).

Vedolizumab

Vedolizumab (Entyvio®) is a humanized IgG1 mAb produced in Chinese hamster ovary cells. Vedolizumab specifically binds to the α4β7 integrin and blocks the interaction of α4β7 integrin with mucosal addressing cell adhesion molecule-1 (MAdCAM-1) and inhibits the migration of memory T-lymphocytes across the endothelium into inflamed gastrointestinal parenchymal tissue. Vedolizumab does not bind to or inhibit function of the α4β1 and αEβ7 integrins and does not antagonize the interaction of α4 integrins with VCAM-1. The interaction of the α4β7 integrin with MAdCAM-1 has been implicated as an important contributor to the chronic inflammation that is a hallmark of CD and UC (Entyvio®, US prescribing information 2022).

The efficacy and safety of vedolizumab have been assessed in three randomized, double-blind, placebo-controlled Phase III studies (Trials CD-1, CD-II, and CD-III) in adult patients with moderately to severely active CD (CDAI ≥220 and ≤450). Trials CD-I and CD-II both assessed induction regimens. A higher number of patients who had an inadequate response, loss of response, or intolerance to one or more TNF blockers were enrolled in CD-II when compared to CD-II (76% vs. 46%). Concomitant stable dosages of aminosalicylates, corticosteroids, and immunomodulators were allowed in both CD-1 and CD-II. In Study CD-I, a statistically significantly higher proportion of patients treated with vedolizumab at the recommended dose regimen (300 mg intravenous infusion at Weeks 0 and 2) achieved clinical remission (defined as CDAI ≤150) as compared to placebo at Week 6 (15% vs. 7%, $p = 0.041$). In Study CD-II, among patients who had an inadequate response, loss of response, or intolerance to one or more TNF blockers, no statistically significant difference was shown in the propor-

tion of patients achieving clinical remission between the 300 mg vedolizumab group and the placebo group at Week 6 (15% vs. 12%). Study CD-III evaluated the maintenance regimen. Among clinical responders ((≥70 decrease in CDAI score from baseline) at Week 6, 39% of patients who received maintenance treatment with 300 mg vedolizumab intravenous infusion every 8 weeks demonstrated clinical remission (defined as the proportion of patients in this subgroup that discontinued corticosteroids by Week 52 and were in clinical remission at Week 52), compared to 22% of patients in the placebo group.

The efficacy and safety of vedolizumab have also been assessed in two randomized, double-blind, placebo-controlled Phase III studies (Trials UC-1 and UC-II) in adult patients with moderately to severely active UC (Mayo score of 6–12 with endoscopy subscore of 2 or 3). UC-I assessed the induction therapy and UC-II assessed the maintenance therapy. Concomitant stable dosages of aminosalicylates, corticosteroids, and immunomodulators were allowed in both studies. Six-weeks treatment with vedolizumab (300 mg intravenous infusion at Week 0 and Week 2) compared to placebo resulted in a greater proportion of patients achieved clinical response at Week 6 (47% vs. 26%, defined as reduction in complete Mayo score of ≥3 points and ≥30% from baseline with an accompanying decrease in rectal bleeding subscore of ≥1 point or absolute rectal bleeding subscore of ≤1 point), clinical remission at Week 6 (17% vs. 5%, defined as complete Mayo score of ≤2 points and no individual subscore >1 point), and improvement of endoscopic appearance of the mucosa at Week 6 (41% vs. 25%, defined as Mayo endoscopy subscore of 0 [normal or inactive disease] or 1 [erythema, decreased vascular pattern, mild friability]). Among clinical responders at Week 6, 42% of patients who received maintenance treatment with 300 mg vedolizumab intravenous infusion every 8 weeks demonstrated clinical remission at Week 52, compared to 16% of patients in the placebo group.

Among IBD indications, vedolizumab is indicated for treatment of adult patients with moderately to severely active CD and UC who have had an inadequate response with, lost response to, or were intolerant to a TNF blocker or immunomodulator; or had an inadequate response with, were intolerant to, or demonstrated dependence on corticosteroids (Entyvio®, US prescribing information 2022).

Asthma

Asthma is a complex chronic inflammatory syndrome of the airways and is characterized by variable symptoms of cough, breathlessness, and wheezing. These episodes may be punctuated by periods of more severe and sustained deterioration in control of symptoms, termed exacerbations, which can result in potentially life-threatening bronchospasm. Asthma affects almost 20 million individuals in the USA, six million of which are children. Pharmacotherapeutic management of the disease has progressed but is suboptimal for a subset of moderately to severely affected patients. Treatment with inhaled corticosteroids and short- and long-acting β-adrenoceptor agonists is considered the standard of care and is generally effective at attenuating symptoms, particularly in mild-to-moderate asthma; however, these therapeutic modalities do not necessarily address the underlying pathology of the disease. A subset of patients with moderate-to-severe asthma remain symptomatic despite treatment with corticosteroids, suggesting persistent inflammation of the airways.

The limitations of existing asthma therapies justify continued research into novel interventions, particularly those that modify disease processes. To that aim, a number of therapeutic proteins targeting cytokines linked to the underlying pathology of the disease have been developed, and five biologic agents have been approved for use in asthma, including omalizumab (Xolair®), a humanized mAb against IgE; mepolizumab (Nucala®) and reslizumab (Cinqair®), two humanized mAbs against IL-5; Dupilumab (Dupixent®) a humanized mAb against IL-4 receptor; and benralizumab (Fasenra®), a humanized mAb against IL-5 receptor.

Anti-IgE Biologic Agent

Omalizumab

Omalizumab (Xolair®) is a recombinant human IgG1κ mAb that selectively binds IgE and inhibits the binding of IgE to the high-affinity IgE receptor (FcεRI) on the surface of mast cells and basophils. IgE plays a central role in increasing allergen uptake by dendritic cells, activated mast cells, and basophils. Reduction in surface-bound IgE on FcεRI-bearing cells limits the degree of the allergic response. Omalizumab also reduces the number of FcεRI receptors on basophils in atopic patients. Omalizumab is a first-in-class selective IgE inhibitor approved for use by patients with allergic asthma inadequately controlled by inhaled corticosteroids (Xolair®, US prescribing information 2021).

The safety and efficacy of omalizumab in treatment of asthma have been evaluated in in three randomized, double-blind, placebo-controlled, multicenter trials in patients 12–76 years old, with moderate-to-severe persistent asthma for at least 1 year, and a positive skin test reaction to a perennial. Omalizumab dosing was based on body weight and baseline serum total IgE concentration. All patients were required to have a baseline IgE between 30 and 700 IU/mL and body weight not more than 150 kg. Patients were treated according to a dosing table to administer at least 0.016 mg/kg/IU (IgE/mL) of omalizumab or a matching volume of placebo over each 4-week period. The maximum omalizumab

dose per 4 weeks was 750 mg. Two of the Phase III trials (Trials 1 and 2) were conducted in patients with concomitant controller medications of inhaled corticosteroids (ICS) and short-acting β2-agonists. Both trials included a run-in period followed by randomization to omalizumab or placebo. Patients received omalizumab for 16 weeks with an unchanged corticosteroid dose unless an acute exacerbation necessitated an increase. Patients then entered a 12-weeks ICS reduction phase during which ICS dose reduction was attempted in a stepwise manner. Omalizumab efficacy was based primarily on the reduction of asthma exacerbations, which were defined as a worsening of asthma that required treatment with systemic corticosteroids or a doubling of baseline-inhaled corticosteroid dose. In Trial 1 and Trial 2, the number of exacerbations per patient was reduced in patients treated with omalizumab compared with placebo in both the 12-weeks steroid-free period (asthma exacerbation frequencies of 0, 1, and ≥2 per patient in 86–87%, 11–12%, and 1–2% of patients receiving omalizumab, and in 70–77%, 17–25%, and 5–7% of patients receiving placebo) and the 16-week steroid reduction period (asthma exacerbation frequencies of 0, 1, and ≥2 per patient in 79–84%, 14–19%, and 1–2% of patients receiving omalizumab, and in 68–70%, 26–28%, 3–4% of patients receiving placebo). Trial 3 had a similar design to that of Trials 1 and 2, but long-acting β2-agonists were allowed. In Trial 3, the number of exacerbations in patients treated with omalizumab was similar to that in placebo-treated patients. The absence of an observed treatment effect in Trial 3 may be related to differences in the patient population compared with Asthma Trials 1 and 2, study sample size (lower in Trial 3), or other factors.

The safety and efficacy of omalizumab in treatment of asthma have also been evaluated in children 6 to <12 years of age with moderate-to-severe asthma who had a positive skin test or in vitro reactivity to a perennial aeroallergen. Omalizumab-treated children had a statistically significant reduction in the rate of asthma exacerbations versus the placebo group.

An anaphylaxis Black-Box warning was placed on the omalizumab (Xolair®) label based on clinical evidence. Anaphylaxis has been reported to occur after administration of omalizumab in pre-marketing clinical trials and in post-marketing spontaneous reports. In pre-marketing clinical trials in patients with asthma, anaphylaxis was reported in 3 of 3507 (0.1%) patients. In post-marketing spontaneous reports, the frequency of anaphylaxis attributed to omalizumab use was at least 0.2% of patients based on an estimated exposure of about 57,300 patients from June 2003 through December 2006. Anaphylaxis has occurred after the first dose of omalizumab but also has occurred beyond 1 year after beginning treatment. Therefore, omalizumab should be available only in a healthcare setting by healthcare providers prepared to manage anaphylaxis that can be life-threatening (Xolair®, US prescribing information 2021).

Omalizumab is indicated for moderate-to-severe persistent asthma in patients 6 years of age and older with a positive skin test or in vitro reactivity to a perennial aeroallergen and symptoms that are inadequately controlled with inhaled corticosteroids. Omalizumab is not indicated for acute bronchospasm or status asthmaticus (Xolair®, US prescribing information 2021).

Anti-IL-5 Biologic Agents

Two anti-IL-5 biologic agents have been approved for asthma, mepolizumab (Nucala®) and reslizumab (Cinqair®). Mepolizumab is a humanized Ig1κ mAb while reslizumab is a humanized IgG4κ mAb. IL-5 is the major cytokine responsible for the growth and differentiation, recruitment, activation, and survival of eosinophils. Mepolizumab or reslizumab binds to IL-5, inhibiting the bioactivity of IL-5 by blocking its binding to the α chain of the IL-5 receptor complex expressed on the eosinophil cell surface. Mepolizumab or reslizumab, by inhibiting IL-5 signaling, reduces the production and survival of eosinophils; however, their mechanism of action in asthma has not been definitively established (Nucala®, US prescribing information 2022; Cinqair®, US prescribing information 2020).

Mepolizumab

Mepolizumab (Nucala®) is a humanized IgG1κ mAb against IL-5, which is produced by recombinant DNA technology in Chinese hamster ovary cells (Nucala®, US prescribing information 2022).

The safety and efficacy of mepolizumab as add-on treatment of severe asthma have been evaluated in patients aged 6 years and older who had asthma despite regular use of high-dose inhaled corticosteroid (ICS) plus additional controller(s), or use of daily oral corticosteroids (OCS) in addition to regular use of high-dose ICS plus additional controller(s). These patients had markers of eosinophilic airway inflammation and continued their background asthma therapy throughout the duration of the trials. In a 32-week double-blind, randomized, placebo-controlled Phase III trial, patients receiving add-on mepolizumab treatment at the recommended dose regimen (100 mg administered subcutaneously every 4 weeks) in combination with background therapy, compared with placebo (plus background therapy), experienced significantly fewer (52% reduction) exacerbations (defined as worsening of asthma requiring use of oral/systemic corticosteroids and/or hospitalization and/or emergency department visits). Additionally, compared with placebo (plus background therapy), there were fewer exacerbations requiring hospitalization and/or emergency department visits and exacerbations requiring only in-patient hospitalization with mepolizumab add-on treatment. In a 24-week OCS-reduction trial, effect of mepolizumab on reducing the use of maintenance OCS was evaluated. Compared with placebo (plus background therapy), add-on mepolizumab at the recommended dose regimen for 24 weeks resulted in greater reductions in daily maintenance OCS dose, while maintaining asthma control. Twenty-three percent (23%) of patients in the add-on mepolizumab treatment group versus 11% in the placebo group (plus background therapy) had a 90–100% reduction in their OCS dose. Thirty-six per cent (36%) of patients in the add-on mepolizumab group versus 56% in the placebo group (plus background therapy) were classified as having no improvement for OCS dose (i.e., no change or any increase or lack of asthma control or withdrawal of treatment). Additionally, 54% of patients receiving add-on mepolizumab treatment achieved at least half reduction in the daily prednisone dose compared with 33% of subjects who received placebo (plus background therapy).

In the Phase III trials for asthma, a total of 28 adolescents aged 12–17 years with severe asthma were enrolled. These adolescent patients showed a reduction in the rate of exacerbations that trended in favor of mepolizumab. The adverse event profiles in adolescents were generally comparable to the overall population in the Phase III studies.

Mepolizumab pharmacokinetics following subcutaneous administration in subjects aged 6–11 years with severe asthma was investigated in the initial 12-week treatment phase of an open-label clinical trial. Exposures, areas under the curve (AUC), following subcutaneous administration of either 40 mg (for children weighing <40 kg) or 100 mg (for children weighing ≥40 kg) were 1.32 and 1.97 times higher, respectively, compared with those observed in adults and adolescents receiving 100 mg. Based on these results, simulation of a 40-mg subcutaneous dose every 4 weeks in children aged 6–11 years, irrespective of body weight, resulted in predicted exposures similar to those observed in adults and adolescents.

Mepolizumab is indicated for add-on maintenance treatment of patients with severe asthma aged 6 years and older, and with an eosinophilic phenotype. Mepolizumab is not recommended for relief of acute bronchospasm or status asthmaticus (Nucala®, US prescribing information 2022).

Reslizumab

Reslizumab (Cinqair®) is a humanized IL-5 antagonist IgG4κ mAb produced by recombinant DNA technology in murine myeloma non-secreting 0 (NS0) cells (Cinqair®, US prescribing information 2020).

The safety and efficacy of reslizumab as add-on treatment of asthma have been evaluated in adult and adolescent patients aged 12 years and older who had at least 1 asthma exacerbation requiring systemic corticosteroid use over the past 12 months. Most patients had markers of eosinophilic airway inflammation, and all patients continued their background asthma therapy throughout the duration of the trials.

While patients aged 12–17 years were included in these trials, reslizumab is not approved for use in this age group. The safety and effectiveness of reslizumab in pediatric patients (aged 17 years and younger) have not been established.

In two 52-week double-blind, randomized, placebo-controlled Phase III trials, patients receiving add-on reslizumab treatment at the recommended dose regimen (3 mg/kg administered intravenously every 4 weeks), compared with placebo (with background therapy), had significant reduction (50–59% reduction) in the rate of all asthma exacerbations (defined as worsening of asthma requiring use of systemic corticosteroids or twofold increase in the use of ICS for 3 or more days, and/or asthma-related emergency treatment including visit to their healthcare professional for nebulizer treatment or other urgent treatment to prevent worsening of asthma symptoms, or a visit to the emergency room or hospitalization). Exacerbations requiring the use of a systemic corticosteroid such as OCS as well as exacerbations resulting in hospitalization or an emergency room visit were also reduced with add-on reslizumab treatment. In two other 16-week double-blind, randomized, placebo-controlled Phase III trials, add-on reslizumab treatment at the recommended dose regimen resulted in improvements in lung function (as assessed by FEV1 [forced expiratory volume in 1 s]) 4 weeks following the first dose of reslizumab and maintained through Week 52.

An anaphylaxis Black-Box warning was placed on the reslizumab (Cinqair®, US prescribing information 2020) label. Anaphylaxis occurred with reslizumab infusion in 0.3% of patients in placebo-controlled Phase III studies. Patients should be observed for an appropriate period of time after reslizumab infusion; healthcare professionals should be prepared to manage anaphylaxis that can be life-threatening.

Reslizumab is indicated for add-on maintenance treatment of patients with severe asthma aged 18 years and older, and with an eosinophilic phenotype. Reslizumab is not indicated for treatment of other eosinophilic conditions, and not for relief of acute bronchospasm or status asthmaticus (Cinqair®, US prescribing information 2020).

Anti-IL-5 Receptor Biologic Agent

In addition to directly neutralizing the IL-5 cytokine, blockage of IL-5 receptor has also been explored as an alternative mechanism of action to target the IL-5 pathway for the treatment of asthma, such as benralizumab.

Benralizumab

Benralizumab (Fasenra®) is a humanized afucosylated IL-5 receptor antagonist IgG1κ mAb produced by recombinant DNA technology in Chinese hamster ovary cells. The IL-5 receptor is expressed on the surface of eosinophils and basophils. Inflammation is an important component in the pathogenesis of asthma. Multiple cell types (e.g., mast cells, eosinophils, neutrophils, macrophages, lymphocytes) and mediators (e.g., histamine, eicosanoids, leukotrienes, cytokines) are involved in inflammation. Benralizumab, by binding to the α chain of IL-5 receptor, reduces eosinophils through antibody-dependent cell-mediated cytotoxicity (ADCC); however, its mechanism of action in asthma has not been definitively established (Fasenra®, US prescribing information 2021).

The safety and efficacy of benralizumab have been evaluated in patients aged 12 years and older who had at least 2 asthma exacerbation requiring systemic corticosteroid use over the past 12 months. Most patients had markers of eosinophilic airway inflammation, and all patients continued their background asthma therapy throughout the duration of the trials. In a 48-week double-blind, randomized, placebo-controlled Phase III trial, patients receiving add-on benralizumab treatment at the recommended dose regimen (30 mg administered subcutaneously at Weeks 0, 4, and 8, followed by every 8 weeks), compared with placebo (plus background therapy of high dose ICS), had significant reduction (35% vs. 51%) in the rate of all asthma exacerbations (defined as a worsening of asthma requiring use of oral/systemic corticosteroids for at least 3 days, and/or emergency department visits requiring use of oral/systemic corticosteroids and/or hospitalization). Exacerbations requiring hospitalization or an emergency room visit and exacerbations requiring hospitalization were also reduced with add-on benralizumab treatment. Compared to placebo (plus background therapy), add-on benralizumab treatment at the recommended dose regimen also provided consistent improvements over time in the mean change from baseline in lung function, as assessed by FEV1.

Benralizumab is indicated for add-on maintenance treatment of patients with severe asthma aged 12 years and older, and with an eosinophilic phenotype. Benralizumab is not indicated for treatment of other eosinophilic conditions, and not for relief of acute bronchospasm or status asthmaticus (Fasenra®, US prescribing information 2021).

Anti-thymic Stromal Lymphopoietin (TSLP) Biologic Agent

Airway inflammation is an important component in the pathogenesis of asthma. Multiple cell types (e.g., mast cells, eosinophils, neutrophils, macrophages, lymphocytes, ILC2 cells) and mediators (e.g., histamine, eicosanoids, leukotrienes, cytokines) are involved in airway inflammation. Blocking TSLP reduces biomarkers and cytokines associated with inflammation including blood eosinophils, airway submucosal eosinophils, IgE, fractional exhaled nitric oxide (FeNO), IL-5, and IL-13; however, the mechanism of action of anti-TSLP agents in asthma has not been definitively established yet.

Tezepelumab-ekko

Tezepelumab-ekko (Tezspire®), a thymic stromal lymphopoietin (TSLP) blocker, is a human monoclonal antibody immunoglobulin G2λ (IgG2λ) produced in Chinese hamster ovary (CHO) cells by recombinant DNA technology. Tezepelumab-ekko can bind to human TSLP and blocks its interaction with the heterodimeric TSLP receptor. TSLP is a cytokine mainly derived from epithelial cells and occupies an upstream position in the asthma inflammatory cascade (Tezspire®, US prescribing information 2022).

The efficacy of Tezepelumab-ekko was evaluated in two randomized, double-blind, parallel group, placebo-controlled clinical trials (PATHWAY and NAVIGATOR) of 52 weeks duration. In the PATHWAY study, adult patients received treatment with tezepelumab-ekko 70 mg subcutaneously every 4 weeks, tezepelumab-ekko 210 mg subcutaneously every 4 weeks, tezepelumab-ekko 280 mg subcutaneously every 2 weeks, or placebo subcutaneously. In the NAVIGATOR study, adult and pediatric patients 12 years of age and older with severe asthma received treatment with tezepelumab-ekko 210 mg subcutaneously every 4 weeks or placebo subcutaneously every 4 weeks. In both, PATHWAY and NAVIGATOR, patients receiving tezepelumab-ekko had significant reductions in the annualized asthma exacerbations rate of 0.2 and 0.93, respectively, compared to placebo of 0.72 and 2.1, respectively. There were also fewer exacerbations requiring emergency room visits and/or hospitalization in patients treated with tezepelumab-ekko (0.03 and 0.06 in PATHWAY and NAVIGATOR, respectively) compared with placebo (0.18 and 0.28 in PATHWAY and NAVIGATOR, respectively). In NAVIGATOR, patients receiving tezepelumab-ekko experienced fewer exacerbations than those receiving placebo regardless of baseline levels of blood eosinophils or FeNO, and the time to first exacerbation was longer for the patients receiving tezepelumab-ekko compared with placebo in NAVIGATOR. Similar findings were seen in PATHWAY. Change from baseline in FEV1 was assessed as a secondary endpoint in PATHWAY and NAVIGATOR. Compared with placebo (−0.06 and 0.1 in two studies, respectively), tezepelumab-ekko provided clinically meaningful improvements in the mean change from baseline in FEV1 in both trials (0.08 and 0.23, respectively).

Tezepelumab-ekko is indicated for the add-on maintenance treatment of adult and pediatric patients aged 12 years and older with severe asthma (Tezspire®, US prescribing information 2021).

Anti-IL-4 Receptor Agents

Dupilumab

In addition to the treatment of AD, dupilumab was also approved as an add-on maintenance treatment of asthma (Dupixent®, US prescribing information 2022).

The efficacy and safety of dupilumab for patients aged 12 years and older were assessed in three randomized, double-blind, placebo-controlled, parallel-group, multicenter trials (DRI12544, QUEST, and VENTURE). In the primary analysis population of DRI12544 and QUEST, subjects receiving either dupilumab 200 mg or 300 mg Q2W had significant reductions in the rate of asthma exacerbations compared to placebo. In the overall population in QUEST, the rate of severe exacerbations was 0.46 and 0.52 for dupilumab 200 mg Q2W and 300 mg Q2W, respectively, compared to matched placebo rates of 0.87 and 0.97. The rate ratio of severe exacerbations compared to placebo was 0.52 (95% CI: 0.41, 0.66) and 0.54 (95% CI: 0.43, 0.68) for dupilumab 200 mg Q2W and 300 mg Q2W, respectively. Elevation of FeNO can be a marker of the eosinophilic asthma phenotype when supported by clinical data. Prespecified subgroup analyses of DRI12544 and QUEST demonstrated that there were greater reductions in severe exacerbations in subjects with higher baseline blood eosinophil levels (≥150 cells/µL) or FeNO (≥25 ppb). In QUEST, reductions in exacerbations were significant in the subgroup of subjects with baseline blood eosinophils ≥150 cells/µL. In subjects with baseline blood eosinophil count <150 cells/µL and FeNO <25 ppb, similar severe exacerbation rates were observed between dupilumab and placebo. In QUEST, the estimated rate ratio of exacerbations leading to hospitalizations and/or emergency room visits versus placebo was 0.53 (95% CI: 0.28, 1.03) and 0.74 (95% CI: 0.32, 1.70) with dupilumab 200 mg and 300 mg Q2W, respectively.

Dupilumab is used as an add-on maintenance treatment of adult and pediatric patients aged 6 years and older with moderate-to-severe asthma characterized by an eosinophilic phenotype or with oral corticosteroid-dependent asthma (Dupixent®, US prescribing information 2022).

CAR T Cell-Induced Cytokine Release Syndrome

T lymphocytes can be genetically modified to target tumors through the expression of a chimeric antigen receptor (CAR). Most notably, CAR T cells have demonstrated clinical efficacy in hematologic malignancies with more modest responses when targeting solid tumors. However, CAR T cells also have the capacity to elicit expected and unexpected toxicities, including neurologic toxicity, "on target/off tumor" recognition, anaphylaxis, and the life-threating cytokine release syndrome (CRS) (Bonifant et al. 2016).

Anti-IL-6 Receptor Biologic Agent

Tocilizumab

An overview of tocilizumab, a human anti-IL-6 receptor mAb, has been provided earlier in this chapter (Actemra®, US prescribing information 2021). In addition to use of this therapeutic protein for the treatment of RA, pJIA, and sJIA (and giant cell arteritis as described later in this chapter), tocilizumab is also indicated for treatment of CAR T cell-induced severe or life-threatening CRS in adults and pediatric patients 2 years of age and older (Actemra®, US prescribing information 2021).

The efficacy of tocilizumab for the treatment of CRS has been assessed in a retrospective analysis of pooled outcome data from clinical trials of CAR T-cell therapies for hematological malignancies. Evaluable patients had been treated with intravenous tocilizumab 8 mg/kg (12 mg/kg for patients <30 kg) with or without additional high-dose corticosteroids for severe or life-threatening CRS; only the first episode of CRS was included in the analysis. The study population included 24 male and 21 female subjects of median age 12 years (range, 3–23 years). The median time from start of CRS to first dose of tocilizumab was 4 days (range, 0–18 days). Resolution of CRS was defined as lack of fever and off vasopressors for at least 24 h. Patients were considered responders if CRS resolved within 14 days of the first dose of tocilizumab, no more than 2 doses of tocilizumab were needed, and no drugs other than tocilizumab and corticosteroids were used for treatment. Thirty-one patients (69%) achieved a response. Achievement of resolution of CRS within 14 days was confirmed in a second study using an independent cohort that included 15 subjects (range: 9–75 years old) with CAR T cell-induced CRS.

Chronic Idiopathic Urticaria

Chronic idiopathic urticaria (CIU) or chronic spontaneous urticaria (CSU) is a common autoimmune skin condition characterized by spontaneously recurring hives, occurring either intermittently or continuously for 6 weeks or longer (Vestergaard and Deleuran 2015). A significant association of CIU is a deeper localized swelling called angioedema, which is observed in about one-third of patients. This leads to remarkable psychosocial morbidity with a negative impact on overall quality of life. CIU occurs largely in young women between 20 and 40 years of age. The exact prevalence of CIU is difficult to determine. A recently published article indicated a point prevalence of at least 0.5% in the general population for CIU (Maurer et al. 2011). Conventional treatment for CIU prescribes a stepwise approach with non-sedating non-impairing antihistamines as first-line agents followed by

increasing to four times the licensed doses as second-line treatment. However, close to half of CIU patients do not achieve adequate symptom relief with H1 antihistamines alone. Third-line treatments include cyclosporine, which is associated with toxicity and requires frequent monitoring.

An autoimmune mechanism is thought to mediate the disease process in up to 50% of patients with CIU. Autoantibody to the α chain of the high affinity IgE receptor and/or intrinsic IgE immune modulation may play an important role. Toward this, omalizumab, a humanized anti-IgE antibody, is developed as an alternative treatment option for patients with CIU. Omalizumab (Xolair®) is the first and only biologic agent currently approved for the treatment of CIU. Rituximab, a humanized anti-CD-20 mAb, has been reported in a few cases to be effective in the treatment of CIU, whereas other cases did not show any effect (Mallipeddi and Grattan 2007). Antibodies against TNFα have also been used in the treatment of CIU but with variable success and only in small numbers of patients (Cooke et al. 2015).

Anti-IgE Biologic Agents

Omalizumab

An overview of omalizumab, a humanized anti-IgE mAb, has been provided earlier in this chapter (Xolair®, US prescribing information 2021). In addition to asthma, the efficacy and safety of omalizumab have been evaluated in patients with CIU. The mechanism of action has been described earlier as an antibody targeting IgE; however, the specific mechanism by which omalizumab results in an improvement of CIU symptoms has not been fully defined.

Two randomized, placebo-controlled, multiple-dose clinical trials have been conducted for omalizumab in adult and adolescent patients 12 years of age and older with CIU. Disease severity was measured by a weekly urticaria activity score (UAS7, range 0–42), which is a composite of the weekly itch severity score (range 0–21) and the weekly hive count score (range 0–21). In both trials, patients who received subcutaneous omalizumab 150 mg or 300 mg every 4 weeks in addition to their baseline H1 antihistamine therapy had greater decreases from baseline in itch severity scores and hive count scores than patients receiving placebo (plus background therapy) at Week 12. In one trial, the change from baseline to Week 12 were −3.63, −6.66, and −9.40 for placebo, omalizumab 150 mg and 300 mg, respectively, in itch severity score, and were − 4.37, −7.78, −11.35 for placebo, omalizumab 150 mg and 300 mg, respectively, in hive count score. Similar results were shown in the other trial.

In addition to asthma, omalizumab is also indicated for chronic idiopathic urticaria in adults and adolescents 12 years of age and older who remain symptomatic despite H1 antihistamine treatment. Omalizumab is not indicated for other allergic conditions or other forms of urticaria (Xolair®, US prescribing information 2021).

Chronic Rhinosinusitis with Nasal Polyps

Chronic rhinosinusitis with nasal polyps (CRSwNP) is an important clinical entity diagnosed by the presence of both subjective and objective evidence of chronic sinonasal inflammation. Symptoms include anterior or posterior rhinorrhea, nasal congestion, hyposmia, and/or facial pressure or pain that last for greater than 12 weeks duration. Nasal polyps are inflammatory lesions that project into the nasal airway, are typically bilateral, and originate from the ethmoid sinus. Males are more likely to be affected than females, but no specific genetic or environmental factors have been strongly linked to the development of this disorder to date. The dysregulation of the host immune system has also been extensively evaluated in CRSwNP.

Originally, this disease was categorized by a type-2 inflammatory response with enhanced tissue eosinophilia. Levels of eosinophilic granule proteins (e.g., eosinophil cationic protein (ECP)), the eosinophilic survival factor (IL-5), and eosinophil chemotactic proteins (e.g., Eotaxin-1, Eotaxin-2, Eotaxin-3, MCP-4) are all elevated in CRSwNP nasal polyps compared to healthy controls. Studies have also reported CRSwNP to have increased numbers of basophils, innate type-2 lymphoid cells, and mast cells. Additionally, type-2 cytokines including IL-5 and IL-13 as well as the epithelial cell-derived thymic stromal lymphopoietin (TSLP) are elevated in CRSwNP. While the inflammatory environment in CRSwNP has been extensively characterized, the specific events and signals that initiate this response are not well defined (Stevens et al. 2016). Recently, biologic agents such as dupilumab (anti-IL-4 receptor antibody) and mepolizumab (anti-IL-5 antibody) are considered promising therapeutic alternatives for CRSwNP.

Anti-IL-4 Receptor Agents

Dupilumab

An overview of duplilumab, a human anti-IL-4 mAb, has been provided earlier in this chapter. In addition to AD and asthma, duplilumab is also indicated for the treatment of CRSwNP (Dupixent®, US prescribing information 2022). The safety and efficacy of dupilumab for the treatment of CRSwNP were assessed in two randomized, double-blind, parallel-group, multicenter, placebo-controlled studies (SINUS-24 and SINUS-52). Patients with CRSwNP received 300 mg duplilumab or placebo administered subcutaneously once every 2 weeks. Statistically significant efficacy was observed in SINUS-52 with regard to improvement in bilateral endoscopic nasal polyp score at Week 24 and Week 52. At Week 52, the least square mean (LSM) difference for nasal congestion in the dupilumab group versus placebo was −0.98 (95% CI: −1.17, −0.79). In both studies, significant improvements in nasal congestion were observed as early as the first assessment at Week 4. The LSM difference for nasal

congestion at Week 4 in the dupilumab group versus placebo was −0.41 (95% CI: −0.52, −0.30) in SINUS-24 and −0.37 (95% CI: −0.46, −0.27) in SINUS-52. A significant decrease in the Lund-Mackay computed tomography sinus score was observed. The LSM difference at Week 24 in the dupilumab group versus placebo was −7.44 (95% CI: −8.35, −6.53) in SINUS-24 and −5.13 (95% CI: −5.80, −4.46) in SINUS-52. At Week 52, in SINUS-52, the LSM difference in the dupilumab group versus placebo was −6.94 (95% CI: −7.87, −6.01). Dupilumab significantly improved the loss of smell compared to placebo. The LSM difference for loss of smell at Week 24 in the dupilumab group versus placebo was −1.12 (95% CI: −1.31, −0.93) in SINUS-24 and −0.98 (95% CI: −1.15, −0.81) in SINUS-52. At Week 52, the LSM difference for loss of smell in the dupilumab group versus placebo was −1.10 (95% CI −1.31, −0.89). In both studies, significant improvements in daily loss of smell severity were observed as early as the first assessment at Week 4. Dupilumab significantly decreased sino-nasal symptoms as measured by the Sinu-Nasal Outcome Test (SNOT-22) compared to placebo. The LSM difference for SNOT-22 at Week 24 in the dupilumab group versus placebo was −21.12 (95% CI: −25.17, −17.06) in SINUS-24 and −17.36 (95% CI: −20.87, −13.85) in SINUS-52. At Week 52, the LS mean difference in the dupilumab group versus placebo was −20.96 (95% CI -25.03, −16.89).

In addition to the treatment of AD and asthma, dupilumab is indicated as an add-on maintenance treatment in adult patients with inadequately controlled CRSwNP (Dupixent®, US prescribing information 2022).

Anti-IL-5 Receptor Agents

Mepolizumab

An overview of mepolizumab, a humanized anti-IL-5 mAb, has been provided earlier in this chapter. In addition to asthma, mepolizumab is also indicated for the treatment of CRSwNP (Nucala®, US prescribing information 2022). In a randomized, placebo-controlled, multicenter study, patients with CRSwNP received 100 mg mepolizumab or placebo administered subcutaneously once every 4 weeks while continuing nasal corticosteroid therapy. Patients who received mepolizumab 100 mg had a statistically significant improvement (decrease) in bilateral nasal polyp score at Week 52 (mean change compared to baseline: −0.87 versus placebo of 0.06) and nasal obstruction visual analogue scale score from Weeks 49–52 at the end of the 52-week treatment period. The proportion of patients who had surgery was significantly reduced by 57% (hazard ratio: 0.43, 95% CI: 0.25, 0.76) in the group treated with mepolizumab 100 mg compared with the placebo group. By Week 52, 18 (9%) patients who received mepolizumab 100 mg had surgery compared with 46 (23%) patients in the placebo group.

In addition to severe asthma and EGPA, mepolizumab is indicated for the add-on maintenance treatments of adult patients with inadequately controlled CRSwNP (Nucala®, US prescribing information 2022).

Cryopyrin-Associated Periodic Syndromes

Cryopyrin-associated periodic syndrome (CAPS) comprises three genetic autoinflammatory disorders including familial cold autoinflammatory syndrome (FCAS), Muckle-Wells syndrome (MWS), and neonatal-onset, multisystem, inflammatory disorder (NOMID) (Kubota and Koike, 2010). These three phenotypically distinct disorders are recognized as a clinical continuum of the same disease in an increasing order of severity since all three disorders are associated with mutations in the NACHT, LRR, and PYD domains-containing protein 3 (NLRP3) gene that encodes the cryopyrin protein. NLRP3 mutations result in overactivation of caspase-1, the enzyme that cleaves precursors of IL-1β, IL-18, and IL-33 into active cytokine forms. CAPS is rare with a prevalence of one in about one million people; however many cases of this disease are believed to be undiagnosed.

Therapeutic treatments of CAPS include nonsteroidal anti-inflammatory drugs, colchicine, immunosuppressants, corticosteroids, and the recent addition of anti-IL-1β biologic therapy (Kubota and Koike 2010). Anti-IL-1β therapy is very effective for the treatment of CAPS since it exerts pharmacologic action directly against the underlying cause of the disease. Clinical evidence suggests that use of anti-IL-1β agents can achieve rapid and complete control of both clinical manifestations and laboratory parameters.

Anti-IL-1β Biologic Agents

Currently, there are two long-acting anti-IL-1β therapeutic proteins (canakinumab [Ilaris®] and rilonacept [Arcalyst®]) that have been approved for the treatment of CAPS, although a short-acting IL-1 receptor antagonist, anakinra, is also effective for this disease (Hawkins and Lachmann 2003). Excessive production of IL-1β is the central pathophysiology of CAPS. Both canakinumab and rilonacept were generally well tolerated in the Phase III trials with infections being a commonly reported adverse event due to the immunosuppressant effect of anti-IL-1β therapy.

Canakinumab

An overview of canakinumab, a human anti-IL-1β mAb, has been provided earlier in this chapter (Ilaris®, US prescribing information 2020). In addition to the use of this therapeutic protein for the treatment of sJIA, canakinumab has also been evaluated in patients with CAPS. Canakinumab binds to human IL-1β and neutralizes its activity.

The efficacy and safety of canakinumab for the treatment of CAPS have been assessed in a double-blind, placebo-controlled, randomized withdrawal trial in patients 9–74 years of age with the MWS phenotype of CAPS. This study consisted of three parts. Part 1 was an 8-week open-label, single-dose period where all patients received canakinumab. Patients who achieved a complete clinical response and did not relapse by Week 8 were randomized to receive either placebo or canakinumab every 8 weeks for 24 weeks in Part 2 of the study. Patients who completed Part 2 or experienced a disease flare entered Part 3, a 16-week open-label active treatment phase. Throughout the trial, patients weighing more than 40 kg received subcutaneous canakinumab 150 mg and patients weighing 15–40 kg received canakinumab 2 mg/kg. In Part 1, a complete clinical response was observed in 71% of patients 1 week following initiation of treatment and in 97% of patients by Week 8. In the randomized withdrawal period, a total of 81% of the patients randomized to placebo flared compared to none (0%) of the patients randomized to canakinumab treatment. In a second trial, patients 4–74 years of age with both MWS and FCAS phenotypes of CAPS were treated in an open-label manner. Treatment with canakinumab resulted in clinically significant improvement of signs and symptoms in majority of patients within 1 week.

Among inflammatory diseases, in addition to AOSD and sJIA, canakinumab is also indicated for the treatment of CAPS, including FCAS and MWS, in adults and children 4 years of age and older (Ilaris®, US prescribing information 2020).

Rilonacept

Rilonacept (Arcalyst®) is a dimeric fusion protein consisting of the ligand-binding domains of the extracellular portions of the human IL-1 receptor component (IL-1RI) and IL-1 receptor accessory protein (IL-1RAcP) linked in-line to the Fc portion of human IgG1. Rilonacept blocks IL-1β signaling by acting as a soluble decoy receptor that binds IL-1β and prevents its interaction with cell surface receptors. Rilonacept also binds IL-1α and IL-1 receptor antagonist (IL-1RA) with reduced affinity (Arcalyst®, US prescribing information 2021).

The safety and efficacy of rilonacept for the treatment of CAPS have been assessed in a randomized, double-blind, placebo-controlled Phase III trial in patients with FCAS and MWS phenotypes of CAPS. After 6 weeks of treatment, a higher proportion of adult patients in the rilonacept group at the recommended subcutaneous dose regimen (320 mg initially followed by 160 mg every week onward) experienced improvement from baseline in a composite CAPS disease score by at least 30% (96% vs. 29% of patients), by at least

50% (87% vs. 8%), and by at least 75% (70% vs. 0%) compared to the placebo group. Six pediatric patients (12–17 years of age) were enrolled directly into the open-label extension phase of this study, and improvements in symptoms were shown in these patients following rilonacept treatment.

Taking rilonacept with TNF inhibitors is not recommended because this may increase the risk of serious infections. An increased incidence of serious infections has been associated with administration of an IL-1 blocker in combination with TNF inhibitors. IL-1 blockade may interfere with the immune response to infections. Serious, life-threatening infections have been reported in patients taking rilonacept. In an open-label extension study, one patient developed bacterial meningitis and died (Arcalyst®, US prescribing information 2021).

Rilonacept is indicated for the treatment of CAPS, including FCAS and MWS, in adults and children 12 years of age and older (Arcalyst®, US prescribing information 2021).

Deficiency of IL-1 Receptor Antagonist (DIRA)

Deficiency of IL-1 receptor antagonist (DIRA) is a monogenic autoinflammatory disease caused by loss of function mutations in the gene encoding IL-1 receptor antagonist (IL-1Ra) that result in unopposed signaling of the proinflammatory cytokines IL-1α and IL-1β through the IL-1 receptor. DIRA presents with multifocal osteomyelitis and pustulosis and can escalate to life-threatening inflammation, with the development of systemic inflammatory response syndrome (SIRS) and organ failure. DIRA patients are homozygous or compound heterozygous for loss-of-function mutations in *IL1RN*, encoding IL-1Ra. Most mutations are nonsense or frameshift mutations that lead to either no expression of protein or expression of nonfunctional protein. To date, 10 different disease-causing mutations in *IL1RN* have been found, including 4 nonsense mutations, 1 in-frame deletion, 3 frameshift deletions, and a 22-kb and a genomic 175-kb deletion on chromosome 2. The 175-kb deletion seen in the DIRA patients of Puerto Rican descent encompasses IL1RN as well as 5 adjacent genes, including antagonist IL36RN, agonists IL36A, IL36B, and IL36G of the IL-36 receptor, and IL38. DIRA can be and has historically been misdiagnosed as infectious osteomyelitis with pustulosis and systemic inflammation, and therefore treatment with antibiotics is often initiated. However, therapeutic intervention with corticosteroids or, more recently, since the discovery of the genetic cause, IL-1–blocking treatment with anakinra can be lifesaving. IL-1 blockade leads to rapid and durable clinical and inflammatory remission in DIRA patients (Garg et al. 2017).

DIRA has been diagnosed in only a very small number of children. It presents at birth or in the first days of life. Cases have been identified from families originating in Puerto Rico, Newfoundland (Canada), the Netherlands, and Lebanon. DIRA is an autosomal recessive genetic disorder, meaning each child of two carrier parents has a 25% chance of suffering from DIRA.

Anti-IL-1β Biologic Agent

Rilonacept

In addition to CAPS and recurrent pericarditis, rilonacept is also indicated for the maintenance of remission of DIRA in adult and pediatric patients.

The safety and efficacy of rilonacept for the maintenance of remission of DIRA was demonstrated in a 2-year, open-label study of six pediatric patients who previously experienced clinical benefit from daily injections of an IL-1 receptor antagonist, anakinra. The study population included patients with loss-of-function IL1RN mutations. Patients had a median age at baseline of 4.8 years (range 3.3–6.2), and stopped anakinra treatment 24 h before initiation of rilonacept. Remission was defined using the following criteria: diary score of <0.5 (reflecting no fever, skin rash, and bone pain), acute phase reactants (<0.5 mg/dL CRP, C-reactive protein), absence of objective skin rash, and no radiological evidence of active bone lesions. Following a rilonacept loading dose of 4.4 mg/kg subcutaneously, patients received a once-weekly maintenance dose of 2.2 mg/kg (up to a maximum 160 mg), and were assessed for remission and possible dose escalation. During the first 3 months of rilonacept administration at the 2.2 mg/kg dose, five of six patients exhibited recurrence of pustular rash, and therefore the dose was escalated to 4.4 mg/kg once-weekly (up to a maximum of 320 mg). One patient remained on the 2.2 mg/kg once-weekly dose. All patients met the primary endpoint of the study, remission at 6 months, and sustained the remission for the remainder of the 2-year study. No patient required steroid use during the study.

Rilonacept is indicated for the maintenance of remission of DIRA in adults and pediatric patients weighing at least 10 kg (Arcalyst®, US prescribing information 2021).

Eosinophilic Granulomatosis with Polyangiitis

Eosinophilic granulomatosis with polyangiitis (EGPA) is a rare disease characterized by disseminated necrotizing vasculitis with extravascular granulomas occurring exclusively among patients with asthma and tissue eosinophilia (Greco et al. 2015). EGPA usually manifests between 7 and 74 years of age, with a mean age at onset of 38–54 years. The estimated incidence is approximately 0.11–2.66 new cases per one million people per year, with an overall prevalence of 10.7–14 per one million adults. Although still considered an idiopathic condition, EGPA is generally considered a Th2-mediated disease. Recent evidence also points to B cells and the humoral response as further contributors to EGPA's pathogenesis. EGPA patients without poor-prognosis factors are usually treated with glucocorticoids alone, whereas those with a worse prognosis are recommended glucocorticoids and immunosuppressants. Recently, biologic agents such as mepolizumab (anti-IL-5 antibody) are considered promising therapeutic alternatives for EGPA.

Anti-IL-5 Receptor Biologic Agents

Mepolizumab

An overview of mepolizumab, a humanized anti-IL-5 mAb, has been provided earlier in this chapter (Nucala®, US prescribing information 2022). In addition to the use of this therapeutic protein for the treatment of asthma and CRSwNP, mepolizumab has also been evaluated in patients with EGPA. Eosinophils and their mediators are considered to be critical effectors to EGPA. Mepolizumab, by inhibiting IL-5 signaling, reduces the production and survival of eosinophils; however, its mechanism of action in EGPA has not been definitively established (Nucala®, US prescribing information 2022).

In a randomized, placebo-controlled, multicenter study, patients with EGPA received 300 mg mepolizumab or placebo administered subcutaneously once every 4 weeks while continuing their stable OCS therapy. A greater percentage of the patients receiving 300 mg mepolizumab treatment achieved remission versus placebo within the first 24 weeks and remained in remission for the remainder of the 52-week treatment period (19% vs. 1%).

In addition to severe asthma and CRSwNP, mepsolizumab is indicated for the treatment of adult patients with EGPA (Nucala®, US prescribing information 2022).

Generalized Pustular Psoriasis (GPP) Flares

Generalized pustular psoriasis (GPP) is characterized by intermittent flares involving the sudden and widespread formation of sterile pustules with associated erythema and pustules often coalescing to form lakes of pus. Generalized pustular psoriasis flares often occur with systemic symptoms, such as fever, chills, malaise, anorexia, nausea, and severe pain. Moreover, pustules during and after a GPP flare can last for months and lead to life-threatening complica-

tions, often requiring emergency care. Generalized pustular psoriasis can present in patients with a history of plaque psoriasis. However, the pathophysiology and genetic factors that are associated with GPP are distinct from those associated with plaque psoriasis. Most cases of GPP with no history of plaque psoriasis are caused by variants of the *IL36RN* gene, whereas cases of GPP with a prior plaque psoriasis diagnosis are associated with variants of the *CARD14* or *AP1S3* genes.

Flares are a hallmark of GPP and may occur de novo or be provoked by triggers, including withdrawal of systemic corticosteroids, infections, stress, pregnancy, and menstruation. Severity of fares varies widely between patients, and between fares in an individual patient. Significant flares are often accompanied by systemic symptoms, notably fever, general malaise, and extracutaneous manifestations such as arthritis, uveitis, and neutrophilic cholangitis (Choon et al., 2022). Currently spesolimab is the only approved biologic for the treatment of flares with GPP in the USA and EU.

Anti-IL-36 Receptor Biologic Agent

As of November 2022, spesolimab-sbzo (Spevigo®) is the only anti-IL-36 receptor (IL-36R) biologic agent approved for the treatment of generalized pustular psoriasis flares in adults in the USA and EU. Binding of spesolimab-sbzo to IL-36R prevents the subsequent activation of IL-36R by cognate ligands (IL-36 α, β and γ) and downstream activation of pro-inflammatory and pro-fibrotic pathways. The precise mechanism linking reduced IL-36R activity and the treatment of flares of GPP is unclear.

Spesolimab-sbzo

Spesolimab-sbzo (Spevigo®) is a humanized monoclonal IgG1 antibody against human IL-36R produced in CHO cells by recombinant DNA technology (Spevigo®, US prescribing information 2022).

The safety and efficacy of spesolimab-sbzo was assessed in a randomized, double-blind, placebo-controlled study in adult patients with flares of generalized pustular psoriasis (GPP). Subjects received a single intravenous dose of 900 mg spesolimab-sbzo or placebo during the double-blind portion of the study. The primary endpoint of the study was the proportion of subjects with a Generalized Pustular Psoriasis Physician Global Assessment pustulation subscore of 0 (indicating no visible pustules) at Week 1 after treatment. The result showed that 54% subjects achieving a subscore of 0 versus the placebo of 6% subjects.

Spesolimab-sbzo is indicated for the treatment of generalized pustular psoriasis flares in adults (Spevigo®, US prescribing information 2022).

Giant Cell Arteritis

Giant cell arteritis (GCA) is an inflammatory vasculopathy that typically occurs in medium and large arteries with well-developed wall layers and adventitial vasa vasorum (Pradeep and Smith 2018). It is the most common systemic vasculitis in the elderly with an estimated incidence of 27 cases in 100,000 people in those over 50 years old with peak incidence at 70–80 years of age. GCA is a medical emergency which, if left untreated, can result in vision loss. Current standard of care is prompt initiation of glucocorticoid treatment when there is a suspicion of GCA. In most patients with GCA, administration of glucocorticoids can improve signs and symptoms; however, glucocorticoid-related morbidity is a common treatment challenge. When glucocorticoids are tapered, disease flares may occur frequently (an average of one to two episodes per person-year). Recent findings suggested a fundamental failure of T regulatory cell function as a main contributor to GCA's pathogenesis. This represents an opportunity for novel therapeutic medicines as possible glucocorticoid-sparing agents, such as abatacept (a lymphocyte activation inhibitor by targeting CTLA-4) (Langford et al. 2017) and tocilizumab (an IL-6 receptor antagonist). In 2017, tocilizumab (Actemra®) was approved for GCA by FDA (and EMA). This is the first FDA-approved therapy specific to this disorder.

Anti-IL-6 Receptor Agents

Tocilizumab

An overview of tocilizumab, a human anti-IL-6 receptor mAb, has been provided earlier in this chapter. In addition to use of this therapeutic protein for the treatment of RA, pJIA, sJIA, and CRS, tocilizumab is also indicative for the treatment of GCA in adult patients (Actemra®, US prescribing information 2021).

In a randomized, double-blind, placebo-controlled Phase III study in patients with active GCA, treatment with subcutaneous tocilizumab 162 mg weekly and 162 mg every other week (in combination with 26 weeks prednisone taper) resulted in 56 and 53% of GCA patients, respectively, achieving sustained remission (defined as absence of GCA signs and symptoms from Week 12 through Week 52, along with normalization of erythrocyte sedimentation rate [ESR], normalization of C-reactive protein [CRP], and adherence to the prednisone taper regime), while the control groups (placebo with 26- or 52-week prednisone taper) had only 14–18% of patients with sustained remission.

Hidradenitis Suppurativa

Hidradenitis suppurativa (HS) is a chronic, inflammatory, recurrent, debilitating skin disease of the hair follicle that usually presents with painful, deep-seated, inflamed lesions in the apocrine gland-bearing areas of the body, most commonly the axillae, inguinal, and anogenital regions. The prevalence of HS is approximately 1–4% in general population, with an onset after puberty and before age of 40, peaking in the second and third decades of life. HS has the highest impact on patients' quality of life among the overall dermatological diseases, and is associated with multiple comorbidities, including obesity, metabolic syndrome, inflammatory bowel disease, and spondyloarthropathy. The general approach to HS includes non-medical interventions, topical and systemic medications, and surgery. Until now, surgery remains the first-line therapy for HS patients with extensive disease. However, even with extensive surgical intervention HS often recurs. This led to the recent interest in the use of adjunctive biologic therapy in the management of HS (Shanmugam et al. 2017).

The role of immune system in the pathophysiology of HS is being increasingly recognized, and several therapeutic proteins are under investigation. The TNF antagonists of infliximab and adalimumab are considered to be efficacious in the treatment of moderate-to-severe HS, as demonstrated in case series, retrospective studies, and randomized controlled trials (Shanmugam et al. 2017). Some benefits of ustekinumab, an anti-IL12/IL-23 mAb, in HS have been indicated in case reports and a prospective open-label study (Blok et al. 2016). Use of IL-1 receptor antagonists (such as anakinra) in HS had mixed results with some studies showing lack of efficacy, but a recent randomized controlled trial (Tzanetakou et al. 2016) has shown efficacy. IL-17 antagonist therapy is also being investigated in the clinic with mAbs such as bimekizumab and secukinumab. Recently, adalimumab (Humira®) was approved by FDA and EMA for the treatment of moderate-to-severe HS. This is the first biologic medicine approved for HS.

Anti-TNFα Biologic Agents

Adalimumab

An overview of adalimumab, a human anti-TNFα mAb, has been provided earlier in this chapter. In addition to arthritides, psoriasis, CD, and UC (and uveitis as described later in this chapter), adalimumab is also indicated for the treatment of moderate-to-severe HS in adult patients and pediatric patients 12 years or older (Humira®, US prescribing information 2022). Elevated levels of TNFα, an inflammatory cytokine, are seen in blood and skin lesions of patients with HS.

The safety and efficacy of adalimumab for the treatment of HS have been assessed in two randomized, double-blind, placebo-controlled Phase III studies in adult patients with moderate-to-severe HS with Hurley Stage II or III disease and with at least 3 abscesses or inflammatory nodules. Hurley staging system is a three-stage classification system developed for assessing extent and severity of HS, with Stage 0 being no active HS and Stages I–III associated with increased severity. All patients used topical antiseptic wash daily in these two studies. Concomitant oral antibiotic use was allowed in Study HS-II but not Study HS-I. Hidradenitis Suppurativa Clinical Response (HiSCR) was used to assess the treatment effect, which was defined as at least a 50% reduction in total abscess and inflammatory nodule count with no increase in abscess count and no increase in draining fistula count relative to baseline. Twelve-week treatment with subcutaneous adalimumab at the recommended dose regimen (160 mg initial dose, 80 mg 2 weeks later, followed by 40 mg weekly dosing thereafter) resulted in 42 and 59% of HS patients achieving HiSCR in Study HS-I and Study HS-II, respectively, while the control group had only 26 and 28% of patients with HiSCR response in Study HS-I and Study HS-II, respectively.

Lupus Nephritis

Lupus nephritis (LN) is a form of glomerulonephritis and constitutes one of the most severe organ manifestations of SLE. LN is histologically classified into six distinct classes that represent different manifestations and severities of renal involvement in SLE. Most patients with SLE who develop LN do so within 5 years of an SLE diagnosis, although it is not uncommon for LN to develop at later times. SLE is characterized by the development of autoantibodies against nuclear and non-nuclear material, which bind to their targets in the extracellular space and form immune complexes in situ or in the circulation. Some of these antibodies are closely associated with different histological classes of LN. For example, proliferative LN is frequently associated with antibodies against U1 small nuclear ribonucleoprotein or against the podocyte antigens α-enolase and annexin A1. Autoantibodies against proteins that are involved in clearing extracellular nuclear material, such as pentraxin 3 or C-reactive protein, can exacerbate systemic autoimmunity and thus also LN. These autoantibodies promote the persistence of nuclear material in the extracellular space, a process that augments autoantigen presentation and drives the activation of autoantigen-specific adaptive immunity in the kidneys and in other organs (Anders et al., 2020).

LN is typically treated with immunosuppressive drugs, such as glucocorticoids, and cyclophosphamide or mycophenolate mofetil (MMF). However, conventional immunosuppressive treatments are not uniformly effective, and even in patients who respond, 35% may relapse. Furthermore, 5–20% of patients with LN develop end-stage kidney disease (ESKD) within 10 years from the initial event, and drug-induced toxicity remains a concern. Belimumab (Benlysta®), a B-lymphocyte stimulator (BLyS) neutralizing agent, is currently the only approved biologic therapy for LN.

Anti-BLyS Biologic Agent

Belimumab

An overview of belimumab has been provided earlier in this chapter (Benlysta®, US prescribing information 2022). In addition to SLE, the safety and effectiveness of belimumab have been assessed in patients aged 5 years and older with active LN.

The safety and effectiveness of belimumab 10 mg/kg administered intravenously over 1 h on Days 0, 14, 28, and then every 28 days plus standard therapy have been evaluated in a 104-week, randomized, double-blind, placebo-controlled trial in adult patients with active proliferative and/or membranous lupus nephritis. The clinical endpoint Primary Efficacy Renal Response at Week 104 of belimumab plus stand therapy was 43% versus placebo plus stand therapy of 32%, and the Complete Renal Response at Week 104 of belimumab plus stand therapy was 30% versus placebo plus stand therapy of 20%, respectively. In this trial, subjects who received belimumab were significantly less likely to experience a renal-related event or death compared with placebo. In descriptive subgroup analyses of time to renal related event or death, results were consistent with the overall endpoint regardless of induction therapy (mycophenolate or cyclophosphamide), biopsy class (Class III or IV, Class III + V or Class IV + V, or Class V; post-hoc analysis), and baseline proteinuria (<3 g/g or ≥3 g/g; post-hoc analysis). The treatment difference was primarily driven by the renal worsening and renal-related treatment failure components of the endpoint.

Use of belimumab in pediatric patients with active LN is based on the extrapolation of efficacy from the intravenous study in adults with active LN, and supported by pharmacokinetic data from intravenous studies in adults with active LN and from pediatric patients with SLE. Estimated belimumab exposures for pediatric patients were comparable to adults with active LN.

Belimumab is indicated for patients aged 5 years and older with active lupus nephritis who are receiving standard therapy (Benlysta®, US prescribing information 2022).

Multiple Sclerosis

Multiple sclerosis (MS) is a chronic demyelinating disease of the CNS. The main pathological findings in MS are inflammation, demyelination, and axonal degeneration. Inflammation and demyelination are responsible for the acute symptoms of the disease, and axonal degeneration is the underlining cause of the progressive disability associated with MS. Although the etiology of MS remains undetermined, it is considered to be an autoimmune disorder. Blood autoreactive T and B lymphocytes, once activated against myelin constituents, migrate across the blood–brain barrier and initiate inflammatory and demyelinating processes within the CNS leading to MS lesions.

Currently, two therapeutic proteins have been approved for the treatment of MS, including an α4β1/α4β7 integrin antagonist, natalizumab (Tysabri®), and a CD20-directed cytolytic agent, ocrelizumab (Ocrevus®).

Anti-Integrin Biologic Agents

Natalizumab

An overview of natalizumab, a α4β1/α4β7 integrin antagonist, has been provided earlier in this chapter (Tysabri®, US prescribing information 2021). In addition to the use of this therapeutic protein for the treatment of CD, natalizumab (Tysabri®) was the first mAb developed for the treatment of MS. The mechanism of action has been described earlier as an antibody targeting integrins; however, the specific mechanism(s) by which natalizumab exerts its effects in MS has not been fully defined (Tysabri®, US prescribing information 2021).

The safety and efficacy of natalizumab in treatment of MS have been evaluated in two randomized, double-blind, placebo-controlled, multicenter Phase III studies in patients with relapsing forms of MS who had not received any interferon (IFN)-β or glatiramer acetate (Study MS1) (Polman et al. 2006) or patients who had experienced relapses despite INF-β-1a treatment (Study MS2) (Rudick et al. 2006). In Study MS1, 2-years treatment with intravenous natalizumab monotherapy at the recommended intravenous dose regimen (300 mg every 4 weeks), when compared to the placebo treatment, lowered the proportion of patients with increased disability (17% vs. 29%) and the annualized relapse rate (68% reduction). Natalizumab also suppressed the formation of new gadolinium enhancing lesions and reduced the mean number of active lesions based on magnetic resonance imaging (MRI) assessment. In Study MS2, natalizumab or placebo was evaluated in combination with IFN-β-1a. The clinical and MRI-associated efficacies associated with natalizumab treatment were similar to those observed in Study MS1. However, Study MS2 ended 1 month earlier than planned because of the occurrence of PML in two patients receiving natalizumab plus IFN-β-1a.

PML is a demyelinating infectious disease of the CNS caused by reactivation of the John Cunningham virus (JCV). PML may be fatal or result in severe disability. Risk factors for the development of PML include duration of therapy, prior use of immunosuppressants, and presence of anti-JCV antibodies. These factors should be considered in the context of expected benefit when initiating and continuing treatment with natalizumab (Tysabri®, US prescribing information 2017). As of July 2001, 145 cases of PML have been reported among 88,100 patients treated with natalizumab worldwide in the post-marketing setting (Laffaldano et al. 2011).

In addition to CD, natalizumab is also indicated as monotherapy for the treatment of patients with relapsing forms of multiple sclerosis (MS). It is generally recommended for patients who have had an inadequate response to, or are unable to tolerate, an alternate MS therapy. Because of the risk of PML, in the USA, natalizumab is available only through a restricted program under REMS called the TOUCH® Prescribing Program. Natalizumab must be given as monotherapy, and any prior immunomodulator or immunosuppressive therapy must be discontinued prior to use (Tysabri®, US prescribing information 2021).

Anti-CD20 Cytolytic Agents

Ocrelizumab

Ocrelizumab (Ocrevus®) is a humanized CD20-directed cytolytic IgG1 antibody. The precise mechanism by which ocrelizumab exerts its therapeutic effects in MS is unknown, but is presumed to involve binding to CD20, a cell surface antigen present on pre-B and mature B lymphocytes. Following cell surface binding to B lymphocytes, ocrelizumab results in antibody-dependent cellular cytolysis (ADCC) and complement-mediated lysis (Ocrevus®, US prescribing information 2022).

The safety and efficacy of ocrelizumab have been evaluated in patients with relapsing or primary progressive forms of MS. In two double-blind, randomized, active comparator-controlled Phase III trials, treatment of ocrelizumab at the recommended dose regimen (initial treatment of two 300 mg IV infusions administered 2 weeks apart, and subsequent doses administered as a single 600 mg IV infusion every 24 weeks) in patients with relapsing forms of MS resulted in significantly lower annualized relapse rate compared with placebo (47% reduction, as well as the proportion of patients with confirmed disability progression (40% reduction). In a double-blind, randomized, placebo-controlled Phase III trial, treatment of ocrelizumab at the recommended dose regimen in patients with primary progressive form of MS resulted in significantly lower proportion of patients with confirmed disability progression (24% reduction).

Ocrelizumab is indicated for the treatment of patients with relapsing or primary progressive forms of MS (Ocrevus®, US prescribing information 2022).

Neuromyelitis Optica Spectrum Disorder

Neuromyelitis optica spectrum disorder (NMOSD) is an uncommon antibody-mediated disease of the central nervous system. Early diagnosis and treatment are important to reduce the risk of long-term disability and death.

AQP4 is the most widely expressed water channel in the brain, spinal cord, and optic nerves. Experimental data suggest that anti-AQP4 antibodies induce IL-6 production in astrocytes expressing AQP4, and that IL-6 signaling to endothelial cells reduces blood–brain barrier function. Once bound to the extracellular domain of the AQP4 receptor, anti-AQP4 antibodies result in complement- and cell-mediated astrocytic damage, in addition to internalization of the glutamate transporter EAAT-2. The astrocyte is subsequently rendered powerless, ultimately culminating in withdrawal of support for surrounding cells such as oligodendrocytes and neurons. Granulocyte infiltration ensues, matched by oligodendrocyte damage and demyelination.

Currently, two therapeutic proteins have been approved for the treatment of NMOSD who are anti-AQP4 antibody positive, including an IL-6R antagonist, satralizumab-nwge (Enspryng®), and a CD19-directed cytolytic agent, inebilizumab-cdon (Uplizna®).

Anti-IL-6 Receptor Biologic Agents

Satralizumab-nwge

Satralizumab-mwge (Ensprynd®) is a recombinant humanized anti-human IL-6 receptor mAb based on a human IgG2 framework. The precise mechanism by which satralizumab-mwge exerts therapeutic effects in NMOSD is unknown but is presumed to involve inhibition of IL-6-mediated signaling through binding to soluble and membrane-bound IL-6 receptors.

The efficacy of satralizumab-mwge for the treatment of NMOSD in adult patients was established in two placebo-controlled studies. Patients with anti-AQP4 antibody positive and negative were randomized in the studies. In Study 1, the time to the first clinical endpoint committee-confirmed relapse was significantly longer in satralizumab-mwge treated patients compared to patients who received placebo (risk reduction 55%). In the anti-AQP4 antibody

positive population, there was a 74% risk reduction. There was no evidence of a benefit in the anti-AQP4 antibody negative patients. In Study 2, the time to the first confirmed relapse was significantly longer in patients treated with satralizumab-mwge compared to patients who received placebo (risk reduction 62%). In the anti-AQP4 antibody positive population, there was a 78% risk reduction. There was no evidence of a benefit in the anti-AQP4 antibody negative patients.

Satralizumab-mwge is indicated for the treatment of NMOSD in adult patients who are anti-AQP4 antibody positive (Ensprynd®, US prescribing information 2022).

Anti-CD19 Cytolytic Agents

Inebilizumab-cdon

Inebilizumab-cdon (Uplizna®) is a humanized CD20-directed cytolytic IgG1 antibody. The precise mechanism by which inebilizumab-cdon exerts its therapeutic effects in NMOSD is unknown but is presumed to involve binding to CD19, a cell surface antigen present on pre-B and mature B lymphocytes. Following cell surface binding to B lymphocytes, inebilizumab-cdon results in antibody-dependent cellular cytolysis.

The safety and efficacy of inebilizumab-cdon have been evaluated in a randomized (3:1), double-blind, placebo-controlled trial. The patients with anti-AQP4 antibody positive and negative were enrolled in this study. The time to the first adjudicated relapse was significantly longer in patients treated with inebilizumab-cdon compared to patients who received placebo (relative risk reduction 73%; hazard ratio: 0.272). In the anti-AQP4 antibody positive population, there was a 77.3% relative reduction (hazard ratio: 0.227). There was no evidence of a benefit in patients who were anti-AQP4 antibody negative. Compared to placebo-treated patients, patients treated with inebilizumab-cdon who were anti-AQP4 antibody positive had reduced annualized rates of hospitalizations (0.11 for inebilizumab-cdon versus 0.50 for placebo).

Inebilizumab-cdon is indicated for the treatment of NMOSD in adult patients who are anti-AQP4 antibody positive (Uplizna®, US prescribing information 2022).

Recurrent Pericarditis

Recurrent pericarditis (RP) is the most common complication of acute pericarditis and is defined as a repeat episode after a symptom-free interval of at least 4–6 weeks. The recurrence rate after an initial episode of acute pericarditis is approximately 15% to 20% and may increase to as much as

50% after a first recurrent episode. Moreover, 5–10% of patients with recurrent pericarditis develop multiple recurrences that are particularly difficult to treat and often refractory to conventional anti-inflammatory drugs.

Based on current evidence, pericarditis should be considered an inflammatory reaction to various stimuli, including chemical/physical and infectious, with viral infection being a common etiology. Interaction of pathogens or irritants with toll-like receptors leads to an increased transcription of pro-inflammatory genes, including those needed for assembly of the nucleotide-binding domain, leucine-rich-containing family, pyrin domain-containing 3 (NLRP3) inflammasome, which leads to activation of IL-1 (autoinflammatory pathway). This pathogenetic scheme has been confirmed indirectly by the beneficial effect of both colchicine (an indirect inhibitor of the NLRP3 inflammasome) and specific IL-1 blockers in patients with recurrent pericarditis.

Anti-IL-1β Biologic Agents

Rilonacept

An overview of rilonacept, a human anti-IL1β mAb, has been provided earlier in this chapter (Arcalyst®, US prescribing information 2021). In addition to CAPS and DIRA, rilonacept is also indicated for the treatment of recurrent pericarditis in adults and pediatric patients 12 years and older.

The efficacy and safety of rilonacept was evaluated in the Phase III study RHAPSODY, a double-blind, placebo-controlled, randomized withdrawal study. The study consisted of a 12-week run-in followed by a double-blind, placebo-controlled, randomized withdrawal period. In the run-in period, adult patients received a loading dose of rilonacept 320 mg followed by 160 mg weekly. Patients between 12 and 17 years of age received a loading dose of rilonacept 4.4 mg/kg (up to 320 mg) followed by 2.2 mg/kg (up to 160 mg) weekly. During the run-in period, patients tapered and discontinued standard of care therapies. In the withdrawal period, patients were randomized 1:1 to remain on rilonacept 160 mg weekly or to receive placebo.

The mean duration of disease was 2.4 years with a mean of 4.4 pericarditis events per year including the qualifying pericarditis event (0–10 point Numerical Rating Scale [NRS] ≥4 and CRP ≥1 mg/dL). Mean qualifying NRS pain score was 6.2, and mean qualifying CRP level was 6.2 mg/dL. The primary efficacy endpoint was time to first adjudicated pericarditis recurrence (based on pain, CRP, and clinical signs) in the event-driven withdrawal period. Of 61 randomized, 23 patients (74%) in the placebo arm had a recurrence compared with 2 patients (7%) in the rilonacept arm who temporarily discontinued treatment for 1–3 doses. The median time to recurrence on rilonacept could not be estimated because too few events occurred and was 8.6 weeks (95% CI: 4.0, 11.7) on placebo with a hazard ratio of 0.04. Rilonacept reduced the risk of recurrence by 96%.

Rilonacept is indicated for the treatment of recurrent pericarditis and reduction in risk of recurrence in adults and pediatric patients 12 years and older (Arcalyst®, US prescribing information 2021).

Uveitis

Uveitis (UV) is a term that describes a heterogeneous collection of diseases including infections, systemic immune-mediated diseases like sarcoidosis, and immune-mediated syndromes confined to the eye like sympathetic ophthalmia. Uveitis is rare with a prevalence of 115.3 cases per 100,000 people; however, it can damage vital eye tissue, leading to permanent vision loss. Most uveitis is anterior in location, which generally permits successful therapy with topical medication alone. Challenge in the treatment of uveitis relates to patients who have inflammation involving the posterior segment, either primarily in the vitreous (intermediate uveitis), the choroid or retina (posterior uveitis), or involving the entire eye (panuveitis). These patients can have refractory uveitis where systemic corticosteroids or other immunosuppressive therapy are required. Weighting of the risk of blindness and the complications related to these drugs need to be carefully planned (Lin et al. 2014).

Uveitis is considered an immune-mediated disease. A recent literature review suggests that anti-TNFα agents, such as infliximab, adalimumab, and golimumab, are reasonably effective for controlling ocular inflammation and sparing patients corticosteroid treatment in non-infectious refractory uveitis (Borrás-Blasco et al. 2015). Adalimumab (Humira®) is currently the first and only FDA-approved non-corticosteroid therapy for uveitis, though other anti-TNFα agents are also being used "off-label."

Anti-TNFα Biologic Agents

Adalimumab

An overview of adalimumab, a human anti-TNFα mAb, has been provided earlier in this chapter. In addition to arthritides, psoriasis, CD, UC, and HS, adalimumab is also indicated for the treatment of moderate-to-severe uveitis (UV) in adult patients and pediatric patients 2 years of age and older (Humira®, US prescribing information 2022).

The safety and efficacy of adalimumab in treatment of uveitis have been assessed in two randomized, double-masked, placebo-controlled Phase III studies (UV I and UV

II) in adult patients with non-infectious intermediate, posterior, and panuveitis despite corticosteroids therapy. The primary efficacy endpoint was "time to treatment failure." Treatment failure was a multi-component outcome defined as the development of new inflammatory chorioretinal and/or inflammatory retinal vascular lesions, an increase in anterior chamber cell grade or vitreous haze grade, or a decrease in best corrected visual acuity. Statistically significant reductions in the time to treatment failure were demonstrated in patients treated with subcutaneous adalimumab at the recommended dose regimen (80 mg initial dose followed by 40 mg every 2 weeks) versus patients receiving placebo, with hazard ratio of 0.50 (95% CI: 0.36–0.70) and 0.57 (95% CI: 0.39–0.84) in Study UV I and Study UV II, respectively.

The use of adalimumab for the treatment of non-infectious uveitis in pediatric patients 2 years of age and older is supported by evidence from the two adequate and well-controlled studies in adults and a 2:1 randomized, controlled clinical study in 90 pediatric patients from 2 to <18 years of age. At the recommended dose regimen (20 mg adalimumab for <30 kg or 40 mg for ≥30 kg SC Q2W), adalimumab significantly decreased the risk of treatment failure by 75% relative to placebo in these pediatric UV patients (hazard ratio = 0.25 [95% CI: 0.12, 0.49]).

Conclusion

The introduction of more than 30 antibody-based biotherapeutics in the last decades has fundamentally changed the treatment paradigm in immune-mediated inflammatory diseases such as RA, IBD, and psoriasis. Though these "targeted biotherapies" are expensive compared to traditional "small molecular" therapies such as methotrexate, they have provided effective treatment alternatives with highly specific targeted novel mechanisms of action. Some of these biotherapeutics can not only provide relief of symptoms but also offer an opportunity to modify or even reverse the course of these diseases, as has been demonstrated in RA. Notably, despite the remarkable clinical improvement in the treatment of inflammatory diseases using antibody-based biotherapeutics, there is still unmet medical need in achieving "permanent cure" for these complex multifactorial disorders. Inflammatory diseases such as RA and IBD are not the products of dysfunction in isolated individual entities or linear pathways; they arise from perturbations in complex dynamic networks that shift from patterns representing normal function to others that give rise to disease. Therefore, treatment of the disease is unlikely to be resolved using a one-size-fits-all approach by targeting a single target. It is important to understand how individual biological components interact for a given patient, which could then be used to guide personalization of therapies and the segmentation or stratification of treatment populations. With the further advance in

protein engineering technology and better understanding of the etiology and disease progression of immune-mediated inflammatory diseases, equipped with more predictive and diagnostic biomarkers, it is reasonable to anticipate that more effective and safer "targeted biotherapeutics" tailored to the individual patients' needs will be added to the therapeutic armory to successfully treat inflammatory diseases. It should also be noted that "new" small molecule therapies in immune-mediated inflammatory disease areas are evolving in the last few years, such as Janus kinase inhibitors (JAKi), Tyrosine Kinase 2 inhibitors (TKY2i), etc. Compared with therapeutic proteins, small molecule drugs have often the advantages of more convenient routes of administration (oral) and being more economically sustainable due to lower costs. However, potential off-target effects remain a concern. In this regard, antibody drug conjugates could be a good approach, in particular for the "difficult-to-treat" patient populations.

Self-Assessment Questions

Questions

1. Are targeted biologic therapies for autoimmune diseases only to be used once drugs like corticosteroids and methotrexate have had an adequate trial of use and have failed to control the patient's symptoms?
2. What is the primary clinical concern with the immunogenicity of biologic therapies?
3. What is the most likely explanation for why a patient who receives a dose of omalizumab might have an increase in their total serum IgE level for many weeks after the first dose?
4. Why do some cell subsets in the peripheral blood increase after dosing with natalizumab?
5. If a trial reports an ACR70 of 20% on active drug, what does that mean?
6. What does PASI 75 mean?
7. Given that there are currently five anti-TNFα biotherapeutic agents on the market, how would you compare and contrast them?
8. What are the key differences in the indication for use of rituximab vs. abatacept in RA?
9. What is the mechanism of action for guselkumab in treating plaque psoriasis?

Answers

1. Though the standard of care in diseases like RA is still to start with older DMARDs like methotrexate, the decision of when to start or switch therapies is complex and impacted by individual issues linked to clinical response like tolerance/adherence to a particular therapeutic regimen, severity and course of disease and its progression, and concomitant medications and medical issues. It is

likely that the standard of care will continue to change and incorporate earlier use of biologic therapies that can modify the disease course with fewer generalized side effects.

2. If a biologic therapy is highly immunogenic, there is a concern that an increasing number of patients exposed to the drug, particularly upon repeated exposure after a hiatus, because their antidrug antibodies could sometimes neutralize the majority of the drug and they would not likely get the full dose or effect. Though less likely there are also rare examples of antidrug antibodies resulting in an autoimmune or allergic-type reaction.

3. Therapeutic monoclonal antibodies that target soluble molecules like IgE form complexes. Though immune complexes are typically cleared from the blood more quickly than monomeric IgG, soluble target molecules typically have a shorter serum half-life than IgG. So an assay detecting the soluble target (in this case IgE) that can detect target even when it is bound to the drug (which is typically a longer-lived IgG) will show more target present in the serum post-dosing as compared to baseline. This is called a carrier effect. Assuming the drug neutralized the bound target, the test detecting the target can be misleading, because the target, though present, is effectively inactive.

4. Natalizumab is a monoclonal antibody that blocks lymphocyte movement between the blood and tissues ("trafficking"); when this movement is effectively blocked in one direction (from the blood into the tissues), an apparent increase in the peripheral lymphocyte population will be evident on assessment by flow cytometry (or perhaps even on a CBC with differential) post-dosing.

5. An ACR70 of 20% means the 20% of the patients had a 70% improvement in their RA disease.

6. The PASI score stands for Psoriasis Area and Severity Index. This tool allows researchers and dermatologists to put an objective number on what would otherwise be a very subjective idea: how bad is a person's psoriasis. The PASI evaluates the degree of erythema, thickness, and scaling of psoriatic plaques and estimates the extent of involvement of each of these components in four separate body areas (head, trunk, upper, and lower extremities).

   If in a clinical study a certain proportion of patients experienced a 75% reduction in their PASI scores, it is reported as a percentage of people achieving "PASI 75."

7. Although the five anti-TNFα biologics have broadly similar efficacy and safety profiles in RA, there are significant differences in the five anti-TNFα agents particularly with respect to dosing characteristics and also in the details of the approved indications for use.

Infliximab is the first approved anti-TNFα agent given intravenously, has the longest dosing interval, and is the first FDA-approved anti-TNFα agent for IBD indication. All the other anti-TNFα agents are administered by subcutaneous administration. It is a chimeric monoclonal antibody that neutralizes TNF-α and has approvals in the most indications (rheumatoid arthritis, psoriatic arthritis, ankylosing spondylitis, adult and pediatric ulcerative colitis, adult and pediatric Crohn's disease, and psoriasis).

Etanercept is a dimeric soluble fusion protein and has approvals for use in several indications (rheumatoid arthritis, polyarticular juvenile rheumatoid arthritis, psoriatic arthritis, ankylosing spondylitis, and psoriasis). It is used as weekly injection.

Adalimumab is a human monoclonal antibody that neutralizes TNF-α and has FDA approvals for use in patients with rheumatoid arthritis, psoriatic arthritis, ankylosing spondylitis, psoriasis, Crohn's disease, pediatric Crohn's disease, ulcerative colitis, hidradenitis suppurative, and uveitis. It is used at a frequency of every week or every other week.

Golimumab is a human monoclonal antibody that neutralizes TNF-α and has FDA approvals for use in patients with rheumatoid arthritis, psoriatic arthritis, and ankylosing spondylitis, and ulcerative colitis via subcutaneous route of administration. It is used at a frequency of every month. It can also be given via intravenous route of administration with a frequency of every 8 weeks for the treatment of rheumatoid arthritis.

Certolizumab pegol is a recombinant, humanized antibody Fab fragment, with specificity for human tumor necrosis factor alpha (TNFα), conjugated to an approximately 40-kDa polyethylene glycol (PEG2MAL40K). It has been approved by FDA for the treatment of rheumatoid arthritis and Crohn's disease.

8. Rituximab in combination with methotrexate is indicated for the treatment of adult patients with moderate-to-severe rheumatoid arthritis who have had an inadequate response to one or more TNF antagonist therapies. Abatacept is indicated for use as monotherapy or in combination with DMARDS in patients with moderate-to-severe active rheumatoid arthritis who have had an inadequate response to DMARDs or TNF antagonists.

9. IL-23 is a naturally occurring cytokine that is involved in inflammatory and immune responses, such as natural killer cell (NK) activation and CD4+ T-cell differentiation and activation. Guselkumab, a human IgG1λ monoclonal antibody that binds with high affinity and specificity to the p19 protein subunit used by IL-23, can prevent human IL-23 from binding to the IL-23 (IL-12Rβ1/23R) receptor complexes on the surface of NK and T cells.

References

Actemra (tocilizumab) (2021) US prescribing information. Genentech Inc., South San Francisco, CA

Adbry (tralokinumab-idrm) (2022) US prescribing information In: LEO Pharma Inc., Madison, NJ

Adedokun OJ, Xu Z, Marano CW et al (2017) Pharmacokinetics and exposure-response relationship of golimumab in patients with moderately-to-severely active ulcerative colitis: results from phase 2/3 PURSUIT induction and maintenance studies. J Crohns Colitis 11:35–46

Airoldi I, Di Carlo E, Cocco C et al (2005) Lack of Il12rb2 signaling predisposes to spontaneous autoimmunity and malignancy. Blood 106:3846–3853

Anders HJ, Saxena R, Zhao MH et al (2020) Lupus nephritis. Nat Rev Dis Primers 6(1):7

Arcalyst (rilonacept) (2021) US prescribing information. Regeneron Pharmaceuticals Inc., Tarrytown, NY

Baumgart DC, Sandborn WJ (2007) Inflammatory bowel disease: clinical aspects and established and evolving therapies. Lancet 369:1641–1657

Benlysta (Belimumab) (2022) US prescribing information. Human Genome Sciences, Inc., (a subsidiary of GlaxoSmithKline), Rockville, MD

Blauvelt A, Reich K, Tsai TF et al (2017) Secukinumab is superior to ustekinumab in clearing skin of subjects with moderate-to-severe plaque psoriasis up to 1 year: results from the CLEAR study. J Am Acad Dermatol 76(60–69):e9

Blok JL, Li K, Brodmerkel C, Horvátovich P, Jonkman MF, Horváth B (2016) Ustekinumab in hidradenitis suppurativa: clinical results and a search for potential biomarkers in serum. Br J Dermatol 174:839–846

Boguniewicz M (2017) Biologic therapy for Atopic dermatitis: moving beyond the practice parameter and guidelines. J Allergy Clin Immunol Pract 5:1477–1487

Bongartz T, Sutton AJ, Sweeting MJ et al (2006) Anti-TNF antibody therapy in rheumatoid arthritis and the risk of serious infections and malignancies: systematic review and meta-analysis of rare harmful effects in randomized controlled trials. JAMA 295:2275–2285

Bonifant CL, Jackson HJ, Brentjens RJ, Curran KJ (2016) Toxicity and management in CAR T-cell therapy. Mol Ther Oncolytics 3:16011

Borrás-Blasco J, Casterá DE, Cortes X, Abad FJ, Rosique-Robles JD, Mallench LG (2015) Effectiveness of infliximab, adalimumab and golimumab for non-infectious refractory uveitis in adults. Int J Clin Pharmacol Ther 53:377–390

Brown SL, Greene MH, Gershon SK et al (2002) Tumor necrosis factor antagonist therapy and lymphoma development: twenty-six cases reported to the Food and Drug Administration. Arthritis Rheum 46:3151–3158

Chaudhary R, Butler M, Playford RJ, Ghosh S (2006) Anti-TNF antibody induced stimulated T lymphocyte apoptosis depends on the concentration of the antibody and etanercept induces apoptosis at rates equivalent to infliximab and adalimumab at 10 micrograms per ml concentration. Gastroenterology 130(Suppl 2):Abstract A696

Choon SE, Navarini AA, Pinter A (2022) Clinical course and characteristics of generalized pustular psoriasis. Am J Clin Dermatol 23(Suppl 1):21–29

Christl LA, Woodcock J, Kozlowski S (2017) Biosimilars: the US regulatory framework. Annu Rev Med 68:243–254

Christophers E (2001) Psoriasis—epidemiology and clinical spectrum. Clin Exp Dermatol 26:314–320

Cimzia (certolizumab pegol) (2019) US prescribing information. UCB Inc., Smyrna, GA

Cinqair (reslizumab) (2020) US prescribing information. Teva Pharmaceutical Industries Ltd., Frazer, PA

Cooke A, Bulkhi A, Casale T (2015) Role of biologics in intractable urticaria. Biologics 9:25–33

Cosentyx (secukinumab) (2021) US prescribing information. Novartis Pharmaceuticals Corporation, East Hanover, NJ

Davis JC, Mease PJ (2008) Insights into the pathology and treatment of spondyloarthritis: from the bench to the clinic. Semin Arthritis Rheum 38:83–100

D'Haens G, Baert F, van Assche G et al (2008) Early combined immunosuppression or conventional management in patients with newly diagnosed Crohn's disease: an open randomised trial. Lancet 371:660–667

Dunleavy K (2021) The top 20 drugs by worldwide sales in 2021. Fierce Pharm. fiercepharma.com/special-reports/top-20-drugs-worldwide-sales-2021

Dupixent (dupilumab) (2022) US prescribing information. Sanofi-aventis U.S. LLC/Regeneron Pharmaceuticals, Inc., Bridgewater, NJ/Tarrytown, NY

Ehrlich P (1891) Experimentelle untersuchungen über immunität. I Ueber Ricin. Dtsch Med Wochenschr 17:976–979

Ellerin T, Rubin RH, Weinblatt ME (2003) Infections and anti-tumor necrosis factor alpha therapy. Arthritis Rheum 48:3013–3022

Enbrel (etanercept) (2022) US prescribing information. Immunex Corporation, Thousand Oaks, CA

Ensprynd® (satralizumab-mwge) (2022) US prescribing information In: Genentech, Inc., South San Francisco, CA

Entyvio (vedolizumab) (2022) US prescribing information. Takeda Pharmaceuticals, Deerfield, IL

Fasenra (benralizumab) (2021) US prescribing information. AstraZeneca Pharmaceuticals LP, Wilmington, DE

Feldman SR, Krueger GG (2005) Psoriasis assessment tools in clinical trials. Ann Rheum Dis 64(Suppl 2):ii65–ii68

Garg M, de Jesus AA, Chapelle D et al (2017) Rilonacept maintains long-term inflammatory remission in patients with deficiency of the IL-1 receptor antagonist. JCI Insight 2(16):e94838

Gold R, Jawad A, Miller DH et al (2007) Expert opinion: guidelines for the use of natalizumab in multiple sclerosis patients previously treated with immunomodulating therapies. J Neuroimmunol 187:156–158

Gomez-Garcia F, Epstein D, Isla-Tejera B et al (2017) Short-term efficacy and safety of new biological agents targeting the interleukin-23-T helper 17 pathway for moderate-to-severe plaque psoriasis: a systematic review and network meta-analysis. Br J Dermatol 176:594–603

Greco A, Rizzo MI, De Virgilio A et al (2015) Churg-Strauss syndrome. Autoimmun Rev 14:341–348

Griffiths CE, Barker JN (2007) Pathogenesis and clinical features of psoriasis. Lancet 370:263–271

Griffiths CE, Strober BE, van de Kerkhof P et al (2010) Comparison of ustekinumab and etanercept for moderate-to-severe psoriasis. N Engl J Med 362:118–128

Gura T (2002) Therapeutic antibodies: magic bullets hit the target. Nature 417:584–586

Haldar P, Brightling CE, Hargadon B et al (2009) Mepolizumab and exacerbations of refractory eosinophilic asthma. N Engl J Med 360:379–384

Hauser SL, Waubant E, Arnold DL et al (2008) B-cell depletion with rituximab in relapsing-remitting multiple sclerosis. N Engl J Med 358:676–688

Hawkins PN, Lachmann HJ (2003) Interleukin-1-receptor antagonist in the Muckle–Wells syndrome. N Engl J Med 348:2583–2584

Hochhaus G, Brookman L, Fox H et al (2003) Pharmacodynamics of omalizumab: implications for optimised dosing strategies and clinical efficacy in the treatment of allergic asthma. Curr Med Res Opin 19:491–498

Hoentjen F, van Bodegraven AA (2009) Safety of anti-tumor necrosis factor therapy in inflammatory bowel disease. World J Gastroenterol 15:2067–2073

Humira (adalimumab) (2022) US prescribing information. AbbVie Inc., North Chicago, IL

Ilaris (canakinumab) (2020) US prescribing information. Novartis Pharmaceuticals Corporation, East Hanover, NJ

Ilumya (tildrakizumab-asmn) (2020) US prescribing information In: Sun Pharmaceutical Industries, Inc., Cranbury, NJ

Jabbar-Lopez ZK, Yiu ZZN, Ward V et al (2017) Quantitative evaluation of biologic therapy options for psoriasis: a systematic review and network meta-analysis. J Invest Dermatol 137:1646–1654

Kabashima K, Irie H (2021) Interleukin-31 as a clinical target for pruritus treatment. Front Med (Lausanne) 8:638325

Kevzara (sarilumab) (2018) US prescribing information. Sanofi-aventis U.S. LLC/Regeneron Pharmaceuticals, Inc., Bridgewater, NJ/Tarrytown, NY

Khattri S, Brunner PM, Garcet S et al (2017) Efficacy and safety of ustekinumab treatment in adults with moderate-to-severe atopic dermatitis. Exp Dermatol 26:28–35

Kraft M, Worm M (2017) Dupilumab in the treatment of moderate-to-severe atopic dermatitis. Expert Rev Clin Immunol 13:301–310

Krueger GG (2003) Clinical response to alefacept: results of a phase 3 study of intravenous administration of alefacept in patients with chronic plaque psoriasis. J Eur Acad Dermatol Venereol 17(Suppl 2):17–24

Kubota T, Koike R (2010) Cryopyrin-associated periodic syndromes: background and therapeutics. Mod Rheumatol 20:213–221

Laffaldano P, Lucchese G, Trojano M (2011) Treating multiple sclerosis with natalizumab. Expert Rev Neurother 11:1683–1692

Langford CA, Cuthbertson D, Ytterberg SR et al (2017) A randomized, double-blind trial of abatacept (CTLA-4Ig) for the treatment of Giant cell arteritis. Arthritis Rheumatol 69:837–845

Langley RG (2012) Effective and sustainable biologic treatment of psoriasis: what can we learn from new clinical data? J Eur Acad Dermatol Venereol 26(Suppl 2):21–29

Langley RG, Elewski BE, Lebwohl M et al (2014) Secukinumab in plaque psoriasis—results of two phase 3 trials. N Engl J Med 371:326–338

Lebwohl M, Christophers E, Langley R et al (2003) An international, randomized, double-blind, placebo-controlled phase 3 trial of intramuscular alefacept in patients with chronic plaque psoriasis. Arch Dermatol 139:719–727

Leonardi CL, Kimball AB, Papp KA et al (2008) Efficacy and safety of ustekinumab, a human interleukin-12/23 monoclonal antibody, in patients with psoriasis: 76-week results from a randomised, double-blind, placebo-controlled trial (PHOENIX 1). Lancet 371:1665–1674

Lin P, Suhler EB, Rosenbaum JT (2014) The future of uveitis treatment. Ophthalmology 121:365–376

Mallipeddi R, Grattan C (2007) Lack of response of severe steroid-dependent chronic urticaria to rituximab. Clin Exp Dermatol 32:333–334

Maurer M, Weller K, Bindslev-Jensen C et al (2011) Unmet clinical needs in chronic spontaneous urticaria. A GA2LEN task force report. Allergy 66:317–330

McInnes IB, Schett G (2011) The pathogenesis of rheumatoid arthritis. N Engl J Med 365:2205–2219

Modigliani R, Mary JY, Simon JF et al (1990) Clinical, biological, and endoscopic picture of attacks of Crohn's disease: evolution on prednisolone. Gastroenterology 98:811–818

Molodecky NA, Soon IS, Rabi DM et al (2012) Increasing incidence and prevalence of the inflammatory bowel diseases with time, based on systematic review. Gastroenterology 142(46–54):e42

Nestle FO, Kaplan DH, Barker J (2009) Psoriasis. N Engl J Med 361:496–509

Nucala (mepolizumab) (2022) US prescribing information. GlaxoSmithKline LLC, Philadelphia, PA

O'Dell JR (2004) Therapeutic strategies for rheumatoid arthritis. N Engl J Med 350:2591–2602

Orencia (abatacept) (2021) US prescribing information. Bristol-Myers Squibb Company, Princeton, NJ

Ocrevus (ocrelizumab) (2022) In: South San Francisco, CA (ed) US prescribing information. Genentech, Inc. A Member of the Roche Group

Papp KA, Langley RG, Lebwohl M et al (2008) Efficacy and safety of ustekinumab, a human interleukin-12/23 monoclonal antibody, in patients with psoriasis: 52-week results from a randomised, double-blind, placebo-controlled trial (PHOENIX 2). Lancet 371:1675–1684

Papp K, Gulliver W, Lynde C et al (2011) Canadian guidelines for the management of plaque psoriasis. J Cutan Med Surg 15:210–219

Polman CH, O'Connor PW, Havrdova E et al (2006) A randomized, placebo-controlled trial of natalizumab for relapsing multiple sclerosis. N Engl J Med 354:899–910

Pradeep S, Smith JH (2018) Giant cell Arteritis: practical pearls and updates. Curr Pain Headache Rep 22:2

Reich K, Burden AD, Eaton JN, Hawkins NS (2012) Efficacy of biologics in the treatment of moderate to severe psoriasis: a network meta-analysis of randomized controlled trials. Br J Dermatol 166:179–188

Remicade (infliximab) (2021) US prescribing information. Janssen Biotech Inc., Horsham, PA

Rituxan (rituximab) (2021) US prescribing information. Genentech Inc., South San Francisco, CA

Rudick RA, Stuart WH, Calabresi PA et al (2006) Natalizumab plus interferon-β-1a for relapsing multiple sclerosis. N Engl J Med 354:911–923

Ruzicka T, Hanifin JM, Furue M et al (2017) Anti-interleukin-31 receptor a antibody for atopic dermatitis. N Engl J Med 376:826–835

Saeki H, Kabashima K, Tokura Y et al (2017) Efficacy and safety of ustekinumab in Japanese patients with severe atopic dermatitis: a randomised, double-blind, placebo-controlled, phase II study. Br J Dermatol 177:419–427

Saif H, Dan W, Maneesh B et al (2019) Neuromyelitis optica spectrum disorders. Clin Med (Lond) 19(2):169–176

Salliot C, Finckh A, Katchamart W et al (2011) Indirect comparisons of the efficacy of biological antirheumatic agents in rheumatoid arthritis in patients with an inadequate response to conventional disease-modifying antirheumatic drugs or to an anti-tumour necrosis factor agent: a meta-analysis. Ann Rheum Dis 70:266–271

Sandborn WJ, Feagan BG, Stoinov S et al (2006) Certolizumab pegol administered subcutaneously is effective and well tolerated in patients with active Crohn's disease: results from a 26-week, placebo-controlled phase III study (PRECiSE 1). Gastroenterology 130:A-107

Saphnelo (anifrolumab-fnia) (2022) US prescribing information In: AstraZeneca Pharmaceuticals LP, Wilmington, DE.

Shanmugam VK, Zaman NM, McNish S, Hant FN (2017) Review of current immunologic therapies for hidradenitis suppurativa. Int J Rheumatol 2017:8018192. https://doi.org/10.1155/2017/8018192

Siliq (brodalumab) (2020) US prescribing information. Valeant Pharmaceuticals North America LLC, Bridgewater, NJ

Simponi (golimumab) (2021) US prescribing information. Janssen Biotech Inc., Horsham, PA

Simponi Aria (golimumab) (2021) US prescribing information. Janssen Biotech Inc., Horsham, PA

Silverberg J, Cork M, Wollenberg A, et al (2020) Early changes in patient-relevant endpoints in three tralokinumab pivotal phase 3 trials (ECZTRA 1-3) in adults patients with moderate-to-severe atopic dermatitis. SKIN Cutan Med 4(6):S98. https://doi.org/10.25251/skin.4.supp.98

Skyrizi (risankizumab-rzaa) (2022) US prescribing information In: AbbVie, Inc., North Chicago, IL.

Smedby KE, Askling J, Mariette X et al (2008) Autoimmune and inflammatory disorders and risk of malignant lymphomas–an update. J Intern Med 264:514–527

Soler D, Chapman T, Yang LL, Wyant T, Egan R, Fedyk ER (2009) The binding specificity and selective antagonism of vedolizumab, an anti-alpha4beta7 integrin therapeutic antibody in development for inflammatory bowel diseases. J Pharmacol Exp Ther 330:864–875

Spevigo® (spesolimab-sbzo) (2022) US prescribing information In: Boehringer Ingelheim Pharmaceuticals, Inc., Ridgefield, CT

Stelara (ustekinumab) (2022) US prescribing information. Janssen Biotech Inc., Horsham, PA

Stevens WW, Schleimer RP, Kern RC (2016) Chronic rhinosinusitis with nasal polyps. J Allergy Clin Immunol Pract 4(4):565–572

Taltz (ixekizumab) (2022) US prescribing information. Eli Lilly and Company, Indianapolis, IN

Taylor PC, Steuer A, Gruber J et al (2004) Comparison of ultrasonographic assessment of synovitis and joint vascularity radiographic evaluation in a randomized, placebo-controlled study of infliximab therapy in early rheumatoid arthritis. Arthritis Rheum 50:1107–1116

Tezspire® (tezepelumab-ekko) (2021) US prescribing information In: Amgen, Inc., Thousands Oaks, CA

Tracey D, Klareskog L, Sasso EH et al (2008) Tumor necrosis factor antagonist mechanisms of action: a comprehensive review. Pharmacol Ther 117:244–279

Tremfya (guselkumab) (2022) US prescribing information. Janssen Biotech Inc., Horsham, PA

Tysabri® (natalizumab) (2021) US prescribing information. Biogen Idec Inc, Cambridge, MA

Tzanetakou V, Kanni T, Giatrakou S et al (2016) Safety and efficacy of anakinra in severe hidradenitis suppurativa: a randomized clinical trial. JAMA Dermatol 152:52–59

Uplizna® (inebilizumab-cdon) (2022) US prescribing information In: Horizon Therapeutics, Deerfield, IL

Van den Brande JM, Braat H, van den Brink GR et al (2003) Infliximab but not etanercept induces apoptosis in lamina propria T-lymphocytes from patients with Crohn's disease. Gastroenterology 124:1774–1785

Vestergaard C, Deleuran M (2015) Chronic spontaneous urticaria: latest developments in aetiology, diagnosis and therapy. Ther Adv Chronic Dis 6:304–313

Wollenberg A, Blauvelt A, Guttman-Yassky E, et al (2021) Tralokinumab for moderate-to-severe atopic dermatitis: results from two 52-week, randomized, double-blind, multicentre, placebo-controlled phase III trials (ECZTRA 1 and ECZTRA 2). Br J Dermatol. https://doi.org/10.1111/bjd.19754

Xolair (omalizumab) (2021) US prescribing information. Genentech Inc., South San Francisco, CA

Further Reading

Lagassé HA, Alexaki A, Simhadri VL, Katagiri NH, Jankowski W, Sauna ZE, Kimchi-Sarfaty C (2017) Recent advances in therapeutic protein drug development. F1000Res 6:113. https://doi.org/10.12688/f1000research.9970.1 eCollection 2017

Murphy K, Weaver C (2016) Janeway's immunology. Garland Science. 9th Edition. ISBN: 978-0815345053

Zhou H, Theil F-P (2015) ADME and translational pharmacokinetics/pharmacodynamics of therapeutic proteins. John Wiley and Sons, Inc, Hoboken, NJ. ISBN: 978-1-118-89864-2

Monoclonal Antibodies in Solid Organ Transplantation

Nicole A. Pilch, Holly B. Meadows, and Rita R. Alloway

Introduction

Administration of targeted immunosuppression, in the form of genetically engineered antibodies, is commonplace in solid organ transplantation. Polyclonal antibodies, such as rabbit antithymocyte globulin, offer global immunosuppression by targeting several cell surface antigens on B and T lymphocytes. However, secondary to their broad therapeutic targets, they are associated with infection, infusion-related reactions, inter-batch variability, and posttransplant malignancies. Nevertheless, polyclonal antibodies are still commonly administered for induction and treatment of allograft rejection and offer an important role in current solid organ transplantation, which is beyond the scope of this chapter.

In an attempt to target solid organ transplant immunosuppression, monoclonal antibodies directed against key steps in specific immunologic pathways were introduced. The first agent, muromonab-CD3 (OKT3), was initially introduced in the early 1980s for the treatment of allograft rejection (Morris 2004). The use of monoclonal antibodies has evolved and expanded over the past three decades, and today monoclonal antibodies are routinely included as part of the overall immunosuppression regimen. Both the innate and adaptive immune systems have multiple components and signal transduction pathways aimed at protecting the host from a foreign body, such as transplanted tissue. The ultimate goal of posttransplant immunosuppression is tolerance, a state in which the host immune system recognizes the foreign tissue but does not react to it. This goal has yet to be achieved under modern immunosuppression secondary to immune system redundancy as well as the toxicity of currently available agents. Therefore, monoclonal antibodies are used to provide targeted, immediate immunomodulation aimed at attenuating the overall immune response. Specifically, monoclonal antibodies have been used to (1) decrease the inherent immunoreactivity of the potential transplant recipient prior to engraftment, (2) induce global immunosuppression at the time of transplantation allowing for modified introduction of other immunosuppressive agents (calcineurin inhibitors or corticosteroids), (3) spare exposure to maintenance immunosuppressive agents, and (4) treat acute allograft rejection. Monoclonal antibody selection, as well as dose, is based on patient-specific factors, such as indication for transplantation, type of organ being transplanted, and the long-term immunosuppression objective. To understand the approach that the transplant clinician uses to determine which agent to administer and when, it is necessary to briefly describe how immunoreactivity can be predicted and review the immunological basis for the use and development of monoclonal antibodies in solid organ transplantation.

Immunologic Targets: Rational Development/Use of Monoclonal Antibodies in Organ Transplant

The rational use of monoclonal antibodies in transplantation is focused on the prevention of host immune recognition of donor tissue. There are two ways in which allograft tissue can be immediately impaired secondary to the host immune response: complement-dependent antibody-mediated cell lysis (antibody-mediated rejection) and T-cell-mediated parenchymal destruction leading to localized allograft inflammation and arteritis (cellular-mediated rejection) (Halloran 2004). Pre-transplant screening for antibodies against donor tissues has significantly reduced

N. A. Pilch (✉)
Department of Pharmacy Practice and Outcomes Sciences, Medical University of South Carolina, Charleston, SC, USA
e-mail: weimert@musc.edu

H. B. Meadows
Department of Pharmacy and Clinical Sciences, South Carolina College of Pharmacy, Medical University of South Carolina, Charleston, SC, USA

R. R. Alloway
Division of Nephrology, Department of Internal Medicine, University of Cincinnati, Cincinnati, OH, USA
e-mail: rita.alloway@uc.edu

© The Author(s), under exclusive license to Springer Nature Switzerland AG 2024
D. J. A. Crommelin et al. (eds.), *Pharmaceutical Biotechnology*, https://doi.org/10.1007/978-3-031-30023-3_25

the incidence and severity of antibody-mediated rejection. However, as will be discussed, preferential destruction of cells that produce these antibodies using monoclonal technology, such as rituximab, prior to transplant has become an option for recipients with preformed alloantibodies. Thus, the use of monoclonal antibodies to target B-cells has become a focus of transplant clinicians as a way to overcome traditional immunologic barriers pre and posttransplant. However, prevention and treatment of cellular-mediated rejection has traditionally been the main focus of maintenance immunosuppression and the rationale for use of monoclonal antibodies in the posttransplant period. Cellular-mediated rejection is characterized by initial recognition of donor tissue by T cells. This leads to a complex signal transduction pathway traditionally described as three signals (Halloran 2004):

- Signal 1: Donor antigens are presented to T cells leading to activation, characterized by T-cell proliferation.
- Signal 2: CD80 and CD86 complex with CD28 on the T-cell surface activating signal transduction pathways (calcineurin, mitogen-activated protein kinase, protein kinase C, nuclear factor kappa B) which leads to further T-cell activation, cytokine release, and expression of the interleukin-2 (IL2) receptor (CD25).
- Signal 3: IL-2 and other growth factors cause the activation of the cell cycle and T-cell proliferation (Halloran 2004).

Monoclonal antibodies have been developed against various targets within this pathway to prevent propagation and

lymphocyte proliferation providing profound immunosuppression (Table 25.1). Monoclonal antibodies that were originally developed for treatment of various malignancies have also been employed as immunosuppressant agents in solid organ recipients. Use of these agents must be balanced with maintenance immunosuppression to minimize the patient's risk of infection or malignancy from over-immunosuppression. Table 25.2 describes common adverse effects associated with maintenance immunosuppressant medications. Table 25.3 summarizes trends regarding the use of monoclonal antibodies for induction immunosuppression in solid organ transplantation. A plethora of factors influence the trends towards more frequent induction therapy use with T-cell depleting agents all focused on increasing accessibility to transplantation and improving long-term outcomes.

Monoclonal Antibodies Administered Pre-transplant

Immunologic barriers to solid organ transplantation are common. Improved management of end-stage organ disease has increased the number of potential organ recipients and produced a significant shortage of organs available for transplant in comparison to the growing demand. Therefore, clinicians have sought to transplant across previously contraindicated immunologic barriers. In addition, more patients are surviving through their first transplant and are now waiting for a subsequent transplant. Monoclonal antibodies are now being employed prior to transplant to desensitize the recipient's immune system. Desensitization is a strategy where immunosuppression is administered prior to transplant to prevent hyperacute or early rejection in patients who

Table 25.1 Use of monoclonal antibodies in solid organ transplantation

Monoclonal antibody	Molecular weight	Animal epitope	Molecular target	Target cells	Use	Currently available
Alemtuzumab (Campath-1H®)	150 kDa	Murine/human	CD52	Peripheral blood lymphocytes, natural killer cells, monocytes, macrophages, thymocytes	Induction	Yes[a]
					Antibody-mediated rejection	
Daclizumab (Zenapax®)	14.4 kDa	Murine/human	CD25 alpha subunit	IL2-dependent T-lymphocyte activation	Induction	No
Basiliximab (Simulect®)	14.4 kDa	Murine/human	CD25 alpha subunit	IL2-dependent T-lymphocyte activation	Induction	Yes
Muromonab-OKT3 (Orthoclone-OKT3®)	75 kDa	Murine	CD3	T lymphocytes (CD2, CD4, CD8)	Treatment of polyclonal antibody-resistant cellular-mediated rejection	No
Rituximab (Rituxan®)	145 kDa	Murine/human	CD20	B lymphocytes	Desensitization	Yes
					Antibody-mediated rejection	
					Focal segmental glomerulosclerosis	
Belatacept (Nulojix®)	90 kDa	Humanized	CD80 and CD86	T lymphocytes	Maintenance immunosuppression	Yes
Eculizumab (Soliris®)	148 kDa	Murine/human	C5	Block formation of membrane attack complex	Desensitization	Yes[a]
	148 kDa				Antibody-mediated rejection	
Tocilizumab (Actenra®)		Murine/human Murine/human	IL-6 receptor	Blocks proinflammatory effects of IL-6	Hemolytic uremic syndrome Desensitization	Yes

[a] Providers must register in the Risk Evaluation and Mitigation Strategy (REMS) to facilitate prescribing

Table 25.2 Complications of current maintenance immunosuppressants

	Hypertension	Hyperlipidemia	Hyperglycemia	Hematologic	Renal dysfunction	Dermatologic
Corticosteroids	+	++	++	−	−	++
Cyclosporine	+++	+++	++	+	+++	++
Tacrolimus	+++	+++	+++	+++	+++	++
Mycophenolate mofetil[a]	−	−	−	+++		−
Sirolimus	++	+++	−	+++	+	+++
Everolimus	++	+++	−	+++	+	+++
Belatacept	−	−	−	−	−	−

Incidence based on manufacturer package insert clinical trial approval reports, + < 1%, ++ 1–10%, +++ > 10%

[a]Adverse effects reported for mycophenolate mofetil (CellCept®) are based on clinical trials using this agent in combination with cyclosporine or tacrolimus and corticosteroids, values modified to account for concurrent agents

Table 25.3 Current trends of monoclonal antibody induction use in solid organ transplantation

Adult induction immunosuppression trends based on scientific registry of transplant recipients 2016 annual report				
Organ	Who receive induction (%)	T-cell Depleting[a] (%)	Basiliximab (%)	None (%)
Kidney	95	75	20	5
Pancreas	92.4	84.8	7.6	7.6
Heart	52.2	22.2	30	47.8
Lung	76	15	61	24
Liver	38	18	20	62
Intestine	69.1	52.9	16.2	30.9

Based on estimates from OPTN/SRTR 2016 Annual Reports for Kidney, Pancreas, Heart, Lung, Liver and Intestine (Hart et al. (2017); Kandaswamy et al. (2017); Colvin et al. (2017); Valapour et al. (2017); Kim et al. (2017); Smith et al. (2017)

[a]Includes poly and monoclonal T-cell depleting

are known to have circulating antibodies against other human antigens. This strategy is generally reserved for patients who are "highly sensitized" during their evaluation for transplant. As the long-term significance of these sensitizing events is better understood, varying degrees of "desensitization" therapy are initiated based upon varying levels of sensitization. The long-term impact of this empirical therapy is yet to be defined. Specifically, as a patient develops end-stage organ disease, their medical and immunologic profiles are characterized. Blood samples from these potential recipients are screened for the presence of antibodies against the major histocompatibility complexes (MHC) on the surface of other human cells, specifically human leukocyte antigens (HLA). Potential recipients who have received blood products, previous organ transplants, or have a history of pregnancy are at higher risk for the development of antibodies against HLA. In addition, all humans have preformed IgG and IgM antibodies against the major blood group antigens (A, B, AB, and A1) (Reid and Olsson 2005). These antibodies will recognize donor tissue and quickly destroy (hyperacute rejection) the implanted organ if the tissue contains previously recognized HLA within minutes to hours following transplant. Therefore, it is necessary to evaluate the presence of preformed circulating antibodies against HLA in the potential organ recipients. Some centers will implement desensitization, which incorporates monoclonal antibodies prior to transplant to diminish the production of antibodies against a new organ, allowing for transplant across this immunologic barrier.

Monoclonal Antibodies Administered at the Time of Transplant

Current maintenance immunosuppression is aimed at various targets within the immune system to halt its signal transduction pathway. Available agents, although effective, are associated with significant patient and allograft adverse effects, which are correlated with long-term exposure (Table 25.2). The leading cause of death in noncardiac transplant recipients is a cardiovascular event. These cardiovascular events have been linked to long-term corticosteroid exposure. In addition, chronic administration of calcineurin inhibitors (cyclosporine and tacrolimus) is also associated with acute and chronic kidney dysfunction leading to hemodialysis or need for a kidney transplant. Monoclonal antibodies given at the time of transplant (induction) have been used to decrease the need for corticosteroids and allow for the delay or a reduction in the amount of calcineurin inhibitor used. Determination of the solid organ transplant recipient's immunologic risk at the time of transplant is necessary to determine which monoclonal antibody to use in order to minimize the risk of early acute rejection and graft loss. Recipients are stratified based on several donor, allograft, and recipient variables to determine their immunologic risk. Patients at high risk for acute rejection or those in which maintenance immunosuppression is going to be minimized should receive a polyclonal or monoclonal antibody that provides cellular apoptosis, for example, alemtuzumab or rabbit antithymocyte globulin. Recipients at low risk for acute rejection may receive a monoclonal antibody which provides immunomodulation without lymphocyte depletion, such as basiliximab.

Several important pharmacokinetic parameters must be considered when these agents are administered to the various organ transplant recipients. The volume of distribution, biological half-life, and total body clearance can differ significantly from a kidney transplant recipient to a heart transplant recipient. Clinicians must consider when to administer monoclonal antibodies in different transplant populations to maximize efficacy and minimize toxicity. For example, heart and liver transplant recipients tend to lose large volumes of blood around the time of transplant; therefore, intraoperative administration may not be the optimal time to administer a monoclonal antibody since a large portion may be lost during surgery. Monoclonal antibodies are also removed by plasma exchange procedures, such as plasmapheresis, which may be performed during the perioperative period in solid organ transplant recipients (Nojima et al. 2005).

Monoclonal Antibodies Administered Following Transplant

Monoclonal antibodies given following transplantation are used to treat allograft rejection and as maintenance immunosuppressants. Administration of these agents is mainly reserved for severe allograft rejection in which the immunologic insult must be controlled quickly. Under normal homeostatic conditions the humoral immune system provides immediate control of infectious pathogens through secretion of antibodies. Cell-mediated immunity, in addition to fighting infections, provides surveillance against the production of mutant cells capable of oncogenesis. Interruption of either of these immune systems through the use of monoclonal antibodies places these patients at significant risk for infection and malignancy. Careful post-administration assessment of infection and posttransplant malignancy is commonplace. While those monoclonal antibodies employed as maintenance immunosuppressants have been developed to decrease the toxicity of long-term exposure to traditional agents such as calcineurin inhibitors, which can lead to chronic kidney damage in all organ transplant recipients, the use of these monoclonal antibodies is not without their own risks.

Specific Agents Used in Solid Organ Transplant

Muromonab

Muromonab was the first monoclonal antibody used in solid organ transplantation. Muromonab is a murine monoclonal antibody directed against human CD3 receptor, which is situated on the T-cell antigen receptor of mature T cells, inducing apoptosis of the target cell (Wilde and Goa 1996). Cells which display the CD3 receptor include CD2-, CD4-, and CD8-positive lymphocytes (Ortho Biotech 2004). Other investigators suggest that muromonab may also induce CD3 complex shedding, lymphocyte adhesion molecule expression causing peripheral endothelial adhesion, and cell-mediated cytolysis (Buysmann et al. 1996; Magnussen et al. 1994; Ortho Biotech 2004; Wilde and Goa 1996; Wong et al. 1990). Muromonab is approved for the treatment of kidney allograft rejection and steroid-resistant rejection in heart transplant recipients (Ortho biotech 2004). Muromonab was initially employed as an induction agent for kidney transplant recipients, in conjunction with cyclosporine, azathioprine, and corticosteroids. When compared to patients who received no muromonab induction, the rate of acute rejection was lower and the time to first acute rejection was substantially greater (Wilde and Goa 1996). Liver recipients with renal dysfunction at the time of transplant who received muromonab induction were also able to run their posttransplant cyclosporine levels lower without an increased incidence of acute rejection (Wilde and Goa 1996). Therefore, administration of OKT3 enabled preservation of renal function in the setting of reduced calcineurin inhibitor exposure when compared to those who did not receive muromonab (Wilde and Goa 1996). The use of OKT3 as an induction agent is nearly extinct with the introduction of newer agents that have more favorable side effect profiles.

Today, muromonab is of historical value as it is no longer being manufactured. Although prior to its withdrawal from the market, it was reserved for treatment of refractory rejection. Muromonab is extremely effective at halting most corticosteroid as well as polyclonal antibody-resistant rejections. These rejections are treated with 5 mg of muromonab given daily for 7–14 days (Ortho Biotech 2004). The dose and duration of therapy is often dependent on clinical or biopsy resolution of rejection or may be correlated with circulating CD3 cell concentrations in the serum.

Most patients who are exposed to OKT3 will develop human anti-mouse antibodies (HAMA) following initial exposure. These IgG antibodies may lead to decreased efficacy of subsequent treatment courses, but premedication with corticosteroids or antiproliferative agents during initial therapy may reduce their development (Wilde and Goa 1996). Following administration, in vitro data indicates that a serum concentration of 1000 µg/L is required to inhibit cytotoxic T-cell function (Wilde and Goa 1996). In vivo concentrations near the in vivo threshold immediately (1 h) following administration but diminish significantly by 24 h (Wilde and Goa 1996). Steady-state concentrations of 900 ng/mL can be achieved after three doses, with a plasma elimination half-life of 18 h when used for treatment of rejection and 36 h when used for induction (Ortho Biotech 2004; Wilde and Goa 1996).

Muromonab administration is associated with significant acute and chronic adverse effects. Immediately following administration, patients will experience a characteristic OKT3 cytokine release syndrome. The etiology of this syndrome is characterized by the pharmacodynamic interaction the OKT3 molecule has at the CD3 receptor. Muromonab will stimulate the target cell following its interaction with the CD3 receptor prior to inducing cell death. Consequently, CD3 cell stimulation leads to cytokine production and release, which is compounded by acute cellular apoptosis leading to cell lysis and release of the intracellular contents. The cytokine release syndrome associated with muromonab manifests as high fever, chills, rigors, diarrhea, capillary leak, and in some cases aseptic meningitis (Wilde and Goa 1996). Capillary leak has been correlated with increased tumor necrosis factor release leading to an initial increase in cardiac output secondary to decreased peripheral vascular resistance, followed by a reduction in right heart filling pressures which leads to a decrease in stroke volume (Wilde and Goa 1996). Sequelae of this cytokine release syndrome can occur immediately, within 30–60 min, and last up to 48 h following administration (Ortho Biotech 2004; Wilde and Goa 1996). This syndrome appears to be the most severe following the initial dose when the highest inoculum of cells is present in the patient's serum or when preformed antibodies against the mouse epitope exist. Subsequent doses appear to be better tolerated, though cytokine release syndrome has been reported after five doses, typically when the dose has been increased or the CD3-positive cell population has rebounded from previous dose baseline (Wilde and Goa 1996). Pretreatment against the effects of this cytokine release is necessary to minimize the host response: specifically, corticosteroids to prevent cellular response to cytokines, nonsteroidal anti-inflammatory agents to prevent sequelae of the arachidonic acid cascade, acetaminophen to halt the effects of centrally acting prostaglandins, and diphenhydramine to attenuate the recipient's response to histamine.

In addition to immediate adverse effects, the potency of muromonab has been associated with a high incidence of posttransplant lymphoproliferative disease and viral infections. For all patients, the 10-year cumulative incidence of posttransplant lymphoproliferative disease is 1.6% (Opelz and Dohler 2004). Review of large transplant databases revealed that deceased donor kidney transplant recipients who received muromonab for induction or treatment had a cumulative incidence of posttransplant lymphoproliferative disease that was three times higher than those who did not receive muromonab or other T-cell depleting induction (Opelz and Dohler 2004). This observation may be multifactorial. It is well known that posttransplant lymphoproliferative disease may be induced secondary to Epstein-Barr viral B-cell malignant transformation. Muromonab's potent inhibition of T lymphocytes over a sustained period of time diminishes the immune system's normal surveillance and destruction of malignant cell lines, consequently leading to unopposed transformed B-cell proliferation and subsequent posttransplant lymphoma (Opelz and Dohler 2004).

Early use and development of muromonab in solid organ transplantation was beneficial for the novel development and use of newer monoclonal agents. The immunodepleting potency of muromonab, combined with the significant risk for malignancy, has made its use obsolete in the setting of modern transplantation. However, this agent still serves as a template for treatment of severe allograft rejection and the use of monoclonal antibodies posttransplant.

Interleukin-2 Receptor Antagonists
Interleukin-2 antagonists were the next monoclonal antibodies to be used and were specifically developed for use in solid organ transplantation. As previously mentioned, monoclonal antibody use and development in solid organ transplantation is rational. The IL-2 receptor was targeted for several reasons. Interleukin-2, the ligand for the IL-2 receptor, is a highly conserved protein, with only a single gene locus on chromosome 4 (Church 2003). Animal IL-2 knockout models have decreased lymphocyte function at 2–4 weeks of age and early mortality at 6–9 weeks of age (Chen et al. 2002). These models also display significantly diminished myelopoiesis leading to severe anemia and global bone marrow failure (Chen et al. 2002). This observation confirms the significant role that IL-2 and the IL-2 receptor complex play in immunity. The function and biological effect of IL-2 binding to the IL-2 receptor was first reported by Robb and colleagues in 1981 (Robb et al. 1981). This in vitro study evaluated murine lymphocytes and found that the IL-2 receptor is only present on activated cells (CD4+ and CD8+) (Church 2003). Uchiyama et al. (1981) reported one of the first monoclonal antibodies developed against activated human T cells. This compound displayed in vitro preferential activity against activated T cells, including terminally mature T cells, but did not exhibit activity against B cells or monocytes (Uchiyama et al. 1981). Later it was determined that this antibody actually bound to the alpha subunit of the activated T-cell receptor, CD25 (Church 2003). The actual T-cell receptor is made up of three subunits, alpha, beta, and gamma. When the beta and gamma subunits combine, they can only be stimulated by high concentrations of IL-2; however, in conjunction with the alpha subunit, the receptor shows high affinity for IL-2 and can be stimulated at very low concentrations. The expression of IL-2 and the IL-2 receptor alpha region is highly regulated at the DNA transcription level and is induced following T-cell activation (Shibuya et al. 1990). The alpha subunit is continuously expressed during allograft rejection, T-cell-mediated autoimmune diseases, and malignancies (Church 2003). The beta

and gamma subunits, however, have constitutive expression, resulting in low levels of expression in resting T lymphocytes (Vincenti et al. 1997, 1998). There is no constitutive expression of IL-2 or the alpha receptor subunit (Noguchi et al. 1993; Shibuya et al. 1990). Both, the beta and gamma subunits, have similar molecular structures and are members of the cytokine receptor superfamily, but are structurally dissimilar to the alpha subunit (Noguchi et al. 1993). Therefore, the alpha subunit (CD25) became a rational target for monoclonal development since it is only expressed on activated T cells. Blockade of the CD25 receptor was to halt the activity of IL-2, thereby decreasing proliferation and clonal expansion of T cells when activated by foreign donor antigens.

Daclizumab

In 1997, daclizumab became the first anti-CD25 monoclonal antibody approved for use in the prevention of allograft rejection in kidney transplant recipients, when combined with cyclosporine and corticosteroids. Daclizumab was the first "humanized" monoclonal antibody approved in the United States for human administration (Tsurushita et al. 2005). The daclizumab molecule is a humanized IgG1 adapted from a mouse antibody against the alpha portion of the IL-2 receptor (Uchiyama et al. 1981). Daclizumab was developed as an alternative to the initial mouse antibody developed against the IL-2 receptor. The mouse antibody led to the development of human anti-mouse antibodies (HAMA) and inability to administer subsequent doses. Although daclizumab bound with one-third the affinity for the T-cell receptor site when compared to the original mouse molecule, it was still able to exhibit a high-binding capacity ($Ka = 3 \times 10^9 \text{ M}^{-1}$) (Queen et al. 1988; Tsurushita et al. 2005). A daclizumab serum concentration of 1 µg/mL is required for 50% inhibition of antigen-induced T-cell proliferation (Junghans et al. 1990). Early, phase I clinical trials in kidney transplant recipients, who received corticosteroids in combination with cyclosporine and azathioprine, used five doses of daclizumab (Vincenti et al. 1997). Pharmacokinetic studies revealed a mean serum half-life of 11.4 days, a steady-state volume of distribution of 5 l, and displayed weight-dependent elimination. There was no change in the number of circulating CD3-positive cells following administration. Five doses of 1 mg per kg body weight given every other week were required to produce the serum concentrations needed to achieve 90% inhibition of T-cell proliferation for 12 weeks. One patient did develop neutralizing antibodies against the daclizumab molecule after receiving weekly doses for 2 weeks. Saturation of the IL-2 receptor did not change. Intravenous doses were well tolerated with no infusion-related reactions. No infection or malignancies were reported up to 1 year following daclizumab administra-

tion. The authors concluded that daclizumab stayed within the intravascular space and doses should be based on patient weight at the time of transplant (Vincenti et al. 1997). Subsequent premarketing clinical trials confirmed these results and dosing schematic and were able to show that daclizumab administration reduced the incidence of acute rejection by 13% in low-risk kidney transplant recipients (Vincenti et al. 1998). Following daclizumab's approval, several trials have been conducted using various dosing regimens and immunosuppression combinations within various solid organ recipients. Secondary to low utilization in solid organ transplant, however, its manufacturing for transplant has ceased, and outcomes and observations from its use are historical in value.

Basiliximab

Basiliximab was developed as a more potent anti-IL-2 receptor antagonist when compared to daclizumab and may have several logistical advantages. Basiliximab, in combination with cyclosporine and corticosteroids, was approved for the prevention of acute allograft rejection in renal transplant recipients in May of 1998. Basiliximab is a murine/human (chimeric) monoclonal antibody directed against the alpha subunit of the IL-2 receptor on the surface of activated T lymphocytes. The antibody is produced from genetically engineered mouse myeloma cells. The variable region of the purified monoclonal antibody is comprised of murine hypervariable region, RFT5, which selectively binds to the IL-2 receptor alpha region. The constant region is made up of human IgG1 and kappa light chains (Novartis Pharmaceuticals 2005). Since the variable region is the only portion with a nonhuman epitope, there appears to be low antigenicity and increased circulating half-life associated with its administration (Amlot et al. 1995). Following administration, basiliximab rapidly binds to the alpha region of the IL-2 receptor and serves as a competitive antagonist against IL-2. The estimated receptor binding affinity (Ka) is $1 \times 10^{10} \text{ M}^{-1}$, which is three times more potent than daclizumab (Novartis Pharmaceuticals 2005). Complete inhibition of the CD25 receptor occurs after the serum concentration of basiliximab exceeds 0.2 µg/mL and inhibition correlated with increasing dose (Kovarik et al. 1996; Novartis Pharmaceuticals 2005). Initial dose finding studies of basiliximab were similar to daclizumab. Basiliximab, combined with cyclosporine and corticosteroids, was administered to adult kidney transplant recipients for the prevention of acute cellular rejection.

Kovarik et al. (1997) performed a multicenter, open-label pharmacodynamic analysis evaluating basiliximab dose escalation in adult patients undergoing primary renal transplantation. Patients received a total of 40 or 60 mg of basiliximab in combination with cyclosporine, corticosteroids, and azathioprine. Thirty-two patients were evaluated and were primarily young (34 ± 12 years), Caucasian (29/32) males (23/32). Basiliximab infusions were well tolerated without changes in blood pressure, temperature, or hypersensitivity reactions. Thirty patients underwent pharmacokinetic evaluation. Basiliximab blood concentrations showed biphasic elimination with an average terminal half-life of 6.5 days. Significant intra- and interpatient variability in observed volume of distribution and drug clearance was observed. This could not be corrected through body weight adjustment. Gender did not appear to influence the pharmacokinetic parameters of basiliximab; however, this cohort contained only a small number of female recipients that may have limited the detection of a difference.

Results also indicated that the use of basiliximab with a combination of cyclosporine, corticosteroids, and azathioprine may be an inadequate immunosuppression regimen to prevent acute rejection, especially if cyclosporine initiation is delayed posttransplant. A total of 22 patients had an acute rejection episode, 16 patients in the 40 mg groups and 6 in the 60 mg group. These rejections appeared within the first 2 weeks following transplantation with a mean time to rejection of 11 days. The study was designed for cyclosporine to begin on day 10 posttransplant. Also, three patients experienced graft loss, two of which were immunologically mediated. There was no difference in the basiliximab serum concentration in the patients who experienced rejection versus those who did not. The authors concluded that increased cyclosporine concentrations, which would inhibit IL-2 production, within the first few days posttransplant may increase the efficacy of basiliximab when used for induction (Kovarik et al. 1996).

The clinical efficacy of basiliximab has been confirmed in several prospective post-marketing trials. Currently, the recommended basiliximab dosing regimen is a total dose of 40 mg, with 20 mg administered 2 h prior to transplanted organ reperfusion and a subsequent 20 mg dose on postoperative day 4.

IL-2 receptor antagonists are currently used in all solid organ transplant populations for induction (Table 25.3), but are only approved for use in kidney transplant recipients. Administration does not reduce the total number of circulating lymphocytes or the number of T lymphocytes expressing other markers of activations, such as CD26, CD38, CD54, CD69, or HLA-DR (Chapman and Keating 2003). Consequently, it is necessary that additional immunosuppressive agents, such a calcineurin inhibitors and antiproliferative agents, be administered as soon as possible to decrease the risk of early acute rejection.

The advantage of IL-2 receptor antagonists is that they confer a decreased risk of infusion-related reactions, posttransplant infection, and malignancy when compared to immunodepleting agents. The use of these agents has increased since the introduction of more potent maintenance immunosuppressant agents, although their preference for induction has fluctuated over the years depending on organ transplant type. Although these agents have been evaluated in organ recipients who are at high risk for acute rejection, they are mainly reserved for patients who are at low to moderate risk. Also, these agents are still being evaluated for use in immunosuppression protocols which withdraw or avoid corticosteroids or calcineurin inhibitors. Contemporary use of IL-2 receptor antagonists is gaining more favor in combined kidney and another organ (e.g., heart) transplant to minimize calcineurin inhibitor early posttransplant. This is typically decided on a patient-by-patient basis and evaluates the predicted function of the kidney when employed.

There may be an increased risk of anti-idiotypic IgE anaphylactic reaction in patients who receive repeat courses of IL-2 receptor antagonists. Two published case reports describe patients who had been previously exposed to an IL-2 receptor antagonist and upon subsequent exposure developed dyspnea, chest tightness, rash, and angioedema. Therefore, caution may be warranted in patients who receive a dose of an IL-2 antagonist without concomitant corticosteroids following previous exposure in the past 6 months when circulating antibodies are expected to be present.

Alemtuzumab

Alemtuzumab is a recombinant DNA-derived, humanized, rat IgG1κ monoclonal antibody targeting the 21–28 kDa cell surface protein glycoprotein CD52, which is produced in a Chinese hamster ovary cell suspension (Genzyme Corporation 2009; Kneuchtle et al. 2004). Initially, the first anti-CD52 antibodies were developed from rat hybrid antibodies that were produced to lyse lymphocytes in the presence of complement (Morris and Russell 2006). Campath-1M was the first agent developed. This molecule was a rat IgM antibody which produced little biological effect. In contrast, the rat IgG (Campath-1G) produced profound lymphopenia (Morris and Russell 2006). In order to prevent the formation of antibodies against the rat IgG, the molecule was humanized and called alemtuzumab or Campath-1H (Morris and Russell 2006). The biologic effects of alemtuzumab are the same as Campath-1G and include complement-mediated cell lysis, antibody-mediated cytotoxicity, and target cell apoptosis (Magliocca and Knechtle 2006). The CD52 receptors account for 5% of lymphocyte surface antigens (Morris and Russell 2006). Cells which express the CD52 antigen include T and B lymphocytes, natural killer cells, monocytes, macrophages, and dendritic cells (Bloom et al. 2006; Genzyme Corporation 2009). However, plasma cells and memory type cells appear to be unaffected by alemtuzumab (Magliocca and Knechtle 2006). Following administration, a marked decrease in circulating lymphocytes is observed. Use in the hematology population indicates that this effect is dose dependent (see Chap. 23). However, single doses of 30 mg or two doses of 20 mg are currently used in the solid organ transplant population.

The plasma elimination half-life after single doses is reported to be around 12 days, and the molecule may be removed by posttransplant plasmapheresis (for more details, please see Chap. 23) (Magliocca and Knechtle 2006). The biological activity of alemtuzumab, however, may last up to several months. One in vivo study of kidney transplant recipients aimed to observe the recovery and function of lymphocytes following administration of 40 mg of alemtuzumab (Bloom et al. 2006). Authors reported a 2-log reduction in peripheral lymphocytes following administration. Absolute lymphocyte counts at 12 months remained markedly depleted, falling below 50% of their original baseline. Monocytes and B lymphocytes were the first cell lines to recover at 3–12 months post-administration. T lymphocytes returned to 50% of their baseline value by 36 months.

Alemtuzumab was primarily FDA approved for the treatment of B-cell chronic lymphocytic leukemia. The first report of alemtuzumab use in solid organ transplantation appeared in 1991. Friend et al. (1991) published a case series on the use of alemtuzumab to reverse acute rejection in renal transplant recipients. Shortly thereafter, Calne et al. (1999) issued the first report of alemtuzumab use as an induction agent. The authors reported the results of 31 consecutive renal transplant recipients. Patients received two 20 mg doses of alemtuzumab; the first dose was given in the operating room and the second dose was given on postoperative day 1. Patients were initiated on low-dose cyclosporine monotherapy 72 h after transplant, with a target trough range of 75–125 ng/mL. Six patients experienced corticosteroid responsive rejection (20%). Three of these were maintained on corticosteroids and azathioprine following rejection, while the other three remained on cyclosporine monotherapy. Allografts remained functional in 94% (29/31) of patients at 15–28 months posttransplant (Calne et al. 1999).

The largest multicenter randomized controlled trial assessing alemtuzumab induction in low- and high-risk renal transplant recipients showed that biopsy-confirmed acute rejection was reduced in low-risk patients receiving alemtuzumab when compared to basiliximab after 3 years of follow-up. In high-risk renal transplant patients, alemtuzumab and Thymoglobulin® appeared to have similar efficacy. However, patients who received alemtuzumab had increased rates of late rejections (between 12 and 36 months) when compared to conventional therapies (8% versus 3%, $p = 0.03$). All patients were withdrawn from steroids by postoperative day 5. Adverse effects were similar with more leukopenia observed in the alemtuzumab group (54%) compared to basiliximab (29%), and more serious adverse effects related to malignancy were seen with alemtuzumab (5%) when compared to a composite of all basiliximab- and

Thymoglobulin®-treated patients (1%). However, overall adverse events related to malignancy were similar between treatment groups (Hanaway et al. 2011).

Most data on the use of alemtuzumab in solid organ transplantation are with kidney transplant recipients, and literature suggests that it is used in 13% of all kidney transplants (Serrano et al. 2015). However, alemtuzumab is currently being used for induction and for treatment of rejection in other organs as well (Morris and Russell 2006). In a review of immunosuppression trends in the United States, alemtuzumab use markedly increased from 2001 to 2004 and has been lumped in with other T-cell depleting induction in most recent reports, with use primarily limited to induction of immunosuppression (see Table 25.3).

In 2004, alemtuzumab was the predominant agent used for induction in both pancreas and intestinal transplant recipients (Meier-Kriesche et al. 2006). Use in liver transplant has been limited but has appeared in a couple of published trials. Specific findings from these trials indicate that patients without hepatitis C were able to tolerate lower levels of calcineurin inhibitors which corresponded to lower serum creatinine levels at 1-year posttransplant (Tzakis et al. 2004). In contrast, administration of alemtuzumab positively correlated with early recurrence of hepatitis C viral replication (Marcos et al. 2004).

Alemtuzumab in heart transplantation has been rarely reported in the literature with only 2% of heart transplant patients receiving alemtuzumab for induction in 2004 (Meier-Kriesche et al. 2006). Teuteberg and colleagues published a retrospective study on 1-year outcomes on the use of alemtuzumab for induction in cardiac transplantation at a single center. Freedom from rejection was higher in the alemtuzumab group (versus no induction); however, survival at 1 year was similar between groups with more adverse effects in the alemtuzumab group (Teuteberg et al. 2010). Despite this publication, there remains a paucity of data in the cardiac transplant population regarding alemtuzumab for induction immunosuppression, which has resulted in limited use in this population. The use of alemtuzumab in lung transplant is also under investigation with some initial analysis of registry data suggesting that it may be beneficial in preventing bronchiolitis obliterans syndrome, a form of chronic rejection, when used for induction and as a rescue (Ensor et al. 2017; Furuya et al. 2016).

A retrospective review of 5-year outcomes on the use of alemtuzumab induction in lung transplant recipients at a single center showed an improvement in patient and graft survival with alemtuzumab compared to no induction or daclizumab induction and higher rates of freedom from cellular rejection than no induction or Thymoglobulin® or daclizumab induction (Shyu et al. 2011). The results of the previous study are consistent with another retrospective study that showed decreased rejection rates with alemtuzumab induction in comparison to Thymoglobulin® and daclizumab in lung transplant patients (McCurry et al. 2005). In 2004, 3% of lung transplant recipients received alemtuzumab for induction (Meier-Kriesche et al. 2006); however, this number may be increasing as more data emerges regarding alemtuzumab use in the lung transplant population.

Alemtuzumab induction has allowed for early withdrawal of corticosteroids in several clinical trials, thereby decreasing long-term steroid exposure. This may lead to improved clinical outcomes since the use of steroids has been correlated with an increased incidence of cardiovascular disease, endocrine, and metabolic side effects. However, the long-term benefit of steroid withdrawal after alemtuzumab induction requires further study. Several trials have also shown success with using low-dose calcineurin inhibitors with alemtuzumab induction. However, in early trials in which calcineurin inhibitor avoidance was initiated, the rate of early acute antibody-mediated rejection was 17% compared to 10% under traditional immunosuppression which included calcineurin inhibitors (Magliocca and Knechtle 2006).

The infusion of alemtuzumab is well tolerated. In general, induction doses are administered immediately preceding reperfusion of the transplanted allograft. Pretreatment with corticosteroids, diphenhydramine, and acetaminophen is generally advised to prevent sequelae from cellular apoptosis. However, cytokine release associated with alemtuzumab is insignificant in comparison to other agents (Morris and Russell 2006).

Initially, there were few published experiences detailing long-term outcomes in patients who received alemtuzumab induction (Magliocca and Knechtle 2006), and clinicians were concerned that the profound lymphodepletion that was observed following administration would lead to a significant increase in the number of severe infections. Therefore, lymphocyte response to donor antigens following alemtuzumab administration was also evaluated in vitro (Bloom et al. 2006). Lymphocytes from patients treated with alemtuzumab were able to respond to donor antigens and cytokines. However, a small subset of patients was hyporesponsive, which is similar to the control patients observed in this study (Bloom et al. 2006). In addition, several reports detailing the use of alemtuzumab thus far suggest that both infection and malignancy rates are minimal when compared to other agents used for the same indication (Magliocca and Knechtle 2006; Morris and Russell 2006). These findings are confirmed with the prospective 3-year data published by Hanaway et al. in kidney transplant recipients as well as the retrospective 5-year data published by Shyu et al. in lung transplant recipients (Hanaway et al. 2011; Shyu et al. 2011). Later, analysis of kidney registry data also indicates that long-term outcomes with alemtuzumab induction are similar to that of Thymoglobulin® induction (Serrano et al. 2015).

At present, a concern associated with alemtuzumab administration is an increased incidence of autoimmune diseases. The exact incidence and etiology of autoimmune diseases following alemtuzumab administration in solid organ transplant is currently unknown, although the most well-designed trial with 3-year follow-up to date did not report autoimmune diseases developing in kidney transplant recipients receiving alemtuzumab for induction (Hanaway et al. 2011). Initial reports of autoimmune diseases associated with alemtuzumab administration came from the multiple sclerosis population. A single center observed the development of Grave's disease in 9 out of 27 patients who received alemtuzumab (Coles et al. 1999). Thyroid function in all patients was normal prior to alemtuzumab, and the mean time to development of autoimmune hyperthyroidism was 19 months (range 9–31 months) (Coles et al. 1999). Autoimmune hyperthyroidism was first reported in a kidney transplant recipient who received alemtuzumab induction 4 years earlier (Kirk et al. 2006). Watson et al. (2005) published a 5-year experience with alemtuzumab induction, in which they reported a 6% (2/33) incidence of autoimmune disease development following administration. One patient developed hyperthyroidism in the early posttransplant period, and one patient developed hemolytic anemia, which was refractory to corticosteroids. With the increased use of alemtuzumab in solid organ transplantation, it is important to continually assess the risk of autoimmune disease development in this population.

Alemtuzumab became such a popular option in solid organ transplant and for other non-transplant indications that the pharmaceutical company removed it from the market in 2012 (Enderby and Keller 2015). Similar to other medications mentioned the drug has been repurposed as a medication to treat multiple sclerosis under the name Lemtrada®. Transplant centers were able to obtain alemtuzumab for transplantation since then but under scrutinized conditions and limited qualities associated with a distribution program (Serrano et al. 2015).

Rituximab

Rituximab is a chimeric murine/human IgG1 monoclonal antibody directed at the CD20 cell surface protein (Tobinai 2003). Rituximab is currently FDA approved for the CD20-positive forms of non-Hodgkin's lymphoma and chronic lymphocytic leukemia (CLL) and Wegener granulomatosis, microscopic polyangiitis, and refractory rheumatoid arthritis (see Chaps. 23 and 24) (Genentech 2011). The CD20 antigen is a 35-kDa phosphoprotein expressed on B cells, from pre-B cells to mature B cells. This protein is not expressed on hematopoietic stem cells, plasma cells, T lymphocytes, or other tissues (Tobinai 2003). The CD20 protein is a calcium channel and is responsible for B-cell proliferation and differentiation (Tobinai 2003). Early monoclonal antibodies developed against CD20 revealed that antibody binding did not result in modulation of activity or shedding of the surface protein, making the development of a humanized anti-CD20 antibody rational (Tobinai 2003). Rituximab was originally developed to treat B-cell lymphomas, as the vast majority of malignant B cells express the CD20 receptor. Following continuously infused, high doses of engineered anti-CD20 monoclonal antibodies clearance of CD20-positive cells occurred within 4 h of administration (Press et al. 1987). Circulating B-cell clearance was immediate; however, lymph node and bone marrow B-cell clearance were dose dependent.

Rituximab was initially used in solid organ transplant recipients to treat posttransplant lymphoproliferative disorder (PTLD). PTLD is a malignancy that develops following exposure to high levels of T-cell depleting immunosuppression (see section "Immunologic Targets: Rational Development/Use of Monoclonal Antibodies in Organ Transplant"). Under normal physiologic conditions, both the humoral and cellular immune systems work in concert to fight infection. In addition, cytotoxic T lymphocytes survey the body for malignant cells. Current immunosuppression and induction therapy are focused on decreasing communication and proliferation of T lymphocytes, which may lead to unopposed B-cell proliferation. The most significant risk factors for the development of PTLD are the use of potent T-cell depleting therapies as well as an Epstein-Barr virus (EBV) negative recipient serostatus. Approximately 60–70% of PTLD cases are associated with EBV. Certain B cells that are infected with EBV or other viruses may go into unopposed cellular differentiation leading to PTLD (Evens et al. 2010).

This disorder was first reported in five living donor renal transplant recipients in 1969 with four of the five patients dying from their disease. The fifth patient survived following radiation and reduction in immunosuppression (Penn et al. 1969). The incidence of posttransplant malignancy, specifically PTLD, increased as the number of solid organ transplants increased. Specific agents linked to the development of PTLD included OKT3 and rabbit antithymocyte globulin (Evens et al. 2010; Swinnen et al. 1990). The initial treatment for PTLD is a reduction in maintenance immunosuppression, to allow T-cell surveillance to resume and aid in the destruction of malignancy causing cells. However, pharmacotherapeutic agents have been used successfully in patients who fail to respond to decreased immunosuppression. Rituximab is the most studied medication for the treatment of PTLD and can be considered in patients with CD20-positive tumors. Rituximab was initially used in the 1990s to target B-cell-specific forms of PTLD that did not involve the central nervous system (Cook et al. 1999; Davis and Moss 2004; Faye et al. 1998). The molecular size of rituximab generally precludes its use for central nervous system tumors with <5% of rituximab penetrating the blood-brain barrier,

although some reports have shown success with rituximab for the treatment of CNS PTLD (Jagadeesh et al. 2012; Kordelas et al. 2008; Patrick et al. 2011). Administration of rituximab in patients with peripheral lymphomas resulted in clearance of malignant B cells for up to 12 months (Davis and Moss 2004). Currently, rituximab is reserved for patients with CD20-positive PTLD who fail to respond to reduction in maintenance immunosuppression. Rituximab can be used alone or in combination with chemotherapy in patients with severe or refractory PTLD.

Rituximab has also been employed as a desensitizing agent (see section "Monoclonal Antibodies Administered Pre-transplant") prior to solid organ transplant. Doses of 375 mg per m^2 administered prior to transplant enabled transplantation across ABO incompatible blood types and transplantation of highly sensitized patients. Often rituximab is given in combination with other immunosuppressants to halt the production of new B lymphocytes and prevent the formation of new plasma cells. Desensitization protocols involve administration of pooled immunoglobulin followed by plasmapheresis to remove donor-specific antibody complexes. Rituximab is administered following the course of plasmapheresis for two reasons: (1) rituximab is removed by plasmapheresis and (2) rituximab only targets B lymphocytes, not the plasma cells currently secreting antibody. Therefore, timing of administration is crucial to the success of the desensitization protocol (Pescovitz 2006). Several protocols with various outcomes exist. One example is a study by Vo and colleagues where they evaluated intravenous immunoglobulin with and without rituximab for desensitization in a double-blind placebo-controlled trial. This trial was halted early secondary to serious adverse events in the placebo arm however did suggest some benefit of rituximab-based regimens in preventing rebound of antibodies (Vo et al. 2014).

Following transplant, rituximab is also used for the treatment of acute, refractory antibody-mediated rejection. Antibody-mediated rejection is characterized by host recognition of donor antigens followed by T-cell proliferation and antigen presentation to B cells. B cells then undergo clonal expansion and differentiation into mature plasma cells, which secrete anti-donor antibody. This immune process may occur before or after transplantation. Often the presence of antibodies against donor tissue is discovered prior to transplant, during final crossmatch, thus preventing hyperacute rejection. In some cases, low levels of antibody or memory B cells exist which can facilitate antibody-mediated rejection within the first several weeks following transplant. Rituximab, therefore, is used to induce apoptosis of the B cells producing or capable of producing antibodies against the allograft. Unfortunately, the CD20 receptor is absent on mature plasma cells; therefore, rituximab can only stop new B cells from forming. Plasmapheresis is necessary to remove antibodies produced by secreting plasma cells. It is important to remember that rituximab may be removed by plasmapheresis and timing of administration is necessary to ensure optimal drug exposure. The optimal number of doses and length of therapy necessary to suppress antibody-mediated rejection is unknown (Pescovitz 2006; Stegall and Gloor 2010).

In 2005 and 2006, rituximab was shown to improve the clinical course of renal transplant patients with recurrent focal segmental glomerulosclerosis (FSGS) in patients who were receiving rituximab for the treatment of PTLD (Nozu et al. 2005; Pescovitz et al. 2006). A subsequent study described seven pediatric patients who had a relapse of proteinuria after transplantation and who failed to respond to initial plasmapheresis. After failure of plasmapheresis, patients received rituximab for treatment of refractory FSGS. Three patients had complete resolution of proteinuria; urine protein decreased by 70% in one patient and by 50% in one patient. One patient failed to respond to therapy and one patient was unable to tolerate the rituximab infusion. This study confirmed that rituximab is a possible treatment option for recurrent FSGS (Strologo et al. 2009). Additional studies are needed to further delineate the role of rituximab in the treatment of recurrent FSGS.

Eculizumab

Eculizumab is a recombinant-humanized IgG2/4 monoclonal antibody with murine complementarity-determining regions grafted onto the framework of the human antibody on the light- and heavy-chain variable regions. Eculizumab binds with specificity and with high affinity to C5, a complement protein. By binding to C5, eculizumab prevents cleavage of C5 to C5a and C5b, which prevents the formation of the membrane attack complex. Currently, eculizumab is approved for use in the treatment of paroxysmal nocturnal hemoglobinuria and atypical hemolytic uremic syndrome (Alexion Pharmaceuticals 2011; McKeage 2011).

Because antibody-mediated rejection (AMR) is associated with complement activation evidenced by C4d$^+$ staining on biopsy, the use of eculizumab for the prevention and treatment of AMR holds promise (Stegall and Gloor 2010). The first case describing the use of eculizumab for the treatment of severe AMR was published in 2009. The patient was a highly sensitized kidney transplant recipient who received desensitization therapy before and after transplant. However, he became anuric with a biopsy that was positive for AMR approximately 8 days after transplant. After clinical failure of plasmapheresis and intravenous immunoglobulin, eculizumab was initiated. Intravenous immunoglobulin was also given in order to decrease donor-specific antibodies, and rituximab was given in order to prevent B-cell proliferation. Donor-specific antibodies did not decrease initially; however, C5d-9 staining was reduced on biopsy, and AMR was completely resolved on follow-up biopsies (Locke et al. 2009).

The use of eculizumab for the prevention of AMR has also been reported. In one study, patients with a positive crossmatch to their living kidney donor received plasmapheresis and eculizumab preoperatively and were compared to a historical control who received only plasmapheresis pre- and postoperatively. The treatment group also received eculizumab posttransplant for at least 4 weeks. Treatment continued in patients who did not have a decrease in donor-specific antibody. The incidence of AMR at 3 months was 7% in the eculizumab group compared to 41% in the historical control group (Stegall and Gloor 2010).

Complement activation is involved in the development of hemolytic uremic syndrome. There have been a few case reports that show that eculizumab can improve the outcomes of patients who develop hemolytic uremic syndrome after renal transplant (Van den Hoogen and Hilbrands 2011). Later, case series detailing use posttranspant in patients either prophylactically, with known hemolytic uremic syndrome, or who developed hemolytic uremic syndrome indicated good response without a significant increase in opportunistic infections (de Andrade et al. 2017).

There is limited data, mainly case reports and series, on the use of eculizumab in solid organ transplantation at this time. However, it is likely that its role in the prevention and treatment of AMR, hemolytic uremic syndrome after transplantation, and other possible indications will be more clearly defined by the next decade.

Tocilizumab

Tocilizumab is a humanized monoclonal antibody directed at the IL-6 receptor and is approved in the United States to treat various forms of arthritis. Tocilizumab is a novel mAb therapy that competitively inhibits the binding of IL-6 to its receptor. Inhibiting the entire receptor complex prevents IL-6 signal transduction to inflammatory mediators that summon B and T cells. IL-6 signaling can be inhibited by suppressors of cytokine signaling, such as antibodies directed against IL-6R. In order to signal, IL-6 and its receptors form a four-part complex at the cell surface that comprises IL-6, an IL-6R, and two gp130 proteins. The signal is then transduced through several members of the Janus-activated kinase-signal transducer and activator of transcription (JAK-STAT) pathway. The signal ultimately leads to the transcription of genes with IL-6 response elements. The most commonly known members of the JAK-STAT pathway are the acute-phase proteins, which are a collection of macromolecules that flood the circulation during certain inflammatory disorders (Sebba 2008).

As attempts to reduce HLA antibody levels in sensitized transplant recipients have not been universally effective, many drugs with favorable mechanisms of action have been tested in desensitization. Since IL-6 is a pleiotropic cytokine that has powerful stimulatory effects on B cells and plasma cells and its function is responsible for normal antibody production, it is reasonable to evaluate its efficacy for desensitization. In November 2016, the first published experience with tocilizumab in patients undergoing desensitization awaiting kidney transplant was reported. Tocilizumab was added to a common intravenous immunoglobulin (IVIG) based desensitization protocol and resulted in decrease in the meantime to transplant and outcomes free of biopsy proven antibody mediated rejection. These early results encourage further studies with tocilizumab and other agents developed outside of transplant indications which possess a novel approach to reducing antibody production (Vo et al. 2015).

Belatacept

In an effort to achieve the "immunotolerant" state posttransplant, research has been focused in the area of co-stimulation blockade. Simplistically, when a T cell is exposed to an antigen particle expressed on an antigen presenting cell through the T-cell receptor, additional co-stimulation is required for full activation of the T cell (Wekerle and Grinyo 2012). If co-stimulation is blocked, then the T cell becomes unresponsive and in essence tolerant. CD28 is expressed on human T cells and is upregulated on activated T cells, while its ligands, on the surface of the antigen presenting cell, are CD80 and CD86 (Wekerle and Grinyo 2012). Cytotoxic T-lymphocyte antigen-4 (CTLA-4) was identified as a compound that would bind the same ligands as CD28 but to a much higher affinity (Wekerle and Grinyo 2012). A modification of CTLA-4, giving it higher binding affinity for CD80/86, was fused with a mutated (no longer able to fix complement) human IgG1, yielding belatacept (Wekerle and Grinyo 2012). Therefore, belatacept binds to CD80 and CD86 with high affinity, blocking their interaction with CD28 on T cells. An artifact of belatacept is that it also blocks intrinsic CTLA-4, which normally acts as an inhibitory ligand on the surface of activated T cells, responsible for limiting the proliferation of the immune response (Wekerle and Grinyo 2012). Blockade of CTLA-4 could prevent tolerance from being achieved when administered posttransplant; however, phase II trials indicate that the synthesis of CD4+ CD25+ regulatory T cells is not interrupted following belatacept exposure (Gupta and Womer 2010). Belatacept is an intravenous infusion, dosed based on actual body weight, and is unaffected by renal or hepatic function, which is administered frequently during the first 1–3 months posttransplant then monthly thereafter (Martin et al. 2011).

Belatacept has been mainly studied and demonstrated efficacy in kidney transplant recipients in combination with basiliximab induction and mycophenolate mofetil/prednisone maintenance immunosuppression. Belatacept has been touted as calcineurin inhibitor sparing and therefore more renal protective posttransplant. The 3-year results of the BENEFIT study were published detailing the safety and efficacy of belatacept versus cyclosporine in combination with mycophenolate mofetil and prednisone (Vincenti et al. 2012). The

BENEFIT trial evaluated 663 kidney transplant recipients who received low intensity (0–3 months; 10 mg/kg on days 1 and 5, 10 mg/kg on weeks 2, 4, 8, 12, 3–36 months 5 mg/kg every 4 weeks; $n = 226$), moderate intensity (0–6 months) 10 mg/kg on days 1 and 5, 10 mg/kg on weeks 2, 4, 6, 8, 10, 12, 16, 20, and 24; 7–36 months 5 mg/kg ($n = 219$) belatacept or cyclosporine ($n = 221$) in combination with mycophenolate mofetil and prednisone. Graft survival at 3 years was 92% in the low- and moderate-intensity groups and 89% in the cyclosporine group. A total of 6 patients died, 2 in each group, and 9 patients lost their graft (4 in the low intensity, 3 in the moderate intensity, and 2 in the cyclosporine group). Calculated glomerular filtration rate was 66 ± 27 mL/min/1.73 m^2 in the low intensity, 65 ± 26 mL/min/1.73 m^2 in the moderate intensity, and 44 ± 24 mL/min/1.73 m^2 in the cyclosporine group, $p < 0.0001$. Acute rejection mainly occurred in the first-year posttransplant with a cumulative rate of 17% in the low intensity and 24% in the moderate intensity versus 10% in the cyclosporine group. There were no new rejections in the belatacept treatments arms between years 2–3. Donor-specific antibody production was reduced by 50% in the belatacept-treated patients. PTLD occurred in five patients who received belatacept versus one patient in the cyclosporine group (Vincenti et al. 2012). In 2016, 7 years after transplantation of this same cohort, patient and graft survival and the mean eGFR were significantly higher with belatacept than with cyclosporine. A 43% reduction in the risk of death or graft loss was observed as compared with the cyclosporine regimen with equal contributions from the lower rates of death and graft loss. The mean estimated glomerular filtration rate increased over the 7-year period with belatacept, but declined with the cyclosporine regimen (Vincenti et al. 2016). Similar results were found at 3 and 7 years in extended criteria kidney transplant recipients (Durrbach et al. 2016; Pestana et al. 2012). When more intensive belatacept dosing was used in combination with mycophenolate mofetil ($n = 33$) or sirolimus ($n = 26$) versus tacrolimus with mycophenolate mofetil ($n = 30$) following rabbit antithymocyte globulin and early corticosteroid withdrawal (4 days), acute rejection rates were low (12% belatacept-mycophenolate, 4% belatacept-sirolimus, and 3% in the tacrolimus-mycophenolate). Graft survival was 100% at 1 year in the tacrolimus group versus 91% in the belatacept-mycophenolate group and 92% in the belatacept-sirolimus group; however, graft function was roughly 8 mL/min/1.73 m^2 higher in the belatacept groups. However, less than 80% of patients in the belatacept groups remained steroid-free at 12 months versus 93% in the tacrolimus group (Ferguson et al. 2011). Patients 6–36 months post-kidney transplant were also enrolled in a conversion trial in which they were randomized to continue their current immunosuppression or be converted to belatacept to evaluate if an improvement in renal function could be obtained following discontinuation of a calcineurin inhibitor (Rostaing et al. 2011). An average improvement in glomerular filtration rate was noted in the belatacept group

(7 mL/min versus 2.1 mL/min, $p = 0.0058$) at 12 months following conversion. Six patients did develop acute rejection following their conversion to belatacept, but these rejections did not result in graft loss (Rostaing et al. 2011).

Evidence for the use of belatacept is currently lacking in nonrenal transplant recipients and high immunologic risk and non-Caucasian organ recipients. Additionally, patients who are EBV positive are at high risk of developing post-transplant lymphoproliferative disease in the central nervous system. This observation warranted a black box warning to be issued in the belatacept package insert detailing that the use of belatacept is contraindicated in patients who are EBV negative (Bristol Myers Squibb Company 2011). As a result of this warning, a risk evaluation and mitigation strategy (REMS) for belatacept was originally approved by the Food and Drug Administration (FDA) on June 15, 2011. On May 9, 2017, the FDA determined that a REMS is no longer required for belatacept and has eliminated the REMS requirement citing current labeling is adequate to address the serious and significant public health concerns to ensure the benefits of the drug outweigh the risks.

Other Monoclonal Antibodies Used in Transplant Recipients

Monoclonal antibodies to treat other comorbid conditions are commonplace in prospective transplant candidates. The use of these antibodies posttransplant becomes a discussion, and the therapeutic impact on overall immunosuppression is unknown. Careful consideration of each monoclonal antibody used in combination with transplant immunosuppression is needed. The patient and clinical team need to weigh the risks and benefits of continuing the monoclonal antibody and potentially see if symptoms resolve for conditions previously treated with maintenance immunosuppression. If monoclonal antibodies used for non-transplant indications after transplant are needed, the team should consider timing based on immunosuppression given at the time of transplantation, and if additional monitoring for opportunistic infections and malignancies is needed.

Conclusion

Currently, there are several challenges remaining in solid organ transplantation. These challenges may be grouped as follows. One challenge is optimizing patient-specific immunosuppression based on risk factors for acute rejection. Monoclonal antibodies provide targeted immunosuppression that when used in conjunction with specific maintenance, immunosuppressants may allow more specific therapy. Another challenge is preventing over-immunosuppression, which may lead to infection and malignancy. Although monoclonal antibodies provide targeted therapy, the toxicity and potency must be balanced with over-immunosuppression. Consideration of the mechanism of action of both the monoclonal antibody and maintenance immunosuppression must be evaluated to ensure that appropriate antimicrobial prophylaxis and malignancy screening tools are utilized to minimize the patient's risk. Finally, increasing patient and graft survival through reducing the incidence of adverse effects associated with long-term exposure to maintenance immunosuppression, such as cardiovascular events or kidney dysfunction, is necessary. Monoclonal, along with polyclonal antibodies, may allow for withdrawal or minimization of specific maintenance immunosuppressants that lead to the increased incidence of these long-term adverse effects. Oftentimes the use of specific monoclonal antibodies in institutional protocols is driven by cost or changing availability (Table 25.4) with careful consideration of the goals of therapy.

Table 25.4 Per dose cost comparison estimates between monoclonal antibodies currently used in solid organ transplantation

Monoclonal antibody	Dose[a]	US cost per course (AWP)[b]
Alemtuzumab	30 mg × 1	$~81,216
Basiliximab	20 mg × 2	$9911
Rituximab	375 mg/m² weekly × 4 doses	$605
Belatacept	10 mg/kg days 1 and 5	$31,793 for the first year
	10 mg/kg after 2 and 4 weeks	$19,076 subsequent years
	10 mg/kg after 8 and 12 weeks	
	5 mg/kg after 16 weeks and every 4 weeks thereafter	
Eculizumab	1200 mg × 1[c]	$3130
Tocilizumab	600 mg × 1 then	$1444
	600 mg weekly × 3	
	8 mg/kg on day 15, then monthly for 6 months [d]	

[a] Based on 70 kg dosing weight, height 165 cm rounded to nearest vial size
[b] Actual wholesale price (AWP) Adapted from Medi-span; Wolters Kluwer (accessed 8/24/22)
[c] Dosing is based on Stegall et al. (2011) study. Adequate dose for transplantation has not yet been established
[d] Dosing is based on Vo et al. (2015) study. Adequate dose for desensitization has not yet been established
[e] Only available through specialized distribution program ~available as 12 mg/1.2 mL, calculation based on 36 mg

Self-Assessment Questions

Questions

1. Monoclonal antibodies are used for several reasons in solid organ transplantation. What benefit do they provide over polyclonal antibodies?
2. The rational development and use of monoclonal antibodies in solid organ transplantation is focused on the prevention of host recognition of donor tissue (rejection). What are the two ways in which the host immune system recognizes donor tissue and may cause tissue damage?
3. Which of the monoclonal antibodies used in solid organ transplant do not cause depletion in cell lines but only target molecular signaling?
4. What are the molecular targets for monoclonal antibodies currently used in solid organ transplantation?
5. Monoclonal antibodies are used at various times in solid organ transplantation. Describe the reasons why a monoclonal antibody would be administered before transplant, at the time of transplant, or following transplant?
6. There are several important pharmacokinetic parameters that must be considered when administering monoclonal antibodies to solid organ transplant recipients. What are these pharmacokinetic parameters?
7. Which of the monoclonal antibodies used in solid organ transplant is considered a maintenance immunosuppressant?
8. There are several benefits, as well as several risks associated with the use of monoclonal antibodies in solid organ transplantation. What are these benefits and risks?

Answers

1. Monoclonal antibodies provide targeted immunosuppression. The advantage monoclonal antibodies offer over polyclonal antibodies is that the receptor target is known. Polyclonal antibody development involves the introduction of human lymphocytes into an animal host immune system. The animal will then develop polyclonal antibodies directed against human lymphocyte cell surface targets. As a consequence, inter-batch variability and potency may vary. Although significant outcome data exists with the use of polyclonal antibodies, monoclonal antibodies have a known target allowing for in vivo and in vitro pharmacokinetic and pharmacodynamic data to aid incorporation into novel immunosuppression regimens.
2. The two ways in which the host immune system recognizes donor tissue: Complement-dependent antibody-mediated rejection occurs when the host (recipient) develops or has preformed antibodies against the donor tissue. Preformed antibodies will aggregate to the implanted tissue and initiate the complement cascade,

which facilitates cell lysis. The majority of these antibodies are usually directed against the major histocompatibility complexes (MHC) located on the surface of the donor tissue. An absolute contraindication to transplantation is the presence of preformed antibodies against MHC complex I, which is located on the surface of all nucleated cells. The second way in which the host immune system attacks donor tissue is through T-cell-mediated rejection. This occurs when the donor tissue is recognized as foreign by host antigen presenting cells. Antigen presenting cells present donor tissue antigens to the T cells which stimulates T-cell proliferation and graft infiltration leading to inflammation and arteritis.

3. IL-2 receptor antagonists and belatacept do not cause cellular depletion but rather target intercellular signaling to suppress communication in the immune system. However, they must be administered in combination with other immunosuppressants to provide adequate immunosuppression to prevent organ rejection.
4. Alemtuzumab (Campath-1H®) targets the CD52 receptor, located on peripheral blood lymphocytes, natural killer cells, monocytes, macrophages, and thymocytes.

   Daclizumab (Zenapax®) targets the CD25 alpha subunit of the IL-2 receptor, located on activated T lymphocytes.

   Basiliximab (Simulect®) targets the CD25 alpha subunit of the IL-2 receptor, located on activated T lymphocytes.

   Muromonab-OKT3 (Orthoclone-OKT3®) targets the CD3 receptor located on CD2-, CD4-, and CD8-positive lymphocytes.

   Rituximab (Rituxan®) targets the CD20 receptor located on B lymphocytes.

   Eculizumab (Soliris®) targets C5 in the complement pathway.

   Tocilizumab targets the IL-6 receptor.
5. The administration of monoclonal antibodies prior to transplant is called desensitization. This strategy is reserved for "highly sensitized" patients, meaning they have high titers of circulating antibodies against donor-specific antigens. Monoclonal antibodies that target cells which produce these antibodies are employed, in conjunction with plasmapheresis and pooled human immune globulins. Removal of these antibodies may facilitate successful transplantation across this immunologic barrier.

   Monoclonal antibodies administered at the time of transplant are called induction. Induction is provided at the time of transplant to decrease the ability of the host immune system to respond to implantation of foreign tissue. In addition, monoclonal antibodies which provide profound T-cell depletion given at the time of transplant may facilitate the need for certain maintenance immunosuppressants.

Following transplantation, monoclonal antibodies may be used to treat cell-mediated or antibody-mediated rejection. Cell and antibody infiltrates found in biopsy specimens in correlation with the clinical status of the patient will dictate the type, dose, and duration of the monoclonal antibody chosen.

6. The volume of distribution, biological half-life, and total-body clearance can differ significantly between solid organ transplant recipients. Careful consideration of these pharmacokinetic parameters must be employed to maximize the efficacy and minimize the toxicity associated with administration of these agents. For example, weight-based dosing in obese patients must be carefully considered, and biological markers of efficacy should be evaluated to determine the appropriate dose and dosing schedule. In addition, monoclonal antibodies are also removed by plasma exchange procedures, such as plasmapheresis, which may be performed during the perioperative period. Therefore, it would be prudent to administer the monoclonal antibody following the plasma exchange prescription to avoid removal of the drug and avoid a possible decrease in efficacy.

7. Belatacept is a co-stimulatory blocker that targets CD80 and CD86 and is considered a maintenance immunosuppressant. It is used in combination with other immunosuppressants to prevent organ rejection. The agent must be given on a monthly basis after the induction period to prevent rejection.

8. Benefits include targeted immunosuppression, no batch variability, and low antigenicity in humanized products. The risks associated with any type of immunosuppression include an increased risk for infection, as well as malignancy. Patients who receive monoclonal antibodies which specifically target a cell line, such as muromonab, are associated with a significantly increased risk of posttransplant lymphoproliferative disease. Appropriate antimicrobial prophylaxis and vigilant screening for posttransplant malignancy may allow for safe and effective use of these monoclonal antibodies in solid organ transplantation.

References

Alexion Pharmaceuticals (2011) Eculizumab (Soliris) package insert. Alexion Pharmaceuticals, Cheshire

Amlot PL, Rawlings E, Fernando ON, Griffin PJ, Heinrich G, Schreier MH, Castaigne JP, Moore R, Sweny P (1995) Prolonged action of a chimeric interleukin-2 receptor (CD25) monoclonal antibody used in cadaveric renal transplantation. Transplantation 60:748–756

Bloom DD, Hu H, Fechner JH, Knechtle SJ (2006) T-lymphocyte alloresponses of Campath-1H treated kidney transplant patients. Transplantation 81:81–87

Bristol Myers Squibb Company (2011) Belatacept (Nulojix) package insert. Bristol Myers Squibb, Princeton, NJ

Buysmann S, Bemelman FJ, Schellekens PT, van Kooyk Y, Figdor CG, ten Berge IJ (1996) Activation and increased expression of adhesion molecules on peripheral blood lymphocytes is a mechanism for the immediate lymphocytopenia after administration of OKT3. Blood 87:404–411

Calne R, Moffatt SD, Friend PJ, Jamieson NV, Bradley JA, Hale G, Firth J, Bradley J, Smith KG, Waldmann H (1999) Campath IH allows low-dose cyclosporine monotherapy in 31 cadaveric renal allograft recipients. Transplantation 68:1613–1616

Chapman TM, Keating GM (2003) Basiliximab: a review of its use as induction therapy in renal transplantation. Drugs 63:2803–2835

Chen J, Astle CM, Harrison DE (2002) Hematopoietic stem cell functional failure in interleukin-2-deficient mice. J Hematother Stem Cell Res 11:905–912

Church AC (2003) Clinical advances in therapies targeting the interleukin-2 receptor. QJM 96:91–102

Coles AJ, Wing M, Smith S, Coraddu F, Greer S, Taylor C, Weetman A, Hale G, Chatterjee VK, Waldmann H, Compston A (1999) Pulsed monoclonal antibody treatment and autoimmune thyroid disease in multiple sclerosis. Lancet 354:1691–1695

Colvin M, Smith JM, Hadley N, Skeans MA, Carrico R, Uccellini K, Lehman R, Robinson A, Israni AK, Snyder JJ, Kasiske BL (2017) OPTN/SRTR 2016 annual report: heart. Am J Transplant 18:291–362

Cook RC, Connors JM, Gascoyne RD, Fradet G, Levy RD (1999) Treatment of post-transplant lymphoproliferative disease with rituximab monoclonal antibody after lung transplantation. Lancet 354:1698–1699

Davis JE, Moss DJ (2004) Treatment options for post-transplant lymphoproliferative disorder and other Epstein-Barr virus-associated malignancies. Tissue Antigens 63:285–292

de Andrade LGM, Contti MM, Nga HS, Bravin AM, Takase HM, Viero RM, da Silva TN, Chagas KN, Palma LMP (2017) Long-term outcomes of the atypical hemolytic uremic syndrome after kidney transplantation treated with eculizumab as first choice. PLoS One 12:e0188155

Durrbach A, Pestana JM, Florman S, Del Carmen RM, Rostaing L, Kuypers D, Matas A, Wekerle T, Polinsky M, Meier-Kriesche HU, Munier S, Grinyó JM (2016) Long-term outcomes in belatacept-versus cyclosporine-treated recipients of extended CriteriaDonor kidneys: final results from BENEFIT-EXT, a phase III randomized study. Am J Transplant 16:3192–3201

Enderby C, Keller CA (2015) An overview of immunosuppression in solid organ transplantation. Am J Manag Care 21:s12–s23

Ensor CR, Rihtarchik LC, Morrell MR, Hayanga JW, Lichvar AB, Pilewski JM, Wisniewski S, Johnson BA, D'Cunha J, Zeevi A, McDyer JF (2017) Rescue alemtuzumab for refractory acute cellular rejection and bronchiolitis obliterans syndrome after lung transplantation. Clin Transplant 31. https://doi.org/10.1111/ctr.12899

Evens AM, Roy R, Sterrenberg D, Moll MZ, Chadburn A, Gordon LI (2010) Post-transplantation lymphoproliferative disorders: diagnosis, prognosis, and current approaches to therapy. Curr Oncol Rep 12:383–394

Faye A, Van Den Abeele T, Peuchmaur M, Mathieu-Boue A, Vilmer E (1998) Anti-CD20 monoclonal antibody for post-transplant lymphoproliferative disorders. Lancet 352:1285

Ferguson R, Grinyo J, Vincenti F, Kaufman DB, Woodle ES, Marder BA, Citterio F, Marks WH, Agarwal M, Wu D, Dong Y, Garg P (2011) Immunosuppression with belatacept-based, corticosteroid-avoiding regimens in de novo kidney transplant recipients. Am J Transplant 11:66–76

Friend PJ, Waldmann H, Hale G, Cobbold S, Rebello P, Thiru S, Jamieson NV, Johnston PS, Calne RY (1991) Reversal of allograft rejection using the monoclonal antibody, Campath-1G. Transplant Proc 23:2253–2254

Furuya Y, Jayarajan SN, Taghavi S, Cordova FC, Patel N, Shiose A, Leotta E, Criner GJ, Guy TS, Wheatley GH, Kaiser LR, Toyoda Y (2016) The impact of alemtuzumab and basiliximab induction on patient survival and time to bronchiolitis obliterans syndrome in double lung transplantation recipients. Am J Transplant 16:2334–2341

Genentech (2011) Rituximab (Rituxan) package insert. Genentech, Inc., San Francisco, CA

Genzyme Corporation (2009) Alemtuzumab (Campath) package insert. Genzyme Corporation, Cambridge, MA

Gupta G, Womer KL (2010) Profile of belatacept and its potential role in prevention of graft rejection following renal transplantation. Drug Des Devel Ther 4:375–382

Halloran PF (2004) Immunosuppressive drugs for kidney transplantation. N Engl J Med 351:2715–2729

Hanaway MJ, Woodle ES, Mulgaonkar S, Peddi VR, Kaufman DB, First MR, Croy R, Holman J (2011) Alemtuzumab induction in renal transplantation. N Engl J Med 364:1909–1919

Hart A, Smith JM, Skeans MA, Gustafson SK, Wilk AR, Robinson A, Wainright JL, Haynes CR, Snyder JJ, Kasiske BL, Israni AK (2017) OPTN/SRTR 2016 annual report: kidney. Am J Transplant 18:18–113

Jagadeesh D, Woda BA, Draper J, Evens AM (2012) Post transplant lymphoproliferative disorders: risk, classification, and therapeutic recommendations. Curr Treat Options in Oncol 13(1):122–136

Junghans RP, Waldmann TA, Landolfi NF, Avdalovic NM, Schneider WP, Queen C (1990) Anti-Tac-H, a humanized antibody to the interleukin 2 receptor with new features for immunotherapy in malignant and immune disorders. Cancer Res 50:1495–1502

Kandaswamy R, Stock PG, Gustafson SK, Skeans MA, Curry MA, Prentice MA, Fox A, Israni AK, Snyder JJ, Kasiske BL (2017) OPTN/SRTR 2016 annual report: pancreas. Am J Transplant 18:114–171

Kim WR, Lake JR, Smith JM, Schladt DP, Skeans MA, Harper AM, Wainright JL, Snyder JJ, Israni AK, Kasiske BL (2017) OPTN/SRTR 2016 annual report: liver. Am J Transplant 18:172–253

Kirk AD, Hale DA, Swanson SJ, Mannon RB (2006) Autoimmune thyroid disease after renal transplantation using depletional induction with alemtuzumab. Am J Transplant 6:1084–1085

Kneuchtle SJ, Fernandez LA, Pirsch JD et al (2004) Campath-1H in renal transplantation: the University of Wisconsin experience. Surgery 136:754–760

Kordelas L, Trenschel R, Koldehoff M, Elmaagacli A, Beelan DW (2008) Successful treatment of EBV PTLD with CNS lymphomas with the monoclonal anti-CD20 antibody rituximab. Onkologie 31:691–693

Kovarik JM, Rawlings E, Sweny P, Fernando O, Moore R, Griffin PJ, Fauchald P, Albrechtsen D, Sodal G, Nordal K, Amlot PL (1996) Pharmacokinetics and immunodynamics of chimeric IL-2 receptor monoclonal antibody SDZ CHI 621 in renal allograft recipients. Transpl Int 9:S32–S33

Kovarik J, Wolf P, Cisterne JM, Mourad G, Lebranchu Y, Lang P, Bourbigot B, Cantarovich D, Girault D, Gerbeau C, Schmidt AG, Soulillou JP (1997) Disposition of basiliximab, an interleukin-2 receptor antibody, in recipients of mismatched cadaver renal allografts. Transplantation 64:1701–1705

Locke JE, Magro CM, Singer AL, Segev DL, Haas M, Hillel AT, King KE, Kraus E, Lees LM, Melancon JK, Stewart ZA, Warren DS, Zachary AA, Montgomery RA (2009) The use of antibody to complement protein C5 for salvage treatment of severe antibody-mediated rejection. Am J Transplant 9:231–235

Magliocca JF, Knechtle SJ (2006) The evolving role of alemtuzumab (Campath-1H) for immunosuppressive therapy in organ transplantation. Transpl Int 19:705–714

Magnussen K, Klug B, Moller B (1994) CD3 antigen modulation in T-lymphocytes during OKT3 treatment. Transplant Proc 26:1731

Marcos A, Eghtesad B, Fung JJ, Fontes P, Patel K, Devera M, Marsh W, Gayowski T, Demetris AJ, Gray EA, Flynn B, Zeevi A, Murase N, Starzl TE (2004) Use of alemtuzumab and tacrolimus monotherapy for cadaveric liver transplantation: with particular reference to hepatitis C virus. Transplantation 78:966–971

Martin ST, Tichy EM, Gabardi S (2011) Belatacept: a novel biologic for maintenance immunosuppression after renal transplantation. Pharmacotherapy 31:394–407

McCurry KR, Iacono A, Zeevi A, Yousem S, Girnita A, Husain S, Zaldonis D, Johnson B, Hattler BG, Starzl TE (2005) Early outcomes in human lung transplantation with thymoglobulin or Campath-1H for recipient pretreatment followed by posttransplant tacrolimus near-monotherapy. J Thorac Cardiovasc Surg 130:528–537

McKeage K (2011) Eculizumab: a review of its use in paroxysmal nocturnal haemoglobinuria. Drugs 71:2327–2345

Meier-Kriesche HU, Li S, Gruessner RWG, Fung JJ, Bustami RT, Barr ML, Leichtman AB (2006) Immunosuppression: evolution in practice and trends, 1994–2004. Am J Transplant 6:1111–1131

Morris PJ (2004) Transplantation–a medical miracle of the 20th century. N Engl J Med 351:2678–2680

Morris PJ, Russell NK (2006) Alemtuzumab (Campath-1H): a systematic review in organ transplantation. Transplantation 81:1361–1367

Noguchi M, Adelstein S, Cao X, Leonard WJ (1993) Characterization of the human interleukin-2 receptor gamma gene. J Biol Chem 268:13601–13608

Nojima M, Yoshimoto T, Nakao A, Itahana R, Kyo M, Hashimoto M, Shima H (2005) Sequential blood level monitoring of basiliximab during multisession plasmapheresis in a kidney transplant recipient. Transplant Proc 37:875–878

Novartis Pharmaceuticals (2005) Basiliximab (Simulect) package insert. Novartis Pharmaceuticals Corporation, East Hanover

Nozu K, Iijima K, Fujisawa M, Nakagawa A, Yoshikawa N, Matsuo M (2005) Rituximab treatment for posttransplant lymphoproliferative disorder (PTLD) induces complete remission of recurrent nephritic syndrome. Pediatr Nephrol 20:1660–1663

Opelz G, Dohler B (2004) Lymphomas after solid organ transplantation: a collaborative transplant study report. Am J Transplant 4:222–230

Ortho Biotech (2004) Muromonab (Orthoclone) package insert. Ortho Biotech, Raritan

Patrick A, Wee A, Hedderman A, Wilson D, Weiss J, Govani M (2011) High-dose intravenous rituximab for multifocal, monomorphic primary central nervous system posttransplant lymphoproliferative disorder. J Neuro-Oncol 103:739–743

Penn I, Hammond W, Brettschneider L, Starzl TE (1969) Malignant lymphomas in transplantation patients. Transplant Proc 1:106–112

Pescovitz MD (2006) Rituximab, an anti-CD20 monoclonal antibody: history and mechanism of action. Am J Transplant 6:859–866

Pescovitz MD, Book BK, Sidner RA (2006) Resolution of recurrent focal segmental glomerulosclerosis proteinuria after rituximab treatment. N Engl J Med 354:1961–1963

Pestana JOM, Grinyo JM, Vanrenterghen Y, Becker T, Campistol JM, Florman S, Garcia VD, Kamar N, Lang P, Manfro RC, Massari P, Rial MD, Schnitzler MA, Vitko S, Duan T, Block A, Harler MB, Durrbach A (2012) Three year outcomes from BENEFIT-EXT: a phase III study of belatacept versus cyclosporine in recipients of extended criteria donor kidneys. Am J Transplant 12(3):630–639

Press OW, Appelbaum F, Ledbetter JA, Martin PJ, Zarling J, Kidd P, Thomas ED (1987) Monoclonal antibody 1F5 (anti-CD20) serotherapy of human B cell lymphomas. Blood 69:584–591

Queen C, Schneider WP, Selick HE, Payne PW, Landolfi NF, Duncan JF, Avdalovic NM, Levitt M, Junghans RP, Waldmann TA (1988) A humanized antibody that binds to the interleukin 2 receptor. Proc Natl Acad Sci U S A 86:10029–10033

Reid ME, Olsson ML (2005) Human blood group antigens and antibodies. In: Hoffman R, Benz EJ (eds) Hematology: basic principles and practice, 4th edn. Churchill Livingstone, Philadelphia, PA, pp 2370–2374

Robb RJ, Munck A, Smith KA (1981) T cell growth factor receptors: quantitation, specific and biological relevance. J Exp Med 154:1455–1474

Rostaing L, Massari P, Garcia VD, Mancilla-Urrea E, Nainan G, del Carmen RM, Steinberg S, Vincenti F, Shi R, Di Russo G, Thomas D, Grinyo J (2011) Switching from calcineurin inhibitor based regimens to a belatacept based regimen in renal transplant recipients: a randomized phase II study. Clin J Am Soc Nephrol 6:430–439

Sebba A (2008) Tocilizumab: the first interleukin-6-receptor inhibitor. Am J Health Syst Pharm 65:1413–1418

Serrano OK, Friedmann P, Ahsanuddin S, Millan C, Ben-Yaacov A, Kayler LK (2015) Outcomes associated with steroid avoidance and alemtuzumab among kidney transplant recipients. Clin J Am Soc Nephrol 10:2030–2038

Shibuya H, Yoneyama M, Nakamura Y, Harada H, Hatakeyama M, Minamoto S, Kno T, Doi T, White R, Taniguchi T (1990) The human interleukin-2 receptor beta-chain gene: genomic organization, promoter analysis and chromosomal assignment. Nucleic Acids Res 18:3697–3703

Shyu S, Dew MA, Pilewski JM, Dabbs AJD, Zaldonis DB, Studer SM, Crespo MM, Toyoda Y, Bermudez CA, McCurry KR (2011) Five-year outcomes with alemtuzumab induction after lung transplantation. J Heart Lung Transplant 30:743–754

Smith JM, Weaver T, Skeans MA, Horslen SP, Harper AM, Snyder JJ, Israni AK, Kasiske BL (2017) OPTN/SRTR 2016 annual report: intesting. Am J Transplant 18:254–290

Stegall MD, Gloor JM (2010) Deciphering antibody-mediated rejection: new insights into mechanisms and treatment. Curr Opin Organ Transplant 15:8–10

Stegall MD, Diwan T, Raghavaiah S, Cornell LD, Burns J, Dean PG, Cosio FG, Gandhi MJ, Kremers W, Gloor JM (2011) Terminal complement inhibition decreases antibody-mediated rejection in sensitized renal transplant recipients. Am J Transplant 11:245–2413

Strologo LD, Guzzo I, Laurenzi C, Vivarelli M, Parodi A, Barbano G, Camilla R, Scozzola F, Amore A, Ginevri F, Murer L (2009) Use of rituximab in focal glomerulosclerosis relapses after renal transplantation. Transplantation 88:417–420

Swinnen LJ, Costanzo-Nordin MR, Fisher SG, O'Sullivan EJ, Johnson MR, Heroux AL, Dizikes GJ, Pifarre R, Fisher RI (1990) Increased incidence of lymphoproliferative disorder after immunosuppression with the monoclonal antibody OKT3 in cardiac-transplant recipients. N Engl J Med 323:1723–1728

Teuteberg JJ, Shullo MA, Zomak R, Toyoda Y, McNamara DM, Bermudex C, Kormos RL, McCurry KR (2010) Alemtuzumab induction prior to cardiac transplantation with lower intensity maintenance immunosuppression: one-year outcomes. Am J Transplant 10:382–388

Tobinai K (2003) Rituximab and other emerging antibodies as molecular target-based therapy of lymphoma. Int J Clin Oncol 8:212–223

Tsurushita N, Hinton PR, Kumar S (2005) Design of humanized antibodies: from anti-Tac to Zenapax. Methods 36:69–83

Tzakis AG, Tryphonopoulos P, Kato T, Nishida S, Levi DM, Madariaga JR, Gaynor JJ, De Faria W, Regev A, Esquenazi V, Weppler D, Ruiz P, Miller J (2004) Preliminary experience with alemtuzumab (Campath-1H) and low-dose tacrolimus immunosuppression in adult liver transplantation. Transplantation 77:1209–1214

Uchiyama T, Border S, Waldmann TA (1981) A monoclonal antibody (anti-Tac) reactive with activated and functionally mature human T cells. J Immunol 126:1393–1397

Valapour M, Lehr CJ, Skeans MA, Smith JM, Carrico R, Uccellini K, Lehman R, Robinson A, Israni AK, Snyder JJ, Kasiske BL (2017) OPTN/SRTR 2016 annual report: lung. Am J Transplant 18:363–433

Van den Hoogen MWF, Hilbrands LB (2011) Use of monoclonal antibodies in renal transplantation. Immunotherapy 3:871–880

Vincenti F, Lantz M, Birnbaum J, Garovoy M, Mould D, Hakimi J, Nieforth K, Light S (1997) A phase I trial of humanized anti-interleukin 2 receptor antibody in renal transplantation. Transplantation 63:33–38

Vincenti F, Kirkman R, Light S, Bumgardner G, Pescovitz M, Halloran P, Neylan J, Wilkinson A, Ekberg H, Gaston R, Backman L, Burdick J (1998) Interleukin-2-receptor blockade with daclizumab to prevent acute rejection in renal transplantation. Daclizumab triple therapy study group. N Engl J Med 338:161–165

Vincenti F, Larsen CP, Alberu J, Bresnahan B, Garcia VD, Kothari J, Lang P, Urrea EM, Massari P, Mondragon-Ramirez G, Reyes-Acevedo R, Rice K, Rostaing L, Steinberg S, Xing J, Agarwal M, Harler MB, Charpentier B (2012) Three-year outcomes from BENEFIT, a randomized, active-controlled, parallel-group study in adult kidney transplant recipients. Am J Transplant 12:210–217

Vincenti F, Rostaing L, Grinyo J, Rice K, Steinberg S, Gaite L, Moal MC, Mondragon-Ramirez GA, Kothari J, Polinsky MS, Meier-Kriesche HU, Munier S, Larsen CP (2016) Belatacept and long-term outcomes in kidney transplantation. N Engl J Med 374:333–343

Vo AA, Choi J, Cisneros K, Reinsmoen N, Haas M, Ge S, Toyoda M, Kahwaji J, Peng A, Villicana R, Jordan SC (2014) Benefits of rituximab combined with intravenous immunoglobulin for desensitization in kidney transplant recipients. Transplantation 98:312–319

Vo AA, Choi J, Kim I, Louie S, Cisneros K, Kahwaji J, Toyoda M, Ge S, Haas M, Puliyanda D, Reinsmoen N, Peng A, Villicana R, Jordan SC (2015) A phase I/II trial of the interleukin-6 receptor-specific humanized monoclonal (tocilizumab) + intravenous immunoglobulin in difficult to desensitize patients. Transplantation 99:2356–2363

Watson CJ, Bradley JA, Friend PJ, Firth J, Taylor CJ, Bradley JR, Smith KG, Thiru S, Jamieson NV, Hale G, Waldmann H, Calne R (2005) Alemtuzumab (CAMPATH 1H) induction therapy in cadaveric kidney transplantation—efficacy and safety at five years. Am J Transplant 5:1347–1533

Wekerle T, Grinyo JM (2012) Belatacept: from rational design to clinical application. Transpl Int 25:139–150

Wilde MI, Goa KL (1996) Muromonab CD3: a reappraisal of its pharmacology and use of prophylaxis of solid organ transplant rejection. Drugs 51:865–894

Wong JT, Eylath AA, Ghobrial I, Colvin RB (1990) The mechanism of anti-CD3 monoclonal antibodies. Mediation of cytolysis by inter-T cell bridging. Transplantation 50:683–689

Willebrordus Petrus Johannes van Oosterhout

Introduction

Migraine is a common, chronic primary headache disorder characterized by attacks of headache and accompanying symptoms, with a significant personal and societal burden, requiring safe and effective, acute, and preventive treatment options in a substantial part of the patient group. The classes of conventional pharmacotherapy used to prevent migraine attacks include beta-blockers, angiotensin receptor blockers, calcium channel antagonists, antiepileptic drugs, serotonin receptor antagonist, tricyclic antidepressants, serotonin norepinephrine reuptake inhibitors, and onabotulinumtoxinA. These all have in common that they were not developed for migraine treatment, they have moderate efficacy, and that adherence is hampered by unfavorable tolerability profiles.

Calcitonin gene-related peptide (CGRP) is a 37-amino acid neuropeptide abundant in the central nervous system, but also in the gastrointestinal tract, vascular system, and in non-neuronal cells. From the 1980s pre-clinical and clinical studies have increasingly recognized it as a pivotal neuropeptide in migraine pathophysiology. This led to the development of fully human and humanized monoclonal antibodies (mAbs) targeting CGRP or its receptor to block the CGRP pathway. Anti-CGRP mAbs have been introduced globally in clinical practice in the past 5–10 years, and further clinical studies are still ongoing. In migraine, these large, high molecular weight mAbs have a presumed site of action outside the blood-brain barrier (BBB), most likely within sensory ganglia such as the trigeminal ganglion (TG). Their use results in a marked reduction of monthly migraines with favorable safety and tolerability profiles.

Four anti-CGRP mAbs are available worldwide, with monthly to quarterly dosing strategies via subcutaneous

W. P. J. van Oosterhout (✉)
Department of Neurology, Zaans Medisch Centrum,
Zaandam, The Netherlands

Department of Neurology, Leiden University Medical Center,
Leiden, The Netherlands
e-mail: Oosterhout.r@zaansmc.nl

(three mAbs) or intravenous (one mAb) administration. After subcutaneous administration, mAbs are distributed via lymph and blood. Interaction with the neonatal Fc receptor (FcRn; Chap. 8) reduces the clearance and results in long half-lives. Drug-to-drug interactions are not a major concern as mAbs are not metabolized via the CYP450 system. The immunogenicity profiles and neutralizing anti-drug antibodies of the four mAbs have shown no effect on efficacy or safety. Future studies will help identifying classes of migraine patients who are expected to benefit best from anti-CGRP-mAb therapies, and to elucidate the potential utility of the mAbs in combination with emerging other acute and prophylactic anti-CGRP-pathway treatment strategies in migraine.

Calcitonin Gene-Related Peptide (CGRP): Structure and Isoforms

CGRP and CGRP Isoforms

α-Calcitonin gene-related peptide (CGRP) is a 37-amino acid neuropeptide produced by tissue-specific alternative RNA splicing of the calcitonin gene (Fig. 26.1). It was discovered in 1982. Its association with migraine (and other head and facial pain) was first postulated in 1985 (Edvinsson 2022).

CGRP Receptor Complex

The CGRP receptor is a unique G protein-coupled receptor comprising three subunits: the calcitonin-like receptor (CLR), the receptor activity-modifying protein 1 (RAMP1), and the receptor component protein (RCP), together giving it its functionality and specificity (Fig. 26.2).

The CGRP action is mainly mediated by adenylate cyclase coupled with the cAMP pathway. CGRP binds strongly to both the CGRP receptor and the AMY1 receptor (formed by

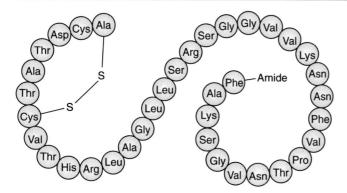

Fig. 26.1 CGRP is a 37 amino acid peptide with two isoforms. The alfa isoform (α-CGRP) is shown here. The human beta isoform (β-CGRP) differs at three amino acid sites in the peptide. α-CGRP mRNA is detected in the sensory ganglia, primary spinal afferent fibers, and motor neurons. β-CGRP is predominantly expressed in the enteric nervous system and in the human pituitary gland. (Adapted from Edvinsson et al. (2018))

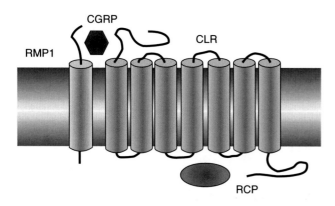

Fig. 26.2 The CGRP receptor consists of three units: the calcitonin-like receptor (CLR), the receptor-activity modifying protein (RAMP1), and the RCP protein to enable signaling via adenylate cyclase through the cAMP pathway. (Adapted from Raddant and Russo (2011))

the calcitonin receptor and RAMP1). RCP facilitates the coupling of CLR to Gs alpha subunit (Gαs). CGRP coupling to Gαs induces activation of adenylyl cyclase and protein kinase A (PKA) via cyclic AMP (cAMP)(Cohen et al. 2022; Edvinsson 2021, 2022).

CGRP Physiology

CGRP receptors are found in high concentrations in the cerebrum (striatum, amygdala, pineal gland, colliculi, trigeminal nucleus caudalis, cortex), cerebellum, and smooth muscle layer of cerebral arteries. They can also be found on ganglionic satellite ganglion cells (SGCs), perivascular mast cells, and immunocytes. CGRP receptors are primarily located on cell bodies and dendrites of second-order neurons and occasionally on axon terminals of first-order neu-

rons. CGRP receptors are not co-localized with CGRP in dural afferents: CGRP receptors are generally found on Aδ fibers and CGRP in C fibers (Edvinsson 2021, 2022). Whereas calcitonin is primarily produced in the thyroid gland with functions in bone and calcium metabolism, CGRP is found throughout the nervous system and implicated in the vascular and somatosensory system. CGRP is the most potent vasodilator of the calcitonin family that also includes the neuropeptides amylin, adrenomedullin-2 (intermedin), and adrenomedullin. The two isoforms α-CGRP and β-CGRP differ by three amino acids (Fig. 26.1) and are encoded by two different genes: the CALC I and CALC II genes. Although the isoforms share similar biological properties, they differ in regional potency and regulation. Both α-CGRP and β-CGRP are found in the nervous system, gastrointestinal tract, vascular system, and in non-neuronal cells, such as endothelial cells, immune cells, and skin cells. α-CGRP is the most abundant in the sensory nervous system, while β-CGRP is more abundant in the enteric plexus and pituitary. CGRP is found widely in neuronal regions subserving a sensory, ingestive, vascular, and nociceptive function. CGRP is a neuropeptide impermeant to the blood-brain barrier (BBB) with a serum half-life of 7–10 min and unknown tissue half-life, although CGRP-injection triggered skin flares lasting over 12 h suggest a slow tissue clearance. CGRP diffuses widely from the site of release and is degraded by mast cell tryptase, neutral endopeptidase, and matrix metalloproteinase II (Cohen et al. 2022; Edvinsson 2021, 2022).

CGRP in the Nervous System and the Trigeminovascular System (cf. Section of Etiology)

CGRP-containing neurons are found in various central nervous system nuclei. In the peripheral nervous system, CGRP is distributed in sensory and motor nerves and ganglia (Cohen et al. 2022; Edvinsson 2021, 2022).

In the trigeminal system, CGRP is found mainly in the small-sized C fibers (less in Aδ and Aβ fibers) innervating the meninges, cerebral vasculature, and other cranial pain-sensitive structures. Most cerebrovascular CGRP-containing nerve fibers originate in the trigeminal ganglion (TG). The TG is a pivotal structure in migraine pathophysiology and a potential therapeutic site of action due to its localization outside of the BBB. CGRP is released from dural fibers upon TG stimulation. Within the TG, CGRP and CGRP receptors are not co-localized at the neuronal level: within the TG, 50% of neurons express CGRP (in so-called C-fibers), whereas only 30% express CGRP receptors (in so-called Aδ fibers), suggesting a paracrine regulation (cf. Chap. 5, Table 5.8).

Pharmacology

Inhibiting the CGRP Pathway in Migraine

There are two different drug classes inhibiting the function of CGRP: small-molecule antagonists (so-called *gepants*) and monoclonal antibodies (mAbs) targeting CGRP or its receptor. In this section, we will only discuss the mechanisms of action and pharmacokinetic profiles of the CGRP-pathway targeted mAb therapies.

CGRP-Pathway Targeted mAbs

The development of mAbs only recently offered new treatment options for migraine. The last breakthrough (introduction of the triptans) dated from over 30 years ago. To date,

four mAbs are available globally (Table 26.1), and their local use is largely depending on national approval and reimbursement decisions. Anti-CGRP mAbs were first studied and approved in chronic migraine (CM)(cf. section on Epidemiology), with increasing evidence and global approval for use in episodic migraine (EM), including implementation in international guidelines (Edvinsson 2022; Nissan et al. 2022; Sacco et al. 2022; Yuan et al. 2017).

These antibodies are cloned isotypes of immunoglobulin G (IgG), and by binding either the CGRP ligand or the CGRP receptor complex they inhibit or prevent CGRP activity (Fig. 26.3).

In general, IgG antibodies have a long half-life of around 21 days, even longer for mAbs targeting CGRP (receptor) (Cf. Table 26.2), allowing low-frequency administration and enabling use for preventive treatment strategies (Edvinsson 2021, 2022; Yuan et al. 2017).

Table 26.1 Calcitonin gene-related peptide (CGRP) monoclonal antibodies

Monoclonal antibody	Molecular weight	Ig serotype	Structure	Molecular target	Target cells	Use
Erenumab (AMG334)	150 kDa	IgG$_2$	Human	CGRP receptor	Sensoric ganglion cells, TG	Blocking CGRP-induced neuro-inflammation, nociceptive transmission
Fremanezumab (TEV48125)	148 kDa	IgG$_2$	Humanized murine	α-, β- CGRP ligand	Not applicable	Blocking CGRP-induced neuro-inflammation, nociceptive transmission
Galcanezumab (LY2951742)	144 kDa	IgG$_4$	Humanized murine	α-, β- CGRP ligand	Not applicable	Blocking CGRP-induced neuro-inflammation, nociceptive transmission
Eptinezumab (ALD403)	143 kDa	IgG$_1$	Humanized murine	α-, β- CGRP ligand	Not applicable	Blocking CGRP-induced neuro-inflammation, nociceptive transmission

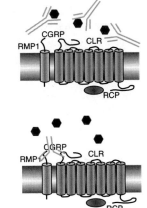

Fig. 26.3 CGRP (receptor) antibodies. Eptinezumab has FDA approval for the preventive treatment of migraine. Fremanezumab, galcanezumab, and erenumab have FDA and EMA approval for the preventive treatment of migraine. For abbreviations, see Fig. 26.2. (Adapted from Raddant and Rosso (2011))

Antibodies against CGRP or its receptor:

■ **Eptine**zumab
Fremanezumab
Galcanezumab
binding to peptide CGRP

■ **Erenumab**
binding to CGRP receptor

Table 26.2 Pharmacokinetics of calcitonin gene-related peptide (CGRP) monoclonal antibodies

Monoclonal antibody	Target	Route	$T_{1/2}$	T_{max}	Dose	Frequency
Erenumab	CGRP receptor	SC	28 days	6 days	70 mg / 140 mg	every 4 weeks/monthly / every 4 weeks/monthly
Fremanezumab	α-, β- CGRP ligand	SC	32 days	5 days	225 mg / 675 mg	every 4 weeks/monthly / every 3 months
Galcanezumab	α-, β- CGRP ligand	SC	27 days	5 days	120 mg[a]	every 4 weeks/monthly
Eptinezumab	α-, β- CGRP ligand	IV	27 days	1-3 h	100 mg / 300 mg	every 3 months / every 3 months

$T_{1/2}$ half-life, T_{max} time to peak drug concentration, *SC* subcutaneous, *IV* intravenous
[a]Start with 240 mg loading

Distribution and Site of Action of Anti-CGRP-Pathway MABS

The site of action of mAbs is largely determined by the distribution, the type of vascular endothelium (continuous, fenestrated, sinusoidal), and the type of transport (receptor mediated vs. passive vs. transcellular vs. paracellular). Considering the large size of these mAbs (approximately 150 kDa; 7-10 nm), they barely cross the healthy blood-brain barrier (BBB), but can cross passively via paracellular extravasation or "leaky" vascular endothelium. Meninges, epidural capillaries, and veins have no BBB. The tissue concentration in the skin, lung, kidney, liver, spleen, or heart after intravenous injection of the IgGs is estimated to be 10–15% of the plasma concentration, but only 0.35% in brain parenchyma and 0.1% in cerebral spinal fluid (CSF). IgG molecules in the CSF are rapidly transported back to the circulation via neonatal Fc receptors in the choroid plexus. There is no clear evidence that the BBB is disrupted in migraine or in migraine attacks, although cortical spreading depolarization induced pinocytosis might serve as a mechanism for transcellular transport. Although in rat models repeated inflammatory dural inflammation induces BBB permeability in the trigeminal nucleus caudalis, there is no evidence that there is any BBB disruption relevant for mAb penetration in (chronic) migraine. Brain-nerve-barrier permeability is comparable to normal BBB permeability. Therefore, the site of action for mAbs is likely to be outside the brain parenchyma. Sensory ganglia, such as the trigeminal ganglion (TG), have fenestrated capillaries, thereby allowing mAb passage. Anti-CGRP mAbs therefore can passively accumulate in sensory ganglia such as the TG. Anti-CGRP receptor mAbs may actively accumulate at the target site due to receptor binding. Non-target mAb distribution (e.g., in spleen, liver, circulation, endocrine glands) has not resulted in any major adverse events (AEs) in clinical trials, open label extension (OLE), or real-world data (RWD) studies up to date. Taken together, anti-CGRP-pathway mAbs are likely to act outside the BBB: in the meninges, circumventricular regions, neural sensory ganglia (such as the TG), and trigeminal vasculature, thereby minimizing central nervous system actions and adverse events (Edvinsson 2022; Joshi et al. 2021; Szkutnik-Fiedler 2020; Yuan et al. 2017).

Efficacy of Anti-CGRP MAB Therapies in Migraine: Clinical Studies

The efficacy of these targeted mAb therapies was proven in several pivotal studies in episodic and chronic migraine, showing a significant reduction in monthly migraine days (MMDs) compared to placebo (for an overview, see Nissan et al. 2022). Various clinical studies have shown favorable safety profiles for all four mAbs, with only minor side effects and no serious adverse events. When stratified for treatment history, these mAbs also were reported effective both in mAb monotherapy as well as concurrently with another prophylactic strategy in patients who had previously failed up to >4 preventive treatments,. RCTs (randomized clinical trials) show a MMD reduction of ~2 days as compared to placebo, with MMD reduction of >5 to >10 days reported in OLE and RWD studies. Furthermore, RCTs show significant improvement in use of acute anti-migraine medication, monthly medication use days, headache-related quality of life (HIT-6), and depressive symptoms (Patient Global Impression Change assessment)(Cohen et al. 2022; Edvinsson 2021, 2022; Joshi et al. 2021; Nissan et al. 2022).

Head-to-head comparison trials with existing preventive anti-migraine drugs have not been performed, with one exception. Erenumab demonstrated a favorable tolerability and efficacy profile compared to topiramate (Reuter et al. 2022).

Dosing and Pharmacokinetics

The anti-migraine monoclonal antibody-based therapies target the CGRP receptor (erenumab) or the ligand (galcanezumab, fremanezumab, eptinezumab). The mAbs are administered parenterally via intravenous (eptinezumab) or subcutaneous injection (Table 26.2).

After subcutaneous administration, the lymphatic system plays an active part in mAb absorption and is the main route of transport to the blood compartment. Therefore, a key factor affecting the bioavailability of s.c mAbs is the transit time of mAbs within the lymphatic system. Prolonged mAb absorption from the injection site within the lymph after subcutaneous administration is the cause for lower concentrations and a longer time to reach peak plasma concentrations (2–8 days after administration) compared to the intravenous administration route.

The bioavailability of subcutaneously administered mAbs is on the average 82% for erenumab and 66% for fremanezumab. For the bioavailability of galcanezumab, a range of 50–100% has been reported (Kielbasa and Helton 2019). Due to their size and polarity, mAbs are transported into tissues by convective transport. mAbs are distributed mainly by lymph and blood; redistribution through bile or saliva is not observed. mAbs penetrate poorly into tissues, resulting in low distribution volumes for the anti-CGRP antibodies ranging from 3.7 to 7.3 L. However, determining the volume of distribution of mAbs is complex as elimination occurs at tissue sites that are not in rapid equilibrium with plasma. High-affinity binding and target-mediated elimination contribute to an underestimation of the volume of distribution using non-compartmental analysis (Joshi et al. 2021; Szkutnik-Fiedler 2020).

Metabolism and Excretion

Most mAbs are metabolized and cleared via specific target-mediated clearance and non-specific elimination by the reticuloendothelial system (RES; mononuclear phagocyte system [MPS]) Cf. Chap. 8). These mechanisms occur in parallel and depend on antibody concentration. Clearance rates are high with low antibody concentrations, and low with high antibody concentrations since the target-mediated clearance pathway gets saturated at high concentrations, and clearance at those concentrations only occurs via the RES/MPS. Binding of mAbs to neonatal Fc receptors (FcRn) in the RES/MPS slows down the clearance process and contributes to their long half-life, allowing for 1–2 months dosing intervals (cf. Chap. 8).

Drug-Drug Interactions (DDI)

Most DDIs can be predicted when the drug is metabolized via a CYP450 isoenzyme type, although not all those "theoretical" DDIs may be clinically relevant. DDIs may become significant in patients with renal or hepatic impairment. Since therapeutic mAbs are not metabolized by CYP450 and its isoenzymes, DDI-drug/mAb are not a major concern when incorporating monoclonal antibodies into a polytherapy regimen. The notion that there was no clinically relevant DDI found between erenumab and sumatriptan, or combined oral contraceptive (ethinyl estradiol/norgestrel), is specifically important given the young and feminine migraine population. Antibody-antibody interactions are also unlikely: IgG clearance is dependent on FcRn saturation, but at typical therapeutic dosing regimens of mAb FcRn saturation is not likely to occur. No DDI information is available for mAb use with gepants (Al-Hassany and MaassenVanDenBrink 2021; Edvinsson 2022; Joshi et al. 2021).

Individual MABS

Erenumab

Erenumab (AIMOVIG®, solution for subcutaneous injection in a pre-filled syringe or pre-filled pen, 70 mg and 140 mg) is a fully human IgG$_2$ monoclonal antibody produced using recombinant DNA technology in Chinese hamster ovary cells. The recommended dose is 70 mg or 140 mg s.c. once a month. Clinical improvement is usually achieved within 3 months. Erenumab's bioavailability upon s.c. administration is 82%, maximum plasma concentrations are reached after 4–6 days (from 3 to 14 days), steady state is achieved after 12 weeks, and its biological half-life is 28 days. Erenumab is eliminated in two phases: at low concentrations

by saturable binding to the target CGRP receptor, and at higher concentrations mainly by non-specific proteolysis. At therapeutic doses, erenumab elimination is mainly via non-specific proteolysis. There is no interaction with sumatriptan or oral contraceptives (ethinyl estradiol/norgestrel) in healthy volunteers, nor is there any relevant food-drug interaction (Al-Hassany and MaassenVanDenBrink 2021; Cohen et al. 2022; Szkutnik-Fiedler 2020).

Fremanezumab

Fremanezumab (AJOVY®, solution for subcutaneous injection, 225 mg in pre-filled syringes) is a humanized IgG$_2$ monoclonal antibody derived from a murine precursor. The fremanezumab mechanism of action involves targeting both α and β isoforms of CGRP ligands. It can be used according to two dosing schedules: 225 mg once per month, or 675 mg every 3 months (three 225 mg injections). The maximum plasma concentration is reached after 5–7 days (from 3 to 20 days), half-life is 32 days, and absolute bioavailability is 55% (225 mg) and 66% (900 mg). Steady state is achieved within 168 days. Based on population pharmacokinetic modeling and simulation, higher body weight was associated with a lower fremanezumab exposure due to increased central clearance and distribution volume. During the clinical studies, concomitant use of acute and preventive antimigraine medication did not affect fremanezumab's pharmacokinetics (Cohen et al. 2022; Szkutnik-Fiedler 2020).

Galcanezumab

Galcanezumab (EMGALITY®, solution for subcutaneous injection, 120 mg in pre-filled pen) is a recombinant humanized IgG$_4$ monoclonal antibody produced in Chinese hamster ovary cells. The dosing schedule is one subcutaneous injection at a dose of 120 mg once a month with the first loading dose of 240 mg. Maximum serum concentration is reached at 5 days (from 7–14 days), its half-life is 27 days, and distribution volume is 7.3 L. Galcanezumab exposure increased proportionally with dose. The population pharmacokinetic analysis showed that absorption rate, apparent clearance, and apparent volume of distribution were independent of the dose (range 5–300 mg) (Cohen et al. 2022; Szkutnik-Fiedler 2020).

Eptinezumab

Eptinezumab (VYEPTI®, 100 mg ampules) is the only drug in its class to be administered intravenously for migraine attacks. The recommended dosage is 100 mg in an i.v.

infusion over 30 min or a maximum of 300 mg every 3 months. Eptinezumab exhibits linear pharmacokinetics, with exposure increasing proportionally with after intravenous administration (100–300 mg). Steady-state plasma concentrations are achieved with the first dose, distribution volume is 3.7 L, and half-life is 27 days. There is no pharmacokinetic interaction with sumatriptan (Cohen et al. 2022; Szkutnik-Fiedler 2020).

Immunogenicity and Anti-drug Antibodies (ADA)

In general, repeated administration of mAbs is known for its potential to be immunogenic, as is manifested in the production of anti-drug antibodies (ADAs). ADAs can alter pharmacokinetics, and ADAs depend on many patient-related factors (genetic background, disease state, comorbidity, and co-treatment) and the drug itself (dose, frequency, route of administration, impurities, formulation, post-translational modifications, antibody origin, mAb target). For general consideration on immunogenicity and anti-drug antibodies, see Chap. 7.

Anti-CGRP mAbs

Immunogenicity profiles of all four anti-CGRP-pathway mAbs (erenumab, fremanezumab, galcanezumab, and eptinezumab) have been reported. No effect of ADA formation on the mAbs' efficacy, treatment outcome, or safety was found (Cohen et al. 2021; Edvinsson 2021, 2022; Joshi et al. 2021; Nissan et al. 2022). ADAs didn't show any effect on the pharmacokinetics of galcanezumab. Reported immunogenicity rates so far were relatively low (see Table 26.3), and the clinical significance of neutralizing antibodies is not known yet (1, 5).

Table 26.3 Incidence of anti-drug antibodies and neutralizing antibodies[a] across phase 2/3 trials of calcitonin gene-related peptide (CGRP) monoclonal antibodies

Mechanism	3–6 months phase 2/3 studies				≥1-year phase 3 studies	
	Anti-drug antibodies in CM (%)	Neutralizing antibodies in CM (%)	Anti-drug antibodies in EM (%)	Neutralizing antibodies in EM (%)	Anti-drug antibodies (%)	Neutralizing antibodies (%)
Erenumab	2–6	0	3–8	0–1	6–10[b]	<1[b]
Fremanezumab	<1	0	<1	<1	2[b]	<1[b]
Galcanezumab	3	2	3–11	3	12[b]	12[b]
Eptinezumab	16–18	6–7	14–18	10	18–19[c]	8[c]

CM chronic migraine, EM episodic migraine
[a]Clinical significance of neutralizing antibodies is not known
[b]EM and CM patients
[c]CM patients

Adverse Events

Clinical studies have confirmed the favorable safety profile of all four anti-CGRP-pathway mAbs. The most common side effects are skin reactions (rash, itching) at the injection site and constipation. Angioedema and anaphylactic reactions, although rare, have also been reported. Hypersensitivity reactions can occur within minutes of drug administration but also more than 1 week after treatment. If a hypersensitivity reaction occurs, consideration should be given to discontinuing the use of the monoclonal antibody. AE rates for the therapeutic mAbs have varied greatly between OLE of RCTs, prospective studies, and real-world data (RWD), ranging from ~20% to >90%; for an overview see ref. (Cohen et al. 2022).

Clinical Practice Migraine

Migraine

Migraine is a common, chronic, primary headache disorder that is typically characterized by recurrent disabling attacks of headache and accompanying symptoms, including aura. Headache attacks are typically unilateral, pulsating, and last 4–72 h. Associated symptoms are, among others, photophobia, phonophobia, osmophobia, and nausea and vomiting. Migraine is classified according to the *International Classification of Headache Disorders*, 3th edition (IHCD3), as migraine without aura, the most common form, and migraine with aura, with aura being defined as fully reversible neurological symptoms which precede or accompany the headache (Goadsby et al. 2017; Headache Classification Committee of the International Headache Society (IHS) The International Classification of Headache Disorders, 3rd edition 2018).

Etiology

Migraine is a multifactorial, neurovascular disorder. The entire chronology of signs and symptoms associated with the different phases of the migraine attack – the premonitory, aura, headache, and post-dromal phases – results from a complex brain disorder affected by environmental and genetic factors. Pathophysiological research has identified the trigeminovascular system or pathway as an important mechanism in the development of the characteristic headaches, whereas the premonitory (non-headache) phase is linked to hypothalamic dysfunction and the migraine aura is suggested to result from a wave of cortical spreading depolarization. The trigeminovascular network is comprised of cranial vessels and their innervation by the fifth cranial nerve (trigeminal nerve) arising from the brainstem. Trigeminal afferents innervate pain-sensitive structures in the head, including skin, meninges, cerebral arteries and venous sinuses, periosteum, and calvarial sutures. The second-order neurons are located in the trigeminal ganglion (TG). The trigeminocervical complex (TC or TCC) spans the brainstem and upper cervical spinal cord and projects ascending pathways to other brainstem nuclei, neural ganglia, and cerebellum and cerebrum; see Fig. 26.4. The trigeminal afferents, long axons with cell bodies in the TG, are a combination of $A\delta$, $A\beta$, and δ fibers. Upon proalgesic stimulation, trigeminal afferents release a variety of neuropeptides (including CGRP, tachykinin, bradykinin, adenylate cyclase-activating polypeptides, such as somatostatin, and neuropeptide Y) which play a major role in trigeminal signaling and sensory modulation in migraine (Edvinsson 2022; Ferrari 2008; Goadsby et al. 2017). Furthermore, the input from the afferents is relayed via the TCC before nociceptive information reaches higher cortical structures, eliciting headache symptoms.

The trigeminovascular input from the meningeal vessels passes through the trigeminal ganglion and synapses in the trigeminocervical complex. The peripheral release of molecules results in vasodilation, plasma leakage, and mast cell degranulation within the dura mater, while central release can cause activation of second order neurons leading to the pain of migraine attacks. Trigeminal activation can mediate parasympathetic responses via the pterygopalatine ganglion. (From Goadsby et al. (2002)).

Migraine attacks are associated with trigeminal release of the potent vasodilatory neuropeptide calcitonin gene-related peptide (CGRP), which is the most abundant neuropeptide in the perivascular nerve endings of the trigeminal nerve. This release does not directly elicit pain via dural nociceptors, but induces cerebral arterial vasodilation, enhanced trigeminal nociception, and inflammatory responses via interaction with neurons, astroglial cells, mast cells, immunocytes, and meningeal blood vessels, resulting in peripheral and central sensitization. Peripheral sensitization sustains neurogenic inflammation and activation of primary neurons, while central sensitization lowers the activation threshold of central neurons, resulting in allodynia and central pain. Despite earlier beliefs, vasodilation nowadays is considered only a secondary phenomenon, due to the aforementioned processes, and vasoconstriction is not essential for therapeutic anti-migraine efficacy (Edvinsson 2022; Goadsby et al. 2017).

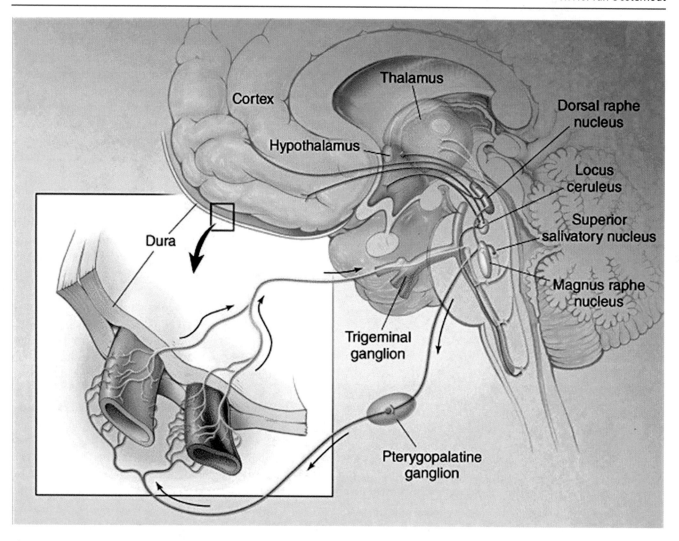

Fig. 26.4 Anatomical structure of trigeminocervical complex, trigeminal ganglion, and trigeminal afferents with regard to migraine pathophysiology (From Goadsby et al. (2002), with permission)

Epidemiology

An average of 11–12% of the population in Europe and North America suffer from migraines, and 75% are women. Prevalence rates show age and gender-specific distributions, being the highest among women in their fertile lifespan, and declining with age. Based on monthly headache days, migraine is clinically defined as episodic migraine (EM; <15 migraine days per month) or chronic migraine (CM; ≥15 headache days per month, of which ≥8 migraine days, for more than 3 months). CM occurs through chronification of EM, with increasing attack frequency and change of nature of the headache into less migraine reminiscent (Headache Classification Committee of the International Headache Society (IHS) The International Classification of Headache Disorders, 3rd edition 2018; Launer et al. 1999).

Migraine Burden

Migraine has a significant burden, both on a personal and societal level, with over 42% of patients reporting a moderate to severe degree of functional disability, 8% unemployment rate, and 6.8% on-disability rate, based on a recent epidemiological study. In the USA, for example, the annual direct healthcare cost for patients with migraine is estimated at US$ 22,364 per person, and the total indirect cost estimated at US$ 19 billion. The Global Burden of Disease Survey 2019 has reported migraine as the second highest cause of years lived with disability, being first among women under 50 years of age. This dramatic socioeconomical impact warrants ongoing efforts in better understanding of migraine pathophysiology and facilitating the development of effective treatment options (GBD2019-Collaborators 2020; Steiner et al. 2018).

Acute and Preventive Treatment

Acute migraine management includes analgesics or NSAIDs for mild attacks and triptans ($5HT_{1B/1D}$ receptor agonists) for moderate or severe attacks. The use of ergots to abort attacks has nearly vanished globally, because of cardiovascular safety concerns, unreliable efficacy, and tolerability issues. Small molecule CGRP receptor antagonists (*gepants*) and lasmiditan, a selective $5HT_{1F}$ receptor agonist, have recently been introduced as effective acute treatments.

Up to 40% of migraine patients are eligible for preventive treatment, as they suffer from ≥2 monthly or very long-lasting migraine attacks. Over the years, various classes of drugs have been used for migraine prevention: beta-blockers, angiotensin receptor blockers, calcium channel antagonists, antiepileptic drugs, serotonin receptor antagonists, tricyclic antidepressant, serotonin norepinephrine reuptake inhibitors, and onabotulinumtoxinA. None of these medications were specifically developed for the prevention of migraine, and none directly target the underlying pathomechanisms. The anti-CGRP-pathway monoclonal antibodies are the first prophylactic drugs specifically designed based on our knowledge of the migraine pathophysiology.

Concluding Remarks

The birth of the monoclonal antibodies targeting the CGRP pathway in migraine has brought new insights in the field of preventive migraine treatment, with proven efficacy in reducing monthly migraine days, in combination with, in general, low-frequency, patient-friendly subcutaneous administration options. Notably, their favorable tolerability and safety profiles as seen in RCTs, OLE, and RWD studies, so far, are promising when compared with the known migraine prophylactics that have low adherence mainly due to adverse events.

In the nearby future, the practical use of anti-CGRP mAbs in migraine is expected to be globally widely accepted and accessible for selected patient groups in both chronic and episodic migraine. Their use may be restricted by national reimbursement policies. Further studies are needed to identify those patients who are most likely to benefit from these treatment strategies and to elucidate the potential utility of the mAbs in combination with other, also emerging, pharmacological migraine treatment options.

Self-Assessment Questions and Answers

1. Where are the cranial CGRP receptors localized to which the mAbs bind and sort their effect?
   (a) Within the blood-brain barrier.
   (b) Outside the blood-brain barrier.
   (c) Both within and outside the blood-brain barrier.
   Answer: b. Cranial CGRP receptors are localized throughout the cerebrum (striatum, amygdala, pineal gland, colliculi, trigeminal nucleus caudalis, cortex), cerebellum, and smooth muscle layer of cerebral arteries. The anti-CGRP mAbs exert their action via the CGRP receptor on ganglionic cells in the trigeminal ganglion, where the lack of blood-brain barriers allows the large-molecule mAbs to be in the direct vicinity of the receptor.

2. Via what mechanism are anti-CGRP mAbs eliminated?
   (a) Via renal mechanisms.
   (b) Via hepatic mechanisms.
   (c) Via specific target cell-mediated clearance and nonspecific elimination by the RES.
   Answer: c. mAbs are not cleared renally or hepatically, and therefore renal or hepatic impairment or dysfunction does not affect mAb pharmacokinetics. mAbs are eliminated by target-cell mediated degradation (in their lysosomes) or via the reticulo-endothelial system.

3. What is a pivotal factor explaining the anti-CGRP mAb long half-life?
   (a) Non-specific elimination via the RES is inhibited as the mAb itself degrades the RES.
   (b) Interaction of anti-CGRP mAbs with the neonatal Fc receptors reduces clearance.
   Answer: b. The interaction with the neonatal Fc receptor is assumed to be the key factor explaining the long half-life of anti-CGRP mAbs as it reduces the clearance rate (cf. Chap. 8).

4. What statement on drug-drug interactions (DDI) is true?
   (a) Since the anti-CGRP mAbs are metabolized via the CYP450 system, theoretically DDIs are very unlikely.
   (b) There is no major concern on DDIs except in the case of using oral anticonceptives.
   (c) DDIs between anti-CGRP mAb and new, emerging anti-CGRP treatment strategies in migraine are to be awaited.
   Answer: c. Anti-CGRP mAbs are not metabolized via CYP450, and therefore DDIs are very unlikely, also in the case of oral anticonceptives. As emerging anti-CGRP strategies also affect the CGRP pathway, those DDIs could be relevant. At this moment in time, there is not much known about these possible DDIs.

5. What are the immunogenicity rates of anti-CGRP mAbs and what is its clinical importance?
   (a) Immunogenicity rates range from 9.8% to 12.5%, and warrant biochemical and clinical monitoring, likely resulting in switching between different anti-CGRP mAb strategies.
   (b) Immunogenicity rates range from 45 to 70% and only neutralizing antibodies are of clinical importance.

(c) The percentage of anti-drug antibody (ADA) formation in patients reaches 18%, and so far ADA formation has not been shown to affect efficacy or safety of anti-CGRP mAbs.

Answer: c. At present, reported ADA rates vary between 0% and 18% (Table 26.3) without reported effects on efficacy or safety.

References

Al-Hassany L, MaassenVanDenBrink A (2021) Drug interactions and risks associated with the use of triptans, ditans and monoclonal antibodies in migraine. Curr Opin Neurol 34(3):330–338. https://doi.org/10.1097/WCO.0000000000000932

Cohen JM, Ning X, Kessler Y, Rasamoelisolo M, Campos VR, Seminerio MJ, Krasenbaum LJ, Shen H, Stratton J (2021) Immunogenicity of biologic therapies for migraine: a review of current evidence. J Headache Pain 22(1):3. https://doi.org/10.1186/s10194-020-01211-5

Cohen F, Yuan H, Silberstein SD (2022) Calcitonin gene-related peptide (CGRP)-targeted monoclonal antibodies and antagonists in migraine: current evidence and rationale. BioDrugs 36(3):341–358. https://doi.org/10.1007/s40259-022-00530-0

Edvinsson L (2021) CGRP and migraine: from bench to bedside. Rev Neurol (Paris) 177(7):785–790. https://doi.org/10.1016/j.neurol.2021.06.003

Edvinsson L (2022) Calcitonin gene-related peptide (CGRP) is a key molecule released in acute migraine attacks-successful translation of basic science to clinical practice. J Intern Med 292(4):575–586. https://doi.org/10.1111/joim.13506

Edvinsson L, Haanes KA, Warfvinge K, Krause DN (2018) CGRP as the target of new migraine therapies - successful translation from bench to clinic. Nat Rev Neurol 14(6):338–350. https://doi.org/10.1038/s41582-018-0003-1

Ferrari MD (2008) Migraine genetics: a fascinating journey towards improved migraine therapy. Headache 48(5):697–700. http://www.ncbi.nlm.nih.gov/pubmed/18471114

GBD2019-Collaborators (2020) Global burden of 369 diseases and injuries in 204 countries and territories, 1990-2019: a systematic analysis for the global burden of disease study 2019. Lancet 396(10258):1204–1222. https://doi.org/10.1016/S0140-6736(20)30925-9

Goadsby PJ, Holland PR, Martins-Oliveira M, Hoffmann J, Schankin C, Akerman S (2017) Pathophysiology of migraine: a disorder of sensory processing. Physiol Rev 97(2):553–622. https://doi.org/10.1152/physrev.00034.2015

Goadsby PJ, Lipton RB, Ferrari MD (2002) Migraine--current understanding and treatment. N Engl J Med. 24;346(4):257–70

Headache Classification Committee of the International Headache Society (IHS) The International Classification of Headache Disorders, 3rd edition (2018) Cephalalgia 38(1):1–211. https://doi.org/10.1177/0333102417738202

Joshi S, Tepper SJ, Lucas S, Rasmussen S, Nelson R (2021) A narrative review of the importance of pharmacokinetics and drug-drug interactions of preventive therapies in migraine management. Headache 61(6):838–853. https://doi.org/10.1111/head.14135

Kielbasa W, Helton DL (2019) A new era for migraine: pharmacokinetic and pharmacodynamic insights into monoclonal antibodies with a focus on galcanezumab, an anti-CGRP antibody. Cephalalgia 39(10):1284–1297. https://doi.org/10.1177/0333102419840780

Launer LJ, Terwindt GM, Ferrari MD (1999) The prevalence and characteristics of migraine in a population-based cohort: the GEM study. Neurology 53(3):537–542. http://www.ncbi.nlm.nih.gov/pubmed/10449117

Nissan GR, Kim R, Cohen JM, Seminerio MJ, Krasenbaum LJ, Carr K, Martin V (2022) Reducing the burden of migraine: safety and efficacy of CGRP pathway-targeted preventive treatments. J Clin Med 11(15):4359. https://doi.org/10.3390/jcm11154359

Raddant AC, Russo AF (2011) Calcitonin gene-related peptide in migraine: intersection of peripheral inflammation and central modulation. Expert Rev Mol Med 13:e36. https://doi.org/10.1017/S1462399411002067

Reuter et al. (2022) Renumab versus topiramate for the prevention of migraine - a randomised, double-blind, active-controlled phase 4 trial. Cephalalgia. 42(2):108–118.

Sacco S, Amin FM, Ashina M, Bendtsen L, Deligianni CI, Gil-Gouveia R, Katsarava Z, MaassenVanDenBrink A, Martelletti P, Mitsikostas DD, Ornello R, Reuter U, Sanchez-Del-Rio M, Sinclair AJ, Terwindt G, Uluduz D, Versijpt J, Lampl C (2022) European headache federation guideline on the use of monoclonal antibodies targeting the calcitonin gene related peptide pathway for migraine prevention - 2022 update. J Headache Pain 23(1):67. https://doi.org/10.1186/s10194-022-01431-x

Steiner TJ, Stovner LJ, Vos T, Jensen R, Katsarava Z (2018) Migraine is first cause of disability in under 50s: will health politicians now take notice? J Headache Pain 19(1):17. https://doi.org/10.1186/s10194-018-0846-2

Szkutnik-Fiedler D (2020) Pharmacokinetics, pharmacodynamics and drug-drug interactions of new anti-migraine drugs-lasmiditan, gepants, and calcitonin-gene-related peptide (CGRP) receptor monoclonal antibodies. Pharmaceutics 12(12):1180. https://doi.org/10.3390/pharmaceutics12121180

Yuan H, Lauritsen CG, Kaiser EA, Silberstein SD (2017) CGRP monoclonal antibodies for migraine: rationale and Progress. BioDrugs 31(6):487–501. https://doi.org/10.1007/s40259-017-0250-5

Index